THE PHYSICS OF ELECTRONIC AND ATOMIC COLLISIONS

AIP CONFERENCE PROCEEDINGS 205

THE PHYSICS OF ELECTRONIC AND ATOMIC COLLISIONS

XVI INTERNATIONAL CONFERENCE

NEW YORK, NY 1989

EDITORS:

A. DALGARNO
HARVARD-SMITHSONIAN CENTER
FOR ASTROPHYSICS

R. S. FREUND
AT&T BELL LABORATORIES

P. M. KOCH
STATE UNIVERSITY OF NEW YORK
at STONY BROOK

M. S. LUBELL
CITY UNIVERSITY OF NEW YORK,
CITY COLLEGE

T. B. LUCATORTO
NATIONAL INSTITUTE OF STANDARDS
AND TECHNOLOGY

AIP

American Institute of Physics   New York

Authorization to photocopy items for internal or personal use, beyond the
free copying permitted under the 1978 US Copyright Law (see statement
below), is granted by the American Insitute of Physics for users registered
with the Copyright Clearance Center (CCC) Transactional Reporting Service, provided that the base fee of $2.00 per copy is paid directly to CCC, 27
Congress St., Salem, MA 01970. For those organizations that have been
granted a photocopy license by CCC, a separate system of payment has been
arranged. The fee code for users of the Transactional Reporting Service is:
0094-243X/87 $2.00.

© 1990 American Institute of Physics.

Individual readers of this volume and non-profit libraries, acting for them,
are permitted to make fair use of the material in it, such as copying an article
for use in teaching or research. Permission is granted to quote from this
volume in scientific work with the customary acknowledgment of the source.
To reprint a figure, table or other excerpt requires the consent of one of the
original authors and notification to AIP. Republication or systematic or
multiple reproduction of any material in this volume is permitted only under
license from AIP. Address inquiries to Series Editor, AIP Conference Proceedings, AIP, 335 E. 45th St., New York, NY 10017.

L.C. Catalog Card No. 90-53183
ISBN 0-88318-390-0
DOE CONF 890708

Printed in the United States of America.

Contents

Preface xi

ICPEAC Executive Committee 1987–1989 xiii
ICPEAC General Committee 1987–1989 xiv
XVI ICPEAC Local Committee xv

KEYNOTE LECTURE

On the Utility and Ubiquity of Atomic Collision Physics 2
    S. Datz

PLENARY LECTURES

Chaos in Atoms? 16
    K. H. Welge
Transition State Spectroscopy of Hydrogen Transfer Reactions 33
    D. M. Neumark
Low-Energy Molecular Collisions with Applications to Interstellar Cloud Problems 49
    K. Takayanagi

REVIEW TALKS, PROGRESS REPORTS, AND HOT TOPICS

ELECTRONS

Electron–Atom Collision Theory 68
    C. J. Joachain
Giant Resonances in Double Ionization of Atomic Ions 82
    H. Suzuki, T. Hirayama, and T. Takayanagi
Determinations of the Products of Dissociative Recombination Reactions 90
    N. G. Adams, C. R. Herd, and D. Smith
R-Matrix Calculations of Electron Molecule Scattering 96
    L. A. Morgan
Spin-Polarization, Orientation, and Alignment in Electron–Atomic Collisions 103
    M. H. Kelley
Near-Threshold Studies of Atomic Hydrogen 115
    J. F. Williams
Simultaneous Electron–Photon Excitation Experiments 122
    W. R. Newell
Relativistic (e, $2e$) Processes on Inner Shells of Heavy Atoms 130
    J. Bonfert, H. Graf, and W. Nakel
The Theory of Electron–Ion Collisions 137
    A. E. Kingston
Theoretical Calculations of Elastic and Inelastic Scattering of Electrons from Hydrogen 149
    D. H. Madison
Electron Impact Ionization of U^{88+}–U^{91+} 157
    N. Claytor, B. Feinberg, H. Gould, C. E. Bemis, Jr., J. Gomez del Campo, C. A. Ludemann, and C. R. Vane

PHOTONS

Synchrotron Radiation Experiments on Atoms and Molecules 162
    U. Becker

Synchrotron-Radiation Experiments with Recoil Ions 176
    J. C. Levin

Atoms in Intense Radiation Fields 184
    M. H. Mittleman

Photoionization of Positive Atomic Ions: A Review of Our Present Understanding 189
    S. T. Manson

Polarizational Radiation in Electronic and Atomic Collisions ("Atomic" Bremsstrahlung) 201
    M. Ya. Amusia

Correlation Effects in Electron Impact- and Photon-Induced Two-Electron Transitions in Rare Gases 215
    K.-H. Schartner

Photoionization of Laser Excited Atoms by Synchrotron Radiation 224
    D. Cubaynes, J. M. Bizau, B. Carré, and F. J. Wuilleumier

Photodetachment Collisions 233
    D. J. Pegg

Double Photoionization near Threshold 241
    V. Schmidt

IONS AND ATOMS

Theoretical Collision Physics of Highly Charged Ions 246
    A. Bárány

Optical Potentials in Ion–Atom Collisions 258
    R. M. Dreizler, H. J. Ast, A. Henne, H. J. Lüdde, and C. Stary

Distorted Wave Models for Ionization in Atomic Collisions 264
    R. D. Rivarola, P. D. Fainstein, and H. Ponce

Electron Capture and Energy-Gain Spectroscopy 273
    K. Taulbjerg

Correlation in Atomic Scattering 280
    J. H. McGuire and J. C. Straton

Ion–Ion Collisions: Charge Transfer and Ionization 290
    E. Salzborn

Multiply Differential Ionization Probabilities in Small Impact Parameter Ion–Atom Collisions 299
    G. Schiwietz, B. Skogvall, N. Stolterfoht, D. Schneider, V. Montemayor

Energy Loss and Charge Exchange Processes of High Energy Heavy Ions Channeled in Crystals 309
    J. C. Poizat, S. Andriamonje, R. Anne, N. V. de Castro Faria, M. Chevallier, C. Cohen, J. Dural, B. Farizon-Mazuy,
    M. J. Gaillard, R. Genre, M. Hage-Ali, R. Kirsch, A. L'hoir, J. Mory, J. Moulin, Y. Quéré, J. Remillieux,
    D. Schmaus, and M. Toulemonde

Dynamics of Inelastic Collisions of Electronically Excited Rare Gas Atoms 317
    H. C. W. Beijerinck

SIFT and FALP Determinations of Ionic Reactions Rate Coefficients 325
    D. Smith and N. G. Adams

Position Sensitive Detection with Laser Induced Fluorescence 337
    L. Hüwel, A. M. Wodtke, P. Andresen, and H. Voges

Optical Spectroscopic Studies on Penning Ionization and Ion–Molecule Reactions at Thermal Energy by Using Flowing Afterglow 342
    M. Tsuji

Multi-Charged Ion/Slow Electron Collisions in Cold Plasmas 350
    L. Shmaenok

Spectral Distribution of Electrons Emitted into the Continuum of Fast Projectiles: Theoretical Approaches of Higher Order in Comparison with Experiment 358
    D. H. Jakubassa-Amundsen

Multiple Electron Capture in Close Ion–Atom Collisions 366
    A. S. Schlachter, J. W. Stearns, K. H. Berkner, E. M. Bernstein, M. W. Clark, R. D. DuBois, W. G. Graham,
    T. J. Morgan, D. W. Mueller, M. P. Stockli, J. A. Tanis, and W. T. Woodland

Momentum Transfer between Projectile and Recoil Ion in Fast Ionizing Proton–Helium Collisions 372
    J. Ullrich, R. E. Olson, R. Dörner, and H. Schmidt-Böcking

First Atomic Physics Experiments with Cooled Stored Ion Beams at the Heidelberg Heavy-Ion Ring TSR 378
    A. Wolf, V. Balykin, W. Baumann, J. Berger, G. Bisoffi, P. Blatt, M. Blum, A. Faulstich, A. Friedrich, M. Gerhard,
    C. Geyer, M. Grieser, R. Grieser, D. Habs, H. W. Heyng, B. Hochadel, B. Holzer, G. Huber, E. Jaeschke, M. Jung,
    A. Karafillidis, G. Kilgus, R. Klein, D. Krämer, P. Krause, M. Krieg, T. Kühl, K. Matl, A. Müller, M. Music,
    R. Neumann, G. Neureither, W. Ott, W. Petrich, B. Povh, R. Repnow, S. Schröder, R. Schuch, D. Schwalm, P. Sigray,
    M. Steck, R. Stokstad, E. Szmola, M. Wagner, B. Wanner, K. Welti, and S. Zwickler

State-Selective Angular-Differential Single-Electron Capture in Very Slow Ar^{4+}–Ar Collisions 384
    C. Biedermann, H. Cederquist, J. C. Levin, R. T. Short, L. Liljeby, L. R. Andersson, H. Rothard, K.-O. Groeneveld,
    C-S. O, C. R. Vane, J. P. Gibbons, S. B. Elston, and I. A. Sellin

l-State Selective Charge Exchange Cross Sections for Collisions of He^{2+} on Atomic and Molecular Hydrogen 390
    R. Hoekstra, F. J. de Heer, and R. Morgenstern

MOLECULES

Chaos in Molecular Rydberg States 398
    M. Lombardi, P. Labastie, M. C. Bordas, and M. Broyer

Recent Developments in Molecular Ion Recombination Research 404
    J. B. A. Mitchell

High-Resolution Electron–Molecule Scattering Using Synchrotron-Generated Electron Beams 410
    D. Field, D. W. Knight, S. Lunt, G. Mrotzek, J. Randell, and J. P. Ziesel

GENERAL

Resonant Recombination and Autoionization in Electron–Ion Collisions 418
    A. Müller

Collisions in, with and on Clusters 430
    J. McCombie and G. Scoles

Quantum Mechanical Reactive Scattering Theory for Simple Chemical Reactions: Recent Developments in Methodology and Applications 442
    W. H. Miller

Harpooning and Chemistry at Surfaces 451
    A. W. Kleyn

Wavepacket Dynamics as a Tool for Calculation of Averaged Photoionization Cross Sections 458
    W. P. Reinhardt and D. R. Kerner

Collisions of Rydberg Atoms with Neutral Particles 466
    V. S. Lebedev

Collisionally Induced Stochastic Dynamics of Fast Ions in Solids 476
    J. Bürgdorfer

Observations of Excited H^0 Atoms Produced by Relativistic H^- Ions in Carbon Foils 487
    A. H. Mohagheghi, P. G. Harris, C. Y. Tang, H. C. Bryant, J. B. Donahue, C. R. Quick, R. A. Reeder, H. Sharifian,
    W. W. Smith, J. E. Stewart, H. Toutounchi, T. C. Altman, and D. C. Rislove

Classical Ghosts in Quantal Microwave Ionization 492
    D. Richards and J. G. Leopold

Recent Progress in Above-Threshold Ionization 499
    P. H. Bucksbaum

Generation of Very High Harmonics of Optical Radiation in Rare Gases 505
    A. L'Huillier, L. A. Lompré, M. Ferray, and G. Mainfray

Photodetachment in Strong Oscillating Fields 513
    D. J. Larson, P. S. Armstrong, M. C. Baruch, T. F. Gallagher,
    and T. Olsson

Atomic Physics in Surface Studies: An Overview 519
    F. B. Dunning
Hydrogen-Surface Electron Transition Rates 529
    P. Nordlander and J. C. Tully

SYMPOSIUM: CORRELATED TRANSFER/EXCITATION AND AUTOIONIZATION

General Considerations 538
    J. A. Tanis
Transfer and Excitation with Heavy Projectiles and Targets 544
    W. G. Graham
Resonant Processes in Atomic Collisions and a Unified View 550
    Y. Hahn
Recombination between Free Electrons and Multiply Charged Ions 556
    P. Hvelplund
RTE of Hydrogen-like and Lithium-like Ions 562
    R. Schuch, E. Justiniano, M. Schulz, P. H. Mokler, S. Reusch, S. Datz, P. F. Dittner, J. Giese, P. D. Miller,
    H. Schoene, T. Kambara, A. Müller, Z. Stachura, R. Vane, A. Warzcak, and G. Wintermeyer
Observation of Electron–Electron Interaction in Collisions of O^{5+} Ions with H_2 Targets 568
    T. J. M. Zouros, D. H. Lee, and P. Richard

SYMPOSIUM: COLLISIONS WITH COLD PARTICLES

Measurements on Very Low-Energy Ion/Atom–Molecule Collisions 574
    G. H. Dunn, M. M. Schauer, and S. R. Jefferts
Theory of Ultracold Atomic Collisions in Optical Traps 580
    P. S. Julienne
Theoretical Treatment of the Associative Ionization Reaction between Laser Excited Sodium Atoms: Energy
Dependence and Anisotropy Effects 586
    A. Henriet and F. Masnou-Seeuws
Collisional Loss Mechanisms in Light-Force Atom Traps 593
    T. G. Walker, D. W. Sesko, C. Monroe, and C. Wieman
Atomic Hydrogen: Gas and Surface Collisions for $T \to 0$ 599
    J. T. M. Walraven
Experiments in Cold and Ultracold Collisions 607
    J. Weiner

SYMPOSIUM: COLLISIONS INVOLVING POSITRONS

Introduction 614
    J. W. Humberston
The Workshop on Annihilation in Gases and Galaxies 616
    R. J. Drachman
Low Energy Positron Hydrogen Atom Scattering Using CCA 622
    A. S. Ghosh, M. Mukherjee, and M. Basu
Measurements of Positron and Electron Total and Elastic Scattering by Atoms 627
    W. E. Kauppila and T. S. Stein
Positron-Impact Ionization of Atomic Hydrogen 633
    W. Raith, B. Olsson, G. Sinapius, W. Sperber, and G. Spicher
Slowing Down of Positrons in Solids 639
    P. J. Schultz, L. R. Logan, W. N. Lennard, and G. R. Massoumi
Theoretical Calculations of Positron Collisions with Atoms 645
    A. D. Stauffer, K. Bartschat, R. I. Campeanu, M. Horbatsch, R. P. McEachran, L. A. Parcell, and S. J. Ward

SYMPOSIUM: SUPERCOMPUTATIONAL COLLISION PHYSICS

Supercomputers and the Future of Computational Atomic Scattering Physics 652
    S. M. Younger
Electron Correlation in the Continuum 658
    C. Bottcher and M. R. Strayer
Ion–Metal and Ion–Atom Collisions: Instant Replays and Mean-Field Theories 663
    J. D. Garcia, N. H. Kwong, and K. J. Schafer
Large Scale Calculations of Electron–Atom/Ion Processes 668
    K. T. Taylor

Author Index 675

Preface

The Sixteenth International Conference on the Physics of Electronic and Atomic Collisions was held in New York from July 25 to August 1, 1989, thirty-one years after the first conference in the series which also took place in New York. The series was initiated in 1958 by Benjamin Bederson and Sidney Borowitz. It attracted 65 participants from 5 countries and a total of 47 papers. The Sixteenth Conference attracted 800 participants from 35 countries and 838 contributed papers, in addition to the invited lectures, progress reports, and review talks. Seven satellite conferences were conducted in association with the Conference.

This volume contains the written versions of most of the invited lectures, review talks, progress reports, and contributed papers selected as hot topics. The keynote lecture was given by Sheldon Datz. Other plenary lectures were given by R. McCray, D. M. Neumark, K. Takayanagi, and K. H. Welge. Symposia were organized on Correlated Transfer/Excitation and Autoionization, Collisions with Cold Particles, Collisions involving Positrons and Supercomputational Collision Physics.

The Sixteenth International Conference on the Physics of Electronic and Atomic Collisions gratefully acknowledges the assistance it has received from a number of foundations, agencies, corporations, and individuals. Specifically the Conference:

- Has received sponsorship and financial support from:

    The International Union of Pure and Applied Physics (IUPAP)
    The National Institute of Standards and Technology

- Has received substantial financial agency assistance from:

    The U.S. National Science Foundation
    The U.S. Department of Energy
    The U.S. Office of Naval Research
    The U.S. Air Force Office of Scientific Research
    The U.S. Defense Advanced Research Projects Agency
    The U.S. Information Agency—Institute of International Education

- Acknowledges financial contributions from:

    Academic Press, Inc.
    Associated Universities Incorporated—Brookhaven National Laboratory
    Balzers
    Brandeis University
    Candela Laser Corp.
    City College of the City University of New York
    Columbia University
    Cornell University
    GTE Laboratories
    Harvard–Smithsonian Center for Astrophysics
    IBM Research Division—T. J. Watson Research Center
    Le Croy Instruments
    McGraw-Hill Book Company, New York, NY
    Newport Corporation
    New York University
    Princeton University
    Questek, Inc.
    Schlumberger–Doll Research
    Sony USA, Inc.
    Spectra-Physics
    State University of New York—Stony Brook
    UHV Instruments
    United Technologies Research Center
    University of Connecticut
    University of Massachusetts
    Vanderbilt University
    Wesleyan University
    Xelon Instrument Sales
    Yale University

The Sixteenth International Conference on the Physics of Electronic and Atomic Collisions expresses appreciation for the important assistance provided by the National Institute of Standards and Technology, the City College of CUNY, and AT&T Bell Laboratories, especially by Ms. Jane Donovan, Ms. Rhina Herrera, and Ms. Beverly Smith.

The Conference also wishes to thank Ms. Nicki Adler and Mr. James Eschinger for the design and production of the official XVI ICPEAC poster and Ms. Marion Shore of the Harvard–Smithsonian Center for Astrophysics for her assistance in the preparation of the Book of Invited Papers.

A. Dalgarno R. S. Freund
P. M. Koch M. S. Lubell
T. B. Lucatorto

March 15, 1990

INTERNATIONAL CONFERENCE ON THE PHYSICS OF ELECTRONIC AND ATOMIC COLLISIONS

ORGANIZATION 1987–1989

EXECUTIVE COMMITTEE

CHAIRMAN
Eugen Merzbacher — USA

VICE CHAIRMAN
Franco A. Gianturco — Italy

SECRETARY
J. S. Risley — USA

TREASURER
Reinhard Morgenstern — The Netherlands

MEMBERS

A. Dalgarno — USA
B. Fastrup — Canada
R. S. Freund — USA
H. B. Gilbody — UK
M. S. Lubell — USA
T. B. Lucatorto — USA
I. E. McCarthy — Australia
W. Mehlhorn — FRG
F. H. Read — UK
I. A. Sellin — USA
M. C. Standage — Australia
N. Stolterfoht — FRG
F. J. Wuilleumier — France

GENERAL COMMITTEE

ARGENTINA
J. E. Miraglia

AUSTRALIA
M. C. Standage

AUSTRIA
H. Winter

BELGIUM
C. J. Joachain

CANADA
L. J. Dubé

DENMARK
N. O. Andersen
B. Fastrup

FEDERAL REPUBLIC OF GERMANY
F. G. Bosch
U. Beck
H. Ehrhardt
H. O. Lutz
M. Mehlhorn
N. Stolterfoht
U. Wille

FRANCE
A. D. Chetioui
F. Masnou-Seeuws
A. Pesnelle
F. J. Wuilleumier

HUNGARY
D. Berényi

INDIA
D. Mathur
A. N. Tripathy

ITALY
V. Aquilanti
F. A. Gianturco

JAPAN
Y. Itakawa
N. Kobayashi

PEOPLE'S REPUBLIC OF CHINA
Zhao-Yong Wang

SPAIN
L. F. Errea

THE NETHERLANDS
R. Morgenstern
B. van Linden van den Heuvell

UNITED KINGDOM
D. S. F. Crothers
H. B. Gilbody
G. King
W. R. Newell
F. H. Read

USA
C. Bottcher
R. J. Celotta
A. Dalgarno
T. F. Gallagher
C. H. Greene
H. Helm
M. S. Lubell
T. B. Lucatorto
J. H. McGuire
E. Merzbacher
F. W. Meyer
J. S. Risley
I. A. Sellin
N. H. Tolk
R. L. Watson

USSR
S. V. Bobashev
Y. N. Demkov
P. Shevelko

YUGOSLAVIA
R. K. Janev

XVI ICPEAC LOCAL COMMITTEE

CHAIRMEN

R. S. Freund
AT&T Bell Laboratories

M. S. Lubell
CUNY, City College

T. B. Lucatorto
National Institute of Standards and Technology

TREASURER

D. Mariani
Schlumberger-Doll

MEMBERS

F. W. Byron, Jr., University of Massachusetts
K. F. Canter, Brandeis University
J. P. Doering, Johns Hopkins University
M. S. Feld, Massachusetts Institute of Technology
W. Happer, Princeton University
S. Hartmann, Columbia University
K. Jones, Brookhaven
P. M. Koch, SUNY Stony Brook
V. O. Kostroun, Cornell University
W. L. Lichten, Yale University
H. H. Michels, United Technologies
T. Morgan, Wesleyan University
A. Pesnelle, Saclay
E. Pollack, University of Connecticut
L. Rosenberg, New York University
W. W. Smith, University of Connecticut
N. N. Tolk, Vanderbilt
P. Vicharelli, GTE Laboratories
J. J. Wynne, IBM Research, NY

Keynote Lecture

ON THE UTILITY AND UBIQUITY OF ATOMIC COLLISION PHYSICS

Sheldon Datz

Physics Division, Oak Ridge National Laboratory, Oak Ridge, Tennessee 37831-6377, USA

> This paper is divided into three parts. In the introduction, we discuss the history and makeup of ICPEAC. In the second part, we discuss the extent of applicability of atomic collision physics. In the third part, we chose one subject (dielectronic excitation) to show the interrelationship of various sub-branches of atomic collision physics

I. INTRODUCTION

From the Preface to Book of Abstracts for ICPEAC II (Boulder, Colorado, 1961):

"This conference is the second in a series of informal meetings organized by a group of workers in the general field of electronic and atomic collisions. The first such meeting was held at New York University in 1958, and we will probably continue to meet at irregular intervals in the future ...."

Benjamin Bederson
Conference Secretary

This is the thirtieth anniversary, actually the thirty first, of ICPEAC. It all began in New York in 1958 with 70-80 people getting together at New York University for two days to discuss their mutual concerns. There were 47 papers, a book of abstracts, a cocktail party at the Fifth Avenue Hotel, and a good time was had by all. Since that time, the conference has grown by more than a factor of ten.

To refresh your memories, the conferences occurred as follows:

I	New York, NY	1958
II	Boulder, CO	1961
III	London, UK	1963
IV	Quebec, Canada	1965
V	Leningrad, USSR	1967
VI	Cambridge, MA	1969
VII	Amsterdam, the Netherlands	1971
VIII	Beograd, Yugoslavia	1973
IX	Seattle, WA	1975
X	Paris, France	1977
XI	Kyoto, Japan	1979
XII	Gatlinburg, TN	1981
XIII	W. Berlin, FRG	1983
XIV	Stanford, CA	1985
XV	Brighton, England	1987
XVI	New York, NY	1989

Committees can vote on policies and invited papers, but participants vote with their interest, their airplane tickets, and their contributed papers. Fig. 1 shows the growth of contributed papers **ab ovo**. The growth is almost continuous with jogs developing at **ca** 1980, because of the accelerated growth of the field in Europe. In its present form, one might expect an asymptote at ~900 - 1000 total contributed papers.

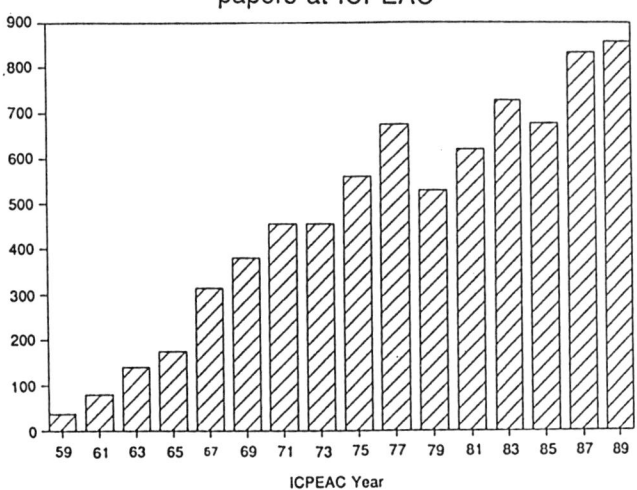

FIG. 1. Total number of contributed papers at ICPEAC.

In Fig. 2, we show the trends in subfields over the last six ICPEACs. Taking an even more detailed look at sub-disciplines and assuming no radical changes in ICPEAC policy or government funding profiles, one might safely predict for example:

- continued but declining dominance of ion-atom collisions
- steady decline in ion/atom-molecule fraction
- growth in high temperature plasma related fields, e.g., ion-ion and electron-ion collisions
- growth in collisions of "exotic" e^+, $\bar{p}$, $\mu^\pm$, etc. species
- possible growth in clusters and solid interactions
- <u>continued</u> growth in photon-atom/molecule collisions.

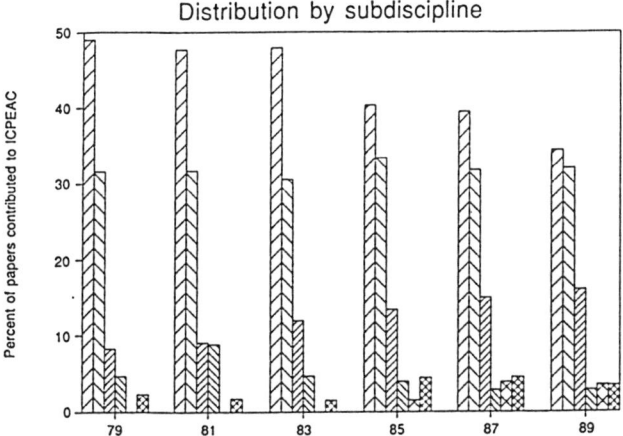

FIG. 2. Distribution of contributed papers among the sub-fields of ICPEAC over the last six conferences. ion/atom-atom/mol; e^--atom/mol.; $h\nu$-atom/mol.; Rydbergs and field assisted; exotic species; solids and clusters.

This latter point is especially worth noting. Since it's humble beginning as two contributed papers in 1965, photon interactions have grown to 130 contributions (16%) at the present meeting. This growth, shown in Fig. 2, is correlated with the development of the new tools, lasers and synchrotron light sources. It is further noteworthy that almost a third of this work involves the interaction of photons and with molecules; neither of these collision partners is mentioned in the present conference title. Evidently the practitioners in this area feel a strong identification with the community of collision physicists, rather than spectroscopists.

II. THE APPLICABILITY OF ATOMIC COLLISION PHYSICS

The need for detailed knowledge of atomic collision physics permeates many fields of science and technology. In nuclear physics, for example, there is a need for information on stopping powers and on the states of energetic ions penetrating solids and gases.[1] In this latter case, the need arises in the study of hyperfine interactions by, e.g., perturbed angular correlations.

A. Condensed Matter Physics

In condensed matter physics, the applications are very broad. They include methods of surface and near surface characterization and materials modification. A number of these are listed in Table I.

You will note that the range of energies used in the atomic probes of solids is quite broad. This breadth is a characteristic of atomic collision physics. In fact, the range exceeds that found in any other branch of science. This is illustrated in Fig. 3 where we show atomic collision physics related branches of science and technology as a function of energy; the range covers 16 orders of magnitude in energy!

Actually, if one wishes, one could include even lower energy interactions such as those involving hyperfine state changes in atomic hydrogen which occur at 1420 MHz = 6×10^{-6} eV (0.1 cal./mol.) in the interstellar medium. But, instead, I have taken, as a lower base, the mean temperature of the interstellar medium 10°K ~ 10^{-3} eV. Since the recent proposal to build THC, a heavy-ion (hadron) collider at CERN, we could extend the upper end to ~10^{15} eV (Pb at 4 TeV/c/amu).

B. Astrophysics

The range of atomic collision energies involved in astrophysics covers the entire extent shown in Fig. 3 and, as was pointed out in Alex Dalgarno's ICPEAC XII paper,[3] *"Most of the knowledge we have about the universe resides in the form of photons. To interpret the message they bring in their journey to us, we must reconstruct the events in which they participated... The processes which produce the photons and the processes which modify them belong usually to the domain of electron, atomic, and molecular physics."*

In Fig. 4, a picture of the sun taken with radiation at 173 Å (±1 Å) is shown. The light emanates mostly from Fe IX and Fe X in the solar corona at ~10^6 degrees (~10^2 eV).[4] Many features in the corona structure (loops, levels, etc.) are visible. It is noteworthy here that a discrepancy of a factor of two in the temperature of the solar corona as measured by the density of gradient of the corona and the Doppler widths of the spectral lines and by abundances of Na-like Fe was settled by Burgess,[5] who introduced dielectronic recom-

Table I. Some Applications of Atomic Collision Physics
to Condensed Matter Science and Technology.[2]

Low Energy Ion Scattering and Atomic Diffraction (0.01 eV - 1 keV)	Top surface composition and structure.
High Energy Ion Scattering (100 keV - 10 MeV)	Atomic composition of surface and near surface layers. Lattice location of impurities in single crystals.
Ion-Induced X-Ray Emission (1 - 100 MeV)	Proton induced X-ray analysis. Heavy ion induced X-ray spectra, atomic environment at lattice site.
Secondary Ion Mass Spectrometry (100 eV - 10 keV)	Sputtered ions are analyzed; depth profiling possible.
Photoelectron and Auger Spectroscopy (50 eV - 1 keV)	Surface and near surface composition.
X-ray Fluorescence Analysis (100 eV - 100 keV)	Atomic composition in near surface region.
Material Modification; Ion-Implantation (100 keV - 10 MeV)	Modification of mechanical, chemical, and electronic properties.
Microfabrication (10 - 100 keV)	Photo-, X ray-, electron beam-, and ion beam-lithography
Radiation Biology (1 eV - 10 GeV)	Interactions of ionizing radiation and particles with biological systems (also heavy ion therapy).

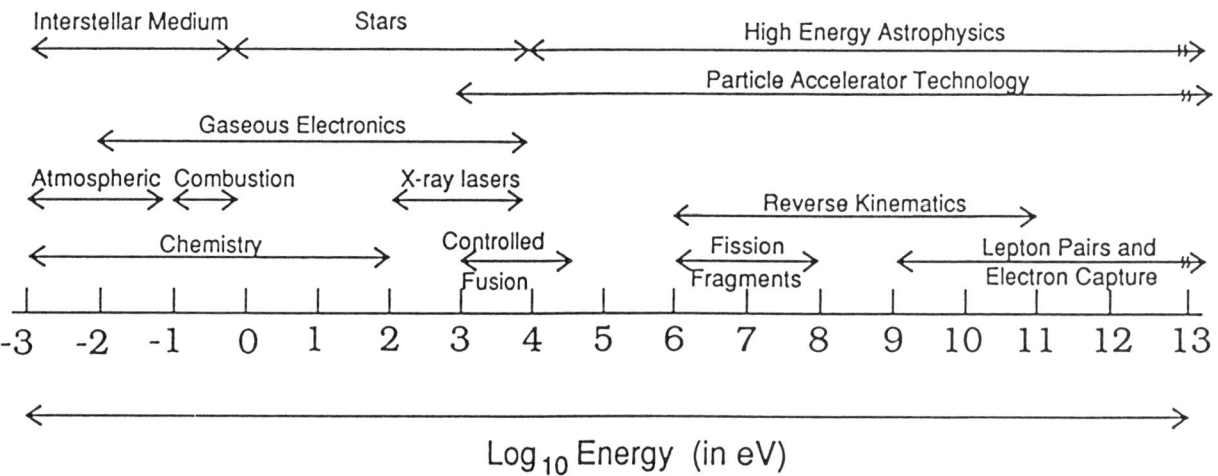

Fig. 3. Some of the sciences and technologies which require atomic collision physics input as a function of the relevant energy range.

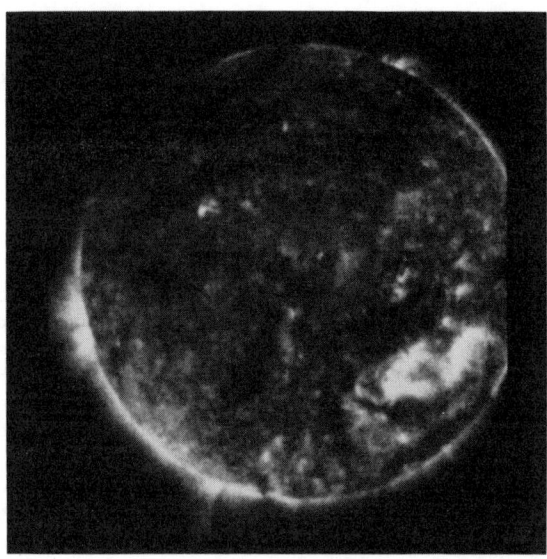

Fig. 4. The solar corona as photographed in the emission of the resonance lines of Fe IX at 171 Å and Ge X at 174.5 Å. The dark bands visible at northern latitudes are due to overlying cool prominences. The image corresponds to coronal structures from $\sim 0.8 \times 10^6$ to 1.4×10^6 K.

bination as an additional mechanism to reduce the mean ionic charge.

A picture of the Cygnus loop,[6] taken with soft (8 — 80 Å) X rays, is shown in Fig. 5. The Cygnus loop is a supernova remnant which has expanded to the point where we can look at detailed structure. Another area is revealed in a photograph (Fig. 6) taken with visible light coming from a cooler shell. In the visible, the oxygen rich supernova remnant will look very green in a color photograph because of the large contribution from the 4959 Å and 5007 Å O III lines. Information of the sort presented in these figures is vital to all theories of Stellar evolution.

At ICPEAC XVI, we shall have two plenary talks on collision physics in relation to astrophysics. At the low energy end, a lecture on "Low Energy Molecular Collisions with Applications to Interstellar Cloud Problems" by K. Takayanagi, and a lecture covering a recent very notable higher energy event "Atomic and Molecular Processes in Supernova 1987A" by R. A. McCray.

C. Chemically Reactive Collisions

Chemically reactive collisions, and here I include ion-molecule reactions, also covers a broad range on a logarithmic scale of energies.

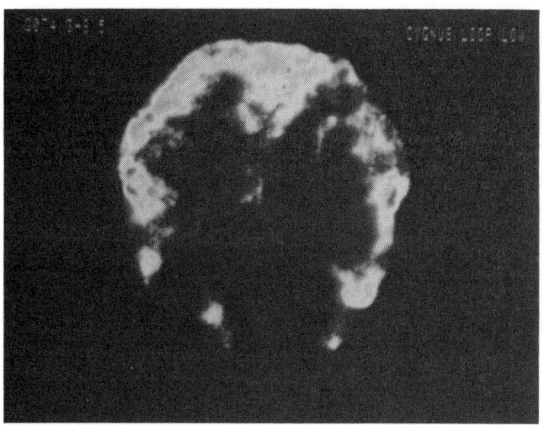

FIG. 5. The Cygnus loop photographed using its emitted soft X-ray radiation.

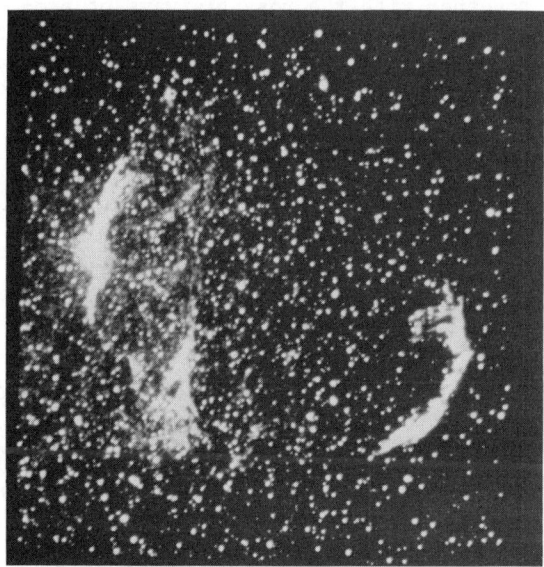

FIG. 6. The Cygnus loop photographed using its emitted green O^+ light.

The coldest to my knowledge being the observation of the reactions

$$He^+ + H_2 \rightarrow He + H + H^+$$
$$\rightarrow He + H_2^+ + h\nu$$

at 10°K (10^{-3} eV) in an ion trap by a group at JILA.[7] It is noteworthy that the reaction of $He^+ + H_2$ has been studied at energies up to 1 MeV, hence 9 orders of magnitude. An entire symposium on "Collisions with Cold Particles," chaired by H. Metcalf, will be presented at this conference.

The upper portion of the energy range belongs to hot atom chemistry and ion molecule reactions.

The subject of chemical kinetics was liberated from the bulb and the test tube in the mid 1950's with the successful application of molecular beam techniques.[8] Two papers on the subject appeared at ICPEAC II in 1961. One was on "Reactive Scattering of Velocity Selected K Atoms" by D. Beck, E. F. Greene, and J. Ross,[9] and the other entitled "Reactive Scattering in Crossed Molecular Beams" by G. H. Kwei, J. A. Norris, J. L. Kinsey, and D. R. Hershbach[10] discussed the reactions of CH_3I + K, Rb, and Cs. This was, in fact, the first publication from the Hershbach group.

By the mid-1970's, the scope of the technique was already permitting the investigation of state-to-state chemically reactive collisions as illustrated in Fig. 7, which is taken from a 1973 review by Hershbach.[11] It illustrates the range of reactions available for study at the time together with the possible ways of preparing specific reagent states and analysis of product states. The enormous progress that has been made since these early days is exemplified at this conference by the plenary paper of Y. T. Lee on "Molecular Beam Studies of Chemical Reactions" and the review paper by W. H. Miller on "Recent Developments in the Theory and Application of Quantum, Scattering Theory for Chemical Reactions."

D. Atmospheric Physics

Atomic collision processes control the composition of the upper atmosphere and, to a growing extent, also the lower atmosphere. One has to be totally illiterate not to be aware of the threats to the lower atmosphere by the massive production of polutants such as NO_x and O_3 and threats to the upper atmosphere by the even increasing CO_2 concentration with its attendant "Greenhouse Effect" and by the disappearance of the protective ozone layer which is being attacked by free chlorine atoms photolytically produced from chloro-fluoro carbons. The potential ozone disaster was first described by S. Rowland, a hot-atom chemist from the

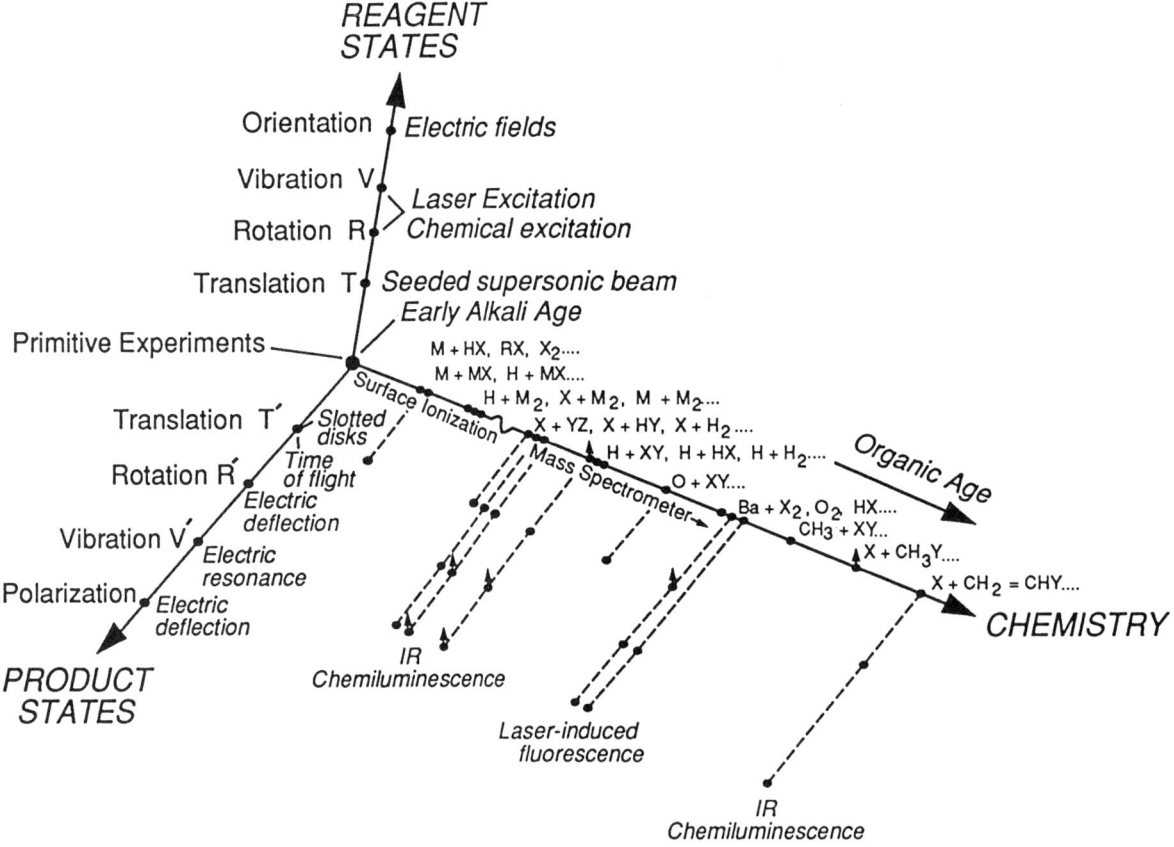

FIG. 7. Domains accessible to molecular beam chemistry, circa 1973.

University of California at Irvine almost twenty years ago. The effect is explained by a simple series of binary atom molecule collisions. To determine whether chloro-fluro carbons were truly a threat or whether competing natural processes would mitigate the effect required careful measurements of a whole series of reaction cross sections. Needless to say, the impact of these many threats on the chemical industry, the worlds economy, methods of energy generation, conservation, etc., are potentially enormous and contributions to knowledge by members of our craft may do much to help find solutions to alleviate these problems.

On a more benificent note, we can enjoy the collisional interactions of electrons exciting the oxygen red line at 6300 Å as they swirl in the earth's magnetic field[12] and create the red aurora borealis via the $2p^4\ ^1D \rightarrow 2p^4\ ^3P$ transition (Fig. 8).

Fig. 8. The red aurora.

E. X-ray Lasers

As we move up in energy, we enter the region of incompletely ionized plasmas and hence the area of gaseous electronics. This is an area of enormously broad utility ranging from such mundane things as gas discharge lighting of our streets at the low energy end to the presently exotic lasers at the higher end.

As is the case with any normal gas laser, the photons arise from the stimulation of transitions of outer shell electrons with inverted populations.[13] Since the energy levels of highly ionized species are further apart, the product photons are in the X-ray region. The population inversion needs pumping, and, because of the short radiative lifetimes involved, this creates special problems. In general, X-ray laser schemes utilize atomic collision processes to pump the inversion. The two inversion schemes presently in use are recombination to inverted states in a cooling plasma and free-electron collisional pumping. In particular, I should like to note that it was necessary to develop a dielectronic recombination theory to explain the kinetics responsible for the strong amplification in the $3p \rightarrow 3s$ $(J = 2 \rightarrow 1)$ line of neon-like selenium[14] (see below). A review talk on the "Atomic Physics of Soft X-Ray Lasers" will be presented at this conference by A. V. Hazi.

F. Fission Fragments, Tracks and Stopping Powers

More complex is the understanding of the interaction of energetic heavy ions, such as fission fragments with the solids into which they recoil and, in general, the physics of heavy particle tracks. Very recent studies by Schmidt-Böcking and co-workers and by Ron Olson on the details of inelasticity and secondary electron production in such collisions now show great promise to finally understand the phenomena on an atomic basis.[15]

G. Fusion

Even a cursory discussion of the need for atomic collision data in the development of magnetically confined controlled fusion plasmas, as well as inertially confined plasmas, could easily take up this entire talk. Numerous reviews and conferences have been devoted to just this subject. Suffice it to say here that any and all information on collisional processes of electrons, hydrogen atoms, hydrogen molecules, He ions, totally and partially stripped heavy ions in the energy range of a few eV (plasma edge) to tens of keV (plasma center) are of immediate interest to this extremely important technology of the future. Immediately relevant invited papers at this conference include those by A. Báránýi, P. Hvelplund, R. Schuch, E. Salzborn, and L. Shmaenok.

H. Accelerator Technology

Particle accelerators are instruments with a much broader range of uses than is generally attributed to them (i.e., nuclear and particle physics). One need only peruse the proceedings of the latest conference on the "Application of Accelerators in Research and Industry"[16] to find in addition to 52 papers on accelerator technology; 95 papers on atomic physics and related phenomena; 20 papers on proton induced X-ray analysis (PIXE) and ion microprobes; 36 papers on materials analysis facilities, accelerator mass spectroscopy, Rutherford backscattering and channeling, nuclear reaction analysis, resonant ionization spectroscopy; and 13 papers on radiation therapy, neurosurgery with ion beams (this is only a partial list of papers).

Applications of atomic collision physics to accelerator technology actually begin with ion sources of which many are properly in the domain of gaseous electronics. Negative ion production is important for tandem Van de Graaffs[17] as is multicharged ion production for heavy particle accelerators. In this regard, a number of these multicharged ion sources, such as Electron Cyclotron Resonance Ion Sources (ECR), Electron Beam Ion Sources (EBIS), and the newly developed Electron Beam Ion Trap (EBIT) are themselves powerful tools with application to atomic collision physics.

Accelerated ions are often stripped of electrons by collisions in either gaseous or thin solid foils as in tandem electrostatic accelerators for further acceleration at higher charge states, or to achieve lower magnetic rigidity for injection into boosters. A knowledge of the electron capture and loss processes, and equilibrium charge states at the relevant energies is vital for the design parameters of these systems.[18]

Especially in synchrotrons where ions travel enormous pathlengths and especially during the acceleration phase, particles can be lost because of ionization or charge capture with background gas. Hence the requisite vacuum conditions and their associated costs are determined from a knowledge of charge-transfer cross sections.

I. Pair Production and Electron Capture

As we accelerate to even higher energies, the standard charge-changing cross sections decrease rapidly, but if we wish to take advantage of the available currents, and the kinematics of storage ring colliders, such as the proposed Relativistic Heavy Ion Collider (RHIC) at Brookhaven or the Hadron Collider at CERN, we encounter some new atomic physics phenomena namely lepton pair creation (e.g., electron-positron pairs) and negative lepton capture from the negative continuum.

To quote from a recent paper by Bottcher and Strayer[19] *"One of the most useful modern probes of hadronic matter is the associated production and decay of lepton pairs during a collision. In such collisions, lepton-hadron final state interactions are usually small, and hence the leptons carry direct information on the space-time region of creation. Historically, lepton pair production has been an important tool in collider experiments, in part, because of the special relationship between deep inelastic lepton-hadron scattering and the large mass Drell-Yan processes provide complementary information on quantum chromodynamics (QCD) in the asymptotic regime. From these experiments have arisen new ideas and phenomena: scaling, chiral and flavor symmetry, charm, and a quantitative understanding of a rich meson and baryon spectroscopy."*

The problem is that the cross section for lepton creation by non-nuclear electromagnetic (i.e., atomic physics) processes is larger than that for the desired process; hence a detailed knowledge of the cross section for their production, energy and angular distributions. The process is simply described as the formation of a virtual photon which decays by pair production. The venerable and elegant Weizaker-Williams approach to the problem is both approximate and perturbative and does not directly yield angular and energy distributions. More exact perturbative calculations have recently been carried out[19] and they, of course, differ somewhat from the Weizaker-Williams result. More disturbing, however, are the results of a non-perturbative calculation,[20] (see Fig. 9) which predicts differences of as much as factors of, e.g., 100 at RHIC energies (100 GeV/nucleon).

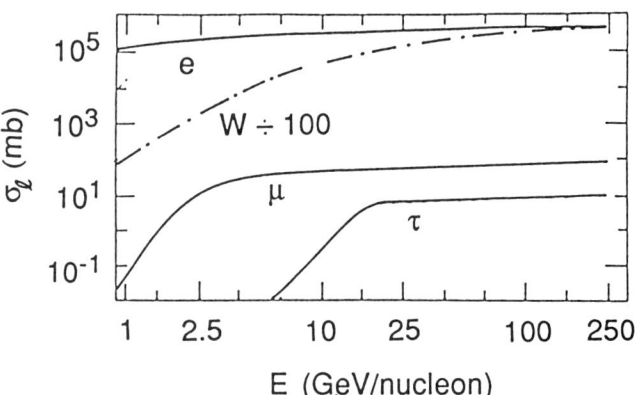

FIG. 9. Predicted cross sections (millibarns) for $e^{\pm}$, $\mu^{\pm}$, and $\tau^{\pm}$ pair production for $U^{92+} + U^{92+}$ collisions vs γ. For RHIC, $\gamma = 100$ for CERN SPS (200 GeV/c/amu) on a fixed target $\gamma = 10$. The solid line for $e^{\pm}$ from nonperturbative calculation is compared with the Weizaker-Williams calculation (W ÷ 100).

Since the pair is created in the immediate vicinity of the projectile nucleus, there is a strong possibility that the electron (negative lepton) may be captured. Hence the ionic charge changes and, if this occurs in a storage ring, the particle is lost. In fact, this process may be the limiting factor for containment times in such devices. Some detailed, but perturbative calculations have recently been carried out (Fig. 10),[21] but thus far no experiments have been performed. Clearly, this possible poison for particle physics poses a very interesting problem in atomic physics and proposals have been mounted for experiments on $e^\pm$ pair production and electron capture at SPS (200 GeV/amu - CERN), the AGS (20 GeV/amu - Brookhaven), and Bevelac (1 GeV/amu - Berkeley).

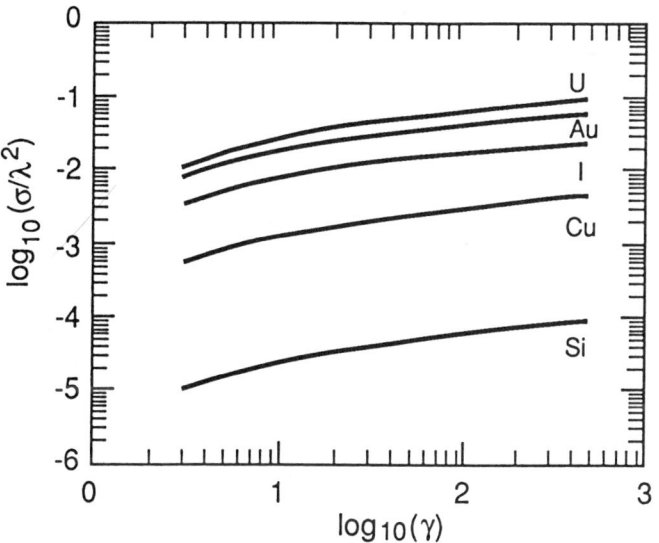

FIG. 10. Capture cross sections for symmetric $A_z + A_z$ collisions, scaled with respect to $\lambda^2 = 1.49$ Kb. Curves correspond as labeled, to the ions $A(z) = Si(14)$, $Cu(29)$, $I(53)$, $Au(79)$, and $U(92)$, e.g., for U^{92+} at RHIC energies $\sigma_c \sim 100$ b.

J. Reverse Kinematics

The remaining bar on Fig. 3 "reverse kinematics" is not a science or a technology, but a name applied by nuclear physicists to a technique in which one uses high laboratory energy techniques to study collisions at lower center-of-mass energies; for example, an accelerated carbon ion is shot at a hydrogen target rather than a proton at a carbon target. Such kinematic tricks are in extensive use by atomic collision physicists as we shall see below.

III. DIELECTRONIC PROCESSES IN ELECTRON-ION, ION-ATOM, AND ION-SOLID COLLISIONS

Up to this point, we have been discussing the general utility of atomic collision physics in other fields of endeavor. At this point, I would like to become more specific and investigate the interaction between various and seemingly diverse branches of atomic collision physics in investigating a single problem, i.e., dielectronic excitation.

First, we will consider dielectronic recombination which occurs in collisions of ions with free electrons. This process is the dominant recombination mechanism in hot heavy-ion plasmas and, as we have indicated above is important to understand for such diverse applications as X-ray lasers and coronal temperatures. Second, we will examine resonant electron transfer and excitation (RTE) which occurs when an ion collides with an almost free electron weakly bound to an atom. Finally, we will take a look at recent studies of dielectronic excitation in crystal channels which have been shown to behave as dense Fermi electron targets.

Dielectronic recombination (DR) is initiated when a continuum electron excites a previously bound electron and in so doing loses just enough energy to be captured itself into a bound state $(n\ell)$. The latter process results in a doubly excited ion (dielectronic excitation) in the next lower charge state which may either auto-ionize or emit a photon resulting in a stabilized recombination (Fig. 11). Thus, for an ion A of charge state q in initial state α, the DR process may be written:

$$A^{q+}(\alpha) + e^-(k,\ell') \rightleftarrows [A^{(q-1)+}(\beta,n\ell)]^{**} \rightarrow [A^{(q-1)+}(\alpha,n\ell)]^* + h\nu. \quad (1)$$

It should be noted here that the first step, i.e., the formation of the dielectronically excited state is the reverse of the Auger process. Experimental results in DR were first reported in 1983 at the Berlin ICPEAC in a "hot topic" symposium. Work has proceeded apace since then but two techniques which are giving qualitative improvements in the data have just come into fruition; these will be reported in invited talks by P. Hvelplund, R. Schuch, and A. Wolf. Both techniques utilize a merged electron-ion beam technique,[22] one in a single pass experiment[23] and the second as part of a cooled heavy-ion storage ring.[24]

The Aarhus experimental setup is described in a paper by P. Hvelplund in this volume. In the example we discuss here,[23] a beam of 20-MeV O^{6+} ions is merged with an electron beam

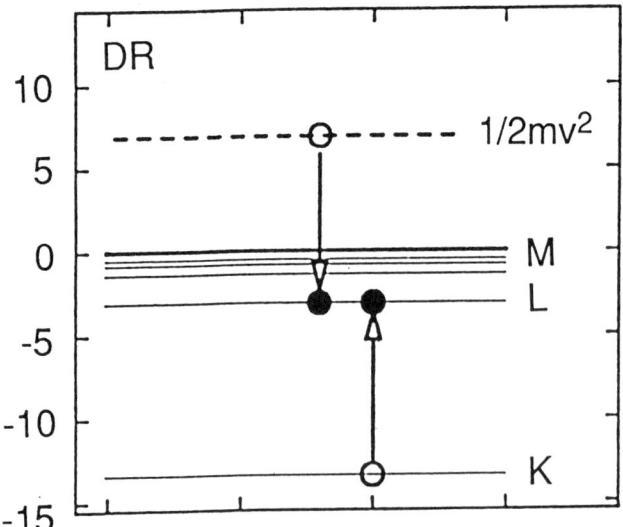

FIG. 11. Dielectronic excitation of a hydrogenic ion to a KLL state. Radiative relaxation leads to dielectronic recombination.

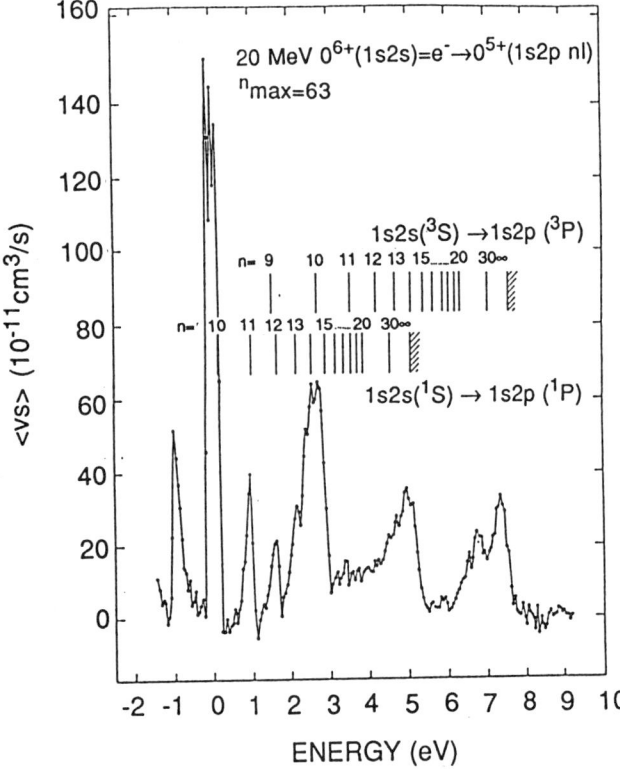

Fig. 12. Dielectronic recombination spectrum of O^{5+} (1s2s) obtained in a single pass merged electron-ion beam experiment (see Ref. 23).

whose velocity is varied to give relative velocities in the 0 — 20 eV range. The resultant recombination (to form O^{5+}) spectrum shown in Fig. 12 arises from the 1s2s metastable component of the beam via the transitions 1s2s(^{1}S) → 1s2p(^{1}P) and the 1s2s(^{3}S) → 1s2p(^{3}P) and (^{1}P) states. Note that although the ion beam energy in the laboratory is 20 MeV, the center-of-mass energy ranges from 0 — 10 eV with a resolution of ~0.15 eV.

The completion of heavy ion storage ring projects will lead to many new and dramatic advances in heavy ion atomic physics. The merged electron beam exists as an integral part of the ring where it is used for "cooling" the stored beam.[24] One advantage of the storage ring is the enormous increase in effective current and attendant luminosity obtained by circulating the same particles through the thin target at frequencies of up to a megahertz. The first results on DR obtained from the Heidelberg ring[23] (the first of the completed projects) will be discussed in invited papers by R. Schuch and by A. Wolf.

Another experiment which gives related results involves electron transfer plus excitation (TE) in ion-atom collisions. In this case, instead of a truly free electron, we use an atomic target which has weakly bound (almost free) electrons with orbital velocities $v_e \ll v_i$, where v_i is the ion velocity. This process has been dubbed "Resonant Transfer and Excitation" (RTE).[25] Here, e.g.,

$$A^{q+}(\alpha) + He \to [A^{(q-1)+}(\beta,n\ell)]^{**} + He^+ \qquad (3)$$

the doubly excited state can then relax via Auger or radiative decay

$$[A^{(q-1)+}(\beta,n\ell)]^{**} \to A^{q+}(\alpha) + e_A$$

$$\to [A^{(q-1)+}(\alpha,n\ell)]^* + h\nu_1 . \qquad (4)$$

Further, the singly excited state can relax with the emission of a second X ray

$$[A^{(q-1)+}(\alpha,n\ell)]^* \to A^{(q-1)+}(\alpha,\alpha') + h\nu_2 . \qquad (5)$$

The end results are either an Auger electron which contains information on the $n\ell$ state of the originating doubly excited state, or an ion of decreased charge and two photons. Experiments have been carried out by measuring the energy dependence of the Auger electron spectra (RTEA); charge capture in coincidence with X rays (RTEX); or two X rays in coincidence (RTEXX). In all cases, the relative collision energy is scanned by varying the energy of the ion beam. RTE will be discussed in papers by Graham, Hahn, and Schuch in a symposium chaired by J. Tanis, who pioneered this technique.

As an example, consider the results of one experiment carried out at the Berkeley Super HILAC by Graham et al.[26] shown in Fig. 13. The ion used was He-like Ca^{18+} and the target was H_2. Here we can see a significant effect of RTE on the total electron capture cross section ($\sigma_{q,q-1}$). The effect is much enhanced in the measurement of a K-X ray in coincidence with the formation of Ca^{17+}. The lower energy bump, in this case, corresponds to a KLL excitation. The width of the peak is the result of a fold of the momentum distribution (Compton profile) of the bound electrons in the H_2 molecular target with the KLL resonance lines.

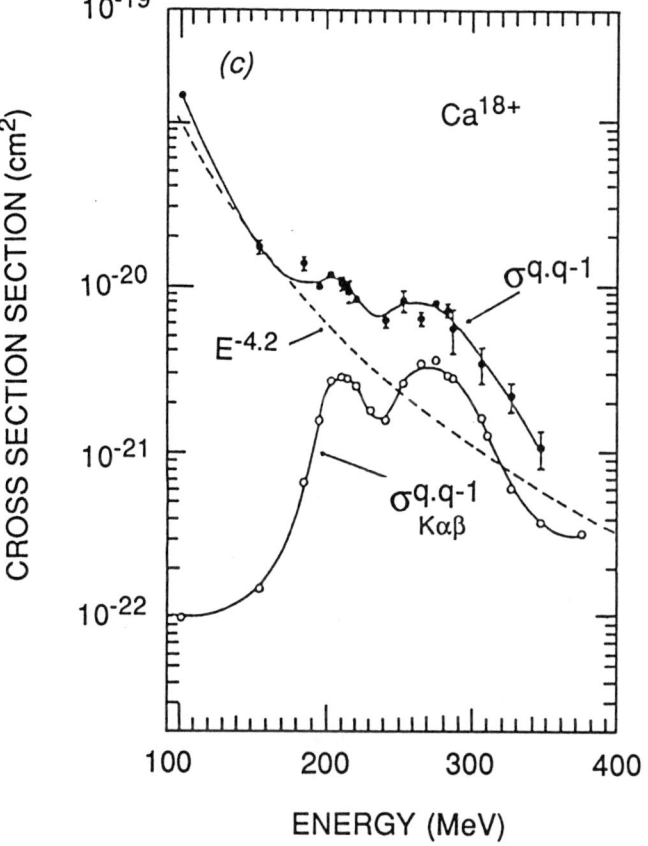

FIG. 13. Upper curve, measured capture cross section for Ca^{18+} on H_2 showing contributions from RTE. Lower curve, capture measured in coincidence with calcium K X ray (see Ref. 26).

Dielectronic excitation processes have now also been studied in crystal channels. Energetic ions traveling through crystals at small angles to low index directions may be steered to avoid small impact parameter collisions with the atomic cores of the lattice atoms ("channeling) and interact directly only with loosely bound electrons.[27] For ion velocities $v_i \gg v_f$, where v_f is the velocity of the target Fermi electrons, the penetrating ion may be viewed as being bombarded by a flux of electrons moving at velocity v_i.

Two invited papers at the present conference deal with consequences of this effect: one by J. C. Poizat on "Energy Loss and Charge Exchange Processes of High Energy Heavy Ions Channeled in Crystals" and the second by N. Claytor on the "Measurement of Electron Impact Ionization of $U^{88+} - U^{91+}$" in which ~500 MeV/amu ions are made to collide with the electrons in a crystal channel ("reverse kinematics"). If this assumption is quantitatively correct, ions traveling through this medium at velocities equivalent to the electron velocities required for sharply varying processes, such as dielectronic or direct excitation of an electron bound to the ion, should experience events similar to those in a hot dense ($\gtrsim 10^{23}$ electrons/cm^3) plasma, but with a relatively narrow electron energy distribution, i.e., a Fermi distribution as against a Maxwell-Bolzman distribution at the temperature necessary to carry out these excitation processes.

In vacuum under single collision conditions, the doubly excited state formed by dielectronic excitation would decay, as above, either by an Auger process or by radiative stabilization via two photons. The final result being the production of an ion of reduced charge state and two photons. However, if the state is created in a dense electron medium (i.e., a dense plasma or a crystal channel), secondary collisional processes leading to further excitation and ionization can come into play and may even dominate.

Take, as an example, the dielectronic excitation of a hydrogenic ion to a $[2p^2]^{**}$ KLL resonance (see e.g., Fig. 11). This state could be collisionally ionized to $[2p]^*$ or it could be excited to a $[2p3\ell]^{**}$ state. The excitation cross section from a given n state to an (n + 1) state is larger than that for direct ionization, but the ionization cross sections for high n states are so large that for n>3 in the cases under consideration,[28] the electron can be considered as removed from the ion by either excitation or ionization. The remaining 2p electron can then either radiate to 1s or undergo a similar collisional excitation or ionization. Thus, one possible path leads the ionization to form a bare nucleus at single collision energies below the first excitation potential. As an example, the results for Ca^{19+} ions (150 – 330 MeV) are shown in Fig. 14. The Ca^{19+} ion beam passed through the ⟨110⟩ axis of a 1.2 μm thick silicon

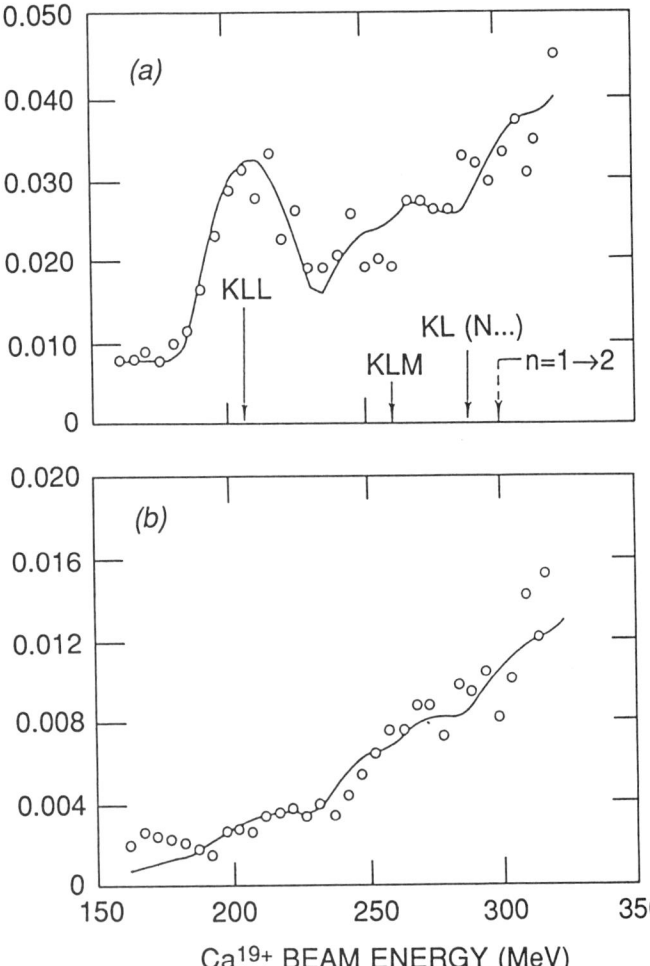

FIG. 14. (a) Yield of H-like K_α calcium X rays and (b) charge fraction of Ca^{20+} ions as a function of Ca^{19+} ion energy incident on a ⟨100⟩ channel in Si (1.2 μm thick) the smooth curves are calculated from the appropriate rates.

crystal. Measurements were made of (a) the emerging charge-state distributions using electrostatic deflection and a solid-state position-sensitive detector, and (b) the X-ray spectra using a Si(Li) detector which can resolve H-like K_α from the He-like K_α lines.

In Fig. 14a, the yield of H-like K_α X rays is shown as a function of ion energy. The features correspond to dielectronic excitation to KLL, KLM, KLN, etc., and at 300 MeV to direct 1s → 2p excitation. The calculated absolute magnitudes of these contributions are obtained from the appropriate rate equations containing radiative and Auger rates and collision cross sections. They are then folded with the appropriate Fermi distribution. Using the same rate coefficient as those used for the X-ray yields shown in Fig. 14a, the yield of bare Ca^{20+} ion is calculated and compared with the data in Fig. 14b. (Dielectronic excitation in channels is discussed in contributed papers 137 and 138).

IV. CODA

In this paper, I have tried to point out not only some features of the utility and ubiquity of atomic collision physics, but also the response of the ICPEAC community in meeting the challenges created by the needs of the general scientific and technological community. The great utility of atomic collision physics is, however, both a strength and a weakness; the weakness being the perception of our field as an adjunct of other disciplines toward which its results are applicable rather than as a separate discipline in and of itself.

The object of the latter part of the talk was to point out the close inter-relationship between seemingly disparate branches of our science, i.e., ion-atom, electron-ion, and ion-solid collisions. All of these are presently represented at ICPEAC. This demonstration was, in part, intended as a response to those who believe that ICPEAC is now too large and ought to be broken up into separate subdiscipline conferences .... When I was a lad growing up in New York, the New York Yankees' baseball team was by far the best in the world habitually winning the "American League Pennant" and the "World Series." Fans who favored other teams were heard to shout "Break up the Yankees!" Break up the Yankees? .... Never!

This research was sponsored by the U.S. Department of Energy, Office of Basic Energy Sciences, Division of Chemical Sciences under Contract No. DE-AC05-84OR21400 with Martin Marietta Energy Systems, Inc.

REFERENCES

1. See e.g., Atomic Physics in Nuclear Experimentation in Annals of the Israel Physical Society 1, B. Rosner and R. Kalish, eds., 1977.

2. Applied Atomic Collision Physics, H. S. W. Massey, E. W. McDaniel, and B. Bederson, eds., Vol. 4 Condensed Matter, S. Datz, ed., Academic Press, New York, 1983.

3. A. Dalgarno, "Applications of Atomic Collision Physics to Astrophysics" in Physics of Electronic and Atomic Collisions, S. Datz, ed., North-Holland Press (1981), pp.

4. A. B. Walker, T. W. Barbee, R. B. Hoover, and J. K. Lindblom, Science 241, 1781 (1988).

REFERENCES (contd)

5. A. Burgess, Astrophys. J. 139, 776 (1964).

6. W. P. Blair, J. C. Raymond, J. Danziger, and F. Matteucci, Astrophys. Journal 338, 812 (1989).

7. M. Schaer, S. Jefferts, S. Barlow, and G. Dunn, J. Chem. Phys. (in press).

8. E. H. Taylor and S. Datz, J. Chem. Phys. 23, 1711 (1955).

9. D. Beck, E. F. Greene, and J. Ross, Proceedings, Second Int'l Conf. on the Physics of Electronic and Atomic Collisions, Books of Abstracts, W. A. Benjamin and Company (1961), p. 94.

10. G. H. Kwei, T. A. Norris, I. L. Kinsey, and D. R. Herschbach, Proceedings, Second Int'l Conf. on the Physics of Electronic and Atomic Collisions, Book of Abstracts, W. A. Benjamin and Company (1961), p. 98.

11. D. R. Herschbach, Farad. Dis. Chem. Soc. 55, 233 (1973).

12. See e.g., D. R. Bates, "Airglow and Auroras" in Applied Atomic Collision Physics, H.S.W. Massey, E. W. McDaniel, and B. Bederson, eds., Academic Press 1, 152-200 (1982). [Photo by Al McNeal, University of Alaska]

13. See e.g., P. Jaeglé, J. Physique 48, C9 (1987).

14. B. L. Whitten, A. U. Hazi, M. H. Chen, and P. L. Hagelstein, Phys. Rev. A 33, 2171 (1986).

15. See e.g., U. Ramm, S. Schmidt, H. Schmidt-Böcking, G. Kraft, and R. E. Olson, GSI Scientific Report No. ISSN 0174-0184, p. 205 (1988).

16. Nucl. Instrum. Meth. Phys. Res. B 40/41, (1989).

17. G. D. Alton, c.f. Ref. 2, pp. 44-171.

18. H.-D. Betz, c.f. Ref. 2, pp. 2-43.

19. C. Bottcher and M. Strayer, Phys. Rev. D 39, 1330 (1989).

20. C. Bottcher and M. R. Strayer, Nucl. Instrum. Meth. Phys. Res. B 31, 122 (1988).

21. M. Rhoades-Brown, C. Bottcher, and M. R. Strayer, Phys. Rev. A (in press).

22. P. F. Dittner, Physica Scripta T22, 65 (1988).

23. L. H. Andersen, H. Knudsen, P. Hvelplund, and P. Kvistgaard, Phys. Rev. Lett. 62, 2656 (1989).

24. R. Schuch, Nucl. Instrum. Meth. Phys. Res. B 24/25, 11 (1987).

25. J. A. Tanis, E. M. Bernstein, W. G. Graham, M. Clark, S. M. Shafroth, B. M. Johnson, K. W. Jones, and M. Meron, Phys. Rev. Lett. 47, 1325 (1982).

26. W. G. Graham, E. M. Bernstein, M. W. Clark, J. A. Tanis, K. H. Berkner, P. Gohil, R. J. McDonald, A. S. Schlachter, J. W. Stearns, R. H. McFarland, T. J. Morgan, and A. Müller, Phys. Rev. A 33, 3591 (1986).

27. "Heavy Ion Channeling," S. Datz and C. D. Moak, in Heavy Ion Science, Vol. 6, D. A. Bromley, ed., Plenum Press, 1985, pp. 169-240.

28. S. Datz, C. R. Vane, P. F. Dittner, J. P. Giese, J. Gomez del Campo, N. L. Jones, H. F. Krause, P. D. Miller, M. Schulz, H. Schöne, and T. M. Rosseel, Phys. Rev. Lett. (in press).

Plenary Lectures

CHAOS IN ATOMS ?

K.H. WELGE

Fakultät für Physik, Universität Bielefeld,

D-4800 Bielefeld, FRG

INTRODUCTION

Stimulated by the discovery of the so called quasi-Landau resonances in the absorption cross section of earth alkalide atoms by Garton and Tomkins in 1969 [1], first explained theoretically by Edmonds [2], the highly excited atoms in external static magnetic fields of laboratory strength has been a subject of much theoretical [3-9] and experimental [3, 4, 10-14] research. However, even the hydrogen atom, with its pure Coulomb field the natural prototype case, has remained for a long time an unsolved problem, theoretically and experimentally.

In uniform magnetic fields the atom is described by the Hamiltonian

$$H = \frac{1}{2} p^2 - \frac{1}{r} + \frac{1}{2} \gamma L_z + \frac{1}{8} \gamma^2 \rho^2 \qquad (1)$$

in cylindrical coordinates with the magnetic field axis along the z-direction and with $\gamma = B/B_0$, $B_0 = 2.35 \times 10^5$ T, the magnetic field parameter. Since the angular momentum L_z ($= \hbar m$) is a constant of motion the Schrödinger equation is non-separable (and non-integrable) in two degrees of freedom (ρ, z). A quantum mechanical solution of the problem has been presented by Clark and Taylor [15] in the perturbative l- and n-mixing regime at B = 4,7 T and energies up to $E \leq -30$ cm^{-1}. However, in the most important and challenging strong field mixing (or quasi-Landau) regime around the ionization limit, where the Coulomb field ($-1/r$) and the diamagnetic interaction ($1/8 \gamma^2 \rho^2$) are the diamagnetic hydrogen atom with two non-integrable degrees of freedom is the known Poincare-surface-of-section method.

With the finding of the diamagnetic hydrogen atom behaving classically chaotic this system moved at the center of the longstanding and much debated question of "quantum chaos" concerning the correspondence between quantum and classical mechanics when the quantum system is chaotic in the classical limit. A definition of chaos in quantum mechanics is at present not available - and may principally not be possible. What one is left with instead is the question of whether there exist characteristic signatures and phenomena in the semiclassical or quantum mechanics of a system when its classical counterpart changes from regular to irregular dynamics. Signatures of "quantum chaos" can be found, if at all, in properties of the quantum system that is in the present case the spectrum of the atom in the strongly field mixed quasi-Landau region.

What causes the particular interest in and the significance of the magnetized hydrogen atom is the fact that it constitutes a real and, with its two non-integrable degrees of freedom, at the same time a most simple system that is amenable to detailed theoretical as well as experimental investigations. Another important property is that the degree of irregularity can be externally controlled and monitored through its dependence on the excitation energy and the field strength. On the other hand, other real systems like molecules, nuclei or solids are, with their many degrees of freedom, to complex for detailed studies of their classical and quantum mechanics. Also, they cannot be manipulated with respect to their irregularity by easily controllable external parameters. Simple chaotic systems with few degrees of freedom, like the stadium or Sinai billiards [22], are purely theoretical models.

of comparable magnitude, the problem has remained unsolved theoretically until most recently, when large-scale numerical solutions of the Schrödinger equation have been carried out at higher and higher energies [6, 8], finally up to the ionization limit E = 0 [16]. Also, no experiments had been performed with hydrogen atoms, which was thought to be the basically important prototype system [17, 18]. In fact, until about the mid 80's it was generally accepted that the structure and dynamics of atoms in the quasi-Landau regime was essentially represented and governed by the Garton-Tomkins-Edmonds (GTE) quasi-Landau resonance pattern only.

In addition to being a fundamentally important quantum mechanical problem of one-electron atoms the strongly perturbed highly excited Rydberg hydrogen atom in uniform laboratory magnetic fields has gained additional attention well beyond atomic physics after it became known in the early 80's that the motion of the electron in the combined Coulomb and diamagnetic fields turns more and more irregular as the excitation energy approaches the ionization limit [19, 20]. This was revealed by classical trajectory calculations performed first by Edmonds and Pullen [21], Robnik [19], and Harada and Hasegawa [20] and in later years confirmed by extensive and detailed further theoretical studies [6, 7, 8]. The essential result is that, for example in a field of 6 T, the regular regime extends up to an energy of about -100 cm^{-1} from where the fraction of phase space with irregular motion gradually increases reaching, roughly speaking, complete chaos at about -20 cm^{-1}. In units of scaled energy, $\tilde{E}$ (see later), the corresponding limits are $\tilde{E} \sim -0.5$ and $\tilde{E} \sim -0.1$.

Chaos is well defined in classical physics, for instance in mechanics by the exponential divergence of nearby trajectories in phase space ($\sim \exp(\lambda t)$) where the degree of irregularity is given by the Liapunov index $0 \leq \lambda$. Another, qualitative representation used to picture chaotic motion of a system like

In this lecture we present and discuss some of the major results of spectroscopic studies with the hydrogen atom in strong magnetic fields. Since many theoretically oriented accounts have been given (examples of most recent reviews are ones by Friedrich and Wintgen [23]; Wunner, Hasegawa, and Robnik [24]) we report on experimental developments achieved during the last few years. Special attention is paid to the question what phenomena and features are experimentally observed in the quantum system, i.e. in the spectrum of the atom when it evolves, classically, from the regular to the chaotic regime.

EXPERIMENTAL

Like in previous works on the spectroscopy of Rydberg atoms in strong magnetic fields [10] the experiments have been carried out with tunable laser light in a crossed laser - atom-beam arrangement (Fig. 1). The hydrogen atoms pro-

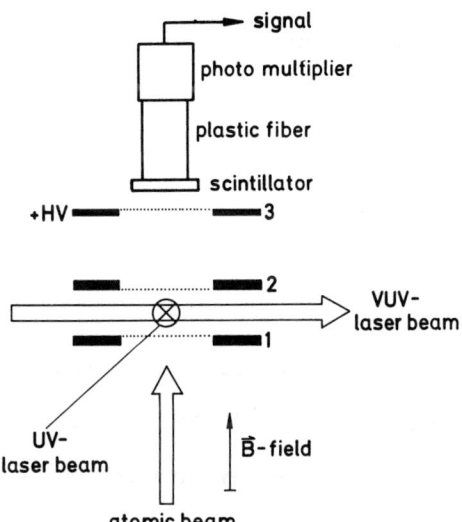

Fig. 1: Experimental crossed laser - atom-beam setup. 1, 2, 3 denotes fine-mesh metal grids.

duced in a microwave discharge source and well collimated to a beam are excited in two steps, $H(1s) + h\nu(vuv) \longrightarrow H(2p) + h\nu(uv) \longrightarrow H^*$, using pulsed dye lasers with ~16 ns pulse length. The two laser beams and the atomic beam cross each other perpendicularly at the center of the magnetic field which is kept electrically field-free by metal grids. The atomic beam is well collimated parallel to the field axis, an important condition to avoid motionally induced electric fields and line broadening by the Doppler effect. In the first step a sublevel of the Paschen-Back manifold with defined quantum number m_i = 0 or ±1 is selectively excited with the vuv laser light appropriately linearly polarized parallel or perpendicular to the magnetic field. In the second step final state spectra are taken by scanning the uv laser wavelength, also appropriately polarized to select individual states with magnetic quantum number m_f = 0, ±1 or ±2. Limited by the bandwidth of the u laser available, the highest resolution achieved in our hydrogen experiments is ~1 GHz. Excited H^* atoms either ionize spontaneously or pass on with the atomic beam and are field ionized behind the fine-mesh metal grids in the region with high electric field. Resulting electrons are detected by means of a scintillator detector connected via a fiber light guide with the photomultiplier.

SPECTRA

Figure 2 shows, just for illustration, a spectrum with m_f = 0 taken at B = 6 T in an energy range well below the ionization limit (E = 0). It covers the region from the perturbative l-mixing regime with quadratic Zeeman splitting of separated n-manifolds up into the lower part of the strong field-mixing quasi-Landau regime. Figure 3a shows a spectrum taken at B = 6 T, m_f = 0 with a resolution of ~1 GHz in the important and most interesting energy range around the ionization limit that is where the atom evolves into the classically irregular regime.

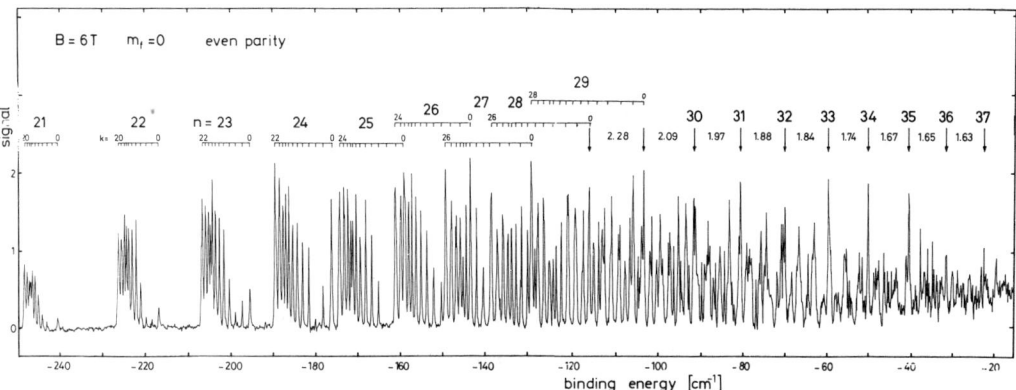

Fig. 2: Low-resolution (~10 GHz) experimental photoabsorption spectrum for $\Delta m = 0$ Balmer transitions to $m = 0$ even parity final states in a magnetic field of 6 T. It covers the range from the perturbative l-mixing regime over the n-mixing regime into the lower part of the strong field-mixing quasi-Landau regime.

The experimental spectra have been compared with the ones recently obtained by Wintgen and Friedrich [3] and by Wunner [4] through exact solution of the Schrödinger equation. Within the limits of the precision of the experiments and up to the experimental resolution limit (at 6 T about -25 cm^{-1}) there is good (or satisfactory) agreement between experiment and theory.

QUASI-LANDAU RESONANCES

Although the spectra grow more and more complex and unorderly with increasing energy this of course cannot be taken as signature of "quantum chaos". Phenomena and features related to the evolving irregularity can however be found in statistical properties of spectral structure, such as fluctuations (e.g.

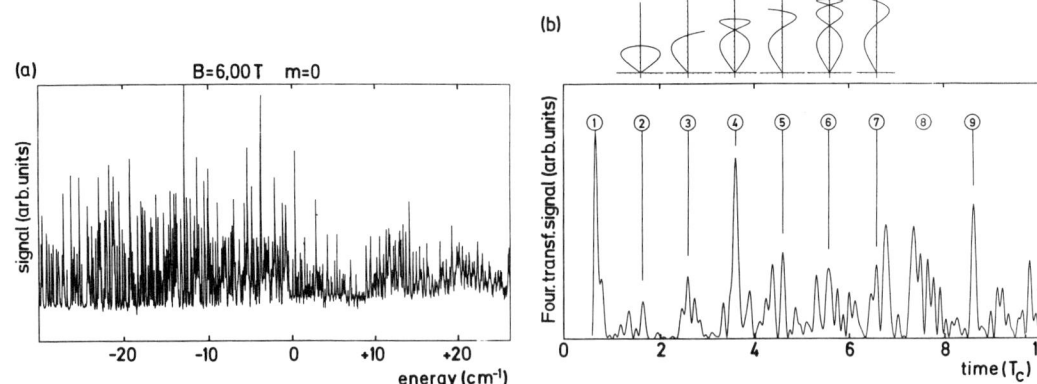

Fig 3:   a) High-resolution (~1 GHz) spectrum around the ionization limit in the classically irregular regime.
b) Fourier-transformed time spectrum of a). Plotted is the absolute value squared. Time scale normalized to cyclotron period T_c. Corresponding to the so-called "fundamental" resonances (2) to (7) calculated closed classical orbits of electron motion are shown in (ρ,z)-projection (z-axis vertically).

nearest-neighbour spacing distribution, spectral rigidity) of the energy level and oscillator strength distribution, a subject which has been extensively investigated theoretically in recent years [6, 7, 8]. Investigations of spectral statistics from experimental spectra has not been possible (or only of limited value) so far due to the relatively low line densities and/or to the limited experimental resolution and precision. However, signatures of quantum chaos contained as quasi-Landau resonances in the long-range structure of the energy spectrum are experimentally observed.

Figure 4 shows one of the first spectra of the magnetized H-atom around the ionization limit [25] taken with relatively very low resolution (~10 GHz). It clearly exhibits the oscillatory structure of quasi-Landau resonances of the same kind as the ones discovered previously by Garton and Tomkins [1] and Edmonds [2]. The energy spacing of the quasi-Landau oscillations can be

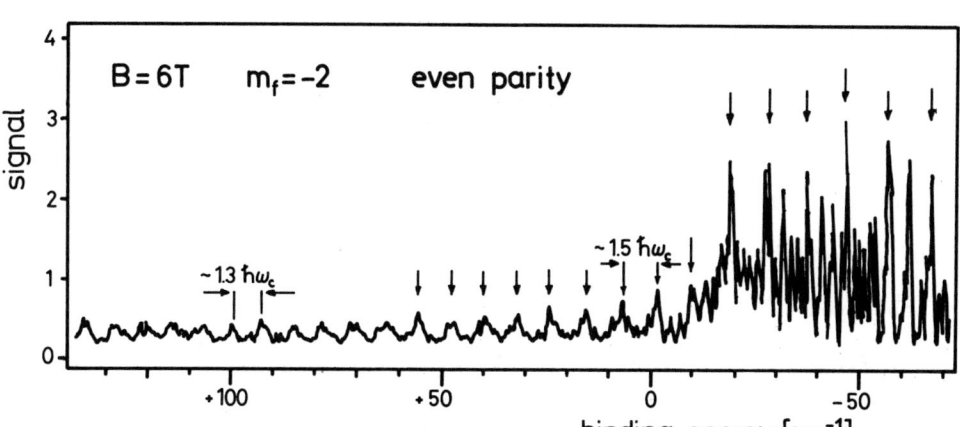

Fig. 4: Low-resolution quasi-Landau spectrum at 6 T with final states of $m_f = -2$ and even parity.

generally represented by $\Delta E = \lambda(E) \cdot \hbar\omega_c$ with $\omega_c = e/m \cdot B$ the cyclotron frequency. In the case of the GTE-resonances the energy dependent factor $\lambda(E)$ has the characteristic value $\lambda(E=0) = 3/2$ at the ionization limit ($E = 0$). This has been previously theoretically explained in several works by classical trajectory calculation [2, 21] and WKB-treatment [26]. Classically, the GTE-resonance type is correlated to a two-dimensional periodic orbit of the electron in the (z=0)-plane through the proton as origin. Related by $\Delta E \cdot T = h$ the period (or recurrence time) of the GTE-orbit is $T_{GTE} = 2/3 \cdot 2\pi/\omega_c$ at $E = 0$.

The disappearance of the oscillatory structures in high-resolution discrete-line spectra like the one shown in Figure 3a came as a total surprise when it was first observed [11]. However, as it is now well understood the oscillations of the low-resolution spectrum are still present in the high-resolution discrete-line spectrum in the form of modulations of the line density and/or oscillator strength distribution. They can be recovered by Fourier transformation of the energy to the time-domain spectrum, a general procedure to analyse spectral frequency distributions. A periodic modulation with spacing (or "wavelength") ΔE_i in the energy spectrum is uniquely related to a resonance at time T_i in the Fourier spectrum according to the general relation $\Delta E_i \cdot T_i = h$. Plotted on a relative time scale T/T_c normalized to the cyclotron period of the electron, $T_c = 2\pi/\omega_c$, Figure 3b shows the Fourier transform of the high-resolution energy spectrum. Unexpected at the time when first observed [11], it does not consist of one peak only, the one to be expected from the GTE-resonance, but shows a great number of additional resonances. The GTE-resonance corresponds to the first peak at shortest time, $T/T_c = 0.67$ with $T_c = 6$ ps at $B = 6$ T. The question how the richness of new quasi-Landau structures can be explained has been a key problem extensively investigated during the last few years in many theoretical works.

PERIODIC ORBITS

Like the GTE-resonance is related to the specific orbit in the (z=0)-plane, any oscillatory modulation of spacing ΔE_i in the quantum spectrum correlated to the corresponding resonance T_i in the Fourier spectrum can generally be related to a closed classical orbit of some type i. While closed orbits not need to go through the nucleus it is physically suggestive that in the case of the spectra excited in the experiments the orbits originate close to the proton, an assumption based on the small initial state quantum numbers ($|m| \leq 1$) at which the electron is localized initially very close to the proton. Figure 3 shows as an example a family of periodic orbits calculated at $E = 0$ and $B = 6$ T through the proton as origin, that is with the choice of initial conditions $\rho(t=0) = 0$, $z(t=0) = 0$ according to the excitation condition [5]. These orbits are unambigiously related to the series of resonances indicated in Figure 3b by integer numbers. For reasons given later we call this special series "fundamental". Proof for the correlation between the orbits and the observed resonances is derived from the fact that the periods or recurrence times of the calculated orbits agree with resonances in the Fourier spectrum.

The general correlation between the long-range oscillatory structures of the quantum spectrum (and the corresponding resonances in the time-domain spectrum) and periodic classical orbits is based on a theory previously developed by Gutzwiller [27], Balian and Bloch [28], and Berry [29]. This periodic-orbit (PO) theory, extended in recent years and specifically applied to the diamagnetic hydrogen atom [30], provides a semiclassical correspondence between the structure of the quantum spectrum, i.e. oscillatory modulation, and the classical motion of the electron on periodic orbits. The point is, that the theory applies not only to the regular regime with stable orbits, but provides a semiclassical correspondence in the irregular regime, where orbits are unstable. Quantitatively, the PO-theory represents the average level density $n(E)$ of a quantum system at energy E, here the highly excited hydrogen atom perturbed by an external magnetic field, by the relation

$$n(E) = n_o(E) + \sum_r \sum_j a_{rj}(E) \cdot \cos\{j[S_r(E)/\hbar - \mu_r]\} \qquad (2)$$

where $n_o(E)$ is the level density without field, $S_r(E)$ is the classical action of a principle periodic orbit of type r, j denotes the recurrence number (or harmonic) of the orbit, a_{rj} is the amplitude of the orbit and μ_r is the phase. The amplitude a_{rj} depends on the stability of a given orbit and is thus a direct indicator for the irregularity of the classical system. The theory has been extended by Delos and coworkers [9] to the average oscillator strength density $\overline{Df(E)}$ as function of the action $S_r(E)$, the phase φ_r, and the amplitude A_{rj}

$$\overline{Df(E)} = \overline{Df_o(E)} + \sum_r \sum_j A_{rj}(E) \cdot \cos \{j[S_r(E)/\hbar + \varphi_r]\}. \qquad (3)$$

A_{rj} now depends on the quantum mechanical probability of the transition to the final states, in addition to the degree of stability of the classical orbits. With the energy dependence of the action $S_r(E)$ given by $\delta S_r/\delta E = T_r$ the relation $\Delta E_r \cdot T_r = h$ is obtained. Based on eqn. 3 Delos and coworkers have calculated Fourier spectra in satisfactory agreement with the experiments [9].

SCALED ENERGY SPECTROSCOPY

With the discovery of a multitude of new quasi-Landau resonances arose the fundamental question: What is the "entire" quasi-Landau spectrum for final states of given m^π quantum number and what is its structure as it evolves from the regular to the irregular regime? The spectroscopy carried out as function of the energy at constant field strength possesses the problem that the quasi-Landau oscillations (and the related orbits) depend on both the energy and the field strength so that the Fourier transformation yields resonances only when the energy dependence is sufficiently weak within the transformation limits. This difficulty is overcome by scaled-energy spectroscopy based on the scaling property of the Hamiltonian (1), which accounts for the fact that the motion of the electron remains unchanged when the Coulomb and Lorentz forces change such that the so called scaled energy $\tilde{E} = E \cdot \gamma^{-2/3}$ remains constant [31]. With scaled variables defined by the transformations $\tilde{\vec{r}} = \gamma^{2/3} \vec{r}$, $\tilde{\vec{p}} = \gamma^{-1/3} \vec{p}$ the resonance condition for a given closed orbit i is

$$\frac{1}{2\pi\hbar} \oint_i (\tilde{p}_z d\tilde{z} + \tilde{p}_\rho d\tilde{\rho}) = S_i(\tilde{E}) = n_i \gamma^{1/3} \qquad (4)$$

where the action $S_i(\tilde{E})$ = constant for $\tilde{E}$ = constant. Thus, a spectrum taken at $\tilde{E}$ = constant and plotted on a scale $\gamma^{-1/3}$ will consist of series of equidistant lines with each series uniquely related to an orbit type i. Consequently, the Fourier transformation to conjugate coordinates $n\gamma^{1/3}$ will contain one resonance only for each i at the corresponding action $S_i(\tilde{E})$.

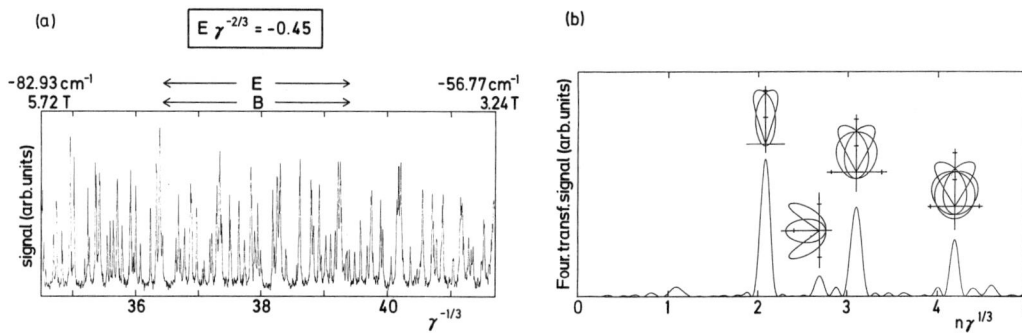

Fig. 5: a) Scaled-energy spectrum at $\tilde{E}$ = -0.45 as linear function of $\gamma^{-1/3}$.
b) Fourier-transformed action spectrum of a); closed orbits correlated to respective resonances in (ρ,z)-projection; z-coordinate vertically.

Figure 5a shows, as an example, a constant-scaled-energy spectrum at $\tilde{E}$ = -0.45 plotted vs $\gamma^{-1/3}$ and Figure 5b the corresponding Fourier spectrum as function of the action. Every peak in the action spectrum is correlated to one or (within the finite peak width) possibly more than one periodic orbit. Such action spectra have been investigated in the range from $\tilde{E}$ = -0.5 to $\tilde{E}$ = +0.2 that is just the regime where the system evolves from regularity ($\tilde{E} \leq -0.5$) to complete irregularity ($\tilde{E} \geq -0.1$). Figure 6a shows in concise overlay form the most important part from $\tilde{E}$ = -0.3 to $\tilde{E}$ = 0.0 at intervalls $\Delta\tilde{E}$ = 0.01.

To understand the remarkably structured distribution of resonances a systematic search of all closed orbits through the proton as origin has been made

28 Chaos in Atoms

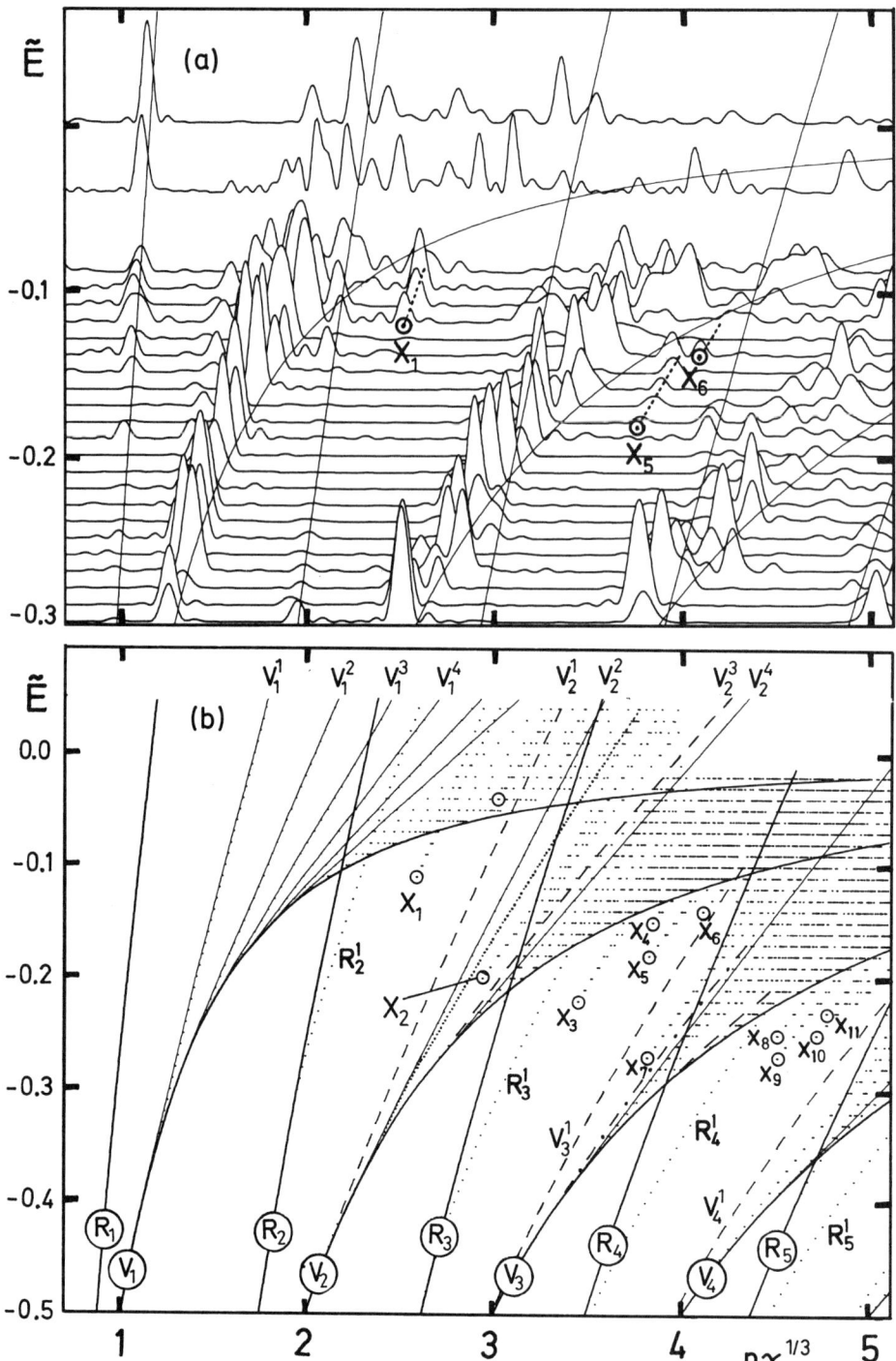

Fig. 6: a) Experimental quasi-Landau resonance action spectrum as a function of $\tilde{E}$ in overlay form. Even-parity, $m = 0$ final state.
b) Semiclassically calculated $(\tilde{E}, C)$-spectrum of quasi-Landau resonances correlated to closed classical orbits through origin.

by classical trajectory calculation. The result is shown in Figure 6b, where each dot represents the position of a closed orbit with $\tilde{E}$ in steps of 0.01. There is a clear similarity in overall distribution of observed and calculated resonances which allows to identify the nature of the most prominent ones excited.

Summarizing a few results: Resonances labelled by R_μ ($\mu = 1, 2, ...$) correlate with two-dimensional orbits in the (z=0)-plane (basic "rotators") where R_1 is the GTE-resonance. Resonances labelled V_μ ($\mu = 1, 2, ..$) correlate with one-dimensional orbits (basic "vibrators") along the z-axis, i.e. the magnetic field direction. The rotators and vibrators with $\mu \geq 2$ are exact harmonics of the principle ones ($\mu = 1$). We identify μ with the recurrence number j in eqn. 2 and 3. From the basic resonances originates, through bifurcation, a multitude of higher generation resonances, i.e. orbits, increasing in number with increasing scaled energy and action.

The basic rotator R_1, i.e. the GTE-resonance does not bifurcate. V_1 bifurcates into a first generation family of resonances V_1^υ ($\upsilon = 1, 2 ...$) which is just the "fundamental" one identified in Fig. 3. This family is found again as exact harmonics bifurcating from the higher basic vibrators. Higher order basic resonances ($\mu \geq 2$) bifurcate in rapidly increasing number into higher generations without any apparent order. Not directly originating from the rotators and vibrators, another kind of resonances occurs, examples of which are indicated in the figure by X_μ. These "exotics" originate suddenly and apparently uncorrelated to each other or other resonances at singular points. Furthermore, they bifurcate right at the point of origin. Exactly how the exotic resonances arise is not yet known.

CONCLUSIONS

During the last few years theoretical and experimental studies with highly excited diamagnetic hydrogen atoms have led to substantial progress in the physics of atoms in the classically chaotic quasi-Landau regime. Phenomena have been observed and investigated in the quantum spectrum that can be correlated to chaotic motion of the electron in the classical limit. A major discovery are new long-range periodic structures – quasi-Landau resonances – in the spectrum of the atom around the ionisation limit that correspond to and arise from periodic orbits, a clearly classical object. Instability and bifurcation, that are signatures of irregular motion, have been observed.

Extending the hydrogen studies, most recent experiments carried out with lithium atoms [14] at much higher resolution have yielded fully resolved quantum spectra around the zero-field ionisation limit. Concerning the long-range quasi-Landau structures with associated periodic orbits essentially the same results have been obtained as with the hydrogen atom indicating that these phenomena are primarily determined by the electron motion at long distances from the nucleus. Another physically appealing confirmation of the periodic orbit picture has been recently provided by theoretical wave packet treatment of the electron motion in three dimension [32, 33].

I wish to express my special thanks to my coworkers, particularly A. Holle, G. Wiebusch, J. Main, and H. Rottke who have made invaluable contributions to the works on the hydrogen atom. I also wish to thank my colleagues D. Kleppner, G. Wunner, and P. Zoller who made results of their works for this lecture available.

REFERENCES

[1] W.R.S. Garton and F.S. Tomkins, Astrophys. J. 158, 839 (1969)

[2] A.R. Edmonds, J. de Physique (Paris) 31, Colloque C4, 71 (1970)

[3] D. Wintgen, A. Holle, G. Wiebusch, J. Main, H. Friedrich, and K.H. Welge, J. Phys. B 19, L557 (1986)

[4] A. Holle, G. Wiebusch, J. Main, K.H. Welge, G. Zeller, G. Wunner, T. Ertl, and H. Ruder, Z. Phys. D 5, 279 (1987)

[5] J. Main, G. Wiebusch, A. Holle, and K.H. Welge, Z. Phys. D 6, 295 (1987)

[6] D. Wintgen and H. Friedrich, Phys. Rev. Lett. 57, 571 (1986) and Phys. Rev. A 36, 131 (1987)

[7] D. Delande and J.C. Gay, Phys. Rev. Lett. 57, 2006 (1986) and 59, 1809 (1987)

[8] G. Wunner, U. Woelk, I. Zech, G. Zeller, T. Ertl, F. Geyer, W. Schweizer, and H. Ruder, Phys. Rev. Lett. 57, 3261 (1986)

[9] M.L. Du and J.B. Delos, Phys. Rev. Lett. 58, 1731 (1987) and Phys. Rev. A 38, 1896 (1988)

[10] A. Holle, G. Wiebusch, J. Main, B. Hager, H. Rottke, and K.H. Welge, Phys. Rev. Lett. 56, 2594 (1986)

[11] J. Main, G. Wiebusch, A. Holle, and K.H. Welge, Phys. Rev. Lett. 57, 2789 (1987)

[12] A. Holle, J. Main, G. Wiebusch, H. Rottke, and K.H. Welge, Phys. Rev. Lett. 61, 161 (1988)

[13] J. Neukammer, H. Rinneberg, K. Vietzke, A. König, H. Hieronymus, M. Kohl, H.-J. Grabka, and G. Wunner, Phys. Rev. Lett. 59, 2947 (1987)

[14] G.R. Welch, M.M. Kash, C. Iu, L. Hsu, and D. Kleppner, Phys. Rev. Lett. 62, 1975 (1989)

[15] C.W. Clark and K.T. Taylor, J. Phys. B 15, 1175 (1982)

[16] G. Wunner, private communication

[17] D. Kleppner, M.G. Littmann, and M.L. Zimmermann, in Rydberg States of Atoms and Molecules, eds. R.F. Stebbings and F.B. Dunning, Cambridge University Press (1983)

[18] J.C. Gay and D. Delande, Comments At. Mol. Phys. 13, 275 (1983)

[19] M. Robnik, J. Phys. A 14, 3195 (1981)

[20] A. Harada and H. Hasegawa, J. Phys. A 16, L259 (1983)

[21] R.A. Pullen, D. Phil. thesis, Imperial College London (1981) (unpublished)

[22] M. Shapiro and G. Gollman, Phys. Rev. Lett. 53, 1714 (1984)

[23] H. Friedrich and D. Wintgen, Phys. Rep. (in press)

[24] H. Hasegawa, M. Robnik, and G. Wunner, Suppl. Prog. Theor. Phys. (to be published)

[25] A. Holle and K.H. Welge, in Proceedings of the Seventh International Conference on Laser Spectroscopy, eds. T.W. Hänsch and Y.R. Shen, Springer, New York (1985)

[26] A.F. Starace, J. Phys. B 6, 585 (1973)

[27] M.C. Gutzwiller, J. Math. Phys. 8, 1979 (1967), and 10, 1004 (1969), and 11, 1791 (1970), and 12, 343 (1971)

[28] R. Balian and C. Bloch, Annals of Physics 85, 514 (1974)

[29] M. Berry and K.E. Mount, Rep. Prog. Phys. 35, 315 (1972)

[30] D. Wintgen, Phys. Rev. Lett. 61, 1803 (1988)

[31] H. Hasegawa, S. Adachi, and H. Harada, J. Phys. A 16, L503 (1983)

[32] G. Alber, Z. Phys. D 14, 307 (1989)

[33] P. Zoller, private communication

TRANSITION STATE SPECTROSCOPY OF HYDROGEN TRANSFER REACTIONS

Daniel M. Neumark

Department of Chemistry, University of California, Berkeley, CA 94720

The transition state region for several chemical reactions of the type A + HB → HA + B has been investigated via photodetachment of the stable negative ion AHB⁻. The photoelectron spectra of BrHBr⁻ and BrHI⁻ show resolved vibrational structure attributable to the transition state region of the Br + HBr and Br + HI reactions, respectively. The photoelectron spectra of IDI⁻ and the cluster ion IDI⁻(N$_2$O) are compared; these spectra probe different parts of the I + HI potential energy surface. Finally, a high resolution threshold photodetachment spectrum of one of the IHI⁻ features shows evidence for reactive resonances in the I + HI reaction.

INTRODUCTION

In a hydrogen transfer reaction of the type A + HB → HA + B, the reactants pass through a short-lived [AHB] collision complex or transition state en route to products. One of the fundamental goals in chemical reaction dynamics is to understand the nature of the potential energy surface on which a reaction occurs in the vicinity of the transition state (the transition state region).[1] This will provide a detailed probe of the interatomic forces which govern chemical bond formation and cleavage in a reaction.

During the last 10-15 years, the most popular approach to this problem has been state-to-state chemistry.[2] In a typical experiment, the product state distribution of a chemical reaction is characterized as completely as possible as a function of reactant initial conditions. For example, one might measure the product angular and internal energy distributions at several reactant collision energies. This information has considerable value in its own right and often provides valuable insight into the mechanism of a reaction. However, the extraction of the properties of the transition state region from these asymptotic measurements is very difficult, even for reactions which have been characterized in detail.

This situation has motivated the development of 'transition state spectroscopy' experiments.[3,4] The goal of these studies is to use spectroscopic means to directly probe the short-lived complex formed in a reaction. The initial experiments in this area were performed by Polanyi, Brooks, and their co-workers in 1980.[5,6] Since then, several techniques involving frequency and time-domain spectroscopy have been applied to this problem.[7-15] However, the short-lived nature of the transition state has limited the successful application of these methods to a relatively small number of systems.

We have developed an

alternative approach to the spectroscopy of the transition state in bimolecular reactions.[16-19] We study the unstable [AHB] complex formed in the A + HB hydrogen transfer reaction by photodetaching the stable, hydrogen-bonded negative ion AHB$^-$. If the ion geometry is similar to that of the neutral transition state, then photodetaching the ion will probe the transition state region of the A + HB potential energy surface. Even though the [AHB] complex is unstable, the photoelectron spectrum of AHB$^-$ can exhibit resolved vibrational structure which yields considerable insight into the spectroscopy and dissociation dynamics of the [AHB] complex. Hydrogen transfer reactions are particularly appealing because, in many cases, the [AHB] transition state for the reaction has good geometric overlap with the strongly hydrogen-bonded ion AHB. In addition, most hydrogen transfer reactions are 'heavy + light-heavy' reactions, in which a hydrogen atom is transferred between two much heavier species. The dissociation dynamics of the [AHB] complex with this mass combination favors the observation of resolved vibrational structure in the AHB$^-$ photoelectron spectrum. In essence, we observe the fast vibrational motion of the light H atom as the complex slowly dissociates.

This experiment provides a high degree of control over important parameters of a chemical reaction. Photodetachment initiates the A + HB reaction with the atoms in the same geometry as they are in the ion AHB$^-$, thereby providing excellent control over the reactant orientation. In addition, since the ions are generated in a source which produces rotationally cold species, we can limit the total angular momentum available to the reaction. The latter restriction is particularly significant since it facilitates comparison with theoretical simulations of our results.

The simplest reactions amenable to this technique are those in which a hydrogen atom is exchanged between two identical halogen atoms. Consider the reaction

$$Br + HBr \rightarrow HBr + Br, \quad (1)$$

which is studied by photodetaching BrHBr$^-$.[19] This ion is predicted to be a linear and centrosymmetric species with an interbromine equilibrium distance $R_e = 3.43 \pm 0.1$ Å.[20,21] Model potential energy surfaces for the Br + HBr reaction[22,23] predict a linear minimum energy path with a barrier < 10 kcal/mol. Thus, reasonable geometric overlap is expected between the ion and neutral transition state.

We have also studied simple asymmetric reactions[24] such as

$$Br + HI \rightarrow HBr + I, \quad (2)$$

as well as more complex reactions involving polyatomic reactants[18] such as

$$F + CH_3OH \rightarrow HF + CH_3O. \quad (3)$$

Finally, the photoelectron spectra of cluster ions such as IDI$^-$(N$_2$O) have been obtained as a first step towards understanding what happens when a reaction is initiated by photodetaching a cluster of known size. Reactions (1) and (2) will be discussed below, along with the cluster ion results. Results obtained with a higher resolution instrument on the I + HI reaction will also be discussed.

EXPERIMENTAL

Two types of negative ion photodetachment experiments are used in this work. The instruments used in these experiments are described in detail elsewhere.

Most of the results to date were obtained with 'fixed-frequency' negative ion photoelectron spectroscopy.[25,26] Our apparatus is a time-of-flight photoelectron spectrometer similar to that described in Ref. 26. Internally cold negative ions are generated by crossing a pulsed, free-jet expansion of an appropriate mixture of gases with a 1 keV electron beam. The ions are mass-selected via time-of-flight and photodetached with a pulsed, fixed-frequency laser. The data shown below were taken using either the 4th or 5th harmonic of a Nd:YAG laser at 266 nm (4.66 eV) or 213 nm (5.82 eV), respectively. The kinetic energy of the ejected photoelectrons is determined with a time-of-flight analyzer. The resolution of the energy analyzer is 5-8 meV at low electron kinetic energy (0.5 - 0.6 eV), but this degrades approximately as $E^{3/2}$ at higher electron kinetic energy.

Higher resolution studies were performed on a threshold photodetachment spectrometer.[27] This instrument has a similar ion source to the fixed-frequency apparatus and also uses time-of-flight mass selection. However, the ions are photodetached with a tunable pulsed dye laser. At a given laser wavelength, only electrons produced with nearly zero-kinetic energy are detected. The zero-kinetic energy spectrum plotted as a function of laser wavelength consists of a series of peaks, each corresponding to an ion→neutral transition. The width of the peaks is determined by the ability of the instrument to discriminate against photoelectrons produced with high kinetic energy. By adapting the methods developed by Schlag and co-workers for threshold photoionization of neutrals,[28] we have achieved a resolution of 3 cm^{-1} (0.37 meV) with this instrument. In the experiments performed with this instrument (Section 4, below), tunable laser light in the range of 300 nm was required. This was obtained by frequency doubling the output of an excimer pumped dye laser in a β-barium borate crystal. The laser repetition rate was 50 Hz. With Rhodamine B as the dye, 2 mJ/pulse of frequency-doubled light was obtained.

RESULTS AND DISCUSSION

1) BrHBr$^-$ Photoelectron Spectrum

The photoelectron spectra of BrHBr$^-$ and BrDBr$^-$ taken at 213 nm are shown in Figure 1. The spectra are discussed and analyzed in considerable detail in Ref. 19; only the salient features are summarized here. Each spectrum shows several resolved peaks of varying widths. In these spectra, the electron kinetic energy E is given by

$$E = h\nu - E_b^- - E_{int}^0 + E_{int}^-. \quad (4)$$

Here $h\nu$ = 5.82 eV is the photon energy, E_b^- = 4.27 eV[29,30] is the binding energy of the electron to BrHBr$^-$, that is, the energy necessary to remove an electron from the ground state of BrHBr⁻ to form Br + HBr(v=0), and E^-_{int} and E^0_{int} are the internal energies of the ion and neutral [BrHBr] complex, respectively. E^-_{int} is measured relative to Br + HBr(v=0). Equation (4) shows that if E^-_{int} = 0, then the peaks in a photoelectron spectrum at <u>lower</u> electron kinetic energy correspond to <u>higher</u> values of internal energy in the complex. Note that all the peaks occur at electron kinetic energies less than hν - E^-_b = 1.55 eV. This is the electron kinetic energy that would result from forming Br + HBr(v=0) from ground state BrHBr⁻. Hence, the observed peaks in each spectrum correspond to states of the complex that lie <u>above</u> Br + HBr(v=0), that is, states of the complex which can dissociate.

The three peaks at highest electron kinetic energy in the BrHBr⁻ spectrum are approximately evenly spaced, suggesting a vibrational progression. This is confirmed by the BrDBr⁻ spectrum, in which the highest energy peak is unchanged, but the spacing between the four highest energy peaks is considerably less than in the BrHBr⁻ spectrum. This isotope shift indicates that we are observing a vibrational progression of the [BrHBr] complex in a mode primarily involving H atom motion. The progression is assigned to the v_3 antisymmetric stretch mode, in which the H atom vibrates between the two essentially stationary halogen atoms. Since the highest energy peak occurs at the same energy in the two spectra, it is assigned to the v_3'' = 0 → v_3' = 0 ion → neutral transition. There is no evidence for 'hot band' transitions in the spectrum originating from excited v_3'' levels of the ion. The v_3 mode is not totally symmetric, so only even v_3' levels of the neutral are accessible from the v_3'' = 0 level of the ion. The peaks in this progression in the two spectra are labelled by their v_3' quantum number. The broad peaks at lower electron energy labelled A' and B' are assigned to an excited electronic state of the [BrHBr] complex and are discussed in more detail in Ref. 19.

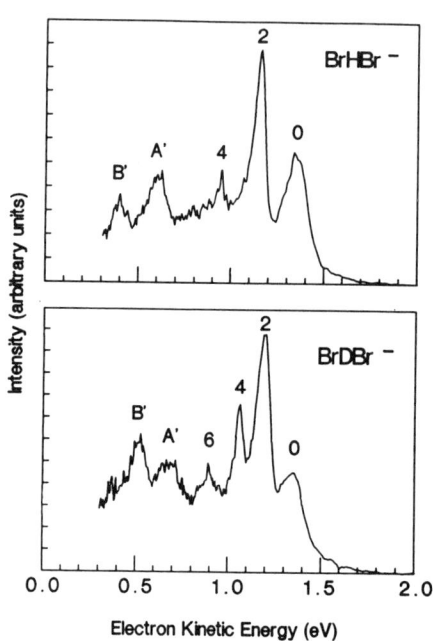

Figure 1. Photoelectron spectra of BrHBr⁻ and BrDBr⁻ obtained at 213 nm.

An important feature of these spectra is that the spacing between the peaks in the v_3 progression is considerably less than the vibrational frequency in diatomic HBr. The $v_3' = 0$ and $v_3' = 2$ peaks in the BrHBr⁻ spectrum are separated by 1700 cm⁻¹, while the HBr fundamental is at 2560 cm⁻¹. This shows that our experiment probes the transition state region of the Br + HBr potential energy surface where the H atom is expected to interact strongly with both Br atoms. In effect, we are probing the region of the surface where the strong bond in diatomic HBr has been replaced by two weak bonds in the [BrHBr] complex.

In a heavy + light-heavy reaction such as Br + HBr, the v_3 mode of the [BrHBr] complex in the transition state region is poorly coupled to dissociation of the complex. Thus, although the peaks in the photoelectron spectra correspond to states of the [BrHBr] complex with enough energy to dissociate, they correspond to a progression in a vibrational mode which is not the dissociation coordinate of the complex. Similar effects have been seen in electronic spectra involving transitions to dissociative states of polyatomic molecules.

The peaks widths in the BrHBr⁻ and BrDBr⁻ spectra are a measure of the dissociation dynamics of the [BrHBr] complex. The peaks become progressively narrower as v_3' increases; the $v_3' = 0$ peak in the BrHBr⁻ spectrum is 150 meV wide, whereas the $v_3' = 4$ peak is only 20 meV wide. This is at first glance a counter-intuitive result; it implies that states of the complex with higher internal energy (in the v_3 mode) have longer lifetimes. However, previous calculations on heavy + light-heavy reactions[31] predict the existence of levels of the complex which are quasi-bound along the dissociation coordinate (the symmetric stretch coordinate in this case) and which become more pronounced for more highly excited antisymmetric stretch levels of the complex. These quasi-bound states are responsible for the sharp resonance structure[32] observed in reactive scattering calculations on these systems. The narrow $v_3' = 4$ peak in the BrHBr⁻ spectrum and $v_3' = 6$ peak in the BrDBr⁻ spectrum are assigned to transitions to these quasi-bound [BrHBr] levels.

The ultimate goal of this experiment is to learn about the Br + HBr potential energy surface in the vicinity of the transition state. This is done by constructing a flexible functional form for the Br + HBr potential energy surface and varying the parameters of the surface until the BrHBr⁻ and BrDBr⁻ photoelectron spectra are successfully reproduced in spectral simulations. The accurate simulation of these spectra requires calculating the Franck-Condon overlap between the ion vibrational wavefunction and the three-dimensional scattering wavefunctions supported by the Br + HBr surface over a wide energy range. These calculations have been performed by Schatz[33] and Bowman[34] on a model Cl + HCl surface in order to simulate the ClHCl⁻ photoelectron spectrum. Miller[35] has done this type of calculation on a model F + H₂ surface in a simulation of the FH₂⁻ photoelectron spectrum. However, these calculations are too complex to use in our iterative analysis. We have therefore developed an approximate method to simulate our spectra.

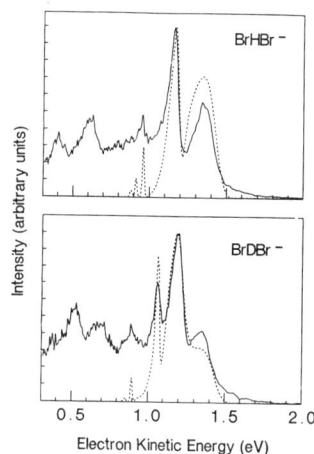

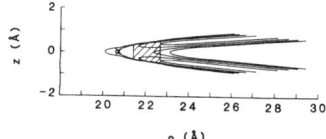

Figure 2. (bottom) Best-fit 'effective collinear potential energy surface for Br + HBr reaction. Shaded area shows region that has best Franck-Condon overlap with ground state of BrHBr⁻.
(top) Simulations of ground state progressions in the BrHBr⁻ and BrDBr⁻ spectra superimposed on the experimental spectra.

We construct an 'effective' collinear potential energy surface[36] which is a collinear Br + HBr surface with the zero-point bending energy of the [BrHBr] complex implicitly included at every point. The surface that provides the best fit to our spectra is shown at the bottom of Figure 2. The shaded region shows the part of the surface which has good Franck-Condon overlap with the ground state of BrHBr⁻. The simulations of the spectra are very sensitive to small changes in this region (the Franck-Condon region) of the surface. This region of the surface plays a key role in the dynamics of the Br + HBr reaction. The simulated spectra obtained with this surface are shown superimposed on the experimental spectra at the top of Figure 2. We see that reasonable agreement is obtained between experimental and simulated peak positions, intensities, and widths for the ground state progression.

In summary, the photoelectron spectra of BrHBr⁻ and BrDBr⁻ are a sensitive probe of the transition state region of the Br + HBr potential energy surface. The peaks in the spectra reveal the spectroscopy and dissociation dynamics of the [BrHBr] complex, and strongly suggest the existence of reasonably long-lived (approximately 100 fs) quasi-bound levels of the complex. The spectra have been analyzed to yield an approximate potential energy surface for the reaction.

2) BrHI⁻ Photoelectron Spectrum

The transition state region for the reaction Br + HI → HBr + I is accessible through photodetachment of the asymmetric bihalide ion BrHI⁻. The BrHI⁻ and BrDI⁻ photoelectron spectra taken at 213 nm are shown in Figure 3. Each spectrum shows two progressions of evenly spaced peaks. The peak widths are between 100-150 meV. The origins of the two progressions occur at the same electron energies in each spectra, and the peak spacing within each progression is noticeably smaller in the BrDI⁻ spectrum. Each progression is assigned to the ν_3 mode of the [BrHI] complex. The two progressions in each spectrum are assigned to two electronic states of the complex. The splitting between the origins of the progressions is 7300 cm⁻¹, which is very close to the splitting between the $^2P_{3/2}$ ground state and the $^2P_{1/2}$ spin-orbit excited state of atomic iodine. Hence the progression at higher electron energy is assigned to the ground electronic state of the [BrHI] complex, while the progression at lower electron energy is likely due to an excited state of the complex which correlates asymptotically to HBr + I*($^2P_{1/2}$).

The scale at the top of Figure 3 shows that the peak spacing in both progressions in the BrHI⁻ spectrum is only slightly less than the vibrational frequency in diatomic HBr. This is in sharp contrast to the much larger 'red shift' seen in the BrHBr⁻ spectrum. One can understand this result by considering the geometry of the BrHI⁻ ion. The proton affinity of Br⁻ is 0.47 eV higher than that of I⁻. Hence, the H atom in the ion should be significantly closer to the Br atom; the ion can be pictured as I⁻·HBr. This means that photodetachment of the ion will primarily access the I + HBr product valley on the neutral potential energy surface. In this region, the nascent HBr bond is nearly complete. We therefore expect to observe a vibrational frequency in the photoelectron spectrum characteristic of this nearly-formed bond, which is indeed the experimental result.

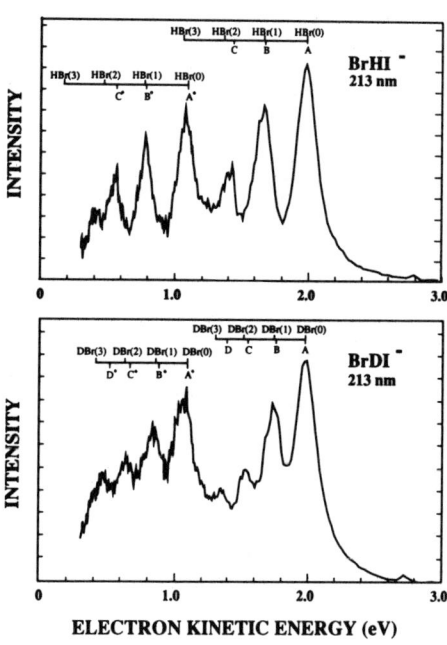

Figure 3. BrHI⁻ and BrDI⁻ photoelectron spectra obtained at 213 nm.

We have simulated the ground state progression in the BrHI⁻ spectrum using the time-dependent wave-packet formalism developed by Heller,[37] Kosloff,[38] and others.[39,40] A model potential energy surface proposed by Broida and Persky[41] for the Br + HI reaction is used in the simulations. This surface is of the London-Eyring-Polanyi-Sato (LEPS) functional form.[42] The procedure for estimating the BrHI⁻ potential energy surface is described by us elsewhere. This is a two-dimensional simulation restricted to collinear geometries.

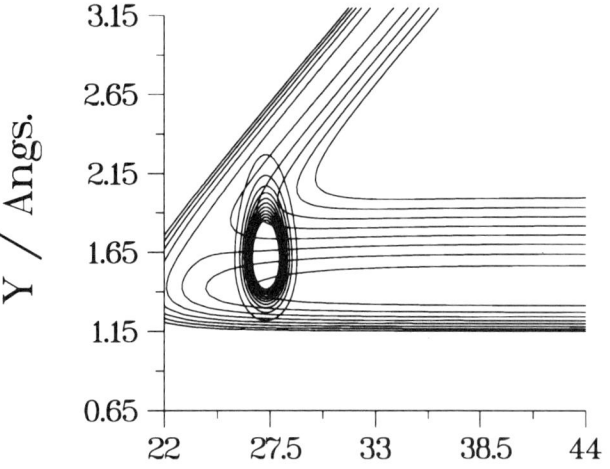
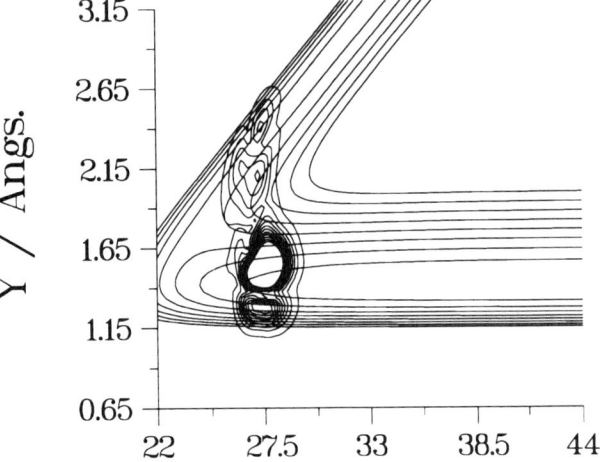
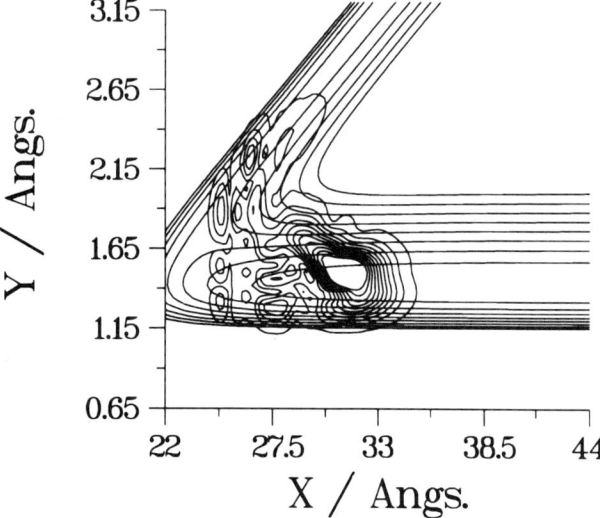

Figure 4. Wavepackets at t=0 (top), t = 40 fs (middle), and t = 160 fs (bottom), on Br + HI potential energy surface resulting from BrHI⁻ photodetachment. The surface is plotted using the mass-scaled coordinates y = $R_{H\text{-}Br}$ and x ≈ $(\mu_{I,HBr}/\mu_{HBr})^{1/2}(R_{I\text{-}Br})$.

The procedure is as follows. The initial wavepacket $\psi(0)$ on the Br + HI surface is obtained by projecting the ion ground state wavefunction onto the neutral surface, assuming the Franck-Condon principle is valid. This wavepacket is then propagated in time on the reactive surface using the method described by Kosloff and Kosloff.[38a] The autocorrelation function $C(t) = \langle\psi(0)|\psi(t)\rangle$ is calculated, and the Fourier transform of C(t) yields the simulated photoelectron spectrum.

Figure 4 shows the initial wavepacket on the Br + HI surface, as well as the wavepacket at the later times t=40 fs and t=160 fs. Figure 5 shows a plot of |C(t)|, and the simulated spectrum is superimposed on the experimental spectrum in Figure 6. Figure 4 shows that the initial wavepacket is localized in the I + HBr product valley, as discussed above. The plot at t= 160 fs shows that photodetachment of BrHI⁻ leads primarily to I + HBr products; very little of the wavepacket ends up in the Br + HI valley.

The plot of the autocorrelation function provides some interesting physical insight into our experiment. The most prominent feature in Figure 5 is a high-frequency oscillation which is damped after about 30 fs. This high frequency motion is the ν_3 mode of the [BrHI] complex, and the rapid decay of this oscillation is due to dissociation of the complex. The frequency of the oscillation determines the peak spacing in the photoelectron spectrum, while the time scale for decay determines the peak widths. Thus the structure in the BrHI⁻ photoelectron spectrum shows the [BrHI] complex vibrating along the v_3 coordinate as it falls apart.

result from deficiencies in the LEPS surface used in the simulations. A surface which is more repulsive in the I + HBr valley will yield broader peaks in the simulated photoelectron spectrum. In addition, if the minimum energy path on the true surface were bent, then a more accurate simulation not restricted to collinear geometries would probably yield broader peaks.

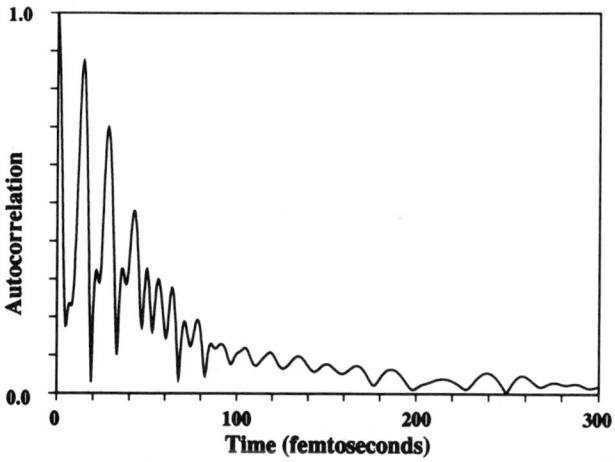

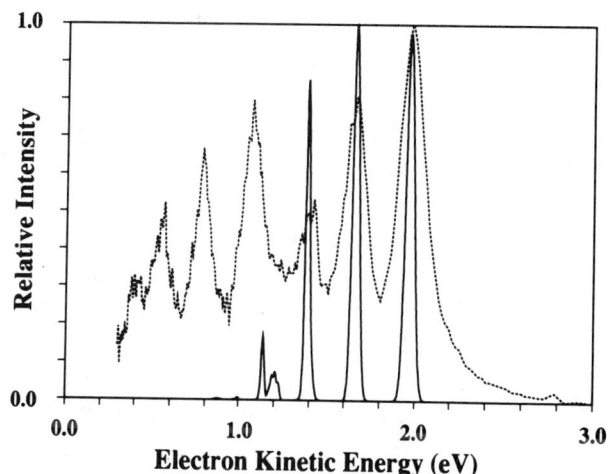

Figure 5. Modulus of autocorrelation function $|C(t)|$ vs. t for propagation of wavepacket shown at top of Figure 4.

The simulated spectrum shows four nearly evenly spaced peaks and a small, partly resolved clump of peaks at 1.20 eV. The positions of the first three peaks agree reasonably well with the three highest energy peaks in the experimental spectrum, although the simulated widths are too narrow. The overly narrow widths may

Figure 6. Simulated BrHI⁻ photoelectron spectrum obtained from Fourier transform of C(t) of Figure 5 superimposed on experimental spectrum. Only the ground state progression was modelled.

The clump of peaks in the simulation at 1.2 eV is from transitions to a series of [BrHI] resonances supported by the LEPS surface localized largely in the Br +

HI reactant valley. The contribution from these states can be seen in Figure 5; they lead to the persistence of $|C(t)|$ at long times. Due to the low intensity of these transitions in the simulation, our experiment does not completely rule out their existence. We therefore cannot say if a more accurate potential energy surface would support these resonances.

3) $IDI^-(N_2O)$ Photoelectron Spectrum

The photoelectron spectroscopy of negative cluster ions has become a very active research field in recent years.[43-47] This class of experiments, particularly those of Bowen and co-workers,[43b] has led us to pursue a new direction in our own efforts. Suppose we measure the photoelectron spectrum of a cluster ion in which a bihalide ion is surrounded by a known number of solvating species, i.e. $IDI^-(S)_n$. There is no ambiguity concerning the number of solvating species since the ions are mass-selected prior to photodetachment. Photodetachment of the cluster ion generates a neutral collision complex surrounded by the solvating species; such an experiment may serve as a probe of condensed phase reaction dynamics. As a first step to studying larger clusters, we have obtained the photoelectron spectrum of $IDI^-(N_2O)$, shown in Figure 7 below the photoelectron spectrum of the bare IDI^- ion. These spectra were taken at 266 nm.

A comparison of the two spectra indicates that the addition of an N_2O molecule has perturbed but not destroyed the structure seen in the IDI^- spectrum. Both spectra show three peaks. These are assigned to even members of the v_3

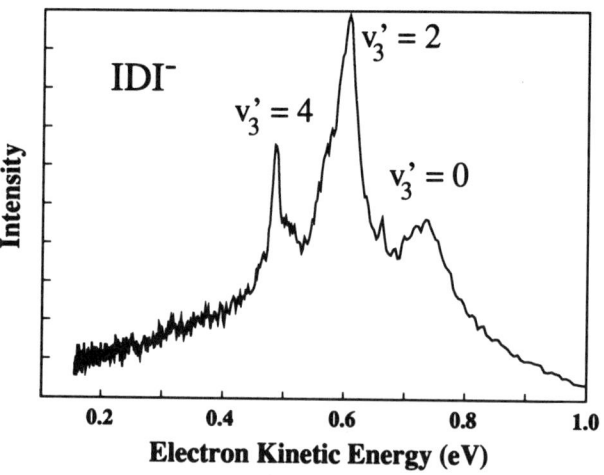

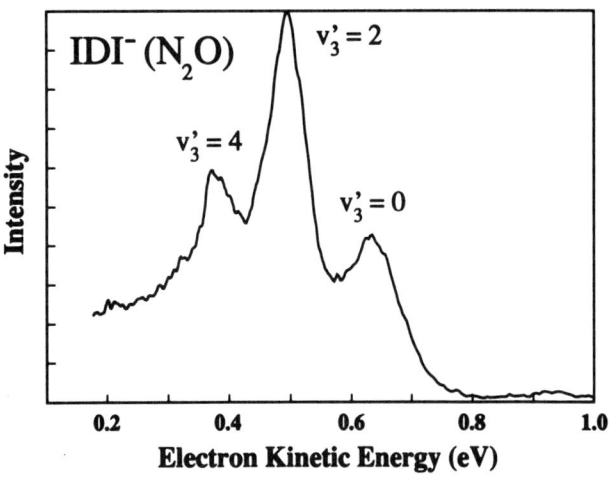

Figure 7. Photoelectron spectrum of IDI^- (top) and $IDI^-(N_2O)$ (bottom) taken at 266 nm.

progression in the neutral complex just as in the BrHBr⁻ spectrum. The 0→0 transition is shifted toward lower electron energy by 0.100 eV in the cluster ion spectrum. Assuming that N_2O interacts much more strongly with IDI⁻ than with the [IDI] complex, then this shift yields the binding energy of N_2O with IDI⁻. Shifts of this magnitude were seen in Bowen's photoelectron spectroscopy study of the series NO⁻, NO⁻(N_2O), and NO⁻(N_2O)$_2$.[43b]

A important difference between the bare and clustered IDI⁻ spectra is that the spacing between the three peaks is slightly different. The v_3' = 0 and v_3' = 2 peaks are more widely spaced by 0.013 eV in the IDI⁻(N_2O) spectrum, and the v_3' = 2 and v_3' = 4 spacing is 0.007 eV less than in the bare ion spectrum. These differences are small but significant and were not seen in Bowen's spectra of the NO⁻ series. We believe this variation in the peak spacing results because photodetachment produces an unstable collision complex rather than a stable molecule in our experiment.

A plausible explanation for our observation is as follows. In the cluster ion IDI⁻(N_2O), some charge transfer is expected from the IDI⁻ ion to the N_2O molecule. This can distort the IDI⁻ geometry, most likely by lengthening the interiodine distance since the extra electron is what holds the ion together in the first place. On the other hand, subsequent to photodetachment, it is a reasonable approximation to ignore the interaction between the [IDI] complex and neighboring N_2O molecule. Thus, the cluster ion photoelectron spectrum is, to first order, equivalent to the photoelectron spectrum of slightly distorted IDI⁻.

Figure 8 shows a model collinear potential energy surface for the I + HI reaction.[48] The horizontal and vertical axes are proportional to the symmetric and antisymmetric stretch normal coordinates, respectively. In our earlier work on the IDI⁻ photoelectron spectrum,[17] we estimated the equilibrium interiodine distance in IDI⁻ to be R_e = 3.88 Å. This distance corresponds to the dashed vertical line drawn through the surface in Figure 8. A cut through the surface at this interiodine distance yields a double minimum potential which is, to a good approximation, the antisymmetric stretch potential for the [IDI] complex with an interiodine distance of 3.88 Å. This potential is shown in Figure 9a, along with the eigenvalues for the first few even v_3' levels. Suppose that the interiodine distance in IDI⁻ (N_2O) is 0.05 Å greater than in the bare ion. The antisymmetric stretch potential for the [IDI] complex for this geometry is obtained by taking the cut through the surface indicated by the dotted line in Figure 8. This potential and its first few even eigenvalues are shown in Figure 9b. The major difference between the two potentials is that the barrier between the two minima is higher at larger interiodine distance. While this pushes up all the eigenvalues, the v_3' = 2 level is the most strongly affected since it lies very close to the barrier. The result is that in the antisymmetric stretch potential associated with the cluster ion, the v_3' = 0 and v_3' = 2 levels are further apart, and the v_3' = 2 and v_3' = 4 levels are closer together. This is consistent with the experimentally observed peak spacings.

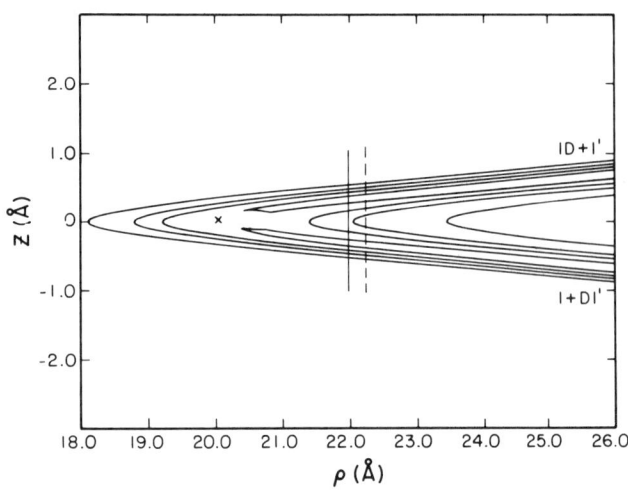

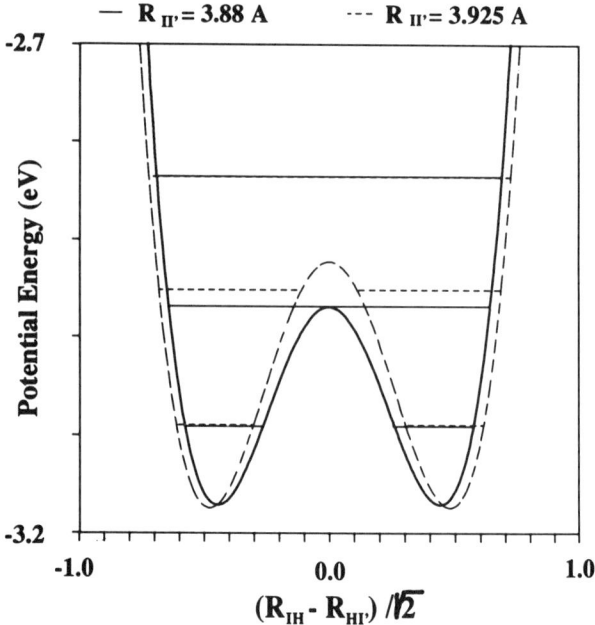

Figure 8. Model collinear potential energy surface for I + DI reaction. Solid vertical line corresponds to estimated interiodine distance R_e = 3.88 Å in IDI$^-$. The dashed vertical line corresponds to estimated R_e = 3.93 Å in IDI$^-$(N$_2$O).

Thus, our simple picture qualitatively explains the experimental results. The photoelectron of a cluster ion such as IDI$^-$(N$_2$O) allows one to probe different regions of the I + DI potential energy surface and also gives an indication of the nature of the interaction in the cluster ion. We plan to study the photoelectron spectra of similar cluster ions in the near future as a function of the type and number of solvating species.

Figure 9. Antisymmetric stretch potentials and v_3 = 0, 2, and 4 energy levels for [IDI]complex obtained by cuts through I + HI potential energy surface at interiodine distances of 3.88 Å (solid lines) and 3.93 Å (dashed lines).

4) Threshold Photodetachment Spectrum of IHI$^-$

Figure 10 shows the IHI$^-$ photoelectron spectron previously obtained by us.[17] The I + HI reaction is predicted to exhibit the most pronounced resonance effects due to quasi-bound IHI states supported by the potential energy surface.[31,49] This results from both the large m_I/m_H mass ratio and the low barrier expected for the reaction. Simulations of the IHI⁻ photoelectron spectrum[17,34,50] predict that some of the v_3' peaks should exhibit underlying structure at higher resolution due to a progression in quasi-bound symmetric stretch levels of the [IHI] complex; these levels are quasi-bound along the dissociation coordinate of the complex. These simulations were performed on a model LEPS surface for the I + HI reaction on which the barrier is probably too low. Nonetheless, the prediction of more structure under the peaks in the IHI⁻ photoelectron spectrum prompted an investigation of this system with our threshold photodetachment spectrometer[27] which, as discussed above, has considerably higher resolution than the 'fixed-frequency' instrument used to obtain the the spectrum in Figure 10.

Figure 11 shows a scan of the $v_3' = 2$ peak (the middle peak in Figure 10) obtained with the threshold photodetachment spectrometer. This spectrum shows three partly resolved peaks which decrease in intensity toward lower wavelength. The peaks are spaced by about 100 cm⁻¹. This is in the range expected for the symmetric stretch frequency for quasi-bound states of the [IHI] complex. Although the narrow peaks in the BrHBr⁻ and BrDBr⁻ spectra were assigned to resonances, the observation of this low frequency progression in the IHI⁻ threshold photodetachment spectrm provides even more compelling evidence for the existence of these quasi-bound levels. The observation of these levels is of considerable interest not only because it confirms long-standing theoretical predictions, but because it also provides more detailed spectroscopic information on the transition state region of the I + HI potential energy surface. Further investigation of the IHI⁻ threshold photodetachment spectrum is currently in progress in order to substantiate the claim that we are observing [IHI] resonances. We plan to take high resolution spectra of the other v_3' peaks in the IHI⁻ and IDI⁻ photoelectron spectra in order to determine which of them, if any, exhibit underlying structure. Meanwhile, the spectrum in Figure 11 can be regarded as very promising result.

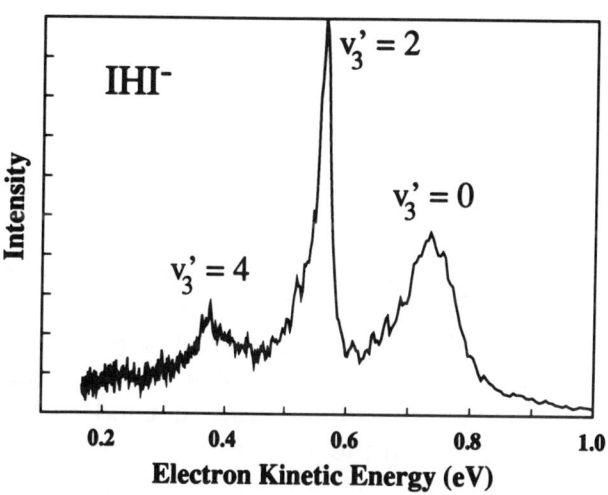

Figure 10. Photoelectron spectrum of IHI⁻ taken at 266 nm.

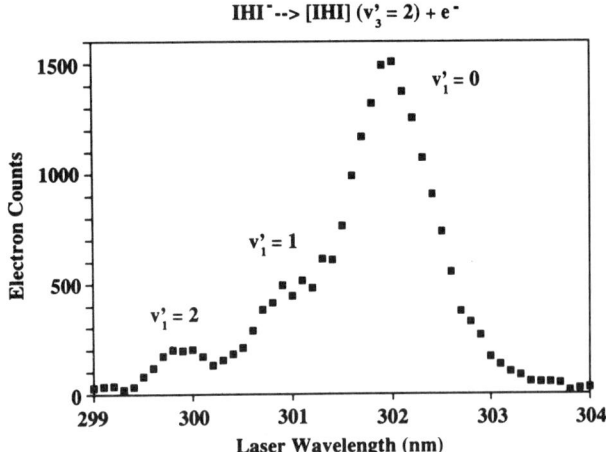

Figure 11. Threshold photodetachment spectrum of [IHI] $v_3 = 2$ peak.

SUMMARY

Negative ion photodetachment can provide a detailed probe of the transition state region of a chemical reaction. The examples above show the application of this method to several simple reactions. Future directions include the investigation of more complex chemical reactions, reactions initiated by photodetaching a cluster ion (Section 3, above), and higher resolution studies with our threshold photodetachment spectrometer (Section 4, above).

ACKNOWLEDGMENTS

The photoelectron spectroscopy work is supported by the Air Force Office of Scientific Research under Grant No. AFOSR-87-0341. The threshold photodetachment studies are supported by the Office of Naval Research Young Investigator Program under Grant No. N0014-87-K-0495.

The results shown here were obtained with the aid of my postdoctoral fellow, Dr. Irene Waller, and my graduate students Don Arnold, Steven Bradforth, Doug Cyr, Theo Kitsopoulos, Jennifer Loeser, Ricardo Metz, and Alexandra Weaver.

REFERENCES

1. R. D. Levine and R. B. Bernstein, <u>Molecular Reaction Dynamics and Chemical Reactivity</u> (Oxford University Press, New York, 1987), pp. 396-510.

2. S. R. Leone, Annu. Rev. Phys. Chem. **35**, 109 (1984).

3. P. R. Brooks, Chem. Rev. **88**, 407 (1988).

4. A. H. Zewail, Science **242**, 1645 (1988).

5. P. Arrowsmith, F. E. Bartoszek, S. H. P. Bly, T. Carrington, Jr., P. E. Charters, and J. C. Polanyi, J. Chem. Phys. **73**, 5895 (1980).

6. P. Hering, P. R. Brooks, R. F. Curl, Jr., R. S. Judson, and R. S. Lowe, Phys. Rev. Lett. **44**, 687 (1980); T. C. Maguire, P. R. Brooks, R. F. Curl, J. H. Spence, and S. J. Ulvick, J. Chem. Phys. **85**, 844 (1986).

7. H. P. Grieneisen, H. Xue-jing, and K. L. Kompa, Chem. Phys. Lett.

82, 421 (1981).

8. P. D. Kleiber, A. M. Lyyra, K. M. Sando, V. Zafiropulos, and W. C. Stwalley, J. Chem. Phys. **85**, 5493 (1986).

9. A. Benz and H. Morgner, Molec. Phys. **57**, 319 (1986).

10. B. A. Collings, J. C. Polanyi, M. A. Smith, A. Stowlow, and A. W. Tarr, Phys. Rev. Lett. **59**, 2551 (1987).

11. J.-C. Nieh and J. J. Valentini, Phys. Rev. Lett. **60**, 519 (1988).

12. S. Buelow, G. Radhakrishnan, J. Catanzarite, and C. Wittig, J. Chem. Phys. **83**, 444 (1985).

13. W. H. Breckenridge, C. Jouvet, and B. Soep, J. Chem. Phys. **84**, 1443 (1986).

14. D. G. Imre, J. L. Kinsey, R. W. Field, and D. H. Katayama, J. Phys. Chem. **86**, 2564 (1982).

15. M. Dantus, M. J. Rosker, and A. H. Zewail, J. Chem. Phys. **87**, 2395 (1987); R. M. Bowman, M. Dantus, and A. H. Zewail, Chem. Phys. Lett. **156**, 131, (1989).

16. R. B. Metz, T. Kitsopoulos, A. Weaver, and D. M. Neumark, J. Chem. Phys. **88**, 1463 (1988).

17. A. Weaver, R. B. Metz, S. E. Bradforth, and D. M. Neumark, J. Phys. Chem. **92**, 5558 (1988).

18. S. E. Bradforth, A. Weaver, R. B. Metz, and D. M. Neumark, <u>Advances in Laser Science - IV Proceedings of the 1988 International Laser Science Conference</u> (in press).

19. R. B. Metz, A. Weaver, S. E. Bradforth, T. N. Kitospoulos, and D. M. Neumark, J. Phys. Chem. (in press- to be published 2/90).

20. B. J. Ault, Acc. Chem. Res. **15**, 103 (1982).

21. A. B. Sannigrahi and S. D. Peyerimhoff, J. Mol. Struct. **165**, 55 (1988).

22. I. Last and M. Baer, J. Chem. Phys. **80**, 3246 (1983).

23. J. M. White and D. L. Thompson, J. Chem. Phys. **61**, 719 (1974).

24. S. E. Bradforth, A. Weaver, D. W. Arnold, and D. M. Neumark, manuscript in preparation.

25. D. G. Leopold, K. K. Murray, and W. C. Lineberger, J. Chem. Phys. **81**, 1048 (1984);

26. L. A. Posey, M. J. DeLuca, and M. A. Johnson, Chem. Phys. Lett. **131**, 170 (1986)

27. T. N. Kitsopoulos, I. M. Waller, J. G. Loeser, and D. M. Neumark, Che. Phys. Lett. **159**, 300 (1989).

28. K. Muller-Dethlefs, M. Sander, and E. W. Schlag, Z. Naturforsch. **39a**, 1089 (1984).

29. H. Hotop and W. C. Lineberger, J. Phys. Chem. Ref. Data, **14**, 731 (1985).

30. G. Caldwell and P. Kebarle, Can. J. Chem. **63**, 1399 (1985).

31. E. Pollak, J. Chem. Phys. **78**,

1228 (1983); D. K. Bondi, J. N. L. Connor, J. Manz, and J. Romelt, J. Mol. Phys. **50**, 467 (1983).

32. D. G. Truhlar and A. Kuppermann, J. Chem. Phys. **52**, 3841 (1970); S.-F. Wu and R. D. Levine, Mol. Phys. **22**, 881 (1971).

33. G. C. Schatz, J. Chem. Phys. **90**, 3582 (1989).

34. J. M. Bowman and B. Gazdy, J. Phys. Chem. **93**, 5129 (1989); B. Gazdy and J. M. Bowman, J. Chem. Phys. (in press).

35. J. Zhang and W. H. Miller, J. Chem. Phys. (submitted).

36. J. M. Bowman, Adv. Chem. Phys. **61**, 115 (1985).

37. E. J. Heller, J. Chem. Phys. **68**, 3891 (1978); Acc. Chem. Res. **14**, 368 (1981).

38. D. Kosloff and R. Kosloff, J. Comput. Phys. **52**, 35 (1983); R. Kosloff and D. Kosloff, J. Chem. Phys. **79**, 1823 (1983); R. Kosloff, J. Phys. Chem. **92**, 2087 (1988).

39. R. H. Bisseling, R. Kosloff, and J. Manz, J. Chem. Phys. **83**, 993 (1985).

40. N. E. Henriksen, J. Zhang, and D. G. Imre, J. Chem. Phys. **89**, 5607 (1988).

41. M. Broida and A. Persky, Chem. Phys. **133**, 405 (1989).

42. S. Sato, Bull. Chem. Soc. Jpn., **28**, 450 (1955).

43. J. V. Coe, J. T. Snodgrass, C. B. Friedhoff, K. M. McHugh, and K. H. Bowen, J. Chem. Phys. **83**, 3169 (1985); J. Chem. Phys. **87**, 4302 (1987).

44. D. G. Leopold, J. Ho, and W. C. Lineberger, J. Chem. Phys. **86**, 1715 (1987).

45. O. Cheshnovsky, S. H. Yang, C. L. Pettiette, Y. Liu, and R. E. Smalley, Chem. Phys. Lett. **138**, 119 (1987).

46. M. J. Deluca, B. Niu, and M. A. Johnson, J. Chem. Phys. **88**, 5857 (1988).

47. G. Gantefor, K. H. Meiwes-Broer, and H. O. Lutz, Phys. Rev. Lett. **37**, 2716 (1988).

48. J. Manz and J. Romelt, Chem. Phys. Lett. **81**, 179 (1981).

49. J. Manz, R. Meyer, and H. H. R. Schor, J. Chem. Phys. **80**, 1562 (1984).

50. G. C. Schatz, J. Chem.Phys. **90**, 4847 (1989).

LOW-ENERGY MOLECULAR COLLISIONS
- with applications to interstellar cloud problems -

Kazuo Takayanagi

Institute of Space & Astronautical Science, Yoshinodai, Sagamihara
Kanagawa-ken 229 Japan

Some commonly used theoretical approaches to the low-energy molecular collision dynamics are briefly reviewed. The present status of theoretical calculations of intermolecular potential is summarized. Taking examples from the systems relevant to the physics and chemistry of the interstellar molecular clouds, recent calculations of the rotational and vibrational cross sections and rate constants are surveyed. Efforts of finding scaling and fitting laws are briefly mentioned. Recent studies of vibrationally excited H_2 molecules, including collision-induced dissociation, are reviewed. Finally, works on ion-molecule collisions are discussed.

I. Introduction

The title of this talk includes atom-molecule and molecule-molecule encounters, where one of the colliding pair may be an ion. By low energy, I mean the energy range lower than a few eV. In the energy range under consideration, the elastic collision, the rotational or vibrational transitions or both in a single collision are common processes. Transitions in inversion doublet as in NH_3 and transitions in Λ doublet as in OH radical may be included as special cases. In some systems, chemical reaction or electron transfer process is also possible. When the collision energy is higher, dissociation of molecule becomes possible. Radiative association may also take place in the very low energy region. There are other type of collision processes including those accompanying electronic transitions. However, I would like to confine myself to the rotational and vibrational transitions in collisions. I will choose some topics from problems on interstellar molecular clouds.

In addition to the astrophysical problem, which I am particularly interested in, there are many other application fields of the molecular collision physics. Some examples are: anomalous ultrasonic absorption and dispersion, excitation of gas laser, transport properties of gases, collision broadening of spectral lines, etc. In early days, these phenomena are also the major experimental sources of molecular collision data. In more recent years, molecular beam technique and laser technique are used to acquire state-selected collision data experimentally. However, in most molecules, particularly molecules without containing hydrogen atoms, the rotational and vibrational level spacings are relatively small, so that even in low energy collisions there are often a large number of excited states open energetically. It is not an easy task to obtain a set of cross sections or rate constants for all the relevant elementary processes. This is the reason why many scientists are trying to find out empirical formula, scaling laws, and other relations between different cross sections for the same collision system or between cross sections for the same type of process for different systems.

Since early 1970's, there have been a large number of theoretical papers published on the subject. Many of them are to propose and test a new idea of theoretical approach in a simple model system. Study of vibrational transitions is often limited to the colliear configuration of the atom-diatom system or to the breathing sphere (averaging over the molecular orientation) approximation. Recently more and more efforts are made by people to calculate cross sections and rate constants for real systems with realistic potential functions. The results of these calculations can be compared directly with some experimental data and the accuracy of the basic theory and numerical calculations are tested. Also the results are very valuable for those who are studying interstellar molecular physics and chemistry and many other application fields.

In this article, I do not want to give too much technical details of the theoretical calculations, but I will describe briefly some frequently used theoretical approaches to the molecular collision dynamics. Also I will mention briefly of the intermolecular potential used in calculations. Then some examples of the calculated cross sections or rate constants will be summarized. Finally, a few specific problems will be discussed. Namely, I will mention a little about collision-induced dissociation and ion-molecule collisions.

II. Theoretical approaches to the Collision Dynamics

Here, only a brief sketch of more commonly used approximation methods will be given. For more details of these and other theoretical approaches please refer to appropriate books or review articles.[1-9] In the interstellar clouds, there are about 70 different molecular species detected so far. Collision partners are either H or He atoms or H_2 molecules, because the other particles have much less abundances. In this section, however, let us consider atom-diatom collisions, for simplicity. The diatom and the atom are assumed to be in a $^1\Sigma$ and a ^1S state, respectively.

Let $\vec{r}$ be the internuclear distance vector of the diatom and $\vec{R}$ the position of the atom relative to the center-of-mass of the diatom. The angle between these two vectors will be denoted by θ. We have to solve, under the appropriate boundary conditions, the Schrödinger equation of the following form:

$$(H - E)\Psi = 0, \quad (1)$$

$$H = -\frac{\hbar^2}{2\mu}\nabla_R^2 - \frac{\hbar^2}{2m}\nabla_r^2 + V(r) + V(\vec{r},\vec{R}). \quad (2)$$

The two kinetic energy terms can be decomposed into the radial part and the angular part:

$$\frac{\hbar^2}{2\mu}\nabla_R^2 = -\frac{\hbar^2}{2\mu R}\frac{\partial^2}{\partial R^2}R + \frac{\vec{\ell}^2}{2\mu R^2}, \quad (3)$$

$$\frac{\hbar^2}{2m}\nabla_r^2 = -\frac{\hbar^2}{2mr}\frac{\partial^2}{\partial r^2}r + \frac{\vec{j}^2}{2mr^2}. \quad (4)$$

The wave function Ψ of the whole system is now expanded in terms of the functions

$$y_{j\ell}^{JM}(\hat{r}, \hat{R})$$

$$= \sum_{m_j}\sum_{m_\ell} C(j\ell J; m_j m_\ell M) Y_{jm_j}(\hat{r}) Y_{\ell m_\ell}(\hat{R}), \quad (5)$$

which represents a state with a definite magnitude of the total angular momentum J and its z-component M. The quantity $C(j\ell J; m_j m_\ell M)$ is the usual Clebsch-Gordan coefficient. The expansion coefficients are further expanded in terms of the eigenfunction $\chi_n^j(r)$ of the molecular vibration. Finally, a set of differential quations for the radial function F's are obtained as follows:

$$[\frac{d^2}{dR^2} + k_{n'j'}^2 - \frac{\ell'(\ell'+1)}{R^2}] F_{n'j'\ell'}^{J\,nj\ell}(R)$$

$$= \frac{2\mu}{\hbar^2}\sum_{n''j''\ell''}\langle n'j'\ell'J|V|n''j''\ell''J\rangle F_{n''j''\ell''}^{J\,nj\ell}(R), \quad (6)$$

$$k_{n'j'}^2 = \frac{2\mu}{\hbar^2}(E - \varepsilon_{n'j'}), \quad (7)$$

where $\varepsilon_{n'j'}$ is the rovibrational energy eigenvalue, and the unprimed set (n,j,ℓ) indicates the incident channel. This set of equations is truncated and then solved numerically. This is the usual CC(close coupling or coupled channel) method.

Various integration methods have been proposed for numerical solution of the CC equations (6).[10,11]

So far the formulation is similar to that of electron-atom collision problems. In the molecular collision problem, however, we have much more energy levels involved, because of the small level spacings for vibration and rotation. Furthermore, a large number (say, hundred) of partial waves appreciably contribute to the inelastic collision. It is extremely laborious to solve the coupled equations (6) including all the relevant energy levels and for all the relevant partial waves. Various decoupling approximations have been introduced to reduce the burden of computation.[1,2]

There are two most frequency used approximations. They are

(i) CS approximation: $\dfrac{\vec{\ell}^2}{2\mu R^2} \rightarrow \dfrac{\hbar^2\bar{\ell}(\bar{\ell}+1)}{2\mu R^2}$ (8)

(ii) ES approximation: $\dfrac{\vec{j}^2}{2mr^2} \rightarrow \dfrac{\hbar^2\bar{j}(\bar{j}+1)}{2mr^2},$ (9)

where $\bar{\ell}, \bar{j}$ are the representative orbital and rotational quantum numbers, respectively. CS means the "(j_z-conserving) coupled states" or "centrifugal sudden", while ES stands for "energy sudden". The CS approximation is often introduced after the collision dynamics is formulated in the body-fixed frame where the z-axis is along the position vector $\vec{R}$. By introducing the CS approximation, the Coriolis coupling which gives rise to change in j_z in the body-fixed frame is nelgected, thus giving the name "j_z-conserving".

In the equation (6) in the space-fixed frame the above approximations lead to the following replacements:

(i) $\dfrac{\ell'(\ell'+1)}{R^2} \rightarrow \dfrac{\bar{\ell}(\bar{\ell}+1)}{R^2},$ (8a)

(ii) $k_{n'j'}^2 \rightarrow k_{n'\bar{j}}^2.$ (9a)

These approximations (with a slight difference) were briefly suggested in my article published in 1963.[12] In those days, however, electronic computers were not available at Saitama University, where I was teaching, and also at most of the other Japanese universities. I had only an electric calculator. Therefore, I could not apply the idea to solve any coupled equations. In 1970, Tsien and Pack[13] applied the combined CS and ES approximations to a molecular inelastic collision problem and their results were promising. Since then, the method has gradually become popular among the molecular collision scientists. The combination of CS and ES is now called the Infinite Order Sudden (IOS) approximation, since it includes the effect of the interaction potential to the infinite order. For more refined derivation of IOS approximation and also concerning the nature of this approximation method, please refer to Secrest,[14] Parker and Pack,[15] and references therein.

A less drastic approximation is obtained if only CS or ES is introduced. Validity and usefulness of these and other approximation methods are usually examined by comparing the resulting data with the corresponding full CC results. Kouri and Fitz,[16] for instance, have discussed in this way the validity and its limitation of the CS and IOS approximations.

The ES approximation in the form (9a) is equivalent to the neglect of the energy level spacings. For rotational transitions, where level spacings are usually small, this approximation and, furthermore, the IOS approximation often gives an encouraging result. However, this does not necessarily apply to the vibrational transitions where the energy quantum is usually much larger. Namely, we cannot further replace $k_{n'\bar{j}}$ by $k_{\bar{n},\bar{j}}$. Therefore, in many published calculations, sudden approximation is applied only to the rotational degrees of freedom, while the vibrational degree(s) of freedom is still treated according to the full CC method or, to save the computation time, the CS method. In this case, the relevant coupled equations are solved for the vibration, coupled with translation, for a fixed molecular orientation θ. Such a calculation is repeated at a sufficiently large number of different orientations to derive the scattering amplitude as the function of the molecular orientation:

$$f(n \rightarrow n'; \theta). \qquad (10)$$

The final scattering amplitude for the simultaneous vibrational-rotational transition is obtained by taking the matrix element of (10) with respect to the initial and the final rotational wave functions. This method is called VCC/RIOS or simply VCC/IOS method.

Foe some nonlinear molecules, Clary[17] introduced a new, intermediate approximation method. For symmetric-top molecules, for instance, the rotational Hamiltonian is of the following form

$$H_{rot} = (A - B)j_a^2 + B\vec{j}^2, \qquad (11)$$

where A and B are the usual rotational constants and j_a is the projection of the angular momentum operator of the molecular rotation along the symmetry axis. In some molecules, B is small but A is fairly large. Then it is a good approximaton to treat the term $B\vec{j}^2$ according to the IOS approximation, while the $(A - B)j_a^2$ term is treated on the same footing as the vibrational degrees of freedom. That is, the rotation around the molecular axis, together with the vibrational motion, are treated in the CC or CS method. When the CC method is applied, Clary called this scheme as AVCC/IOS method. "A" meaning "azimuthal". He has applied this method to the excitation of C_2H_2, D_2CO, and CH_3CN.

In early days when the electronic computers were not powerful enough to solve a large set of coupled equations, the breathing sphere approximation was often adopted to study the vibrational transitions. That is, the orientation-averaged interaction is used in the calculation of the cross sections. By doing so, the possibility of the rotational transitions are entirely suppressed. The vibrational transitions through anisotropic potentials are completely neglected. Thus, this approximation may give a considerable error.

So far, all the degrees of freedom are treated quantum-mechanically. Sometimes, the relative translational motion of the colliding pair is treated classically. Classical trajectory is determined either by assuming a more or less arbitrarily chosen spherical potential or by approximately taking account of the coupling between the translation and internal motion of the molecule. Billing,[8] for instance, proposes to use the expectation value of the intermolecular force potential with respect to the instantaneous wave function of the internal motion. Sometimes, both the translation and the molecular rotation are described classically and only the vibration is treated as quantized degree of freedom.

Because of its complexities, the vibrational transitions are often studied within the framework of the collinear collisions. Results of such calculations cannot be compared directly with experimental data. Three-dimensional treatment of vibrational transitions are often performed in purely classical manner, i.e., by the classical trajectory (CT) Monte Carlo method. When the potential surface has been calculated quantum-mechanically, this approach is sometimes called the quasiclassical trajectory calculation. The initial condition will be an ensemble of rotational-vibrational states corresponding to a specified quantum state. The final state distribution obtained in this method will be a continuous one, so that one has to introduce a discretization method to derive the state-to-state transition probability.

For low-velocity collisions, there are another group of approaches based on the use of adiabatic basis functions. An example of this type of formulation will be given when I discuss the ion-molecule collisions later on.

III. Interaction Potentials

Before we solve the dynamical part of a collision problem, we need the reliable potential functions for both intermolecular and intramolecular interactions. This is not a trivial part of the job. Reasonably accurate potential data are available only for a small number of systems.

Let us take the simplest system, $H_2 + H$, as an example. Margenau[18] was one of the earliest to calculate the intermolecular potential. He calculated the short-range force by the valence bond method and then added to it the long-range dispersion force. Mason and Hirschfelder[19] made another calculation for the same system. There was a considerable difference between the results of these two calculations.

In 1964, Porter and Karplus[20] calculated the potential surface for the H_3 system over a wider range of nuclear configuration. Their emphasis was placed on the saddle point region which is important for atom-transfer chemical reaction. Their result has been improved by Liu and Siegbahn.[21] Analytical fitting to the new result has been made by Truhlar and Horowitz.[22] The result is usually called LSTH surface. This surface has been used often to calculate the rotational and vibrational excitation cross sections as well as the chemical reaction rates. More recently, Varandas, et al.[23] have made further calculations and refit an analytic function to the numerical results. Somewhat earlier, Lewchenko, et al.[24] have made a near-Hartree-Fock SCF calculation. There are also some empirical potentials. One of them is by McGuire and Krüger.[25] They adopted for the spherical part of the potential the empirical potential from the beam experiment by Gengenbach, Hahr and Toennies,[26] while the Porter-Karplus potential was adopted for the nonspherical part. As is expected, the discrepancy among the potentials obtained recently, either by ab initio calculations or semiempirically, is much less than the discrepancy among the early calculations.

For this and for many other atom-homonuclear diatom systems, the analytic expression of the potential often takes the following form:

$$V(r,R,\theta) = V_0(r,R) + V_2(r,R) P_2(\cos \theta). \quad (12)$$

In the rigid rotor approximation, r is taken to be a constant. Calculations (see, for instance, Lewchenko, et al.[24]) show that the dependence on the internuclear distance r is very small for V_0, while it is fairly large for V_2 in $H_2 + H$, at least in the short-range repulsive force.

For another simple system, $H_2 + He$, a larger number of potential energy calculations have been made and analytical expressions have been proposed by fitting either to theoretical surface or to experimental data. The total and differential scattering cross sections, the rotational and vibrational excitation cross sections from beam experiments provide us a good guideline in selecting an empirical potential. Furthermore, the transport properties of H_2-He gas mixture, the NMR spin-lattice relaxation, the ultrasonic absorption and disperion, the line broadening, etc. are also used in determination of the potential. One of the most frequently used analytic potentials for this system is the one proposed by Gordon and Secrest,[27] based on their SCF calculation with configuration interaction. It is a short-range repulsive potential of the exponential form and has no attractive part. Shafer and Gordon[28] have derived an empirical potential with long-range dispersion part. Further theoretical calculations with configuration mixing have been made by Tsapline and Kutzelnigg[29] and then extended by Raczkowski and Lester.[30] Their results are fitted to an analytical form with $V_4(r,R) P_4(\cos \theta)$ added to (12). Probably, the most accurate ab initio potential surface available today for this system is the one by Meyer, Hariharan & Kutzelnigg.[31] The analytic form suggested by them also has the $V_4 P_4$ term. However, the magnitude of this term and the corresponding term from the TKRL calculations do not agree well. In any case, this term is more than an order of magnitude smaller than the V_0 amd $V_2 P_2$ terms and does not affect the rotational cross sections in any important way. However, Alexander[32] has shown that this small term gives an non-negligible contribution to the vibrational excitation.

As we proceed to more complex systems, the ab initio calculation becomes more and more difficult. In principle, the SCF method plus the configuration interaction calculations with a large basis set will give us a reliable short-range potential. At sufficiently large distances, the interaction consists of permanent multipole-multipole interactions, permanent multipole-induced multipole interactions, and the dispersion potential. The last-mentioned can be calculated by the perturbation method. In the intermediate distances near the van der Waals minimum, however, the perturbation method cannot be used. This is an important region to the low-energy collisions which we often encounter in the interstellar molecular clouds. The cancellation between the repulsive and attractive forces makes it difficult to accurately calculate the potential in this region. Sometimes, the ab initio short-range potential plus the damped long-range potential ("damped" means "with an artificial cut-off factor") is adopted and the damping factor or adjustable parameters in this factor are chosen empirically. For instance, Bačić, Schinke and Diercksen[33] used the following potential for the $CO + H_2$ collisions.

$$V = \sum_{\lambda_1\lambda_2} [V^{SCF}_{\lambda_1\lambda_2}(r_1,R) + V^{dis}_{\lambda_1\lambda_2}(r_1,R) f(R)]$$

$$\times P_{\lambda_1}(\cos \gamma_1) P_{\lambda_2}(\cos \gamma_2), \quad (13)$$

$$f(R) = \exp[-\gamma(\frac{D}{R} - 1)] \quad \text{for} \quad R \leq D$$
$$= 1 \quad \text{for} \quad R \geq D.$$

The H_2 is regarded as a rigid rotor, r_1 is the internuclear distance of CO, γ_1 and γ_2 are the orientation angles of the CO and H_2 molecules relative to the intermolecular distance vector $\vec{R}$. The two terms V^{SCF} and V^{dis} are the short-range potential and the asymptotic form of the dispersion potential, respectively. The two adjustable parameters had been fixed by comparing the rotationally inelastic beam experiment and a coupled state calculations at 87.2 meV for the system of $D_2 + CO$. The values chosen in this way were $D = 4.46$ Å and $\gamma = 6$.

In early 1970's, when the detailed ab initio calculation of the short-range potential was still too difficult for most systems, a simple, approximate approach to estimate the intermolecular potential was introduced. That is the free-electron-gas model.[34,35] When there is no reliable interaction potential available, people often use this method to obtain the potential for cross section calculations. In this approach, it is assumed that the electron distributions in the colliding atoms or molecules (in the isolated condition) are known from, say, the SCF calculations. These charge distributions will be denoted by ρ_A and ρ_B. Then it is assumed that when the two atoms or molecules approach each other, the total charge density is approximately given by the sum of the undeformed density distributions: $\rho \cong \rho_A + \rho_B$. Now the Coulomb energy (static force potential) can be calculated for this charge distribution. Then, the energy densities for the kinetic energy, for electron exchange effect, and the correlation effect are calculated using the formulae for the uniform free-electron gas. For instance, the kinetic energy density is given by a known constant factor times $\rho^{2/3}$. In this way, the total energy of the system and thus the intermolecular interaction potential can be calculated. Improvements and extensions of this method have been proposed by some people.[36]

In Table III in a later section, where some theoretical studies of rotational transitions in polyatomic targets are listed, the potential function used is also indicated. It is seen that the free-electron-gas model potetnail has been used in many systems.

IV. Molecular Processes in Interstellar Clouds

A. Interstellar Molecules

Before discussing some examples of rotational and vibrational cross sections and rate constants, a brief introduction to the interstellar medium will be given. Forty years ago, when I got interested in this field of astrophysics, the interstellar medium was roughly classified into H I and H II regions. This classification was based on the state of existence of hydrogen, which is the most abundant element in the universe. In H I regions, hydrogen is in the neutral atomic form, while they are almost completely ionized in the H II regions. In both H I and H II regions, the portion of the medium where gas density is much higher than the surroundings is called the interstellar clouds. The H I clouds known in those days have typical density of ten atoms per cm^3 and representative temperature is 100 K. These clouds are now called the diffuse clouds in order to make distinction from much denser dark clouds. In early days, the dark clouds were known only as the obstacles to stop the stellar radiations from behind and these clouds are outside the investigations of atomic and molecular processes. In the diffuse H I clouds, only three molecular species (CH, CH^+, and CN) were detected through optical observations.

The H II regions are kept ionized either because of the extremely low density (where the electron-ion recombination rate is very low) or because of the UV and X-rays illumination by the nearby star(s). In the latter case, the typical temperature is 10^4K. These ionized clouds are usually seen as emission nebulae. Here we cannot expect any molecular species to exist.

Now the dark clouds with higher gas density and with dust particles can be studied by observing the microwave and the millimeter-wave emission lines arising from the rotational transitions of the interstellar molecules. In more recent years, the infrared radiations are also providing us valuable information concerning molecular species (and also the dust particles) in these clouds. Beginning with the discovery of the OH radicals in early 1960's, about 70 different molecular species (isotpe variations disregarded) have been detected so far. In the dark clouds, hydrogen is in the molecular form. The H_2 molecule is now the most abundant species. Relative to this molecule, the abundance of the He atom is about 10^{-1}. The second most abundant molecular species is CO. The relative abundance of this species is about 10^{-4}. A partial list of detected interstellar molecules is given in Table I. The relative abundances of the complex organic species are of the order of $10^{-8}-10^{-10}$ or less.

Table I A Partial List of Detected Interstellar Molecules

Simple molecules:
H_2, CO, CN, CS, NS, PN, CH, OH, SiO, SO, HCl, HCO, HCN, HNO, OCS, SiC_2, H_2O, H_2S, NH_3, H_2CO, H_2CS, HNCO, HNCS, C_2H_2, etc.

Long chains:
C_nH (n = 1-6), C_2S (n = 1-3), $HC_{2n+1}N$ (n = 0-5)

Other larger molecules:
HCOOH, CH_3OH, CH_3CN, NH_2CHO, CH_3NH_2, CH_3CHO, CH_3C_3N, $HCOOCH_3$, CH_3CH_2OH, etc.

Ring molecules:
C_3H, C_3H_2

Maser action found in
OH, H_2O, SiO, CH_3OH

Ions:
CH^+, HCO^+, HOC^+, HCS^+, $HOCO^+$, etc.

These dark clouds, often called the interstellar molecular clouds, have temperature much lower than the diffuse H I clouds, because of the large optical depth for the UV radiations from outside and the efficient cooling due to the rotational transitions of molecules, emitting infrared radiations. In the core of the clouds, 10-20 K are the representative temperature, while the surface layer is often warmer. The number density of the molecular clouds varies from cloud to cloud, but typically $10^3 - 10^6$ particles/cm^3. The relative abundance of the atoms and molecules tells us that we need to consider only H_2 and He as the collision partners when we discuss the rotational and vibrational transitions of molecular species in these clouds. Near the surface of the molecular clouds and inside the diffuse H I clouds, however, hydrogen is mostly in atomic form. Therefore, collisions with H atoms are also important.

Even in the dark clouds, some UV photons with longer wavelengths which cannot ionize the major constituents, H_2 and He, will be found particularly in the surface layer. These photons ionize particles with lower ionization energies, such as the carbon atom. The cosmic ray particles can penetrate through the dark clouds. These energetic particles produce some ionization throughout the cloud. Therefore, there are small number of free electrons and ions in the dark clouds although the number density of these charged particles is very low.

In Table II, low-energy molecular processes are listed. Some of them, e.g. (I,4), are not important in the interstellar clouds. In this review talk, as was already mentioned, I would like to concentrate to the rotational and vibrational transitions and ion-molecule reactions.

Molecular collision processes in interstellar clouds have been reviewed recently in two articles, one by Black[37] and the other by Flower[9].

Table II Low-Energy Molecular Processes

I. Molecular formation:
   (I,1) $A^{(+)} + B \rightarrow AB^{(+)} + h\nu$
       e.g., $H^+ + H$; $CH_3^+ + H_2$
   (I,2) $A^- + B \rightarrow AB + e^-$
       e.g., $H^- + H$
   (I,3) $A^* + B \rightarrow AB^+ + e^-$
   (I,4) $A + B + C \rightarrow AB + C$
   (I,5) surface reactions

II. Molecular destruction:
   (II,1) $AB + h\nu \rightarrow A + B$
   (II,2) $AB + e^- \rightarrow A^- + B$
   (II,3) $AB^+ + e^- \rightarrow A + B$
   (II,4) AB + ion, such as
       $CO + He^+ \rightarrow C^+ + O + He$
   (II,5) cosmic-ray particle impact

III. Rotational and vibrational transitions

IV. Charge transfer

V. Chemical reaction

B. Rotational and Vibrational Transitions of Interstellar Diatomic Molecules

Cross sections for collision-induced transitions between low-lying rotational and vibrational levels of H_2 and CO were summarized in my previous review article.[38] In H_2, the rotational excitation from j=0 to j=2 (note that only Δj = even is allowed in the homonuclear diatomic molecules) has the cross section maximum of about several Å^2 in the energy region around 1 eV, while in CO the cross section peaks for $j=0 \rightarrow j=1$ and $0 \rightarrow 2$ are in the energy range of a few meV, immediately above the respective excitation threshold. The peak values are about several tens of $Å^2$. The difference in the maximum cross sections for H_2 and CO may be due to the fact that H_2 is small and nearly spherical (internuclear distance is small), while CO is not. The difference in the location of the cross section maxima in the two cases can be understood, at least qualitatively, by the Massey's criterion. The Massey's parameter representing the adiabaticity is

$$w = a\, \Delta E/\hbar v, \qquad (14)$$

where a is the range of the relevant interaction, ΔE is the energy transfer between the translation and the internal degrees of freedom, $2\pi\hbar = h$ is the Planck's constant and v is the velocity of relative motion. The criterion says that if $w \gg 1$, the cross section for the inelastic collision under consideration should be negligibly small. Very often the cross section maximum comes in the velocity region where $w \sim 1$. Thus the difference in the level spacing ΔE explains the large difference between H_2 and CO in the location of cross section maxima.

Multiplying the velocity and taking average over the Maxwellian distribution, we obtain the rate constant as a function of the gas temperature. The rate constant for an excitation usually increases rapidly first and then have a broad maximum, while the rate constant for deexcitation has much less T-dependence. In the rotational transitions of CO in collision with, say, H_2 for T < 100K, there is an approximate selectrion rule, i.e. the rate constants for $\Delta j = \pm 1, \pm 2$ are comparable and the rates for larger change in j are considerably smaller. As T increases further, collision rates for $\Delta j = \pm 3$ and ± 4 become comparable with those for $\Delta j = \pm 1$ and ± 2. [It is noted that in the review article by Flower[9] the theoretical rate constants for the rotational transitions of H_2, OH, CO and NH_3 in collision with H_2 are tabulated.]

The CO molecule has a small, but finite dipole moment. Therefore, the radiative transitions from j to j-1 can take place. The rotational level population is thus determined by the balance between the collisional and the radiative transitions. If the number density of the major component, H_2, of the clouds is 10^6/cm^3 or more, the collisional transition rates are much larger than the radiative transition rates at least for the levels below j = 5. [The level population in the higher levels is very small for the typical cloud temperature which is below 20K. Therefore, we need not worry about those higher levels.] Thus the Boltzmann distribution will be maintained, at least approximately. However, if the number density is less than 10^4/cm^3, the collisional transition rates for the levels j > 3 are less than the radiative decay rates, so that the Boltzmann distribution cannot be maintained. In this way, the relative population of the rotational levels of CO (and other molecules also) reflects the number density (and of course also temperature) of the gas, provided that the gas density is within an appropriate range. The absolute intensity of the emission lines then gives us the column density of the molecular species under condieration. It may be added that CO is a rather stable molecule and can be found in many parts of interstellar clouds, so that its emission lines, particularly the line at the wavelength of 2.6 mm arising from j=1 → 0 transition is convenient to probe the structure and motion of the interstellar clouds.

Another widely distributed diatomic species in interstellar molecular clouds is the OH radical. In this radical, the lowest electronic state is $^2\Pi_{3/2}$, j = 3/2. Because of the Λ doubling, this lowest state (as well as other rotational states) is split into two energy levels with opposite parity. The transition between these two levels in the ground state produces the radiowave with wavelength of 18 cm. To say more precisely, this spectral line is further split into four hyperfine components. It was in this group of lines where the astropohysical maser action was discovered for the first time. Theoretical calculations of Λ doublet and hfs transitions, see Dewangan, Flower and Alexander[39] and Corey and Alexander[40] and references therein.

In this and other cases where the molecules are not in $^1\Sigma$ state, the theoretical formulation of the dynamical problems requires some modification. General formulation of collision dynamics with proper angular momentum coupling for molecules in $^2\Sigma$, $^2\Pi$, $^3\Sigma$, etc. interacting with either an atom in a ^1S state, or, more generally, an open shell target, has been discussed in many papers.[41-44]

In addition to H_2, CO, OH, there are some other diatomic interstellar species for which theoretical studies of the rotational transitions have been made. (CH, CS, SiO, etc.)

Except for H_2 and CO, theoretical studies of the vibrational transitions in diatomic species are very little. For hydrogen molecules, the absorption of UV photon followed by fluorescence may produce the vibrationally excited states,[45-47] to which I will come back later.

C. Rotational and Vibrational Transitions of Polyatomic Molecules

For polyatomic molecules, some theoretical studies of rotational transitions are listed in Table III. NH_3 and H_2CO have been more frequently studied than the other molecules. The ammonia molecule has a large dipole moment and level spacings are also large, so that the rates for the radiative transitions are large. Even when the number density of the cloud is as high as

Table III Some Theoretical Studies of Collision-Induced
Rotational Transitions in Polyatomic Molecules*)

Molecule	Authors	Interaction; Calculation Method; E (or T) Range
HCN	Green, Thaddeus (1974, ApJ)	e-gas potential; CC; 5-100K
	Monteiro, Stutzki (1986, MN)	Green-Thaddeus pot.; CC; hfs taken account; $\leq$30K
H_2O	Green (1980, ApJSuppl)	by He, $H_2(j=0)$; e-gas potential
	Palma, Green, DeFrees, McLean (1988, ApJSuppl)	SCF-correlation pot.; CS; below 1400K
SO_2	Palma (1987, ApJSuppl)	e-gas pot.; IOS; 25-125K
OCS	Green, Chapman (1978, ApJSuppl)	by $H_2(j=0)$; e-gas pot.; CT; 10-100K
H_2CO	Garrison, Lester, Miller, Green (1975 ApJL); Garrison, Lester, Miller (1976, JCP)	CI potential; CC; 20-95K
	Garrison, Lester (1977, JCP)	CI potential; CS; 25-95K
	Green, Garrison, Lester, Miller (1978, ApJSuppl)	CI potential; CS; 10-80K
	Bocchetta, Gerratt, Guthrie (1988, JCP)	SCF+correlation pot.; R-matrix; 18-120K
NH_3	Green (1976, JCP)	e-gas pot.; CC, CS, EP; 100, 190, 250 cm^{-1}
	Davis, Boggs (1978, JCP)	e-gas pot.; SC; 50-1000 cm^{-1}
	Green (1980, JCP)	SCF+semiempirical long-range pot.; CS; 15-300K
	Billing, Poulsen, Diercksen (1985, CP)	SCF+correlation(MBPT); SC; 50-300K
	Danby, Flower, Kochanski, Kurdi, Valiron, Diercksen (1986, JPhysB); Danby, Flower, Valiron, Kochanski, Kurdi, Diercksen (1987, JPhysB)	by H_2; SCF+correlation(MBPT); CC; 15-300K
	Billing, Diercksen (1987, CP)	by $H_2(j=0,1)$; SCF+corrlation(MBPT); SC-CS; 50-300K
HC_3N	Green, Chapman (1978, ApJSuppl)	e-gas potential; CT; 10-80K
C_3H_2	Green, DeFrees, McLean (1987, ApJS)	SCF+correlation pot.; IOS; 30-120K
	Avery, Green (1989, ApJ)	SCF+correlation pot.; CS; 10-30K
CH_3OH	Billing (1986, CP)	CH_3-torsion excitation; sum of atom-atom potentials; SC; 5-50 kJ/mol
CH_3CN	Green (1986, ApJ)	by $H_2(j=0)$; e-gas potential for CH_3CN+He system used; IOS; 20-140K
	Clary, Green (1987, CP)	e-gas potential; ACC/IOS; 60-150 cm^{-1}
HCO^+	Monteiro (1984, MN)	SCF potential; CC; 5, 10K
	Monteiro (1985, MN)	by $H_2(j=0)$; SCF+CI; CC; below 110 cm^{-1}
HCS^+	Monteiro (1984, MN)	SCF potential; CC(5, 10K), CS(10-60K)

*) Collision partner is He unless otherwise stated. ▽ Journal abbreviations: ApJ=Astrophys.J.; AA= Astron. & Astrophys.; CP=Chem.Phys.; JCP=J.Chem.Phys.; MN=Mon.Not.; MP=Molecular Phys. ▽ Calculat. Method: CC=close coupling; CS=coupled states; CT=classical trajectory; SC=semiclassical, MBPT= manh-body perturbation theory, etc.

$10^7/cm^3$, the radiative decay rate is much larger than the collisional transition rates. Thus, the molecule in an excited level will radiate and come down quickly to the lowest level for the given K quantum number. [K represents the projection of the rotational angular momentum vector along the molecular symmetry axis. The radiative selectrion rule is $\Delta K = 0$.] Then, a molecule in (j,K) = (2,2) state will jump to (1,1) state slowly by collisions and from (3,3) the molecule will go mainly to (1,0), while from (4,4) the result of collision is either (2,1) or (1,1).[9] Since para-ortho transitions are forbidden, NH_3 molecules will remain mainly in (0,0) and (1,1) states. This is far from the Boltzmann distribution. Only in the regions with high density and high temperature, the finite population in nonmetastable levels, such as (2,1), (2,2), (3,3), etc. can be detected.[48] Sometimes interstellar molecules are found in fairly high rotational states. These are explained usually by collisional excitations. For instance, Loren, et al.[49] detected in Orion Molecular Cloud OMC-1 region the cyanoacetylene molecules (HC_3N) distributed up to the level j = 31. This can be explained by assuming the gas kinetic temperature of 125K and the number density around $10^7/cm^3$. Sometimes the vibrationally excited molecules are detected. These are usually interpreted as due to the pumping by infrared radiations from nearby stars, protostars or warm dust particles.

Theoretical studies of vibrational excitations of polyatomic molecules in collision are much less than for rotational transitions. Exception is the molecule CO_2.[50] However, the existence of this molecule in the interstellar clouds has not been confirmed yet. Clary[51] has studied the system SO_2 + He, and Peet and Clary[52] studied D_2CO + He in the AVCC/IOS approximation.

D. Scaling and Fitting

In pure rotational transitions and also rotational transitions associated with vibrational transitions, there are often a large number of energetically open channels. Therefore, a large array of state-to-state inelastic cross sections at many values of total energy of the system or a large array of state-to-state rate constants at various temperatures are required in order to interpret the astrophysically or laboratory observed data or to make modeling calculations. In most theoretical works, however, only the excitations from the ground state are reported. Experimental data are also fragmentally. Therefore, it is desirable to find out useful relations among the cross sections or rate constants for the same system so that we can complete the state-to-state transition array from the available portion of it. Since all the cross sections and all the rate constants come from the common interaction potential for a given system, they are not entirely independent. The simplest relation of this kind has been derived from the sudden approximation. For a linear molecule colliding with an atom, for instance, there holds a relation between the rotational cross sections:[53]

$$\sigma(j \to j'; E) = \sum_{j_t=|j-j'|}^{j+j'} \sqrt{(E'/E_t)} \times [C(jj_tj';000)]^2 \, \sigma(0 \to j_t; E), \quad (15)$$

where E in the two cross sections (σ's) is the energy of relative motion before $j \to j'$ or $0 \to j_t$ transition and E', E_t are the final energies in the respective transitions. The quantity $C(:::)$ is the usual Clebsch-Gordan coefficient. This formula is useful when the excitation cross section from the ground state $\sigma(0 \to j_t)$ becomes rapidly small as j_t increases. Then all the state-to-state cross sections are represented by only a small number of cross sections. Of course, the sudden approximation is not always applicable. Various corrections to this simple relation have been proposed. One of these approaches is the so-called "energy-corrected sudden" (ECS) approximation. Richard and DePristo[54] applied this approach to the nonlinear molecule collision, i.e., to NH_3 + He with encouraging results. The ECS theory has been applied also to vibrational transitions. See, for instance, Clary and DePristo,[50] where CO_2 + He collisions are analyzed, and references therein.

There are another type of scaling method where experimental data are fitted to a parametrized function. This empirical function is then used to predict rate constants which have not been measured directly. In the so-called "statistical power gap law", the rate constant is assumed to be proportional to some statistical weight factor and a power $|\Delta E|^{-\gamma}$ of the change ΔE of the relative kinetic energy. In the "exponential gap law", an exponential factor $\exp(-\theta|\Delta E|)$, where θ is a parameter, is introduced instead of the power. Information theoretic arguments are often invoked. These approaches and applications to some real systems are discussed in the review article by Brunner and Pritchard.[55] See also the "AON" (A on N) procedure introduced by Whitaker and Brechignac.[56]

Usually, the parametrization procedure is applied to the rate constant or cross section. Recently, a new scaling method has been proposed by Kreutz and Rabitz[57] who seek for an intermediate level parametrization of the collision dynamics. The scattering matrix is expressed as

$$S = \exp(-i\underline{B}). \quad (16)$$

In strong coupling cases, the behavior of the S matrix is often rather complicated. Even then, the behavior of the B matrix will be much smoother. Therefore, the parametrization of the latter will be more efficient way of scaling than the direct parametrization of the S matrix or the cross section. The authors of the above-mentioned work have obtained satisfactory result for the system of HD + He.

V. Processes Involving H_2 Molecules in Highly-Excited Vibrational States

The H_2 molecule, the most abundant molecule in the dense clouds, does not have the dipole moment, so that the rotational and vibrational transitions are optically forbidden. However, because of the quadrupole moment, the radiative transitions are possible although the spontaneous emission rates are very small. In recent years, the interstellar H_2 molecules have been detected through the infrared emissions due to the vibrational-rotational transitions. The regions producing these infrared emission lines are either shock-heated region or the region illuminated by strong ultraviolet radiations from nearby star(s). In the former, the H_2 emission is the thermal emission from the hot gases (a few to several thousand degrees). In the latter, if the gas density is sufficiently high and the UV radiation is sufficiently strong, the gas is heated up to more than a thousand degree, so that thermal emission is again possible. However, there is another type of emission coming from the UV absorption followed by the fluorescence. The absorption of UV photons in appropriate wavelength region around 1000 Å leads to the electronic excitation to $B^1\Sigma_u^+$, $C^1\Pi_u$ states. The fluorescence does not necessarily return H_2 in the vibrationally ground level in the $X^1\Sigma_g^+$ state. Generally, the H_2 molecule will return to one of the vibrationally excited states and about 10 % of the flourescence will lead to the continuum of the electronically ground state, i.e., the dissociation of the molecule.[58] In this way, the highly excited H_2 can be produced without involving collision processes. The apparent vibrational temperature can become several thousand degrees. Of course, as the gas density increases, the collisional de-excitation of these highly excited molecules must be taken into account. In both shock-heated regions[59] and UV-illuminated regions,[60-64] if there are an appreciable number of H_2 in the highest vibrational level (v=14), even the low energy collision with another particle or the absorption of low-energy photon can produce the dissociation of the molecule. Therefore, the collisional transitions among the highly-excited vibrational levels and also the collision-induced dissociation in H_2 are the important problems in the interstellar clouds.

A. Collision-Induced Transitions of H_2 Among the Highly-Excited Vibrational States

Most of the theoretical works in the past are limited to the study of the vibrational excitation from the ground state (v=0) to the first or the second excited states. Exceptions are the work by Garcia and Lagana[65] and a more recent work by Cacciatore, Capitelli and Billing.[66] The former is the classical trajectory calculation for H_2 + H using the LSTH potential surface and the rate constants of the following processes have been derived at 300K:

$$H_2(v<10) + H \rightarrow H_2(v-\Delta v) + H, \quad \Delta v = 1-4. \quad (17)$$

The latter authors studied

$$H_2(v) + H_2(0) \rightarrow H_2(v-1) + H_2(0) \quad (18)$$

$$H_2(1) + H_2(v) \rightarrow H_2(0) + H_2(v). \quad (19)$$

The molecular rotation and the relative motion were described classically, while the vibration was treated quantum mechanically. Both in (18) and (19), the cross sections increase rather rapidly with increasing v. However, they did not calculate the cross sections for the process with $|\Delta v| > 1$ and also for the simultaneous transitions such as

$$H_2(v) + H_2(0) \rightarrow H_2(v-1) + H_2(1), \quad (20)$$

which are expected to be important to accelerate the vibrational relaxation.

Experimentally, the vibrational relaxation time derived from laboratory experiments gives primarily the rate constant for collisional transition between the v=0 and v=1 states. As far as I know, there is no experimental data for transitions among the highly-excited vibrational states of H_2.

Therefore, we still need a large number of other rate constants in, for instance, the modeling of the interstellar shocks. Usually, some cross sections are tentatively assumed. In some work, it is assumed that only the transitions to the nearest neighbor states are important and use such an expression as

$$k(v \rightarrow v-1; T) = k(1 \rightarrow 0; T) f(v,T). \quad (21)$$

The simplest choice of the conversion factor $f(v,T)$ is the Landau-Teller relation

$$f(v,T) = v, \quad (22)$$

corresponding to the harmonic oscillator colliding with an atom through the interaction which is linear in terms of the internuclear distance of the molecule. Sternberg and Dalgarno[62] used this approximation (21), (22). In the paper by Draine, Roberge and Dalgarno[59] discussing the molecular processes in interstellar shocks, they are using the rate constant for the multiquantum transitions as follows:

$$k(v \to v-\Delta v) = k(1 \to 0) \left[\frac{1 + 1.64x}{1+1.64x/\Delta v}\right]^{3/2}$$

$$\times \exp\left[\frac{1.5\Theta}{\Theta + T}(1-\Delta v)\right], \quad (23)$$

where $\Theta = 5987K$ and $x = |\Delta E_{vib}|/\kappa T \cong \Delta v \cdot \Theta/T$, κ is the Boltzmann constant. For collisions with He and with another H_2, the rate constant $k(1\to 0)$ is adopted from the experiment by Dove and Teitelbaum.[67] In a more recent paper by Sternberg and Dalgarno,[62] the cross section for $H_2(v,j) + H$ is arbitrarily chosen to be

$$\sigma(vj \to v'j') = \left(\frac{2j'+1}{2j+1}\right)\frac{6 \times 10^{-16}}{(v-v')^2} \text{ cm}^2$$

$$\text{with } \Delta j = -2 \Delta v. \quad (24)$$

The propensity that the simultaneous transitions $\Delta j = -\alpha \Delta v$ are favorable is known in some collision systems in laboratory experiments.[68] However, the functional form in (24) must be checked by theoretical studies.

Onda[69] has recently calculated vibrational cross sections for highly-excited H_2 in collision with H. The potential surface adopted is "Double Many Body Expansion" potential calculated by Varandas, et al.[23] His approach to the dynamical problem is VCC/IOS, i.e. the close coupling method for the vibrational degree of freedom and the IOS approximation for the molecular rotation. The preliminary results show that the de-excitation rate $k(1 \to 0; T)$ is not far from that adopted in the work by Draine, Roberge and Dalgarno,[59] but the rates $k(2 \to 1;T)$ and $k(2 \to 0; T)$ are more than a factor of two smaller than the expression in the reference 59.

In the temperature range from 500 to 5000K, all the de-excitation cross sections from the initial state $v_i = 6$ are close to each other (within a factor of two). That is, the multiquantum transitions are as fast as the single quantum transition, indicating the strong coupling nature of the problem.

For initial states above about $v = 5$, the good convergence of the CC calculation cannot be achieved unless the continuum states are also included in the set of basis functions. Because of the difficulties handling the continuum wave functions, it is extremely laborious to calculate the reliable cross sections, for, say, $v_i > 10$.

B. Collision-Induced Dissociation of H_2

Dalgarno and Roberge[70] discussed the dissociation of H_2 (and CO) in the interstellar clouds. Because of the low gas density, collision frequency is very low in these gases. For this reason, vibrationally excited molecules may radiate infrared photons and go down to lower energy levels before further excitation, de-excitation, or dissociation takes place by collision. Even for H_2, which has a long lifetime (around 10^6s) against the infrared photon emission, the radiative transitions can not be ignored when the gas density is low. Dalgarno and Roberge have estimated that the rate of H_2 dissociation will drastically decrease when the gas number density becomes below about 10^4 cm^{-3}. In order to make this and other estimations more conclusive, the reliable cross section data for collision processes are required. Since there is no reliable cross section data for collision-induced dissociation, it is often assumed that the molecule in its highest vibrational state (v=14 for H_2) will get dissociated as soon as another atom or molecule collide with the excited H_2. Some people assume that the dissociation cross section for $H_2(v=14)$ is the same as the cross section for excitation from $v = 13$ to $v = 14$.[71] In any case, we need the accurate excitation and de-excitation cross sections before we can estimate the dissociation rate.

For the dissociation in the system $H_2 + He$, the classical trajectory calculations are available. Dove and Raynor with their coworkers studied this system in a series of papers.[72-73] The calculations are mainly classical, but for the transitions to the nearest neighbor levels, they also applied the distorted wave method. They derived the analytic expression of the overall rate constant for dissociation in the low pressure as well as high pressure conditions. In the low pressure limit, only the dissociation from the ground state $v = 0$ is important. The rate constant they obtained is

$$\log k(\text{cm}^3/\text{s}) = 3.801 \log T -27.029$$
$$-29487/T. \quad (25)$$

In the high pressure limit, on the other hand, the radiative decays can be neglected. The rate constant becomes

$$\log k(\text{cm}^3/\text{s}) = -1.75 \log T - 2.729$$
$$-23474/T. \quad (26)$$

At T = 2000K, for instance, there are nine orders of magnitude difference between these two limits.

For $H_2 + H$ collisions, Blais and Truhlar[74] made the classical trajectory calculation for the dissociation for the initial molecular states v = 0, 1, 2, 6, 12, and 14 with selected rotational states. The LSTH potential was used. More recently, Dove and Mandy[75] calculated, again in classical treatment, the dissociation rate for $H_2(v=j=0) + H$ in the temperature range of 10^3–10^5 K. Here the excitation to the quasi-bound states is interpreted as the indirect dissociation.

In addition to these, there are the dissociation cross section calculation by the distorted wave method by Sakai[76] and another work by an impulse approximation by Koizumi.[77] Onda[69] has extended his calculations to the collision-induced dissociation of highly vibrationally excited H_2 by H atom. He discretizes the continuum of H_2 by assuming that the molecular wave function vanishes at and beyond $r = 18\, a_o$ (a_o = the Bohr radius = 0.529 Å). This artificial limitation of the continuum wave functions should not affect the dissociation cross section drastically because the transition matrix elements of the intermolecular potential are appreciable only within about $r < 10\, a_o$. The excitation of discretized levels above the dissociation threshold is regarded as the dissociation. In this way he has obtained some dissociation cross sections.

The rotational excitation is also important because the centrifugal force is generally to assist the dissociation by pushing up the vibrational levels. Furthermore, the centrifugal force, combined with the attractive dispersion force, produces the potential barrier, which gives rise to some quasibound (predissociating) states above the dissociation energy.

The collision-induced dissociation of diatom is closely related to the three-body recombination of atoms.

$$H + H + M \rightarrow H_2^* + M \quad (27)$$

When the gas pressure is relatively low, the chance of three atoms coming together into interacting region simultaneously should be very small. Therefore, the main recombination process in low pressure gases will be those via the quasibound states:

$$H + H \rightarrow H_2^*, \quad (28)$$

where H_2^* is a quasibound state. If H + H and H_2^* are in statistical equilibrium, the recombination rate can be obtained by calculating the probability of collisional stabilization of H_2^* into a really bound state. Recent work of this kind include the one by Orel[78] for H + H + H and the one by Schwenke[79] for H + H + H_2. In the former, the details of the calculation is not shown, except that the accurate SLTH potential has been used. In the latter, the classical trajectory approach was used to calculate the stabilization cross section. When the gas pressure increases, the real three-body recombination without passing through quasibound states will become non-negligible. On the other hand, if the gas pressure is very low as in the interstellar space, the quasibound state will decay into H + H before collisional stabilization, so that the three-body recombination process does not work.

VI. Ion-Molecule Collisions

In diffuse interstellar clouds, ultraviolet radiations with photon energy above 13.6 eV are immediately absorbed in the surface layer by the H atoms. However, the photons with energies below 13.6 eV can still ionize some atoms and molecules, such as the C atom (ioniz. energy = 11.26 eV), the Si atom (I.E. = 8.15 eV), the CH radical (I.E. = 11.13 eV), etc. From this kind of consideration, we expect that the number density of free electrons and that of positive ions are respectively about 3×10^{-4} times the number density of hydrogen atoms, which are the most abundant particles in these clouds. In dense clouds, the radiation with photon energy above 15.4 eV is absorbed by H_2 and there are also dust particles which prevent radiations penetrating into deep inside the clouds. In the core part of these dense clouds, the main source of ionization is the cosmic ray particle impact. The ionization degree in the cloud core will be less than in diffuse clouds. In spite of this low ionization degree, the existence of some ions has significant effects in the interstellar chemistry. It is well known that most neutral-neutral chemical reactions have rate constant rapidly decreasing with decreasing temperature below about room temperature. Therefore, in the core of dense clouds, where the typical temperature is of the order of 10 K, the neutral-neutral reactions are practically forbidden. On the other hand, the ion-molecule reactions have usually no potential barrier along the reaction path and the rate constant remains appreciable down to the very low temperatures as long as the reactions are exothermic. For this reason, the interstellar chemistry is controlled primarily by the ion-molecule reactions.

Because of the long-range interactions characteristic to the ion-molecule systems, the rotationally inelastic cross sections are also large in these systems as compared with neutral-neutral collisions. However, because of the low number density of ions, the neutral-neutral collisions are probably more important in the rotational excitations of interstellar molecules.

In theoretical studies of the low-energy ion-molecule collisions, it is convenient to use the adiabatic basis functions. More than ten years ago, I got interested in the collisions between two polar molecules, such as HF + HF. Because of the strong dipole-dipole interaction, the molecule will tend to take a preferable orientation during collision and this will have some influence on the vibrational transitions. In order to study the relative orientation during collision, I immediately adopted the PSS

(Perturbed Stationary State) approach. The PSS approach was originally introduced into the ion-atom collisions by Massey and Smith[80] in 1933. We used this idea in the collisions between two polar molecules and calculated the rotational cross sections.[81] Later, one of my colleagues tried to study V-V transfer problem in the adiabatically deformed rotational state. But this work has not completed yet. Then, I applied the perturbed rotational state approach to the simpler problem, i.e., the rotational transitions in ion-polar molecule collisions. I have already reported this topic in the 12th ICPEAC,[82] so that only a brief outline of the theory will be given in the next subsection and then review the more recent developments.

A. Ion-Molecule Collisions. Use of Adiabatic Basis Functions.

The formulation was semiclasical in the sense that the relative motion of the colliding pair was assumed to follow a classical trajectory $\vec{R}(t)$. The rotational wave function of the polar molecule, which is regarded as a rigid linear rotor, is obtained by solving an equation of the following form:

$$i\hbar \frac{\partial \psi}{\partial t} = [H_{rot} + V(\hat{r}, \vec{R})]\psi, \quad (29)$$

where $\hat{r}$ is the unit vector along the molecular axis. First, we solve the eigenvalue problem:

$$[H_{rot} + V(\hat{r}, \vec{R})]\chi(\hat{r}; \vec{R}) = \varepsilon(R)\chi(\hat{r}; \vec{R}) \quad (30)$$

to obtain the PRS (Perturbed Rotational State) wave functions χ and the corresponding eigenvalues $\varepsilon(R)$. At sufficiently large distances, each eigenfunction should tend to a single spherical harmonic function $Y_{jm}(\hat{r})$, representing the free rotation. This set of quantum numbers (j, m) may be used to distinguish different PRS functions χ's down to shorter distances. In solving (30), it is convenient to choose $\vec{R}$ as the z-axis (rotating coordinate system !). Now the wave function in (29) is expanded in terms of the PRS functions:

$$\psi = \sum_{jm} C_{jm}(t) \chi_{jm}(\hat{r}; \vec{R}(t)) \exp[-i\int \varepsilon_{jm}(t')dt'/\hbar]. \quad (31)$$

Substituting this into (29), we obtain a set of differential equations for the coefficients $C_{jm}(t)$. After trunction, this set of equations will be solved and the transition probability per collision is obtained. By integrating over the impact parameter, the effective cross section is obtained. The calculated cross sections are normally very large. This reflects the fact that the main contributon to the cross section comes from distant collision, supporting the simple straight clasical path approximation.

When the same problem is treated in a fully quantum mechanical way, it is convenient to add the angular part of the orbital motion to H_{rot} in (30), thus deriving the adiabatic basis functions including the whole angular coordinates ($\hat{r}$ and $\vec{R}$) of the system,[83-84] leaving only the radial functions in the coupled equations.

The use of the adiabatic basis functions in the rotational and vibrational states in molecular encounters is not new. The PSS approach has been applied to the vibrational transitions in collinear atom-diatom collisions.(Takayanagi[85]; Nyeland and Hunding[86]), while Child[87] used the rotationally adiabatic basis functions and studied the vibrational transitions in H_2 + He.

Mullaney and Truhlar[83] have shown that even the neutral-neutral collisions, such as HF + He, where the long-range force is no more dominant, the use of the adiabatic basis functions gives a much faster convergence of the results than the usual unperturbed basis.

B. Ion-Molecule Reactions.

During the past fifteen years or so, a large number of models have been proposed to explain relative abundances of various chemical species in dense interstellar clouds. 100-200 species (neutral atoms and molecules, positive and negative ions and electrons) are included and more than 1000 readtions are taken into account in some big computations. Some of them study equilibrium state, while the others study time-dependent problems. Many of these models are listed in a recent review article by Herbst.[88] Anicich and Huntress[89] compiled ion-molecule bimolecular reaction rates from laboratory experiments published during 1965-1985. 135 ion species are more than 1200 ion-neutral pairs are included in this compilation. And yet, many of these data are for temperatures at and above the room temperature and only for a limited number of reactions data are available down to 80 K or lower temperatures. In recent years, some experimental rate constants for ion-molecule reactions down to 10-20K region have been reported from a few laboratories. (See the review articles, such as articles in *Rate coefficients in Astrochemistry*,[90] and a report by Barlow, Luine and Dunn.[91]

The behavior of the rate constant of many ion-molecule reactions was first interpreted as due to the polarization interaction, which is the attractive force between the ion charge qe and the induced-dipole moment of the molecule (polarizability α). The effective interaction at large distances is

$$V_{eff}(R) = -\frac{\alpha q^2 e^2}{2R^4} + \frac{\vec{\ell}^2}{2\mu R^2}. \quad (32)$$

This interaction has a barrier at $R = R_c \equiv 2\alpha\mu \times q^2e^2/\vec{\ell}^2$. When the velocity v of the relative motion is very low, the orbital angular momentum squared $\vec{\ell}^2$ is small even when the impact parameter (in classical description) is fairly large ($|\vec{\ell}| \cong \mu vb$). This tells us that R_c is far outside the size of molecules where we can safely ignore the detailed interaction between the ion and the molecule. Once the system passes over the potential barrier, the colliding pair will get accelerated and hit each other. The event is usually called "capture" but I prefer "hitting collision", meaning that the colliding particles come to close contact to each other, but it does not necessarily mean the amalgamation of the particles into a compound state. The above consideration gives the well-known Langevin cross section (an upper limit to the reaction cross section):

$$\sigma_L(v) = 2\pi qe \sqrt{(\alpha/\mu)}/v, \qquad (33)$$

and the corresponding rate constant

$$k_L = 2\pi qe \sqrt{(\alpha/\mu)}, \qquad (34)$$

which is independent of temperature T.

C. Ion-Polar Molecule Collisions

For ion-polar molecule reactions, the experimental rate constant is sometimes larger than the Langevin rate constant k_L. In order to explain these findings, the ion-dipole interaction must be introduced. The effective potential is now

$$V_{eff} = \frac{Dqe}{R^2}(\hat{r}\cdot\hat{R}) - \frac{\alpha q^2 e^2}{2R^4} + \frac{\vec{\ell}^2}{2\mu R^2}, \qquad (35)$$

where D is the dipole moment of the molecule. For simplicity, the molecule is again assumed linear and $\hat{r}$ is the unit vector along the molecular axis. The locked-dipole model [$(\hat{r}\cdot\hat{R}) = -1$ in (35)] studied by Moran and Hamill,[92] however, gave too large a rate constant. Dugan and his coworkers[93] performed a series of classical trajectory Monte Carlo calculations for ion-polar molecule collisions and succeeded in explaining the observed reaction rates to some extent. Su and Bower[94] proposed a simple approach called the ADO (Average Dipole Orientation) theory. The original ADO theory could reproduce many rate constants at room temperature, but it was not so successful in explaining the temperature dependence of those constants. Bates[95] tried to refine the ADO theory by introducing the adiabatic invariance approximation.

The PRS theory which I had introduced to study the molecular rotational state during the low-energy collision processes can be used to calculate the hitting cross section. Since we are considering the slow collisions and since the dipole interaction is of the long range, the first term in (35) changes very slowly as a function of the time t. Therefore, we can expect that the molecular rotation will be almost adiabatically deformed during collision. If we neglect nonadiabatic transitions between different PRS eigenstates, we can replace the first term of (35) combined with the rotational Hamiltonian H_{rot} by one of the PRS energies $\varepsilon_{jm}(R)$. Then the hitting cross section can be easily derived by the procedure similar to that used to derive the Langevin cross section for the simple polarization interaction. Sakimoto and myself[96] applied this idea to calculate the hitting cross section and rate in 1980 and found that the reaction rates of some interstellar reactions can be more than ten times larger than the widely-used Langevin values. Later, Clary[97] has followed a similar line of approach. In his formulation of the problem, he introduced the centrifugal sudden approximation (CSA), so that he calls his approach ACCSA (Adiabatic Capture Centrifugal Sudden Approximation). He has applied this method to calculate rate constants for various realistic systems and compared the results with experimental data when they are available. In ion reactions with HCN and HCl, his theoretical results are in excellent agreement with the laboratory data down to 200K by SIFT (selected ion flow tube technique).[98] The rates calculated by the ADO theory are too small in the low temperature region as compared with experimental values. Unfortunately, laboratory experiments for polar molecules in very low temperatures are difficult because of the possibility of dimer formation. Thus the available data in the lowest temperature range are mostly for nonpolar molecules.

Su and Chesnavitch[99] have found, through their classical trajectory Monte Carlo calculations, that the hitting rate in unit of Langevin value is a function of a parameter

$$x = (D^2/2\alpha\kappa T)^{1/2} \qquad (36)$$

alone, at least approximately. They proposed an empirical formula

$$\bar{k}/k_L = 0.4767 x + 0.62. \qquad (37)$$

They obtained this relation for the limited range of x: $7 \gtrsim x \gtrsim 2$. Sakimoto,[100] applying the PRS method, has confirmed that the empirical reaction (37) is a good approximation for a wider range of x. For x = 32, for instance, he found that

$$\bar{k}/k_L = 17.6 \quad \text{for} \quad \kappa T/B = 97.86$$
$$= 15.1 \qquad \qquad = 0.4883.$$

Namely, for 200 times difference in T/B (B is the

rotational constant), the rate constant changes only 1.5 %. The empirical relation (37) gives for x = 32 the value $\bar{k}/k_L$ = 15.87.

It is noted that Sakimoto[100] applied the Bohr-Sommerfeld quantization procedure to derive the adiabatic potential curves quickly, without solving the wave equation (30).

Sakimoto[101] also studied the reaction of symmetric-top molecule with an ion. When the molecule is in a rotational state where both the quantum numbers K and M are nonvanishing (K is the projection of the rotational angular momentum $\vec{j}$ along the molecular axis, while M is the projection onto $\vec{R}$), the adiabatic potential will contain a long-range (R^{-2}) interaction. However, when he takes average over various orientation of the molecule, effects of this interaction become unnoticeable.

At very low temperatures, only the j = 0 state contributes to the rate constant. In this case, the quantum effect, i.e., the discreteness of the rotational state, becomes significant and the Su-Chesnavich scaling formula is no more applicable. On the other hand, because of j = 0, even the symmetric-top molecule becomes practically the same as a linear molecule.

It is interesting to note that Marquette, et al.[102] have recently measured the rates for He^+, C^+, and N^+ ions reacting with NH_3 and H_2O down to 27K. Most of the rates obtained are somewhat below the theoretical prediction (37), but the rates for C^+, N^+ + H_2O at the lowest temperature exceed the theoretical value. This may be due to the breakdown of the classical trajectory prediction. On the other hand, He^+ reactions with NH_3 and H_2O have rates considerably less than the theoretical prediction. This may be due to the small reaction probability per hitting collision.

Recently, Morgan and Bates[103] made extensive calculations to parameterize the cross section and rate constant. The relevant parameters are polarizability, dipole moment, reduced mass and temperature.

As was pointed out earlier, the gas density in the interstellar clouds is so low that the rotational level population is not necessarily in the Boltzmann distribution. Thus the rate constant for each individual value of the quantum number should be used for accurate estimation of the rate constants.

D. Ion-Quadrupolar Molecule Collisions

So much for the ion-polar molecule collision problem. When the molecule does not have the dipole moment, the next possible long-range interaction is the quadrupole interaction.

$$\frac{Qqe}{R^3} P_2(\hat{r}\cdot\hat{R}), \qquad (38)$$

where Q is the quadrupole moment of the molecule and P_2 is the Legendre polynomial. Hyatt and Stanton[104] used the first-order Stark effect approximation to calculate the effective potential out of (38):

$$V = \frac{j(j+1) - 3m^2}{2(2j-1)(2j+3)} \frac{Qq}{R^3}, \qquad (39)$$

from which they calculated the hitting cross section. Su and Bower[105] proposed the AQO (Average Quadrupole Orientation) method similar to the ADO method for polar molecule reactions. I gave the PRS formulation for the collision problem within the semiclassical framework.[106] Sakimoto[107] discussed the effect of potential ridge in the quadrupole interaction. One important point to be noted in (38) is that the sign of Qq is very important. The different sign of this product gives entirely different interaction. Bates and Mendas[108] applied the Bohr-Sommerfeld quantization procedure similar to the work of Sakimoto for ion-polar molecule collision. Kosmas' work[109] was based on a simplified version of AQO method. With further simplifying assumptions, he obtained the average interaction potential. He further took average of this potential over the Bolzmann distribution before calculating the hitting cross section. The derived effective interaction is

$$V = -0.023 \frac{q^2Q^2}{\kappa TR^6}. \qquad (40)$$

This is no more dependent on the sign of Qq. He suggested an approximate formula for $\bar{k}/k_L$, but the reliability of results derived from the potential (40) is doubtful.

Bhowmik and Su[110] made classical trajectory calculations for ion-quadrupole interaction. They took the isotropic polarization potential plus the point-quadrupole interaction. The calculations were compared with the experimental D^+ transfer rate constant in various D_3^+ + molecule reactions. They also studied the fictious reaction in D_3^- + C_6F_6 and compared the result with the one calculated for D_3^+ + C_6F_6. It is found that the effect of quadrupole moment in the rate constant is larger in the Qq = positive case than in the negative case. They also found that the rate for the quadrupole molecule depends significantly on the moment of inertia of the molecule.

E. Other Topics

So far, I have not mentioned the effect of the anisotropy of the polarization interaction. It can be easily included in the formulation of the collision dynamics (see, for instance, ref. 106). However, when the molecule has a dipole moment or a large quadrupole moment, the anisotropic part of the polarization force will have only a minor role. Schelling and Castleman[111] used the anisotropic polarization potential and made classical trajectory calculations. They added to this potential a isotropic short-range repulsive potential of the form c/R^{12}. They calculated the collision duration, or the lifetime of the collision complex, and concluded that the anisotropic part of the polarization potential has a large influence on the lifetime obtained. However, it must be noted that they have neglected the anisotropy of the short-range repulsive potential. In the real systems, there might be some cancellation between the anisotropic parts of the short-range and the polarization potentials in the important range of R. If that is the case, the effect of the anisotropic polarization force has been overestimated in their calculations. It is desirable that more realistic potential surface is used to do this sort of investigations.

Rowe and his coworkers[112] measured the rate constants for the systems

$$He^+ + N_2, O_2, CO$$
$$N^+ + O_2, CO, CH_4$$

down to 8K. The measured reaction rate for N^+ + CO was the same at 8K as at 88K and its value was close to the Langevin value. All the other rates at 8K are close to the corresponding values at 300K and somewhat smaller than the Langevin value. There is no indication of the influence of the quadrupole interaction and the anisotropic part of the polarization interaction.

Clary[97] suggested that even in neutral-neutral reactions, long-range interactions sometimes enhance the reaction rate, particularly in the radical-radical collisions. Recently, Sims and Smith[113] measured the reaction rate for the system of CN + O_2 down to 99K. The rate constant was found rising monotonically as the temperature was decreasing. Although the Clary's calculation predicted this increase, the experimental rate increased more rapidly than the theoretical prediction.

All the theories so far introduced calculate the rate of the colliding system passing over the barrier of the effective potential. At this stage, the arguments are classical. In the very low energy reactions, it may be necessary to calculate the quantum mechanical transmission probability for the barrier crossing, both in the energy ranges above and below the barrier height. Another weak point of the theories in the past is that there has been no investigation of the reaction dynamics after the potential barrier crossing. In order to obtain the final understandings of the ion-molecule reactions, the barrier crossing dynamics should be combined with the rest part of the collision problem.

Finally, it must be mentioned that the radiative ion-molecule association reactions, such as

$$CH_3^+ + H_2 \rightarrow CH_5^+ + h\nu \qquad (41)$$

have been investigated in detail by Bates[115] and Herbst.[116] Dunn and coworkers at JILA in Colorado measured the rate of (41) at 13K using a trapped-ion technique.[117]

An Additional Note:

In V-A, I discussed the transitions in vibrationally excited H_2 molecule. Most of the calculations for the H_2 + H system are for the direct transitions among the excited states. However, in this particular system, there is another possibility, i.e., the atom exchange reactive collisions. When H_2 is in a low vibrational state and when the gas temperature is also low, the reactive collisions have very small cross sections. However, when H_2 is in high vibrational state, the reactive collision is not rare. This will contribute to the rotational and vibrational transitions. Also, the exchange of atoms can result in the para-ortho conversion in the gas phase. However, the relevant cross section or rate constant data are still fragmentary. They are mainly from classical trajectory calculations.[118] Bowers, et al.[119] have used the coupled channel distorted wave method in this problem.

References

1. R.B.Bernstein, ed., *Atom-Molecule Collision Theory- A Guide for Experimentalist* (Plenum, New York and London, 1979).
2. A.D.Dickinson, Computer Phys. Commun.17,51 (1979).
3. F.A.Gianturco, *The Transfer of Molecular Energies by Collision: Recent Quantum Treatments* (Springer-Verlag, Berlin, 1979)
4. A.E.DePristo and H.Rabitz, Adv. Atom. Mol. Phys. 42, 271 (1980).
5. T.Yardley, *Introduction to Molecular Energy Transfer* (Academic Press, Boston & San Diego, 1980).
6. A.S.Dickinson and D.Richards, Adv. Atom. Mol. Phys. 18, 165 (1982).
7. J.M.Bowman, ed., *Molecular Collision Dynamics* (Springer-Verlag, Berlin, 1983).
8. G.D.Billing, Computer Phys. Rept.,1, 237 (1984).
9. D.R.Flower, Phys. Rept.174, 1 (1989).
10. L.D.Thomas, M.H.Alexander, B.R.Johnson, W.A. Lester,Jr., J.C.Light, K.D.McLenithan, G.A. Parker, M.J.Redmon, T.G.Schmalz, D.Secrest, and R.B.Walker, J.Computer Phys.41,407 (1981).
11. A.C.Allison, Adv. Atom. Mol. Phys.25, 323 (1988)
12. K.Takayanagi, Prog. Theor. Phys. Suppl.25, 1 (1963).
13. T.P.Tsien and R.T.Pack, Chem. Phys. Letters

6, 54 (1970).
14. D.Secrest, J. Chem. Phys. 62,710 (1975).
15. G.A.Parker and R.T.Pack, J. Chem. Phys.68, 1585 (1978).
16. D.J.Kouri and D.E.Fitz, J. Phys. Chem. 86, 2224 (1982).
17. D.C.Clary, J. Chem. Phys. 81, 4466 (1984); A.C.Peet and D.C.Clary, Mol. Phys. 59, 529 (1986); D.C.Clary and S.Green, Chem. Phys. 112, 15 (1987).
18. H.Margenau, Phys. Rev. 63, 385 (1943).
19. E.A.Mason and J.O.Hirschfelder, J. Chem. Phys. 26, 756 (1957).
20. R.N.Porter and M.Karplus, J. Chem. Phys.40, 1105 (1964).
21. B.Liu, J. Chem. Phys. 58, 1925 (1973); P. Siegbahn and B.Liu, J. Chem. Phys. 68, 2457 (1978).
22. D.G.Truhlar and C.J.Horowitz, J. Chem. Phys. 68, 2466 (1978), 71, 1514 (1979).
23. A.J.C.Varandas, F.B.Brown, C.A.Mead, D.G. Truhlar and N.C.Blais, J. Chem. Phys. 86, 6258 (1987).
24. V.Lewchenko, G.C.Hancock and P.R.Certain, J. Chem. Phys. 76, 3119 (1982).
25. P.McGuire and H.Krüger, J. Chem. Phys. 63, 1090 (1975).
26. R.Gengenbach, Ch.Hahr and J.P.Toennies, J. Chem. Phys. 62, 3620 (1975).
27. M.D.Gordon and D.Secrest, J. Chem. Phys.52, 120 (1973).
28. R.Shafer and R.G.Gordon, J. Chem. Phys. 58, 5422 (1973).
29. B.Tsapline and W.Kutzelnigg, Chem. Phys. Letters 23, 173 (1973).
30. A.W.Raczkowski and Wm.A.Lester,Jr., Chem. Phys. Letters 47, 45 (1977).
31. W.Meyer, P.C.Hariharan and W.Kutzelnigg, J. Chem. Phys. 73, 1880 (1980).
32. M.H.Alexander, J. Chem. Phys. 66, 4608 (1977).
33. Z.Bacic, R.Schinke and G.H.F.Diercksen, J. Chem. Phys. 82, 236, 245 (1985).
34. V.K.Nikulin, Zhur. Tekhnicheskoi Fiz. 41, 41 (1971); Sov.Phys.-Tech. Phys. 16, 28 (1971).
35. R.G.Gordon and Y.S.Kim, J. Chem. Phys. 56, 3122 (1972).
36. A.I.Rae, Chem. Phys. Letters 18, 574 (1973); M.Waldman and R.G.Gordon, J. Chem. Phys. 71, 1340 (1979).
37. J.H.Black, Adv. Atom. Mol. Phys. 25, 477 (1988).
38. K.Takayanagi, in *Astrochemistry* - Proc. of IAU Symposium 120, M.S.Vardya and S.P.Tarafdar eds. (D.Reidel, Dordrecht, 1987) p.31.
39. D.P.Dewangan, D.R.Flower and M.H.Alexander, Mon.Not.Roy.Astr.Soc. 226, 505 (1987).
40. G.C.Corey and M.H.Alexander, J. Chem. Phys. 88, 6931 (1988).
41. H.Klar, J. Phys. B(Atom.Mol.Phys.) 6,2139 (1973).
42. M.H.Alexander, J. Chem. Phys. 76, 3637, 5974 (1982); 78, 1625 (1983).
43. T.Orlikowski, Mol. Phys. 56, 35 (1985).
44. G.C.Corey, M.H.Alexander and P.J.Dagdigian, J. Chem. Phys. 84, 1547 (1986).
45. R.Gould and M.Harwit, Astrophys. J. 137, 694 (1963).
46. J.H.Black and A.Dalgarno, Astrophys. J.203, 132 (1976).
47. J.H.Black and E.F. van Dishoeck, Astrophys. J. 322, 412 (1987).
48. R.Güsten and D.Fiebig, Astron. & Astrophys. 204, 253 (1988).
49. R.B.Loren, N.R.Erickson, R.L.Snell, L.Mundy and J.H.Davis, Astrophys. J. Letters 244, L107 (1981).
50. D.C.Clary, J. Chem. Phys. 75, 209 (1981); G.D.Billing and D.C.Clary, Chem. Phys. Letters 90,27 (1982). D.C.Clary and A.E.DePriston, Chem. Phys. 79, 2206 (1983); A.J.Banks and D.C.Clary, J. Chem. Phys.86, 802 (1987) and references therein.
51. D.C.Clary, J. Chem. Phys. 75, 209, 2899(1981).
52. A.C.Peet and D.C.Clary, Mol. Phys. 59, 529 (1986).
53. I.Shimamura, Z. Physik A309, 107 (1982) and references therein.
54. A.M.Richard and A.E.DePristo, Chem. Phys. 69. 273 (1982).
55. T.A.Brunner and D.Pritchard, in *Dynamics of the Excited State*, K.P.Lawley, ed. (Wiley, New York, 1982) p.589.
56. B.J.Whitaker and Ph.Brechignac, Chem. Phys. Letters 95, 407 (1983); Ph.Brechignac and B.J. Whitaker, J. Chem. Phys.84, 2101 (1986).
57. Th.G.Kreutz and H.Rabitz, J. Chem. Phys.90, 1701 (1989).
58. For the relevant radiative transition rates, see A.C.Allison and A.Dalgarno, Atomic Data 1, 289 (1970) for b-b transitions and T.L.Stephens and A.Dalgarno, J. Quantit. Spectrosc.Radiat. Transfer 12, 569 (1972) for dissociation.
59. For collision processes in interstellar shock waves, see, e.g., B.T.Draine, W.G.Roberge and A.Dalgarno, Astrophys. J.264, 485 (1983).
60. For fluorescent excitation of H_2, see K. Takayanagi, K.Sakimoto and K.Onda, Astrophys. J.Letters 318,L81 (1987) and the following two articles and also references therein.
61. J.H.Black and E.F.van Dishoeck, Astrophys. J. 312,412(1987).
62. A.Sternberg and A.Dalgarno, Astrophys. J. 338,197(1989).
63. For observation of fluorescent excitation of interstellar H_2, see I.Gatley and N.Kaifu, in *Astrochemistry* - Proc. of IAU Symposium 120, M.S.Vardya and S.P.Tarafdar, eds. (D.Reidel, Dordrecht, 1987) p.153.
64. Ten infrared lines of H_2 from NGC2023 reported by T.Hasegawa, I.Gatley, R.Garden, P.Brand, M.Ohishi, J.Lightfoot, M.Hayashi and N.Kaifu, Astrophys. J.Letters 318, L77 (1987).
65. E.Garcia and A.Lagana, Chem. Phys. Letters 123, 365 (1986).

66. M.Cacciatore, M.Capitelli and G.D.Billing, Chem. Phys. Letters **157**, 305 (1989).
67. J.E.Dove and H.Teitelbaum, Chem. Phys. **6**, 431 (1974).
68. For instance, P.D.Magill, T.P.Scott, N.Smith and D.E.Pritchard, J. Chem. Phys. **90**, 7195 (1989) studied the rovibrational transition in $Li_2(v_i=9, j_i)$ + He, Ne, etc. at $600^{\circ}C$ and found that as a function of Δj for a fixed Δv the rate constant becomes maximum at $\Delta j = -4 \Delta v$ although the energy transfer to translation becomes minimum at about $\Delta j = -6 \Delta v$.
69. K.Onda, to be published.
70. A.Dalgarno and W.G.Roberge, Astrophys. J. **233**, L25 (1979); W.G.Roberge and A.Dalgarno, Astrophys. J. **255**, 176 (1982).
71. A.W.Yau and H.O.Pritchard, J. Phys. Chem. **83**, 134 (1979).
72. J.E.Dove and S.Raynor, Chem. Phys. **28**, 113 (1978); J. Phys. Chem. **83**, 127 (1979); J.E. Dove, S.Raynor and H.Teitelbaum, Chem. Phys. **50**, 175 (1980).
73. J.E.Dove, A.C.M.Rusk, P.H.Cribb and P.G. Martin, Astrophys. J. **318**, 379 (1987).
74. N.C.Blais and D.G.Truhlar, in *Potential Energy Surfaces and Dynamics Calculations*, ed. D.G.Truhlar (Plenum, New York, 1981), p.431; Astrophys. J. **258**, L79 (1982); J. Chem. Phys. **78**, 2388 (1983).
75. J.E.Dove and M.E.Mandy, Astrophys. J. **311**, L93 (1986).
76. K.Sakai, Univ.California Riverside, Master's Degree Thesis (1986), unpublished.
77. H.Koizumi, to be published.
78. A.E.Orel, in *XV ICPEAC, Abstracts of Contributed Papers*, p.702 (1987).
79. D.W.Schwenke, J. Chem. Phys. **89**, 2076(1988).
80. H.S.W.Massey and R.A.Smith, Proc. Roy. Soc. (London) **A142**, 142 (1933).
81. M.Hashi, S.Tsuchiya and K.Takayanagi, J.Phys. Soc. Japan **49**, 1486 (1980); see also T.Wada, ibid **53**, 3362 (1984); K.Takayanagi and T.Wada, ibid **54**, 2122 (1985).
82. K.Takayanagi, in *Physics of Electronic and Atomic Collisions*, ed. S.Datz (North-Holland, Amsterdam, 1982) p.343.
83. N.A.Mullaney and D.C.Truhlar, Chem. Phys. **39**, 91 (1979).
84. K.Sakimoto, Chem. Phys. **79**, 137 (1983).
85. K.Takayanagi, Sci. Rep. Saitama Univ. **A4**, No.3, 87 (1963)
86. C.Nyeland and A.Hunding, Chem. Phys. Lett. **5**, 143 (1970).
87. M.S.Child, Proc. Roy. Soc. **A315**, 259 (1970).
88. E.Herbst, Rev. Mod. Astronomy **1**, 114 (1988).
89. V.G.Anicich and W.T.Huntress, Astrophys. J. Suppl. **62**, 553 (1986).
90. T.J.Millar and D.A.Williams, eds. *Rate Coefficients in Astrochemistry*, particularly the article by D.K.Rowe (p.135) and the one by D.Smith and N.G.Adams (p.153). (Kluwer, Dordrecht, 1988).
91. S.E.Barlow, J.A.Luine and G.H.Dunn, Intern.J. Mass Spectr.Ion Processes **74**, 97 (1986).
92. T.F.Moran and Wm.H.Hamill, J. Chem. Phys. **39**, 1413 (1963).
93. J.V.Dugan,Jr. and J.L.Magee, J. Chem. Phys. **47**, 3103 (1967); **58**, 5816 (1973); J.V.Dugan, Chem. Phys. Letters **21**, 476 (1973).
94. T.Su and M.T.Bower, J. Chem. Phys. **58**, 3027 (1973); L.Bass, T.Su, W.Chesnavich and M.T.Bower, Chem. Phys. Letters **34**, 119 (1975).
95. D.R.Bates, Proc. Roy. Soc. **A384**, 289 (1982); Chem. Phys. Letters **97**, 19 (1983).
96. K.Sakimoto and K.Takayanagi, J. Phys. Soc. J. **48**, 2076 (1980).
97. D.C.Clary, Mol. Phys. **53**, 3 (1984); **54**, 605 (1985); see also ref.90, p.1.
98. D.C.Clary, D.Smith and N.G.Adams, Chem. Phys. Letters **119**, 320 (1985).
99. T.Su and W.J.Chesnavich, J. Chem. Phys. **76**, 5183 (1982).
100. K.Sakimoto, Chem. Phys. **85**, 273 (1984).
101. K.Sakimoto, Chem.Phys.Letters **116**, 86(1985).
102. J.B.Marquette, B.R.Rowe, G.Dupeyrat, G. Poissant and C.Rebrion, Chem. Phys. Letters **122**, 431 (1985).
103. Wm.L.Morgan and D.R.Bates, Astrophys. J. **314**, 817 (1987).
104. D.Hyatt and L.Stanton, Proc. Roy. Soc. (London) **A318**, 107 (1970).
105. T.Su and M.T.Bower, Int.J. Mass Spectrom. Ion Phys. **17**, 309 (1975).
106. K.Takayanagi, J. Phys. Soc. Japan **51**, 3337 (1982).
107. K.Sakimoto, J. Phys. Soc. Japan **51**, 2657 (1982).
108. D.R.Bates and I.Mendas, Proc. Roy. Soc. (London) **A402**, 245 (1985).
109. A.M.Kosmas, J. Physique Lett. **46**, L799 (1985).
110. P.K.Bowmik and T.Su, J. Chem. Phys. **84**, 1432 (1986).
111. F.J.Schelling and A.W.Castleman,Jr., Chem. Phys. Letters **111**. 47 (1984).
112. B.R.Rowe, J.B.Marquette, G.Dupeyrat and E.E.Ferguson, Chem. Phys. Letters **113**, 403 (1985).
113. I.R.Smith and I.W.M.Smith, Chem. Phys. Lett. **151**, 481 (1988).
114. D.R.Bates, Astropohys. J. **270**, 564 (1983); **298**, 382 (1985); **312**, 363 (1987).
115. D.R.Bates, in *Recent Studies in Atomic and Molecular Processes*, ed. A.E.Kingston (Plenum, New York, 1987) p.1.
116. E.Herbst and D.R.Bates, Astrophys. J. **329**, 410 (1988) and references therein.
117. S.E.Barlow, G.H.Dunn and M.Schauer, Phys. Rev. Letters 902 (1984).
118. N.C.Blais and D.G.Truhlar, in *Potential Energy Surfaces and Dynamics Calculations* ed. D.G.Truhlar (Plenum,New York, 1981) p.431.
119. M.S.Bowers, B.H.Choi and K.T.Tang, Chem. Phys. Letters **136**, 145 (1987).

Electrons

ELECTRON-ATOM COLLISION THEORY

C.J. Joachain

Physique Théorique, Faculté des Sciences, Université Libre de Bruxelles
and Institut de Physique Corpusculaire, Université de Louvain, Belgium

A review is given of recent advances in the theory of electron-atom collisions. The first part deals with elastic and inelastic scattering and the second part with ionization. Positron-atom collisions are also considered.

1. INTRODUCTION

In this talk I would like to give a survey of recent progress in the theory of electron-atom collisions. I shall begin by discussing elastic and inelastic collisions, first in the low-energy region and then at intermediate and high energies. The second part of the talk will be devoted to (e,2e) reactions. The emphasis will be on simple systems for which the collision dynamics can be studied in detail. Comparisons between results obtained for incident electrons and positrons will also be made.

2. ELASTIC AND INELASTIC COLLISIONS

We start by considering elastic or inelastic collisions

$$e^- + A(i) \rightarrow e^- + A(f) \tag{1}$$

where $A(i)$ is an atom or ion in state $|i\rangle$ and $A(f)$ denotes this atom or ion in state $|f\rangle$

2.1 The low-energy region

Let us assume that the incident electron energy E_i is low enough so that only a few target states can be excited. We neglect for the moment all relativistic effects, which restricts our treatment to light atoms and ions. The Schrödinger equation describing the collision of an electron with a target atom or ion containing N electrons and with nuclear charge Z is

$$H_{N+1}\Psi = E\Psi \tag{2}$$

where E is the total energy of the electron-atom (ion) system and the $(N+1)$-electron Hamiltonian H_{N+1} is given by

$$H_{N+1} = \sum_{i=0}^{N}\left(-\frac{1}{2}\nabla_i^2 - \frac{Z}{r_i}\right) + \sum_{i>j=0}^{N}\frac{1}{r_{ij}} \tag{3}$$

where $r_{ij} = |\mathbf{r}_i - \mathbf{r}_j|$ and the index (0) labels the unbound electron. We also introduce the target eigenvectors $|n\rangle$ and corresponding eigenenergies w_n such that

$$H_N|n\rangle = w_n|n\rangle \tag{4}$$

where the target Hamiltonian H_N is given by eq. (3) with the summation starting at the index 1. We shall denote by $\mathbf{x}_i \equiv (\mathbf{r}_i, \sigma_i)$ the space and spin coordinates of the electron i, by $\mathbf{X} \equiv \mathbf{x}_1, \mathbf{x}_2, \ldots \mathbf{x}_N$ the ensemble of target coordinates and by $\psi_n(\mathbf{X}) \equiv \langle \mathbf{X}|n\rangle$ the target eigenfunctions.

In order to obtain the scattering amplitude for a transition from the initial target state $|i\rangle$ to the final target state $|f\rangle$, Schrödinger's equation (2) must be solved subject to appropriate boundary conditions. Most of the approximation methods of solution which have been used in the low-energy region may be described in terms of an expansion of the total wave function of the form[1,2]

$$\Psi = \mathcal{A}\sum_{i=1}^{M} F_i(\mathbf{x}_0)\psi_i(\mathbf{X}) + \mathcal{A}\sum_{i=M+1}^{M+P} F_i(\mathbf{x}_0)\bar{\psi}_i(\mathbf{X}) + \sum_i c_i\chi_i(\mathbf{x}_0,\mathbf{x}_1,\ldots\mathbf{x}_N) \tag{5}$$

where $\mathcal{A}$ is the antisymmetrization operator. The first expansion goes over a limited number of target eigenstates ψ_i. If only this expansion is retained, with all the open channels included, then one obtains the familiar close-coupling expansion. The second expansion contains a certain number of pseudo-states $\bar{\psi}_i$, which are often chosen to optimize the representation of long-range polarization effects. The third expansion goes over a set of quadratically integrable functions χ_i which are individually antisymmetric with respect to exchange of the coordinates of all the $(N+1)$ electrons.

By substituting the expansion (5) of Ψ into Schrödinger's equation (2), projecting onto the target functions ψ_i, the pseudo-states $\bar{\psi}_i$ and the L^2 functions χ_i, and using the Feshbach projection operator formalism[3], one obtains a set of coupled partial integro-differential equations for the functions $F_i(\mathbf{x}_0)$. These equations can then be partial-wave analyzed by using the fact that the Hamiltonian H_{N+1} given by eq. (3) commutes with the operators $\mathbf{L}^2, \mathbf{S}^2, L_z, S_z$ and Π, where $\mathbf{L}$ is the total orbital angular momentum, $\mathbf{S}$ the total spin and Π the parity operator. In this way, one obtains a set of coupled integro-differential equations for the radial functions $F_i^\Gamma(r_0)$, corresponding to the channel index $\Gamma \equiv LSM_L M_S \Pi$. By solving these equations subject to the appropriate boundary conditions, the scattering amplitudes and cross sections can be calculated.

Most of the recent results emerging from the coupled integro-differential equations have been obtained by using one of four methods : the R-matrix method developed by Burke et al [4,5], the reduction of these equations to a system of linear algebraic equations [6], the non-iterative integral equations method[7], and the matrix variational method[8]. These methods and the associated computer programs have been reviewed by Burke and Eissner[9].

As an example, we show in Fig. 1 the differential cross section for the $1^1S \to 2^3S$ transition in electron-helium collisions, at a scattering angle $\theta = 54°$, as a function of the incident electron energy. The agreement between the 19 state R-matrix calculation of Fon et al [10], convoluted with a 25 meV Gaussian, and the experimental data of Bass [11] is seen to be very good.

In the case of electron-ion scattering it is often convenient to consider a Rydberg series of resonances. This is one of the motivations behind the development of multichannel quantum defect theory[12]. In this theory the properties of an electron in the field of a positive ion are expressed in terms of analytical functions of the energy. In particular, if the K-matrix elements are calculated from the coupled integro-differential equations at a few energies above threshold, they can be extrapolated yielding an infinite series of resonances below threshold. The theory can also be used to fit experimental data above and below threshold in terms of a few parameters.

Before leaving the low-energy domain, let us briefly consider relativistic effects. For target atoms or ions with small or intermediate values of Z, these effects are small, so that the calculations can in first approximation be performed in LS coupling using the non-relativistic Hamiltonian defined by eq. (3); the corresponding K-matrices are subsequently recoupled to yield transitions between fine structure levels[13]. On the other hand, for atoms or ions with high Z values, relativistic terms must be kept in the Hamiltonian. This can be done by including in the Hamiltonian the Breit-Pauli correction[14-16]. R-matrix calculations based on the Breit-Pauli Hamiltonian have been performed for electron-mercury scattering by Scott et al [17] and Bartschat et al[18]. The integrated Stokes parameters calculated by Bartschat et al[18] for the $6s6p\ ^3P_1^o \to 6s^2\ ^1S_0^e$ transition excited by polarized electrons [19,20] are in good agreement with the experimental results of Wolcke et al [21]. An alternative way of including relativistic terms in the Hamiltonian is to use the Dirac Hamiltonian[22]. Calculations using this approach have been performed for hydrogen-like ions[23,24] and preliminary work has been carried out to develop a general code[25,26], but so far the only significant results for heavy atoms have been obtained by using the Breit-Pauli Hamiltonian.

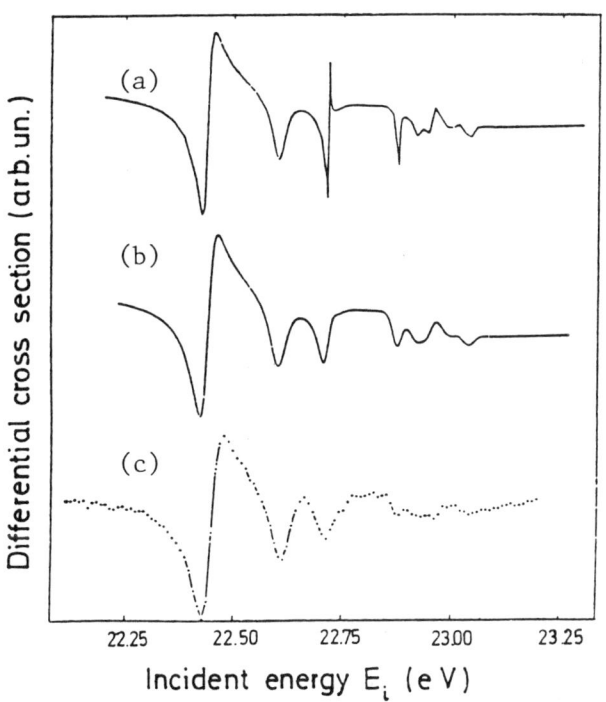

Figure 1. The differential cross section (in arbitrary units) for the excitation of the 2^3S state of helium by electron impact, at a scattering angle $\theta = 54°$, as a function of the incident electron energy E_i. (a) Theoretical 19 state R-matrix calculation of Fon et al [10], (b) calculated cross section convoluted with a 25 meV Gaussian, (c) experimental results of Bass [11].

2.2 The region of intermediate and high energies

Let us now turn to elastic and inelastic electron-atom collisions at impact energies larger than the ionization energy of the target. In this case an infinite number of channels are open, and we shall examine various theoretical methods which have been proposed to deal with this situation. For simplicity we shall restrict our attention to non-relativistic collisions, and assume that the target is a neutral atom ($N = Z$); the modifications required when the target is an ion are straightforward. The case of incident positrons will be treated as well as that of incident electrons.

We begin by considering the high-energy domain which extends from several times the ionization threshold upwards. It is then reasonable to try an approach based on perturbation theory. We shall first discuss direct elastic and inelastic collisions. We denote respectively by $\mathbf{k}_i$ and $\mathbf{k}_f$ the initial and final momentum of the projectile, $\mathbf{\Delta} = \mathbf{k}_i - \mathbf{k}_f$ being the momentum transfer. The free motion of the colliding particles before the collision is described by the direct arrangement channel Hamiltonian $H_d = K + H_N$ where K is the kinetic energy operator of the projectile and H_N is the target Hamiltonian introduced in eq. (4), having eigenenergies w_n and eigenvectors $|n\rangle$. The full Hamiltonian H of the system (which coincides with the $(N+1)$-electron Hamiltonian H_{N+1} when the projectile is an electron) can be decomposed in the direct arrangement channel as $H = H_d + V_d$, where the direct interaction V_d between the projectile and the target is

$$V_d = \frac{ZQ}{r_0} - Q \sum_{j=1}^{Z} \frac{1}{r_{0j}} \qquad (6)$$

with $Q = -1$ for incident electrons and $Q = +1$ for incident positrons.

The Born series for the direct scattering amplitude may be written formally as

$$f = \sum_{n=1}^{\infty} \bar{f}_{Bn} \qquad (7)$$

where the nth Born term $\bar{f}_{Bn}$ contains n times the interaction V_d and $(n-1)$ times the direct Green's operator $G_d^{(+)} = (E - H_d + i\epsilon)^{-1}$, $\epsilon \to 0^+$.

The first term $\bar{f}_{B1}$ is the familiar first Born amplitude, which has been evaluated for many transitions. The second Born term, $\bar{f}_{B2}$, plays a central role in a number of methods which have been proposed to obtain systematic improvements over the first Born approximation. For the direct transition $|\mathbf{k}_i, i\rangle \to |\mathbf{k}_f, f\rangle$, where $|i\rangle$ and $|f\rangle$ denote respectively the initial and final target states, we have[27]

$$\bar{f}_{B2} = 8\pi^2 \sum_n \int d\mathbf{q} \frac{\langle \mathbf{k}_f, f|V_d|\mathbf{q}, n\rangle \langle \mathbf{q}, n|V_d|\mathbf{k}_i, i\rangle}{q^2 - k_i^2 + 2(w_n - w_i) - i\epsilon}, \epsilon \to 0^+ \qquad (8)$$

A large amount of work has been devoted to the study of $\bar{f}_{B2}$. In a previous lecture[28] at the 10th ICPEAC, I analyzed the main properties of $\bar{f}_{B2}$, and at the 15th ICPEAC Walters[29] reviewed recent developments.

Detailed discussions of the second Born term may be found in the review articles of Byron and Joachain[30] and of Walters[31].

It is worth stressing that while the first Born term $\bar{f}_{B1}$ gives the dominant contribution to the scattering amplitude at large k_i for direct elastic scattering at all momentum transfers and for direct inelastic scattering at momentum transfers $\Delta < 1$, it is the second Born term $\bar{f}_{B2}$ which for large k_i dominates the Born series in the case of direct inelastic scattering at large momentum transfers. The dominance of the second Born term for direct inelastic collisions at large Δ may be understood by remembering that large momentum transfer collisions can only take place if the projectile electron or positron collides with the much heavier atomic nucleus. Now, for inelastic scattering, the $e^\pm$-nucleus interaction term ZQ/r_0 does not contribute to the first Born term $\bar{f}_{B1}$ because of the orthogonality of the initial and final target states, $\langle f|i\rangle = 0$. As a result, $\bar{f}_{B1}$ falls off rapidly at large Δ and the Born series is dominated by the second Born term $\bar{f}_{B2}$. The fact that $\bar{f}_{B2}$ falls off more slowly than $\bar{f}_{B1}$ for inelastic collisions at large Δ is due to the possibility of off-shell elastic scattering in the intermediate states $|i\rangle$ and $|f\rangle$ (the initial and final target states of the inelastic transition), where the projectile electron or positron can experience the singular Coulomb potential ZQ/r_0 of the nucleus.

The accurate evaluation of the higher Born terms $\bar{f}_{Bn}(n \geq 3)$ is beyond the limit of present expertise. Nevertheless, Byron and Joachain[30] have obtained useful information about these terms in the large k_i limit. For example, at large momentum transfers, they showed that the terms $\bar{f}_{Bn}(n \geq 2)$ for direct elastic or inelastic scattering are dominated by processes in which off-shell elastic scattering can occur in intermediate states; this again emphasizes the importance of the singular $e^\pm$-nucleus interaction at large Δ. Byron and Joachain[30,32] also made a detailed comparison of the terms $\bar{f}_{Bn}$ with the corresponding terms $\bar{f}_{Gn}$ of the Glauber series (obtained by expanding the Glauber amplitude f_G in powers of the interaction V_d) with the aim of obtaining a consistent expansion of the direct scattering amplitude in powers of k_i^{-1}. This analysis leads to the eikonal-Born series (EBS) direct amplitude[32]

$$f_{EBS} = \bar{f}_{B1} + \bar{f}_{B2} + \bar{f}_{G3} \qquad (9)$$

which is correct through order k_i^{-2}.

The terms $\bar{g}_{Bn}$ of the Born series for the exchange amplitude are much more difficult to analyze than the direct Born terms $\bar{f}_{Bn}$. We simply recall that for large k_i the term $\bar{g}_{B2}$ falls off more slowly than $\bar{g}_{B1}$, except for elastic exchange scattering at small Δ, where the Bonham-Ochkur amplitude[33,34], which is the leading piece of $\bar{g}_{B1}$, is of order k_i^{-2}. Thus, in the EBS theory of elastic scattering, the Bonham-Ochkur exchange amplitude must be taken into account in order to perform a consistent calculation of the differential cross section through order k_i^{-2}.

The EBS method is very successful when perturbation theory converges rapidly, that is for fast $e^\pm$ collisions with light atoms, at not too large angles. If such conditions are not met, improvements are necessary. These can be obtained by constructing methods which include terms from all orders of perturbation theory in order to ensure unitarity. The first method of this kind, proposed by Byron and Joachain[35], consists in a simple modification of the EBS approximation, based on the use of the EBS' direct amplitude (also called subsequently the modified Glauber approximation[36])

$$f_{EBS'} = f_G - \bar{f}_{G2} + \bar{f}_{B2}$$
$$= \bar{f}_{B1} + \bar{f}_{B2} + \bar{f}_{G3} + \sum_{n=4}^{\infty} \bar{f}_{Gn} \quad (10)$$

This amplitude also gives a consistent picture of the direct scattering amplitude through order k_i^{-2}. Moreover, upon comparison with eq. (9), we see that the EBS' direct amplitude extends the EBS direct amplitude by the addition of an infinite number of terms of perturbation theory (of order $n \geq 4$) calculated in the Glauber approximation. This addition improves the situation at large momentum transfers, where the singular $e^\pm$-nucleus Coulomb interaction ZQ/r_0 plays a dominant role.

A more elaborate way of extending the third order perturbative EBS direct amplitude of eq. (9) to infinite order is to use the unitarized eikonal-Born series ($UEBS$) amplitude of Byron, Joachain and Potvliege[37–39], which is constructed in the following way. First, the potential scattering Wallace amplitude[40], which is an improved eikonal amplitude, is generalized to the case of multi-channel $e^\pm$-atom scattering[37–39]. This amplitude, which we call f_W, includes terms from all orders of perturbation theory. In fact, at large Δ, f_W contains the two leading terms (in powers of k_i^{-1}) of each order of perturbation theory summed to all orders, and hence is more accurate in this angular region then the Glauber amplitude f_G. However, the multi-channel Wallace amplitude f_W suffers, like the Glauber amplitude f_G, from important small angle deficiencies in second order due to the fact that in obtaining these amplitudes the average excitation energies in both the initial and final channels have been set equal to zero. The amplitude f_W is therefore corrected in second order to yield the $UEBS$ amplitude[37–39]

$$f_{UEBS} = f_W - \bar{f}_{W2} + \bar{f}_{B2} \quad (11)$$

where $\bar{f}_{W2}$ is the second order term in the expansion of f_W in powers of the interaction V_d. At small momentum transfers the $UEBS$ amplitude (11), expanded through order k_i^{-2}, reduces to the EBS amplitude (9). At large Δ, f_{UEBS} differs negligibly from the multi-channel Wallace amplitude f_W. Thus, the direct amplitude f_{UEBS} combines all the advantages of the EBS method at small and intermediate angles with those of the multi-channel Wallace amplitude f_W at large Δ. In addition, Byron, Joachain and Potvliege[38,39] have obtained a non-perturbative expression for the exchange amplitude, which is appropriate to use with the direct $UEBS$ amplitude (11) for electron-atom collisions.

As an example, we compare in Fig. 2 the $UEBS$

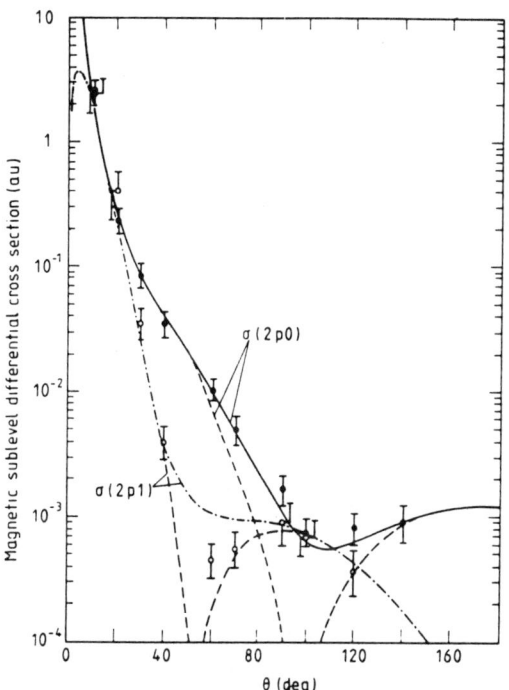

Figure 2. The differential cross sections (in atomic units au) for the excitation of the $2pm$ states of atomic hydrogen by electron impact at an incident electron energy $E_i = 54.4 eV$. Curve ——— : $UEBS$ results of Byron, Joachain and Potvliege[39] for the excitation of the $2p0$ state; curve - · - : $UEBS$ results for the excitation of the $2p1$ (or $2p-1$) state; curves - - - : $UEBS$ results without exchange; • experimental data of Williams[41] for the excitation of the $2p0$ state; ○ experimental data of Williams[41] for the excitation of the $2p1$ state.

magnetic sublevel differential cross sections for excitation of the $2pm$ states of atomic hydrogen by incident electrons having an energy $E_i = 54.4 eV$ with the experimental data of Williams[41] (obtained by performing electron-photon coincidence measurements). The agreement between theory and experiment is seen to be very satisfactory. We note from Fig. 2 that at this relatively low impact energy exchange effects are very significant in certain angular regions.

Another approach which is capable of including terms from all orders of perturbation theory is that based on the optical potential[27]. Let us begin by considering $e^{\pm}$-atom direct elastic scattering, the corresponding direct part of the optical potential being denoted by V_{opt}^d. Formal expressions for V_{opt}^d are readily derived by using the Feshbach projection operator formalism[3]. At intermediate and high energies it is appropriate to make a perturbative expansion of V_{opt}^d in powers of V_d. That is,

$$V_{opt}^d = V^{(1)} + V^{(2)} + V^{(3)} \qquad (12)$$

where $V^{(1)} = \langle i|V_d|i\rangle$ is the static potential, which is real and of short range, and hence does not account for polarization or absorption effects. However, for small values of the projectile coordinate r_0, the static potential $V^{(1)}$ correctly reduces to the Coulomb interaction ZQ/r_0 acting between the projectile and the target nucleus, and hence gives a good account of large angle direct elastic scattering.

The second and higher order terms of the direct optical potential are in general complicated, non-local, complex operators[27]. However, at sufficiently high energies local approximations to them can be obtained[27,28,42-44]. For example, Byron and Joachain[44] converted the lowest order terms of perturbation theory for the direct elastic amplitude, calculated by using the EBS method, into an ab-initio optical potential. The second order part $V^{(2)}$ of the direct optical potential may then be written approximately as

$$V^{(2)} = V_{pol} + iV_{abs} \qquad (13)$$

where V_{pol} and V_{abs} are real and central but energy-dependent. The term V_{pol}, which falls off like r_0^{-4} at large r_0 accounts for dynamic polarization effects and iV_{abs} for absorption effects. The leading contribution to the third order potential $V^{(3)}$ has also been obtained for $e^{\pm} - H(1s)$ elastic scattering[44], and more recently for $e^{\pm} - H(2s), e^{\pm} - He(2^1S)$ and $e^{\pm} - He(2^3S)$ elastic collisions[45].

Having obtained a local approximation for the direct optical potential V_{opt}^d, exchange effects (which must be included for the case of incident electrons) may be taken into account by using a local exchange pseudo-potential V_{opt}^{ex}, which is usually determined within the framework of the static-exchange approximation[46-49]. The full optical potential V_{opt} containing the direct and exchange parts (for incident electrons) or the direct part only (for incident positrons) is then treated in a unitary, essentially exact manner by using the partial wave method. It is worth noting that in performing such an exact treatment of the (approximate) optical potential one generates approximations to all terms of perturbation theory. In particular, the static potential $V^{(1)}$ is taken care of exactly, a feature which is an important advantage for large angle scattering.

As an illustration of this ab-initio optical model theory, we compare in Fig. 3 the theoretical predic-

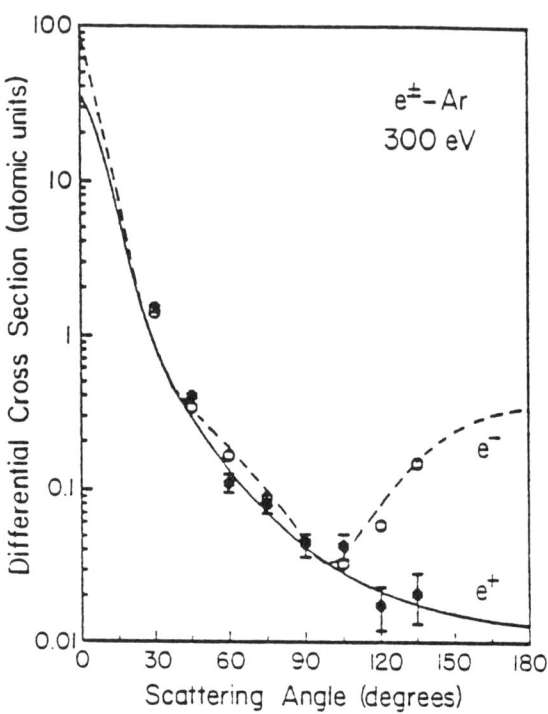

Figure 3. The differential cross section (in au) for electron (- - -) and positron (——) elastic scattering by argon at 300 eV, calculated by Joachain et al[28,50] from an ab-initio optical model theory. The experimental data of Hyder et al[51] are : ○ electron scattering, ● positron scattering.

tions of Joachain et al[28,50] with the differential cross sections measured subsequently by Hyder et al[51] for $e^{\pm}$-argon elastic scattering at an incident energy of 300 eV. The agreement is seen to be very good. A discussion of the angular distributions for $e^{\pm}$-argon elastic scattering has been given recently by Joachain and

Potvliege[52].

The optical model formalism can readily be generalized to the analysis of transitions between a certain number of target states, in which case the optical potential becomes a potential matrix. In particular, the second order potential (SOP) method of Bransden and Coleman[53] is an approximation such that this optical potential matrix is treated to second order in the projectile-target interaction V_d. Among the various applications of the SOP method, I would like to mention the study of elastic scattering, $1s - 2s$ and $1s - 2p$ excitation in $e^- - H(1s)$ collisions performed by Bransden et al[54]. McCarthy et al[55] have studied approximations to the optical potential matrix going beyond second order; their method is often called the coupled-channel optical model ($CCOM$).

We now turn to distorted-wave treatments, which are characterized by the fact that the interaction is broken in two parts, one which is treated exactly and the other which is handled by perturbation theory[27]. Many kinds of distorted wave methods have been applied to electron or positron collisions with atoms or ions. Detailed reviews have been written by Bransden and Mc Dowell[56], Henry[57], Walters[31] and Itikawa[58]. We mention in particular the distorted wave Born approximation[27] ($DWBA$), the distorted wave second Born approximation[59-61] ($DWSBA$) and the multi pseudo-state close coupling plus distorted wave second Born approximation[62] ($MPCC$).

The most elaborate methods discussed above can usually be applied down to relatively low impact energies, but as the projectile energy is decreased to be only somewhat larger then the ionization energy of the target, other approximation schemes are necessary. Instead of pushing down the validity of high-energy methods, one can try to extend the low energy approaches discussed in Section 2.1 to that "sub-intermediate" energy region. For example, Burke and Webb[63] and Callaway and Wooten[64] have generalized the pseudo-state approach to the case where the pseudo-states can carry away flux into the open channels. This method presents some undesirable features due to the unphysical pseudo-resonances associated with the pseudo-states, although Burke et al[65] have shown that useful information can be obtained by performing a suitable averaging of the T-matrix elements over the pseudo-resonances.

There has also been much interest in methods which attempt to approximate the wave function in a limited region of configuration space by a combination of square integrable (L^2) functions, thus avoiding the explicit representation of the scattering boundary conditions. These L^2 methods have been reviewed by Reinhardt[66] and Broad[67]. Bransden and Stelbovics[68] have also proposed to expand the closed-channel part of the optical potential on an L^2 basis.

Finally, we remark that Burke et al[69,70] have recently proposed a new R-matrix method with the aim of obtaining accurate scattering amplitudes in the "sub-intermediate" energy region. In this new R-matrix method the calculation of the scattering of an electron by an $(N + 1)$-electron target is carried out in two steps. In the first one a complete set of $(N + 1)$-electron bound and continuum basis states are determined by considering the collision of an electron with an N-electron target. In the second step these states are then used as target states for the $(N + 2)$-electron system. In this new version of the R-matrix method the partitioning of configuration space is such that the internal region is divided into two sub-regions, as shown in Fig. 4. The radius a_1 of the inner sub-

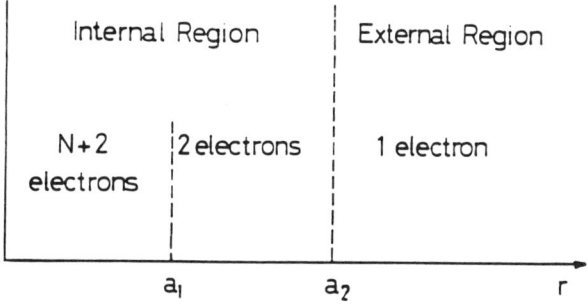

Figure 4. Partitioning of configuration space in the new R-matrix method.

region is chosen to just envelope the bound states of the N-electron target considered in the first step of the calculation. The radius a_2 of the outer sub-region is chosen to just envelop the bound states of the $(N+1)$-electron target retained in the second step. In the region $a_1 \leq r \leq a_2$ only the valence electron of the $(N + 1)$-electron target and the scattered electron are present; both of these are represented by continuum wave functions. This new version of the R-matrix theory has the additional possibility of representing ionization processes as well as elastic and inelastic ones.

To conclude this section, I shall discuss some stringent tests of the theory. A first one is to make detailed comparisons between electron and positron scattering. I recall in this connection that simple methods like the first Born or the Glauber approximations give identical cross sections for electron or positron impact, so that more sophisticated theories are required to analyze the differences between electron and positron collisions[71]. Fortunately, it has become feasible in recent years to perform positron-atom differential scattering measurements[72], as illustrated in Fig. 3. Since higher current positron beams are now available at various facilities[73], such differential measurements will soon be considerably extended.

A second severe test of the theory is provided by experiments using polarized electron beams and targets[19]. In particular, the development of improved polarized electron and hydrogen atom beams[74,75] has led to the first measurements[74] of the asymmetry between spin-parallel ($\uparrow\uparrow$) and spin-antiparallel ($\uparrow\downarrow$) scattering in $e^- - H$ collisions. This asymmetry can be characterized for a transition from target state $|i\rangle$ to $|f\rangle$ by the parameter

$$A_{fi} = \frac{\sigma_{fi}(\uparrow\downarrow) - \sigma_{fi}(\uparrow\uparrow)}{\sigma_{fi}(\uparrow\downarrow) + \sigma_{fi}(\uparrow\uparrow)} \quad (14)$$

where $\sigma_{fi}(\uparrow\downarrow)$ and $\sigma_{fi}(\uparrow\uparrow)$ are respectively the cross sections (differential or integrated) corresponding to the process $i \to f$, for antiparallel ($\uparrow\downarrow$) and parallel ($\uparrow\uparrow$) spins of the projectile electrons and the atoms. For example, Mc Dowell et al[76] have studied the angular distribution of the elastic and inelastic asymmetry parameter in $e^- - H$ collisions, using several theoretical methods. The agreement with the experimental results[74] is good.

A third exacting test of theoretical calculations is provided by electron-photon coincidence experiments, as illustrated in Fig. 2 for the electron impact excitation of the $2pm$ states of atomic hydrogen. Comprehensive reviews of this subject have been written by Slevin[77] and Andersen et al[78]. In particular, the angular correlation parameters[77] λ, R and I for the excitation of the $2pm$ states of atomic hydrogen by electron impact have been calculated by using several elaborate theoretical methods such as the $UEBS$ theory[39], the second order potential (SOP) method[54], the distorted wave second Born approximation[61] ($DWSBA$) and the multi pseudo-state close coupling ($MPCC$) approach[62]. The comparison with the measurements[41,79,80] at an incident electron energy of 54.4 eV shows that there is still room for improvement.

The complete determination of the density matrix describing the excitation of the $n = 2$ manifold of atomic hydrogen also requires the knowledge of the off-diagonal matrix elements characterizing the coherence between the $1s \to 2s$ and $1s \to 2pm$ transitions. Back et al[81] reported the first observation of $s - p$ coherence in an electron-photon coincidence experiment, using an external electric field to mix the $2s$ and $2p$ states. Their data yielded values of certain combinations of state multipoles[82], which were found to be in excellent agreement with the theoretical values obtained from the $UEBS$ method[83] or the $MPCC$ approach[84]. Unfortunately, the data of Back et al[81] do not provide a stringent test of the theory. Recently, however, Williams and Heck[85] have measured all the state multipoles describing the excitation of the $n = 2$ states of atomic hydrogen by incident electrons, at an energy of 350 eV and a scattering angle of 3°. Their data for the state multipoles describing the coherence between the $1s \to 2s$ and $1s \to 2pm$ transitions are compatible with the theoretical $UEBS$ and $MPCC$ results.

3. IONIZATION

The second part of this review deals with the ionization of atoms by electron impact. This is certainly one of the most challenging collision processes to analyze, since it exhibits the difficulties of several particle problems, coupled with delicate features due to the infinite range of the Coulomb interaction. I shall limit the discussion to the most basic electron impact ionization process, namely the single step $(e, 2e)$ reaction

$$e^- + A(i) \to A^+(f) + 2e^- \quad (15)$$

where $A(i)$ is a neutral atom in state $|i\rangle$ and $A^+(f)$ is the final ion left in state $|f\rangle$. A recent review of this subject is that of Byron and Joachain[86].

3.1 Kinematics

Let us consider the $(e, 2e)$ reaction (15), in which an electron of momentum $\mathbf{k}_i$ and non-relativistic energy $E_i = k_i^2/2$ is incident on a target atom A in the eigenstate $|i\rangle$ corresponding to the eigenenergy w_i. In the final state, two electrons emerge with momenta $\mathbf{k}_A$ and $\mathbf{k}_B$ and corresponding energies $E_A = k_A^2/2$ and $E_B = k_B^2/2$, the remaining ion being left in the eigenstate $|f\rangle$, with eigenenergy w_f. If $\mathbf{Q}$ denotes the recoil momentum of the ion, momentum conservation requires that $\mathbf{k}_i = \mathbf{k}_A + \mathbf{k}_B + \mathbf{Q}$. The energy conservation condition gives $E_i + w_i = E_A + E_B + w_f$, where the recoil energy of the ion has been neglected since the ion mass is much larger than unity. We shall denote by $\mathbf{\Delta} = \mathbf{k}_i - \mathbf{k}_A$ the momentum transfer, or more precisely the momentum lost by the faster (also called "scattered") electron A, with $E_A \geq E_B$. The slower electron B is usually called the "ejected" electron.

The most detailed information presently available about single ionization reactions of the type (15) has been obtained by analyzing triple differential cross sections ($TDCS$), measured in $(e, 2e)$ coincidence experiments. Such experiments were first performed by Ehrhardt et al[87] and Amaldi et al[88]. Since then, many $(e, 2e)$ coincidence measurements have been carried out; they have been reviewed recently by Ehrhardt et al[89]. The $TDCS$ is a measure of the probability that in an $(e,2e)$ reaction an incident electron of momentum $\mathbf{k}_i$ and energy E_i will produce on collision with the target two electrons having energies E_A and E_B and momenta $\mathbf{k}_A$ and $\mathbf{k}_B$, emitted respectively into the solid angles $d\Omega_A$ and $d\Omega_B$ centered about the directions (θ_A, ϕ_A) and (θ_B, ϕ_B). The $TDCS$ is usually denoted by the symbol $d^3\sigma/d\Omega_A d\Omega_B dE$. For unpolarized incident electrons and targets, which is the case considered here, it is a function of the quantities E_i, E_A (or E_B), θ_A, θ_B and $\phi_A - \phi_B$. By integrating the $TDCS$ over $d\Omega_A$, $d\Omega_B$ or dE one can form various double and single differential cross sections. Finally, the total ionization cross section is obtained by integrating over all outgoing electron scattering angles and energies, and depends only on E_i, the incident electron energy.

It is useful when studying $(e,2e)$ experiments to distinguish between several kinematical arrangements, since these have important implications on the theoretical analysis of the collision. In coplanar geometries the momenta $\mathbf{k}_i$, $\mathbf{k}_A$ and $\mathbf{k}_B$ are in the same plane (so that $\phi_A = 0$ and $\phi_B = 0$ or π) while in non-coplanar geometries the momentum $\mathbf{k}_B$ is out of the $(\mathbf{k}_i, \mathbf{k}_A)$ reference plane. Another useful distinction can be made between asymmetric and symmetric geometries. I shall soon return to this point in analyzing the $TDCS$ for $(e,2e)$ reactions with fast incident electrons. Before doing this, however, I shall briefly discuss the threshold behaviour of ionization cross sections.

3.2 Threshold behaviour of ionization cross sections

The behaviour of ionization cross sections just above threshold has been the subject of many investigations. In particular, threshold laws for ionization, which give the energy dependence of the total ionization cross section near threshold, but not its actual magnitude, have attracted much interest. Since threshold laws only provide limited information, it is not unreasonable to expect that they should be obtainable without knowing the full solution of the collision problem. This was emphasized in a fundamental paper by Wigner[90], who pointed out that the derivation of threshold laws does not require a detailed knowledge of the collision dynamics in the "reaction zone", where all the particles are close together. Wigner then applied this idea to various cases in which the reaction product consists of two particles escaping from each other.

Wannier[91] has extended Wigner's theory to the single ionization of atoms or ions by electron impact, in which the final number of particles is three : two electrons and a positive ion. If the three particles escaping from each other would interact only via short range forces, then by using phase space considerations alone it is a simple matter to show that the corresponding total "break-up" cross section would obey a quadratic threshold law, $\sigma \sim E^2$, where E is the excess energy above threshold. When long-range forces are present, as in the case of the $(e,2e)$ reaction with a positive ion and two electrons in the final state, phase space considerations alone are not sufficient to determine the threshold law. Indeed, one must take into account (i) the attractive Coulomb forces acting between each electron and the residual ion and (ii) the repulsive Coulomb force between the two electrons. Using hyperspherical coordinates, assuming that for large values of the hyper-radius R the Coulomb potential varies sufficiently slowly for classical mechanics to be valid, and assuming also that the distribution in phase space of the two electrons is quasi-ergodic when they leave the "reaction zone", Wannier[91] obtained for a system of total angular momentum equal to zero the total ionization cross section threshold law

$$\sigma \sim E^m, \quad m = \frac{1}{4}\left[\left(\frac{100Z - 9}{4Z - 1}\right)^{1/2} - 1\right] \quad (16)$$

where Z denotes the charge of the positive ion. We note that $m = 1.127$ when $Z = 1$. In the limit $Z \to \infty$ we have $m \to 1$, corresponding to a linear threshold law. This last result is to be expected since when $Z \to \infty$ the electron-electron interaction can be neglected, so that the final state can be described as a product of two Coulomb wave functions for the two electrons, giving a linear threshold law[92]. The Wannier result (16) has been confirmed by Peterkop[93] and Rau[94], who used the WKB approximation. Experimental evidence[95-101] is also consistent with the Wannier threshold law.

As I already mentioned above, the original classical calculations of Wannier[91] were performed for a system of total angular momentum equal to zero. Likewise, in the semi-classical calculations of Peterkop[93] and Rau[94], only the particular set $^1S^e$ of the quantum numbers $(LS\Pi)$ of the two electrons was considered. Several authors[102-107] have examined how the Wannier theory can be extended to other $(LS\Pi)$ values. In particular, Klar and Schlecht[103] concluded that all singlet states ($S = 0$) have the original Wannier exponent m while all triplet states ($S = 1$) have a different, higher exponent m' so that they tend to be suppressed at threshold. This result would lead to a value $A_I = 1$ of the ionization asymmetry parameter at threshold. Now, as seen from Fig. 5, the experiments[75,108] using polarized electron beams and targets show that the threshold value of A_I differs from unity. Upon re-examination of this problem, it was found[104-107]

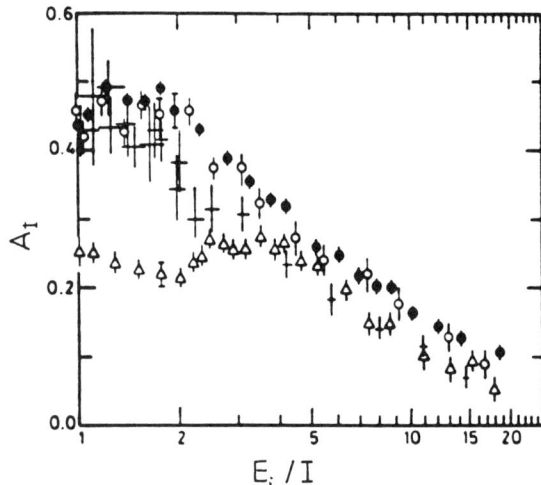

Figure 5. The ionization asymmetry parameter A_I as a function of the incident electron energy E_i expressed in units of the ionization potential I. The experimental data are from Fletcher et al[75] for atomic hydrogen (+) and from Baum et al[108] for lithium (•), sodium (○) and potassium (△).

that the threshold behaviour is the same for all $(LS\Pi)$ quantum numbers, except in the case of the $^3S^e$ and $^1P^e$ states, for which the Pauli principle requires the wave function to vanish at the Wannier saddle point. A review of this subject, together with other aspects of the Wannier theory, has been given by Rau[109].

Crothers[110] has recently developed a semi-classical theory of threshold ionization which yields absolute differential and total ionization cross sections near threshold. His theoretical values of the total and double differential cross sections agree well with the data of Pichou et al[98]. Crothers[110] has also estimated the contribution to the total $e^- - H$ ionization cross section arising from the Coulomb-dipole configurations, which according to Temkin[111] are the dominant ones close to threshold. Crothers found that this contribution is only of the order of 5% of that coming from configurations considered in the Wannier theory.

Finally, we note that several $(e,2e)$ coincidence experiments[100,101,112,113] giving $TDCS$ have been performed near threshold. Semi-empirical fits to the observed $TDCS$ have been obtained by using a method developed by Altick et al[114–116], but so far no ab-initio theoretical calculation has been able to reproduce the measured $TDCS$ in a satisfactory way.

3.3 Theory of coplanar asymmetric $(e,2e)$ reactions

Let us now turn to $(e,2e)$ reactions in which a fast (but non-relativistic) electron is incident on a neutral atom. I shall only give a survey of the situation concerning triple differential cross sections $(TDCS)$, which provide the most severe test of the theory. Detailed discussions may be found in the reviews of Joachain et al[86,117,118].

Let us begin by considering the Ehrhardt coplanar asymmetric geometry[87,89] where, for a given energy E_i of the fast incident electron, a fast electron A is detected in coincidence with a slow electron B. The scattering angle θ_A of the fast electron is fixed and small, while the angle θ_B of the slow electron is varied. It is worth noting that in this geometry the magnitude $\Delta = |\mathbf{k}_i - \mathbf{k}_A|$ of the momentum transfer is small.

A number of measurements of $TDCS$ have been performed in the asymmetric coplanar geometry[89]. The basic features emerging from these experiments are illustrated in Fig. 6, which shows the absolute $TDCS$ measured by Ehrhardt et al[119] for the $(e,2e)$ reaction in atomic hydrogen

$$e^- + H(1s) \rightarrow H^+ + 2e^- \qquad (17)$$

for the case $E_i = 250 eV$, $E_B = 5 eV$ and $\theta_A = 3°$. As seen from Fig. 6, there is a strong angular correlation

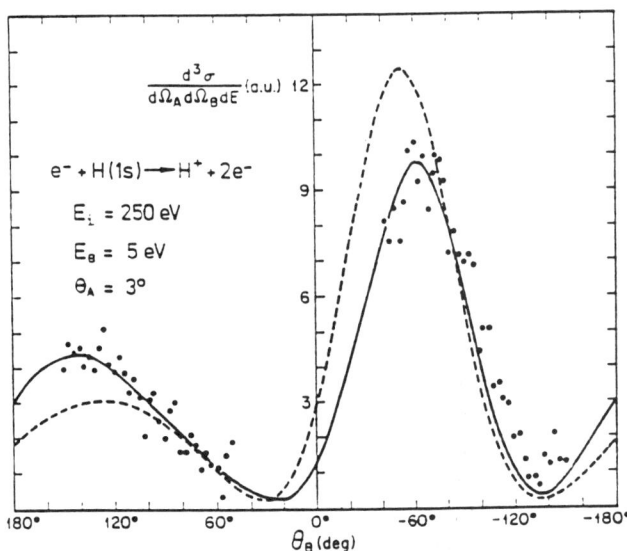

Figure 6. The $TDCS$ (in au) for the ionization of atomic hydrogen from the ground state by electron impact, for the case $E_i = 250 eV$, $E_B = 5 eV$ and $\theta_A = 3°$, as a function of the ejected electron angle θ_B. Theoretical curves : - - - first Born approximation, —— eikonal Born series (EBS) results of Byron, Joachain and Piraux[122]. Experimental data : • absolute measurements of Ehrhardt et al[119].

between the scattered and ejected electrons, characterized by two peaks : the forward or "binary" peak and the backward or "recoil" peak. According to the first Born approximation, also shown in Fig. 6, the maximum of the forward peak should occur in the direction $\hat{\Delta}$ of the momentum transfer, and that of the backward peak should be in the opposite direction $-\hat{\Delta}$. However, we see from Fig. 6 that the experimental peak is shifted towards larger angles $|\theta_B|$ with respect to the first Born prediction, and that the measured recoil peak is shifted by an even larger amount, also towards larger angles. Moreover, the ratio of the intensity of the binary peak to that of the recoil peak is considerably reduced with respect to the first Born prediction.

These features have been illustrated for the $(e, 2e)$ reaction (17) in atomic hydrogen since in that case the wave function describing the target in the initial $(1s)$ state is known exactly, as is the wave function corresponding to the final, unbound, target state $(H^+ + e^-)$. As a result, the comparison with absolute experimental data is unambiguous. In the case of $(e, 2e)$ reactions in other atoms, for example in helium,

$$e^- + He(1^1S) \rightarrow He^+(1s) + 2e^- \quad (18)$$

difficulties arise in describing accurately the target initial ground state and especially its final continuum state consisting of an ion and an unbound electron. Although interesting in themselves, these problems introduce additional complications in the interpretation of data on $TDCS$. Nevertheless, the experiments[89] clearly show that at intermediate incident electron energies there are significant shifts of the forward and recoil peaks away from the directions $\hat{\Delta}$ and $-\hat{\Delta}$ predicted by the first Born approximation, and that the relative magnitude of the forward and recoil peaks is not given correctly by the first Born approximation.

The experimental work performed since 1969 on coplanar asymmetric $(e, 2e)$ reactions was steadily accompanied by theoretical treatments of the problem. The early calculations, which were all of first-order in the electron-electron interaction, failed to account for the experimental data. The first theoretical treatment in which all the characteristic features of the measurements were reproduced was a second Born calculation performed by Byron, Joachain and Piraux[120] for the $(e, 2e)$ reaction (17) in atomic hydrogen, within the framework of the EBS method[30,32]. For an unpolarized $e^- - H$ system the $TDCS$ is given by

$$\frac{d^3\sigma}{d\Omega_A d\Omega_B dE} = \frac{k_A k_B}{k_i} \left[\frac{1}{4}|f+g|^2 + \frac{3}{4}|f-g|^2 \right] \quad (19)$$

where f and g are the direct and exchange ionization amplitudes, respectively. According to the EBS method, the direct amplitude is given by eq. (9). Now, in the Ehrhardt asymmetric geometry, where the magnitude Δ of the momentum transfer is small, the third order Glauber term $\bar{f}_{G3}$ is unimportant. Moreover, since one electron emerges with a high velocity and the other with a low velocity, exchange effects are also small. A good approximation to the $TDCS$ in the Ehrhardt asymmetric geometry is therefore provided by the second Born expression

$$\frac{d^3\sigma_{B2}}{d\Omega_A d\Omega_B dE} = \frac{k_A k_B}{k_i} |\bar{f}_{B1} + \bar{f}_{B2}|^2 \quad (20)$$

It is this expression which was evaluated by Byron, Joachain and Piraux[120] for the reaction (17), using the closure approximation to obtain $\bar{f}_{B2}$. More recently, these authors have performed full EBS calculations[121,122] for the reaction (17). As seen from Fig. 6, the predictions of the EBS theory (which are very close to the second Born values) have been fully confirmed by the measurements[119]. Further confirmation of the EBS results has also come recently from the $UEBS$ calculations of Joachain et al[123], the EBS' results of Balyan and Srivastava[124], the coupled pseudo-state calculations of Curran and Walters[125] and the results of Brauner et al[126].

Before leaving the subject of coplanar asymmetric ionization reactions in atomic hydrogen, it is interesting to see what happens when the incident electron is replaced by a positron. This is illustrated in Fig. 7 for the case $E_i = 250 eV$, $E_B = 5 eV$ and $\theta_A = 3°$. The $TDCS$ for incident positrons, calculated according to the EBS method[117], is compared with the corresponding one for incident electrons[122] and with the first Born $TDCS$. We see from Fig. 7 that all the characteristic features for the case of incident positrons are reversed with respect to those corresponding to incident electrons. In particular, the peaks are shifted in opposite directions, and the ratio of the intensity of the forward peak to that of the recoil peak is increased for incident positrons while it is decreased for incident electrons with respect to the first Born value.

Let us now turn to the analysis of asymmetric coplanar $(e, 2e)$ reactions in other atoms. As an example, we shall consider the reaction (18) in helium, which has been the subject of detailed investigations. In this case, the second Born $TDCS$ of eq. (20) has been calculated by Byron, Joachain and Piraux[127,128], using a Hartree-Fock wave function for the helium ground state and a symmetrized product of the $He^+(1s)$ wave function times a Coulomb wave (orthogonalized to the ground state Hartree-Fock orbital) to describe the final $He^+(1s) + e^-$ continuum state. Furtado and O'Mahony[129] have extended the work of Byron, Joachain and Piraux by using improved wave

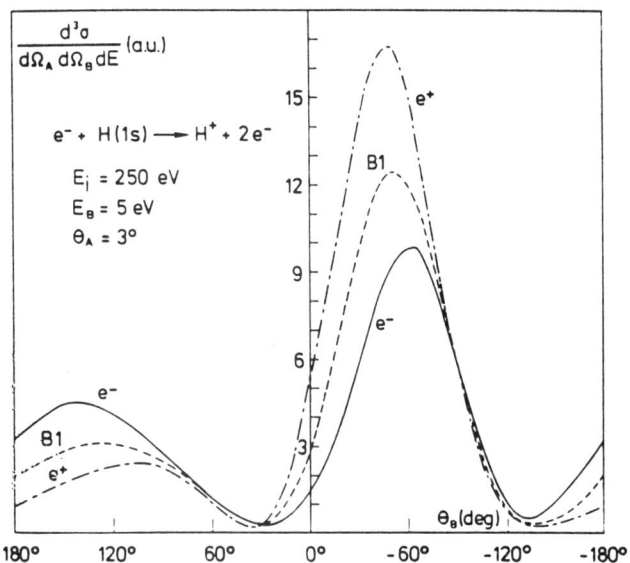

Figure 7. The $TDCS$ (in au) for the ionization of atomic hydrogen from the ground state by electron and positron impact, for the case $E_i = 250 eV$, $E_B = 5 eV$ and $\theta_A = 3°$. Curve - - - : first Born approximation. Curve —— : EBS results for incident electrons. Curve - · - : EBS results for incident positrons. From Joachain and Piraux[117].

functions for the helium ground state and for the final continuum state. The agreement with experiment is then nearly as good as that shown in Fig. 6 for the case of atomic hydrogen.

The second Born and EBS calculations discussed so far concern the ionization of atoms initially in the ground state. Recently, the $TDCS$ for ionization of atomic hydrogen from the metastable $2s$ state has been obtained in the coplanar asymmetric geometry by Vucic et al[130] using the second Born approximation, and subsequently by Potvliege et al[131] using the EBS method. These $TDCS$ differ markedly from those corresponding to ionization from the ground state. In particular, interference effects between the first and second Born terms distort the first Born $TDCS$ in a way different from that corresponding to the ionization of $H(1s)$.

3.4 Theory of coplanar symmetric $(e, 2e)$ reactions

Symmetric geometries are such that $E_A \simeq E_B$ and $\theta_A \simeq \theta_B$. The first $(e, 2e)$ symmetric coincidence experiments were performed by Amaldi et al[88] and since then many measurements of this type have been performed. As an example, I shall consider a coplanar, fully symmetric geometry with $E_A = E_B$, $\theta_A = \theta_B (= \theta)$, $\phi_A = 0$ and $\phi_B = \pi$. I shall assume that the incident electron is fast, so that the magnitude of the momentum transfer is given approximately by $\Delta \simeq k_i (3/2 - \sqrt{2} \cos \theta)^{1/2}$, and we see that Δ is never small in this geometry. The recoil momentum of the ion is $Q = |2k_A \cos \theta - k_i| \simeq k_i |\sqrt{2} \cos \theta - 1|$.

Two angular regions can therefore be distinguished. For $\theta < 70°$, Δ is relatively large while Q remains small or moderate. In this case (as well as in non-coplanar symmetric geometries with similar values of Δ and Q), one has impulse-type collisions from which the electron momentum density distribution of the target atom can be obtained[132]. This important property has given rise to the field of $(e, 2e)$ spectroscopy[133-135], and I shall call the angular domain in which Δ is large and Q small or moderate the $(e, 2e)$ spectroscopy region. The most elaborate calculations in this region have been performed by using the distorted wave impulse approximation[133-135] $(DWIA)$. It is worth noting that in the region for which $Q \simeq 0$ (corresponding to $\theta \simeq 45°$) the first Born term is of order k_i^{-2} (for large k_i) while the second Born term is of order k_i^{-3}, and hence is only a correction to $\bar{f}_{B1}$.

Let us now consider the large angle region ($\theta > 70°$) for which both Δ and Q are large. In this region one expects on the basis of the calculations performed for inelastic (excitation) scattering[30] that the second Born term should be very important, and that this term will be dominated by the contributions of the initial and final target states, acting as intermediate states. This is confirmed by second Born calculations performed for $e^- - H(1s)$ ionization by Byron, Joachain and Piraux[136]. For large k_i and in the case of a coplanar symmetric geometry at large θ it is found that the first Born term $\bar{f}_{B1}$ is of order k_i^{-6} while the contribution $\bar{f}_{B2}(1s)$ of the initial $(1s)$ target state to $\bar{f}_{B2}$ is also of order k_i^{-6}, except near $\theta = 135°$ where it is of order k_i^{-5}. The contribution $\bar{f}_{B2}$ (cont) of the final (continuum) target state to $\bar{f}_{B2}$ can also be estimated for large k_i, and is found[136] to be of order k_i^{-6}. Hence second order effects, particularly those due to the term $\bar{f}_{B2}(1s)$, should be very important at large angles. This analysis is confirmed by an exact (numerical) calculation[136] of the $TDCS$, using $\bar{f}_{B1} + \bar{f}_{B2}(1s)$ as the direct scattering amplitude and taking into account the fact that the collision only occurs in the singlet mode.

A similar striking behaviour of the calculated $TDCS$, due to second order effects, arises for large angle symmetric $(e, 2e)$ reactions in helium (see Fig. 8) and has been observed by Pochat et al[137]. As seen from Fig. 8, the agreement between experiment and theory is considerably better at large angles when the contribution $\bar{f}_{B2}(1^1S)$ of the initial (1^1S) target state to the second Born term is included in the scattering amplitude. This agreement could still be improved by including the contribution $\bar{f}_{B2}$ (cont) of the final (continuum) target state to $\bar{f}_{B2}$. Moreover, since the incident electron energy $E_i = 200 eV$ is relatively low, the contributions to $\bar{f}_{B2}$ arising from other target states acting as intermediate states should also be analyzed, as well as those coming from higher order terms of perturbation theory.

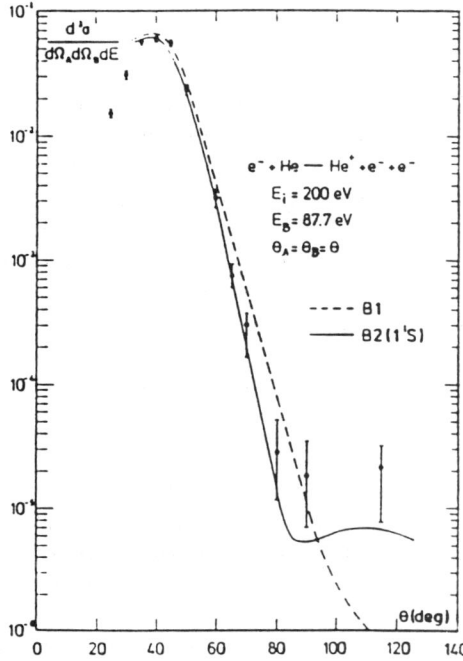

Figure 8. The $TDCS$ (in au) for the $(e, 2e)$ reaction (18), as a function of the angle $\theta = \theta_A = \theta_B$, for a coplanar symmetric energy-sharing geometry, with $E_i = 200 eV$ and $E_A = E_B = 87.7 eV$. The full curve, $B2(1^1S)$, corresponds to a second Born calculation[137] in which $f = \bar{f}_{B1} + \bar{f}_{B2}(1^1S)$. The broken curve, $B1$, corresponds to the first Born approximation. The experimental data (•) are those of Pochat et al[137].

REFERENCES

1. P.G. Burke, in "Fundamental Processes in Atomic Collision Theory", ed. by H. Kleinpoppen and J.S. Briggs (Plenum, New York, 1985).
2. C.J. Joachain, in "Collision Theory for Atoms and Molecules", ed. by F.A. Gianturco (Plenum, New York, 1989).
3. H. Feshbach, Ann. Phys. (N.Y.) 5, 357 (1958); 19, 287 (1962).
4. P.G. Burke, A. Hibbert and D.W. Robb, J. Phys. B4, 153 (1971).
5. P.G. Burke, in "Atomic Physics", ed. by H. Narumi and I. Shimamura, Vol. 10 (Elsevier, Amsterdam, 1987) p. 243.
6. M.J. Seaton, J. Phys. B7, 1817 (1974).
7. E.R. Smith and R.J.W. Henry, Phys. Rev. A7, 1585 (1973); A8, 572 (1973).
8. R.K. Nesbet, "Variational Methods in Electron-Atom Scattering Theory" (Plenum, New York, 1980).
9. P.G. Burke and W. Eissner, in "Atoms in Astrophysics", ed. by P.G. Burke, W.B. Eissner, D.G. Hummer and I.C. Percival (Plenum, New York, 1983), p. 1.
10. W.C. Fon, K.P. Lim, K.A. Berrington, P.G. Burke and A.E. Kingston, to be published.
11. A. Bass, Ph. D. thesis, University of Manchester (1988).
12. M.J. Seaton, Rep. Progr. Phys. 46, 167 (1983).
13. H.E. Saraph, Comp. Phys. Comm. 3, 256 (1972); 15, 247 (1978).
14. M. Jones, Phil. Trans. Roy. Soc.A277, 587 (1975)
15. N.S. Scott and P.G. Burke, J. Phys. B13, 4299 (1980).
16. A.L. Sinfailam, Aust. J. Phys. 33, 261 (1980).
17. N.S. Scott, P.G. Burke and K. Bartschat, J. Phys. B16, L361 (1983).
18. K. Bartschat, N.S. Scott, K. Blum and P.G. Burke, J. Phys. B17, 269 (1984).
19. J. Kessler, "Polarized Electrons" (Springer-Verlag, Berlin, 2d ed., 1985).
20. J. Kessler, in "Electronic and Atomic Collisions", ed. by H.B. Gilbody, W.R. Newell, F.H. Read and A.C.H. Smith (North Holland, Amsterdam, 1988) p. 21.
21. A. Wolcke, K. Bartschat, K. Blum, H. Borgmann, G.F. Hanne and J. Kessler, J. Phys. B16, 639 (1983).
22. I.P. Grant, Comp. Phys. Comm. 17, 149 (1979).
23. G.D. Carse and D.W. Walker, J. Phys. B6, 2529 (1973).
24. D.W. Walker, J. Phys. B7, 97 (1974).
25. J.J. Chang, J. Phys. B8, 2327 (1975).
26. P.H. Norrington and I.P. Grant, J. Phys. B14, L261 (1981).
27. C.J. Joachain, "Quantum Collision Theory" (North Holland, Amsterdam, 3d ed., 1983).
28. C.J. Joachain, in "Electronic and Atomic Collisions", ed. by G. Watel (North Holland, Amsterdam, 1978) p. 71.
29. H.R.J. Walters, in "Electronic and Atomic Collisions", ed. by H.B. Gilbody, W.R. Newell, F.H.

Read and A.C.H. Smith (North Holland, Amsterdam, 1988) p. 147.
30. F.W. Byron, Jr. and C.J. Joachain, Phys. Rep. 34, 233 (1977).
31. H.R.J. Walters, Phys. Rep. 116, 1 (1984).
32. F.W. Byron, Jr. and C.J. Joachain, Phys. Rev. A8, 1267 (1973).
33. R.A. Bonham, J. Chem. Phys. 36, 3260 (1962).
34. V.I. Ochkur, Zh. Eksp. Teor. Fiz 45, 734 (1963) [Sov. Phys. JETP 18, 503 (1964)].
35. F.W. Byron, Jr. and C.J. Joachain, J. Phys. B8, L284 (1975).
36. T.T. Gien, J. Phys. B9, 3203 (1976).
37. F.W. Byron, Jr., C.J. Joachain and R.M. Potvliege, J. Phys. B14, L609 (1981).
38. F.W. Byron, Jr., C.J. Joachain and R.M. Potvliege, J. Phys. B15, 3915 (1982).
39. F.W. Byron, Jr., C.J. Joachain and R.M. Potvliege, J. Phys. B18, 1637 (1985).
40. S.J. Wallace, Ann. Phys. (N.Y.) 78, 190 (1973).
41. J.F. Williams, J. Phys. B14, 1197 (1981).
42. F.W. Byron, Jr. and C.J. Joachain, Phys. Lett. A49, 306 (1974).
43. F.W. Byron, Jr. and C.J. Joachain, Phys. Rev. A15, 128 (1977).
44. F.W. Byron, Jr. and C.J. Joachain, J. Phys. B14, 2429 (1981).
45. S. Vucic, R.M. Potvliege and C.J. Joachain, J. Phys. B20, 3157 (1987).
46. M.H. Mittleman and K.M. Watson, Phys. Rev. 113, 98 (1959); Ann. Phys. (N.Y.) 10, 268 (1960).
47. J.B. Furness and I.E. McCarthy, J. Phys. B6, 2280 (1973).
48. R. Vanderpoorten, J. Phys. B8, 926 (1975).
49. M.E. Riley and D.G. Truhlar, J. Chem. Phys. 65, 792 (1976).
50. C.J. Joachain, R. Vanderpoorten, K.H. Winters and F.W. Byron, Jr., J. Phys. B10, 227 (1977).
51. G.M.A. Hyder, M.S. Dababneh, Y.F. Hsieh, W.E. Kauppila, C.K. Kwan, M. Madhavi-Hezaveh and T.S. Stein, Phys. Rev. Lett. 57, 2252 (1986).
52. C.J. Joachain and R.M. Potvliege, Phys. Rev. A35, 4873 (1987).
53. B.H. Bransden and J.P. Coleman, J. Phys. B5, 537 (1972).
54. B.H. Bransden, T. Scott, R. Shingal and R.K. Raychoudhury, J. Phys. B15, 4605 (1982).
55. I.E. McCarthy and A.T. Stelbovics, Phys. Rev. A22, 502 (1980); J. Phys. B16, 1233 (1983); I.E. McCarthy, B.C. Saha and A.T. Stelbovics, J. Phys. B14, 2871 (1981); B15, L401 (1982); Phys. Rev. A25, 268 (1982).
56. B.H. Bransden and M.R.C. McDowell, Phys. Rep. 30, 207 (1977); 46, 249 (1978).
57. R.J.W. Henry, Phys. Rep. 68, 1 (1981).
58. Y. Itikawa, Phys. Rep. 143, 69 (1986).
59. D.P. Dewangan and H.R.J. Walters, J. Phys. B10, 637 (1977).
60. A.E. Kingston and H.R.J. Walters, J. Phys. B13, 4633 (1980).
61. D.H. Madison, J.A. Hughes and D.S. McGinness, J. Phys. B18, 2737 (1985).
62. W.L. van Wyngaarden and H.R.J. Walters, J. Phys. B19, 929 (1986); B19, 1817 (1986); B19, 1827 (1986).
63. P.G. Burke and T.G. Webb, J. Phys. B3, L131 (1970).
64. J. Callaway and J.W. Wooten, Phys. Lett. A45, 85 (1973); Phys. Rev. A9, 1924 (1974); A11, 1118 (1975).
65. P.G. Burke, K.A. Berrington and C.V. Sukumar, J. Phys. B14, 289 (1981).
66. W.P. Reinhardt, Comp. Phys. Comm.17, 1 (1979)
67. J.T. Broad, in "Electron-Atom and Electron-Molecule Collisions", ed. by J. Hinze (Plenum, New York, 1983) p. 91.
68. B.H. Bransden and A.T. Stelbovics, J. Phys. B17, 1877 (1984).
69. P.G. Burke, C.J. Noble and P. Scott, Proc. Roy. Soc. (London) A410, 289 (1987).
70. T.T. Scholz, M.P. Scott, P.G. Burke and C.J. Noble, in "Electronic and Atomic Collisions", ed. by H.B. Gilbody, W.R. Newell, F.H. Read and A.C.H. Smith (North Holland, Amsterdam, 1988) p. 215.
71. C.J. Joachain, in "Atomic Physics with Positrons", ed. by J.W. Humberston and E.A.G. Armour (Plenum, New York, 1987) p. 71.
72. W. Kauppila and T.S. Stein, in "Atomic Physics with Positrons", ed. by J.W. Humberston and E.A.G. Armour (Plenum, New York, 1987) p. 27.
73. W. Raith and G. Sinapius, Comments At. Mol. Phys. 22, 199 (1989).
74. G.D. Fletcher, M.J. Alguard, T.J. Gay, V.W. Hughes, C.W. Tu, P.F. Wainwright and M.S. Lubell, Phys. Rev. Lett. 48, 1671 (1982).
75. G.D. Fletcher, M.J. Alguard, T.J. Gay, V.W. Hughes, P.F. Wainwright, M.S. Lubell and W. Raith, Phys. Rev. A31, 2854 (1985).
76. M.R.C. McDowell, P.W. Edmunds, R.M. Potvliege, C.J. Joachain, R.Shingal and B.H. Bransden, J. Phys. B17, 3951 (1984).
77. J. Slevin, Rep. Progr. Phys. 47, 461 (1984).
78. N. Andersen, J.W. Gallagher and I.V. Hertel, Phys. Rep. 165, 1 (1988).
79. E. Weigold, L. Frost and K.J. Nygaard, Phys. Rev. A21, 1950 (1980).
80. J.F. Williams, Aust. J. Phys. 39, 621 (1986).

81. C.G. Back, S. Watkin, M. Eminyan, K. Rubin, J. Slevin and J.M. Woolsey, J. Phys. B17, 2695 (1984).
82. K. Blum, "Density Matrix Theory and Applications", (Plenum, New York, 1981).
83. R.M. Potvliege and C.J. Joachain, J. Phys. B18, L585 (1985).
84. W.L. van Wyngaarden and H.R.J. Walters, J. Phys. B18, L689 (1985).
85. J.F. Williams and E.L. Heck, J. Phys. B21, 1627 (1988).
86. F.W. Byron, Jr. and C.J. Joachain, Phys. Rep. (to be published).
87. H. Ehrhardt, M. Schulz, T. Tekaat and K. Willmann, Phys. Rev. Lett. 22, 89 (1969).
88. U. Amaldi, Jr., A. Egidi, R. Marconero and G. Pizzella, Rev. Sci. Instr. 40, 1001 (1969).
89. H. Ehrhardt, K. Jung, G. Knoth and P. Schlemmer, Z. Phys. D1, 3 (1986).
90. E.P. Wigner, Phys. Rev. 73, 1002 (1948).
91. G.H. Wannier, Phys. Rev. 90, 817 (1953).
92. R.K. Peterkop, "Theory of Ionization of Atoms by Electron Impact" (Colorado Univ. Press, 1977).
93. R.K. Peterkop, J. Phys. B4, 513 (1971).
94. A.R.P. Rau, Phys. Rev. A4, 207 (1971).
95. J.W. McGowan and E.M. Clarke, Phys. Rev. 167, 43 (1968).
96. S. Cvejanovic and F.H. Read, J. Phys. B7, 1841 (1974).
97. D. Spence, Phys. Rev. A11, 1539 (1975).
98. F. Pichou, A. Huetz, G. Joyez and M. Landau, J. Phys. B11, 3683 (1978).
99. M.H. Kelley, W.T. Rogers, R. Celotta and S.R. Mielczarek, Phys. Rev. Lett. 51, 2191 (1983).
100. P. Fournier-Lagarde, J. Mazeau and A. Huetz, J. Phys. B17, L591 (1984); B18, 379 (1985).
101. P. Selles, A. Huetz and J. Mazeau, J. Phys. B20, 5195 (1987).
102. T.A. Roth, Phys. Rev. A5, 476 (1972).
103. H. Klar and W. Schlecht, J. Phys. B9, 1699 (1976)
104. C.H. Greene and A.R.P. Rau, Phys. Rev. Lett. 48, 533 (1982); J. Phys. B16, 99 (1983).
105. A.D. Stauffer, Phys. Lett. A91, 114 (1982).
106. R. Peterkop, J. Phys. B16, L587 (1983).
107. J.M. Feagin, J. Phys. B17, 2433 (1984).
108. G. Baum, M. Moede, W. Raith and W. Schröder, J. Phys. B18, 531 (1985).
109. A.R.P. Rau, Phys. Rep. 110, 369 (1984).
110. D.S.F. Crothers, J. Phys. B19, 463 (1986).
111. A. Temkin, Phys. Rev. Lett. 49, 365 (1982); Phys. Rev. A30, 2737 (1984).
112. E. Schubert, K. Jung and H. Ehrhardt, J. Phys. B14, 3267 (1981).
113. P. Schlemmer, T. Rösel, K. Jung and H. Ehrhardt (to be published).
114. P.L. Altick, J. Phys. B18, 1841 (1985).
115. K.D. Shaw and P.L. Altick, J. Phys. B19, 3161 (1986).
116. P.L. Altick and T. Rösel, J. Phys. B21, 2635 (1988)
117. C.J. Joachain and B. Piraux, Comments At. Mol. Phys. 17, 261 (1986).
118. C.J. Joachain, in "Few-Body Problems in Particle, Nuclear, Atomic and Molecular Physics", ed. by J.L. Ballot and M. Fabre de la Ripelle (Springer Verlag, Berlin, 1987) p. 294.
119. H. Ehrhardt, G. Knoth, P. Schlemmer and K. Jung, Phys. Lett. A110, 92 (1985).
120. F.W. Byron, Jr., C.J. Joachain and B. Piraux, J. Phys. B13, L673 (1980).
121. F.W. Byron, Jr., C.J. Joachain and B. Piraux, Phys. Lett. A99, 427 (1983); A102, 289 (1984).
122. F.W. Byron, Jr., C.J. Joachain and B. Piraux, J. Phys. B18, 3203 (1985).
123. C.J. Joachain, B. Piraux, R.M. Potvliege, F. Furtado and F.W. Byron, Jr., Phys. Lett. A112, 138 (1985).
124. K.S. Balyan and M.K. Srivastava, Phys. Rev. A32, 3098 (1985); A33, 2155 (1986).
125. E. Curran and H.R.J. Walters, J. Phys. B20, 337 (1987).
126. M. Brauner, J.S. Briggs and H. Klar (to be published).
127. F.W. Byron, Jr., C.J. Joachain and B. Piraux, J. Phys. B15, L293 (1982).
128. F.W. Byron, Jr., C.J. Joachain and B. Piraux, J. Phys. B19, 1201 (1986).
129. F. Furtado and P.F. O'Mahony, J. Phys. B20, L405 (1987); B21, 137 (1988).
130. S. Vucic, R.M. Potvliege and C.J. Joachain, Phys. Rev. A35, 1446 (1987).
131. R.M. Potvliege, S. Vucic and C.J. Joachain, J. Phys. B20, 4883 (1987).
132. A.E. Glassgold and G. Ialongo, Phys. Rev. 175, 151 (1968).
133. I.E. McCarthy and E. Wiegold, Phys. Rep. 27, 275 (1976).
134. E. Wiegold and I.E. McCarthy, Adv. Atom. Mol. Phys. 14, 127 (1978).
135. A. Giardini-Guidoni, R. Fantoni, R. Camilloni and G. Stefani, Comments Atom. Mol. Phys. 10, 107 (1981).
136. F.W. Byron, Jr., C.J. Joachain and B. Piraux, J. Phys. B16, L769 (1983).
137. A. Pochat, R.J. Tweed, J. Peresse, C.J. Joachain, B. Piraux and F.W. Byron, Jr., J. Phys. B16, L775 (1983).

GIANT RESONANCES IN DOUBLE IONIZATION OF ATOMIC IONS

Hirosi Suzuki[*], Takato Hirayama[**] and Toshinobu Takayanagi[*]

[*]Department of Physics, Sophia University, Chiyada-ku, Tokyo 102 Japan
[**]Department of Physics, Gakushuin University, Toshima-ku, Tokyo 171 Japan

Experimantal studies on giant resonances in heavy atoms and ions by electron impact, carried on in our laboratory at Sophia University, are described. Three topics are discussed; the first one is measurements of the double ionization cross sections of singly charged alkaline-earth ions by means of the crossed electron-ion beams experiment. The second topic is measurements of partial cross sections for the inner shell ionization of rare-gas atoms by means of the Auger electron spectroscopy. The third one is measurements of partial cross sections for the outer s-subshell ionization of rare-gas atoms by means of the VUV spectroscopy. Experimental results are discussed in relation to the giant resonance effect.

1. Preface

Recently, double ionization cross sections of atomic ions by electron impact have been measured in different laboratories and a distinctive gigantic structure has been observed in the cross section curve as a function of impact energy for the ions which have the 4d electrons in their electron configuration.[1,2] Younger proposed the first interpretation of these phenomena as a genuine giant resonance which occurs in the electron impact excitation and ionization of heavy atoms and ions, namely, a shape resonance which appears in the scattering channel.[3,4,5]

In this paper, we describe subsequent experimental results on the giant resonance in the electron impact double ionization of Ba^+ ion,[6] together with a series of measurements of the cross sections of singly charged alkaline-earth ions.

In relation to the giant resonance in double ionization processes, we discuss a pronounced behavior of the partial cross sections of the electron impact 4d ionization in Xe atom, which has been firstly observed in a series of measurements of partial cross sections for the inner ahell ionization of rare-gas atoms.[7,8,9] These results are compared with theoretical cross sections calculated by Younger using the distorted wave Born exchange approximation.[5,10] We discuss further a giant resonance type behavior of the partial cross sections for the outer s-subshell electron ionization in rare-gas atoms by electron impact, which have been observed in a series of experiments in our laboratory.

2. Giant resonances in atomic processes

Typical examples of the giant resonances previously known are the pronounced structure in the photoabsorption spectra of heavy atoms and ions.[11,12] If the angular momentum l of the inner shell electron to be ionized is large (l>2), the potential for its final scattered wave has a barrier between two potential wells by virtue of the addition of the centrifugal repulsive part $l(l+1)/r^2$ to the Coulombic attractive part $V(r)$. Formation of the pseudo-bound-state with the positive energy in the inner potential well results in the shape resonance effect due to a virtual compound state formed for the scattered electron. Younger performed the first calculation of the single ionization of the 4d electron in Cs^+ ion using the distorted wave Born exchange approximation (DWBEA), and found that the phase shift for kf waves for l=3 partial waves undergoes a rappid increase from zero to pi in the vicinity of the electron energy 0.83 a.u., reflecting the transfer of kf orbital density from the outer potential well into the inner well for the f-wave scattering potential.[3,5] Experimental cross sections for double ionization of Cs^+ [1] are well reproduced by the theoretical results for the single ionization of the 4d-electrons,[3,5] suggesting the branching ratio for autoionization is unity.

In order to illustrate an idea of the giant resonances in the electron impact ionization of atomic ions, we show the cross section curves for double ionization of singly charged rare-gas ions and alkali ions in Fig.1 and Fig.2, respectively. In contrast to small cross sections and gently sloping shapes for Kr^+ and Ar^+ curves in Fig.1 and for Rb^+ and K^+ curves in Fig.2, cross sections for Xe^+ and Cs^+ overwhelm in their absolute values, and their curves possess a very distinctive profile. It is a huge peak which rises steeply from the threshold and has a tail towards higher energies. It has generally a very broad feature of the spectrum, suggesting lifetimes of a few orders of magnitude shorter than typical excitation-autoionization processes.

3. Double ionization cross sections of alkaline-earth ions

Single and multiple ionization cross sections of alkaline-earth ions are measured by meams of the crossed electron-ion beams technique, which is essentially the same as that developed by Harrison and Dolder[18]. Schematic diagram of the apparatus is shown in Fig.3. It consists of a surface ionization type ion source and a collision chamber which contains an electron collision region and a charge state analyzer with ion detectors. The collision chamber is evacuated being kept at a pressure of ultrahigh vacuum region. The double beam chopping technique are employed in order to discriminate the true signals from various background noises. Typical signal to noise ratio was from 5 to 10 for double ionization measurements. Measurements of the form factor, which represents a spatial overlap of the electron and ion densities at the crossed region and is essential to deduce the absolute cross section, were performed using a L-shaped probe plate with fine horizontal slits, which was driven vertically by a stepping motor. Overall uncertainty of the absolute cross sections are estimated about ±10% for the 90% confidence level.

Results of the double ioinization cross sections for Ba^+, Sr^+ and Ca^+ are

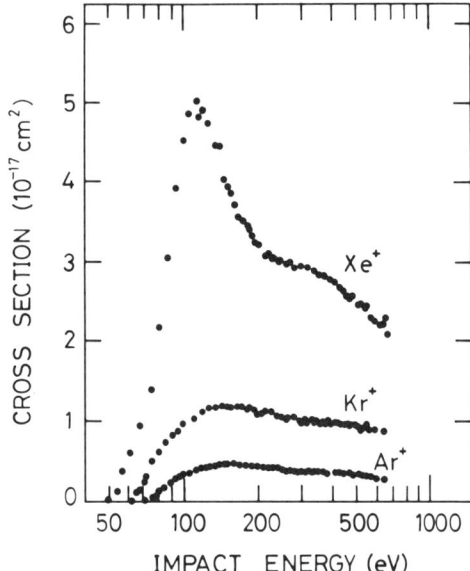

FIG.1. Double ionization cross sections of rare-gas ions, Xe+ (Muller et al.[13]), Kr+ (Tinschert et al.[14]), and Ar+ (Muller et al.[15]) as functions of electron impact energies.

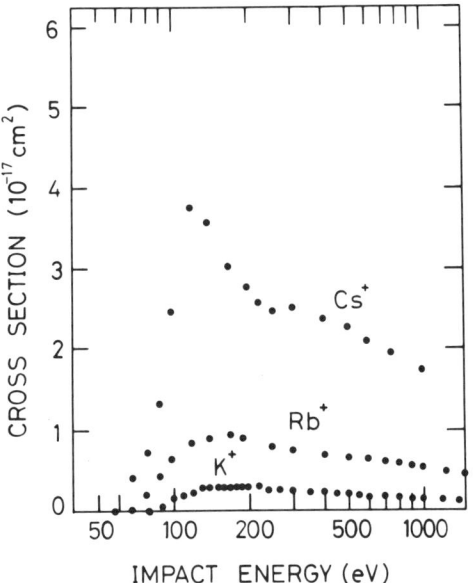

FIG.2. Double ionizaition cross sections of alkali metal ions, Cs+ (Hertling et al.[1]), Rb+ (Hughes et al.[16]), and K+ (Hirayama et al.[17]) as functions of electron impact energies.

shown in Fig.4. In contrast to mild profiles in the ionization curve for Sr+ and Ca+, the curve for Ba+ reveals distinctive features; the cross section increases rather steeply from the threshold (45.5 eV) and show a remarkable increase at around 85 eV of the impact energy. The curve possesses a broad tail towars higher energies with a small shoulder around 200 eV. The huge peak with a typical profile of the giant resonance is attributed to the 4d ionization followed by the autoionization expressed as,

$$Ba^+(4d^{10}5s^25p^66s) + e$$
$$\to Ba^{2+}(4d^95s^25p^66s) + 2e$$
$$\hookrightarrow Ba^{3+}(4d^{10}5s^25p^5) + e \ .$$

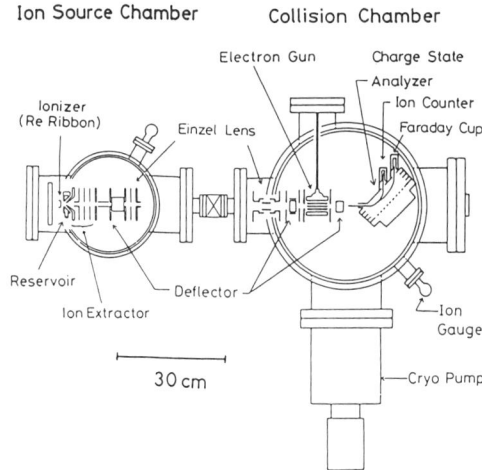

FIG.3. Schematic diagram of the apparatus for the crossed electron-ion beams experiment.

The resonance shape closely resembles those of Xe+ and Cs+, suggesting that these resonances are from the same origin. The cross section curve for Ba+, however, has another feature near the threshold that differs from Xe+ and Cs+. The Ba+ curve possesses a bump ranging from 50 eV to 90 eV, which is

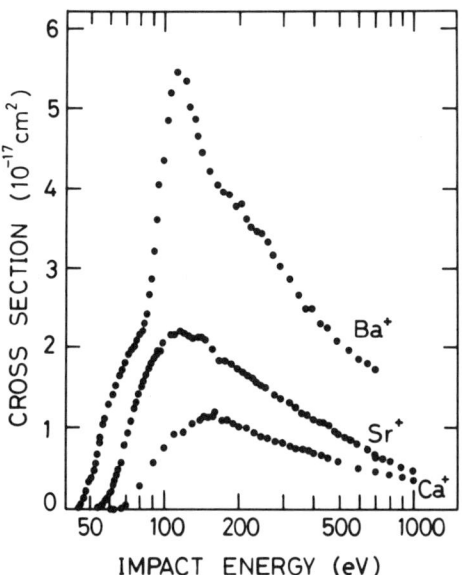

FIG.4. Double ionizaition cross sections of alkaline-earth ions, Ba^+ (Hirayama et al.[6]), Sr^+, and Ca^+ (Ikehara et al.[19]) as functions of electron impact energies.

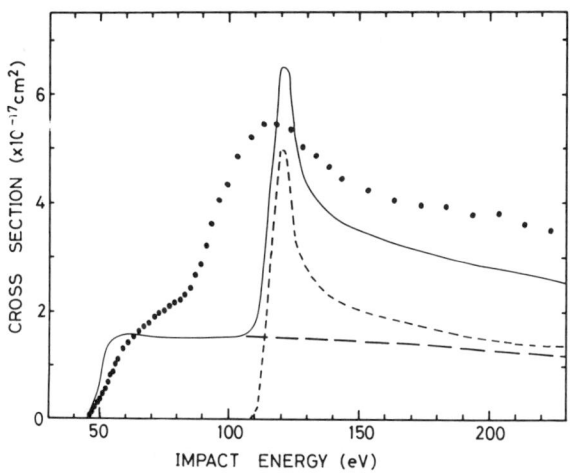

FIG.5. Comparison of the double ionizaition cross sections for Ba^+ with the theoretical calculation of the 5s and 4d-ionizaition cross sections by Younger[20].

attributed to the ionization of the 5s-electron followed by autoionization expressed as,

$Ba^+(5s^25p^66s) + e$
$\rightarrow Ba^{2+}(5s5p^66s) + 2e$
$\hookrightarrow Ba^{3+}(5s^25p^5) + e$.

In Fig.5, results of Younger's DWBEA calculation[20] are compared with the experimantal results. The dashed curve and the dotted curve represent the ionization cross sections for the 5s and 4d electrons, respectively, and the solid line represents the sum of the both contributions.

It is noteworthy that the theory predicts a sharp resonance structure for the 4d-ionization, whereas the experiment indicates a much broader structure. While the Hartree-Fock 4d ionization threshold is 109 eV[20], the onset of the giant peak in the experimental curve appears at 85 eV. This suggests us a significant contribution from the excitation of the 4d-electrons which results in double autoionization processes, such as

$Ba^+(4d^95s^25p^66snl)$
$\rightarrow Ba^{2+}(4d^{10}5s5p^66s) + e$
$\rightarrow Ba^{3+}(4d^{10}5s^25p^5) + e$.

In order to make clear these contributions, the cross section measurements of higher energy resolution may be desired.

4. Partial cross sections for the inner-shell ionization of rare-gas atoms

Partial ionization cross sections for the 4d-electrons in Xe, the 3d-electrons in Kr, and the 2p-electrons in Ar have been measured by means of the Auger-electron spectroscopy. The inner-shell ionization cross sections are deduced directly by the total Auger-electron emission cross sections, because the Auger yield is practically unity for the present case. The total Auger-emission cross sections are composed of that of the mormal Auger process, the double Auger process to the discrete states, and the double Auger process to the continuum state. The sum of the former two processes are measured directly from total Auger-line intensities of the diagram and satellite lines. The total Auger-emission cross sections are obtained by an aid of experimental data by previous authors on the ratio of the continuum double Auger to the sum of the former two processes. Absolute scale of the cross sections are normalized to the differential elastic scattering of electrons with respective atom species.

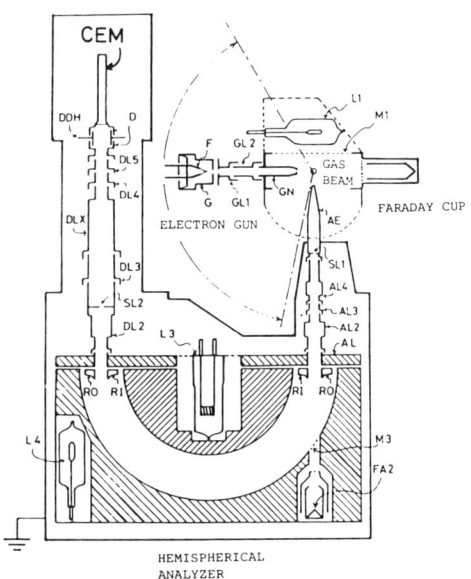

FIG.6. Schematic diagram of the Auger-electron spectrometer for determination of the inner shell ionizaition cross sections.

The electron spectrometer is shown in Fig.6, which is equipped with a conventional hemispherical analyzer, whose meam orbit radius is 5 cm. The analyzer was operated by a constant resolution mode, namely the potential difference between the hemispheres was kept constant and the potential of the entrance and the exit slits was continuously scanned. The potentials on the intermediate lenses before the analyzer was also scanned so that the variation in the transmission efficiency of the lense with respect to the decelerating voltage, i.e. the chromatic aberration, was minimized.

Results of partial cross sections for the 4d-ionization in Xe, the 3d-ionization in Kr, and the 2p-ionization in Ar are shown in Fig.7. Overall uncertainties were estimated to be about 35%, which is not surprising if we consider that the experimental procedures to deduce the absolute cross sections are much complicated. The cross sections for the 4d-ionization in Xe are overwhelmingly greater than those for the 3d-ionization in Kr and the 2p-ionization in Ar. The 4d-ionization cross section in Xe rises abruptly from 80 eV and forms a first maximum at 135 eV, then it decreases once creating a distinct minimum and rises again giving a second maximum at around 500 eV. The profile of the first maximum possesses the typical feature of the giant resonance, while the second maximum appears to dominate the cross section.

We would try to compare the experimental 4d-ionization cross section in Xe with the theoretical calculation by Younger[5] in Fig.8, where one third of the DWBEA including term-dependent ejected f waves and no ground-state correlation is drawn by a solid curve ($\frac{1}{3}Q^I_{4d}$) and one seventh of

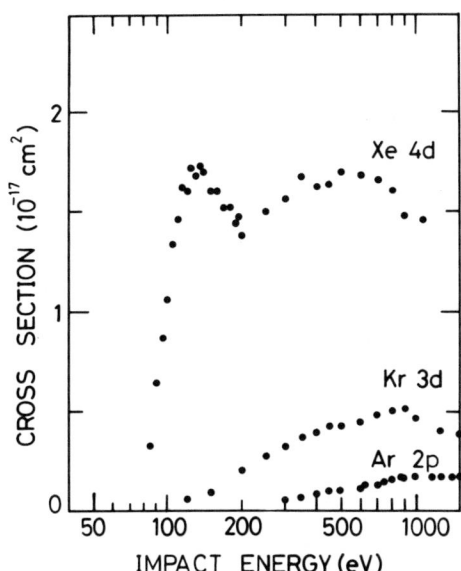

FIG.7. Partial cross sections for the inner shell ionizations[9], 4d-ionization in Xe, 3d-ionization in Kr, and 2p-ionization in Ar.

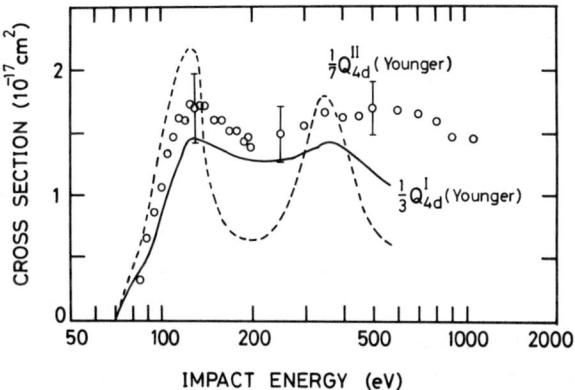

FIG.8. Comparison of the 4d-ionization cross sections in Xe with the theoretical calculation by Younger[5].

the DWBEA with semiclassical exchange, patial wave and ground-state correlation is drawn by a dashed curve ($\frac{1}{7} Q^{II}_{4d}$). Although the absolute scale quite disagrees with experimental results, the distinct feature of the double maximum charactor has been reproduced by the theory. A discrepancy remains, however, with respect of the widths of the resonance structure.

5. Partial cross sections for ionization of the subshell s-electrons in rare-gas atoms

Partial ionization cross sections for the 4s-electrons in Kr,[21] the 3s-electrons in Ar,[22] and the 2s-electrons in Ne[23] have been measured by means of the vacuum ultraviolet (VUV) spectroscopy. These subshell ionizaiton cross sections are determined from the emission cross sections of the VUV light quanta in the transition from the $nsnp^6$ $^2S_{1/2}$ state to the ns^2np^5 $^2P_{3/2, 1/2}$ states, where n are 4 for Kr, 3 for Ar, and 2 for Ne; wavelenths of emitted VUV light are 917 Å and 965 Å for Kr, 920 Å and 932 Å for Ar, and 460.7 Å and 462.4 Å for Ne. Relative efficiencies of the spectrometer and detector assembly as a function of wavelength are calibrated using the relative intensities of the lines in a known spectrum, e.g. H_2 Werner-band, available by previous authors. Absolute scale of the cross sections is normalized in Ar, which is determined by the electron energy-loss spectroscopy.[24] Undesirable effect due to the self-absorption of the resonance lines through the target gas was minimized and corrected by a careful procedure.

The VUV spectrometer used to measure the emission spectra is shown in Fig.9. To realize the polarization-free measurements, the collision

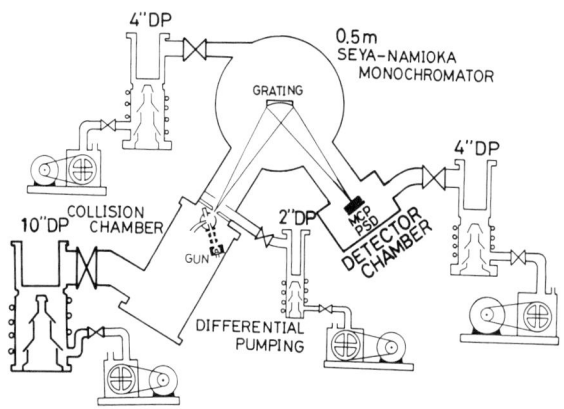

FIG.9. Schematic diagram of the VUV spectroscopic equipment used to determine the sub-shell ionization cross sections.

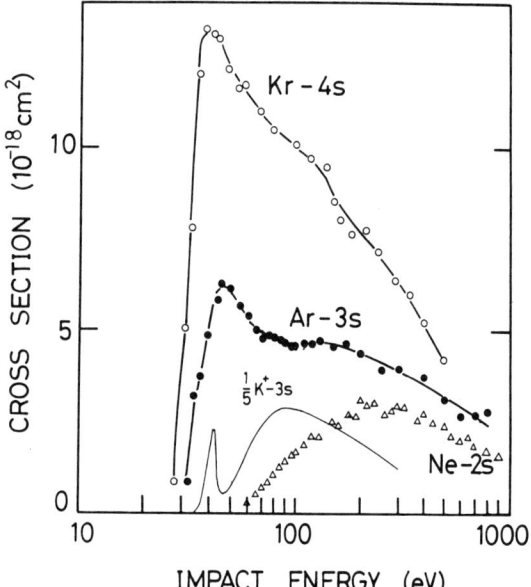

FIG.10. Partial cross sections for the sub-shell ionizations, 4s-ionization in Kr[21], 3s-ionization in Ar[22], and 2s-ionization in Ne (not in absolute scale[23]).

assembly are fixed to the spectrometer so that the angle between the direction of the electron beam and the optical axis of the spectrometer is the magic angle 54.7°, and also the scattering plane containing the electron beam and the optical axis are made directed at 45° with respect to the entrance slit of the spectrometer. A position-sensitive detector assembly is used for a stable photon counting.

Results of partial cross sections for the 4s-ionization in Kr, the 3s-ionization in Ar, and the 2s-ionization in Ne are shown in Fig.10. Overall uncertainties are estimated to be about 20%.

In spite of the simple monotonous shape of the curve for Ne with a broad maximum at around 250 eV, the curves for Kr and Ar show a remarkable increase from threshold and reveal a profile which resemble the giant resonance with a sharp peak around 38 eV and 46 eV, respectively. The authors do not know whether the characteristic features are really due to the shape resonance type effect or not, further experimental and theoretical studies are expected to be performed.

6. Conclusions

Experimantal results on double ionization cross sections of alkaline earth ions as well as partial cross sections for the inner-shell ionization and for the outer sub-shell ionization in rare-gas atoms have been discussed in relation to the giant resonance effect in the electron scattering channel.

We conclude that, in order to attain deeper understanding of the giant resonance phenomena in electron scattering channel, we need further experimantal studies on various kinds of inelastic collision processes of electrons with atoms and ions, especially multiple ionization processes and inner or sub-shell ionization processes by means of different kinds of experimental method including newly devoped ones.

References

1. D.R.Hertling, R.K.Feeney, D.W.Hughes and W.E.Sayle, J. Appl. Phys. 53 5427 (1982).
2. Ch.Achenbach, A.Müller, E.Salzborn and R.Becker, Phys. Rev. Lett. 50 2070 (1983).
3. S.M.Younger, Phys. Rev. Lett. 56 2618 (1986).
4. S.M.Younger, in Giant Resonances in Atoms, Molecules and Solids, NATO ASI Series Vol.151, ed., J.P.Connerade, J.M.Esteva and R.C.Karnatak (Plenum Press, 1986) 237.
5. S.M.Younger, Phys. Rev. 35 2841 (1987).
6. T.Hirayama, S.Kobayashi, A.Matsumoto, S.Ohtani, T.Takayanagi, K.Wakiya and H.Suzuki, J. Phys. Soc. Jpn. 56 851 (1987).
7. H.Suzuki, T.Takayanagi, K.Morita and Y.Iketaki, in Electron-Molecule Collisions and Photoionizaiton Process, ed., McKoy et al. (Verlag Chemie Intermational, Inc., 1983) 43-56.
8. H.Suzuki, Sing. J. Phys. 3 1 (1986).
9. T.Takayanagi et al., in preparation for publication in J. Phys. B.
10. S.M.Younger, Phys. Rev. 35 4567 (1987).
11. R.Haensel, G.Keitel, P.Schreiber and C.Kunz, Phys. Rev. 188 1375 (1969).
12. H.Peterson, K.Radler, B.Sonntag and R.Haensel, J. Phys. B 8 31 (1975).
13. A.Müller, C.Achenbach, E.Salzborn and R.Becker, J. Phys. B 17 1427 (1984).
14. K.Tinschert, A.Müller, R.Becker and E.Salzborn, J. Phys. B 20 1823 (1987).
15. A.Müller, K.Tinschert, C.Achenbach, R.Becker and E.Salzborn, J. Phys. B 18 3011 (1985).
16. D.W.Hughes and R.K.Feeney, Phys. Rev. A 23 2241 (1981).
17. T.Hirayama, K.Oda, Y.Morikawa, T.Ono, Y.Ikezaki, T.Takayanagi, K.Wakiya and H.Suzuki, J. Phys. Soc. Jpn. 55 1411 (1986).
18. K.T.Dolder and B.Peart, Rep. Prog. Phys. 39 697 (1976).
19. W.Ikehara et al., in preparation for publication in J. Phys. Soc. Jpn.
20. S.M.Younger, Private communication.
21. T.Takayanagi et al., in preparation for publication in Phys. Rev. A.
22. G.P.Li, T.Takayanagi, K.Wakiya and H.Suzuki, Phys. Rev. A 38 1831 (1988).
23. Y.Akagi, M.Hayakawa, Y.Morikawa, T.Takayanagi, K.Wakiya and H.Suzuki, Atomic Collision Research in Japan, Progress Report, 10 24 (1984).
24. G.P.Li, T.Takayanagi, K.Wakiya, H.Suzuki, T.Ajiro, S.Yagi, S.S.Kano and H.Takuma, Phys. Rev. A 38 1240 (1988).

DETERMINATIONS OF THE PRODUCTS OF DISSOCIATIVE RECOMBINATION REACTIONS

N G. Adams[a], C.R. Herd[b] and D. Smith

School of Physics and Space Research,
University of Birmingham, Birmingham, B15 2TT, UK

Product distributions for the electron-ion dissociative recombination of HCO^+, O_2H^+, HCO_2^+, N_2OH^+ and H_3O^+ have been determined at 300K using a flowing afterglow/Langmuir probe apparatus. Electron (and thus ion) densities were determined using the Langmuir probe and the OH products of the reactions were detected using laser induced fluorescence spectroscopy. The OH fluorescence was quantitatively related to the OH number density by using the calibration reaction $H+NO_2 \longrightarrow OH+NO$ and by determining the H-atom number density using vacuum ultraviolet absorption spectroscopy. Preliminary determinations of the H atom and O atom contributions to the product distributions are also described.

1. Introduction

The dissociative recombination of electrons with positive ions, i.e.

$$AB^+ + e \longrightarrow A + B \qquad (1)$$

is an important de-ionization process in many plasma situations such as laser plasmas [1], combustion flames [2], interstellar gas clouds [3] and planetary ionospheres [4]. Ion-neutral reactions also occur in these media and, since recombination acts to return the medium to the neutral state, then a change in the neutral composition of the medium results. Such changes are extremely important in situations where neutral species are the main species detected (e.g. in interstellar gas clouds) or where they contribute to a parallel neutral chemistry (e.g. in combustion flames). The mechanisms of recombination reactions are also of great interest, and the identification of the products and their state of excitation can provide information concerning the crossings between the stable ionic potential curve and the repulsive neutral potential curves to products.

---

a) Present address: Dept. of Chemistry, University of Georgia, Athens, GA30602.
b) Present address: Columbian Chemicals Co., PO Box 96, Swartz, LA71281.

Very few experimental data are available concerning the products of recombination and these are only for atomic products (e.g. H, N and O) and for ions in uncertain states of internal excitation (i.e. O_2^+ [5], NO^+ [6], CO_2^+ [7], and H_2O^+ [8]; note N_2^+ ($v = 1$) has also been studied [9]). Some theories which predict the products of recombination are available [10,11], however this is a difficult theoretical problem. These theories require a knowledge of the ionic and neutral potential curves, few of which are available [12], and therefore theories are, at present, in the development phase [13]. Thus there is a great need for experimental data concerning the atomic and molecular products of the recombination of many ground vibronic state ions both for application to real plasma situations and to guide theory. It is a start on such measurements, i.e. the quantitative detection of the OH products of the dissociative recombination of O_2H^+, HCO_2^+, N_2OH^+ and H_3O^+, that is summarized here. These are the first determinations of the molecular products of recombination reactions.

2. Experimental

A schematic diagram of the flowing

afterglow used for these measurements is given in Fig. 1. It illustrates an axially movable Langmuir probe (to measure electron and ion densities), a mass spectrometer (to identify the ion types present in the plasma), laser induced fluorescence (LIF, to detect molecular species, i.e. OH in the present study), and vacuum ultraviolet (VUV) absorption spectroscopy (to measure atom densities). Since the technique has been described in detail elsewhere [14,15], it will be discussed only briefly here. A plasma is created in He by a microwave discharge and flows along a stainless steel tube (1m length, 8cm diameter) due to the action of a Roots pump such that the residence time in the flow-tube is ~20 ms. At a He pressure of 1.6 torr, the He^+ ions are rapidly converted to He_2^+ (see Fig. 2; He_2^+ can also be produced in the discharge in collisions between highly excited He atoms and He). Ar is added downstream to remove the He metastable atoms by Penning ionization [15] and also reacts with He_2^+ generating Ar^+ in both cases, thus producing an Ar^+/e plasma.

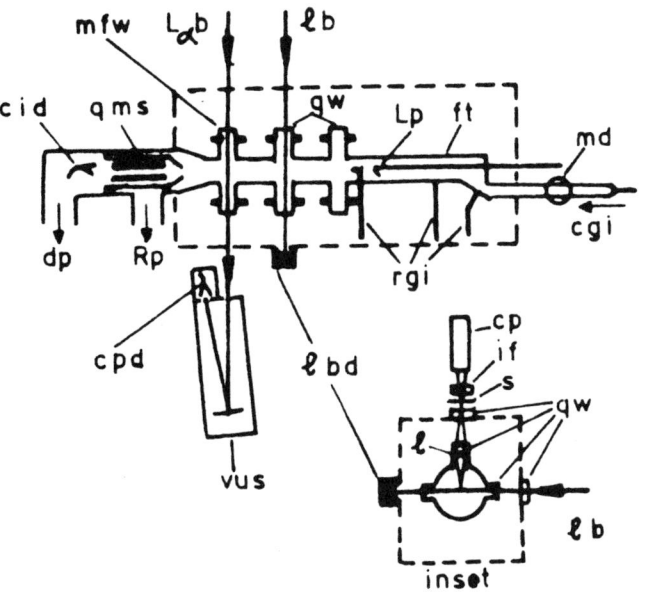

Figure 1. A simplified schematic diagram of the flowing afterglow/Langmuir probe apparatus indicating the Langmuir probe, the mass spectrometer, the laser induced fluorescence facility (also illustrated in cross-section in the inset) and the atomic absorption facility. The symbols have the following meanings: cgi - He carrier gas inlet; md - microwave discharge; ft - flow tube; Lp - Langmuir probe; rgi - reactant gas inlets; lb - laser beam; qw - quartz windows; lbd - laser beam dump; L_αb - Lyman α beam; mfw - magnesium fluoride windows; vus - vacuum ultraviolet spectrometer; cpd - channeltron photon detector; qms - quadrupole mass spectrometer; cid - channeltron ion detector; dp - diffusion pump; Rp - Roots pump. Additional symbols in inset : l - lens; s - slit; if - interference filter; cp - cooled photomultiplier. The dashed line represents the vacuum jacket which thermally insulates the flow tube for high and low temperature measurements.

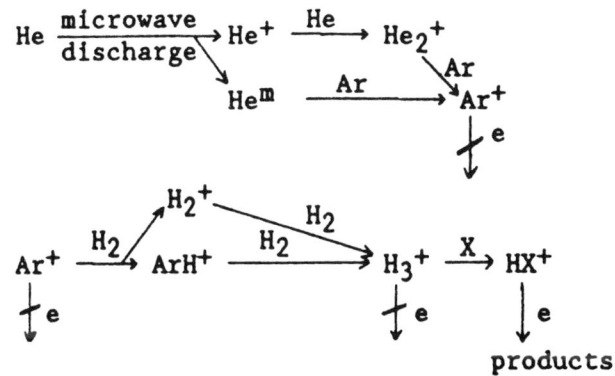

Figure 2. Reaction sequence utilized for producing a recombining HX^+/e plasma.

This Ar^+/e plasma recombines only very slowly by radiative recombination [16]. H_2 is then added to this plasma and the reactions illustrated in Fig. 2 rapidly result in the production of an H_3^+/e plasma which again recombines only very slowly [17]. Vibrationally excited $H_3^+(\nu \geq 3)$, which can recombine at an appreciable rate, is readily quenched in proton transfer reactions with H_2. Since H_2 has a small proton affinity, H_3^+ will readily transfer a proton to many other molecules, X, [18] and the recombination of HX^+ can then be studied [19]. Similar to the case of H_3^+, any internal excitation of the HX^+ ion is quenched by near-resonant proton transfer with X. In this manner, plasmas of electrons with HCO^+, O_2H^+, HCO_2^+, N_2OH^+ and H_3O^+ were separately created and their recombination studied. The energetically allowed product channels of these recombination reactions are listed in Table I from which it can be seen that OH, H and O are all possible products.

Table I. Exothermicities of the dissociative recombination reactions of several protonated ions, HX^+, for production of ground state products (calculated using data from refs. 20 to 24).

Recombining Ion	Reaction Products	Exothermicity(eV)
HCO^+	H + CO	7.5
	OH + C	0.8
O_2H^+	H + O_2	9.2
	OH + O	8.5
	H + 2O	4.1
$HOCO^+$	H + CO_2	8.0
	OH + CO	6.9
	O + H + CO	2.5
	O + HCO	3.8
$NNOH^+$	H + N_2O	7.7
	OH + N_2	10.4
	OH + 2N	0.7
	O + H + N_2	6.0
	NH + NO	6.5
	N + H + NO	2.7
H_3O^+	H + H_2O	6.4
	OH + H_2	5.8
	OH + 2H	1.3
	O + H + H_2	1.5

The OH product was monitored downstream of the position where the gas X was added, using laser induced fluorescence by exciting the $A^2\Sigma(\nu'=1) \leftarrow X^2\Pi(\nu''=0)$ rovibronic transitions with the doubled output (~280nm) of a dye laser pumped by a Nd YAG laser. The fluorescence in the (1,1) band at ~312 nm (see the inset in Fig.1) was collected by a lens, focussed onto a slit and passed through an interference filter to a cooled photomultiplier. Fluorescence photons were counted for a period of 3μsec (equivalent to several times the radiative lifetime of the $A^2\Sigma(\nu'=1)$ state) after each laser shot. These precautions resulted in a signal to noise ratio of 100:1 on the most intense lines. A laser excitation spectrum over the wavelength region 281 to 284.3nm gave the characteristic spectrum of OH [15,25].

The absolute OH $X^2\Pi(\nu''=0)$ number density, [OH], was determined using the well-studied reaction [26,27]

$$H + NO_2 \longrightarrow OH + NO \qquad (2)$$

as a calibration, the H-atom number density, [H], being determined using vacuum ultraviolet absorption spectroscopy at the L_α resonance wavelength [15]. In order for [H] to be determined by this technique, a large H atom concentration was produced by flowing H_2 through the microwave discharge.

To establish that the OH present in the recombining plasma originated from the recombination reaction, [OH] was monitored as a function of the flow rate of the molecules X and of the electron density, [e]. [OH] varied with X in the manner expected for the production of HX^+ via proton transfer from H_3^+ followed by HX^+ recombination. The way in which [OH] is expected to depend on [e] is not obvious, however a simple analysis[15] shows that

$$\frac{[e]_o^2}{[OH]_z} = \frac{[e]_o}{f} + \frac{v}{\alpha_e f z} \qquad (3)$$

where the subscripts o and z represent the axial positions at which no recombination had occurred and at which the laser probed the recombination products respectively, f is the fraction of the recombinations that produce OH $X(\nu''=0)$, v is the plasma flow velocity and α_e is the recombination coefficient. Plots of $[e]_o^2/[OH]_z$ versus $[e]_o$ for the reactions studied are given in Fig.3 and show excellent linearity. The slopes provide the percentages of OH $X(\nu''=0)$ generated in the reactions, and these are listed in Table II. The intercepts

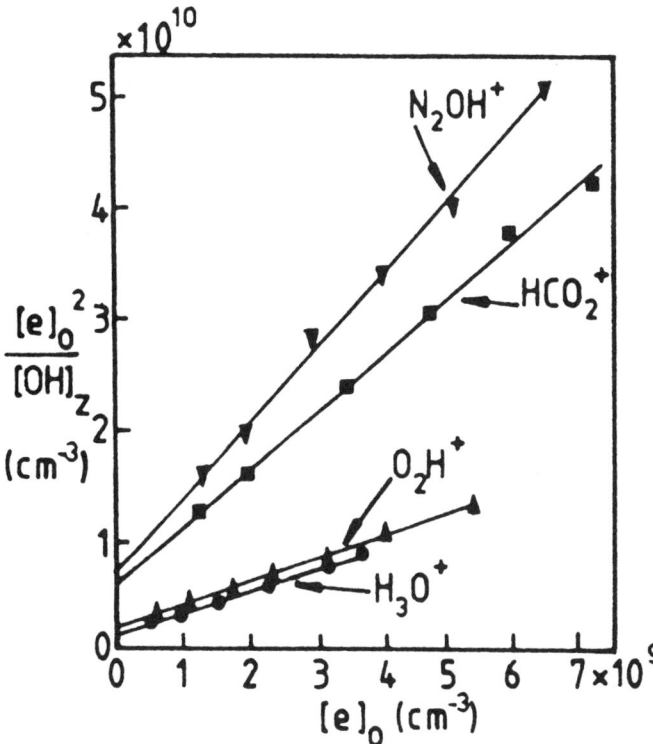

Figure 3. Plots of $[e]_o^2/[OH]_z$ versus $[e]_o$ (see eqn.(2)).

Table II. Experimentally determined product distributions for the dissociative recombination reactions of several protonated ions, HX^+, at 300K. The OH product percentages are for the ground electronic ($X^2\Pi$) state.

Recombining Ion	Reaction Products	Product Distribution(%)
HCO^+	$OH + C$	0
O_2H^+	$OH(\nu''=0)+O$	40
	$OH(\nu''>0)+O$	20
	other products	40
$HOCO^+$	$OH(\nu''=0)+CO$	17
	$OH(\nu''>0)+CO$	17
	other products	66
$NNOH^+$	$OH(\nu''=0)+N_2(2N)$	14
	$OH(\nu''>0)+N_2(2N)$	17
	other products	69
H_3O^+	$OH(\nu''=0)+H_2(2H)$	46
	$OH(\nu''>0)+H_2(2H)$	19
	other products	35

provide values of α_e which agree to within error with those previously deduced from the decrease in [e] with distance z along the flow tube under recombination controlled conditions [19]. The above checks established beyond doubt that OH is produced in the recombination reactions. Also included in Table II are the percentages of OH $X(\nu''>0)$. These data were obtained by adding NO to the flow upstream of the laser excitation point, but downstream of the recombination region, and monitoring the increase in OH $X(\nu''=0)$ due to quenching of all vibrationally excited states to the ground vibrational state [26,28]. Note that in the case of the N_2OH^+ recombination, any NH product will react with NO producing OH with a rate coefficient similar to that for the quenching reaction [29]. Thus the percentage of OH $X(\nu''>0)$ quoted in Table II for this recombination may contain a contribution due to NH production. This could be checked by using LIF to directly detect NH or its companion product NO.

In the recombination reactions studied here it is also energetically possible to produce OH $A^2\Sigma$ directly. So far, weak $A^2\Sigma \longrightarrow X^2\Pi$ emissions have been observed from the recombination of O_2H^+ and of HCO_2^+, however, these were such that the OH $A^2\Sigma$ state did not contribute appreciably to the product distribution.

In order to fully characterize the product distributions, the percentages of H and O are also required (as well as N and NH in the case of the N_2OH^+ recombination). Experiments are underway at the University of Rennes to measure H-atom densities directly using VUV absorption spectroscopy [30], and preliminary data have been obtained in the present study in the Birmingham laboratory using reaction (2) as a monitor of [H] via the very sensitive LIF technique for detecting OH. Using the O_2H^+ recombination as a calibration, the data indicate for the other recombination reactions listed in Table I that, on average, an H-atom is produced when OH is not a product. This is to be expected for the HCO^+ recombination reaction where the other energetically allowed channel producing OH was not observed. However, this observation does indicate that the energetically possible channel producing O+HCO in the HCO_2^+ recombination reaction and that producing OH+2H in the H_3O^+ recombination reaction (see Table I) are not very significant. In the course of this study, H-atoms were also detected from the recombination of N_2H^+ and CH_5^+ at a level of about one H-atom per recombination for N_2H^+ and somewhat greater for CH_5^+. For N_2H^+, this indicates that the only other exothermic channel (NH+N+2.5eV) is not significant. For CH_5^+ there are several exothermic channels resulting in the production of H- atoms

$$CH_5^+ + e \longrightarrow H + CH_4 \quad (4a)$$
$$\longrightarrow 2H + CH_3 \quad (4b)$$
$$\longrightarrow H_2 + CH_3 \quad (4c)$$
$$\longrightarrow H + H_2 + CH_2 \quad (4d)$$
$$\longrightarrow 2H_2 + CH \quad (4e)$$

and thus additional studies will be required to determine the product distribution for the recombination of this ion.

Measurements are also in progress to determine O-atom contributions to the product distributions for recombination using the relatively inefficient reaction

$$O + GeH_4 \longrightarrow OH + GeH_3 \quad (5)$$

to convert O-atoms to OH [31]. The situation here is complicated since the OH product is rapidly lost in a sequential reaction with GeH_4. However, initial studies have indicated that this approach is viable. It has already been established that a smaller percentage of O atoms is generated in the H_3O^+ recombination than in the O_2H^+ and HCO_2^+ recombinations.

3. Conclusions

A great deal of progress has been made in quantitatively identifying the neutral products (both atomic and molecular) of the dissociative recombination of ground vibronic state ions. Already the significance of these measurements to molecular synthesis in interstellar clouds is being assessed [32], and the data are providing impetus to theoreticians to extend their theoretical models of the recombination process [13].

Acknowledgements

Financial support of this work by the USAF and the SERC is gratefully acknowledged. We are grateful to Mark Geoghegan for his assistance with some of the data aquisition.

References

[1] J.B. Laudenslager, in *Kinetics of Ion-Molecule Reactions*, edited by P. Ausloos, (Plenum, New York, 1979) p.405.
[2] J.M. Goodings, N.S. Karellas, and C.S. Hassanali, Int. J. Mass Spectrom. Ion Proc. 89, 205 (1989).
[3] E. Herbst, in *Rate Coefficients in Astrophysics*, edited by T.J. Millar and D.A. Williams (Kluwer, Dordrecht, 1988) p.239.
[4] A. Dalgarno, in *Rate Coefficients in Astrophysics*, edited by T.J. Millar and D.A. Williams (Kluwer, Dordrecht, 1988) p.321.
[5] E.C. Zipf, J. Geophys. Res. 85, 4232 (1980).
[6] D. Kley, G.M. Lawrence, and E.J. Stone, J. Chem. Phys. 66, 4157 (1977).
[7] F. Vallee, B.R. Rowe, J.C. Gomet, J.L. Queffelec, and M. Morlais, Chem. Phys. Lett. 124, 317 (1986).
[8] F. Vallee, J.C. Gomet, B.R. Rowe, J.L. Queffelec and M. Morlais, in *Rate Coefficients in Astrophysics*, edited by T.J. Millar and D.A. Williams (Kluwer, Dordrecht, 1988) p.29.
[9] J.L. Queffelec, B.R. Rowe, M. Morlais, J.C. Gomet, and F. Vallee, Planet. Space Sci. 33, 263 (1985).
[10] D.R. Bates, Ap. J. (Letters).306, L45 (1986); in *Modern Applications of Atomic and Molecular Processes*, edited by A.E. Kingston (Plenum, London, 1987) p.1.
[11] E. Herbst, Ap. J. 222, 508 (1978).
[12] J.B.A. Mitchell, and S.L. Guberman. Eds. *Dissociative Recombination: Theory, Experiment and Applications* (World Scientific, Singapore 1989).
[13] D.R. Bates, Ap. J. in press (1989).

[14] N.G. Adams and D. Smith, in *Techniques for the Study of Ion Molecule Reactions*, edited by J.M. Farrar and W.H. Saunders, Vol.22 (Wiley, New York, 1988) p.165.

[15] N.G. Adams, C.R. Herd, and D. Smith, J.Chem. Phys. 15 July 1989.

[16] D.R. Bates, Adv. At. Mol. Phys., 15, 235 (1979); in *Case Studies in Atomic Physics* Vol 4, edited by M.R.C. McDowell and E.W. McDaniel (North Holland, Amsterdam, 1974) p.57.

[17] N.G. Adams and D. Smith in *Rate Coefficients in Astrochemistry*, edited by T.J. Millar and D.A. Williams (Kluwer, Dordrecht, 1988) p.173.

[18] Y. Ikezoe, S. Matsuoka, M. Takebe, and A. Viggiano, *Gas Phase Ion-Molecule Reaction Rate Constants through 1986* (Maruzen Co. Ltd., Tokyo, 1987).

[19] N.G. Adams, and D. Smith, Chem. Phys. Lett., 144, 11 (1988).

[20] H.M. Rosenstock, K. Draxl, B.W. Steiner, and J.T. Herron, J. Phys. Chem. Ref. Data, 6, Suppl.1 (1977).

[21] S.G. Lias, J.F. Liebman and R.D. Levin, J. Phys. Chem. Ref. Data, 13, 695 (1984).

[22] S.G. Lias, J.E. Bartmess, J.F. Liebman, J.L. Holmes, R.D. Levin and W.G. Mallard, J. Phys. Chem. Ref. Data, 17, Suppl.1 (1988).

[23] K.P. Huber, and G. Herzberg, *Molecular Spectra and Molecular Structure: IV Constants of Diatomic Molecules* (Van Nostrand, New York, 1979).

[24] G. Herzberg, *Molecular Spectra and Molecular Structure: III Electronic Spectra and Electronic Structure of Polyatomic Molecules* (Van Nostrand, Princeton, 1967).

[25] G.H. Dieke, and H.M. Crosswhite, J. Quant. Spectrosc. Radiat. Transfer, 2, 97 (1962).

[26] M.J. Howard, and I.W.M. Smith, Prog. React. Kinet., 12, 55 (1983).

[27] D. Klenerman, and I.W.M. Smith, J. Chem. Soc. Farad. Trans 2, 83, 229 (1987).

[28] I.W.M. Smith, and M.D. Williams, J. Chem. Soc., Farad. Trans. 2, 81, 1849 (1985).

[29] W.H. Brune, J.J. Schwab, and J.G. Anderson, J. Phys. Chem. 87, 4503 (1983).

[30] B.R. Rowe, priv. comm. (1989).

[31] B.S. Agrawalla, and D.W. Setser, in *Gas-Phase Chemiluminescence and Chemi-ionization*, edited by A. Fontijn (Elsevier, Amsterdam, 1985) p.157.

[32] C.R. Herd, N.G. Adams, and D. Smith, Ap. J., (1989) submitted.

R-MATRIX CALCULATIONS OF ELECTRON MOLECULE SCATTERING

Lesley A. Morgan

Computer Centre, Royal Holloway and Bedford New College, Egham, Surrey TW20 0EX, UK

Recent calculations of low energy electron scattering by diatomic molecules, using the R-matrix method, are reviewed. These include the vibrational excitation of HF and HCl, and the electronic excitation of H_2 and N_2. Calculations, using similar methods, of positron scattering by H_2 and HF are discussed briefly.

INTRODUCTION

The R-matrix method has been used extensively in recent years to obtain high quality cross sections for low energy electron scattering by atomic targets. The application of the method to molecular targets is made more complicated, especially at the computational level, by the reduced symmetry and the possibility of vibrational and rotational motion. However, we are now reaching the point where accurate calculations are possible for a wide range of diatomic targets. This review will concentrate on recent work, carried out in the U.K., with particular emphasis on electron impact excitation, vibrational or electronic, dissociative attachment and positron scattering.

METHOD

The multi-centre R-matrix method has been described in detail elsewhere, see for example Gillan et al[1], and we will simply outline the method here.

In the inner region of configuration space we use a fixed nuclei trial wavefunction of the form

$$\psi_k = \mathcal{A} \sum_{i,j} \phi_i(x_1\ldots x_N) u_{ij}(x_{N+1}) a_{ijk} + \sum_i \chi_i(x_1\ldots x_{N+1}) b_{ik} \quad (1)$$

where the ϕ_i are multicentre target wavefunctions constructed from an STO basis and the $u_{ij}(x)$ are numerical continuum orbitals. The χ_i are quadratically integrable functions, constructed from the target real and virtual STOs, which allow for short range polarization effects in the collision. $\mathcal{A}$ is the usual antisymmetrisation operator. The coefficients a_{ijk} and b_{ik} are obtained by diagonalizing the fixed nuclei hamiltonian in the inner region. This is carried out using a modified version of the ALCHEMY molecular structure package[2].

Vibrational motion is included using the non-adiabatic method proposed by Schneider et al[3]. We carry out a sequence of fixed nuclei calculations on a grid of internuclear separations spanning the range of R of interest. The fixed nuclei electronic R-matrix poles then provide the potentials for an R-matrix treatment of nuclear motion. Usually, only the lowest few poles need be retained. We use a trial vibronic wavefuction of the form

$$\theta_i = \sum_{jk} \psi_k(x_1\ldots x_{N+1}; R)\eta_j(R)c_{ijk} \quad (2)$$

where the ψ_k are the electronic wavefunctions defined in Eq. (1) and the η_j are orthogonal polynomials. In order to keep the model entirely *ab initio* we use target vibrational wavefunctions obtained from the potential energy curve of the target wavefunctions ϕ_i used in Eq. (1).

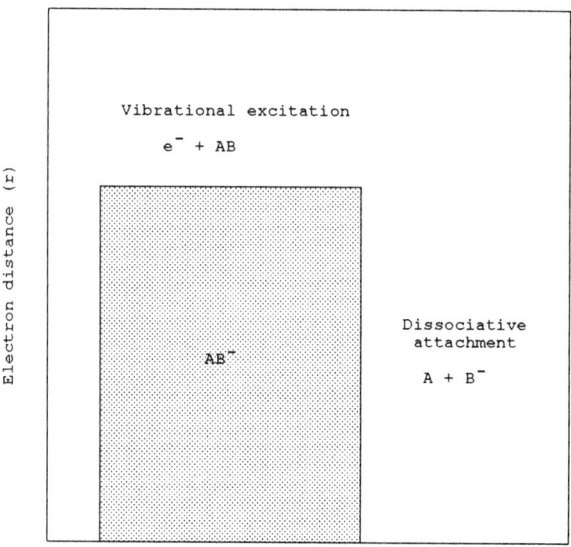

Figure 1. Partition of configuration space

In the outer region of configuration space, short range potentials may be neglected and the scattering equations have the form

$$\left(\frac{d^2}{dr^2} - \frac{l_i(l_i+1)}{r^2} + k_i^2\right)F_i(r) = 2\sum_j V_{ij}(r)F_j(r). \quad (3)$$

where i labels the vibrational and angular momentum quantum numbers and the potentials have the form

$$V_{ij}(r) = \sum_\lambda \frac{a_{ij}^\lambda}{r^{\lambda+1}} \quad (4)$$

We have also considered the dissociative attachment process $e^- + AB \to A^- + B$ using the method of Schneider et al[3]. They give expressions for the R-matrix boundary conditions for coupled vibrational excitation and dissociative attachment on the boundary of a region bounded by the scattered electron coordinate $r = a$ and the nuclear coordinate $R = A$ (Fig. 1).

VIBRATIONAL EXCITATION OF HF AND HCl

The origins of the sharp peaks, first observed by Rohr and Linder[4], in the near threshold vibrational excitation of the hydrogen halides are still the subject of much debate (see for example the recent review by Morrison[5]). Recent high resolution measurements carried out by the Kaiserslautern group[6] have largely confirmed the earlier results and have provided a wealth of new detail. They suggest that, in the case of HCl, there is more structure in the cross sections than was apparent from the earlier measurements. A feature observed in both sets of measurements for HCl, but not for HF, is a broad resonance centred near 3 eV.

We have already used the methods outlined above to carry out a fully *ab initio* calculation for electron scattering by HF[7]. The threshold peaks in this system can be attributed directly to the strong permanent dipole moment of the target. In a fixed nuclei approximation, this by itself is strong enough to bind the electron at all but small internuclear separations. The additional attractive short range forces extend the range of geometries for which HF$^-$ is bound to even shorter bond lengths. When we use the fixed nuclei data to construct a nonadiabatic description of the nuclear motion, the bound state is transformed into a series of 'nuclear excited Feshbach resonances'[8] which arise from the scattered electron being temporarily bound to the tails of the target vibrational wavefunctions. These resonances manifest themselves as sharp peaks just above each vibrational threshold in the scattering cross sections. The peak in our $v = 1$ cross sections (Fig. 2) is significantly narrower than experiment and is very close to the threshold. The major deficiency in our calculations is probably our use of a SCF, rather than CI, target wavefunction which leads to small inaccuracies in the shapes of the potential energy curves and hence in the vibrational energy levels. Rotational effects, omitted from our model, can also be expected to be significant. However, we expect these effects to be quantative rather than qualitative.

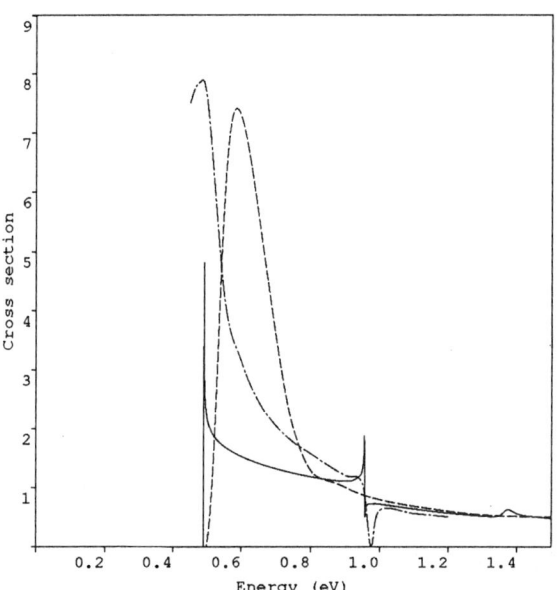

Figure 2. HF integrated cross section $v = 0 \to 1$
——— R-matrix (Morgan and Burke 1988)
—·—·— Gauyacq (1983)
------- Experiment (Knoth et al 1989)

The situation in HCl is not so straightforward, due to the weaker dipole moment and hence the more important role of the short range forces. We have therefore carried out R-matrix calculations using two different models[9]. The first is a static exchange plus short range polarization approximation, very similar to that used in our work on HF. In the second model we add polarized pseudo-states to represent the longer range polarization effects, as in recent work on N_2 [10]. The fixed nuclei results from both models are qualitatively similar. HCl$^-$ is bound only for large internuclear separations. However, as R decreases, the HCl⁻ potential curve moves parallel to the target HCl curve and does not cross it to become a resonance (Fig. 3)

At higher energies (3.5 eV in the equilibrium geometry) we observe a broad resonance, but this is somewhat weaker that that found by previous calculations and does not produce a significant effect on the vibrationally resolved cross sections.

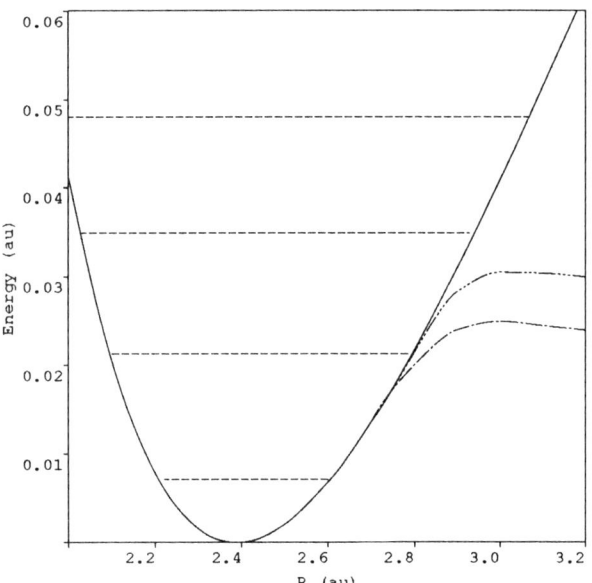

Figure 3. HCl potential energy curves
— HCl ground state
- - - - Vibrational levels
-·-·- Pseudo state model
······· Short range polarization only

Our low energy cross sections for the excitation of the $v = 1$ level are compared with experiment in Fig. 4. (Cross sections are given in cm² $\times 10^{-16}$ throughout this paper.) A major qualitative difference is now apparent between our two sets of results. The cross section from the pseudo-state model has a sharp threshold peak which is entirely absent from the results from the simpler model. These show a small peak just below the $v = 2$ threshold. As the two models differ only in the treatment of long range polarization effects, (they have identical short range polarization terms and long range dipole and quadrupole potentials) we can conclude that these effects are the primary cause of the sharp peak just above threshold.

Two other, quite different, non-adiabatic calculations for HCl have already been published. Our pseudo-state results are in remarkably good agreement (see Fig. 5) with those of Teillet-Billy and Gauyacq[11] who used an effective range approximation. The resonance model results of Domcke and Mundel[12] show a double peak with a much smaller magnitude than either our pseudo-state cross section or that of Teillet-Billy and Gauyacq. They show rather more resemblance to the results from our simpler model.

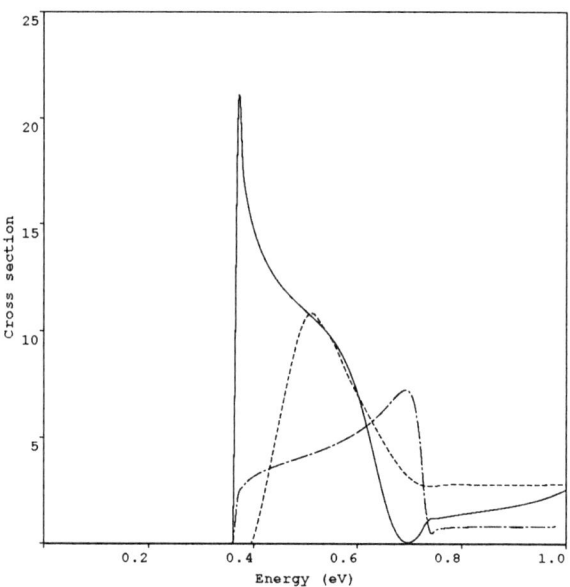

Figure 4. HCl integrated cross section $v = 0 \rightarrow 1$
——— Pseudo-state model
-·-·- Short range polarization only
- - - - Experiment (Knoth et al 1989)

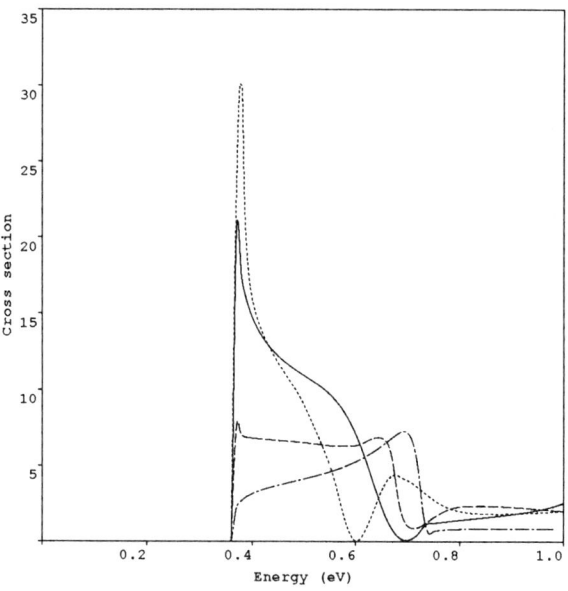

Figure 5. HCl integrated cross section $v = 0 \rightarrow 1$
——— Pseudo-state model
-·-·- Short range polarization only
········ Teillet-Billy and Gauyacq
- - - - Domcke and Mundel

The comparison with experiment suggests that the 'correct' result lies between our two sets of results. This, in turn, suggests that the polarization effects produced by our pseudo-states are too strong. This conclusion is also supported by the fact that the potential produced by our pseudo-state model is sufficiently strong to produce an unmistakable resonance in the vibrationally elastic cross section just below the $v = 1$ threshold. Despite a careful search in this region by Knoth et al[6] no such resonance has been observed. On the other hand there is evidence that our polarization potential is too weak ! Our pseudo-states were not optimized and we know that they give a significant underestimate of the static polarizability of HCl. This is supported by a comparison of our fixed nuclei results with those obtained from models where the polarization potential has been adjusted to give the correct asymptotic form. Unfortunately vibrationally elastic cross sections, which could provide unambiguous evidence for the nature of the sharp threshold structure, have not been reported for previous non-adiabatic calculations.

Cross sections for the excitation of the $v = 2$ level show less sensitivity to the model employed (Fig. 6) and are all in reasonable accord with the earlier measurements of Rohr and Linder[4].

We can use evidence from the non adiabatic eigenphase sum (Fig. 7) and a direct search for S-matrix poles in the complex energy plane to analyse our results. Both models produce a pole in the S-matrix between the $v = 1$ and $v = 2$ thresholds. This can be attributed to a nuclear excited Feshbach resonance associated with the $v = 2$ state of the target. In addition, the sharp peak in the pseudo-state results appears to be due to a second resonance very close to the $v = 1$ threshold, though the evidence for this is not so conclusive. However, from Fig. 3 we can see that the HCl$^-$ curve obtained from the pseudo-state results starts to diverge from the HCl curve just below the $v = 1$ threshold, so it would seem probable that the scattered electron could be bound temporarily to the tail of the $v = 1$ wavefunction.

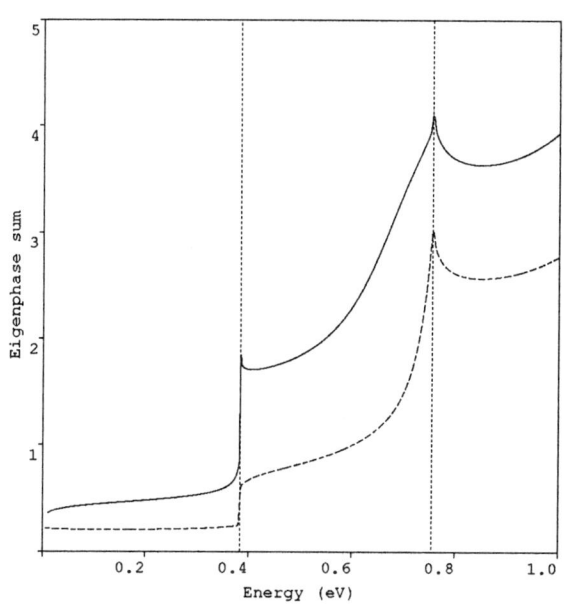

Figure 7. HCl Eigenphase Sum (Nuclear motion included)

———————— Pseudo-state model
- - - - - - - - Short range polarization only
·············· Vibrational thresholds

We have also calculated dissociative attachment cross sections using the formalism of Schneider et al[3]. This is, to the best of our knowledge, the first time that this formalism has been applied to a 'real' system. Our results (Fig. 8) are in reasonable accord with other theory and with experiment. Our use of an SCF target wavefunction is probably the most serious deficiency in our calculations. This manifests itself in several ways, the most obvious being the inaccuracies in our vibrational energy levels. Configuration interaction

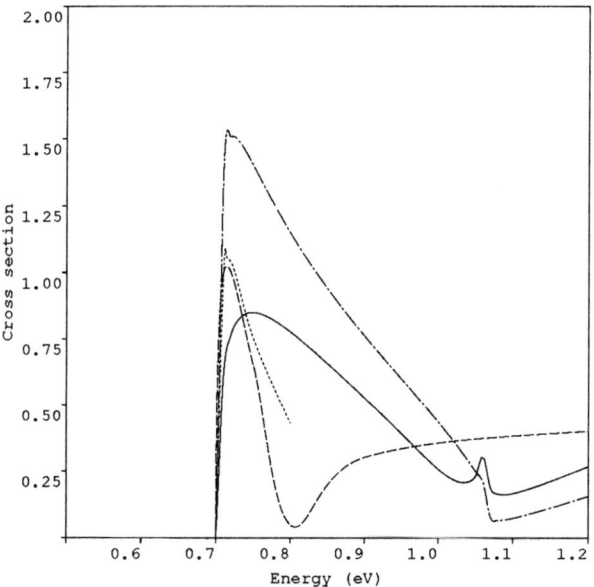

Figure 6. HCl integrated cross section $v = 0 \rightarrow 2$

———————— Pseudo state model
— · — · — · Short range polarization only
·············· Teillet-Billy and Gauyacq
- - - - - - - - Domcke and Mundel

effects can be expected to be important at large R and will probably have a significant effect on the dissociative attachment cross section.

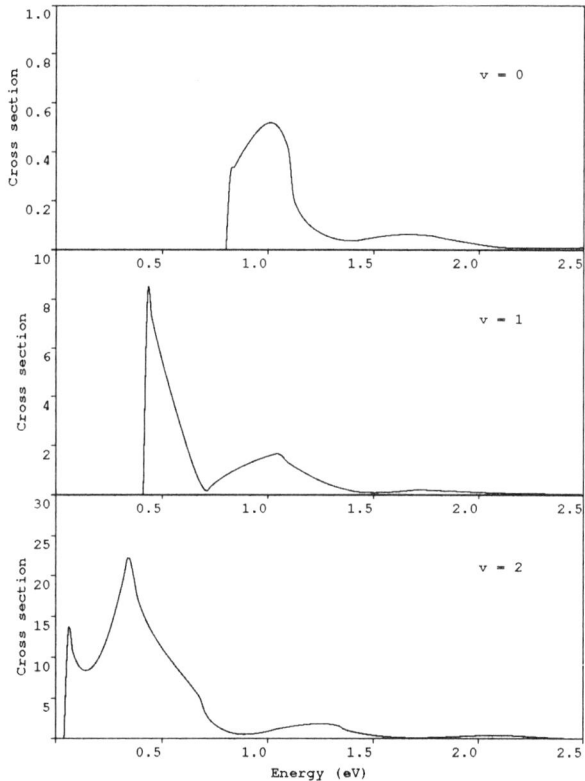

Figure 8. Dissociative Attachment to HCl

ELECTRONIC EXCITATION OF H_2 AND N_2

The scattering of electrons by molecular hydrogen is one of the most fundamental processes in the field of molecular physics. Despite this there are many areas of uncertainty concerning a range of processes including electronic excitation, dissociative recombination and differential cross sections[13]. In particular, even for this apparently simple system, theoretical calculations on the electronic excitation process have been restricted to two states[14]. All of these studies used single configuration wavefunctions to represent the electronic states involved.

Hydrogen, in common with most molecules, has a number of nearby low-lying electronic states. In particular there is experimental data[15] on excitation to the lowest five excited states all of which have vertical excitation thresholds in the range 10 - 13 eV. These studies yield a wealth of data including features which have been associated with several series of resonances. There has been little theoretical work on these resonances and their exact provenance is still controversial. Clearly a thorough theoretical treatment of low energy electronic excitation processes in H_2 must take account of all these states. To do this successfully, it is also necessary to have accurate electronic excitation thresholds for each state included in the calculation. This cannot be achieved by using simple single configuration wavefunctions.

Cross sections for the electronic excitation of the lowest five excited states, b $^3\Sigma_u^+$, a $^3\Sigma_g^+$, B $^1\Sigma_u^+$, c $^3\Pi_u$ and C $^1\Pi_u$, have recently been calculated[16]. The states are represented using a full configuration interaction treatment within a basis of Slater Type Orbitals optimised to give accurate vertical excitation energies. Results have been obtained, for excitation from the ground state, for energies up to 40 eV. Three models have been investigated. A two-state calculation was first carried out for comparison with earlier two-state models. Subsequently, four-state (all the Σ s) and six-state calculations were performed.

The new two-state results show larger cross sections at low energy and are similar to the earlier results at higher energies[14], except for the presence of pseudo-resonances. Both these effects are due to the improved short-range polarization potential used in the new calculations. Comparison with the 4 and 6 state models shows that inclusion of more target states leads to an increase in the cross section near threshold and a decrease at higher energies. This latter effect can be understood in terms of the inclusion of further open channels which compete for the flux.

The six-state model has been used to calculate cross sections for excitation to the higher target states. These cross sections all show sharp strucures near threshold. Similar behaviour has been observed experimentally by Mason and Newell[15]. A full analysis of these effects and the total cross sections will be published shortly[16].

The electron impact excitation of N_2 is another case where there has been very little previous theoretical work. In some respects this is an easier system to tackle than H_2 since its charge cloud is not so diffuse.

Calculations[17] are in progress which include the four lowest states, X $^1\Sigma_g^+$, A $^3\Sigma_u^+$, B $^3\Pi_g^+$, and W $^3\Delta_u^+$. A compact CI basis has been obtained to describe these states which gives excitation energies in excellent agreement with experiment. Preliminary results, for fixed nuclei, have been obtained for scattering energies up to 25 eV. Cross sections for the excitation of the A $^3\Sigma_u^+$ state are shown in Fig. 9. There is very good overall agreement with experiment[18]. The origin of the structure near 19 eV is not clear at the present time, though calculations in progress, which include higher target states, may shed some light on this.

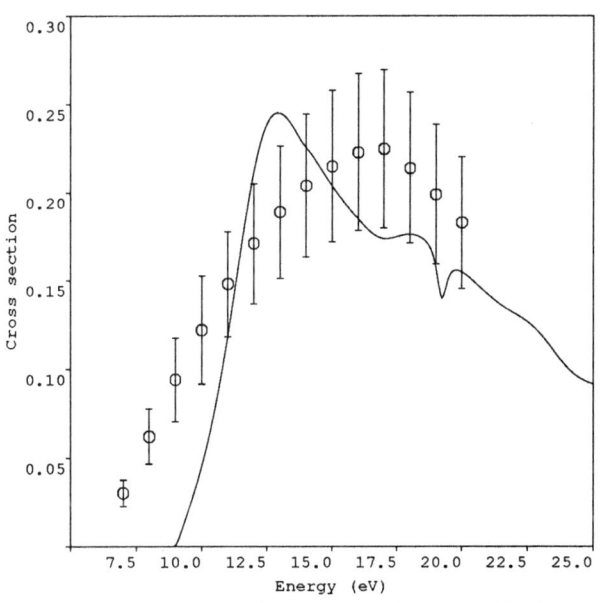

Figure 9. Cross section for the excitation of the lowest triplet state of nitrogen.
——— R-matrix results (Gillan et al 1989)
○ Experiment (Cartwright et al 1977)

POSITRON SCATTERING BY H_2 AND HF

The study of positron - molecule scattering has become an increasingly active area of study due, at least in part, to the increased availability of positron beams. Several theoretical methods have been developed to deal with low energy collisions and these have recently been reviewed by Armour[19]. The R-matrix method can be straightforwardly applied to positron scattering with only minor changes needed to the electron scattering codes.

Some preliminary calculations of e^+ - H_2 scattering have recently been extended by Tennyson and Danby[20]. These use a full CI description of the target and two pseudo-states chosen to reproduce the static polarizability, very similar to those used for electron scattering by Nesbet et al[21]. Both long and short range polarization effects are found to be important, especially in the Σ_g^+ symmetry.

Differential cross sections have been obtained for energies up to 1 Rydberg. In order to obtain converged results, symmetries up to Δ_u were included. The most interesting feature of these cross sections is the prediction of a minimum at about 45° for scattering energies between 0.3 and 0.6 Ryd. A similar behaviour has been predicted for e^+ - He scattering but the larger total cross section of the e^+ - H_2 system makes it a more promising candidate for observation.

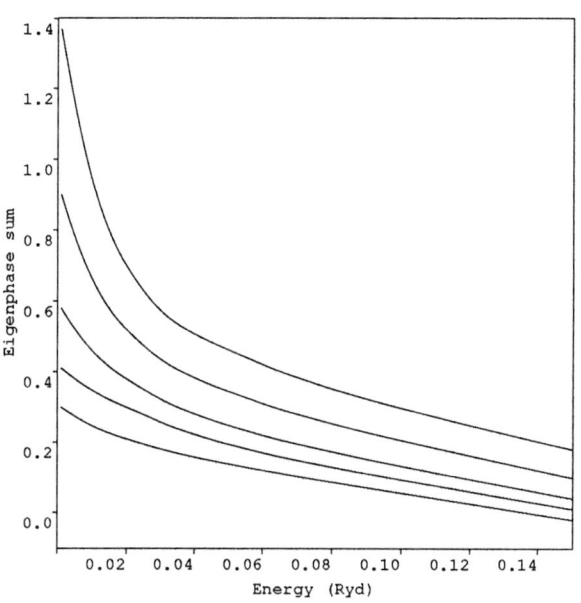

Figure 10. Positron - HF eigenphase sums
Fixed nuclei results for R=1.3 (bottom), R=1.5, R=1.733, R=2.1 and R=2.5 (top)

Danby and Tennyson have also carried out calculations for positron - HF scattering[22]. These were very similar to the calculations described above for electron - HF scattering. The strong, long range, dipole potential is equally attractive to both positrons and electrons, though the short range potential will be quite different. Again, the long range polarization potential is underestimated in these calculations. However the resultant potential is found to be sufficiently attractive to bind the positron at all the geometries considered. The major difference in the results, compared with the electron scattering case, is that the binding energy remains very small even at the larger internuclear separations. The eigenphase sum (Fig. 10) shows a pronounced decrease just above threshold, characteristic of the existance of a bound state, at all geometries. As in the electron scattering case, the existance of the bound states was confirmed by directly locating the corresponding poles in the S-matrix.

References

1. C.J. Gillan, O. Nagy, P.G. Burke, L.A. Morgan and C.J. Noble, J. Phys. B: At. Mol. Phys. bf 20 4583 (1987)
2. A.D. McLean in 'Proceedings of the Conference on Potential Energy Surfaces in Chemistry' (Ed. W A Lester Jr., IBM San Jose) 87 (1971). C.J. Noble Daresbury Laboratory Report DL/SCI/TM33T (1982)
3. B.I. Schneider, M Le Dourneuf and P.G. Burke, J. Phys. B: At. Mol. Phys. **12** L365 (1979)
4. K. Rohr and F. Linder, J. Phys. B: At. Mol. Phys. **13** 2521 (1976)
5. M.A. Morrison, Adv. At. Mol. Phys. **24** 51 (1988)
6. G. Knoth, M. Rädle, M. Gote, H. Ehrhardt and K. Jung, J. Phys. B: At. Mol. Opt. Phys. **22** 299 (1989). M. Rädle, G. Knoth, K. Jung and H. Ehrhardt, J. Phys. B: At. Mol. Opt. Phys. **22** 1455 (1989)
7. L.A. Morgan and P.G. Burke, J. Phys. B: At. Mol. Opt. Phys. **21** 2091 (1988)
8. J.P. Gauyacq and A. Herzenberg, Phys. Rev. A **25** 2959 (1982)
9. L.A. Morgan, P.G. Burke and C.J. Gillan, submitted to J. Phys. B: At. Mol. Opt. Phys. (1989)
10. C.J. Gillan, C.J. Noble and P.G. Burke, J. Phys. B: At. Mol. Phys. **21** L53 (1988)
11. D. Teillet-Billy and J.P. Gauyacq, J. Phys. B: At. Mol. Phys. **17** 4041 (1984)
12. W. Domcke and C. Mundel, J. Phys. B: At. Mol. Phys. **18** 4491 (1985)
13. J.W. McConkey, S. Trajmar and G.C.M. King, Comments At. Mol. Phys. **22** 17 (1988)
14. K.L. Baluja, C.J. Noble and J. Tennyson, J. Phys. B: At. Mol. Phys. **18** L851 (1985), T.L. Gibson, M.A.P. Lima, V. McKoy and W.M. Huo, Phys. Rev. A **35** 2473 (1987), T.N. Rescigno and B.I Schneider, J. Phys. B: At. Mol. Opt. Phys. **21** L691 (1988)
15. M.A. Khakoo, S. Trajmar, R. McAdams and T.W. Shyn, Phys. Rev. A **35** 2832 (1987). N.J Mason and W.R. Newell, J. Phys. B: At. Mol. Phys. **19** L203 (1986)
16. S.E. Branchett and J. Tennyson, to be published.
17. C.J.Gillan, C.J. Noble and P.G. Burke, to be published.
18. D.C. Cartwright, S. Trajmar, A. Chutjian and W. Williams, Phys. Rev. A **16** 1041 (1977)
19. E.A.G. Armour, Phys. Rep. **169** 1 (1988) and Comments At. Mol. Phys. **22** 173 (1989)
20. J. Tennyson, J. Phys. B: At. Mol. Phys. **19** 4255 (1986), G. Danby and J. Tennyson, submitted to J. Phys. B: At. Mol. Opt. Phys. (1989)
21. R.K. Nesbet, C.J. Noble and L.A. Morgan, Phys. Rev. A **34** 2798 (1986).
22. G. Danby and J. Tennyson, Phys. Rev. Lett. **61** 2737 (1988)

SPIN-POLARIZATION, ORIENTATION, AND ALIGNMENT IN ELECTRON-ATOM COLLISIONS

Michael H. Kelley

Center for Atomic, Molecular, and Optical Physics, NIST, Gaithersburg, MD

The use of state-selection techniques has added greatly to our fundamental understanding of electron–atom collision phenomena. In particular, measurements of atomic orientation and alignment in electron impact excitation, and measurements of spin dependence in both elastic and inelastic scattering, provide substantially more detail about collision processes than conventional measurements of differential cross sections alone. This work reviews recent experiments which, through the use of spin dependence and atomic orientation and alignment, attempt to provide the most complete characterization possible of electron–atom collision phenomena.

1 Introduction

Collision studies which make use of quantum state preparation and detection techniques have led to a considerable advance in our knowledge of scattering phenomena. In particular, for the case of electron-atom collisions, the substantial progress seen in coherence, correlation, coincidence, polarization, optical pumping, and step-wise excitation techniques has made possible experimental investigations which are increasingly detailed in their characterization of scattering processes. In some favorable cases, the truly "complete" or "perfect" measurements envisioned by Bederson[1] are possible. Such measurements provide the most detailed and complete characterization possible of the interactions at work in and scattering dynamics of collisions between electrons and atoms.

Several excellent reviews document the status of experiments utilizing state-selection techniques, including electron–photon coincidence measurements,[2–5] collisions using laser-excited atoms,[6,7] and spin-polarization techniques in electron–atom scattering.[8–11] The breadth of these combined fields of study far exceeds the scope of the present work. We will rather concentrate on several recent experiments which combine elements of each of these fields in order to approach more closely the ultimate goal of "complete" scattering experiments.

We begin in Sec. 2 with the introduction of some general concepts, in the context of inelastic scattering, which are required for the understanding of the experiments to be presented. We follow in Sec. 3 with a discussion of the origin of spin dependence in electron scattering, and describe spin-dependent measurements of elastic electron scattering. Finally, in Sec. 4 we describe several recent experiments which study both the spin dependence and orbital angular momentum alignment in inelastic electron–atom scattering.

2 General Principles of State Selection: S → P Excitation

Let us consider a prototypical electron scattering event in which an incident electron collides with a target atom and scatters to a detector placed at some scattering angle. Our goal is to determine from this measurement as much as possible about this collision process and the interactions at work.

How much could we learn about this collision event? Everything that can be learned about any such collision is contained in the wavefunction of the colliding system, so that's what we should try to determine. One generally describes this wavefunction in terms of complex scattering amplitudes which connect the initial state with each possible final state of the system. Determination of all scattering amplitudes is then equivalent to a determination of the full wavefunction, and provides a complete description of the collision process.

Of course, characterization of the final wavefunction requires knowledge of both the final state of the target atom and the final state of the scattered electron, so we must determine the complete set of scattering amplitudes connecting all initial to all final states. That is, we must study the scattering to all angles, for all combinations of initial and final states of the electron and target atom.

This is a rather ambitious undertaking, but is exactly what is necessary for a "complete" experiment. For essentially all experiments to date, however, one has been content to determine the set of scattering amplitudes for a particular scattering channel, e.g. for elastic scattering or for some particular atomic excitation. While such measurements are in a limited sense "complete," we should not lose sight of the ultimate objective of determining the scattering amplitudes for all processes at work during the collision.

Even in the more restricted sense of "complete" experiments for a single resolvable scattering channel, there are two immediate conceptual problems with the wavefunction determination as proposed above. First, it is not possible to determine directly the phase of the complex valued wavefunction. It would thus seem that at least the phases of the scattering amplitudes would be inaccessible. Second, it is generally not possible to ensure that either the initial or final state of the system is a single, well-defined quantum state, as there are typically degenerate magnetic sublevels which cannot be resolved by conventional techniques. The goal of much of the effort invested in coherence, correlation, coincidence, polarization, and step-wise excitation techniques has been to develop methods with which to overcome these difficulties.

Let us take the illustrative example of electron impact excitation from an S to a P state, ignoring any effects of electron spin. The S state has only the single $M_L = 0$ magnetic sublevel, but the P state has three degenerate sublevels, $M_L = \pm 1, 0$. Thus, three scattering amplitudes, f_{+1}, f_{-1}, and f_0, are required to describe this excitation. However, because of the overall positive reflection symmetry about the scattering plane (as defined by the incident and scattered electron momenta), these three are not all independent and only two need to be determined.

How these three amplitudes reduce to two is dependent upon the choice of coordinate system. Because one result of the reflection symmetry is that angular momentum transferred in the collision must be perpendicular to the scattering plane, it is convenient to choose the normal to the scattering plane as the quantization axis for the atomic angular momentum. In this coordinate system, the so-called "natural" frame, excitation of the $M_L = 0$ level is forbidden, so the f_0 amplitude vanishes exactly. Only the amplitudes f_{+1} and f_{-1} need to be determined.[†]

It is in principle straightforward to determine the magnitudes of f_{+1} and f_{-1}. This can be done in alkali atoms, for example, by collisional de-excitation of atoms maintained in the first excited state by laser optical pumping. One knows from time reversal symmetry that the same amplitudes must describe both the excitation and the de-excitation processes. By pumping with circularly polarized light incident perpendicular to the scattering plane, one can prepare excited atoms in fully oriented, pure magnetic sublevels of the P state, either

---

[†]Another common coordinate system is one in which the incident direction of the projectile electron serves as the quantization axis, the so-called "collision" frame. In this system, the $M_L = \pm 1$ sublevels cannot independently be excited and the corresponding scattering amplitudes have a fixed relationship with each other, i.e. $\mathcal{F}_{+1} = -\mathcal{F}_{-1}^*$. One thus needs to determine the two amplitudes $\mathcal{F}_0$ and $\mathcal{F}_1$.

$M_l = +1$ or $M_L = -1$. From the scattering cross section for collisional de-excitation of these P atoms, which is simply proportional to $|f_{+1}|^2$ or $|f_{-1}|^2$, one determines the magnitude of the amplitudes, $|f_{+1}|$ and $|f_{-1}|$.

Rather than determine $|f_{+1}|$ and $|f_{-1}|$ separately, it is generally preferable to determine the conventional differential cross section, proportional to $|f_{+1}|^2 + |f_{-1}^2|$, and a relative measure of the difference between $|f_{+1}|$ and $|f_{-1}|$. One example of such relative measurements is $L_\perp$, the net orbital angular momentum which would be transferred to the P state via electron impact excitation from the ground state, defined by

$$L_\perp = \frac{I_{+1} - I_{-1}}{I_{+1} + I_{-1}} \qquad (1)$$

where $I_{\pm 1}$ refer to the scattering signals for exciting or de-exciting the $M_L = \pm 1$ sublevels. Because the experimental signals appear both in the numerator and denominator of this expression, experimental factors such as target densities, detector efficiencies, and beam fluxes are completely canceled out. From a determination of the conventional differential cross section and $L_\perp$, or R, one can determine both $|f_{+1}|$ and $|f_{-1}|$.

It is in principle also straightforward in this case to determine the phase difference between these two scattering amplitudes from a study of the collisional de-excitation of atoms pumped with linearly polarized light, again incident normal to the scattering plane. Atoms excited with linearly polarized light have a "peanut"-shaped charge distribution, and the cross section for de-excitation depends on the alignment of this "peanut" with respect to the direction of incidence of the incoming electron. A determination of the alignment angle which maximizes (or minimizes) the cross section gives a direct measure of the phase difference between the two scattering amplitudes.

The connection between this alignment angle and the relative phase of the scattering amplitude is readily apparent if one thinks of the "peanut"-shaped P state as a linear superposition of the $M_L = \pm 1$ sublevels. That is, one thinks of the "peanut" as a standing wave generated by the interference between the two counter-rotating "doughnut"-shaped charge clouds of the $M_L = \pm 1$ sublevels. The relative phase of the two "doughnuts" determines the alignment angle of the "peanut".

The scattering from each of the $M_L = \pm 1$ sublevels contributes coherently to the observed scattering signal, generating an "interference pattern" at the detector which, in this picture, gives rise to the sinusoidal dependence of the scattering cross section on the alignment angle. The phase of this "interference pattern", i.e. the angle at which the maximum scattering occurs is determined by both the relative phase between the two initial states and the phase difference between the two scattering amplitudes. Consequently, determination of phase of the interference pattern gives one a direct measurement of the phase difference between the two scattering amplitudes.

It is important to point out, however, that only through the lack of degeneracy in the final S state was a complete determination of the amplitudes possible with no explicit analysis of the atomic state after scattering. In the more general case, where multiple initial and final states exist, state selection is required both before and after the collision in order to make a full determination of scattering amplitudes.

This principle of separately determining the amplitudes and phases of the scattering amplitudes, as described above, applies equally well to any other scattering process and is readily generalized to collision systems involving state degeneracy for both the initial and final states of the electron and target atom. Independent of what the scattering process might be, one chooses a suitable coordinate system and basis set with which to describe the collision system. Then, one only needs to measure the scattering cross section for each pair of incident and final states, and to determine, in a pair-wise fashion, the interferences between scattering channels in order to completely determine the magnitudes and phases of all scattering amplitudes, apart from one overall phase.

Of course, this idealized measurement protocol can rarely be achieved in practice. Nevertheless, the conceptual description is very useful and is the fundamental idea underlying the density matrix formalisms commonly used to handle the complicated bookkeeping involved in describing complicated collision processes involving techniques of state preparation and detection. There has been much effort directed towards developing general formulae for relating calculated scattering cross sections to quantities observed in state-selected measurements.[12,13] We will not be concerned here with specific formulae but will refer the reader to the original publications of individual experiments for further details.

For the present we merely note that for essentially all experiments, results are reported in terms of physical parameters, analogous to the angular momentum transfer, $L_\perp$, discussed above, which characterize some physical aspect of the collision or measurement process and are related in a known way to the underlying scattering amplitudes. For the case of electron–photon coincidence studies of atomic excitation, results are generally reported in terms of the shape of the charge cloud for the excited atom ($L_\perp$, P_{lin}, γ, and ρ_{00},[4] or alternatively λ, $\overline{\chi}$, Δ, and ε,[14,15]) or the state multipoles of the density matrix which describes the excited state.[13]

3 Spin-Dependence in Elastic Electron Scattering

The preceding description of inelastic $S \to P$ transitions would be complete if electrons had no spin. Of course they do, and the study of spin dependence in electron scattering has proven valuable for furthering our understanding of fundamental aspects of electron–atom scattering. Besides the additional information about scattering dynamics available from studies of the spin dependence, one can use spin dependence to aid in classification of unknown resonance features in elastic or inelastic scattering and to study directly optically-forbidden, spin-changing transitions which are otherwise difficult to observe.[8,10,11]

In order to interpret the results of any experiment involving polarized electrons, it is useful first to consider separately the two underlying causes of spin dependence in elastic electron scattering — the spin-orbit interaction and exchange. For each, as for the case of inelastic $S \to P$ transitions, one chooses a basis set which simplifies the understanding of the corresponding pure-state to pure-state collision events. A determination of the corresponding cross sections and pairwise interferences then comprises a complete measurement.

3.1 The Spin-Orbit Interaction

The spin-orbit interaction arises from the interaction of the magnetic moment (spin) of an incident electron with the magnetic field observed in the rest frame of that electron due to its motion in the electric field of the scattering target. This effective magnetic field is always normal to the scattering plane determined by the incident and scattered electron momenta, so that electrons at a given impact parameter whose spins are "up", relative to the scattering plane, have an energy which is different from that of electrons whose spin is "down". The scattering of spin "up" and spin "down" electrons is thus determined by different scattering amplitudes, say $f_\uparrow$ and $f_\downarrow$, and hence different scattering cross sections.

This difference between the scattering cross sections leads to both of the primary observables for spin-orbit scattering. First, it causes the left/right scattering asymmetry, S_A, which is exploited in Mott detectors for the determination of electron spin. Second, it causes electrons with no initial spin component normal to the scattering plane to acquire a normal component, S_P, after scattering from heavy atoms. For most cases studied in elastic scattering, where the target atom has no net angular momentum, $S_A = S_P$ so that observation of either asymmetry, along with the spin-averaged differential cross section, is sufficient to determine $|f_\uparrow|$ and $|f_\downarrow|$.

The phase difference between $f_\uparrow$ and $f_\downarrow$ can be determined in a measurement analogous to that use for the $P \to S$ de-excitation described in Sec. 2. One prepares an initial ground state which is a superposition of "up" and "down", i.e. polarized in the scattering plane, and measures the phase of the resulting "interference pattern", i.e. the net precession angle of the spin about the scattering plane normal. This net precession angle is directly related to the phase difference between $f_\uparrow$ and $f_\downarrow$. Thus, complete characterization of elastic spin-orbit scattering is achieved by determining the spin polarization after scattering of an electron initially polarized in the scattering plane.

Rather than using the amplitudes $f_\uparrow$ and $f_\downarrow$ which we have just described for the scattering of "up" and "down" electrons, one conventionally uses the amplitudes f and g for "direct" and "spin-flip" scattering.[8] These pairs of amplitudes are related by

$$f_\uparrow = f - ig \qquad f_\downarrow = f + ig. \qquad (2)$$

There have been many studies of spin asymmetries in elastic scattering which have added significantly to our understanding of scattering from heavy atoms.[8-10] We take as an example work from the Münster group for electron scattering from xenon and mercury in which a complete set of such measurements were made.[16-19]

The essential features of the apparatus used by the Münster group are a source of electrons which are spin-polarized in the scattering plane, an atomic beam scattering target, and an electron spin polarimeter. The spin polarimeter determines only two transverse spin components of the scattered electron, but a Wien filter was optionally used to rotate the longitudinal spin component of the scattered electron into the transverse direction so that a complete determination of all three components could be made.

The measurement of the spin polarization after scattering, together with a measurement of the absolute spin-averaged differential scattering cross section comprises a determination of the complete set of observables for this scattering process. Consequently, Berger, et al. were able to determine directly both the magnitude and relative phase of the two complex scattering amplitudes f and g. We show in Fig. 1 their results for scattering from Xe at 50.0 eV incident energy.

The striking difference between the angular dependences for $|f|$ and $|g|$ stems from the rather fundamental difference between the contributions of f and g to the scattering cross section. The conventional spin-averaged differential cross section is largely determined by f, with g, typically an order of magnitude smaller than f, adding a rather small correction due to the spin-orbit interaction. This interaction is strongest in the high electric field near the nucleus and is thus most important at small impact parameters. Consequently, only the lowest few partial waves should contribute to g, as is reflected by its rather smooth angular dependence.

Also shown in Fig. 1 are results of theoretical calculations from various groups.[20-22] The overall agreement is quite good at this as well as the other incident energies studied. Similar agreement was also found for electron scattering from mercury, testifying to the ability of current scattering calculations to model this scattering process accurately.

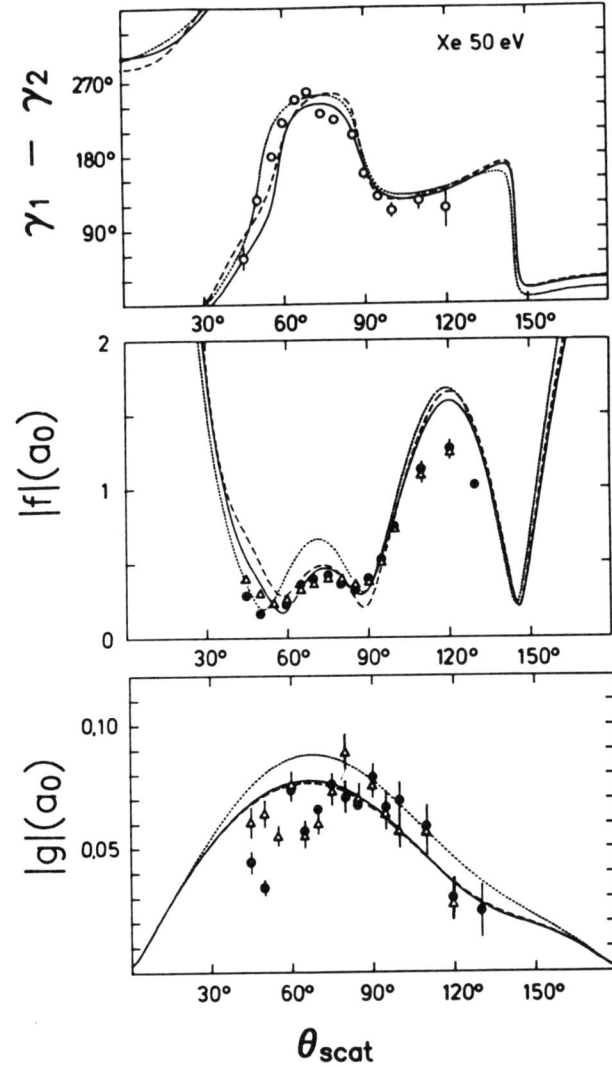

Fig. 1. Scattering amplitudes for Xe at 50 eV incident energy. Experimental data from Berger, et al.[18] For the determination of $|f|$ and $|g|$, the absolute cross section measurements of Register, et al.[23](●) and Mehr[24](△) were used. Theoretical curves are from Haberland, et al.[20](———), McEachran and Stauffer[21](· · · ·), and Awe[22](- - - - -).

3.2 Exchange Scattering

The other source of spin dependence, exchange, differs from the spin-orbit interaction, in that it is not the result of any spin-dependent force at work during the scattering, but is a manifestation of inherent symmetry properties of the wavefunction for spin-$\frac{1}{2}$ particles. The simplest case to consider is the scattering of a polarized electron from a polarized one-electron atom such as hydrogen, or equivalently, an alkali. These systems provide a particularly good example for the study of exchange because the dependence of the scattering cross section on the relative spin orientation of the incident electron and atom gives a direct measure of the role of exchange in the collision.

Neglecting the spin-orbit interaction, there are only three possible non-equivalent collision events:

$$e(\uparrow) + H(\downarrow) \rightarrow e(\uparrow) + H(\downarrow) \quad (3)$$
$$e(\uparrow) + H(\downarrow) \rightarrow e(\downarrow) + H(\uparrow) \quad (4)$$
$$e(\uparrow) + H(\uparrow) \rightarrow e(\uparrow) + H(\uparrow) \quad (5)$$

In the first example, the spins of both particles remain unchanged, so one can say that no "exchange" has occurred, and the scattering is characterized by a "direct" scattering amplitude f.[‡] In the second example, both spins have changed, so "exchange" has clearly taken place, and the scattering is characterized by the "exchange" amplitude $-g$. In the third process, one cannot tell whether the electrons have "exchanged" or not, so both channels contribute and the scattering is described by the amplitude $f - g$.

While this description parallels closely the measurements one can actually make, it does not clearly reflect the inherent symmetry of spin-$\frac{1}{2}$ particles. This symmetry is emphasized by describing the two particle wavefunction in a basis set in which the two spins have been coupled to form singlet and triplet states. Because we have assumed no spin-orbit interaction, the total spin of the colliding system is conserved, and these singlet and triplet states are eigenstates of the scattering Hamiltonian and form independent, non-intermixing scattering channels. Each is described by a complex scattering amplitude, which we label S or T, having the following relationship with the "direct" and "exchange" amplitudes:

$$S = f + g \qquad f = \tfrac{1}{2}(S + T) \quad (6)$$
$$T = f - g \qquad g = \tfrac{1}{2}(S - T) \quad (7)$$

Several possible approaches to the determination of the spin-dependent information are apparent from the scattering scenarios shown in Eqs. 3–5. For example,

---

[‡]We follow the unfortunate historical convention of referring to the two scattering amplitudes relevant for both the spin-orbit interaction and exchange by the same pair of symbols, f and g.

consider experiments in which both incident particles are spin-polarized, but for which no spin analysis of the scattered particles is performed. In this case, two scattering intensities can be measured which are proportional to the scattering cross sections for the incident parallel and antiparallel relative spin orientations. These cross sections are $\sigma_{\uparrow\uparrow} = |T|^2$ and $\sigma_{\uparrow\downarrow} = \tfrac{1}{2}(|S|^2 + |T|^2)$, respectively. One can define an exchange asymmetry, A_{ex}, by

$$A_{ex} = \frac{\sigma_{\uparrow\downarrow} - \sigma_{\uparrow\uparrow}}{\sigma_{\uparrow\downarrow} + \sigma_{\uparrow\uparrow}} = \frac{|S|^2 - |T|^2}{|S|^2 + 3|T|^2} \quad (8)$$

From a determination of the spin-averaged cross section and A_{ex}, one can determine both $|S|$ and $|T|$.

Determination of the phase difference between S and T requires, again in analogy with the discussion of Sec. 2, interference between the singlet and triplet channels. For example, in the antiparallel configuration of the scattering scenarios in Eqs. 4 and 5, the incident spin state is a linear superposition of singlet and triplet. The "interference" determines the relative cross section for reversing the incident electron's spin. Unfortunately, such a measurement would suffice to determine only the magnitude of the phase difference, not its sign. If, however, the incident electron's polarization were orthogonal to that of the incident atom, then the phase difference between S and T could be determined from the electron spin polarization direction after scattering.

Studies of spin dependence due to exchange are not yet as mature as studies of the spin-orbit interaction and there are to date no reports of a complete determination of the two scattering amplitudes. There are, however, several groups which have made partial measurements, from the pioneering work of Collins, et al.[25] and Hils, et al.[26] on scattering from potassium, and Jaduszliwer, et al. on scattering from rubidium,[27] to the determined efforts of the Yale and City College groups to study scattering from atomic hydrogen,[28] to the work in Bielefeld on scattering from lithium[29] and our work at NIST on scattering from sodium.[30]

These experiments represent a rather wide range of interesting experimental approaches, and it is unfortunate that space constraints prevent describing each in detail. We show here as examples the results from two studies of the elastic scattering of spin-polarized electrons from spin-polarized alkali atoms. First we describe measurements made in Bielefeld of elastic scattering from polarized lithium atoms.

Spin-polarized electrons were scattered from a beam of atomic lithium which was spin-polarized in a hexapole magnet. The direction of atomic polarization was determined by diabatic or adiabatic passage through a magnetic field whose direction reverses over a short flight path of the atoms. The spin of neither the scattered electron nor scattered atom was determined. The scattering asymmetry defined above for exchange, A_{ex}, was measured at fixed angles as a function of incident electron energy. These results are shown in Fig. 2. Also shown are the theoretical results of close-coupling[31,32] and modified polarized-orbital calculations.[33] The close-coupling results are in substantially better agreement with the data.

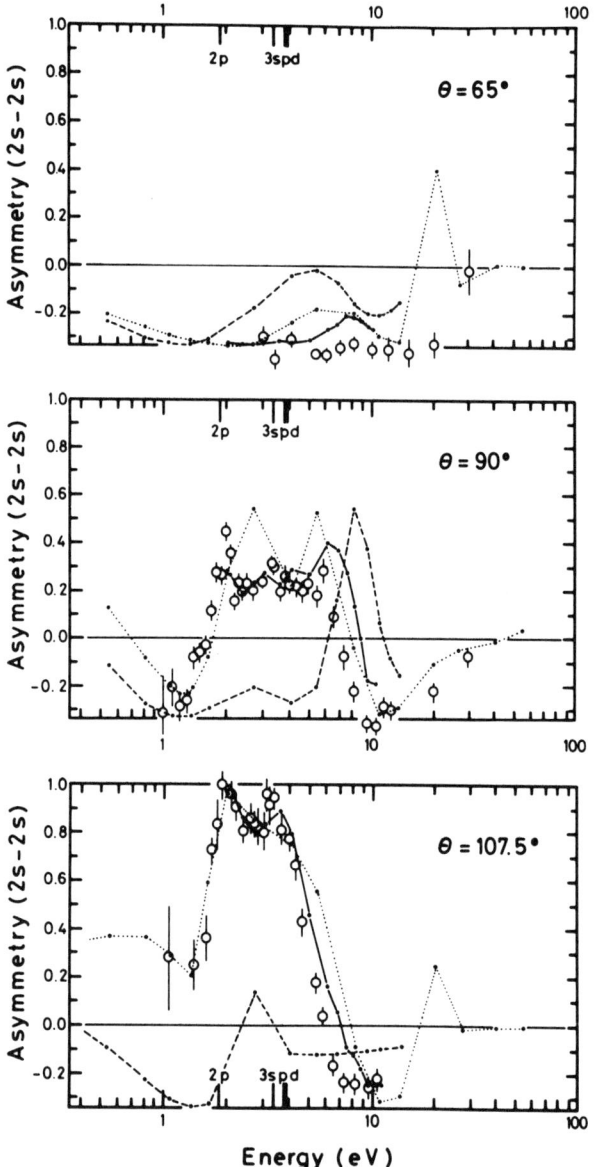

Fig. 2. Elastic spin asymmetry for scattering from Li. Experimental data from Baum et al.[29] Theoretical curves from Moores[32](—), Burke and Taylor[31](· · · ·), and Bhatia et al.[33](- - - -).

Two general features of the scattering are worth noting. First, for lower energies, below and slightly above the ionization threshold, this exchange asymmetry is very large. Limiting asymmetry values of $+1$ and $-\frac{1}{3}$ indicate pure singlet and triplet scattering, respectively. The scattering at 65° is dominated by triplet scattering, while the scattering to 107.5° is strongly dominated by singlet scattering. The second interesting general feature is that exchange continues to play an important role to relatively high incident energy, several times the ionization threshold.

We have made similar measurements at NIST for elastic scattering from sodium. Laser optical pumping was used to prepare ground state atoms polarized normal to a scattering plane. Electrons, also spin polarized normal to the plane, were scattered into angles in the range from 20° to 135°. As in the Bielefeld measurements, we measured the asymmetry A_{ex} for scattering with incident anti-parallel versus incident parallel spins. In Fig. 3 are shown results of our measurements for an incident energy of 54.4 eV. Even at this relatively high energy, about ten times the ionization threshold, exchange is seen to play an important role in the scattering.

Also included in Fig. 3 are results of a two-state close-coupling calculation of Mitroy, et al.[34] A four-state close-coupling calculation of Oza[35] gives near identical theoretical results. Though there are marked differences between the theory and experiment, there is agreement in a qualitative sense. The theory predicts correctly the size of the slight predominance of triplet scattering in the near forward direction, albeit not in the correct angular range. The source of this discrepancy is at present not understood.

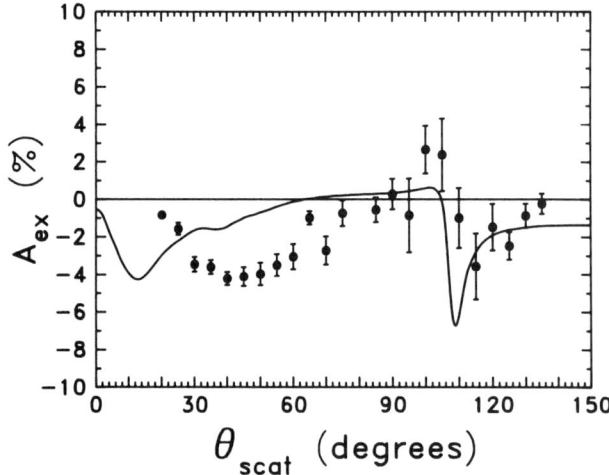

Fig. 3. Elastic spin asymmetry for scattering from Na at 54.4 eV. Experimental data from McClelland et al.[30] Theoretical curve from close-coupling calculation by Mitroy, et al.[34]

One further interesting note from these elastic scattering measurements on Li and Na is that, contrary to conventional wisdom "exchange" is not necessarily most important at larger scattering angles. For example in sodium at an incident energy of 54.4 eV, we find that exchange effects are most apparent for scattering angles between 30° and 60°.[30]

In our discussion thus far of exchange in elastic scattering from the alkalis, we have specifically neglected two important effects. We have neglected any effect of the spin-orbit interaction on the scattering and we have not considered exchange in scattering from targets with no net spin.

3.3 Exchange and Spin-Orbit Scattering

For the case where both the spin-orbit interaction and exchange are important, the formalism is substantially more complicated, with a total of six complex amplitudes required rather than the two complex amplitudes required for a description of either exchange or the spin-orbit interaction alone.[36] Though it is possible to devise a set of eighteen measurements which would be sufficient to determine this full set of amplitudes,[37] there are to date no reports of such a complete measurement.

One interesting feature of the simultaneous working of exchange and the spin-orbit interaction is a cooperative effect which should occur if both are present, but must vanish in the absence of either.[38,39] One should look for a left/right scattering asymmetry in the elastic scattering of unpolarized electrons from atoms polarized normal to the scattering plane. Equivalently, one would look for an asymmetry in the scattering of unpolarized electrons from "up" and "down" atoms. Neither the spin-orbit effect nor exchange could alone generate such an asymmetry.

In our measurements of spin dependence in elastic scattering from sodium at 54.4 eV, we have observed both an exchange and a spin-orbit asymmetry near the cross section minimum at about 110° scattering angle.[30] From those measurements we were also able to determine this joint asymmetry. However, within our present experimental uncertainty, we have not yet found evidence for this interesting cooperative effect.

3.4 Singlet → Triplet Excitation

The other important effect ignored in the above simplified description of exchange in electron–atom scattering is exchange with a target which has closed electronic shells and hence no net spin. In collision with such a target, an exchange event which changes the spin of the electron must also change the spin of the atom. As a result of the Pauli exclusion principle, this is not possible for elastic scattering. This is not to say that exchange with the spin-zero core of an alkali or with any spin-zero atom is not important — exchange certainly does contribute to the scattering cross-section and must be treated correctly before agreement with experiments can be reached. It does say that observable spin dependence in scattering from spinless atoms can appear only in conjunction with excitation to higher electronic states.

For example, excitation of triplet P levels in helium can proceed only via exchange and provides a unique opportunity to study exchange in electron atom collisions without use of spin-polarization techniques. While triplet excitation provides a direct measurement of an "exchange" process, there is no simple comparison with the analogous "direct" process, which would result in excitation of a singlet state in helium. Because of the energy difference between the singlet and triplet levels, far more than exchange determines the difference between singlet and triplet excitation amplitudes. Nevertheless, the results of such measurements contribute greatly to our overall understanding.

Excitation of the He(3^3P) level has been studied by groups in Stirling,[40] Belfast,[41] Utrecht,[42] and Perth.[43] We use the work of the Stirling group as an example. Electrons which excite ground state helium to the 3^3P state were detected in coincidence with the photon subsequently emitted from the excited state. The circular polarization of photons emitted normal to the scattering plane and linear polarization of photons emitted parallel to the scattering plane was determined. From these polarized coincidence measurements the three parameters λ, χ, and γ, which completely characterize the scattering process, could be determined.[5] Results of the measurement are shown in Fig. 4. Also shown are results of a first-order many-body theory calculation.[44] The general agreement of the model calculation with the experimental results is encouraging, but because of the extreme difficulty of the measurements and the consequently large experimental uncertainties, one cannot yet draw firm conclusions about agreement or disagreement between the theory and experiment.

4 Spin Dependence in Inelastic Electron Scattering

Having set the stage in the separate discussions of excitation and spin-dependent scattering, we now generalize the ideas of the previous sections and discuss the spin dependence observed in orientation and alignment studies of atomic excitation. As in the case of elastic scattering, we discuss separately those scattering processes in which the spin-orbit interaction and exchange play the central role.

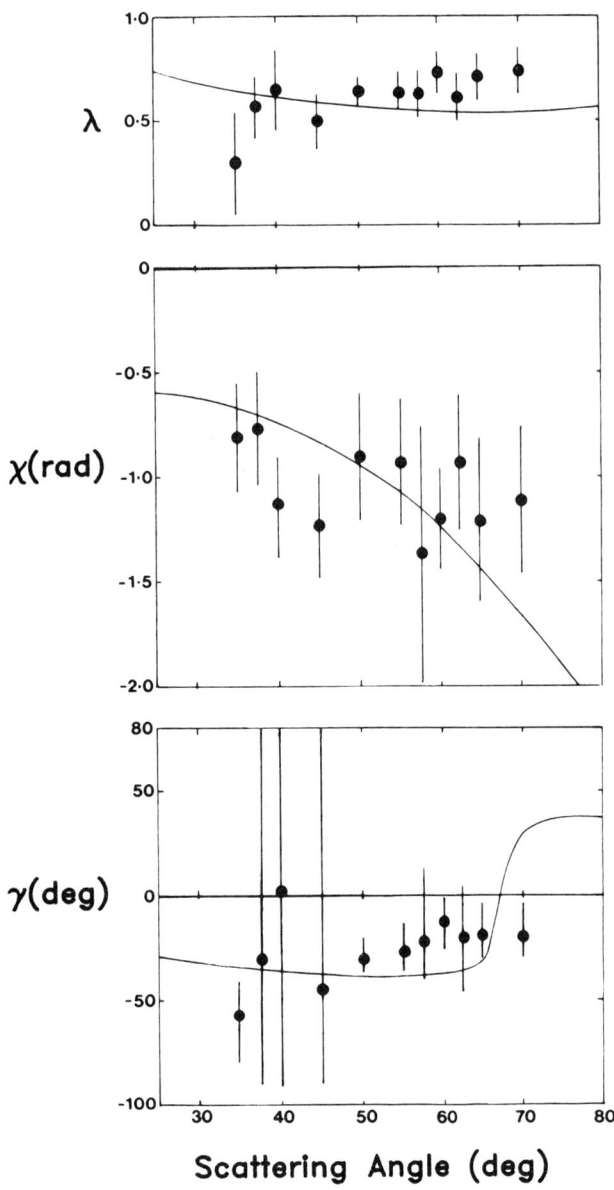

Fig. 4. The scattering parameters λ, χ, and γ from the Stirling measurements of He(3^3P) excitation at 60 eV.[40] Theoretical curve from FOMBT calculation of Cartwright and Csanak.[44]

4.1 Excitation of Hg(6^3P_1)

Electron impact excitation of heavy atoms is a good testing ground for state-of-the-art calculations of electron scattering. One must contend with the simultaneous importance of both exchange and the spin-orbit interaction, with transitions between multiple magnetic sublevels, with a rather complex atomic structure for which LS-coupling is a poor approximation, and with an often rich resonance structure for the colliding system.

Because all of these aspects of the collision interact in a complicated way, they can be very difficult to disentangle in order to extract an intuitive understanding of the collisional process. Methods of spin polarization and state selection can prove extremely valuable aids in attempts to unravel the details of this complicated collision system.

We take as an example the work of the Münster group for excitation of the 6^3P_1 level in mercury.[45] A detailed theoretical analysis of this system is too lengthy for the present, except to mention that a total of six complex amplitudes are required for a complete characterization of this collision system. While current experimental techniques do not allow a complete determination of this full set of amplitudes, the results of this measurement provide one of the most detailed characterizations yet of any collisional process.

Electrons, transversely polarized normal to the scattering plane, excited the target atoms and were detected in coincidence with fluorescence photons subsequently emitted perpendicular to the incident electron's direction, both in and perpendicular to the scattering plane. Excitation of the different magnetic sublevels, assumed quantized in the "collision" frame, was distinguished by the linear polarization of the photons. Photons from the $M_L = 0$ level are polarized parallel to the quantization axis while those from the $M_L = \pm 1$ are polarized perpendicular to it.

From the coincidence signals were derived six spin asymmetries. For each magnetic sublevel were determined:

$S_A(M_J)$ – the left/right scattering asymmetry for excitation by an initially polarized electron, averaged over the emission angle of the photon

$S_P(M_J)$ – the spin polarization after scattering of initially unpolarized electrons, also averaged over the emission angle of the photon.

The corresponding asymmetries, S_A and S_P, averaged over excitation of the different magnetic sublevels, were also determined.

Shown in Fig. 5 are the measurement results for these six asymmetries. Also shown in Fig. 5 are results of an R-matrix calculation from Bartschat *et al.*[46] The overall good quality of the agreement between theory and experiment is testimony to the current state-of-the-art for such scattering calculations.

The work of the Münster group also included similar studies of 6^3P_1 excitation at an incident energy of 15 eV. At the higher energy, the theory is not yet capable of accurately modeling the collision data.

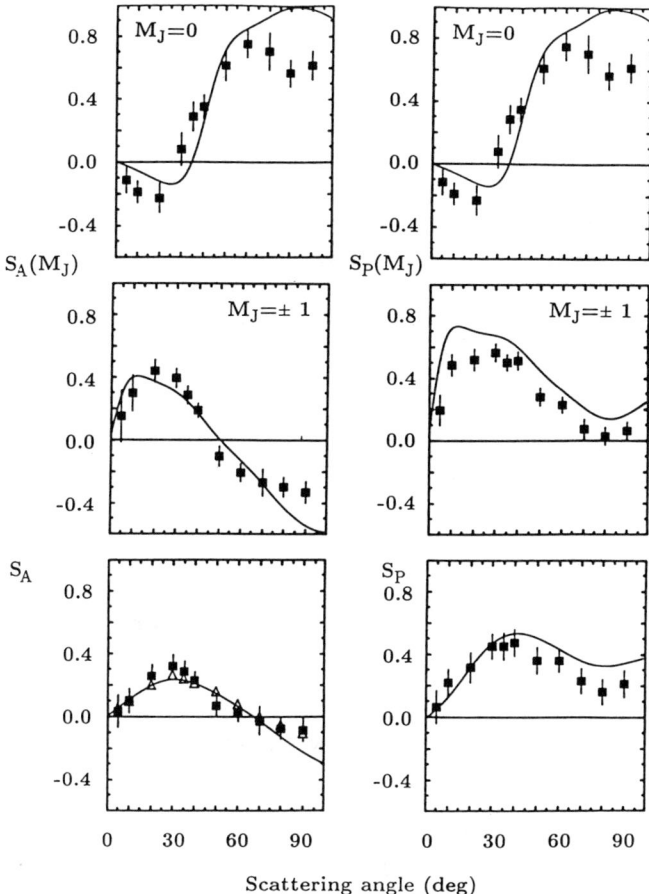

Fig. 5. Spin asymmetry functions $S_A(M_J)$, $S_P(M_J)$, S_A, and S_P at incident energy of 8 eV. Experimental results from Goeke, et al.[45] Theoretical curve from the R-matrix calculation of Bartschat, et al.[46]

4.2 Spin Dependence in Na 3P

We take as our final topic the spin dependence in $S \to P$ excitation of one-electron atoms. As noted in Sec. 3.2, electron scattering from alkali atoms provides a particularly good system for the study of exchange because an observation of the dependence of some scattering process on the relative spin orientations of the incident particles provides a direct measure of the effect of exchange on that collision.

Because a thorough treatment of the theoretical details necessary for an accurate description of this scattering process are available elsewhere,[47,48] we provide here only a brief introduction to aid in understanding of the experimental results. We take as the starting point the simplified description of $S \to P$ excitation provided in Sec. 2. Two complex scattering amplitudes were required for that case, one for excitation of each of two allowable magnetic sublevels. But now, as in the case of elastic scattering, there is a different amplitude for each relative orientation of the incident spins, or equivalently, each of the two incident composite spin states, singlet and triplet. This gives a total of four complex amplitudes to be determined, $f_{\pm 1}^S$ and $f_{\pm 1}^T$.

It is also, of course, possible to describe this collision process in the "collision" frame and to use uncoupled wavefunctions for the spins of the electron and atom. In this case, one would use, perhaps, the amplitudes f_0^d, f_1^d, f_0^e, and f_1^e for "direct" and "exchange" excitation of the $M_L = 0$ and $M_L = 1$ sublevels.

We will discuss two rather different approaches for the determination of the amplitudes for spin-polarized excitation in alkali atoms. The first is the method used at JILA by Han, et al.[49] to measure directly the excitation of the four M_J sublevels of the $3P_{\frac{3}{2}}$ excited state in sodium, starting from the $M_J = +\frac{1}{2}$ ground state. Excitation of these four sublevels corresponds rather closely to collision events in which the orbital angular momentum M_L either does or does not change by 1 during collisions in which the electrons either do or do not "exchange". Han et al. denote the partial cross sections for these events by $Q_{|\Delta M_L|}^{|\Delta m_S|}$, with $|\Delta m_S|$, $|\Delta M_L| = 0$ or 1. Here $|\Delta M_L| = 0$ and 1 refer to excitation of the $M_L = 0$ and 1 sublevels in the collision frame. $|\Delta m_S| = 1$ and 0 refer respectively to changing or not changing the atomic spin in the collision.

The experimental technique was based on optical pumping to prepare a pure $M_J = +\frac{1}{2}$ initial state, and a spectroscopic determination of the M_J level excited by electron impact. A magnetic field of about 220 Gauss was applied to the collision region to split the magnetic sublevels of the excited state so that they could be resolved spectroscopically. This magnetic field had no fundamental effect on the electron impact excitation process.

The relevant energy level diagram is shown in Fig. 6. Laser optical pumping prepares spin polarized atoms in the upper of the two ground state sublevels. Electrons incident along the direction of the applied magnetic field collisionally excite the polarized atoms and the populations in the four $3^2P_{\frac{3}{2}}$ magnetic sublevels are determined by tuning a laser through the $3P_{\frac{3}{2}}(M_J) \to 5S_{\frac{1}{2}}(m_S)$ transition manifold and detecting the $4P \to 3S$ cascade fluorescence from the $5S$ state.

From the observed sublevel populations, Han et al. were able to extract the four partial cross sections mentioned above. Measurement results for these partial cross sections are shown in Fig. 7. Also shown are the results of a four-state close-coupling calculation from Moores and Norcross[50] which are overall in good qualitative agreement. It is unfortunate that this useful technique has very limited angular resolution and lacks a generalization

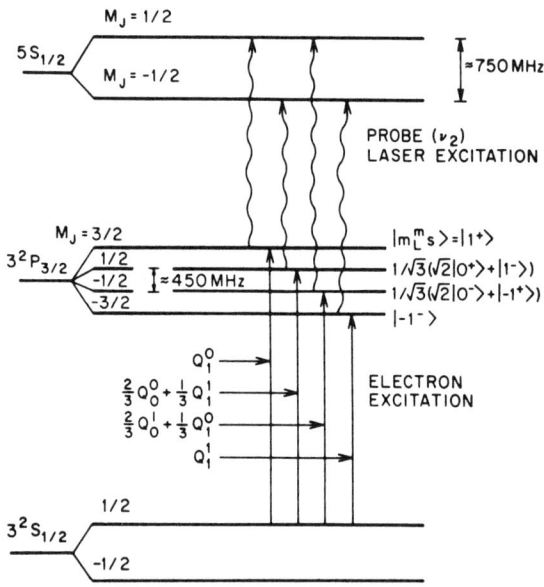

Fig. 6. Energy level diagram showing principle of technique used by Han et al..[49]

for determination of the phase relationship between the underlying scattering amplitudes.

Conventional crossed-beams experiments which include spin polarization of both the incident electrons and atoms in angle resolved collision studies have also been carried out. At NIST, we have studied this process using electron impact de-excitation of laser-excited sodium atoms. Similar measurements have been performed in Bielefeld using excitation of spin-polarized lithium atoms by spin-polarized electrons.[51]

The apparatus at NIST for these measurements is the same as is used for the elastic scattering measurements described in Sec. 3.2 except that the optical pumping laser illuminates the scattering volume directly to maintain a population of excited and optically pumped atoms. The primary advantages of optical pumping over magnetic state selection and electron-photon coincidence methods are that it allows studies of either elastic or inelastic transitions with the same apparatus, and that the inelastic measurements have a much larger scattering signal than the equivalent electron-photon coincident experiments.

At each incident energy and scattering angle, four count rates were recorded, corresponding to the four relative spin orientations of the incident electrons and atoms. From these four count rates were derived three relative quantities, the ratio between the triplet and singlet scattering cross sections, $R = \frac{|T|^2}{|S|^2}$, and angular momentum transferred normal to the scattering plane, separately for the singlet and triplet channels, $L_\perp^S$ and

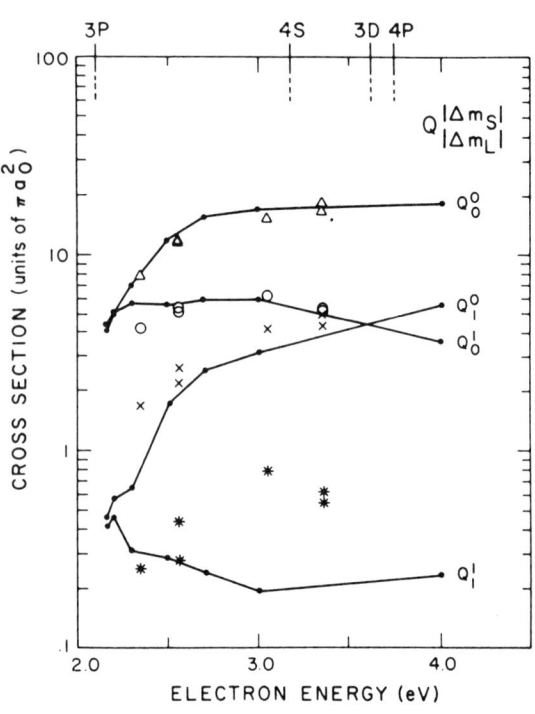

Fig. 7. Partial cross sections $Q_{|\Delta M_L|}^{|\Delta m_S|}$ measured by Han et al.,[49] compared to close-coupling calculation of Moores and Norcross.[50]

$L_\perp^T$. Because no spin analysis is performed after collision, no information is accessible about the phase differences between the scattering amplitudes, except for the spin-averaged phase difference determined from scattering from atoms excited with linearly polarized light as discussed in Sec. 2.

In Fig. 8 we show results of our measurements of these quantities at an incident energy of 2.0 eV. Also shown, in Fig. 8a, is the spin-averaged $L_\perp$ both from our measurements and from Hermann, et al.[52] The theoretical results of Moores and Norcross[50] are included in Fig. 8 for comparison. Several features are worth noting. First, the generally good agreement apparent in Fig. 8a for the unpolarized $L_\perp$ at smaller scattering angles is misleading. The substantial disagreement for the singlet channel angular momentum observed in Fig. 8b is masked in the unpolarized data by the much larger contribution of the triplet state to the unpolarized measurements due to its degeneracy factor of three. Because the triplet channel shows good agreement at small angles, the disagreement for singlet scattering is masked.

The second interesting feature is that while the theoretical results for the triplet channel are in better agreement at smaller scattering angles, the situation is reversed at larger angles. At larger angles the calculation is better for the singlet channel and drastically overestimates the triplet channel angular momentum transfer for angles larger than about 90°. The marked discrepancy apparent in Fig. 8a for scattering larger than about 110° comes entirely from the triplet channel. Over the intermediate angular range from about 45° to 110° the discrepancy is shared more or less equally between the two channels.

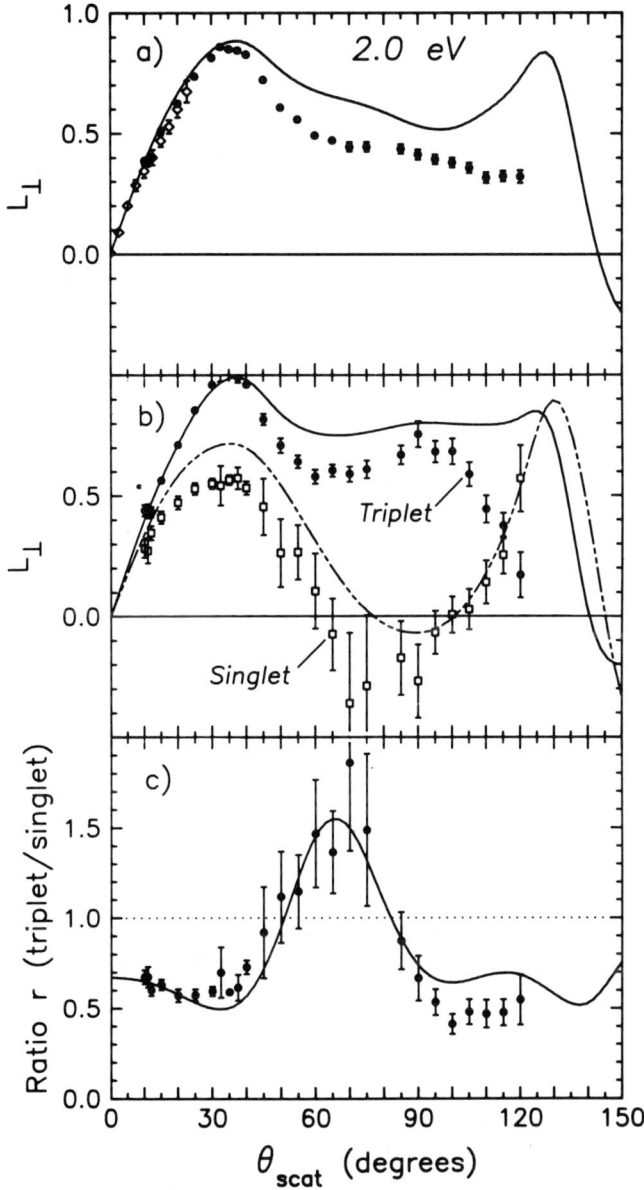

Fig. 8. $L_\perp$, $L_\perp^S$, $L_\perp^T$, and R measured at NIST for Na($3P \rightarrow 3S$) de-excitation at 2.0 eV.[53] Theoretical curves are close-coupling results from Moores and Norcross.[50]

Finally, we note that the calculation reproduces very well the ratio of triplet to singlet scattering show in Fig. 8c. Over much of the angular range, singlet scattering is dominant by roughly a factor of two. In a rather narrow range, from about 55° to 80°, triplet scattering becomes about 50% larger than singlet.

In the spirit of "complete" measurements, as put forward in Sec. 2 of this work, we have been attempting to make detailed measurements of both the elastic and the dominant inelastic channel for each incident energy. We have completed measurements of elastic and inelastic scattering at an incident energy of 54.4 eV,[53-55] and are just completing a similar series of measurements at 20.0 eV.[56]

5 Summary

We have seen that great strides have been made toward a quantitative understanding of electron–atom collisions, and that state-selection techniques continue to play a very important role in the advancement of our knowledge. The extension of alignment and orientation studies to include spin dependence has brought us closer to realization of the goal of "complete" measurements for electron–atom collisions. Nevertheless, there is much work yet to be done before the experimental characterization of these collisions is complete.

Acknowledgements

This work is supported in part by the U. S. Dept. of Energy, Office of Basic Energy Sciences, Division of Chemical Sciences.

References

1. B. Bederson, Comments At. Mol. Phys. **1**, 41–44 (1969).
2. K. Blum and H. Kleinpoppen, Physics Reports **52**, 203–61 (1979).
3. J. Slevin, Rep. Prog. Phys. **47**, 461–512 (1984).
4. N. Andersen, J.W. Gallagher, and I.V. Hertel, Physics Reports **165**, 1–188 (1988).
5. H. Kleinpoppen, In *Fundamental Processes of Atomic Dynamics*, edited by J.S. Briggs, H. Kleinpoppen, and H.O. Lutz (Plenum Press, New York, 1988), page 393.
6. I.V. Hertel and W. Stoll, In *Advances in Atomc and Molecular Physics*, Vol. 13, edited by D.R. Bates and B. Bederson (Academic Press, New York, 1978), page 113.
7. W.R. MacGillivray and M.C. Standage, In *Coherence in Atomic Collision Physics*, edited by H.J. Beyer, K. Blum, and R. Hippler (Plenum Press, New York, 1988), page 103.

8. J. Kessler, *Polarized Electrons* (Springer-Verlag Berlin, 1985), 2nd edition.
9. G.F. Hanne, Physics Reports **95**, 95–165 (1983).
10. G.F. Hanne, In *Coherence in Atomic Collision Physics*, edited by H.J. Beyer, K. Blum, and R. Hippler (Plenum Press, New York, 1988), page 41.
11. W. Raith, In *Fundamental Processes of Atomic Dynamics*, edited by J.S. Briggs, H. Kleinpoppen, and H.O. Lutz (Plenum Press, New York, 1988), page 429.
12. K. Blum, In *Fundamental Processes in Atomic Collision Physics*, edited by H. Kleinpoppen, J.S. Briggs, and H.O. Lutz (Plenum New York, 1985), page 103.
13. K. Blum, *Density Matrix Theory and Application* (Plenum Press, New York, 1981).
14. F.J. da Paixão, N.T. Padial, GY. Csanak, and K. Blum, Phys. Rev. Lett. **45**, 1164–7 (1980).
15. K. Blum, F.J. da Paixão, and G. Csanak, J. Phys. B: At. Mol. Phys. **13**, L257–61 (1980).
16. O. Berger, J. Kessler, K.J. Kollath, R. Möllenkamp, and W. Wübker, Phys. Rev. Lett. **46**, 768–770 (1981).
17. R. Möllenkamp, W. Wübker, O. Berger, K. Jost, and J. Kessler, J. Phys. B: At. Mol. Phys. **17**, 1107–21 (1984).
18. O. Berger and J. Kessler, J. Phys. B: At. Mol. Phys. **19**, 3539–57 (1986).
19. G. Holtkamp, K. Jost, F.J. Peitzmann, and J. Kessler, J. Phys. B: At. Mol. Phys. **20**, 4543–69 (1987).
20. H. Haberland, L. Fritsche, and J. Noffke, Phys. Rev. A **33**, 2305 (1986).
21. R.P. McEachran and A.D. Stauffer, J. Phys. B: At. Mol. Phys. **19**, 3523–38 (1986).
22. B. Awe, F. Kemper, F. Rosicky, and R. Feder, Phys. Rev. Lett. **46**, 603 (1983).
23. D.F. Register, L. Vušković, and S. Trajmar, J. Phys. B: At. Mol. Phys. **19**, 1685 (1986).
24. J. Mehr, Z. Phys. **198**, 345 (1967).
25. R.E. Collins, B. Bederson, and M. Goldstein, Phys. Rev. A **3**, 1976–1987 (1971).
26. D. Hils, M.V. McCusker, H. Kleinpoppen, and S.J. Smith, Phys. Rev. Lett. **29**, 398–401 (1972).
27. B. Jaduszliwer, N.D. Blasker, and B. Bederson, Phys. Rev. A **14**, 162–68 (1976).
28. G.D. Fletcher, M.J. Alguard, T.J. Gay, V.W. Hughes, P.F. Wainwright, M.S. Lubell, and W. Raith, Phys. Rev. A **31**, 2854–84 (1985).
29. G. Baum, M. Moede, W. Raith, and U. Sillmen, Phys. Rev. Lett. **57**, 1855–8 (1986).
30. J.J. McClelland, M.H. Kelley, and R.J. Celotta, Phys. Rev. Lett. **58**, 2198–2200 (1987).
31. P.G. Burke and A.J. Taylor, J. Phys. B: At. Mol. Phys. **2**, 869 (1969).
32. D.L. Moores, J. Phys. B: At. Mol. Phys. **19**, 1843–1851 (1986).
33. A.K. Bhatia, A. Temkin, A. Silver, and E.C. Sullivan, Phys. Rev. A **18**, 1935 (1978).
34. J. Mitroy, I.E. McCarthy, and A.T. Stelbovics, J. Phys. B: At. Mol. Phys. **20**, 4827–4850 (1987).
35. Dipak H. Oza, Phys. Rev. A **37**, 2721–2723 (1988).
36. P.G. Burke and J.F.B. Mitchell, J. Phys. B: At. Mol. Phys. **7**, 214–228 (1974).
37. S.M. Khalid and H. Kleinpoppen, Phys. Rev. A **27**, 236–242 (1983).
38. P.S. Farago, J. Phys. B: At. Mol. Phys. **7**, L28–L31 (1974).
39. D.W. Walker, J. Phys. B: At. Mol. Phys. **7**, L489–L492 (1974).
40. H.A. Silim, H.-J. Beyer, A. El-sheikh, and H. Kleinpoppen, Phys. Rev. A **35**, 4454–7 (1987).
41. B.P. Donnelly, P.A. Neill, and A. Crowe, J. Phys. B: At. Mol. Phys. **21**, L321–5 (1988).
42. J.P.M. Beijers, S.J. Doornenbal, J. van Eck, and H.G.M. Heideman, J. Phys. B: At. Mol. Phys. **20**, 5529–5540 (1987).
43. J.F. Williams and I. Humphrey, In *Electronic and Atomic Collisions, Contributed papers of ICPEAC XIV* (Palo Alto, CA, July 1985), edited by M.J. Coggiola, D.L. Huestis, and R.P. Saxon (North-Holland, Amsterdam, 1985), page 112.
44. D.C. Cartwright and G. Csanak, J. Phys. B: At. Mol. Phys. **19**, L485–91 (1986).
45. J. Goeke, G.F. Hanne, and J. Kessler, J. Phys. B: At. Mol. Phys. **22**, 1075–1093 (1989).
46. K. Bartschat, N.S. Scott, K. Blum, and P.G. Burke, J. Phys. B: At. Mol. Phys. **17**, 269–77 (1984).
47. I.V. Hertel, M.H. Kelley, and J.J. McClelland, Z. Phys. D **6**, 163–183 (1987).
48. S.M. Khalid and H. Kleinpoppen, J. Phys. B: At. Mol. Phys. **17**, 243–258 (1984).
49. X.L. Han, G.W. Schinn, and A. Gallagher, Phys. Rev. A **38**, 535–8 (1988).
50. D.L. Moores and D.W. Norcross, J. Phys. B: At. Mol. Phys. **5**, 1482–1505 (1972).
51. G. Baum, L. Frost, W. Raith, and U. Sillmen, J. Phys. B: At. Mol. Phys. **22**, 1667–1677 (1989).
52. H.W. Hermann, I.V. Hertel, and M.H. Kelley, J. Phys. B: At. Mol. Phys. **13**, 3465–3479 (1980).
53. J.J. McClelland, M.H. Kelley, and R.J. Celotta. To be published in Phys. Rev. A, 1989.
54. J.J. McClelland, M.H. Kelley, and R.J. Celotta, J. Phys. B: At. Mol. Phys. **20**, L385–8 (1987).
55. J.J. McClelland, M.H. Kelley, and R.J. Celotta, Phys. Rev. Lett. **58**, 2198–200 (1987).
56. S.J. Buckman, J.J. McClelland, M.H. Kelley, and R.J. Celotta, In *Electronic and Atomic Collisions, Abstracts of Contributed Papers, ICPEAC XVI* (New York, 1989), page 149.

NEAR-THRESHOLD STUDIES OF ATOMIC HYDROGEN

J F Williams

Physics Department, The University of Western Australia
Nedlands, Perth, Australia 6009

Detection of the Lyman alpha, Lyman beta and Balmer alpha photons and near-threshold energy loss electrons enables the determination of excited state lifetimes, total and differential excitation functions and various angular and polarization correlation parameters. Data are given for the summed n=2 and 3 states as well as the separate 2P, 3S, 3P and 3D states.

Introduction

Electron scattering from atomic hydrogen is a central part of both theoretical and experimental studies of quantum scattering and atomic structure, see for example, Massey et al[1]. The most recent review of King et al[2] discusses resonance phenomena below the n=2 and 3 thresholds in both the elastic and 2S and 2P decay channels, differential elastic and 2S and 2P inelastic cross sections and angular correlation parameters at energies from 54 to 350 eV. This progress report primarily concerns coincidence studies just above the n=2 and 3 thresholds and gives an indication of measurements in progress using very low energy electrons.

An indication of the thrust of some recent work in Perth follows from Table 1 which shows the binding energy of the lowest states of atomic hydrogen as well as their energy separations and from fig 1 which shows decay schemes of the n=3 and 2 levels. Electron energy loss techniques should be able to separate states with principle quantum number up to about n=14 since electron energy resolutions down to 30 meV are routinely attainable and of the order of 5 meV are possible. The response of most photomultipliers limits photon detection to transitions of the Balmer series (656.3 nm and shorter wavelengths) and part of the Paschen series (from n=7 to 3 at 820.4 nm and higher n) while the Lyman transitions are detected by channeltrons. For the n=3 to n=2 Balmer-alpha radiation of 656.3 nm wavelength transmission filters and polarizers are available. The n=3 to n=1 Lyman beta radiation (102.6 nm) is not readily filtered with a narrow bandpass filter but may be separated from Lyman alpha (121.6 nm) by a cut-off filter. Also linear and circular reflection-type polarizers are not as efficient as their visible wavelength counterparts. The different angular momentum states for n=3 are not separable using simple electron energy loss or spectroscopic techniques, however they may be separated by their lifetimes which are about 158, 5.3 and 15.5 nsec for the 3^2S, 3^2P and 3^2D states respectively. It follows from the results of Humphrey et al[3] that radiation from states whose lifetimes differ by at least a factor of about three may be separable with an accuracy of about 98% under normal conditions in an electron-photon coincidence measurement. The separation via lifetime measurement requires an event to start the timing measurement and this is usually accomplished by either pulsing the incident electron beam or detecting the scattered electron with an energy loss

Table 1: Energy levels, and their separations, for atomic hydrogen.

Principal quantum number n	Binding energy eV	Separation energy meV
14	13.529	
		11
13	13.518	
		13
12	13.595	
		18
11	13.486	
		24
10	13.463	
		32
9	13.431	
		45
8	13.386	
		65
7	13.321	
		100
6	13.221	
		166
5	13.055	
		306
4	12.749	
		661
3	12.088	
		1889
2	10.199	

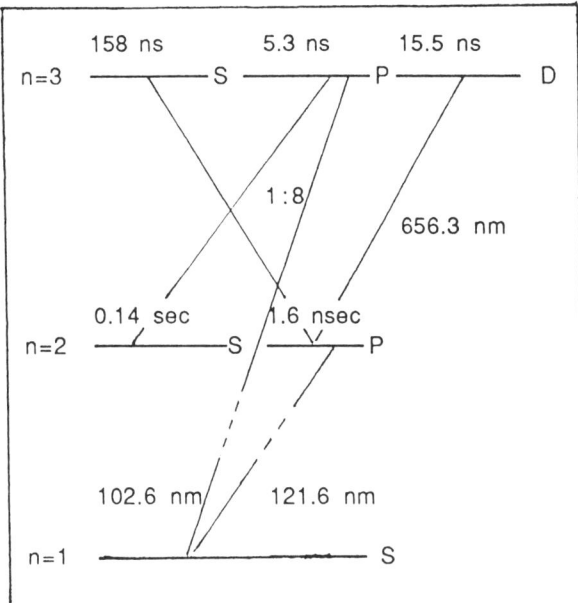

Fig 1. Decay scheme of lowest excited states of atomic hydrogen

equal to the binding energy of the parent excited state. The latter method is preferred because it eliminates cascade effects independently of the incident electron energy and it permits the significant advantages of the angular and polarization correlation techniques to be used to determine the state multipoles of the excited state. Measurements of Balmer-alpha radiation are difficult because in crossed electron-hydrogen atom experiments the atoms originate from a thermal oven or high-frequency discharge source which produces large amounts of Balmer-alpha radiation and because photomultiplier tubes at Balmer-alpha wavelengths have a non-negligible dark count which may be in the range of 10^2 to 10^4 counts/second even when cooled. A full description of the apparatus used for the present work and experimental problems is given in Williams and Heck[4] and earlier papers.

2. Integral excitation cross sections

The collection of all energy loss electrons from a given state presents many problems except in the special case when the incident electron energy is just above a threshold, that is when the energy loss electrons have low energies. The technique has been adequately discussed by Cvejanovic and Read[5]. In the present apparatus potentials are applied to the separate mesh sides of a cube to create a field distribution which enhances the extraction and collection efficiency of very low energy electrons. If the collection efficiency can be estimated, then a measure is obtained of the total excitation cross section. The method appears to be valid over a small energy interval from threshold up to an energy at which the overall efficiency of detection becomes too small; it is easily justified up to about 30 meV above threshold and has been extended up to 0.6eV above threshold but that limit is very dependent on the field value and distribution. Details will be given in a paper in preparation.

Figure 2 shows a representative threshold energy loss spectrum for an incident electron energy resolution of 12 meV, an electric field across the interaction volume of about 1.0 meV/mm and a displacement of the potential at the interaction region of about 10 meV. Peaks arising from all principal quantum number n states up to n=15 are seen. The peak for n=2 was more clearly defined than that for n=3 but is not shown because of the change of energy scale required. The raw data were smoothed to obtain this diagram. Similar spectra have been obtained over a wide range of experimental conditions which generally indicate both a broadening and displacement of the threshold energy loss peak as the potential and field at the interaction region increased.

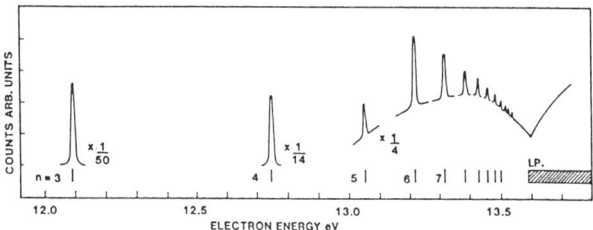

Fig 2. A threshold electron energy loss spectrum for an energy resolution of 12 meV.

The areas under the n=2 energy loss threshold peak at energies of 30, 50, 100, 150, 170, 290, 450 and 560 meV above threshold were calibrated by normalizing the 30 meV peak area to the total n=2 excitation cross section[6] of 0.64 πa_0^2. The 50 meV peak area then coincided within 5% with the total cross section value of 0.44 πa_0^2 while the ratio of the other areas to the cross section uniformly decreased to 560 meV above threshold. With that calibration the total n=3 excitation cross section was measured to be 0.112, 0.121, 0.104 and 0.130 πa_0^2 at 12.27, 12.38, 12.55 and 12.65 eV respectively. The data are further discussed in section 4.

The total n=3 excitation cross section could also be measured by detection of the Balmer-alpha and

Lyman-beta decay radiation. The radiation from the P state has an angular distribution

$$I(\theta) = I_{total} \frac{3}{4\pi} (100 - P \cos^2 \theta)/(300 - P)$$

where P is the percentage polarization. The intensity measured at an angle where $\cos^2\theta = 1/3$ is independent of polarization, and proportional to the total 3^2P cross section. The S state radiation gives an isotropic angular distribution but no simple way exists for separating the D state radiation by virtue of angular distribution, so the radiation would need to be collected over at least $\pi/2$ steradians to avoid polarization effects. Since the incident electron energy is kept below the n=4 threshold, cascade is not possible. The use of a parabolic mirror to collect radiation over a large solid angle and focus it onto the cathode of a photomultiplier tube is a standard technique, for example, in lifetime measurements[7]. The method could be used here for the total Balmer alpha radiation. Allowance for the 8:1 Ly β to Bα branching ratio must be made in determining the total cross section and the method is attractive for incident energies below the n=4 threshold when cascade radiation from higher states does not occur. This type of measurement will be attempted in future work.

3. n=2 differential cross sections

In the absence of electric fields in the interaction region the low energy electrons are scattered and drift into the energy analyzer. It was possible to detect electrons with energies as low as 28 meV, which corresponded to those electrons which had excited the large ^1P doubly excited resonance at its maximum value. Those scattered electrons were then detected in coincidence with the 2P decay photons at $\theta_e = \pi/2 = \phi_\gamma$ when the coincidence count rate is directly proportional to $\sigma(2P)$, the 2P differential cross section[8]. In this way the data for $\sigma(2P)$ at energies of 10.227, 10.336 and 10.744 (i.e. k^2 = 0.752, 0.76 and 0.79 a.u.) were obtained as shown in figure 3. The P resonance gives rise to a large peak near 90° and the forward angle scattering is rising strongly at 0.5 eV above threshold. The measured data are in good agreement over most of the ranges with the recent pseudostate close-coupling values of Callaway[9].

Subsequently at 10.227 and 10.744 eV in-plane angular correlations of the n=2 energy loss electrons

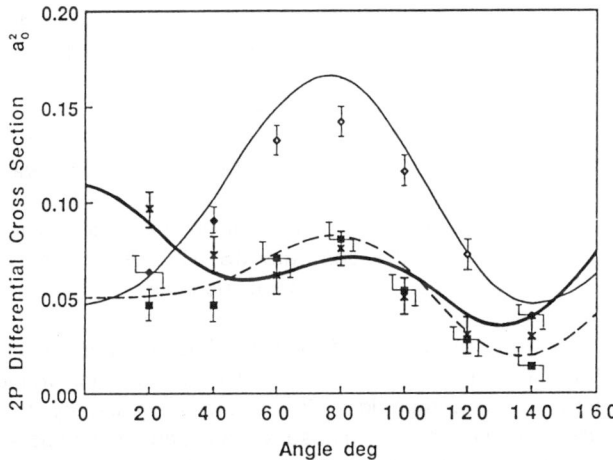

Fig 3. The differential cross section for exciting the 2P state as a function of electron scattering angle at incident energies of 10.227 eV (open squares), 10.336 eV (solid squares) and 10.744 eV (crosses). The lines are the pseudostate close coupling values of Callaway.

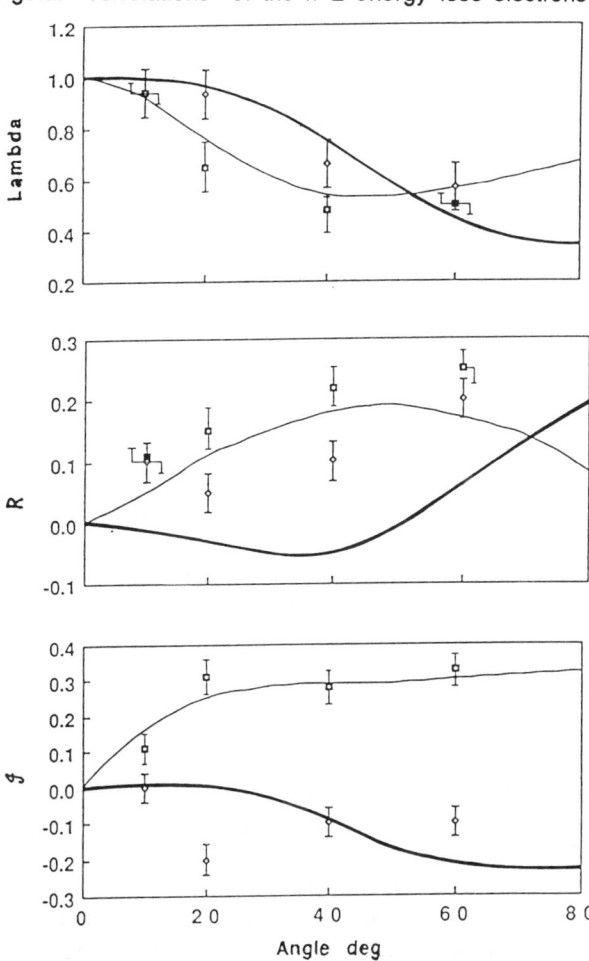

Fig 4. The correlation parameters λ, R and ɟ are shown as a function of electron scattering angle for incident energies of 10.227 and 10.744 eV. The lines are the pseudostate close coupling values of Callaway (1989).

and the 2P decay photons were made from which the parameters λ and R were deduced. The circular polarization normal to the scattering plane was measured to give the 𝟅 correlation parameter. These data are shown in figure 4 over the electron scattering angular range from 10 to 60° and are seen to be indicating similar trends to the close-coupling values of Callaway. The predicted very small negative values of R are however not observed. These data are the first with sufficient energy resolution to see the effects of a resonant state on the angular correlation parameters. Any effects would have been more clearly shown by a measurement of the parameters at a fixed scattering angle and sweeping the electron energy, however the time to accumulate data with acceptable statistical accuracy would be prohibitively long. Nevertheless over the limited angular range studied it seems that the 𝟅 parameter is more sensitive than either the λ or R parameters to the resonance and as well it changes sign. 𝟅 is negative similar to its behaviour at higher energies however at the peak of the resonance it is positive. With the electron energy resolution of 15 meV and an uncertainty in the energy scale of about 5 meV over the data compared with the width of the resonance of 22 meV there is considerable averaging of the effects of the resonance. A more detailed study is required before further explanation can be offered.

4. Excitation cross sections for the separate 3S, 3P and 3D states

The total 3P cross section was also measured by collecting the Lyβ photons at 54.7°. The effectiveness of an aluminium oxide film of 300Å thickness as a low pass filter to pass 120.6 nm photons but not 121.6 nm photons is readily measured by detecting energy loss electrons in coincidence with all photons radiated into the detector and stepping the incident electron energy across the n=2 and 3 thresholds which was done at energies of 10.1, 11.0 and 12.4 eV. At 10.0 eV there is only background molecular radiation, at 11.0 eV only Lyα and at 12.4 eV Lyα and Lyβ. The excitation threshold for the 3^3P state was clearly seen and the relative Lyα/Lyβ transmissions measured. The time-dependence of the Bα separates the 3S and 3D and 3P states and was measured with collection of the n=3 energy loss electrons, as in section 2, to act as time zero for the decay time. Allowance for the 8:1 Lyβ to Bα branching ratio was made.

The measured excitation functions of the 3S, 3P and 3D states were calibrated through the 3P measurements. The 3P state excitation function was measured both by detecting Lyman beta and Balmer alpha photons and calibrated by simultaneous measurements of the n=2 excitation function which is well known at these energies[6]. Figure 5 shows these data and the theoretical values of Callaway[9]. The close-coupling calculations were made variationally and used a large basis set of 10 exact states 1s to 4f and 18 pseudostates. The relative measured values are in good agreement with theory although the experimental uncertainty is large. Also there is a clear indication of the presence of resonances although they are too close to be separated.

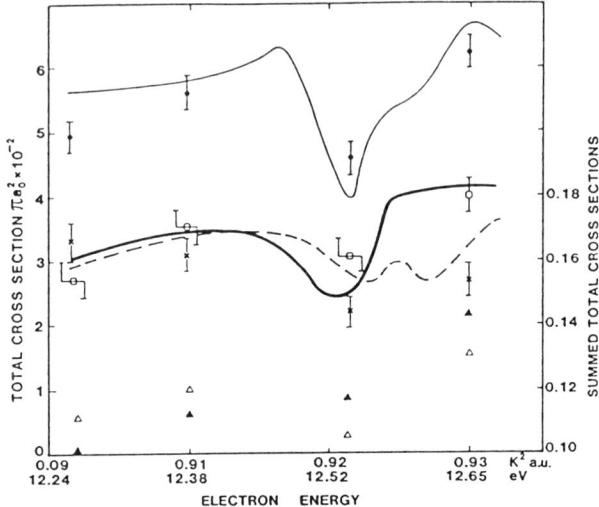

Fig 5. The cross sections for excitation of the separate 3S, 3P and 3D states (left hand scale) and the total n=3 excitation cross (the sum of the 3S, 3P and 3D state cross sections) (right hand scale) as a function of electron energy. The measured data are 3P state , 3D state and 3S state while the lines 3P , 3D and 3S are the calculated values of Callaway. The measured total cross section for n=3 excitation are given by solid triangles and the sum of the separate 3S, 3P and 3D cross sections given by open triangles.

With the threshold electron detection method the total n=3 cross section was measured and calibrated against the threshold n=2 cross section as reported above in section 2. These values can be compared with the sum of the 3S, 3P and 3D total cross sections measured above as shown in figure 5. It is seen that there is consistency between the values deduced from the two methods which gives confidence to the extension of this general approach to higher states.

The decay scheme in fig 1 also indicates that the coincident detection of the n=3 energy loss electrons with the total Balmer alpha radiation is equivalent to the coincident detection of the n=3 energy loss electrons with 2P decay radiation plus the Lyman beta radiation plus the quenched 2S state radiation. This measurement can be made since it requires only two photon detectors, one without lithium fluoride filter to detect the prompt 3P and 2P cascade radiations and a second to detect the quenched 2S radiation. Placement of those detectors at an angle of 54.7° gives a signal proportional to the total n=3 radiation intensity, which can be calibrated to give the total n=3 excitation cross section. It is also seen that separation of the 2S quenched radiation from the Lyman beta radiation permits a measurement of the branching ratio of the two decay modes of the 3P state. This branching ratio is well predicted by quantum theory and measurement, (see Moisewitsch and Smith[10] for a review) and there is little reason to doubt existing values. However the measurement does serve as a check on the accuracy and validity of the present methods.

An alternative method of determining the 3P cross section is seen from fig 1. If all the Balmer-alpha photons are collected then the 3S and 3D components can be used as a coincidence gate to veto their cascade 2P radiation. All the remaining true coincidence counts detected by a UV photon detector with an appropriate filter will be 120.6 nm photons. It is easy to construct and use a parabolic mirror to collect Balmer-alpha radiation over a solid angle of precisely one quarter 4π steradians (or similar large fraction) and to locate its focus at an appropriate distance from the interaction region so that the radiation is focussed onto the photocathode of a photomultiplier tube. Future work will explore this method.

5. Angular distributions for the separate 3S, 3P and 3D states

The n=3 energy loss electrons are detected with a rotatable electron energy analyzer to measure the n=3 differential angular distributions. The separation of the degenerate angular momentum contributions requires, as above, the detection in coincidence of a photon signal which is independent of polarization. For the 3^2P state this requires the photon detector to be located at either 54.7° in the scattering plane or at $(\theta,\phi) = (\pi/2, \pi/2)$, that is normal to the scattering plane. The 3^2S state photon decay is isotropic. The 3^2D_j photon distribution is given, for example, in the paper by Percival and Seaton[11]. A single Bα detector was placed normal to the scattering plane. A coincidence time window at least three lifetimes long is required to observe most of the radiation from a given state. For the 3S state with a lifetime of 158 nsec, the time window would be large enough to make integration times inconveniently long. A time window of 0.3 times the lifetime was used. The angular distribution for the P state decay (5.3 nsec) decreased by a factor of 3 over five angles equally spaced from 15° to 75°. There are no theoretical values for comparison.

6. State multipoles of the 3S, 3P and 3D states.

A detailed review of scattered electron-radiated photon angular and polarization theory pertaining to the state multipoles of the excited state has been given by Blum[12]. Specific applications to the n=3 states of atomic hydrogen have been given by Blum et al[13] and Heck and Gauntlett[14] and measurements of scattered-electron, cascade-photon angular correlations were reported by Chwirot and Slevin[13] for the n=3 P and D states of atomic hydrogen. Their analysis used the results of van Linden van den Heuvell[16], obtained for the 3^1D state of helium to parameterize the ^2D angular correlations of H and used the formulation of Blum[12] for the ^2P state correlation of H. From that work it is apparent that the hyperfine interaction can be neglected to a first approximation in the time evolution of the excited states. Fine-structure effects can be allowed for by the use of the appropriate time perturbation coefficients, which implies that either the coincidence resolving time can be made long compared with the fine-structure characteristic time so that an average over any fine-structure effects occurs or the coincidence time is made relatively short so that fine structure effects may be studied. Such effects are being considered for the n=3 states but are not reported here.

Information about the 3D state multipoles could be sought by measurements of the total intensity of the Balmer-alpha radiation in coincidence with the n=3 energy loss electrons but there are nine multipoles to be deduced from the angular correlations[13] which would be difficult. It is easier to determine just the $3^2D_j - 2^2P_j$ contribution to the Balmer alpha radiation from angular correlations of the Lyman alpha cascade radiation and the n=3 energy loss electrons. Chwirot and Slevin expressed the angular correlation of the L$_\alpha$ fluorescence in the scattering plane ($\phi_\gamma = 0$ in the collisional reference frame) as

$N_c(\theta_\gamma) \approx 1 + 0.3\ S_{20}(2P) + 0.75\ S_{22}(2P) - 1.5\ S_{21}(2P)\sin2\theta_\gamma + 0.3[3\ S_{20}(2P) - \sqrt{6}\ S_{22}(2P)]\cos2\theta_\gamma$

and in the plane perpendicular to the scattering plane ($\theta_\gamma = \pi/2$ in the collisional reference frame) by

$N_c(\theta_\gamma) \approx 1 - 0.6\ S_{20}(2P) + 1.5\ S_{22}(2P)\cos2\theta_\gamma$.

In these expressions the $S_{20}(2P)$ multipoles refer only to the ensemble of atoms populated via spontaneous decay of the $3^2S_{1/2}$ and 3^2D_j states. They are approximately given by $S_{2q}(2P) = 0.2\ S_{2q}(3D)$ since the 3S atoms contribute isotropically to the coincidence signal. In the above forms the angular correlations are similar to those for direct excitation of the 2^2P_j states but do not reduce to just two independent parameters because of the linear combinations of the $S_{2q}(2P)$ required to describe the data. Initial measurements are shown in fig. 6 for an incident electron energy of 12.27 eV and an electron scattering angle of 20 degrees. A time window of 50 nsec, which is three times the lifetime of the 3D state and about 0.3 times the 3S state lifetime, was used for convenient data accumulation times and a reduction of the isotropic background counts from the 3S state. The data were limited to an angular range of the photon detectors of less than 90° which did not give a good fit of the above angular correlation expressions to the data. Modifications to the apparatus are in progress.

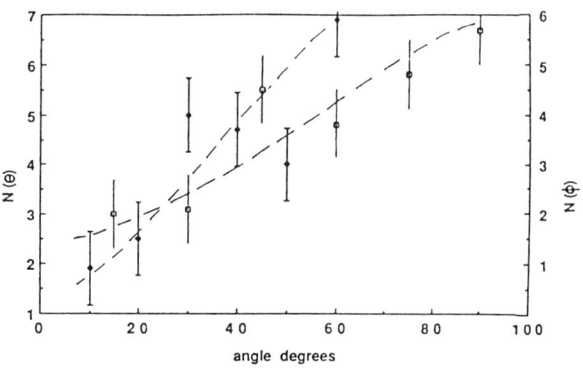

Fig 6. Angular correlations for the detection of the n=3 energy loss electrons in coincidence with Lyman alpha radiation. The incident electron energy was 12.27 eV and the electron scattering angle 20 degrees. The solid points are $N(\phi)$ data.

The angular correlation function for the excited 3^2P_j states of hydrogen giving rise to Lyman beta fluorescence is analogous to the expression for 2^2P_j states of hydrogen.

$N_c(\theta_\gamma) \approx 4 + 3\lambda(1 - 2\lambda)\cos^2\theta_\gamma - 3\sqrt{2}\ R\sin2\theta_\gamma$

yielding values for the parameters λ and R where $\lambda = \sigma_0/\sigma$ and $R = \text{Re}<f_0 f_m^*>/\sigma$. Here f_m are the scattering amplitudes for the excitation of the mth magnetic sublevel of the P state, $\sigma_m = <f_m f_m^*>$ and the total differential cross section $\sigma_0 + 2\sigma_1$. Also measurements of the circular polarization yield the I parameter where $\text{I} = \text{Im}<f_1 f_0^*>/\sigma$. Initial measurements of the in-plane angular correlations at an incident energy of 12.27 eV and an electron scattering angle of 10° have yielded values of $\lambda = 0.96 \pm 0.11$ and $R = 0.04 \pm 0.08$ while a measurement of the circular polarization normal to the scattering plane has yielded $\text{I} = +0.0 \pm 0.1$. These values are about what one expects for a P to S transition at small scattering angles however they have demonstrated the feasibility of further detailed study of these parameters in an energy region close to threshold.

In the quest to determine all the elements of the density matrix for the n=3 state, the study of photon-photon correlations becomes interesting. An experiment detecting coincidences between one energy loss electron and one radiated photon without external fields will give information[12] on state multipoles of rank $0 \leq K \leq 2$. For n=3 excitation L = 0,1 and 2 and K ≤ 4 so that the coincidence detection of the n=3 energy loss electron with both of the sequential cascading photons (n=3→2 and n=2→1) is required for completely specifying the reduced density matrix[14]. As a small step in this direction one can detect just the two photons in coincidence. The initial state of the excited atoms is axially symmetric with respect to the incident electron beam and only the angle between the emission directions of the two photons is of concern. Only the zeroth components of the state multipoles are non zero where

$<T(00)_{00}> = \int \sigma_{00}\ d\Omega$
$<T(20)_{20}> = \int <f_{20}f_{00}^*>d\Omega$
$<T(22)_{00}> = 1/\sqrt{5} \int (2\sigma_{22} + 2\sigma_{21} + \sigma_{20})\ d\Omega$
$<T(22)_{20}> = \sqrt{2}/\sqrt{7} \int (2\sigma_{22} - \sigma_{21} + \sigma_{20})\ d\Omega$
$<T(22)_{40}> = \sqrt{2}/\sqrt{35} \int (\sigma_{22} - 4\sigma_{21} + 3\sigma_{20})\ d\Omega$

and where the integration is over the solid angle of detection. In principle a (γ_1,γ_2) measurement will yield a value of $<T(22)_{40}>$. Such a measurement has been attempted for an incident electron energy of 12.27 eV at which cascading radiation will not obscure the n=3 information. A sinusoidal variation of the coincident count rate as a function of the angle between the two photons was observed however a least squares fit of the quantum expression yielded a value for $<T(22)_{40}>$ of 0.0 ± 0.2. Further data need to be collected in order to reduce the statistical uncertainty of the data.

In conclusion, a variety of measurement schemes have been discussed and used to obtain more information than previously available, particularly in the threshold region of the n=2 and 3 states. The collection of 'threshold electrons' has led to quantitative measurements on the n=2 and 3 total cross sections and the technique can be extended to higher states. The various coincidence detection schemes using energy loss electrons, direct decay and cascade photons and the lifetime of the excited states has led to separation of the degenerate angular momentum components of the n=3 level and the technique may be extended at least to the n=4 level.

This work was supported by The University of Western Australia and the Australian Universities Research Grants Scheme.

References

1. H S W Massey, E Burhop and H B Gilbody *Electronic and Ionic Impact Phenomena*, Oxford Press (1969)
2. G C King, S Trajmar and J W McConkey in preparation (1989)
3. I Humphrey, J F Williams and E L Heck *J Phys B: At Mol Phys* **20** 367 (1987)
4. J F Williams and E L Heck *J Phys B: At Mol Phys* **21** 1627 (1988)
5. S Cvejanovic and F H Read *J Phys B: At Mol Phys* **7** 1841 (1974)
6. J F Williams *J Phys B: At Mol Opt Phys* **21** 2107 (1988)
7. A Adams *Proc 7th Yugoslav. Summer School on the Physics of Ionized Gases*, 35 (1974)
8. J F Williams *J Phys B: At Mol Phys* **14** 1197 (1981)
9. J Callaway *Phys Rev A* **37** 3692 (1988) and to be published
10. B L Moiseiwitsch and S J Smith *Rev Mod Phys* **40** 238 (1968)
11. I C Percival and M J Seaton *Phil Trans Roy Soc (London)* **A251** 113 (1958)
12. K Blum *Density matrix theory and applications*, New York: Plenum (1980)
13. K Blum, E E Fitchard and H Kleinpoppen *Zeit Phys* **A287** 137 (1978)
14. E L Heck and J P Gauntlett *Phys B: At Mol Phys* **19** 3633 (1986)
15. S Chwirot and J Slevin *J Phys B: At Mol Phys* **18** L881(1985); **20** 3885 (1987), 6139 (1988)
16. H B van Linden van der Heuvell, E M van Gasteren, J van Eck and H G M Heideman *J Phys B: At Mol Phys* **16** 1619, 2667 (1983)

SIMULTANEOUS ELECTRON-PHOTON EXCITATION EXPERIMENTS

W.R. Newell

Department of Physics and Astronomy,
University College London, Gower Street,
LONDON WC1E 6BT, U.K.

In this paper the progress made in the experimental study of laser-assisted excitation of discrete atomic states is considered. The simultaneous electron photon excitation of the 2^3S level in helium is reviewed together with current progress in the work on other atoms and molecules.

INTRODUCTION

The general area of laser-assisted electron scattering has, in the experimental domain, been restricted to super-elastic electron scattering of sodium and barium which have been pumped to their first excited levels using low-power tunable dye lasers (see Hertel & Stoll[1] and Register et al.[2]). In this type of collision the projectile electron (energy E_i) is scattered from a discrete bound state and the quantum of energy ($h\nu$) transferred to the final electron energy ($E_f = E_i + h\nu$) was firstly resonantly absorbed by the target atom. These state selection experiments are described by the equation

$$e(E_i) + ATOM \to ATOM^* + e(E_i + h\nu) \quad (1)$$

where of course the energy resolution (ΔE) of the electron scattering apparatus must be less than $h\nu$. In a different type of experiment we can measure the transfer of photons to and from electrons while they are undergoing an elastic collision, described by the equation

$$e(E_i) + nh\nu + ATOM \to e(E_i \pm mh\nu) + (n-m)h\nu + ATOM. \quad (2)$$

In this case the atom is not excited by the projectile nor by the photons but the electron does absorb m photons. In this type of scattering the atom acts only as a third body to conserve the energy and momentum in the reaction. These free-free or Inverse and Stimulated Bremmstrahlung transitions have been reported by several authors (eg. Weingartshofer et al.[3]) and require high-power laser intensities of $10^8 \to 10^{12}$ watts/cm^2 in order to achieve a multi-photon (m>1) transfer to the scattered electrons. Single-photon transfer to the scattered electrons is achieved with the lower laser intensity of $10^4 \to 10^5$ watts/cm^2. Andrick & Langhans[4], in a pioneering experiment demonstrated this process for low energy (11eV) electron-argon scattering in the field of a continuous-wave (cw)50W CO_2 laser.

In free-free scattering the atom plays only a passive role but there exists an excitation process in which the incident electron combines with a photon to cause an excitation of the atom. This paper is concerned with a review of the progress made in the experimental demonstration of this simultaneous electron-photon excitation. This particular reaction is described by the equation

$$e(E_i) + nh\nu + He(1^1S) \to \quad (3)$$
$$e(E_j = E_i - \Delta E\ h\nu) + (n-1)h\nu + He(2^2S)$$

in which the excitation of the 2^3S stationary state in helium, which lies at ΔE above the 1^1S ground state, is accomplished by the absorption of one quantum of radiation $h\nu$ from a CO_2 laser field combined with a simultaneous inelastic electron scattering in which the electron provides the energy decrement ($\Delta E - h\nu$) required to excite the 2^3S state. Since no stationary state exists between the 1^1S and 2^3S levels this excitation must proceed by a virtual interaction.

Electron-atom scattering in the presence of an external electromagnetic radiation field was initially studied by Goppert-Mayer[5] in association with her theoretical work on two-photon excitation. However, it is only by combining the more recent development of laser technology and the expertise of high-resolution electron scattering that an experimental study of the simultaneous interactions of bosons and fermions with a discrete atom can be investigated. Such interactions are of practical importance in the heating of plasmas by radiation and laser-induced gas breakdown phenomena, in addition to being of fundamental interest in the understanding of three-body interactions and the relative coupling between radiation fields and particles. Measurements of simultaneous electron-photon excitation (SEPE) are a step towards the understanding of important processes in astrophysics and laser-surface interactions.

THE PHYSICAL PROCESS

Symbolically the problem of electron atom scattering in a laser field can be represented by the diagram in Figure 1 which will apply to any universal treatment. However any real theory of

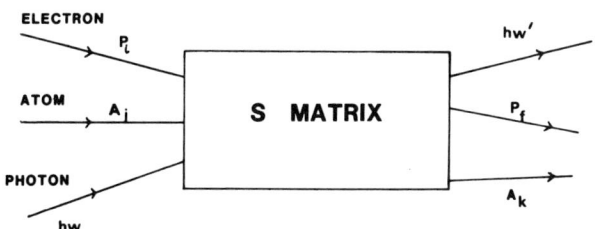

Figure 1

this process must consider all the parameters involved namely, the laser frequency, polarisation and power, the electron energy and spin, and the nature of the atomic target. In addition, we must ask is the scattering instantaneous or time-evolved and in what sense do these particles interact in pairs, ie. (e,hν), (A,hν), (e,A) and is there any heirarchy of pair interactions. A rigorous solution should account for all of this; however even the (e,A) interaction is difficult to treat accurately, especially at the threshold, and the addition of radiation will certainly increase the complexity. A single direct picture of the coupling between the radiation field, electron and atom is given by the four Feynman diagrams in Figure 2 where each diagram represents one particular scattering amplitude.

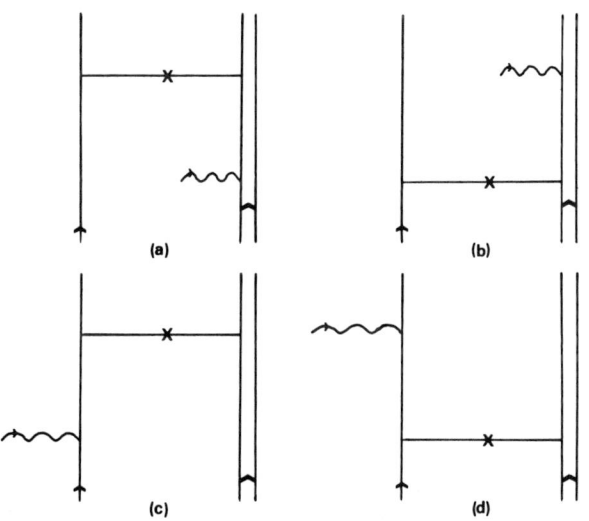

Figure 2

In fact the complete transition amplitude is obtained for a sum of all the Feynman diagrams in Figure 2 that represent first the atom interacting with the radiation field followed by the electron interaction, Figure 2(a), and, alternatively, the electron-atom interaction preceding the interaction of the radiation field, Figure 2(b). In addition Figures 2(c) and (d) represent the class of interactions in which the free incident electron interacts with the radiation field. It is known that, under the conditions of high incident energy (>100eV) the transitions between stationary states of the same parity, the diagrams (c) and (d) can be neglected (Raman & Faisal[6]). The role of these four Feynman diagrams in F-F scattering have been studied by Dubois et al.[7] in atomic hydrogen while the SEPE cross-sections in atomic hydrogen and helium have been calculated by Jetzke et al[8,9] but only at high-laser powers and for relatively high-incident electron energies. In the present work, where we measure the threshold excitation of a spin exchange mechanism, we require detailed close-coupling calculations of the onset of the cross-section. The final transition amplitude is perhaps enhanced in this process for, in exciting the He(2^3S) state, we are also utilising an exchange interaction. A consideration of exchange in free-free processes has only recently been attempted (Ferrante et al.[10], Mandal et al.[11,12]) for the atomic hydrogen 1S-1S transition. In very intense laser fields (>10^8V cm^{-2}) with high-incident electron energies (100eV) there is evidence for the suppression of the exchange amplitude since the incident electron seems to become distinguishable from the target electron but such a suppression is highly dependent on the alignment of the laser field polarisation. Such an influence upon the exchange amplitude is clearly of importance in our study of the simultaneous electron-photon excitation of the He(2^3S) state. However, the present experiments utilise both low-intensity (10^5W cm^{-2}) and high-intensity (10^8W cm^{-2}) laser fields and are performed at low-incident electron (E_i<20eV), and none of the present theories is therefore suitable for describing the role of the laser field in modifying the exchange process studied in these experiments.

Since the photon energy (hν=117meV) is low the experiments are described by the 'soft' photon approximation (Low[13] and Kroll & Watson[14]) in which E_i>>hν and the scattering interaction time (h/E_i) is very much less than the time (ν^{-1}) necessary for photon absorption. Consequently the only type of interactions that are observable are described by Feynman diagrams 2(c) and 2(d) in which the projectile electron is virtually dressed by the laser field. For low-radiation frequencies the atoms are 'transparent' to the laser field. Figure 2 can be generalised to include higher orders of radiation coupling in the electron-radiation interaction. The solution to the Schrödinger equation for the interaction of a plane electromatic wave and a charged particle is the Volkov[15] solution and it can be visualised using Feynman diagrams as in Figure 3 where ▯ represents the Volkov solution. It is clearly seen that this consists of an

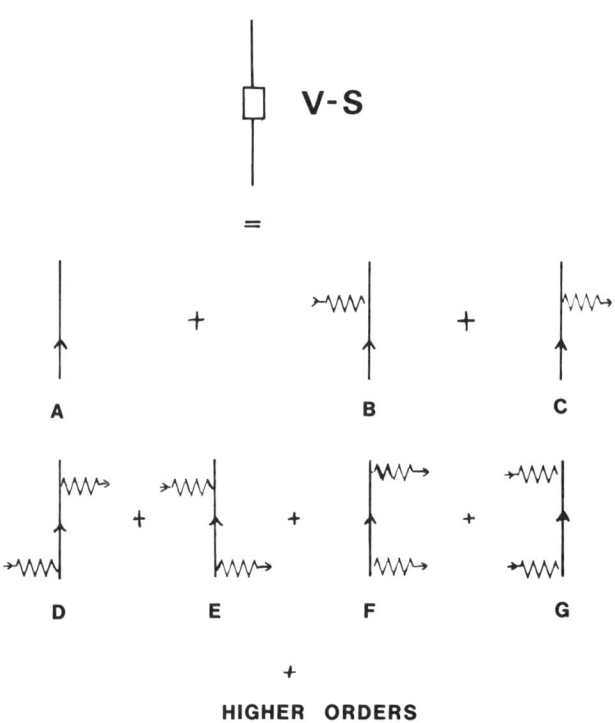

Figure 3

infinite number of amplitudes given in increasing order of photon coupling, ie. A is zero order, B and C first order, D, E, F, G second order etc. Due to the non-conservation of energy and momentum between electrons and photons all the interactions are VIRTUAL. The interactions only become REAL when there is a third body present which will conserve the energy and momentum. The third body is a potential $V(r)$ which of course will also have different orders of interaction with the electron. If we consider the lowest order (ie. First Born) then the solution of the photon-assisted collision can be written as

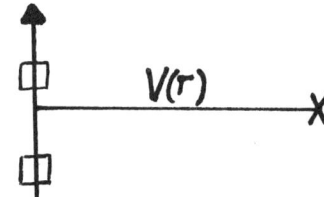

where ▯ is the dressed electron propagator; virtually dressed to all orders in the radiation field. If in the collision N photons are transferred via the real states formed due to the presence of $V(r)$ then the cross-section $(\frac{d\sigma}{d\Omega})^N$ can be written as

$$\left(\frac{d\sigma}{d\Omega}\right)^N = \frac{P_f}{P_i} J_N^2 \left(\frac{E \cdot Q}{w^2}\right) \left(\frac{d\sigma}{d\Omega}\right)^{BORN} \quad (4)$$

where P_f is final electron momentum P_i is initial electron momentum and Q the momentum transfer. E is electric field strength and w the frequency of the radiation and J is a Bessel function. It should be noted that this formula is correct to all orders in radiation field with no restriction placed on laser frequency or intensity. The separation of the scattering process and the influence of the radiation field is a consequence of the soft photon approximation. Consequently providing $(E \cdot Q / w^2) < N$ then Bessel function can be written as $J_N(y) \sim \frac{1}{N!} (\frac{1}{2}y)^N$ which gives the correct low field result, ie. identical to the lowest order treatment. The only restriction implied in equation (4) is on the order of the scattering which in this case is the First Born approximation.

In order to improve on equation (4) we must include the higher-order terms of $V(r)$. Inclusion of the higher-order Born terms can easily be included in schematic form as given in Figure 4(A). Mathematical summation of this series

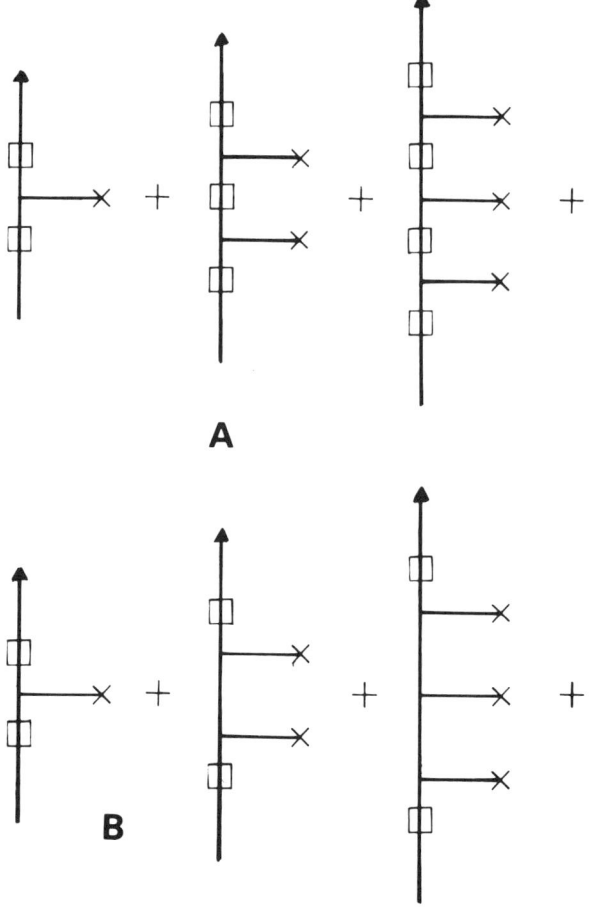

Figure 4

would give an exact result but it is difficult. However, by replacing the dress electron propagators (other than the initial and final operators) by free electron propagators in Figure 4(B) we obtain

$$\left(\frac{d\sigma}{d\Omega}\right)_N = \left(\frac{P_f}{P_i}\right) J_N^2 \left(\frac{E \cdot Q}{w^2}\right) \left(\frac{d\sigma}{d\Omega}\right)^{EXACT} \quad (5)$$

where equation (5) differs from (4) in that we now have the EXACT scattering cross-section but one not associated with photon transfer. The physical condition implied in replacing Figure 4(A) by Figure 4(B) is that w is small.

Consequently equation (4) is correct to all orders in photon transfer but only the lowest order in scattering; whereas equation (5) is correct to all orders in scattering but only lowest orders in photon transfer.

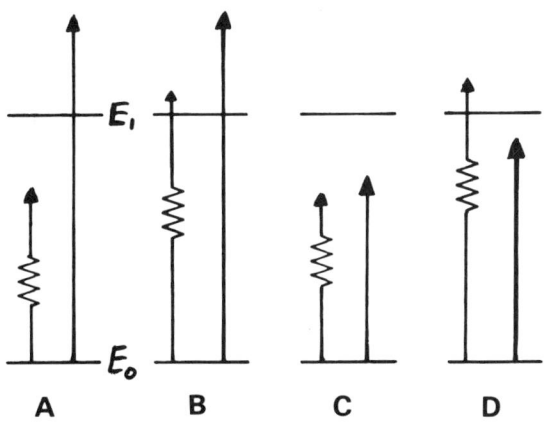

Figure 5

Figure 5 illustrates the non-resonant process which we can consider as an extension of the free-free scatterkng given by equations (4) and (5). In the reaction, equation (3), it is possible for the transition to occur without the electron losing the energy $\Delta E = (E_1 - E_0)$ providing the photon(s) can be used to 'bridge' the energy gap. In Figure 5 no energy, neither $nh\nu$ nor E_0, is resonant with E_1. In 5(A) and 5(B) the electron can separately excite E_1 but in no case can $nh\nu$. In 5(B) and 5(D) the electron is superelastically scattered whereas in 5(D) and 5(B) the electron loses energy. Only in the case of 5(C) does it require both the electron (E_0) and the photons ($nh\nu$) to excite E_1. This is the process of simultaneous electron photon excitation.

ELECTRON SCATTERING APPARATUS

The main requirement of the electron scattering apparatus is that the energy spread ($\Delta\epsilon$) in the electron beam is $\Delta\epsilon \ll h\nu$ in order to resolve the effects. This is easily accomplished using a filtered electron gun (L1→L6) and hemispherical monochromator, Figure 6, followed by an accelerator lens stack (L10-L16) which sets the electron beam energy and focusses it onto the intersection of the gas and laser beams (see Mason & Newell[16]).

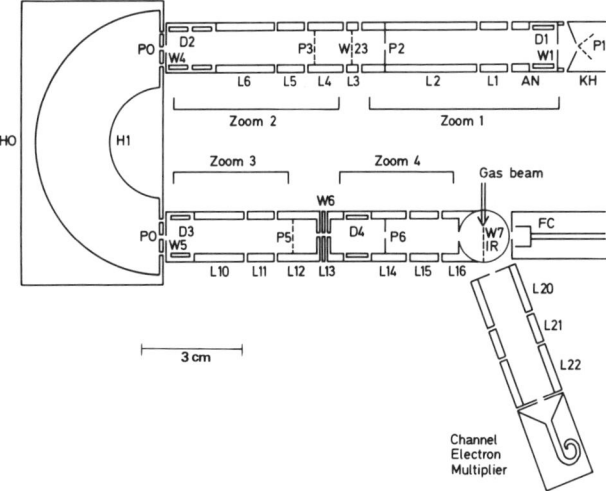

Figure 6

Electron Scattering Apparatus

Detection of the SEPE can be made by observing the scattered electrons which should consist of satellite peaks at $19.817 \pm n(0.117)$eV about the inelastic energy-loss peak at 19.817eV associated with the direct excitation of the He(2^3S) state. However, it should be noted that at the threshold when $E_i < 19.817$eV but $E_i + h\nu > 19.817$eV there is no inelastic peak and the simultaneous signal should appear superimposed upon the background. No experiment demonstrating this excitation process using energy-loss spectroscopy has as yet been performed. A more appealing prospect is to detect the excited atom itself since the detection of metastable flux is a measure of the total excitation cross-section not just of the differential cross-section (Mason & Newell[17,18]). Consequently the signal observed will be 2-3 orders of magnitude higher than for any angular-selected electron scattered signal ($\theta = 0°$ to $180°$). This is easily accomplished by pulsing the electron beam and analysing the scattered particle flux (electrons, photons and metastable atoms) using the time of flight detector (L20-L22). This makes it possible to examine the metastable flux independently of any photons and electrons present thus improving its signal-to-noise ratio and so reducing the total counting times required to detect the simultaneous electron-photon signal.

Observation of metastable He(2^3S) atoms is particularly favourable because He(2^3S) atoms have a large stored energy (19.817eV) and hence are easily detected by Penning ionisation of surfaces with a high efficiency (Allison[19]) using a channel electron multiplier. Since this metastable state is very long-lived ($\sim 10^4$s) lifetime) there is no appreciable decay during the flight time ($\sim 40\mu$s) to the detector. The detection method has also subsequently been used by Wallbank et al.

LASER

In all the experiments to date Mason & Newell[20] and Wallbank et al.[21] a CO_2 laser source of radiation has been employed. In the case of Mason & Newell it was a c.w. 300W laser whereas Wallbank et al. employed a pulsed CO_2 source with an intensity of $\sim 10^8$ W/cm^2. A typical arrangement is shown schematically in Figure 7. An early attempt at the experiment by Khakoo et al.[23] employed a pulsed Nd. laser with an intensity of 10^{10} W/cm^2.

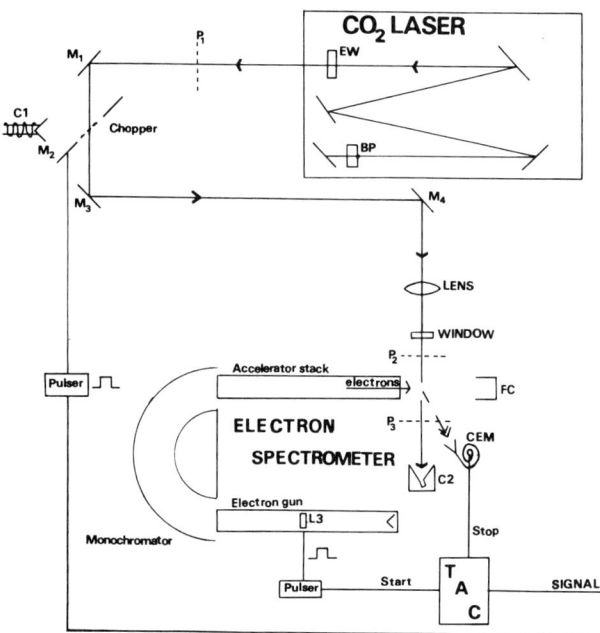

Figure 7
Schematic of Experiment

Observation of the SEPE process in the total-cross-section channel will be more easily detected at the threshold for the excitation of the He(2^3S) state when the SEPE process will produce metastable atoms even when the incident electron energy E_i is less than E_{ex}(19.817eV). Such an observation is assisted by a sharp threshold in the cross-section and the He(2^3S) forbidden excitation cross-section rises quickly to a maximum in 450meV. Consequently there is a gain in the metastable production given by the ratio of the cross-section (σ_S) at $E_i + h\nu$ to that (σ_D) at E_i where σ_S is the cross-section for simultaneous excitation and σ_D is the cross-section for direct excitation. This gain effect arises when the effective incident electron energy is increased by $h\nu$(=117meV) and that fraction of the incident electron beam which absorbs the photon then produces metastable atoms at a higher point of the cross-section curve than they would had they not absorbed a photon.

The effective cross-section gain is then σ_S/σ_D, which in the ideal case (where the background $S_B=0$) is infinite, but in the current measurement where $S_B=0.05 \times 10^{-18}$ cm^2 the gain is is 42. Consequently observation of SEPE is enhanced where this gain factor is greatest and where the direct cross-section has its largest rate of change with electron energy, and this is maximum in the threshold region.

The interaction region was defined by the overlap of the electron, laser and gas beams as viewed by the time-of-flight section of the apparatus. The general method of signal acquisition used was by TOF spectra of the metastable species in helium produced at some fixed incident electron energy E_i. However, in order to distinguish between He(2^3S) atoms produced by direct excitation (by electron scattering only) and by simultaneous excitation (by electron and photon scattering), it is also necessary to pulse the laser beam. This was accomplished by Mason & Newell by moving the mirror M_2 (with a period of 15S) whereas the laser employed by Wallbank et al. had a natural pulse repetition rate of 10pps. The pulsing was also used to route the T.O.F. signal and the difference in these TOF spectra gives a direct measure of the simultaneous electron-photon excitation.

The signal difference due to SEPE was carefully investigated by Mason & Newell to establish that it was not due to an experimental artifact. The SEPE signal was only detected when all three beams were present and detailed tests were performed using different combinations of the three beams to understand in details all possible sources of spurious signal. These included contact potential charges, effects of gas beam heating, laser and electron beam stability and beam overlap geometry.

RESULTS

To date SEPE experiments have been reported by two groups of workers; Mason & Newell and Wallbank et al. In essence the technique initially used by Mason & Newell was also employed by Wallbank et al., namely T.O.F. detection of the excited helium atoms which recoil through an angle of $\sim 17°$ from the primary gas beam direction. The main difference between the experiments is in the laser intensities employed with Mason & Newell using a single mode c.w. 300W CO_2 laser (intensity $\sim 10^4$ W/cm^2) polarised along the electron scattering angle of $\theta = 135°$ whereas Wallbank el al. used a pulsed multimode CO_2 laser (intensity $\sim 10^8$ W/cm^2) with polarisation and pulse repetition rate unspecified. Consequently the two experiments measure single photon and multi-photon effects respectively.

The simultaneous electron-photon excitation cross-section measurements (Mason & Newell) of the 2^3S state are shown in Figure 8, together with the direct cross-section (σ_D). The threshold

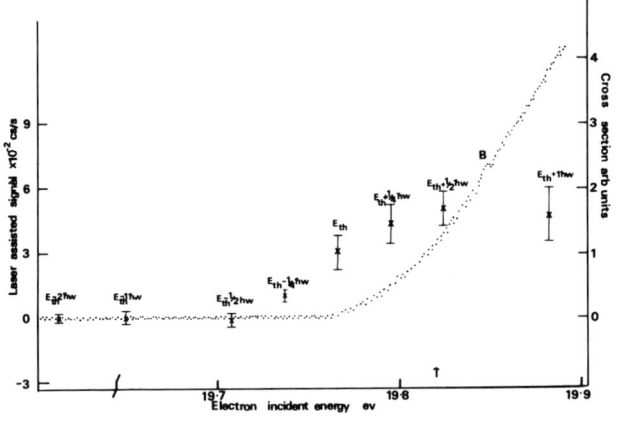

Figure 8

SEPE signal (X) compared with the direct excitation curve (....)

($E_{th} = 19.757$ eV) of the 2^3S state was taken as the point where the direct signal was first observed and is used as a reference point for the incident electron energies employed. The experiment was performed at incident electron energies E_i of $2h\nu$, $1h\nu$, $0.5h\nu$ and $0.25h\nu$ below E_{th}, at E_{th} and at $0.25h\nu$, $0.5h\nu$ and $1h\nu$ above E_{th}.

A completely resolved threshold for SEPE is reported between $E_{th} - 0.5h\nu$ and $E_{th} - 0.25h\nu$, not as might be expected at $E_{th} - 1h\nu$. The failure to observe a threshold at $E_i - 1h\nu$ is not ascribed to any physical process (eg. a Stark shift) but is purely a consequence of the signal-to-noise ratio. At an incident energy $E_i - E_{th} - 2h\nu$ it is readily understood that one-photon absorption will produce no SEPE of He(2^3S) state and that the laser 'on' - laser 'off' difference should be zero; this is as observed in Figure 8. At $E_i = E_{th} - 1h\nu$ one-photon absorption will produce a SEPE signal but with a cross-section indiscriminate from the 'noise' concurrent with the background and only

when $E_i - E_{th} - 0.5h\nu$ and $E_i = E_i - 0.25h\nu$ is the SEPE cross-section such that it is sufficient to overcome the background 'noise' of the apparatus. A similar dependence is also reported by Wallbank et al. Figure 9, only in this case the influence of multiphoton transfer to the projectible electron shows a SEPE cross-section rising at $E_{th} - 4h\nu$. This multiphoton SEPE curve has a sharp maximum at E_{th} followed by a rapid fall to 'zero' at $E_{th} + h\nu$. In contrast, Figure 8 shows a more gradual increase with a smooth maximum at $E_{th} + 0.5h\nu$ followed by a slight turnover. Mason & Newell estimate the simultaneous to direct excitation to be (1.4×10^{-3}) at an incident energy of $E_{th} + 0.25h\nu$. Such a value is perhaps larger than that expected from free-free experiments, but the present experimental method is expected to give an upper limit to the cross-section. This value contrasts with the value of 210^{-2} given by Wallbank et al. which indicates the effects of non-linear features present in their experiment.

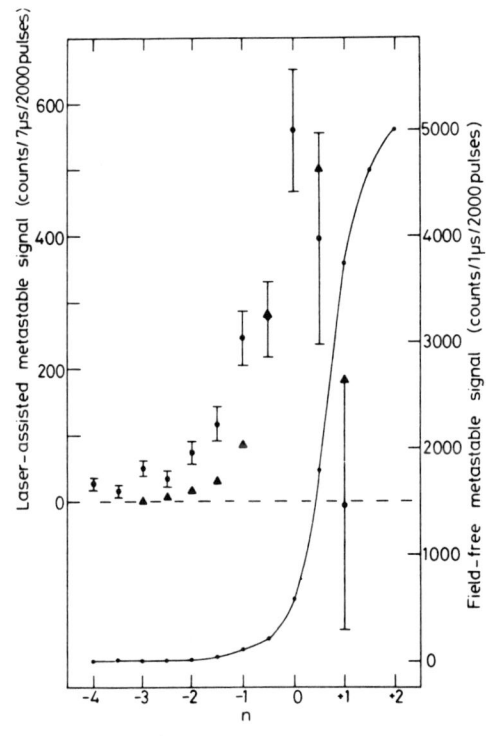

Figure 9

SEPE signal (ϕ) compared with the direct signal (—•—•—). Note that the zero is displaced for clarity.

It is supposed that this SEPE process proceeds totally without any target dressing effects but the process may be influenced by the presence of resonance effects close to the threshold. There is evidence of a small resonance feature labelled B in Figure 8 in the direct cross-section and, although the nature of this feature remains unclear, the presence of a resonance (particularly a wide-shaped resonance) could affect the SEPE cross-section in a more dramatic manner than the gain from the direct cross-section. However, the influence of the laser polarisation and the rotation of the momentum transfer vector must be fully understood before the influences of resonance can be studied.

Using an extension of the Kroll & Watson[14] formulae and the field-free total electron impact excitation cross-section of the He(2^3S) given by Fon et al.[22], Geltman & Maquet[24] have calculated the laser-assisted excitation cross-sections for the conditions prevailing in both experiments. In order to provide an accurate comparison the calculated SEPE curves were averaged over a Maxwell distribution function of FWHM of 35meV (Mason & Newell) and 45meV (Wallbank et al.) In the case of Wallbank et al. the calculated SEPE cross-section is lower than the direct cross-section above $E_{th}+1h\nu$ and this derives from the large argument in the Bessel function produced by the high-laser intensity. Consequently several orders of Bessel functions are involved. In contrast, the lower laser power used by Mason & Newell only invokes the use of the zero-order Bessel function and the argument is such that no zeros are accessed. In this case there is only a smooth charge in the calculated SEPE cross-section.

Extension of this work to atoms other than helium has been undertaken for neon by Wallbank et al. in the excitation of the $^1S \to {}^3P$ transition. No SEPE signal was observed. Mason & Newell have also attempted to detect SEPE in the hydrogen excitation $^1\Sigma \to {}^3\pi$; again no signal was detected. In both these cases the direct cross-section (σ_D) did not rise as quickly as in helium which would result in a reduced gain. However it is probable that a selection rule is operative and although both transitions are singlet → triplet neither is of an even-even parity change as is the case in helium.

An attempt has been made (Mason & Newell) to measure SEPE in doubly-excited negative ion resonances ($2s2p^2$) ^2D and ^2P in helium which occur at 58.3 and 57.2eV respectively. These resonances decay into the metastable flux in which they are detected as features on the direct signal. A positive signal has been obtained but the statistics are poor due to the low gain of 1.1 which applies. This work is perhaps more suitable to a pulsed laser system.

FUTURE

A detailed investigation of the polarisation dependence of the scattering dynamics is now required as this is the main reason for investigating SEPE in the soft photon limit. In addition, it is important to change the nature of the atomic target to systems which have a large dynamic polarisability or permanent dipole moment. This will enhance the coupling of the projectile electron to the target and consequently increase the recoil coupling. It is now timely to employ shorter wavelengths in these experiments eg. 1μm although the reduced signal (10^{-4}) inherent in such a change from CO_2 radiation will require a different experimental approach.

1. I.V. Hertel and W. Stoll, Adv.At.Mol.Phys. 13, 113 (1977).
2. D.F. Register, S. Trajmar, G. Csanak, S.W. Jensen, M.A. Fineman and R.T. Poe, Phys.Rev.A. 28, 151 (1983).
3. A. Weingartshofer, E.M. Clarke, J.K. Holmes and C. Jung, Phys.Rev. 19, 2371, (1979).
4. D. Andrick and L. Langhans, J.Phys.B: At.Mol. Phys. 18, 4135, (1985).
5. M. Goppert-Mayer, Ann.Phys.,Lpz. 9, 273 (1931).
6. N.K. Raman and F.H.M. Faisal, J.Phys.B: At. Mol.Phys. 9, L275 (1976).
7. A. Dubois, A. Maquet and S. Jetzke, Phys. Rev. A 34, 1888 (1985).
8. S. Jetzke, J. Broad and A. Maquet, J.Phys. B: At.Mol.Phys. 20, 2887 (1987).
9. S. Jetzke, F.H.M. Faisal, R. Hippler and M.O. Lutz, Z. Phys. A 315, 271-6 (1984)
10. G. Ferrante, G. Leone and F. Trombetta, J.Phys.B: At.Mol.Phys. 15, L475 (1982).
11. S.K. Mandal, M. Basu and A.S. Ghosh, J.Phys. B: At.Mol.Phys. 19, 3333 (1986).
12. S.K. Mandal, M. Basu, P.S. Majumdar and A.S. Ghosh, J.Phys.B: At.Mol.Phys. 18, 3339 (1985).
13. F.E. Low, Phys.Rev. 110, 974 (1958).
14. N.M. Kroll and K.M. Watson, Phys.Rev. A. 8, 804 (1973).
15. D.M. Volkov, Z.Physik, 94, 250 (1935).
16. N.J. Mason and W.R. Newell, J.Phys.E: Sci. Instrum. 19, 722 (1986).
17. N.J. Mason and W.R. Newell, J.Phys.B: At. Mol.Phys. 20, 1357 (1987).
18. N.J. Mason and W.R. Newell, J.Phys.B: At. Mol.Phys. 20, 3913 (1987).
19. W. Allison, Ph.D. Thesis University of London (1978).

20. N.J. Mason and W.R. Newell, J.Phy.B: At.Mol. Opt.Phys. 22, 777 (1989).
21. B. Wallbank, J.K. Holmes, L. Le Blanc and A. Weingartshofer, Z.Phys. D 10, 467 (1988).
22. W.C. Fon, K.A. Berrington, P.G. Burke and A.E. Kingston, J.Phys.B: At.Mol.Phys. 14, 2921 (1981).
23. M.A. Khakoo, W.R. Newell, W.T. Toner and R.W. Eason, Rutherford Appleton Laboratory Annual Report To The Laser Faculty Committee pp608-11 (1982).
24. Private Communication.

Relativistic (e,2e) processes on inner shells of heavy atoms

J. Bonfert, H. Graf, and W. Nakel

Physikalisches Institut, Universität Tübingen
D-7400 Tübingen, West Germany

The investigation of electron impact ionization by the electron-electron (e,2e) coincidence technique has been until recently restricted to the nonrelativistic energy region (typically: 10 eV to 10 keV). We report here on the continuation of an (e,2e) experiment using a 500 keV electron beam impinging upon thin-foil targets. Absolute triply differential cross sections and angular distributions for K-shell ionization of silver and gold were measured and are compared with theoretical predictions.

I. Introduction

Experiments on electron impact ionization of atoms in which both the scattered and the ejected electron are detected in coincidence after angular and energy analysis have provided an increasing understanding of this fundamental collision process. Since the first coincidence measurement of Ehrhardt and co-workers[1] and of Amaldi et al.[2] the investigation in this field of so-called (e,2e) processes has expanded rapidly (for a review see Ehrhardt[3] and McCarthy and Weigold[4]). However, regarding the incident electron energies and the binding energies of the atomic electrons, until recently only the nonrelativistic region has been studied (typically in the order of 10 eV to 10 keV). Work in the relativistic region began in 1982 with coincidence measurements of Schüle and Nakel[5] and later of Ruoff and Nakel[6] at an incident electron energy of 500 keV. There it was possible to measure the absolute triply differential cross section for K-shell ionization of atoms with higher atomic numbers (Z=47 and 73). Inner shells of heavy atoms are particularly suitable for studying the collision dynamics, being comparatively little influenced by problems of atomic structure. In this paper, we report on the continuation of these (e,2e) experiments at 500 keV with improved technical facilities. Absolute triply differential cross sections and angular distributions for K-shell ionization of silver and gold were measured and compared with theoretical predictions.

The triply differential cross section $d^3\sigma/d\Omega_1 d\Omega_2 dE$ (TDCS) represents the probability that in a single ioniza-

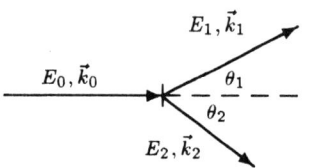

FIG. 1. Schematic diagram of the coplanar geometry showing the energies, momenta, and angles of the incoming and the two outgoing electrons, respectively, indexed 0, 1 and 2.

tion collision an incident electron of energy E_0 and momentum $\vec{k}_0$ will produce two outgoing electrons having energies E_1 and E_2 and momenta $\vec{k}_1$ and $\vec{k}_2$, emitted, respectively, into the solid angles Ω_1 and Ω_2 centered about the directions θ_1 and θ_2 (Fig. 1). In our experiments the electron directions are coplanar so the azimuthal angle is zero.

Since the collision is fully determined with respect to the kinematics of all particles (except for the spins), the measurements provide a stringent test of theoretical predictions without averaging over unobserved parameters. In the TDCS measurements one selects particular electron shells by resolving their ionization energies. The energy conservation in the collision process demands that $E_0 - E_B = E_1 + E_2$ where the excess energy $E_0 - E_B$ can be shared between the kinetic energies E_1 and E_2 of the two outgoing electrons (E_B is the binding energy). The ion recoil energy is negligible because of the smallness of the electron to ion mass ratio. The expected kinematical curves are shown in Fig. 2. The position of the lines of the particular shells are given by the respective binding energies. The ability to separate the lines exper-

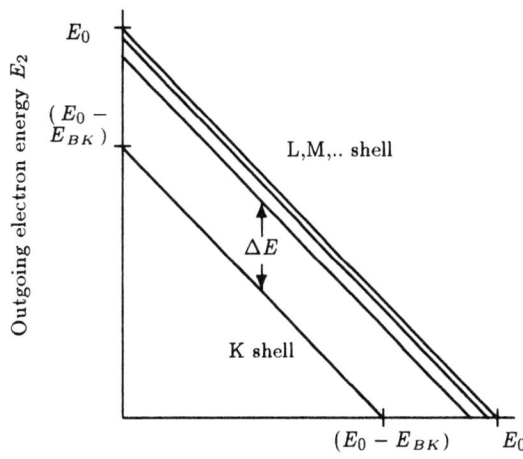

FIG. 2. Kinematics of the ionization process as a function of the energies E_1 and E_2 of the outgoing electrons (arbitrary units, not to scale). The position of the lines of the particular shells is given by the respective binding energies (E_0: incident energy, E_{BK}: binding energy of the K shell). For atomic number $Z = 79$, $\Delta E = 66$ keV; for $Z = 47$, $\Delta E = 22$ keV.

imentally depends on the energy resolution of the detectors.

Since the cross section depends not only on a description of the ionization mechanism but also on the structure of the target, there is another category of electron-electron coincidence experiments which aim to study target electronic structure. Under selected kinematical conditions the electron momentum distributions for the individual orbitals are attainable. This so-called wave function mapping is well established in the nonrelativistic energy region.[4] With higher incident electron energies it should be possible to study inner shells of heavy elements and the pertinent relativistic effects.

A theoretical treatment of electron impact ionization of inner shells at higher energies has to use relativistic interaction, as well as relativistic wave functions. This introduces an additional challenge because, again, in the nonrelativistic region, there currently exists no rigorous theory of the electron impact ionization. The reason for this is found in the three-body nature of the electron ionization process, where in the final state one has three particles (ion and two electrons) interacting via the long-range Coulomb potential.

Additionally, the problem is complicated by the fact that in testing the theory it is usually unknown to what extent disagreement between theory and experiment results either from an inapplicable collision model or from a not adequate description of the target atom (except for hydrogen). Increasing understanding is achived by an iterative interaction of experiment and theory.

To date, there is no fully relativistic calculation available for the TDCS. We compare our measurements with calculations of Jakubaßa-Amundsen[7] and of Das and Konar.[8]

II. Experimental

II.1. Electron beam

The primary electron beam of 500 keV is produced by a van de Graaff generator and focused to a 1 mm diameter spot on the target foil placed at the center of a vacuum chamber (80 cm diameter). The beam divergence is about ±0.2°. The unscattered part of the beam is collected by a Faraday cup.

II.2. Electron spectrum analyzers

Since the outgoing electrons of interest arise in one ionization process so they arrive with a well defined time difference at their respective detectors. Standard coincidence techniques are used to select the (e,2e) events from the many other uncorrelated electrons. However, the electrostatic spectrum analyzers used to date in nonrelativistic (e,2e) measurements (see for instance Lahmam-Bennani[9]) are not suitable for energies in the order of 100 keV. Therefore, in our laboratory two types of electron spectrum analyzers have been developed which are suitable for the special experimental problems (and described in detail in a paper of Ruoff et al.[10]).

The first type consists of a Si surface barrier (SB) detector combined with a nondispersive and doubly focusing magnet i.e. a triply focusing magnet. The magnet was inserted between the defining aperture and the detector in order to eliminate the numerous elastically scattered electrons but to transmit the inelastically scattered electrons within a broad energy range. These spectrometers had been used in energy sharing (e,2e) experiments[5,6] using a two parameter coincidence technique. However, the moderate energy and time resolution of the SB detectors and the low energy tails in the response function are disadvantageous for certain applications. Therefore, we have developed a second type of spectrum analyzers which consists of a magnet for the energy analysis and a plastic scintillation detector for a good time resolution. These analyzers have been used for the measurements presented here and will be described in brief.

The magnet consists of a doubly focusing homogenous sector field shaped by an iron core with a deflection angle of 141°(Fig. 3). Its medium plane is identical with the

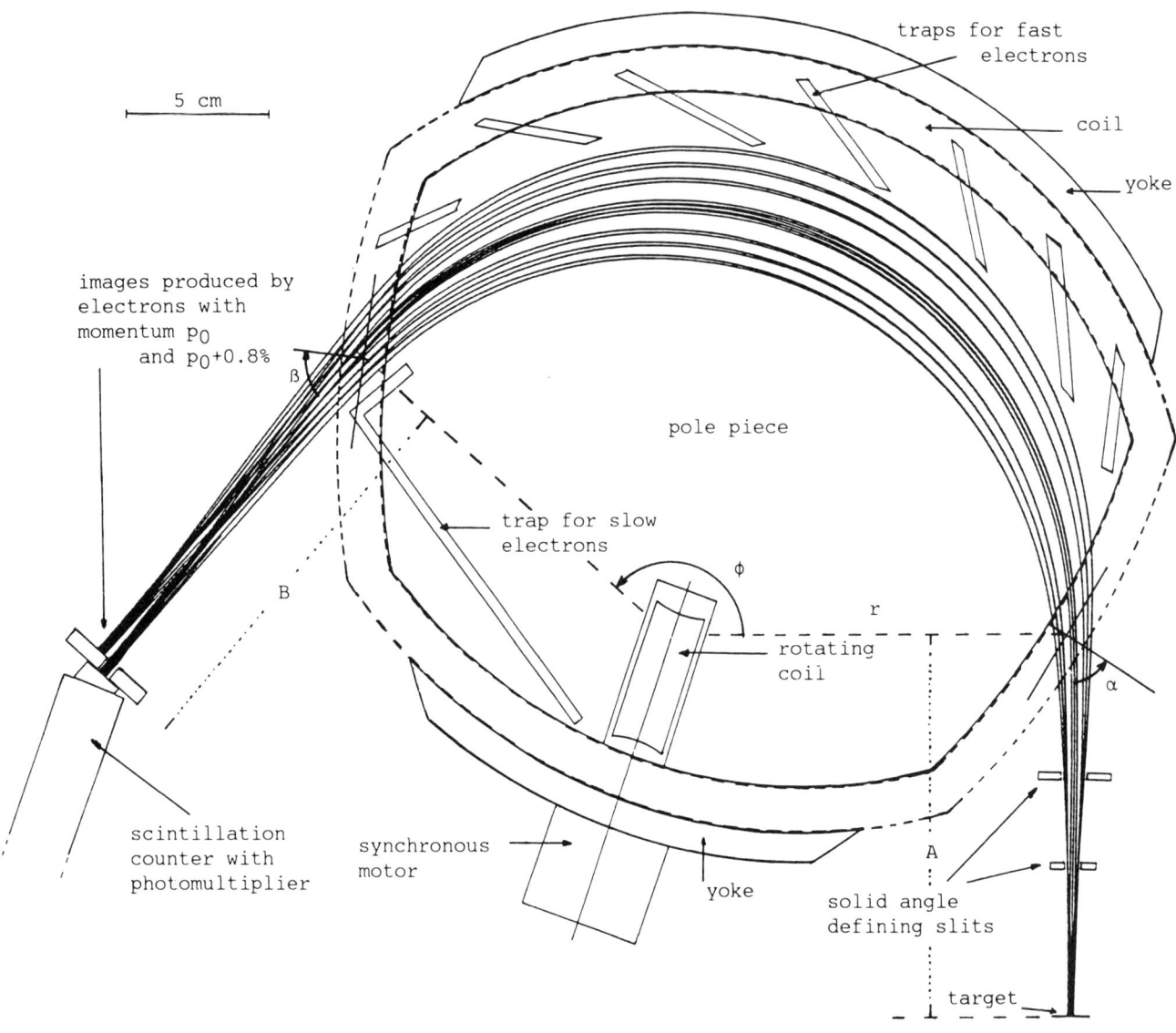

FIG. 3. Schematic diagram of one of the two the magnets mounted inside the scattering chamber (Tab. I). The horizontal focusing properties are shown: trajectories of electrons with a difference in momentum of 0.8% have been calculated by use of the ray-tracing program[15] for a target spot diameter of 2 mm and a horizontal aperture of $\pm 2.3°$.

scattering plane. The fringing field is responsible for vertical focusing. Entrance and exit angles of 57° determine that the vertical focus is at the same place as the horizontal one. They also cause relatively large object and image distances and a large focus length. Thus the target and scintillator with photomultiplier are outside the fringing field of the magnet. The large focus length permits one to have several slits in front of the energy defining slit in order to hold back electrons which do not have the desired momentum. To eliminate the numerous elastically scattered electrons there are also several traps inside the gap of the magnet. Fig. 3 shows a schematic diagram of the whole spectrometer. As can be seen the large entrance angle offers two further advantages: firstly acute-angled positioning with other detection systems is possible; secondly a wide range of scattering angles is attainable.

Since in our device the position of the energy defining slit is fixed the magnetic field has to be measured to an acurracy much better than the momentum resolution of the magnet (0.6%). We do this by use of a rotating coil driven by a synchronous motor to an accuracy of about 0.05%. The induced AC-signal from the coil is fed into

Momentum resolution $(\Delta p/p)_{FWHM}$ for target spot size 1 mm and with:	0.6 %
- Maximum accepted solid angle Ω	6.5 msr (3.5° hor. × 6.0° ver.)
- Apparatus transmission	100 %
Object distance A	125 mm
Image distance B	125 mm
Deflection angle Φ	141°
Entrance angle α	57°
Exit angle β	57°
Radius of central ray r	148 mm
Radius of convex pole piece curvature	268 mm
Air gap d	22 mm
Field index n	0

TAB. I. Data of the sector-field magnet

a special electronic network and further processed by a microcomputer. This device allows one to select desired momentum values exactly and controls the coil currents in the right way. The change of the scattering angle is done by means of a microprocessor controlled stepping motor rotating the spectrometer around the target at the center of the chamber. The special features of the spectrometer are given in Tab. I.

We dispose of two such magnets, a left handed and a right handed version. Both have been tested by observing the image of the target spot produced by elastically scattered electrons on a fluorescent screen at the place of the energy slit. With a SB detector behind the energy slit the electron spectra for different energies selected by the magnet were recorded. In this way the apparatus transmission and the suppression of undesired electrons could be checked. In order to obtain the experimental value of the momentum resolution the peak of the elastically scattered electrons was measured by increasing the magnetic field in small steps by a special computer control routine.

II. 3. Targets

Solid targets (thicknesses between 40 and 70 $\mu g/cm^2$) are used in our experiments whereas most parts of the non-relativistic measurements on TDCS have been made with gas targets.[3,4] At these lower energies, due to the higher scattering cross sections, the incoming and outgoing electrons are unlikely to leave a solid target without further interaction with the target itself. The electrons suffer multiple scattering disturbing the measurement of the elementary ionization process. At such high energies as in our experiments due to the low cross sections the multiple scattering is small and the use of thin-foil targets becomes possible. Thereby we have the great advantage of obtaining absolute cross sections easily. Moreover, the higher density of solid targets relative to the density of gas targets partially compensates for the decrease of the cross section with increasing energy.

In spite of the high electron energies there remains a certain probability that the electron will be scattered (Mott, Møller) before and/or after the ionization process. Especially the outgoing electron of smaller energy is affected. The influence of plural scattering can be checked by using targets of different thickness.

The target thicknesses were determined from measurements of the energy loss of α particles passing through the foil. The conversion factor was taken from the tables of Barkas and Berger.[11] Moreover, the product of target thickness and beam current was checked by measuring the cross section for elastic scattering and comparing it with the theoretical values. For silver targets we used a further possibility. With the total cross section for K-shell ionization of silver,[12] the fluorescence yield[13] and the beam current, we calculated the respective target thickness.

II. 4. Absolute cross section determination

The absolute triply differential cross section for K-shell ionization was determined from:

$$\frac{d^3\sigma}{d\Omega_1 d\Omega_2 dE}(E_0, \theta_1, \theta_2, E_1) = \frac{N_e}{N_0 n_t \Delta\Omega_1 \Delta\Omega_2 \Delta E_c}$$

with:

- N_e number of true coincidences
- N_0 number of primary electrons
- n_t number of target atoms per unit area
- $\Delta\Omega_1$, $\Delta\Omega_2$ solid angles of the spectrometers
- ΔE_c energy width for registration of coincident events (FWHM)

III. Results and comparison with theory

Some results of the experiment are shown in Figs. 4–8. We have plotted the absolute triply differential cross section $d^3\sigma/d\Omega_1 d\Omega_2 dE$. The error bars represent the standard deviations. The systematic error of the absolute values was estimated to be ±15%.

The full curves (JA1) and the circles and triangles (JA2) are calculations of Jakubaßa-Amundsen[7] while the dashed curves (DK) are calculations from a formula given

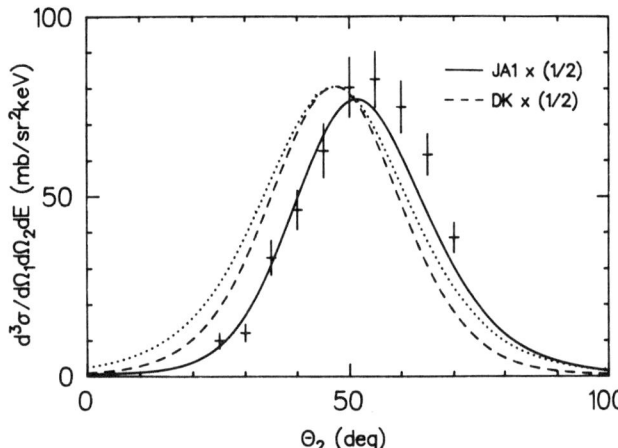

FIG. 4. Absolute triply differential cross section for K-shell ionization of silver vs scattering angle θ_2 of the slow outgoing electron ($E_0 = 500\,\text{keV}$, $\theta_1 = -15°$, $E_1 = 375\,\text{keV}$, $E_2 = 100\,\text{keV}$). The error bars represent one standard deviation. The systematic error of all points, i.e. the error in the absolute scale, is estimated to be $\pm 15\%$. The solid line (JA1) is a recent calculation given by Jakubaßa-Amundsen.[7] The dashed line is a calculation according to a theory of Das and Konar.[8] The dotted curve represents the momentum distribution $\rho(q)$ of the K electrons[14] in arbitrary units.

by Das and Konar.[8] Both calculations apply the first-order Born approximation and a relativistic interaction Hamiltonian.

The JA1 theory uses relativistic plane waves for the incident and the high-energy (scattered) electrons. Darwin approximate relativistic wave functions are used for the atomic electrons in the K shell and for the low-energy (ejected) electrons. In the JA2 calculations Coulomb waves are used instead of plane waves. In the JA1 calculations the exchange interaction is correctly accounted for, whereas in the JA2 calculations this could not be achieved for all parameters (circles with exchange, triangles without exchange).

The DK theory uses relativistic plane waves for the incident and the high-energy outgoing electrons, a non-relativistic hydrogen wave function for electrons in the K shell and a relativistic Sommerfeld-Maue wave function for the low-energy outgoing electrons. The exchange interaction was included via the Ochkur approximation.

Fig. 4 and 5 show angular distributions of 100 keV electrons emerging from K-shell ionization on silver ($E_B = 25.5\,\text{keV}$) and gold ($E_B = 80.7\,\text{keV}$) for 500 keV electron impact. The coincident fast outgoing electrons (375 keV for silver and 319 keV for gold) are detected at $\theta_1 = -15°$.

All calculations overestimate the absolute cross section considerably. For silver (Fig. 4) there are no JA2 calcu-

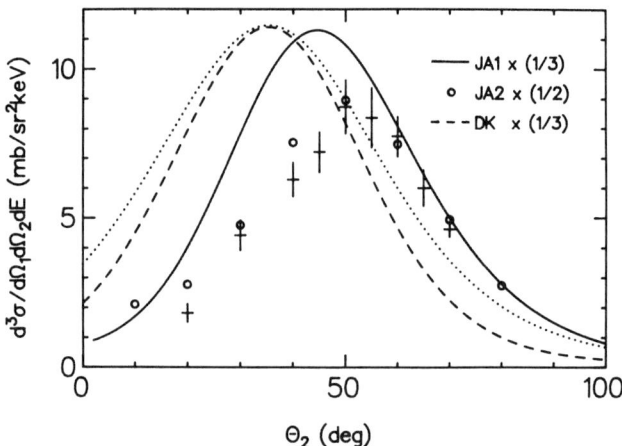

FIG. 5. Same as Fig. 4 ($E_0 = 500\,\text{keV}$, $\theta_1 = -15°$, $E_2 = 100\,\text{keV}$), but for K-shell ionization of gold, resulting in $E_1 = 319\,\text{keV}$. In addition, the JA2 calculation is shown (circles: exchange interaction has been accounted for).

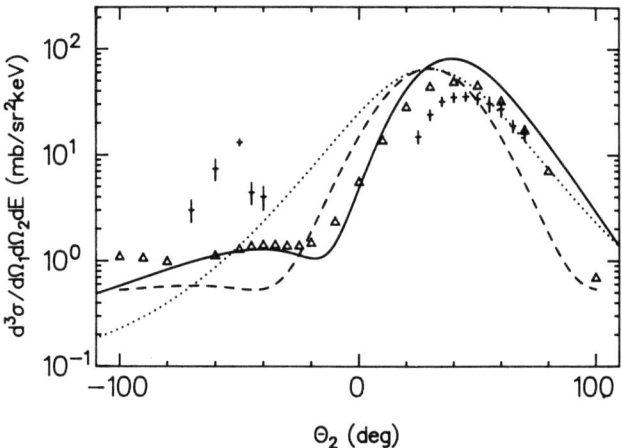

FIG. 6. Same as Fig. 4. (silver, $E_0 = 500\,\text{keV}$, $E_1 = 375\,\text{keV}$, $E_2 = 100\,\text{keV}$) but for smaller scattering angle $\theta_1 = -7°$. In addition the JA2 calculation is shown (triangles: exchange interaction not included).

lations presently available.

The angular distributions are lobes called binary peaks since they are mostly due to binary collisions between incident electrons and atomic electrons. Indeed, the shapes of the lobes mainly depend on the momentum distribution $\rho(q)$ of the bound electrons before their ejection. This shows that the influence of the dynamical conditions is small. The measured shapes are predicted quite well by both theories even though in the angular positions of the lobes there are differences.

Important kinematic parameters in the collision are the momentum transfer which is the difference of the momenta of the incoming and the fast outgoing electron ($\vec{K} = \vec{k}_0 - \vec{k}_1$), and the ion recoil momentum $\vec{k}_{ion} = \vec{k}_0 - \vec{k}_1 - \vec{k}_2$. If the nucleus acts only as a spectator of the collision, the momentum $\vec{q}$ of the K electron in its bound state is equal and opposite to $\vec{k}_{ion}$ and becomes minimal for $\vec{q} \parallel \vec{K}$, i.e. $\vec{q} \parallel \vec{k}_2$. Since the momentum distribution $\rho(q)$ of K electrons rises monotonicly with decreasing $|\vec{q}|$, the maximum of the binary peak is expected to occur in the direction of $\vec{K}$. In Fig. 4 and 5 the maximum of the DK calculations coincides with the direction of $\vec{K}$. The JA1 calculation shows a shift of the maximum towards larger angles nearly reaching the position of that of the measurements. For gold (Fig. 5), the JA2 calculation shows a good agreement with the measured angular position.

In a collision of the incoming electron with an unbound electron initially at rest (Møller scattering) the outgoing electron would emerge at a still larger angle (58.6° for the present parameters).

In Fig. 6 a measurement on silver is shown with the same parameters as in Fig. 4 but with a smaller scattering angle of the fast outgoing electron ($\theta_1 = -7°$) resulting in a smaller amount of momentum transfer. Now, in addition to the dominant binary peak, a smaller peak appears. This so-called recoil peak (well known from nonrelativistic (e,2e) experiments) is essentially the result of interactions with the ion. For our parameters it is observed in the same quadrant as the fast outgoing electron. All calculations underestimate the recoil peak considerably.

In the other category of (e,2e) experiments[4] the coincidence technique is used as a tool to investigate electron momentum distributions in atoms. As mentioned above, under selected kinematical conditions (at high enough energies and for high momentum transfer) an incoming electron interacts with essentially a single electron of the target atom, thereby removing it. It holds $\vec{k}_{ion} = -\vec{q}$. Since the two-electron collisions are well understood, the measurement provides direct information on the electron momentum distribution. The triply differential cross sections obtained are proportional to the momentum distributions of the atomic electrons if the probed momentum space volume is constant. This is given in our experiments using *magnetic* spectrometry.

One of the suitable geometries frequently used is the coplanar symmetric geometry. This means that the momenta of the incoming and the two outgoing electrons are coplanar and that $\vec{k}_1$ and $\vec{k}_2$ are equal in magnitude and have the same angle on either side of $\vec{k}_0$. The different values of $\vec{q}$ selected by varying the angles are all parallel (or antiparallel) to the direction of $\vec{k}_0$.

We used this geometry for a measurement on the K shell of gold. The result is shown in Fig. 7. Also plotted

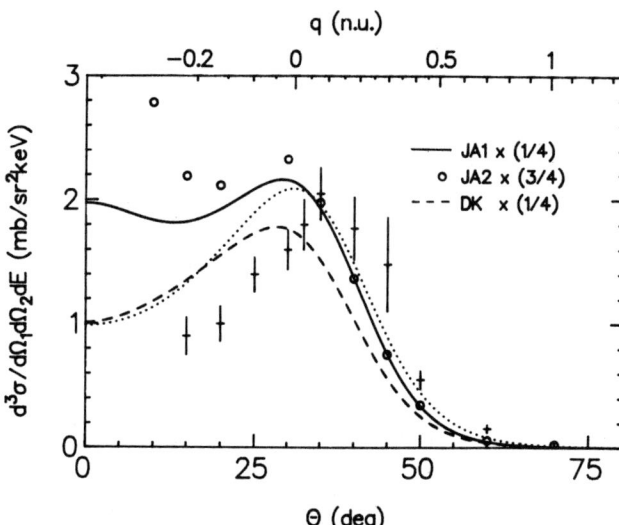

FIG. 7. Absolute triply differential cross section for K-shell ionization of gold vs scattering angle θ in the coplanar symmetric geometry ($E_0 = 500\,\text{keV}$, $\theta_1 = -\theta_2 = \theta$, $E_1 = E_2 = 210\,\text{keV}$). For theories and error bars see Fig. 4 and 5. The dotted curve represents the momentum distribution $\rho(q)$ of the K electrons[14] in arbitrary units. The top scale gives the values of q in natural units. Negative values mean that $\vec{q}$ is parallel to $\vec{k}_0$.

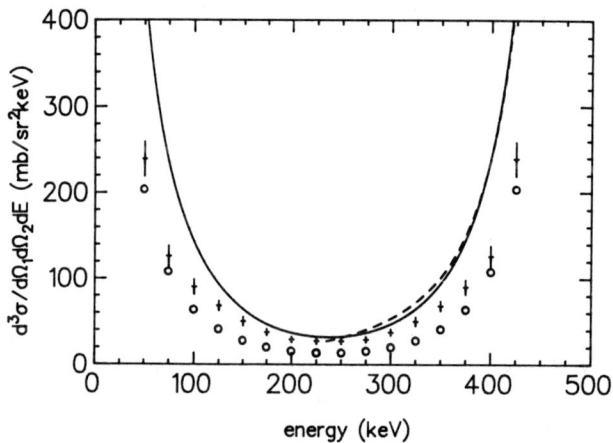

FIG. 8. Absolute triply differential cross section for K-shell ionization of silver vs the energy of one outgoing electron. ($E_0 = 500\,\text{keV}$, $\vec{k}_{ion} = \text{const.}$). Angles and energies were varied to get the cross section under different kinematical conditions subject to the constraint that the ion recoil momentum $|\vec{k}_{ion}| = 51\,\text{keV}/c$ (0.1 n.u.) parallel to the direction of the incoming electron. For theories and error bars see Fig. 4 and 5.

is $\rho(q)$ according to the relativistic hydrogenic wave function for Z=79 (dotted curve) given by Mukoyama.[14] The shape of the measured curve is nearly the same as that of the calculated momentum distribution. However, there is a significant shift in the position of the measurement also in this symmetric case. All theories peak at about the same angular position as $\rho(q)$. On the right side the slope is predicted quite well, whereas on the left side the discrepancies are considerable. The absolute values are overestimated by all theories (see factors in Fig. 7).

In contrast to the procedure of varying $|\vec{k}_{ion}|$ in order to extract momentum distributions from (e,2e)-experiments, it is also possible to change the angles and energies of the outgoing electrons in a way that keeps the vector $\vec{k}_{ion}$ constant. In this way one can study the collision process at a fixed value and direction of $\vec{k}_{ion}$ under different kinematical conditions. In the case of $\vec{q} = -\vec{k}_{ion}$ the influence of the theoretical description of the momentum distribution $\rho(q)$ cancels out when comparisons are made of the measured and calculated shapes of the cross sections. This kind of experiment has been performed on the K shell of silver. Angles and energies were varied subject to the constraint that $|\vec{k}_{ion}| = 51 \,\text{keV}/c$ (0.1 n.u.), parallel to the direction of the incoming electron. The result is shown in Fig. 8. Plotted is the absolute triply differential cross section vs the energy of one outgoing electron. The region where the cross section is nearly independent of the kinematical parameters is the region of wavefunction mapping. The overall curve resembles that obtained in the case of free electron-electron scattering (Møller scattering), characterized by $q = 0$. It is seen that the JA2 predictions are close to the measured values for asymmetric energy sharing, the JA1 theory for symmetric energy sharing. Since in the DK theory the exchange interaction is accounted for only approximately, good agreement there can be expected for asymmetric cases only.

In conclusion, most of our measurements show appreciable discrepancies with regard to the theoretical predictions especially for the absolute cross sections. We hope that better understanding will be achieved by a further iterative interaction between experiment and theory.

Acknowledgments

We would like to thank Professor G.J. Wagner for his support of the work, Dr. D.H. Jakubaßa-Amundsen for helpful discussions and for computing the theory behind our parameters. The financial support of the Deutsche Forschungsgemeinschaft (Na 102/6-3) is gratefully acknowledged.

References

[1] H.Ehrhardt, M.Schulz, T.Tekaat, and K.Willmann, Phys.Rev.Lett. **22**, 89 (1969)

[2] U.Amaldi, A.Egidi, R.Marconnero, and G.Pizzella, Rev.Sci.Instrum. **40**, 1001 (1969)

[3] H.Ehrhardt, K.Jung, G.Knoth, and P.Schlemmer, Z.Phys.D **1**, 3 (1986)

[4] I.E.McCarthy and E.Weigold, Rep.Progr.Phys. **51**, 299 (1988)

[5] E.Schüle and W.Nakel, Phys.B.:At.Mol.Phys. **15**, L639-41 (1982)

[6] H.Ruoff and W.Nakel, J.Phys.B:At.Mol.Phys. **20**, 2299 (1987)

[7] D.H.Jakubaßa-Amundsen, Z.Phys.D **11**, 305 (1989)

[8] J.N.Das and A.N.Konar, J.Phys.B:At.Mol.Phys. **7**, 2417 (1974)

[9] A.Lahmam-Bennani, H.F.Wellenstein, A.Duguet, and M.Lecas, Rev.Sci.Instrum. **56**, 43 (1985)

[10] H.Ruoff, E.Schüle, J.Bonfert, H.Graf, and W.Nakel, Rev.Sci.Instrum. **60**, 17 (1989)

[11] W.H.Barkas and M.J.Berger, Natl Acad. Sci.-Natl Res. Council Publ. **1133**, 103 (1964)

[12] H.Kolbenstvedt, J.Appl.Phys. **38**, 4785 (1967)

[13] M.O.Krause, J.Phys.Chem.Ref.Data **8**, 307 (1979)

[14] T.Mukoyama, J.Phys.B:At.Mol.Phys. **15**, L785 (1982)

[15] H.A.Enge and S.B.Kowalski, in proceedings of the 3rd International Conference on Magnet Technology (Hamburg, 1970)

THE THEORY OF ELECTRON-ION COLLISIONS

A E Kingston

Department of Applied Mathematics and Theoretical Physics
The Queen's University of Belfast, Belfast BT7 1NN, UK

Abstract

The results of recent large scale R-matrix calculations are used to study the convergence of the close coupling approximation for the calculation of electron excitation cross sections for atoms and ions. For the electron excitation of atoms these studies suggest that to obtain reliable cross sections below the ionization threshold it is necessary to include in the calculations a number of bound states which lie above the state under consideration. If the incident electron energy is just above the ionization threshold it is found that, even if we include a large number of bound states in the calculation, the cross section results do not converge to the true result and it is necessary to take account of continuum states in the close coupling expansion. However for the electron excitation of ions the convergence of the close coupling expansion is more rapid than that for atoms and the rate of convergence appears to increase as the nuclear charge of the ion increases.

1 Introduction

An accurate knowledge of the rates for electron excitation of ions is important in the understanding of hot gaseous plasmas. Over the last ten to twenty years the theoretical atomic physics group at Queen's University, Belfast, under the leadership of Phil Burke, have carried out extensive calculations on the electron excitation rates of a large number of atoms and ions. The results of these calculations have been used by us and other workers in the modelling of stellar plasmas, fusion plasmas and laser produced plasmas.

In this work we have used the close coupling approximation and since about 1975 we have used the R-matrix programmes to solve the relevant equations [1]. The current suite of programmes is remarkably versatile, it can be used to obtain excitation rates for any atom or ion. The numerical techniques used are also very stable. The only real restriction on the codes both now and in the past has been the size and speed of the available computers.

Since we usually try to do the largest calculations that we can do with the available computers, developments in R-matrix calculations tend to follow new developments in computer power. For example our work on the electron excitation of helium started with a 5-state calculation in 1975 on an IBM 360/195 [2]. This was followed by an 11-state calculation in 1985 on a CRAY-IS [3], then a 19-state calculation in 1987 on a CRAY-XMP/14 [4], and our latest calculation which is reported at this conference is a 29-state calculation which used the four processors on the CRAY-XMP/48 [5].

These increasingly complex calculations not only give us estimates of the cross sections between more and more levels but they are now beginning to give us information on the convergence of the cross sections as the number of states included in the calculations increases. In this talk I would like to discuss what we have discovered about the convergence of the close coupling approximation over the last few years.

Theory

Almost every approximation used in the calculation of electron excitation cross sections is based on an expansion of the collision complex in terms of the wave functions of the target atoms or ions. For example if we were to consider the scattering of electrons by atomic hydrogen we basically take the total wave function of atom plus electron to be of the form

138 Theory of Electron Ion Collisions

$$\Psi(\underline{r_1},\underline{r_2}) = \psi_{1s}(\underline{r_1})F_{1s}(\underline{r_2}) + \psi_{2s}(\underline{r_1})F_{2s}(\underline{r_2})$$
$$+ \psi_{2p}(\underline{r_1})F(\underline{r_2}) + \psi_{3s}(\underline{r_1})F_{3s}(\underline{r_2}) + \cdots \quad (1)$$
$$+ \int \psi_E(\underline{r_1})F_E(\underline{r_2}).$$

Here $\psi_{1s}, \psi_{2s}, \psi_{2p}\ldots$ are the wave functions of the bound states of hydrogen and the F's represent the motion of the free electron. The final term in eq.(1) contains the free electron wave function ψ_E for the hydrogenic system. If we could take account of the infinite number of bound states and the free electron continuum this type of expansion would give exact results.

For ease of explanation I will use eq.(1) throughout this talk but in a proper formulation we have of course to ensure that during the collision the orbital L and spin S angular momentum as well as the parity are conserved and in the R-matrix method we expand the total wave function in the internal region in terms of the R-matrix basis

$$\Psi_k = A \sum_{ij} a_{ijk} \Phi_i u_i + \sum_j b_{jk} \phi_j \quad (2)$$

The Φ_i are channel functions formed from the atomic states and the angular and spin functions for the scattered electron. The ϕ_j are functions formed from the atomic orbitals alone and are included for completeness. A is the usual anti-symmetrization operator. The coefficients a_{ijk} and b_{jk} are obtained by diagonalizing the electron-atom Hamiltonian in this basis

$$\langle \Psi_k | H | \Psi_{k'} \rangle = E_k \delta_{kk'} \quad (3)$$

where the integrals are taken over the internal region. The R-matrix on the boundary can then be obtained immediately and the K-matrix and cross section calculated by solving the asymptotic equations in the outer region at each required energy.

2 Electron excitation of atomic hydrogen and hydrogenic ions

Hydrogen and the hydrogenic ions are unique as their wave functions are known exactly. Hence these systems are ideal for testing ideas in collision theory, such as the convergence of the close coupling approximation, as any difficulties can be attributed to the approximations in the collision theory. For other systems such as atomic helium as we increase the accuracy of the collision theory we also tend to increase the accuracy of the bound state target wave functions and it is diffucult to say if an improvement in a calculation is due to better target state wave functions or a better treatment of the collision.

2.1 Electron excitation of Atomic Hydrogen

We have recently carried out a series of R-matrix calculations on atomic hydrogen [6] in which we calculated the electron excitation cross sections from the ground state to the 2s and 2p states. In the first calculation, the 3-state calculation we included the 1s, 2s and 2p states in the close coupling expansion eq.(1). The 6-state calculation added the 3s, 3p and 3d states to the three lower states while the 10-state calculation also included the four states 4s, 4p, 4d, 4f. The largest calculation carried out, the 15-state includes the fifteen states 1s, 2s, 2p, 3s, 3p, 3d, 4s, 4p, 4d, 4f, 5s, 5p, 5d, 5f, 5g in eq.(1). This calculation is one of the largest which we have carried out and it is approaching the limit of what can be done with a CRAY-XMP/48 but we are now considering 21-state calculations which will include the n=6 states.

The cross section for excitation of the ground state of hydrogen to the 2s and 2p states for the four R-matrix calculations are displayed in Figures 1 and 2. Unlike other atoms, the excitation cross sections of atomic hydrogen have infinite series of resonances. These resonances are important but they tend to confuse comparisons between different calculations. Hence in the figures we have attempted to plot points which are not affected by resonances. A full discussion of the R-matrix calculations on resonances converging to the n=2, 3 and 4 thresholds may be found in an earlier paper [7].

Figure 1 a. compares the four R-matrix calculations between the n=2 and n=3 thresholds. Apart from the resonance regions close to the n=2 and the n=3 thresholds, the simple 3-state calculation is in good agreement with 6, 10 and 15-state calculations. The 6, 10 and 15-state results are in good agreement with each other and the addition of more bound states in eq.(1) would not change the cross section.

Between the n=3 and 4 thresholds Figure 1 b. there is an important change in the pattern, for the 3-state is now about 50% greater than the other R-matrix results. The 6-state results are a little different from the 10 and 15-state results and they lack the resonances close to the n=4 threshold. However in this region the addition of further bound state in eq.(1) should only modify the 15-state cross section slightly.

The pattern again changes between the n=4 and 5 thresholds (Figure 1 c). Again the 3-state results are much higher than those for the 15-state calculation but the 6-state

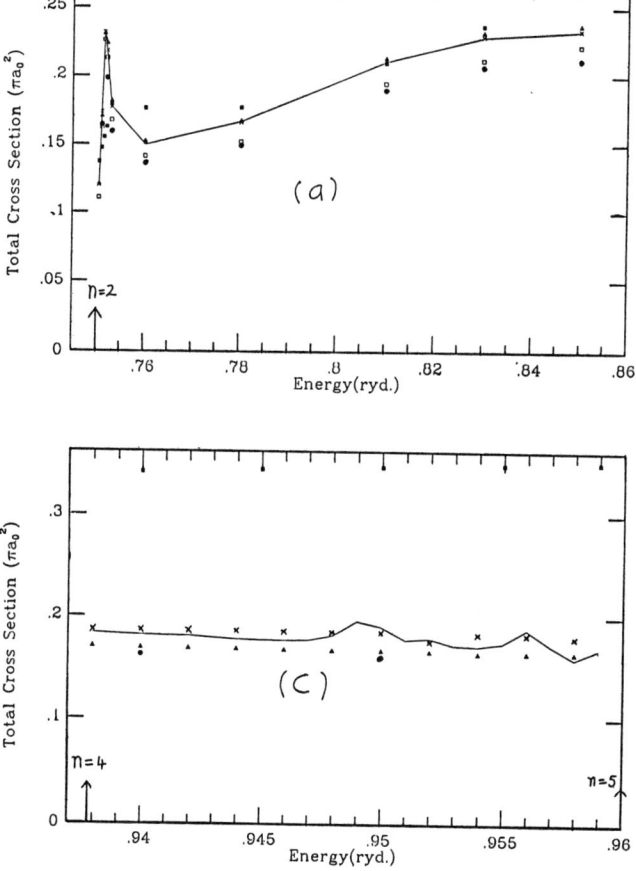
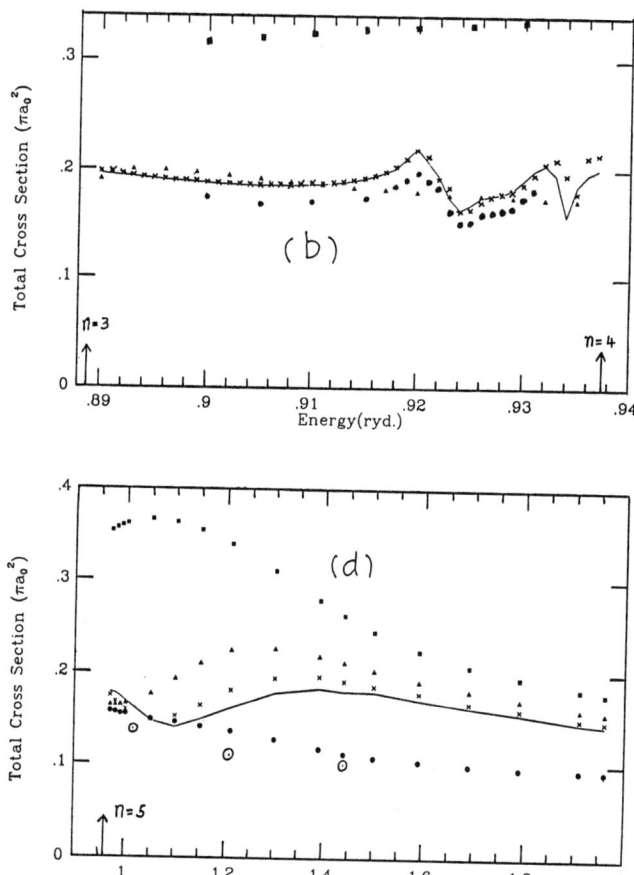

Figure 1. The cross section for excitation of the ground state of hydrogen to the 2s state. R-matrix 3-state calculation ■ ; 6-state calculation ▲ ; 10-state calculation X and 15-state calculation ——— . Other calculations, Taylor and Burke [8] □; Callaway [9, 10, 11] ● ; Scott et al [12] ○.

(a) From n=2 to n=3 thresholds,
(b) from n=3 to n=4 thresholds,
(c) from n=4 to n=5 thresholds and
(d) above the ionization potential.

results lie a little below the 15-state cross section while the 10-state calculations, in general, are greater that 15-state results. This oscillating pattern would suggest that, as for the other two lower energy regions, that the inclusion of further bound states in eq.(1) will not improve on the 15-state calculation to a significant extent.

Perhaps the most important results of this work are given in Figure 1 d, which compares the four R-matrix results for electron energies above the ionization potential of hydrogen. In this energy region the 3-state results are very much greater than the 15-state results and although there is a definite trend in the cross section as we increase the number of bound states, it is clear that the addition of further bound states in eq.(1) will change the cross section significantly.

In our R-matrix calculations we only use bound states in the expansion of the total scattering wave eq.(1) and hence we can say nothing about the importance of the final term in eq.(1) which gives the contribution from the hydrogenic continuum wave functions. However there have been a number of calculations which have taken account of this term, these include the correlation wave function work of Taylor and Burke [8], the pseudo state calculations of Callaway [9, 10, 11] and the intermediate energy R-matrix (IERM) theory of Burke [12].

The results of these calculations are compared with our R-matrix results for the electron excitation of the 2s state of hydrogen in Figure 1. For electron energies for the n=2 to the n=5 the cross section from the bound states in (1) have converged to the present 15-state results. There is a difference of

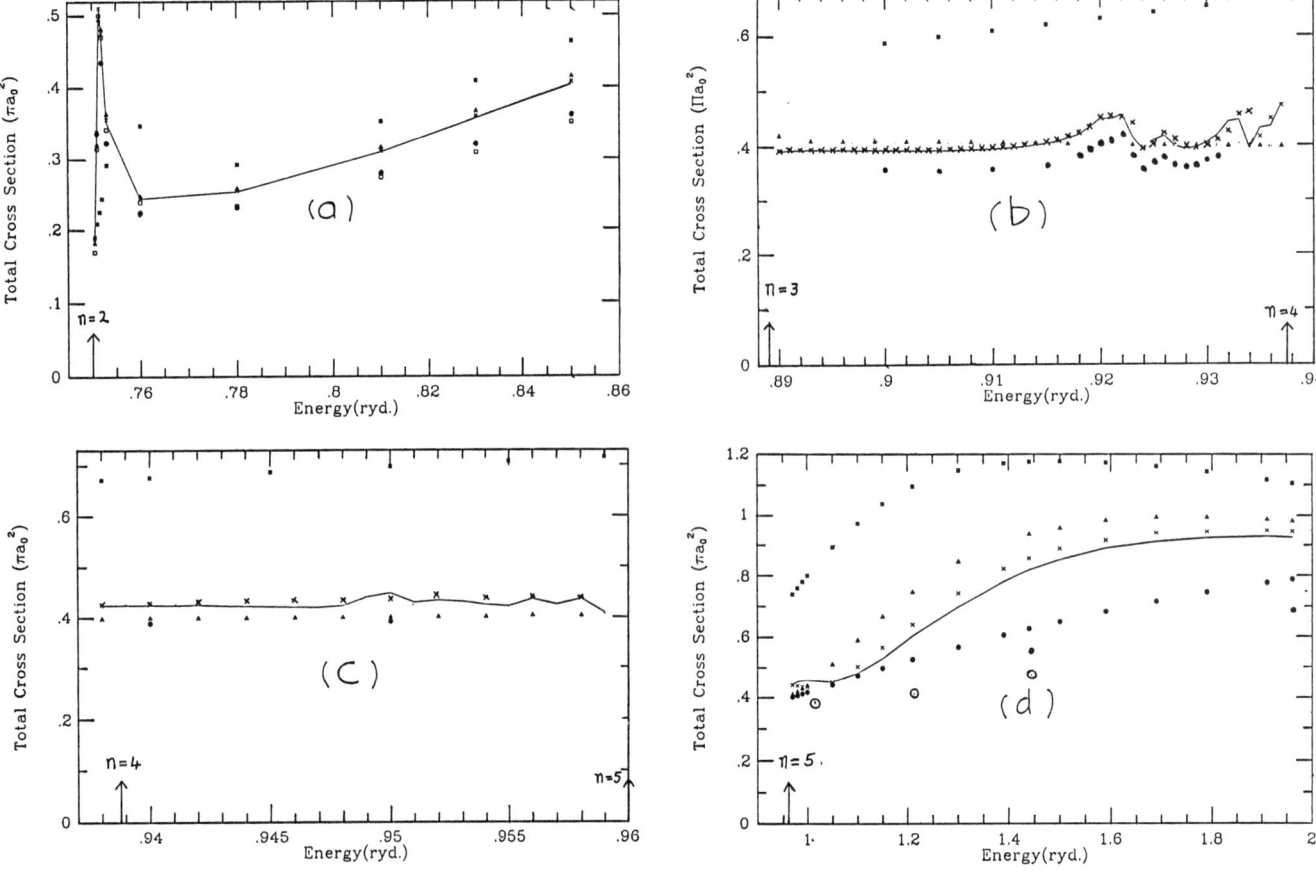

Figure 2. The cross section for excitation of the ground state of hydrogen to the 2p state. R-matrix 3-state calculation ■ ; 6-state calculation ▲ ; 10-state calculation X and 15-state calculation —— . Other calculations Taylor and Burke [8] □ ; Gallaway [9, 10, 11] ● ; Scott et al [12] ○ .

(a) From n=2 to n=3 thresholds
(b) from n=3 to n=4 thresholds
(c) from n=4 to n=5 thresholds and
(d) above the ionization potential

about 10% between these results and the more accurate results which can be attributed to the final term in eq.(1). Above the ionization potential the situation is different. Here our best 15-state results are almost a factor of two too large. This error can be attributed to a neglect of higher bound states and also a neglect of the continuum states. By noting that the contribution to the cross section for states with principle quantum number n is approximately proportional to n^{-2}.

We have been able to estimate that the n > 5 bound states would reduce the cross section by about 20% at 1.2 ryd, 13% at 1.4 ryd and by about 6% from 1.5 to 2 ryd. This suggests that the final term in eq.(1) will reduce to converged bound state cross section by about 20% at 1.2 ryd and by about 40% from 1.4 to 2 ryd. Clearly in this region it is essential to take account of the final term in eq.(1).

Figure 2 compares our four R-matrix calculations for the electron excitation from the grand state of hydrogen to the 2p state. The pattern of the results is similar to that found in Figure 1 for excitation of the 2s state. Between the n=2 and 3 thresholds, Figure 2 a, the 3-state results are in somewhat less good agreement with the 15-state results, as is found for the 1s-2s results but again, it is clear that the addition of further bound states will not change the 1s-2p results significantly. Above the n=3 threshold the 3-state results are greatly in error but again from Figures 2 b and 2 c we see that the addition of further bound states will not change the cross section greatly. For electron energies greater than the ionization potential the pattern is similar to that found for the 1s-2s excitation. The 3-state results are much greater than the 15-state results but the close coupling results have not converged with 15 states and the addition of further bound states in eq.(1) will change the cross section significantly.

We can again estimate the effect of including the final term in eq.(1) by comparing with the calculations which include this effect. Between the n=2 and n=5 thresholds our 15-state results are about 10% higher than the other more accurate calculations and this 10% can be attributed to the final term in eq.(1). Above the ionization potential there is some difference between the most recent results of Callaway and Burke but this difference is small compared with the difference between them and the 15-state results. Using the 6, 10 and 15-state results we estimate that the higher bound states n > 5 will reduce the 15-state cross section by about 10% at 1.2 ryd, 9% at 1.4 ryd and by less than 5% between 1.5 and 2 ryd. This suggests that in the region from 1.2 to 2 ryd the continuum term in eq.(2) reduces the 1s-2p cross section by between 25 and 35 %.

Summarizing the results for both the excitation of the 2s and 2p states it is found that we must consider two distinct energy regions.

A. Electron energies from the n=2 to n=5 thresholds where

1. the close coupling expansion eq.(1) has converged for most practical purposes for n=5.

2. apart from the resonance regions the 6 and 10-state results are quite reliable.

3. the 3-state results are greatly in error above the n=3 threshold, it appears that to obtain reasonable results for excitation to an n=2 state it is important to include the n=3 states.

4. compared with more accurate results the 15-state results are accurate to about 10%, this 10% is due to the neglect of the final term in eq.(2).

B. Electron energies above the ionization potential where

1. the close coupling expansion eq.(1) has not converged for n=5.

2. the contribution from the final term in eq.(1) is large.

Between the n=5 threshold and the ionization potential the R-matrix results have not converged but if we neglect the 3-state results they are in reasonable agreement with more accurate calculations.

2.2 Electron excitation of He^+

As well as calculating 3, 6, 10 and 15-state R-matrix calculations on atomic hydrogen we have also carried out a complete set of calculations of the electron excitation of the ground state of He^+ to the 2s and 2p states. As in the case of atomic hydrogen, the He^+ wave functions are known exactly and any errors in the collision calculations can be attributed to the approximations used in the collision theory.

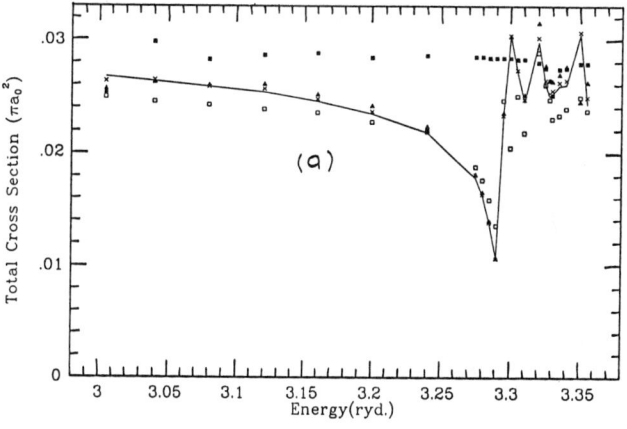

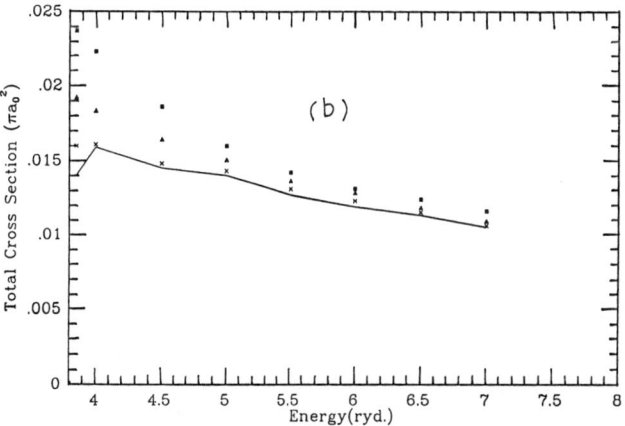

Figure 3. The cross section for excitation of the grand state of He^+ to the 2s state. R-matrix 3-state calculations ■ ; 6-state calculations ▲ ; 10-state calculations X and 15-state calculations — . Other calculations, Taylor and Burke [13] □ .
(a) From n=2 to n=3 thresholds
(b) above the ionization potential.

Figure 3 and 4 give the results of some of our calculations for He^+. We only display the results for electron energies between the n=2 and 3 thresholds and for energies above the ionization potential.

If we consider Figure 3 a. it is seen that for the 1s → 2s excitation between the n=2 and n=3 thresholds the convergence of the form R-matrix cross section to the 15-state result is rather similar to that for atomic hydrogen (Figure 1 b). Between the n=3 and n=5 thresholds the convergence to the 15-state results is also good but the 3-state results are between 30 and 40% greater than the 15-state results, however for atomic hydrogen this difference was almost twice as large as for He^+. Above the ionization potential the 3-state results (Figure 3 b) are much larger than the others and the series does not appear to have converged indeed the pattern is similar to that for the 1s-2s excitation of hydrogen (Figure 1 d) but is less marked. For example if we consider the ratio of the 3-state calculation to the 6, 10 and 15-state calculation at an electron energy of about 1.75 ΔE_{1s-2s} for He^+ the ratios are 0.96, 0.92, 0.89 respectively but for H the ratios are 0.73, 0.63, 0.57 respectively.

The cross section for the 1s-2p excitation of He^+ (Figure 4) follows a similar pattern for the excitation to the 2s state (Figure 3). Between the n=2 and n=3 thresholds there is quite rapid convergence for the four R-matrix calculations. From the n=3 to n=5 thresholds the 3-state results are about 30% greater than the 15-state cross section; this compares with a factor of 60% greater in the 1s-2p excitation of hydrogen. Figure 4 b gives the results above threshold, clearly the four R-matrix results have not converged. The ratio of the 3-state calculation to the 6, 10 and 15-state calculation at an electron energy of about 1.75 ΔE_{1s-2p} for He^+ are 0.93, 0.90 and 0.88 respectively while the same ratios for H are 0.74, 0.65 and 0.61.

For both the 1s-2s and the 1s-2p transitions in He^+ accurate correlation wave function calculations are available for energies between the n=2 and n=3 thresholds. These suggest that for He^+ the contribution from the last term in eq.(1) is much less than the contribution in hydrogen. For example at an electron energy of about 1.03 ΔE_{1s-2p} the continuum term only reduces the 1s-2s cross section in He^+ by 4% but for H the reduction is 10%. Similarly for the 1s-2p transition at this energy the final term in eq.(1) reduces the He^+ cross section by 5% and the H cross section by 10%.

Summarizing the results for He^+, it is clear that the general pattern is similar for that of H. For electron energies between

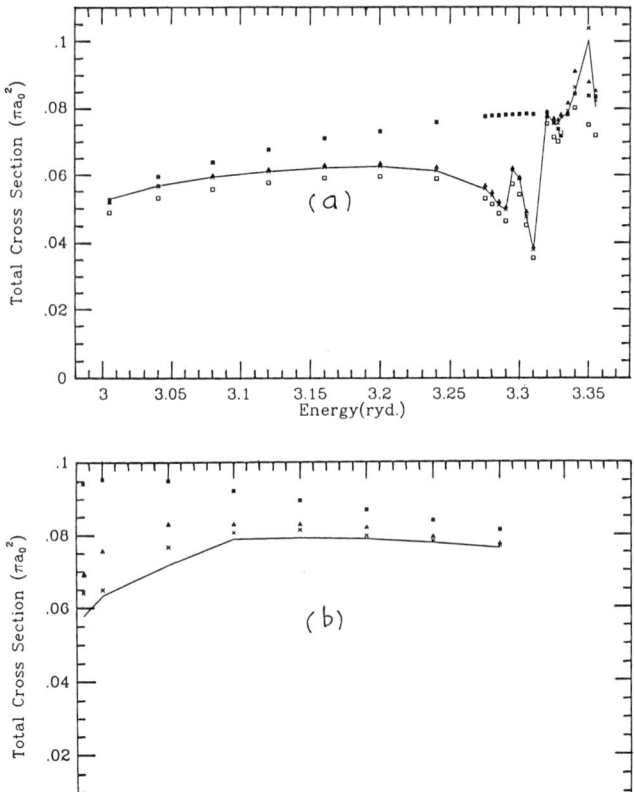

Figure 4. The cross section for excitation of the ground state of He^+ to the 2p state. R-matrix 3-state calculation ■ ; 6-state calculation ▲ ; 10-state calculation X and 15-state calculation — . Other calculations, Taylor and Burke [13] □ .
(a) From n=2 to n=3 thresholds,
(b) above the ionization potential.

the n=2 and n=5 thresholds the 15-state results will not be modified significantly by including more bound state terms in eq.(1). The 3-state cross sections are significantly bigger than the other calculations but the 6 and 10-state results are in general close to the 15-state results. Above the ionization potential the bound state calculations have not converged. However it is clear that for He^+ the convergence is much more rapid than for H.

2.3 Electron excitation of C^{5+}

In order to explore the convergence of the close coupling approximation for more highly charged ions we have started a series of R-matrix calculations on hydrogenic Carbon. The results are still preliminary but they suggest that for highly charged ions the close coupling approximation converge very rapidly. In Table 1 we give for H, He$^+$ and C^{5+} percentage decreases in the 1s-2s and 1s-2p cross section as we increase the number of terms in eq.(1) from 6-states to 10-states at an energy of 1.75 ΔE_{1s-2s}

Table 1 Percentage decrease in cross section in a 10-state calculation compared with a 6-state calculation.

	1s-2s	1s-2p
H	14%	12%
He$^+$	1.5%	2.0%
C^{5+}	0.3%	0.5%

Clearly coupling to higher states is less important for highly charged ions.

3 Photoionization of He-like atoms and ions

To an experimental physicist photoionization and electron excitation are somewhat different subjects but to a theoretical physicist the photoionization cross section is obtained from a matrix element of a bound state wave function and a free state wave function. Hence photoionization cross sections can shed light on the convergence of the close coupling expansion but in this work we have the added complication that the results depend on the accuracy of the bound state of the atom or ion which is photoionized.

3.1 Photoionization of He

The R-matrix method can also be used to obtain photoionization cross sections. In this approach the free state wave function, of an electron in the field of He$^+$, is generated in the same way as in a scattering calculation and the same orbitals are used to generate the bound state of the ion plus electron. The free electron wave function is similar to eq.(1) and as in the work on excitation cross sections we can increase the accuracy of our results by increasing the number of terms in the expansion.

In this photoionization work we carried out a series of four calculations; in the first calculation we included only the 1s state, then we did a 3-state calculation with 1s, 2s and 2p target states, this was followed by a 6-state calculation with 1s, 2s 2p, 3s, 3p and 3d states and finally we added the 4s, 4p, 4d and 4f states to give a 10-state calculation.

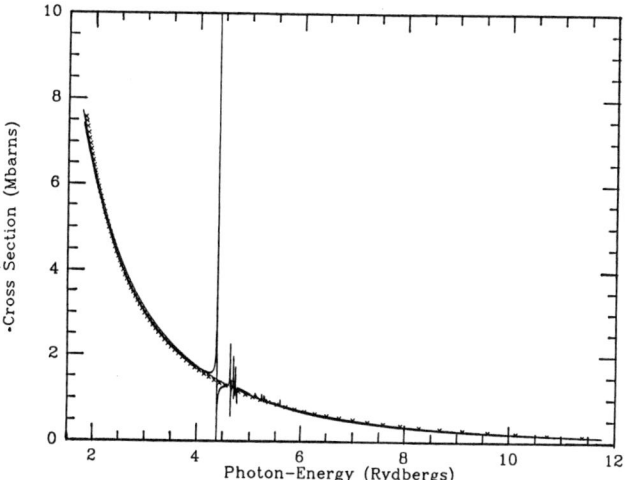

Figure 5. Photoionization cross section for He. R-matrix 1-state calculations ······ ; 3, 6, 10-state calculations are not distinguishable ——— . Experimental results of West [14] xx.

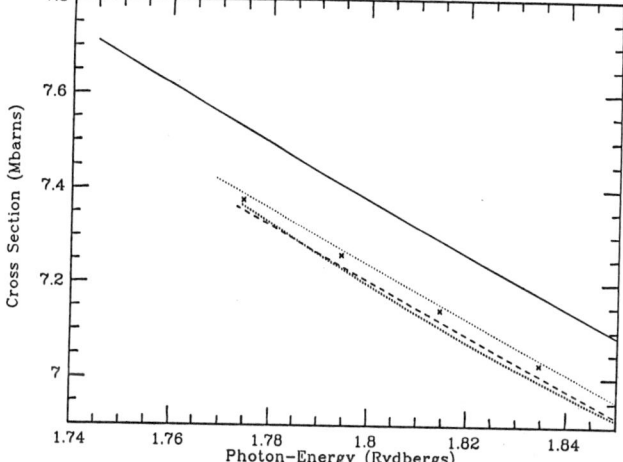

Figure 6. Photoionization cross section for He. R-matrix 1-state calculations ——— ; 3-state calculations ····· ; 6-state calculations — — —; 10-state calculations ━ ━ ━. Accurate theoretical calculations of Stewart [15] xx.

Figure 5 gives an over view of these results. On this scale it is not possible to distinguish the difference between the various R-matrix results but it is seen that they are all in reasonably good agreement with the experimental data [14]. A more detailed graph of the results at the photo-ionization threshold is given in Figure 6. Here we note that the threshold for each calculation is different as the bound state wave function is also dependant on the number of states used in the wave function expansion. At the threshold of the 10-state calculation the 1-state results are only 2% greater than the 10-state results and the other results agree to better than 0.5%. The results are also in excellent agreement with the accurate calculations of Stewart [15]. This region however corresponds to the very low electron scattering in He$^+$ and it is perhaps not surprising that there is rapid convergence in this region.

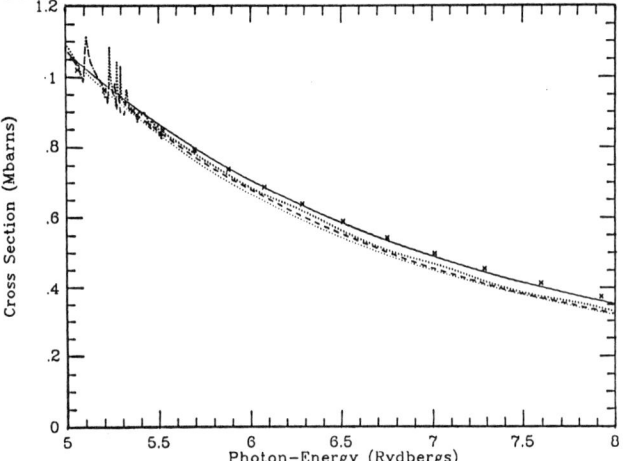

Figure 7. Photoionization cross section for He. R-matrix 1-state calculations ——— ;
3-state calculations ·····;
6-state calculations – – – ;
10-state calculations - - - - ;
Experimental results of West [14] xx.

A more stringent test of the convergence of the series of R-matrix calculations is given in Figure 7 where the results are given in detail for ejected electron energies of close to and above the He$^+$ ionization potential. Here it is seen that the R-matrix results are converging to a cross section which lies between the 6-state and 10-state results but at the largest energies this result can be as much as 10% below the experimental results. More accurate measurements in this region could tell if this descrepancy is due to our neglect of the final continuum term in our total wave function expansion.

3.2 Photoionization of He-like ions

Calculations are now underway to calculate the photoionization cross section of a number of He-like ions. Preliminary results suggest that, as for electron excitation, convergence is very rapid for highly ionized systems. For example Figure 8 shows the close agreement for the results of a 1 and a 3-state calculation for the photoionization of C^{4+}.

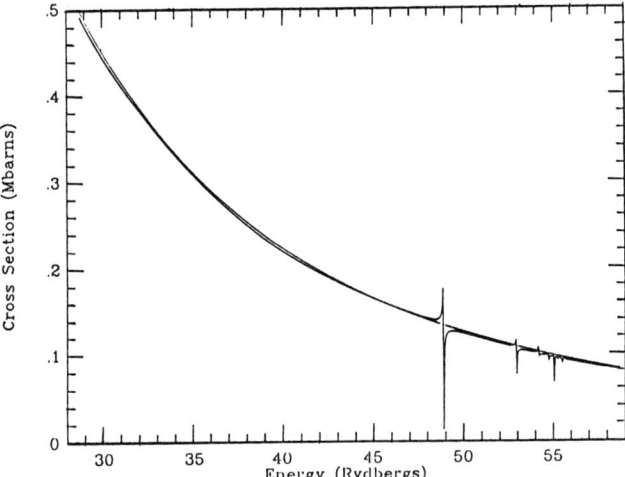

Figure 8. Photoionization of C^{4+}. R-matrix calculation 1-state calculations ····· ;
3-state calculations ——— .

4 Electron excitation of atomic helium and helium like ions

For non-hydrogenic atoms and ions target state wave functions are not known exactly and it is difficult to assess whether improvements in calculations are due to improvements in the target state functions or in the scattering of proximation. However in practice improvements in the target state wave functions go hand in hand with improvements in the collision approximation.

4.1 Electron excitation of atomic helium

Helium is the second most abundant element in the universe and it is also much easier to observe in the laboratory hence it has been studied extensively by both experimental and theoretical physicists. We have also made extensive studies of it using the R-matrix codes. As I mentioned earlier this series of calculations shows how machine dependant these calculations are. In our first calculation on the electron excitation of helium we used the following target states, 1^1S, 2^3S, 2^1S, 2^3P and 2^1P. This was carried out in 1975 on an IBM 360/195 [2]. Ten years later [3] with a CRAY-IS we were able to add the six 3^1S, 3^3S, 3^1P, 3^3P, 3^1D and 3^3D states and do an 11-state calculation. When we got time on a CRAY-XMP/14 [4] we could include a further eight n=4 target states to give a 19-state calculation. By using all four processors on the CRAY-XMP/48 [5] we have now been able to include 29 target states in the expansion of the wave function eq.(1). This calculation includes all helium states up to n=5 and is the largest calculation which

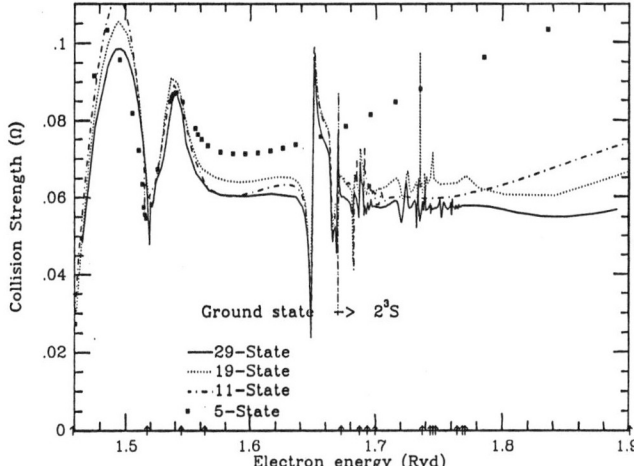

Figure 9. Collision strength for the electron excitation of the ground state of helium to the 2^3S state. R-matrix calculations 5-state results ■ ; 11-state results —·—; 19-state results ····; 29-state results ——.

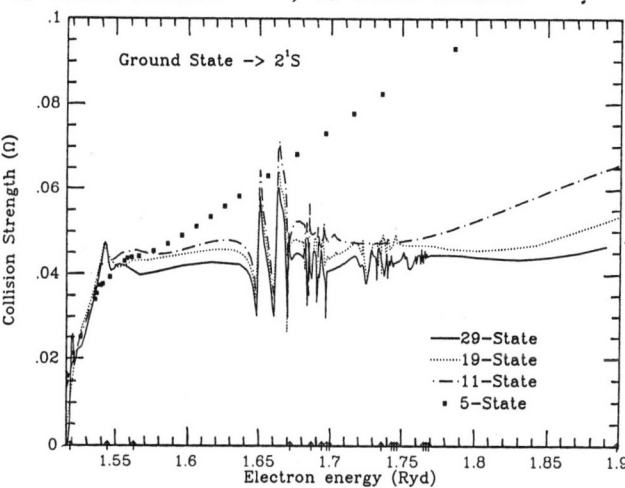

excitation of the ground state of helium to the 2^1S state. R-matrix calculations, 5-state results ■ ; 11-state results —·—; 19-state results ····; 29-state results ——.

Figures 9 and 10 display the results of our four R-matrix calculations for the collision strength for the electron excitation of the 2^3S and 2^1S states of helium. For both of these transitions the collision strength is dominated by resonances which confuse the pattern of convergence. However Figure 9 suggests that for energies up to the n=5 thresholds the 29-state results cannot be greatly in error. An important feature of the calculations is the fact that the 5-state calculation which includes the 1^1S, 2^3S, 2^1S, 2^3P and 2^1P states, are in good agreement with the other calculations up to the excitation energy of the 2^1P state and then it deviates from them. We also note that the 11, 19, and 29-state calculations are all in quite good agreement with each other up to about the ionization threshold when they begin to deviate from each other.

The pattern of convergence of the collision strengths for the 2^1S state, Figure 10, is similar to that for the 2^3S state. All four calculations are in good agreement up to the 2^1P excitation threshold, above this energy the 5-state results deviate from the other calculation. The other three calculations are in quite good agreement up to about the ionization threshold but above this they begin to disagree with each other.

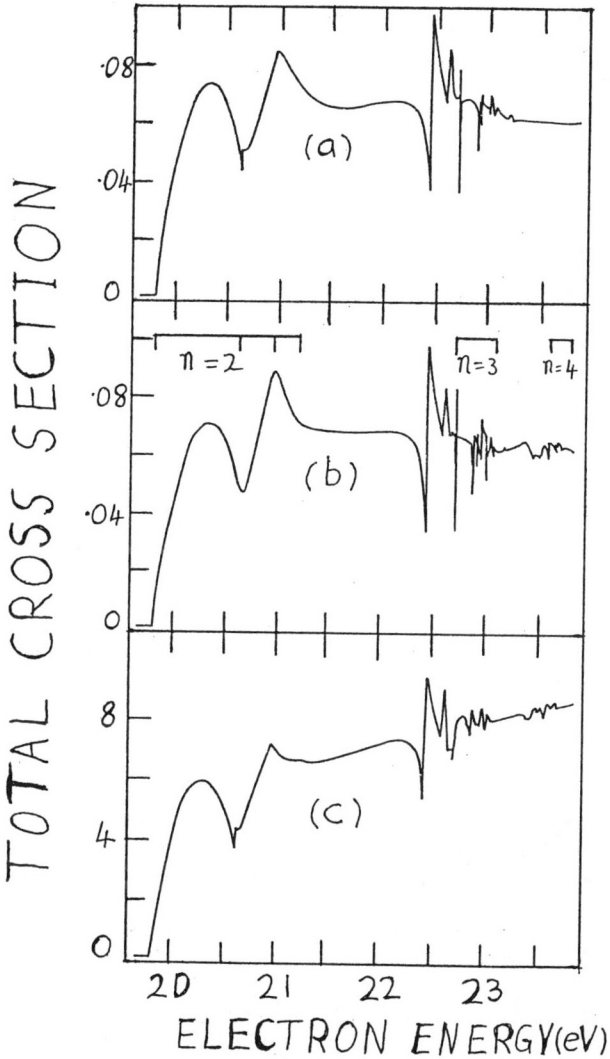

Figure 11. Cross section for metastable atom formation. R-matrix calculation of sum of $1^1S \rightarrow 2^3S + 2^1S$ cross section in (a) 11-state calculation (b) 19-state calculation. Experimental measurements of metastable atom yield in arbitrary units [16, 17]

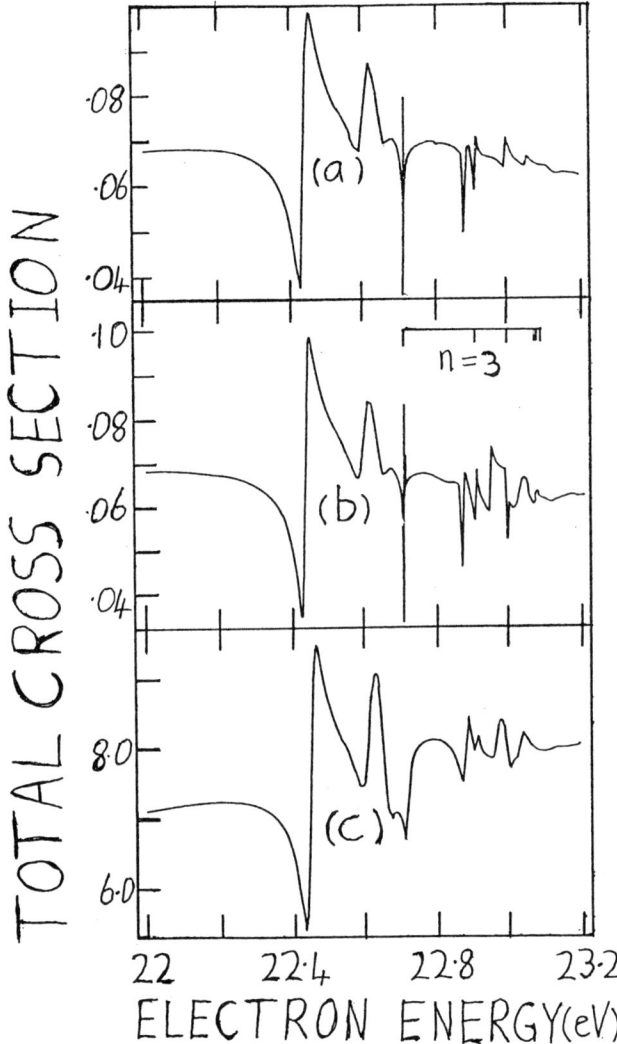

Figure 12 Cross section for metastable atom formation. The legend is the same as for Figure 10.

good overall agreement with the 11 and 19-state calculations. The maxima and minima of the cross sections are all at about the same energies. There is a small feature at the first minima which appears in the 11-state calculation and the experiment but not in the 19-state calculations. However this feature reappears in our most recent 29-state calculations.

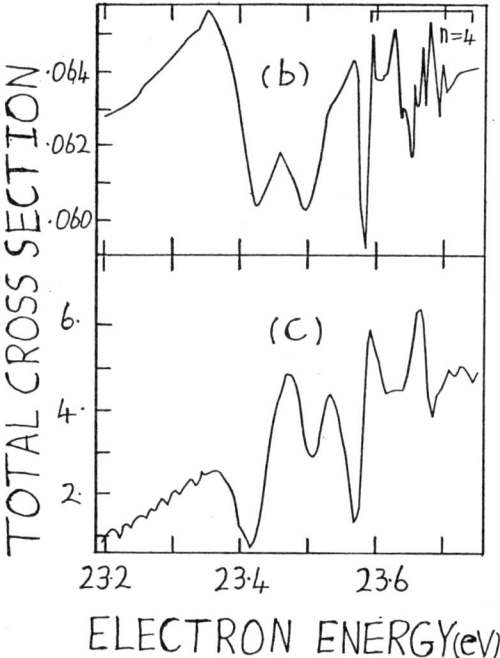

Figure 13 Cross section for metastable atom formation. The legend is the same as for Figure 10.

These two cross sections have exactly the same pattern of convergence as we found for the excitation of the 2s and 2p states of hydrogen and it would appear that this pattern may hold for other atoms.

For the excitation of the 2^3S and 2^1S states of helium experimental work by the Manchester group provides a check on the calculations [16, 17]. In this work they measure the $2^3S + 2^1S$ yield for the electron excitation of ground state helium. The results are not absolute and the fraction of 2^3S to 2^1S produced could vary with energy, hence we would not expect numerical agreement with them but we would expect quantative agreement.

Figure 11 highlights the region close to the n=2 thresholds, the experiments are in

The n=3 threshold region is given in Figure 12 again the maxima and minima all appear at the same energy for the two calculations and the experiments except at about 23ev where the 19-state calculations have a small broad feature not given by the other calculation and experiment. The good qualatative agreement between theory and experiment is maintained up to about 23.5ev (Figure 13) but above that energy the agreement is much less good.

4.2 Electron excitation of helium-like ions

A number of R-matrix calculations have been carried out on the electron excitation of helium-like ions [18]. A comparison of 5-state and 11-state calculations has shown that resonances converging to the n=3 thresholds can have important effects in the electron excitation cross sections to the n=2 levels. Systematic studies including 19 or 29-states are now possible and they would provide useful information on the convergence of the close coupling approximation.

5. Photoionization of lithium-like ions

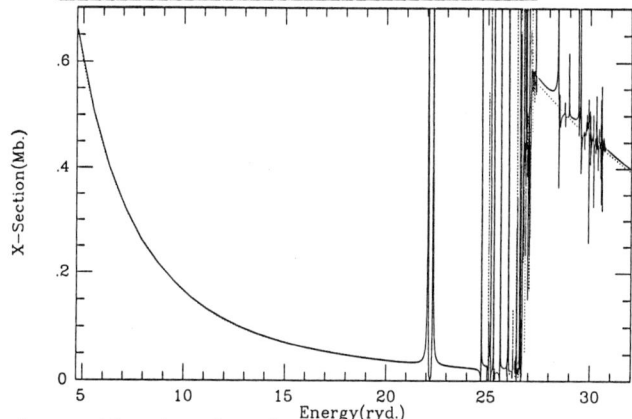

Figure 14 The photoionization cross section for C^{3+} calculated using the R-matrix method, with a 5-state calculation ···· and with an 11-state calculation ——.

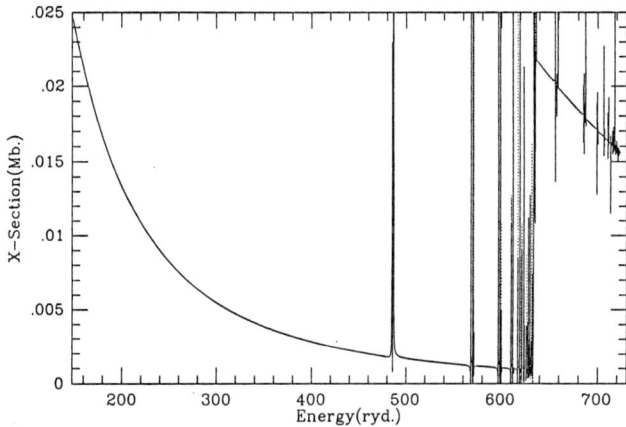

Figure 15 The photoionization cross section for Fe^{23+} calculated using the R-matrix method, with a 5-state calculation ···· and with an 11-state calculation ——.

The helium like ions discussed above form the basis for calculations on the photo-ionization of lithium-like ions. Photo-ionization calculations are in general much easier to carry out than excitation calculations so they can provide information on convergence quite easily.

Figures 14 and 15 give R-matrix calculations for the photoionization of C^{3+} and Fe^{23+} carried out using 5 target states. For both of these calculations it is clear that the cross section have converged.

Acknowledgement

I would like to thank all my colleagues in Belfast for their help and encouragement and for providing me with much unpublished data in time for ICPEAC.

References

1. K A Berrington, P G Burke, M le Dourneuf, W D Robb, K T Taylor and Vo Ky Lan, Comput.Phys.Commun. 14 (1978) 367.

2. K A Berrington, P G Burke and A L Sinfailam, J.Phys.B 8 (1975) 1459.

3. K A Berrington, P G Burke, L C G Freitas and A E Kingston, J.Phys.B 18 (1985) 4135.

4. K A Berrington and A E Kingston, J.Phys.B 20 (1987) 6631.

5. P M J Sawey, K A Berrington, P G Burke and A E Kingston, 16th ICPEAC (1989) Abstract 214.

6. A Pathak, K A Berrington and A E Kingston, J.Phys.B (in preparation) 1989.

7. A Pathak, A E Kingston and K A Berrington, J.Phys.B. 21 (1988) 2939.

8. A J Taylor and P G Burke, Proc.Phys.Soc. 92 (1967) 336.

9. J Callaway, Phys Rev A26 (1982) 199.

10. J Callaway, Phys Rev 32A (1985) 775.

11. J Callaway, 1988 Private communication.

12. M P Scott, T T Scholz, H R J Walters and P G Burke, J.Phys.B (1989) (in press).

13. P G Burke and A J Taylor, J.Phys.B 2 (1969) 44.

14. J B West and G V Marr Proc.Roy.Soc. A349 (1976) 397.

15. A L Stewart, J.Phys.B. 11 (1978) 2449.

16. S J Buckman, P Hammond, F H Read and G C King. J.Phys.B 16 (1983) 4039.

17. A Bass, Private communication (1988). Ph.D. thesis. Manchester University, England, 1988.

18. S S Jayal and A E Kingston J.Phys.B 17 (1984) L145.

THEORETICAL CALCULATIONS OF ELASTIC AND INELASTIC SCATTERING OF ELECTRONS FROM HYDROGEN

D. H. Madison

Department of Physics
University of Missouri-Rolla, Rolla, MO 65401 U.S.A.

Abstract

The electron-hydrogen scattering problem is one of the fundamental problems in atomic physics. Surprisingly, agreement between experiment and even the most sophisticated of theoretical calculations is not good. In this paper, we will discuss our recent attempts at understanding the cause for this disagreement. We have examined the problem both from a perturbation series approach and from the close coupling approach. For the perturbation series method we have calculated exact (no approximations) second order distorted wave amplitudes including second order exchange. For the close coupling method, we have used the optical potential approach and have evaluated the optical potential including the continuum states exactly for the first time. Results will be shown for elastic scattering and for excitation of hydrogen.

I. Introduction

One of the major embarrassments to theoretical atomic physics is the fact that theory and experiment do not agree for scattering of the most elementary projectiles from the most elementary atoms. The electron-hydrogen scattering problem is probably the best example since there would appear to be no reason to not be able to do this problem properly. We believe that we understand the forces of interaction and how to treat the quantum mechanics. There is no evidence which would indicate that the lack of agreement between experiment and theory stems from some new force that has yet to be discovered. For the case of proton scattering, one can at least make the argument that part of the problem lies in numerical difficulties since typical wavelengths are too short for a proper numerical solution of the Schroedinger equation. However, this argument cannot be applied to electron scattering since the wavelengths for typical experiments present no particular numerical difficulties. Consequently, we have a situation where we think we know the physics but at the same time even the most sophisticated theoretical calculations do not agree with experiment.

In this paper, I will describe our attempts to try to understand this problem. If one wishes to perform theoretical calculations of quantum mechanical scattering amplitudes, there are two basic types of approaches which have been used--the perturbation series approach and the close coupling approach. We have been studying this problem from both approaches. I will first discuss the perturbation series results and then the close coupling results.

II. Perturbation Series

The hamiltonian for the electron-hydrogen scattering problem is given by

$$H = T_0 + V + T_1 + V_1 \qquad (1)$$

where T_i is the kinetic energy operator for particle i, V_i is the interaction of particle i with the nucleus and V contains the sum of the interaction of the projectile electron with the nucleus plus the electron-electron interaction. We label the initial projectile electron as "0" and the atomic electron as "1". The first three terms of the perturbation series T-matrix is given by

$$T_{fi} = 2 \langle x_f^-(0)\psi_f(1)|V-U_f|A\psi_i(1)x_i^+(0)\rangle$$

$$+ 2\langle x_f^-(0)\psi_f(1)|U_f|A\psi_i(1)\beta_i(0)\rangle$$

$$+ 2\sum_N \langle x_f^-(0)\psi_f(1)|(V-U_f)A\psi_N(1)$$

$$*\frac{1}{E_N^+ - T_o - U_g}\psi_N^*(1)A(V-U_i)|\psi_i(1)x_i^+(0)\rangle \quad (2)$$

where ψ_i, ψ_f and ψ_N are the initial, final, and intermediate state atomic wavefunctions, A is the antisymmetrizing operator, β_i is a plane wave, and x_i and x_f are initial and final state distorted waves which are obtained from

$$(T_K + U_K - E_K^\pm)x_K^\pm = 0 \quad (3)$$

In (3), the subscript K may be either i or f and $U_i(U_f)$ are the initial or final state distorting potentials. The positive or negative superscripts designate the usual outgoing or incoming wave boundary conditions. Lastly, it is important to note that the distorting potential in the Green's function U_g is an arbitrary spherically symmetric distorting potential for the projectile electron which is independent of the initial and final state distorting potentials U_i and U_f.

The first two terms of (2) represent the first order distorted wave approximation and the third term is the second order term.

For the first order term, the atomic electron makes a single step from the initial state to the final state. In the first order exchange term, the incident electron makes a single step to the final atomic state while the atomic electron is ejected into the final continuum state.

For the second order term, there is an intermediate state involved in the transition. For the direct second order term, the atomic electron makes a transition from the initial state to the intermediate atomic state followed by a transition to the final atomic state. During this time, the projectile electron makes a transition from its initial state to a corresponding intermediate projectile state to the final projectile state. For the second order term, there are three exchange possibilities. In two of these terms, the projectile electron is left in the atom and in the remaining term the projectile electron is finally back in the continuum. The three possibilities are the following. For the first case, the atomic electron makes a transition to the intermediate atomic state followed by a transition to the final projectile state while the projectile makes the corresponding transition from its initial state to the intermediate projectile state and then to the final atomic state. For the second case, the atomic electron makes a transition to the intermediate projectile state and then to the final projectile state while the projectile makes a transition to the intermediate atomic state followed by a transition to the final atomic state. For the last possibility, the atomic electron makes a transition to the intermediate projectile state followed by a transition back into the final atomic state while the projectile makes the corresponding transition to the intermediate atomic state followed by a transition back to the final projectile state. Since these intermediate states are virtual, one must sum over all discrete possibilities and integrate over all continuum possibilities.

When we started this work a few years ago,[1] the second order term was typically either not calculated at all or the direct part of this term was calculated after several approximations had been made. A typical state-of-the-art second order calculation at that time made the following approximations:

a) $U_i = U_f = U_g = 0$
b) The sum over intermediate states was performed using closure

While these second Born calculations (PWB2) with closure seemed to improve agreement between experiment and theory, the agreement certainly was not satisfactory. We initially thought that the problem might simply lie in the various approximations producing a poor representation of the second order term. We were particularly suspicious of the closure

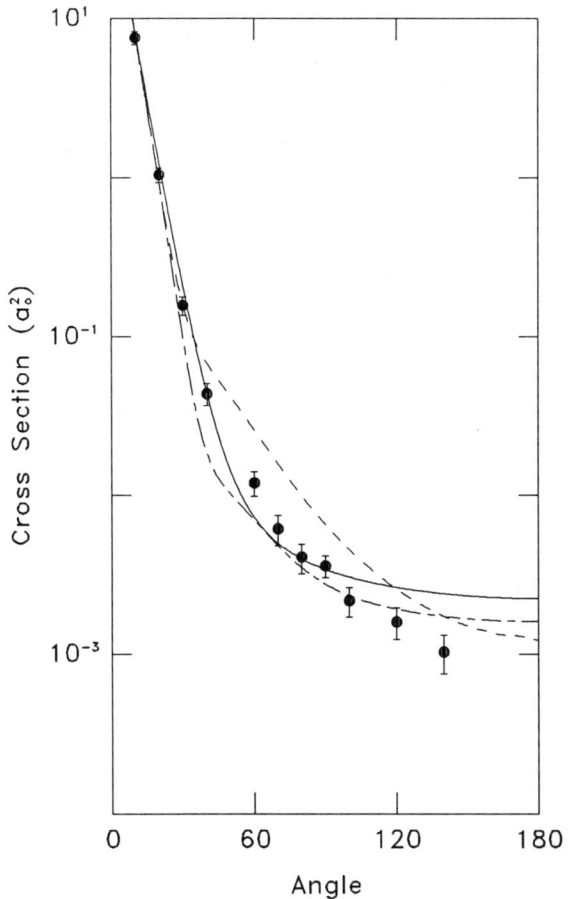

Fig. 1 First and second order distorted wave results for 54.4 eV excitation of the 2p state of hydrogen. The experimental data are those of Williams.[4] The theoretical curves are ——— DWB1, — · — PWB2 and — — — DWB2d(g_o).

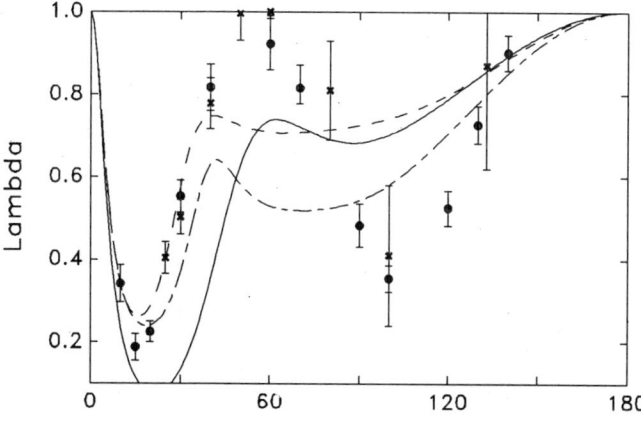

Fig. 2 Same as fig. 1 except for the λ-parameter. The experimental data are: ● Williams[4] and × Weigold et al.[5]

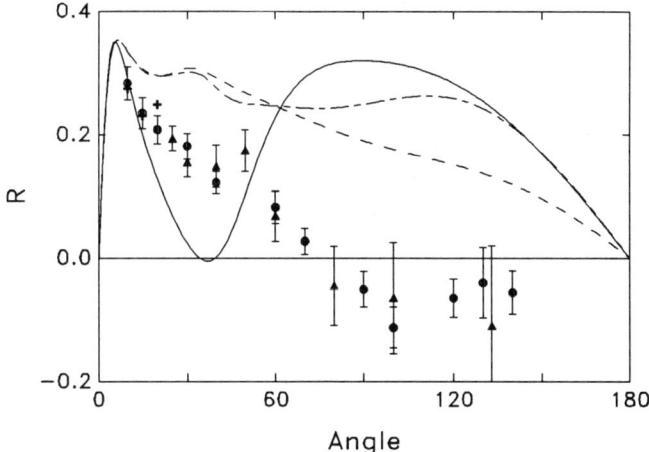

Fig. 3 Same as fig. 1 except for the R-parameter. The experimental data are: ● Williams,[4] ▲ Weigold et al.[5] and + Slevin et al.[6]

approximation since there was solid evidence indicating that the continuum intermediate states should be important and it was not clear that these states would be well represented by closure. With the exception of a PWB2 calculation for elastic scattering and a similar calculation for 2s excitation by Ermolaev and Walters,[2,3] no second order calculation had properly included continuum intermediate states. Consequently, we decided to develop the necessary numerical techniques which would allow one to calculate the second order direct amplitude without making any approximations. This meant that we had to be able to sum over all contributing discrete states and integrate over all contributing continuum intermediate states and we wanted to be able to perform calculations for arbitrary distorting potentials. It took us almost two years to accomplish this objective.

Second order results including the direct term only are compared with first order results in figs. 1-3 for the differential cross section (DCS), λ-parameter and R-parameter respectively. These figures contain the first order distorted wave approximation (DWB1), the second order plane wave Born approximation PWB2 ($U_i = U_f = U_g = 0$), and the second order distorted wave approximation DWB2d(g_o) (U_i = initial state potential, U_f = final state potential, $U_g = 0$ and direct term only). All calculations include first order exchange. There are two striking aspects of these figures:

a) Second order effects are very important
b) Second order calculations including the direct term only and no distortion in the Green's function are not in good agreement with experiment.

The next question concerns the effect of the second order exchange terms. Initially it was believed that these terms should not be important since first order exchange was relatively unimportant. The customary logic is that, at least for the distorted wave series, the perturbation series should be converging and each succeeding term should be smaller than the proceeding one. This is observed for the second order direct DWB2 term which, although important, is smaller than the first order term. This logic would indicate that second order exchange should be almost negligible since first order exchange is small. However, results from close coupling calculations and ionization calculations indicated that second order exchange might be more important than previously thought. Consequently, we decided to include these terms in our calculations. This, of course, greatly increased the length and complexity of the calculation since a full second order calculation had to be performed for each of the three different exchange terms. While exchange terms are normally expected to be one to two orders of magnitude more difficult to evaluate than the direct terms, we have found that it is possible to cast the three terms in such a manner that they can be evaluated with approximately the same difficulty as the direct term.

Second order results including second order exchange are shown in figs. 4-6. Each figure contains the direct result DWB2d(g_o) and the corresponding calculation including exchange DWB2e(g_o). From these figures we see that not only is second order exchange very important, it also significantly improves agreement between experiment and theory for the DCS and λ-parameter. For the DCS, the second order exchange results are in very good agreement with experiment over the entire angular range. While the agreement is not as good for the λ-parameter, the effect of second order exchange is still dramatic. For the R-parameter, there is little similarity between experiment and theory even with second order exchange.

The next question concerns the importance of the distorting potential in the Green's function U_g. Prior to our work, all theoretical second order calculations were performed with $U_g = 0$; i.e. the free particle Green's function was used in the second order interaction operator. However, in our original work for the direct term,[7,8] it was shown that the results were sensitive to the choice for this distorting potential. It can be shown that this potential plays the role of an effective average potential for all the intermediate states. As a result this average distorting potential U_g is roughly analogous to the average energy associated with closure calculations. Since it had been previously assumed that the results would be insensitive to this potential and further that $U_g = 0$ would be adequate, we initially performed calculations with various choices for U_g to determine the effect of this potential.

As a first exercise, we performed calculations with U_g equal to the distorting potential of one of the bound states of hydrogen. This exercise revealed that the results were fairly sensitive to U_g and further that setting U_g equal to a lower bound state tended to give results closer to the experimental data than the potentials for the higher bound states. Since U_g is an effective average potential for all the intermediate states, it would seem more appropriate to use some weighted average of the distorting potentials for all the intermediate states than the potential for a single intermediate state. While it is not immediately clear how to assign weights for various intermediate states, one sensible possibility would be to weight a particular state according to the total integrated cross section for exciting that state. Consequently, we have formed an average distorting potential U_g which is the weighted average of the

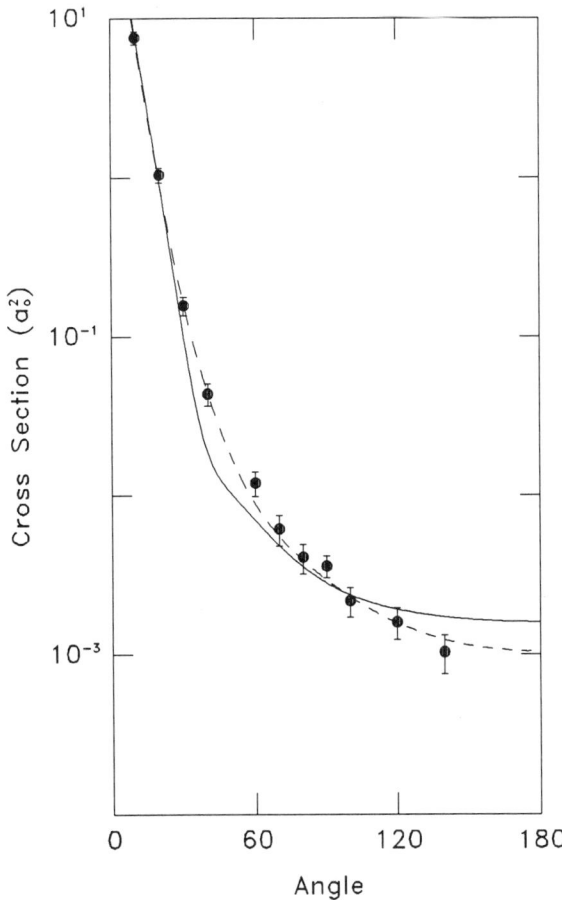

Fig. 4 Second order calculations including the direct term only ——— DWB2d(g_o) and the direct plus exchange terms - - - DWB2e(g_o) for 54.4 eV excitation of the 2p state of hydrogen. The experimental data are those of Williams.[4]

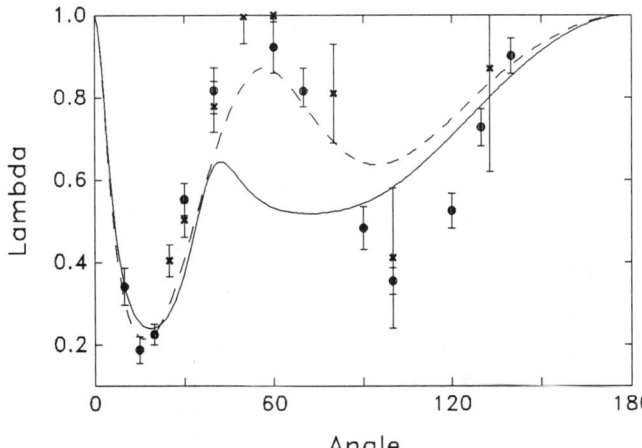

Fig. 5 Same as fig. 4 except for the λ-parameter. The experimental data are the same as fig. 2.

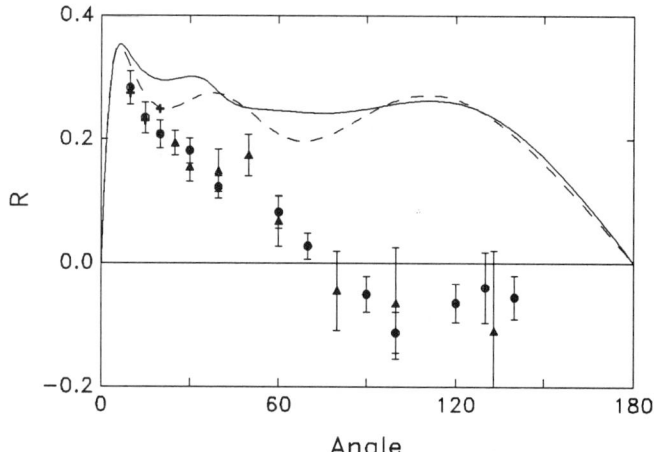

Fig. 6 Same as fig. 4 except for the R-parameter. The experimental data are the same as fig. 3.

distorting potentials for the seven states 1s, 2s, 2p, 3s, 3p, 4s and 4p with the weights determined by the individual integrated cross sections. Figures 7-9 show the results using this average distorting potential for U_g, which we label as DWB2e(g_{ave}), results when U_g is the distorting potential for the 1s state DWB2e(g_{1s}) as well as the DWB2e(g_o) results.

From these figures, it is seen that the angular correlation parameters are more sensitive to the form of the distorting potential in the Green's function than is the differential cross section. Both of the non-zero choices for U_g tend to improve agreement between experiment and theory. For the DCS, the choice of $U_g = U_{1s}$ is in even better agreement with experiment than the original DWB2e(g_o) results. For the λ-parameter, both non-zero choices give improved agreement at small scattering angles and the U_{1s} results are in fairly good agreement with experiment over the entire angular range. For the R-parameter, distortion in the Green's function significantly improves agreement with experiment with the U_{1s} results being closer to the data. The large angle R-parameter continues to be baffling in that experiment and theory don't even have the same sign.

We are still in the process of trying to come to a better understanding of the role of distortion in the Green's function so these results

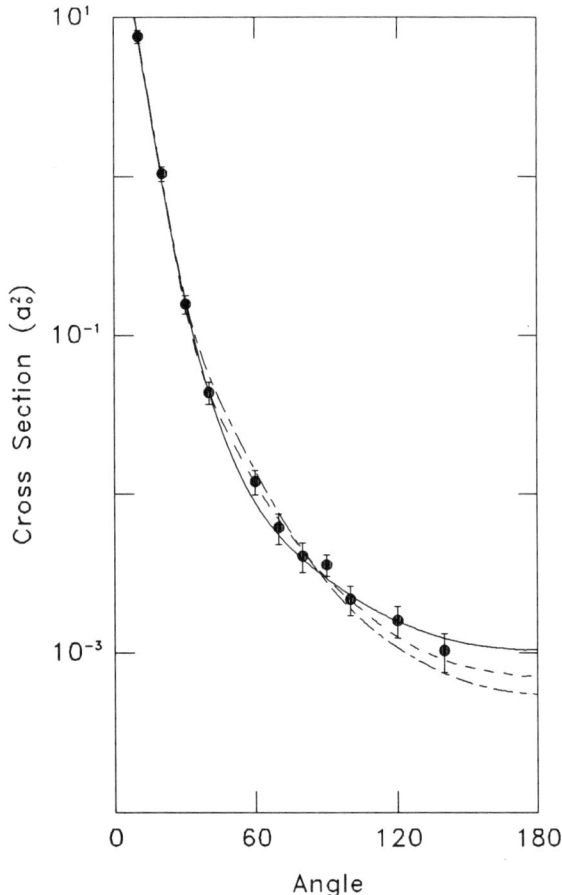

Fig. 7 Second order exchange calculations with different distorting potentials in the Green's function for 54.4 eV excitation of the 2p state of hydrogen. The experimental data are those of Williams.[4] The theoretical curves are: ——— DWB2e(g_o); - - - DWB2e(g_{1s}); and — - — DWB2e(g_{ave}).

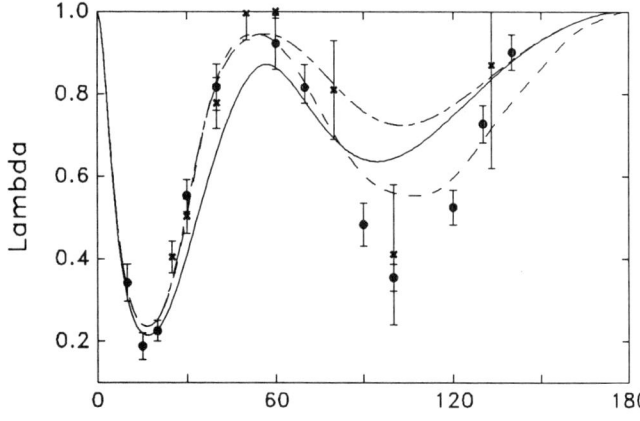

Fig. 8 Same as fig. 7 except for the λ-parameter. The experimental data are the same as fig. 2.

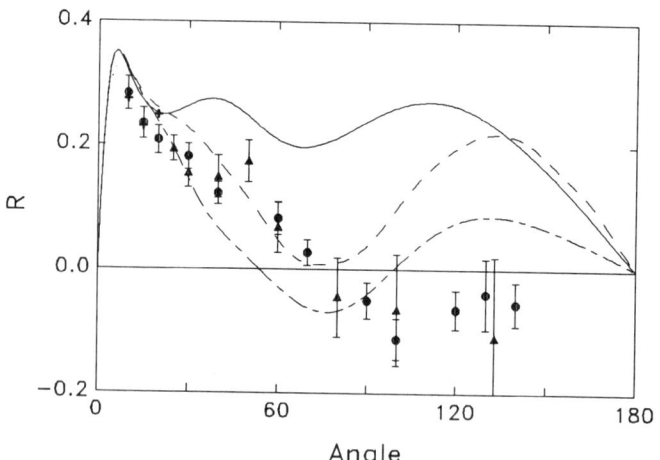

Fig. 9 Same as fig. 7 except for the R-parameter. The experimental data are the same as fig. 3.

should be regarded as preliminary. So far we have not included any continuum contribution in the formation of U_g. This is probably not very realistic since the continuum states play an important role in the second order process. As a result, we plan to modify our codes to allow for long range Coulomb potentials so that we can take the continuum states into account. It is also possible that it is inappropriate to try to represent the distortion of all the intermediate states by a single average potential in the Green's function. Consequently, we are also examining the possibility of setting U_g equal to the distorting potential for each individual intermediate state. We hope to be able to address these issues more definitively in the near future.

III. Close Coupling Approach

In the close coupling approach, one expands the scattering wavefunction in terms of a complete set of target states. This complete set will contain contributions from both discrete states and continuum states. Unfortunately, it is not presently possible to explicitly include continuum states in this expansion. However, our perturbation series work shows that the continuum states are important and must be taken into account. The two current methods for including the continuum are either through the use of pseudostates or through the use of optical potentials.

Madison and Callaway[9] have shown, however, that pseudostates do not accurately represent the continuum for a perturbation series calculation so it is highly questionable if pseudostates can accurately represent the continuum for close coupling calculations. Consequently, the optical potential method would appear to provide the most promise for including the continuum in close coupling calculations. In this method, a subset of the discrete states are treated explicitly in the close coupling expansion and the remaining discrete and continuum states are treated through the optical potential.

In the optical potential approach, the Schroedinger equation for the scattering problem can be written[10]

$$P(E-T_1-V_1-T_0-V^1-V^Q)P\Psi_i^+ = 0 \quad (4)$$

where P is the operator that projects the states to be included explicitly in the close coupling expansion, Ψ is the scattering wavefunction, and V^1 is defined to be

$$V^1 = V_0 + V_{01} + (-)^S(H-E)P_{01} \quad (5)$$

Here P_{01} is the operator that interchanges particles 0 and 1, V_{01} is the electron-electron interaction and V^Q is the optical potential which is given by

$$V^Q = V^1 + V^1 Q \frac{1}{Q(E^+-H)Q} QV^1 \quad (6)$$

Here Q is the operator that projects the states which are not explicitly included in the close coupling expansion. The optical potential (6) is similar to the interaction operator in the second order term of the perturbation series in (2). Consequently, we can evaluate the optical potential using the same technique we used to evaluate the second order T-matrix if we approximate

$$H \approx T_1 + V_1 + T_0 + U_g \quad (7)$$

The net effect of this approximation is to neglect off diagonal couplings in Q-space. All couplings in P-space and between P-space and Q-space are retained however.

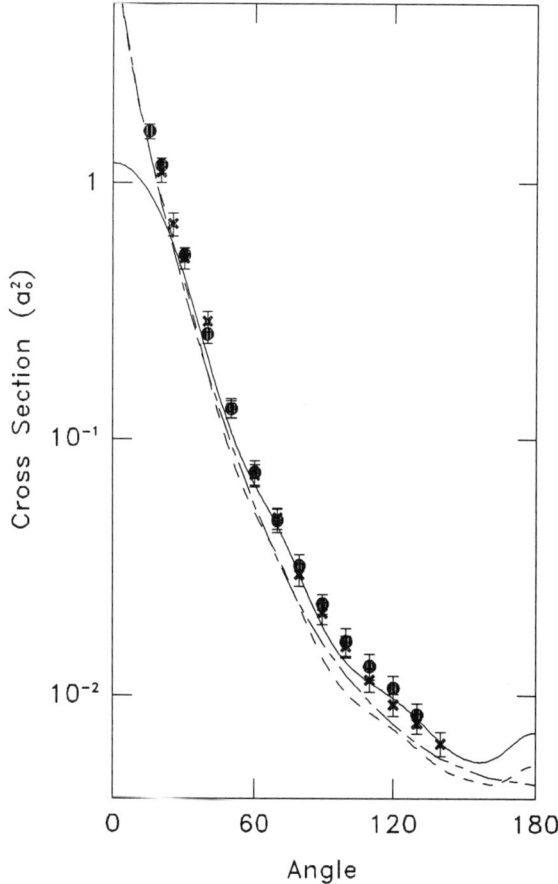

Fig. 10 Differential cross sections for 100 eV elastic scattering of electrons from hydrogen. The experimental data are: x Williams[15] and ● van Wingerden et al.[16] The theoretical calculations are: ———— DWB1; — — — DWB2d(g_{1s}); and — - — - CCd(g_{1s}).

We have developed computer codes to evaluate the optical potential making no other approximation than (7). All contributing continuum states are explicitly integrated over and the codes are designed for arbitrary distorting potentials U_g. This work represents the first calculation to explicitly include the continuum states without approximations in a close coupling calculation. Previous works have either made simplifying approximations[10] or have used pseudostates to represent the continuum.[11-1]

The optical potential contains one direct and three exchange terms just as the second order T-matrix. To date, we have included the direct term in our close coupling computer code and are currently putting in the exchange terms. For these calculations, P-space was the 1s state, Q-space was everything else, and the Green's function distorting potential $U_g = U_{1s}$. Results for 100 eV incident electrons are compared with perturbation series results and with experiment in fig. 10. In this figure, the close coupling results obtained from the direct term of the optical potential $CCd(g_{1s})$ are compared with the DWB1 and $DWB2d(g_{1s})$ results.

There are several interesting aspects of fig. 10. First, while the DWB1 results are clearly too small for small scattering angles, they are in reasonable agreement with data at the larger angles. When the second order direct term is added, the results for small scattering angles are increased to bring them more into line with experiment, but the large angle results are decreased away from experiment. The close coupling calculation is a more complete calculation than the DWB2. If the close coupling T-matrix is expanded, it can be seen that the close coupling results contain the DWB2 plus higher order terms in the perturbation series. However, there is not much difference between the $DWB2d(g_{1s})$ results and the $CCd(g_{1s})$ results. A detailed comparison of the two calculations reveals that they lie within about 10% of each other. This suggests that the higher order terms in the perturbation series contained in the CC results represents about a 10% effect. It remains to be seen if the exchange terms in the optical potential will increase the close coupling results to the level of the experimental data. For the DWB2 calculation, exchange did increase the large angle cross section and brought experiment and theory into better accord. Unfortunately, however, the small angle cross sections were lowered away from experiment.

In conclusion, we are now in a position to evaluate second order amplitudes for perturbation series and optical potentials for close coupling calculations without making simplifying approximations. Probably the most significant advancement represented by this work lies in the fact that we can now include continuum states in these calculations exactly. The most significant recent results are the fact that second order exchange terms are important and dramatically improve agreement between experiment and theory. In spite of all this work, we still have not achieved satisfactory agreement between experiment and theory but there are positive indications in that direction.

Acknowledgements

This work was performed in collaboration with I. Bray and I. E. McCarthy who deserve much of the credit for the results presented here. This work was supported by the NSF.

References

1. D. H. Madison, Phys. Rev. Lett. 53, 42 (1984).
2. A. M. Ermolaev and H. R. J. Walters, J. Phys. B12, L779 (1979).
3. A. M. Ermolaev and H. R. J. Walters, J. Phys. B13, L473 (1980).
4. J. F. Williams, J. Phys. B14, 1197 (1981).
5. E. Weigold, L. Frost, and K. J. Nygaard, Phys. Rev. A21, 1950, (1980).
6. J. Slevin, M. Eminyan, J. M. Woolsey, G. Vassilev, and H. Q. Porter, J. Phys. B13, L341 (1980).
7. D. H. Madison and K. H. Winters, J. Phys. B20, 4173 (1987).
8. D. H. Madison and K. H. Winters, J. Phys. B22, 1651 (1989).
9. D. H. Madison and J. Callaway, J. Phys. B20, 4197 (1987).
10. I. E. McCarthy and A. T. Stelbovics, Phys. Rev. A28, 2693 (1983).
11. J. Callaway, Phys. Rept. 45, 89 (1978).
12. L. A. Morgan, J. Phys. B15, 4247 (1982).
13. J. Callaway and D. H. Oza, Phys. Rev. A34, 965 (1986).
14. W. L. van Wyngaarden and H. R. J. Walters, J. Phys. B19, 929 (1986).
15. J. F. Williams, J. Phys. B8, 2191 (1975).
16. B. van Wingerden, E. Weigold, F. J. de Heer, and K. J. Nygaard, J. Phys. B 10, 1345 (1977).

ELECTRON IMPACT IONIZATION OF U^{88+}–U^{91+}[*]

Nelson Claytor[†,a], B. Feinberg[†], Harvey Gould[†],
Curtis E. Bemis, Jr.[‡], Jorge Gomez del Campo[‡], Carl A. Ludemann[‡], and Charles R. Vane[‡]

[†]Lawrence Berkeley Laboratory, University of California, Berkeley, California 94720

[‡]Oak Ridge National Laboratory, Oak Ridge, Tennessee 37831

We channeled 405 MeV/nucleon uranium ions in Si single crystals to determine the electron impact ionization cross sections for hydrogenlike–berylliumlike uranium ions by 222 keV electrons. Our cross sections are 3.9, 11.0, 16.0, and 31.0 barns (+100%,–50%) respectively, for ionizing 1s, $1s^2$, 2s, and $2s^2$ electrons. Our 1s and $1s^2$ results disagree with present theory; our 2s and $2s^2$ results are not accurate enough to distinguish between theories.

I. Introduction

We report here a measurement of electron impact ionization cross sections for few–electron uranium ions. Until the work described here, there was no way to make such measurements. The measurements require, in addition to the few–electron uranium ions, a dense electron gas of known thickness.

We obtain very high charge state ^{238}U ions from the Lawrence Berkeley Laboratory's Bevalac. We obtain a dense electron gas by channeling[1),2] the ions through a single crystal of silicon. In a single crystal of Si, the atoms are arranged in a periodic structure. Within this structure are "channels," along which there are no nuclei. If the crystal is precisely aligned with the incident beam, the incident ions travel in these channels, guided by electromagnetic interactions with the nuclei of the target crystal. Ions traveling in the channels make only large–impact–parameter collisions with the Si nuclei, since a small–impact–parameter collision with a nucleus knocks the ion out of the channel. Since for channeled ions the collisions with the nuclei are distant, the channeled ions do not acquire enough energy in these collisions to ionize their tightly bound electrons. The ionization of channeled ions is due to collisions with the (essentially free) valence electrons of the solid. We measure the thickness of the electron gas by comparing the measured radiative electron capture (REC) cross sections to REC cross sections measured in a previous experiment[3] for unchanneled ions.

II. Experimental details

For each combination of ion, crystal, and beam energy there is a maximum transverse energy beyond which the ions will be deflected out of the channel. For our experimental conditions (few–electron 405 MeV/nucleon uranium incident on the <110> axis of Si), this maximum energy is reached when the angle between the beam and the crystal is about 0.01°. In order to achieve the necessary small transverse energy, we collimate the uranium beam with circular apertures 0.30 and 0.15cm in diameter, separated by a distance of 10.6m. This defines a maximum beam divergence half-angle of 0.012°, and also decreases the beam intensity by a factor of about 5000. We thus obtain a count rate of about one channeled ion per second.

We use 405 MeV/nucleon uranium in this experiment. Seen in the rest frame of the uranium, the electrons in the Si crystal have a kinetic energy of 222 keV. The binding energy of U^{91+} is about 133 keV. We measure ionization cross

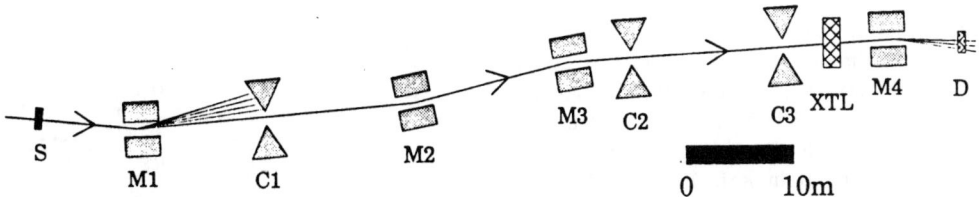

FIGURE 1. Diagram of the beam line and experimental apparatus. M1–M4 are dipole bending magnets and C1–C3 are collimators. Six quadrupole doublets used for focusing are not shown. The spacing of the components along the beam line is to scale, but everything else is shown schematically. 407 MeV/nucleon U^{40+} from the Bevalac is stripped at S and the resulting charge states are analyzed by M1. C1 selects a single uranium charge state and C2 and C3 collimate the beam to achieve the necessary small beam divergence. After the beam passes through the crystal XTL, the resultant charge state distribution is analyzed by M4 and detected by the position sensitive proportional counter D.

sections for incident charge states from hydrogen-like U^{91+} through beryllium-like U^{88+}. These incident charge states (and bare U^{92+}) are produced by stripping 407 MeV/nucleon U^{40+} at the exit of the Bevalac and magnetically separating the resulting charge states (see Figure 1). The ions lose about 2 MeV/nucleon in the stripper. The ions are then collimated (C2 and C3 in Figure 1) and channeled along the <110> axis of either a 0.11 or a 0.37mm thick Si single crystal (XTL in FIgure 1). We use thick crystals because at relativistic energies the charge-changing cross sections are small. Thick crystals also allow us to ignore the effects of thin layers of dirt, oxides, and dislocated atoms on the surface of the crystal.

After the ions exit the Si crystal, the resultant charge state distribution is magnetically analyzed (M4 in Figure 1), and the ions are detected by a position sensitive proportional counter (D in Figure 1). The raw data used for the determination of charge-changing cross sections are the relative charge state fractions of the ions exiting the crystal.

At 405 MeV/nucleon the cross section for ionization of uranium ions by collisions with target nuclei is much larger than that for collisions with target electrons. The purpose of channeling the ions is to reduce or completely eliminate these collisions with the Si nuclei. Figure 2(a) shows that the percentage of incident U^{89+} ions which exits the 0.11mm-thick Si crystal without changing charge state increases from about 7% to 50% as the crystal's <110> axis is rotated into alignment with the beam. Figure 2(b) shows a similar effect for U^{89+} incident on the 0.37mm-thick Si crystal. Figures 3(a) and 3(b) show our raw data for charge state distributions exiting the 0.37mm-thick crystal for incident U^{89+} ions traversing the crystal in a random direction [Figure 3(a)] and along the <110> axis [Figure 3(b)]. Figures 3(a) and 3(b) show the dramatic change in the charge state distribution when channeling occurs relative to a random orientation of crystal and beam; the Figures also show that the U^{89+} and U^{90+} exiting the crystal have lost less energy when they channel in the crystal than when they traverse the crystal in a random direction. This reduction in energy loss is the usual signature for channeling of heavy ions.

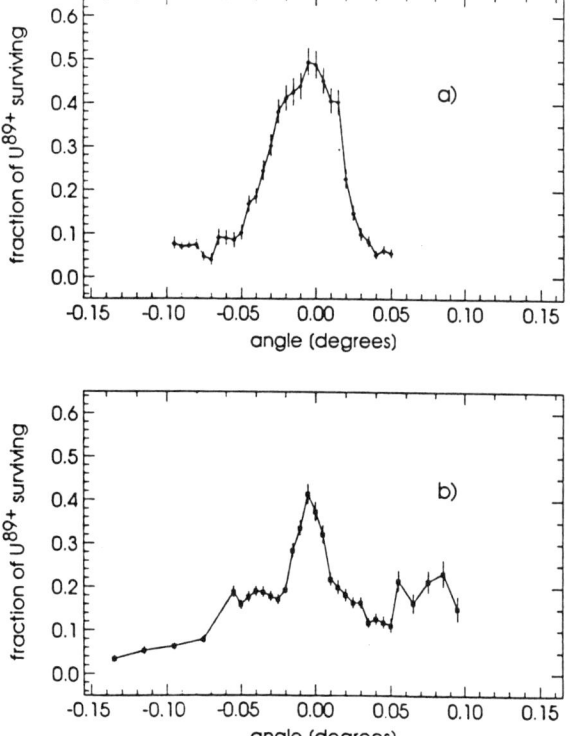

FIGURE 2. Rocking curves, showing the fraction of U^{89+} observed surviving passage through (a) the 0.11-mm Si crystal and (b) the 0.37-mm Si crystal as a function of angle between the U^{89+} beam and the <110> axis of the crystal. The expected channeling half-angle (including thermal vibrations), calculated from formulas in Ref. 1, is 0.011° (0.19 mrad). The half-angle in (a) is 0.025°, possibly because the crystal is strained in its mount. The channeling half-angle of the central peak in (b) is 0.011°, consistent with the predicted value. The high fraction of U^{89+} at larger angles may be due to planar channeling effects. When the 0.37-mm crystal was moved to a fully random orientation, the fraction of U^{89+} was less than 2%.

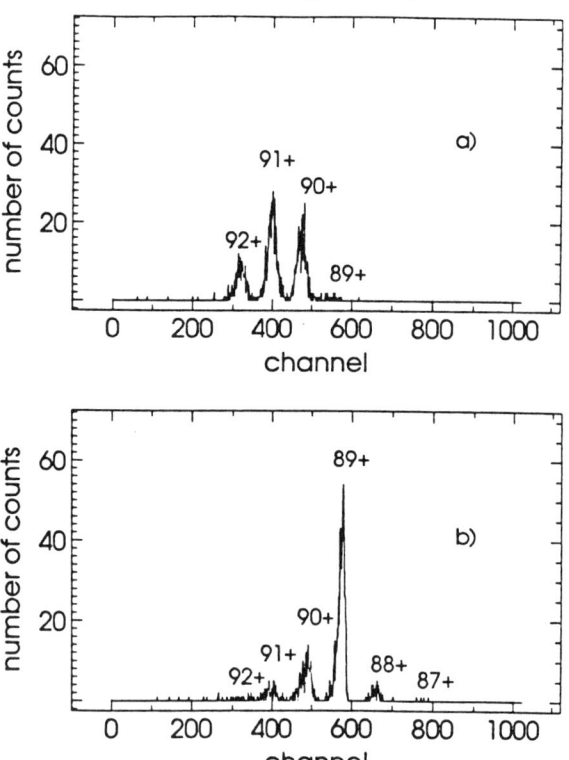

FIGURE 3. Observed charge state distributions from 405 Mev/nucleon U^{89+} exiting the 0.37-mm thick Si crystal. (a) The ions pass through the crystal in a random direction relative to the <110> axis of the crystal. (b) The incident ions are aligned with the <110> axis of the crystal. Approximately 80% of the ions in (b) have channeled.

III. Data analysis

Even if an ion has small transverse energy and is aligned with the crystal, it may not channel. This can occur if, for instance, the ion enters the crystal too close to a row of nuclei. Ions which do not channel present a large background of ionization from ion–nucleus collisions; we subtract them from our measurement. Because of the considerably larger ionization probability for ions traversing the crystal in the random direction then for ions which channel, we assume that all ions which have lost several electrons have not channeled. Thus in Figure 3(b), the U^{91+} and U^{92+} come from U^{89+} ions which did not channel.

When we compare the fractions of U^{91+} in Figures 3(a) and 3(b), we find that 80% of the U^{89+} aligned with the <110> axis channeled in the 0.37mm-thick crystal. When we include data from all incident charge states and from both crystals, we find that 79%±2% of the incident ions channeled in the 0.37mm-thick crystal, and that 38%±2% of the incident ions channeled in the 0.11mm-thick crystal. We think that the smaller channeling fraction for the 0.11mm-thick crystal is a result of the thinner crystal being more strained in its mount than the thicker 0.37mm-thick crystal. This is also reflected in a much larger acceptance angle for the 0.11mm-thick crystal [Figure 2(a)] than for the 0.37mm-thick crystal [Figure 2(b)].

Previous experiments[2),4),5)] have seen large differences between charge state distributions for channeled and unchanneled ions and have measured total charge changing cross sections[5)] for oxygen ions. However, ours is the first experiment to measure the electron density seen by the channeled ions. This measurement allows us to extract electron impact ionization cross sections from our data.

We measure the electron density integrated along the paths of the channeled ions by comparing a cross section for electron capture by the channeled ions with a previously measured[3)] capture cross section for ions in the random direction. What makes this comparison possible is that the only capture process involved in either case is REC, which to a good approximation involves only the target electrons and not the target nuclei. (REC is the process in which a free or loosely bound electron is captured by the ion with the simultaneous emission of a photon–the inverse of photoionization.) Thus REC scales linearly with the electron density in the target. For relativistic ions in low-Z targets, REC has been shown to be the dominant capture mechanism both for channeled ions[6)] and for ions in random directions (amorphous materials), where cross sections have been measured,[3),7)–9)] and agree with theory.[3),7)–9)] The dominance of REC over small-impact-parameter processes like nonradiative capture is intuitively obvious for channeled ions.

Production of U^{88+} by REC can easily be seen in Figure 3(b) [but not in Figure 3(a) because of competition from the large probability of ionization]. Comparing capture using incident charge states U^{89+}–U^{92+}, we find the average integrated electron density in the <110> channel of Si. to be 0.44±0.01 of the electron density in the bulk material, or approximately 6.2 electrons per Si nucleus. The quoted uncertainty in the measurement is purely statistical—it does not reflect the much larger systematic uncertainties in our measurements and in the measurements in Ref. 9.

We determine cross sections for capture and ionization by a least-squares fit[10)] to the curves of charge state yield versus target thickness. We estimate the uncertainty in the cross sections to be a factor of two (from 50% smaller to 100% larger), because of systematic uncertainties in combining a large number of measurements made with only two target thicknesses, uncertainties in determining the channeling fractions, other possible effects not included in our analysis, uncertainties in previously measured capture cross sections, and uncertainties in the limits of validity of our approximations.

To determine electron impact ionization cross sections from our measured ionization cross sections, we multiply the cross sections from our least-squares fit by the relative electron density in the channel. Further, to compare our ex-

TABLE I. Electron impact ionization cross sections (barns).

Ion	State	Expt.	PMG[a)]	PB[b)]	Sco[c)]	Younger	Lotz[g)]
U^{91+}	1s	3.9	1.1		1.5	0.8[d)]	0.7
U^{90+}	$1s^2$	11.0	2.2		3.0	1.7[e)]	1.4
U^{89+}—U^{90+}	2s	16.0		13.0	29.0	9.4[d)]	12.0
U^{88+}—U^{90+}	$2s^2$	31.0		26.0	57.0	19.8[f)]	24.0

a) From Ref. 17.
b) Extrapolated from Ref. 16.
c) Extrapolated from Ref. 11.
d) Extrapolated from Ref. 12.
e) Extrapolated from Ref. 13.
f) Extrapolated from Ref. 14.
g) Extrapolated from Ref. 15.

perimental results with calculations of L–shell ionization, we subtract our cross section for ionization of U^{90+} (two K–shell electrons) from our measured cross section for ionization of U^{89+} (one L–shell electron) and for ionization of U^{88+} (two L–shell electrons).

IV. Results

Our electron impact ionization cross sections for hydrogenlike U^{91+}—berylliumlike U^{88+} by 222–keV electrons are listed in Table I and are compared with theory. Our cross sections for ionization of U^{88+}—U^{91+} are compared to calculations of K–shell and L–shell ionization by Scofield[11], Younger[12-14], and Lotz[15]; K–shell ionization alone by Pindzola, Moores, and Griffin[17]; and L–shell ionization alone by Pindzola and Buie[16]. Our U^{90+} and U^{91+} (K–shell) cross sections of 11.0 and 3.9 barns respectively (Table I), determined to a factor of two, are not in agreement with the K–shell ionization cross sections extrapolated from Scofield,[11] or with the calculations of Younger,[12-14] or with the calculations of Pindzola, Moores, and Griffin,[17] or with the formula of Lotz.[15] We do no think that ionization of excited states, populated by electron excitation, makes a significant contribution to our measured U^{91+} or U^{90+} cross sections. This is because the mean free time between ionizing collisions is considerably longer than the radiative lifetime of all the low–lying states of U^{91+} and U^{90+} except the $1s2s\,^3P_0$ state of U^{90+}, which is not easily populated.[18]

As noted above, in order to facilitate comparison of our experimental results with calculations of L–shell cross sections, we subtract our experimental result of 11.0 barns for ionization of U^{90+} from our measured ionization cross sections of 27.0 barns for U^{89+} (one L–shell electron) and 42.0 barns for U^{88+} (two L–shell electrons). Our L–shell results are not sufficiently accurate to distinguish between the different calculations.[11-16]

We thank Paul Luke and Jack Walton for preparing the crystals, Joseph Jaklevic, Steve Withrow, and Ray Zuhr for advice and assistance in crystal prealignment, George Kalnins for accurately calculating the beam optics, Richard Leres for data acquisition software support, and Lynette Levy for editorial assistance. We especially thank the operators, staff, and management of the Bevalac for making experiments with relativistic uranium possible. This work was supported by the Director, Office of Energy Research, Office of Basic Energy Sciences, Chemical Sciences Division; and Office of High Energy and Nuclear Physics, Nuclear Sciences Division, U.S. Department of Energy under Contract No. DE–AC–03–76SF00098 (LBL); and the Office of High Energy and Nuclear Physics, Nuclear Sciences Division of the U.S. Dpeartment of Energy under Contract No. DE–AC–05—84OR21400 with Martin Marietta Energy Systems, Inc. (ORNL).

*) N.Claytor, B.Feinberg, H. Gould, C.E. Bemis Jr.,J. Gomez del Campo, C.A. Ludemann, and C.R. Vane, Phys. Rev. Lett. **61**, 2081 (1988).

a) Present address: University of Pennsylvania Physics Dept., 209 S. 33rd. St., Philadelphia, PA 19104.

1) For a general review of channeling, see D.S. Gemmel, Rev. Mod. Phys. **46**,129 (1974).

2) For a review of heavy ion channeling, see for example S. Datz and C.D. Moak, in Treatise on Heavy Ion Science, edited by D.A. Bromley (Plenum, New York, 1984), Vol. 6, p. 169.

3) R. Anholt, W.E. Meyerhof, X.–Y. Xu, H. Gould, B. Feinberg, R.J. McDonald, H. Wegner, and P. Thieberger, Phys. Rev. A **37**, 1586 (1987).

4) H.O. Lutz, S. Datz, C.D. Moak, T.S. Noggle, and L.C. Northcliffe, Bull. Am. Phys. Soc. **11**, 177 (1966); F.W. Martin, Phys. Rev. Lett. **22**, 329 (1969); M. Kaminsky, in Recent Developments in Mass Spectroscopy, Proceedings of the International Conference on Mass Spectroscopy, Kyoto, Japan, 1969, edited by K. Ogata and T. Hayakawa (University Park Press, Baltimore, MD, 1970), p.1167; H.O. Lutz, S. Datz, C.D> Moak, and T.S. Noggle, Phys. Lett. **33A**, 309 (1970); T. Andersen, S. Datz, P. Hvelplund, and G. Sorenson, Phys. Lett **33A**, 121 (1970).

5) S. Datz, F.W. Martin, C.D. Moak, B.R. Appleton, and L.B. Bridwell, Radiat. Eff. **12**, 163 (1972).

6) S.A. Andriamonje, M. Chevallier, C. Cohen, J. Dural, M.J. Gaillard, R. Genre, M. Hage–Ali, R. Kirsch, A. L'Hoir, B. Mazuy, J. Mory, J. Moulin, J.C. Poizat, J. Remillieux, D. Schmaus, and M. Toulemonde, Phys. Rev. Lett. **59**, 2271 (1987).

7) H. Gould, D. Greiner, P. Lindstrom, T.J.M. Symons, and H. Crawford, Phys. Rev. Lett. **52**, 180, 1654(E) (1984).

8) R. Anholt, S.A. Andriamonje, E. Morenzoni, Ch. Stoller, J.D. Molitoris, W.E. Meyerhof, H. Bowman, J.–S. Xu, Z.–Z. Xu, J.O. Rasmussen, and D.H.H. Hoffmann, Phys. Rev. Lett. **53**, 234 (1984).

9) W.E. Meyerhof, R. Anholt, J. Eichler, H. Gould, Ch. Munger, J. Alonso, P. Thieberger, and H. Wegner, Phys. Rev. A **32**, 3291 (1985).

10) S. Datz, H.O. Lutz, L.B. Bridwell, C.D. Moak, H.D. Betz, and L.D. Ellsworth, Phys. Rev. A **2**, 430 (1970).

11) J.H. Scofield, Phys. Rev. A **18**, 963 (1978).

12) S.M. Younger, Phys. Rev. A **22**, 111 (1980), and private communication.

13) S.M. Younger, Phys. Rev. A **22**, 1425 (1980), and private communication.

14) S.M. Younger, Phys. Rev. A **24**, 1278 (1981), and private communication.

15) W. Lotz, Z. Phys. **216**, 241 (1968).

16) M.S. Pindzola and M.J. Buie, Phys. Rev. A **37**, 3232 (1988).

17) M.S. Pindzola, D.L. Moores, and D.C. Griffin, submitted to Phys. Rev. A.

18) C.T. Munger and H. Gould, Phys. Rev. Lett. **57**, 2927 (1986).

Photons

SYNCHROTRON RADIATION EXPERIMENTS ON ATOMS AND MOLECULES

U. Becker

Institut für Strahlungs- und Kernphysik,
Technische Universität Berlin,
Hardenbergstr. 36, D-1000 Berlin 12

Some representative synchrotron radiation experiments on atoms and molecules are described. Among other topics such as electron correlations and resonances special emphasis is put on angle resolved experiments both in electron and ion spectroscopy. The results are discussed within the more unifying approach of angular momentum transfer theory and related symmetry arguments.

1. INTRODUCTION

The large increase of publications concerned with synchrotron radiation experiments on atoms and molecules shows clearly what a remarkable resurgence of interest in atomic and molecular photoionization phenomena has taken place during the last few years. This development was mostly driven by the fast progress made in instrumentation, particularly on the side of synchrotron radiation sources as an initiating moment to plan and perform new and more sophisticated experiments. Parallel to this development, advances in theoretical methods made it possible to meet the growing expectations on theory to interpret the new and more accurate data on a highly sophisticated level. This opened a period of fruitful interplay between theory and experiment which is still ongoing. In addition, beyond the more fundamental aspects of the majority of these studies, first applications such as VUV and soft X-ray lasers, quantitative surface analysis and photoinduced processes in the field of soft X-ray lithography appeared at the horizon, giving rise to still increasing activity. This review covering the fundamental aspects of atomic and molecular photoionization, will try to show how the various experiments, although concerned with very different aspects of this subject, contribute to the general progress in our understanding of photoionization phenomena particularly in terms of unifying concepts.

The development in the field of experimental atomic and molecular photoionization progressed along three directions: (a) extended energy regions at an improved level of resolution, (b) increased requirements on the completeness of the experimental data and (c) use of more sophisticated targets to study complex and prepared systems.

Activity along direction (a) is part of the development in synchrotron radiation instrumentation mentioned above whereas progress in direction (b) and (c) is carried on by the different research groups performing the experiments. As a natural consequence of the experimental requirements at the different levels of sophistication, the most highly differential experiments cannot be performed in all wavelengths regions, for all transitions and arbitrary targets. Therefore, spin polarized detection of photoelectrons became a parallel, yet independent subject within the larger field of photoionization, which was reviewed by U. Heinzmann [1] within the series of these conferences. The same is true for complex targets such as clusters and larger molecules, both constitute also a subject by their own, the latter being reviewed by Irene Nenner [2]. Therefore, results from these fields will not, or only be slightly, referred to here. More specialized topics such as double-ionization near threshold [3], two-electron resonances and photoionization studies of laser excited atoms [4] are presented within this conference and will be remarked only in passing. Main emphasis is put on four subjects, (1) extended measurements of both, partial cross sections and angular-distribution asymmetry parameters, (2) threshold behavior and resonances, including analysis of complete electron spectra and angular distributions of resonantly emitted electrons, (3) "complete" experiments via fragment analysis and target alignment and finally, (4) angle resolved fragmentation studies of molecules.

2. EXTENDED PHOTON ENERGY PHOTOIONIZATION STUDIES

Most previous experimental and theoretical studies on the photoionization of atoms and molecules were restricted to relatively limited wavelengths ranges, concentrating on photoionization phenomena in the rare gases. In addition, the only theoretical study over a wide range of energies (0-1keV) was the one by Kennedy and Manson performed within the independent particle model [5]. M.O. Krause [6] and other more recent review articles [7] showed the results of these calculations as a showcase for atomic innershell photoionization. However, the comparison of all calculated partial cross sections with experimental data was restricted to an absorption curve representing the total photoionization cross section. This was the only curve available at this time for the large energy range covered by these calculations. This situation has changed dramatically during the last couple of years. For example, different groups have measured partial cross sections and angular distribution asymmetry parameters for most of the Xe subshells between 10-1000 eV [8-14]. The new finding is that all classical photoionization "single particle" phenomena such as Cooper minima, delayed onset and shape resonances being exhibited in the 4d partial cross section, are transferred to most other subshells via interchannel coupling. This demonstrates clearly part of their "multielectron" or "collective" character [15]. Motivated by this set of

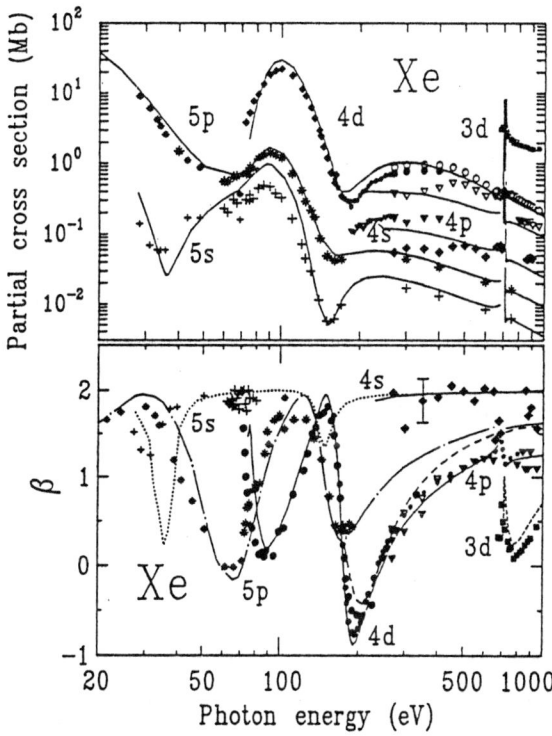

Figure 1: σ and β for the subshell photoionization of Xe up to 1 keV. The different exp. data sets are from:
σ: 5p *[8], *[9] ; 5s +[8], +[9] ; 4d ●, ○[9], •[10] ; 4p ▼, ▽[9], ▼[10] ; 4s ◆[9] ; 3d ■[11] ;
β: 5p *[12] , *[13]; 5s +[14], +[13] ; 4d ●[13], •[10], ...[11] ; 4p ▼, ▽[9] ; 4s ◆[9] ; 3d ■[11].

The different theoretical curves represent extended RRPA calculations [16].

Xe photoionization data, extended photon energy calculations using the relativistic random phase approximation (RRPA) were performed [16]. Most of the cross sectional behaviour is well described. The largest deviations are exhibited for subshells with strong correlation effects such as 4s and 4p, it is expected that inclusion of these correlations will improve the agreement. Fig. 1 shows the set of experimental partial cross sections and β parameters together with the RRPA results of Kutzner, Radojević and Kelly [16].

For molecular photoionization a similar, but slightly less complete level of data coverage up to 1000 eV was achieved recently for CO [17]. However, corresponding calculations, particularly for angular- distribution asymmetry parameters, are in this case, not available.

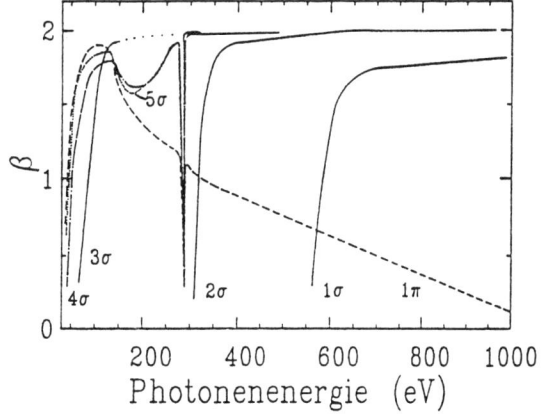

Figure 2: Angular-distribution asymmetry parameters for the photoionization of the different orbitals of CO [17].

Figure 2 shows β for the various molecular orbitals of CO. The β values in this survey figure are displayed for clarity by lines representing an average of the different separated data points. The dip in β at 287 eV for all valence orbitals is due to the strong $\pi - \pi^*$ discrete resonance just below the carbon K edge.

3. THRESHOLD BEHAVIOR AND RESONANCES

3.1 Photoelectron satellites

Threshold behavior and resonances attracted much interest with an increasing number of groups being involved. Here we will concentrate on three aspects only: photoelectron satellites, two-step processes and angular distributions of resonant Auger electrons.

The threshold behaviour of photoelectron satellites will be discussed for neon because all types of characteristic behaviour are represented there. Satellite lines in photoelectron spectra represent two-electron transitions caused by electron correlations. Examination of satellite intensities at different photon energies led to the assumption that the threshold and near-threshold behaviour of these satellites may be characteristically associated with certain correlations which contribute to their intensity. To illustrate this Fig. 3 shows three Ne valence satellite spectra taken at very different energies: a zero-volt kinetic energy spectrum [18], a spectrum at 150 eV photon energy [19] and one with Al K_α radiation at the sudden limit [20]. The comparison between these spectra reveals quite a change in relative intensity for many satellites in going from threshold to the sudden limit. Bridging the energy gap between the two extremes was the main purpose of many experiments, some of them outlined here.

Studying threshold behaviour requires an electron detection system with high transmission for electrons with low ki-

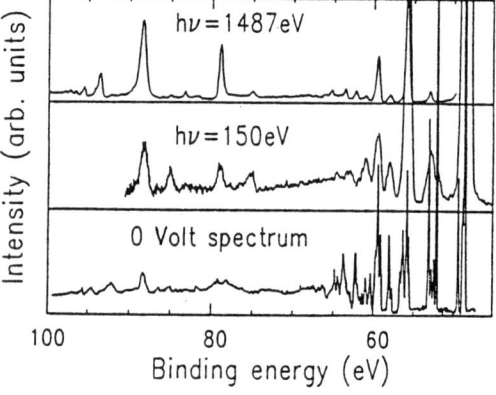

Figure 3: Three Ne valence satellite spectra taken at different energies [18-20].

netic energies. A method particularly suited for such purpose, is time-of-flight detection. Figure 4 shows for example the schematic set-up of an angle resolved photoemission experiment using synchrotron radiation as an excitation source. Two rotatable analyzers allow simultaneous partial cross section and angular distribution measurements, as well as evaluation of the degree of polarization of the monochromatized synchrotron radiation.

The results of near threshold studies of Ne valence and K-shell satellites are displayed in Fig. 5. This figure exemplifies four types of characteristic behaviour: constant fractional intensity, decreasing and increasing intensity towards threshold respectively, and structures due to resonance enhancement. These different kinds of behaviour were found for many other atoms and molecules, both for inner- and outer-shell photoionization [25]. It is assumed that the different dominant electron correlations give rise to this variety of threshold behaviours making it possible to differentiate to a first approximation among them. The correlations can be subdivided into two classes depending on their threshold behaviour: "intrinsic" and "dynamic" correlations respectively. Complementary information may be gained from angular distribution measurements. First ab initio calculations support this phenomenological concept to iden-

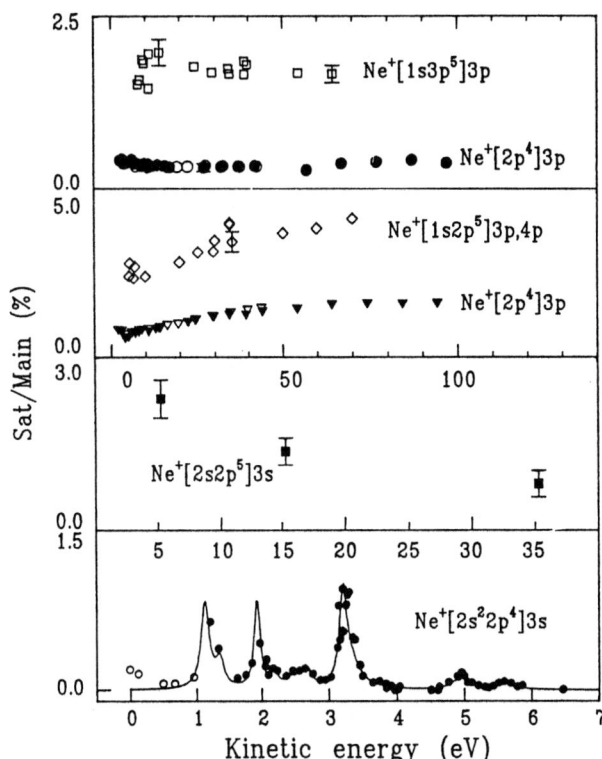

Figure 5: Examples of different satellite threshold behavior in neon. Data: $\bigcirc,\nabla$[21]; $\square,\diamond$[22]; $\bullet,\blacktriangledown,\bullet$[23]; o[18]; $\blacksquare$[24]

tify the dominant correlations contributing to satellite intensities. Although satellite intensities are usually regarded as of minor importance compared to the total photoionization cross section there are examples where they contribute significantly. Recent experiments on Ba, for example, indicate 38% of the total 4d intensity is due to the satellites. The origin of this strong satellite fraction is still unknown. There are "intrinsic" correlations such as configuration interaction in the initial (ISCI) and final ionic state (FSCI) [26] in the discussion, but additional "dynamic" contributions due to conjugate shake-up into the $4d^9 5s^2 5p^6 4f$ final ionic-states have to be considered. Near threshold measurements of the individual satellite intensities may help to disentangle the various interactions involved in the photoionization of Ba. Figure 6 shows the partitioning of the Ba total cross section into the 4d, $4d_{sat}$ and valence components. Part b) and c) show

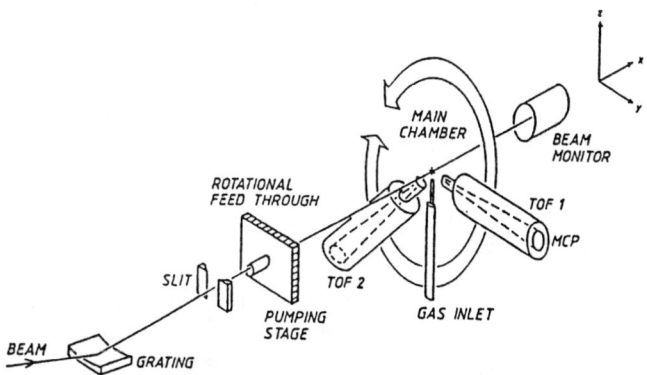

Figure 4: Scheme of an angle resolved photoemission experiment with synchrotron radiation.

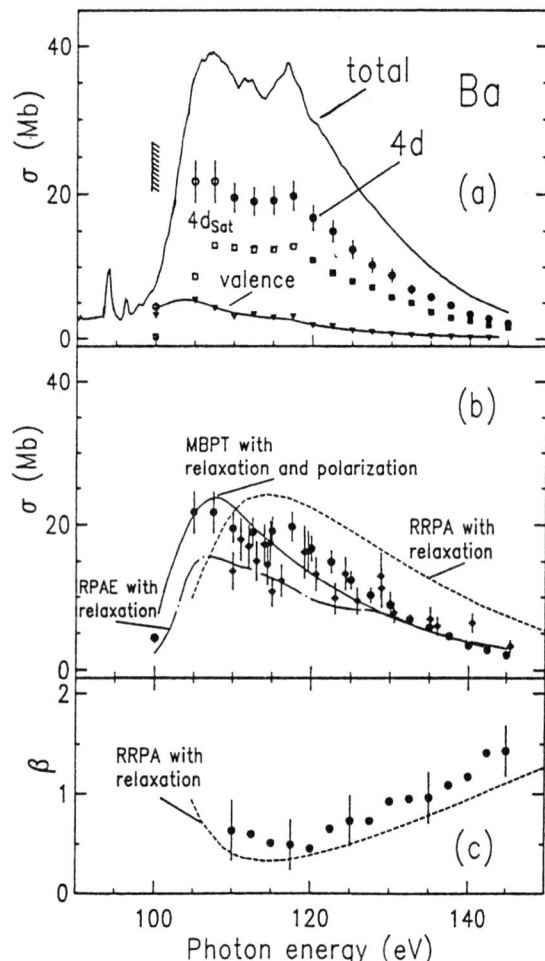

Figure 6: (a) Ba 4d, $4d_{sat}$ and valence partial cross sections in the vicinity of the giant resonance. The total cross section is from ref. [27]. (b) σ_{4d} and (c) β_{4d} results ● [28], ♦ [29] in comparison with theory MBPT [31], RRPA [32], RPAE [33].

the 4d σ and β measurements compared to different theoretical results including many body perturbation theory (MBPT). The β variation of the satellite lines is similar to that of the 4d main line, although the absolute values are higher [28,30].

3.2 Two-step processes

Multielectron processes such as satellite production but also two-electron emission via shake-off are regarded as simultaneous processes which have to be treated within a one-step model. However, the occurrence of the simultaneous processes is superimposed by other, so-called, two-step processes. A typical process of two-step character is Auger decay and its counterpart i.e. two-step autoionization. Both effects have to be considered regarding near threshold behavior and resonances. An example is Auger decay affecting threshold behavior in the valence double ionization of Ne. In contrast to inner-shell ionization where Auger decay is one of the main sources of multiple ionization, valence double-ionization was regarded as due to simultaneous two-electron emission via shake-off. However, a very recent study exposed that part of the double ionization rate results from Auger decay of valence satellite states [23]. Figure 7 shows a "complete" valence photoelectron spectrum of Ne taken at 120 eV exhibiting many so-called participator Auger lines in the low kinetic energy part of the spectrum. These lines are superimposed on the continuously distributed shake-off electrons resulting from the one-step process of simul-

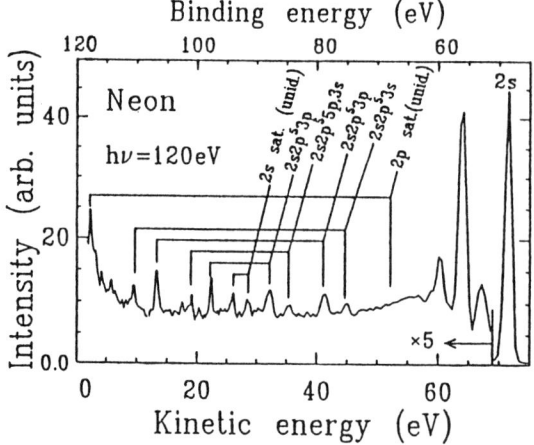

Figure 7: "Complete" valence photoelectron and Auger spectrum at 120 eV [23].

taneous two-electron ejection. The same effect was also observed for other rare gases, e.g. Ar, and for small molecules, e.g. CO. Comparing the middle part of Fig. 3 with Fig. 7, both showing Ne photoelectron spectra at intermediate energies, reveals that the search for valence Auger lines requires recording of "complete" electron spectra down to 0 eV kinetic energy or as close as possible.

The effect of two-step autoionization on the decay of resonances is exemplified for the Xe 4d → 6p resonance. This extensively studied resonance decays via different mechanisms such as spectator decay, associated shake-up and shake-down, and resonant shake-off, resulting in doubly charged ions. The question of the leading decay mode is still open because an exact partition of all decay channels requires knowledge of how much line intensity stems from second-step transitions [34]. This question is even more important for 0 eV measurements because second-step transitions may be accidentally located in the vicinity of a threshold giving rise to electrons with zero kinetic energy. Quantitative information on this problem was obtained for the first time by an electron-electron coincidence experiment by v.Raven et al. [35] First and second step processes could be correctly related and fractional intensities of two-step resonant Auger transitions could be derived. A decay spectrum of the Xe 4d → 6p resonance showing all types of decay transitions including resonant shake-off and second-step resonant Auger is shown in Fig. 8. The tick marked lines connecting first-step processes with second-step ones are taken from the electron-electron coincidence experiment. Again, as in the nonresonant case of Ne, the low energy part of the spectrum is the most interesting regarding the superposition of one- and two step processes.

3.3 Angular distributions of resonant Auger electrons

Complementary information on the character of a transition may be obtained by measuring angular distributions not only for satellite transitions but also for transitions occurring in the decay of a resonance. Resonantly enhanced satellites are a link between the

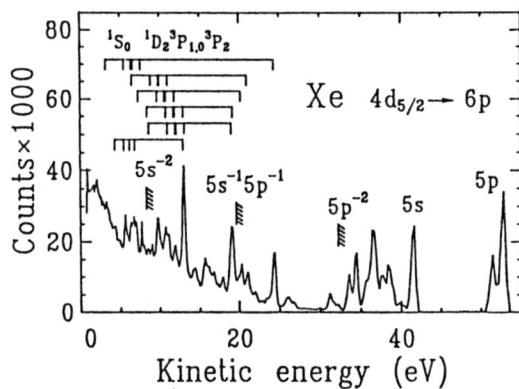

Figure 8: Decay spectrum of the Xe 4d → 6p resonance with second step transitions [35] marked as indicated.

two cases. Carlson et al. [36] found the first manifestation of an unusual degree of angular anisotropy in the resonant Auger spectrum of Kr. Specifically, unusually low values, close to -1, for the angular-distribution asymmetry parameter β, were seen for two bands of resonant Auger lines. The same effect was observed also for the other rare gases [37]. To explain this effect angular momentum transfer theory was employed [38]. The Ar 2p → 4s resonance transition is particularly suited for a straight forward theoretical explanation due to the high experimental resolution that can be achieved. This is because the excited electron has no angular momentum. Consequently the spectator electron causes splitting in the final ionic state into two spin-coupled components. Two highly resolved decay spectra of this resonance taken under two different angles are displayed in Fig. 9. The two spectra are very similar to the one published by Carlson et al. [39], the only difference is, that they are free of higher order contributions of the monochromatized light. The β-values obtained are drawn in the lower part of the figure. The two spin-coupled components ^2P and ^4P exhibit β-values with opposite sign. This can be explained by simple symmetry arguments based on the antiparallel and parallel coupling schemes of the corresponding spins and angular momenta. Cooper [39] has evaluated an extended angular momentum transfer formalism to calculate these β-values. His results are principally in good agreement with the experimental data. These calculations are equivalent to calculations performed

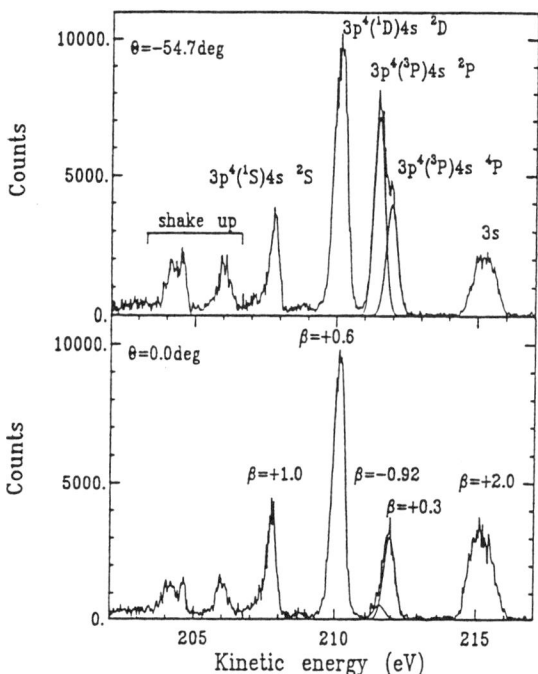

Figure 9: High resolution decay spectra of the Ar 2p → 4s resonance taken at two different angles.

within Auger and alignment theory [40] if one uses Fano-Macek's definition for the alignment of the excited state [41]. The alignment parameter A_{20} is given by:

$$A_{20} = \frac{\sum_M \sigma(J,M)[3M^2 - J(J+1)]}{J(J+1) \sum_M \sigma(J,M)},$$

where $\sigma(J,M)$ is the population of the (J,M_J) magnetic sublevel, equal to the cross section for production of that level. An indication for the complete analogy of the two different methods is the fact that both theories predict a β-value of 1 for the coupling- and Auger matrix-element independent transition to the ^2S state. This value was also verified experimentally. Because of

the well defined Fano-Macek alignment in the excited state angular-distribution measurements of resonant Auger electrons may in turn be used to determine these Auger matrix-elements experimentally. The latter is of substantial interest for nonresonant measurements in achieving a "complete" description of the photoionization process.

4. "COMPLETE" EXPERIMENTS

One of the fundamental problems in photoionization is the fact that the partial cross section σ_{if} is related to a coherent superposition of several partial waves:

$$\sigma_{if}(h\nu) = \frac{4\pi^2 \alpha a_0^2}{3} h\nu \sum_{i',f'} | <f' | \sum_\mu r_\mu | i'> |^2$$

f' and i' being the degenerated final and initial states. Treating photoionization nonrelativistically reduces the number of partial waves emitted by the photoionization of a given subshell to two, which are only coupled by a relative phase. Several approaches were choosen to deal with this "intrinsic" degeneracy of the photoioinization process, the most sophisticated being spin- and angle resolved photoelectron spectroscopy [1]. However, this method has to cope with low detection efficiencies and depends on the availability of circularly polarized light, a requirement still restricted to very limited energy regions. Transitions with small partial cross sections, such as correlation satellites and extended studies over wider energy ranges have to overcome the degeneracy problem by other methods, some of which are described in the following.

4.1 Fragment analysis

Photoionization of an electron with an angular momentum j > 1/2 leaves the ion in an aligned state. This alignment corresponds to the alignment created by photoexcitation with the only difference that the "excited electron", which contributes to the system alignment, is a coherent superposition of continuum waves. The population of the magnetic sublevels of the remaining ion reflects, therefore the strengths of the different partial wave contributions and is uneffected by their relative phase. This is the reason that fragment analysis regarding the degree of alignment of the ionic state is one way to achieve further information on the photoionization process [40,42]. The alignment analysis may be performed by studying the degree of polarization of subsequently emitted fluorescence radiation or alternatively by investigating the nonisotropic angular distribution of the Auger electrons. Both methods are completely equivalent. However, for inner shell processes, angle-resolved Auger measurements are more easily performed than angular-distribution or polarization-degree measurements of soft X-ray fluorescence. Therefore, first "complete" experiments following this approach were performed by determining angular-distributions of Auger electrons in addition to those of the corresponding photoelectrons of which the partial cross section had also to be evaluated. Hausmann et al. [43] used this approach to determine the transition amplitudes of the outgoing ϵs and ϵd waves including their relative phase for 2p photoionization in atomic magnesium. They compare their measurement, performed at 80 eV, with presently available theoretical calculations for this system, claiming inadequancy of most of the methods. However, it should be noted that the choosen approach is an approximation, particularly if one neglects parity unfavoured terms as done in the data evaluation of this experiment. The importance of those terms is still only brought to light via spin polarization measurements as was shown very recently on the photoionization of 4f electrons in Yb [44]. Nevertheless, for simple closed shell systems the alignment method should yield enough additional information to decide conclusively if the theoretical approaches are adequate to describe the photoionization of a given subshell qualitatively and also, are quantitatively correct. This is illustrated by the 2p photoionization of argon. The relevant quantities are the same as in the case of the 2p photoionization of Mg. The only difference is, that we have choosen the dipole matrix element description instead of using scattering amplitudes which contain part of the correlations. This is done because we have to neglect the corresponding parity unfavoured transitions due to the limited number of experimental quantities determined in our experiment. Figure 10 summarises the result of the Ar 2p study together with the derived dipole matrix elements $R_{\epsilon s}$ and $R_{\epsilon d}$ of the ϵs and ϵd partial waves and their relative phase Δ to each other. The bulk of the β_{2p} values is taken from Lindle et al. [45]. The σ, β and A_{20} values are compared to different theoretical curves of which the "non single-particle" methods such as MBPT and RRPA fit the data best. For the energy dependent alignment a simple Hartree-Slater calculation was used as this was the only one available in the literature. The interesting point is, however, that if one derives dipole matrix elements and relative phases from the σ^{MBPT}, β^{RRPA} and A_{20}^{HS} curves, no consistent set of $R_{\epsilon s}$, $R_{\epsilon d}$ and Δ can be achieved. This is because the larger dip in β around 250 eV which is correctly described by the RRPA requires in our approximation a larger s-wave contribution, otherwise the superposition of the two partial waves cannot make up for the low β values. As a result of this conflict, the relative phase $\cos(\Delta)$ reaches unphysical values larger than 1. This means that a correct description of the Ar 2p photoionization has to reproduce the alignment increase towards threshold as seen by the experiment. In summary our alignment values and the measured σ_{2p} and β_{2p} values yield physically meaningful dipole matrix elements and relative phases in the whole energy range.

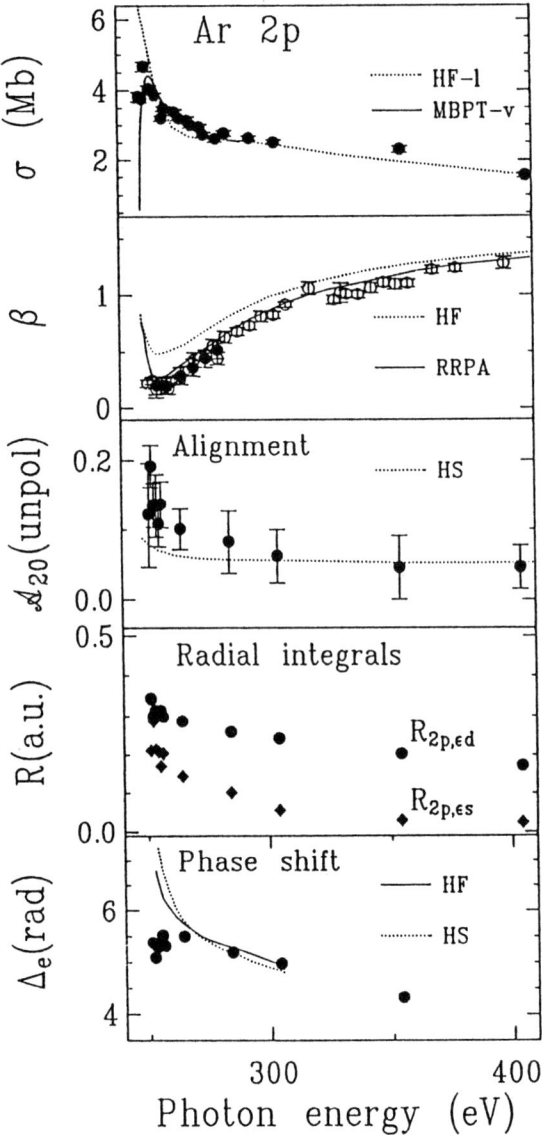

Figure 10: σ_{2p}, β_{2p} and ionic alignment A_{20} left by photoionization of the Ar 2p subshell. Open circles represent β_{2p} values from Lindle et al.[45]. Theoretical curves are: HF, ref. [5]; HS, ref. [46]; RRPA, ref. [45] and MBPT, ref. [47]. The lower part of this figure shows dipole matrix elements and relative phases derived from these data together with the theoretical phases from ref. [5].

4.2 Target alignment

As mentioned in the preceeding chapter, one of the main shortcomings of the approach is the insufficient number of independent experimental quantities which can be principally determined. The reason for this deficiency is the use of randomly oriented targets for photoionization. Following Yang's [48] theorem, only second order harmonics contribute to the differential cross section formula for photoelectrons, because other harmonics cancel due to symmetry arguments. However, if aligned targets are used for photoionization, more information is carried over in the description of the angular-distribution of photoelectrons. These additional terms, specifically their coefficients, give rise to a more complete description of the photoionization process approaching in certain cases a "perfect" experiment containing all necessary quantities to disentangle the matrix elements and their phases completely [49]. One way to align an atom is photoexcitation as the first step. Photoionization of these excited aligned atoms yields photoelectrons exhibiting non β-like angular distributions containing higher order harmonics such as P_{21}, P_{22}, P_{41} and P_{42}. The first angle integrated experiment, in this direction was performed by Meyer et al. [50] using dye laser excitation of Li atoms as the first step and then ionizing these aligned atoms in a second step by synchrotron radiation. This experiment concentrated on the population of autoionizing states showing the effect of the polarization direction of the first step photons (laser) on the intensity of the photoelectrons emitted via autoionization. The intensity changes are the direct result of the different alignment created by the first step excitation, if one chooses a parallel or antiparallel orientation of the polarization vectors of the two photons used for excitation and ionization, respectively. The upper part of Fig. 11 indicates a scheme for the two-step excitation of the autoionizing state and implies that the observed intensity changes, shown in the lower part of this figure, is dependent on the alignment or sublevel-population. The factor of 4 ob-

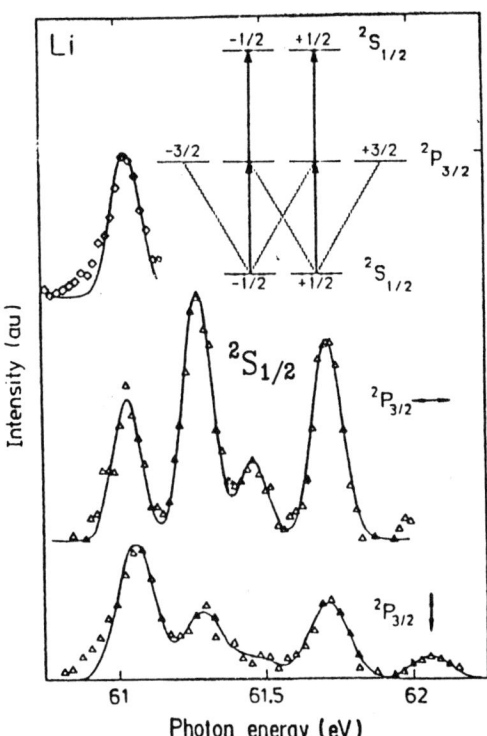

Figure 11: Two-step excitation of autoionizing states in Li. Parallel and perpendicular orientations of the two electric vectors with respect to each other are marked by corresponding arrows [50]

served for the $^2P_{3/2} \to {}^2S_{1/2}$ transitions (line 6,7) is easily derived from the m dependent partial cross sections as shown schematically. This effect appears in the integrated partial cross section.

To obtain more differential information from photoionization of aligned systems angle resolved measurements are necessary. First experiments in this direction were performed on Yb targets using a VUV lamp instead of synchrotron radiation, although synchrotron radiation measurements would be, in principle, the same. Figure 12 shows the geometry chosen by C. Kerling [51] for this experiment. Laser and VUV radiation are collinear whereas the electric vectors of the two photon beams and the direction of electron observation are in the plane perpendicular to this direction. Two angles are varied, the

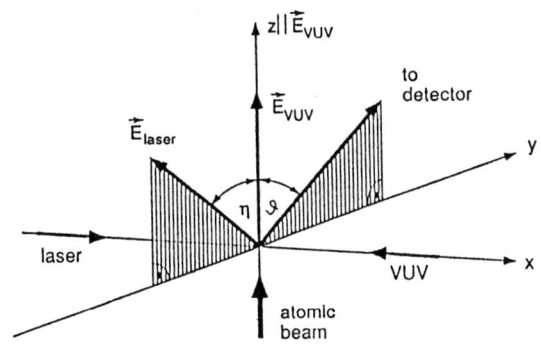

Figure 12: Geometry of an angle-resolved photoionization experiment on aligned atoms.

angle ϑ between the electric vector of the VUV photons and the observation direction, and the angle η between the two electric vectors. The differential cross section of the emitted electrons is given by [52]:

$$I(\vartheta) = A_{00}(1 + A_{20}\ P_{20}(\cos\vartheta) + A_{40}\ P_{40}(\cos\vartheta) + \\ + A_{21}\ P_{21}(\cos\vartheta) + A_{41}\ P_{41}(\cos\vartheta))$$

In addition to the coefficient A_{20} corresponding to β, three other coefficients A_{21}, A_{40} and A_{41} may be obtained. The resultant emission pattern for the two-step photoionization of Yb atoms are shown in Fig. 13 for different angles η. Future progress in intense synchrotron radiation sources will certainly give rise to such experiments as a function of photon energy.

5. ANGLE RESOLVED MOLECULAR FRAGMENTATION

In contrast to atoms, molecules exhibit additional symmetries with respect to the molecular axis. Whereas, in atoms photoexcitation creates aligned states regarding magnetic sublevel population, photoexcitation in molecules produces alignment with respect to the molecular axis. In a similar way as magnetic sublevel dependent partial cross sections are responsible for the alignment in an excited atomic state, the dependence of the partial cross sections on the orien-

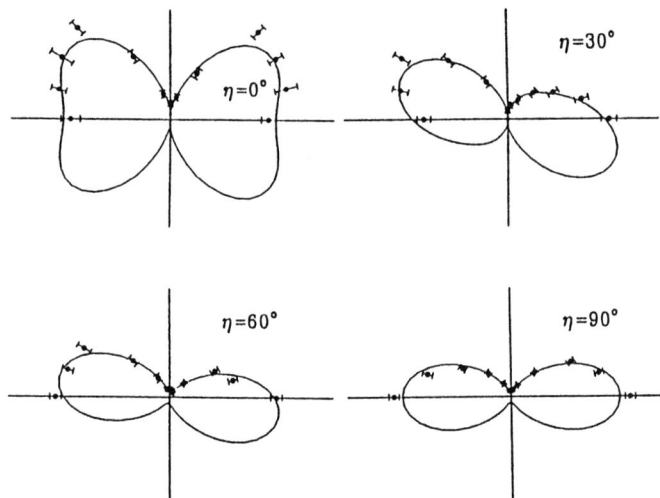

Figure 13: Photoelectron emission pattern resulting from two-step photoionization of Yb atoms [51].

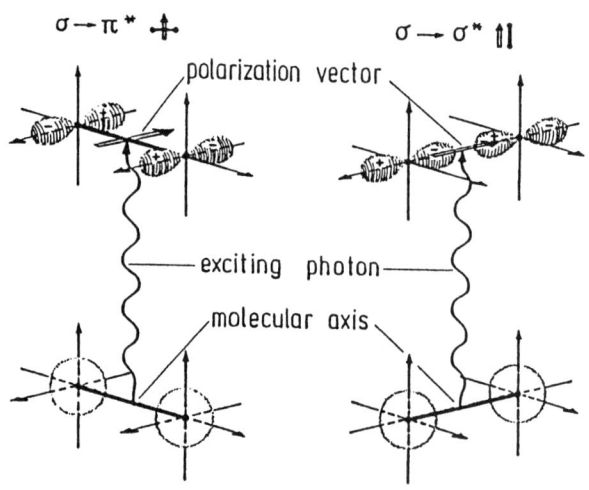

Figure 14: Schematic representation of a "perpendicular" $\sigma \to \pi$ and a "parallel" $\sigma \to \sigma$ transition in a diatomic molecule.

tation of the molecular axis with respect to the electric vector is the reason for molecular alignment in a photoexcited state. This effect is most pronounced for $\sigma \to \pi$ and $\sigma \to \sigma$ transitions analogous to an atomic S $\to$ P transition. Those transitions are called perpendicular and parallel transitions, respectively, because in a $\sigma \to \pi$ transition the electric vector has to be perpendicular to the molecular axis, whereas for $\sigma \to \sigma$ transitions it has to be parallel. Figure 14 shows this situation schematically. Probing the resultant molecular alignment or anisotropy may yield information on the symmetry of unknown excited molecular states.

A specific feature of molecular photoionization is the existence of shape resonances in K-shell photoionization due to back scattering of outgoing waves in the anisotropic molecular potential [53]. These so called molecular shape resonances are related to unbound continuum orbitals of the molecule. These quasi-orbitals are also of a certain symmetry and are in some cases of unknown symmetry. Molecular inner-shell photoionization or excitation is, in most cases, followed by fast dissociation of the core excited molecule or molecular ion. In case of diatomic molecules this fragmentation occurs along the molecular axis making angle resolved fragment spectroscopy of molecular ions a probe for molecular alignment and symmetries. The ionic fragments have to have some kinetic energy in order to be detected without an extraction field, which otherwise destroys the angular resolution. CO and N_2 are classic examples for symmetry dependent K-shell excitations into both bound and unbound continuum orbitals. In specific, K-shell excitations into bound π^* and unbound σ^* shape resonances are of interest. In the first angle resolved fragmentation study on N_2 two groups obtained evidence for strong molecular alignment due to photoabsorption [54,55]. They measured the molecular anisotropy parameter β_m a quantity defined in a similar way as the angular-distribution asymmetry parameter of photoelectrons. However, this quantity represents a static distribution of molecular axis with respect to the electric vector, similar to A_{20}, causing the angular distribution of the fragments, measured in the experiments. Figure 15 shows the results for β_m from Yagishita et al. [58] The

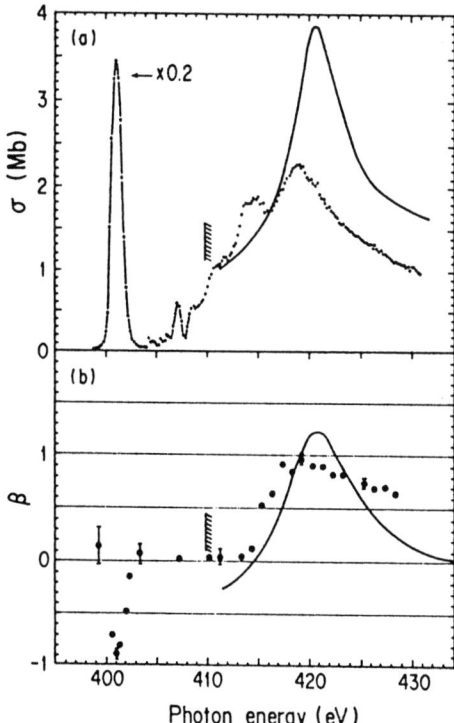

Figure 15: Molecular anisotropy parameter β_m obtained from an angle resolved fragmentation study [55] in comparison with theoretical predictions [56].

data are compared to theoretical predictions of Wallace et al. [56] which are confirmed, particularly within the shape resonance, a result not obvious from simple symmetry arguments. Further studies in this direction for other molecules are underway. Angle resolved coincidence measurements between ejected electrons [57] and photofragments will open the field of angular distribution studies of photo- and Auger electrons from oriented molecules. Theoretical calculation again predict non-β-like angular distributions as in the case of aligned atoms [58].

6. SUMMARY

The present state of synchrotron radiation experiments on atoms and molecules shows trends towards more extended and more complete measurements. Measurements covering wide energy ranges in order to prove the general capability of theory to describe the photoionization process are complemented by specific studies regarding threshold behavior and resonances. The intention to obtain a set of experimental quantities as complete as possible, to disentangle the principle degeneracy of the photoionization process, is realized along different approaches using angle resolved and coincidence techniques. Studies of aligned targets and their dependence on the excitation energy become feasible.

7. ACKNOWLEDGEMENT

The author likes to thank Cordula Kerling for making her recently aquired data available to him. He is grateful to David Kilcoyne for critical reading of the manuscript and all others who have contributed to this work. He is indebted to the Deutsche Forschungsgemeinschaft for a research grant and to the Bundesminister für Forschung und Technologie who funded the project under contract No. 05.314 EX B2.

REFERENCES

1. U.Heinzmann, in: Electronic and Atomic Collisions, ed. by D.C.Lorentz, W.E.Meyerhof, and J.R.Peterson (North-Holland, Amsterdam 1986) pp. 37 and references therein
2. I.Nenner, in: Electronic and Atomic Collisions, ed. by H.B.Gilbody, W.R.Newell, F.H.Read, and A.C.H.Smith (North-Holland, Amsterdam 1988) pp. 517 and references therein
3. V.Schmidt, in: Electronic and Atomic Collisions, ed. by R.S.Freund, M.S.Lubell, and T.B.Lucatorto, to be published (1990)
4. K.Schartner, ibid., and D.Cubaynes, ibid.
5. D.J.Kennedy and S.T.Manson, Phys.Rev. A **5**, 227 (1972)
6. M.O.Krause, in: Synchrotron Radiation Research, ed. by H.Winick and S.Doniach (Plenum, New York, 1980), p. 104

7. B.Crasemann and F.Wuilleumier, in: *Atomic Inner-Shell Physics*, ed. by B.Crasemann (Plenum, New York, 1985), p. 281
8. A.Fahlmann, M.O.Krause, M.A.Carlson, and A.Svensson, Phys.Rev.A **30**, 812 (1984)
9. U.Becker, D.Szostak, H.G.Kerkhoff, M.Kupsch, B.Langer, R.Wehlitz, A.Yagishita, and T.Hayaishi, Phys.Rev.A **39**, 3902 and references therein (1989)
10. D.W.Lindle, T.A.Ferrett, P.A.Heimann, and D.A.Shirley, Phys.Rev.A **37**, 3808 (1988)
11. U.Becker, H.G.Kerkhoff, M.Kupsch, B.Langer, D.Szostak, and R.Wehlitz, J.Phys. (Paris) Colloq. **48**, C9-497 (1987)
12. M.O.Krause, T.A.Carlson, and P.R.Woodruff, Phys.Rev.A **24**, 1374 (1981)
13. S.Southworth, U.Becker, C.M.Truesdale, Ph.H.Kobrin, D.W.Lindle, S.Owaki, and D.A.Shirley, Phys.Rev.A **28**, 261 (1983)
14. A.Wachter, R.Malutzki, and V.Schmidt, in: *Abstracts of the International Conference on Vacuum Ultraviolet Radiation Physics*, ed.by P.O.Nilsson, Vol. I, 33 (1986)
15. M.Ya.Amusia and V.K.Ivanov, Sov. Phys.Usp. **30**, 449 (1987)
16. M.Kutzner, V.Radojevic, and H.P.Kelly, Phys.Rev.A (1989) in press
17. O.Hemmers, Diplomarbeit, Technische Universität Berlin (1988), unpublished
18. P.A.Heimann, U.Becker, H.G. Kerkhoff, B.Langer, D.Szostak, R.Wehlitz, D.W.Lindle, T.A. Ferrett, and D.A.Shirley, Phys. Rev. A **34**, 3782 (1986) and P.Lablanquie, Thèse d'Etat, Université de Paris-Sud, Centre d'Orsay (1989)
19. U.Becker, R.Wehlitz, O.Hemmers, B.Langer, and A. Menzel, Phys. Rev. Lett. **63**, 1054 (1989)
20. S.Svensson, B.Eriksson, N.Martensson, G.Wendin, and U.Gelius, J.Electron Spectrosc.Phenom. **47**, 327 (1988)
21. P.A.Heimann, C.M.Truesdale, H.G.Kerkhoff, D.W.Lindle, T.A. Ferrett, C.C.Bahr, W.D.Brewer, U.Becker, and D.A.Shirley, Phys. Rev.A **31**, 2260 (1985)
22. U.Becker, T.A.Ferrett, P.A.Heimann, H.K.Kerkhoff, D.W. Lindle, and D.A.Shirley, to be published in Phys.Rev.A
23. U.Becker, R.Hölzel, H.G.Kerkhoff, and B.Langer, Phys. Rev.Lett. **56**, 1120 (1986)
24. U.Becker, O.Hemmers, B. Langer, A.Menzel, J.Viefhaus, and R.Wehlitz, in: *Abstracts of the Ninth International Conference on Vacuum Ultraviolet Radiation Physics*, ed. by S.Fadley and D.A.Shirley, 47 (1989)
25. U.Becker and D.A.Shirley, to be published in Physica Scripta (1989) and references therein
26. W.Mehlhorn, B.Breuckmann, and D.Hausamann, Physica Scripta **16**, 177 (1977)
27. M.Richter, M.Meyer, M.Pahler, T.Prescher, E.v.Raven, B.Sonntag, and H.E.Wetzel, Phys. Rev.A **39**, 5666 (1989)
28. A.Sivasli, Diplomarbeit, Technische Universität Berlin (1989), unpublished
29. J.Bizeau, PhD thesis, Université de Paris-Sud, Centre d'Orsay (1987), unpublished
30. H.G.Kerkhoff, Dissertation, Technische Universität Berlin (1989), unpublished
31. M.Kutzner, Z.Altun, and H.P.Kelly, to be published in Phys.Rev.A (1989)
32. V.Radojevic, M.Kutzner, and H.P.Kelly, Phys.Rev.A **40**, 727 (1989)
33. M.Ya.Amusia, private communication
34. U.Becker, D.Szostak, M.Kupsch, H.G.Kerkhoff, B.Langer, and R.Wehlitz, J. Phys. B**22**, 749 (1989)
35. E.v.Raven, M.Meyer, M.Pahler, and B.Sonntag, to be published in J.Electron Spectrosc. Relat. Phenom. (1989)
36. T.A.Carlson, D.R.Mullins, C.E.Beall, B.W.Yates, J.W.Taylor, D.W.Lindle, B.P.Pullen, and F.A.Grimm, Phys.Rev.Lett. **60**, 182 (1988)
37. T.A.Carlson, D.R.Mullins, C.E.Beall, B.W.Yates, J.W.Taylor, D.W.Lindle, and F.A.Grimm, Phys. Rev.A **36**, 1170 (1989)
38. U.Fano and D.Dill, Phys.Rev.A **6**, 185 (1972)
39. J.W.Cooper, Phys.Rev.A **39**, 3714

(1989)
40. E.G.Berezhko, N.M.Kabachnik, and V.S.Rostovsky, J.Phys.B **11**, 1749 (1978)
41. U.Fano and J.H.Macek, Rev.Mod.Phys. **45**, 553 (1973)
42. Greene and Zare, Phys.Rev.A **25**, 2031 (1982)
43. A.Hausmann, B.Kämmerling, H.Kossmann, and V.Schmidt, Phys. Rev.Lett. **61**, 2669 (1988)
44. M.Müller, Dissertation, Universität Bielefeld (1989), unpublished
45. D.W.Lindle, L.J.Medhurst, T.A.Ferrett, P.A.Heimann, M.N.Piancastelli, S.H.Liu, D.A.Shirley, T.A.Carlson, P.C.Deshmukh, G.Nasreen, and S.T.Manson, Phys.Rev.A **38**, 2371 (1988)
46. E.G.Berezhko, N.M.Kabachnik, and V.V.Sizov, J.Phys.B **11**, 1819 (1978)
47. C.Pan and H.P.Kelly, Phys.Rev.A **39**, 6232 (1989)
48. C.N.Yang, Phys.Rev. **74**, 764 (1948)
49. H.Klar and H.Kleinpoppen, J.Phys.B **15**, 933 (1982)
50. M.Meyer, B.Müller, A.Nunnemann, Th.Prescher, E.v.Raven, M.Richter, M.Schmidt, B.Sonntag, and P.Zimmermann, Phys.Rev.Lett. **59**, 2963 (1987)
51. C.Kerling, N.Böwering, and U.Heinzmann, in: <u>Book of Abstracts of the 3rd ECAMP</u>, ed.by A.Salin, Part II, p.579 (1989) and C.Kerling, private communication (1989)
52. J.C.Hansen, J.A.Duncanson, R.L.Chien, and R.S.Berry, Phys.Rev.A **21**, 222 (1980)
53. J.L.Dehmer and D.Dill, Phys.Rev.Lett. **35**, 213 (1975)
54. N.Saito and I.H.Suzuki, Phys.Rev.Lett. **61**, 2740 (1989)
55. A.Yagishita, H.Maezawa, M.Ukai, and E.Shigemasa, Phys.Rev.Lett. **62**, 36 (1989)
56. D.Dill, J.R.Swanson, S.Wallace, and J.L.Dehmer, Phys.Rev.Lett. **45**, 1393 (1980)
57. U.Becker, R.Hölzel, H.G.Kerkhoff, B.Langer, D.Szostak, and R.Wehlitz, Phys.Rev.Lett. **56**, 1455 (1986)
58. W.Davenport, Phys.Rev.Lett. **36**, 945 (1976)

SYNCHROTRON-RADIATION EXPERIMENTS WITH RECOIL IONS

Jon C. Levin

Department of Physics, University of Tennessee, Knoxville, Tennessee 37996-1200
and Oak Ridge National Laboratory, Oak Ridge, Tennessee 37831-6377

Studies of atoms, ions and molecules with synchrotron radiation have generally focused on measurements of properties of the electrons ejected during, or after, the photoionization process. Much can also be learned, however, about the atomic or molecular relaxation process by studies of the residual ions or molecular fragments following inner-shell photoionization. Measurements are reported of mean kinetic energies of highly charged argon, krypton, and xenon recoil ions produced by vacancy cascades following inner-shell photoionization using white and monochromatic synchrotron x radiation. Energies are much lower than for the same charge-state ions produced by charged-particle impact. The results may be applicable to design of future angle-resolved ion-atom collision experiments. Photoion charge distributions are presented and compared with other measurements and calculations. Related experiments with synchrotron-radiation produced recoil ions, including photoionization of stored ions and measurement of shakeoff in near-threshold excitation, are briefly discussed.

I. INTRODUCTION

Beams of low-energy, high-charge-state ions, typically formed from recoils produced by swift heavy-ion impact, have long been used to study charge-changing reactions with gas targets. In recent years, interest in the fundamental process of electron transfer in atomic collisions, in ion-atom collisions in cold interstellar clouds, and in the spectroscopy of multiply charged ions, has resulted in ever-lower center-of-mass energies. Production of highly ionized and excited ions at low energies poses a difficult challenge, since typical sources of such ions, such as stellar, fusion, and laser plasmas typically involve energies in the 1 to 100 keV energy range.[1] Fast-beam sources which achieve similar ionization-excitation states typically involve beam velocities $v/c \approx 0.1$.[2] In both cases, Doppler broadening limits spectroscopic precision. Similarly, the emittance of plasma and fast beam sources is not suited for high-charge state ion-atom collision experiments at eV to keV energies under conditions where good energy and angular definition of high-charge-state projectile ions are important.

Use of fast, heavy-ion impact on target atoms to produce excited, high-charge state recoil ions ("hammer" method) permits achievement of four orders of magnitude lower energy than plasma sources in which similar ionization states have comparable abundance, as well as three orders of magnitude in v/c relative to fast beam sources. As beam energies are pushed lower,[3] the inherent recoil energy resulting from the primary ionization may begin to limit the resolution of the low-energy secondary beam. Our measurements of energies of recoil ions indicate that highly charged ions produced by synchrotron radiation have kinetic energies one to two orders of magnitude less than do similar charge states of recoil ions produced by charged-particle impact, providing up to six orders of magnitude lower energies and four orders of magnitude advantage in v/c compared to plasma sources. The results may be applicable to development of a very-cold ion source featuring low energy spread and good angular definition.

Use of synchrotron radiation to produce a low emittance, sub-nanosecond pulsed ion source may permit improved coincidence experiments in such areas as:

o angle-resolved high-charge-state ion-atom and ion-molecule collisions in the few eV to several hundred eV range;

o interaction of tunable monochromatic radiation simulating stellar radiation with cold ions and molecules such as are found in interstellar clouds;

o precision spectroscopy of few-electron ions, following trapping and further cooling made easier by production of already cold ions;

o study of low-energy ion-surface and molecule-surface interactions.

In recent years, development of electron cycloctron resonance (ECR) ion sources, and a new generation of cryogenic electron-beam ion sources (such as CRYSIS at MSI in Stockholm and CRYEBIS at Kansas State's MacDonald Laboratory) has resulted in production of usable numbers of ions in very-high charge states. The high charge states accessible with such machines cannot be produced by creation of a single inner-shell vacancy, where the charge state reached is limited by the subsequent vacancy cascade, including shakeoff processes. Existing synchrotron-radiation sources are not sufficiently intense to permit multiple ionization without storage of the ions in the photon beam. An experimental program aimed at observing sequential inner-shell photoionization of ions stored in a Penning trap will be discussed.

II. RECOIL ENERGIES

Synchrotron x radiation from the Stanford Synchrotron Radiation Laboratory (SSRL) was used to produce K-shell vacancies in neon and argon, and L-shell vacancies in krypton and xenon. The subsequent vacancy cascades produced multiply charged ions which were detected by time-of-flight (TOF) techniques. Examination of the resultant charge-state peaks permitted extraction of information about the mean kinetic energy of each charge state.

Beams of white or monochromatic x rays from an eight-pole wiggler on SSRL beam line IV-II operated at 15 kG were focused by a toroidal mirror and collimated to a 1 mm diameter x-ray beam at the position of a thin gas target (≈ 0.2 mTorr). The critical energy of the radiation was 4 keV (corresponding to the 2 GeV electron-beam energy) and was attenuated below 3 keV by beryllium windows. The gas target was viewed by a vertically mounted TOF anaylzer through a 3-mm aperture located just above the x-ray beam. X-ray intensity was monitored by ion chambers positioned both upstream and downstream of the vacuum system housing the analyzer. Typical flux through the target was $\approx 10^{12}$ photons mm^{-2} s^{-1}. The TOF analyzer was designed so that ion flight times were less than the 780 ns between bursts of photons characteristic of SSRL timing mode operation. Each pulse of light had a measured full width at half maximum (FWHM) of ≈ 300 picoseconds[4] and was synchronous with a start signal to a CAMAC time-to-digital converter (TDC). Ions detected by a dual channel-plate detector after traversing the ≈ 10 cm analyzer provided the stop signal to the TDC. Timing resolution was better than 1 ns, due to the stability of the fast-rise-time TOF start pulses derived from the storage-ring rf electronics, and to low-noise time-pickoff techniques applied to pulses from the detector. Typical counts rates were from 1 to 5 kHz.[5] The apparatus used is shown schematically in Fig. 1.

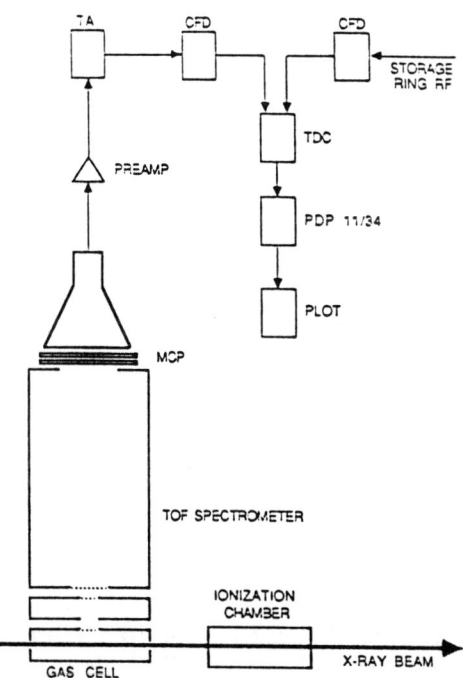

FIG. 1. *Schematic of time-of-flight apparatus and associated electronics used to measure recoil-ion charge distributions and mean energies.*

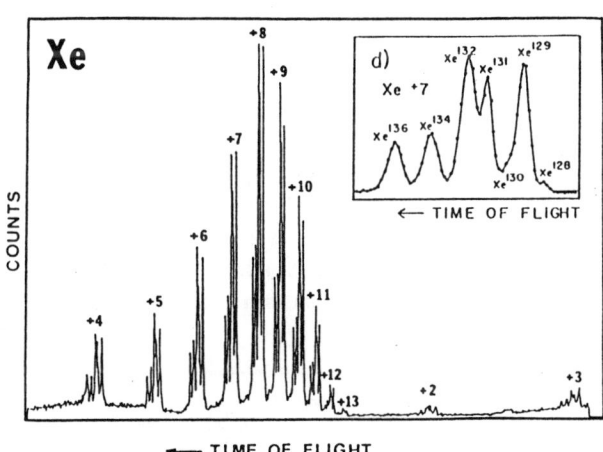

FIG. 2. *Xenon time-of-flight spectrum obtained with unmonochromatized synchrotron radiation. The charge distribution is the result of a single vacancy created, most frequently, in the L shell. Expanded view of Xe^{7+} shows resolution of xenon isotopes. Peak widths are ≈ 2.2 nsec. From ref. 6.*

The timing resolution was sufficient to obtain adjacent-mass isotopic resolution of xenon photoionized, primarily in the L shell, by unmonochromatized synchrotron radiation (Fig. 2). This fact is qualitative indication of low ion kinetic energy—otherwise the TOF spectra would be smeared. The peaks shown for Xe^{7+} in Fig. 2 have areas corresponding closely to known isotopic abundances and have FWHM of ≈ 2.2 nanoseconds.

Mean recoil-ion energies were determined for each charge state q through study of the FWHM of the corresponding TOF peaks, obtained by a least-squares gradient-search fitting procedure.[7] The ratios of the electric fields across the four spectrometer field regions provide for first-order focusing in flight time of ions created at different distances from the detector across the 1 mm diameter source region. The magnitudes of the electric fields largely determine flight time. For each charge state q, however, TOF peak widths are the result of several contributions. These peak widths are assumed to be the result of three contributions to variations in flight time: a constant electronic timing contribution, denoted by α, independent of spectrometer field strength E and charge q; a contribution arising from field fringing and imperfect compensation for finite source width (only first-order "time focusing" is expected) which scales as $\sqrt{(1/qE)}$, represented by β; and a third contribution resulting from the vector velocity distribution of the ions at creation, denoted by γ. Ions of mass it m originally headed away from the detector must be turned around, and consequently arrive after those originally headed towards the detector. This difference in time is

$$(\Delta t)_{re} = 2\sqrt{2mU_0}/(qE) = \gamma/(qE), \quad (1)$$

indicating that the initial energy (U_0) of the ions produces a width contribution which scales as $1/(qE)$.[8] Each contribution to peak width has a different functional dependence on the product qE; addition in quadrature of these three components results in the expression:

$$(FWHM)^2 = \alpha^2 + \frac{\beta^2}{(qE)} + \frac{\gamma^2}{(qE)^2}. \quad (2)$$

By varying the absolute magnitude of the electric fields in the extraction and acceleration regions of the spectrometer, while preserving their ratio to maintain space focusing, a set of spectra can be obtained whose peak widths (FWHM) can be fitted to the quadratic function of Eq. 2. Mean recoil-ion energy U_0 can thus be extracted from the fitted quadratic term using Eq. 1. Monte-Carlo simulations of our spectrometer using principles first outlined by Wiley and McClaren[8] have confirmed the general validity of this procedure.[9]

The clustering of data along quadratic curves is illustrated in Fig. 3 for recoil ions Ar^{6+}, Kr^{4+}, and Xe^{8+}, and is representative of the quadratic behavior exhibited by other charge states. The influence of quadratic terms is evident, and the near-zero intercept is consistent with electronic timing resolution better than 1 ns. Extraction of recoil energy from fitted quadratic terms γ using Eq. 1 resulted in the energies summarized in Table 1.

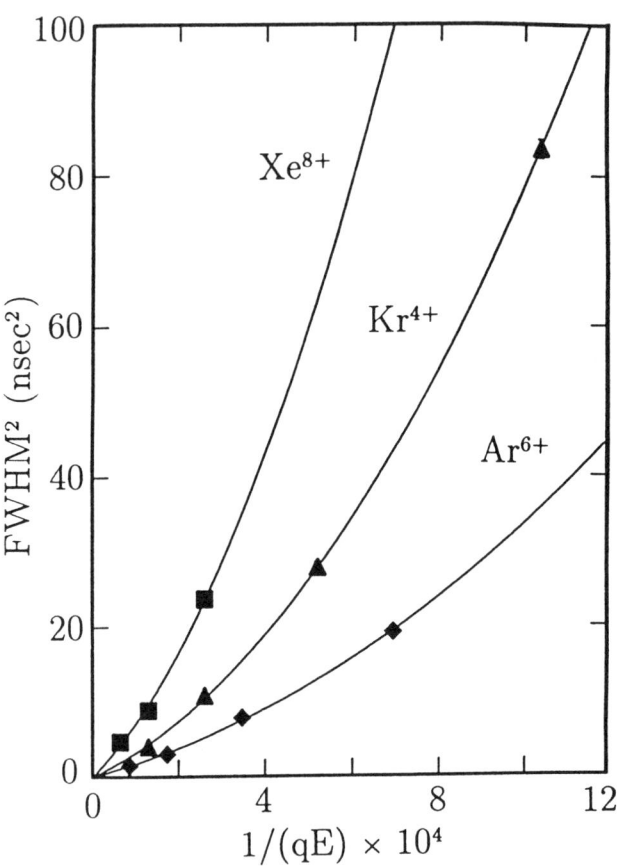

FIG. 3. *Quadratic behavior of $(FWHM)^2$ as a function of $1/qE$ for Ar^{6+}, Kr^{4+}, and Xe^{8+}, from which kinetic energy is extracted.*

We can make some simple estimates of kinetic energies expected of ions created by photoionization. Ion kinetic energy is the cumulative result of several effects: initial thermal energy, recoil from the photoelectron, and recoil from the several Auger and Coster-Kronig electrons ejected during the vacancy cascade which follows the primary ionization. Only velocity associated with thermal energy is isotropic. Angular momentum and parity conservation restrict the form of the angular distribution of electrons ejected by electric dipole excitation.[10] The differential cross section for photoelectrons ejected from a randomly oriented target by plane-polarized incident radiation, such as synchrotron radiation, is given by

$$\frac{d\sigma(\epsilon)}{d\Omega} = \frac{\sigma(\epsilon)}{4\pi}[1 + \beta(\epsilon)P_2(cos\theta)], \quad (3)$$

where the Legendre polynomial P_2 is evaluated at the angle θ between the electron ejected with energy ϵ and the photon polarization vector. The $\beta(\epsilon)$ asymmetry parameter is constrained to be between -1 and +2 and oscillates strongly as a function of energy.[11] Similarly, the Auger-electron distribution is not isotropic and depends on photon energy.[11] Since the energies in Table 1 were determined for photoionization with unmonochromatized synchroton radiation, for which photon energies range from the photoionization threshold to the 10 keV cutoff of the focusing mirror, the direction of the recoil-ion velocity vector will depend in a complicated way on photon energy. Without taking these angular and energy effects, or the effect of different decay channels into account, the simple predictions of recoil energy in Table 1 were obtained by the addition, in quadrature, of thermal energy, recoil from the photoelectron produced by photons at the SSRL critical energy, and recoil from the most energetic Auger event.

The possible use of synchrotron-radiation-produced highly charged ions in low-energy ion-atom collision experiments can be illustrated by comparing the ion energies in Table 1 with energies of the same charge state ions of Ar produced by fast charged-particle beams. We have used TOF techniques similar to those already described to detect Ar recoil ions created by beams of 23 MeV/u and 27.6 MeV/u Cl^{5+} produced by the Oak Ridge National Laboratory EN Tandem accelerator.[12] The charged-changed projectiles, separated by charge state in a parallel-plate electrostatic field, provided the start pulse to a time-to-amplitude converter (TAC). Recoil ions detected by the TOF spectrometer stopped the TAC. High-charge-state recoil ions produced in this manner can have kinetic energies 1-2 orders of magnitude higher than when created by synchrotron radiation (Fig. 4). This is because recoil energies associated with charged-particle production are the result of the internuclear Coulomb repulsion between projectile and target, which typically are separated, at the distance of closest approach, by less than 0.5 a.u. At such small internuclear separations, near the radius of both projectile and target L-shell electrons, the effective charges due to reduced screening are much larger than the final, asymptotic charges detected experimentally. Hartree-Fock calculations indicate that at the distance of closest approach, 0.35 a.u.[12], effective target and projectile charges for the $Cl^{5+} + Ar \rightarrow Cl^{8+} + Ar^{9+}$ system are 12.6 and 11.9, respectively, much higher than the asymptotic charges 9 and 8.

Charge	Ar(K)	Kr(L)	Xe(L)
3	0.031(13)	0.052(13)	
4	0.038(18)	0.066(17)	
5	0.037(23)	0.083(22)	0.076(11)
6	0.046(29)	0.094(27)	0.089(14)
7	0.036(82)	0.099(32)	0.125(16)
8	0.023(99)	0.059(34)	0.121(47)
9			0.196(21)
10			0.145(23)
estimate	0.05	0.03	0.03

Table 1. *Kinetic energies, in eV, of Ar, Kr, and Xe ions produced by vacancy cascades following inner-shell vacancy creation, primarily in the shell indicated, by synchrotron-radiation photoionization. Statistical uncertainties in the last digits are indicated in parentheses.*

The recoil energy produced by a charged-particle projectile depends on beam energy. As indicated by Fig. 4, slower projectiles have more time to interact with the target, producing more recoil energy. The results of Fig. 4 are consistent with data obtained at 18 MeV and 33 MeV, but not shown on Fig. 4. Shell effects are suggested by the much larger kinetic energies associated with production of Ar^{9+} in coincidence with triple-electron loss from the Cl^{5+} projectile. To achieve this final state, it is necessary that both projectile and target be ionized into the L shell. The small impact parameters necessary to effect this degree of ionization produce the large recoil energy measured.

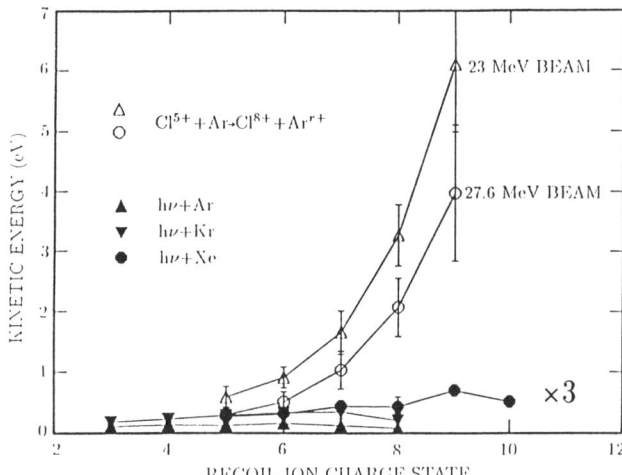

FIG. 4. *Comparison of kinetic energies of Ar ions produced by charged-particle impact and by synchrotron radiation. The synchrotron-radiation data have been multiplied by 3 for clarity. From ref. 13.*

II. CHARGE DISTRIBUTIONS

It is well known that the atomic rearrangement process following inner-shell vacancy production can lead to highly ionized states of the residual atom. The theoretical estimation of the charge distributions which result from this vacancy cascade must include relative cross sections for the various radiative, Auger, and Coster-Kronig processes which fill the initial, and subsequent, vacancies until the atom reaches a stable state in which no more transitions are possible. Immediately following ejection of an electron, the remaining ensemble of electrons are not in eigenstates of the ion; the sudden change in the effective charge can then result in "shake" phenomena, in which one or more electrons may be excited to unoccupied bound states ("shakeup") or into the continuum ("shakeoff"). Shakeoff probabilities can be quite large, and when the effect is multiplied during a vacancy cascade, shakeoff can contribute strongly to the formation of high-charge-state ions. Following creation of a single L_2 vacancy in argon, for example, calculations show that shakeoff results in 14.8% abundance of Ar^{3+}, and 0.2% of Ar^{4+}; in the absence of shakeoff, the result is 100% Ar^{2+}.[14]

FIG. 5. *Comparison of measured Xe charge distributions following creation of an inner-shell vacancy. The deepest shell ionized is indicated. Solid bars are data from our group and open bars are theory from ref. 16.*

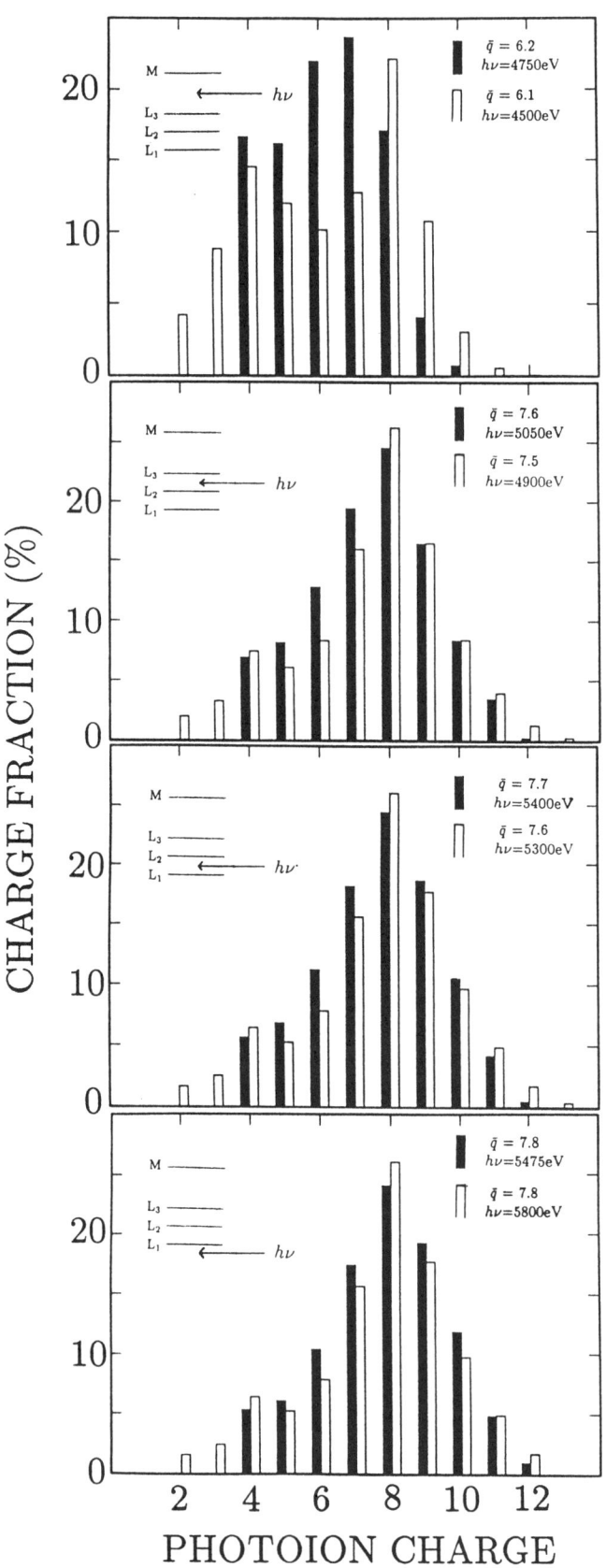

We have examined shell effects by tuning synchrotron radiation with a Si(111) double-crystal monochromator (bandpass 1×10^{-4}) to interleave the L_1, L_2, and L_3 edges of Xe. The resultant charge distributions, which are shown in Fig. 5, can be compared with early mass-spectroscopic measurements made by Carlson et al,[15] using x-ray guns and filters, and recent measurements by Tonuma et al[16] with monochromatic x rays at the Photon Factory. The agreement between our results and the measurements of Tonuma et al is very good, resolving a discrepancy we observed[5] with Carlson et al in the mean-charge-state increment measured for Xe as the L_1 edge is crossed. Consideration of Coster-Kronig yields[17] for transfer of an L_1 vacancy to the $L_{2,3}$ subshells and of relative photoionization cross sections of the $L_{1,2,3}$ levels[18], leads to an estimated shift in mean charge state similar to that seen by our group and that of Tonuma et al.

Calculations of charge distributions following L-shell ionization have been performed by Tonuma et al[16] using estimates of shakeoff made by Carlson et al[15] in 1966. These estimates do not include the relaxation of the atomic-electron ensemble that occurs during the vacancy cascade, the decrease in Auger transition rates which occurs because of the depletion of outer-shell electrons, or the energy dependence of shakeoff probabilities across the photoionization threshold before the asymptotic value is reached. The result of the first effect is that some Auger decays become energetically impossible as a result of the increased electron binding energies due to decreased screening as the number of inner-shell vacancies increases. Nevertheless, the agreement between measurement and theory is reasonable for the distributions following L-vacancy production, but not so good for the M-shell results (Fig. 5.)

Our group plans to measure the threshold energy dependence of shakeoff by determining the photoion charge distribution in coincidence with Auger electrons from particular inner shells as synchrotron radiation is tuned in small increments from below the photoionization threshold up toward the asymptotic limit. This approach should complement the more usual measurements of shake phenomena made by examining photoelectron and Auger-electron satellites,[19] but may offer higher count rates due to the high collection efficiency for low-energy photoions.

Studies of charge distributions following inner-shell photoionization may help elucidate more complicated multiionization phenomena in ion-atom collisions. The independent-electron-ejection model has long been employed, with substantial success, to describe multiple-vacancy production in target gases by beams of energetic charged projectiles. In this model, whose validity requires that the collisional velocity be much larger than the orbital velocity of the electrons being ionized, electron ejection is described by a binomial distribution of outer-shell vacancies. Extension of the model to include inner-shell ionization is accomplished by introducing an independent binomial distribution for each inner shell. In most treatments, the effect of vacancy cascades following inner-shell ionization on the final recoil-ion charge distribution is thus neglected. We have made measurements of argon recoil-ion charge-state distributions, in coincidence with charge-changed $Cl^{5+,8+,10+}$ projectiles at 0.7 MeV/u, whose interpretation requires the assumption of Auger vacancy cascades filling target L-shell vacancies. Good agreement with the recoil-ion charge spectra cannot be obtained by assuming only the customary binomial distribution of M-shell ionization.[20] Rather than losing its electrons sequentially, from the most weakly bound to the more tightly bound, the target argon atom achieves its final charge state through a superposition of outer-shell stripping with an L-shell vacancy cascade. Similar results have been reported for oxygen and fluorine projectiles at 1 MeV/u.[21]

III. SEQUENTIAL PHOTOIONIZATION

A research program to study sequential photoionization of multiply charged ions with unmonochromatized synchrotron radiation has been initiated by D. A. Church of Texas A&M University on National Synchrotron Light Source beamline X-26C. Motivations include the possible development of a high-charge-state, high-brightness ion beam with very-low energy for use in secondary ion-atom collisions. In addition, atomic inner-shell phenomena are very sensitive to relativistic and quantum-electrodynamical effects. Precision spectroscopic studies of a range of ion charges q for a fixed Z, or of a range of isoelectronic states, could yield information on parity-nonconserving interactions[22] and hyperfine structure.

The trap used at NSLS is of the Penning type, composed of a cylindrical ring electrode and two end caps shaped to produce a dominant quadrupole potential when the end caps are at zero potential and a negative potential is applied to the ring. The trap was mounted in a vertical uniform magnetic field $B \approx 0.7T$. The target gas was pulsed into the trap in unison with the opening of a fast shutter which permitted synchrotron radiation, focused through a slot in the trap ring electrode by a platinum-coated cylindrical mirror, to enter the trap. The pulsed-gas system permitted attainment of very-low background pressures ($\leq 10^{-9}$ Torr), thus increasing storage times. The stored ions were detected using a resonantly excited tuned circuit.[23] Due to the low energies of the multiply charged ions produced, the confining well depth required was ten times shallower than for recoil ions of similar charge states produced by fast heavy-ion impact.[24]

Stored xenon ions in charge states as high as q=11+ have been observed with this apparatus. Note the similarity between the stored-ion charge distribution (Fig. 6) and the distribution of xenon ions created by a similar distribution of "white" photons (Fig. 2). The trapped ions show a shift to lower charge as a result of interactions between the stored thermal multicharged ion gas and the residual neutral atoms, mainly xenon, in the trap. By varying the delay time between the close of the photon shutter and the start of the detection cycle, it is possible to measure rate coefficients for electron transfer from the neutral Xe to Xe^{q+}. Rate coefficients have been obtained for argon[23] and are presented in a poster at this conference (Wed 136).

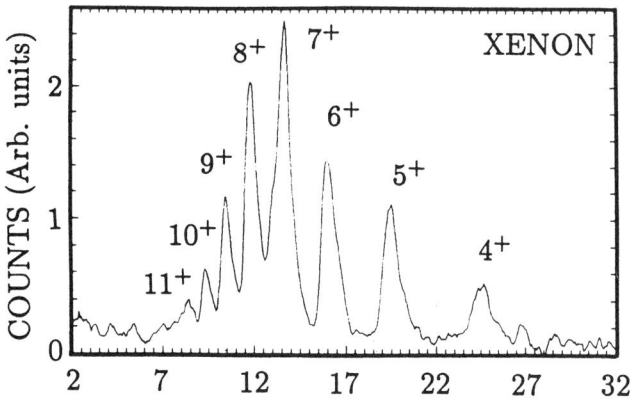

FIG. 6. *Signal of multicharged Xe photoions produced by unmonochromatized synchrotron radiation and stored in a Penning ion trap. Compare with Fig. 2.*

A recent measurement (June 1989) aimed at observing sequential photoionization of argon ions is undergoing analysis. Signal intensity is currently limited by decay of stored ions as a result of charge transfer and by the flux of bending-magnet radiation that passes through the beryllium windows separating the experimental chamber from ring vacuum. In the future, it may be possible to remove these windows, permitting greater flux at low energies where the cross section for M-shell photoionization is large.

The next generation of dedicated synchrotron-radiation facilities, such as the Advanced Photon Source (APS) at Argonne and Advanced Light Source (ALS) at Berkeley, will offer many orders of magnitude greater flux than are presently available, due, in part, to increased reliance on wiggler and undulator insertion devices. These machines should make possible detailed photoionization studies of stored ions. In addition, the new high-flux light sources may permit production and extraction of highly charged photoions in sufficient quantities to form a beam useful in secondary angle-resolved ion-atom collision studies at very-low energy.

Acknowledgements

The work conducted at SSRL was performed in collaboration with R. T. Short, C.-S. O, and I. A. Sellin from the University of Tennessee (UT) and ORNL, L. Liljeby from the Manne Siegbahn Institute (MSI) in Stockholm, D. A. Church from Texas A&M University, and S. Huldt, S.-E. Johansson, and E. Nilsson from Lund University, Lund, Sweden. The EN Tandem experiments at ORNL involved R. T. Short, C. Biedermann, C.-S. O, S. B. Elston, J. P. Gibbons, and I. A. Sellin from UT and ORNL and H. Cederquist, now at MSI. The Penning trap experiments at NSLS are led by D. A. Church and S. Kravis from Texas A&M and include B. Johnson, M. Meron, and K. Jones of Brookhaven National Laboratory, R. T. Short, C.-S. O, and I. A. Sellin from UT and ORNL, G. Berry and Y. Azuma of Argonne National Laboratory, and M. Druetta of the University of Lyon.

The work of the researchers from UT/ORNL is supported in part by the National Science Foundation and by the U.S. Department of Energy, Office of Basic Energy Sciences, Division of Chemical Sciences, under Contract No. DE-AC05-84OR21400 with Martin Marietta Energy Systems, Inc. SSRL is supported by the U.S. Department of Energy, Office of Basic Energy Sciences, and the National Institutes of Health, Biotechnology Resource Program, Division of Research Resources.

References

[1] E. Källne, Atomic Physics **10**, 395 (1987).

[2] For example, see *Beam Foil Spectroscopy*, S. Bashkin, ed., Springer-Verlag, Berlin (1976); and Proceedings of the 1987 Symposium on Atomic Spectroscopy and Highly-Ionized Atoms, H. G. Berry, ed., (1988).

[3] H. Cederquist, C. Biedermann, J. C. Levin, C.-S. O, I. A. Sellin and R. T. Short, Nucl. Instr. and Meth. **B34**, 243(1988).

[4] M. G. White, R. A. Rosenberg, G. Gabon, E. D. Poliakoff, G. Thornton, S. H. Southworth, and D. A. Shirley, Rev. Sci. Instrum. **50**, 1269(1979).

[5] R. T. Short, C.-S. O, J. C. Levin, I. A. Sellin, L. Liljeby, S. Huldt, S.-E. Johansson, E. Nilsson, and D. A. Church, Phys. Rev. Lett. **56**, 2614 (1986).

[6] I. A. Sellin, J. C. Levin, C.-S. O, H. Cederquist, S. B. Elston, R. T. Short, and H. Schmidt-Böcking, Physica Scripta **T22**, 178(1988).

[7] P. Bevington, *Data Reduction and Error Analysis for the Physical Sciences*, McGraw-Hill, New York, 1969.

[8] W. C. Wiley, and L.H. McLaren, Rev. Sci. Instr. **26**, 1150(1955).

[9] J. C. Levin, R. T. Short, and I. A. Sellin, to be published.

[10] C. N. Yang, Phys. Rev. **74**, 764(1948); M. Peshkin, Adv. Chem. Phys. **18**, 1(1970); V. L. Jacobs, J. Phys. B **5**, 2257(1972).

[11] S. H. Southworth, P. H. Kobrin, C. M. Truesdale, D. Lindle, S. Owaki, and D. A. Shirley, Rapid Communications, Phys. Rev. A **24**, 2257(1981).

[12] J. C. Levin, R. T. Short, C.-S. O, H. Cederquist, S. B. Elston, J. P. Gibbons, I. A. Sellin, and H. Schmidt-Böcking, Phys. Rev. A. **36**, 1649(1987).

[13] J. C. Levin, C. Biedermann, H. Cederquist, C.-S. O, R. T. Short, and I. A. Sellin, Nucl. Instr. and Meth. **B40/41**, 402(1989).

[14] T. Mukoyama, Bull. Inst. Chem. Res., Kyoto Univ. **63**, 373(1985).

[15] T. A. Carlson, W. E. Hunt, and M. O. Krause, Phys. Rev. **151**, 41(1966).

[16] T. Tonuma, A. Yagishita, H. Shibata, T. Koizuma, T. Matsuo, K. Shima, T. Mukoyama, and H. Tawara, J. Phys. **B20**, l31(1987).

[17] M. H. Chen, B. Crasemann, and H. Mark, Phys. Rev. **A24**, 177(1981).

[18] W. H. McMaster, N. Kerr Del Grande, J. H. Mallet, and J. H. Hubbel, Compilation of x-ray cross sections, UCRL-50174 Sec. II Rev. 1, Lawrence Radiation Laboratory, University of California, Livermore, 1969.

[19] G. B. Armen, T. Åberg, K. R. Karim, J. C. Levin, B. Crasemann, G. S. Brown, M. H. Chen, and G. E. Ice, Phys. Rev. Lett. **54**, 182(1985).

[20] J. C. Levin, C.-S. O, H. Cederquist, C. Biedermann, and I. A. Sellin, Rapid Communications, Phys. Rev. A **38**, 2674(1988).

[21] O. Heber, G. Sampoll, B. B. Bandong, R. J. Maurer, E. Moler, R. L. Watson, I. Ben-Itzhak, J. L. Shinpaugh, J. M. Sanders, L. Hefner, and P. Richard, Rapid Communications, Phys. Rev. A **39**, 4898(1989).

[22] S. A. Blundell, W. R. Johnson, and J. Sapirstein, Phys. Rev. A **38**, 4961(1988).

[23] D. A. Church, S. D. Kravis, I. A. Sellin, C.-S. O, J. C. Levin, R. T. Short, M. Meron, B. M. Johnson, and K. W. Jones, Rapid Communications, Phys. Rev. A **36**, 2487(1987).

[24] D. A. Church, R. A. Kenefick, W. S. Burns, I. A. Sellin, M. Breinig, S. B. Elston, S. Huldt, R. Holmes, J.-P. Rozet, and S. Berry, Phys. Scr. **T3**, 173(1983).

ATOMS IN INTENSE RADIATION FIELDS

Marvin H. Mittleman

Physics Department, The City College of New York

The state of the field is reviewed and some speculations about the future of very-intense-radiation physics are given.

Historically, intense field (defined here as greater than about 10^{13} W/cm^2) physics really started with the experiment by Agostini et al.[1] in which electrons which absorbed more than the minimum number of photons were first observed. This yielded an electron energy spectrum with peaks separated by a photon energy. They were however, either explicitly or implicitly predicted in earlier theories.[2] A more quantitative understanding emerged from the experiment of Kruit et al.[3] in which the suppression of the lowest energy peak of the electron spectrum was observed. This was explained[3,4] by the idea that the ionization energy of the atom inside the field was shifted upward. This came as a surprise because the role of the ponderomotive "potential" was not well understood. It works in the following way: The creation of a zero energy "free" electron is not possible in the field since the electron must have a quiver energy. The time average of this quiver energy is the ponderomotive "potential," $U_p = e^2E^2/4m\omega^2$, and it can be envisaged as acting to shift the continuum states upward by this amount. The field also acts to shift the initial bound state, usually downward, by a much smaller energy and so the difference, the ionization energy, is increased. The ponderomotive potential also acts to expel the "free" electron from the radiation field so the electron gains the ponderomotive energy on its way out of the field to the detector.

Therefore, it is deprived of the increased ionization potential energy by the field and has it essentially totally restored by the expulsion from the field so that it is detected with an energy which is almost unshifted (from the zero field case) outside the field.

Is it possible that the cancellation occurs because we are looking at the same effect in two different ways? This was answered negatively first theoretically[5] and then experimentally[6] with short pulse lasers. The almost total cancellation of the increased ionization potential and the expulsion energy, yielding an unshifted electron energy at the detector, depends critically upon the electron being expelled from the field while the field amplitude is constant. If the field is switched off before the electron escapes the spatial region of the field then it only recovers part of the ponderomotive energy during its expulsion and the cancellation described above is incomplete. Pico-second pulses are short enough to demonstrate this phenomena. Evidently the kinetic energy of the emerging electron sets the time scale required to observe it and so high energy electrons, which escape quickly, will show a better cancellation of the ponderomotive potential than do low energy electrons. That is, the degree of cancellation of the ponderomotive energy, will depend upon the electron energy itself and so the peaks in the spectrum will no longer be exactly separated by the photon energy.[5,6]

Another novel application of these ideas by Freeman and co-workers[6] exploits the spatial dependence of the field and its effect on resonance transitions. The states of the atom in the field are shifted by the field and so a particular (multiphoton) transition will be resonant at some spatial points and not others. These resonances will enhance the ionization rate. But in a long pulse, where the cancellation described above is complete, the electron energy spectrum will not depend on the ponderomotive potential and will only show the peaks which depend on the number of photons absorbed. Then the resonance enhancement will not be observable. However, in the short pulse experiments each of those peaks will be split into subpeaks which present intermediate resonances in the multiphoton absorption process. Two new and interesting results have emerged.[6] First, the resonance enhancement of the non-resonant reaction rate is large, a factor of five or so, and second it is possible to unfold the experimental results and trace out the dependence of the resonance energy on the field. The ground state field dependence is known so this is a measurement of the excited state field dependence. It has also been found that excited state energies, for surprisingly low lying states, are linear in the intensity with a coefficient given by the __ponderomotive energy__. After some preliminary disagreement, this last effect has been experimentally confirmed.[7] Neither of these effects have as yet been incorporated into any theory.

Other recent experiments[8] on angular distributions of electrons produced by multiphoton ionization by elliptically polarized lasers have led to further constraints on theories. The electrons were observed in the plane of polarization of the laser and found to exhibit only a symmetry of reflection through the origin in that plane. All previous existing multiphoton theories[2] showed an __additional__ symmetry of reflection __in__ this plane through the line of the major axis of the polarization and the experimental breaking of this additional symmetry proved to be a large effect.[8] However, two photon absorption described be second order perturbation theory[9] showed only the symmetry observed in the experiment. The earlier multiphoton theories[2] are first Born approximations, in a rearrangement collision sense,[10] treating either the electron-core potential or electron-field potential as a perturbation in lowest order. Ordinary second order perturbation theory is not described in this way. It can be shown[11] that a multiphoton theory which correctly gives the angular distribution can not be a first order theory in the sense described above and so must be rather complex. And to repeat, experiment shows[8] that this is not a small effect.

Trying to extrapolate to ultra-intense (10^{16} W/cm^2 or greater) radiation fields requires some theoretical courage in view of the unexpected phenomena revealed by experiment at lower intensities. We proceed anyway.

Perhaps the most obvious question to ask concerns the existence of metastable states of atoms in ultra-intense fields. So far only hydrogen has been considered.

A rather complex and obscure calculation by Janjusevic and Mittleman[12] gives a qualified yes an answer but a recent much more transparent series of papers by Gavila and co-workers[13] seems to give an unqualified yes as an answer. They use the Kramers[14] representation in which the Hamiltonian of the atom in the field is (in dipole approximation)

$$H = \frac{p^2}{2m} + V(\vec{r} - \vec{\alpha}(t)) \qquad (1)$$

where

$$\vec{\alpha}(t) = \int^t dt' \frac{e}{m}\vec{A}(t') \quad (2)$$

If the frequency of the field is high compared to the frequency of motion of the bound electron then a time average of the potential can be performed[15] and the eigenfunctions and eigenvalues of that time independent Hamiltonian are a good approximation to the true quantities. The smearing of the (time averaged) position of the nucleus weakens the potential and thereby lowers the binding energy and this effect is increased as the field strengthens. Thus the frequency of the motion of the bound electron decreases as the field increases. Gavrila et al.[13] realized the critical fact that the validity of the method required that the field frequency be large compared to the <u>field dependent</u> electron frequency not compared to the larger field independent electron frequency and this recognition made this a high intensity theory as much as a high frequency theory. In their theory for the binding energy the field frequency ω, enters only through the parameter entering in (2), $\alpha_0 = eE/m\omega^2$ and ω can be considered another independent parameter. The binding energy is a function only of α_0 (proportional to $\alpha_0^{-1/3}$ for linear polarization and high intensity)[16] and they show that the width is proportional to the other independent parameter ω^{-1}. Then, mathematically speaking, there is some range of the parameters α_0 and ω for which the width is small compared to the binding energy and so a metastable state exists. Details of this are not yet available.

The theory is non-relativistic and in the dipole approximation and it is not clear how the relaxation of these restrictions will affect the results. The problem for more than one electron is a challenging one. In the Kramers representation electron-electron interactions are unchanged but electron-nucleus interactions are weakened. This may result in a Coulomb explosion for some multi-electron atoms but not necessarily for all.

A variety of novel high intensity phenomena have been suggested by Rhodes[17] and others[18] which I can only touch upon here. The possibility of multiphoton pair creation (near a nucleus) was raised[19] but the cross section is so small as to be negligible. This is fortunate for someone who is trying to propagate an intense pulse through anything but a perfect vacuum. The reason for the small cross section is the action of the ponderomotive potential. A very intense laser is required to have a dense enough field of photons to make up the energy of pair creation. But in such an intense field the energy of pair creation is at least $2(mc^2+U_p)$ so that as the field gets more intense the number of photons required for pair creation rises. Detailed calculation[19] shows that the cross section is always small.

It has been suggested by Becker[20] et al. that nuclear β decay rates could be significantly altered by the presence of an intense field. Their mechanism is the following: If the maximum energy of the emerging electron is small then the phase space available to it is small and so is the decay rate. The presence of the radiation field allows the electron to absorb photons, increase its energy and therefore its phase space and so the decay rate is increased. A significant increase in the decay rate of tritium was predicted for field at 10^{17}W/cm^2 intensity. Again, the inclusion of the ponderomotive potential eliminates this effect since the energy required to emit the electron is raised by the ponderomotive potential.

Another phenomenon that has been demonstrated[21] and is theoretically understood[22] for moderate fields is high order harmonic generation. Harmonic numbers in the thirties[21] (only odd harmonics occur because of parity conservation) have been observed in dense gases. Kulander's theory[22] of this phenomena relies on the wave function of a model single particle atom in a field. He obtains $\psi(\vec{r},t)$ numerically by a solution of the resulting Schrodinger equation. The wave function yields a current density which acts as a source for the higher harmonics in the gaseous medium. Excellent agreement with experiment results for intensities of the order of 10^{13} W/cm^2. For higher intensities ionization is more rapid and the numerical problem becomes more difficult. For ultra-intense fields ionization is essentially instantaneous and the electrons are described as "free" particles in a field. Sarachik and Schappert[23] in, a classic paper, have given a simple description of the relativistic electron motion and the high harmonic generation resulting from its anharmonic motion. Very high harmonics are generated, (as high as 10^4 for an intensity of 10^{21} W/cm^2) in a tight cone centered about the field propagation direction. The intensity of radiation from a single electron is of course small but a (spatially) small sample of N electrons will move coherently and radiate a factor of N^2 larger than a single electron. However, if the density is too large the plasma becomes opaque at the laser frequency and the effect is quenched. Clearly there is some density at which the harmonic generation is optimized but this formidable problem has not been solved.

There have been several suggestions of exotic effects induced by ultra-intense laser fields, laser induced fission[17] laser induced fusion,[18] laser induced inner atomic shells excitation[17] and others.[18] In all of them collective effects resulting from the laser ionized medium are crucial and these have been only incompletely treated so conclusions reached are only provisional. In addition the motion of a free electron in these ultra-intense fields has some surprising effects which must be carefully treated. For example, when an ultra-intense pulse overtakes an electron at rest it is given the usual relativistic oscillatory motion in the perpendicular direction, the details of which depend upon the field polarization. But it is also given a much more energetic forward motion by the pulse and, if no collisions occur, it comes back to rest after the passage of the pulse. If an inelastic collision occurs inside the pulse the electron is propelled forwards after the passage of the pulse. This has some consequences in the process of inner shell excitation. The idea proposed[17] was that the ultra-intense pulse would immediately strip away the outer electrons which would oscillate coherently and so constitute a current density which after half a period would pass close to the parent ion and so cause inner-shell excitation. This neglects the forward motion (at essentially the speed of light) of the electrons which, after half a period, are half a wave length away. They therefore must collide with another ion, not the parent. Again this requires the description of a laser induced relativistic plasma.

Preliminary calculations of laser driven electrons with a potential prechosen to mock up the plasma potential have been performed.[24] The results are too complex to try to review here, they suggest the possibility of an enhancement of energy transfer from the field to the electron gas due to this "plasma potential." If this is true for the real case then the laser induced "exotic" effects may indeed occur.

This work was supported by a grant from the U.S. National Science Foundation.

References

1. Agostini, Fabre, Mainfrey, Petite and Rahman Phys. Rev. Lett. 42 1127 (1979).
2. L.V. Keldysh, JETP 20 1307 (1965); F.H.M. Faisal, J. Phys. B 6 L89 (1973); J. Gersten and M.H. Mittleman, Phys. Rev. A 10 74 (1974), A.J. Pert, J. Phys. B 8 L173 (1975); H.R. Reiss, Phys. Rev. A 22 1786 (1980).
3. Kruit, Kimman, Muller and van der Wiel Phys. Rev. A 28, 248 (1983).
4. M.H. Mittleman, Phys. Rev. A 29, 2245 (1984); A. Szoke, J. Phys. B 21, L125 (1988); J. Kupersztych, Europhys. Lett. 4, 23 (1987).
5. E. Fiordilino and M.H. Mittleman, J. Phys. B 18, 4425 (1985); Agostini, Kupersztych, Lompre, Petite and Yergeau, Phys. Rev. A 36, 4111 (1987).
6. Freeman, Bucksbaum, Milchberg, Darack, Schumacher and Geusic, Phys. Rev. Lett. 59, 1092 (1987). See also the last ref. 5.
7. H. Muller, private communications, June 1989.
8. M. Bashkansky, P.H. Bucksbaum and D.W. Schumacher, Phys. Rev. Lett. 60, 2458 (1988).
9. See for example, M. Crance and M. Aymar, J. Phys. (Paris) 46, 1887 (1985).
10. See for example, M.L. Goldberger and K.M. Watson, "Collision Theory," J. Wiley (N.Y.) 1964.
11. P. Krstic and M.H. Mittleman, to be published.
12. M. Janjusevic and M.H. Mittleman, J. Phys. B 21, 2279 (1988).
13. See M. Pont, N.R. Walet, M. Gavrila and C.W. McCurdy, Phys. Rev. Lett. 61, 939 (1988) for previous references.
14. H.A. Kramers, "Collected Scientific Papers," North-Holland, Amsterdam 1956.
15. J.I. Gersten and M.H. Mittleman, J. Phys. B 9, 2561 (1976).
16. Radiactive Distortion of the Hydrogen Atom in Superintense, High-Frequency Fields of Linear Polarization, M. Pont, N.R. Walet and M. Gavrila, preprint 1989.
17. K. Boyer and C.K. Rhodes, Phys. Rev. Lett. 54, 1490 (1985).
18. M. Perry, private communciation.
19. V.P. Yakovlev, JETP 22, 233 (1966); M.H. Mittleman, Phys. Rev. A 35, 4624 (1987).
20. W. Becker, W.H. Louisell, J.D. McCullen and M.O. Scully, Phys. Rev. Lett. 47, 1269 (1981).
21. Ferray, L'Huiller, Li, Lompre, Mainfrey and Manus, J. Phys. B 21, L31 (1987).
22. K.C. Kulander, Phys. Rev. A 38, 778 (1988) and K.C. Kulander and B.W. Shore, preprint (1989).
23. E.S. Sarachik and G.T. Schappert, Phys. Rev. D 1, 2738 (1970).
24. Relativistic Dynamics of Electrons in Intense Laser Fields, J.N. Bardsley, B.M. Penetrante and M.H. Mittleman, Phys. Rev. A, in press.

PHOTOIONIZATION OF POSITIVE ATOMIC IONS: A REVIEW OF OUR PRESENT UNDERSTANDING

Steven T. Manson

Department of Physics and Astronomy, Georgia State University, Atlanta, Georgia 30303, USA

Our present understanding of the phenomenology of ionic photoionization is reviewed. Particular emphasis is placed on the broad theoretical predictions which have been made which have not yet been tested by experiment.

I. INTRODUCTION

The absorption of electromagnetic radiation by matter is one of the fundamental processes in the universe. A basic understanding of the energetics of the photoionization process was given by Einstein almost a century ago.[1] The years since have seen much progress in our understanding of the details of photoionization of neutral atoms.[2] This progress has been largely the result of the interplay between theory and experiment.[2-4] Thus, while there are still many gaps in our understanding of atomic photoionization,[5] it is also true that we have significant knowledge of a number cases.[2-5]

The situation is otherwise for photoabsorption by positive atomic ions. Despite their importance in a number of areas, there has been relatively little work on ionic photoionization to date. Laboratory measurements have been hampered by the difficulty in obtaining ions in sufficient quantity to perform meaningful photoionization experiments. Consequently the only extant measurements are for six singly-charged ions at low photon energies[6] (and some rough measurements for two others[7,8]), along with some measurements of Ar^+ and multi-charged helium-like ions each at a single energy,[9,10] all of these measurements are for ground states of ions; as far as we are aware no experiments involved excited states have been performed. Thus, in most ways, theory is ahead of experiment in the area of photoionization of positive ions.

In this paper, therefore, the status of the theory shall be reviewed, along with the available experiment. Before doing that, however, it is worthwhile to point out some general features of positive ions; this is given in the next section. The following section presents a selection of results of investigations of ionic photoionization with an eye to making our present level of understanding clear. Finally, the last section presents a summary along with some remarks related to where the field is going.

II. POSITIVE ATOMIC IONS

Aside from its particular state, a positive atomic ion can be characterized by (Z, N), the nuclear charge and the number of electrons respectively. Thus, ionic data can be viewed in at least three alternative pictures: the <u>isonuclear</u> picture where Z is held constant; the <u>isoelectronic</u> picture, where N is held constant; and the _isoionic_ picture, where Z-N is held constant. Looking at ionic properties in each of these pictures brings out different aspects of the systematics of the properties than others.[12] The Hamiltonian for an N-electron atomic ion of nuclear charge Z is given (non-relativistically) by

$$H = \Sigma(P_i^2/2m - Ze^2/r_i) + \Sigma e^2/r_{ij} \quad (1a)$$

$$= \Sigma P_i^2/2m - Ze^2[\Sigma 1/r_i - Z^{-1}\Sigma 1/r_{ij}] \quad (1b)$$

The term in brackets in Eq. (1b) is the total potential energy of the ion. If an isoelectronic sequence is considered, increasing Z (while keeping N fixed) clearly lessens the importance of the second term in the brackets owing to the Z^{-1} factor. This term is just the interelectron interactions which represents the deviation of the ionic Hamiltonian from hydrogenic. Thus, with increasing Z along an isoelectronic sequence, we would expect the photoionization to become more hydrogenic. Furthermore, we would expect that theoretical calculations would improve, with increasing Z, because the term becoming less important in the Hamiltonian is just the one being approximated. But, it does _not_ follow that the approach to simple hydrogenic behavior, with increasing Z, is monotonic or rapid. Nevertheless, this idea, that ions are simpler and less interesting than neutral atoms, has led to a _relative_ lack of theory for ions. The fallacy of this idea shall be discussed.

III. PHOTOIONIZATION RESULTS

A great deal of theoretical work has been done on inner shell photoionization of ions. These calculations have pointed up an important simplification on the inner shell cross sections, but only when scrutinized in an isonuclear sequence and plotted vs. photon energy, $h\nu$. In particular, it has been found that the removal of outer shell electrons has no significant effect on the cross sections of inner shells, shells with smaller principal quantum number, except for a shift of threshold to higher photon energy.

As an example, the photoionization of 2p electrons, calculated using Hartree-Slater (HS) central-field wave functions, in the iron isonuclear sequence[13] is shown in Fig. 1 where it is seen that the cross section from neutral iron to Fe^{+16} lies along the same curve, to roughly the thickness of the line. Since Fe^{+16} is neon-like, it means that all of the n=3 and n=4 electrons have been removed. Removing the

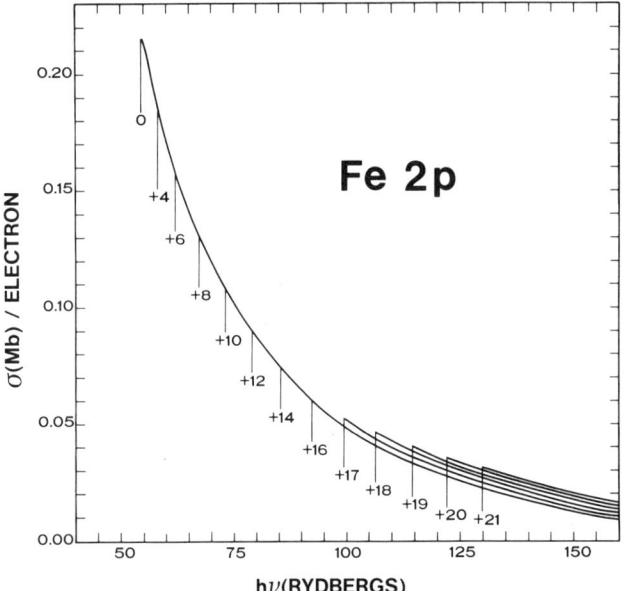

Figure 1. Photoionization cross section per electron for the 2p subshell of the Fe isonuclear sequence calculated in the Hartree-Slater approximation. The vertical lines are the theoretical thresholds for the given stage of ionization.

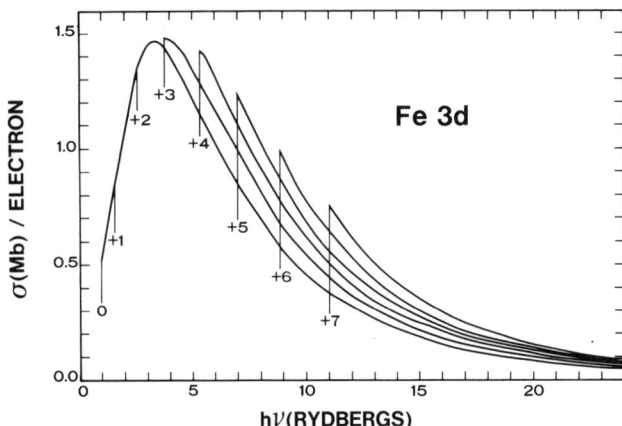

Figure 2. Photoionization cross section per electron for the 3d subshell of the Fe isonuclear sequence calculated in the Hartree-Slater approximation. The vertical lines are the theoretical thresholds for the given stage of ionization.

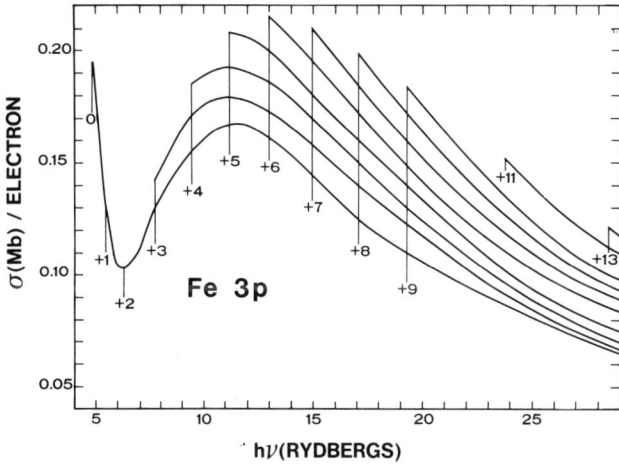

Figure 3. Photoionization cross section per electron for the 3p subshell of the Fe isonuclear sequence calculated in the Hartree-Slater approximation. The vertical lines are the theoretical thresholds for the given stage of ionization.

next electron, a 2p electron, and going to Fe^{+17}, it is clear from Fig. 1 that the cross section does then change. Thus, in this case, it is clear that removing electrons with higher n only increases the threshold energy but otherwise leaves the 2p cross section unchanged.

The 2p cross section is a simple monotone decreasing function of photon energy. To investigate whether or not this effect remains for more complicated cross sections, the HS situation for Fe 3d is displayed[13] in Fig. 2 where a delayed maximum,[14] caused by the centrifugal barrier in the f-wave potential of the 3d → f transition, occurs. It is clear from Fig. 2 that up to Fe^{+2}, where only 4s electrons are removed, the constancy of the inner shell cross section is seen. Likewise for the Fe 3d subshell,[13] which displays a Cooper minimum,[15] i.e., a zero in the 3p → d matrix element, the removal of outer shell electrons has no effect on inner shell cross sections, except for a threshold shift.

This phenomenon can be explained by noting that <r> for a given subshell depends strongly upon principle quantum number n, but only very weakly upon angular momentum l. For iron atoms and ions, <r> for n = 1, 2, 3, and 4 is $~0.06a_0$, $~0.25a_0$, $~0.9a_0$, and $~3a_0$, respectively.[12] Thus the spatial extend of wave function is much smaller than that of any of the other electrons of higher n. The nl electrons, therefore, "see" the higher-n electrons roughly as spherical shells of negative charge concentrated at their respective radii. Then, since these radii are well beyond the point where the nl wave function is effectively zero, and since spherical shells of charge produce no force (constant potential) within the shell, the only effect of their removal is to lower the potential in the inner region by a constant amount without otherwise affecting the nl wave function. This leads to a σ_{nl} which is constant as a function of photon energy along with a shift of the threshold to higher energy.

Based upon this explanation, the phenomenon would occur at high-Z as well,[17] which Fig. 4 for Hg 4f in the HS approximation clearly demonstrates. It has also been shown for relativistic calculations in Fe[18]; and for Th (Z = 90) the cross sections have been found to be equal within a few percent for the neutral and for 80 times ionized Ne-like Th.[19] Many-body perturbation theory also shows the same effects.[20] Further, the situation should not be substantially altered when exchange and correlation are included in the calculation. To test this, relativistic-random-phase approximation[21] (RRPA) calculations have been performed.[22] The results for the 2s cross section in the Ar isonuclear sequence is given in Fig. 5 for the neutral, Ne-like and Be-like ions; the agreement among the cross sections and the shift of threshold to higher energy is clearly seen. Finally, there is some experimental indication of the validity of this phenomenon. Cross sections for both neutral[23] and singly ionized[24] potassium have been measured. The results, shown in Fig. 6, indicate that the quality of the data, particularly for the atom, is such that all that can be inferred is that data is not inconsistent with this inner shell effect, but the error bars are so large that it can hardly be said that the effect is confirmed experimentally.[25]

The importance of this inner shell effect is two-fold. First, it allows the transfer of significant portions of photoionization data for neutral atoms to ions. Second, since at high energies the total photoionization cross section for an atom (or ion) is dominated by the in-

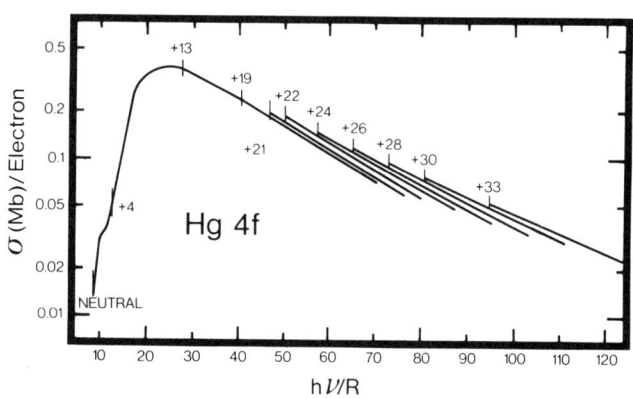

Figure 4. Photoionization cross section per electron for the 4f subshell of the Hg isonuclear sequence calculated in the Hartree-Slater approximation. The vertical lines are the theoretical thresholds for the given stage of ionization.

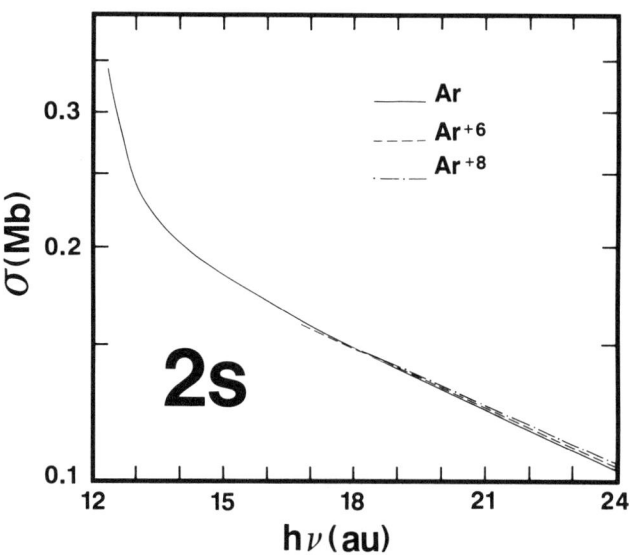

Figure 5. Photoionization cross section for the 2s subshell of Ar, Ar^{+6} and Ar^{+8} calculated in the relativistic-random-phase approximation.

ner shells,[5] total atomic and ionic total cross section should approach equality at high energies. Note too, that to observe the phenomenon, the ionic cross sections must be examined in an isonuclear sequence as a function of photon, not photo-

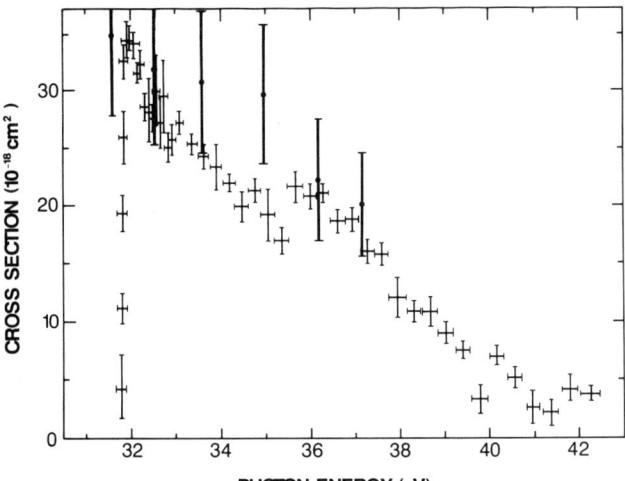

Figure 6. Comparison of experimental photoionization cross section for K^+ (crosses) and neutral K.

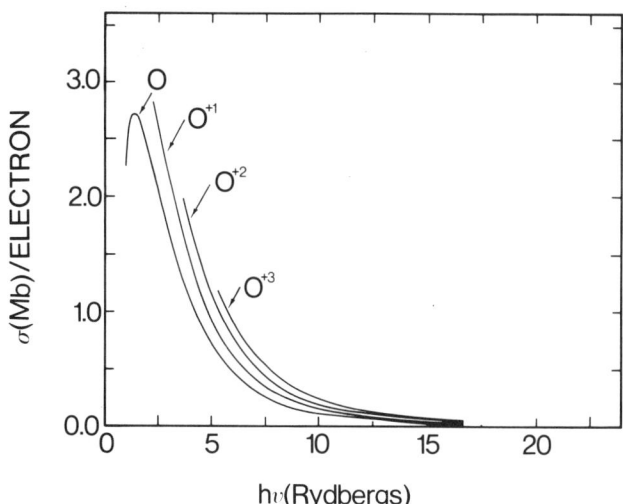

Figure 7. Photoionization cross section per electron for the 2s subshell of oxygen isonuclear sequence calculated in the Hartree-Slater approximation.

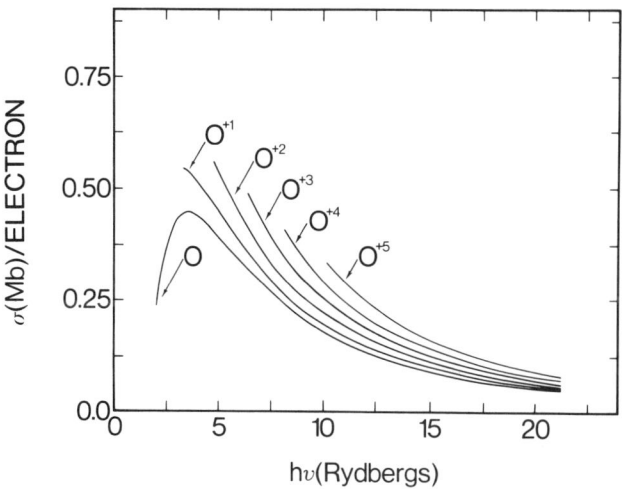

Figure 8. Photoionization cross section per electron for the 2p subshell of oxygen isonuclear sequence calculated in the Hartree-Slater approximation.

electron, energy.[26,27]

Looking at the inner shell cross sections when the inner shells have been broken into shows a consistent pattern in the isonuclear sequences given in Figs. 1-4. In each case the cross section per electron in the subshell is shown, to account for the depletion of the given subshell. In any case, these results show that the removal of electrons from a given shell increases the threshold energy of the remaining electrons in that shell, due to decreased screening, and increases the cross section per electron, as a function of photon energy, in the threshold region. This behavior occurs irrespective of the shape of the cross sections, as Figs. 1-4 clearly show. Physically this increase in cross section occurs because the decrease in screening, which means that the remaining electrons "see" a larger effective charge, causes the electron wave functions to contract, thus giving better overlap between initial and final states, near threshold, thereby increasing the matrix element and the cross section. By this reasoning, then, it would be expected that at energies quite far above threshold, this ordering will reverse itself, just as in the hydrogenic calculation.[28] This reversal is hinted at in Fig. 2. These figures also suggest the possibility of simply scaling the cross section from on ionization state to the next along an isonuclear sequence since the various cross sections along the sequence, are nearly parallel. This turns out to be somewhat more difficult than it looks; some attempts, however, have been made.[29,30]

Turning our attention to outer shells, it is found that, of course, removing electrons change the cross sections. The change in cross section along an isonuclear sequence is quite similar for outer shells as it was for inner shells, as discussed above. For example, the results of HS calculations are shown in Figs. 7 and 8 for 2s and 2p cross sections respectively in the oxygen isonuclear sequence.[31] From these figures it is seen that removal of electrons increases the ionization potential, of course, and shifts the cross section per electron upward. The cross sections for the various members of the isonuclear sequence are much less parallel than the inner shell results, however.

As discussed in the previous section, ionic cross sections can be scrutinized in other ways, not just along isonuclear sequences; just because isonuclear sequences are useful for inner shells, whereby the significant simplification of the inner shell effect is brought out, it does not follow that they are necessarily the most revealing for outer shells.

Looking at the isoelectronic sequence for the 3p shell of Ar[32] in Fig. 9, a somewhat more complex picture is presented than that seen in isonuclear sequences. Most notably, for lowly-ionized members of the sequence, the curves cross. Roughly speaking, the cross sections per electron increased moving along an isonuclear sequence; the curves for succeeding members of an isoelectronic sequence appear

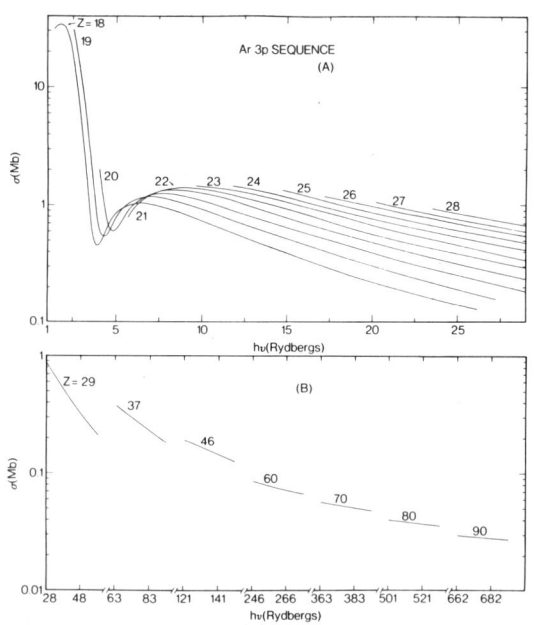

Figure 9. Photoionization of the 3p subshell of the Ar isoelectronic sequence calculated in the Hartree-Slater approximation.

to move to increasing energy, as seen in Fig. 9, thus causing crossing in energy regions where the cross section is increasing with increasing energy. The shift in energy is seen to be a small fraction of the increase in ionization threshold along the isoelectronic sequence. Similar results have been found for other isoelectronic sequences, for both outer shells[32,33] and inner shells.[33]

Before continuing, it is worthwhile to mention that ionic cross sections can also be studied in an isoionic picture, i.e., a sequence of the same stage of ionization. This, of course, has been done for neutrals for some time and has yielded some useful insights into how to interpolate between measured elements to obtain cross sections for unmeasured ones.[34,35] For ions, particularly highly charged ones, we find that an isoionic sequence is the least

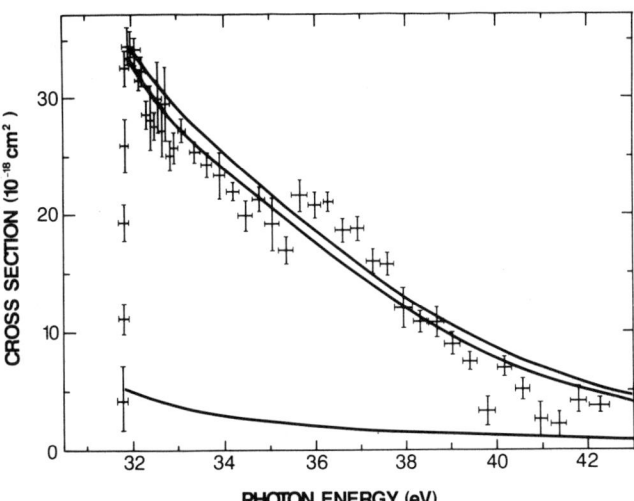

Figure 10. Photoionization cross section of K^+; the points are experimental, the upper pair of curves are from relativistic-random-phase approximation calculations (upper 5-channel, lower 14-channel) and the lower curve from a Hartree-Slater calculated.

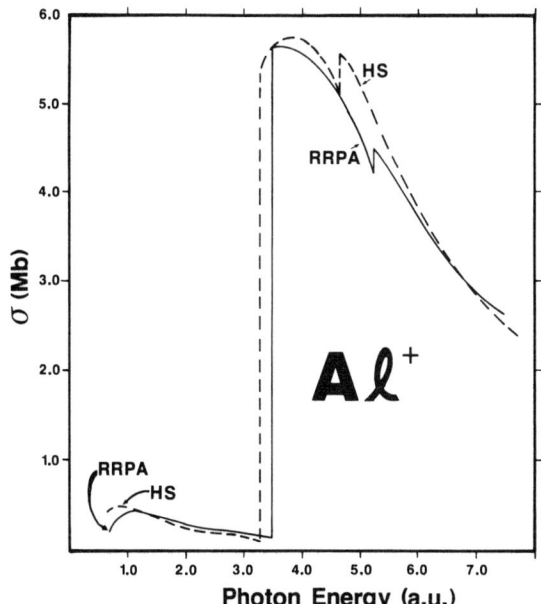

Figure 11. Photoionization cross section of Al^+ calculated in relativistic-random-phase approximation (RRPA) and Hartree-Slater (HS) approximation.

useful for revealing properties.

Now that the phenomenology based on theory has been laid out, it is of importance to inquire as how good these results are. Ultimately, the only appeal is to experiment, which is scarce but not non-existent, to determine the utility of theoretical calculation. As an example, the situation for K^+ in the threshold region is shown in Fig. 10. The measured cross section[24] is seen to be in quite good agreement with the RRPA calculations[36] while the simple central-field HS calculation[37] is in rather poor agreement. It is encouraging that the RRPA agrees so nicely (except for the region at about 36eV where experiment shows a structure which may be an artifact), but the poor agreement of HS is cause for concern since most of the theory that has been done is at the HS level.[37] It turns out that this is a worst (and unusual) case since the K^+ photoionization cross section, which is all from the 3p subshell in this energy region, contains a 3p → d Cooper minimum, which does not show up as a minimum in the cross section due to the 3p → s channel. In any case, the HS calculation predicts the minimum at much too low an energy (by about 6 eV) thoroughly destroying agreement with experiment in this narrow region. A Hartree-Fock (HF) calculation[38] (not shown) with a better placement of the minimum gives much better agreement with experiment.

Lest this give the impression that the HS results are quantitatively useless, one can point to good agreement with rough measurements[7,8] in Li^+ and Na^+ along with helium-like C^{+5} measurements[9] where excellent agreement with HS results[37] were found; the experimental threshold cross section was 0.47 ± 0.05 Mb compared with the theoretical value of 0.49 Mb.

Furthermore, since the RRPA result agreed so well with

experiment, even a Cooper minimum region, it seems likely that RRPA results would make a good benchmark. A comparison of HS and RRPA[39] for Al$^+$ is given in Fig. 11 where it is seen that, in contrast to the K$^+$ case, agreement is quite good. The basic difference in the Al$^+$ case is that the cross section is dominated by the 2p subshell which exhibits no minima. The largest discrepancies are at threshold, where the 3s cross section appears to have a minimum just below threshold, and in the vicinity of the opening of the 2s channel, above 5 a.u., where the difference in theoretical threshold energies, along with interchannel coupling, are important.

For another example, the comparison between HS and multi-configuration Hartree-Fock (MCHF) for 3p photoionization in several members of the argon isoelectronic sequence[32] is shown in Fig. 12. In neutral Ar (Z = 18), large differences are seen in the Cooper minimum region, by Z = 21, the minimum is just below threshold and significant differences are seen near threshold, but by Z = 23, with no minimum in evidence, agreement is found to be quite good.

Thus, it appears that HS calculations will be quite reasonable for ions except in the neighborhood of Cooper minima. For light and medium ions, all of our experience is that the minima move well below threshold fairly rapidly, with increasing stage of ionization, at least for ground states. For very heavy elements, certain zeros do not seem to move below threshold with increasing stage of ionization[40] due to the relativistic effects.[41] Thus it appears likely that for ground states of highly charged low and medium nuclear charge ions, the HS photoionization cross section

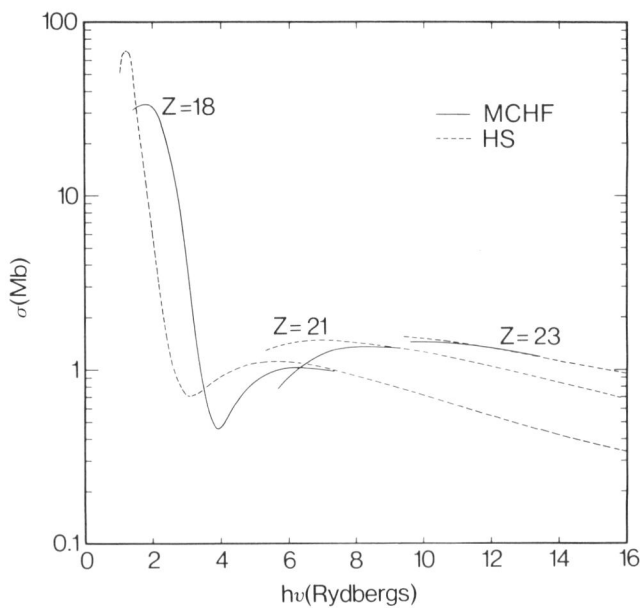

Figure 12. Photoionization cross sections for members of the Ar 3p isoelectronic sequence calculated in muticonfiguration Hartree-Fock (solid lines) and Hartree-Slater (dashed lines) approximations.

should be good indeed.

Up to this point no mention has been made of photoionization of excited states of ions. There is no experiment and few calculations in this area that we are aware of. One theoretical study has produced a particularly interesting result, a study of nf excited state photoionization in the Cs isoelectronic sequence.[42] The cross sections for neutral Cs are shown in Fig. 13 where it is seen that they simply fall off monotonically with energy; at low energy the cross sections are essentially hydrogenic since both states in the dominant f → g transition have vanishingly small phase shift (quantum defect). At higher energies, the final g-state exhibits a shape resonance,[43] giving rise to a non-zero phase shift and, thereby, the change in slope seen.

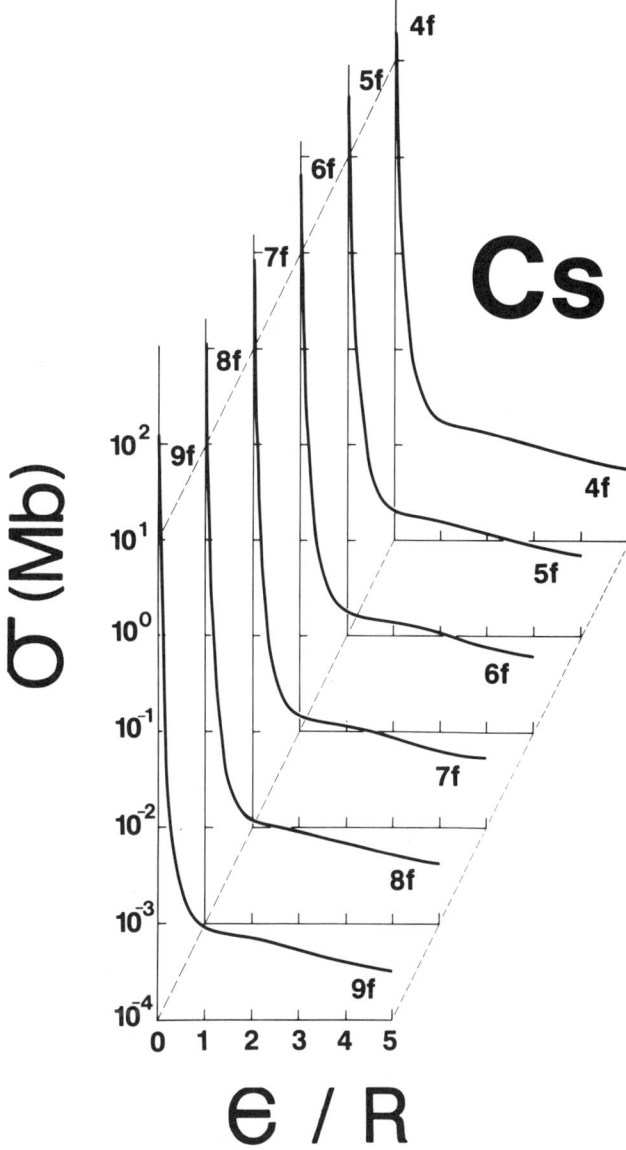

Figure 13. Photoionization cross sections for nf excited states of neutral Cs vs photoelectron energy in Rydbergs calculated in the Hartree-Slater approximation.

It might be thought that in going to Ba$^+$, the next member of the isoelectronic sequence, these cross sections would become even simpler and more hydrogenic. This, however, is not the case. Looking at Fig. 14, the photoionization cross sections of all of the excited nf states of Ba$^+$ show very dramatic Cooper minima.[15] To understand these results, we note

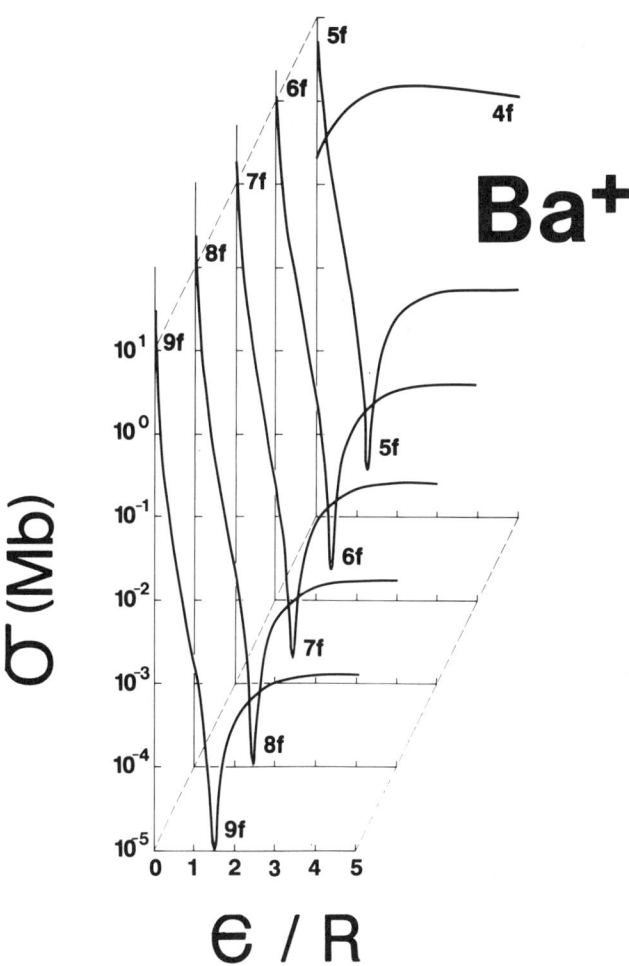

Figure 14. Photoionization cross sections for nf excited states of neutral Ba$^+$ vs photoelectron energy in Rydbergs calculated in the Hartree-Slater approximation.

that the effective f-wave potential (electrostatic attraction plus centrifugal repulsion) is double welled in Cs, but the nf states are bound in the outer well.[44] This means that the nf wave functions in Cs do not really "feel" the inner well and are essentially hydrogenic as discussed above. Going to Ba$^+$, however, the added electrostatic attraction causes the nf states to "collapse" and to be bound in the inner well, making them non-hydrogenic and giving them each a quantum defect close to unity (except for the 4f which becomes a

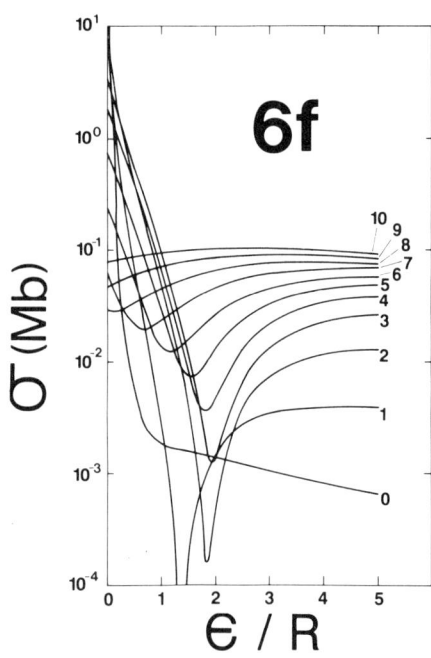

Figure 15. Photoionization cross sections for the 6f excited state of the first 11 members of the Cs isoelectronic sequence vs. photoelectron energy in Rydbergs calculated in the Hartree-Slater approximation.

core level). Then, since the continuum g-state still have near-zero phase shifts near threshold, there is a relative phase-shift difference between initial and final states in the nf → εg transitions in Ba$^+$ of about π. From previous work,[45-48] this difference implies a Cooper minimum in each, as seen in Fig. 14.

But Ba$^+$ is no fluke. Looking at the 6f cross section for the first 11 members of the isoelectronic sequence (Fig. 15), it is clear that the existence of the Cooper minimum dominates the cross section in the threshold region even where it has moved below threshold, e.g., the cross section for Tb^{+10} is rising from threshold. Thus, although going far enough along the isoelectronic will likely produce simple cross sections, it clearly does not occur rapidly or monotonically. This means that using neutral atom cross section data to infer ionic data can only be reasonably done when the physics of the situation is understood as Fig. 15 dramatically shows.

IV. FINAL REMARKS

It is evident from the discussion of the previous section that ionic photoionization has been the subject of quite a bit of theoretical scrutiny over the years; a variety of systematic trends have been uncovered. Experiment, however, is urgently required to put this theory on a firm footing.

This paper has not dealt with the auto-ionizing resonances in the photoionization cross section of ions because, while there is some experiment,[6] there is very little theory,[49] not enough to even begin to understand systematic behavior. We can infer that, based on the inner shell effect, that oscillator strength moves into the (auto-ionizing) discrete region with the removal of outer electrons as threshold moves to higher energies (c.f., Figs. 1-5). Thus, it appears that positive ions will have more of their oscillator strength in auto-ionization than do the corresponding neutral atoms. This inference needs to be checked experimentally.

Finally, we note that information can be gleaned about the resonant (auto-ionization) and non-resonant ionic photoionization cross section from the inverse processes, dielectronic and radiative recombination respectively. Experimental setups to do such measurements are in operation[50] and others are to come on-line in the future.

ACKNOWLEDGEMENTS

This work was supported by the U.S. Army Research Office and the National Science Foundation.

REFERENCES

1. A. Einstein, Ann. Phys. (Leipzig) Ser. IV 17, 132 (1905).
2. See, for example, A. F. Starace, in Corpuscles and Radiation in Matter I, Vol. 31 of Handbuch der Physik, edited by W. Mehlhorn (Springer-Verlag, Berlin, 1982), pp. 1-122.
3. J. A. R. Samson, in Corpuscles and Radiation in Matter I, Vol. 31 of Handbuch der Physik, edited by W. Mehlhorn (Springer-Verlag, Berlin, 1982), pp. 123-213.
4. J. Berkowitz, Photoabsorption, Photoionization and Photoelectron Spectroscopy (Academic Press, N.Y., 1979).
5. E. B. Saloman, J. H. Hubbell and J. H. Scofield, Atomic Data Nuc. Data Tables 38, 1 (1987).
6. For a review of these measurements, see K. Dolder, in Electronic and Atomic Collisions, edited by H. B. Gilbody, W. R. Newell, F. H. Read and A. C. H. Smith (North-Holland, Amsterdam, 1988), pp. 549-556.
7. T. B. Lucatorto and T. J. McIlrath, Phys. Rev. Letters 37, 428 (1976).
8. T. J. McIlrath and T. B. Lucatorto, Phys. Rev. Letters 38, 1390 (1977).
9. E. Jannitti, in SPIE Vol. 911, X-Ray and Vacuum Ultraviolet Data Bases, Calculations, and Measurements, edited by N. K. Del Grande, P. Lee, J. A. R. Samson and D. Y. Smith (SPIE, Bellingham, WA, 1988), pp. 11-18.
10. R. B. Cairns and G. L. Weissler, Bull. Am. Phys. Soc. 7, 129 (1962).
11. C. E. Theodosiou, S. T. Manson and M. Inokuti, Phys. Rev. A 34, 943 (1986).
12. S. T. Manson, M. Inokuti, and C. E. Theodosiou, in Dynamic Processes of Highly Charged Ions (Fuji, Japan, 1986), pp. 39-45.
13. R. F. Reilman and S. T. Manson, Phys. Rev. A 18, 2124 (1978).
14. S. T. Manson and J. W. Cooper, Phys. Rev. 165, 126 (1968).
15. J. W. Cooper, Phys. Rev. 128, 681 (1962).
16. J. B. Mann, Atomic Structure Calculations. II. Hartree-Fock Wave Functions and Radial Expectation Values: Hydrogen to Lawrencium, Los Alamos Scientific Report LA-3691, 1968.
17. K. D. Chao and S. T. Manson, Phys. Rev. A 24, 2481 (1981).
18. D. J. Botto, J. McEnnan and R. H. Pratt, Phys. Rev. A 18, 580 (1978).
19. W. Ong, S. T. Manson, H. K. Tseng and R. H. Pratt, Phys. Letters 69A, 319 (1979).
20. T. N. Chang and Y. S. Kim in X-Ray and Atomic Inner-Shell Physics-1982, edited by B. Crasemann (American Institute of Physics, New York, 1982), pp. 633-643.
21. W. R. Johnson, Adv. Atomic Molec. Phys. 25, 375 (1988) and references therein.
22. G. Nasreen, S. T. Manson and P. C. Deshmukh, Phys. Rev. A 40, xxxx (1989).
23. R. D. Driver, J. Phys. B 9, 817 (1976).
24. B. Peart and I. C. Lyon, J. Phys. B 20, L673 (1987).
25. S. T. Manson, Phys. Rev. A 38, 506 (1988).
26. M. Lamoureux and F. Combet Farnoux, J. Phys. (Paris) 35, 205 (1974).
27. F. Combet Farnoux and M. Lamoureux, J. Phys. B 9, 897 (1976) and references therein.
28. H. Hall, Rev. Mod. Phys. 8, 358 (1936).
29. W. D. Barfield, G. D. Koontz and W. F. Huebner, J. Quant. Spectrosc. Rad. Transf. 12, 1409

(1972).
30. W. F. Huebner, M. F. Arao and L. D. Ohlsen, J. Quant. Spectrosc. Rad. Transf. 19, 93 (1978).
31. D. W. Missavage, S. T. Manson and G. R. Daum, Phys. Rev. A 15, 1001 (1977).
32. A. Msezane, R. F. Reilman, S. T. Manson, J. R. Swanson and L. Armstrong, Jr., Phys. Rev. A 15, 668 (1977).
33. G. Nasreen and S. T. Manson, Bull. Am. Phys. Soc. 34, 1413 (1989).
34. B. L. Henke, P. Lee, T. J. Tanaka, R. L. Shimabukuro and B. J. Fujikawa, Atomic Data Nuc. Data Tables 27, 1 (1982).
35. E. M. Henry, C. L. Bates and W. J. Veigele, Phys. Rev. A 6, 2131 (1972).
36. G. Nasreen, P. C. Deshmukh and S. T. Manson, J. Phys. B 21, L281 (1988).
37. R. F. Reilman and S. T. Manson, Ap. J. Supp. 40, 815 (1979).
38. M. J. Seaton, Mon. Not. Roy. Astron. Soc. 110, 248 (1950).
39. G. Nasreen and S. T. Manson, Phys. Rev. A 38, 504 (1988).
40. R. Yin and R. H. Pratt, Phys. Rev. A 32, 125 (1985).
41. S. T. Manson, C. J. Lee, R. H. Pratt, I. B. Goldberg, B. R. Tambe and A. Ron, Phys. Rev. A 28, 2885 (1983).
42. J. Lahiri and S. T. Manson, Phys. Rev. A 37, 1047 (1988).
43. U. Fano and J. W. Cooper, Rev. Mod. Phys. 40, 441 (1968).
44. A. R. P. Rao and U. Fano, Phys. Rev. 167, 7 (1970).
45. J. Lahiri and S. T. Manson, Phys. Rev. A 33, 3151 (1986).
46. S. Wane and M. Aymar, J. Phys. B 20, 2657 (1987).
47. A. Z. Msezane and S. T. Manson, Phys. Rev. Lett. 48, 473 (1984).
48. J. Lahiri and S. T. Manson, Phys. Rev. Lett, 48, 614 (1984).
49. M. J. Seaton, in *Electronic and Atomic Collisions*, edited by H. B. Gilbody, W. R. Newell, F. H. Read and A. C. H. Smith (North-Holland, Amsterdam, 1988), pp. 41-56 and references therein.
50. R. E. Marrs, M. A. Levine, D. A. Knapp and J. R. Henderson, in *Electronic and Atomic Collisions*, edited by H. B. Gilbody, W. R. Newell, F. H. Read and A. C. H. Smith (North-Holland, Amsterdam, 1988), pp. 209-214.

POLARIZATIONAL RADIATION IN ELECTRONIC AND ATOMIC COLLISIONS ("ATOMIC" BREMSSTRAHLUNG)

Amusia M.Ya.

A.F. Ioffe Physical-Technical Institute
Academy of Sciences, Leningrad K-21, USSR 194021

The new mechanism of formation of continuum spectrum radiation in electronic and atomic collisions is discussed. This mechanism leads to a number of peculiarities of this radiation. Specific features of this radiation, frequency dependence, angular distributions, are discussed and demonstrated is the prominent difference of it from ordinary bremsstrahlung (OB). This new, "atomic", radiation (AB) dominates over OB in a wide region of atomic frequences. The results of numerical calculation of AB intensity in electron-atom and atom-atom collisions are presented and compared to experimental data demonstrating a satisfactory agreement.

The AB accompanying nuclear processes, α- and β- decay, are also discussed.

1. The main source of continuum spectrum radiation, both in nature and laboratory, is the bremsstrahlung (BS). Usually this process is treated as a deceleration of a charged particle under the action of the target field, mainly Coulombic. This process must be accompanied by electromagnetic wave radiation emitted by the projectile. For fast incoming particles it is described by the following diagrams:

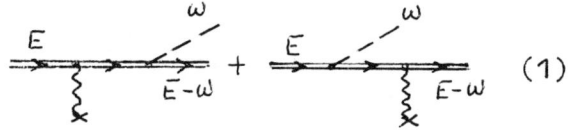 (1)

Here the solid doubled line represents the free charged particle with initial energy E, the dashed line denotes a photon, while the wavy line with a cross at the end represents the external field in which the projectile is decelerated. At low projectile velocity the amplitude of such a process called in this paper ordinary bremsstrahlung (OB) is depicted in the following way:

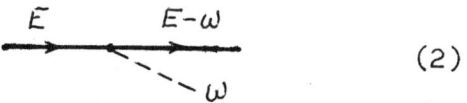 (2)

where the heavy line represents the decelerating charge particle in the external field. But in many natural and laboratory circumstances both, the projectile and target, have internal structure, internal degrees of freedom which may excited either really or virtually in the process of collision. This excitation or polarization during the collision leads also to radiation. In what follows it is called polarizational or "atomic" bremsstrahlung (AB) which as is shown below is quite different from ordinary bremsstrahlung (OB), i.e. from the radiation of a charged particle in the static target field. For the structureless incoming particle AB is exemplified by such a diagram:

$$\text{(3)}$$

In (3) the solid line with an arrow to the right represents an excited (or continuum spectrum) target electron E, that with an arrow to the left is the vacancy i formed in the target atom due to its interaction denoted by wavy line, with the projectile. The electron-vacancy excitation is the simpliest microscopical version of the target polarization by the projectile. Of course, (3) is only an example, althoug very important, and the real target polarization is described by infinite number of other diagrams.

2. The OB and AB have a number of different features, so one can consider AB as a new type of radiation. As is well known [1], for OB: a) its intensity is proportional inversely to the square of the projectile mass m_p: it is more difficult to decelerate a heavy particle than a light one by a given field;

b) its amplitude is proportional, at least for high velocities, to the second degree of the projectile charge e_p, e_p coming from the interaction with the target field, and the other from the photon emission; c) it has a very simple frequency ω dependence: ω^{-1}. As is also well known, OB dominates the radiation spectrum in the low frequency region [1].

The main features of AB reviewed in [2,3] directly follow from the mechanism of its formation. Let us consider at first AB in a collision of a structureless particle of charge $e_p < 0$ and mass m_p with an atom. In collision the incoming particle repulses the atomic electrons and attracts the nucleus. So, the gravity center of an electron cloud is shifted off the nucleus. The atom becomes polarized and the dipole moment being induced. The direction of this moment rotates in space trying to follow the projectile. The space variation of the dipole moment leads to emission of radiation or to AB. So, in AB the radiation comes from the polarized target atom. From the AB mechanism described above it follows straightforwardly that its intensity: a) is almost independent on m_p, because the dipole moment induced in the target atom during the collision process depends on the ability of the projectile to polarize the target. It is even larger for heavy particles; b) for high velocities its amplitude is proportional only to e_p, coming from the interaction of the projectile with the target atom; c) has a rather complicated frequency dependence being determined by target atom dipole polarizability $\alpha_d(\omega)$. It describes both the reaction of the target atom upon the projectile field and the emission spectrum of the target.

From the description of the AB mechanism it is evident that the projectile need not to be charged: even being neutral but interacting differently with electrons and nucleus, it polarizes the atom, leading to AB. If the projectile is structureful, it will be polarized in its turn, becoming a source of AB. For a fast projectile, the AB in collision of structureful perticles is represented by four diagrams:

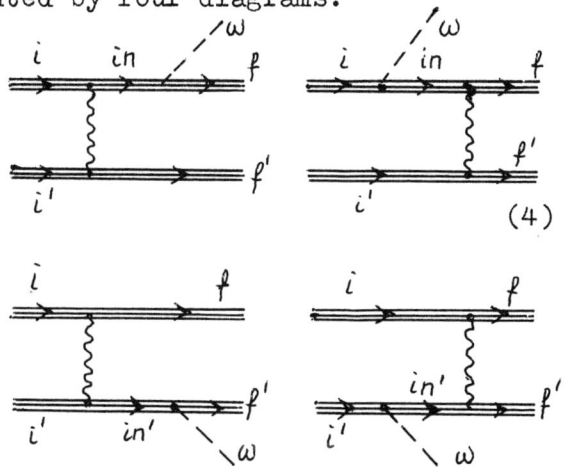
(4)

Here i (i') and f (f') denote the initial i (i') and final f (f') states of the projectile (target). If after photon ω emission the colliding particles are not excited, the only difference between i (i') and f (f') is in the total momenta of the particles which is altered in the collision process. AB in this case comes from electron cloud polarization of the projectile and target, while the source of radiation itself is the same as in (3). They are the virtually excited atomic electrons.

It is seen that the AB features are so different fron OB that AB may be considered as another kind of radiation.

3. It is essential to know how large is the AB contribution to the total radiation emission spectrum. As was mentioned above, the differential in frequence OB cross section is $d\sigma^{OB}/d\omega \sim \omega^{-1}$, while for AB it is [4,5] $d\sigma^{AB}/d\omega \sim \omega^3 |\alpha_d(\omega)|^2$. So, with the growth of ω the role of AB increases. In fact, AB becomes dominant, where $\alpha_d(\omega)$ is large, it is for ω of the order of atomic subshell ionization potential. Of course, the region of AB dominance is large for heavy projectiles. But even for electrons, AB becomes predominant for small-angle scattering. To understand this it is essential to have in mind that $d\sigma^{OB}/d\omega \sim R_a^2$, where R_a is the target atom radius, while $d\sigma^{AB}/d\omega$ is determined by long-range dipole interaction between the colliding particles, and therefore is much large. The relative intensity of AB and OB in a fast electron-atom collision may be estimated from the formula [6]: *)

$$\omega \frac{d\sigma}{d\omega} = \frac{16}{3} \frac{Z^2}{c^3 v^2} \ln 2vR_a + \qquad (5)$$
$$\frac{16}{3} \frac{1}{c^3 v^2} \omega^4 |\alpha_d(\omega)|^2 \ln \frac{v}{R_a},$$

v being the projectile velocity.

In (1) the first term describes the OB, while the second one AB contributions. For high ω the fast increase of ω^4 is compensated by decrease of $\alpha_d \approx -Z/\omega^2$, so the contribution of both terms are almost the same. For lower ω, even not to close to the atomic excitation levels, where $\alpha_d(\omega)$ has

―――――
*) Here and below the atomic system of units is used: $e = \hbar = m_e = 1$.

poles, the second term dominates.

4. If only OB is taken into account in electron-atom collision, then BS proceeds on a screened nucleus. However this result is incorrect at least for very high frequencies. For ω exceeding all ionization potentials the atomic electron may be considered as unbounded, so the incoming one together with them cannot have a variating dipole moment and are unable to radiate. Only the inclusion of AB leads to a correct result: the total radiation coinsides with that of an incoming electron upon the bare nucleus [7]. The atomic shells are well separated in energy. Therefore with increase of ω from the first ionization potential the BS spectrum $\omega\, d\sigma/d\omega$ is not constant, contrary to the prediction of OB but increases step by step, the role of one by another electron shells becoming negligible. Finally the $\omega\, d\sigma/d\omega$ must reach the

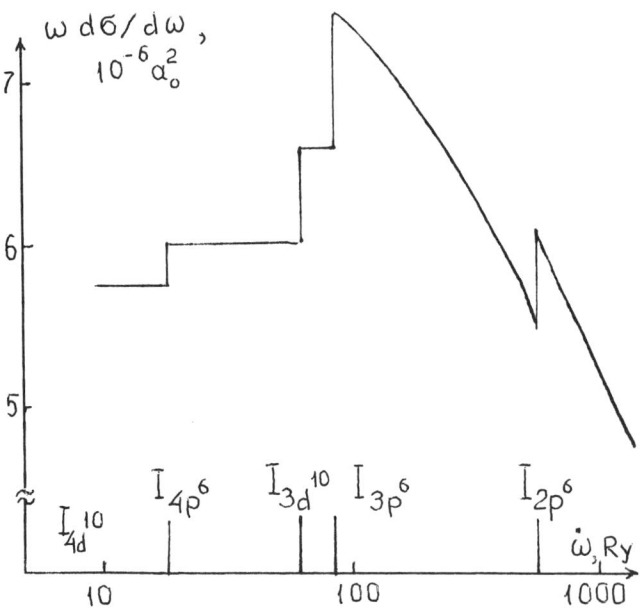

Fig. 1. Shell-by-shell decreening of Xe atom in the bremsstrahlung arising due to fast electron-Xe atom collision (shematical representation). I_q denotes the ionization potential of different Xe subshells.

value which $\omega\, d\sigma/d\omega$ has on a bare atom. This process is called nuclear striptease off the electron shells [8]. This limit is reached only if the incoming electron energy is much higher than the 1s ionization potential of the target atom, in Fig. 1 - Xe. But for not too high electron energies when the emitted photon energy is of the order of that for the incoming particle E, $\omega\, d\sigma/d\omega$ starts to decrease, reaching zero for $\omega = E$.

The schematic picture of AB depicted in Fig. 1 does not take into account the $\omega\, d\sigma/d\omega$ variation in the near threshold regions. To describe this, the calculations are necessary which allow for the dynamical polarizability of the target atom and the polarizability dependence of the transferred momentum. The results of these calculations are presented in Fig.2. For the long time experimental investigations were performed of bremsstrahlung in intermediate energy electron-atom collisions. At first, a prominent maximum was observed in the radiation emission spectrum of 1 keV electrons colliding with lanthanum metal. Then this maximum was interpreted as AB coming from polarization of lanthanum $4d^{10}$ subshell [6]. Extensive measurements of radiation spectrum in electron-xenon collision demonstrate prominent maxima in the region above $4d^{10}$ subshell photoioniza-

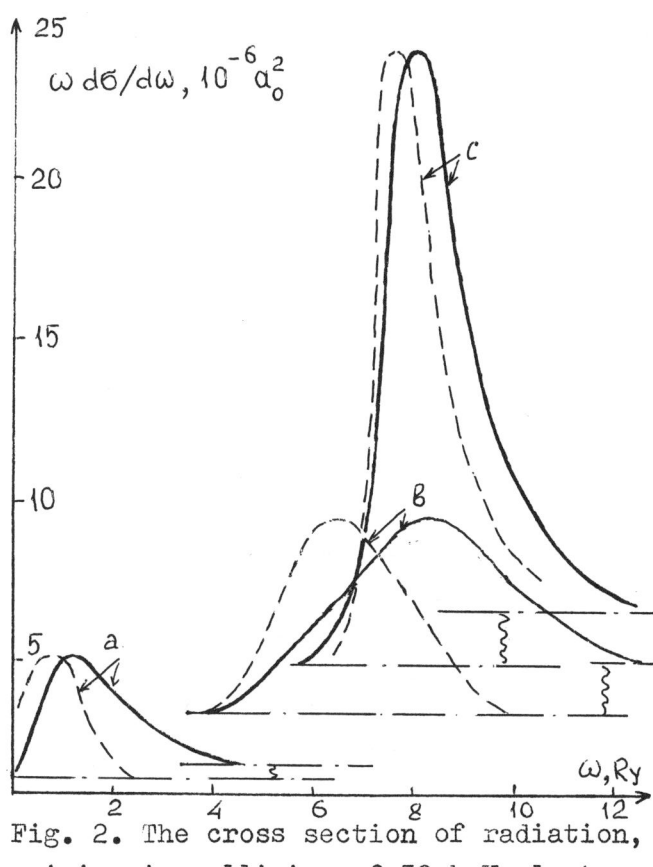

Fig. 2. The cross section of radiation, arising in collision of 50-keV electrons with Ar, Xe, and La atoms. a - Ar, b - Xe, c - La, the solid line is $\omega d\sigma/d\omega$, the dashed line is $\omega\sigma_j(\omega)$ normalized to the $\omega d\sigma/d\omega$ maximum. Dashed-dotted line is the schematical representation of the jump in $\omega d\sigma/d\omega$, see Fig. 1.

tion threshold. The first measurements [9] demonstrate the prominent role of AB in electron-atom collisions, giving the radiation intensity proportional to $|\alpha_d(\omega)|^2$. Now these investigations are continuing [10] and demonstrate not only the very existence of the maximum in the radiation spectrum above $4d^{10}$-subshell ionization threshold but also some more fine details, for instance the systematical shift of the position of the maximum to lower energies with increase of incoming electron energy, see Fig. 3.

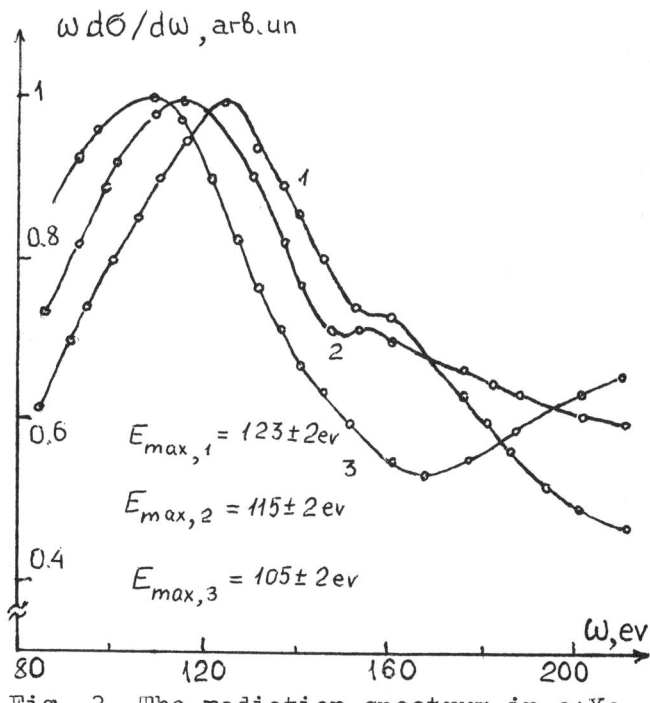

Fig. 3. The radiation spectrum in e+Xe collisions. The magnitude $d\sigma/d\omega$ is in arbitrary units. The electron energies for curves 1,2,3 are 0.3; 0.6, and 0.9 keV, respectively, the maximum position being at 123±2 eV, 115±2 eV, and 105±2 eV.

The results of measurement are in satisfactory agreement with our recent calculations.[*] The comparison is presented in Fig. 4, where the experimental data are normalized to the calculated curve. It is seen that a general agreement exists in both the form of the curves, the position of maximum, and the AB relative role.

5. The role of AB prominently increases if the projectile becomes relativistic. In this case the virtual photon by which the projectile exchanges with the target atom becomes almost real leading to the cross section increase [11]:

---

[*] Performed together with A.V. Korol' and L.V. Chernysheva.

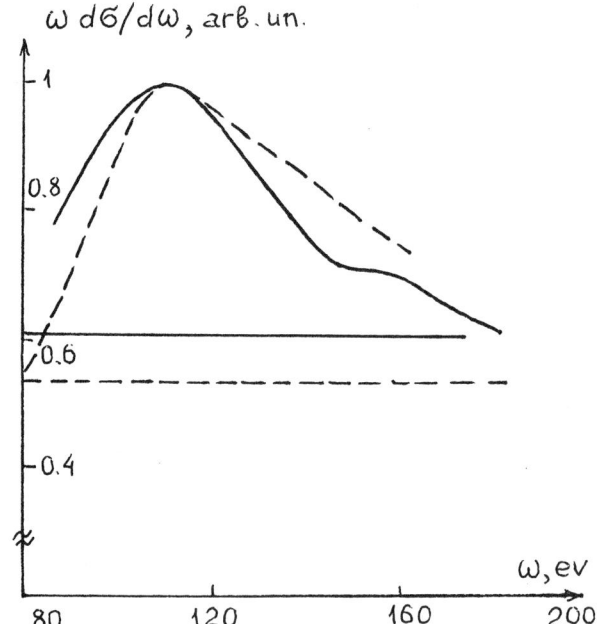

Fig. 4. The radiation spectrum of 0.6 keV electrons on Xe. The solid line is the results of measurements [10], while the dashed one presents calculational data. Background is the OB radiation.

$$\omega \frac{d\sigma^{AB}}{d\omega} = \frac{16}{3} \frac{\omega^4}{c^3 v^2} |\alpha_d(\omega)|^2 \ln\left[\frac{v}{R_a} \frac{E}{c^2}\right]. \quad (6)$$

The factor $E/c^2 \gg 1$ is the relativistic correction to the AB spectrum. Now so high energies are available that $\ln E/c^2$ reaches 10.

The relativistic effects are seen more prominently in the angular distribution of radiation: OB is concentrated within a narrow cone along the projectile momentum direction, with an apex angle $\gamma^{-1} = (1-v^2/c^2)^{-1/2}$, while AB being nonrelativistic in essence, has an ordinary dipole radiation angular distribution. The angular distribution of the bremsstrahlung in relativistic electron-atom collision is presented in Fig. 5.

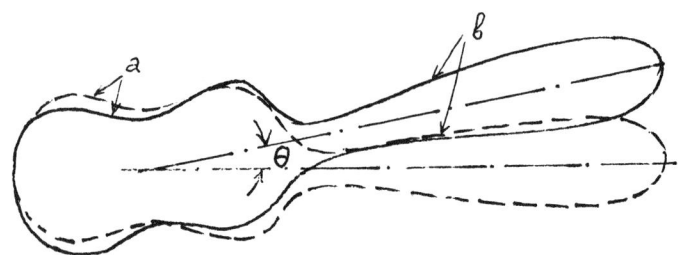

Fig. 5. The angular distribution of the radiation in the relativistic electron-atom collisions: a - AB, b - OB, θ is the scattering angle. Solid line is the radiation after the electron scattering on the atom, dashed line is the same before the electron scattering.

At ultrarelativistic conditions at high ω the phenomenon opposite to striptease takes place: projectile radiation amplitudes on atomic electron and on nuclear proton in this case almost cancel, so the total radiation is in fact that of atomic electron in projectile field.

6. For heavy projectiles or ions only AB exists, the OB being suppressed by the factor $(M_e/M_a)^2$, where M_a is atom's or ion's mass. If projectile velocities are high, the BS spectrum may be calculated in the first Born approximation which leads to a very simple and transparent formula [12]:

$$\omega \frac{d\sigma}{d\omega} = \frac{16 \omega^4}{3c^3 v^2} \int_{\omega/v}^{2\mu v} \frac{dq}{q} \left| [Z_2 - Q^{(2)}(q)] \alpha_d^{(1)}(\omega,q) - [Z_1 - Q^{(1)}(q)] \alpha_d^{(2)}(\omega,q) \right|^2 \quad (7)$$

where 1 and 2 denote the projectile and target, respectively, μ is the reduced mass of colliding particles, Z_i being their charge and $Q^{(i)}(q) = \int \rho_i(r) \times \exp(-i\vec{q}\vec{r})d\vec{r}$ - a so-called form-factor. Here $\rho_i(r)$ is the electron density. In (3) $\alpha_d^{(i)}(\omega,q)$ denotes the generalized dipole polarizability which reduces to $\alpha_d^{(i)}(\omega)$ at $q = 0$, and q is the momentum transferred in the collision. The formula (3) is transparent: it describes the AB amplitude of one atom polarized by the action of the distributed charge $[Z-Q(q)]$ of the other and vice versa. These amplitudes interfere destructively. This is demonstrated by the sign (-) which separates their contributions. As it is seen from (1), there is no dipole radiation in the collision of identical particles. This is clear, because such a pair has no variating dipole moment. The intensity of AB for atomic particles is of the order of AB in electron-atom collision with the same velocity. While in the latter case OB and AB are of the same order, the heavy particle radiates due to polarizational mechanism at least not less intensively than an electron. This statement was confirmed by direct calculations of BS of electrons, α-particle and the atom on Xe. The results of this calculations are presented in Fig. 6. It is seen that for frequencies above $4d^{10}$-subshell ionization threshold, where $\alpha_{Xe}(\omega)$ has a huge maxima resembling that in photoabsorption cross section, $d\sigma/d\omega$, for an electron only several times exceeds that for neutral He [12]. It proved to be that both terms in (3) must be taken into account,

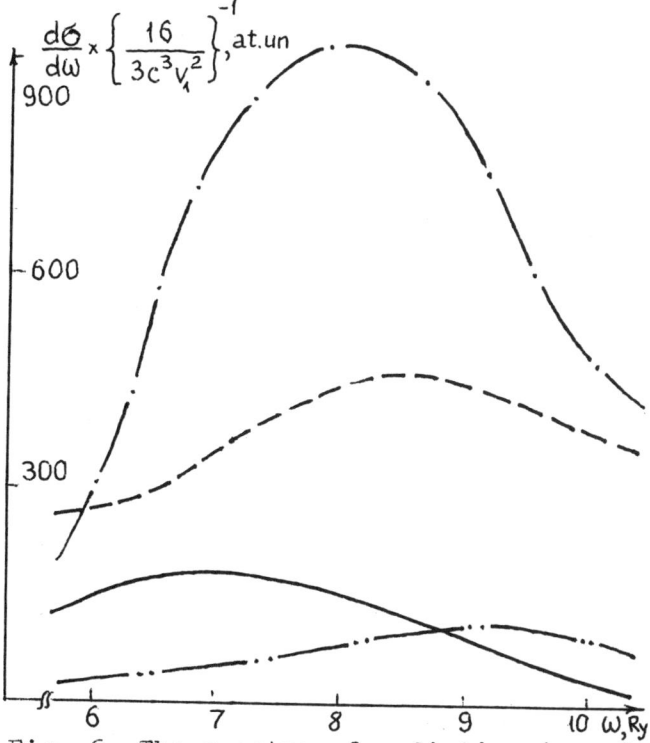

Fig. 6. The spectra of radiation in collisions of Xe with fast particles: electrons, α-particle, He. The projectiles are: dashed line is an electron, dashed-dotted line -particle, dash-dotted-dotted line the He without its polarizability, solid line He, allowing for its polarizability. The projectile velocity is $v_1 = 5$ at.un.

although helium polarizability is small. The reason is that Xe distributed charge is large. The cross sections of BS at maximum above the $4d^{10}$ threshold for collisions with e, p, α-particle and He are 4, 2, 8, 2 of 10^{-22} cm^2, respectively.

7. AB of heavy atomic particles permit to suggest two experiments in which it leads to macroscopically observable effects. a) Suppose a closed volume of homogeneous gas which cools by radiating. According to (3), the dipole AB is forbidden, the quadrupole being much less intensive, so the cooling

proceeds rather slowly. But if to connect two volumes of different gases at the same temperature, then they mix. The AB becomes allowed which leads to much more intensive (by four orders!) radiative cooling. b) The light beam may accelerate even neutral particles, leading to an ordered motion of them. This may be due to the light pressure which amplitude is of the second order in the fine structure constant. But another mechanism also exists: the bremsstrahlung of light in atomic particles collisions, a process which is reverse to AB. In this process not only the energy but the photon momentum may be acquired which leads to ordered motion of the gas. The amplitude of this process is proportional to the gas density and only to the first degree of the fine structure constant. Estimations show that even for low radiation frequences this mechanism becomes more effective starting from the densities as low as $10^{12} cm^{-1}$ [13].

8. Up to now we consider BS as a process in which only a photon is emitted, the inner states of both the target and projectile remain unaltered. In this sence it is an elastic bremsstrahlung process (EBS). However, for colliding particles with inner structure, the photon emission may be accompanied by their excitation or ionization. Moreover, the emission of these secondary particles, electrons, must be accompanied by both OB on colliding atoms and AB of them. These processes are exemplified by the following diagrams:

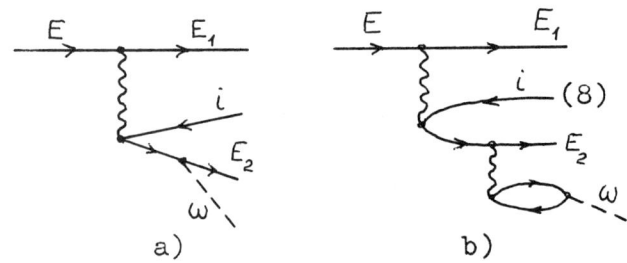
(8)

Here (8a) describe the OB of a secondary electron in its own target field, while (8b) presents AB for the same electron. Of course, the secondary target electron may radiate in the static field of the projectile and by polarizing the projectile electron cloud. The same mechanisms act for secondary electron emitted off the projectile and moving not only in its own field but also in that of the target. This complicated bremsstrahlung process is called inelastic BS (IBS). Then a natural question arises which is more intense, mainly in heavy atomic particle collisions, the EBS in this case entirely coming from AB or IBS. In IBS the process of the AB of colliding particles may be a large extend or even completely masked by OB of the secondary electrons. Then the experimental observation of AB in atomic collisions would be very difficult. Fortunately enough, this is not the case [14]: if exists a broad frequency region in which elastic atomic bremsstrahlung contribution is larger than that of inelastic bremsstrahlung, the ratio of them being of the order of N_p, the number of electrons in the shell which polarization mainly contributes to AB at the considered frequency. This fact is possible to explain qualitatively: in AB all electrons of the mostly polarized shell are pushed simultaneously, the elastic AB amplitude being proportional to N_p. Therefore, the cross section is proportional to N_p^2. On contrary, in inelastic AB the total cross section is a sum of processes with different final states, so it is proportional only to N_p, thus being N_p times smaller than the elastic AB contribution. The effect of coherency is seen quite clearly from comparison of (3) and (8): the summation over all vacancy states i is performed in the amplitude, leading to N_p factor in it and to N_p^2 in the cross section. On the contrary, the contribution of (8) is summed over i only in the cross section leading there to the factor N_p.

As was demonstrated in [15,16,17] by direct calculation of AB in proton-atom and ion-atom collision in MeV per nucleon energy region, the account of AB in a number of cases increases the total radiation intensity by an order of magnitude, leading to good agreement with experimental data. It proved to be that AB contribution is large for frequencies higher than 1s-shell ionization threshold I_{1s}. In Fig. 7 the results of calculation of AB (and other possible radiation mechanisms) are presented for 1 MeV proton-aluminium scattering. Recently the comparison of AB contribution with that of other mechanisms was performed in [18]. However, for $\omega \gg I_{1s}$ the atomic electron do not feel its nucleus and it cannot accept an energy larger than $2v^2$, v being the projectile velocity. So for $\omega > 2v^2$ the nucleus-nucleus radiation will dominate. It is confirmed by calculations [19].

If the incoming ion's charge is higher than that of target atom or ion,

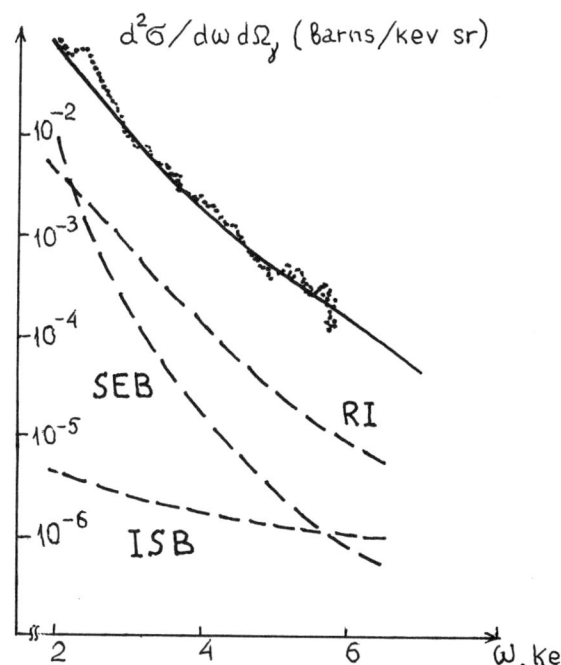

Fig. 7. Radiation emission at 90° to the 1 MeV proton beam on Al target[17]. solid line is AB, dashed lines: radiative ionization (RI), i.e. the radiation accompanied by ionization, secondary electron bremsstrahlung (SEB), their radiation in the field of other target atoms, and the internuclear ISB that of proton plus aluminium nucleus radiation. The experimental points are given from [15].

the radiation comes mainly due to virtual intercept of target electron to the projectile. The entire AB picture looks as radiation of an unbound electron in the projectile's field [20]. Note that in this case also a broad frequency region exists where elastic AB dominates.

9. An interesting problem is the AB of heavy relativistic atomic particles. It is not pure theoretical, because very high -factors are new achieved: $\gamma = (1-v^2/c^2)^{-1/2}$ reaches 10^2. The

process was recently investigated [21]. It determined, as in the nonrelativistic case, by mutual polarization of colliding particles. But in the laboratory frame the projectile spectrum is modified due to Doppler effect which alters the energy distribution and by angular alteration which alters the angular distribution of radiation.

The AB cross section differential in photon energy and the scattering angle are given by the following formulae [21]:

$$\frac{d\sigma^t}{d\omega d\Omega} = \frac{\omega^3(1+\cos^2\theta)}{\pi v^2 c^3} \int_0^\infty \frac{dq}{q} |[Z^{(1)} - Q^{(1)}(q)]\alpha_d^{(2)}(\omega,q)|^2, \quad (8)$$

$$\frac{d\sigma^p}{d\omega d\Omega} = \frac{3(1+\cos^2\theta^p)}{\pi v^2 c^3} \left(\frac{\omega^p}{\omega}\right)^2 \int_0^\infty \frac{dq}{q} |[Z^{(1)} - Q^{(1)}(q)]\alpha_d^{(1)}(\omega,q)|^2, \quad (9)$$

$$\frac{d\sigma^i}{d\omega d\Omega} = -\frac{2\omega^3(1+\cos\theta\cos\theta^p)}{\pi v^2 c^3} \frac{\omega^p}{\omega}\int_0^\infty \frac{dq}{q} [Z^{(1)} - Q^{(1)}(q)][Z^{(2)} - Q^{(2)}(q)] \cdot \text{Re}[\alpha_d^{(1)}(\omega^p,q)\alpha_d^{(2)*}(\omega,q)]. \quad (10)$$

The indices t and p refer to radiation of the target and projectile, respectively. The latter term is due to the interference of the target and projectile contributions, q is not the total but transverse transferred momentum and

$$\omega^p = \omega\gamma[1-(v/c)\cos\theta], \quad \cos\theta^p = [\cos\theta-(v/c)]/[1-(v/c)\cos\theta]. \quad (11)$$

It is seen from (8),(9) and (11) that the difference between target and projectile radiation results from Doppler effect ($\omega \to \omega^p$) and angular abberation $\theta \to \theta^p$.

It is interesting to note that for very low frequencies for which $\alpha_d^{(1)}(\omega^p)$ is frequency independent the angular distribution projectile radiation is not concentrated within a narrow angle γ^{-1}, being almost isotropic:

$$d\sigma^p/d\omega d\Omega \sim \{[1-(v/c)\cos\theta]^2 + [\cos\theta-(v/c)]^2\}. \quad (12)$$

Only for high frequencies the projectile radiation angular distribution is concentrated within the cone with apex angle γ^{-1}. Note that the radiation of discrete frequency ω_0 in the projectile frame transforms into a broad band $\omega_0/\gamma \leq \omega \leq \gamma\omega_0$ in the laboratory frame. Taking into account Doppler effect and angular aberation leads to nonzero dipole radiation intensity even for identical particle collision. Instead of having two equal amplitudes with opposite signs (see (7) for (1) = (2)), the projectile amplitude is shifted in frequency and angle, thus destroying the compensation. This leads in the nonrelativistic collision to a small as to the case of different particles, of the order of $(v/c)^2$, dipole radiation spectrum:

$$\frac{d\sigma}{d\omega} = \frac{16}{3}\frac{\omega^3}{c^5}\int_0^\infty \frac{dq}{q}[Z-Q(q)]^2\{|\alpha_d(\omega,q)|^2 + \omega \text{Re}\,\alpha_d^*(\omega,q)\frac{d}{d\omega}\alpha_d(\omega,q) + \frac{2}{5}\omega^2|\frac{d}{d\omega}\alpha_d(\omega,q)|^2\}. \quad (13)$$

10. AB appears also if the atomic nucleus is shifted off its normal position in the center of nucleus. Then

also a time-dependent dipole moment appears which radiates. Thus, AB accompanies the neutron-atom collision in which the nucleus acquires some recoil momentum [22]. The radiation spectrum of atomic electrons up to rather high neutron energies may be presented as a product of two factors. One determines the probability for the nucleus to acquire a momentum in its collision with the neutron which is given by the neutron-nucleus scattering cross section σ_{nN}. The other describe the probability for a photon to be emitted by the electron cloud due to its polatization which is determined by $\alpha_d(\omega)$. The entire expression for the AB spectrum [22] is

$$\frac{d\sigma}{d\omega} = \frac{32}{3} \frac{\omega^3 E^2}{c^3 v^2 M_N^2} \frac{\sigma_{nN}}{4\pi} |\alpha_d(\omega)|^2 \times \quad (14)$$

$$(1-\frac{\omega}{E_n})^{1/2}(2-\frac{\omega}{E_n}),$$

where E_n is the incoming neutron energy, $E_n = NM_n v^2/2$. For example, if $\omega = 0.5$ and $E_n = 5$, then in neutron-hydrogen collision (9) gives a cross section $\approx 5 \cdot 10^{-33}$ cm^2 at the interval $\Delta\omega$ 0.1. This is about 20 times higher than the radiation cross section for the same ω in neutron proton collisions. As in other cases, the AB spectrum is expressed via the atomic shell polarizability. In the region of resonances in σ_{nN}, just as the resonances in $\alpha_d(\omega)$, the radiation intensity may be prominantly amplified. For not too high frequencies the AB is more intensive than the direct radiation of the heavy nucleus turned into motion due to collision with the neutron. But for ω exceeding atomic ionization potentials the electrons finding in atom is negligible, so they cannot radiate dipolarly, and the entire radiation comes from the nucleus recoil.

The atomic electron shells may be polarized also by the neutron magnetic moment. This results in an expression [22]:

$$\frac{d\sigma}{d\omega} = \frac{16\omega^3}{3c^3 v^2}|\alpha_d(\omega)|^2 \mu_n^2 \left(\frac{\omega}{c}\right)^2 \times \quad (15)$$

$$\ln\frac{1+(1-\omega/E)^{1/2}}{1-(1-\omega/E)^{1/2}} + 4(\frac{E}{c})^2(1-\frac{\omega}{2E})(1-\frac{\omega}{E})^{1/2}.$$

Here μ_n is the neutron magnetic moment. Its interaction with an electron is comparatively long-range, and $\alpha_d(\omega,q)$ decreases fastly with the increase of the transferred momentum q. So, the main contribution to (10) comes from small scattering angles, while (14), depending only upon short-range forces, collect contributions from a large region of q. The numerical estimation shows that for light atom (15) is more essential, while for the heavy (14) becomes negligible. In neutrino-atom collisions AB is also generated. It may be demonstrated[*] that although neutrino interacts with both electrons and nucleons, the main contribution comes from nucleus recoil which leads to AB amplification as compared to that coming from direct electron shell polarization by neutrino, by a factor $(M_N/m_e)^2$, M_N being the nucleus mass. The AB cross section is about 10^{-46} cm^2. If, as is now discussed, neutrino would have a prominent magnetic moment, it might increase the AB cross section.

11. The nuclear processes proceed normally much faster than the atomic ones, so they may be treated as a fast disturbance for the electron shells. This dis-

---

[*] These results are obtained together with A.V. Korol.

turbance leads to AB. A general consideration of AB which comes from fast external perturbation was performed recently [23]. A number of nuclear processes such as α- and β-decays and fission are accompanied by electron shell radiation. Being in general rather complex, the investigation of these processes is in a number of cases considerably simplified. Recently, in the general framework of the approach discussed in this short review the AB accompanying α- and β-decay were investigated [24,25]. In β-decay the electron emission by nucleus is instant for atomic shells. The radiation coming from the change of nuclear charge is determined by the atomic shell polarizability. However, it is suppressed due to smallness of the nuclear radius as compared to the atomic one: for point nucleus the variation of its charge does not lead to radiation. The other source of radiation in β-decay is the internal BS of β-electron which is created and then moves via the atomic cloud in the nucleus field. The simpliest diagram describing the bremsstrahlung of β-electron is the following [24]:

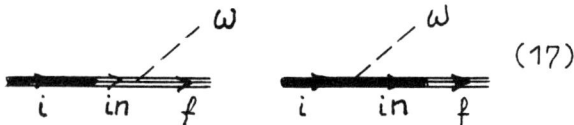
(16)

The notation $r = 0$ emphasizes that the β-electron originates from the nucleus. Even if the action of the nucleus upon the β-electron is neglected, the contribution of (16) is nonzero: originated from $r = 0$ electron, contrary to the free one, may emit a photon. So, even a fast β-electron is a source of both OB and AB.

Although being suppressed, the contribution of nuclear charge variation is also essential. This radiation amplitude is presented by the following diagram [24]:

(17)

Here the solid thick line represents the state of the atom with the nuclear charge Z, while the triple line denotes the states of the atom after β-decay with the nuclear charge Z+1. Most essential is the contribution of (17) for i≠f, while for i=f it is suppressed. The radiation of the β-electron on atomic cloud is represented by the following diagrams:

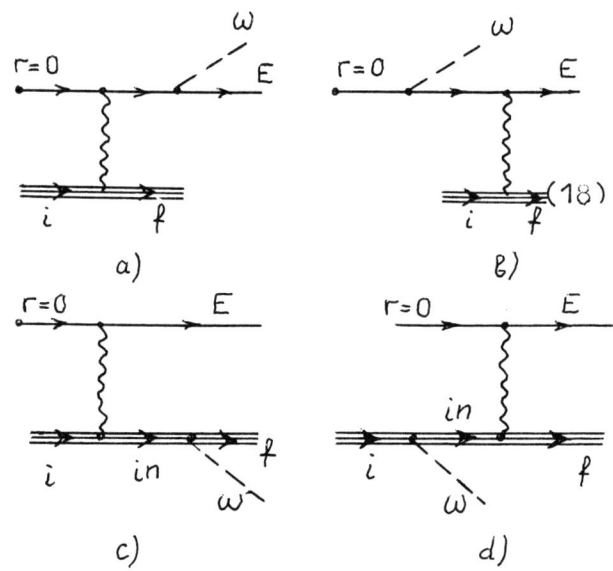
(18)

Here the triple line denotes the atom, its initial and final states being i and f, respectively, while "in" denotes the intermediate states of the atom over which summation (and integration) is performed. The diagrams (18 a,b) represent the OB of β-electron in the field of atomic electrons, while (8 c,d) describe the AB in the β-decay, both elastic and inelastic.

While in general the β-electron is fast and even relativistic, the α-particle moves rather slowly as compared to inner and intermediate shell atomic electrons. This essentially simplifies the consideration and the compact expression for AB can be given, expressed also via atomic shell polarizability. In both cases in α- and β-decay processes the AB due to nucleus recoil may be essential.

Entirely, the radiation of atomic shells due to nuclear processes is a large domain of investigation where a lot of work must be done.

12. The main attention was given above to pairvise collisions. However, it is clear that some cooperative phenimena in connection with Atomic Bremsstrahlung must be essential. Namely, if the incoming particles beam is dense, then they may polarize the target atom coherently, so that AB amplitude is amplified by a factor proportional to projectile beam density. This factor may be very large. The other possibility for cooperative phen mena comes from AB in dense targets, where the projectile polarizes simultaneously not a single atom but a number of them, leading to AB intensity proportional to the square of the total induced momentum, which is much larger than the sum of AB intensities coming from each of the target atoms.

I am grateful to my collaborators A.V. Korol', M.Yu. Kuchiev, and A.V. Soloviev for many fruitful discussions on AB problems and K.L. Tsemekhman for help in preparation of this talk.

References

1. V.B. Berestetsky, E.M. Lifhits, and L.P. Pitaievsky, Quantum Electrodynamics, Nauka, Moscow, 1980.
2. M.Ya. Amusia, Physics Reports, 162, 5, 249 (1988).
3. Polarizacionnoe izluchenie atomov i chastits (polarizational radiation of atoms and particles), ed. V.N. Tzitovich and Oiringel, Nauka, Moscow, 1987 (in Russian).
4. V.M. Buimistrov and L.I. Trakhtenberg, Zh. Eksp. Teor. Fiz. (Sov. Phys. JETP) 69, 108 (1975) (in Russian).
5. M.Y. Amusia, A.S. Baltenkov, and A.A. Paiziev, Pis'ma Zh. Eksp. Teor. Fiz., 24, 366 (1976) (Sov. Phys. JETP. Lett., 24, 332 (1976).
6. M.Ya. Amusia, T.M. Zimkina, and M.Yu. Kuchiev, Zh. Tech. Fiz. (Sov. Phys.-Tech. Phys.) 52, 1424 (1982) (in Russian).
7. V.M. Buimistrov and L.I. Trakhtenberg, Zh. Eksp. Teor. Fiz. (Sov. Phys. JETP) 73, 3(9), 850-853 (1977) (in Russian).
8. M.Ya. Amusia, N.B. Avdonina, L.V. Chernysheva, and M.Yu. Kuchiev, J.Phys. B 18, L 791 (1985).
9. E.T. Verkhovtseva, E.V. Gnatenko, and P.S. Pogrebnjak, J. Phys. B:At.Mol. Phys. 16, L 613-616 (1983), 19, 2089 (1986).
10. E.T. Verkhovtseva, E.V. Gnatchenko, and A.A. Tkachenko, Abstracts of papers, This conference, p.   (1989).
11. M.Ya. Amusia, A.V. Korol', M.Yu. Kuchiev, and A.V. Soloviev, Zh. Eksp. Teor.

Teor. Fiz. (Sov. Phys. JETP) $\underline{88}$, 383 (1985) (in Russian),

12. M.Ya. Amusia, M. Yu. Kuchiev, and A.V. A.V. Soloviev, Zh. Eksp. Teor. Fiz. (Sov.Phys. JETP) $\underline{89}$, 1512 (1985). (in Russian).

13. M.Ya. Amusia, A.S. Baltenkov, Pis'ma v Zh. Tekh. Fiz. (Letters to Sov. Phys. Tekh. Phys.) $\underline{12}$, 18, 1123 (1986) (in Russian).

14. M.Ya. Amusia, A-V- Korol', and A.V. Soloviev, Pis'ma v Zh. Tekh. Fiz. (Sov. Phys.-Tekh. Phys.) $\underline{12}$, 12, 705 (1986) (in Russian).

15. K. Ishii and S. Morita, Phys. Rev., A 30, 2278 (1984).

16. K. Ozawa, J.H. Chang, Y. Yamamoto, and S. Morita, Phys. Rev. A $\underline{33}$, 5, 3018 (1986).

17. A.D. Gonzales, J.E. Miraglia, and C.R. Garibotti, Phys. Rev. A $\underline{34}$, 2834 (1986).

18. A.D. Gonzales, M.C. Pacher, and J.E. Miraglia, Phys. Rev. A $\underline{37}$, 4974 (1988).

19. M.C. Pacher and J.E. Miraglia, Phys. Rev. A $\underline{39}$, 6, 2905 (1989).

20. M.Ya. Amusia and A.V. Soloviev, J. Phys. B, in press, 1989.

21. M.Ya. Amusia, M.Yu. Kuchiev, and A.V. Soloviev, Zh. Eksp. Teor. Fiz. (Sov. Phys. JETP) $\underline{94}$, 1, 74 (1988) (in Russian).

22. M.Ya. Amusia, A.S. Baltenkov, and A.V. Soloviev, Zh. Eksp. Teor. Fiz. (Sov. Phys. JETP), 1988 (in Russian).

23. M.Ya. Amusia and A.V. Soloviev, Abstracts of papers, this conference, p.   (1989).

24. A.V. Soloviev, Abstracts of papers, this conference, p.   (1989).

25. A.V. Soloviev, Abstracts of papers, this conference, p.   (1989).

CORRELATION EFFECTS IN ELECTRON IMPACT- AND PHOTON-INDUCED TWO-ELECTRON TRANSITIONS IN RARE GASES

K.-H. Schartner

I. Physikalisches Institut der Justus-Liebig-Universität, Giessen, FRG

The application of VUV fluorescence spectroscopy in studies of one- and two-electron processes in the outer shells of the rare gases, induced by photon impact or by fast electron impact, is described. Results for ionization of Ar-3s electrons, for additional 3p-electron ionization and for satellite production are discussed in a comparison with recently published results from photoelectron spectroscopy. Cross sections for electron impact ionization into the Ar^{++}-$3s3p^5$ $^{1,3}P$ states are compared with cross sections for photoionization. The experiments have been carried out to study electron correlations in connection with external perturbations of atomic systems.

I. Introduction

The single-electron model, i. e. the use of one-electron wavefunctions in the cross section formulas, cannot be applied to two-electron processes induced by photon- or high-energy electron-impact. This follows from the fact that the photon-electron interaction and the electron-electron interaction at small momentum transfer are described by one-electron operators [1,2]. Within this context experimental and theoretical studies of two- and more-electron processes are of great interest. They are sensitive to improvements of the single-electron model by a proper consideration of the neglected parts of the many-electron interactions. Thus they can provide a deeper understanding of the atomic structure and its dynamics under the influence of external perturbations. So it is only natural, that more and more work is devoted to multi-electron processes revealing the influence of electron correlations. These are studies of ionization accompanied by a simultaneous excitation of a second electron or of double ionization. The powerful synchrotron radiation facilities, which are available nowadays, moved the interest from the higher energies of earlier studies [3] into the threshold range of the ionization processes [4].

Rare gases are still advantageous for fundamental research in experiment and theory [5]. They offer a variation of the number of atomic electrons and have a closed-shell structure. Photoelectron spectroscopy (PES) and ion mass spectroscopy are the dominant experimental methods in photoionization studies. Only recently, photon induced fluorescence spectroscopy (PIFS) in the VUV has been introduced[6-10]. It is this method which will be treated with most detail in the following. On the contrary, fluorescence spectroscopy has a long tradition in electron impact excitation [11], as has electron loss spectroscopy [12]. Detailed information about the dynamics of electron impact ionization can be obtained by electron momentum spectroscopy (EMS). In the limits of small momentum transfer, this method has close connections to photoionization [13].

II. Photon- and Electron Impact-Induced Fluorescence Spectroscopy

Fluorescence spectroscopy is based on a quantitative analysis of the fluorescence radiation emitted during the decay of the level of interest i, excited from a lower state 0, which is mostly but not necessarily the ground state of an atom or molecule (fig 1).

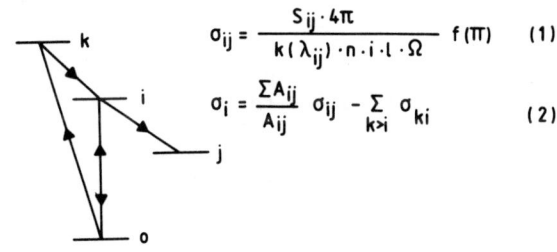

$$\sigma_{ij} = \frac{S_{ij} \cdot 4\pi}{k(\lambda_{ij}) \cdot n \cdot i \cdot l \cdot \Omega} f(\pi) \quad (1)$$

$$\sigma_i = \frac{\Sigma A_{ij}}{A_{ij}} \sigma_{ij} - \sum_{k>i} \sigma_{ki} \quad (2)$$

Fig. 1: Population and decay scheme of the level of interest i

In general, the determination of an emission cross section σ_{ij} for inducing the transition $i \rightarrow j$ precedes the derivation of the excitation cross section σ_i. For a measurement of σ_{ij}, the signal rate S_{ij} obtained from the observation of a part ℓ of the photon or electron beam under the solid angle Ω has to be converted into an intensity by the quantum efficiency $k(\lambda_{ij})$ of the particular monochromator-detector combination (equ. 1). The target density n and the flux i of impinging particles have to be measured. The function $f(\Pi,\Theta)$ takes care of an eventually anisotropic pattern of the fluorescence radiation [14].

Main experimental difficulties arise from $k(\lambda)$, especially in the VUV, where the development of suitable intensity transfer standards is still not completed [15]. For photon impact, the derivation of i from secondary electron currents from e. g. Al- or Au-cathodes, serving as flux monitors, is problematic. Here Schottky type photodiodes are promising, at least as long as higher order contaminations of an impinging photon beam are suppressed [16]. Due to the sketched experimental difficulties, often normalizing procedures are applied. The 1st Born approximation for high enough electron energies offers such a possibility. In the same connection, well known emission cross sections have an important application in radiometry [15, 17].

The excitation cross section σ_i follows from σ_{ij} by subtraction of cascade processes, represented by σ_{ki} (equ. 2). τ_i and A_{ij} are the lifetime of level i and the transition probability for the decay mode $i \rightarrow j$.

PIFS and PES can be regarded as alternative and complementary methods. When total cross sections are measured, they are equivalent (fig. 2). On the other hand, a number of basic differences exists:

1. Multiple ionization events can be observed without coincidence techniques.

2. The bandwidth of the exciting photons and the resolution of the fluorescence radiation are not coupled.

3. Thresholds are passed without eventual transmission problems caused by the strongly varying photoelectron energies.

A further equivalence exists between the polarization of the fluorescence radiation and the angular distribution of photoelectrons. Both provide a more detailed information about the photoionization processes than the total cross sections [18].

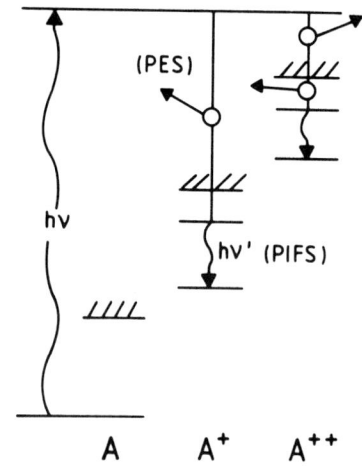

Fig. 2: Equivalence of photon induced fluorescence spectroscopy (PIFS) and photoelectron spectroscopy (PES) for single and double ionization by photons of energy $h\nu$.

A disadvantage of PIFS is certainly the before mentioned cascade correction.

As an optical method, PIFS is in principle capable of a high resolution, reaching values for $\lambda/d\lambda$ of a few thousand. This fundamental advantage is momentary hampered by the comparatively small quantum efficiency of optical spectrometers. Moreover, much progress has been made in PES with respect to resolution. Present studies of satellite phenomena reach resolutions close to 1000 at 30 eV [19, 20, 21].

III. Schematic Level Scheme of the Heavier Rare Gases

The rare gases except He have the same general level scheme, shown in fig. 3. The single terms are denoted by their dominant configurations. It is well known that electron correlations are important and are responsible for strong configuration interactions, already present for Ar [22].

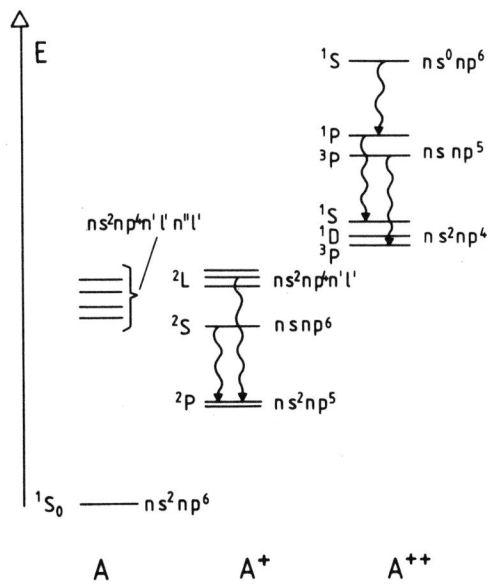

Fig. 3: Schematic term diagram for ionization and excitation of the outer shell electrons of the heavier rare gases (n = 2, ..., 5)

Ionization of the outer s-electron is observable through the ns^2np^5 $^2P_{1/2, 3/2}$ - $nsnp^6$ $^2S_{1/2}$ doublet. The $nsnp^6$ $^2S_{1/2}$ level is the lowest excited state of the singly ionized rare gas atom. Higher excited ion states are the upper levels of satellite transitions ns^2np^5 ^2P - $ns^2np^4n'l'$. Additional ionization of a p-electron forms the $nsnp^5$ $^{1,3}P$ levels. Transitions from the doubly ionized ns^0np^6 ^1S state cascade to the $nsnp^5$ ^1P level. For all rare gases, transitions to the ionic ground states lead to fluorescence in the VUV.

IV. Experimental set-up for Photon Induced Fluorescence Radiation

Experiments described in the following have been carried out at the 1 MeV van-de-Graaff accelerator of the University of Giessen and at the synchrotron radiation facility BESSY at Berlin. In the accelerator target, proton or electron impact ionization processes have been studied. The proton beam could be replaced in situ by an electron beam. Details of the experimental set up have been published [23]. Here it is only mentioned that a 1m normal incidence monochromator equipped with a one-dimensional position sensitive channelplate detector has been used to monochromatize the fluorescence radiation.

Fig. 4 is a schematic view of the apparatus used in the experiments at BESSY, carried out in a collaboration between research groups at Giessen and at Kaiserslautern [6, 9, 10]. Synchrotron radiation, monochromatized by a normal incidence monochromator or alternatively by a toroidal grating monochromator, is focussed into a narrow photon beam. It passes a differentially pumped target cell and is monitored by secondary electron emission from an Al cathode. An open photodiode calibrated at the National Bureau of Standards, serves as flux standard. The transmission of the target cell can be controlled by two diaphragms.

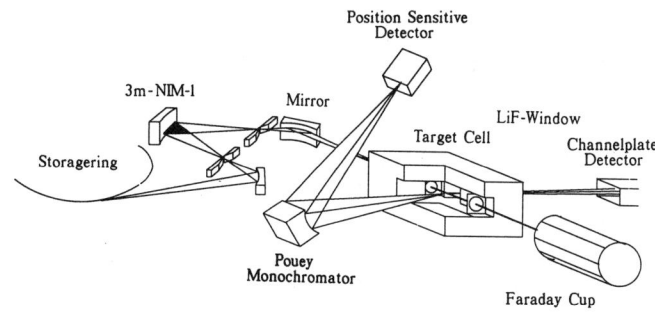

Fig. 4: Schematic view of the experimental set-up used in photoionization studies

Fluorescence radiation can be registered either undispersed by a channelplate with or without LiF-window in front, or dispersed by a high-luminosity monochromator of the Pouey-type [24] with a two-dimensional position sensitive channelplate detector [25].

In fig. 5 and 6 we compare fluorescence spectra registered from Ar, bombarded by 3 keV-electrons or by photons with different energies, respectively. The $3s^23p^5$ $^2P_{1/2, 3/2}$-$3s3p^6$ $^2S_{1/2}$-doublet lines are in both cases dominating. It is evident that the multiplet splitting resolved in fig. 5 is unresolved in fig. 6. This resolution difference is not of fundamental nature, but points out an experimental difficulty of PIFS, which is the low impinging photon flux of $5 \cdot 10^{11}$/s with respect to 10^{15}/s of electron beams. As a consequence, a high luminosity monochromator - with a loss of resolution as the price - and a large detection efficiency for the fluorescence radiation are stringent. They are achieved by:

1. Using the photon beam as entrance slit,

2. A large grating of 110 mm x 110 mm size in 400 mm distance,

3. And by a multiplex detection mode.

A quantum efficiency of $5 \cdot 10^{-3}$ at 92 nm was measured for the described system, using electron impact induced line radiation as intensity standard [15]. This number might be improved a little, but the essential feature of the detection system is the multiplex detection mode. A scan mode would increase the accumulation time of the single spectra in fig. 6, which is 10 min, to a few hours.

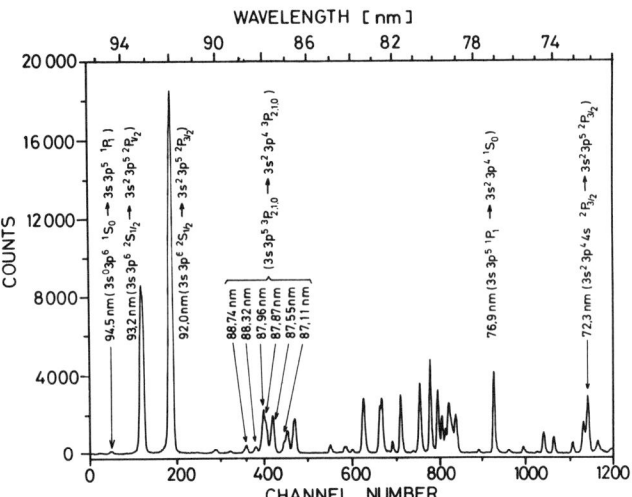

Fig. 5: Fluorescence spectrum of Ar induced by 3 keV-electron impact

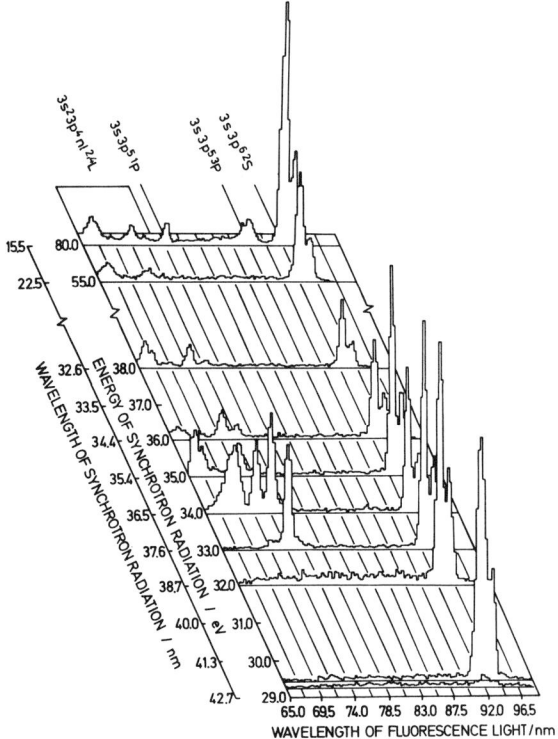

Fig. 6: Fluorescence spectra of Ar with energy and wavelength of the exciting photons as parameters

V. Electron Impact Induced Two-Electron Processes

The photoionization studies at BESSY were initiated by experiments with fast electrons as ionizing particles [26], which on the other hand were carried out for comparison with fast protons. The results for

$$e^-, H^+ + Ar \rightarrow e^-, H^+ + Ar^{++} \left\{ \begin{array}{l} 3s3p^{5\ 1,3}P \\ 3s^o3p^{6\ 1}S \end{array} \right\}$$

are displayed on an equal velocity scale in fig. 7 [15]. Their derivation is complicated by a strong cascade, which has to be subtracted according to equ. 2, resulting from a 2p-electron ionization followed by an Auger transition [27]. Fig. 7 shows that the cross sections for the fast electrons have a slightly weaker velocity dependence than the cross sections for the slower protons.

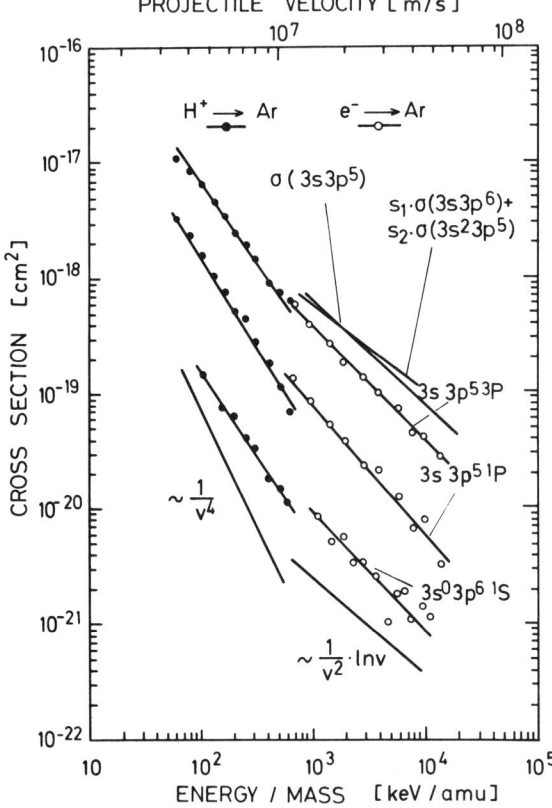

Fig. 7: Cross sections for double ionization of the Ar-M-shell with contributions of the s-electrons by fast electrons and protons as function of the specific projectile energy or the velocity. —●— H^+-impact, —○— e^--impact. For explanation of the full lines see text

The observed difference is taken as a manifestation of electron correlations contributing to double ionization. This interpretation is based on modelling the electron- and proton-impact induced double ionization by a superposition of a two step mechanism and of the influence of electron correlations [29].

The effect of electron correlations for fast-electron impact is described by the Bethe approximation [28], yielding the expression

$$\sigma \sim \frac{Z^2}{v^2} \cdot M^2 \cdot \ln cv^2 \quad (3)$$

with

$$M^2 = \int_I^\infty \frac{df}{dE} \cdot \frac{R}{E} dE \quad (4)$$

and

$$\frac{df}{dE} = \lim_{K \to 0} \frac{df(K,E)}{dE} = \frac{E}{Ra_0^2} \cdot |\langle \phi_f | \sum_j \hat{r}_j | \phi_i \rangle|^2 \quad (5)$$

It is the limit $K \to 0$, which links ionization by fast electron with the optical oszillator strength and consequently makes it sensitive to deviations from the single-electron model through the matrix element in (5). Provided a shake-off mechanism is a valid description of the electron correlations in this case, we would expect a velocity dependence for the simultaneous ejection of a s- and a p-electron, which is in accordance with (3), and which is superimposed on a $1/v^4$ dependence of the two-step ionization mechanism [29]. Both v-dependences have been inserted in the lower part of fig. 7. When compared with the experimental data, neither mechanism fits the experiment. Nevertheless there is also no contradiction to the suggestion, that electron correlations are the dominant mechanism for simultaneous ionization of Ar-3s-3p-electrons by fast electrons.

A quantitative analysis by the shake-off model would derive cross sections for ionization into the $3s3p^5$-configuration from

$$\sigma^{++}(3s3p^5) = S_1 \cdot \sigma^+(3s3p^6) + S_2 \cdot \sigma^+(3s^23p^5)$$

S_1 and S_2 are probabilities for shake-off of a 3p-electron after removal of a 3s-electron and vice versa. Using values for S_1 and S_2 from [30] and cross sections for single ionization from [11, 31] one obtains the measured σ^{++}, as shown in the upper part of fig. 7.

This agreement surprises from the viewpoint of the applicability of the sudden approximation to electron impact with respect to photon impact: The high velocity limit for the ratio of double to single ionization by charged particles should be smaller than the photon value [33], which is used for S_1 and S_2.

It should be noted that we observe a common velocity dependence for proton and electron impact for a single-electron transition in agreement with (3). This is demonstrated in fig. 8 for the Ar-3s ionization and for the production of the Ar^+-ion in the $3s^23p^43d\,^2S$ state [27]. The latter is strongly mixed with the $3s3p^6\,^2S$ state [22]. Considerable experimental and theoretical efforts have been undertaken over the past to study this admixture [33] (see also the inset of fig. 8).

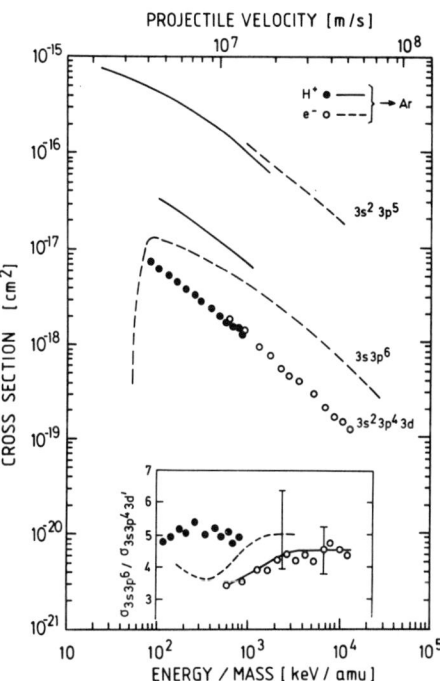

Fig. 8: Cross sections for Ar-3s ionization by He^+-, H^-- and e^--impact, and for excitation of Ar to the Ar^+-$3s^23p^43d\,^2S$-state as function of the specific projectile energy or projectile velocity. --- from [11] but renormalized [15]. Inset: Cross section ratios, --- Tan et al.[53], but $\sigma(3s3p^6)$ renormalized [15]

VI. Double Photoionization of Ar into the $3s3p^5$ $^{1,3}P$-States

Fig. 6 shows the $3s^23p^4$ ^1S-$3s3p^5$ ^1P- and the $3s^23p^4$ ^3P-$3s3p^5$ ^3P-transitions at 77 nm and at 88 nm for $h\nu = 80$ eV. Fig. 9 and fig.10 reproduce the cross sections which have been measured so far. The singlet- and the triplet-cross sections have been added in fig. 9 for comparison with the only existing calculation of this particular double ionization process, based on many-body-perturbation theory [34].

Shake-off values for $\sigma(3s3p^5)$, calculated in the same way as in the preceding section, have been included in fig. 9, with published photoionization cross sections used for the derivation of the three inserted values [36,37]. In contrast to electron impact ionization, the shake-model leads to smaller cross sections with respect to experiment. The conclusion might be, that the calculated shake values [30] for the Ar-M-shell are to small. Larger values for S_1 and S_2 are needed to match experiment and shake-off values for photoionization. They would lead to too large shake-off values for electron impact. This again would fit into the physics of particle impact induced double ionization [29].

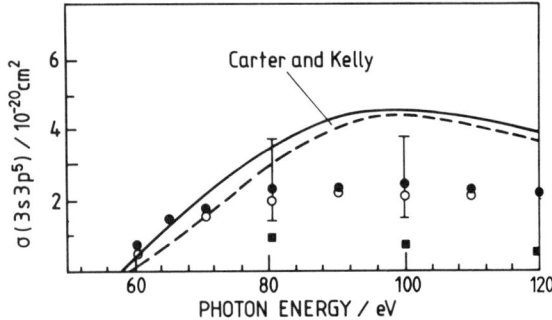

Fig. 9: Cross sections for photoionization into the Ar-$3s3p^5$ configuration as function of impinging photon energy.
○● data from two runs,
─ ─ ─ MBPT calculations [34],
■ shake-off values [30, 36, 37].

Fig. 9 should be interpreted at the moment as an illustration for the capability of PIFS to measure cross sections in the 10^{-20} cm^2 range. A judgement of the validity of the calculations would certainly gain from a reduction of the error bars, coupled to the availability of more intense photon beams. Progress is made in this direction, as documented by fig. 10, where the threshold range of the 3s-3p-double ionization cross sections is displayed [35]. The data in fig. 10 have been obtained at a wiggler/undulator beam line at BESSY, providing a higher photon flux with a bandwidth of 0.5 eV in this experiment.

Both the ^3P- and ^1P-component of the $3s3p^5$-configuration are populated even close to threshold. The ^3P-cross section is structured at the ^1P-threshold. The result in fig. 10 is the third one devoted to tests of a final-state selectivity of double ionization of rare gases by a single photon. Lablanquie et al., using coincidence techniques, observed a selection of the ^3P-state of the Ar^{++}-$3s^23p^4$ configuration with respect to the ^1D- and ^1S-state for a impinging photon energy of 56 eV, i. e. 12 eV above threshold [38]. Their result is in qualitative agreement with theoretical predictions of the threshold behaviour of two-electron processes, based on symmetry arguments of the electron-pair wave functions [39].

Recently, a different behaviour of the ground state of Xe^{++} was reported [40]: All three states of the $5s^25p^4$-configuration were observed to be populated at 40.8 eV photon energy, i. e. 8 eV above threshold. Within the context of the mentioned studies, fig. 10 is certainly of great interest, since the symmetry arguments cited above would predict a preferential population of the ^1P-level, which is not observed. From an experimental viewpoint, the result of fig. 10 is promising for motivations of wiggler/undulator beamlines.

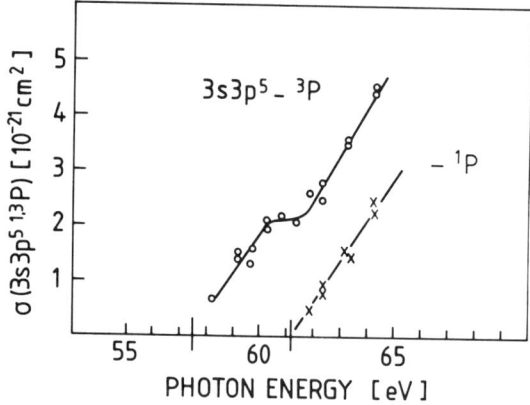

Fig. 10: Cross sections for photoionization of Ar into the ^1P- and ^3P-levels of the $3s3p^5$-configuration as function of photon energy [35], ionization thresholds marked.

VII. Resonances in Correlation Satellites and in the Ar-3s Photoionization

Progress in PES has been made [19], and especially recently [20, 21], in detecting the strong contributions of atomic two-electron excitations to the production of correlation satellites of the rare gases. Equivalent results have been obtained for Ar by PIFS [9] (fig. 11). Strong satellite transitions $3s^23p^5$ ^2P - $3s^23p^4nl$ are only observed in a limited photon energy range (see also fig. 6). A number of resonances, which enhance the satellite intensities, have been identified by absorption experiments [41]. It becomes evident, that a satisfactory theoretical reproduction of correlation satellite intensities will have to include many two-electron states, more than have been considered in the recent MBPT calculations, which concentrate on the region above 35 eV photon energy [42].

The photoionization of the Ar-3s electron is an often studied process. Its energy dependence is discussed in nearly all reviews on photoionization [3]. As a weak channel with respect to the Ar-3p ionization, it is sensitive to electron correlations, e. g. electron-hole interactions [43]. The published data points are on the other hand still scarce, and more experimental efforts are needed to improve their quality for this "benchmark" cross section [44-46]. The contribution of PIFS to the present status is shown in fig. 12, covering as a first measurement the threshold range. Among the calculated cross sections [42, 43, 47-49], the MBPT [42] shows the best overall agreement with the experimental results. It is evident that progress in the reduction of the experimental error bars is needed. Efforts in this direction do exist [51, 52].

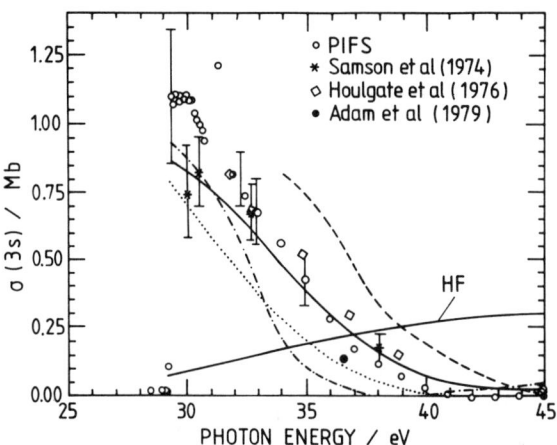

Fig. 12: Ar-3s photoionization cross section between threshold and 45 eV. present values from PIFS [35], PES data [44, 45, 46]. Calculations: HF [47], —·— RPAE [48], ···· SRPAE [43], — — R matrix [49], —— MBPT [42].

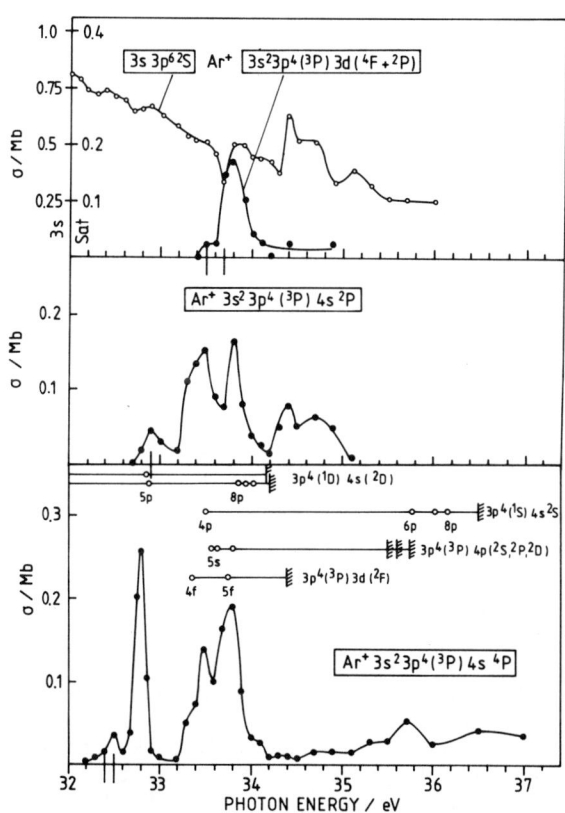

Fig. 11: Resonance structures in the photon energy dependence of correlation satellites. Corresponding thresholds marked [9]. Classification of doubly excited atomic states from absorption spectra [41]. σ(3s3p⁶) for comparison inserted in the upper part.

The PIFS data in fig. 12 contain a few points which are part of rich and prominent resonance structures in the photon energy dependence of σ(3s). This influence of autoionization resonances could have been anticipated from the early highly resolved absorption spectra [41], but was detected only recently by PIFS [9] and independently by PES [50], moreover predicted simultaneously by the MBPT [42] (though here not yet for photon energies between 29.24 eV and 35 eV). Fig. 13 demonstrates these resonancelike features for Ar (see also fig. 11). The classification in fig. 13 stems from absorption spectra [41].

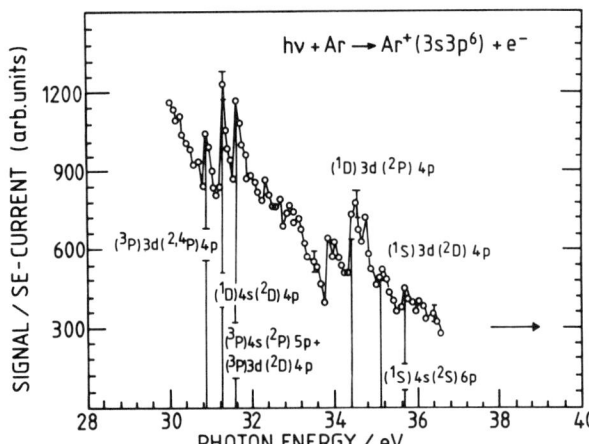

Fig. 13: Autoionization resonances in the energy dependence of the Ar-3s photoionization process [9]

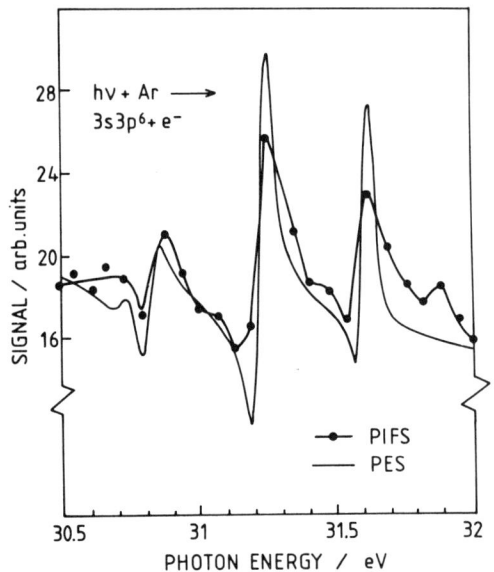

Fig. 14: Comparison of resonance structures in the Ar-3s photoionization cross section, observed by PIFS [9] and by PES [20].

Fig. 14 compares the prominent resonances between 30.5 eV and 32 eV with recent data from a PES study [20]. Good agreement in position is seen, differences in shape are due to a better resolution of the PES data (70 meV with respect to 150 meV for PIFS). From the absorption features [41], even narrower resonances should be expected.

Resonances in the ns-photoionization cross sections are present for all rare gases. Their influence seems to become more complex with increasing Z number [10, 52].

Summary and Conclusions

Applications of fluorescence spectroscopy - as an alternative and complementary experimental method to electron spectroscopy - to studies of total cross sections for electron impact- or photon-induced two-electron processes are described. Results for Ar are surveyed and discussed, partly with the intention to compare electron impact with photon impact and partly to demonstrate the capabilities of photon induced VUV-fluorescence spectroscopy, introduced to photoionization studies of the rare gas outer s-shells. Progress in the knowledge about the process of the Ar-3s photoionization is pointed out, and the need for further experimental efforts to improve the quality of absolute cross sections is underlined.

Acknowledgements

The presented results have been obtained in a collaboration with B. Kraus, P. Lenz, B. Magel, B. Möbus, University of Giessen, and with H. Schmoranzer and M. Wildberger, University of Kaiserslautern. Support by the Deutsche Forschungsgemeinschaft and by the Bundesministerium für Forschung und Technologie under contract numbers 05 352 AX IO and 05 347 AA I5 is greatly acknowledged.

References

[1] J. W. Cooper, in Photoionization and other Probes of Many-Electron Interactions, edited by F. J. Wuilleumier (Plenum Press, New York, 1976) p. 31

[2] M. Inokuti, in Photoionization and other Probes of Many-Electron Interactions, edited by F. J. Wuilleumier (Plenum Press, New York, 1976) p. 165

[3] J. A. R. Samson, in Handbuch der Physik XXXI, edited by S. Flügge and W. Mehlhorn (Springer Verlag) p. 123

[4] U. Becker, B. Langer, H. G. Kerkhoff, M. Kupsch, D. Szostak, R. Wehlitz, P. A. Heimann, S. H. Liu, D. W. Lindle, T. A. Ferrett, and D. A. Shirley, Phys. Rev. Lett. **60**, 1490 (1988)

[5] V. Schmidt, Z. Physik **D2**, 275 (1986)

[6] K.-H. Schartner, P. Lenz, B. Möbus, H. Schmoranzer, and M. Wildberger, Phys. Lett. **128**, 374 (1988)

[7] J. A. R. Samson, Y. Chang, and E. M. Lee, Phys. Lett. **A127**, 171 (1988)

8. B. M. Johnson, M. Meron, A. Agagu, and K. W. Jones, Nucl. Instr. Methods **B24/25**, 391 (1987)
9. K.-H. Schartner, P. Lenz, B. Möbus, H. Schmoranzer, M. Wildberger, Phys. Rev. Lett. **24**, 2744 (1988)
10. K.-H. Schartner, P. Lenz, B. Möbus, H. Schmoranzer, M. Wildberger, J. Phys. B: At. Mol. Opt. Phys. **22**, 1573 (1989)
11. B. F. J. Luyken, F. J. de Heer, and R. C. Baas, Physica **61**, 20 (1972)
12. C. E. Brion, in Physics of Electronic and Atomic Collisions, edited by S. Datz (North Holland, Amsterdam, 1982), p. 579
13. E. Weigold, in Electronic and Atomic Collisions, edited by D. C. Lorents, W. E. Meyerhof, J. R. Petersen (North Holland, Amsterdam, 1985), p. 125
14. J. van den Bos, G. J. Winter, and F. J. de Heer, Physica **40**, 357 (1988)
15. K.-H. Schartner, B. Kraus, W. Pöffel, and K. Reymann, Nucl. Instrum. Methods **B 27**, 519 (1987)
16. M. Krumrey, E. Tegeler, J. Barth, M. Krisch, F. Schäfers, and R. Wolf, Appl. Opt. **27**, 4336 (1988)
17. P. J. M. van der Burgt, W. B. Westerveld, and J. S. Risley, 1989 to be published
18. H. Klar, J. Phys. B: At. Mol. Phys. **13**, 2037 (1980)
19. U. Becker, R. Hölzel, H. G. Kerkhoff, B. Langer, D. Szostak, and R. Wehlitz, Phys. Rev. Lett. **56**, 1120 (1986)
20. R. I. Hall, L. Avaldi, G. Dawber, P. M. Rutter, and G. C. King, J. Phys. B: At. Mol. Opt. Phys. **22**, ... (1989)
21. A. A. Wils, A. A. Cafolla, F. J. Currell, J. Comer, A. Svensson, M. A. MacDonald, J. Phys. B: Atom. Mol. Opt. Phys. **22**, ... (1989)
22. H. Smid and J. Hansen, J. Phys. B: At. Mol. Phys. **16**, 3339 (1983)
23. R. Hippler and K.-H. Schartner, J. Phys. B **7**, 618 (1974)
24. H. Schmoranzer, K. Molter, T. Noll, and J. Imschweiler, Nucl. Instrum. Methods in Physics Research A **246**, 485 (1986)
25. B. Kraus, K.-H. Schartner, F. Folkmann, A. E. Livingston, P. H. Mokler, EUV, X-Ray and Gamma-Ray Instrumentation for Astronomy and Atomic Physics, Proc. SPIE **1159** (1989), to be published
26. M. Eckhardt and K.-H. Schartner, Z. Physik A **312**, 321 (1983)
27. B. Kraus, Diplomarbeit, Univ. Giessen 1986
28. H. Bethe, Ann. Physik **5**, 325 (1930)
29. J. H. McGuire, Phys. Rev. Lett. **49**, 1153 (1982)
30. T. A. Carlson and C. W. Nestor, Phys. Rev. **A8**, 2887 (1973)
31. P. Nagy, A. Skulartz, and V. Schmidt, J. Phys. B: At. Mol. Phys. **13**, 1249 (1980)
32. J. H. McGuire, J. Phys. B: At. Mol. Phys. **17**, L779 (1984)
33. C. E. Brion, A. O. Bawagan, and K. H. Tan, Can. J. Chem. **66**, 1877 (1988)
34. Carter and H. P. Kelly, Phys. Rev. A **16**, 1525 (1977)
35. B. Möbus, B. Magel, K.-H. Schartner, H. Schmoranzer, M. Wildberger, private communication
36. M. Y. Adam, F. Wuilleumier, S. Krummacher, V. Schmidt, W. Mehlhorn, J. Electron. Spectrosc. **15**, 211 (1979)
37. J. B. West and G. V. Marr, Proc. R. Soc. Lond. A **349**, 397 (1976)
38. P. Lablanquie, J. H. Eland, I. Nenner, P. Morin, J. Delwiche, and M. J. Hubin-Franskin, Phys. Rev. Lett. **58**, 992 (1987)
39. C. H. Greene and A. R. Rau, J. Phys. B: At. Mol. Phys. **16**, 99 (1983)
40. S. D. Price and J. H. D. Eland, J. Phys. B: At. Mol. Opt. Phys. **22**, L153 (1989)
41. R. D. Madden, D. L. Ederer, and K. Codling, Phys. Rev. **177**, 136 (1969)
42. W. Wijesundera and H. P. Kelly, Phys. Rev. A **39**, 634 (1989)
43. C. D. Lin, Phys. Rev. **A9**, 171 (1974) and references therein
44. J. A. R. Samson and J. L. Gardner, Phys. Rev. Lett. **33**, 671 (1974)
45. R. G. Houlgate, J. B. West, K. Codling, and G.V. Marr, J. Electr. Spectr. **9**, 205 (1976)
46. M. Y. Adam, F. Wuilleumier, S. Krummacher, V. Schmidt, and W. Mehlhorn, J. Electr. Spectr. **15**, 211 (1979)
47. D. J. Kennedy, S. T. Manson, Phys. Rev. **A5**, 227 (1972)
48. M.Ya. Amusia, V. Ivanov, N.A. Cherepkov, L. V. Chernysheva, Phys. Lett. **A40**, 361 (1972)
49. P. G. Burke, K. T. Taylor, J. Phys. **B8**, 2620 (1975)
50. M. Y. Adam, P. M. Morin, and G. Wendin, Lure Activity report, 33 (1983-1985)
51. U. Becker, private communication
52. K.-H. Schartner, P. Lenz, B. Magel, B. Möbus, H. Schmoranzer, M. Wildberger, BESSY-Jahresbericht 1988, 110
53. K.-H. Tan and J. W. McConkey, J. Phys. B: At. Mol. Phys. **7**, L183 (1974)

PHOTOIONIZATION OF LASER EXCITED ATOMS BY SYNCHROTRON RADIATION

D. Cubaynes, J.M. Bizau, B. Carré* and F.J. Wuilleumier

Laboratoire de Spectroscopie Atomique et Ionique and
L.U.R.E.Université Paris-Sud, Bat 350, 91405 Orsay Cedex, France

*Service de Physique des Atomes et des Surfaces, CEN Saclay, 91191
Gif sur Yvette, France

ABSTRACT

The present status of the experimental studies in the photoionization of excited atoms using laser and synchrotron radiations is briefly surveid. Recent improvements in the experimental capabilities are summarized. Examples of new experiments are given. In particular, a full study of the photoionization of an excited atom, using electron spectrometry, is presented more in details.

Introduction :

The photoionization process is useful for the characterization of the dynamical and geometrical properties for the correlated motion of the electrons in atoms. This characterization occurs via several parameters which are the partial photoionization cross section σ_i from all the differents subshells, the total cross section which is the sum of all the partials cross sections, the asymmetry parameter of each subshell wich characterizes the angle at wich the photoelectrons from the ith subshell are emitted, and the spin parameters. These parameters are predicted by the theory and are measured by the experiment. For a best comparison, they must be determined on the largest energy scale avalaible. The synchrotron radiation, wich is continuous in wavelength from the infra-red to the X-ray, is a good tool for the studies of photoionization.

From the experimental point of view, as soon as the photon energy is higher than the second ionization threshold (which corresponds to the first satellite state,i.e. to the lower excited final state of the ion), photoabsorption spectroscopy or ionic spectroscopy are unable to separate the contributions of the different partial photoionization cross sections to the total cross section. Only electron spectrometry is able to give access to all differents exit channels.

The first synchrotron radiation photoionization experiments were carried out on the closed shell species, mostly the rare gases[1]. These first pioneering studies on the atomic photoionization have shown the importance of the relativistic and correlation effects. The agreement with the best theoretical calculations at that time (RPAE and MBPT) was good (often within 10%). Starting in 1980, with the developement of new high flux troughput monochromators, it became possible to study in more details the spectroscopy of ground state atoms. The increase of the photon flux was used, for example, to improve the results previously obtained on the rare gases[2,3,4] : satellite states, which are produced via electron correlations, are presently extensively studied at threshold and under resonant excitation conditions. It was also possible to start the study of new atomic systems such as open shell atoms[5,6,7] : the coupling of at least two open shells in the final state of the ion gives rise to a multiplet structure wich is more difficult to account for theoretically. It was also possible to start study on the photoionization of laser excited atoms[8,9]. Most recently,the first experiments on the photoionization of ions[10,11] have been reported on.

Photoionization of excited atoms :

Photoionization of laser excited atoms are studied since 1981[12]. Data avalaible on this subject are useful for astrophysics or for plasma-physics. For the atomic physicist, laser excited atoms provide with a particularly interesting case : the wavefunction of the excited electron extends often on a radius larger than in the ground state atom. Therefore, in the photoionization of a laser excited atom, the atomic potential is probed over a larger space range. On the other hand, the parity of a laser excited state is opposed to the one of the ground state : a new full set of states become avalaible in photoionization. These states may have a different behavior than the ground state : theoretical calculations predict the occurence of several new features[13,14,15]. Also, the initial state of an atom must often be represented by a linear combinaison of several configurations; with the laser, it is possible to remove in principle this degeneracy, and to prepare an atom in a state with well defined quantum numbers.

Two recent revues have covered the field of photoionization of laser excited atoms[16,17]. To summarize the situation in 1987, the first experimental studies were on resonant processes. Since the direct photoionization cross sections are usually lower than a resonant cross section, their experimental studies started later, when the photon flux has been significantly improved. Briefly, the feasibility experiments on photoionization of excited atoms were achieved at L.U.R.E. in 1981 on sodium[12] and in 1982 on barium[18] atoms and at BESSY in 1985 on barium atoms[19].

Resonant photoionization studies have firstly been made on laser excited sodium atoms. The transitions between a laser excited $2p^63p\ ^2P_{3/2}$ state and the autoionizing levels with $2p^53s3p$ configuration were systematically investigated[8]. Assuming that the radiative decay of these states is negligible, values of the oscillator strength associated to the excitation of the different possible autoionizing states were determined. The sum of all these strength was found to be 0.22(4), assuming no difference in the 2p photoionization cross section for the ground state and for the laser excited atom. Similar experiments[20] were performed later for the $2p^63p\ ^2P_{1/2}$ state. Studies of the 5p resonances in laser excited $5p^66s5d\ ^{1,3}D_2$ barium atom were also undertaken[17]. They showed that the increase in the number of intermediate channels in the excited case spread the oscillator strength over a large number of resonances.

Photoionization cross section for outer and inner-shell direct ionization in these excited atoms have also been determined between 20 eV and 140 eV photon energy range[9,21]. This range covers several atomic core thresholds, and especially the 4d ionization threshold. One of the mains result of these experiments is that the 5d, 5p and 5s photoionization cross sections are strongly enhanced via intershell correlation with the 4d subshell.

New experiments were also started on the resonant photoionization of highly excited $2p^64d$ and $2p^64p$ sodium atoms produced by two laser beams: transitions to autoionizing states of configuration $2p^53s4d$ and $2p^53s4p$ were observed and some oscillator strength could be determined[22].

Several new studies were proposed in a recent revue[16]. The use of the high flux storage ring such as BESSY and of the new generation of synchrotron ring such as Super-A.C.O. allowed to make new advances in this field.

Experimental apparatus :

As outlined in reference 16, the determination of partial cross sections over an extended wavelength range requires the use of synchrotron radiation and electron spectroscopy.

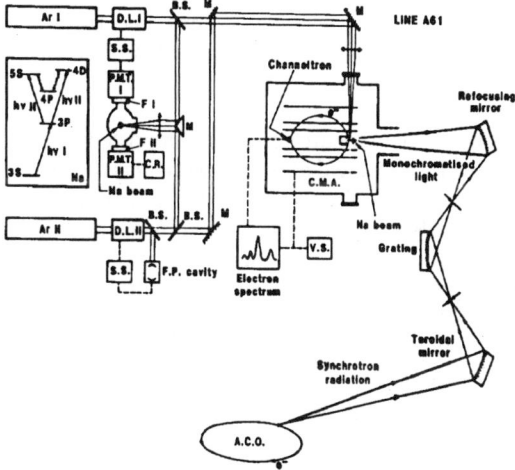

Fig 1. Experimental apparatus used at L.U.R.E. for photoionization of laser excited atoms

Figure 1 shows the example of the apparatus used at L.U.R.E. for photoionization of laser excited atoms in the highest extension it has reached. Electrons (A.C.O. case), or positrons (Super-A.C.O. case) rotating in the storage ring emit the synchrotron radiation. This white light is selected in wavelength by a toroidal monochromator, and then focussed in the source volume of an electron analyzer (cylindrical mirror analyser : C.M.A.) onto an atomic beam. The electrons emitted at the magic angle in the photon-atom interaction are energy analyzed and counted by a channeltron. A multichannel analyzer and a micro-computer are then used to record and to analyze the spectra. A furnace mounted on the axis of the C.M.A. produces an effusive atomic beam whose density is limited to 10^{13} atoms/cm^3 to avoid inelastic electron collisions and collisional depletion of excited atoms. The laser beam is sent in the source volume of the analyzer and crosses the atomic beam at right angle, to minimize Doppler effect. This laser radiation is produced by a commercial c.w. ring dye laser pumped by an argon ionic laser. A small amount of this photon beam is sent into an auxiliary furnace whose fluorescence is recorded with a P.M. and is used to lock the wavelength to the desired fine structure transition. For the experiments on more highly excited atoms, a second c.w. dye laser is added to the first one. It achieves the second excitation step $3p\,^2P_{3/2}$ -> $nl\,^2L_{1/2,5/2}$ (nl = 4d or 5s). Its wavelength is locked with an external F.P. cavity when the 4p--> 3s U.V. fluorescence is observed. At the atomic density we use in the source volume of the analyzer, no polarisation effects occurs; thus, the differential photoionization cross sections are proportional to the subshell cross sections .

As previously mentionned, one important problem for the laser excited atom experiment is the electron background. A high amount of low energy electron is produced by collisional phenomena, and diffuse thermally throughout the analyser. Thus, these electrons are no more energy analysed, and are counted at any analyzing voltage. One major improvement consists in mounting a negatively polarized grid in front of the channeltron to prevent these low energy electrons to reach the detector.

The most important improvement since the first experiments mentionned earlier comes from the use of high flux storage rings. The increase of photon flux combined with the diminution in the size of the positron bunch allow on one hand to reduce the monochromator band pass and, on the other hand to reduce the counting time for a given spectrum. The plasma potential due to collisional ionization in the source volume of the analyser is very sensitive to any change of the lasers, or of the atomic density and drift slowly with the time. Thus the resolution of an electron spectrum decreases with the increase of the counting time. High flux storage rings allow to study some second order phenomena by reducing the monochromator band pass and the counting time.

New experimental capabilities :

The use of high flux storage ring such as Super-A.C.O. (or A.C.O. in its last time) or B.E.S.S.Y. allowed to perform some importants progress. Most of them were sugested in ref. 17. Briefly, it was proposed to improve the resolution, to work with a lower atomic density, to pump with a lower laser power, to excite a core electron on a Rydberg state level or to study the angle of emission of the photoelectrons.

It is now possible to work with some densities of excited atom lower than in the first experiments. So, the usuable tunability range of c.w. dye laser has been extended. The loss of laser power when these laser are tuned to the blue-side can be compensated with a higher synchrotron flux : one spectrum on resonant photoionization of laser excited calcium atoms has been obtained[23].

We show in figure 2 one example of such an improved spectrum obtained with a reduced monochromator band pass, and the grid in front of the channeltron to reduce the background, in the case of photoionization of the $2p^64d$ highly excited

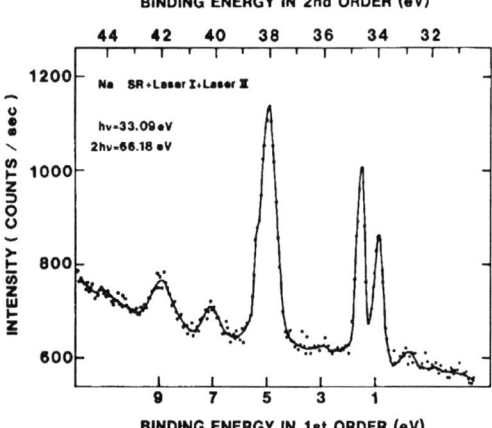

Fig 2. Photoelectron spectrum for ionization of $2p^6nl$ (n=4) sodium atoms recorded for 33.09 eV photon energy. The autoionization structure is clearly visible for photoelectron of around 1eV binding energy. The structures observed at lower kinetic energy arise from direct photoionization in 2p subshell by photons diffracted in second order (from Ref. 24).

states in Na. The general behavior of this spectrum has already been explained[22,24]. Briefly, one can see mainly two structures. The doublet around 1 eV binding energy in first order arises from the resonant photoionization of the 4d and the 4p electrons. The structures seen at higher binding energy come from photoionization in 2p subshell by second order diffracted photons. This spectra is to be compared with the one shown in ref.22 . The signal to noise ratio has been improved by one order of magnitude. The contribution of the 4d resonant photoionization can now be separated from the 4p one. Moreover, the increase of flux allows to see the resonant photoionization of the 4s excited state populated by radiative decay from the 4p. The density of 4s atoms is around 1% of the total atomic density. These new results are summarized in a recent paper[24].

Another possibility offered by the increase of the photon flux is to decrease the global atomic density. In this case, for atomic densities lower than 10^{11} atoms/cm^3, collisional process and radiation trapping do not destroy the alignment induced by the laser polarization. In this case, the angle of 54°44' is no

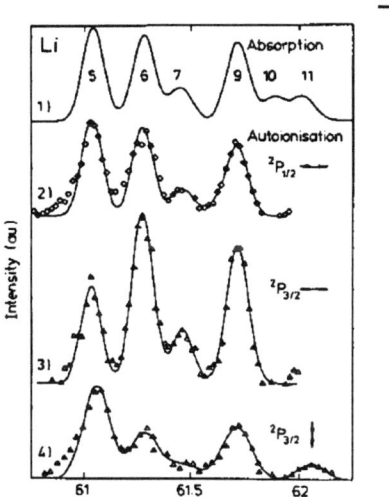

Fig 3. Laser polarization dependence of autoionization in lithium atoms (from Ref. 25).

more magical : the intensity and the angular distribution of photoelectrons depend of the relative orientation of the polarisation vector of the laser and of the synchrotron radiation. One example, obtained on autoionization of excited lithium atoms with B.E.S.S.Y.'s radiation is shown in fig. 3[25]. Taking into account the variation of density of excited state with the laser transition, this can be used to easely recognise the S symetry levels : the electron emission of such a level is isotropic and do not depend on the polarization of the laser radiation. This has been used for the example shown in fig. 3 to demonstrate than the peak labelled 9 was issued of a S symetry level instead of a D one. Similar results have been reported on sodium atoms.

The first experiments on laser excited atoms have started on Super-A.C.O. on february 1989. The higher photon flux and the lower emittance of this ring allowed to really start the spectroscopy of excited states, in the few cases wich are allowed by the tunability of the present c.w. dye lasers. These experiments where performed on sodium atoms[26]: the photoionization in the 2p and the 2s subshells in excited sodiums atoms, especially the behavior of the correlation satellites were investigated. In the following, we present a summary of these new experimental results.

Full electron spectroscopy of an excited atom : the sodium 2p case

a) Single photoionization : the main lines

We show in figure 4 two photoelectron spectra recorded in the 2p binding energy range with the apparatus shown in figure 1. They have been recorded at 75.6 eV photon energy. The first spectrum, in the upper pannel, has been obtained for $2p^63s\ ^2S_{1/2}$ ground state sodium atoms, the second comes from a medium composed of atoms in the ground state and in the $2p^63p\ ^2P_{3/2}$ excited state. The ground state spectrum exhibits three main structures : the most intense one is for electrons of 38.0 eV binding energy wich is the 2p binding energy. It is called main line. The two other structures, observed at higher binding energy, are the satellite lines: the first structure is observed for binding energy of nearly 46 eV, wich corresponds to $2p^54s$ and $2p^53d$ final ionic states, and are due to shake up transitions (the excited electron change its orbital quantum number by 0 or 2 units). The second one, detected at 42 eV binding energy, is produced in conjugate shake up transitions ($\Delta l=1$), and corresponds to $2p^53p$ final ionic state. When the laser tuned to the 3s $^2S_{1/2}$---> 3p $^2P_{3/2}$ transition irradiates the vapor, up to 30% of the sodium atom can be tranfered into the 3p-excited state. The vapor probed by the

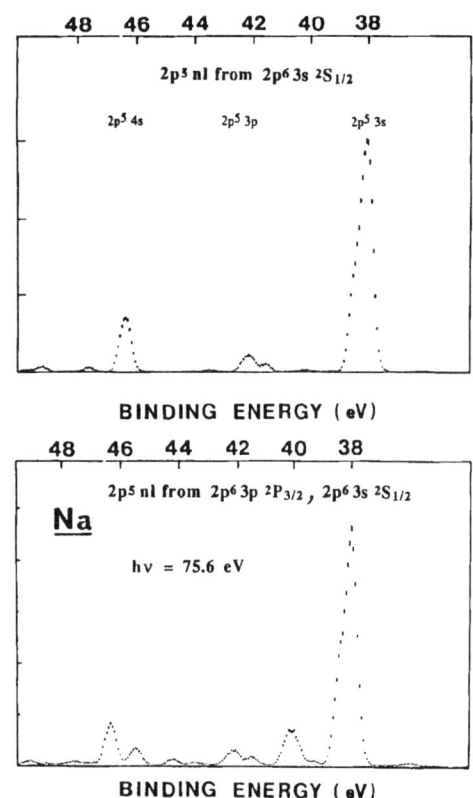

Fig 4. Two photoelectron spectra recorded at 75.6 eV photon energy. Atoms are in the ground state for the upper pannel and in both ground and excited states for the lower pannel.

synchrotron radiation is then a mixture of atoms in ground and in excited states. Thus the electron spectrum must show the structures of the ground and of the excited states. This is actually the case for the spectrum shown in the lower pannel of Fig.4 which has been recorded while some atoms in the excited state were present in the vapor. In addition to the electron lines due to photoionization of the ground state atoms, which have been previously described, this spectrum exhibits several surnumerary electron structures. The most intense ones are at 40 eV binding energy. These peaks come from the excited atom. Their mean kinetic energy is 2.11 eV higher (energy brought by the laser) than the one of the conjugate shake up satellites in the ground state, and thus correspond to the same final state configuration :$2p^5 3p$. Therefore, this peak at 40 eV binding energy is the main line of the excited state. As previously mentionned, the structures observed on the high binding energy side of the main line in the excited state arise from satellite states. One notes that the binding energy of the main line in the excited state is higher than in the ground state. This means that the 2p subshell is less screened by the outer electron in the excited state than in the ground state.

A theoretical study of photoionization in the 2p subshell of these sodium atoms with outer electron either in its ground state or in the first resonant state has predicted that this 2p cross section was independent of the state that the valence electron occupies[27]. One of the conclusions of the first experiment on laser excited states was that, within a 20% uncertainty, there were no large discrepancies between these two cross sections. A significant increase of the photon flux was needed to allow for decreasing the monochromator band pass , in order to improve and to extend these results on a broader scale (the two electron main lines for 2p photoionization in excited and ground state atoms are only distant of 2 eV). This has been made possible with the improved photon flux of the new storage ring Super-A.C.O. where the whole apparatus shown in figure 1 has been tranferred. We have measured the behavior of the ratio between the intensity of the main lines in the excited and in the ground state over a wavelength range extending from 50 eV to 110 eV.The variation of this ratio is shown in Figure 5. One can observe that this ratio is roughly constant and equal at 0.78 (0.05) over the whole scale range. The higher value of the data at 99 eV photon energy can be explained by the

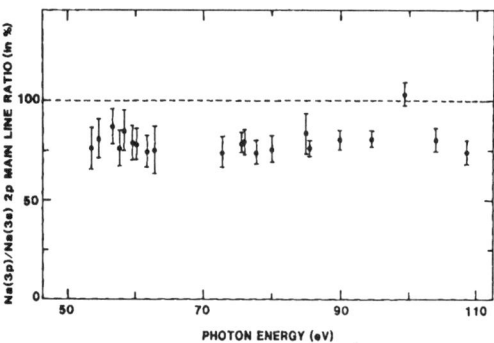

Fig 5. Variation with the photon energy of the branching ratio between the main line in the excited state and in the ground state.

coincidence of this photon energy with Rydberg series converging towards the $2p^4 3s$ double ionization threshold. The value of the ratio, clearly lower than 1, means that exciting the outer electron removes some oscillator strength out of the 2p subshell main line. This clearly proves the existence of some coupling between the valence electron and the 2p core subshell.

b) Two electron transitions in photoionization: the satellite case

One powerful source of information for the photoionization studies is the behavior of the satellite lines. As a matter of fact, satellite lines involve a change of at least two electrons even though the interaction photon-atom is classically described by a one electron operator. Thus, satellite lines cannot been described by the classical frozen core central field model. The occurence of satellite states is then a direct result of electron-electron correlations in the atom, and their study is of fundamental importance for the understanding of correlation effects in atoms. Numerous experimental studies have been reported on satellite states in ground state atoms. None have been published yet for atoms prepared in an excited state. In the simultaneous excitation of two electrons by one photon, both electrons can be tranferred to discrete states (resonant excitation), one of the electron is excited and the other is ionized (shake up process) or both electrons can be ionized (shake off process).

1) Resonant excitation

As an example of the resonant case, we choose to show the first satellite resonance observed for two electron excitation in 3p-excited sodium. The oscillator strengths for the excitation transition of one 2p electron to the 3s orbital in the $2p^63p\ ^2P_{3/2}$ excited state ($2p^63p \rightarrow 2p^53s3p$) have been previously studied. A satellite transition of these resonances consists in exciting a second electron in the 2p -> 3s transition. The first parity allowed one is the 3p -> 4p transition. We have observed the decay of the $2p^53s4p$ autoionizing states populated directly from the $2p^63p\ ^2P_{3/2}$ state by absorption of a single photon and we have measured the oscillator strength of this transition. We show in figure 6 typical electron spectra recorded in the energy range of this resonance. They have been taken at 34.23 eV photon energy. Photons of twice this energy (68.46 eV) are also transmitted in second order by the monochromator. The spectrum displayed in the upper pannel is issued from only ground state atoms. Two main structures can be seen : the first one, at higher kinetic energy, is observed for 38 eV binding energy in second order (upper scale) and comes fom photoionization in 2p subshell by 68.46 eV photons. The second one is observed for 5.14 eV binding

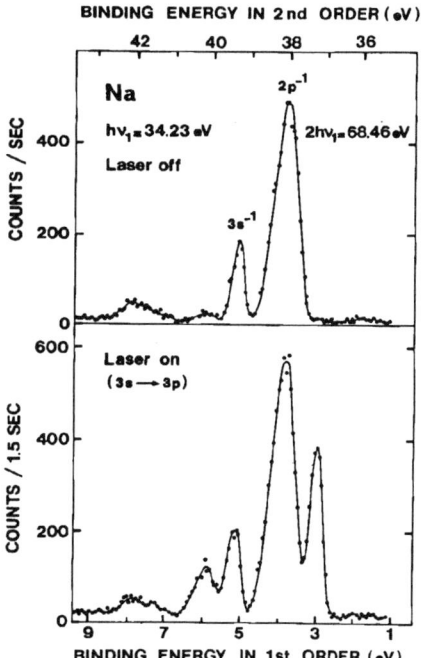

Fig 6. Two sodium photoelectron spectra recorded at 34.23eV photon energy. The resonant ionization via the $2p^53s4p$ autoionizing state of the 3p outer electron is observed in the lower pannel at 3eV binding energy.

energy in first order and comes from photoionization of ground state atoms in 3s subshell by 34.23 eV photons. The structure observed at lower kinetic energy are some satellites produced in the 2p photoionization and have already been described. The spectrum shown in the lower pannel was recorded with the laser and the synchrotron radiation on. As previously mentionned, the vapor probed by the synchrotron radiation is a mixture of ground and excited atoms, and the electron spectrum shows structures arising from photoionization in these two states. One can see two more lines coming from the excited state in this spectrum : the highest one is observed at 3 eV binding energy in first order and comes from 3p resonant ionization of the excited atom through a $2p^53s4p$ autoionizing state. When the monochromator is tuned off the resonant region, the 3p electron line disappears because the 3p cross section is very low in this energy region. The 3p line, in the figure, is clearly more intense than the 3s photoionization of the ground state in spite of a much lower atomic population, wich confirms the resonant character of this line. The second additional line, observed at 40 eV binding energy in second order, has previously been described and comes

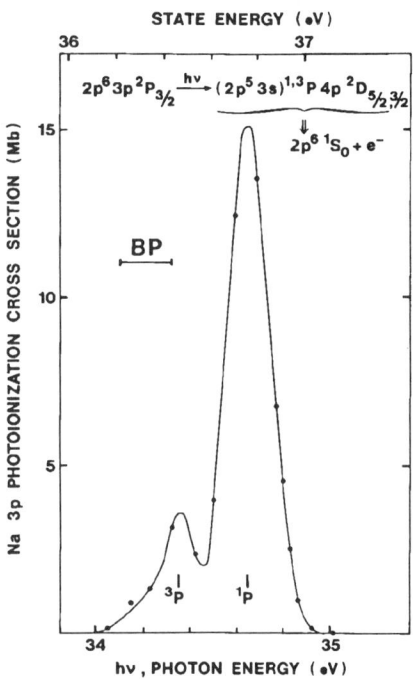

Fig 7. Excitation function for the resonant ionization of the 3p-electron. The resonant $2p^53s4p$ intermediate sate is populated via shake up transition.

from 2p subshell photoionization in laser excited atoms. We have measured the behavior, with the photon energy, of the 3p resonant photoionization peak. Corrected for the analyser tansmission, the synchrotron photon flux, the atomic density and the laser efficiency, we obtain the excitation function diplayed in Figure 7: it is the result of the convolution of the atomic spectrum with the band pass of the monochromator. A previous photoabsorption experiment[28] has determined the energy position of the differents intermediate states and a HF calculation was used to identify them. Two peaks can clearly be seen in figure 7. The most intense corresponds to several states with a ^1P intermediate coupling for the core. The energy of the resonance that we measure is in excellent agreement with the previous photoabsorption results. These transitions have an associated oscillator strength of 0.029(0.007) assuming the radiative decay of the autoionizing state to be negligible. Unpublished calculations of this oscillator strength in the HF approximation[29] gives a value of 0.036 for the $^2D_{5/2}$, 0.0097 for the $^2D_{3/2}$ and 0.011 for the $^2S_{1/2}$ states, respectively. The lower peak in Fig. 7 correspond to ^3P intermediate coupling and is associated with a 0.007(0.002) oscillator strength under the same assumption (the calculated values are 0.006 and 0.0025 for the $^2D_{3/2}$ and the $^2S_{1/2}$ states, respectively). The P final symetry states cannot autoionize for parity reasons. Thus, the oscillator strength corresponds to $2p^63p\ ^2P_{3/2} \longrightarrow 2p^53s4p\ ^2D,^2S$. The previous studies of the $2p^53s3p$ resonances and of resonant photoionization of $2p^64d$ and $2p^64p$ atoms led us to think that the major contribution to the oscillator strength comes from ^2D states. Because of the change of core coupling between the $2p^53s3p$ states and the $2p^53s4p$ shake up states, it is difficult to directly compare the satellite line to the main line. However, the sum of the oscillator strength for these two electron excitation lines can be compared to the sum of all the oscillators strength determined[8] for the one electron transition $2p^63p\ ^2P_{3/2} \longrightarrow 2p^53s3p$, leading to a ratio of 17%. It is worthwhile to note that this value is very close to the asymptotic value obtained for the relative intensity of the shake up satellites in direct photoionization of ground state atoms.

2) The direct case : continuum photoionization

A new improvement of these experiments was to succesfully study the same effect in the non resonant case. This has been recently made possible for the first time by the higher photon flux of Super-A.C.O.. We have seen previously that the sodium 2p- partial cross section (main line) decreased in the excited state while theoretical calculations[27] predicted that the total photoionization cross section in 2p subshell was independent of the orbital of the valence electron. Let us see now the satellite behavior. We show in the lower pannel of the Figure 8 a full electron spectra which is the difference of the two spectra shown in Figure 4. These two spectra are shown again for visual comparison in the upper pannels of the figure. All spectra have been corrected for the analyser transmission. This is a pure spectrum of 3p-excited sodium atoms : the contribution of the ground state atoms has been substracted. The importance of the satellite lines is clearly visible in this spectrum. Except the main line wich is observed around 40.1 eV binding energy, all the others lines are satellite lines. Satellite lines produced by shake up transitions correspond now to $2p^54p$ and $2p^54f$ final ionic configurations ($\Delta l = 0,2$) while, in the ground state spectrum, the shake up lines were coincident ($2p^53d$ and $2p^54s$ configurations). They are separed in the excited state : the former is observed at 47 eV binding energy and the latter (the most intense) around 45.5 eV. The satellite line coming

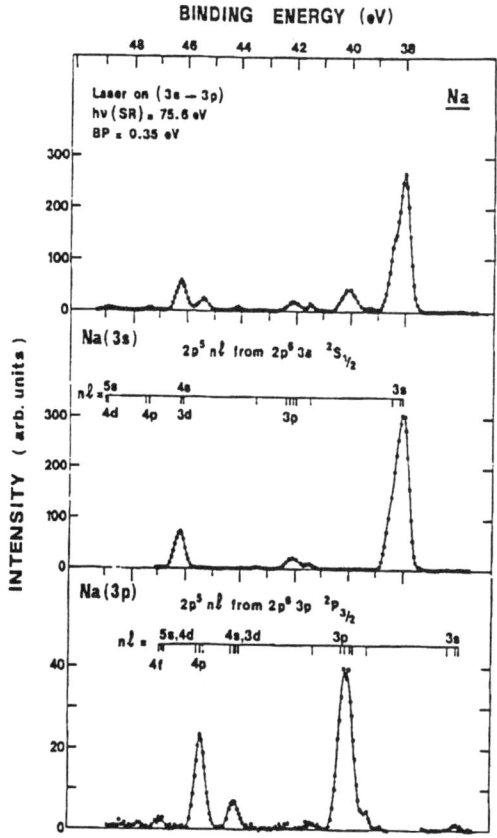

Fig 8. Three photoelectrons spectra at 75.6 eV photon energy, corrected for the transmission of the analyzer. The spectrum in the lower pannel is the normalized difference of the two upper, previously shown in Fig. 4, and is the pure photoelectron spectrum from the 3p excited state.

from conjugate shake up transitions appears around 44 eV ($2p^5 3d$ and $2p^5 4s$ configurations). It is also possible in the excited atom to observe a shake down satellite ($2p^5 3s$ configuration) which is detected at lower binding energy than the main line. We have determined the variation of the satellite to main line ratio with the photon energy for all the outgoing channels. As an example, we show in figure 9 the difference of behavior between the shake up to main line branching ratio for the ground state atom and for the excited atom (only the $2p^5 4p$ configuration in this last case). The two curves exhibit the normal behavior of a shake up satellite : increasing at threshold with the photon energy to reach a constant value as soon as the kinetic energy of the outgoing electron is equal to a few time the excitation energy of the satellite. The amazing point is that the asymptotic value of the satellite intensity is two times higher in the excited atom than in the ground state atom : the electron correlations are twice more intense in the excited state. This hudge effect is connected to the loss of oscillator strength in the 2p main line of the excited state : it corresponds to a tranfer of oscillator strength to the satellite states. When one adds the satellite contributions in the ratio between the two 2p photoionization cross sections, this ratio increases to 0.92(0.07), wich brings the theory and the experiment in good agreement. This ratio,wich is still sligthly lower than one, will increase if the shake off contribution is taken into account. This process, which corresponds to double ionization, is not directly observed in photoelectron spectroscopy. No data are yet avalaible for its contribution.

This example underlines that the study of satellites from excited atoms gives some very important results for the comprehention of the correlated motion of the electrons in the atom.

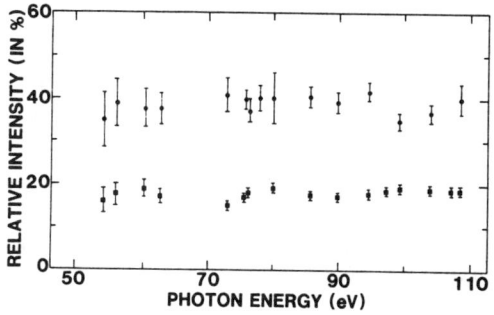

Fig 9. Variation with the photon energy of the behavior of the shake up satellites of sodium atoms produced in photoionization of 2p subshell in the ground state (■) and in the 3p-excited state (•).

Conclusion :

We have shown in this paper the progress that have been made since the first experiments on excited atoms. The powerful tool that is the study of excited atoms has been demonstrated by the number of phenomena which can now be investigated. The next improvement will come from the use of undulators. The increase of photon flux and of the band pass will allow a great number of new experiments such as the determination of the angular parameter ß, or the study of double Rydberg atoms. In addition to the specific experiments on excited states, the excited atoms will offer all the experimental possibilities of the ground state atoms. One next step will be to use some pulsed laser : it will be possible to study the dynamics of the excitation making some stroboscopy betwen the pulsed synchrotron and the pulsed laser radiation.

/1/ F.J. Wuilleumier, Atomic Physics, 7, 481, (1981).

/2/ H. Kossmann, B. Krässig, V. Schmidt and J.E. Hansen, Phys. Rev. Lett., 58, 1620, (1987).

/3/ S. Svensson, N. Martensson and U. Gelius, Phys. Rev. Lett., 58, 2639, (1987).

/4/ P. Lablanquie, J.H.D. Eland, I. Nenner, P. Morin, J. Delwiche and M.J. Hubin_Franskin, Phys. Rev. Lett., 58, 992, (1987).

/5/ E. Schmidt, H. Schröder, B. Sonntag, H. Voss, H.E. Wetzel, J. Phys. B 16, 2961, (1983); ibid 18, 79, (1985).

/6/ M.O. Krause, A. Svensson, A. Fahlman, T.A. Carlson and F. Cerrina, Z. Phys. D 2, 327, (1986).

/7/ J.M. Bizau, P. Gérard, F.J. Wuilleumier and G. Wendin, Phys. Rev. Lett., 38, 63, (1984).

/8/ J.M. Bizau, F.J. Wuilleumier, D.L. Ederer, J.C. Keller, J.L. LeGoüet, J.L. Picqué, B. Carré and P.M. Koch, Phys. Rev. Lett., 55, 1281, (1985).

/9/ J.M. Bizau, D. Cubaynes, P. Gérard, F.J. Wuilleumier, J.L. Picqué, D.L. Ederer and G. Wendin, Phys. Rev. Lett., 57, 306, (1986).

/10/ I.C. Lyon, B. Peart, J.B. West and K. Dolder, J.Phys B 19, 4137, (1985).

/11/ I.C. Lyon, B. Peart, K. Dolder and J.B. West, J.Phys B 20, 1471, (1987).

/12/ J.M. Bizau, F.J. Wuilleumier, P. Dhez, D.L. Ederer, J.L. Picqué, J.L. LeGoüet, and P. Koch, "Laser techniques for the extreme ultra-violet spectroscopy", ed. T.J. McIlrath and R. Freeman, American Institute of Physics Conference Proceeding n° 90, (American Inst. of Phys. New-York),(1982), p. 331.

/13/ A. Msezane and S.T. Manson, Phys. Rev. Lett, 35, 364, (1975).

/14/ A. Msezane and S.T. Manson, Phys. Rev. Lett, 48, 473, (1982).

/15/ J. Lahiri and S.T. Manson, Phys. Rev. Lett., 48, 614, (1982).

/16/ F.J. Wuilleumier, D.L. Ederer and J.L. Picqué, Adv. Atom. Mol. Phys. 23, 197, (1986).

/17/ D. Cubaynes, J.M. Bizau, F.J. Wuilleumier, D.L. Ederer, J.L. Picqué, B. Carré, M. Ferray and F. Gounand, J. Physique 48, C9-453, (1987).

/18/ F.J. Wuilleumier, in " X- Ray and Inner-shell Physics", ed. B. Crasseman, AIP Conference Proceeding Series n°94, American Institute of Physics, New York, 1982, p. 615.

/19/ A. Nunneman, T. Prescher, M. Richter, M. Schmidt, B. Sonntag and P. Zimmermann, J.Phys B 18, L337, (1985).

/20/ T. Prescher, Ph.D. Thesis, University of Hamburg (1988).

/21/ D. Cubaynes, J.M. Bizau, T.J. Morgan, F.J. Wuilleumier, M. Ferray, F. Gounand, P.d Oliveira et P.R. Fournier, J. Physique 48, C9-513, (1987).

/22/ M. Ferray, F. Gounand, P.d Oliveira, P.R. Fournier, D. Cubaynes, J.M. Bizau, T.J. Morgan and F.J. Wuilleumier, Phys. Rev. Lett. 59, 2040, (1987).

/23/ M. Meyer, T. Prescher, E. v. Raven, M. Richter, B. Sonntag, B.R. Müller, W. Fiedler and P. Zimmerman, J. Physique 48, C9-547, (1987).

/24/ B. Carré, P. d'Oliveira, M. Ferray, P. Fournier, F. Gounand, D. Cubaynes, J.M. Bizau and F.J. Wuilleumier, submitted to Z. Phys. D., 1989.

/25/ M. Meyer, B. Müller, A. Nunnemann, T. Prescher, E.v.Raven, M. Richter, M. Schmidt, B. Sonntag and P. Zimmerman, Phys. Rev. Lett. 59, 2963, (1987).

/26/ D. Cubaynes, J.M. Bizau, F.J. Wuilleumier, B. Carré and F. Gounand, submitted to Phys. Rev. Lett.

/27/ T.N. Chang and Y.S. Jim, J. Phys. B 15, L835, (1982).

/28/ J. Sugar, T.B. Lucatorto, T.J. Mc Ilrath and A.W. Weiss, Opt. Lett. 4, 109, (1987).

/29/ A. Weiss, unpublished results, communicated by T.B. Lucatorto.

PHOTODETACHMENT COLLISIONS

David J. Pegg

Department of Physics, University of Tennessee, Knoxville, Tennessee 37996-1200, USA
and
Physics Division, Oak Ridge National Laboratory, Oak Ridge, Tennessee 37831-6377, USA

A crossed-beams apparatus has been used to measure the cross sections (angular differential and integral) for photodetaching electrons from metastable He$^-$ and stable B$^-$ ions at λ = 696.2 nm. Kinematic effects were exploited, in a novel way, to enhance the precision of the measurements. The partial (angular integral) cross sections for He$^-$ photodetachment are $\sigma(2^3S)$ = 22.9 ± 1.0 Mb and $\sigma(2^3P)$ = 10.0 ± 0.6 Mb. The corresponding cross section for B$^-$ is $\sigma(2^2P)$ = 16.5 ± 2.1 Mb. The calculated cross section for the photodetachment of D$^-$ was used to establish an absolute scale for the relative measurements. Radiative attachment cross sections are derived from the photodetachment results using the principle of detailed balance.

I. INTRODUCTION

Interest in photon-ion interactions has grown rapidly over the past few decades, due primarily to the advent of lasers and synchrotron light sources. At the Brighton ICPEAC, for example, Dolder[1] reported on elegant experiments designed to measure cross sections for photoionization of positive atomic ions. In this progress report, I shall attempt to complement that contribution by describing measurements of cross sections for the photodetachment of negative atomic ions made recently at the Oak Ridge National Laboratory.

The pioneering work on photodetachment began at NBS in the 1950's and early 1960's with the development, by Branscomb and Fite,[2] of a crossed photon-negative ion beams apparatus. In these early experiments, an arc lamp was used as the light source. The absolute cross sections that were measured have been reviewed by Branscomb[3] and Smith.[4] A second generation of crossed-beams experiments began at JILA in the late 1960's and 1970's, when the conventional light source was replaced by a laser. This modification not only greatly enhanced available photon fluxes but also made it possible, for the first time, to measure photoelectron angular distributions with relative ease. Most of the experiments of this period were designed primarily, however, to measure electron affinities and determine relative changes in cross sections, particularly in the vicinities of thresholds and resonances. This work has been reviewed by Lineberger.[5] Somewhat suprisingly perhaps, only a few absolute photodetachment cross section measurements have been reported to date. Photodetachment cross sections are not only of intrinsic interest but they frequently form the most accessible path to the determination of cross sections for the far less probable process of radiative attachment.

The present paper describes the use of energy- and angle-resolved photoelectron detachment spectroscopy to determine partial cross sections (angular differential and integral) for the photodetachment of the metastable He$^-$ ion at a wavelength of λ = 696.2 nm. Brehm et al.[6] reported the first energy-resolved studies of the photodetachment of He$^-$. In this experiment, a crossed-beams technique, which employed a fixed-frequency laser, was used to determine the electron affinity of He(2^3S). As a by-product of these measurements, the investigators were also able to estimate, within a factor of two, the cross sections for photodetaching He$^-$ at λ = 514.5 nm. The large quoted uncertainties reflect the fact that no information on the photoelectron angular distributions was obtained in the experiment. Total cross sections for the photodetachment of He$^-$ (via two competing unresolved exit channels) have been reported by Compton et al.[7] and Hodges et al.[8] The cross section for photo-

© 1990 American Institute of Physics

detachment of B⁻ at λ = 696.2 nm has also been measured in the present work for the first time.

II. THEORETICAL CONSIDERATIONS

The He⁻ ion is unstable as a result of being formed in the spin-aligned $(1s2s2p)^4P$ state (see Fig. 1) when an electron is captured by an excited He $(1s2s\,^3S)$ atom. The quartet state, unable to decay via allowed radiative or autodetachment processes, autodetaches via forbidden processes induced by the relatively weak spin-dependent interactions. The varying strengths of these interactions result in a differential metastability among the fine structure levels. Lifetimes of the J = 1/2, 3/2, and 5/2 levels have been measured[9] to be 10, 16, and 500 μsec, respectively. Metastable He⁻ ions are therefore sufficiently long lived that they undergo very little exponential depletion by autodetachment prior to photodetachment in the present crossed-beams apparatus. Small corrections (~2%) to the measured yields are, however, made to account for the unstable nature of this ion.

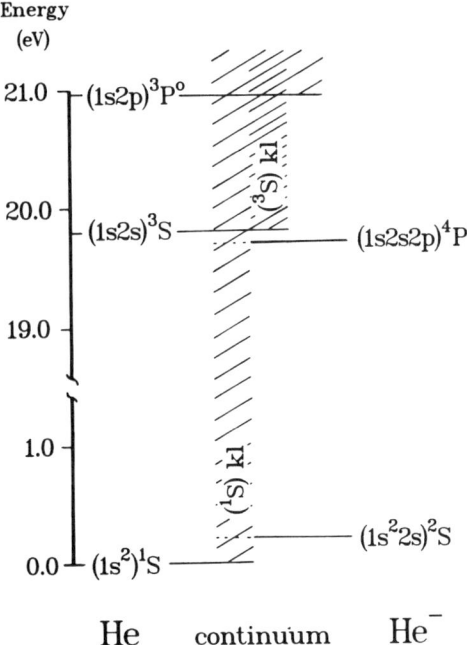

FIG. 1. Partial energy level diagram for the He and He⁻ systems. The He⁻ ion exists in the metastable $(1s2s2p)^4P$ state. The $(1s^22s)^2S$ ground state is unbound.

When the He⁻(2^4P) ion interacts with visible radiation, two competing photodetachment channels to the continuum are open:

$$h\nu + He^-(2^4P) \rightarrow He(2\,^3S) + e^-(ks,d) \quad (1)$$

$$h\nu + He^-(2^4P) \rightarrow He(2\,^3P) + e^-(kp). \quad (2)$$

In this paper, photodetachment processes will frequently be labeled by their exit channels. In an independent electron description of process (1), a p-orbital electron is ejected, resulting in an outgoing electron represented by s- and d-waves. In process (2), a s-orbital electron is ejected, leading to a pure outgoing p-wave in the independent electron model. The spectral dependences of the asymmetry parameters for both of these processes have recently been measured by Thompson et al.[10] Correlation effects are found to play a significant role in determining the shape of the electron emission patterns following photodetachment.

The B⁻ ion is stable, being formed in the $(1s^22s^22p^2)^3P$ state when an electron is captured by a ground state B(^2P) atom. At visible wavelengths, only one photodetachment channel is open when B⁻(^3P) interacts with radiation

$$h\nu + B^-(^3P) \rightarrow B(2\,^2P) + e^-(ks,d) . \quad (3)$$

Since a p-orbital electron is ejected, the outgoing electron is characterized by s- and d-partial waves in the independent electron model.

In the present work, the partial cross sections, $\sigma(^3S)$ and $\sigma(^3P)$, for the photodetachment processes shown in Eqns. (1) and (2) have each been measured relative to the cross section, $\sigma(^2S)$, for the photodetachment of D⁻$(1\,^1S)$ ions via the process

$$h\nu + D^-(1\,^1S) \rightarrow D(1\,^2S) + e^-(kp). \quad (4)$$

In this process, the emission is well described by an independent electron model and the photoelectron is represented by a pure p-wave resulting in a $\cos^2\theta$ distribution.

III. EXPERIMENTAL CONSIDERATIONS

The technique of energy- and angle-resolved photoelectron spectroscopy has been used to investigate the interaction between negative ions and radiation. Cross sections for photodetachment are determined from measurements of the yields and angular distributions of photoelectrons ejected from the ions at the interaction of crossed laser and negative ion beams. Fig. 2 is a schematic of the crossed-beams apparatus. A fast moving (v ~ 10⁸ cm/sec) and tenous beam (ρ ~ 10³ ions/cm³) of negative ions is intersected perpendicularly by an energy-resolved and linearly-polarized beam of photons from a pulsed dye laser. As a result of the photon-ion interaction, precisely known amounts of energy and angular momentum are transferred to the ion. Following photodetachment events, electrons ejected from the field-free interaction region in the direction of motion of the ion beam (forward-directed electrons) are collected, energy analyzed, and detected. The angular distribution of the detached photoelectrons can be determined by measuring their yield as a function of the angle between the electric field vector of the linearly polarized laser beam and the collection direction. This is achieved in the present apparatus by keeping the collection fixed in the forward direction and rotating the laser polarization vector, using a $\lambda/2$ phase retarder (double Fresnel rhomb). Fig. 3 shows the result of an angular distribution measurement. A typical photoelectron spectrum, taken in this case with the polarization vector parallel to the collection direction, is shown in Fig. 4.

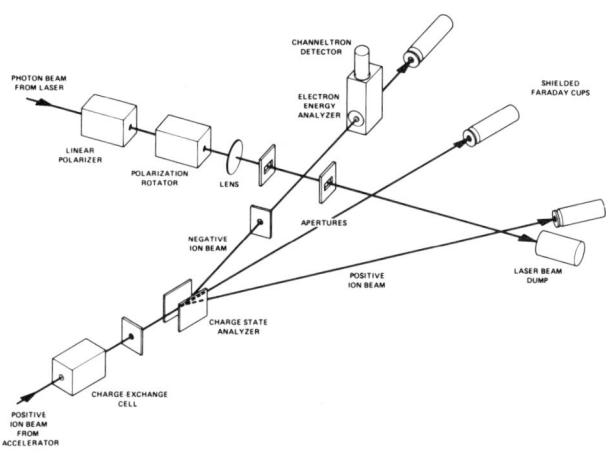

FIG. 2. A schematic view of the crossed laser-ion beams apparatus used in the energy- and angle-resolved photoelectron spectroscopy measurements.

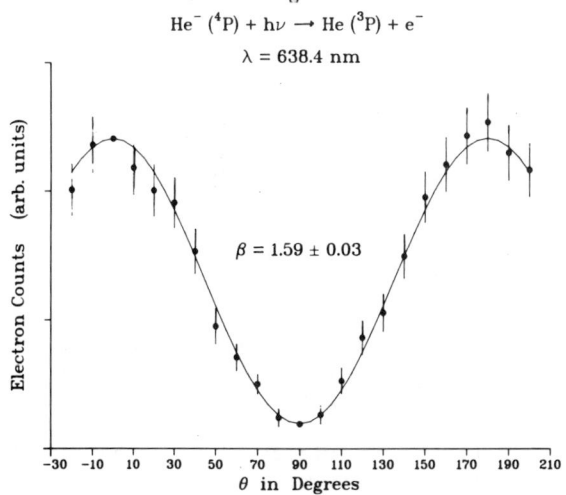

FIG. 3. Angular distribution of electrons photodetached from a 40 keV beam of He⁻ ion at $\lambda = 638.4$ nm.

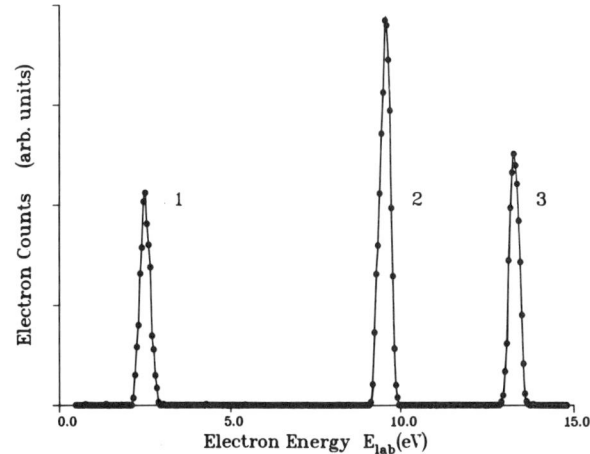

FIG. 4. Energy spectrum of electrons photodetached from a 40 keV beam of He⁻ ions at $\lambda = 689.5$ nm.

The beam that is the source of negative ions in these measurements is produced by double charge transfer (sequential electron capture) when a momentum-selected positive ion beam from an accelerator is passed through a Li vapor charge-exchange cell. After a delay of a few microseconds, the beam exiting the cell is charge state analyzed. The negative ion component ($\lesssim 1\%$) is then deflected by 10° into a beam line containing an electron spectrometer consisting of a spherical-sector energy analyzer and a channel-electron multiplier detector. The analyzer is operated in the fixed pass energy, preacceleration mode. About 2 cm in front of the entrance to the electron spectrometer, the ion beam is crossed with the laser beam. Sets of apertures are used to ensure that the overlap of the two beams remains unaltered during the measurements. The flashlamp-pumped pulsed dye laser used in this investigation has a maximum repetition rate of 10 Hz and a pulse duration of 2.2 μsec. The operating laser bandwidth was 0.2 nm. The output power of the laser was carefully chosen to avoid saturation on each of the photodetachment processes studied. A 10-cm focal length injection lens focussed the laser beam onto the ion beam in the interaction region. The whole apparatus is surrounded by three mutually orthogonal sets of Helmholtz coils designed to reduce the effect of stray magnetic fields on the electron trajectories.

The data is stored in a CAMAC-based multichannel analyzer data acquisition system. A synchronous detection scheme is used to extract the photoelectron signal from a rather large electron background. The time structure of the detection scheme is illustrated in Fig. 5. Four multichannel scalers (MCS) are used to record the data. MCS1 is triggered by the laser pulse via a fast photodiode detector. It is gated on for 4 μsec, after a 1 μsec delay, to record the electron signal associated with the laser being on. The delay is set to allow for the finite transit time of electrons through the spectrometer and to avoid noise associated with the "firing" of the laser. MCS2, which is operated with the same gate width as MCS1, is triggered on 2 μsec after MCS1 is turned off. The purpose of MCS2 is to record, with the laser off, the background associated with electrons generated by ion impact with apertures and, to a lesser extent, electrons produced by detachment following collisions between the ions and the residual gas. In the case of metastable ions, a small contribution arises from autodetachment. Subtraction of the contents of MCS1 and MCS2 allows one to effectively discriminate against the background contributions. MCS3 and MCS4 are used to monitor the ion and laser beam intensities for normalization purposes. The ion beam current is collected and integrated in a Faraday cup situated directly downstream of the electron spectrometer. The ion beam intensities were chosen so that approximately the same background of electrons was always incident on the detector. This made it less likely that pulse pile-up effects and count rate-dependent gain changes would affect the relative measurements. The intensity of the photon beam is recorded by monitoring the light scattered from the front face of the injection lens with a photodiode detector. Further details of the apparatus can be found in the paper by Pegg et al.[11]

As a consequence of the fast moving and unidirectional nature of the source, the kinetic energies, yields, energy distributions, and angular distributions of the photoelectrons ejected from the moving ions will be modified by kinematics. The combination of high-energy beams and forward-directed collection used in the present experiment permits one to effectively exploit kinematic effects such as spectral peak shifting and doubling. Previous cross section measurements using crossed beams have employed ion beam energies an order of magnitude lower than those used in the present work and collection has been perpendicular to the ion beam direction. The kinematically-shifted energy of a forward-directed ($\theta_L = 0$) electron in the laboratory frame, E_L, is related to its energy in the ion frame, E_C, via the relationship $E_L = ME_C$, where M represents a kinematic "magnification" factor, $M = (1 \pm \sqrt{\varepsilon/E_C})^2$. Here, ε is the energy of an electron moving with the same velocity as the ions of the beam. The positive sign in the factor M corresponds to emission of electrons in the forward direction ($\theta_C = 0$) in the ion frame and the negative sign corresponds to backward-directed ($\theta_C = \pi$) emission in the ion frame. If the ion beam velocity exceeds the electron velocity in the ion frame, both the forward and backward electrons will be swept forward in the laboratory frame. Kinematic peak doubling arises since E_L becomes double valued in this case. The separation of the twin peaks is $\Delta E_L = 4\sqrt{\varepsilon E_C}$. Details of these and other kinematic modifications have been discussed previously by Pegg et al.[12]

IV. RELATIVE CROSS SECTIONS

The angular integral cross section, $\sigma(x)$, for photodetachment of an ion via an exit channel labeled x can be expressed in terms of the yield of photoelectrons, $Y(x)$, ejected in the direction of the linear polarization vector of the light beam in the following manner:

$$\sigma(x) = \frac{4\pi Y(x)}{[1+\beta(x)]g(x)\phi(x)k(x)\rho(x)V\Delta\Omega GT} . \quad (5)$$

In this expression, β is the asymmetry parameter characterizing the shape of the emission pattern, g is the frame-transformed solid-angle ratio, k represents the photoelectron collection and detection efficiency, ρ is the ion beam density, ϕ is the photon flux, V is the interaction volume, $\Delta\Omega$ is the solid angle defined by the collection geometry, G measures the spatial overlap of the two crossed beams, and T is the integration time for each yield measurement. In the present work, relative cross sections are obtained by comparing, under identical geometrical conditions, the yields of electron produced in the photodetachment of the ions of interest, $Y(x)$, and a beam of reference ions, $Y(r)$, whose cross section is known. Thus the relative cross sections for the two photodetachment processes can be written as

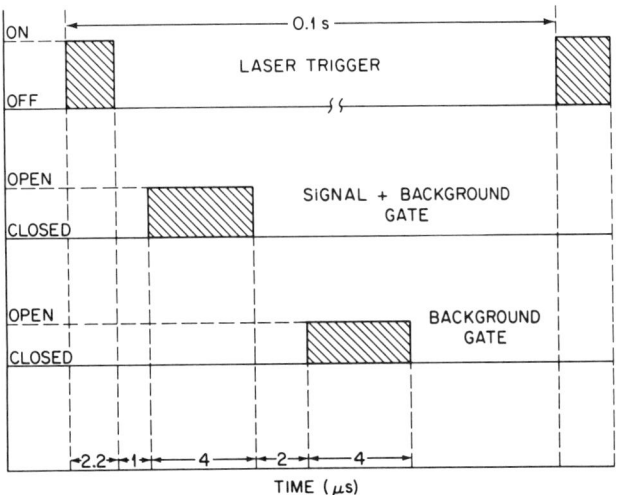

FIG. 5. The time structure of the synchronous detection scheme.

$$\frac{\sigma(x)}{\sigma(r)} = \frac{Y(x)\,\rho(r)\,\phi(r)\,g(r)\,[1+\beta(r)]}{Y(r)\,\rho(x)\,\phi(x)\,g(x)\,[1+\beta(x)]} \quad . \quad (6)$$

Here the measured yield ratio is multiplied by several measured factors in order to account for the different frame-transformed solid angles, photon fluxes, ion densities, and asymmetry parameters associated with photodetaching electrons from the two beams. All geometric factors, being equal, cancel out. In this work it is arranged by appropriate choice of beam energies, that the photoelectron peaks in each spectra are kinematically shifted so that they coincide in energy in the laboratory frame. The efficiency factors, $k(x)$ and $k(r)$, associated with the collection and detection of photoelectron at the same energy from the ions of interest and the reference ions will then be equal and hence cancel out. For example, Fig. 6 shows three photoelectron spectra in which the electron peaks all occur at ≈ 11 eV in the laboratory frame. The top and bottom peaks are associated with the photodetachment of He$^-$ via the He(^{3}S) and He(^{3}P) exit channels, respectively. The peak of the middle spectrum arises from the photodetachment of D$^-$ ions in the reference beam via the D(^{2}S) exit channel. To ensure that the electrons from these three different photodetachment channels are analyzed at the same energy in the laboratory frame, the ion beam energies for the top, middle, and bottom spectra were chosen to be 29.19, 19.17, and 50.47 keV, respectively.

The ion beam densities, ρ, and the solid-angle transformation factors, g, both depend on the velocity of the ion beam. This quantity can be determined very precisely ($\sim 0.1\%$), <u>in situ</u>, by simply analyzing the photoelectron spectra. In Fig. 4, the backward-directed peak 1 ($\theta_c = \pi$) and the forward-directed peak 2 ($\theta_c = 0$) form a kinematically-doubled pair, both being associated with the photodetachment process that leaves the He atom in the $2\,^3$P state. Their separation, <u>in the laboratory frame</u>, is $\Delta E_L(12) = 4\sqrt{\varepsilon E_c(1)}$. The quantity E_c, the energy of the electron in the ion frame, can be determined from a measurement of the photon wavelength and the use of the precise theoretical value of the electron affinity of He($2\,^3$S) calculated by Bunge et al.[13] Thus by measuring $\Delta E_L(12)$, one can determine the reduced ion beam energy, ε, and hence the ion beam energy itself. Another way of determining the ion beam energy is by measuring the separations of peaks 2 and 3. Peak 3 is due to the forward-directed electrons associated with the competing process via the He($2\,^3$S) exit channel. In the ion frame, peaks 2 and 3 are separated by the $2\,^3$S $- 2\,^3$P transition energy, E_e, known precisely from optical spectroscopy. The separation of peaks 2 and 3 in the laboratory frame is given by

$$\Delta E_L(23) = E_e + 2\sqrt{\varepsilon}\left[\sqrt{E_c(2)+E_e} - \sqrt{E_c(2)}\right]. \quad (7)$$

In a similar manner, the velocities of beams of other ions, such as D$^-$ and B$^-$, can be precisely determined by measuring the separation between the peaks of their photoelectron spectra and the peaks of the He$^-$ spectrum.

Angle-resolved measurements are made to determine the asymmetry parameter, β. The yield of a photoelectron peak is measured as a

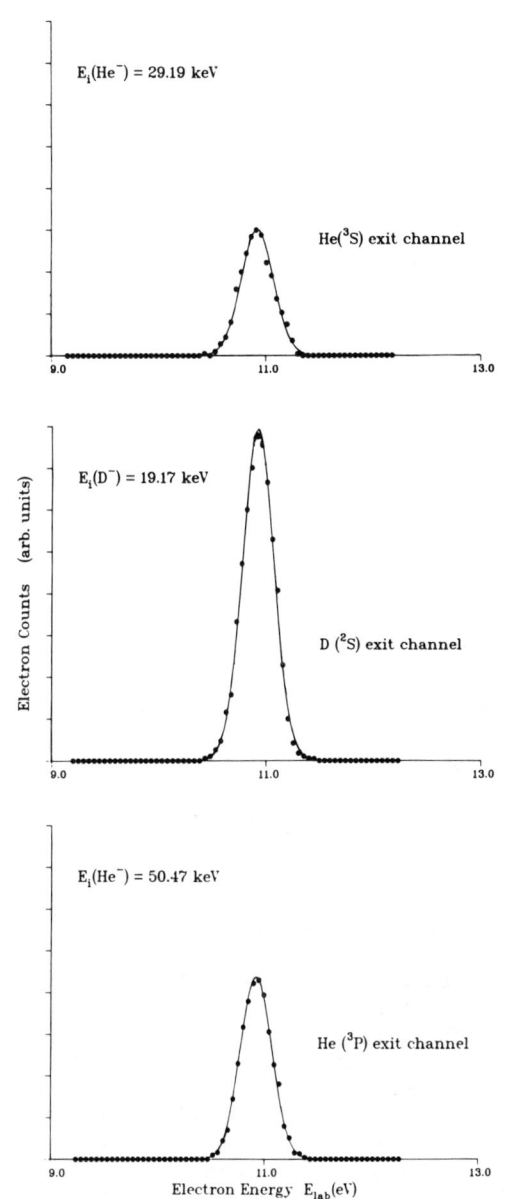

FIG. 6. Three spectral peaks kinematically shifted to the same laboratory-frame energy by an appropriate choice of ion beam energy. The peak yields were used in the measurement of relative cross sections for photodetachment of He$^-$ and D$^-$ ions.

function of the angle, θ, between a fixed collection direction (in the present case, the direction of motion of the ion beam) and the variable direction defined by the electric field vector of the linearly polarized laser beam. For the case of plane polarized radiation in the electric dipole approximation (assuming an independent electron model), the angular correlation between the outgoing photoelectron and the incoming photon should take the form, $f(\theta) = 1 + \beta P_2(\cos\theta)$. Here $P_2(\cos\theta)$ is the second-order Legendre polynomial. Multiphoton absorption or an ion state that is prepared anisotropically prior to photoabsorption would produce higher (even) order polynomials in the fit. It has been demonstrated that both processes make a negligible contribution in the present work. The angular distribution measurement technique has been tested, at several different wavelengths in the visible, by photodetaching beams of D⁻ and Li⁻ ions. The predicted $\cos^2\theta$ ($\beta = 2$) distribution was obtained in all cases.

Angular differential cross sections, $I_c(\theta)$, can most easily be determined from the measured angular integral cross sections, σ, and the asymmetry parameters, β, using the expression

$$I_c(\theta) = \frac{\sigma}{4\pi}[1 + \beta P_2(\cos\theta)] . \qquad (8)$$

V. RESULTS

A. Partial Cross Sections for He⁻ Photodetachment

The partial (angular integral) cross sections, $\sigma(^3S)$ and $\sigma(^3P)$, for photodetaching electrons from He⁻(2^4P) via the He(2^3S) and the He(2^3P) exit channels, respectively, have been measured relative to the cross section, $\sigma(^2S)$, for photodetaching D⁻ via the D(1^2S) exit channel. The ratio of the $\sigma(^3S)$ to $\sigma(^3P)$ cross sections has also been measured. At a wavelength of λ = 696.2 nm, the following results were obtained: $\sigma(^3S)/\sigma(^2S) = 0.60 \pm 0.02$, $\sigma(^3S)/\sigma(^3P) = 2.28 \pm 0.10$, and $\sigma(^3P)/\sigma(^2S) = 0.27 \pm 0.02$. The uncertainties quoted on these values are at the level of two standard deviations of the weighted mean obtained from about ten data sets, each consisting of a number of individual spectra. The major source of statistical error in the ratios obtained from the individual data sets stems from measurements of the photoelectron yield (~4%) and asymmetry parameter (~3%) ratios. Uncertainties in the measurement of other factors in Eqn. (6) are negligibly small. Systematic errors are also estimated to be within the quoted limits. The asymmetry parameters characterizing the photoelectron angular distributions have also been measured at λ = 696.2 nm. The results are shown in Table I along with the angular integral cross section results. An absolute scale for the He⁻ partial cross sections has been established by assuming a theoretical value for the $\sigma(^2S)$ cross section. The calculations of Stewart,[14] for example, which are estimated to have an uncertainty of better than 3%, produce a result of $\sigma(^2S) = 38.1$ Mb at λ = 696.2 nm. Broad and Reinhardt[15] calculate a value within 1% of this result. Combining the theoretical result (assuming a 3% uncertainty) with the measured ratios produces the partial (angular integral) cross sections shown in Table I. Table I indicates that at a wavelength of 696.2 nm, the probability that a given He⁻ ion photodetaches into the He(^3S) + e⁻(ks,d) continuum is 70% (i.e. the branching ratio, $R(^3S) = 0.70 \pm 0.06$). At the present time, there are no other experimental or theoretical results to compare with these measured partial cross sections. The sum of the partial cross sections, σ, can, however, be compared to the total cross section calculations of Hazi and Reed.[16] The scatter plot shown in Fig. 7 indicates that there is excellent agreement between the present result of $\sigma = \sigma(^3S) + \sigma(^3P) = 32.9 \pm 2.4$ Mb and the theoretical result of σ = 33.4 Mb (length form) and σ = 29.9 Mb (velocity form). Also shown in Fig. 7 are the experimental results of Compton et al.[7] and Hodges et al.[8] These less precise values are also seen to be in agreement with the present result.

The angular integral cross sections, σ, and the asymmetry parameters, β, can be conveniently combined, using Eqn. (7) to determine the angular differential cross sections, $I_c(\theta)$. The results are shown in Fig. 8.

TABLE I. Measured Angular Integral Photodetachment Cross Sections and Asymmetry Parameters at λ = 696.2 nm.

Exit Channel	σ(Mb)	β
He(^3S)	22.9 ± 1.0	1.15 ± 0.02
He(^3P)	10.0 ± 0.6	1.52 ± 0.04
B(^2P)	16.5 ± 2.1	0.18 ± 0.04
D(^2S)	38.1 ± 1.1[a]	2.00 - 0.02

[a]Reference 14.

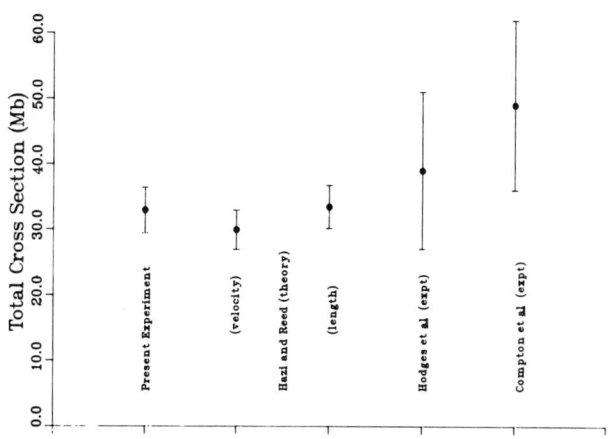

FIG. 7. Scatter diagram showing different experimental and theoretical results for the total (angular integral) cross section for photodetachment of the He⁻ ion at λ = 696.2 nm.

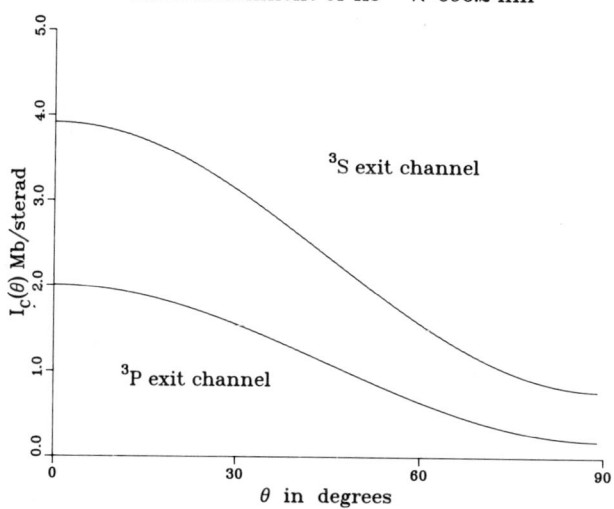

FIG. 8. Measured angular differential cross sections for the photodetachment of the He⁻ ion at λ = 696.2 nm.

B. Cross Section for B⁻ Photodetachment

The cross section, $\sigma(^2P)$, for the photodetachment of B⁻ via the B($2\,^2P$) exit channel has been measured relative to the cross section, $\sigma(^3P)$, for the photodetachment of He⁻ via the He($2\,^3P$) exit channel at a wavelength of λ = 696.2 nm. The B⁻ and He⁻ ion beam energies were chosen to be 51.04 and 30.18 keV, respectively, so that the peaks in the photoelectron spectra coincided in energy in the laboratory frame. This ensured that photoelectrons from the two beams were collected and detected with equal efficiency. The result of the measurement is $\sigma(^2P)/\sigma(^3P) = 1.65 \pm 0.19$. The asymmetry parameter for this process has been measured and the result is shown in Table I along with angular integral cross section. In this case, an absolute scale for the $\sigma(^2P)$ cross section has been established by assuming the value of the $\sigma(^3P)$ cross section that had previously been measured relative to the $\sigma(^2S)$ cross section of D⁻, i.e. $\sigma(^3P) = 10.0 \pm 0.6$ Mb. There are no other existing results, experimental or theoretical, to compare with the present measurement.

Again, the angular differential cross section, $I_c(\theta)$, is conveniently determined from the measured angular integral cross section, σ, and the asymmetry parameter, β. The results are shown in Fig. 9.

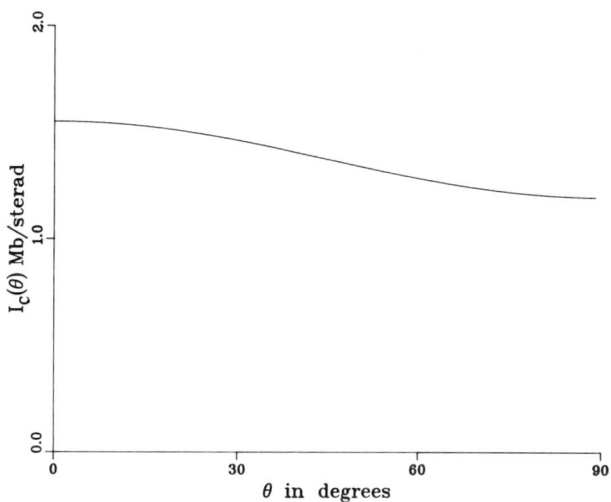

FIG. 9. Measured angular differential cross section for photodetaching the B⁻ ion at λ = 696.2 nm.

C. Radiative Attachment Cross Sections

The measured cross sections for photodetaching He⁻ and B⁻ ions can be used to indirectly determine, using detailed balance arguments, the cross sections, σ_a, for the far more improbable inverse process of radiative attachment. The processes are the inverse of those shown in Eqns. (1), (2), and (3). The derived radiative attachment cross sections (in barns) at λ = 696.2 nm are shown in Table II.

TABLE II. Derived Radiative Attachment Cross Sections at λ = 696.2 nm.

Entrance Channel	σ_a(b)
He(^{3}S)	166 ± 7
He(^{3}P)	74.2 ± 4.5
B(^{2}P)	51.0 ± 6.5

VI. SUMMARY

A crossed-beams apparatus that employs energy- and angle-resolved photoelectron spectroscopy has been used to measure angular differential and integral cross sections for the photodetachment of He$^-$ and B$^-$ ions at a wavelength of 696.2 nm. An absolute scale has been established by making relative photoelectron yield and angular distribution measurements between beams of the ions of interest and D$^-$ reference ions whose photodetachment cross section is known from theory to better than 3%.

Absolute cross sections are difficult to determine accurately. The relatively high precision (~5% in favorable cases) of the present measurements reflects our ability to exploit, in a novel way, certain kinematic effects associated with a fast moving source of ions and the collection of ejected photoelectrons in the forward direction. The measurements are currently being continued at several discrete wavelengths across the visible spectrum. It is hoped that these new results will stimulate further calculations of photodetachment cross sections and asymmetry parameters.

ACKNOWLEDGMENTS

I wish to acknowledge my fellow collaborators, past and present, at the University of Tennessee (J. S. Thompson and J. Dellwo) and the Oak Ridge National Laboratory (G. D. Alton, R. N. Compton, and T. J. Kvale). Special thanks go to Jeff Thompson. This research was supported by the U.S. Department of Energy, Office of Basic Energy Sciences, Division of Chemical Sciences through the University of Tennessee. Oak Ridge National Laboratory is operated by Martin Marietta Energy Systems, Inc. under Contract No. DE-AC05-84OR21400 with the U.S. Department of Energy.

REFERENCES

1. K. Dolder, in Electronic and Atomic Collisions (invited papers of the XV ICPEAC), ed. by H. B. Gilbody, W. R. Newell, F. H. Read, and A.C.H. Smith (North-Holland, Amsterdam, 1988), p.549.
2. L. M. Branscomb and W. L. Fite, Phys. Rev. 93, A651 (1954).
3. L. M. Branscomb, in Atomic and Molecular Processes, ed. by D. Bates (Academic, New York, 1962), p. 100.
4. S. J. Smith, in Methods of Experimental Physics, ed. by B. Bederson and W. Fite (Academic, New York, 1968), Vol. 7, p. 179.
5. W. C. Lineberger, in Applied Atomic Collision Physics, ed. by H.S.W. Massey, E. W. McDaniel and B. Bederson (Academic, New York, 1982), Vol. 5, p. 239.
6. B. Brehm, M. A. Gusinow, and J. L. Hall, Phys. Rev. Lett. 19, 737 (1967).
7. R. N. Compton, G. D. Alton, and D. J. Pegg, J. Phys. B 13, L651 (1980).
8. R. V. Hodges, M. J. Coggiola, and J. R. Peterson, Phys. Rev. 23, 59 (1981).
9. L. M. Blau, R. Novick, and D. Weinflash, Phys. Rev. Lett. 24, 1269 (1970).
10. J. S. Thompson, D. J. Pegg, R. N. Compton, and G. D. Alton, J. Phys. B (to be published).
11. D. J. Pegg, J. S. Thompson, R. N. Compton, and G. D. Alton, Phys. Rev. Lett. 59, 2267 (1987).
12. D. J. Pegg, J. S. Thompson, R. N. Compton, and G. D. Alton, Nucl. Instrum. Meth. B 40/41, 221 (1989).
13. A. V. Bunge and C. F. Bunge, Phys. Rev. A 19, 452 (1979).
14. A. L. Stewart, J. Phys. B 11, 3851 (1978).
15. J. T. Broad and W. P. Reinhardt, Phys. Rev. A 14, 2159 (1976).
16. A. U. Hazi and K. Reed, Phys. Rev. A 24, 2269 (1981).

DOUBLE PHOTOIONIZATION NEAR THRESHOLD

V. Schmidt

Fakultät für Physik, Universität Freiburg, 7800 Freiburg, FRG

Experimental and theoretical investigations of double photoionization in rare gases brought a deep understanding for the highly correlated motion of the atomic electrons. A straightforward but challenging extension is provided by investigation of the double photoionization process in the simplest molecular two-electron system H_2. Very recent experiments on H_2 yield an interesting view of the angular distributions in the four-particle break-up of H_2 and give a quite consistent picture for the single and double ionization cross sections of H_2 and its atomic counterpart He. However, the mechanisms underlaying the described observations still await theoretical clarification.

In the last decades, much attention has been directed to such photoinduced two-electron transitions in free atoms as double excitation, ionization with excitation and double ionization, because they are strongly influenced by the correlated motion of the electrons in the initial and final state of the photoprocesses. Naturally, helium was a unique candidate for the study of such fundamental processes, and considerable progress has been achieved due to the stimulating interplay between theory and experiment. Of particular interest in the present context are three phenomena related to the near threshold double photoionization in helium: First, the magnitude of the double photoionization cross section itself,[1-7] second, its energy dependence in the vicinity of the threshold[8-13] and third, the angular distribution for photoelectron emission leaving the helium ion in a highly excited state.[14-18]

To extend the knowledge obtained from the studies on helium, two directions can be followed: i) investigations in other atoms, e.g. in the p shells of rare gases where double and even higher ionization processes occur and different fine structure components of the remaining ion state have to be taken into account (for recent experiments compare Ref. [19-21]). ii) Analogous studies in non-atomic systems where the hydrogen molecule as the simplest molecular two-electron system provides again an ideal case to start with.

The present communication deals with double photoionization in H_2 studied by angle-resolved detection of the pair (H^+, H^+) of charged fragment ions in coincidence. Of special interest are the magnitude of the double photoionization cross section, the angular distribution of the electron-pair released from the double ionization and of the proton-pair from the dissociation process and, more generally, the comparison of ionization processes in the molecular two-electron system H_2 with similar ionization processes in the atomic counterpart helium. These points are discussed in a forthcoming publication[22] and are related to two other articles.[23,24] In the present communication only the results shall be summarized.

One of the mentioned publications[22] considers double photoionization processes in H_2 and presents two unexpected results. Firstly, the double photoionization cross section $\sigma(H_2^{++}) = \sigma(H^+,H^+)$ exposes a remarkable discrepancy to data which could be expected on theoretical grounds.[25,26] Such a conclusion could not be drawn from a former investigation[27] because of the larger experimental uncertainty. Secondly, the fragment ions show a pronounced non-isotropic angular distribution which was not observed before. Towards the ionization threshold, the proton-pair from the dissociation process flies apart preferentially perpendicular to the electric vector of the light (sideways dissociation). Transferring the sideways double-electron emission results of helium[14,15] to H_2, one arrives at a situation which is close to a proposed[28] angular distribution pattern in the threshold region: sideways and orthogonal emission of the proton and electron pair in the four-particle break-up.

The second publication[23] describes the determination of the single photoionization cross section $\sigma(H_2^+)$ leading to H_2^+ in its ground state. Special attention was given to disturbances of the ion signal caused by impurities of the monochromatized synchrotron light (higher order components and stray light contributions). The absolute $\sigma(H_2^+)$ and $\sigma(H^+,H^+)$ values then allow a direct comparison with the corresponding single and double photoionization cross sections of helium: Whilst the shape of the cross sections $\sigma(H_2^+)$ and $\sigma(He^+)$ is quite different, the ratios of double to single photoionization are practically the same in both cases.

Finally, the third publication[24] concerns electron impact data for single and double ionization in H_2 where the absolute cross sections $\sigma_e(H_2^+)$ and $\sigma_e(H^+,H^+)$ were determined. Special emphasis was given to necessary calibration procedures needed also for the photoionization experiment, e.g. information on the transmission and detection efficiency of the ion analyzer. Our $\sigma_e(H_2^+)$ cross section values are in excellent agreement with data from Edwards et al.,[29] but our $\sigma_e(H^+,H^+)$ values are higher by approximately a factor of 8. As a consequence, our ratio of double to single ionization in H_2 gets close to the corresponding ratio value of helium. This result is similar to and can be expected from the observation obtained for photon impact.

The presented experimental data provide already a detailed and consistent view on double photoionization in H_2 with its four-particle break-up. It is hoped that they will stimulate further theoretical work on this fundamental two-electron, two-center system. Eventually, an understanding of the underlying dynamics may be achieved as it is reached already for the atomic counterpart helium.

It is a pleasure for me to thank the members of my research-group, especially H. Kossmann, for their great engagement in the described experiments. The financial support by the German Federal Minister for Research and Technology (BMFT) under Contract No. 05372AA and 05472AA is gratefully acknowledged

References
1. V. Schmidt, N. Sandner, H. Kuntzemüller, P. Dhez, F. Wuilleumier and E. Källne, Phys.Rev. A13, 1748 (1976).
2. G.R. Wight and M.J. Van der Wiel, J.Phys. B9, 1319 (1976).
3. D.M.P. Holland, K. Codling, J.B. West and G.V. Marr, J.Phys. B12, 2465 (1979).
4. F.W. Byron and C.J. Joachain, Phys.Rev. 164, 1 (1967).
5. R.L. Brown, Phys.Rev. A1, 586 (1970).
6. S.L. Carter and H.P. Kelly, Phys.Rev. A24, 170 (1981).
7. S.N. Tiwary, J.Phys. B15, L323 (1982).

8. G.H. Wannier, Phys.Rev. 90, 817 (1953).
9. R. Peterkop, J.Phys. B4, 513 (1971).
10. A.R.P. Rau, Phys.Rev. A4, 207 (1971).
11. M.J. Van der Wiel, Phys.Lett. 41A, 389 (1972).
12. H. Kossmann, V. Schmidt and T. Andersen, Phys.Rev.Lett. 60, 1266 (1988).
13. G.C. King, M. Zubek, P.M. Rutter, F.H. Read, A.A. MacDowell, J.B. West and D.M.P. Holland, J.Phys. B21, L408 (1988).
14. C.H. Greene, Phys.Rev.Lett. 44, 869 (1980).
15. C.H. Greene, J.Phys. B20, L357 (1987).
16. H.R. Sadeghpour and C.H. Greene, Phys.Rev. A39, 115 (1989).
17. P.A. Heimann, U. Becker, H.G. Kerkhoff, B. Langer, D. Szostak, R. Wehlitz, D.W. Lindle, T.A. Ferrett and D.A. Shirley, Phys.Rev. A34, 3782 (1986).
18. D.W. Lindle, P.A. Heimann, T.A. Ferrett and D.A. Shirley, Phys.Rev. A35, 1128 (1987).
19. P. Lablanquie, J.H.D. Eland, I. Nenner, P. Morin, J. Delwiche and M.J. Hubin-Franskin, Phys.Rev.Lett. 58, 992 (1987).
20. J.A.R. Samson and G.C. Angel, Phys.Rev. Lett. 61, 1584 (1988).
21. S.D. Price and J.H.D. Eland, J.Phys. B22, L153 (1989).
22. H. Kossmann, O. Schwarzkopf, B. Kämmerling and V. Schmidt (1989) submitted to Phys. Rev.Lett.
23. H. Kossmann, O. Schwarzkopf, B. Kämmerling, W. Braun and V. Schmidt, accepted for publ. in J.Phys. B (1989).
24. H. Kossmann, O. Schwarzkopf and V. Schmidt (1989) submitted to J.Phys. B.
25. H. LeRouzo, J.Phys. B19, L677 (1986).
26. H. LeRouzo, Phys.Rev. A37, 1512 (1988).
27. G. Dujardin, M.J. Besnard, L. Hellner and Y. Malinovitch, Phys.Rev. A35, 5012 (1987).
28. J.M. Feagin and R.E. Stevens, to be published.
29. A.K. Edwards, R.M. Wood, A.S. Beard and R.L. Ezell, Phys.Rev. A37, 3697 (1988).

Ions and Atoms

THEORETICAL COLLISION PHYSICS OF HIGHLY CHARGED IONS

Anders Bárány

Manne Siegbahn Institute of Physics, S-104 05 Stockholm, Sweden

We review the main aspects of some of the theoretical models that have been developed during the last years to describe highly charged ion-atom collisions in the low-energy regime. These models are the products of a fruitful mixing of experiment and theory and provide us with a general, albeit still only rough, picture of many-body processes still to be described in detail by more elaborate physical theories.

1. Introduction. Collision processes involving highly charged ions have been observed and studied with varying precision and intensity since the beginning of this century. At that time Thomson[1] observed traces of mercury ions of charge states as high as eight capturing from one to seven electrons on their passage through a cathode ray tube. Having already formulated a successful theoretical model for electron-impact single ionization, he now went on to discuss electron capture using a rough velocity-matching argument based on this model. He also argued that the highly charged ions must have been produced in head-on collisions between singly charged ions and mercury atoms, in what in modern language could be called a shake-off process. But since the acceleration voltages only were of the order of 1 kV, that particular conclusion seems somewhat doubtful. Instead electron-impact multiple ionization probably was the mechanism active in creating the eight times ionized mercury atoms.

In the wake of the discovery of nuclear fission at the end of the 1930's, a number of questions concerning the interaction of highly charged fission fragments with matter needed to be studied. Since the fragments typically were created inside solid matter, direct observation of the primary processes of excitation, ionization and capture was difficult. That so many of the relevant phenomena still could be described in simple formulae by Bohr[2] attests to the usefulness and predictive power of theoretical modelling, so far as it is based on a sound understanding of the relevant physics.

During the 1950's a number of studies on total cross sections for electron capture by charge- and energy-selected beams of highly charged ions probed in more detail processes of the type observed by Thomson[1]. Many of these studies were summarized by Hasted[3] and Fedorenko[4] at a 1959 ICPIG (not to be mixed up with the first ICPEAC!) conference in Uppsala, Sweden. Production of beams of krypton ions in charge states up to seven were reported by Hasted[3], who also found indications of double capture and transfer ionization, i.e. capture plus ionization, in collisions between the krypton ions and a neon target. Fedorenko[4] showed total cross sections for rare gas ions with somewhat lower charges capturing up to three electrons from a selection of rare gas targets. He also reported low-resolution angular differential cross sections associated with such captures. (That summer, totally unaware of the magic of highly charged ion-atom collisions, the author of this review paper was a junior assistant in the workshop of the Physics Department, about 500 m away from the conference!)

The theoretical interpretations used in those days at first relied heavily on a modified Massey-parameter argument, but were later based on rather sophisticated two-state Landau-Zener calculations using quantum defect wave functions as developed by Bates, Moiseiwitsch, Dalgarno and co-workers[5]. Good agreement with experiments were only found after adjusting the locations of the cross section maxima to the energy regions determined experimentally.

The importance of highly charged ions in astrophysical and fusion plasmas led during the 1970's to an increasing interest from the collision community, eventually reflected in a series of specialized ICPEAC symposia starting 1979 in Kyoto. At that time, among others, Olson[6] and Salzborn[7] reviewed electron capture from one- and multielectron targets, respectively, with special emphasis on scaling relations. A broad general theory review by Janev and Presnyakov[8] appeared in 1981 (eventually to become extended to a monograph[9]) and at the 1981 ICPEAC in Gatlinburg[10] there were actually two symposia on highly charged ions, one for electron-ion and one for ion-atom collisions. Also reflecting the increasing interest in highly charged ions, a biennial series of topical conferences was started through the organizing of the first one in Stockholm 1982. At that conference around 50 invited and contributed papers were presented, among them Donets[11] description of an EBIS capable of producing helium-like xenon ions (Xe^{52+}). Taken together these three conferences in 1979, 1981 and 1982 together with the monograph by Janev et al [9] give a rather complete picture of the field at the beginning of this decade, and form a convenient starting point for the present review.

Within the last few years there has been a dramatic development in the field of ion-atom collisions with highly charged ions, a development which in its experimental aspects in a sense parallels that of general heavy particle collisions in the 1960's and 70's. Techniques then perfected, such as coincident-charge-state analysis, energy-gain spectroscopy and high-resolution angular differential measurements, all have their counterparts with highly charged ions in the 1980's. There were also a number of successful theoretical methods for general heavy particle collisions developed at that time, as the new data became available. These methods have to a certain extent survived the projection onto the highly charged ion-atom collision system and some are still used today.

In the present review I will try to describe some of the developments that have been made in recent years, not so much within pure theory as in the theoretical understanding of the important physical mechanisms that makes it possible for us to predict with some certainty the *gross* outcome of encounters between highly charged ions and atoms. I stress the word "gross" because there have been quite a lot of "surprises" also (usually found in the Hot Topic sessions of the ICPEAC). But such surprises are, of course, what really makes physics interesting!

The scope of the review will be restricted to low-energy collisions, since higher energies were well covered at the last ICPEAC by Olson[12]. "Low-energy" is here somewhat loosely defined as the domain where the kinetic energy carried by the projectile is of the order of its recombination energy or smaller. (The recombination energy for an ion of charge state q is taken as the sum of the first q ionization potentials.) In this energy regime potential effects start to dominate over kinetic ones, which means that quasi-static theoretical arguments usually suffice to interpret the experimental results. Very often the terminology "slow collisions" is used, in the sense that collision velocities are much smaller than the active electronic velocities, but since this terminology also includes inner-shell phenomena at higher energies, which are not reviewed here (see Schuch[13] for an excellent review), we will stick to "low-energy". Fig.1 shows the recombination energy of Xe^{q+} as function of charge state q. Also shown is a typical kinetic energy favoured by many experimentalists, $1q$ keV. We note that the kinetic energy dominates up to a rather high charge state. If we define a parameter d as the ratio between recombination

energy and kinetic energy, the collision systems discussed here typically have d in the range 0.1-1.5. What is meant by "highly charged" usually depends to a large extent on the person using the expression, but I will try to justify its use in this review by the choice of collision systems.

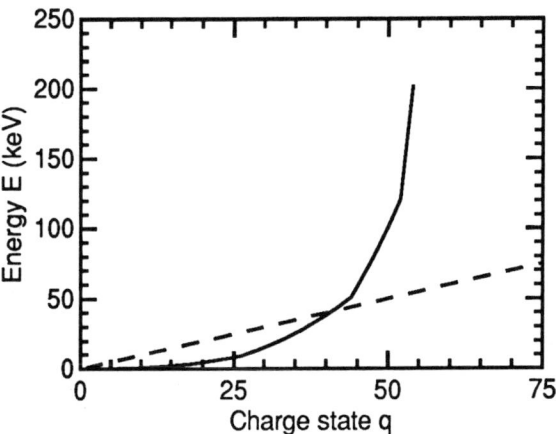

Figure 1. The recombination energy of Xe^{q+} (full-drawn) and the energy $1q$ keV as functions of charge state q.

2. Capture and ionization Within the present decade a number of groups have measured cross sections σ_{qr}^{0s} for the general reaction

$$A^{q+} + B \rightarrow A^{r+} + B^{s+} + (r+s-q)e^- \quad (1)$$

through coincident charge state analysis of the highly charged "fast" projectiles and "slow" (recoil) product ions in the low-energy domain.

First out with the new technique was the KSU group[14], using recoil ions, the Giessen group[15], using an EBIS, and the Stockholm-Aarhus collaboration[16], again using recoil ions. Eventually a rather large amount of absolute cross section data has become available[17,18] and is still being extended towards higher charge states[19]. This data base has been very important in providing a testing ground for theoretical models, through which a first order understanding of the dominating reaction mechanisms has been reached.

Since up to around five electrons have been observed to leave the target even in these low-energy collisions, the theoretical method used to describe the collisions cannot for practical reasons be a quantum mechanical *ab initio* theory nor is any known perturbation scheme of practical use. Instead concepts from classical and semiclassical mechanics have come into the foreground and have led to the formulation of manageable collision models and a framework in which we can grasp the experimental results.

Before concentrating on this framework we note in passing the two statistical theories presented at the XIII ICPEAC by Åberg[20] and Müller[21]. These give charge state fractions of the recoil ions that agree well with experiments, but the physical picture given for the production of the recoils (a target left in a state of multiple excitation from which electrons are ejected through Auger processes) seems not to be corroborated by observations of Doppler shifts in electron spectra from many-electron processes, as is evident, e.g., in the recent work of the Toulouse group[22].

Formulating the classical equations of motion for a non-relativistic many-body system is not very difficult[23], what is difficult is of course to choose the initial conditions properly, solve the equations and interpret the solutions evolving from the chosen conditions. At first glance the nCTMC method as developed and applied to reaction (1) by Olson seems to be ideal[12], but for computational reasons it is so far only practical at higher energies (or when the collision time is smaller than or of the order of the relevant electronic periods[24]). As so many times before in the history of physics, we have to rely on qualitative model arguments to be confronted with experimental results. Even with supercomputers around it still seems to be more feasible to do the experiments than the *ab initio* calculations - a good reason for further experiments!

Taking into account that the relevant collision energies here typically are of the order of a few keV per nucleon, it is clear that a quasimolecular description of the collision system is the adequate one, i.e. that to a first approximation the electronic and nuclear motions can be adiabatically decoupled. It is also clear that the timescales are such that decay of the highly excited transient quasimolecule generally does not occur during the actual collision (which could be the case at lower energies), but only after separation of projectile and target. Except in the well localized regions of non-adiabaticity that drive the reaction (1), the electronic and nuclear motions can thus be treated separately in the ordinary Born-Oppenheimer way.

Considering first the electronic motion, the general physical guiding principle is the one embodied in the so called classical overbarrier model for one-electron targets. For a reaction to occur in this model, the projectile has to come at least so close to the target that the potential barrier separating the target potential well from the projectile one does not hinder the classical trajectories of the electron to wander into the latter. Since the electron is moving around rapidly, there will always be time for it to find its way into the projectile well. Having once done so it will then oscillate back and forth during the collision time in the quasimolecular double-well potential formed by the two positively charged cores.

When these cores start to move away from each other the probability that the electron is left on the projectile will just equal the fraction of time spent in the projectile well. In early applications of the model to symmetric charge transfer[25], this fraction was taken to be 0.5, a factor which at first was used even for highly charged ions[26], but which was later changed simply to unity[27]. Considering that the period for a classical Kepler orbit at fixed energy scales as the force constant, the probability factor is just $q/(q+1)$, a factor which rapidly becomes insignificantly close to unity as the charge q increases. There is also the question about the quantization of the energy of the captured electron, which may be used to infer[26] a cross section with an oscillatory q-dependent structure. But since the final capture states in highly charged ions typically are Rydberg states with n-values of the order of the charge state q, the oscillations are damped out for high charge states q and a simple continuum approximation[27] is generally appropriate.

For the nuclear motion the dominating force will be a weak attractive polarization on the way into the collision followed, for those trajectories which lead to electron capture, by a strong Coulomb repulsion on the way out. For simplicity the polarization force is neglected, leaving a straight line trajectory before the collision. Thus a simple parallel projection of the overbarrier critical internuclear distance onto the impact parameter plane leads to the geometrical cross section πR_1^2, where the critical radius R_1 is given by

$$R_1 = (2q^{1/2} + 1)/I_1, \qquad (2)$$

and where I_1 is the ionization potential of the target. (The relevant formulae for the one-electron case have been collected by Bárány and Danared[28].)

For a true classical adiabatically treated one-electron system the effective overbarrier radius is diminished because of symmetry conditions[9] not allowing the classical trajectory of the electron to traverse the barrier at the saddle point with unlimited small kinetic energy. The breaking of these symmetries by, e.g., a second electron also would lead to the possibility of chaotic electron trajectories. Neglecting tunneling phenomena both effects would imply that the cross section should be depleted in true one-electron systems, an effect which has yet to be seen in highly charged ion collisions with hydrogen atoms.

Before discussing multi-electron processes it would be nice to compare the model described above with some experimental measurements. Because of the scarcity of cross section data for highly charged ions colliding true atomic hydrogen targets, the comparison will be made with cross sections taken with He targets. An important point then has to be taken into account, namely that transfer ionization (two-electron capture followed by autoionization) may lead to a projectile which apparently has captured just one electron. This process, and also double capture into bound states, will compete with single capture in sharing the intensity of the incoming beam.

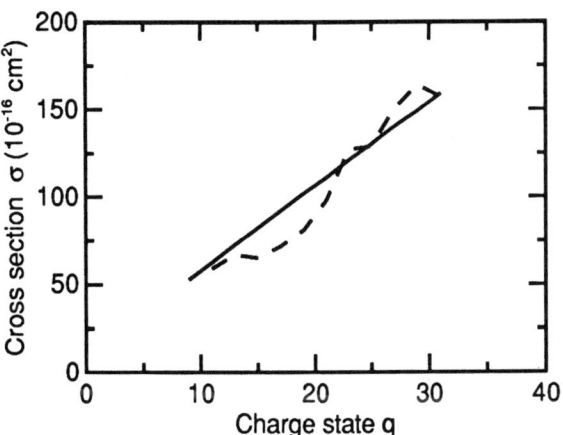

Figure 2. Absolute cross sections for $4q$ keV Xe^{q+} capturing one or two electrons from He. The solid line is the overbarrier model cross section πR_1^2 and the dashed line has been drawn through the experimental values[19] for odd charge states q.

We meet this difficulty by invoking a very general argument by Janev and Presnyakov[8], namely that in theoretical models of this "absorbing-sphere" type, the estimated geometrical one-electron cross section should be looked upon as containing also the sum of all multielectron processes. That this argument is very reasonable is clearly seen as soon as one realizes that multi-electron processes typically occur at smaller impact parameters than those involving just one electron. The comparison should thus be made between the one-electron classical overbarrier model cross section πR_1^2 (using of course the first ionization potential of helium) and cross sections $\sigma_{qq-1}^{01} + \sigma_{qq-1}^{02} + \sigma_{qq-2}^{02}$ that include transfer ionization and double capture. Such a comparison is made in Fig.2, which shows measurements by Andersson et al [19] taken at the CRYEBIS in Stockholm using Xe ions of charge states 11 through 31. The agreement between the model theory and the experiment is excellent.

The two-electron case was actually treated in somewhat more detail by Janev and Presnyakov[8] by assuming that the potential overbarrier criterion could be used also for the second electron, but now with the change that the second ionization potential takes the place of the first and that both the charge q of the highly charged ion and the doubly charged core of the target are screened by the first electron. This latter assumption, about target screening, was criticized by Bárány et al [29] who were the first to extend the model to an arbitrary number of electrons and to realize that the higher-order criteria in themselves only lead to estimates of the target loss of 2 or more electrons, a point not mentioned by Janev and Presnyakov[8] and apparently leading to their excellent agreement with experiment.

Bárány et al [29] assumed screening of the highly charged ion, where the first electron spends most of its time, but no screening of the target core. The assumption about the screening of the highly charged ion was then criticized by Roncin et al [30] and independently by Niehaus[31], who both argued for neglecting also that part of the screening. The different assumptions can be summarized in the expression for the critical barrier radius of the second electron,

$$R_2 = (2(q-\alpha)^{1/2}(2-\beta)^{1/2} + (2-\beta))/I_2, \quad (3)$$

where I_2 is the second ionization potential of the target, α and β the screening parameters. The alternatives used are determined through $\alpha = 1, \beta = 1$ (full screening[8]), $\alpha = 1, \beta = 0$ (screening of projectile only[29]) and $\alpha = \beta = 0$ (no screening[30,31]). It is immediately clear that the effect of the screening of the target core is rather important, while the question of screening of the highly charged projectile is of more academic interest, at least for these partial cross sections.

Using similar arguments about the nuclear trajectory as in the one-electron case, the geometrical cross section πR_2^2 is now to be compared to the sum of the two-electron processes, i.e. to the sum $\sigma_{qq-1}^{02} + \sigma_{qq-2}^{02}$. This is done in Fig.3, again for the Xe^{q+} − He collisions of Andersson et al [19]. We note that the experimental results are nicely bracketed by the model values for full screening ($\alpha = \beta = 1$) and those for screening of the projectile only ($\alpha = 1, \beta = 0$). This particular comparison does not give a clear answer on the best screening assumption for two-electron processes with a He target.

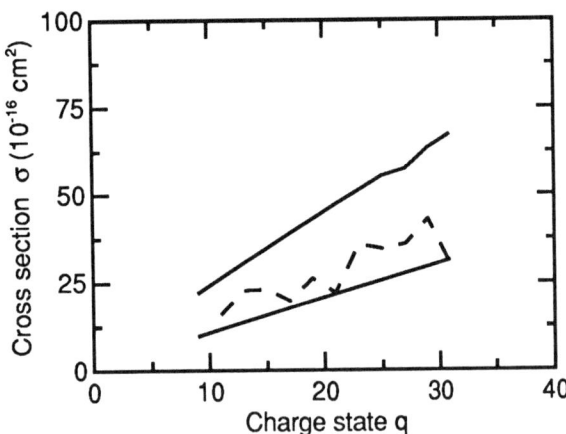

Figure 3. Absolute cross sections for $4q$ keV Xe^{q+} capturing both electrons from He into bound or autoionizing projectile states. The solid lines are the extended overbarrier model values πR_2^2, with full screening (lower) and with screening of the projectile only (upper). The dashed line has been drawn through the experimental points[19] for odd charge states q.

The extension of the classical model described above to the general case of an arbitrary number of electrons was presented by Bárány et al[29] and Liljeby et al[17]. The first paper[29] contains a rather straight-forward extension of the arguments given above, using for the general critical barrier radii the expressions ($m = 1, 2, ..$)

$$R_m = (2(q-\alpha)^{1/2}m^{1/2} + m)/I_m \qquad (4)$$

with $\alpha = m - 1$ and with I_m the mth ionization potential of the target. The paper by Liljeby et al[17] contains a much more elaborated version including, in particular, the assumption that all electrons lost from the target end up in a multiply excited state formed on the projectile. This state is then assumed to autoionize as far as is energetically possible according to a simple model estimate of the excitation energy.

Niehaus[31-33] has also presented a highly elaborated version of the classical model, a version which is more detailed and flexible, but which also is somewhat more difficult to apply in a straight-forward parameter-free fashion. But since his version can accomodate different assumptions about the decay of excited states one can, under suitable conditions, possibly draw conclusions about branching ratios between autoionization and radiative decay after comparing model cross sections with experimental ones through a fitting procedure. The expressions for the critical barrier radii used are those of Eq.(4) with $\alpha = 0$, which means that the predictions for the cross sections shown in Figs. 1 and 2 agree with the ones given there (upper curve in Fig.2). The model also takes into account that capture of inner electrons might screen the outer ones, leading to a new set of critical radii to be used when the two cores separate and determining the division of the excitation energy between target and projectile. Thus this version also allows for a number of higher-order processes that were left out by Bárány et al[29] and Liljeby et al[17], such as re-capture by the target and capture of inner-shell electrons. Niehaus' model[31-33] has been compared to the one of Liljeby et al[17] for different charge-changing processes in Ar^{q+} - Ar collisions by Cederquist et al[34]. Fig. 4 shows a comparison at somewhat higher charge states, using the cross sections taken by Danared et al[18] at the ECR source in Louvain-la-Neuve.

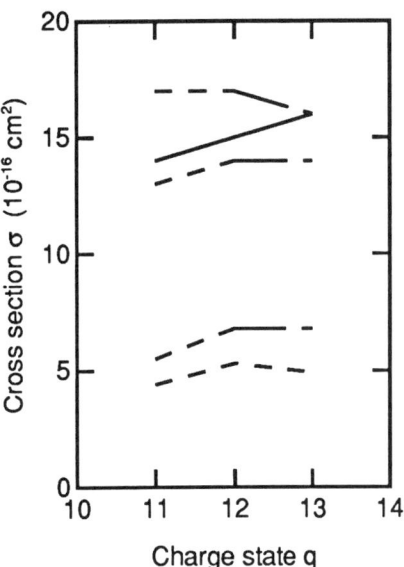

Figure 4. Absolute cross sections for $1.8q$ keV Ar^{q+} colliding with Ar and capturing one target electron while three are lost to the continuum. The dashed line is drawn through the experimental points, the solid line is the model values from Ref.17 and the chain lines those from Ref.31 for three different assumptions about the relaxation of excited states: Full autoionization in projectile only (upper); full autoionization in projectile and target (middle); reduced autoionization in projectile and none in target (lower)[18].

Very recently Mank et al [35] have combined the extended overbarrier model[29] with statistical arguments similar to the ones used by Müller et al [21]. Thereby the physical picture of a multiply excited state formed on the projectile, from which electrons are "boiled off" in a statistical way, is for the first time introduced into the statistical models. The theoretical estimates compare well with experiments[35], even though there is only a rather limited number of electrons among which the statistical sharing of the excitation energy is assumed.

3. Formation of excited states.

Another experimental technique which has been successfully applied to the general multi-electron reaction (1) during the present decade is energy-gain spectroscopy. This technique leads, through kinetic energy analysis of the projectile, to a determination of the potential energy of the excited state formed in the collision. A large number of groups have been active studying low-energy collisions with multiply charged ions, but with the restriction to somewhat higher charge states, energy-gain experiments have been run notably by the NICE collaboration[36], the KSU group[37], the Aarhus-Stockholm collaboration [38] and the Orsay group[39]. The information gained on the excited states formed in reaction (1) under potential energy dominated conditions, has had a major influence on the development of the theoretical models described in the last section, thereby helping us to form a more detailed picture of the dominating reaction mechanisms.

Before going in more detail into this development we note that also other techniques have been used to probe the state-selectivity of one- and multi-electron processes in the lower charge state region. Photon spectra have been the major source of information on the state-selectivity of single-electron processes[40] and have thereby in particular furthered the development of close-coupling *ab initio*-type theories[41,42]. A few cases of two-electron processes have been observed in high resolution optical[43] and soft X-ray spectra[44], but since the multiply excited states generally decay non-radiatively (at least at lower charge states) electron spectroscopy have instead been extensively used[45–47].

In a first attempt to understand the state selectivity of the multi-electron process along the classical mechanical lines used by Bárány et al [29], the model was developed to predict also energy defects for the target loss of an arbitrary number of electrons, as described by Hvelplund et al [48]. The general idea is already present in the one-electron continuum version of the classical overbarrier model, namely that the only energy exchange between the electronic and the nuclear motion is the minimal one which is necessary for the electron to adiabatically adjust to the changing potentials, eventually to end up in a state on the projectile which is bound by the original binding energy plus the minimal energy defect Q. Within the model all reactions are exothermic, and this seems to apply also to the real collision systems (1) under the conditions of potential energy dominance.

From the first development of the model for multi-electron processes[48] one thus derived minimum energy defects

$$Q_1^{min} = (q-1)/R_1, Q_2^{min} = Q_1^{min} + (q-3)/R_2, .. \quad (5)$$

that in an energy gain spectrum of $A^{(q-1)+}$ could be compared to the energy gains of single capture, double capture followed by autoionization, triple capture followed by double autoionization, etc. A problem immediately arises when comparing the model Q-values with experimental energy-gain spectra, namely the fact that even though the intensity of the scattered projectiles might be confined to small forward angles, because of the strong Coulomb repulsion in the final state, the peak intensity is generally at a non-zero scattering angle[49]. With an apparatus having a very small acceptance angle the intensity becomes very low in the forward direction, while positioning the slit at a non-zero angle or using a larger slit gives rise to a kinematic shift of the experimental peaks. This shift depends on the angular scattering characteristics of the inelastic collision system, i.e. on a trajectory effect not originally included in the model. At first[38] the shift was calculated by estimating the peak intensity simply through the so-called "half-Coulomb" angle, i.e. the angle obtained for a trajectory passing through the critical barrier radius, not suffering any scattering on the way into the collision and pure Coulomb scattering on the way out[49]. The results were rather encouraging[48] and led to a further development of the model ideas[50,51], which we will describe before showing some comparisons with experiments.

The basic ingredient in this latest development is the introduction of an assumption which leads to the occurrence of a maximal energy defect for each n-electron process. This is a multi-step reaction window concept, in which it is assumed that the electrons transfer one at the time within spherical shells defined by the critical potential barrier radii. A justification for this assumption can be found in multichannel Landau-Zener calculations for one-electron problems, since these show the appearance of similar reaction windows[52]. Even in a purely classical calculation, which so far has not been performed, one would probably find a similar effect, since electron correlation could keep an electron around the target even if the barrier criterion is fulfilled. The maximal energy defects are given by expressions similar to Eq.(5), but with the number j of each radius R_j increased by unity. Also considered in this development[50,51] is a more detailed description of the nuclear trajectories, leading to more realistic estimates of the scattering angles and the kinematic shifts.

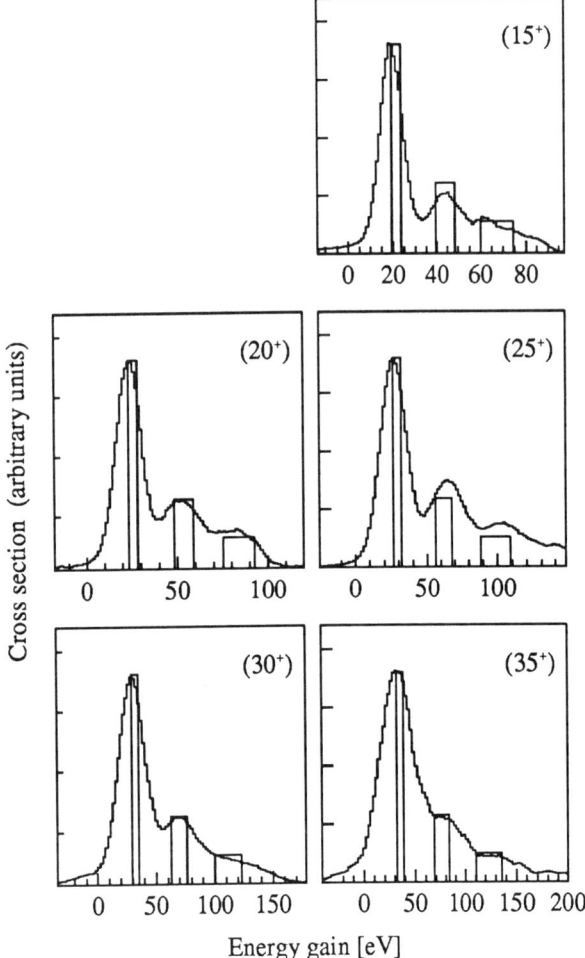

Figure 5. Experimental energy-gain spectra for $4q$ keV Xe^{q+} colliding with Xe and ending up with one electron captured[53]. Model energy-gain spectra[50] for the same collision system are shown. The three peaks of the model spectra belong to single capture and transfer ionization involving two and three electrons, respectively. The model single-capture peaks have been normalized in height with the experimental ones.

Fig.5 shows a comparison between the model predictions and the very recent experimental results of Cederquist et al [53], who used the Stockholm CRYEBIS to furnish beams of Xe ions in charge states up to 35 at an energy of $4q$ keV and collided these with Xe. Of special interest for these collisions is the question whether or not q has become large enough for radiative decay of multiply excited states to start competing with autoionization[8,54]. The conclusion made by Cederquist et al [53] is that this is the case for q larger than about 25, which also ties in with the fact that the model seems to overestimate transfer ionization at the higher q values.

Also Niehaus' more detailed version of the classical model[31-33] can be used to predict energy defects of reaction (1). Because of the different screening assumptions the Q-values are slightly different, the more so the higher the number of active electrons and the lower the charge state q. For capture of both electrons from He the model energy defect predictions differ by less than 1 eV from the ones of Hvelplund et al [48] already at $q = 10$, and the difference becomes even smaller as q increases. For collisions in which $\Delta q/q$ is small, the difference between the two assumptions leads to negligibly small effects in the energy gains.

A major difference between the models, though, is Niehaus' use of a dynamic width given to the peaks in the energy-gain spectrum[31-33]. This width is taken to derive from the time-energy uncertainty relations, and probably makes his version better applicable to somewhat higher velocities, where also the kinematic shifts (not considered by him) can be neglected. Since each electron is kept track of individually, the model also predicts energy defects of higher-order processes such as capture of an inner-shell electron (made possible by re-capture of the outer shell electrons). In a non-precise way it seems possible to look upon the extended overbarrier model presented by Bárány and Hvelplund et al [29,48,50,51] as a high charge and low-velocity limit of the more detailed version by Niehaus[31-33].

Several interesting extensions and applications of the state-selective predictions of the classical models have been made recently. Among them are the angular-momentum distributions given by Burgdörfer et al [55,56], basically derived on the assumption that the angular momentum of a captured s-electron is conserved in the reference frame of the highly charged ion. Subdividing the impact parameter plane into annular regions corresponding to different electronic angular momenta, Burgdörfer et al find an l-distribution with an average that agrees quite well with photon spectroscopic observations of single-electron capture by light ions. At first glance this speaks against the quasi-static model interpretations that have been used so far, but for the particular ions and energies considered the potential energy is not really dominating over the kinetic energy, which probably explains the success of this dynamic extension of the quasi-static model. Recently these ideas have also been applied to two-electron processes by Meyer et al[57], who found large angular momentum states $2pnl$ populated in two-electron capture by O^{6+} from He. These states were explained by assuming that the two electrons were transferred as a pair, both acquiring angular momenta with the same sense of rotation. Through electron correlation the total angular momentum is then divided up, resulting in high l-states for the outer electron. Again the collision system is not quasi-static and it seems somewhat early to evaluate this particular extension of the model ideas. Finally, Mack et al[58] recently discussed the formation of specially correlated doubly excited rotor states in terms of a time-dependent screening of the high-charge core. Some justification for this hypothetical mechanism seems to be present in electron spectra.

4. Angular scattering effects. During the later part of the present decade a number of experimental groups have begun to produce direct or indirect measurements of angular differential cross sections. While Barat et al[59], Danared et al[60,18] and Schmidt-Böcking et al[61] have made direct registrations of angular differential cross sections for general multi-electron processes (1) at keV energies, Cocke[62] and co-workers studied simpler collision processes (single and double capture) at lower energies. Also at low energy the Aarhus-Stockholm collaboration made indirect observations of angular scattering effects in energy-gain spectra[63-66]. Very recently Cederquist et al[67] have constructed a new set-up capable of measuring both energy-gain spectra and angular differential cross sections at very low energies. This development will be presented as a Hot Topic at this ICPEAC, and will not be reviewed here. All these measurements of angular differential cross sections and observations of angular scattering effects in energy-gain spectra probe the screening conditions in the quasimolecule and have helped clarify questions on details of the reaction mechanisms active in the highly charged quasimolecular complex, in particular questions about the reaction pathways realized for double electron capture[39].

We first note that the angular scattering properties of the general multi-electron reaction (1) can be analyzed in a gross way along the lines described in the last section, i.e. using classical scattering trajectories related to the barrier reaction window criteria formulated there. In this way both a maximal and a minimal scattering angle is associated with each particular target electron loss. Here the question of the particular screening assumption used acquires more relevance, since the main part of the scattering occurs at the smallest internuclear distances and is quite sensitive to the details of the force between the two cores. Barat et al[39] reached the conclusion that the unscreened version seemed to describe the data better than the screened one. For more highly charged projectiles, as studied by Danared et al[60,18], the differences seem not to be significant. We note, though, that Barat et al[39] are not using the same screened version as Danared et al[60,18], since the latter are applying the equations of Hvelplund et al[50].

The classical model by Niehaus[31-33] described above can not only be used to produce upper and lower bounds for the scattering angles, but also to generate differential cross sections. This is done by following the trajectories for all the different possible reaction paths leading to a given final state. Since there are so many reaction paths, the results can be smeared out into angular scattering distributions. This was realized by Danared et al[18], who found reasonably good agreement with experimental distributions, cf Fig.6.

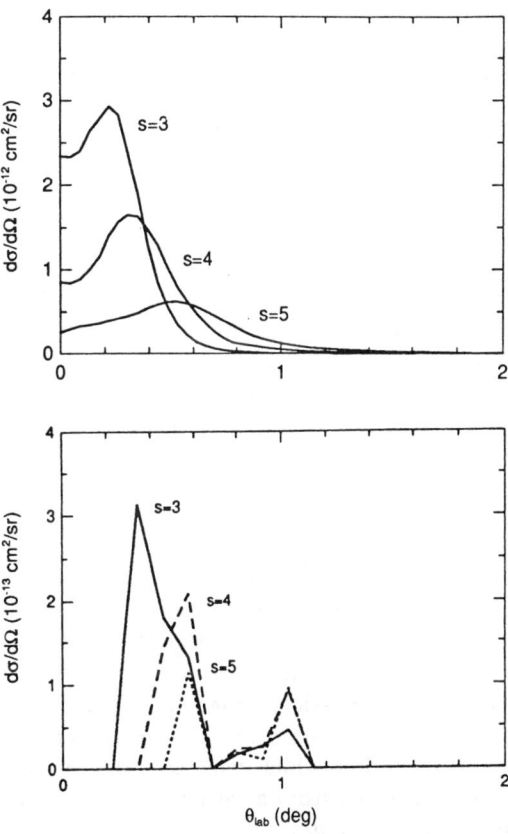

Figure 6. A comparison between the experimental (top) and the model (bottom) cross sections for 16 keV Ar^{9+} capturing two electrons from Ar and leaving the target in charge states s=3,4 and 5. Both measurement and calculation are absolute[18].

Before reviewing the latest developments in the field of angular scattering effects, we should consider that almost all theoretical arguments so far have been of a purely classical nature (even though they sometimes have been founded on experiences gained through quantum mechanics). In one particular collision system, which is not very highly charged, quantum interference effects were strong enough to be observed, making the classical arguments insufficient. This is the collision system C^{4+}-He, which at low energies has the peculiarity that double electron capture dominates over single capture and which also is the almost ideal two-state system. The observation was first made through angular scattering effects in energy-gain spectra[63], but were later confirmed in direct angular differential cross section measurements[37,68] and explained[68–70] in terms of Stueckelberg oscillations of the kind first observed in the 1960's for singly charged ions.

A rather detailed theoretical study performed by Danared and Bárány[69] also discussed the rainbow scattering phenomena that are generally present in such collision systems and that are well known from earlier work with singly charged ions. A similar investigation was performed for the collision system Ne^{4+}-He by Tan and Lin[71], for Ar^{2+}-He by Koslowski et al [72] and, at lower energy, by Friedrich et al [73].

An interesting question is to what extent similar interference effects will be seen in collision systems with highly charged ions, where generally several excited states are populated at large internuclear distances, leading to complex and rapidly varying phases. In our view the rainbow phenomena will be the more dominating features and a simple formalism to calculate angular differential cross sections and energy gain spectra for one-electron capture in such a way that interference effects are averaged out, but rainbows are kept, has recently been developed by Andersson et al [74,65]. The method used can best be described as a semiclassical multichannel Landau-Zener theory which includes trajectory effects. For each possible reaction pathway through the curve-crossing network, an impact-parameter dependent probability and a classical deflection function is calculated. These are then used to build up the angular differential cross section, which generically shows a rainbow singularity separated from a rather broad maximum for each final state populated[74,65].

In Fig.7 we show energy-gain spectra of Ar^{6+} capturing one electron from He at 200 eV impact energy. The experiment was performed by Hvelplund et al [65] at Aarhus, using recoil ions. For the semiclassical calculations three final states for the captured electron were kept open: $4s, 4p, 4d$. Since $4p$ dominates the capture at this energy, we compare with a two-state semiquantal calculation[69,75] which only keeps the $4p$ for the capture state, but also with a very recent full quantum mechanical close-coupling calculation which keeps six final channels open. This latter calculation was performed by Andersson et al [76], using molecular basis states generated by the model potential technique and a formulation of the dynamical equations incorporating translational coordinates. One observes a rather pronounced dark region separating the rainbow peak from the broad maximum. Since interference effects are averaged out in the first calculation, this max-min structure has nothing to do with Stueckelberg oscillations (cf Olson and Kimura[49]).

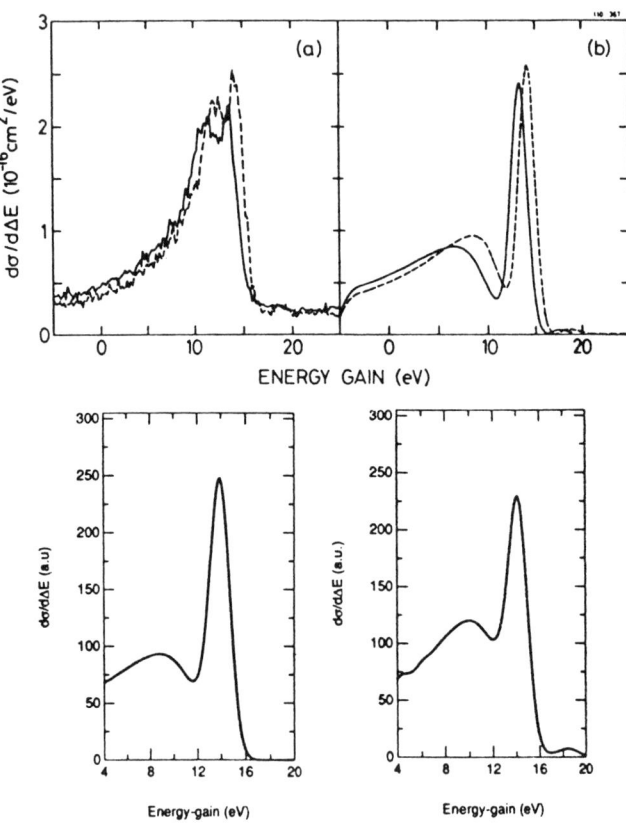

Figure 7. Experimental (upper left)[65] and theoretical energy-gain spectra of Ar^{6+} one-electron capture from He at 200 eV. In the upper part the fulldrawn curves relate to ^{3}He as target. Three levels of sophistication are used: Upper right is a semiclassical calculation[65], lower left a semiquantal[69,75] and lower right is a full quantum mechanical calculation[76].

The full quantum mechanical close-coupling method has also been used by Andersson et al [77] to calculate the angular differential cross section for electron transfer in O^{5+}-H collisions at eV energies. Even though no direct measurement of this cross section has been made so far, questions about the angular scattering characteristics of low-energy collision systems involving multiply charged ions and atomic hydrogen have recently gained actuality. One reason for this is the start-up of the powerful new merged-beams apparatus at ORNL[78]. In estimating the angular collection efficiency, Havener et al [78] used the "half-Coulomb" angle of Olson and Kimura[49], which at very low energies can be a serious underestimation[77]. At extremely low energies there might even be an orbiting phenomenon[6] involving resonantly enhanced probabilities for electron transfer[79], the angular scattering properties of which are still unknown.

Returning now to somewhat higher energies, a very recent development is the use by Taulbjerg and co-workers[80] of large-scale atomic basis expansion calculations of scattering effects in energy-gain spectra. These are presented at this ICPEAC by Taulbjerg and will not be further reviewed here. There has also been a recent attempt by Barat et al [39] to model also multi-electron processes in a similar way. The impact parameter equations are solved for a set of simplified potentials and couplings. The resulting impact parameter probabilities and state-selected differential cross sections are then used to discuss details of the reaction mechanisms. In particular the conclusion is reached that two-electron captures may occur as a single-step process without invoking electron correlation effects[81].

5. The shape of things to come. We are now reaching the end of this review and would like to conclude with some speculations on the future developments within the field of ion-atom collisions with highly charged ions at low energies.

One trend is of course self-evident, and that is the evolution towards higher and higher charge states. The developments of the new generation ion sources (EBIS, ECRIS) are by no means over yet, and since the production of recoil ions is connected to the general advancement of heavy ion accelerators, there are indications of a rapid progress in the production of highly charged ions. In particular the construction of electron-cooled heavy-ion storage rings[82] might, among many other atomic physics applications[82,83], very well lead to extremely powerful sources of low-energy highly charged recoil ions[84].

This trend towards ion sources for higher charge states, though, does not in itself act as a strong incitement for theoretical developments. We have seen that for the kind of physics considered in this review, the high charge states actually tend to make the processes increasingly simpler. But with the techniques for beam-handling, ion cooling and storage, particle detection, and the whole field of ion traps rapidly becoming more sophisticated[82], we can look forward to experimental investigations showing finer and finer details of the collision processes. This now becomes of major theoretical interest, since the highly-charged ion-atom collision system resides in the border-region of many-particle classical and quantum physics, where still many fascinating questions remain. These questions could *a priori* be answered by extensive use of very large computers, but a continuation of the inspiring coupling between experiment and theory, which has guided us so far, we hope will be possible also in the future. Let me end with the words of Max Born[85] "My advice to those who wish to learn the art of scientific prophecy is not to rely on abstract reason, but to decipher the secret language of Nature from Nature's documents, the facts of experience".

Acknowledgements. In my attempts to "collect and decipher the facts of experience" for highly charged ion-atom collisions 1981-89 I wish to acknowledge the help from all members of the Scandinavian collaboration, in particular from Preben Hvelplund, Jens Olaf Pedersen, Leif Liljeby, Håkan Danared, Henrik Cederquist, Håkan Andersson and Lars Andersson. Without their help this review would never have been realized. Many inspiring interactions with Ingmar Bergström, Derrick Crothers and Reinhold Schuch are also gratefully acknowledged. Ronald Olson gave some decisive general comments on how to write a review paper for the ICPEAC, but all faults of the present one reside, of course, with me.

References.

[1] J.J. Thomson, *Rays of Positive Electricity and Their Application to Chemical Analyses* (Longmans, Green and Co, London, 1913), p.46.

[2] N. Bohr, Dan. Mat. Fys. Medd. **18**, no.8 (1948).

[3] J.B. Hasted, J.T. Scott, and A.Y.J. Chong, in *Ionization Phenomena in Gases*, edited by N.R. Nilsson (North-Holland, Amsterdam, 1960), p.34.

[4] N.V. Fedorenko, I.P. Flaks, L.G. Filippenko, and E.S. Solov'ev, *ibidem*, p.41.

[5] (a) D.R. Bates and B.L. Moiseiwitsch, Proc. Phys. Soc. A **67**, 805 (1954), (b) A. Dalgarno, Proc. Phys. Soc. A **67**, 1010 (1954), (c) T.J.M. Boyd and B.L. Moiseiwitsch, Proc. Phys. Soc. A **70**, 809 (1957).

[6] R.E. Olson, in *Electronic and Atomic Collisions*,

edited by N. Oda and K. Takayanagi (North-Holland, Amsterdam, 1980), p.391.

[7]E. Salzborn and A. Müller, *ibidem*, p.407.

[8]R.K. Janev and L.P. Presnyakov, Phys. Rep. **70**, 1 (1981).

[9]R.K. Janev, L.P. Presnyakov, and V.P. Shevelko, *Physics of Highly Charged Ions* (Springer-Verlag, Berlin Heidelberg, 1985).

[10]*Physics of Electronic and Atomic Collisions*, edited by S. Datz (North-Holland, Amsterdam, 1982).

[11]E.D. Donets, Phys. Scr. **T3**, 11 (1983).

[12]R.E. Olson, in *Electronic and Atomic Collisions*, edited by H.B. Gilbody, W.R. Newell, F.H. Read, and A.C.H. Smith (Elsevier Science Publishers B.V., 1988), p.271.

[13]R. Schuch, in *Electronic and Atomic Collisions*, edited by D.C. Lorents, W.E. Meyerhof, and J.R. Peterson (Elsevier Science Publishers B.V., 1986), p.297.

[14](a)C.L. Cocke, R. DuBois, T.J. Gray, E. Justiniano, and C. Can, Phys. Rev. Lett. **46**, 1671 (1981), (b)E. Justiniano, C.L. Cocke, T.J. Gray, R.D. DuBois, and C.Can, Phys. Rev. A **24**, 2953 (1981).

[15](a)W. Groh, A. Müller, C. Achenbach, A.S. Schlachter, and E. Salzborn, Phys. Lett. **85A**, 77 (1981), (b)W. Groh, A. Müller, A.S. Schlachter, and E. Salzborn, J. Phys. B **16**, 1997 (1983).

[16](a)G. Astner, A. Bárány, H. Cederquist, H. Danared, P. Hvelplund, A. Johnson, H. Knudsen, L. Liljeby, and L. Lundin, Phys. Scr. **T3**, 163 (1983), (b)G. Astner, A. Bárány, H. Cederquist, H. Danared, S. Huldt, P. Hvelplund, A. Johnson, H. Knudsen, L. Liljeby and K.-G. Rensfelt, J. Phys. B **17**, L877 (1984).

[17]L. Liljeby, G. Astner, A. Bárány, H. Cederquist, H. Danared, S. Huldt, P. Hvelplund, A. Johnson, H. Knudsen, and K.-G. Rensfelt, Phys. Scr. **33**, 310 (1986).

[18]H. Danared, H. Andersson, G. Astner, P. Defrance, and S. Rachafi, Phys. Scr. **36**, 756 (1987).

[19]H. Andersson, G. Astner, and H. Cederquist, J. Phys. B **21**, L187 (1988).

[20]T. Åberg, A. Blomberg, and O. Goscinski, in *Electronic and Atomic Collisions*, edited by J. Eichler, I.V. Hertel, and N. Stolterfoht (Elsevier Science Publishers B.V., 1984), p.213.

[21]A. Müller, W. Groh, and E. Salzborn, *ibidem*, p.371.

[22]P. Benoit-Cattin, A. Bordenave-Montesquieu, M. Boudjema, A. Gleizes, S. Dousson, and D. Hitz, J. Phys. B **21**, 3387 (1988).

[23]H. Goldstein, *Classical Mechanics* (Addison-Wesley Publishing Co, 1980).

[24]R.E. Olson, J. Phys. B **13**, 483 (1980), and private communication (1989).

[25]D.R. Bates and R.A. Mapleton, Proc. Phys. Soc. **87**, 657 (1966).

[26]H. Ryufuku, K. Sasaki, and T. Watanabe, Phys. Rev. A **21**, 745 (1980).

[27]F. Folkmann, R. Mann, and H.F. Beyer, Phys. Scr. **T3**, 88 (1983).

[28]A. Bárány and H. Danared, Nucl. Instrum. Methods Phys. Res. B **23**, 1 (1987).

[29]A. Bárány, G. Astner, H. Cederquist, H. Danared, S. Huldt, P. Hvelplund, A. Johnson, H. Knudsen, L. Liljeby, and K.-G. Rensfelt, Nucl. Instrum. Methods Phys. Res. B **9**, 397 (1985).

[30]P. Roncin, M.N. Gaboriaud, M. Barat, and H. Laurent, Europhys. Lett. **3**, 53 (1987).

[31]A. Niehaus, J. Phys. B **19**, 2925 (1986).

[32]A. Niehaus, Nucl. Instrum. Methods Phys. Res. B **23**, 17 (1987).

[33]A. Niehaus, Nucl. Instrum. Methods Phys. Res. B **31**, 359 (1988).

[34]H. Cederquist, G. Astner, A. Bárány, H. Danared, A. Johnson, L. Liljeby, and K.-G. Rensfelt, Nucl. Instrum. Methods Phys. Res. B **24/25**, 43 (1987).

[35]G. Mank, R. Völpel, T. Grewe, K. Huber, and E. Salzborn, J. Physique **50**, C1-151 (1989).

[36]M. Kimura, in *Electronic and Atomic Collisions*, edited by D.C. Lorents, W.E. Meyerhof, and J.R. Peterson (Elsevier Science Publishers B.V., 1986), p.471.

[37]C.L. Cocke, J.P. Giese, L.N. Tunnell, W. Waggoner and S.L. Varghese, *ibidem*, p.453.

[38](a)E.H. Nielsen, L.H. Andersen, A. Bárány, H. Cederquist, P. Hvelplund, H. Knudsen, K.B. MacAdam, and J. Sørensen, J. Phys. B **17**, L139 (1984), (b)E.H. Nielsen, L.H. Andersen, A. Bárány, H. Cederquist, J. Heinemeier, P. Hvelplund, H. Knudsen, K.B. MacAdam, and J. Sørensen, J. Phys. B **18**, 1789 (1985).

[39]M. Barat, H. Laurent, M.N. Gaboriaud, L. Guillemot, and P. Roncin, in *Electronic and Atomic Collisions*, edited by H.B. Gilbody, W.R. Newell, F.H. Read, and A.C.H. Smith (Elsevier Science Publishers B.V., 1988), p.613.

[40]R.K. Janev and H. Winter, Phys. Rep. **117**, 265 (1985).

[41]W. Fritsch, Nucl. Instrum. Methods Phys. Res. B **23**, 9 (1987).

[42]C. Harel and A. Salin, in *Electronic and Atomic Collisions*, edited by H.B. Gilbody, W.R. Newell, F.H. Read, and A.C.H. Smith (Elsevier Science Publishers B.V., 1988), p.631.

[43]S. Martin, A. Salmoun, Y. Ouerdane, M. Druetta, J. Désesquelles, and A. Denis, Phys. Rev. Lett. **62**, 2112 (1989).

[44]J.-E. Rubensson, J. Nordgren, A. Bárány, M.G. Suraud, S. Bliman, D. Hitz, and E.J. Knystautas, J. Physique **50**, C1-321 (1989).

[45]A. Bordenave-Montesquieu, P. Benoit-Cattin, M. Boudjema, and A. Gleizes, in *Electronic and Atomic*

Collisions, edited by H.B. Gilbody, W.R. Newell, F.H. Read, and A.C.H. Smith (Elsevier Science Publishers B.V., 1988), p.643.

[46]F.W. Meyer, D.C. Griffin, C.C. Havener, M.S. Huq, R.A. Phaneuf, J.K. Swenson, and N. Stolterfoht, ibidem, p.673.

[47]M. Mack, Nucl. Instrum. Methods Phys. Res. B 23, 74 (1987).

[48]P. Hvelplund, L.H. Andersen, A. Bárány, H. Cederquist, J. Heinemeier, H. Knudsen, K.B. MacAdam, E.H. Nielsen, and J. Sørensen, Nucl. Instrum. Methods in Phys. Res. B 9, 421 (1985).

[49]R.E. Olson and M. Kimura, J. Phys. B 15, 4231 (1982).

[50]P. Hvelplund, A. Bárány, H. Cederquist, and J.O.K. Pedersen, J. Phys. B 20, 2515 (1987).

[51]A. Bárány and P. Hvelplund, Nucl. Instrum. Methods Phys. Res. B 23, 40 (1987).

[52]K. Taulbjerg, J. Phys. B 19, L367 (1986).

[53]H. Cederquist, H. Andersson, G. Astner, P. Hvelplund, and J.O.P. Pedersen, Phys. Rev. Lett. 62, 1465 (1989).

[54]J.E. Hansen, J. Physique 50, C1-603 (1989).

[55]J. Burgdörfer, R. Morgenstern, and A. Niehaus, J. Phys. B 19, L507 (1986).

[56]J. Burgdörfer, R. Morgenstern, and A. Niehaus, Nucl. Instrum. Methods Phys. Res. B 23, 120 (1987).

[57]F.W. Meyer, D.C. Griffin, C.C. Havener, M.S. Huq, R.A. Phaneuf, J.K. Swenson and N. Stolterfoht, Phys. Rev. Lett. 60, 1821 (1988).

[58](a)M. Mack, J.H. Nijland, P. v.d.Straten, A. Niehaus, and R. Morgenstern, Phys. Rev. A 39, 3846 (1989),(b)J.H. Nijland, M. Mack, P. v.d.Straten, A. Niehaus, J.H. Posthumus, and R. Morgenstern, J. Physique 50, C1-341 (1989).

[59]P. Roncin, M. Barat, and H. Laurent, Europhys. Lett. 2, 371 (1986).

[60]H. Danared, H. Andersson, G. Astner, A. Bárány, P. Defrance and S. Rachafi, J. Phys. B 20, L165 (1987).

[61]H. Schmidt-Böcking, M.H. Prior, R. Dörner, H. Berg, J.O.K. Pedersen, C.L. Cocke, M. Stockli, and A.S. Schlachter, Phys. Rev. A 37, 4640 (1988).

[62]C.L. Cocke, J. Physique 50, C1-19 (1989).

[63]H. Cederquist, L.H. Andersen, A. Bárány, P. Hvelplund, H. Knudsen, E.H. Nielsen, J.O.K. Pedersen, and J. Sørensen, J. Phys. B 18, 3951 (1985).

[64]J.O.P. Pedersen, P. Hvelplund, J.P. Bangsgaard, A. Bárány, and L.R. Andersson, J. Physique 50, C1-145 (1989).

[65]L.R. Andersson, J.O.P. Pedersen, A. Bárány, J.P. Bangsgaard, and P. Hvelplund, J. Phys. B 22, 1603 (1989).

[66]J.P. Bangsgaard, P. Hvelplund, J.O.P. Pedersen, L.R. Andersson, and A. Bárány, Phys. Scr. T28, 91 (1989).

[67]H. Cederquist, L. Liljeby, C. Biedermann, J.C. Levin, C.-S. O, H. Rothard, K.-O. Groeneveld, C.R. Vane and I.A. Sellin, Phys. Rev. A 39, 4308 (1989).

[68]A. Bárány, H. Danared, H. Cederquist, P. Hvelplund, H. Knudsen, J.O.K. Pedersen, C.L. Cocke, L.N. Tunnell, W. Waggoner, and J.P. Giese, J. Phys. B 19, L427 (1986).

[69]H. Danared and A. Bárány, J. Phys. B 19, 3109 (1986).

[70]J. Tan, C.D. Lin, and M. Kimura, J. Phys. B 20, L91 (1987).

[71]J. Tan and C.D. Lin, Phys. Rev. A 37, 1152 (1988).

[72]H.R. Koslowski, B.A. Huber, and V. Staemmler, J. Phys. B 21, 2923 (1988).

[73](a)B. Friedrich, S. Pick, L. Hládek, Z. Herman, E.E. Nikitin, A.I. Reznikov, and S.Ya. Umanskii, J. Chem. Phys. 84, 807 (1986), (b)E.E. Nikitin, A.I. Reznikov, and S.Ya. Umanskii, Sov. Phys. JETP 64, 937 (1986).

[74]L.R. Andersson, H. Danared, and A. Bárány, Nucl. Instrum. Methods Phys. Res. B 23, 54 (1987).

[75]L.R. Andersson and A. Bárány, in Electronic and Atomic Collisions, Abstracts of Contributed Papers, edited by J. Geddes, H.B. Gilbody, A.E. Kingston, C.J. Latimer, and H.J.R. Walters (Brighton, 1987), p.546.

[76]L.R. Andersson, M. Gargaud, J.P. Hansen and R. McCarroll, contributed paper to XVI ICPEAC.

[77]L.R Andersson, M. Gargaud, and R. McCarroll, contributed paper to XVI ICPEAC.

[78]C.C. Havener, M.S. Huq, H.F. Krause, P.A. Schulz, and R.A. Phaneuf, Phys. Rev. A 39, 1725 (1989).

[79]M. Rittby, N. Elander, E. Brändas, and A. Bárány, J. Phys. B 17, L677 (1984).

[80](a)J.P. Hansen and K. Taulbjerg, J. Phys. B 21, 2459 (1988), (b)J.P. Hansen, L. Kocbach, and K. Taulbjerg, J. Phys. B 22, 885 (1989), (c)J.P. Hansen and L.R. Andersson, J. Phys. B 22, L285 (1989).

[81]M. Barat, Comments At. Mol. Phys. 21, 307 (1988).

[82]Workshop and Symposium on the Physics of Low-Energy Stored and Trapped Particles, edited by A. Bárány, A. Kerek, M. Larsson, S. Mannervik, and L.O. Norlin, Phys. Scr. T22 (1988).

[83]R. Schuch, A. Bárány, H. Danared, N. Elander, and S. Mannervik, Nucl. Instrum. Methods Phys. Res. B, in press (1989).

[84]J. Ullrich, C.L. Cocke, S. Kelbch, R. Mann, P. Richard, and H. Schmidt-Böcking, J. Phys. B 17, L785 (1984).

[85]M. Born, Experiment and Theory in Physics (Cambridge Univ. Press, Cambridge, 1943)

OPTICAL POTENTIALS IN ION-ATOM COLLISIONS

R.M. Dreizler, H.J. Ast, A. Henne, H.J. Lüdde and C. Stary

Institut für Theoretische Physik
Universität, Frankfurt, West Germany

The time dependent extension of the Feshbach formalism allows the formulation of a set of nonunitary coupled channel equations in a given finite model space based on the assumption that consecutive interactions occur simultaneously. For a variety of collision systems we demonstrate that a microscopic optical potential model yields reasonable results for excitation capture and total ionisation at comparatively small basis sets.

1. Motivation

The theoretical discussion of ion-atom collision processes in the intermediate energy range would be much simpler, if one did not have to deal with ionisation. A typical example for the variation of the total reaction cross sections with energy is indicated in Fig. 1. It shows the summed capture, excitation and ionisation cross sections for the one electron collision system He^{2+} + H(1s) in the energy range from 1 to 10^3 keV/amu. At lower energies capture is the dominant process, at intermediate energies all reaction channels are in competition, at still higher energies only excitation and ionisation remain.

The standard theoretical method for the calculation of reaction cross sections is the close coupling method in the semiclassical approximation (SCA). The representation of the bound state channels, though not necessarily a simple matter, does not represent a major difficulty. The representation of the continuum channels is, and in most cases they are neglected. A wellknown though somewhat extreme example is the collision of neutral alkali atoms with rare gas atoms. The ionisation of the loosely bound, active alkali electron is a dominant process, even at lower energies. If one neglects ionisation channels in the theoretical discussion, one finds for instance that excitation cross sections (Fig.2: 2p excitation in Li for Li+Ne) are grossly overestimated.

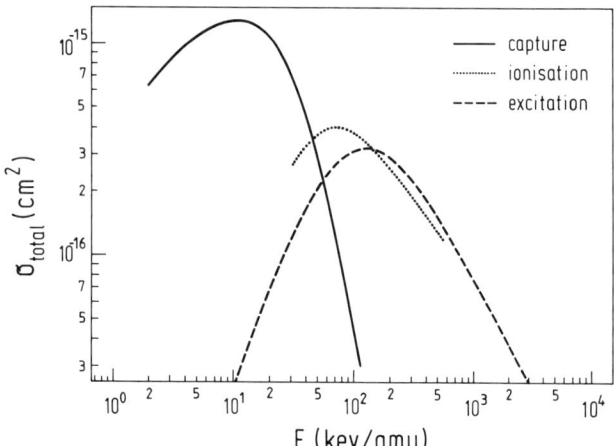

Figure 1 He^{2+} + H: capture, ionisation and excitation

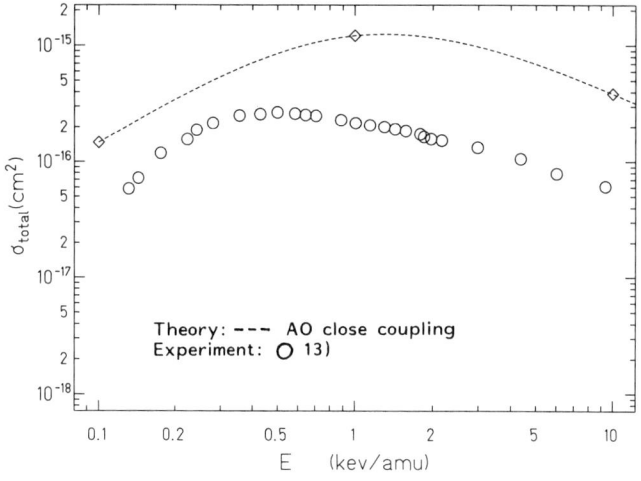

Figure 2 Li + Ne: excitation into Li(2p)

The message thus is: One has to deal with continuum states in order to improve the theoretical description. Continuum states enter into the theoretical discussion at two stages

(i) For energies, for which ionisation does not contribute strongly to the total reaction cross section, they still feature as intermediate states in the representation of the collision system around the point of closest approach.
(ii) If ionisation is a dominant process, one definitely has to account for the probability flux into these channels.

There are essentially three options to cope with ionisation channels
(i) Use a discretised representation of the continuum states. This is possible for one centre situations as eg. the description of laser excitations. It is much more involved for two centre situations, that one encounters in ion-atom collisions.
(ii) Include suitable pseudostates in the basis set and hope that they model ionisation channels. This approach has been used with some success, it is however open to the criticism that it does not represent a systematic way to deal with the problem.
(iii) The optical potential approach, where one eliminates the channels not explicitly included in the representation space via a projection formalism. This method is not easily put into a viable form either, but we chose to make the attempt, as it seemed to offer a more systematic approach towards the resolution of the difficulties indicated.

2. Theory

The microscopic optical potential model was first introduced by Feshbach[1] for the stationary formulation of the collision problem. The corresponding formulation for the time dependent collision problem in the SCA can be stated as follows:
Consider the solution of the explicitly time dependent Schrödinger equation

$$i\partial_t |\psi(t)\rangle = \hat{H}(t)|\psi(t)\rangle = (H_0 + V(t))|\psi(t)\rangle$$

by expansion in terms of a complete set of basis states

$$|\psi(t)\rangle = \sum_{j=1}^{N} a_j(t)|j\rangle + \sum_{\alpha=N+1}^{\infty} a_\alpha(t)|\alpha\rangle.$$

The representation space has been divided into the P-space ($\sum_{j=1}^{N}$) and the complementary Q-space ($\sum_{\alpha=N+1}^{\infty}$). From the standard set of coupled channel equations for the P and Q-space amplitudes one can eliminate the Q-space amplitudes a_α and arrive at an exact closed set of equations for the P-space amplitudes alone

$$i\dot{a}_k(t) = \sum_{j=1}^{N} \langle k|\hat{H}(t)|j\rangle a_j(t)$$
$$+ \sum_{j=1}^{N} \langle k|\hat{V}(t)\hat{v}_j(t)|j\rangle$$

The optical potential $\hat{v}_j(t)$ is given by the perturbative expression

$$\hat{v}_j(t) = \sum_{n=1}^{\infty} (-i)^n \int_{t_0}^{t} dt_1 \int_{t_0}^{t_1} dt_2 \cdots \int_{t_0}^{t_{n-1}} dt_n$$
$$Q H(t_1) Q H(t_2) \cdots Q H(t_n) a_j(t_n)$$

with $\hat{Q}$ being the projector onto Q-space

$$\hat{Q} = \sum_{\alpha=N+1}^{\infty} |\alpha\rangle\langle\alpha| = (1 - \hat{P}).$$

The potential is complex (describing the loss of flux from the P-space) and can be said to involve memory effects. The nonlocal term of the integro-differential equation is determined by the prehistory of the system and not (as the first term) by the status of the system at time t. The optical model coupled channel equations can not be solved as they stand. The simplest option, a first order optical potential, has been investigated by Bransden and Coleman[2] in 1972 and has since been applied to electron-atom[3] and to a lesser extent to ion-atom[4] scattering. The first order theory does, however, not describe the coupling to the Q-space correctly and fails completly with respect to the coupling among Q-space states. One is thus forced to go beyond first order theory, eg. by partial (and approximate) resummation of the perturbation expression. The resummation scheme, that we have investigated, is the scaled local approximation[5,6]. If one assumes that the interaction time is short, one can replace two consecutive interactions in the perturbative expression by their instantaneous limit

$$\hat{H}(t)\hat{Q}\,\hat{H}(t') \to f\,\hat{H}(t)\hat{Q}\,\hat{H}(t')\,\delta(t-t')$$

This approximation can be rationalised by a variety of arguments, which then show that the scaling factor depends on the external parameters of the collision process

$$f = f(v, Q_P, Q_T, b)$$

but not on time. One finds for instance that the dependence on the relative velocity goes as

$$f = \frac{\text{const}}{v}.$$

Intuitively one should be willing to accept that this approximation is adequate for higher impact energies and the question to be answered by explicit calculation is: At what energies does it start to fail. We note in particular that memory effects are suppressed by this approximation. With this approximation the perturbation series for the optical potential can be resummed and the coupled channel eqs. become

$$i\dot{a}_k(t) = \Sigma_j <k|\hat{V}(t) \frac{1}{(1+i f Q H(t))} + \hat{H}_0|j> a_j(t)$$

The P-space matrix elements can be evaluated if one accepts a closure approximation. As an excerpt of the further theoretical points, which I am also not able to present in detail, I would like to mention the following:

(i) The formulation indicated above refers to the situation that the P-space is composed of target states only. The Q-space then contains all capture and ionisation channels. If the P-space contains target and projectile boundstates, a formulation[7] that differs in some details is required. One has to ensure that P- and Q-space orthogonality is maintained and that double counting of states is avoided. The Q-space is in this case essentially composed of ionisation channels only.

(ii) If the P-space becomes complete, then the optical model coupled channel equations go over into the exact representation of the time dependent Schrödinger equation.

(iii) Individual Q-space amplitudes[8] could be calculated from the Q-space amplitudes with some additional labour.

3. Results

We have applied the local optical model to a number of collision systems. We have considered one electron systems with either a neutral hydrogen target or one electron ion-ion processes[9], effective one electron systems as eg. hydrogen atoms in collision with noble gas targets[10] or protons and antiprotons colliding with alkali targets[11], as well as collisions in two electron systems[12]. The energy range covered is between a few keV/amu up to 100 MeV/amu. Most of the calculations are carried through for a P-space, that involves only target basis states, for instance KLM shell states for the case of fully stripped ions colliding with hydrogen or correlated He-states in the case of protons colliding with He. In this case the dominant Q-space channels are capture and ionisation. Besides target excitation, we can then calculate total electron loss cross sections via the loss of flux from the P-space

$$\sigma_{loss} = \int b^2 db (1 - N_p(b)),$$

with

$$N_p(b) = \Sigma_{j=1}^N |a_j(b, t \to \infty)|^2.$$

For some cases we have also used a two centre AO basis, as eg. for protons on Li or Na. The loss cross section can in this case be identified with the total ionisation cross section. A few explicit examples of our results are shown in the following figures. We first look at ionisation and capture for the one electron collision system p + H (Fig.3). We compare the results of a two centre optical model calculation with experiment in the energy

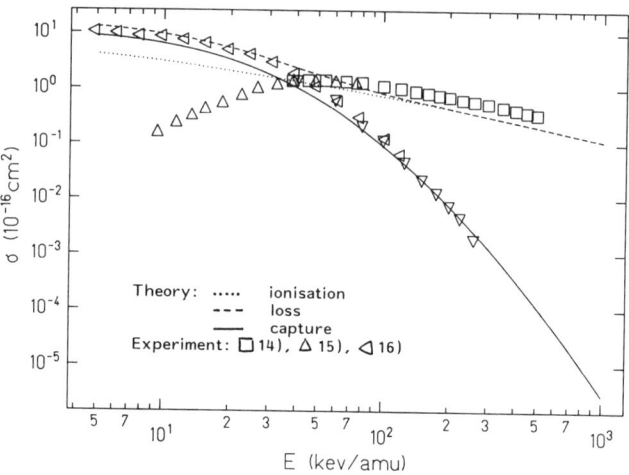

Figure 3 $H^+ + H$: ionisation, capture and loss

range between 5 and 1000 keV/amu. The P-space is rather modest consisting of the KLM-shells of the target and projectile systems. One finds that this optical model calculation reproduces the total loss quite well over the complete energy range, but one also notes that the separate capture (P-space channels) and ionisation (Q-space channels) are not given correctly at energies below 50 keV/amu. The conclusion is: The local approximation is not able to describe the resonant charge exchange mechanism in symmetric collision systems for lower energies. This point is emphasised if one looks at the corresponding excitation cross sections. In Fig. 4 we compare the 2s excitation cross section obtained in a one and a two centre optical model calculation. The one centre results are obtained with a P-space consisting of the KLM-shells of the target alone. Local coupling to the Q-space (ionisation and capture channels) is obviously deficient at lower energies.

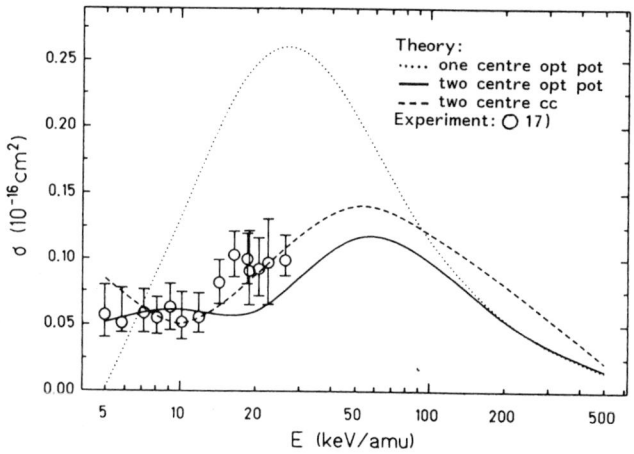

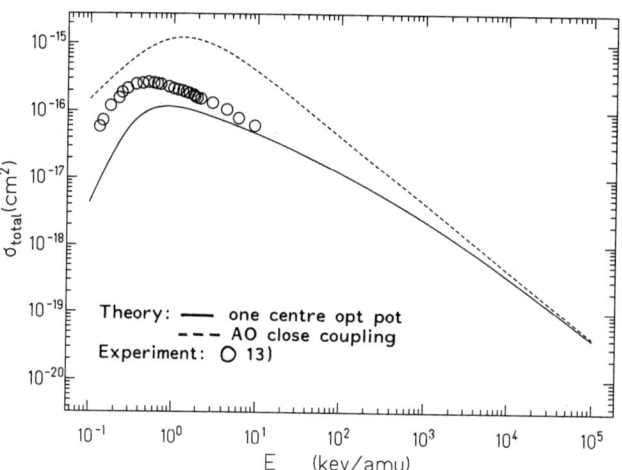

Figure 4 H$^+$ + H: excitation into H(2s)

In the two centre calculation the direct coupling to the capture channels is treated correctly and results of reasonable quality are obtained. As charge exchange proceeds, however, in part via intermediate continuum states, the assumption of local coupling to the Q-space is still not sufficiently adequate at low energies.

The situation is different for asymmetric collision systems. As an example I show the

Figure 6 Li + Ne: excitation into Li(2p)

(Fig.6) the optical model cures the overestimation (too strongly below 5 keV/amu due to the local approximation) and for larger energies goes (together with the results of a truncated close coupling calculation) over into the correct perturbation (Born) limit.

A corresponding situation is found for the neutral collision system H on noble gas targets. I show the 2s and 2p excitation

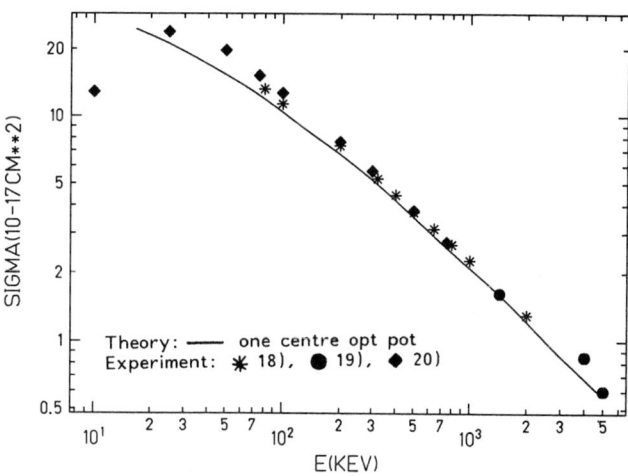

Figure 5 H$^+$ + He: total electron loss

loss cross section for the two electron system p + He, (Fig.5) which is obtained with a one centre calculation with a P-space consisting of He-configurations. The local approximation fails only below 20 keV/amu. For the 2p excitation cross section of the active electron in Li + Ne

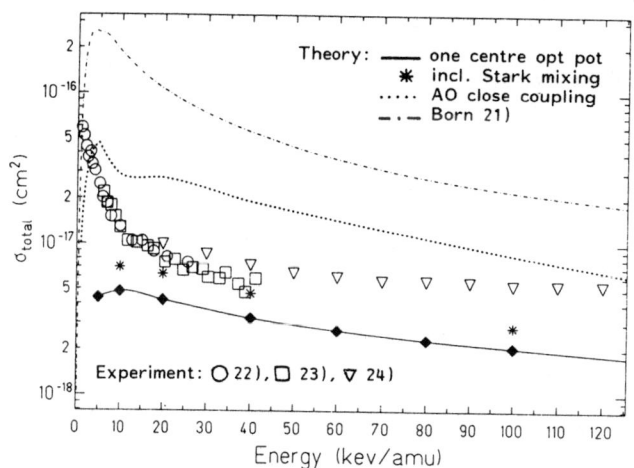

Figure 7 H + Ne: a)excitation into H(2s)

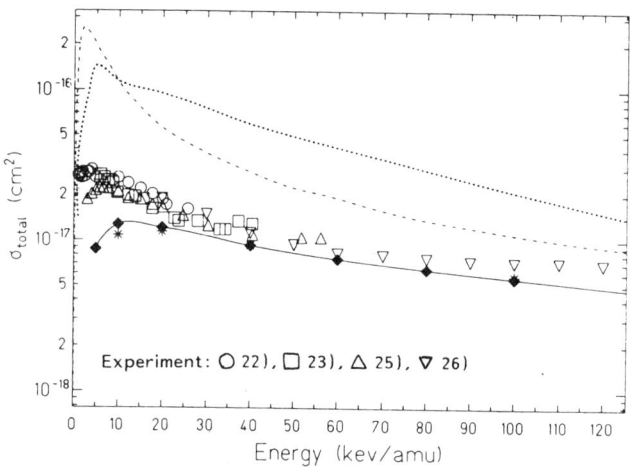

b) excitation into H(2p)

cross sections for H + Ne, comparing Born, standard close coupling and one centre optical potential results with experiment (Fig. 7). In this case we describe the noble gas target by a frozen effective HF-potential. This means that we exclude all doubly inelastic processes. If these are included (in a rather summary fashion) one obtains the results indicated by an asterix. Notwithstanding the fact that some more work is required to deal with a satisfactory description of the doubly inelastic channels, we note again that the local approximation fails below about 20 keV/amu.

I finally show some results for the system p + Na, which is notorious in the sense, that rather large scale pseudostate calculations (eg. 63 two centre pseudostates in the work of Shingal and Bransden) are required to obtain adequate excitation cross sections. I show one and two centre optical potential results for the 3s-3p excitation in Na, obtained with a comparatively modest basis set (Fig. 8). The corresponding results for capture and ionisation (no experiment) are given in Fig.9. We note that capture is adequately described

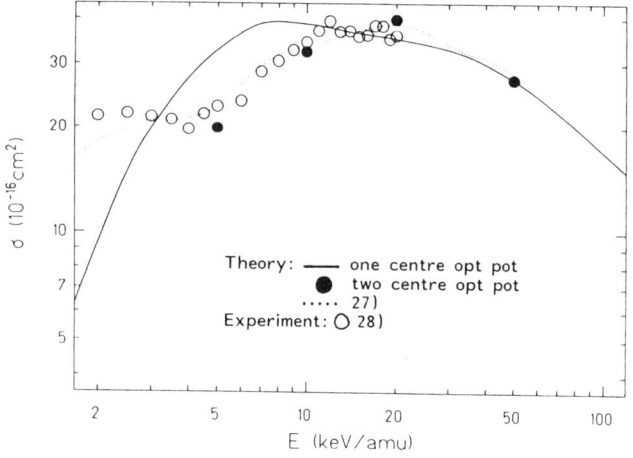

Figure 8 H$^+$ + Na: excitation into Na(3p)

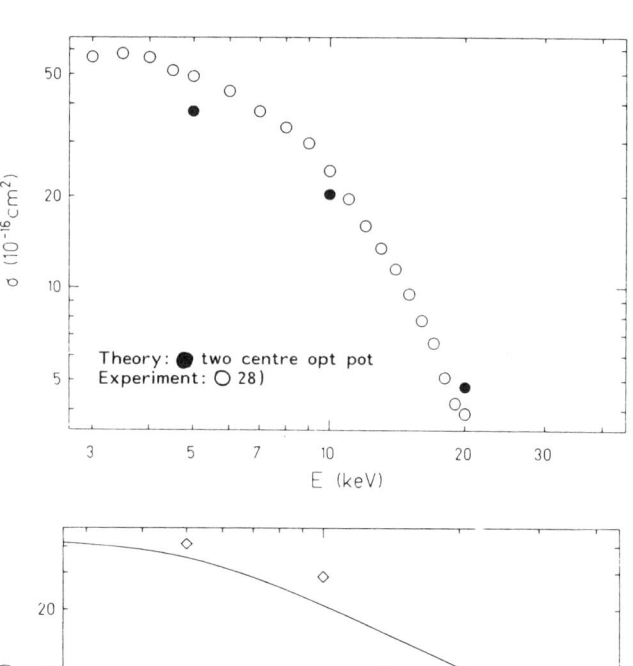

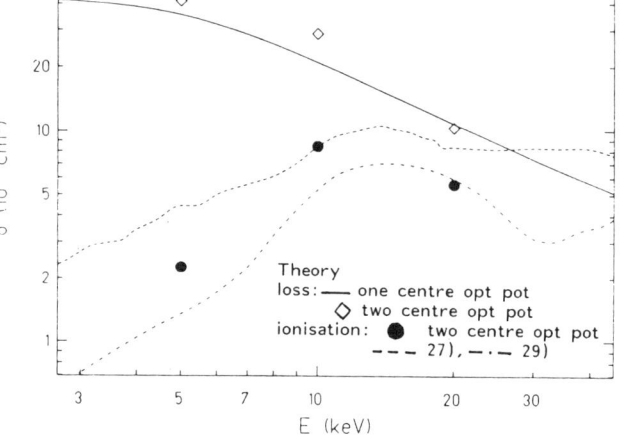

Figure 9 H$^+$ + Na: a) total capture
b) ionisation and loss

by the two centre optical model. There is little to say about ionisation at present. Our results interpolate, so to speak, between the pseudostate calculations of Shingal and Bransden and of Fritsch.

4. Conclusion

As a brief summary I offer the following remarks. The local approximation is not satisfactory at lower energies. The critical energy below which it starts to fail is roughly characterised by the statement that the corresponding impact velocity is of the order of the orbital velocity of the active electron(s). On the other hand the local approximation is able to bridge the gap between the critical energy and the energies, for which perturbative Born results become valid. The optical model results converge to the Born limit in all cases.

The results reported above are all obtained with the scaling factor

$$f = \frac{v_0}{v},$$

which seems to be adequate for the discussion of total cross sections. More work is needed to tie down the impact parameter dependence of f for the discussion of differential cross sections.

For impact energies below the critical energy, improvements over the local approximation are required. One option is: use the first order approximation in order to restore memory effects and use the local approximation for the remainder of the perturbation series

$$\Sigma_j <k|\hat{V}(t)\hat{v}_j(t)|j>$$

$$\rightarrow \Sigma_j \int_{t_0}^{t} dt_1 \{<k|\hat{V}(t)\hat{Q}\hat{V}(t_1)|j>$$

$$(1-f\delta(t-t_1))a_j(t_1)\} + \Sigma_j <k|\hat{V}(t)v_j^{loc}(t)|j>$$

This is not the only option, one can envisage. It remains to be seen which contribution of perturbative and resummed scheme can cover the entire intermediate energy range.

References

1. H. Feshbach, Ann Phys 5, 357 (1958)
2. B.H. Bransden and J.P. Coleman, J Phys B5, 537 (1972)
3. H.R.J. Walters, Phys Rep 116, 1 (1984)
4. R.M. Potvliege, F. Furtado and C.J. Joachain, J Phys B20, 1771 (1987)
5. H.J. Lüdde, H. Ast and R.M. Dreizler, Phys Lett A125, 197 (1987)
6. H.J. Lüdde, H. Ast and R.M. Dreizler, J Phys B21, 4131 (1988)
7. H.J. Lüdde and R.M. Dreizler, J Phys B accepted for publication
8. H.J. Lüdde, A. Henne and R.M. Dreizler, J Phys B submitted for publication
9. H. Ast, H.J. Lüdde and R.M. Dreizler, J Phys B21, 4143 (1988)
10. H. Ast, H.J. Lüdde and R.M. Dreizler, J Phys B submitted for publication
11. C. Stary, H.J. Lüdde and R.M. Dreizler, J Phys B submitted for publication
12. A. Henne, H.J. Lüdde and R.M. Dreizler, J Phys B submitted for publication
13. J. Østgaard Olson, N. Andersen and T.J. Andersen, J Phys B10, 1723 (1977)
14. M.B. Shah and H.B. Gilbody, J Phys B14, 2361 (1981)
15. M.B. Shah, D.S. Elliot and H.B. Gilbody, J Phys B20, 2481 (1987)
16. H. Tawara, T. Kato and Y. Nakai, IPPJ-AM30 Nagoya, 21 (1983)
17. T.J. Morgan, J. Geddes and H.B. Gilbody, J Phys B6, 2118 (1973)
18. M.B. Shah and H.B. Gilbody, J Phys B18, 899 (1985)
19. H. Knudsen, L.H. Andersen, P. Hveplund, G. Astner, H. Cerderquist, H. Danared, L. Liljeby and K.-G. Rensfelt, J Phys B17, 3545 (1984)
20. R.D. DuBois, L.H. Toburen and M.E. Rudd, Phys Rev A29, 70 (1984)
21. H. Levy II, Phys Rev 185, 7 (1969a)
22. J.H. Birely and R.J. McNeal, Phys Rev A5, 257 (1972)
23. A.L. Orbeli, E.P. Andreev, V.A. Ankudinov and V.M. Dukelskii, Sov Phys JETP 30, 63 (1970)
24. R.M. Hughes and S.S. Choe, Phys Rev A5, 1758 (1972b)
25. V. Dose, R. Gunz and V. Meyer, Helv Phys Acta 14, 269 (1968)
26. R.H. Hughes and S.S. Choe, Phys Rev A5, 656 (1972a)

DISTORTED WAVE MODELS FOR IONIZATION IN ATOMIC COLLISIONS

R. D. Rivarola*, P. D. Fainstein** and . H. Ponce**

*Instituto de Física Rosario (Conicet-UNR)
Av. Pellegrini 250, 2000 Rosario, Argentina
**Centro Atómico Bariloche and Instituto Balseiro, 8400 Bariloche, Argentina

A progress report on the use of distorted wave models to represent single electron ionization in ion-atom collision is given here. The work is focused on the Continuum Distorted Wave-Eikonal Initial State approximation. Evidence is given that electron ejection must be described as a two center reaction at intermediate and high collision energies. Experimental data confirm this asseveration.

1. INTRODUCTION

In the present work we study the single electron ionization of atomic targets by impact of bare ions at intermediate and high collision velocities. It has been proved that this process must be interpreted at least as a three-body one[1-4]. The fields of the projectile and of the residual target act simultaneously on the ejected electron. Thus, electron excitation to the continuum of the target or electron capture to the continuum of the projectile are only partial images of the two center reactions. So, first-order Born approximations (5,6) give incomplete representation of the process and appear to be inadequate to describe double differential cross sections (as a function of the scattering angle and energy of the ejected electron) even at high enough collision velocities. The failure of these first order theories is extended also for total cross sections at high collision energies when the projectile charge Z_p increases[4] and at intermediate ones when binding and polarization effects play an important role[3,7].

New experimental facilities related to the obtaining of a few hundred keV-antiproton beams (8) have permitted us to study two center effects for a combined field of repulsive and attractive potentials[9-12].

Different distorted wave theoretical models have been developed since the pioneering work of Salin (13). We focus our attention on the Continuum Distorted Wave-Eikonal Initial State (CDW-EIS) one.

This distorted wave model was introduced by Crothers and McCann (14) to study K-shell single electron ejection for monoelectronic systems. Extensions of the CDW-EIS model for ionization from any initial state has been given by us for monoelectronic[15] as well as for multielectronic[4,16] targets. Multiple scattering terms associated with the active electron-projectile interaction are introduced in the initial distorted wavefunction χ_i^+ by means of an eikonal phase which is multiplied by the electron bound wave function. So, χ_i^+ is normalized and satisfies the correct boundary conditions for $t \to -\infty$ (the collisional time is taken as zero for the position of closest approach between the nuclei). In the exit channel, the two center-continuum state of the electron in the simultaneous presence of the projectile and target fields is simulated by a double product between two continuum factors and a plane wave. The continuum factors take account of the active electron-projectile and active electron-residual target (target nucleus plus passive electrons) interactions.

Double, single and total CDW-EIS cross sections are presented and compared with experimental data in order to determine the adequacy of the theoretical model to describel the single ionization reaction.

Atomic units will be used unless otherwise stated.

2. THEORY

The straight-line version of the impact parameter approximation, $R=\rho+vt$ (where R is the internuclear vector, ρ is the impact parameter and v is the relative velocity between the nuclei is used. Without loss of generality, we describe the reaction from a framework fixed on the target nucleus.

Let us consider that a bare projectile of nuclear charge Z_P hits on a target of nuclear charge Z_T, which is ionized. We indicate with $x(s)$ the position vector of the active electron referred to the target (projectile) nucleus. To reduce the description of the collision to a one-electron-type process, we neglect the passive electrons dynamics and assume that they remain frozen in their initial state during the collision. So, following Fainstein et al[4], it is possible to prove that the scattering amplitude in its prior form can be written as a function of the impact parameter as

$$A_{if}(\underline{\rho}) = -i \exp\left(-i \int_{-\infty}^{+\infty} V_s(R)\, dt\right)$$

$$\times \int_{-\infty}^{+\infty} dt \langle \Psi_f^-(\underline{x},t) | H_a - i\partial/\partial t$$

$$|\chi_i^+(\underline{x},t) \exp(-i\varepsilon_i t)\rangle \quad (1)$$

with

$$H_a = -\frac{\nabla^2}{2} - \frac{Z_T}{x} + V_{ap}(x) - \frac{Z_P}{x} \quad (2)$$

where $V_{ap}(x)$ is the active electron-passive electrons interaction averaged on the passive electrons-wavefunction. In expression (1), Ψ_f^- is the incoming exact solution of the time dependent Schroedinger equation constructed with the Hamiltonian H_a, ε_i is the active electron-initial orbital energy and $V_s(R)$ is the potential coming from the interaction between the projectile and the residual target averaged on the passive electrons-wavefunction. The distorted wavefunction χ_i^+ is given by

$$\chi_i^+ = \phi_i(\underline{x}) \exp(-i(Z_p/v)\ln(vs + \underline{v}\cdot\underline{s})) \quad (3)$$

where ϕ_i is the stationary active electron-initial solution corresponding to the Hamiltonian $(H_a - Z_P/s)$. The function ϕ_i is taken as a Roothaan-Hartree-Fock or a hydrogenic wave for multi-electronic or monoelectronic targets respectively.

In a first order calculation of the scattering amplitude, Ψ_f^- is replaced by the final distorted wavefunction χ_f^-, given by

$$\chi_f^- = \phi_f(\underline{x},t) N^*(\zeta)\,_1F_1(-i\zeta; 1; -ips - i\underline{p}\cdot\underline{s}) \quad (4)$$

where ϕ_f represents the continuum state of the electron in the field of the residual target, $N(a) = \exp(\pi a/2)\,\Gamma(1-ia)$, $z = Z_P/P$ and p is the momentum of the electron referred to the projectile nucleus.

If in the exit channel, the Hamiltonian H_a is simplified by assuming that the electron-residual target interaction may be replaced by a coulomb potential with an effective charge Z_T^*:

$$H_a \simeq \frac{\nabla^2}{2} - \frac{Z_T^*}{x} - \frac{Z_P}{s} \quad (5)$$

Following the prescription used by Belkic et al (17) for electron capture, we choose $Z_T^* = (-2n^2\varepsilon_i)^{1/2}$ with $\underline{n}$ the principal quantum number. It is obvious that for a monoelectronic case $Z_T = Z_T^*$. Thus we obtain

$$\phi_f = (2\pi)^{-3/2} \exp(i\underline{k}\cdot\underline{x}) \exp\left(i\frac{k^2}{2}t\right) N^*$$

$$(\xi)\,_1F_1(-i\xi; 1; -ikx - i\underline{k}\cdot\underline{x}) \quad (6)$$

with k the momentum of the electron referred to the target nucleus and $x = Z_T^*/K$. Finally, the continuum state is assumed to be described by a product of Coulomb continuum factors associated with the projectile and residual target times a plane wave. The main idea is that in this first order approximation the two center continuum state is represented by the product given by expression (4).

Replacing in equation (1) the operator $(H_a - i\partial/\partial t)$ by $H_a - i\partial/\partial t)^\dagger$ and Ψ_f by χ_f given by expression (4), we obtain also the first order approximation of the scattering amplitude in its post form.

Using the relation

$$A_{if}(\underset{\sim}{\rho}) = \exp\left(-i \int_{-\infty}^{+\infty} V_s(\underset{\sim}{R}) dt\right) a_{if}(\underset{\sim}{\rho}) \quad (7)$$

the scattering amplitude $R(\eta)$ as a function of the transverse momentum transfer η can be then obtained by the transform

$$R(\underset{\sim}{\eta}) = (2\pi)^{-1} \int d\underset{\sim}{\rho} \exp(i\underset{\sim}{\eta}\cdot\underset{\sim}{\rho}) a_{if}(\underset{\sim}{\rho}) . \quad (8)$$

Analytical forms of $R(\eta)$ have been obtained for monoelectronic (15) as well as for multielectronic (4,16) targets by using the post form of $A(\rho)$ and the approximation (5).

The double differential cross section associated with electronic degrees of freedom is defined as

$$\frac{d^2\sigma}{dE_k d\Omega_k} = k \int d\underset{\sim}{\eta} |R(\underset{\sim}{\eta})|^2 \quad (9)$$

with E_K- $K^2/2$ and Ω_k the solid angle of the ejected electron. The single differential cross sections $d\sigma/d\Omega_k$ and the total cross section σ can be obtained by successive integrations.

3. RESULTS AND DISCUSSION

Total cross sections for single ionization of $He(1s^2)$ by the impact of different projectiles with collision energies of 1.4 MeV amu^{-1} (v-7.48 au) as a function of the projectile charge Z_P are shown in figure 1. At this energy, measurements for the impact of Fe^{15+}, Kr^{18+}, Fe^{20+}, U^{36+}, Gd^{37+} and U^{44+} are shown[18]. Measurements from Knudsen et al.[19] for the impact of 1.44 MeV amu^{-1}-projectiles with Z_P varying from 1 to 8 are also included in the figure. CDW-EIS calculations are in close agreement with experimental data showing large deviations from the first order-Born (B1) results when the projectile charge increases. Total cross section obtained by using a Glauber (G) approximation are in good agreement with our calculation. The B1 and G results taken from McGuire et al.[18] have been computed by considering a monoelectronic hydrogenic model with an effective target nuclear charge equal to 1.3. This effective charge has been chosen to fit experimental data for Z_P = 1. It is observed on the figure that the Z_P^2 law predicted by the BI approximation does not work for large Z_P at this relatively high collision energy. So, multiple scattering contributions associated with the projectile potential play a main role in the description of this saturation effect (see also 20).

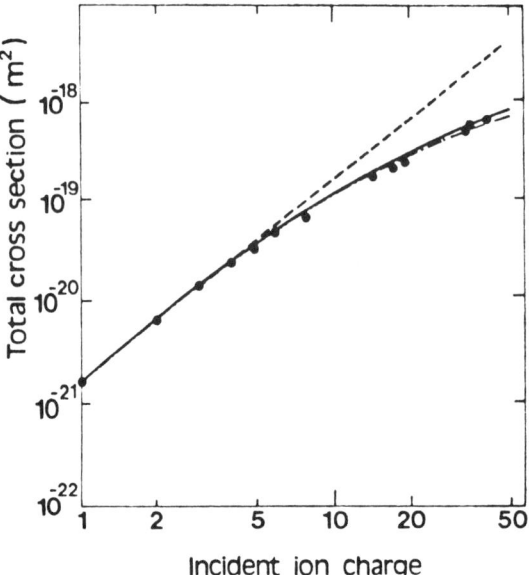

Figure 1. Total cross sections for ionization of He targets as a function of the projectile charge. Theoretical results: _____, CDW-EIS from ref. (4); _ _ _ _, B1; and _._._, Glauber from ref. (18). Experiments from ref. (18) and (19).

However, at high enough collision velocities it could be expected that the B1 approximation could represent the single ionization reaction. Platten and coworkers (2) and Stolterfoht et al.[1] have measured double differential cross section ratios for the impact of 5 MeV amu^{-1} - C^{6+}, O^{8+} and Ne^{10+} and 25 MeV amu^{-1}-Mo^{40+} ions on He atoms respectively. These cross section ratios are obtained between double differential cross sections for impact of an ion of charge Z_P and the corresponding ones for protons on the same target, all divided by a factor Z_P^2, at the same collision velocity. The B1 approximation must give a constant ratio equal to the value one. However, experimental data and the CDW-EIS approximation show a very different behavior from the B1 prediction for all measured cases. In figure 2, results are presented for the Ne^{10+} + He reaction. The ratio is shown as a function of the electron energy for fixed electron scattering angles. It is seen that with respect to proton projectiles the cross sections are enhanced at forward angles and reduced at backward ones. It is a two center effect and is described in our theoretical model because the continuum of the electron is represented in the final channel as a two center function. If, as in an eikonal approximation[21] the initial state is chosen as in CDW-EIS, but the final one only as a continuum of the target, the ratio gives the same constant value for all ejection angles. The electron is ejected in the simultaneous presence of the projectile and residual target and is attracted in the forward direction for the larger projectile charge. In the figure, the two center effect appears to increase with increasing electron energy. It may be understood from the fact that a fast electron feels the two center aspect of a rapidly disintegrating collision system more strongly than an electron, which emerges from the target center. It is expected that the separated projectile-target system is most likely probed by electrons which have a velocity comparable to that of the projectile. Recently, Fainstein et al.[22] have shown that at even higher emission velocities the electrons feel predominantly a combined two center system of effective nuclear charge $Z > Z_P$, Z_T[13]. So the ratio for forward angles reaches its maximum in the region of the CTC peak, decreasing to the value one at an electron energy corresponding to the binary encounter peak and to lower values at higher emission velocities. Unfortunately, measurements have been made for electron energies up to around 1 keV. It would be welcome to have measurements at higher ejection energies or at lower projectile impact velocity to test this new two center effect.

In figure 3, the total cross sections for ionization of He targets by H^+, He^{2+} and Li^{3+} ions calculated with the CDW-EIS model [4] are compared with calculations using the B1 approximation (23) and with experimental data[19,24]. The CDW-EIS results are in good agreement with experimental data and give the correct behavior at the maximum of the cross section. The B1-total cross sections overestimate the experiments at intermediate and low energies. In these energy regions, Z_P^3-polarization and binding effects play a principal role. These effects have been studied in detail by Basbas and coworkers[25,26] by using the perturbed-stationary-states (PSS)

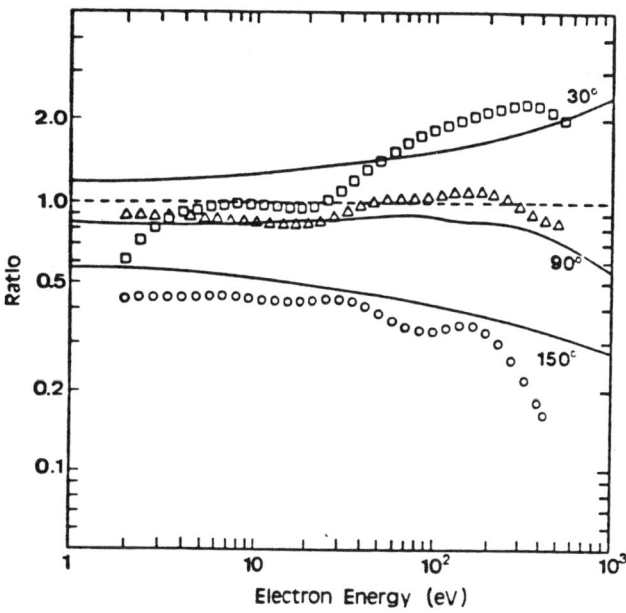

Figure 2. Ratio between total cross sections for 5 meV/amu-Ne^{10} and 5 Me-H^+ ions on He, all divided by Z_P^2 (see the text). Theory: _____, CDW-EIS results from ref. (4). Experiments from ref. 2.

approximation. They showed that when positive charged projectiles are considered (with $Z_P << Z_T$), binding effects decrease the B1-cross section at low energies and polarization effects increase the B1-cross section at intermediate energies. In figure 4 we have plotted the ratio between total cross sections for ionization of He by different projectiles (antiproton, He^{2+} and Li^{3+}) and the total cross section for ionization of He by proton impact. The B1 approximation gives constant values Z_P^2 for the different systems analyzed. Our CDW-EIS calculations are in close agreement with experimental data showing binding effects for the positive ions and anti-binding effects for antiprotons (see also (27)). For antiprotons, polarization effects decrease the B1 predictions and for positive charges, polarization is shadowed by binding effects. However, Martínez et al.[7] have shown that the CDW-EIS approximation represents the effect giving larger total cross sections than the B1 ones for $1 \leq Z_P << Z_T$ in coincidence with the predictions of

Basbas et al.

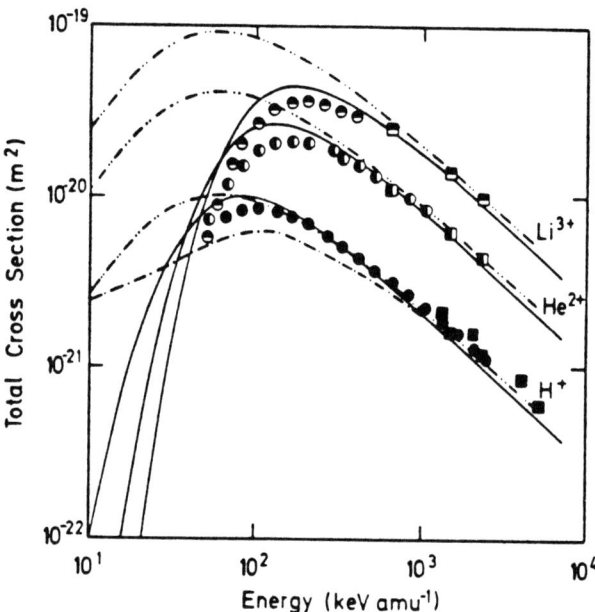

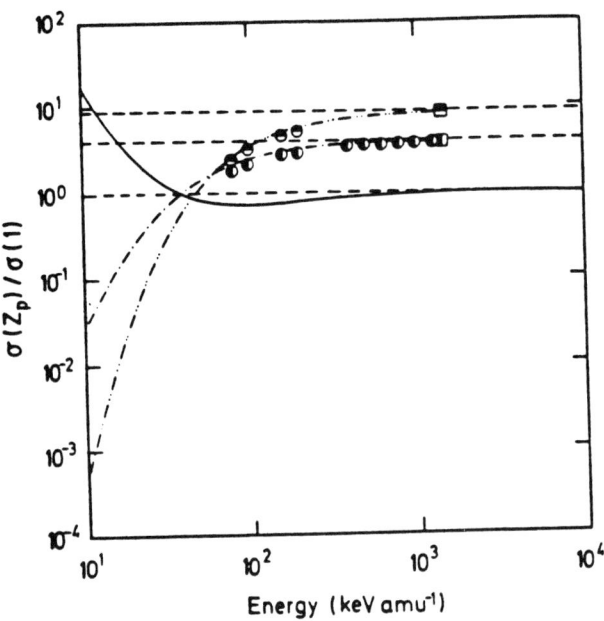

Figure 3. Total cross sections for ionization of He by H^+, He^{2+}, and Li^{3+}. Theory: ___, CDW-EIS; and _._._, Glauber approximations (see ref. 3); and _..._, B1 approximation (ref. 23). Experiments from ref. (19) and (24).

Figure 4. Ratio between total cross sections for ionization of He by bare ion impact (p^-, He^{2+}, Li^{3+}) and by H^+ impact. Theory: ____, CDW-EIS model for p^-; _._, CDW-EIS model for He^{2+}, _.._, CDW-EIS model for Li^{3+} impact; all results from ref. (3). Experiments from ref. (19) and (24).

In order to study ionization from outer shells than the K-one we have calculated the cross sections for impact of protons on Ne targets at collision energies for which the L-shell plays a main role. In figure 5, double differential cross sections for an ejection angle $\theta_K = 0$ degrees and as a function of the final electron energy are presented for impact of 300 KeV-protons on Ne. Contributions coming from the 1s, 2s, $2p_0$, $2p_{\pm 1}$ states are shown (with the z-quantization axis chosen in the direction of v). At small E_k the ionization is dominated by L-shell electrons, in particular those coming from the 2p state. A similar effect is also observed in the CTC peak region. In the binary encounter domain a dip appears for electrons ejected from the $2p_0$ initial state and at higher electron velocities a shoulder is distinguished for the 2s initial state. To see the origin of these structures, we have employed a B1 approximation. If the description is further simplified assuming a plane wave instead of the final target continuum state, we obtain the transition amplitude as a function of the transverse momentum transfer $\underset{\sim}{\eta}$:

$$R(\underset{\sim}{\eta}) = \frac{2\pi i Z_p}{v|\underset{\sim}{q}|^2} \tilde{\phi}_i(\underset{\sim}{K})\Big|_{\underset{\sim}{K}=\underset{\sim}{q}+\underset{\sim}{k}} \quad (10)$$

with $\underset{\sim}{q}$ the total momentum transfer and where $\hat{\phi}_i$ indicates the Fourier transform of the initial state. The Fourier transform $\tilde{\phi}_{2p_0}$ cancels for values of $\underset{\sim}{k}$ situated on the binary sphere, independently of $\underset{\sim}{\eta}$. For the 2s case, the values of $\underset{\sim}{k}$ producing the zeros are functions of $\underset{\sim}{\eta}$. Therefore, when integrating $|R(\underset{\sim}{\eta})|^2$ over $\underset{\sim}{\eta}$ to obtain the double differential cross sections from a 2p initial state, it cancels for values of $\underset{\sim}{k}$ on the binary sphere. For a 2s initial state, the $\underset{\sim}{\eta}$ integration eliminates the possibility of reaching a zero for the cross section at a given value of $\underset{\sim}{k}$. However, since the main contribution to the integral comes from a localized region of $\underset{\sim}{\eta}$, we may expect to have a minima for values of $\underset{\sim}{k}$ connected to the range of $\underset{\sim}{\eta}$ through the equation giving the values of $\underset{\sim}{k}$ and $\underset{\sim}{\eta}$ which cancel $\widetilde{\phi}_{2s}$. The two structures, which can be identified in the Fourier transform of the initial states, are due to the nodes of the corresponding initial bound state wavefunctions[16].

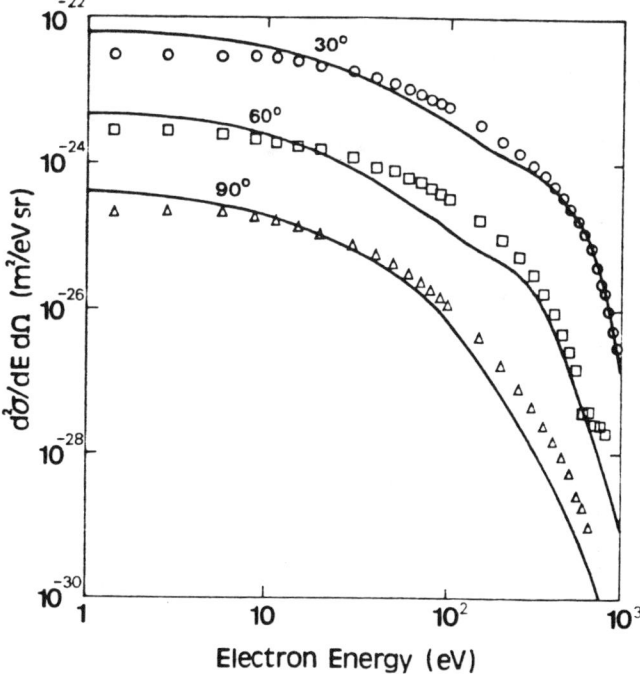

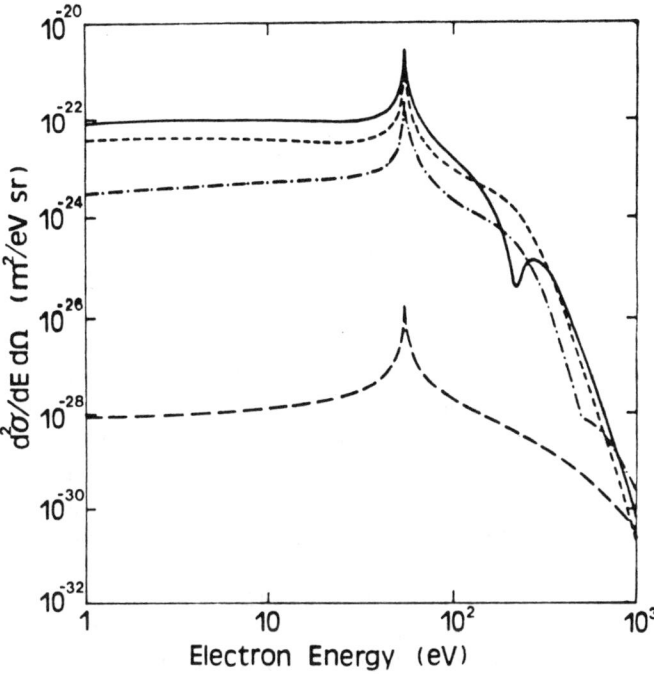

Figure 5. Double differential cross sections for impact of 300 keV-protons on Ne targets, for a 0 degrees-fixed electron scattering angle as a function of the final electron energy. Contributions from: ___, 1s; _._, 2s; ...., $2p_{\pm 1}$; and ___, $2p_0$ initial states (present CDW-EIS calculations).

In figure 6, double differential cross sections for ionization from any initial state are presented for the same system at fixed scattering angles ϕ_k = 30, 60 and 90 degrees. Present CDW-EIS results and experimental data (28) are in reasonable qualitative agreement.

Figure 6. Same as in figure 5, but for fixed scattering angles of 30, 60 and 90 degrees. Theory: ___, present CDW-EIS approximation. Experiments from ref. (28).

Total cross sections for the p+Ne system are shown in figure 7. Contributions coming from the 1s, 2s and 2p initial state are discriminated. For the 1s case the agreement between present CDW-EIS results (see also (29)) and experiments (30,31) is very close. Not so close an agreement is obtained for the 2s initial state, with experiments from Eckhardt and Schartner (32). Theoretical CDW-EIS cross sections and measured data (33) are in good accordance for collision energies higher than 30 keV.

As indicated above, experiments are now possible at CERN with antiproton beams at a few hundred keV- collision energies. Comparisons for antiprotons and protons impacting on the same targets permits us to look for different behavior of the electron in the presence of a combined field of repulsive and attractive potentials or of attractive potentials. It must be also noted that the channels of charge exchange, which can influence the single ionization process at intermediate energies, are not present in the case of antiprotons. In figure 8 we show CDW-EIS double differential cross sections for impact of protons and antiprotons on C targets at 1 Me-collision energy and a fixed angle θ_k = 2 degrees.

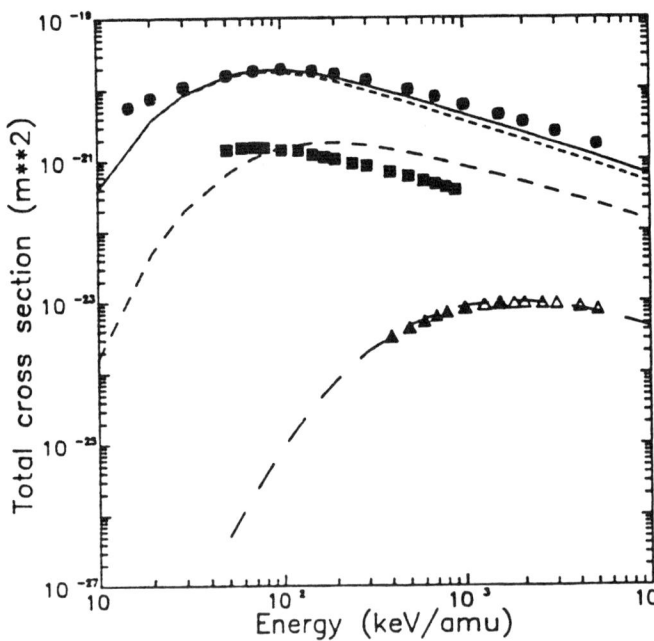

Figure 7. CDW-EIS cross sections for impact of H+ on Ne targets, as a function of the collision energy. Contributions from 1s (long dashed line), 2s (short dashed line) and 2p (shortest dashed line) states and total cross sections (full line). Experiments from ref. (30)-(33).

Contributions coming from initial K- and L-shell electrons are also discriminated. At low energies E_k, in both physical cases, the ionization is dominated by L-shell electrons, but in the CTC region ejection is dominated by K-shell ones. The more significant difference is the presence for antiprotons of an "anti-cusp" located at these electron velocities for which the CTC peak appears for protons. This dip, previously obtained[34] for the p⁻ + H (1s) system, has been recently measured for the system here studied[12] at forward scattering angles, with a 5 degree-angular resolution. Electrons moving with $k \simeq v$ tend to be repelled by antiprotons and attracted by protons. Even when the CTC peak quickly disappears when q_k increases, the dip remains for larger angles. Fainstein et al.[9] have studied this behavior for the ionization of He targets by impact of protons and antiprotons. This effect arises because the density of continuum states corresponding to protons and antiprotons is very different for $k \simeq v$. A dominance of K-shell electrons is also observed for higher electron velocities (which are associated with closer collisions), except in the binary encounter peak. It is associated with the fact that L-shell electrons move slower than K-shell ones, satisfying in a more approximate way the classical condition of binary encounter.

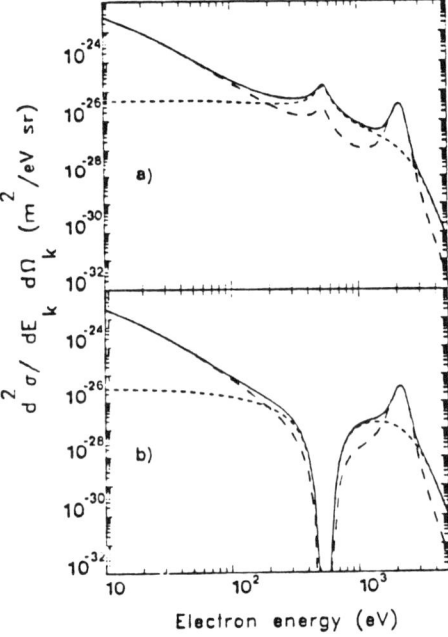

Figure 8. Double differential cross sections for impact of a) protons and b) antiprotons on C targets at a fixed 2 degrees-electron scattering angle. CDW-EIS calculations from ref. (10) : contributions from initial K-shell (short dashed line), L-shell (long dashed line) and K+L-shells (full line) electrons.

Differences between proton and antiproton projectiles can also be observed when single differential cross sectiosn are studied for He targets.[11] In figure 9, these cross sections are shown as a function of the electron scattering angle for a 200 keV-collision energy. For the proton case, a close agreement is obtained between CDW-EIS results and experimental data[33] over all the angular range. No differences with experiments are found at large angles as given by CTMC results.[35] As pointed out by Olson and Gay[34], ejection is enhanced in the forward direction with protons and in the backward direction with antiprotons. This effect is related to the attractive or repulsive potential of the projectile. The good agreement obtained for single differential cross sections between experiments and CDW-EIS calculations remains for total cross sections. The recommended data for protons from Rudd et al.[36] is $(7.0\pm0.7)\times10^{-17}$ cm² while the CDW-EIS calculations give 7.3×10^{17} cm² and 5.9×10^{-17} cm² for proton and antiproton impact respectively. At 500 keV-collision energy, the recommended data results

$(3.7\pm0.4)\times10^{-17}$ cm^2 for protons while the CDW-EIS calculations give 3.7×10^{-17} cm^2 for protons and antiprotons respectively.

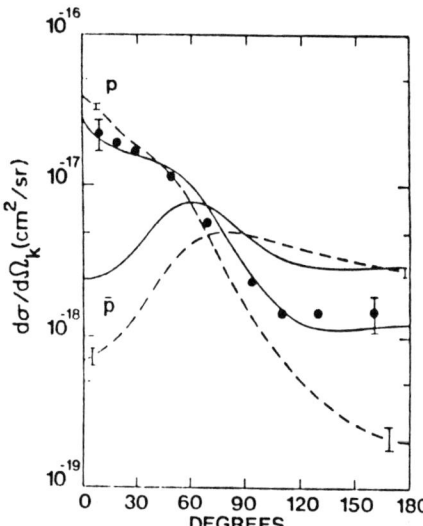

Figure 9. Single differential cross sections as a function of the electron scattering angle for impact of 200 keV-protons and antiprotons on He targets. CDW-EIS calculations from ref. (11) and CTMC results form ref. (35). Experiments from ref. (33).

Other distorted wave models, such as the Continuum Distorted Wave (CDW) and the Symmetric Eikonal (SE) hae been developed. However, the CDW-EIS one appears to give the better representation of the single ionization process. For details on the CDW and SE approximations the reader is referred to the original works[37-39].

4. CONCLUSIONS

The single electron ionization process in ion-atom collisions at intermediate and high impact energies has been studied. Attention has been focused on the use of the CDW-EIS approximation which takes account of the simultaneous action of the projectile and final channels. Evidence is given that ionization is a two center reaction and two center-approximations must be used in order to obtain an adequate represenation of the process. Double, single and total cross sections for ionization of electrons from different shells by impact of bare positive ions and antiprotons are theoretically analyzed and when possible compared with experimental data.

ACKNOWLEDGMENT

This work was partially supported by the International Atomic Energy Agency research contract N° 5365/RB.

REFERENCES

1. N. Stolterfoht, D. Schneider, J. Tanis, H. Altevogt, A. Salin, P. D. Fainstein, R. D. Rivarola, J. P. Grandin, J. N. Scheurer, S. Andriamonje, D. Bertault and J. F. Chemin, Europphys. Lett. 4 (1987) 899.

2. H. Platten, G. Schiwietz, T. Schneider, D. Schneider, W. Zeits, K. Musiol, T. Zouros, R. Kowallik and N. Stolterfoht, Proc. of the 15th Int. Conf. on the Physics of Electronic and Atomic Collisions, Brighton (Amsterdam: North Holland, Abstracts of Contributed Papers (1987) 437.

3. P. D. Fainstein, V. H. Ponce and R. D. Rivarola, Phys. Rev. A. 36 (1987a) 3639.

4. P. D. Fainstein, V. H. Ponce and R. D. Rivarola, J. Phys. B. 21 (1988a) 287.

5. D. R. Bates and G. Griffing, Proc. Roy. Soc. A 66 (1953) 961.

6. K. Dettman, K. G. Harrison and M. W. Lucas, J. Phys. B 7 (1974) 269.

7. A. E. Martínez, G. R. Deco, R. D. Rivarola and P. D. Fainstein, Nucl. Instrum. Meth. B (1989), in press.

8. K. Elsener, Comments At. Mol. Phys. 22 (1989) 263.

9. P. D. Fainstein, R. D. Rivarola and . H. Ponce, Phys. Rev. A. (1989a), in press.

10. P. D. Fainstein, R. D. Rivarola and V. H. Ponce, Phys. Rev. A (1989a) in press.

11. P. D. Fainstein, R. D. Rivarola and . H. Ponce, J. Phys. B (1989b), submitted for publication.

12. H. Knudsen, Invited Lecture at the Tenth Int. Conf. Appl. Accel. Res. Ind., North Texas State University, Denton and private communication (1988).

13. A. Salin, J. Phys. B 2 (1969) 631.

14. D. S. F. Crothers and J. F. McCann, J. Phys. B 16 (1983) 3229.

15. P. D. Fainstein, R. D. Rivarola and . H. Ponce, J. Phys. B (1989c), submitted for publication.

16. P. D. Fainstein, R. D. Rivarola and V. H. Ponce, J. Phys. B 22 (1989d) 1207.

17. Dz. Belkic, R. Gayet and A. Salin, Phys. Rev. 56 (1979) 279.

18. J. H. McGuire, A. Muller, B. Schuch, W. Groh and E. Salzborn, Phys. Rev. A 5 (1987) 2479.

19. H. Knudsen, L. H. Anderson, P. Hvelplund, G. Astner, H. Cederquist, H. Danared, L. Liljeby and K.-G. Rensfelt, J. Phys. B 17 (1984) 3545.

20. A. Salin, Phys. Rev. A 36 (1987) 5471.

21. J. F. Reading and E. Fitchard, Phys. Rev. A 10 (1974) 168.

22. P. D. Fainstein, V. H. Ponce and R. D. Rivarola, J. Phys. (Paris) 50 (1989e) C1-183.

23. K. L. Bell and A. E. Kingston, J. Phys. B 2 (1969) 653.

24. M. B. Shah and H. B. Gilbody, J. Phys. B 18 (1985) 899.

25. G. Basbas, W. Brandt and R. Laubert, Phys. Rev. A 7 (1973) 983.

26. G. Basbas, W. Brandt and R. Laubert, Phys. Rev. A 17 (1978) 1655.

27. W. Brandt and G. Basbas, Phys. Rev. A 27 (1983) 578.

28. L. H. Toburen (1986), private communication.

29. A. E. Martínez, G. R. Deco, R. D. Rivarola and P. D. Fainstein, Nucl. Instrum. Meth. B 34 (1988) 32.

30. C. L. Cocke, R. K. Gardner, B. Curnutte, T. Bratton and T.K. Saylor, Phys. Rev. A 16 (1977) 2248.

31. M. Rodbro, E. Horsdal Pederson, C. L. Cocke and J. R. Macdonald, Phys. Rev. A 19 (1979) 1936.

32. M. Eckhardt and K. H. Schartner, Z. Phys. A 312 (1983) 321.

33. M. E. Rudd, L. H. Toburen and N. Stolterfoht, At. Data Nucl. Data Tables 18 (1976) 413.

34. C. R. Garibotti and J. E. Miraglia, Phys. Rev. A 21 (1980) 572.

35. R. E. Olson and T. J. Gay, Phys. Re. Lett. 61 (1988) 302.

36. M. E. Rudd, Y. K. Kim, D. H. Madison and J. W. Gallagher, Rev. Mod. Phys. 57 (1985) 965.

37. Dz. Belkic, J. Phys. B 11 (1979) 3529.

38. P. D. Fainstein and R. D. Rivarola, J. Phys. B. 20 (1987b) 1285.

39. R. D. Rivarola and P. D. Fainstein, Nucl, Instr. Meth. B 24/25 (1987c) 240.

ELECTRON CAPTURE AND ENERGY-GAIN SPECTROSCOPY

Knud Taulbjerg*

Oak Ridge National Laboratory
Oak Ridge, Tennessee 37831-6373
and
University of Tennessee
Knoxville, Tennessee 37996-1200

The applicability of translation energy spectroscopy as a tool to determine individual reaction cross sections in atomic collisions is analyzed with special emphasis on the electron capture process in highly charged ion collisions. A condition is derived to separate between higher collision energies where translation energy spectroscopy is problem free and lower energies where strong overlap of individual spectra features prohibits an analysis of the total translation energy spectrum by means of a simple deconvolution procedure.

I. INTRODUCTION

It is a well-known experimental technique to employ an energy analysis of scattered projectiles or recoiling target atoms to determine differential cross sections for inelastic processes in ion-atom collisions. As a matter of fact, the fundamental studies of inner shell processes in the sixties[1] that stimulated the development of the Fano-Lichten model for ion-atom collisions were performed using this technique. The method is based on the fact that various inelastic channels at fixed scattering angle are separated in the residual energy of the scattered projectile or in the energy of the recoiling target atom by characteristic amounts uniquely related to the corresponding Q-values.

Modern applications exploit that the resolution power of the technique is strongly increased at forward angles to allow a separation of channels that differ by as little as an electron volt in inelasticity. For processes that are dominated by sufficiently small scattering angles, it is possible to record complete differential cross sections for individual inelastic processes as separated features in the spectrum of residual projectile energy without discrimination against scattering angles. The technique is then commonly referred to as translation energy spectroscopy.

It is the purpose of this paper to discuss the applicability as well as the limitations of translation energy spectroscopy with special emphasis on the electron capture process in highly charged ion collisions, in which case the technique is often referred to as energy-gain spectroscopy. First, we briefly review the basic kinematics needed to provide a unique correspondence between differential cross sections and translation energy spectra in collisions of known inelasticity. Next, a qualitative example is considered to illustrate the conditions under which total translation energy spectra can be resolved into individual components corresponding to different intrinsic states. Quantitative applications are discussed at the end.

II. KINEMATICS

The kinematical relations that are needed to determine the energy of a projectile after a collision of given inelasticity, Q_n, are trivial to express in the center-of-mass frame of the two-body collision system. The transformation to the laboratory frame is more complicated but may be completed in closed form (see, for example, Ref. 2). In this work, we shall only be concerned with small laboratory scattering angles, θ, in which case we find to first order in $\sin^2\theta$,

$$\Delta E_n = E_f - E_i = Q_n - \alpha E_i \sin^2\theta, \qquad (1)$$

where E_i and E_f are projectile energies before and after the collision, and Q_n is the amount of inelasticity in the considered reaction channel n, positive for exothermic and negative for endothermic processes. The parameter α is given by

$$\alpha = M_p/M_T + (Q_n/E), \qquad (2)$$

---

*Permanent address: Institute of Physics, Aarhus University, DK-8000 Aarhus, Denmark.

where M_p and M_T are the mass of the projectile and the target, and where the last term can be ignored for all practical purposes in this work. The parameter α is accordingly unity for symmetric systems but becomes much larger than unity in the important case of collisions of highly charged ions with light target atoms or molecules.

The relation given by Eq. (1) may be used to express the translation energy spectrum pertaining to a specific inelastic process, n, in terms of the corresponding differential scattering cross section. By definition, we write

$$\frac{d\sigma_n}{d\Delta E} = \left.\frac{d\sigma_n}{d\theta} \middle/ \frac{d\Delta E}{d\theta}\right|_{\theta = \theta_n(\Delta E)}. \quad (3)$$

Using Eq. (1), we obtain

$$\frac{d\sigma_n}{d\Delta E} = \left.\frac{1}{\alpha E_i \sin 2\theta} \frac{d\sigma_n}{d\theta}\right|_{\theta = \theta_n(\Delta E)}, \quad (4)$$

which is valid at small scattering angles, or, introducing the solid angle $d\Omega = 2\pi \sin\theta d\theta$,

$$\frac{d\sigma_n}{d\Delta E} = \left.\frac{\pi}{\alpha E_i} \frac{d\sigma_n}{d\Omega}\right|_{\theta = \theta_n(\Delta E)} \quad (5)$$

still only valid at small scattering angles. It is seen that the translation energy spectrum at forward angles terminates at $E_f = E_i - Q_n$ for the nth reaction channel and extends towards lower energies with a profile that uniquely represents the corresponding differential scattering cross section $d\sigma_n/d\Omega$. The width of the profile is, according to Eq. (1), proportional to the parameter α. To estimate how the width varies with projectile energy, it is appropriate to consider the reduced scattering angle

$$\rho = E_i \sin\theta, \quad (6)$$

which covers a range which is expected to be insensitive to energy variations since ρ^{-1} is an approximate measure of impact parameters in the type of collisions considered here. Combining Eqs. (4) and (6), we find

$$\frac{d\sigma_n}{d\Delta E} = \left.\frac{E_i}{\alpha} \frac{d\sigma_n}{d\rho^2}\right|_{\rho^2 = (Q_n - \Delta E)E_i/\alpha}. \quad (7)$$

To the extent that the width of $d\sigma_n/d\rho^2$ is independent of energy, it appears that the width, W, of the corresponding structure in the translation energy spectrum is proportional to the parameter α/E_i

$$W \propto \alpha/E_i. \quad (8)$$

Note that α may be varied independently of other relevant parameters by considering different target isotopes.

III. QUALITATIVE EXAMPLE

To illustrate the kinematic transformations discussed in the previous section, we consider a qualitative example where two inelastic channels may be populated. The differential cross

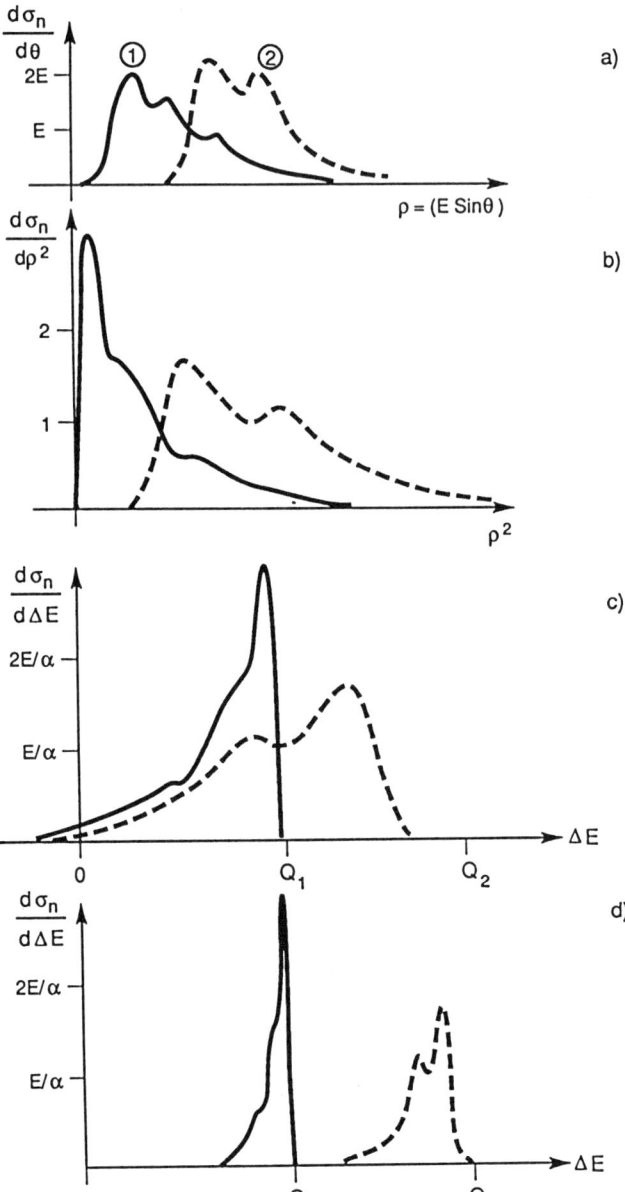

Figure 1. Schematic representation in arbitrary units of differential reaction cross sections and corresponding translation-energy spectra. The mass parameter $\alpha = M_p/M_T$ is a factor of four larger in 1c than in 1d.

sections, $d\sigma_n/d\theta$ are shown schematically in Fig. 1a as a function of ρ. The detailed shape of these curves will, of course, vary with energy, but the range of ρ-values which is covered is expected to be characteristic of the impact parameter range, where the two channels are populated and accordingly to be rather insensitive to energy variations. The cross sections given in Fig. 1a are shown again in Fig. 1b as $d\sigma_n/d\rho^2$ versus ρ^2. Notice how reaction 1 appears with a rather sharp feature at small values of ρ^2, while process 2 gives rise to a broader structure. The compression effect at small scattering angles is essential for the resolution power in translation energy spectroscopy.

According to Eq. (7), we may obtain the translation energy spectra by a reflection of the curves in Fig. 1b combined with a suitable scaling of axes and a shift along the energy axis by an amount which equals the Q-value for the considered reaction channel. This is illustrated in Figs. 1c and 1d for two representative values of the α-parameter which, according to Eq. (7) determines the scale parameters for constant energy. The α-parameter in 1c is four times larger than in 1d. The complete scale parameter is, however, given by (α/E_i), and since the range of ρ-values is expected to be insensitive to energy variations, Fig. 1c and 1d also provide a qualitative picture of typical translation energy spectra for a given system (fixed α) at different energies (up by a factor of four in 1d compared with 1c). The resolution may accordingly be increased by using either a heavier target atom or more energetic projectiles.

In practical experiments it is the total translation energy spectrum; i.e., the sum of individual components that is measured. It is therefore essential for a simple analysis of experimental data that individual components are well separated as in Fig. 1d, in which case Eq. (5) can be used directly to extract differential cross sections for the various reaction channels. Similarly, integrated reaction cross sections are correctly represented by the area under the individual components in the spectrum. In case (c), on the other hand, it is not possible to resolve a measurement of the total spectrum unambiguously into individual components, and it is realized that simple procedures to determine the relative population of the two channels are bound to fail, typically with the result that the part of the cross section for process 2 that appears at energies below Q_1 is misinterpreted.

To provide a qualitative estimation of the width of individual components in translation energy spectra, we may consider electron capture in highly charged ion collisions as an example. Using the Landau-Zener model, the relevant range of impact parameters for capture to a state with a specific Q-value is $b \lesssim Z_p Z_T/Q$ in atomic units. The corresponding range of scattering angles is given as $\theta \gtrsim Q/E_i$, which implies that the ρ-parameter is limited to a range between Q and, say, 2Q. The width of the corresponding structure in the translation-energy spectrum is accordingly

$$W \simeq 3 \frac{\alpha}{E_i} Q^2. \qquad (9)$$

Spectral overlap is avoided if

$$W < \Delta Q = Q_n - Q_{n-1}. \qquad (10)$$

This implies that the condition

$$E_i > 3\alpha \, Q^2/\Delta Q \qquad (11)$$

must be satisfied to ensure a non-overlapping spectrum. In a typical case we may have

$$Q \lesssim 20 \text{ eV} \quad , \quad \Delta Q \gtrsim 3 \text{ eV},$$

which implies that

$$E_i(\text{keV}) \gtrsim \alpha/2$$

to separate the features in the translation energy spectrum. The condition in Eq. (11) is satisfied in most experimental situations for symmetric systems ($\alpha = 1$) but not at lower collision energies in the case of light targets like helium or hydrogen. This conclusion is, of course, somewhat disappointing since studies of highly charged ions in collision with light targets, and especially with atomic hydrogen or its heavier isotopes, are of particular interest. Note, however, that spectral overlap does not mean that translational energy spectroscopy is useless. It only means that some spectral information must be provided to analyze experimental data or that a comparison with theory is to be made directly at spectral levels. A quantitative example is considered in the following section.

IV. QUANTITATIVE RESULTS

The qualitative discussion above will now be substantiated by quantitative results. As an example, we consider experimental translation energy spectra obtained by Giese et al.[3] The data shown in Fig. 2 represent electron capture in collisions of Ar^{+6} with argon and with molecular and atomic deuterium at 3.27 keV. The shape of the experimental spectra is, to a large extent, determined by the experimental resolution which is about ±1 eV, but the effect of the variation of the mass parameter $\alpha = M_p/M_T$ is seen in the data. The width parameter is estimated by Eq. (9) as $W \lesssim \alpha/3$ eV. The individual spectral features are accordingly much narrower

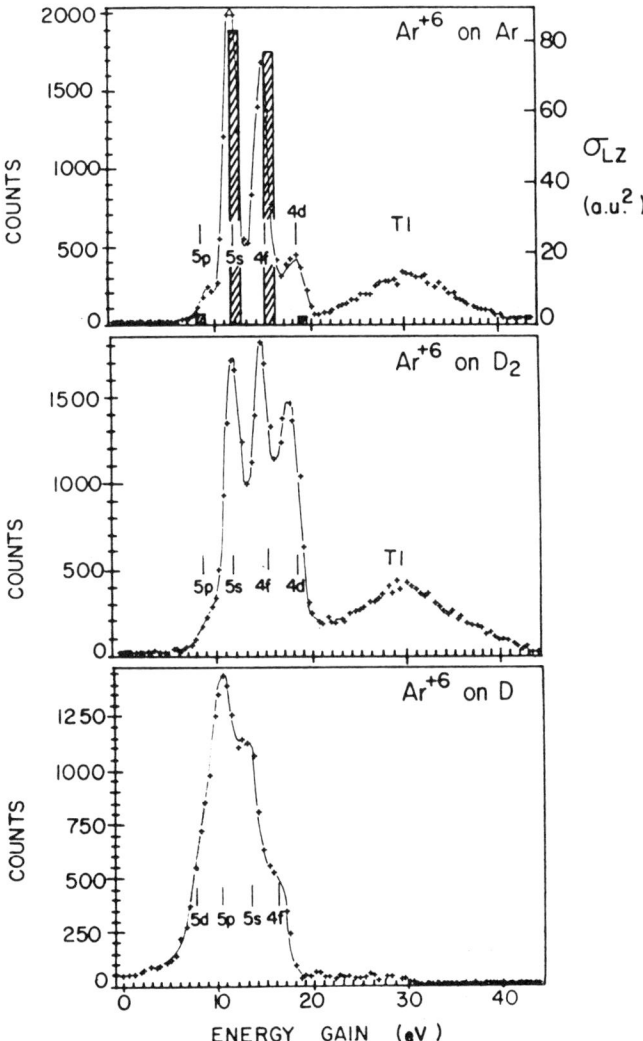

Figure 2. Experimental energy gain spectra in Ar^{+6} electron capture collisions at 3.27 keV impact energy (reproduced from Ref. 2).

simple coupled-channel model developed by Larsen and Taulbjerg,[5] combined with an eikonal transformation to determine theoretical energy-gain spectra for Ar^{+6}-D(1s) collisions. Briefly, the model employs an expansion of the time-dependent Schrödinger equations in a basis of projectile and target states and a Galilean-invariant first-order treatment of electron translation factors. Final states on the projectile are modeled by a quantum defect wavefunction of the Bates-Damgaard type. The parameters of the model are accordingly determined only by the binding energies of the various states. These are known experimentally or may be derived by interpolation methods. The first-order treatment of translation factors allows the coupling elements to be pre-evaluated in a suitable mesh of internuclear separations independent of collision velocity and impact parameter. This is essential to reduce the time consumption in large-scale multi-channel calculations. The approximation is sufficiently accurate (see Fig. 2 of Ref. 5) in the velocity range considered in this work. A further, essential reduction in computation time is gained by employing the pre-orthonormalization procedure described in Ref. 6.

Convergence is tested by varying the size of the basis. In our experience, it is generally important to include complete principal shells to allow appropriately for the combined effect of Stark mixing and rotational coupling. In the Ar^{+6}-D(1s) case, this means inclusion of complete n=4 and n=5 principal shells in the ArVI $1s^2 2s^2 2p^6 3s^2 n\ell$ configuration. This corresponds to a total of 26 reflection symmetric states in the basis. Some of these states are not significantly populated after the collision, and it may be expected that the corresponding states could be eliminated if a molecular expansion was used. The larger atomic basis is, however, a low price to pay to avoid the practical problems in connection with the generation of molecular wavefunctions and coupling elements, and, more importantly, the principal problems with the choice of molecular translation factors.

than the experimental resolution in case of argon targets. The experimental resolution is, however, fine enough to separate between different Q values. A simple fitting procedure determined by the experimental resolution function is accordingly quite adequate to determine individual reaction cross sections or the relative population over final states of the captured electron. In case of atomic deuterium, on the other hand, the mass parameter α equals 20. This implies that individual contributions are expected to be spread out by up to 7 eV or so. More detailed information about the shape of the spectral contributions is therefore required to analyze the data. This is clearly a task for theory.

To provide the required theoretical information, Hansen and Taulbjerg[4] have used the

The coupled channel calculations provide electron capture amplitudes $a_f(\underline{b})$ for each of the considered states in a suitable mesh of impact parameters $\underline{b}$. Capture probability functions, defined as sums over magnetic substates,

$$P_f(b) = \sum_{m_f} | a_f(\underline{b}) |^2$$

and partial cross sections

$$\sigma_f = 2\pi \int_0^\infty b\, db\, P_f(b)$$

are then readily obtained. Capture probability curves for the most important final states are shown in the lefthand panels of Fig. 3. These P(b) curves are characteristic of the primary

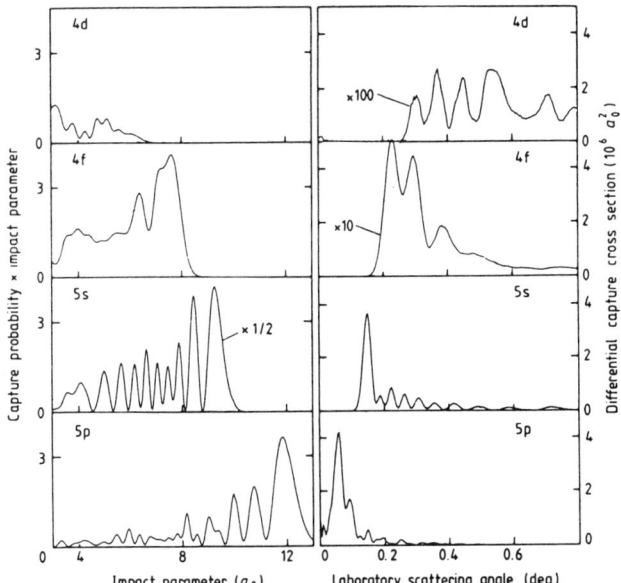

Figure 3. Results from the 26-state calculation of capture in Ar^{6+}-D collisions at 3.27 keV. The lefthand side presents the dependence on impact parameter for the four dominant channels. The righthand side presents the corresponding differential cross sections in the eikonal approximation.

potential curve crossing mechanism for electron capture in highly charged ion collisions. This mechanism controls the regular phase interference oscillations and determines the effective cut-off in the P(b) curves at large impact parameters.

To determine the differential capture cross sections, the capture amplitudes $a_f(\underline{b})$ are amended appropriately by phase factors due to elastic scattering potentials not included in coupled-channel calculations with straight-line trajectories and Bessel transformed to obtain corresponding scattering amplitudes in the eikonal approximation.[7] The resulting differential cross sections are shown in the panels to the right in Fig. 3. Note that the phase interference oscillations also appear in the inelastic scattering cross sections.

Now it is a simple matter to derive theoretical energy gain spectra by aid of the exact relation in Eq. (3). These are shown in Fig. 4. We notice that the dominant lines exhibit a characteristic asymmetric shape with a tail to the low-energy side. The tail is relatively more important in final states with higher Q-values, plainly reflecting that these states are populated in more close encounters. Note, in particular, that the 4d state gives rise to a very broad feature without any reminiscence of a line structure. Generally, we observe that the tails of the spectra extend below the threshold of the adjacent line. The corresponding

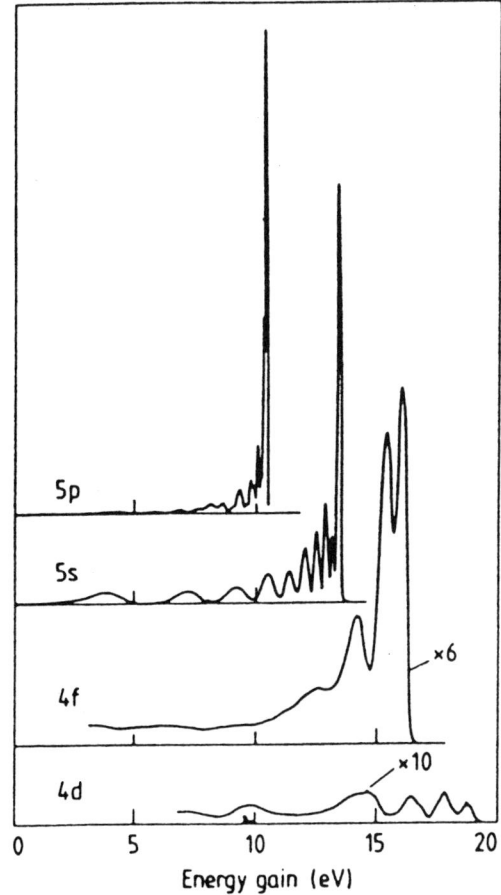

Figure 4. Calculated (26-state) energy-gain spectra for the four dominant capture channels in Ar^{6+}-D collisions at 3.27 keV.

fractions of the reaction cross sections are likely to be misinterpreted in a simple analysis of the summed spectrum. The 4d state, in particular, would be entirely misrepresented since only a small fraction of the 4d cross section appears above the rather sharp threshold for the 4f state in the energy-gain spectrum.

The complete energy-gain spectrum is obtained by summation of the individual components in Fig. 4. To compare with the experimental data, we have modeled the experimental resolution function by a Gaussian with a standard deviation of 1 eV. The theoretical spectrum folded with this resolution function is shown in Fig. 5 in comparison with the experimental data. Apart from the deviation below 5 eV, which originates from the small impact parameter contribution to the 5s channel, the comparison is well within the experimental uncertainties. The theoretical result for the distribution over final states

$$4d:4f:5s:5p = 8:29:44:18 \qquad (12)$$

is accordingly also strongly supported by the

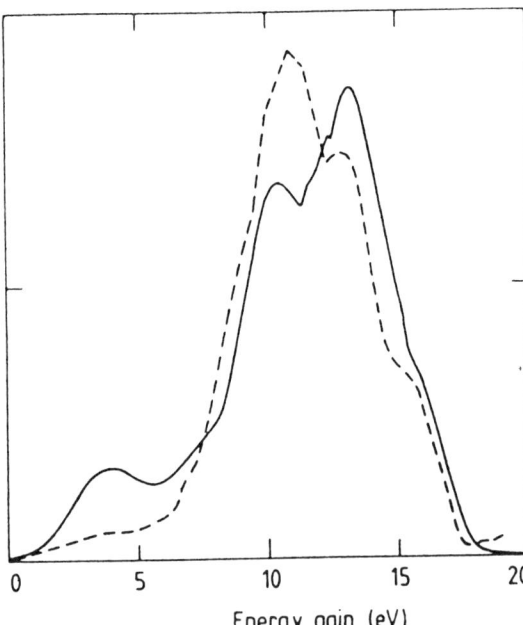

Figure 5. Experimental (broken curve) and theoretical (full curve) energy-gain spectra in Ar^{6+}-D collisions at 3.27 keV. The theoretical spectrum has been folded with a Gaussian resolution function with a standard deviation of 1 eV.

The effect of overlapping features is not peculiar to hydrogen or deuterium targets. This is illustrated in our final example where we consider single electron capture in Ar^{+6}-He collisions. Theoretical energy-gain spectra are shown in Fig. 6 for two impact energies. Our qualitative considerations in Section III are clearly confirmed by the observed energy dependence of the width of the individual spectral features. The overlap between the spectral components is not insignificant in Fig. 6a. As before, this implies that a careful analysis of experimental data is needed to derive the experimental data. This is particularly interesting since a fitting procedure based on Gaussian profiles provides the following distribution:[3]

$$4f:5s:5p:5d = 12:28:41:9. \quad (13)$$

This distribution is similar to the one in Eq. (12) but it is shifted one unit in the final state assignment. This plainly just reflects the fact that the 4d state is spread out over the whole spectrum and therefore is lost on the high energy side of the structure and that the tails of the strongly populated lines accumulate on the low energy side in the region normally assigned to the 5d level.

We have expanded our calculations to a wider range of Ar^{+6} impact energies in collisions with atomic hydrogen and compared with the experimental data by Afrosimov et al.[8] on the distribution over final states, derived from energy-gain spectra, presumably by a simple deconvolution procedure. There is a consistent departure at lower energies (E < 20 keV) between these distributions and the calculated ones. The trend is similar to the departure described in detail above. At higher energies (E = 20-40 keV) there is good agreement between experiment and theory. This is quite understandable from the spectral shapes in Fig. 4, considering that the width of the individual features is expected to be reduced by a factor of 5 at E ~ 30 keV.

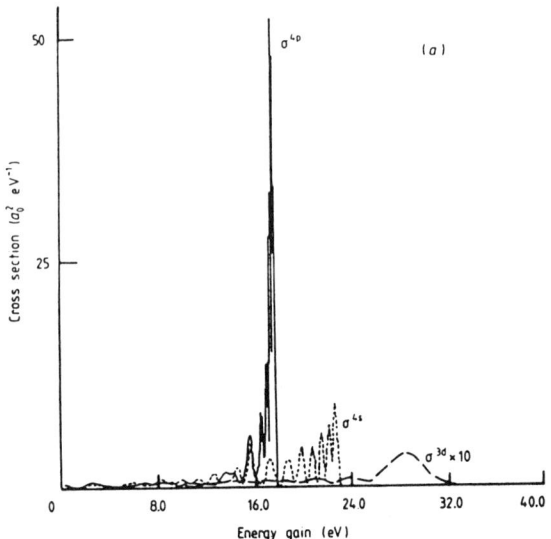

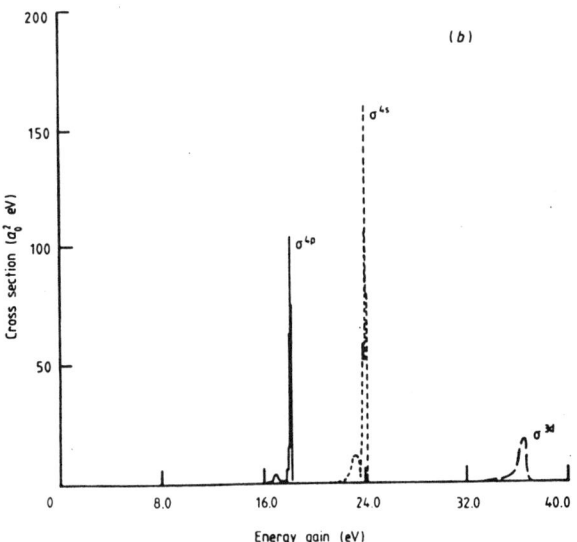

Figure 6. (a) Calculated theoretical energy-gain spectra for the three dominant channels in Ar^{6+}-He collisions at 3.27 keV. (b) Calculated theoretical energy-gain spectra for the three dominant channels in Ar^{6+}-He collisions at 40 keV.

distribution over final states. Generally, cross sections with high Q-values tend to be underestimated in a simple experimental analysis.

The general conclusion of the quantitative computations and of the comparison with available experimental data is in good accord with the qualitative considerations in Section III. In particular, it appears that the condition

$$E \geq 3M_p \cdot Q^2/M_T \, \Delta Q \qquad (14)$$

has been substantiated to separate between the region at higher impact energies where translation energy spectroscopy is essentially problem free from the region at lower energies where a very careful treatment is needed to provide an unambiguous analysis of experimental spectra.

In the low-energy region it is best to accept theoretical results for the distribution over final states until a sound analysis of experimental data has been performed.

ACKNOWLEDGEMENTS

This research was sponsored by the U.S. Department of Energy, Office of Basic Energy Sciences, Division of Chemical Sciences, under contract No. DE-AC05-84OR21400 with Martin Marietta Energy Systems, Inc.

REFERENCES

1. V. V. Afrosimov, Yu. S. Gordeev, M. N. Panov, and N. V. Fedorenko, Sov. Phys. Tech. Phys. 9, 1248 (1965); Q. C. Kessel and E. Everhart, Phys. Rev. 146, 16 (1966); E. Everhart and Q. C. Kessel, ibid. 146, 27 (1966).
2. E. H. Nielsen et al., J. Phys. B18, 1789 (1985).
3. J. P. Giese et al., Phys. Rev. A34, 3770 (1986).
4. J. P. Hansen and K. Taulbjerg, J. Phys. B21, 2459 (1988).
5. O. G. Larsen and K. Taulbjerg, J. Phys. B17, 4523 (1984).
6. J. P. Hansen and K. Taulbjerg, Comp. Phys. Comm. 51, 317 (1988).
7. R. McCarroll and A. Salin, J. Phys. B1, 163 (1968).
8. V. V. Afrosimov, A. A. Basalaev, K. O. Lozhkin, and M. N. Panov, JETP Lett. 46, 107 (1987).

CORRELATION IN ATOMIC SCATTERING

J.H. McGuire and Jack C. Straton

Department of Physics
Kansas State University
Manhattan, KS 66506 USA

We review electron correlation in atomic scattering. The uncorrelated limit, or independent electron approximation, is our starting point. Correlation, or interdependence of electrons, may occur in either the asymptotic or the scattering region. Examples of various kinds of correlated processes are considered.

I. Introduction

Concepts in atomic physics provide understanding useful in other areas of physics, some of chemistry, and a little of biology. In other words atomic physics gives a conceptual basis for understanding our human environment. A problem common to physics, chemistry and biology is the interdependency of atoms, molecules and other particles. Correlation is how interacting individuals affect each other (somewhat like members in a family). In this sense correlation is a conceptual bridge from the properties of individuals to the properties of groups.

Because the term "correlation" is used a little differently by various people working in various fields, we shall briefly consider some of the meanings of correlation. We concentrate on electron-electron correlation in atomic interactions, and use the uncorrelated limit, or independent electron approximation, as a starting point for understanding few and many electron effects in atomic collisions. Static correlation is relatively well understood for two electron systems. We apply existing knowledge of static correlation to asymptotic states of colliding atomic systems. Correlation in the collision region is not so well understood and is a focal point of this review.

II. Correlation

In a general sense correlated means interdependent; that is, correlated particles influence one another. Electron-electron correlation occurs because electrons interact with each other. In a technical sense there are different ways to specify this kind of correlation. It is possible to specify that electron correlation is any effect arising from the Coulomb interaction between electrons. However, it is also possible to define correlation as a deviation from a product of electron probability densities. This is a sensible definition of correlation because the probability for any reaction may be written as a product of individual single electron probabilities if the electron-electron interaction is ignored. This latter definition is also sensible since it corresponds to the definition of correlation used in stochastics and statistical mechanics.[1-5] To the extent that understanding macroscopic properties in terms of microscopic atomic properties is interesting, it is sensible to choose the broader stochastic definition of correlation since statistics are ultimately useful in dealing with large numbers of atoms.

As an introduction to the use of correlation in atomic scattering we review briefly the origins and applications of correlation in stochastics and statistical mechanics. Correlation is often introduced[1,2] as a generalization of the notion of statistical standard deviation of a distribution,

$$\sigma^2 = \langle (x - \langle x \rangle)^2 \rangle. \quad (1)$$

If σ^2 is not zero then the distribution is not localized at a single value $\langle x \rangle$, but rather the distribution is spread out. This leads to the idea that correlated functions are also spread out, i.e., not confined to a single x_i (denoting a single particle), but connect x_i and x_j. Consider

$$f_1(x_i) = \rho(x_i) \ (= \psi^*\psi) \quad (2a)$$

one particle distribution function.

Then the two particle distribution function is uncorrelated if $f_2(x_1, x_2) = f_1(x_1)f_1(x_2)$. In general, however, f_2 is correlated and may be written as

$$f_2(x_1, x_2) = f_1(x_1)f_1(x_2) + g_2(x_1, x_2). \quad (2b)$$

g_2 is called the two particle correlation function. For three particles one may generally write

$$f_3(x_1, x_2, x_3) = f_1(x_1)f_1(x_2)f_1(x_2)$$
$$+ f_1(x_1) g_2(x_2, x_3) + f_1(x_2)g_2(x_1, x_3)$$
$$+ f_1(x_3)g_2(x_1, x_2) + g_3(x_1, x_2, x_3) \quad (2c)$$

and so forth for $f_4, f_5, \ldots f_N, \ldots$ This is called a cluster expansion of the distribution function f_N in terms of the jth order correlation functions g_j.

Correlation has been defined and studied in both space and time.[1-4] In this review we shall consider spatial correlation. Examples of time correlation include time ordering effects and memory. Mostly studies have been confined to two particle or pair correlation since experimental studies of higher order correlation are often difficult. We do note, however, that the N particle distribution function can be related to the N+1 particle distribution function via the two particle correlation potential $v(x_i - x_j)$. This is referred to as the BBGKY hierarchy. Stochastic applications of correlation include molecular clusters, phase transitions, the Fokker-Planck Equation and Pauli correlation for quantum systems of identical particles.

We also wish to mention the generalized correlation coefficient τ introduced recently into atomic structure by Christensen-Dalsgaard.[6] If P is a one particle operator then for a two electron system

$$\tau_f = \frac{\iint g_2(x_1, x_2) \ P(r_1) \ P(r_2) \ dr_1 dr_2}{(\int f_1(r) \ P^2(r) dr) - (\int f_1(r) \ P(r) dr)^2} \quad (4)$$

such that (i) $\tau = 0$ if the electrons are uncorrelated, and (ii) $|\tau| < 1$. Most importantly, τ is independent of choice of basis functions.

In atomic scattering a variety of effective methods have been developed and used to evaluate and study few and many electron effects including correlated coupled channel,[7-14] time dependent Hartree-Fock,[15-17] density functional,[18] statistical energy deposition,[19] classical,[20] stochastic,[21] and hydrodynamic techniques.[22] Our stream of thinking about few and many electron effects will tend, however, to flow along the following course: we are guided by[23]

Exact = Uncorrelated + Correlated Contributions

= Independent Electron Approximation + Asymptotic plus Scattering Correlation

This exact quantum probability amplitude is given by

$$a = \langle \phi_f | U - 1 | \phi_i \rangle \qquad (4)$$

where asymptotic correlation is contained in the asymptotic states ϕ and and scattering correlation is carried by the scattering operator U-1. This is is illustrated in the scattering diagram in Figure 1.

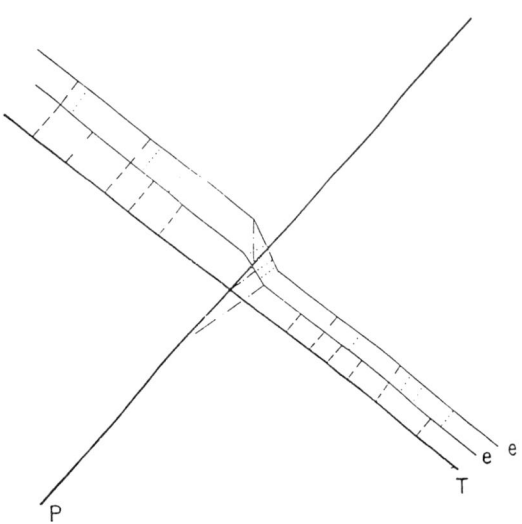

Fig. 1 Scattering diagram for collision of a projectile (P) with a target (T) with two electrons (e). Time increases in the upward direction. The e-e correlation interaction, represented by... may be asymptotic (static) correlation before or after the collision or scattering correlation occurring during the collision.

Solution of the few and many electron problem is incomplete. For Coulomb collisions the asymptotic wavefunction for three particles is not even fully known (although for collisions with neutral particles or collisions where charges are asymptotically screened this is in principle, perhaps, not crucially important). And in the relatively simple three body reaction e + H => e + H* experiment and theory are in fundamental disagreement[24-26] at the present time. Thus we are lead to consider either approximate or numerical solutions of Eqn.(4). This brings us to a further difficulty: neither the scattering problem nor the problem of electron-electron correlation is fully understood for three particles; and we must deal with both. Since the static three body problem is the better understood, we concentrate on solving the scattering part using what is known about electron correlation.

As a beginning we consider collisions with a two electron target (e.g. p + He) and expand Eqn.(4) in terms of the scattering potential V assuming that the effects of the correlation potential v are reasonably well known, namely[23]

$$\langle \phi_f | U-1 | \phi_i \rangle = -i \int \langle \phi_f | V_I(t) | \phi_i \rangle dt$$

$$+ (-i)^2 \iint dt\, dt' (\langle \phi_f | V_I(t) V_I(t') | \phi_i \rangle$$

$$+ \ldots \qquad (5)$$

The first term is linear in Z/v_p where Z and v_p are the charge and velocity of the projectile, the second term is quadratic in Z/v_p, and so forth. For a two particle transition the first term is zero if there is no correlation because interacting with only one electron is not sufficient for a double transition unless the electrons influence each other. To lowest order in v one may think of the term linear in Z (or V) as a term which must be truly a second order term bi-linear in V and v in order to be non-zero. It is the second order term in Z/v_p that contains the lowest order term in the independent electron approximation where the projectile interacts once with each of the two electrons.

If we explicitly expand the first order term in Z in powers of the correlation interaction v we obtain a variety of terms (since V_I is a many electron operator if $v \neq 0$), for example,[23]

$$V_I = e^{-i(H_0+v)t} V e^{i(H_0+v)t}$$
$$= e^{-iH_0 t} V e^{iH_0 t}$$
$$+ (vt) e^{-iH_0 t} V e^{iH_0 t}$$
$$+ \tfrac{1}{2} [[H_0,v],V] + \ldots \quad (6)$$

It may be possible to associate such terms with a variety of terms obtained using diagrams from many body perturbation theory (MBPT)[27,28] such as the first order terms shown in Figure 2. However, separating the contribution of each of these terms amy not be simple. Furthermore, what is correlated and what is not correlated may depend on the choice of basis states used as well. For the moment, at least, we shall therefore avoid serious use of terms like shake, TS1 and ground state correlation, although we shall retain and discuss the physical mechanisms associated with these terms. We shall retain the use of terms linear in Z/v_p, quadratic in Z/v_p, etc., when Z/v_p is small.

A qualitative (and speculative) simple overview of mechanisms for many electron transitions in atomic collisions is given in Figure 3.

III. Independent Electron Approximation

For many electron transitions (e.g., multiple ionization, multiple electron capture), the probability may be represented by a binomial distribution[29-38] of single electron probabilities[30,31] under certain conditions.[34-37] This method originated in the classical

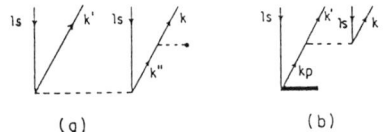
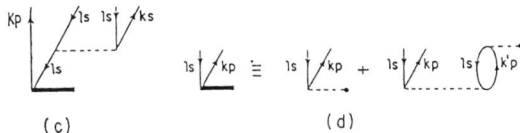

Fig. 2 Many Body Perturbation Theory (MBPT) Diagrams. The ground state (or vacuum), which is not shown, propagates upward until correlation interaction (---) or interaction with the projectile (X——) occurs. Only diagrams first order in correlation are shown: (a) ground state correlation, (b) TS1 or final state correlation, (c) shakeoff, (d) exchange terms.

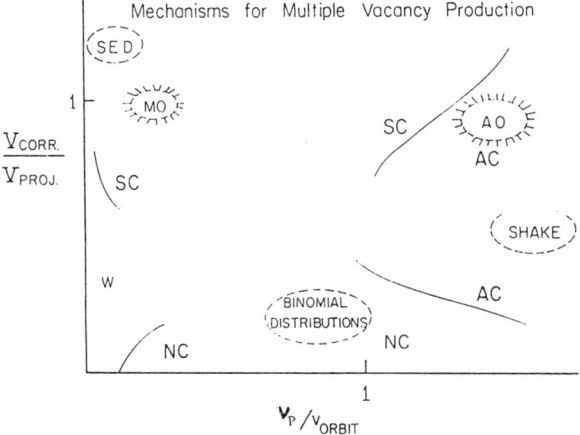

Fig. 3 Overview at Many Electron Transition Mechanisms. $V_{corr}=v$ is the electron electron correlation interaction and V_{proj} is the interaction with the projectile nucleus. If $V_{corr} \to 0$, then there is no correlation (NC) and the independent electron approximation (including binomial distribution) is valid. If the collision velocity v is large the asymptotic correlation (AC) may dominate. If correlation is strong then scattering correlation (SC) may also be significant. AO and MO denote the atomic orbital and molecular orbital regimes. At low v_p the collision time may be long enough to permit a statistical energy distribution (SED) or a Wanner (w) like process.

calculations of Gryzinski,[29] whose success was initially rather surprising. The original picture of Gryzinski is shown in Figure 4. There is now considerable experimental evidence to support the validity of this independent electrons approximation when correlation is small. A recent example[39] is given in Figure 5.

Reading and his co-workers[35-37] have worked out the effects of electron symmetry and Pauli exclusion. This is sometimes referred[1] to as Pauli correlation and sometimes included in the definition of the independent electron approximation. These effects can be large as illustrated in Figure 6. There is, unfortunately, no experimental evidence at present for these effects.

If the projectile charge Z is large then the binomial distribution restricts[38] the maximum value of the single electron probabilities to n/N where n is the degree of ionization and N is

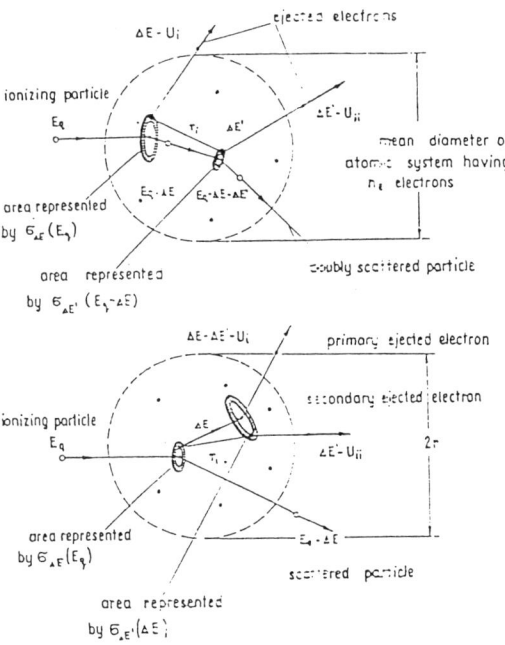

Fig. 4 Gryzinski's original picture of double ionization. The probability was by taking the area $\sigma_{\Delta E}$ divided by the geometric area $4\pi \bar{r}^2$ when $\bar{r}$ is the radius of the sphere denoted by the (---) circle.

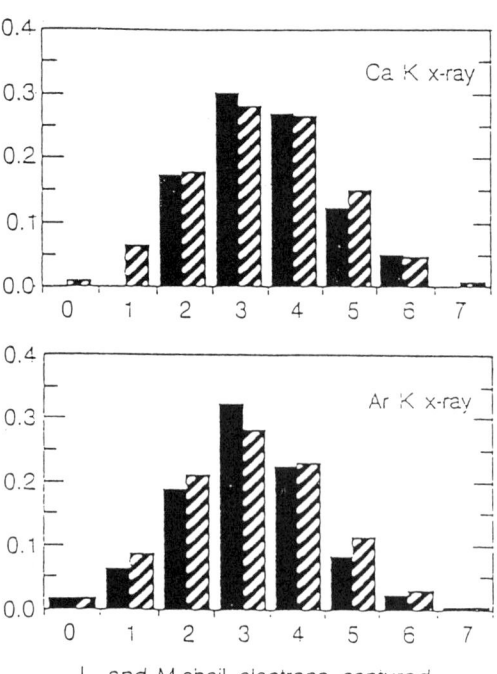

Fig. 5 Typical fit of data (solid) to binomial distribution (hatched) of probabilities given by the independent electron approximation. Taken from Schlacter et al.

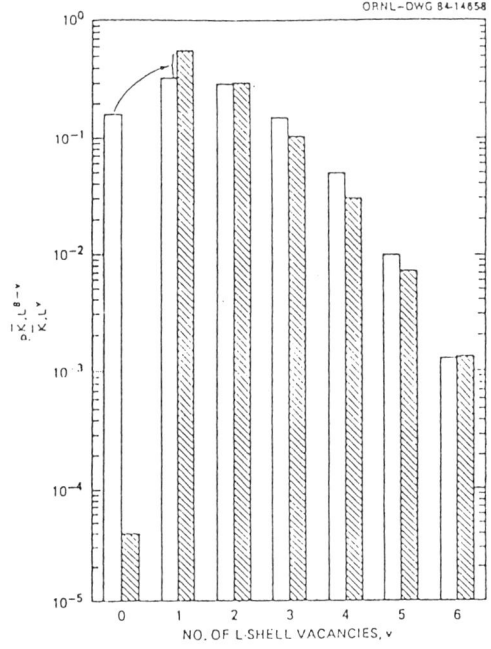

Fig. 6 Pauli correlation distributions (shaded) compared to binomial distributions (unshaded) for $H^+ + Ne$ at 100 keV. Figure taken from Becker, et al. This effect has not yet been verified experimentally.

the total number of equivalent electrons. This means that the useful tabulated SCA values of transition probabilities may sometimes be used even if Z/v_p is not small. Ben-Itzhak has further pointed out[38] that the ratio of double to single ionization is independent of Z and weakly dependent on vp when Z/v_p becomes large.

IV. Asymptotic Correlation

Perhaps the simplest example of asymptotic correlation is shakeoff. Historically double ionization has often been explained[40-43] as a shakeoff process where the second electron was ionized due to a change of screening following removal of a first electron. We also note the possibility of shakeup[44] where the second electron is excited and shakeover[45] where the second electron is captured by the projectile. It has been reasonably argued that shake is not a correlation since it can be explained terms of products of electron density distributions. In the simplest shakeoff picture it was expected that the ratio of double to single ionization in this picture would be independent of the projectile. Recent observations by Cocke's group[46] suggest that there is no single shake limit for various projectile parameters.[46-49] The data shown in Figure 7 suggests rather different values of this ratio for small momentum transfer Q collisions and large Q collisions. However, we feel that the data for photoionization is inconclusive and more is needed. Although Aberg[42] has a good qualitative explanation of this data, good calculations are in our opinion both possible and desirable to determine this relatively simple and fundamental limit, i.e., the behavior of the first order term in Z/v_p of Eqn.(5).

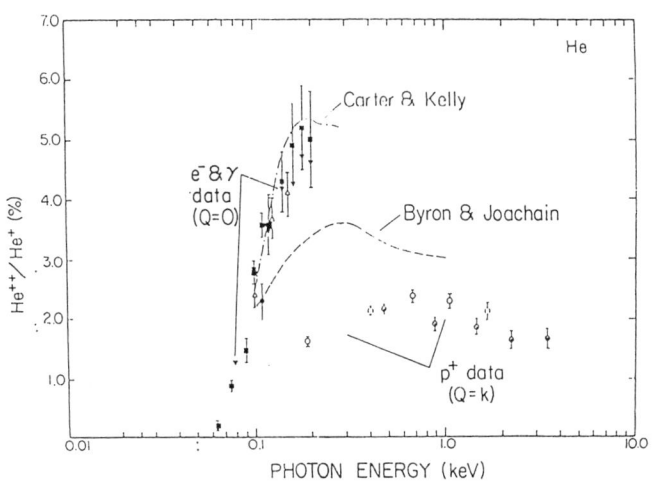

Fig. 7 Different asymptotic limits suggested by observations of Kamber et al. The small momentum transfer (Q=0) ratios seem to differ from the large momentum transfer binary encounter ratios (Q=k = ejected electron momentum) at large (equivalent) photon energies. Observations for Q=0 at large photon energies and calculations for Q=k are needed to complete this picture, representing the simplest example of asymptotic correlation.

V. Scattering Correlation

How correlation works is not at present clear. A number of theorists have attempted to explain a beautiful series of experiments[50-58] done by a growing number of groups who have observed the double ionization of atoms and molecules by ions, electrons, antiprotons and positrons. One sample of these results is given in Figure 8. The theorists all agree that the difference between protons and antiprotons and electrons and antielectrons is due to correlation. However, there is a lot of disagreement[59-63] as to the physical

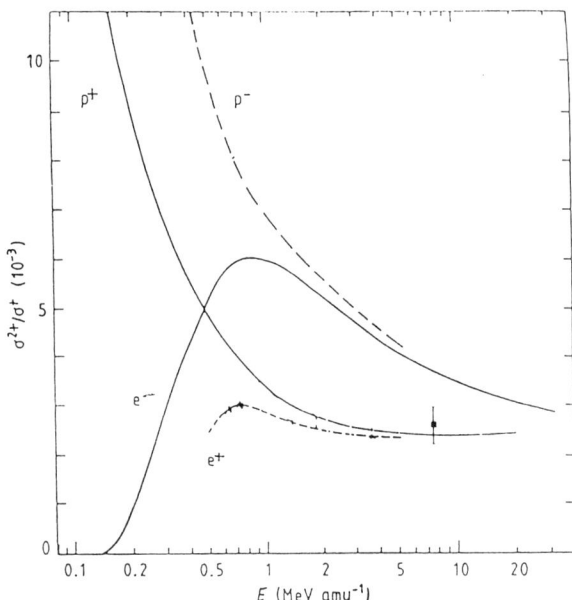

Fig. 8 Observed ratios of double to single ionization of helium by proton (p^+), antiprotons (p^-), electrons (e^-) and positrons (e^+) of the same incident velocity in MeV/amu. Data taken from Charlton, et al. The differences between negative and positive projectiles above 1 MeV/amu are thought to be caused by correlation. But the physical mechanism for this correlation is not clear.

mechanism for correlation. McGuire[58] suggested an interference of quantum amplitudes, but Becker[60] argued that the interference may be restricted by selection rules. Olson[20] suggests that the effects may be explained classically without quantum interference. Vegh suggests[61] that polarization of the charge cloud of the target may be especially important in double ionization. The relatively complete calculations of Reading and Ford[62,11] indicate that correlation must occur between at least two collisions with the projectile. Sorensen[52] suggests that correlation can occur after one collision with the projectile when one electron hits another. Clear understanding remains a challenge.

We continue to encourage studies of double excitation which is easier to understand than double ionization in which the asymptotic three body Coulomb wavefunction is not known. Furthermore, one can look for the effects of possible selection rules more directly in double excitation. Data exists,[64-66] including experiments by Pedersen and Hvelplund[64] and by Giese et al.[65] Calculations of double excitation are now in progress, and the first order term in Z/v_p has been reduced to closed form[67] for double excitation and deexcitation of arbitrary states. The combination of experimental and theoretical studies may give a clearer understanding of the nature of correlation in atomic collisions.

Let us turn for a moment what Stolterfoht[68] calls two center correlation. It is possible to have interactions between projectile and target electrons. In some cases, as Bohr suggested in 1948, the effects of the projectile nucleus and projectile electrons on the target may simply add.[69,70] This is confirmed in the quantum mechanical first Born approximation[71-75] under certain conditions and is confirmed[76,77] as illustrated by the data of Morgan's group shown in Figure 9. In an impact parameter representation the probability amplitude for this process does not factor. Thus it is correlated. However, the wave amplitude for this process in the first Born approximation is a product of independent terms.[71-74] We suggest that use of Wigner distribution functions may resolve this interesting paradox of the meaning of correlation. This has not yet been done in this case.

There are a number of other correlated processes in which pair correlation is being studied[68,13] including transfer excitation,[78-81] double capture,[82,83] ionization excitation,[84,85] transfer ionization[86-88] studies of recoil distributions[89,90] (and other differential cross sections), and molecular fragmentation.[91] Past this lowest order correlation lies the larger expanse of higher order correlation where discovery of new concepts awaits.

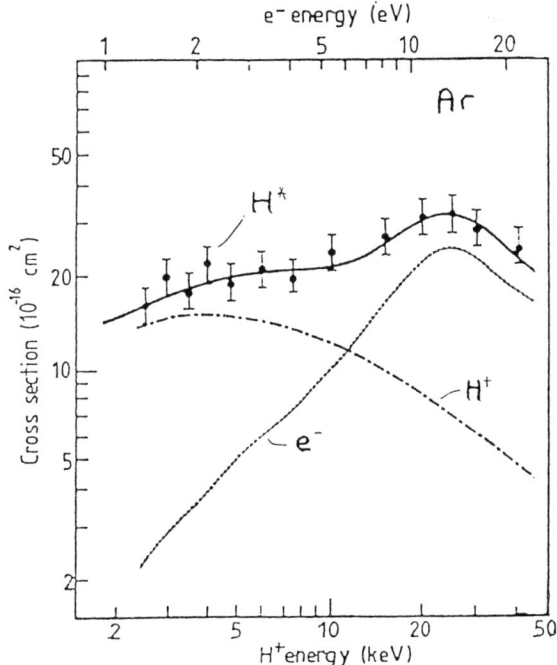

Fig. 9. Effect of projectile electrons. In the free collision model (Bohr, 1948) cross sections for the projectile nucleus and projectile electron add incoherently. This figure shows data from Wang et al. for H^+, e^- and H^* colliding with Ar. The H^* cross sections are equal to the sum of the H^+ and e^- cross sections. In quantum first Born calculations (McGuire et al. 1981) this incoherent limit occurs for collisions with large momentum transfers.

VI. Conclusion

It is quite possible, even likely, that quantum mechanics is too detailed for use by humans (even using supercomputers) to do a complete study of atomic and molecular systems with large numbers of particles. It is, nevertheless, possible for us to use other less detailed methods such as information theory,[21] density functional[18] or other stochastic methods. Common to this effort to understand the nature of both our world and ourselves is the concept of correlation which appears to bridge from what we do understand to what we want to understand.

The difficulty of this endeavor was expressed by Hesse in his story of <u>The Steppenwolf</u>, a person struggling to deal with the gap between total independence and total interdependence:

"Man is not capable of thought in any high degree and even the most spiritual and cultivated of men habitually sees the world and himself through lenses of delusive formulas and artless simplifications."

This problem is a challenge worthy of the very best of us.

VII. Acknowledgment

We thank Professor Chris Sorenson for discussion on stochastic methods. This work was supported by the Division of Chemical Sciences, Office of Energy Research, U.S. Department of Energy.

References

[1] Radu Baleseu, <u>Equilibrium and Non-Equilibrium Statistical Mechanics</u> (John Wiley, NY, 1975).
[2] Kerson Huang, <u>Statistical Mechanics</u> (John Wiley, NY, 1987).
[3] N.G. VanKamper, <u>Stochastic Processes in Physics and Chemistry</u> (North Holland, Amsterdam, 1981).
[4] C.W. Gardiner, <u>Handbook of Stochastic Methods</u> (Springer Verlag, Berlin, 1985).
[5] W.T. Eadie et al. <u>Statistical Methods in Experimental Physics</u> (North Holland, Amsterdam, 1977).
[6] Birte Christensen-Dalsgaard, J. Phys. B21, 2539 (1988).
[7] W. Fritsch and C.D. Lin, Phys. Rev. A27, 3361 (1983).
[8] W. Fritsch and C.D. Lin, J. Phys. B19, 2683 (1986).
[9] M. Kimura, H. Sato and R.E. Olson, Phys. Rev. A28, 2085 (1985).
[10] R. Shingal and B.H. Brandsen, A.M. Ermalov, D.R. Flower, C.W. Newby and C.J. Noble, J. Phys. B19, 309 (1986).
[11] J.F. Reading and A.L. Ford, J. Phys. B20, 3747 (1987).
[12] C. Bottcher, Phys. Rev. Lett. 48, 85 (1982).

[13] P. Rocin, M. Barat, M.N. Gaborland, L. Guillemot, H. Laurant, J. Phys. B$\underline{22}$, 509 (1989).
[14] C. Hertal and A. Salin, J. Phys. B$\underline{13}$, 785 (1980).
[15] K.R. Sandhya Devi and J.D. Garcia, J. Phys. B$\underline{18}$, 4589 (1984).
[16] W. Stich, H.J. Ludde and R.M. Dreizler, J. Phys. B$\underline{18}$, 1195 (1985).
[17] K. Gramlich, N. Grun and W. Scheid, J. Phys. B$\underline{19}$, 1457 (1986).
[18] H. Kohland, R.M. Dreizzler, Phys. Rev. Lett. $\underline{56}$, 1993 (1986).
[19] A. Russek, Phys. Rev. A$\underline{32}$, 246 (1963).
[20] R.E. Olson, Phys. Rev. A$\underline{36}$, 1519 (1987).
[21] A. Blomberg, T. Aberg and O. Goscinski, J. Phys. B$\underline{19}$, 1063 (1986).
[22] M. Horbatsch and R.M. Dreizzler, Z. Physik D$\underline{2}$, 183 (1986).
[23] J.H. McGuire, Phys. Rev. A$\underline{36}$, 1114 (1987).
[24] D.H. Madison, J.A. Hughes and D.J. McGinness, J. Phys. B$\underline{18}$, 2737 (1985).
[25] H.R.J. Walters, Electronic and Atomic Collisions, ed. by H.B. Gilbidy, W.R. Newell, F.H. Read and A.C.H. Smith (North Holland, Amsterdam, 1988) p. 147.
[26] J.F. Williams, Aust. J. Phys. $\underline{39}$, 621 (1986).
[27] S.L. Carter and H.P. Kelly, Phys. Rev. $\underline{24}$, 170 (1981).
[28] T. Ishihara and K. Hino, private communication.
[29] M. Gryzinski, Phys. Rev. $\underline{138}$, A349 (1965).
[30] J.M. Hansteen, A.M. Johansen and L. Kocbach, At. Data Nucl. Data Tables $\underline{15}$, 305 (1975).
[31] D.H. Madison and E. Merzbacher, in Atomic Inner Shell Processes, ed. B. Crasemann.
[32] J.H. Hansteen and O.P. Mosebekk, Phys. Rev. Lett. $\underline{29}$, 1961 (1972).
[33] J.H. McGuire and O.L. Weaver, Phys. Rev. A$\underline{16}$, 41 (1977).
[34] V.A. Sidorovitch and V.S. Nikolaev, J. Phys. B$\underline{16}$, 3743 (1983).
[35] J.F. Reading and A.L. Ford, Phys. Rev. A$\underline{21}$, 124 (1980).
[36] A.L. Ford, J.F. Reading and R.L. Becker, J. Phys. B$\underline{12}$, 2905 (1979).
[37] R.L. Becker, A.L. Ford and J.F. Reading, J. Phys. B$\underline{13}$, 4059 (1980).
[38] I. Ben-Itzhak, T.J. Gray, J.C. Legg and J.H. McGuire, Phys. Rev. A$\underline{37}$, 3685 (1988).
[39] A.S. Schlacter, E.M. Bernstein, M.W. Clark, R.D. DuBois, W.G. Graham, R.H. McFarland, T.J. Morgan, D.W. Mueller, K.R. Stadler, J.W. Stearns, M.P. Stockli and J.A. Tanis, J. Phys. 21, L291 (1988).
[40] M.H. Mittleman, Phys. Rev. Lett. $\underline{16}$, 498 (1966).
[41] T.A. Carlson, Phys. Rev. $\underline{156}$, 142 (1967).
[42] T. Aberg, in "Photoionization and Other Probes of Many Electron Interactions," ed. F. Wuillemier (Plenum Press, NY, 1976) p. 49.
[43] F.W. Byron and C.J. Joachain, Phys. Rev. $\underline{164}$, 1 (1967).
[44] V.L. Jacobs and P.G. Burke, J. Phys. B$\underline{5}$, L67 (1972).
[45] J.H. McGuire, N.C. Deb, Y. Aktas and N.C. Sil, Phys. Rev. A$\underline{38}$, 3333 (1988).
[46] E.Y. Kamber, C.L. Cocke, S. Cheng and S.L. Varghese, Phys. Rev. Lett. $\underline{60}$, 2026 (1988).
[47] V. Schmidt, N. Sander and H. Kuntzemuller, Phys. Rev. A$\underline{13}$, 1748 (1976).
[48] G.R. Wright and M.J. Vander Weil, J. Phys. B$\underline{9}$, 1319 (1976).
[49] D.M.P. Holland, K. Codling, J.B. West and G.V. Marr, J. Phys. B$\underline{12}$, 2465 (1979).
[50] H.K. Haugen, L.P. Anderson, P. Hvelplund, and H. Knudsen, Phys. Rev. A$\underline{26}$, 1950 (1982).
[51] L.H. Andersen, P. Hvelplund, H. Knudsen, H.P. Moller, K. Elsner, K.G. Rensfelt, and E. Uggerhoj, Phys. Rev. Lett. $\underline{57}$, 2147 (1986).
[52] L.H. Andersen, P. Hvelplund, H. Knudsen, S.P. Moller, A.H. Sorensen, K. Elsner, K.G. Rensfeld and E. Uggerhoj, Phys. Rev. $\underline{36}$, 3612 (1987).
[53] M. Charlton, L.H. Andersen, L. Brun-Nielsen, B.I. Deutch, P. Hvelplund, F.M. Jacobsen, H. Knudsen, G. Lariccha, M.R. Poulsen and J.O. Pedersen, J. Phys. B$\underline{21}$, L545 (1988).
[54] M.B. Shah and M.B. Gilbody, J. Phys. B$\underline{18}$, 899 (1985).
[55] M.E. Rudd, Y.K. Kim, D.H. Madison and J.W. Gallagher, Rev. Mod. Phys. $\underline{57}$, 967 (1985).

[56] M. Eckhardt and K. Schartner, Z. Physik 312, 231 (1983).
[57] A.K. Edwards, R.M. Wood and R.L. Ezell, Phys. Rev. A34, 4411 (1986).
[58] A.K. Edwards, R.M. Woods, A.S. Beard, R.L. Ezell, Phys. Rev. A37, 3679 (1988).
[59] J.H. McGuire, Phys. Rev. Lett. 49, 1153 (1982).
[60] R.L. Becker, unpublished but widely quoted; J.H. McGuire and J. Burgdorfer, Phys. Rev. A36, 4089 (1987).
[61] L. Vegh, Phys. Rev. A37, 992 (1988).
[62] J.F. Reading and A.L. Ford, Phys. Rev. Lett. 58, 543 (1987).
[63] DuBois and S.T. Manson, Phys. Rev. A35, 2007 (1987).
[64] J.O.P. Pedersen, P. Hvelplund, Phys. Rev. Lett. 62, 2373 (1989).
[65] J.P. Giese, M. Schultz, J.K. Swensen, H. Schoene, M. Benhennu, S.L. Varghese, C.R. Vane, P.F. Dittner, S.M. Shafroth and S. Datz, private communication.
[66] M.E. Rudd, undergraduate thesis, unpublished.
[67] Jack C. Straton, unpublished.
[68] N. Stolterfoht, conference on the Spectroscopy and Collisions of Few Electron Ions, Bucharest ed. by V. Floreseu and V. Zoran (World Scientific, Singapore, 1989).
[69] L.H. Andersen, L.B. Nielsen, and Jens Sorensen, J. Phys. B21, 1587 (1988).
[70] N. Bohr, K. Danske Vidensk. Selsk. Mat. Fys. Meddr. 18, No. 8 (1948).
[71] J.H. McGuire, N. Stolterfoht, and P.R. Simony, Phys. Rev. A24, 97 (1981).
[72] J.S. Briggs and K. Taulbjerg in Topics in Current Physics, ed. by I. Sellin (Springer Verlag, NY, 1978) Vol. 5, p. 105.
[73] M. Inokuti, Argonne National Laboratory Report No. ANL-76-88, p. 177 (unpublished).
[74] Michio Matsuzawa, J. Phys. B13, 3201 (1980).
[75] R. Anholt, Phys. Lett. A114, 126 (1986).
[76] L.J. Wang, M. King and T.J. Morgan, J. Phys. B19, 623 (1986).
[77] W.E. Meyerhof, private communication.
[78] J.M. Feagin, J. Briggs and T.M. Reeves, J. Phys. B17, 1057 (1984).
[79] J.A. Tanis, E.M. Bernstein, M.W. Clark, R.H. McFarland, T.J. Morgan, A. Muller, M.P. Stockli, K.H. Berkner, P. Gohil, A.S. Schlachter, J.W. Stearns, B.M. Johnson, K.W. Jones, M. Meron, and J. Nason, Electronic and Atomic Collisions, ed. D.C. Lorents, W.E. Meyerhot and J.R. Peterson (North Holland, Amsterdam, 1985) p. 425.
[80] C.P. Bhalla and K.R. Karim, Phys. Rev. A39, 6060 (1989).
[81] T.M. Zouros, D.H. Lee, J.M. Sanders, J.L. Shinpaugh, T.N. Tipping, P. Richard and S.L. Varghese, unpublished.
[82] A. Jain, C.D. Lin and W. Fritsch, Phys. Rev. A39, 1741 (1989).
[83] T.J.M. Zouros, D. Schneider and N. Stolterfoht, Phys. Rev. A35, 1963 (1987).
[84] A.K. Edwards, private communication.
[85] T.J.M. Zouros et al., private communication.
[86] J.H. McGuire, A. Mueller and E. Salzborn, Phys. Rev. A35, 3265 (1987).
[87] S. Datz, R. Hoppler, L.H. Anderson, P.F. Dittner, H. Knudsen, H.F. Krause, P.D. Miller, P.L. Pepmiller, T. Roseel, N. Stolterfoht, Y. Yamazako and C.R. Vane, Nucl. Instr. Meth. A262, 62 (1987).
[88] E. Horsdal-Pedersen and J. Larsen, J. Phys. B12, 4085 (1985).
[89] R.E. Olson, J. Ullrich and H. Schmidt-Bocking, Phys. Rev. A39, 5572 (1989).
[90] J.H. McGuire, Jack C. Straton, W.J. Axmann, T. Ishihara and E. Horsdal, Phys. Rev. Lett. 62, 2933 (1989).
[91] N.B. Mahli, I. Ben-Itzhak, T.J. Gray, J.C. Legg, V. Needham, K. Carnes and J.H. McGuire, J. Chem. Phys. 87, 6502 (1987).

ION-ION COLLISIONS: CHARGE TRANSFER AND IONIZATION

Erhard Salzborn

Institut für Kernphysik, Universität Giessen
D-6300 Giessen, Federal Republic of Germany

This review is concerned with recent crossed-beams studies of electron capture and ionization in collisions between positive ions. Attention is given to the simplest collision systems involving only one and two electrons, respectively. Wherever possible, results involving multiply-charged ions are included.

INTRODUCTION

In atomic collision experiments usually neutral atoms - either free, or bound in solids - are used as targets for incident projectile ions. The first experimental approach to use a target consisting of free ions was reported in 1965 by Guidini et al.[1]. The principal difficulties of such a collision experiment arise from the tenuous thickness provided by the ionic target. However, within the last two decades a few laboratories could successfully meet these problems by means of rather complicated experimental techniques, and since 1977 reliable experimental data on electron capture and ionization in ion-ion collisions have become available.

Ion-ion collisions are fundamental elementary processes in all types of plasmas-in astrophysical objects as well as in laboratory discharges. Therefore, our understanding or modelling of plasmas is subject to reliable information about ion-ion interactions. The impetus for accurate cross section data has increased in recent years due to research on controlled thermonuclear fusion using either magnetically or inertially confined plasmas.

For example, in magnetic fusion there is presently much interest in plasma neutralizers needed for efficient auxiliary heating of next generation fusion devices by injection of multimegawatt neutral $H^o(D^o)$ beams. These plasma neutralizers offer a considerably higher neutralization efficiency than gas targets (85% vs 60%) for the conversion of H^- beams accelerated to injection energies of hundreds of keV. However, design and modelling studies for plasma neutralizers have suffered from the lack of cross section data for single - and double-electron removal from H^- in energetic collisions with positive ions.

In inertial confinement fusion, on the other hand, the intense heavy ion beams, as proposed for igniting a DT pellet, may suffer from severe intensity losses due to charge changing collisions between ions within the beam pulses in the storage rings. In order to obtain an accurate assessment of the expected loss rates, the cross sections both for electron capture and for ionization in collisions between identical heavy ions have to be known as a function of the relative velocity.

More fundamentally, ion-ion collision data are particularly important also because they provide ideal testing grounds for theoretical models. Let us consider, for example, the charge-transfer process in collisions between protons and the hydrogenic ions He^+, Li^{2+}, Be^{3+}, etc. Since the wave functions are known exactly, the accuracy of the calculated cross sections directly reflects the quality of the approximations used in the different theoretical models.

This short review will focus on very recent experimental results obtained by means of the crossed-beams technique. Special attention is given to processes occuring in simple collision systems since they provide the best opportunities to test collision theories.

For more comprehensive discussions the reader is referred to several previous review articles[2-9] on ion-ion collisions.

EXPERIMENTAL TECHNIQUES

All the experiments discussed here have been performed by means of the intersecting-beams technique. The general principles of this method have been described[10] in detail and only an outline of the experimental approach will be given here.

The method, in principle, appears to be straightforward: Two well-collimated ion beams of variable energies are arranged to intersect in an ultra-high vacuum region. The collision products formed in both beams are separated downstream of the interaction region from their parent ion beams and analyzed with respect to their charge states.

Experimentally, inherent difficulties however arise from the inevitably tenuous nature of the ion beams which has to be limited in order to avoid space charge effects. Even at ultra-high vacuum (10^{-10} mbar) in the interaction region the residual gas density exceeds the ion densities within the beams, typically by one or two orders of magnitude. As a result, comparatively low ion-ion signal count rates (typically between 0.1/s and 100/s) are obtained which are completely masked by background counts due to ion interactions with residual gas particles.

Therefore, signal recovery methods have to be employed in order to separate the ion-ion signals from the background events which typically dominate by a few orders of magnitude. Furthermore, carefully designed charge-state analyzers are needed for efficient separation of the tiny fractions of scattered reaction products from the primary ion beams. Here, typically, the intensity ratios are of the order of 10^{-11} to 10^{-13}.

The attainable range of c.m. interaction energies and the energy resolution depend greatly upon the angle θ at which the two beams intersect. Crossing angles used at different laboratories include $\theta = 160°$ (Newcastle), $\theta = 90°$ (Belfast), $\theta = 45°$ (Giessen), $\theta = 8.5°$ (Newcastle) and $\theta = 0°$ (Louvain-la-Neuve). While obtuse angles provide access to high c.m. collision energies, the merged-beams geometry ($\theta = 0°$) allows to study collisions at very low interaction energies with greatly enhanced energy resolution. This can be readily seen from the equation

$$E_{cm} = \frac{M_1 \cdot M_2}{M_1 + M_2} \left[\frac{E_1}{M_1} + \frac{E_2}{M_2} - 2 \left(\frac{E_1 E_2}{M_1 M_2} \right)^{1/2} \cdot \cos\theta \right] \quad (1)$$

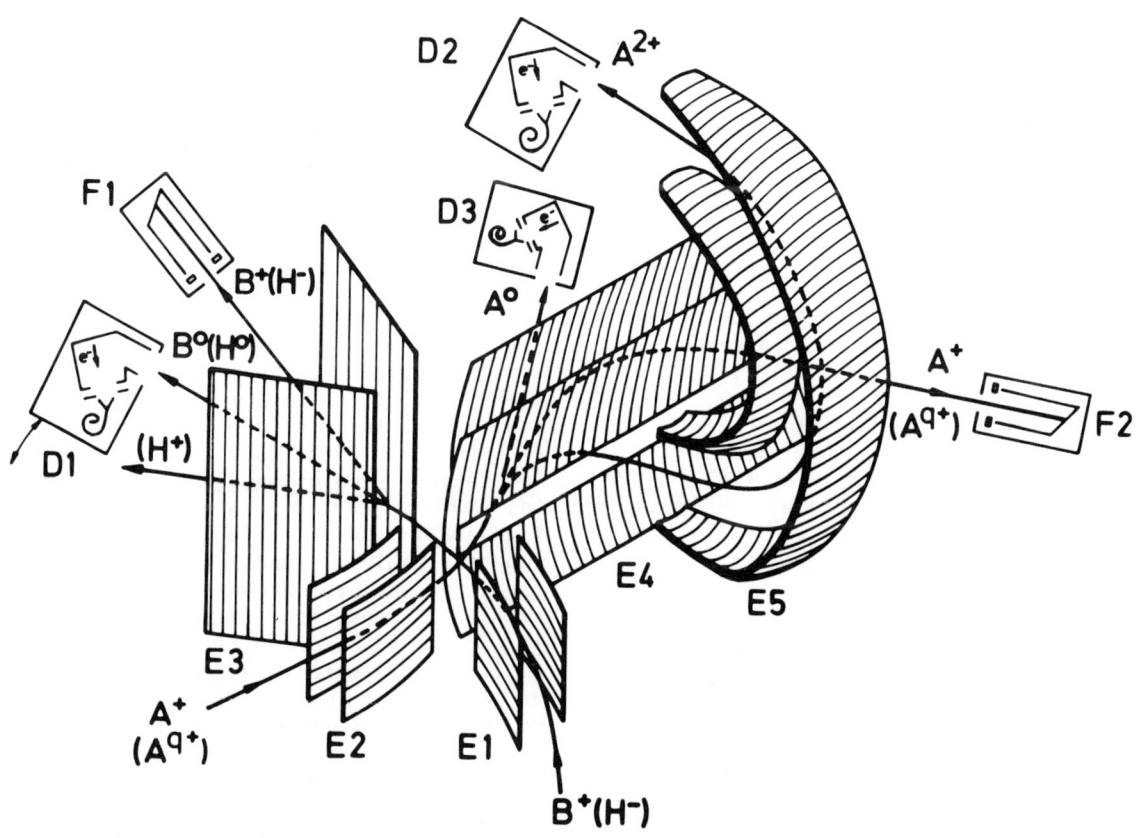

Figure 1. Schematic diagram of the crossed-beams arrangement used in Giessen[11].

where E_{cm} is the interaction energy in the center-of-mass frame and M_1, E_1 and M_2, E_2 denote the masses and laboratory energies, respectively, of the ions of the two beams. The range of interaction energies attained in intersecting-beams experiments spans from 0.1 eV up to hundreds of keV.

Fig. 1 illustrates the essential features of the crossed-beams arrangement used in Giessen. Two momentum-analyzed and well-collimated (< 2 mm diam.) ion beams are arranged to intersect at an angle $\theta = 45°$ in an ultra-high vacuum region. In some experiments the pressure measured with both beams "on" has been as low as $2 \cdot 10^{-11}$ mbar. Both ion beams are cleaned, shortly before intersection, by electrostatic deflectors E1, E2 from particles in other charge states resulting from charge-changing collisions in the residual gas on the path downstream from the analyzing magnets. This precaution is imperative in order to reduce background events. Immediately downstream from the interaction region the reaction products formed in both beams are separated from their parent ion beams. In the fast-beam line (energies up to 400 keV) an electrostatic deflector E3 separates B⁰ atoms from the parent B⁺ ion beams. In case of an incident H⁻ ion beam both reaction products H⁰ and H⁺ are identified. In the slow-beam line (energies up to 20 keV) a two-stage electrostatic analyzer system E4, E5 is used to separate A^{2+} collision products from the A^+ parent beam. The product A⁰ atoms pass undeflected through the ion-optical mirror E4. In this beam line also multiply-charged ions A^{q+} are available from a small 5 GHz ECR-ion source installed recently. All reaction products are counted by channeltron-based single-particle detectors D1, D2, D3 whilst the parent ion beams are recorded by biased Faraday cups F1, F2. The beam dumps are separately pumped in order to avoid degradation of the low pressure in the intersection region.

With this experimental arrangement the following processes have been studied:

electron capture: $A^+ + B^+ \rightarrow A^{2+} + B^0$ (2)
ionization: $A^+ + B^+ \rightarrow A^{2+} + B^+ + e$ (3)
mutual neutralization: $H^+ + H^- \rightarrow H^0 + H^0$ (4)
transfer ionization: $H_a^+ + H_b^- \rightarrow H_a^0 + H_b^+ + e$ (5)
single-electron removal: $H^- + A^{q+} \rightarrow H^0 + ...$ (6)
double-electron removal: $H^- + A^{q+} \rightarrow H^+ + ...$ (7)

In all crossed-beams experiments methods of signal recovery have to be employed in order to discriminate signal counts from background counts which are normally some orders of magnitude more frequent. In measuring the processes (2), (4), and (5) a coincidence technique is used. True ion-ion signal pulses from the two ion/atom-detectors show a fixed time delay whereas there is no time correlation with pulses from background events involved. Thus, in a time spectrum the signals resulting from ion-ion collisions form a well defined peak on top of a flat background due to random coincidences. In measuring the processes (3), (6), and (7) only one detector is operated. In this case a beam modulation technique is used for signal recovery. Both ion beams are chopped by fast electrostatic deflection. The true signal count rate can be either obtained from the difference in the count rates of two properly gated scalers or from the actual time spectrum of recorded detector pulses[12].

It should be noted that the cross section σ_i for ionization (process (3)) cannot be obtained in a direct measurement. The measurements only provide cross sections $\sigma(A^{2+})$ for the production of A^{2+} ions from the combined processes of electron capture (2) and ionization (3). Knowing the cross section σ_c for electron capture from a separate measurement, the ionization cross section is obtained from the difference $\sigma_i = \sigma(A^{2+}) - \sigma_c$.

RESULTS

1. One-electron systems

Experimentally and theoretically, much attention has been paid to H⁺ + He⁺ collisions representing the simplest ion-ion collision system. The reactions

$$H^+ + He^+ \begin{cases} \rightarrow H^0 + He^{2+} & \text{electron capture} \quad (8a) \\ \rightarrow H^+ + He^{2+} + e & \text{ionization} \quad (8b) \end{cases}$$

provide an ideal testing ground for different theoretical models since only a single electron is involved.

The groups in Belfast[13-16], Newcastle[17,18] and Giessen[11,12] have reported measurements of σ_c and $\sigma(He^{2+})$ for this collision system in different but overlapping energy ranges. As can be seen from Fig. 2 the experimental cross sections σ_c for electron capture (into all bound H⁰ states) are in excellent agreement.

In order to facilitate comparison with theoretical predictions, the combined experimental data of Fig. 2 are represented by a solid line in Fig. 3. The available theoretical data can be roughly divided into two groups. Calculations of the first group only describe the cross section σ_c above its maximum at about $E_{cm} = 36$ keV ($v_{rel} \approx 3 \cdot 10^8$ cm/s). Whereas the results of the continuum-distorted-wave approximation by Belkić et al[19], as well as the classical trajectory Monte-Carlo calculations by Olson[20], overestimate the measured cross section, the predictions of the Coulomb-Born approximation by Sinha and Sil[21] are systematically below the data. More recent calculations by Datta et al[22] in the framework of the continuum-intermediate-state approximation seem to converge to the measurements with increasing energy. The results of the 'one-and-a-half-center' expansion by Reading et al[23] agree with experiment above the cross section maximum. The second group of calculations based on coupled-state approaches is in excellent agreement with the measurements over the entire energy range. Basis sets of atomic orbitals have been used by Fritsch and Lin[24], Winter[25,26], and Bransden et al[27]. In the lower energy range the data are also very well reproduced by the molecular-orbital results of Winter et al[28], Errea et al[29], and Kimura and Thorson[30] (not shown).

In conclusion, for electron capture in $H^+ + He^+$ collisions, the experimental data from 3 groups are in excellent accord with each other and also with a number of theoretical predictions.

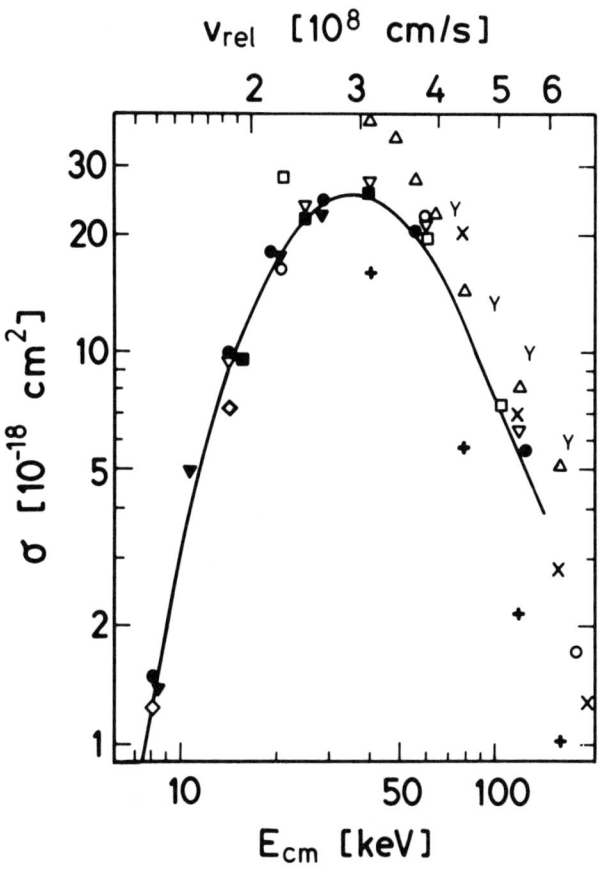

Figure 3. Cross sections σ_c for electron capture in $H^+ + He^+$ collisions: —— combined experimental data of Fig. 2; Calculations: ● Fritsch and Lin[24]; ▽ Winter[25]; ■ Winter[26]; o Bransden et al[27]; ◇ Winter et al[28]; ▼ Errea et al[29]; □ Reading et al[23]; + Sinha and Sil[21]; △ Belkić et al[19]; Y Olson[20]; x Datta et al[22].

In an extension to the $H^+ + He^+$ system, electron capture in collisions between H^+ and the hydrogenic ions Li^{2+}, Be^{3+}, B^{4+}, and C^{5+} has been theoretically studied by Winter[31] within a coupled-Sturmian-pseudostate approach. Fig. 4 shows scaled cross sections $Z^7 \cdot \sigma_c$ plotted versus scaled proton energy $E/(25 \cdot Z^2)=(v/Z)^2$, where v/Z is the proton speed (relative to the target ion) in units of the Bohr velocity of the target ion of charge Z. The scaled curves agree closely with one another showing a peak at the scaled energy $E/(25 \cdot Z^2) \approx 0.5$. Experimental data, however, are not available for hydrogenic target ions beyond He^+.

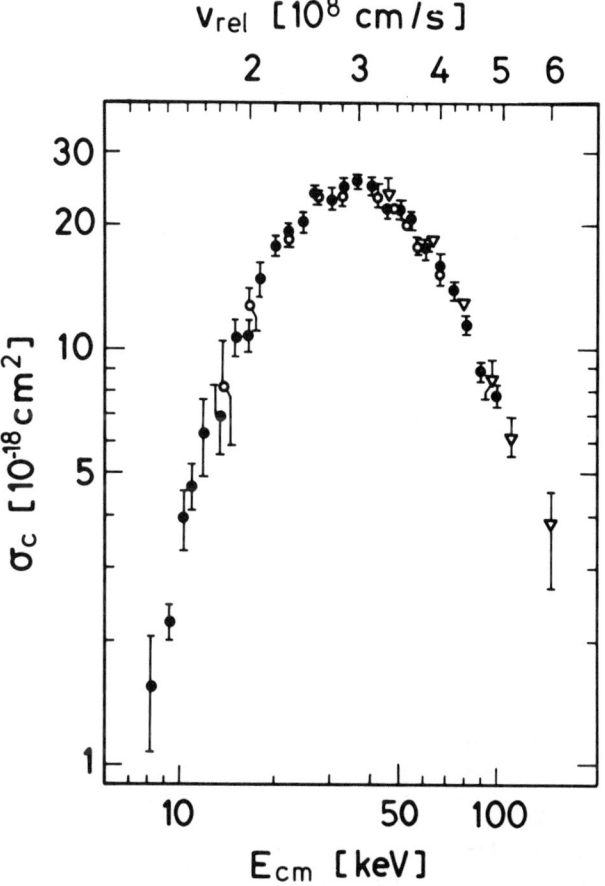

Figure 2. Cross sections σ_c for electron capture (into all bound H^0 states) in $H^+ + He^+$ collisions: ● Rinn et al.[11]; ▽ Watts et al[16]; this data revises earlier data (not shown) by Angel et al[15]; o Peart et al[18]. The error bars represent the 90% (●,o) and 68% (▽), respectively, confidence limit of statistical error.

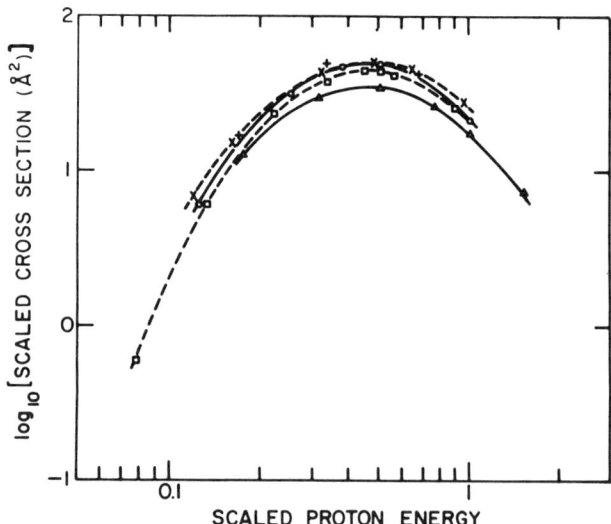

Figure 4. Scaled cross sections $Z^7 \cdot \sigma_c$ versus scaled proton energy $E/(25 \cdot Z^2) = (v/Z)^2$ for electron capture into all states of H^o in collisions between H^+ and the ground-state hydrogenic ions $He^+(\Delta)$, $Li^{2+}(\square)$, $Be^{3+}(o)$, $B^{4+}(x)$, and $C^{5+}(+)$. From Ref. 31

Compared to electron capture, the process of ionization in $H^+ + He^+$ collisions is much less understood. Experimentally, the cross sections σ_i are obtained from the difference of two separate measurements of $\sigma(He^{2+})$ and σ_c, respectively. As a result, the error bars on the experimental σ_i values become increasingly larger towards lower energies where ionization decreases. In Fig. 5 the measurements of σ_i by the groups in Belfast[14,16], Newcastle[18], and Giessen[12] are shown together with the results of theoretical approaches. As can be seen, there is considerable scatter in both the experimental and theoretical data. The Giessen results are somewhat larger in absolute magnitude than the measurements from Belfast and Newcastle. This must be due to enhanced cross sections $\sigma(He^{2+})$ since the three sets of σ_c measurements all agree very well.

The large experimental uncertainties do not permit an accurate assessment of all the theoretical approaches available. None the less, four of them clearly overestimate the ionization cross section at lower collision energies. These include the continuum distorted-wave approximation (CDW) by Belkić[32], a modification of this (MCDW) by Miraglia[33], Salin's[34] approximation (S), and the 'one-and-a-half-center' expansion (POHCE) by Reading et al[23]. The results of the remaining theoretical approaches are largely compatible with the experimental data sets within error bars. These include a calculation by Bates and Griffing[35] based on the first Born approximation (B1), a classical-trajectory Monte

Carlo calculation (open circles) by Olson[20], a close-coupling calculation by Fritsch and Lin[36], and two versions of a multiple-scattering approximation (MSI and MSF) by Garibotti and Miraglia[37] and Miraglia[33], respectively. Very recently, Grozdanov and Solov'ev[38] have calculated cross sections for low-energy ionization using a simple model based on analytic properties of the quasi-molecular adiabatic energy surfaces and the adiabatic theory of inelastic collisions.

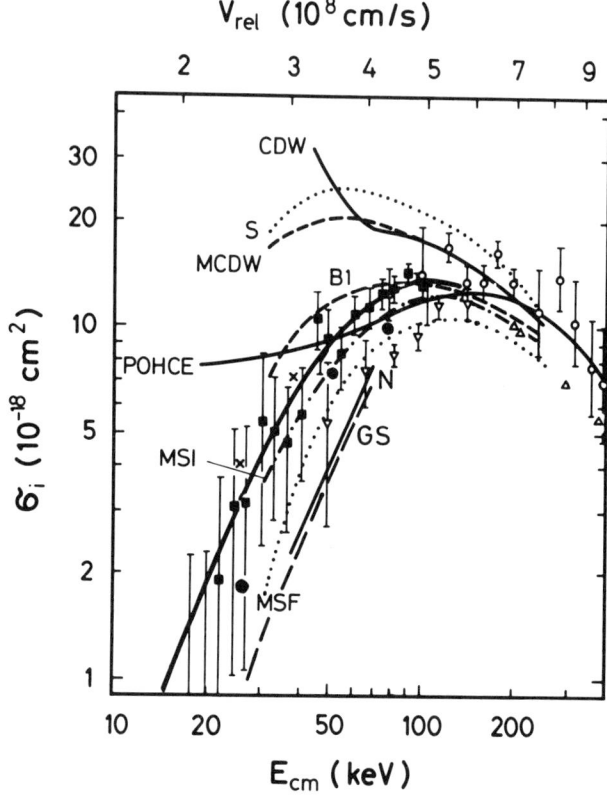

Figure 5. Cross sections σ_i for ionization in $H^+ + He^+$ collisions. Experimental results: ■ measurements with 4-parameter fit curve by Rinn et al[12]; N —— smoothed measurements by Peart et al[18]; ▽ measurements by Watts et al[16] which revise earlier data (not shown) by Angel et al[15]; △ $\sigma(He^{2+})$ data by Angel et al[14] at energies where $\sigma_i \approx \sigma(He^{2+})$. Theoretical results: CDW Belkić[32]; MCDW Miraglia[33]; S Salin[34]; B1 Bates and Griffing[35]; POHCE Reading et al[23]; MSI Garibotti and Miraglia[37]; ··· MSF Miraglia[33]; ● Fritsch and Lin[36]; o Olson[20]; x Winter[26]; – – GS Grozdanov and Solov'ev[38]; CDW-EIS results by Martinez et al[39] almost coincide with the 4-parameter best fit curve to the Giessen measurements and are not shown for clarity.

Their results (GS) are in very good accord with the Newcastle/Belfast data. Two other very recent calculations, on the other hand, are in excellent accord with the Giessen measurements. These are an elaborate triple-center calculation (crosses) by Winter[26] and a continuum-distorted wave-eikonal initial state approach by Martinez et al[39]. The latter results (not shown for clarity) almost coincide with the solid curve which represents a 4-parameter fit function of the Giessen measurements.

In conclusion, ionization in $H^+ + He^+$ collisions is much less understood, both experimentally and theoretically, than electron capture.

Within the coupled-Sturmian-pseudostate approach, Winter[31] has also calculated ionization cross sections σ_i for H^+ collisions with higher-Z hydrogenic targets. Figure 6 illustrates scaled cross sections $Z^4 \cdot \sigma_i$ plotted versus scaled incident proton energy $E/(25 \cdot Z^2)$ for ground-state He^+, Li^{2+},..., C^{5+} targets. The scaled ionization cross sections show a broad maximum at about $E/(25 \cdot Z^2) \approx 1.1$. It is seen that, as for electron capture (Fig. 4), the scaled cross sections generally move upward in a fairly regular way with increasing Z and seem to approach a limit, with the variation, however, being roughly twice as large as for electron capture.

Again, experimental data are only available for the least charged He^+ target.

2. Two-electron system

The simple case of two positive ions which each have only one electron, i.e. $He^+ + He^+$ collisions, is of fundamental theoretical interest because of the additional electron-electron interaction involved. Only very recently, however, the inherent formidable difficulties could be successfully met by theoretical approaches.

Fig. 7 shows cross sections σ_c for electron capture (into all bound He^0 states) in collisions

$$He^+ + He^+ \rightarrow He^0 + He^{2+} \qquad (9)$$

It can be seen that the measurements by Peart et al[40] and by Melchert et al[41] are in excellent agreement.

In Fig. 8 these experimental results are compared with theoretical predictions. Since the calculations often have been performed only at a few discrete energies, the measurements of Fig.7 are represented by a solid line in order to fa-

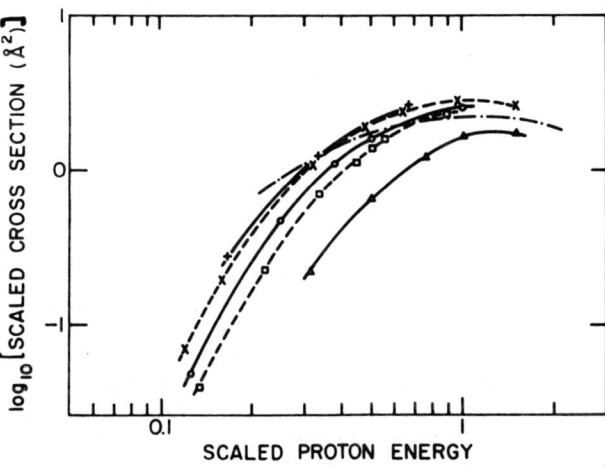

Figure 6. Scaled cross sections $Z^4 \cdot \sigma_i$ versus scaled proton energy $E/(25 \cdot Z^2)$ for ionization in collisions between H^+ and the ground-state hydrogenic ions He^+ (Δ), Li^{2+}(□), Be^{3+}(o), B^{4+}(x), and C^{5+}(+). The dash-dotted curve is the first- Born-approximation result by Bates and Griffing[35]. From Ref.31.

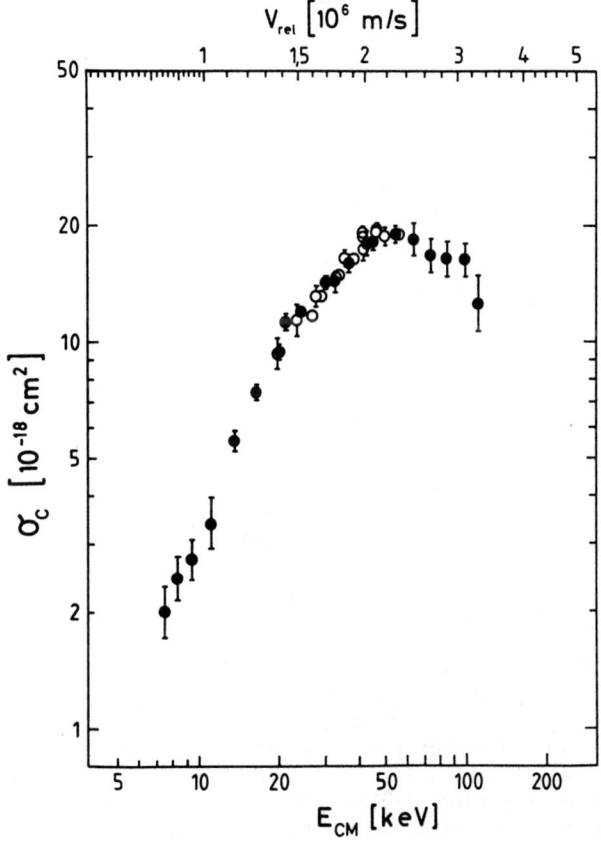

Figure 7. Cross sections σ_c for electron capture (into all bound He^0 states) in $He^+ + He^+$ collisions: ● Melchert et al[41]; o Peart et al[40]. The error bars represent the 90% confidence limit of statistical error.

cilitate comparison. The curve was obtained by a 4 parameter best fit to the set of data combined from both measurements.

As can be seen from Fig.8 the time-dependent Hartree-Fock calculations by Henne et al[42] systematically overestimate the cross section. At the lower collision energies the discrepancy increases up to an order of magnitude. Apparently, the ansatz of a single determinant is not adequate for the given process. The results of classical-trajectory Monte-Carlo calculations by Willis et al[43] using three different types of model potentials exhibit considerable scatter.

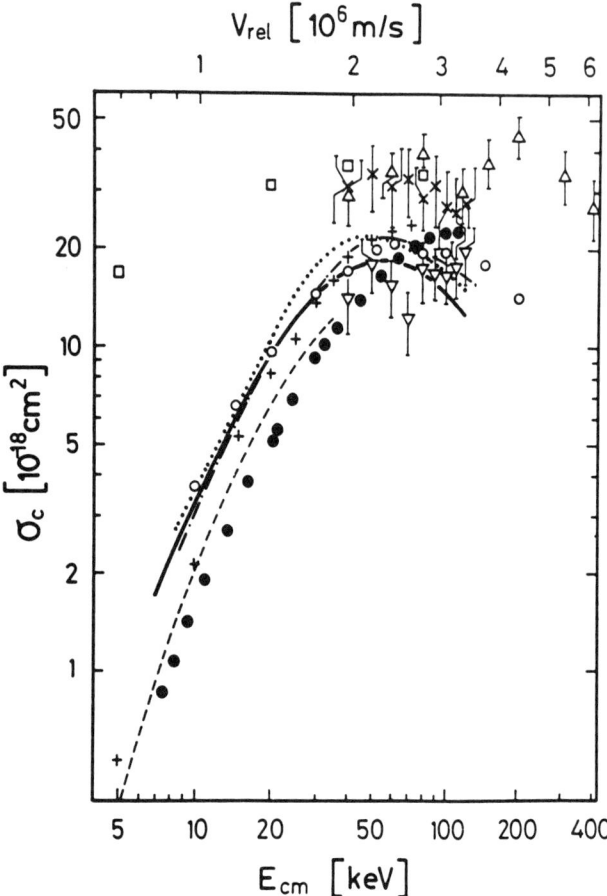

Figure 8. Cross sections σ_c for electron capture in $He^+ + He^+$ collisions: —— combined experimental results of Fig. 7; --- scaled experimental and theoretical data by Afrosimov et al[44] and Harel and Salin[45], respectively, for the inverse reaction $He^{2+}+He(1s^2) \rightarrow He^+(1s)+He^+(1s)$; ▽ CTMCA, △ CTMCB, and × CTMCC calculations, respectively, by Willis et al[43]; □ time-dependent Hartree-Fock calculation by Henne et al[42]; ● two-state MO calculation by Melchert et al[41]; ... AO calculation by Fritsch and Lin[46]; + MO calculation by Allan[48]; -·-·- MO calculation by Kimura[49]; ○ AO calculation by Gramlich et al[47].

Two approaches (CTMCB and CTMCC) also clearly overestimate the measurements. Only the results labelled CTMCA seem to show fair agreement. On the other hand, however, this approach completely fails in predicting the energy dependence of the cross section σ_i for ionization in $He^+ + He^+$ collisions (see Fig. 9).

The σ_c measurements are about a factor of two above the scaled data of the inverse reaction

$$He^{2+} + He(1s^2) \rightarrow He^+(1s) + He^+(1s) \quad (10)$$

which has been studied experimentally and theoretically by Afrosimov et al[44] and Harel and Salin[45], respectively. This indicates that the capture into the ground state of He^o is approximately only 50 % of the total capture cross section. This conclusion is also supported by the good agreement of the scaled data with a simple two-state coupling MO-calculation[41] for ground-state capture.

Only recently, more elaborate coupled-channel calculations using larger basis sets of atomic[46,47] or molecular[48,49] orbitals have became available. All of them are in very good agreement with the measurements below the cross section maximum. It was pointed out[46,47,49] however that close agreement with experiment is obtained only when capture both into spin-singlet and spin-triplet states of He^o is taken into account. The triplet states lead to an almost constant contribution to the total cross section over the considered range of energy. Therefore, the good agreement of the MO-results by Allan[48] is puzzling, since his calculations include only capture into singlet states. It should be noted that the AO-approach by Gramlich et al[47] not only yields an excellent description of electron capture but also provides the first quantum calculation of ionization in $He^+ + He^+$ collisions (see Fig. 9).

In conclusion, only a few recent attempts could successfully cope with the complex difficulties involved in solving the two-electron Schrödinger equation for electron capture in $He^+ + He^+$ collisions.

Finally, in Fig.9 are shown cross sections σ_i for ionization in $He^+ + He^+$ collisions measured by Melchert et al[50]. This data is in very good accord with σ_i values (not shown in Fig. 9 for clarity) which can be deduced from experimental work by Peart et al[40]. It is amazing that only very recently the first theoretical results based on quantum calculations have become available for this process. The coupled-channel AO-calculations by Gramlich et al[47] simultaneously yield cross sections for electron capture and ionization in $He^+ + He^+$ collisions. Their σ_i results (solid line in Fig. 9) are in good agreement with experiment at higher impact energies but tend to overestimate the measurements in the lower energy range. The authors point out that the cross sections for singlet and triplet states are almost the same over the energy range studied. However, the statistical weight is three times larger for triplet than for singlet states. Thus, the triplet states are the dominant ionization channels.

The only other theoretical approach with which the σ_i data can be compared are classical-trajectory Monte-Carlo calculations by Willis et al[43]. These CTMC calculations have been performed in a single-electron picture with three distinct types of model potentials. Only the results labelled CTMCB (upright triangles) appear to be quite good at the higher collision energies. On the other hand, however, this approach (using switching potentials during the collision) predicts electron capture cross sections which cleary overestimate the measurements (cf. Fig. 8). The approaches CTMCA (open circles in Fig. 9) and CTMCC (not shown for clarity) give nearly the same σ_i values. They both fail completely in describing the energy dependence of the ionization cross section.

In conclusion we note that only very recently ionization in $He^+ + He^+$ collisions could be successfully treated in a first theoretical approach based on quantum calculations.

CONCLUSION

In recent years, much progress has been achieved in measuring and calculating the elementary processes of electron capture and ionization in the simplest collision systems, $H^+ + He^+$ and $He^+ + He^+$, involving only one and two electrons, respectively. As compared to electron capture, the process of ionization still appears to be less understood for both systems.

Future innovative experiments may aim at the determination of the final state population in the electron capture reactions or at differential cross section measurements. Another challenge to experimentalists is the unexplored field of collisions involving higher-Z hydrogenic ions.

ACKNOWLEDGEMENTS

The author would like to thank his coworkers, especially Dr.F.Melchert and Dr.K.Rinn. Financial support by the German Federal Minister for Research and Technology (BMFT) under contract 06 GI 658 is gratefully acknowledged.

REFERENCES

1. J.Guidini, C.Manus, T.Sinda, and G.Watel, Proc. 4th ICPEAC, Quebec (1965), Abstracts, p. 450
2. H.B.Gilbody, in Physics of Electronic and Atomic Collisions, ed. S.Datz, North-Holland/Amsterdam (1982) p. 223
3. K.T.Dolder, Comments At.Mol.Phys. 11 (1982) 211
4. K.T.Dolder, in Physics of Electron-Ion and Ion-Ion Collisions, ed. F.Brouillard and J.W. McGowan, Plenum/New York (1983) p. 373
5. K.T.Dolder, in Atomic Processes in Electron-Ion and Ion-Ion Collisions, ed.F.Brouillard, Plenum/New York (1986) p. 313
6. K.T.Dolder, and B.Peart, Rep.Prog.Phys. 48 (1985) 1283
7. K.T.Dolder and B.Peart, in Advances in Atomic and Molecular Physics, Vol. 22 (1985) p. 197
8. K.F.Dunn, in Atomic Processes in Electron-Ion and Ion-Ion Collisions, ed. F.Brouillard, Plenum/New York (1986) p. 333
9. E.Salzborn, Journal de Physique, Colloque C1 (1989) 207
10. K.T.Dolder, in Case Studies in Atomic Collision Physics, Vol. 1, ed. E.W. McDaniel and M.R.C. McDowell, North Holland, Amsterdam 1969, p. 249
11. K.Rinn, F.Melchert, and E.Salzborn, J.Phys. B 18 (1985) 3783
12. K.Rinn, F.Melchert, K. Rink and E.Salzborn, J.Phys.B 19 (1986) 3717

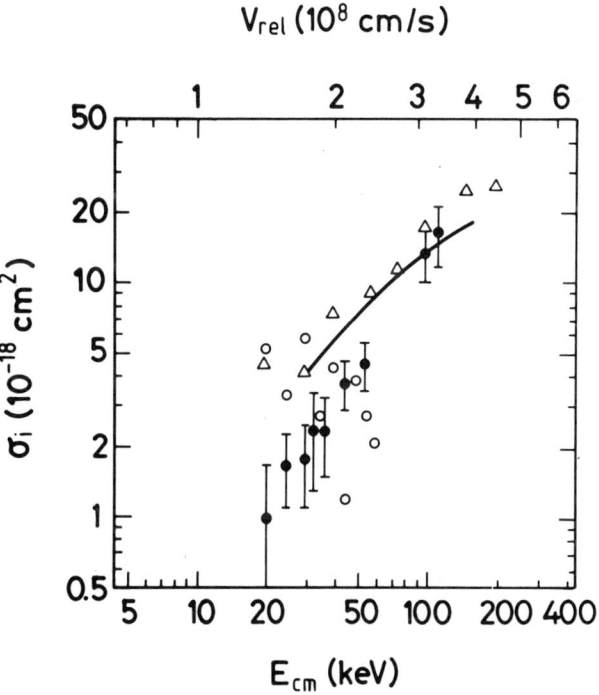

Figure 9. Cross sections σ_i for ionization in $He^+ + He^+$ collisions: ● measurements by Melchert et al[50] (error bars represent the 90% confidence limit of statistical errors); —— coupled-channel AO-calculations by Gramlich et al[47]; o CTMCA and △ CTMCB calculations, respectively, by Willis et al[43].

13. J.B.A.Mitchell, K.F.Dunn, G.C.Angel, R. Browning and H.B.Gilbody, J. Phys. B **10** (1977) 1897
14. G.C.Angel, K.F.Dunn, E.C.Sewell and H.B. Gilbody, J.Phys. B **11** (1978) L 49
15. G.C.Angel, E.C.Sewell, K.F.Dunn and H.B. Gilbody, J.Phys. B **11** (1978) L 297
16. M.F.Watts, K.F.Dunn and H.B.Gilbody, J. Phys. B **19** (1986) L 355
17. B.Peart, R.Grey and K.T.Dolder, J.Phys. B **10** (1977) 2675
18. B.Peart, K.Rinn and K.T.Dolder, J.Phys. B **16** (1983) 1461
19. Dž. Belkić, R.Gayet and A.Salin, Phys.Rep. **56** (1979) 279
20. R.E.Olson, J.Phys. B **11** (1978) L 227
21. C.Sinha and N.C.Sil, Phys.Lett. **71** A (1979) 201
22. S.Datta, C.R.Mandal and S.C.Mukherjee, Can. J.Phys. **62** (1984) 307
23. J.F.Reading, A.L.Ford and R.L.Becker, J. Phys. B **15** (1982) 625
24. W.Fritsch and C.D.Lin, J. Phys. B **15** (1982) 1255
25. T.G.Winter, Phys.Rev. A **25** (1982) 697
26. T.G.Winter, Phys.Rev. A **37** (1988) 4656
27. B.H.Bransden, C.J.Noble and J.Chandler, J.Phys. B **16** (1983) 4191
28. T.G.Winter, G.J.Hatton and N.F.Lane, Phys. Rev. A **22** (1980) 930
29. L.F.Errea, J.M.Gomez-Llorente, L.Mendez and A.Riera, J.Phys. B **20** (1987) 6089
30. M.Kimura and W.R.Thorson, Phys.Rev. A **24** (1981) 3019
31. T.G.Winter, Phys.Rev. A **35** (1987) 3799
32. Dž. Belkić, J.Phys. B **13** (1980) L 589
33. J.E.Miraglia, J.Phys. B **16** (1983) 1029
34. A.Salin, J.Phys. B **2** (1969) 631
35. D.R.Bates and G.Griffing, Proc.Roy.Soc. A **66** (1953) 961
36. W.Fritsch and C.D.Lin, Proc. 13th. Int. Conf. on Physics of Electronic and Atomic Collisions, ed. J.Eichler et al, North Holland/Amsterdam (1983), Abstracts p. 502
37. C.R.Garibotti and J.E.Miraglia, Phys.Rev. A **21** (1980) 572; Phys.Rev. A **25** (1982) 1440
38. T.P.Grozdanov and E.A.Solov'ev, Phys.Rev. A **38** (1988) 4333
39. A.Martinez, R.Rivarola, G.Deco and P.Fainstein, Nucl. Instr. and Meth. B, 1989, in print
40. B.Peart, K.Rinn and K.T.Dolder, J.Phys. B **16** (1983) 2831
41. F.Melchert, K.Rink, K.Rinn, E.Salzborn and N.Grün, J.Phys. B **20** (1987) L 223
42. A.Henne, A.Toepfer, H.J.Lüdde and R.M. Dreizler, J.Phys. B **19** (1986) L 361
43. S.L.Willis, G.Peach, M.R.C.McDowell and J.Banerji, J.Phys. B **18** (1985) 3939
44. V.V.Afrosimov, A.A.Basalev, G.A.Leiko and M.N.Panov, Sov. Phys.-JETP **47** (1978) 837
45. C.Harel and A.Salin, J.Phys. B **13** (1980) 785
46. W.Fritsch and C.D.Lin, Phys.Lett. **123A** (1987) 128
47. K.Gramlich, N.Grün, W.Scheid, J.Phys. B **22** (1989) 2567
48. J.R.Allan, J.Phys. B **19** (1986) L 683
49. M.Kimura, J.Phys. B **21** (1988) L 19
50. F.Melchert, K.Rink, K.Rinn and E.Salzborn, J.Phys. B **20** (1987) L 797

Multiply Differential Ionization Probabilities in Small Impact Parameter Ion-Atom Collisions

G.Schiwietz, B.Skogvall[*], N.Stolterfoht, D.Schneider[+], V.Montemayor[§]

Hahn-Meitner-Institut Berlin, GmbH, Glienicker Str. 100, 1000 Berlin 39, F.R.Germany
[*]*On leave from University of Lund, Lund, Sweden*
[+]*Present address: LLNL, Livermore, California USA*
[§]*Present address: Department of Physics and Astronomy, University of Toledo, Toledo, Ohio USA*

Absolute doubly differential electron emission yields were measured as a function of projectile scattering angle for 30 to 300 keV H^+ + He and Ar^+ + Ar collisions. Coincidence spectra were taken for projectile scattering angles between 0.3° and 3°, electron energies in the range of 5 eV to 1000 eV, and ejection angles from 40° and 110°. In the case of 300 keV H^+ + He collisions, triple coincidences were performed in order to measure electron spectra as a function of the impact parameter and final target charge-state. The importance of higher-order effects in the description of the well-known $4f\sigma$-promotion in 150-300 keV Ar^+ + Ar collisions is investigated. Furthermore, comparisons with experiment show the limits of single-center atomic orbital expansions and of the independent-particle model in the case of H^+ + He collisions.

1 Introduction

The investigation of electron emission in ion-atom collisions has recently gained increasing interest. Electron spectroscopy in connection with the measurement of the projectile scattering angle is an especially sensitive tool for identifying excitation mechanisms, and for testing the limits of even sophisticated collision theories.[1,2]

For low energy Ar^+ + Ar collisions it will be shown in a detailed investigation of electron emission for intermediate impact parameter collisions (b≈0.5 a.u) that a description of the collision process in terms of couplings between individual molecular orbitals may fail. We were able to identify the dominant excitation mechanisms for this 35-electron system by comparison with the molecular orbital model.[2] However, for the well-known rapidly promoting $4f\sigma$-level, it turns out that not only molecular, but also atomic mechanisms influence the excitation process, even when the projectile velocity v_p is much less than the mean electron orbital velocity v_e during the promotion.

Previous investigations of ionization cross sections or probabilities have demonstrated the success of first-order theories for the case of highly energetic light projectile ions.[3-6] For intermediate energies, various corrections, such as a binding correction, projectile Coulomb deflection, and energy loss, tend to improve the agreement between experimental and theoretical results.[7] However, it was shown only recently that the commonly used monopole-like binding correction is not able to describe the measured doubly-differential ionization probabilities for 100 keV H^+ + He collisions at small impact parameters.[8]

Single-center atomic-orbital expansions are probably the most common highest-order theories for the calculation of ionization probabilities.[9,10] In section 5 we will concentrate on the question of under which conditions single-center expansions are appropriate and whether two-center theories[11-13,20] are necessary in order to describe the collision process.

Most of the ion-atom collision theories to date are based on the assumption of the independent motion of the electrons. The corresponding independent-particle model (IPM) successfully describes a large variety of collision systems as long as gross quantities, such as total cross sections or charge-state distributions, are considered.[14-15] However, recent measurements of double capture or transfer ionization, where the final state[16] or a class of two-electron states[17-19] have been selected, indicate strong deviations from the IPM. Here, we will show that measured doubly differential electron emission yields for double ionization in 300 keV H^+ + He collisions also deviate from IPM theories. Comparison of these results will be made with a theory which includes a dynamic electron-electron interaction[23] and with coupled channel results on the basis of the IPM.

2 Experiment

Electrons emitted in ion-atom collisions were measured in coincidence with scattered ions as described elsewhere.[2,8] A Van-de-Graaff generator was used to produce 30 to 300 keV ions. The beam was collimated to a diameter of less than 0.25 mm at a divergency of less than $2 \cdot 10^{-4}$ rad in the target volume. Scattered projectiles were detected with a fast position-sensitive particle detector in the 50Ω-technique[21] and ejected electrons were detected with two types of electron-time-of-flight analyzers. Spectra for up to eight different projectile scattering angles were taken simultaneously. High coincidence rates (10 - 20 cps) with small random coincidence rates were achieved due to the relatively large solide angle of the electron spectrometers as well as the fact that the time-of-flight analyzer energy-analyzes all of the incoming electrons. The product of solid

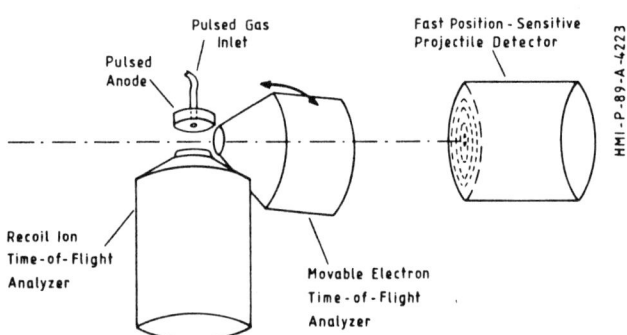

Figure 1. Experimental setup for triple coincidences between electrons, recoil ions and scattered projectiles.

angle, transparency, and channel plate efficiency for the electron spectrometer was either 10 msr or 70 msr in the different experiments.

The experimental setup used for the experiments in section 6 is shown in Fig 1. It displays the position-sensitive projectile detector and the large-solid-angle electron analyzer. Also shown is the newly developed recoil-ion time-of-flight analyzer.[22] This analyzer allows for the determination of the final charge-state of the recoil ion without affecting the spectra of the emitted electrons. The target region is initially kept field free. After an electron is detected, a pulsed field of 150 V/cm is applied to the target region. Thus, the low energy recoil ions are accelerated into the recoil time-of-flight analyzer where they are further accelerated as determined by the recoil charge-state. After passing a drift tube, the recoil ions are detected with a channel plate device. The time difference between pulsing the target region and detecting the ion is a direct measure of the recoil-ion charge-state.

In this paper we present doubly differential electron emission yields (number of emitted electrons per energy interval per electron solid angle per scattered projectile), in some cases as a function of the final charge-state of the target. The data analysis follows closely the description given in ref. 2 and includes the correction for pressure-induced slit scattering. In order to improve the accuracy and to check for random signals, two pulsed gas inlets were installed in the μ-metal shielded chamber. One gas inlet was placed directly above the target region and the other was installed about 30 cm away. Every 10 minutes, the target gas flow was switched from one gas inlet to the other, thus allowing for the determination of the number of scattered projectiles as well as double and triple coincidences produced in the target region.

3 Theoretical Models

Here follows a short overview of the principles of the theoretical ion-atom collision models used for comparison with the experimental results. First it should be mentioned that all theoretical results given in this paper except the dCTMC results[23] are calculated in the independent particle model (IPM). The IPM assumes that one active electron is moving whereas other electrons act as spectators. Furthermore, all theories include curved projectile trajectories and make use of a sophisticated target potential. Except for the AO+ approximation,[24] all target potentials are of single-zeta type as defined by Clementi and Roetti.[25] Thus, deviations between the different theories are due to different treatments of the collision dynamics and not to wave function effects.[20] Therefore we do not compare with calculations based on hydrogen-like states[4] as was done in ref. 8.

The important properties of the considered models are listed in table 1. There are two fundamental ways of classifying the models. One is the distinction between perturbative and non-perturbative theories, as defined by the number of coupled channels (0 corresponds to perturbation theory). The other classification depends on whether the electron motion is treated classically or quantum mechanically. It should be noted that some of the properties discussed above are not commonly attributed to a particular theory, but rather refer to the special version used in this paper.

The semi-classical approximation with separated-atom wavefunctions (SCA) corresponds to an impact-parameter formulation of the Born approximation using frozen continuum wave-packets (see, e.g., ref.10) and dynamic curved projectile trajectories.[26] Thus dynamic screening and polarization effects show up in the relation between impact parameter and projectile scattering angle. Partial waves up to $l=8$ were taken into account, but only small m quantum numbers ($m \leq 3$) were included. The results were checked against the PWBA results of Salin[27,28] who achieved conver-

Table 1 Properties of the considered collision models as described in the text.

Model	number of partial waves (l)	number of coupled channels	dynamic electron-electron interaction	quantum mechanical electron motion
SCA	8 (target)	0		X
AO (single center)	8 (target)	90		X
AO+ (two-center)	2 (target) + 2 (projectile)	40		X
dCTMC	∞	∞	X	

gence in angular momentum space. The agreement was generally better than 20%. At high incident energies and low projectile charges, the SCA as well as the PWBA are known to yield a good description of the reaction dynamics.

The calculation labelled AO is a single-center atomic orbital expansion using exactly the same wavefunctions as in the SCA calculation. However, in this highest-order theory, the couplings between all states were taken into account to all orders.[26] Thus, dynamical binding and polarization effects are incorporated into this model.

The AO + by Fritsch[29] is a two-center coupled channel atomic-orbital expansion (AO) plus united atom orbitals.[24,29,30] This model is non-perturbative in the sense that binding effects, a non-monopole-like polarization, and the formation of molecular orbitals are all taken into account dynamically. However, it is restricted to 40 states, including target, projectile, and continuum, with different partial waves. The continuum is represented by pseudo states[24] allowing also for continuum-continuum interaction.

The dynamic-screening classical trajectory Monte Carlo (dCTMC) model[23] used in this work treats the electron motion and the projectile motion completely classically. But the calculations are non-perturbative, taking all four-body interactions (including dynamic polarization and binding effects) into account. However, the electron-electron interaction is approximated by a spherically-symmetric dynamic-screening function. The dCTMC involves no partial-wave expansions. The density in momentum space and the binding energy are similar to the Hartree-Fock values. The classical density in coordinate space is somewhat higher than the quantum mechanical one in the region of the target nucleus, but there are significant deviations for larger distances because of the classical turning point. Thus, an investigation of small impact parameter (close) collisions will test whether the collision dynamics, but not the initial distribution, can be described classically.

4 Molecular Inner Shell Ionization in $Ar^+ + Ar$ Collisions

The Ar+Ar system has been studied in detail by various groups. The results of these studies have confirmed that the dominant mechanism causing $L_{2,3}$-shell vacancies is electron promotion along the $4f\sigma$ molecular orbital (MO) at impact parameters around 0.5 a.u. (see Refs 1 and 2). A calculated MO correlation diagram is shown in Fig 2. Until very recently, the question of whether the $4f\sigma$-promotion leads to excitation, ionization, or a quasimolecular Auger transition of the two electrons occupying the $4f\sigma$ level could not be answered definitely, although the question was addressed in several investigations.

It will be shown that the molecular $4f\sigma$-promotion mechanism leads to secondary electron emission with a high probability. However, the corresponding electron spectra also exhibit atomic orbital signatures, such as strong high-velocity components resulting from the rapid expansion of the $4f\sigma$ orbital at small internuclear distances. Thus comparison will be made with a MO theory as well as with a coupled-channel model of atomic collisions (AO +).

Figure 3(a) shows electron energy spectra for 200-keV $Ar^+ + Ar$ collisions which were obtained from the corresponding time-of-flight spectra. Doubly differential electron yields are plotted versus electron energy for several projectile scattering angles. The electron yields nearly coincide for electron energies below 20 eV. The spectra for scattering angles smaller than about 1° show an exponentially decreasing continuum. For the larger scattering angles, i.e., for decreasing impact parameters, a strong increase in intensity is observed between about 130 and 220 eV. This structure, with a centroid energy near 180 eV, is due to LMM Auger electron emission from the two excited collision partners. We attribute the spectrum for the projectile scattering angle of 0.50° to direct M-shell ionization since neither line structures corresponding to the decay of doubly excited states nor L Auger lines were observed in this case. From auxilliary measurements we estimated that 0.14±0.08 doubly excited states are formed per collision at impact parameters around 1 a.u.

The spectra in fig 3(b) were obtained by subtracting the spectrum for $\theta_p = 0.50°$ from those in Fig. 3(a). The spectra

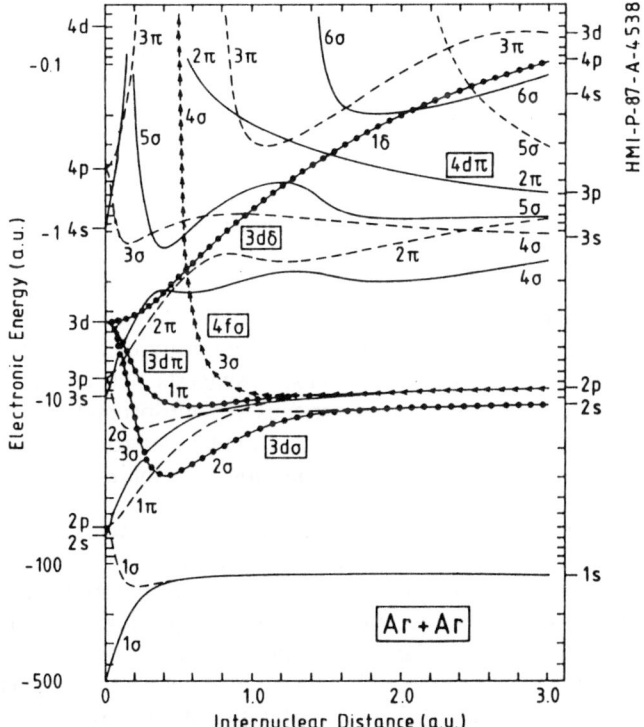

Figure 2. Molecular-orbital correlation diagram for the Ar-Ar collision system taken from Ref 1.

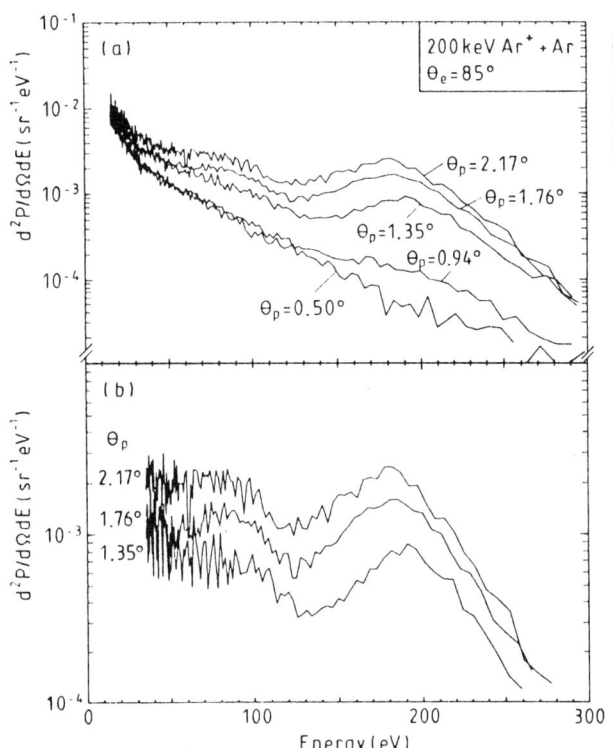

Figure 3. (a) Electron spectra as a function of the projectile scattering angle in 200 keV Ar$^+$ + Ar collisions. The electron observation-angle is 85°. (b) Electron energy spectra after subtraction of the M-shell contribution (corresponding to scattering angles < 1°).

clearly show a flat continuum at electron energies below 110 eV. This continuum, as well as the Auger peak, reaches a maximum intensity at a scattering angle of about $\theta_p = 2°$, which corresponds to an impact parameter of about 0.5 a.u. It has been demonstrated from projectile-energy-dependence measurements and from earlier impact-parameter-dependence measurements, that the internuclear distance at which the $4f\sigma$-promotion occurs is 0.5 a.u. (see Fig 2). That is the internuclear distance at which the promotion causes the production of L vacancies and consequently the Auger electron emission. It should be mentioned that the spectra taken at $\theta_p = 2.6°$ and 3° show no fundamental deviation from the 2.17° spectrum to within experimental uncertainties.

Figure 4(a) shows the Ar L-vacancy production versus distance of closest approach for 50- to 300-keV Ar$^+$ + Ar collisions. The L-vacancy production was calculated by integrating the Auger peak over energy and emission angle. It is seen that the L-vacancy production increases for impact parameters less than 0.7 a.u. and reaches a maximum of 2.0 ± 0.2 L vacancies per collision at about 0.5 a.u., in qualitative agreement with the predictions of the promotion model. The conversion of projectile scattering angle to distance of closest approach was calculated using Molière interaction potentials (see, e.g., Ref 1) with a Lindhard-Nielsen-Scharff screening length.

Auger data of Shanker et al.[31] and inelastic-energy-loss measurements of Fastrup et al.[32] are also given in Fig 4(a). In order to facilitate the comparison between the different measurements, the corresponding projectile scattering angles were transformed into distances of closest approach using the same interaction potentials as above. Such transformations were also done with all other impact parameters and distances of closest approach given in this work.

It has been pointed out earlier[31] that flourescence yields do not affect these results to within the errors usually quoted for Auger-electron-production cross sections. Therefore the number of Ar L vacancies was set equal to the number of L Auger electrons per collision. It should be noted that the measured number of L vacancies per collision for impact parameters below 0.45 a.u. agrees to within ±10% with the prediction of two vacancies from the MO model of $4f\sigma$-promotion. However, from the MO diagram in Fig. 2 and for a collisional broadening of about 1 a.u., we may estimate a distance of closest approach between 0.51 and 0.53 a.u. (shaded area in Fig. 4) for the onset of the $4f\sigma$ vacancy production, which is contrary to the experimental results. The reason for this discrepancy might be a nonadiabaticity of the rapidly promoting $4f\sigma$ level. Furthermore, it is important to note the good agreement between Auger data and inelastic-energy-loss measurements at distances of closest approach below 0.45 a.u. Three-electron Auger processes and also molecular Auger processes would reduce the number of separated-atom Auger electrons, but the probabilities for these two processes are expected to be low, and, hence, would not substantially change the results plotted in Fig. 4(a).

Comparison of the integrated electron intensities for energies less than 110 eV with the integrated Ar L intensities for small impact parameters below about 0.5 a.u. [Fig. 3(b)] shows that these intensities are nearly identical. From the impact-parameter dependence and the electron energy dependence it is inferred that the structure for energies below 110 eV [see Fig. 3(b)] is due to a process which follows the $4f\sigma$-promotion. This rather strong intensity of about two continuum electrons per collision suggests that electrons with energies below 110 eV are caused by $4f\sigma$-promotion into the molecular continuum. This particular electron emission mechanism will be referred to as molecular ionization.

Figure 4(b) displays the ratio between the electron yield from molecular ionization and the number of L vacancies (Ar L Auger yield). Both yields were determined to within an uncertainty of about 20% by integrating the corresponding parts of the time-of-flight spectra over time and angle of ejection. The ratio falls off at about 0.48 a.u. from a level of about 1.3±0.2. The maximum ratio should be about 1.21 corresponding to two ionized $L_{2,3}$-shell electrons, about 0.14 M-shell shake-off electrons, and 0.28 electrons ejected via three-electron Auger transitions, as discussed

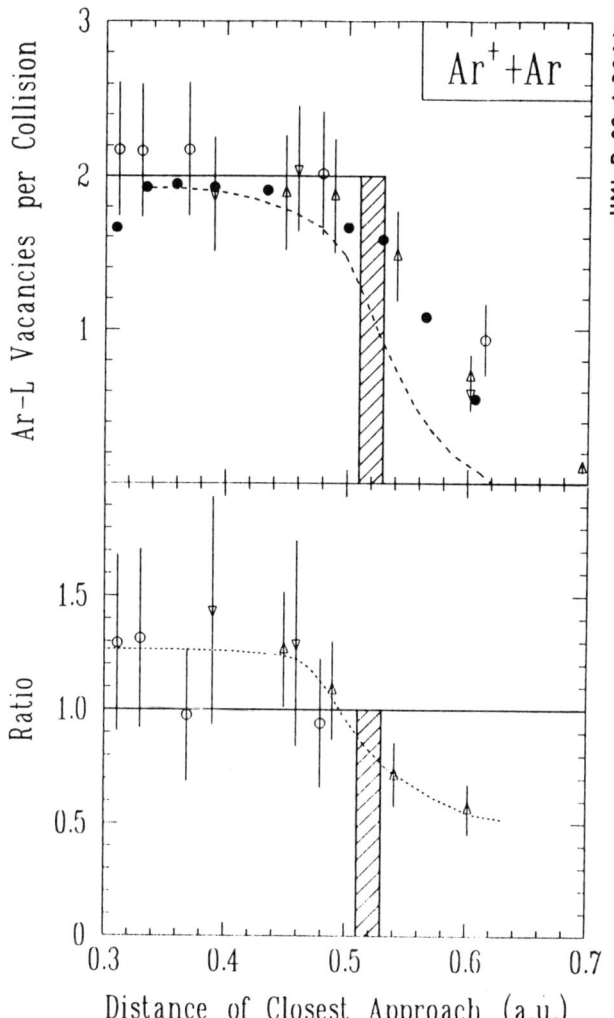

Figure 4 (a) Number of Ar-L shell vacancies per collision as derived from the integrated Auger-peak intensities vs distance of closest approach for incident energies of 150, 200 and 300 keV. This work (open symbols): 150 keV (triangles), 200 keV (inverted triangles), 300 keV (circles). Data for 300 keV taken from Shanker et al (solid circles).[31] Inelastic-energy-loss measurements by Fastrup et al[32] for 50 keV incident energy (dashed line). Solid lines and shaded area correspond to the prediction of the $4f\sigma$-promotion model.
(b) Ratio of the number of electrons (from molecular L-shell ionization) with energies less than 110 eV to the number of Ar-L vaccancies as a function of the distance of closest approach.

elsewhere.[2] At larger impact parameters the ratio approaches 0.5, indicating that electrons are possibly promoted into bound states rather than into the continuum.

The experimental data presented here are in agreement with preliminary results of coupled-channel calculations which are based on an expansion in terms of atomic orbitals (of the separated atoms), united-atom orbitals and pseudostates for the continuum.[29,11] In such a calculation, molecular orbitals are modeled by a coherent superposition of atomic orbitals. The results of the calculation show that two electrons are ionized out of the $4f\sigma$ orbital at small impact pararmters (< 0.3 a.u.).

In order to investigate the spectra in more detail, we performed calculations for direct molecular-orbital ionization according to a model of Briggs,[33] in which the ionization probability is expressed in terms of the first-order amplitudes for a coherent ionization of a static united-atom orbital (at the center of charge) by the projectile and target charges. Since the $4f\sigma$ binding energy decreases by more than 2 orders of magnitude as the collision partners approach one another (cf Fig.2), we treated the binding energy as a fitting parameter.[34,35] With relatively high binding energies (> 50 eV), we could compute spectra of similar shape as those observed experimentally, but then the calculated ionization probability was too low by more than one order of magnitude. Such a discrepancy may be due to the neglect of dynamic effects on the bound-state wave function during the promotion.

A close inspection of the MO diagram in Fig. 2 shows that the mean orbital radius of the $4f\sigma$ level increases more than 20 times faster than the rate at which the collision partners approach each other at internuclear distances below 0.55 a.u., which may result in nonadiabatic processes. At these distances, the $4f\sigma$-level energy is about 50 eV, in agreement with the estimate from the Briggs model. However, dynamic MO-effects are incorporated into the coupled-channel calculation by Fritsch,[29,11] whose results are consistent with our results of two ionized L-shell electrons per collision at small internuclear distances.

In conclusion, our results suggest that two velocity requirements must be met for the promotion model to be valid: the projectile speed as well as the expansion velocity of the molecular wave function should be small compared to the mean electron orbital velocity.

5 Ionization of He in the Presence of Strong Coulomb Fields

In the case of high projectile charges, or when the projectile velocity matches the mean electon orbital velocity, it is known that neither molecular nor atomic first-order theories are appropriate to describe the collision process. This is the domain of higher-order models such as coupled channel theories. Here, we will focus our attention on the question of whether a description in terms of a finite number of target-centered states is sufficient to describe the experimental results, or whether two-center theories are necessary.

In H^+ + He collisions, small impact parameters were selected (small compared to the K-shell radius). At these small impact parameters and energies around 80 keV, the probability for inelastic scattering is nearly 70%. Thus, even the realtively weak interacting protons can produce a high degree of excitation or ionization of He-targets. In order to selectively investigate the influence of the continuum-continuum interactions on the spectrum of directly ionized elec-

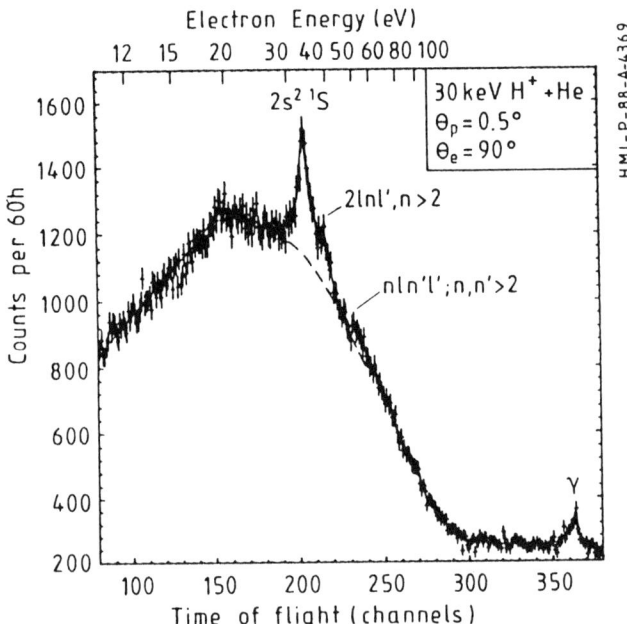

Figure 5. Electron time-of-flight coincidence spectrum for 30 keV H$^+$ + He collisions. The projectile scattering angle is 0.5° and the electron emission angle is 90° with respect to the beam axis.

trons, we also performed noncoincidence measurements for 5 to 25 MeV/u highly charged ions on He.

Fig 5 shows a typical time-of-flight coincidence spectrum for 30 keV H$^+$ + He collisions. It should be emphasized that this spectrum contains roughly 150000 true coincidences. At channel 362, a narrow peak is visible which originates from photon-projectile coincidences (the electron spectrometer has a nonzero detection efficiency for photons with an energy above 5 eV). This peak was used to determine the time-zero for the transformation of the time-of-flight spectra into energy spectra. The broad continuum due to direct ionization will be discussed below. The line structures at 33-54 eV originating from the decay of doubly excited $2l2l$, $2lnl'$, and $nln'l'$ states with n, n' > 2 are discussed elsewhere.[36]

In table 2, experimental yields for ionization, electron capture, and excitation of p-states are given in comparison with theoretical values for H$^+$ and He^{2+} + He collisions.

The total ionization yields were obtained by integration of the measured doubly differential yields over electron energy and angle of ejection. The excitation probabilities were estimated from the photon peak intensities. Since the photon detection efficiency maximizes at 20 eV and the observed photon peak has a line width of less than 3 ns, we can exclude transitions with energies below 8 eV or above 30 eV and lifetimes above 3 ns. Thus, the measured photons are mainly due to He(1s2p → 1s^2), He(1s3p → 1s^2), and H(2p → 1s) transitions, and reflect the excitation of p-states. The total capture probabilities were obtained from the intensities of scattered projectiles with or without electrostatic deflection of charged projectiles behind the target region.

It is seen from table 2 that the results of quantum mechanical (AO+)[29,30] and also classical (CTMC)[12] highest-order theories are in good agreement with the experimental values. It should be noted that ionization probabilities for 33 keV/u He^{2+} + He calculated in the plane-wave Born approximation (PWBA) or in the first-order semiclassical approximation (SCA) overestimate the experimental data by a factor of about 17, corresponding to 4.2 ionized electrons per collision.[27,28,34] This discrepancy between first-order and highest-order theories originates from the high transition probabilities at low velocities and small impact parameters. In a single collision, an electron may be ionized in a first step and then recaptured by either the target or the projectile nucleus in a second step. Such effects, as well as dynamic binding effects, are included in the AO+ and the CTMC theories. Another indication for a breakdown of first-order theories becomes obvious when equal velocity H$^+$ and He^{2+} on He results are compared. All classical and quantum mechanical first-order theories predict the excitation and ionization probabilities to be proportional to the squared projectile charge (Z_p). However, the yields measured for He^{2+} + He are significantly lower than 4 times the H$^+$ + He values.

It should be noted that ionization probabilities for 30 keV H$^+$ + He calculated within a semiclassical approximation (SCA-UA) are in agreement with even measured doubly differential yields[8] when united atom orbitals are used to describe the initial state.[34,4] Therefore it may be concluded that molecular effects are of importance in 30 keV H$^+$ + He collisions. This is to be expected, since the projectile velocity

Table 2 Experimental and theoretical ionization, one-electron capture, and excitation yields (P_I, P_C and P_{Ex}) in percent for 30 and 100 keV H$^+$ + He and for 100 kev He^{2+} + He collisions at a projectile scattering angle of 1°. The experimental excitation probabilities are restricted by the channel-plate efficiency function to the He(1s2p), He(1s3p), and H(2p) states.

	Experiment			AO+			CTMC		SCA	SCA-UA
	P_I	P_C	P_{Ex}	P_I	P_C	P_{Ex}	P_I	P_C	P_I	P_I
30 keV H$^+$ + He	32±13	73±14	3.3±1.6	29	65	2.5[a)]	27	60	107	20
100 keV H$^+$ + He	54±22	26±7	1.4±0.7	44	25	-	50	20	51	36
33 keV/u He^{2+} + He	24±12	-	4.0±2.0	16	62	7.2	25	86	418	-

a) AO+ calculation for He(1s2p) and H(2p) include electron correlation.

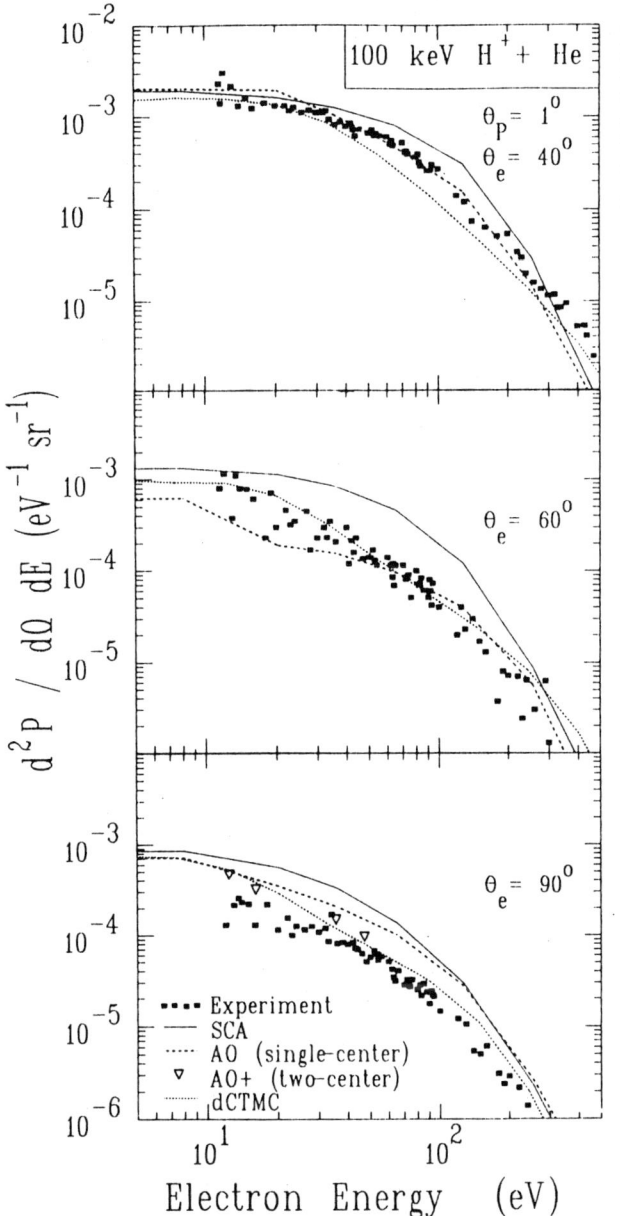

Figure 6. Absolute doubly differential electron emission yields for 40°, 60° and 90° in 100 keV H^+ + He collisions at small impact parameters.

v_p is lower than the mean orbital velocity v_e of the electrons in the K-shell (v_p = 1.1 a.u., v_e = 1.35 a.u.), resulting in nearly adiabatic relaxation of the bound-state wave function. It is noted that the theoretical results given in table 2 are in good agreement with the experimental values for the total ionization yield at 100 keV H^+ + He. In the following, we will demonstrate the ability of multiply differential ionization yields to allow for a judgement about the quality of different theories for direct ionization.

In Fig.6, doubly differential electron emission yields are plotted as a function of electron energy for three ejection angles in 100 keV H^+ + He collisions. The experimental data are compared to results of four different theories (see table 1). It is seen that first-order SCA calculations[26] disagree with experiment. The same holds for SCA calcuations using united-atom orbitals (not shown in the figure). For a discussion of SCA-UA results for 30-300 keV H^+ + He, see Ref. 8.

The single-center coupled channel calculations (AO)[26] are in good agreement with experiment for an electron ejection angle of 40°. However, for 90° there is only a slight improvement compared to the SCA results. It should be emphasized that 90 states (6 bound and 84 continuum states) were included in the calculation. All couplings were taken fully into account. It is noted that the spectra are dominated by single ionization. In principle, a single-center coupled-channel calculation should yield the exact solution of the single-active-electron problem. However, the simulation of projectile-centered states requires an infinite number of target states. Thus, from the discrepancy at 90° we may conclude that two-center effects are important for this collision system. In fact, we measured a total capture yield of 26% at these small impact parameters ($b \approx 0.05$ a.u.). This is also an indication for asymptotic projectile-centered states being important for the collision dynamics.

Asymptotic projectile-centered states are included in the AO+ calculation,[29] which could be evaluated from the target-centered s-continuum only for 90°. At this emission angle there is little influence from other continuum states. The present status of this model does not allow for a treatment of the interfering amplitudes for the different continuum states. However, the AO+ results[29] show an improvement compared to the single-center AO-predictions.[26] The remaining difference between AO+ and experimental data might be a result of the neglected interference or an indication that 40 states are not enough to calculate doubly differential ionization probabilities in 100 keV H^+ + He collisions.

The classical dCTMC results are in good agreement with experiment, except for the binary encounter peak at 40°. This discrepancy might be a consequence of the classical initial distribution. The initial electronic orbits are confined by the classical turning point. This leads to a compression of the radial density distribution, and to an overestimated influence of the target nuclear potential. Thus, ionized electrons are deflected too strongly by the target nucleus, and the binary encounter peak is broadened. However, the overall agreement between experimental and dCTMC values is better than for the other theories. We attribute this to the fact that the dCTMC includes two-center effects and is not restricted by an expansion in terms of stationary states.

Two-center effects also appear in non-coincident investigations when highly-charged projectiles are used. Fig. 7 displays the ratios of experimental doubly differential electron emission cross-sections (DDCS) to DDCS obtained with the plane-wave Born approximation (PWBA). The

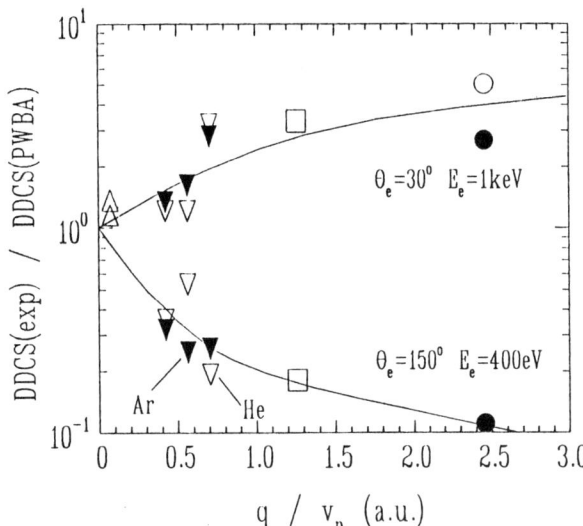

Figure 7. Ratio of experimental to theoretical doubly differential cross sections (DDCS) as a function of q/v_p for ejection angles and energies of 30° and 1keV, and 150° and 400 eV. Open symbols: He targets; closed symbols: Ar targets. Experiments: Triangles are taken from Ref.5, inverted triangles from Ref.37, squares and circles are from Refs. 38 and 39. The solid lines are drawn to guide the eye.

PWBA results by Manson et al[5] include sophisticated Hartree-Fock wavefunctions and partial waves up to l = 16. The experiments were carried out at the VICKSI heavy-ion facility in Berlin,[5,37] at GANIL,[38] and at the Super-HILAC accelerator.[39] High incident energies of 5 to 25 MeV/u were selected in collision with He and Ar atoms. Thus, electron capture, binding effects, and intermediate excitation channels are of minor importance even for projectile charge states up to q = 40.

The displayed enhancement of forward electron ejection and the reduction of backward electron ejection has been referred to as two-center electron emission.[38] Qualitative agreement was found between experimental data and results of the continuum distorted-wave eikonal initial state model by Fainstein and Rivarola.[13,38] In this model, continuum wavefunctions include asymptotic two-center effects. However, the behaviour shown in Fig. 7 might also be explained in terms of a two-step process. In a first step, an electron is ionized according to the description of the PWBA. In a second step, this electron is attracted by the highly-charged projectile and deflected towards small laboratory angles. It is seen that this effect increases with increasing projectile charge or decreasing velocity. However, it is not clear whether the deviation from the PWBA prediction saturates for strong distortions.

In the case of single ionization and large distances from the target nucleus, the continuum wavefunctions for different target atoms are indentical to within a phase shift. Thus, the second step in the two-step process should be nearly independent of the target. This is consistent with the fact that the ratios in Fig. 7 are similar for He and Ar targets. Another indication of the described two-step process is the agreement between experimtent and PWBA results for the singly differential cross-sections (integrated over angle) found in Ref. 37. Thus, the second step leads mainly to a redistribution of the initial angular distribution of ejected electrons. Finally, the range of validity for PWBA calculations may be given as $Z_p/v_p < 0.3$.

6 Double Ionization in Small Impact Parameter $H^+ + He$ Collisions

Most of the theories available to date are based on the independent particle model (IPM). In the IPM, it is generally assumed that only one electron is capable of changing its state while the other electrons act as spectators. Thus, there is no exchange of informations such as energy or angular momentum between the electrons. Consequently, probabilities for multi-electron processes should follow binomial statistics. If an impact parameter is not selected in an experiment, it is hardly possible to make any statements about the validity of the IPM without referring to theories which generally include several additional approximations. Thus, we decided to measure triple coincidences between the projectile, recoil ion, and electrons. This allows for an investigation of electron energy and angular distributions as a function of impact parameter (projectile scattering angle) and final target charge-state.

Fig. 8 displays electron spectra for an impact parameter of 0.02 a.u. and an electron ejection angle of 60° in 300 keV $H^+ + He$ collisions. The spectra are given for both recoil ion charge-states ($q_t = 1,2$) and are compared to the predictions of different theories (see table 1). There is good agreement between experimnetal and nearly all theoretical data for single ionization to within the experimental uncertainties. This was noted also for PWBA calculations by Salin for the same collision system.[27] The dCTMC results overestimate the experimental values for electron energies below 30 eV by a factor of about 2. This discrepancy is probably due to the enhanced initial electron density near the target nucleus, as discussed in section 3. It is noted that the coupled-channel calculations (single-center AO) are similar to the SCA predictions. Thus, two-center effects seem to be of minor importance. This is in accordance with the small ratio of Z_p/v_p (see Fig.7). Therefore, the single-center AO-results should be highly accurate.

In Fig. 8, electron spectra for double ionization ($q_t = 2$) are compared to the same theories as above. Additionally, an upper estimate for the shake-off contribution to double ionization is given. The shake-off spectrum was calculated as the overlap between $He(1s^2)$ and $He^+(\varepsilon s)$ wavefunctions and scaled according to the IPM.[26] Consideration of the velocity distribution of directly ionized electrons would further reduce the shake-off contribution. Thus, shake-off can be neglected in the present case. The measured electron spectra in Fig. 8 show a similar slope as in the case of single

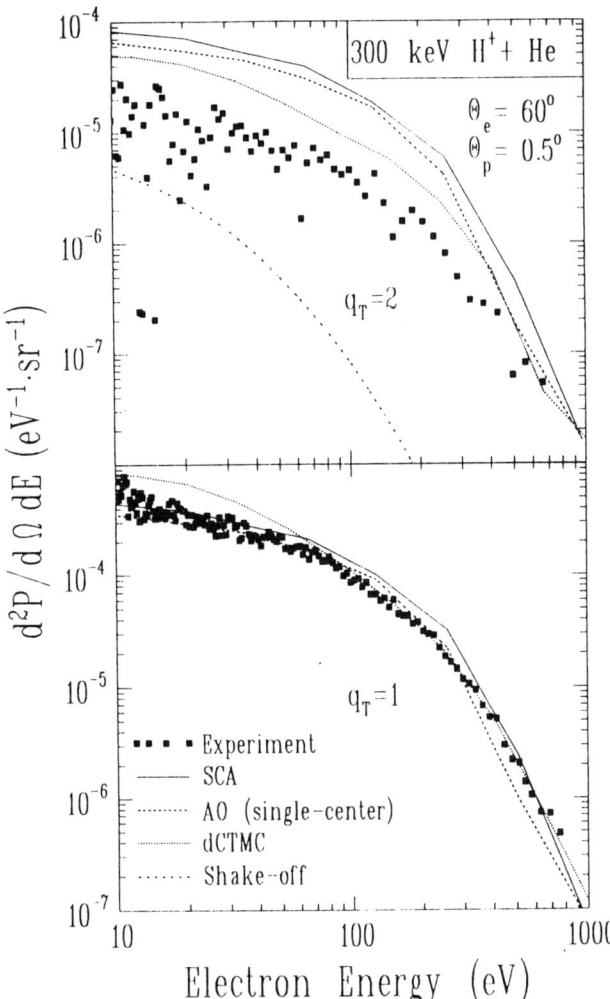

Figure 8. Absolute doubly differential ionization yields as a function of electron energy for two final target charge states ($q_t = 1, 2$) in 300 keV H^+ + He collisions.

ionization, but are reduced by a factor of about 30. Thus SCA and single-center AO results are about a factor of 5 too high. This discrepancy is an indication of the breakdown of the IPM. A similar effect was found in an earlier work by Hippler et al,[40] where the target region was not kept field-free and the impact parameter was not defined. It was argued that a sequential ionization mechanism is responsible for the small ionization cross sections. The "first" electron has to overcome an ionization potential of 24.6 eV. Afterwards, the "second" electron is ionized from a partially relaxed state with increased binding energy. However, two facts are in contradiction with this assumption:

i) In the case of double ionization, the probability of detecting "electron one" or "electron two" at 60° is about equal. Therefore, a contribution of ≈50% in the electron spectrum should stem from "electron two," with enhanced high-velocity components due to the relaxed bound state wavefunction. However, the corresponding enhancement of high energy electrons compared to the single ionization case was not observed.

ii) The dCTMC model includes also effects as discussed in i), since a dynamic electron-electron interaction via a mean field is taken into account. Nevertheless, the dCTMC results overestimate the experimental data by a factor of 2-3.

Because the incident velocity is relatively high and the collision time is short, one can exclude relaxation effects. However, it is conceivable that the dynamic enhancement of the target potential as well as an angular correlation in the initial state are responsible for the deviations between single-center AO and experimental values. It is noted that such an angular correlation is not included in the present dCTMC model and would further reduce the corresponding double ionization probabilities. An enhancement of the double ionization probability is expected if a dynamic electron-electron correlation (during the collision) comes into play. Thus, it is not possible to quantify the effect of the initial state angular correlation on the spectra shown in Fig. 8.

7 Conclusions

Absolute doubly differential electron emission yields were measured as a function of electron energy, ejection angle, and projectile scattering angle. Experiments were performed for low energy Ar^+ + Ar collisions in order to identify the mechanisms responsible for electron ejection. It was found that both inner shell electrons occupying the $4f\sigma$-level are ionized for impact parameters below 0.48 a.u. The onset of the corresponding electron continuum in the range of 0-110 eV starts at 0.6 to 0.7 a.u. The intensity, the onset, and the shape of this continuum are in contradiction with the Briggs model of direct ionization. It is suggested that the rapid expansion of the $4f\sigma$-orbital is responsible for non-adiabatic effects and leads to a breakdown of the MO-model.

For intermediate energy H^+ + He collisions, it was found that most of the collision theories can reproduce total ionization yields at small impact parameters. However, for doubly differential ionization yields in 100 keV H^+ + He collisions, even single-center coupled-channel calculations fail. In this case, agreement with experiment was found only for theories including two-center effects (AO+ and dCTMC).

Non-coincident measurements were performed for high-energy ions (5MeV/u) incident on He. The results suggest that two-step processes are responsible for the deviations between experimental data and results of a plane-wave Born approximation (PWBA). In a first step, an electron is ejected according to the PWBA. In a second step, this electron is attracted by the highly charged projectile. The deviations between PWBA and experimental values increase with increasing projectile charge q, or reduced velocity v_p. A limit of q/v_p for the validity of the PWBA may be extracted from the data.

Triple coincidences were measured for 300 keV H^+ + He. Electron spectra were taken for small impact parameters as a function of the final charge-state of He. The absolute values in the case of double ionization show significant deviations from predictions based on the independent particle model (IPM). Theoretical results including a dynamic electron-electron interaction (dCTMC) are clearly favoured by the experimental data. The remaining diference between experiment and theory might be an indication of correlation effects. Further experimental and theoretical work is needed to establish the influence of correlation processes in the case of double ionization.

References

1) U.Wille and R.Hippler; Phys.Rep.132, 129 (1986)
2) G.Schiwietz, B.Skogvall, J.Tanis, and D.Schneider; Phys.Rev.A38, 5552 (1988)
3) J.Bang and J.M.Hansteen; Kgl.Danske Videnskab.Selskab.,Mat.-Fys.Medd.31 (1959)
4) D.Trautmann, F.Rösel, and G.Baur; Nucl.Instr.Meth. 232, 218 (1984) - F.Rösel, D.Trautmann, and G.Baur; Nucl.Instr.Meth. 192,43 (1982)
5) S.T.Manson, L.H.Toburen, D.H.Madison, and N.Stolterfoht; Phys.Rev.A12, 60 (1975); S.T.Manson (private communication)
6) M.Inokuti; Rev.Mod.Phys.43, 297 (1971)
7) H.Paul; Z.Phys.D4, 249 (1987)
8) G.Schiwietz; Phys.Rev.A36, 370 (1988)
9) R.L.Becker, A.L.Ford, and J.F.Reading, J.Phys.B13, 4059 (1980)
10) U.Heinz, B.Müller, and W.Greiner; Phys.Rev.A23, 562 (1981)
11) W.Fritsch and C.D.Lin; Phys.Rev.A27, 3361 (1983)
12) G.Schiwietz and W.Fritsch, J.Phys.B20, 5463 (1987)
13) P.D.Fainstein, V.H.Ponce, and R.D.Rivarola; J.Phys.B21, 287 (1988)
14) Q.C.Kessel and E.Everhardt, Phys.Rev.146, 16 (1966)
15) R.Gayet and A.Salin; J.Phys.B20, L571 (1987)
16) N.Stolterfoht, C.Havener, R.Phaneuf, J.Swenson, S.Shafroth, and F.Meyer; Phys.Rev.Lett.57, 74 (1986)
17) J.Tanis, G.Schiwietz, D.Schneider, N.Stolterfoht, W.Graham, H.Altevogt, R.Kowallik,A.Mattis, B.Skogvall, T.Schneider, and E.Szmola; Phys.Rev.A39, 1571 (1989)
18) R.Hippler, G.Schiwietz, and J.Bossler, Phys.Rev.A35, 485 (1987)
19) R.Schuch (private communication)
20) V.Montemayor and G.Schiwietz; J.Phys.B22 (in print)
21) G.Schiwietz, U.Stettner, T.Zouros, and N.Stolterfoht; Phys.Rev.A35, 598 (1987)
22) B.Skogvall and G.Schiwietz (in preparation)
23) V.Montemayor and G.Schiwietz; accepted for publication in Phys.Rev.A
24) W.Fritsch and C.D.Lin, J.Phys.B15, 1255 (1982)
25) E.Clementi and C.Roetti; Atomic Data and Nucl. Data Tables 14, 177 (1974)
26) G.Schiwietz (in preparation)
27) A.Salin; submitted to J.Phys.B
28) A.Salin (private communication); see also Ref.8
29) W.Fritsch (private communication)
30) T.Mukoyama, C.D.Lin, and W.Fritsch; Phys.Rev.A32, 2490 (1985)
31) R.Shanker, R.Bilau, R.Hippler, U.Wille, and H.O.Lutz; J.Phys.B14, 997 (1981)
32) B.Fastrup, G.Hermann, and K.J.Smith; Phys.Rev.A3, 1591 (1971)
33) J.S.Briggs, J.Phys.B8, L485 (1975)
34) The calculations were performed using a program written by Trautmann and Rösel; see also Ref.35
35) D.Trautmann, F.Rösel, and G.Baur; J.Phys.B18, 1167 (1985)
36) G.Schiwietz, B.Skogvall, N.Stolterfoht, D.Schneider, V.Montemayor, and H.Platten; Nucl.Instr.Meth.B40, 178 (1989)
37) G.Schiwietz, H.Platten, D.Schneider, T.Schneider, W.Zeitz, K.Musiol, R.Kowallik, and N.Stolterfoht; Hahn-Meitner-Institut report, HMI-B447 (Berlin, 1987, ISSN 0175-8349)
38) N.Stolterfoht, D.Schneider, J.Tanis, H.Altevogt, A.Salin, P.D.Fainstein, R.Rivarola, J.P.Grandin, J.N.Scheurer, S.Andriamonje, D.Bertault, and J.F.Chemin; Europhys.Lett.4, 899 (1987)
39) D.Schneider, A.Schlachter, R.Olson, W.Graham, J.Mowat, R.DuBois, D.DeWitt, D.Loyd, V.Montemayor, and G.Schiwietz; submitted to Phys.Rev.A
40) R.Hippler, J.Bossler, and H.O.Lutz; J.Phys.B17, 2453 (1984)

ENERGY LOSS AND CHARGE EXCHANGE PROCESSES OF HIGH ENERGY HEAVY IONS CHANNELED IN CRYSTALS *

J.C. Poizat [1], S. Andriamonje [2], R. Anne [3], N.V. de Castro Faria [1], M. Chevallier [1], C. Cohen [4], J. Dural [5], B. Farizon-Mazuy [1], M.J. Gaillard [1], R. Genre [1], M. Hage-Ali [6], R. Kirsch [1], A. L'hoir [4], J. Mory [7], J. Moulin [4], Y. Quéré [7], J. Remillieux [1], D. Schmaus [4] and M. Toulemonde [5].

(1) - Institut de Physique Nucléaire de Lyon, IN2P3-CNRS/Université Claude Bernard,
   43, Bd du 11 Novembre 1918, 69622 Villeurbanne Cedex, France
(2) - Centre d'Etudes Nucléaires de Bordeaux (and IN2P3)
   33170 Gradignan, France
(3) - GANIL (IN2P3), 14021 Caen Cedex, France
(4) - Groupe de Physique des Solides de l'Ecole Normale Supérieure
   75251 Paris Cedex 05, France
(5) - Centre Interdisciplinaire de Recherches avec les Ions Lourds
   14040 Caen Cedex, France
(6) - Groupe de Physique Appliquée aux Semiconducteurs, Centre de Recherches Nucléaires (and IN2P3)
   67037 Strasbourg Cedex, France
(7) - Laboratoire des Solides Irradiés, Ecole Polytechnique
   91128 - Palaiseau Cedex, France

The interaction of moving ions with single crystals is very sensitive to the orientation of the incident beam with respect to the crystalline directions of the target. Our experiments show that high energy heavy ion channeling deeply modifies their slowing down and charge exchange processes. This is due to the fact that channeled ions interact only with outershell target electrons, which means that the electron density they experience is very low and that the binding energy, and then the momentum distribution of these electrons, are quite different from the corresponding average values associated to random incidence.

The two experimental studies presented here show the reduction of the energy loss rate for fast channeled heavy ions and illustrate the two aspects of channeling effects on charge exchange, the reduction of electron loss on one hand, and of electron capture on the other hand.

In the first study we have obtained the charge distributions and energy loss spectra of 27 MeV/u Xe^{35+} ions transmitted through a thin silicon single crystal, and deduced electron impact ionization cross sections for Xe^{35+} to Xe^{45+} ions by 14.7 KeV electrons and the transverse energy dependence of energy loss of channeled ions. In the second one, performed with highly stripped Xe incident ions we have observed how the drastic reduction of the electron capture probability allows most of the channeled ions to keep their initial charge state throughout the crystal. The observation of the spectrum of X-rays emitted from the impact area shows how channeling conditions reduce innershell excitation of the incident ions and primary and secondary electron bremsstrahlung. In channeling conditions the X-ray spectra are characterized by the predominance of Radiative Electron Capture

(REC) lines, because the REC process is the only one that can allow channeled ions to capture quasi-free electrons. The shape of the REC lines reflects the momentum distribution of electrons encountered by channeled ions.

At last we briefly present two resonant processes by means of which the capture of quasi-free electrons can be made possible, the Resonant Transfer and Excitation (RTE), and the recently proposed Nuclear Excitation by Electron Capture (NEEC).

I - Introduction

Channeling of positive ions in a crystal occurs when the direction of the incident beam is close to a plane or axis direction of the crystal[1]. In perfect axial alignment for example the ions entering the crystal with an impact parameter ρ_o to an atomic row larger than the thermal vibration amplitude are repelled by the atomic row as a whole and can be considered as submitted to a continuum potential $V(\rho)$ that is the zero-order term of the Fourier series in which the periodic ion-atom potential can be decomposed. A complete discription of axial channeling must take more than one atomic string into account. In figure 1 we show the maps of the electron density averaged along the $< 110 >$ direction of silicon (a), and of the continuum potential for a single charged projectile (b). The electron density has been calculated by means of the electron wave functions in solid silicon, and the potential deduced from the electron density. The entrance conditions, incidence angle to the axis direction and impact location, define what is called the transverse energy of the incident projectile, that can be considered to be constant, at a first approximation, during the passage through a thin crystal. The value of the transverse energy determines the closest distance of approach to the atomic rows. The best channeled particles have the smallest transverse energy and then are maintained quite far from the atomic rows. In particular some particles have such a small transverse energy that they are trapped between the same set of atomic strings all along their pathlength. They are said to be hyperchanneled (for instance, in figure 1b, particles with a transverse energy less than 1.5 eV are confined inside the contour 2).

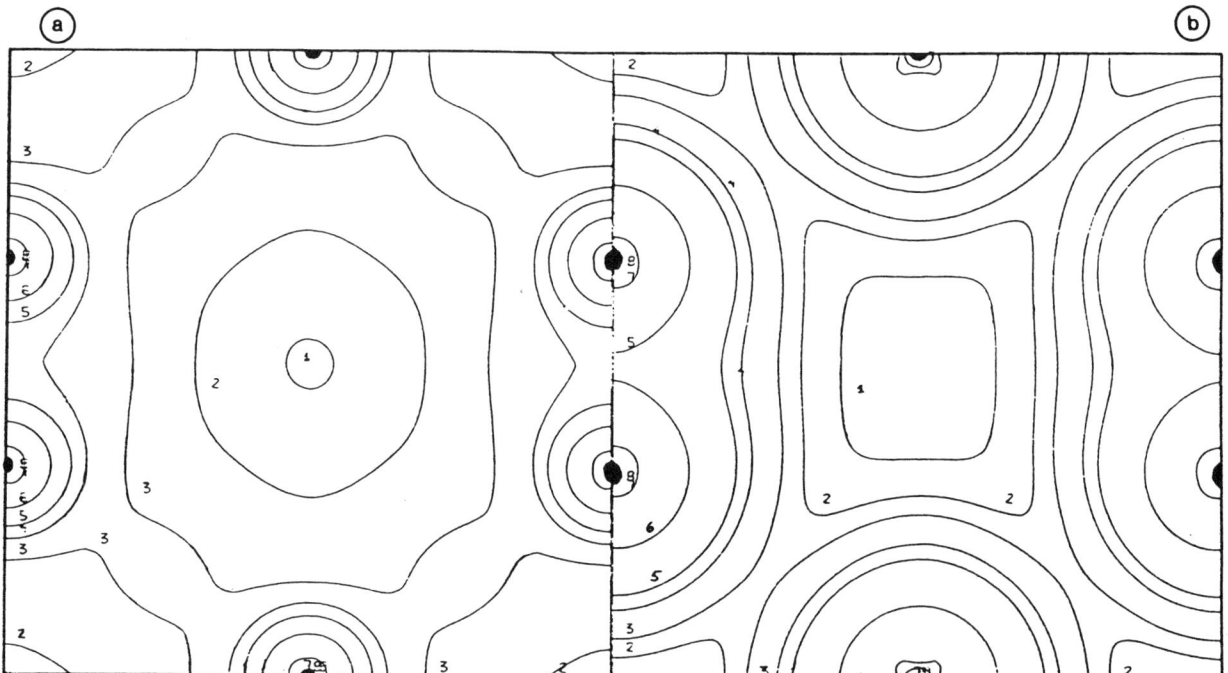

Figure 1 - Map of the electron density in a silicon crystal, averaged along the $< 110 >$ direction. Contour lines 1 to 8 : 0.032, 0.1, 0.24, 0.5, 1, 3, 10 and 30(A^{-3}) (a).

Map of the continuum potential for a unit charge along the $< 110 >$ direction of Si. Contour lines 1 to 8 : 0.5, 1.5, 2.5, 7, 20, 50, 100, 200 (eV) (b).

The main properties of channeled ions can be summarized as follows : as they cannot approach the atomic rows, they do not interact closely with the target nuclei and the innershell target electrons (in particular the suppression of Rutherford scattering is, from a practical point of view, the most useful feature of channeling for low energy positive particles). As a consequence the lowering of the encountered electron density reduces the energy loss rate by a factor up to two or three. We shall see that using highly charged fast ions allows to study channeling energy loss in a very detailed way. Moreover the charge exchange process, dominated in amorphous or randomly oriented materials by the interaction with the target nuclei and core electrons, is there entirely due to the interaction with quasi-free outershell and valence target electrons, which results in a huge reduction of both electron loss and capture. Electron loss and electron capture by channeled heavy ions will be the respective themes of the two experiments we present here, after a brief description of the experimental set-up common to both studies.

II - Experimental set-up

We used the beam line LISE of the GANIL facility (Caen, France), a line specially designed to deliver highly stripped ion beams of excellent space and momentum distributions. The experimental set-up is schematized in figure 2. The beam delivered by the accelerator was made of 27 MeV/u Xenon ions of charge $q = 35$. These Xe^{35+} ions were used as incident projectiles on the crystal for the channeling electron loss study, whereas a thin Be foil could be inserted in order to strip, up to $q = 54$ (bare nuclei) -the charge distribution being centered at $q = 52$-, the transmitted ions used in the second study devoted to channeling effects on electron capture. In all cases the ions with the desired charge state were magnetically selected (M_1) after tight collimation. After passage through a rotating beam chopper, used to monitor the beam intensity (either by the beam current measurement or by counting, in a $Si - Li$ detector, of the X-rays produced in the silver coating of the rotating blades), the beam was sent onto the crystal, a 17 μm thick (111) Si crystal hold by a three-axis goniometer. The crystal was viewed by a hyperpure Ge detector positioned at 90° to the beam. The beam transmitted through the crystal was energy and charge analyzed by means of the magnetic spectrometer M_2 and a wire chamber. The optics was set in such a way that the angular divergence of the outgoing beam had a negligible influence on the wire chamber response. The overall resolution was 45 keV/u and the counting dynamics was $\sim 10^6$. By stepping the current intensity in M_2, the energy loss spectrum associated to each charge state at emergence and the charge state distribution could be obtained. At last the wire chamber could be used to adjust the position of a pair of movable slits, and then could be taken out of the beam, in order to send Xe ions of selected charge state and energy onto a microchannel plate detector (MCP) and to allow coincidence measurements of transmitted ions and X-rays emitted from the crystal.

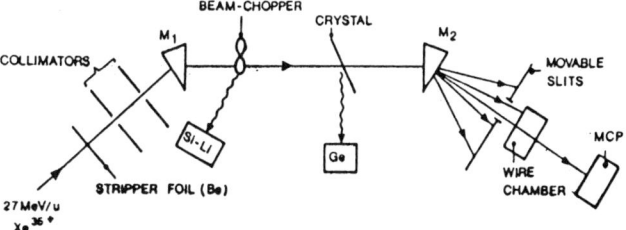

Figure 2 - Experimental set-up.

III - Channeling study with incident Xe^{35+} ions

The charge distributions of Xe ions transmitted after a pathlength of 21 μm through the Si crystal in random conditions and for $< 110 >$ axial alignment are shown in figure 3. The distribution obtained for

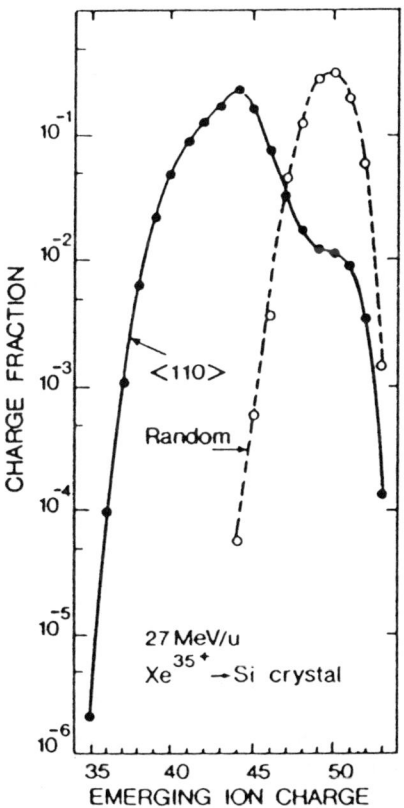

Figure 3 - Charge state distributions after the passage of incident 27 MeV/u Xe^{35+} ions though a 21 μm Si crystal in random and in $< 110 >$ alignment conditions, respectively.

random incidence is quite symmetrical, centered around $q = 49 - 50$, and reflects charge state equilibrium, which means that Xe ions have suffered so many electron loss and capture events that they have lost the memory of their initial charge. On the contrary the distribution obtained in axial alignment extends from $q = 35$ to $q = 53$, which reveals a large dispersion in the charge changing process of aligned projectiles. As it will be confirmed by the analysis of energy losses, this broad distribution results from the transverse energy distribution of the aligned particles. It is clear in particular that the small fraction of ions which have kept their initial charge state are the best channeled ones, with a very small transverse energy, that have travelled through regions of very low electron density. The only part of the distribution that is charge equilibrated is the highest charge side, which contains the nonchanneled component ($\sim 2\%$) of the beam. Another noticeable feature is the sharpness of the distribution at its top ($q = 44$) which is due to a shell effect on the electron loss cross section.

Although a complete study of the orientation dependence of the charge distribution has not been done, the transverse energy dependence of the distribution is well illustrated by the data shown in figure 4 where the orientation dependence of the fraction $q = 36$ around the $< 110 >$ direction is compared with the orientation dependence of the Xe $Ly\alpha$ photon yield given by the Ge detector. Whereas the $Ly\alpha$ yield, resulting from the

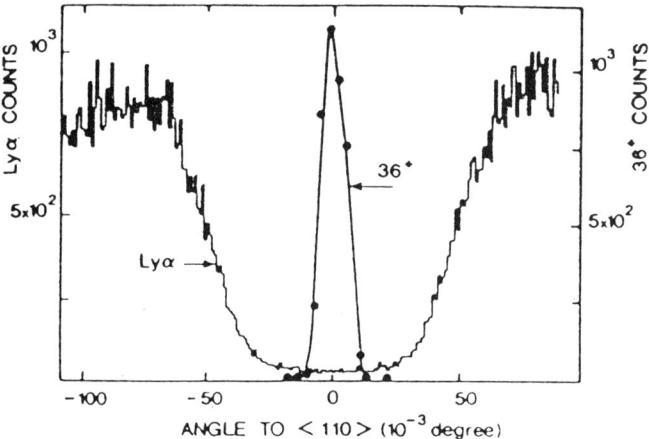

Figure 4 - Orientation dependences of the $Xe\ Ly\alpha$ yield and of the 36^+ fraction in the transmitted beam, for Xe^{35+} ions incident around the $< 110 >$ direction of the Si crystal.

creation of K-shell vacancies in close collisions of Xe projectiles with Si target atoms, presents the usual channeling dip, the 36^+ component presents a sharp peak, that shows that only the very best channeled ions, i.e with a very small transverse energy, are able to lose only one electron during their traversal of the crystal.

A complete analysis of the charge distribution with the aim of obtaining ionization cross sections by collision with target electrons must take the transverse energy distribution of channeled ions into account. But this distribution depends heavily on the angular dispersion of the incident beam, which is too small to be determined directly. However, as energy loss of channeled particles is very sensitive to their transverse energy, we have the very detailed source of information given by the energy spectra associated with each charge state at emergence. The dependence of the mean energy loss of the transmitted ions upon their charge Q at emergence is shown on figure 5. In order to get rid of the charge dependence of

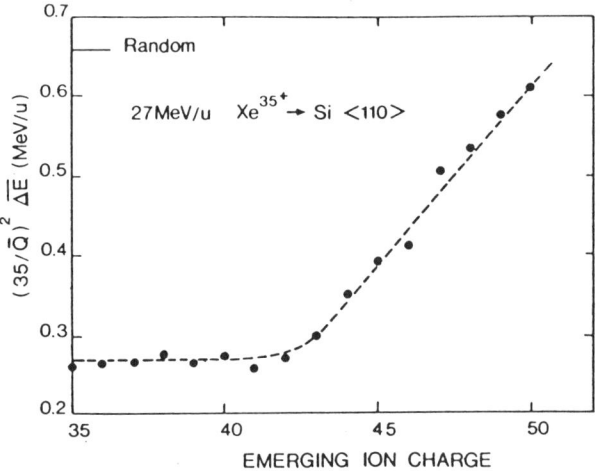

Figure 5 - Dependence of the mean energy loss upon the charge state at emergence for 27 MeV/u Xe^{35+} incident along the $< 110 >$ direction of the Si crystal.

stopping power, each energy loss has been multiplied by $(35/Q_{rms})^2$, where Q_{rms}, that depends on Q, is the root mean square value of the charge over the pathlength of an ion entering the crystal with the charge 35^+ and emerging with charge Q (estimates of Q_{rms} values are readily obtained, but more accurate values have been derived from the simulation described below). Then we obtain a reduced energy loss which depends only on the transverse energy in the crystal of the outcoming particles. As indicated above the charge at emergence essentially reflects the transverse energy of channeled particles. We observe that all the best channeled particles, with charges at emergence from 35 to 42 (about 30% of the incident beam) have the same energy loss rate, 0.4 times the random value, although they do not all experience the same reduced average electron density. This is a nice illustration of the non-local aspect of electronic energy loss of charged projectiles of high velocity in a quasi-free non uniform electron gas.

In order to understand quantitatively our results, both charge distribution and energy spectra associated with the various charge states (not shown here) we have performed a simulation that we summarize as follows : first of all we have been led to consider that the incident beam dispersion is characterized by the sum of two gaussian distributions (the sharpest one is necessary to reproduce the sharp peak of figure 4 but is unable to reproduce the shapes of the energy spectra which are mainly determined by the tail of the beam angular distribution). From the incident beam angular profile and the uniform entrance impact parameter distribution we get the transverse energy distribution. We then start a Monte-Carlo simulation in which we calculate values of electron loss cross sections in collisions with target electrons. We assume that flux statistical equilibrium is rapidly reached and that the electron loss probability is proportional to the mean local electron density experienced by each channeled ion. We calculate successively all the cross sections, starting from the $35 \to 36$ cross section, up to $q = 45$, a safe limit beyond which we may think that charge exchange processes induced by target atom nuclei can come into play for these poorly channeled ions. Using these values we now try to reproduce the experimental energy spectra. For this we need the relationship, for a given charge state in the solid, between the transverse energy of channeled particles and the rate of energy loss. We choose empirically a curve expressing this relationship, which is not too difficult since we know experimentally the rate for near zero transverse energy ions that have been transmitted with their initial charge state, and since particles with a high transverse energy have a "random" type energy loss. Using such a curve and the previous set of electron loss cross sections we obtain simulated energy spectra. The procedure of choosing a beam divergence and a relation between transverse energy and energy loss is repeated until a general agreement is reached. The calculations on channeling energy loss will be published with more details [2,3]. The final electron loss cross sections are given in figure 6. They are compared to values, available in the literature, of 14.7 KeV electron impact ionization cross sections for Xe ions at rest, i.e for the same electron-ion relative velocity. These values are experimental data by Donets [4], computed values from the empirical Lotz formula [5], and also from the simple Thomson formula (with a cut-off for energy transfer below the binding energy of the electron to be ejected). The agreement with the Donets data is rather good except for $q = 35$. As all our values are above Donets values it can be guessed that the difference is due to the fact that, in our case, ion excitation may also occur. This might yield higher loss cross sections if the relaxation does not happen before subsequent ionization in the crystal. Nevertheless the agreement is quite satisfactory if one considers how different the two experiments are, and how indirect our cross section determination is.

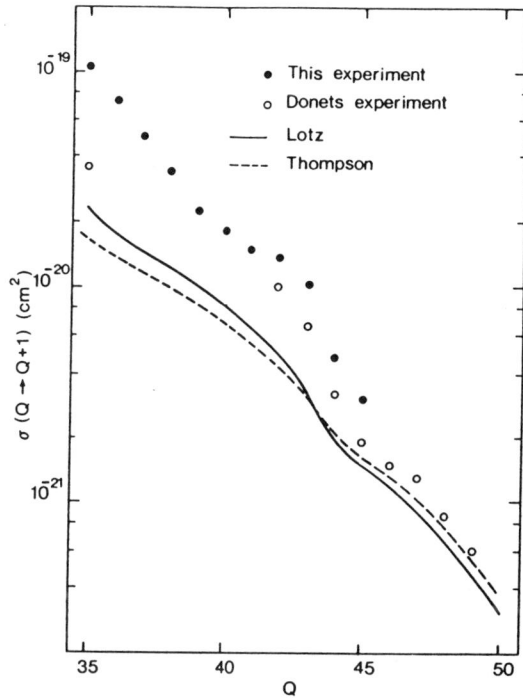

Figure 6 - Electron impact ionization cross sections of Xe^{35+} to Xe^{45+} ions by 14.7 KeV electrons. The values deduced from our measurements are compared to experimental values of Donets, and to the calculations from Lotz and Thomson formula.

IV - Channeling studies with highly stripped Xe ions

Charge distributions of Xe ions transmitted in $<110>$ alignment conditions through the same Si crystal are shown on figure 7. As the various used charge states at incidence, from 50 to 53, result from charge exchange of 27 MeV/u Xe ions in the Be stripper foil, the incident energy is lower (~ 25.5 MeV/u) and the beam intensities are quite smaller. This explains in particular why the 54^+ case is missing in this series. (However we have reported previously on some data obtained with 54^+ incident ions [6]). The charge distributions obtained for random incidence of the various charge states are nearly identical and this distribution, equilibrated around the mean value 49.5, is also shown on figure 7. On the contrary the distributions obtained for axial alignment are quite different from each other and then quite far from being equilibrated. They exhibit the same essential feature, which is the "freezing" of the incident charge state : in the four cases, 75 to 85% of the incident ions have kept their initial charge state. Moreover the detailed observation of the energy spectrum of the transmitted ions shows that "frozen" ions were channeled in the crystal, as revealed by their reduced energy loss. We had already observed this effect in previous studies [6,7] with Xe ions and it must be noted that the first observation of frozen charge states is due to the Oak-Ridge group, with Tandem energy Oxygen ions transmitted through a thin gold crystal [8]. We first describe the distributions obtained with 52^+ and 53^+ ions, i.e He-like and H-like species : they show that these two species cannot lose their K-shell electron(s) if they are channeled, which is easily understood since collisions with quasi-free target electrons, the only collisions allowed to channeled particles, cannot transfer enough energy. Even excitation to L-shell would require at least an incident energy of 55 MeV/u to be possible. In both cases a fraction of the incident ions has been able to capture one electron and again the energy spectrum of these ions shows that most of them are well channeled (cf. below). This fraction is 15% for 53^+ and 12% for 52^+ but we refrain from giving too much importance to the comparison of so close numbers because we know (from the 36^+ data in particular) the large influence of the angular dispersion of the incident beam on the charge exchange process. We know [7] that the capture process involved is Radiative Electron Capture (REC), the only process allowed since the dominant capture process in non crystalline targets, or for random incidence, the Mechanical Electron Capture (MEC), is suppressed for channeled particles, due to their large impact parameters to target atoms. As for the unchanneled part of the beam, it contributes to the low-charge tail of the distributions.

The channeling charge state distributions obtained with 50^+ and 51^+ incident ions are different from the previous ones because the loss of L-shell electrons becomes possible for these channeled ions. The electron loss probability is of course much higher for 50^+ ions than for 51^+ ions. They also may capture electrons by means of REC. The capture probability is seen on figure 7 to be higher for 51^+ ions than for 50^+ ions. Even though these two charge distributions are not equilibrated, the balance between electron loss and electron capture for each of the two incident species indicates that the mean charge state of the equilibrated distribution for channeled ions would be about 50.5, since a mean charge state at equilibrium corresponds to the charge state for which electron loss and capture have the same probability to occur. It is peculiar to observe that channeled and unchanneled ions happen to present nearly the same mean charge at equilibrium, which means that channeling reduces both electron loss and capture, but by nearly the same amount for Xe ions of this energy.

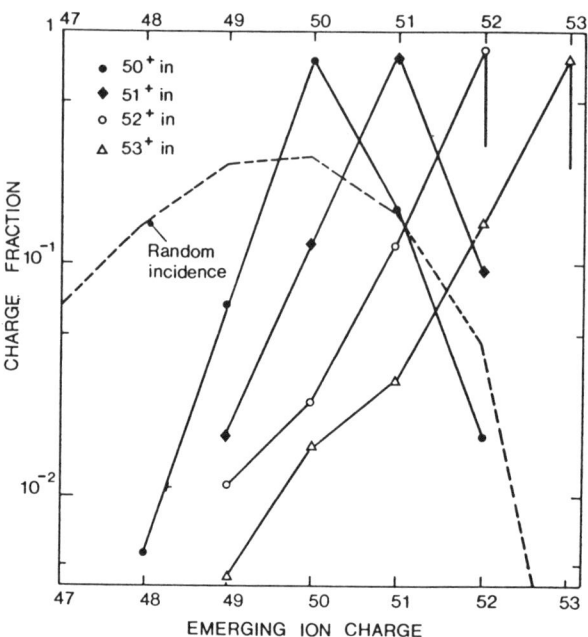

Figure 7 - Charge state distributions obtained with 25.5 MeV/u 50^+ to 53^+ Xe ions after a pathlength of 21 μm through a Si crystal, for $< 110 >$ axial alignment and for random conditions.

In figure 8 we show as an example the energy spectra of 52^+ and 53^+ Xe ions transmitted for $< 110 >$ axial alignment of 53^+ incident ions, as given by the wire chamber. The large 53^+ peak is composed of "frozen" incident ions and the energy loss distribution is quite typical of channeling (the energy spectra obtained with "frozen" 50^+, 51^+ and 52^+ ions are similar), with the low energy tail associated with poorly channeled ions, but also with the shoulder on the high energy side, which corresponds to the best channeled ions that we have seen, in the previous study with incident 35^+ ions, to have the same reduced energy loss rate of about 0.4 times the normal rate. By the way, the comparison between the two sets of results shows very clearly why the use of 35^+ incident ions allows a much more precise study of energy loss processes in channeling conditions. The 52^+ component in figure 8 is also composed of channeled particles, as shown by their energy loss spectrum, but these ions have captured one electron by means of REC and this explains why the 52+ spectrum differs from the 53+ one : the top of the peak and the shoulder are still there but the low energy part is relatively enhanced because the poorly channeled particles experience higher electron densities and then have a higher probability of electron capture.

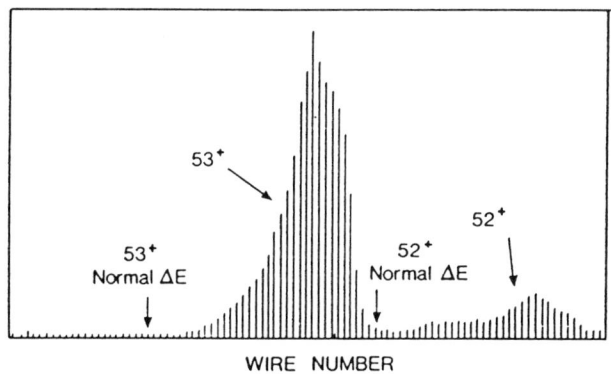

Figure 8 : Energy spectra of 52+ and 53+ transmitted ions for < 110 > axial alignment of incident Xe^{53+} ions (for a given charge state, energy increases from the left to the right).

The charge distributions allow to get a first insight on the capture process but the most precise study of the radiative capture lies in the observation of the X-ray spectrum given by the detector looking at the crystal. Such a spectrum, obtained with Xe^{53+} ions in < 110 > axial alignment is shown in figure 9. As previously observed [6,7], this spectrum is composed of primary and secondary electron bremsstrahlung (two continua extending up to 14 KeV and 60 KeV respectively in our case). The Lyman series, dominated by Ly α, is due to relaxation after electron excitation but also after electron capture, that is known to be most probable into excited states.

These above contributions are due mainly to the unchanneled part of the beam. On the contrary the REC series is mostly due to channeled ions, and is composed of three lines of decreasing intensity from the K-REC and L-REC lines to the M-REC line, in agreement with the theoretical value of the photon energy for capture of an electron at rest, $[E_b + (m/M)E_o]$, where E_b is the electron binding energy in its shell after capture, m and M the electron and projectile masses, respectively, and E_0 the kinetic energy of the projectile. As it can be seen on figure 9, the REC lines are wider than Lyman lines, and this is due to the fact that the target electrons captured by the REC process are not at rest, but have a momentum distribution, usually called the Compton profile. This feature of REC has been studied first by Sohval et al. [9]. The K-REC peak width (FWHM) is $\sim$ 1.25 KeV, in agreement with recent measurements by Vane et al. [10]. The asymmetry of the REC lines is due to the strong (v^{-4}) dependence of the REC cross section upon the relative electron-projectile velocity.

In order to clean up the X-ray spectrum and to isolate the photon emission due to the channeled ions, we have performed coincidence measurements with Xe^{53+} incident ions. The X-ray spectrum of figure 10 is made of photons detected in coincidence with transmitted Xe^{52+} ions with a low energy loss (right part of the 52+

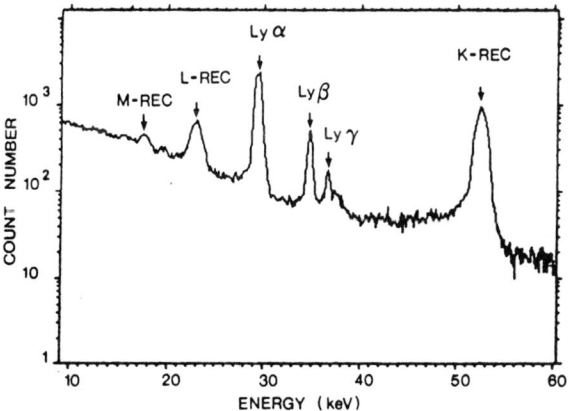

Figure 9 : X-ray spectrum obtained in the Ge detector for 25.5 MeV/u Xe^{53+} ions incident in < 110 > axial alignment on a 21 μm Si crystal.

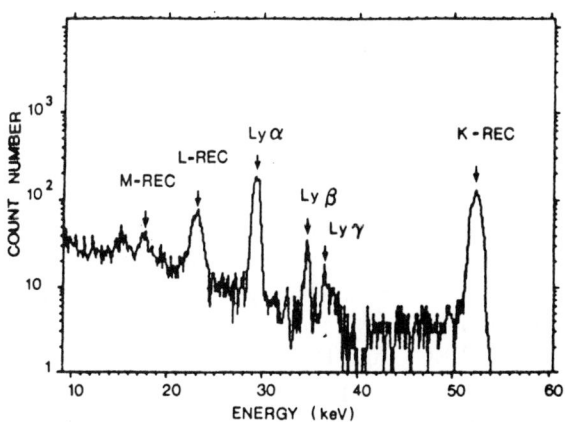

Figure 10 : As in figure 9, but in coincidence with well channeled ions transmitted with the 52+ charge.

spectrum of figure 8), i.e. with well channeled ions that have captured one electron. In this spectrum, to compare to the spectrum of figure 9 obtained simultaneously without coincidence, the REC lines are narrower and less asymmetric (the width of the K-REC line is 1.05 KeV), which is due to the narrower Compton profile of target electrons that can be captured by well channeled electrons. The Lyman lines are here entirely due to deexcitation after radiative capture into excited states. In particular the L-REC and Ly α peaks correspond to the same process and the difference of the peak areas probably shows that the angular distributions of L-REC and Ly α photons are different.

As a conclusion of this study, we must say that crude estimates show that the total REC photon yield is compatible with the measured fraction of channeled ions that have captured one electron (as given by the charge distributions). Estimates of REC cross-sections have already been published [2]. A more detailed analysis will be published.

V - CONCLUSION

These studies on channeling effects on fast heavy ions show that a thin single crystal is quite equivalent, for channeled ions, to a dense electron target composed of electrons nearly at rest. Although the electron density experienced is much lower than the mean electron density in a solid, it is much higher than in a crossed beam experiment.

As a concluding remark we present two other processes that can make possible the capture of quasi-free electrons by fast highly charged ions. They are best defined on fig. 11. As REC is the inverse process of photo-ionization (PI) (fig. 11 a), Resonant Transfer and Excitation (RTE), also called Dielectronic Recombination when it involves free electrons, is the inverse process of Auger Transition (AT) (fig. 11b), and has recently been observed in channeling conditions [11,12] and the recently proposed [13] Nuclear Excitation by Electron Capture (NEEC) (fig. 11c), in which the excess energy is transferred to the projectile nucleus, is the inverse process of Internal Conversion (IC). In the near future Radiative Electron Capture and these two resonant processes should play together a large role in the study of the interaction of fast heavy ions with matter.

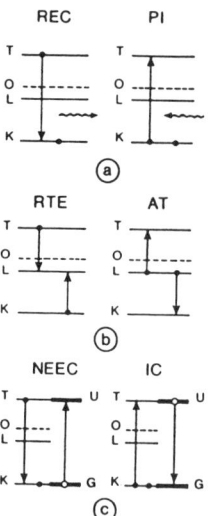

Figure 11 : The processes of target electron capture by a fast heavy ion (viewed from its rest frame) and their inverses : T refers to the target electron state, K, L and O to the projectile's atomic orbitals (with O designating the continuum) in the atomic system, while U and G refer to the excited and ground states, respectively, in the nuclear system.

*) Work performed at GANIL, supported by CNRS-GDR86 and IN2P3.

1) D.S. Gemmell, Rev. Mod. Phys. 46, 129 (1974).
2) A. Andriamonje, M. Chevallier, C. Cohen, J. Dural, M.J. Gaillard, R. Genre, M. Hage-Ali, R. Kirsch, A. L'hoir, B. Mazuy-Farizon, J. Mory, J. Moulin, J.C. Poizat, J. Remillieux, D. Schmaus and M. Toulemonde, Proceedings of the 13th International Conference on Atomic Collisions in Solids, Aarhus, August 1989 (to be published).
3) S. Andriamonje et al. (to be published).
4) E.D. Donets, Physica Scripta T3, 11 (1983).
5) W. Lotz, Z. Physik 216, 241 (1968).
6) S. Andriamonje et al., J. Physique 50, C1-285 (1989).
7) S. Andriamonje et al., Phys. Rev. Lett. 59, 2271 (1987).
8) S. Datz, F.W. Martin, C.D. Moak, B.R. Appleton, C.D. Bridwell, Rad. Eff. 122, 163 (1972).
9) A.R. Sohval et al., J. Phys. B9, L25 (1975).
10) C.R. Vane et al., communication at the First International Symposium of Swift Heavy Ions in Matter (Caen, May 1989) (to be published).
11) E.P. Kanter et al. and S. Datz et al., communications at the Workshop on highly charged ions, Berkeley (March 1989).
12) S. Datz et al., Proceedings of the First International Symposium of Swift Heavy Ions in Matter (Caen, Mai 1989) (to be published).
13) N. Cue, J.C. Poizat and J. Remillieux, Europhys. Lett. 8, 19 (1989).

DYNAMICS OF INELASTIC COLLISIONS OF ELECTRONICALLY EXCITED RARE GAS ATOMS

H.C.W. Beijerinck

Department of Physics, Eindhoven University of Technology,
P.O. Box 513, 5600 MB Eindhoven, The Netherlands

Abstract

Recent advances in the field of reactive and inelastic schattering of electronically excited rare gas atoms in metastable $R^*\{(np)^5(n+1)s\}$ and shortlived $R^{**}\{(np)^5(n+1)p\}$ states are discussed in case studies of the relevant processes: excimer formation, excitation transfer, Penning ionization and intramultiplet mixing. For reactive scattering the fine structure dependence of (KrBr) excimer formation is studied, showing that the propensity for core state conservation depends strongly on the halogen donor used. As a prototype of excitation transfer collisions the $Ar^*(^3P_{0,2})-N_2$ system is analysed with respect to electronic, vibrational and rotational excitation. Vibrational excitation is well understood in terms of a Landau Zener type semiclassical model; the two observed rainbow singularities in the rotational distribution lead to a classical description of scattering on a hard shell with both a P_2- and a P_4-anisotropy. For Penning ionization new insight is given for the well-studied $Ne^*(^3P_{0,2})-Ar$ system. Ab-initio calculations of the auto-ionization width confirm the validity of a simple two-state basis model for the metastable states. Polarisation effects in the ionization cross section for the $Ne^{**}(3p)-Ar$ system are used to establish a practical semiclassical criterion for "locking" of the total electronic angular momentum to the internuclear axis. Finally, intramultiplet mixing is discussed for the $Ne^{**}(3p)-He$ system. The observed polarised-atom cross sections are in good agreement with quantum mechanical coupled channel calculations, using the model potentials of Hennecart and Masnou–Seeuws as input. The results are equally well described with a semiclassical model that incorporates both rotational and radial coupling.

1. Introduction

The field of inelastic collisions of electronically excited atoms was initiated by the experiments of Penning in 1927[1]. Since then, it has been realised that inelastic and reactive collisions play an important role in the physics of gas discharges and laser plasma's. The He*-Ne energy transfer reaction, which lies at the basis of the He Ne laser, is perhaps the best known example.

In this paper we limit ourselves to a discussion of the metastable states $R^*(^3P_{0,2})$ of the $\{(np)^5 (n+1)s\}$ multiplet and the shortlived states $R^*\{\alpha_k\}$ of the $\{(np)^5 (n+1)p\}$ multiplet (with k = 1 through 10 indicating the different fine structure states with decreasing energy) for the heavy rare gases Ne, Ar, Kr and Xe. Depending on the metastable rare gas atom involved and its rare gas collision partner, a variety of inelastic processes can occur[2]. For Ne**-He the only open channel is intramultiplet mixing; when switching to Ne**-Ar, Kr, Xe we have to include Penning ionization. For the Ne**-Ne system only intramultiplet mixing occurs, but the description becomes more complicated because we have to take into account indentical particle symmetry ($^{20}Ne^{**}-^{20}Ne$) or electron inversion symmetry for non-identical nuclei with equal charge ($^{22}Ne^{**}-^{20}Ne$). In the Ar*-Kr system and the Kr*-Xe system excitation transfer is a competing channel. Typical cross section values for intramultiplet mixing, excitation transfer and ionization are $Q_{mix} = 0.1 - 5$ Å^2, $Q_{exc} = 1 - 10$ Å^2 and $Q_{ion} = 10 - 100$ Å^2, respectively.

For molecular targets ionization dominates for Ne*-M systems. For Ar* through Xe* the transfer of electronic energy into electronic, vibrational and rotational degrees of freedom of the molecule is the only possibility. Only for NO with its ionization energy equal to 9.26 eV the ionization channel is open for all metastable states except Xe*($^3P_2$). Other channels with molecular targets are dissociation to excited product states and reactive scattering. For the latter process most effort has been put into collisions with halogen containing molecules, leading to excimer formation with cross sections of the order of $Q_{excimer} = 20 - 100$ Å^2.

The major advances in the last five years are due to scattering experiments with better defined initial and final states. The former is obtained by using beams of state selected and polarised atoms as produced by optical pumping or magnetic devices. Most results are available for Ne*, due to the good match to c.w. dye lasers in the 600 – 650 nm range. For Ar*, Kr*, and Xe* single mode diode lasers in the 760 – 840 nm range have recently been applied for state selection. The final state analysis has been improved by using high intensity beam sources and well designed optical or electronic selection devices for detecting and analyzing the products. The measurement of polarised-atom cross sections has proven to be a valuable source of information. Conversely, the availability of high quality data has triggered a full quantum mechanical and semiclassical analysis in terms of adiabatic potentials, which has resulted in a large increase in insight in the dynamics of the inelastic process involved.

Finally the accessible range of collision energies has been extended considerably by applying a combination of thermal sources[3,4], seeded beams[5] and super thermal sources as the hollow cathode arc[6] and the arc heated supersonic expansion[7], resulting in e.g. $0.05 \leq E(eV) \leq 5$ for the Ne*-Ar system.

2. Analysis of scattering experiments

For the metastable atoms the long range interaction is dominated by the (n+1)s valence electron, resulting in an isotropic interaction potential. Only at short range the orientation of the $(np)^{-1}$ empty core orbital results in an appreciable anisotropy, resulting in an Ω-splitting which is limited to the repulsive branch. The characteristic parameters, like the well depth ϵ and the well position R_m have been determined from elastic scattering experiments[8-11]. As yet, there is no experimental evidence for a fine structure dependence of the potentials in the well area and beyond. The results are interpreted in terms of semi-empirical potentials such as the Ion-Atom-Morse-Morse-Spline-Van der Waals (IAMMSV) potential of Gregor and Siska[8] for the Ne*-Ar system. In general, the repulsive branch is not smooth but shows a "kink" like structure. Evidence for this type of structure can also be derived from the ab-initio calculations of excited state potentials by Spiegelmann[12], which even show a second local minimum in the potential at small internuclear distances for the system.

For the shortlived states with a (n+1)p valence electron the medium range forces are already anisotropic, resulting in significant Ω-splitting in the well region and beyond. Because no elastic scattering experiments are available, all information on the potential surfaces stems from calculations. The model potential method has proven to be highly sucessful at larger values of R, e.g. $R \geq 5a_0$ for the Ne**{(3p)}-He system. Using the model potentials of Hennecart and Masnou-Seeuws[13] as input we observe an excellent agreement of quantum mechanical coupled channel calculations for intramultiplet mixing in the Ne**{(3p)}-He system with the experimental results of Manders et al[14-15]. At smaller values of R the model potential is less reliable. For processes such as Penning[16] ionization, with an important contribution at small distances, we thus observe a more qualitative agreement with experimental results.

Polarisation effects are an important means to obtain detailed insight into the mechanisms governing the collision process[17-19]. In a collision experiment, the excited atom can only be prepared in an initial state $|J,M>_g$ in a space-fixed frame, with the relative velocity g as quantisation axis. The inelastic process, however, is described in a body-fixed frame with quantum numbers $|J,\Omega>$ for the total angular momentum J and its orientation Ω with respect to the internuclear axis. The polarised-atom cross section $Q^{|M|}$ depends on two features. First, we need to know the spatial evolution of the asymptotic initial state $|J,M>_g$ to the local molecular states along the particle trajectory, known as rotational coupling. Second, the inelastic processes have an Ω-dependent transition probability, which in general is the main objective of our research.

In a quantum mechanical calculation both features are taken into account when we include rotational and radial coupling (e.g. for intramultiplet mixing) and the loss of amplitude of the wave function by an imaginary component of the potential. When a large number of states is involved, quantum mechanical coupled channel calculations are more or less a "black box" description of the process. In general, a semiclassical description provides more insight in the dynamics of the collision. Accurate semiclassical models for the Ω-dependent inelastic processes are widely available. For rotational coupling the situation is quite different. The concept of "locking", i.e. coupling of J to the internuclear axis, has been used quite often. A suitable semiclassical recipe, however, is not readily available. For the interpretation of scattering data it is essential to investigate locking in detail[15,16,20].

In classical terms the Ω-dependent potentials, as encountered for Ne**(3p), translate into a torque on J, resulting in a precession with respect to the internuclear axis with frequency

$$\omega_{\text{prec}} = \Delta V^{\Omega,\Omega'}/\hbar, \qquad (1)$$

with $\Delta V^{\Omega,\Omega'}$ the difference potentials for the $|J\Omega>$ and $|J\Omega'>$ adiabatic states. Counteracting this precession, which would result in a well defined quantum number Ω in a body-fixed system, is the rotation of the internuclear axis with an angular velocity

$$\dot{\phi} = (N + \tfrac{1}{2})\,\hbar/\mu R^2, \qquad (2)$$

with N the angular momentum quantum number of nuclear motion and μ the reduced mass.

When $\omega_{prec} \ll \dot{\phi}$, the coupling to the internuclear axis is weak and a space-fixed description of J is favored. On the other hand, when $\omega_{prec} \gg \dot{\phi}$, a body-fixed description of J is correct. Our task is to determine a locking factor f_L, defined as

$$\omega_{prec}(R_L) = f_L \dot{\phi}(R_L), \quad (3)$$

and the corresponding locking radius R_L that can serve as an effective boundary between the two limiting cases. The analysis of polarised atom cross sections for intra-mulitplet mixing for Ne**-He[15] and for Penning ionization for Ne**-Ar[16] result in $f_L = 4 \pm 1$ (sections 5 and 6).

For metastable atom scattering, with the Ω-splitting limited to small values of R, no "locking" will occur. The effective distribution over the molecular Ω-states depends on the impactparameter b and the distance R where the inelastic process occurs, e.g. for b = 0 the initial asymptotic value of M directly results into $\Omega = |M|$. For larger impact parameters we have an effective scrambling of the initial asymptotic M value.

For the shortlived atoms, with their more pronounced Ω splitting, "locking" will be important at low collision energies. At higher energies the criterion for locking is only met for small values of R, resulting in a situation not very different from metastable atom scattering.

3. Excimer formation

To study the propensity of core-conservation we have studied the fine structure dependence of (KrBr) excimer formation[21] in collisions of $Kr^*(^3P_{0,2})$ with Br-containing molecules. The chemiluminescence (170–300 nm) of the B-, C-, and D-states is measured with a compact beam-gas cell apparatus. State selection of the Kr* atoms is achieved by optical pumping with a laserdiode, its beam aligned antiparallel with the atomic beam. Due to the long interaction length (90 mm), the transition is saturated to a line width of 1 GHz at 1 mW laser power, quaranteeing that all isotopes and all velocities of $Kr^*(^3P_0)$ (11% beam population) are equally depopulated.

Both high resolution spectra without state selection and low resolution spectra with state selection have been taken. The former are analyzed in terms of the vibrational state distribution of the excimer products. We limit ourselves to the state selected data, which have been analysed in terms of cross section ratios $(^0Q/^2Q)^{B,C,D}$, with the superscripts indicating the metastable states with J = 0 and 2. For Kr* + Br$_2$ we find $(^0Q/^2Q)^{B,C} = 0.2$, while for the D-state the ratio is equal to $(^0Q/^2Q)^D = 2.5$. The preference for the D-state is even more pronounced for the CH$_2$Br$_2$ halogen donor resulting in $(^0Q/^2Q)^D \approx 10$. Typical data are shown in Fig. 1. For the halogen donors CHBr$_3$ and CF$_2$Br$_2$ we find $(^0Q/^2Q)^D = 5$, without any B-state production. These measurements show a clear preferen-

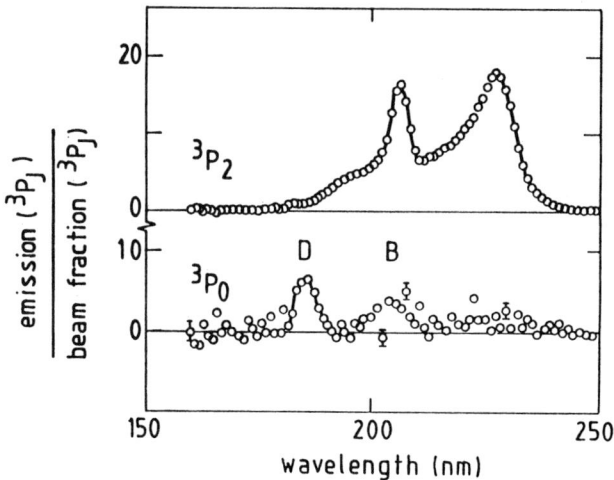

Figure 1. Chemiluminescence of the (KrBr) excimers as produced by $Kr^*(^3P_{0,2})$–CH$_2$Br$_2$ collisions. The signal is normalized to the beam fraction of the $Kr^*(^3P_0)$ and $Kr^*(^3P_2)$ fine structure states and is thus proportional to the reaction cross section. We observe a strong preference for D-state emission produced by $Kr^*(^3P_0)$ atoms.

ce to conserve the core state $^2P_{1/2,3/2}$ of the Kr* atom throughout the reaction, as proposed by Vrakking et al[22]. However, for the halogen donor CBr$_4$ no chemiluminescence at all is produced by the $Kr^*(^3P_0)$ fine structure state. This strong dependence on the halogen donor still has to be explained.

When taking into account the statistical population ratio $Kr^*(^3P_2) : Kr^*(^3P_0) = 5:1$ in a beam experiment, a large fraction of the observed D → X emission is still due to $Kr^*(^3P_2)$: 80% for Br$_2$ and 50% for CHBr$_3$. The absence of D → X emission in previous flowing afterglow experiments[23] on KrBr and XeBr formation thus cannot be explained by the usual assumption of quenching of $Kr^*(^3P_0)$ before it can react.

Recently, the group of Sadeghi[24] has investigated the fine structure dependence of ArCl and ArF formation in flowing afterglows, which opens the possibility of determining the propensity of core-conservation for the whole group of rare gas-halogen donor reactions.

4. Excitation transfer

The collision process

$$Ar^*(^3P_{0,2}) + N_2(X,v=0) \rightarrow Ar(^1S_0) + N_2(C,v',N') \quad (4)$$

is considered as a prototype of exothermal excitation transfer[25]. The exothermicity is $\Delta E = 701$ meV for $Ar^*(^3P_0)$ and final state v' = 0, and $\Delta E = 526$ meV for $Ar^*(^3P_2)$. The process is nonresonant and all accessible vibrational and rotational levels are populated. This system has been studied both in crossed molecular beams and in bulk, with very few state selected results.

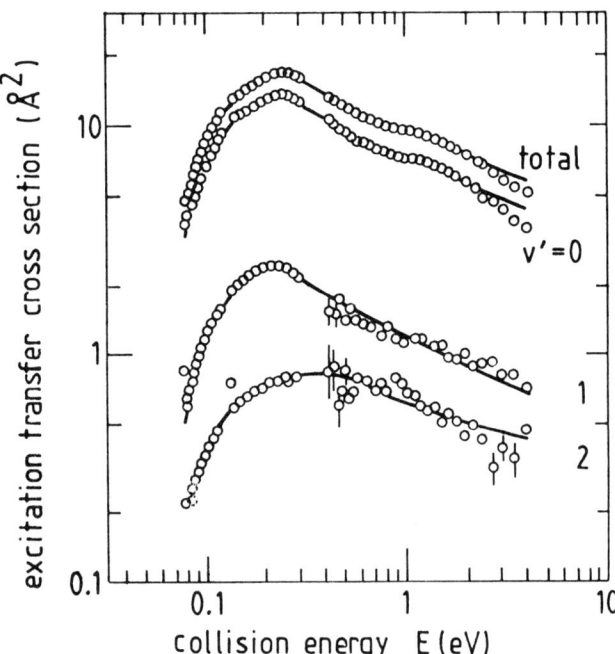

Figure 2. Experimental results of Vredenbregt et al.[31] for the cross section for excitation of the $N_2(C,v')$ final state by collisions with a mixed beam of $Ar^*(^3P_{0,2})$ atoms, for the vibrational states $v' = 0, 1$ and 2 and the total yield of fluorescence. The solid line is a semiclassical model calculation described in the text.

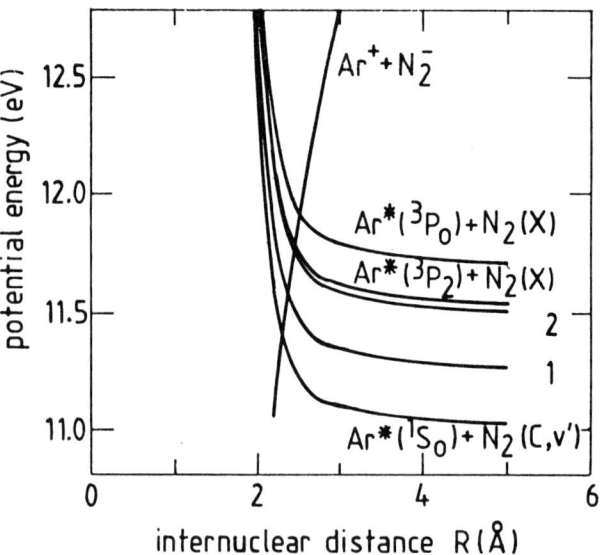

Figure 3. Semi-empirical diabatic potential surfaces for the initial and final states of the $Ar^*(^3P_{0,2}) + N_2(X)$ excitation transfer process, coupled by the ionic $Ar^+-N_2^-$ curve as an intermediate, as suggested by Van Vliembergen et al.[5]

In all cases the $N_2(C) \rightarrow N_2(B)$ radiative decay is analysed in terms of product vibrational and rotational distributions. We have studied both distibutions in detail, resulting in a semiclassical (vibration) or classical (rotation) description of the inelastic collisions. The same approach has also been applied to the endothermic Kr^*-N_2 system with $N_2(C)$ as final state[7,26].

4.1. Vibrational excitation

The experiments have been performed in a crossed beam experiment, using a mixed beam of Ar^* atoms. The fluorescence radiation produced in the $3 \times 8 \times 6$ mm^3 scattering volume is analysed with narrow band interference filters. Their transmission profiles are well matched to the $\Delta v = -2, -1, 0$ and 1 ro-vibrational bands of the $N_2(C) \rightarrow N_2(B)$ transition, taking into account the high degree of rotational excitation of the $N_2(C)$ products. The measured effective cross sections are then analysed in terms of vibrational excitation cross sections Q_{exc}^v for the $v' = 0, 1$ and 2 states (Fig. 2)[31].

The interaction potentials for the system are not well known. However, realistic semi-empirical diabatic potential curves were proposed by Van Vliembergen et al[5] (Fig. 3). The initial and final states are connected through two curve crossings with the ionic $Ar^+ + N_2^-$ potential as an intermediate. Of the latter, only the attractive Coulomb branch is shown. For these potentials the curve crossing with the inital state occurs at an energy $E = 225$ meV with respect to the asymptotic $Ar^*(^3P_2)$ initial state. This value is unrealistic, because the results in Fig. 2 show a threshold at $E = 65$ meV. Other groups find even lower threshold energies[25].

We have analysed the cross sections in terms of a semiclassical Landau Zener description for the two curve crossings involved[31]. Following Bauer[27] we have extended this model with a Franck Condon like factor for the second crossing, which takes into account the wave function overlap of the vibrational states of N_2^- and $N_2(C,v')$. We find a good agreement of our model calculations with the experimental data (solid lines in Fig. 2). For the first crossing of the initial state to the ionic intermediate we find a crossing radius $R_{x,1} = 4.02$ Å and a reference velocity $v_{ref,1} = 4700$ ms^{-1}. The latter is defined by the Landau Zener probability

$$p = 1 - \exp(-v_{ref}/v_{rad}) \qquad (5)$$

for following the adiabatic potential curve, with $v_{ref} = 2\pi H_{kl}^2/\hbar F_{kl}$ in the usual terms of the coupling matrix element H_{kl} and the difference potential derivative F_{kl} and v_{rad} the radial velocity at the crossing. The threshold behaviour of Q_{exc}^v at low energies is thus due to the increasing range of impact parameters of trajectories that can reach the first crossing. Because $v_{rad} \ll v_{ref}$ all particles then follow the adiabatic curve to the ionic intermediate. The second crossing then determines the branching to the different v' states. The crossing parameters are $R_{x,2} = 2.45$ Å (for all v' states) and $v_{ref,2} = 845, 508$ and 3400 ms^{-1} for $v' = 0, 1$ and 2, respectively.

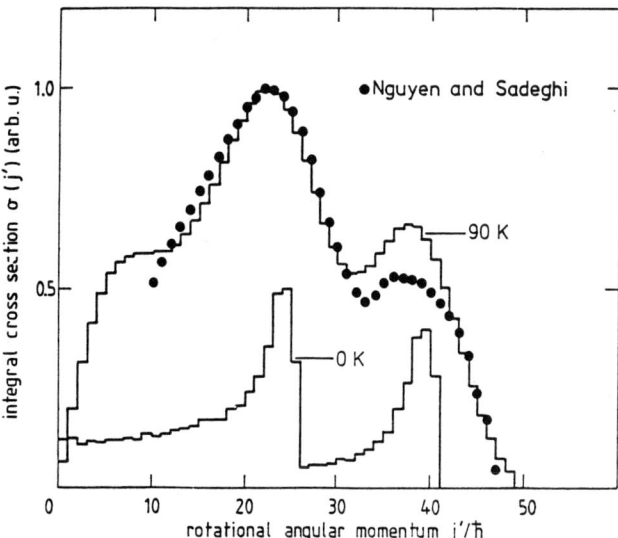

Figure 4. Experimental results for the product rotational distribution of the Ar*($^3P_{0,2}$) + N$_2$(X) excitation transfer process, as measured by Nguyen and Sadeghi[28] at 90 K for the Ar*($^3P_2$) initial state and the N$_2$(C,v' = 0) final state. The solid curves show the result of the hard shell model calculation of Vredenbregt et al[29], both without (lower curve) and with (upper curve) the effect of initial rotation of the N$_2$(X) molecule. The two rainbow singularities are clearly visible.

The threshold of these crossings is located –227, –195 and –720 meV below the Ar*($^3P_2$) initial state. The location of the crossings for v' = 0 and 1 is in surprising good agreement with the potentials of Fig. 3. For v' = 2 the agreement is rather unsatisfactory. In our opinion this is due to resonant effects in this transition, due to the small energy difference ΔE = 30 meV.

4.2. Rotational excitation

In a bulk experiment Nguyen and Sadeghi[28] have studied the product rotational distribution of N$_2$(C,v' = 0, N') with state selected Ar*($^3P_{0,2}$) atoms, at temperatures of 90 K (<E> = 11 meV) and 300 K (<E> = 39 meV). For initial state Ar*($^3P_2$) and 90 K, the product rotational distribution extends to rotational level N' = 48, the highest level that can be reached on grounds of energy conservation. Surprisingly, the distribution has maxima at two different values N_1' = 22 and N_2' = 37 (Fig. 4). For Ar*($^3P_0$) the rotational distribution is similar but shifted to higher N' values by an amount $\Delta N'$ = 6. Nguyen and Sadeghi[28] show that these "bimodal" distributions cannot be explained by common statistical models. The existence of "bimodal" distributions is supported by the more qualitative atomic beam measurements of Van Vliembergen et al[5].

The interaction potentials of Fig. 3 indicate a repulsive energy release which is large compared to the initial collision energy, which leads to a "half-collision" description. In this case rotational rainbows still shine through in the product rotational distributions. Vredenbregt et al[29]. have studied rotational excitation for highly exothermic collisions, using the classical approximation of hard-shell scattering as developed by Beck and co-workers[30]. The Monte Carlo calculations of Vredenbregt et al[29] show that for a hard-shell with both a P_2– and a P_4-anisotropy two rainbow singularities are observed in the product rotational distribution. Using R_0 = 3.0 Å, c_2 = 0.06 and c_4 = 0.11 as hard-shell parameters their model agrees well with the results of Nguyen and Sadeghi (Fig. 4), both with respect to peak positions and general shape. Going to Ar*($^3P_0$) the shift $\Delta N'$ = 6 is correctly predicted and we also observe a good agreement.

It should be realised that these anisotropy parameters belong to the Ar($^1S_0$)–N$_2$(C) final state. A direct comparison with existing results for ground state molecules is thus not possible. However, looking at R_0 = 2.5 Å and c_2 = 0.07 for K–N$_2$, as determined by Beck et al[30], we conclude that our values are not unrealistic. Moreover, it is not unlikely that a P_4-anisotropy exists for the N$_2$(C) state. The excitation of N$_2$(X) to N$_2$(C) requires an anti-bonding N$_2$(2σ_u) orbital to be promoted to a N$_2$(2π_g) orbital[25]. The result is that the charge density of the anti-bonding orbital is shifted to the plane perpendicular to the N–N axis of N$_2$(C). The addition of a P_4-anisotropy also causes a bulge in the direction perpendicular to the N–N axis.

5. Penning Ionization

5.1. Metastable Ne*(3s) atoms

Penning ionization has been investigated in detail by many groups, for a variety of projectiles and an even larger number of target atoms and molecules. The most detailed information is available for Ne*– Ar. Measurements are available for the elastic differential cross section[8,32], the velocity dependence of the total cross section[11], the velocity dependence of the ion yield[33] (including the branching ratio for associative ionization[34]), polarisation effects in the total ionization cross section[35] and the electron energy spectrum[36]. The only remaining problem is the interpretation in terms of suitable, Ω-dependent optical potentials.

Although a large amount of insight has been obtained, some very fundamental problems have not been solved as yet. A striking example is that Ne*($^3P_0$) always has a larger ionization cross section than Ne*($^3P_2$), which holds for all target atoms and molecules investigated. For the Ne*–Ar system Hotop et al[36,37] and Verheijen and Beijerinck[33] find $Q_{ion}(^3P_0)/Q_{ion}(^3P_2)$ = 1.32 at E = 100 meV; with increasing energy this ratio increases to a constant value of 2 at E ≥ 4 eV. Despite all experimental and theoretical effort it has not been decided whether differences in the real or the imaginary part are responsible for this effect[32,38]. In our opinion the imaginary part, the so-called auto-ionization width $\Gamma(R)$, is the prime candidate to explain this effect.

Following Bieniek[39] and Jones and Dahler[40] we have investigated the formal analysis of the process of ionization within the frame work of the Feschbach[41] formalism, with the discrete fine structure states (before ionization) and the continuum states (after ionization) as the two separate sub-spaces of Hilbert space. The result is a non-local coupling of the discrete states to the continuum states. Using the local approximation the auto-ionization width is given by

$$\Gamma_{ii} = 2\pi \sum_f \int dq \, |V_{\Omega fq,i}(R)|^2 \, \delta(E_q - \epsilon(R)), \quad (6)$$

with q and E_q the final wave number and energy of the asymptotic wave function of the free electron, respectively. The latter is determined by the adiabatic difference potential $\epsilon(R)$ of initial Ne*-Ar and final Ne($^1S_0$)-Ar$^+$ state. The index f (final) indicates the quantum number of the free electron, while the index i (initial) describes the quantum numbers of the discrete state. The transition matrix elements are diagonal in the molecular quantum number Ω.

The calculation of the coupling matrix element $V_{\Omega fq,i}(R)$ is the remaining task. In general this expression is a n-electron two-center integral. We assume that only two electrons play an active role in this process and that all other electrons are "frozen". For the initial state these electrons are the Ar(3p) core electron and the Ne*(3s) valence electron; in the final state the free electron is described by a Coulomb wave function with respect to Ar$^+$ while the bound electron is a Ne(2p) core orbital. The transition matrix element then is a linear combination of two-electron two-center integrals for the exchange and the radiative mechanism. Most important for Ne*-Ar is the exchange mechanism, where simultaneously the Ar(3p) electron jumps to the Ne*(2p)$^{-1}$ core hole and the Ne*(3s) valence electron becomes a free electron. The radiative mechanism does not contribute in pure triplet states, due to spin-selection rules.

The usual approximation for the coupling matrix element is the one-electron one-center overlap integral of the Ne*(2p)$^{-1}$ empty core orbital with the Ar(3p) orbital, which has been considered as sufficiently accurate for explaining the Penning ionization data. We have calculated the two-electron two-center integrals on a Cyber 205. As input we have used the Clementi[42] orbital for Ar(3p) and the orbitals of Haberland[43] for Ne*(2p)$^{-1}$ and Ne*(3s). The free electron wave function has been calculated by solving the Schrödiger equation for an effective Ar$^+$ potential. Preliminary results for Γ_Ω for the Ne*($^3P_{0,2}$) states are in good agreement with the predictions of the simple model using a two-state basis $\Gamma_{\sigma'}$ and $\Gamma_{\pi'}$, as discussed by Driessen et al[16]. For the Ne*($^3P_2$) state we have used the resulting auto-ionization widths Γ_Ω as input for a semiclassical model calculation for polarised-atom ionization cross sections $Q_{ion}^{|M|}$ as a function of collision energy E. The calculated behaviour polarisation effects are in good agreement with recent experimental results of Driessen et al[35], which show a switchover from $Q_{ion}^{|M|=0,1}/Q_{ion}^{|M|=2} = 1.3$ at E = 80 meV to a value of 0.7 at E = 1000 meV. At E = 125 meV the polarisation ratio is equal to unity. As yet unexplained by the recent auto-ionization width calculations is the $Q_{ion}(^3P_0)/Q_{ion}(^3P_2)$ ratio and its energy dependence.

5.2. Shortlived Ne**(3p) atoms

In contrast to the metastable states, large polarisation effects have been observed in the ionization cross section of the Ne**(3p)-rare gas systems[16,37,44]. For Ne**{(3p); J = 3}-Ar at E = 100 meV we observe $Q_{ion}^{|M|=0,1}/Q_{ion}^{|M|=3} = 2.5$, decreasing to 1.4 for $E \geq 1$ eV[16]. The strong preference for $|M| = 0$ and 1 at thermal energies reflects the larger ionization probability for the $\Omega = 0$ and 1 molecular states. Due to the large Ω-splitting of the real part of the potential the total angular momentum **J** "locks" to the internuclear axis at a fairly large internuclear separation. The final molecular state with quantum number Ω will thus have nearly the same orbital alignment as the asymptotic inital state $|JM>_g$ with respect to the relative velocity **g**, as produced by excitation with a polarized laser beam. With increasing collision energy the effect of locking will decrease, due to the increasing angular velocity $\dot\phi$ of the internuclear axis. Finally, at high energies, a space-fixed description of **J** is correct, resulting in a scrambling of the asymptotic polarised beam into a wide distribution of Ω-states at the classical turning point where ionization preferentially occurs. The strong Ω dependence of the process of ionization is then no longer visible in the polarised beam experiments.

To determine the locking factor, as described in section 2 and to investigate the Ω-dependence of the process of ionization, we have performed a semiclassical model calculation for the Ne**-Ar system. For the auto-ionization we have used improved one-electron one-center overlap integrals, because the ab-initio results were not yet available at that time. For a locking factor $f_L = 0$, i.e. locking at all R-values, the polarisation effect in the thermal energy range is too large by a factor 2. Moreover, we observe no energy dependence. For $f_L = \infty$, i.e. no locking at all, the polarisation effect nearly vanishes over the whole energy range. For $f_L = 4 \pm 1$ we observe a good agreement between experiment and model calculation. This result is confirmed by Grosser's[45] analysis of the results of Na$^+$ scattering on laser excited Na*(3p) atoms of Hertel et al[17], which also results in $f_L = 4.5 \pm 1$ in our approach.

6. Intramultiplet mixing

In comparison to the available results for the two-electron alkaline earth systems, the description of intramultiplet mixing for the Ne**-He system has reached a high degree of sophistication. On the experimental side, Manders et al[46] have developed a new, compact crossed beam apparatus, with a well defined relative velocity,

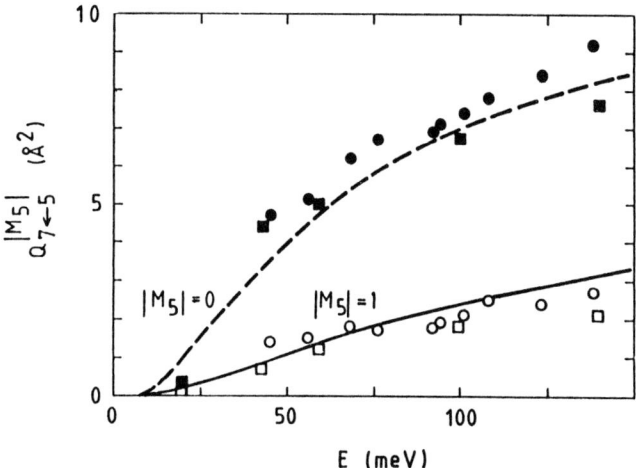

Figure 5. Energy dependence of the polarised–atom cross section $Q_{7-5}^{|1|}$. Comparison of experimental data of Manders et al.[15] (circles) with the result of quantum mechanical calculations (quadrangles) and semiclassical model calculations (lines), using the potentials of Hennecart and Masnou–Seeuws[13] as input.

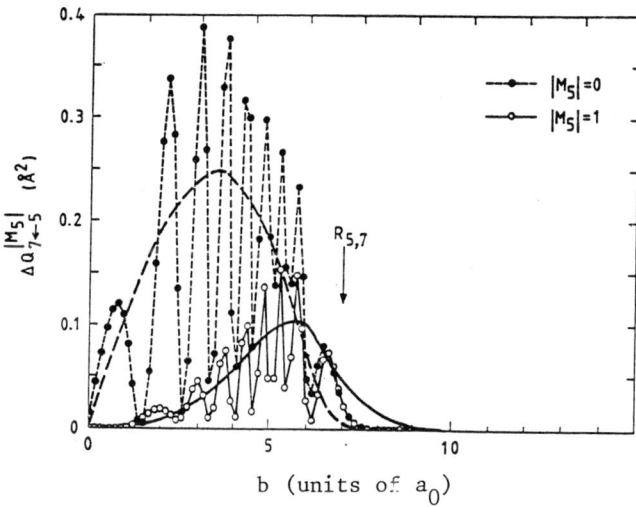

Figure 6. Semiclassical cross section contributions[15] $dQ_7^{|M|}/db$ for the $5 \to 7$ transition as a function of the inpact parameter b (dashed and solid line for $|M| = 0$ and $|M| = 1$, resp.), in comparison with their quantum mechanical analogues $\Delta Q_{7-5}^{|M|}$ (filled and open circles for $|M| = 0$ and $|M| = 1$, resp.). The arrow indicates the crossing radius $R_{5,7}$ in the $\Omega = 0^-$ manifold of adiabatic potentials.

both in magnitude and direction. Both absolute values and polarisation effects can be measured as a function of collision energy with great accuracy. On the theoretical side, quantum mechanical coupled channel calculations have been done for this system with 23 molecular states, using the model potentials of Hennecart and Masnou–Seeuws[13] as input. We observe a good agreement between experiment and theory. Finally, Manders et al have performed a semiclassical analysis of the inelastic collisions, using the extended Landau Zener model of Nikitin[47] for the radial coupling and the locking model of Section 2 for rotational coupling. The results are in good agreement with the quantum mechanical description.

Very strong polarisation effects have been observed, with $0.1 \leq Q_{1\leftarrow k}^{|0|}/Q_{1\leftarrow k}^{|1|} \leq 10$. As a typical example we show the energy dependence of the polarised–atom cross section for the Ne**{α_5} $\to$ Ne**{α_7} transition in Fig. 5, in comparison with quantum mechanical and semiclassical predictions. In Fig. 6 we show the impact parameter dependence of the cross section for this transition, in both the quantum mechanical and the semiclassical description. The agreement is very good.

A detailed analysis of a large number of transitions has been published by Manders et al[14,15,48], including data for the Ne**–Ne[48] and Ne**–Ar systems[49]. We limit ourselves to this brief discussion.

7. Concluding remarks

Going from excimer formation in section 3 to intramultiplet mixing in section 6 we see an increasing degree of sophistication, both with respect to the experimental data available and with respect to the amount of insight and detail in the theoretical description of the collision in terms of potential surfaces. This indicates that the field is open for expansion in new directions. In view of the recent interest in laser cooling of atomic beams, both in transverse direction to increase the intensity by orders of magnitude and in axial direction to obtain translational energies in the (milli)Kelvin range, crossed beam experiments with ultracold atoms or even collisions of two excited atoms with each other seem feasible in the near future. This will open an exciting new field and provide more insight in the collision dynamics involved. For example, at these low energies locking will occur at very large internuclear separations, resulting in very large polarisation effects and/or more rigorous selection rules for the inelastic process. Collisions of two excited atoms will shed more light on correlation effects in inelastic collisions.

In view of the wide scope of this progress report, full referencing to all original papers would require at least 200 references. In view of the limited space, we have chosen to refer to recent papers that will lead to the broad basis underlying this field. We fully realise the limitations of this approach and apologize for all sources not mentioned directly.

References

1. F.M. Penning, Naturwissenschaften 15 (1927) 181.
2. H.C.W. Beijerinck, Comm. At. Mol. Phys. 19 (1987) 227.
3. D.W. Fahey, W.F. Parks and L.D. Shearer, J. Phys. E 13 (1980) 381.
4. M.J. Verheijen et al, J. Phys. E 17 (1984) 904.
5. E.J.W. van Vliembergen et al, Chem. Phys. 114 (1987) 117.
6. P.G.A. Theeuws et al, J. Phys. E 15 (1982) 573.
7. K. Tabayashi and K. Shobatake, J. Chem. Phys. 84 (1986) 4919.
8. R.W. Gregor and P.E. Siska, J. Chem. Phys. 74 (1981) 1078.
9. W. Beyer and H. Haberland, Phys. Rev. A 29 (1984) 2280.
10. E.R.T. Kerstel et al, Chem. Phys. 121 (1988) 211 and references cited.
11. E.R.T. Kerstel et al, Chem. Phys. 119 (1988) 325 and references cited.
12. F. Spiegelmann, private communication as cited by M.C. Castex.
13. D. Hennecart and F. Masnou-Seeuws, J. Phys. B 18 (1985) 657.
14. M.P.I. Manders et al, Phys. Rev. Lett. 57 (1986) 1577; ibid., 57 (1986) 2472; ibid., Phys. Rev. A 37 (1988) 3237.
15. M.P.I. Manders et al, Phys. Rev. A 39 (1989) 4467.
16. J.P.J. Driessen et al, Phys. Rev. Lett. 62 (1989) 2369.
17. I.V. Hertel, H. Schmidt, A. Bähring and E. Meyer, Rep. Prog. Phys. 48 (1985) 375 and references cited.
18. H.A.J. Meyer, H.P. van der Meulen and R. Morgenstern, Z. Phys. D 5 (1987) 299 and references cited.
19. D. Neuschäfer, H.O. Hale, I.V. Hertel and S.R. Leone, in Invited Papers XIV ICPEAC, eds. D.C. Lorentz, W.E. Meyerhof and J.R. Peterson (North Holland, Amsterdam 1986).
20. M.H. Alexander and B. Pouilly, in "Selectivity in Chemical Reactions", ed. J.C. Whitehead (Kluwer, Dordrecht, 1988) p. 265 and references cited.
21. Ch.A. Brau, in "Excimer lasers", ed. Ch.K. Rhodes (Springer, Berlin, 1984) p. 87.
22. M.J.J. Vrakking, K. Tabayashi, and K. Shobatake, to be published.
23. See, for example, C.T. Reffner and J.P. Simons, Faraday Disc. Chem. Soc. 67 (1979) 329.
24. N. Sadeghi, M. Cheaib and D.W. Setser, J. Chem. Phys. 90 (1989) 219.
25. Recent review by G.W. Tyndall, M.S. de Vries, C.L. Cobb and R.M. Martin, J. Chem. Phys. 87 (1987) 5830.
26. R.J.F. van Gerwen et al, Chem. Phys. 118 (1987) 407.
27. E. Bauer, E.R. Fisher and F.R. Gilmore, J. Chem. Phys. 51 (1969) 4173.
28. T.G. Nguyen and N. Sadeghi, Chem. Phys. 79 (1983) 41.
29. E.J.D. Vredenbregt et al, Phys. Rev. A 39 (1989) 5597.
30. W. Schepper, U. Ross and D. Beck, Z. Phys. A 290 (1979) 131; ibid., Phys. Rev. A 19 (1979) 2173; Z. Phys. A 293 (1979) 107; 299 (1979) 97.
31. E.J.D. Vredenbregt, W. Boom and H.C.W. Beijerinck, to be published.
32. D. Hausamann, Ph.D. Thesis, Albert Ludwig University, Freiburg (1986).
33. M.J. Verheijen and H.C.W. Beijerinck, Chem. Phys. 102 (1986) 255 and references cited.
34. C. Weiser and P.E. Siska, J. Chem. Phys. 85 (1986) 4746.
35. J.P.J. Driessen et al, to be published.
36. H. Hotop, J. Lorenzen and A. Zastrow, J. Electr. Spectr. 23 (1981) 347.
37. T. Bregel et al, in Invited papers XIV ICPEAC, eds. D.C. Lorents, W.E. Meyerhof and J.R. Peterson (North Holland, Amsterdam, 1986) p. 577.
38. H. Morgner, J. Phys. B 18 (1985) 251; ibid., Comm. At. Mol. Phys. 11 (1982) 271.
39. R.J. Bieniek, Phys. Rev. A 124 (1978) 392.
40. D.M. Jones and J.S. Dahler, Phys. Rev. A 37 (1988) 2916.
41. H. Feschbach, Ann. of Phys. 19 (1962) 287.
42. E. Clementi, in "Tables of Atomic Functions", Supplement to IBM J. Res. Dev. 9 (1965) 2.
43. H. Haberland, Private comm.
44. W. Buszert et al, J. de Physique 46 (1985) C1-75.
45. J. Grosser, Comm. At. Mol. Phys. 21 (1988) 107.
46. M.P.I. Manders et al, J. Chem. Phys. 89 (1988) 4777.
47. E.E. Nikitin, in "Chemische Elementar Prozesse", ed. H. Hartmann and J. Heidelberg (Springer, Berlin, 1968).
48. M.P.I. Manders et al, Phys. Rev. A 39 (1989) 5021.
49. W. Boom et al., Book of Abstracts, XII Int. Symp. on Mol. Beams, Perugia (1989) p. 394.

SIFT AND FALP DETERMINATIONS OF IONIC REACTIONS RATE COEFFICIENTS

D. Smith and N.G. Adams

School of Physics and Space Research
University of Birmingham, Birmingham, B15 2TT, UK

A brief review is presented of some recent results of studies of ionic reactions at thermal energies using the selected ion flow tube (SIFT) and the flowing afterglow/Langmuir probe (FALP) techniques in our laboratory. From the SIFT results, we summarise the recent data on positive ion-molecule reactions, much of which relate to the synthesis of interstellar molecules. Thus, for example, data are discussed relating to the reactions of some hydrocarbon ions in the series $C_nH_m^+$, and phosphorus-bearing ions in the series PH_n^+, which can lead to interstellar hydrocarbon and phosphorus-bearing molecules. From the FALP results, we summarise the data obtained on the rate coefficients for dissociative recombination, highlighting the recent work on the determination of the products of some dissociative recombination reactions, and discuss recent results on studies of dissociative electron attachment reactions. Reference is also made to our earlier studies of ion-ion mutual neutralization.

I. Introduction

Thermal and near-thermal energy reactions involving atomic and molecular species (neutral or ionized) and electrons occur in most ionized media, thus modifying the chemical composition of the media. In this sense, such ionic reactions are important in naturally-occurring ionized gases including the terrestrial atmospheric regions (the ionosphere, stratosphere and troposphere), other planetary atmospheres and interstellar gas clouds, and in laboratory plasmas such as gas lasers and surface etchant plasmas, and in electron capture detectors and atmospheric pollution monitors. The types of reactions that may occur include: (i) binary and ternary ion-molecule reactions, which convert positive ions and negative ions to chemically different ions and can also produce large cluster ions (especially at high ambient pressures and when polar molecules are present); (ii) electron attachment reactions which convert free electrons to negative ions thus greatly modifying the physical and chemical characteristics of the ionized gas; (iii) positive ion-electron recombination which acts to return the ionized state to the neutral state and also creates new neutral species (often reactive radical species) in the gas; (iv) positive ion/negative ion mutual neutralization, which obviously removes ions of both sign of charge from the media and again generates new neutral species in the gas; (v) associative detachment reactions between negative ions and radicals (atoms or molecules) which release free electrons and form new neutral species. In the media mentioned above, most of these reaction processes will be occurring both sequentially and in parallel, usually amounting to a large number of individual reactions. Thus 'chemical evolution' of the media occur. For example, in the terrestrial stratosphere each of the processes (i) to (v) occur [1]. However, there is yet no convincing evidence that negative ions are involved in the gas phase ion chemistry which results in the production of the molecules observed in interstellar clouds (although there is a possible proviso to this which we mention later in Section 3.2(iii)). It should be mentioned also that reactions between neutral species (especially those involving neutral radical species) also occur in many ionized media and contribute to the gas phase chemical evolution of the media, but we will not be concerned with such reactions here.

The reasons for studying thermal energy ionic reactions is both to understand the fundamentals of the processes involved and also to provide critical data for inclusion in the complex ion-chemical models which have been constructed to describe the chemistry of ionized media. Much has been written about this topic and many comprehensive reviews are available [2]. In this paper, we are concerned specifically with the data obtained relating to the thermal energy ionic reaction processes (i) to (v) above, obtained using the extraordinarily versatile and productive selected ion flow tube (SIFT) [3] and flowing afterglow/Langmuir probe (FALP) [4] apparatuses developed and exploited in our laboratory. In particular, we discuss some of the most recent results we have obtained, most of which relate to reactions believed to occur in the terrestrial atmosphere and in interstellar gas clouds. Interest in these media has provided great stimulus to the study of ionic reactions at thermal energies.

2. The SIFT and the FALP Apparatuses

Both the SIFT and the FALP are fast flow tube experiments. They have been described in detail in some recent reviews [3,4].

The SIFT apparatus was designed for the study of ion-neutral reactions under truly thermalised conditions. The principle of the technique is simple. Ions are created in an ion source, usually by electron impact on an appropriate source gas (however, several types of source have been used including flowing afterglows [5]), mass selected using a quadrupole mass filter and then injected through a venturi inlet into fast-flowing carrier gas (usually helium). These primary ions thermalise in multiple collisions with the carrier gas and are convected downstream where they are sampled via a pinhole orifice into a mass spectrometer/detection system. Reactant gases are introduced at a controlled flow rate into the thermalised swarm of ions and the resulting decrease in the primary ion signal to the downstream detection system and the detection of product ions of the reaction provide values for the rate coefficients and the product ion distributions for the reactions. Measurements can be made at temperatures within the approximate range 80 to 600K. Reactions of positive ions and negative ions can be studied; indeed some thousands have been studied [6,7] involving a wide variety of reactant ions (including cluster ions) and reactant neutrals (including vapours, acids and atomic radicals). A variant on the SIFT, the selected ion flow drift tube (SIFDT), in which an accurately defined drift field is established along the length of the flow tube, allows measurement of rate coefficients for ion-neutral reactions at elevated centre-of-mass energies of the reactants [8]. The SIFDT is providing an increasing amount of data, and comparisons with low energy beam data are being made. The SIFDT data find valuable application in modelling the chemistry of shocked regions of interstellar gas [9].

The FALP apparatus was designed for the study of plasma reaction processes, especially positive ion-electron recombination, ion-ion neutralization and electron attachment [10,11,12]. A flowing afterglow plasma is created by a microwave discharge in the upstream region of a fast flowing carrier gas. By virtue of the flow, ions and electrons are transported downstream and they relax from the suprathermal energies in the discharge to thermal energy in the downstream afterglow plasma. Again, helium is the most commonly used carrier gas and thus He^+ ions, electrons and metastable helium atoms He^m (both 2^1S and

2^3S) are transported into the afterglow. Also He_2^+ ions are created in three-body collisions involving He^+ ions and two He atoms (i.e. $He^+(2He,He)He_2^+$) and by associative ionization involving highly-excited He atoms (i.e. $He^*(He,e)He_2^+$). Chemical modification of the plasma can readily be achieved by the introduction of appropriate gases into the afterglow. For example, the addition of Ar removes both He^m (by the reaction $He^m(Ar,e)Ar^+$) and He_2^+ ions (by $He_2^+(Ar,2He)Ar^+$). A downstream mass spectrometer/detection system (standard to traditional flowing afterglow apparatuses) is included to monitor the ionic species (both positively- and negatively-charged) in the afterglow plasma.

The electron number density, n_e, can be determined at any position, z, along the axis of the afterglow plasma column using a movable Langmuir probe. The axial variation of n_e can be related to the nature and the speed of the loss processes for electrons. Thus, for example, a linear plot of ln n_e versus z is indicative of loss by ambipolar diffusion (as is the case for plasmas comprising atomic ions (He^+, Ar^+) and electrons only)[4]. A linear plot of n_e^{-1} versus z is characteristic of plasmas in which loss of electrons is by dissociative recombination; such can be produced by adding molecular gases to the He^+/Ar^+/electron plasma following which recombining molecular ions are produced by ion chemistry (e.g. $Ar^+(O_2,Ar)O_2^+$). From such observations, dissociative recombination coefficients, α_e, can be obtained [10]. Similarly, the addition of electron attaching gases increases the z-gradient of n_e as electrons are converted to negative ions (e.g. $e(CCl_4,CCl_3)Cl^-$), and thus electron attachment coefficients, β, can be determined [12]. If sufficient electron attaching gas is added to the afterglow plasma, all the free electrons are rapidly converted to negative ions and a positive ion-negative ion plasma is created. Then, using the Langmuir probe to determine the axial variation in the positive ion and/or negative ion number density (n_+ or n_-), ion-ion mutual neutralization rate coefficients, α_i, can be determined [11].

The above briefly illustrates the range of processes that can be studied using the SIFT (ion swarm) and the FALP (afterglow plasma) methods. Using these techniques, most of the ionic and electronic processes that occur in low temperature ionized media can be studied. In combination they have been used to study dissociative attachment (FALP) and the reverse associative detachment (SIFT) reactions (e.g. $e(HBr,H)Br^-$; see Section 3.2(ii)). The value of SIFT experiments in assisting in understanding the ion chemistry occurring in the FALP experiments is great. For example, the ion molecule chemistry which occurs in the creation of specific afterglow plasmas for recombination studies can be understood by exploiting the SIFT apparatus. This has been useful in many ways, not least in the study of the recombination of a variety of protonated ions which are created by proton transfer from H_3^+ ions (e.g. $H_3^+(CO_2,H_2)HCO_2^+$) [13] which we discuss briefly in Section 3.2(i). Finally, it cannot go unmentioned that the recent addition of laser and VUV spectroscopic diagnostics to the FALP experiment is now allowing the neutral products of dissociative recombination reactions to be determined. This major advance is reported briefly in Section 3.2(i).

3. Some Recent SIFT and FALP Studies

As was previously mentioned, much of the SIFT and FALP work in our laboratory has been stimulated by our interest in interstellar chemistry, and a good deal of the new data discussed in this paper refer to interstellar reactions. Thus it is pertinent to mention here the elements of interstellar chemistry that are responsible for the synthesis of the wide variety of molecules that have been detected in interstellar clouds. Several recent reviews discuss this chemistry in some detail [14].

Interstellar molecules are formed in the gas phase by many parallel and sequential binary positive ion-neutral reactions which generate a diversity of polyatomic ions. These ions can then dissociatively recombine with free electrons to form the observed neutral molecules. The initial phases of this chemistry, starting from simple ions such as C^+ and H_3^+, are now generally understood as is demonstrated by the reasonable agreement between the observed relative abundances of many interstellar molecular species and the predicted abundances based on ion- chemical models (see, for example, ref. 15). Such models required laboratory kinetic data for a large number of ion-neutral reactions and many of these data are now available; the lack of data relating to dissociative recombination reactions has been an inhibiting factor in this work until quite recently (see Section 3.2(i)).

Simple ion-molecule reaction sequences such as those which are usually considered to lead to H_2O production e.g.

$$O^+ \xrightarrow{H_2} OH^+ \xrightarrow{H_2} H_2O^+ \xrightarrow{H_2} H_3O^+ \xrightarrow{e} H_2O \quad (1)$$

and NH_3 production in interstellar clouds

$$N^+ \xrightarrow{H_2} NH^+ \dashrightarrow{H_2} NH_4^+ \xrightarrow{e} NH_3 \quad (2)$$

are readily studied using the SIFT method. Generally, it is found that such reactions are rapid at room temperature [6,7]. All three ion-molecule reactions in sequence (1) occur on every collision, i.e. the rate coefficients, k, for the reaction are equal to the collisional rate coefficient k_c [16]. However, this cannot always be assumed to be the case, and indeed the first and the last reactions in sequence (2) are slow at 300K [17]. More importantly, the k for these reactions vary greatly with temperature and when it is appreciated that the temperatures of interstellar clouds are very low (perhaps so low as 5K in some dense clouds, but more typically 20K) then it is important to obtain kinetic data at appropriately low temperatures. Currently, SIFT apparatuses are operated down to 80K and a good deal of valuable data have been obtained at this temperature [3,6]. A very exciting new experiment (the so-called CRESU apparatus [18], which we refer to again below) can be operated down to 8K

and is now providing important data on ion-molecule reactions relevant to interstellar chemistry.

The ion-chemical routes to the production of some interstellar species are uncertain, especially those leading to the larger, polyatomic molecules such as the cyanopolyynes (HC_nN), and the larger hydrocarbons. This acts as the stimulus for further active laboratory studies as does the discovery of new interstellar molecules, such as the recent detection in the interstellar medium of the first phosphorus-bearing molecule, PN.[19] We refer to recent relevant SIFT work below.

As mentioned above, it is only recently with the advent of the FALP that data relating to the dissociative recombination of interstellar molecular ions have become available [13]. Using the FALP, recombination coefficients for reactions such as the last in sequences (1) and (2) are readily obtained. A great uncertainty, however, is in the neutral products of such reactions; we refer briefly to very recent FALP studies on this in Section 3.2(i).

Before discussing particular SIFT and FALP studies, it must be reiterated that a stimulus for much of the work in our laboratory has also been interest in atmospheric reactions and in reactions occurring in laboratory plasmas [20]. The FALP studies of electron attachment reactions and ion-ion mutual neutralization referred to below have been carried out in part with these applications in mind.

3.1 SIFT Results

In a short review such as this, which covers several types of processes, it is only possible to highlight a few of the more interesting recent results that have been obtained. Perhaps none is more interesting and significant to interstellar chemistry than the coordinated theoretical and experimental work that has shown that collisional rate coefficients, k_c, for ionic reactions involving polar reactant neutral molecules increase dramatically with decreasing temperature [21]. The physical reason for this is clear. The minimum potential energy path for ion-molecule reactions is a collinear neutral reactant dipole/ion trajectory and this is more readily achieved as the rotational excitation reduces (as the temperature reduces). Clary [22] devised a procedure for calculating the rate coefficients for such reactions for reactant molecules in specific rotation states (the so-called ACCSA method) and in a co-ordinated SIFT study the proton transfer reactions of H_3^+ with HCN and HCl were studied over the temperature range 205 to 540K [21]. Excellent agreement was obtained between the theoretically predicted k_c and the experimentally measured k. Subsequent measurements of k for several reactions involving polar molecules at temperatures as low as 27K using the CRESU method have further demonstrated the validity of the theory [18]. Thus the general conclusion may be drawn that the k_c for reactions involving polar reactants (e.g. HCN, NH_3, etc.) at the low temperatures of interstellar clouds (~20K) greatly exceed their 300K values (typically $10^{-9} cm^3 s^{-1}$), reaching $~10^{-7} cm^3 s^{-1}$ for very polar reactants at ~20K. This work has been reviewed quite recently by Clary [23]. Recognition of this phenomenon provided the answer to a long standing problem in astrochemistry, i.e. the anomolously high abundance ratio of HCS^+ to CS in interstellar dense clouds [24]. Because of the large dipole moment of CS the reaction $H_3^+(CS, H_2) HCS^+$ is now expected to proceed more rapidly than had previously been thought, thus enhancing the abundance of HCS^+ relative to CS.

The abundance of HCS^+ in interstellar clouds is also enhanced due to the very large proton affinity (PA) of CS radicals [25]. Proton transfer between neutral species is known to be facile for proton transfer reactions $XH^+(Y, X)YH^+$ when PA(Y)>PA(X). It is therefore important to know the PA of a large number of neutral species in order to assess the loss rates of protonated species in real media such as interstellar gas clouds, the terrestrial stratosphere and laboratory plasmas. In this regard, the SIFT technique is especially valuable.

It is a straightforward procedure to inject particular protonated ions into the helium carrier gas, allow them to thermalise and then add a variety of neutral gases or vapours to the ion swarm, and observe whether or not proton transfer occurs from the ion to the reactant neutral. In this way, it can be ascertained if the PA(donor) is greater or less than the PA(acceptor) and thus a 'proton affinity ladder' can be constructed. The PA(CS) was determined to be 188.2 ± 1 kcal mol^{-1} by this method [25], and very recently we have constructed an accurate PA ladder for the molecules CO(PA=141.4 kcal mol^{-1}), Br_2, HBr, N_2O, HCl, Br, CH_4 and CO_2 (128.5 kcal mol^{-1}) [26]. This work was greatly facilitated by studying proton transfer reactions at different temperatures, a feature essential for some of the other SIFT studies mentioned below.

A very large amount of data has been obtained relating to ionic reactions thought to occur in interstellar clouds notably using SIFT and ion cyclotron resonance (ICR) methods (see the data compilation by Ikezoe et al [7]). As a result of the skilled and often inspired work of those who construct and interpret the detailed models of interstellar clouds [15,27], the routes to many of the less complex interstellar molecules have been identified with reasonable certainty. The routes to the more complex molecules are, however, less certain principally because of the lack of laboratory data on reactions involving polyatomic species. Thus work on this goes on unabated. We have recently carried out a detailed survey of the reactions of the hydrocarbon ions in the series $C_nH_m^+$ (n=4,5,6; m=0 to 6) with H_2 and CO (the two most abundant molecules in dense interstellar clouds) [28], which represents the beginning of a determined effort to identify the routes to the production of polyatomic hydrocarbons in dense interstellar clouds. (Important work in this area has also been carried out by Bohme [29] and McElvany [30] and their colleagues). It is observed that generally only the very unsaturated ions in the series (i.e. m < 2) react with H_2,

undergoing H-atom insertion e.g.
$$C_4^+, C_4H^+ + H_2 \longrightarrow C_4H^+, C_4H_2^+ + H \quad (3)$$
So hydrocarbon ions cannot become heavily hydrogenated by such reactions. Thus more saturated polyatomic hydrocarbon ions can only form from simple hydrocarbon ions in reactions with hydrocarbon molecules e.g.
$$C_3H_3^+ + C_2H_2 \longrightarrow C_5H_5^+ + h\nu \quad (4)$$
Reaction (4) is an example of a radiative association reaction, a process which is considered to be important in low temperature interstellar ion chemistry [31]. In laboratory experiments, these reactions are generally observed to proceed via ternary association [6,32] e.g.
$$C_3H_3^+ + C_2H_2 + He \longrightarrow C_5H_5^+ + He \quad (5)$$
Such is the case for the reactions with CO of the highly unsaturated ions in the series $C_nH_m^+$, [28] e.g.
$$C_4H^+ + CO + He \longrightarrow C_4HCO^+ + He \quad (6)$$
and we have shown that the radiative association analogue of these followed by electron-ion recombination could produce C_nO molecules such as C_3O which has been detected in interstellar clouds [33]. (Note it is assumed that the ion C_4HCO^+ formed in (6) could dissociatively recombine with an electron to form C_5O.)

One of the most interesting interstellar observations is that several molecular species are observed to be enriched in deuterium (relative to expectations based on the D/H cosmical abundance ratio) [34]. As a result of much careful SIFT work [35], the suggestion first made by Watson[36] that this enrichment is the result of isotope fractionation in ion-molecule reactions has been verified. Deuterium enhancement occurs in the elementary reactions
$$D^+ + H_2 \rightleftharpoons H^+ + HD \quad (7)$$
$$H_3^+ + HD \rightleftharpoons H_2D^+ + H_2 \quad (8)$$
$$CH_3^+ + HD \rightleftharpoons CH_2D^+ + H_2 \quad (9)$$
$$C_2H_2^+ + HD \rightleftharpoons C_2HD^+ + H_2 \quad (10)$$
All these reactions proceed much faster to the right than to the left at low temperatures by virtue of the zero-point-energy differences between the reactants and products, and so (7) fractionates D into HD and (8),(9) and (10) fractionate D into H_2D^+, CH_2D^+ and C_2HD^+. Hence, the subsequent reactions of these ions can result in other molecules which are enriched in deuterium. The rate coefficients for the forward and reverse of reactions (7),(8),(9) and (10) have been determined over significant temperature ranges using the SIFT and the enthalpy and entropy changes in the reactions have been determined [35]. Very recent SIFT studies in our laboratory of the reactions of H_3^+, CH_3^+ and $C_2H_2^+$ and their deuterated analogues H_2D^+, CH_2D^+ and C_2HD^+ are providing data also on the mechanisms of such ion-molecule reactions at thermal energies (a technique used to great effect also by others [37]). For example, these studies show that in the reaction of CH_2D^+ with CH_4.
$$CH_2D^+ + CH_4 \longrightarrow C_2H_5^+(HD), C_2H_4D^+(H_2) \quad (11)$$
the D and the H atoms are quite equivalent and so the ratio of $C_2H_5^+$ and $C_2H_4D^+$ products are in accordance with simple statistical expectations. However, for the reaction:
$$CH_2D^+ + C_2H_5OH \longrightarrow C_2H_5^+(CH_2DOH)$$
$$\longrightarrow CH_2DOH_2^+(C_2H_4) \quad (12)$$
the D atom remains bonded to the same carbon as it was before the reaction and D/H scrambling does not occur. Thus CH_2DOH is formed in reaction (12) but not CH_3OD. Such experiments are pointers to the deuterated molecules that may be produced and be detectable in interstellar clouds.

SIFT studies of reactions in recognisable series with many reactant gases have made valuable contributions to the understanding of the fundamentals of ion-molecule interactions as well as to interstellar chemistry. Thus we have previously studied in detail the reactions of ions in the series CH_n^+, NH_n^+, H_nCO^+ and H_nS^+ (for n from zero to the value appropriate to saturated ions in each case) [6] and these have indicated important routes to the synthesis of several interstellar molecules and greatly enhanced the understanding of the mechanisms of ion-molecule reactions. Recently we have studied the reactions of the ions derived from phosphine, i.e. the PH_n^+ ions (n = 0 to 4) with several molecular species most of which have been detected in interstellar clouds [38]. This work was stimulated by the recent detection of the first phosphorus-bearing

molecule, PN, in the interstellar medium [19] and the desire to know how it was formed. The PH_n^+ ions are seen to be very reactive with most molecules (except, significantly, with H_2 and CO); the reactions with NH_3 produce PNH_2^+ and PNH_3^+ which on recombination with electrons could result in PN. The reactions of the PH_n^+ ions with hydrocarbons commonly generate organo-phosphorus ions which implies that organo-phosphorus molecules are probably present in interstellar gas (molecules like CP, HCP and HC_3P; note that the analogous nitrogen- bearing molecules CN, HCN and HC_3N are abundant interstellar species). A subsequent study of the reactions of the protonated species PNH^+, POH^+, PSH^+ and H_2CP^+ has enabled the proton affinities of PN, PO, PS and HCP to be determined (by the method of bracketing outlined above) [39]. The surprise result of this study was the large PA of PN ($=191\pm2$ kcal mol^{-1}) which suggests that PNH^+ must surely co-exist with PN in interstellar gas, since PN will accept a proton from most protonated species, e.g.
$H_2CN^+ + PN \longrightarrow PNH^+ + HCN$ (13)
Also, the large dipole moment of PN ($=2.747$ Debye) [19] means that reactions like (13) will have large rate coefficients at low temperatures (again, following the ideas discussed at the beginning of this Section).

The versatility of the SIFT technique is clear; it can be used to study the reactions not only of ground state ions but also of metastable excited ions [8]. Following up earlier SIFT studies [40], we have studied the reactions of the spin-orbit states ($^2P_{3/2}$, $^2P_{1/2}$) of Kr^+ and Xe^+ and the reactions of the molecular ions Kr_2^+ and Xe_2^+ with a variety of molecules [41]. The molecular ions undergo facile switching reactions with many gases, e.g.
$Xe_2^+ + CH_4 \longrightarrow XeCH_4^+ + Xe$ (14)
Further, the molecular product ions of these reactions undergo switching reactions thus:
$XeCH_4^+ + CH_4 \longrightarrow CH_4 \cdot CH_4^+ + Xe$ (15)
generating the dimer ions, M_2^+ (in the above example M is CH_4). Such reactions have been observed also for M = H_2O, COS, C_2H_2, C_2H_6, HCl, N_2O and CO [42]. The interest in these results is clear when it is appreciated that the M_2^+ dimer ions cannot generally be produced in the gas phase by association reactions of the M^+ ions with their parent molecule M, since such reactions usually result in other products, e.g.
$CH_4^+ + CH_4 \longrightarrow CH_5^+ + CH_3$ (16)
However, some of the dimer ions e.g. $H_2O \cdot H_2O^+$, $HCl^+ \cdot HCl$ etc. have been produced by ionizing the neutral dimers and their bond energies have been determined (see references in ref.42)..

All the above examples of recent SIFT studies refer to reactions between translationally-thermalised reactants. The inclusion of the drift field in the SIFT (creating a SIFDT) allows reactions between ions and neutrals to be studied at elevated centre-of-mass energies. SIFDT apparatuses are being exploited increasingly for such studies. A recent review describes some of this work [9]. Of note is the study of the $C^+(H_2, H) CH^+$ reaction [43] which is endothermic by 0.4 eV but which is initiated at elevated energies. This reaction is considered to be important in the shocked regions of interstellar gas [44]. A comparison of the data on this reaction obtained using a SIFDT and low energy ion-neutral beam methods has been made [9,43]. However, it must be remembered that the reactants in a SIFDT are not in thermal equilibrium. This feature has been exploited to study the influence of the rotational energy of the reactant neutral molecules on the efficiency of some ion-neutral reactions [45], studies which are difficult to envisage using other methods.

3.2 FALP Results

As mentioned above, the extraordinarily versatile FALP apparatus is currently being used to study (i) dissociative recombination; (ii) electron atachment and (iii) ion-ion mutual neutralization. We briefly discuss some of the recent studies of these processes in the order given.

(i) Dissociative Recombination
Prior to the FALP work in this area,

the most comprehensive study of dissociative recombination at thermal energies and the determination of recombination coefficients, α_e, had been achieved using the stationary afterglow (SA) technique [46]. Thus the α_e for some important atmospheric ions, including O_2^+ and NO^+, had been determined over appreciable temperature ranges. The initial FALP studies of $\alpha_e(O_2^+)$ agreed well with the SA data both in magnitude and in the variation with temperature $(\alpha_e(O_2^+) \sim T^{-0.7})$ [10]. However, an obvious disagreement between the FALP and SA data was apparent regarding the temperature dependence of $\alpha_e(NO^+)$ and this has since been resolved in favour of the FALP data and the data obtained using the trapped ion method [47].

To date, the greatest value of the FALP in this area has been in the determination of the α_e for a number of interstellar positive ions [13]. Most significant of all is that it has been shown that the most important interstellar ion H_3^+ in its ground vibronic state (from which much of interstellar chemistry begins) does not recombine at a measurable rate in binary collisions with electrons at and below room temperature (and even at 500K). Thus, under such conditions, the $\alpha_e(H_3^+)$ is very small at $\leq 10^{-10} cm^3 s^{-1}$, a result in general accord with theoretical expectations [48]. (Note, however, that vibrationally-excited H_3^+ does indeed recombine rapidly [13]). This quite unexpected result has great import to interstellar chemistry and is, of course, also of fundamental interest. A discussion of this and other $\alpha(H_3^+)$ data has been included in recent reviews [49].

That H_3^+ does not recombine in the FALP plasma is of great practical value. H_3^+/electron plasmas can be readily formed and then converted to plasmas comprising other protonated species, since proton transfer from H_3^+ to most other species is facile (because $PA(H_2)$ is small). Thus the addition of gases like, for example, CO and H_2O convert the H_3^+ plasma to HCO^+ and H_3O^+ plasmas via the proton transfer reactions $H_3^+(CO,H_2)HCO^+$ and $H_3^+(H_2O,H_2)H_3O^+$. In this way the α_e for a wide variety of protonated ions have been determined, many of interstellar importance, some over a range of temperature [13,50]. A significant practical point is that such studies are not possible using the SA technique, but are possible with the FALP because of the extraordinary chemical versatility of this fast flow tube method.

These kinetic data on α_e are vital to the proper understanding of molecular synthesis in interstellar clouds. Equally important is a knowledge of the neutral products of the dissociative recombination reactions of ground state polyatomic ions for which no data existed until quite recently. Now the FALP apparatus has been used in conjunction with LIF and VUV spectroscopic techniques to determine the fraction of OH radicals and H atoms produced in the dissociative recombination of H_3O^+, HCO_2^+, N_2OH^+ and O_2H^+. The details of the experimental method are given in a very recent paper [51], and the detailed results and their interstellar significance is discussed in another recent paper [52]. This exciting FALP/spectroscopy work is just beginning. It will be extended soon to probe the production of CN radicals in the recombination of CN-bearing ions; it also highlights the great value of the combination of the versatile FALP method and spectroscopy in studying ionic processes at thermal energies. The enormous potential of this powerful combination is clear. It is now being vigorously exploited in our laboratory.

(ii) Electron Attachment

Electron attachment is an important process in ionized gases containing electronegative gases. It converts free electrons to negative ions and thus results in a change of the physical characteristics of the media (conductivity, dielectric constant etc.) and the ion chemistry occurring therein. This process is important in the lower terrestrial atmosphere, the negatively charged component of which is negative ions and not free electrons [1]. Electron attachment also plays an important role in laboratory surface etchant plasmas [20]. Two distinct attachment processes have been recognised: (i) dissociative attachment e.g.

$$CCl_4 + e \longrightarrow Cl^- + CCl_3 \quad (17)$$

in which a negative ion and a free radical are formed, and direct attachment, e.g.

$$SF_6 + e \longrightarrow (SF_6^-)^* \longrightarrow SF_6^- \quad (18)$$

in which an excited negative ion of the parent molecule is formed which is relaxed either by radiation emission or collisional deactivation. Both (17) and (18) are fast, occurring essentially with unit efficiency (when $\beta \sim \beta_{max} \sim 3 \times 10^{-7} cm^3 s^{-1}$ at 300K) [12].

The FALP technique has proved to be ideal for the study of these attachment processes at thermal energies and attachment coefficients, β, have been measured over the approximate temperature range 200-600K for many reactions[53]. Generally speaking, the β for dissociative attachment reactions increase with increasing temperature (except when the reactions proceed very rapidly as is the case for reaction (17) and (18)) and then the FALP data provide values for the "activation energies" for the reactions [12,53]. The direct attachment process is not fully understood; occasionally the β actually decrease with increasing temperature for these reactions, as was recognised in the FALP experiment for the C_6F_6 reaction [54]. A very recent FALP discovery has been that the CCl_3 radicals formed in the "primary" dissociative attachment reaction (17) undergo rapid "secondary" dissociative attachment [55] thus :

$$CCl_3 + e \longrightarrow Cl^- + CCl_2 \quad (19)$$

This has implications to the efficiency of electron capture detectors, and may commonly occur for molecular radicals containing halogen atoms (it certainly also occurs for CCl_2Br radicals [55], but note that dissociative attachment to CCl_2 radicals is endothermic).

Dissociative electron attachment in the gas phase is analogous to Bronsted acid behaviour in the liquid phase. Thus strong Bronsted acids undergo rapid dissociative attachment, as a detailed FALP study has shown [56]. For example, the reaction

$$H_2SO_4 + e \longrightarrow HSO_4^- + H \quad (20)$$

is extremely rapid at 300K. Similarly, the 'superacids' FSO_3H and CF_3SO_3H rapidly form the stable FSO_3^- and $CF_3SO_3^-$ negative ions in reaction with thermal electrons. Gaseous HI is similarly 'acidic':

$$HI + e \longrightarrow I^- + H \quad (21)$$

The FALP measurements have shown that the β for reaction (21) is $3 \times 10^{-7} cm^3 s^{-1}$ [56]. The reverse of these dissociative attachment reactions, in which a free electron is released and a new molecule formed, is known as associative detachment. Such reactions can be studied using the SIFT technique, and in a co-ordinated FALP and SIFT study, the forward and reverse reaction rate coefficients for reaction (21) and also for the analogous reactions involving HBr have been measured [57]. This has allowed the enthalpy and entropy changes in these reactions to be determined. The results obtained indicate that reaction (21) is essentially thermoneutral as expected from the known bond energy of HI molecules and the electron affinity of I atoms.

It must be stressed that using the FALP method, the β are detemined under truly thermalised conditions at the particular temperature of the experiment. Thus by increasing the temperature, both the electron (translational) energy and the reactant molecule translational and internal energies are increased. This is not the case for the well-known drift tube and Krypton photoionization methods of determining β (referred to in ref 58) in which only the electron energy is usually varied. Therefore, not surprisingly, comparison of the results for these non-thermal experiments with those from FALP experiments must be approached with caution. We have attempted to clarify the differences to be expected in the measured β for some dissociative attachment reactions of haloethanes by thermal and non-thermal experiments in a recent paper [58].

A final point worthy of note here is that in the study of numerous reactions of halocarbon compounds (such as the Freons), atomic halogen negative ions are seen to be the only products of dissociative attachment reactions. However, in our very recent FALP studies of some dibromoethanes (i.e. $(CF_2Br)_2$, $(CH_2Br)_2$, the molecular ion Br_2^- appears as a significant product (together with Br^-) [59]. This observation bears some consideration and forcibly illustrates the importance of mass identification of the product ions of attachment reactions, a facility often not available in some experiments.

(iii) Positive Ion/Negative Ion Mutual Neutralization

This process, often termed ionic recombination, is important when negative ions form a significant fraction of the negatively-charged species in a plasma. Thus it plays an important role in limiting the ionization in the terrestrial stratosphere and troposphere [1]. At low pressures, the binary process is operative, e.g.

$$NO^+ + NO_2^- \longrightarrow NO + NO_2 \quad (22)$$

but at higher pressures, collision-enhanced mutual neutralization and ternary ionic recombination become apparent e.g.

$$NO^+ + NO_2^- + M \longrightarrow \text{product molecules} + M \quad (23)$$

where M is a third body (in the lower atmosphere M would be N_2 or O_2). The FALP has been used to provide much of the reliable data on the binary process [60]. The binary ionic recombination coefficients, α_i, have been determined for reactions involving (variously) simple and clustered molecular positive and negative ions. (Reaction (22) has been given considerable attention and the neutral products indicated (i.e. NO and NO_2) have been identified using emission spectroscopy [61]). The temperature dependence of the α_i for this reaction has also been studied; thus $\alpha_i \sim T^{-0.5}$ in accordance with theoretical expectations. A summary of some of the FALP work in this area is available [11].

Negative ions are not generally considered to be involved in interstellar chemistry. Recently, however, it has been proposed that negatively-charged polycyclic aromatic hydrocarbons (PAH^-) may be present in dense clouds [62]. If this were so then mutual neutralization reactions with abundant positive ions (e.g. H_3^+, HCO^+) could result in ionization loss and the production of neutral molecules, e.g.

$$PAH^- + H_3^+ \longrightarrow \text{neutral products} \quad (24)$$

At present this is somewhat speculative. It does, however, present another challenge to the FALP to provide data on such interesting reactions as 24).

4. Concluding Remarks

The above is merely intended to illustrate the versatility and productivity of the SIFT and FALP techniques in the study of gas phase ionic processes at or near thermal energies. Whilst much has been achieved using these techniques much more can yet be achieved by their continued exploitation. Currently both experiments operate in the low pressure regime (< 1 Torr). Developments are underway in our laboratory to extend our studies of ionic reactions to much higher pressures (up to atmospheric pressure). Undoubtedly, new reaction phenomena will be identified in higher pressure regimes, and a better understanding of the behaviour of practical devices such as electron capture detectors and plasma chromatography cells will be obtained.

Acknowledgments

We gratefully acknowledge the contributions of Kevin Giles and Charles R. Herd to some of the recent work summarised in this paper. We are also grateful to the SERC and the USAF for financial support of the work.

References

[1] D. Smith and N.G. Adams, in *Topics in Current Chemistry*, edited by S. Veprek and M. Venugopalan (Springer-Verlag, Heidelberg, 1980), Vol. 89, p.1.

[2] See for example the papers in *Swarms of Ions and Electrons in Gases*, edited by W. Lindinger, T.D. Mark and F. Howorka (Springer-Verlag, Vienna, 1984), and the references therein.

[3] D. Smith and N.G. Adams, in *Gas Phase*

Ion Chemistry, edited by M.T. Bowers (Academic, New York, 1979), Vol.1, p.1.
D. Smith and N.G. Adams, in *Advances in Atomic and Molecular Physics*, edited by D. Bates and B. Bederson (Academic, San Diego, 1987), Vol.24, p.1.

[4] D. Smith, N.G. Adams, A.G. Dean, and M.J. Church, J. Phys. D. **8**, 141, (1975).
N.G. Adams, M.J. Church and D. Smith, J. Phys. D. **8**, 1409, (1975).
N.G. Adams and D. Smith, in *Techniques for the Study of Ion/Molecule Reactions*, edited by J.M. Farrar and W. Saunders Jr., (Wiley, New York, 1988), Vol.XX, p.165.

[5] D. Smith and N.G. Adams, J. Phys. D. **13**, 1267, (1980).

[6] N.G. Adams and D. Smith, in *Reactions of Small Transient Species*, edited by M.A. Clyne and A. Fontijn (Academic, London, 1983), p.311.

[7] Y. Ikezoe, S. Matsuoka, M. Takebe and A.A. Viggiano, in *Gas Phase Ion-Molecule Reaction Rate Constants through 1986* (Maruzen Company Ltd., Tokyo, 1987).

[8] W. Lindinger and D. Smith, in *Reactions of Small Transient Species*, edited by M.A. Clyne and A. Fontijn, (Academic, London, 1983) p.387.

[9] D. Smith and N.G. Adams, in *Rate Coefficients in Astrochemistry*, edited by T.J. Millar and D.A. Williams, (Kluwer, Dordrecht, 1988), p.153.

[10] E. Alge, N.G. Adams and D. Smith, J. Phys. B. **16**, 1433, (1983).

[11] D. Smith and N.G. Adams, in *Physics of Ion-Ion and Electron-Ion Collisions*, edited by F. Brouillard and J.W. McGowan (Plenum, New York, 1983), p.501.

[12] D. Smith, N.G. Adams and E. Alge, J. Phys. B. **17**, 461 (1984).

[13] N.G. Adams, D. Smith, and E. Alge, J. Chem. Phys. **81**, 1778, (1984).

[14] See papers in *Rate Coefficients in Astrochemistry*, edited by T.J. Millar and D.A. Williams (Kluwer, Dordrecht, 1988)
D. Smith, Phil. Trans. R. Soc. Lond. A. A323, 269, (1987).
D. Smith and N.G. Adams, Faraday II, (1989) in press.

[15] E. Herbst and C.M. Leung, Ap. J. Suppl., (1989) in press.

[16] T. Su and M.T. Bowers, in *Gas Phase Ion Chemistry*, edited by M.T. Bowers, (Academic, New York, 1979) p.83.

[17] N.G. Adams and D. Smith, Int. J. Mass Spectrom. Ion Proc. **61**, 133 (1984).
J.A. Luine and G.H. Dunn, Ap. J. (Letters) **299**, L67 (1985).

[18] B.R. Rowe, in *Rate Coefficients in Astrochemistry*, edited by T.J. Millar and D.A. Williams (Kluwer, Dordrecht, 1988), p.153.

[19] L.M. Ziurys, Ap. J. (Letters) **321**, L81 (1987).

[20] D. Smith and N.G. Adams, Pure and Appl. Chem. **56**, 175 (1984).

[21] D.C. Clary, D. Smith and N.G. Adams, Chem. Phys. Lett., **119**, 320 (1985).

[22] D.C. Clary, Mol. Phys. **54**, 605 (1985).

[23] D.C. Clary, in *Rate Coefficients in Astrochemistry*, edited by T.J. Millar and D.A. Williams (Kluwer, Dordrecht, 1988) p.1.

[24] T.J. Millar, N.G. Adams, D. Smith, and D.C. Clary, Mon. Not. R. Astron. Soc. **216**, 1025 (1985).

[25] D. Smith, and N.G. Adams, J. Phys. Chem. **89**, 3964 (1985).

[26] N.G. Adams, D. Smith, M. Tichy, G. Javahery, N.D. Twiddy, and E.E. Ferguson, J. Chem. Phys. (1989) in press.

[27] J.H. Black and A. Dalgarno, Astrophys. Letts. **15**, 78 (1973).
T.J. Millar and A. Freeman, Mon. Not. R. Astron. Soc. **207**, 405 (1984).

[28] K. Giles, N.G. Adams and D. Smith, Int. J. Mass Spectrom. Ion Proc., **89**, 303 (1989).

[29] D.K. Bohme, S. Dheandhanoo, S. Wlodek, and A.B. Raksit, J. Phys. Chem., **91**, 2569 (1987).

[30] S.W. McElvany, B.I. Dunlap, A. O'Keefe, J. Chem. Phys. **86**, 715 (1987).

[31] D.R. Bates and E. Herbst, in *Rate Coefficients in Astrochemistry*, edited by T.J. Millar and D.A. Williams (Kluwer, Dordrecht, 1988) p.41.

[32] D. Smith, and N.G. Adams, Int. J.

Mass Spectrom Ion Proc. **76**, 307 (1987).

[33] E. Herbst, D. Smith and N.G. Adams, Astron. Astrophys. L13 (1984).
N.G. Adams, D. Smith, K. Giles and E. Herbst, Astron. Astrophys., (1989) in press.

[34] G. Winewisser, E. Churchwell and C.M. Walmsley, in *Modern Aspects of Microwave Spectroscopy*, edited by G.W. Chantry (Academic, New York, 1979), p.313.

[35] D. Smith and N.G. Adams, in *Ionic Processes in the Gas Phase*, edited by M.A. Almoster-Ferreira (Reidel, Dordrecht, 1984) p.41.

[36] W.D. Watson, in *Topics in Interstellar Matter*, edited by H. van Woerden (Reidel, Dordrecht, 1977) p.135.

[37] C.H. DePuy, in *Ionic Processes in the Gas Phase*, edited by M.A. Almoster-Ferreira (Reidel, Dordrecht, 1984) p.227.

[38] D. Smith, B.J. McIntosh and N.G. Adams, J. Chem. Phys. **90**, 6213 (1989).

[39] Paper in preparation.

[40] N.G. Adams, D. Smith, and E. Alge, J. Phys. B. **13**, 3235 (1980).

[41] K. Giles, N.G. Adams and D. Smith, J. Phys. B. **22**, 873 (1989).

[42] D. Smith, and N.G. Adams, Chem. Phys. Lett. (1989) in press.

[43] N.D. Twiddy, A. Mohebati, and M. Tichy, Int. J. Mass Spectrom. Ion Proc. **74**, 251 (1986).

[44] M. Elitzur, and W. D. Watson, Ap. J. **236**, 172 (1980).

[45] N.G. Adams, D. Smith and E.E. Ferguson, Int. J. Mass Spectrom. Ion Proc. **67**, 67 (1985).
N.G. Adams and D. Smith, Int. J. Mass Spectrom. Ion Proc. **81**, 273 (1987).
A.A. Viggiano, R. A. Morris, F. Dale, J.F. Paulson, K. Giles, D. Smith and T. Su, J. Chem. Phys. (1989) in press.

[46] J.N. Bardsley and N.A. Biondi, Adv. Atom. Mol. Phys. **6**, 1 (1970).

[47] R. Johnsen, Int. J. Mass Spectrom. Ion Proc. **81**, 67 (1987).

[48] H.H. Michels and R.H. Hobbs, Ap. J. (Letters) **286**, L27 (1984).

[49] N.G. Adams and D. Smith in *Rate Coefficients in Astrochemistry*, edited by T.J. Millar and D.A. Williams (Kluwer, Dordrecht, 1988) p.173.
N.G. Adams and D. Smith, in *Dissociative Recombination: Theory, Experiment and Applications*, edited by J.B.A. Mitchell and S.L. Guberman (World Scientific, Singapore, 1989).

[50] N.G. Adams and D. Smith, Chem. Phys. Lett. **144**, 11 (1988).

[51] N.G. Adams, C.R. Herd and D. Smith, J. Chem. Phys. (1989) in press.

[52] C.R. Herd, N.G. Adams and D. Smith, Ap. J. (1989) in press.

[53] E. Alge, N.G. Adams, and D. Smith, J. Phys. B. **17**, 3827 (1984).

[54] N.G. Adams, D. Smith, E. Alge, and J. Burdon, Chem. Phys. Lett. **116**, 460 (1985)

[55] N.G. Adams, D. Smith, and C.R. Herd, Int. J. Mass Spectrom. Ion Proc. **184**, 243 (1988).

[56] N.G. Adams, D. Smith, A.A. Viggiano, J.F. Paulson, and M.J. Henchman, J. Chem. Phys. **84**, 6728 (1986).

[57] D. Smith and N.G. Adams, J. Phys. B. **20**, 4903 (1987).

[58] D. Smith, C.R. Herd and N.G. Adams, Int. J. Mass Spectrom. Ion Proc. (1989) in press.

[59] Paper in preparation.

[60] D. Smith and M.J. Church, Int. J. Mass Spectrom. Ion Proc. **19**, 185 (1976).
D. Smith, M.J. Church and T.M. Miller, J. Chem. Phys. **68**, 1224 (1978).
M.J. Church and D. Smith, J. Phys. D. **11**, 2199 (1978).

[61] D. Smith, N.G. Adams and M.J. Church, J. Phys. B. **11**, 4041 (1978).

[62] A. Omont, Astron. Astrophys. **164**, 159 (1986).
S. Lepp and A. Dalgarno, Ap. J. **324**, 553 (1988).

POSITION SENSITIVE DETECTION WITH LASER INDUCED FLUORESCENCE

L. Hüwel*, A.M. Wodtke#, P. Andresen+ and H. Voges+

*Physics Department, Wesleyan University, Middletown, CT 06457
#Chemistry Department, Univ. of California, Santa Barbara, CA 93106
+Max-Planck-Institut für Strömungsforschung, D3400 Göttingen, FRG

Two dimensional density maps reflecting the spatial distribution of individual quantum states of molecules such as OH, O_2 and NO have been obtained at atmospheric pressure. The experimental technique used in these studies takes advantage of the high spectral brightness of tunable, narrowband excimer lasers to obtain easily detectable fluorescence from short-lived or predissociating states of these molecules. Gated image intensifiers coupled with CCD cameras and appropriate imaging electronics hardware capture and store the fluorescence for further analysis. Because of the short pulse length of the excimer laser of about 10 ns temperature and density profiles of turbulent systems can be obtained.

The quantum state selective detection and preparation of atoms and molecules is one of the most powerful techniques that the laser brought to the field of atomic and molecular physics. In general, it is used to obtain information that is averaged or summed over a finite volume because of geometric constraints in the excitation or probe volume and/or the properties of the detection system. Spatially resolved data can ordinarily only be obtained by systematically changing the position of the excitation source or the detector. The same statement can be made for the detection of charged particles. Recently, however, position sensitive devices have become available that can measure simultaneously the flux distribution of charged particles over a small area. We report here the equivalent development of an optical excitation/detection scheme that allows the instantaneous <u>quantum state resolved</u> measurement of the density of atoms and molecules over an area of a few cm^2.

The experimental approach of this work is summarized in figure 1 for the special case of an atmospheric flame as the object of investigation for which results have been published[1-3]. Briefly, the output of a slightly modified[4], commercial tunable excimer laser (1; numbers in this section refer to figure 1) operating either with KrF or ArF (see also table 1) is directed through a beam shaping cylindrical lens telescope (2) into the desired target (3). The laser beam is transformed into a sheet typically 18 mm wide and 1 mm thick. The energy per pulse is of the order of 100 mJ with a quoted pulse width of 10-15 ns and measured line width of about 1 cm^{-1}. Fluorescence of the excitation process is monitored with a standard monochromator/photomultiplier system (4) which enables us to unambiguously assign the emitting states that the laser is tuned to. Fluorescence emerging in the opposite direction is imaged with high quality optics (5) onto the photocathode of a gated, proximity focussed image intensifier whose phosphor screen is coupled via fiberoptics to a 8 bit CCD camera (6). The camera and the intensifier gate are triggered by an external clock that is suitably synchronized with the excimer laser. The video signal of the CCD camera is captured by a frame-grabber interface board and either continuously stored on a video recorder or transferred to computer memory - which could only hold one image at a time - for immediate analysis. The spatial resolution of the system as well as the maximum object size that can be probed simultaneously depend obviously on the details of the imaging optics and the CCD camera employed. A typical set-up in the present studies integrates about 1000 μm^2 per pixel in a 512x512 array, corresponding to a total observed area of about 250 mm^2.

The origins of the development of two (or even three) dimensional imaging techniques can be traced to the field of combustion diagnostics[5-7]. Laser-induced fluorescence spectroscopy has widely been used in this

	ArF	KrF
Wavelength [nm]	193	248
hν [eV]	6.42	5.00
Tuning Range [cm^{-1}]	265	160
Bandwidth FWHM [cm^{-1}]	1	0.1

Table 1

Output characteristics of tunable ArF and KrF excimer lasers.

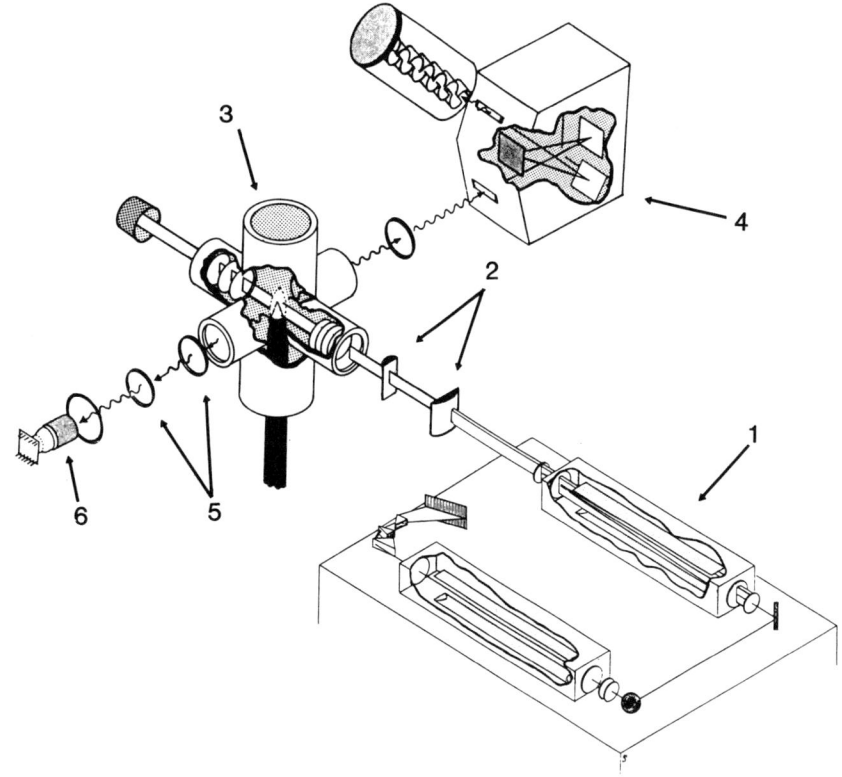

Figure 1 Schematic representation of experimental set-up: (1) tunable injection locked excimer laser, (2) cylindrical lens telescope, (3) flame, (4) scanning monochromator, (5) imaging optics, (6) gated image intensifier and CCD image sensor. Other details are described in the text.

context. One of the outstanding problems in this field is to measure species and state concentration of a variety of atoms and molecules in systems where turbulent motion, steep gradients in the temperature field and in the distribution of constituents make it next to impossible to use standard laser-induced fluorescence methods[8]. A step towards a solution of this problem has been taken by realizing that there are several molecules that have predissociating states within the tuning range of the above mentioned excimer lasers[1-3]. That the loss of fluorescence quantum yield due to predissociation is indeed a blessing in disguise can be understood from the simple equation (1). This relation expresses the local fluorescence intensity $I_F(\mathbf{r})$ that is observable when a laser with pulse energy E excites an absorbing species with local number density $n(\mathbf{r})$

$$I_F(\mathbf{r}) = c\, n(\mathbf{r})\, E\, k_F\, (k_F + k_Q(\mathbf{r}) + k_P)^{-1} \quad (1)$$

where c is a constant incorporating detection and excitation probabilities of the state being pumped and k_F, k_Q, k_P stand for the rates of fluorescence, quenching and predissociation, respectively, out of the state in question. For states and conditions such that $k_P \gg k_Q$ it becomes apparent from equation (1) that the observable fluorescence intensity $I_F(\mathbf{r})$ is related to the desired concentration function $n(\mathbf{r})$ directly via known spectroscopic constants. At atmospheric pressure typical quench cross sections translate into a quench rate of about 10^9 s^{-1}. Therefore, predissociating states with lifetimes of 10 ps - such as the $B^2\Sigma^+$ state of O_2 - can be considered collision free even at atmospheric pressure to an excellent approximation. The chance to fluoresce is drastically reduced, but if a photon is emitted it is from the same state that was pumped by the laser. The high spectral brightness of the tunable excimer lasers overcomes the problem of signal reduction to such an extent that it becomes feasible to excite a large volume of gas and detect the ensuing emission with image intensifiers thus retaining the position information in $I_F(\mathbf{r})$ that is lost in the conventional method with photomultiplier detection. Examples of such intensity maps are presented in figures 2a, 2b, and 2c for O_2 and the intermediate reaction product OH.

The black and white rendering of the original false color images of digital density maps is only meant to give a qualitative impression of the kind of differentiation that is possible with this method. It is obvious that the O_2 state interrogated in figure 2a is located much more to the periphery of the flame than the two different OH lines of figure 2b and 2c. The OH features are the $P_1(8)$ and the $Q_1(11)$ lines after KrF excitation at around 248 nm. The two images are obtained with two separate laser pulses (one image = one pulse, no averaging!). Because the flame is operated under very stable conditions it is possible to extract from such a pair of quantum state maps the corresponding temperature field by relating pixel by pixel the two observed intensities. Work in this direction is in progress in particular to test sophisticated calculations[9] of temperature and partial pressures of stable and turbulent flame front behavior. The oxygen line (figure 2a) was excited with ArF radiation and is due to population in the v=2, N=17 level of O_2. The distribution of this particular vibrational level indicates that it is not a direct combustion product but rather heated ambient gas at the edge of the active flame. It should be pointed out that distributions of O_2 like the one shown in figure 2a can be made with extreme state specificity. By optimizing the laser bandwidth and tuning to the appropriate wavelength region individual spin states of the molecule can be excited[2]. Apart from OH and O_2 several other molecules have transitions in the tuning range of ArF and KrF excimer lasers. Of particular importance for combustion studies are NO^3, H_2O^1 and CO^{10}. It is believed that the above described technique originally developed for combustion diagnostic studies - can be adapted and employed for such atomic and molecular collision studies where a knowledge of the spatial distribution of individual quantum states is important. An example for such a case might be the investigation of photofragmentation where a quantum level imaging technique could conceivably lead to a straightforward determination of state selective differential cross sections.

Figure 2 Two-dimensional density maps of flame constituents. The fluorescence intensity in these images is digitized with a dynamic range of 256. Since there are only 64 different colors available with the current system 4 identical groups of shades can be seen for the range from maximum to zero intensity.

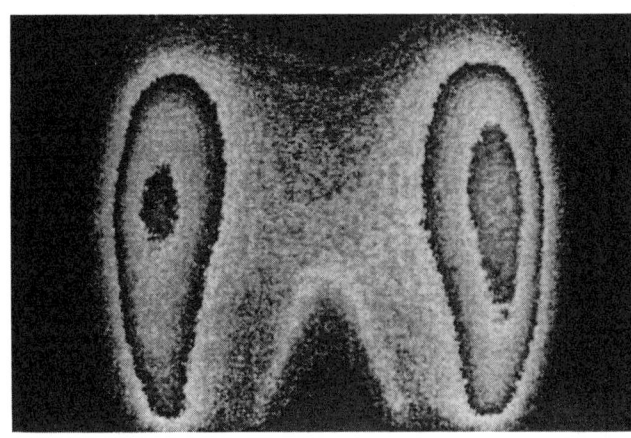

(2a) O_2 in the v=2, N=17 level.

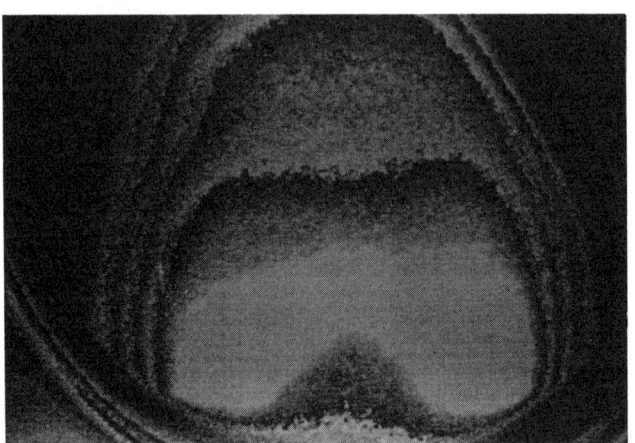

(2b) OH in the v=0, N=8 level.

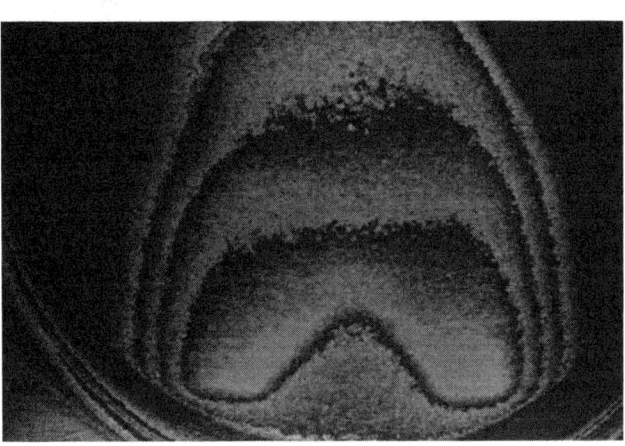

(2c) OH in the v=0, N=11 level.

References

[1] P. Andresen, A. Bath, W. Gröger, H. W. Luelf, G. Meijer, and J. J. ter Meulen, Appl. Opt. **27**, 365 (1988)

[2] A. M. Wodtke, L. Hüwel, H. Schlüter, G. Meijer, P. Andresen, J. Chem. Phys. **89**, 1929 (1988)

[3] A. M. Wodtke, L. Hüwel, H. Schlüter, G. Meijer, P. Andresen, and H. Voges, Opt. Lett. **13**, 910 (1988)

[4] A. M. Wodtke, L. Hüwel, H. Schlüter, and P. Andresen, Rev. Sci. Inst. **60**, 801 (1989)

[5] G. Kychakoff, R. Howe, R. K. Hanson, and J. C. McDaniel, Appl. Opt. **21**, 3225 (1982)

[6] M. P. Lee, P. H. Paul, and R. K. Hanson, Opt. Lett. **11**, 7 (1986)

[7] B. Hiller and R. K. Hanson, Appl. Opt. **27**, 33 (1988)

[8] R. P. Lucht, in: <u>Laser Spectroscopy and it's Applications</u>, L. J. Radziemski, R. W. Solarz, and J. A. Paisner (eds.), p. 623, Marcel Dekker, Inc., New York and Basel (1987)

[9] F. Behrendt, H. Bockhorn, B. Rogg, and J. Warnatz, in: <u>Complex Chemical Reaction Systems</u>, Springer Series in Chemical Physics, vol. **47**, p. 376, Berlin (1987)

[10] G. Meijer, A.M. Wodtke, H. Schlüter, H. Voges, and P. Andresen, J. Chem. Phys. **89**, 2588 (1988)

OPTICAL SPECTROSCOPIC STUDIES ON PENNING IONIZATION AND ION-MOLECULE REACTIONS AT THERMAL ENERGY BY USING FLOWING AFTERGLOW

Masaharu Tsuji

Institute of Advanced Material Study,
Kyushu University, Kasuga-shi,
Fukuoka, 816 Japan

The flowing afterglow apparatus has been coupled with a low pressure chamber by a small orifice for optical spectroscopic studies on Penning ionization and ion-molecule reactions under single collision conditions. Nascent ro-vibrational distributions are determined from a spectral simulation. As selected examples, detailed results are reported about non-Franck-Condon type $He(2^3S)/O_2$ Penning ionization and near-resonant $He^+/H_2O, D_2O$ charge-transfer reactions. A brief summary is also given of our optical spectroscopic studies by using the conventional flowing afterglow apparatus.

I. INTRODUCTION

The flowing afterglow (FA) method was developed in 1963 by Ferguson and his co-workers of the NOAA Laboratories in Boulder, Colorado for the study of ion-molecule reactions in the earth's atmosphere.[1] The main emphasis of their studies was placed upon measurements of formation rate constants and branching ratios of ionic products by using a mass spectrometer. More than a few thousands kinetic data at (near) thermal energy were compiled in the reported tables.[2] In our laboratory the FA coupled with a UV and visible emission detection system has been used for studying the following ionization processes in collisions of metastable atoms (M^*) and ions (M^+) with a molecule (AB):

$$M^* + AB \longrightarrow AB^{+*} + M + e^-,$$
(Penning ionization) (1)

$$\longrightarrow A^{+*} + B + M + e^-,$$
(Dissociative Penning ionization) (2)

$$M^+ + AB \longrightarrow AB^{+*} + M,$$
(Charge-transfer) (3)

$$\longrightarrow A^{+*} + B + M,$$
(Dissociative charge-transfer) (4)

$$M^+ + AB \longrightarrow MA^{+*} + B.$$
(Chemiluminescent reaction) (5)

One purpose of our FA optical study is the detection of new ion fluorescences, which are difficult to excite by photoionization and electron-impact ionization. In Table I are listed ion fluorescences identified in our laboratory and their ionization processes. During spectroscopic studies on new fluorescences of group IV monohalide ions, broad visible bands have been found in reactions of He^* and/or Ne^* with $SiCl_5$,[6,7] $GeCl_4$,[8] and $SiBr_4$.[9] Although these emissions have been ascribed to neutral XY_2^* and/or XY_3^* fragments, they must be reassigned to the $C^2T_2-A^2T_2$ and $C^2T_2-X^2T_1$ bound-free transitions of parent cations (XY_4^{+*}) on the basis of recent optical studies by Tuckett et al. for CF_4^+, SiF_4^+, GeF_4^+, $SiCl_4^+$, and $GeCl_4^+$.[10]

The other purpose is the analysis of internal state distribution for an understanding of ionization dynamics at the microscopic level. Ionization processes studied by our group are listed in Table II. Until our FA optical research, the FA coupled with UV and visible emission spectroscopy has dominantly been used for product state analysis in energy transfer reactions by metastable species such as $Ar(^3P_{0,2})$, $He(2^3S)$, and $N_2(A^3\Sigma_u^+)$. In most cases, ionic active species involved in the discharge flow has been treated as impurities. Combining a microwave discharge source with a high-capacity pumping system (10,000 L/min), the FA optical studies was extended to various type of ion-molecule reactions. The overlap of emissions due to ionic reactions with those due to metastable reactions has often made detailed FA optical studies on ion-molecule reactions difficult. To exclusively detect emission from ionic reactions, we used a pulse modulation technique.[18]

Table I. Ionic emissions identified in Kyushu University.[3-5]

Ions	Transition	Reactions
CS^+	$B^2\Sigma^+-A^2\Pi$	$He^+/CS_2, OCS$
	$B^2\Sigma^+-X^2\Sigma^+$	$He^*, Ne^*/CS$
PN^+	$B^2\Sigma^+-X^2\Sigma^+$	$He^*, Ne^*/PN$
SO^+	$A^2\Pi-X^2\Pi_r$	He^+/SO_2
S_2^+	$A^2\Pi-X^2\Pi_r$	He^*/S_2Cl_2
CCl^+	$A^1\Pi-X^1\Sigma^+$	He^+/CCl_4
	$a^3\Pi_1-X^1\Sigma^+$	He^+/CCl_4
$SiCl^+$	$a^3\Pi_{0^+,1}-X^1\Sigma^+$	$He^+/SiCl_4$
$GeCl^+$	$a^3\Sigma^+-X^1\Sigma^+$	$He^+/GeCl_4$
$SnCl^+$	$a^3\Sigma^+-X^1\Sigma^+$	$He^+/SnCl_4$
CBr^+	$A^1\Pi-X^1\Sigma^+$	He^+/CBr_4
	$a^3\Pi_{0^+,1}-X^1\Sigma^+$	He^+/CBr_4
$SiBr^+$	$a^3\Pi_{0^+,1}-X^1\Sigma^+$	$He^+/SiBr_4$
BBr^+	$A^2\Pi-X^2\Sigma^+$	He^+/BBr_3
GeH^+	$a^3\Pi_{0^+,1}-X^1\Sigma^+$	$He^*, He^+/GeH_4$
SnH^+	$a^3\Pi_{0^+,1}-X^1\Sigma^+$	$He^*, He^+/SnH_4$

Advantages of the FA optical method are as follows.

(1) Because of a high ion density, emission intensity is strong. Therefore, the detection limit of an emission system is low ($\sim 10^{-14}$ cm^{3}s^{-1}), and high resolution measurements that can separate ro-vibrational structures effectively are possible in most cases.

(2) Product state analysis in ion-molecules reactions is possible at thermal energy, where the application of beam experiments is difficult.

A great disadvantage of the FA method is that initial internal state distributions are often modified or lost by collisions with buffer gas, because operating pressures are rather high. This effect is especially severe for molecules with long radiative lifetimes ($> 1\mu s$) and large quenching rate constants. In order to overcome this problem, the FA apparatus has recently been combined with a low pressure cell by a small orifice. This beam apparatus was successfully applied to studying Penning ionization and ion-molecule reactions. Ionization processes that have been investigated by our group are listed in Table III. Among our studies, some selected examples are presented here.

Table II. Ionization processes studied by using conventional FA apparatus[3,11-17]

Reactant	Target	Product ions
Penning Ionization		
He^*	N_2	$N_2^+(B^2\Sigma^+)$
He^*	CO	$CO^+(B^2\Sigma^+)$
He^*, Ne^*	HCl, HBr	$HCl^+, HBr^+(A^2\Sigma^+)$
He^*, Ne^*	CS	$CS^+(B^2\Sigma^+)$
He^*, Ne^*	PN	$PN^+(B^2\Sigma^+)$
He^*	CO_2	$CO_2^+(A^2\Pi, B^2\Sigma^+)$
He^*, Ne^*	N_2O	$N_2O^+(A^2\Sigma^+)$
He^*, Ne^*	OCS	$OCS^+(A^2\Pi)$
He^*, Ne^*	CS_2	$CS_2^+(A^2\Pi)$
He^*, Ne^*	ICN	$ICN^+(A^2\Sigma^+)$
He^*	H_2S	$SH^+(A^3\Pi)$
He^*, Ne^*, Ar^*	$X-(C\equiv C)_2-X$ ($X=H, CH_3, C_2H_5$)	$X-(C\equiv C)_2-X^+(A)$
Charge-transfer reaction		
He^+	N_2O	$N_2O^+(A^2\Sigma^+)$
		$N_2^+(B^2\Sigma^+)$
He^+	CO_2	$CO^+(A^2\Pi)$
He^+, Ne^+	OCS	$CO^+(A^2\Pi)$
He^+	$SiBr_4$	$SiBr^+(a^3\Pi_{0^+,1})$
He^+	$GeCl_4, SnCl_4$	$GeCl^+, SnCl^+(a^3\Sigma^+)$
He_2^+	CO_2	$CO_2^+(A^2\Pi, B^2\Sigma^+)$
He_2^+	S_2Cl_2	$S_2^+(A^2\Pi)$
Ar^+, N_2^+	OCS	$OCS^+(A^2\Pi)$
He_2^+, Ne_2^+	N_2	$N_2^+(B^2\Sigma^+)$
He_2^+	CO	$CO^+(A^2\Pi, B^2\Sigma^+)$
Ar^+	HCl, DBr	$HBr^+, DBr^+(A^2\Sigma^+)$
Ar^+	CS	$CS^+(B^2\Sigma^+)$
Ar^+	PN	$PN^+(B^2\Sigma^+)$
Ar_2^+	$H-(C\equiv C)_2-H$	$H-(C\equiv C)_2-H^+(A^2\Pi)$
Ar_2^+, CO^+	CS_2	$CS_2^+(A^2\Pi)$
Chemiluminescent reaction		
C^+	O_2, CO_2, NO_2, N_2O	$CO^+(A^2\Pi)$
C^+	N_2O	$CN(A^2\Pi)$
C^+	CS_2, OCS	$CS^+(A^2\Pi)$

II. EXPERIMENTAL

The beam apparatus used in this study is shown in Fig. 1. It is composed of two parts, a source chamber and a reaction cell. Rare gas active species were generated by a microwave discharge in the source chamber operated at 0.05-1.0 Torr (1 Torr=133 Pa). After thermalized in the discharge flow, they were expanded into a low-pressure collision chamber through a 2.0 mm diam orifice. The contribution of ionic active species were examined by using an ion-collector grid placed in front of the orifice. Target gases were introduced through a stainless steel nozzle located about 3 mm downstream from the orifice. The partial pressure in the collision chamber was 0.5-1 mTorr for rare gas and 0.5-2 mTorr for target gases. The emission spectrum was dispersed by a 1.0 m scanning monochromator equipped with a cooled photomultiplier.

III. RESULTS AND DISCUSSION

He(2^3S)/O_2 Penning ionization:

Penning ionization of O_2 by metastable He(2^3S) atoms leading to $O_2^+(A^2\Pi_u)$ has captured the attention since Robertson,[27] Richardson et al.,[28] and Richardson and Setser[29] found non-Franck-Condon (FC) vibrational distribution of $O_2^+(A^2\Pi_u)$ in their FA studies using optical spectroscopy (PIOS):

$$He(2^3S) + O_2 \longrightarrow O_2^+(A^2\Pi_u) + He + e^-,$$
$$\Delta H = -2.77 \text{ eV}. \quad (6)$$

Penning ionization (6) consists of the following steps:

$$He(2^3S) + O_2 \longrightarrow He\text{-}O_2^*, \quad (7)$$
$$He\text{-}O_2^* \longrightarrow He\text{-}O_2^+(A) + e^-, \quad (8)$$
$$He\text{-}O_2^+(A) \longrightarrow O_2^+(A) + He, \quad (9)$$
$$O_2^+(A) \longrightarrow O_2^+(X) + h\nu. \quad (10)$$

Penning ionization electron spectroscopy (PIES) method has also been used to the study of He($2^3S, 2^1S$)-O_2 Penning ionization.[30-33] Since the PIES method measures kinetic energy of Penning electrons ejected in process (8), a comparison of the PIES data and FC factors for ionization provides information on an interaction in the entrance channel. On the other hand, the PIOS method detects photoemission from excited ions in process (10), and then gives information on an interaction in the exit channel by comparison with

Table III. Ionization processes studied by using the FA apparatus coupled with a low pressure chamber

Reactant	Target	Product ions	Refs
Penning ionization			
He*	O_2	$O_2^+(A^2\Pi_u)$	19
He*	CO	$CO^+(A^2\Pi)$	20
He*	HCl,HBr	$HCl^+,HBr^+(A^2\Sigma^+)$	21
He*	CS_2	$CS_2^+(A^2\Pi)$	22
Charge-transfer reaction			
He^+	N_2	$N_2^+(D'^2\Pi)$	23
He^+	H_2O,D_2O	$OH^+,OD^+(A^3\Pi)$	24
He^+	C_2H_2	$CH^+(A^1\Pi)$	25
He^+	SiH_4	$SiH^+(A^1\Pi)$	26
He^+	CS_2	$CS^+(B^2\Sigma^+)$	
He^+	OCS	$CO^+(A^2\Pi)$	
He_2^+	CO	$CO^+(A^2\Pi,B^2\Sigma^+)$	
Ar^+	H_2O	$H_2O^+(A^2A_1)$	

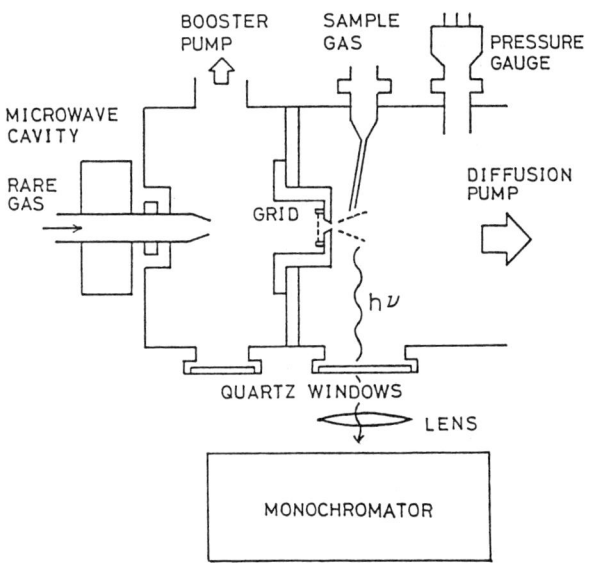

Fig. 1. The flowing afterglow apparatus coupled with a low pressure chamber.

the PIES data. Although a weak PIES peak of $O_2^+(A)$ was identified,[30-33] it was very broad and heavily overlapped with the $a^4\Pi_u$ and $b^4\Sigma_g^-$ bands. Therefore, the vibrational distribution and the peak shift have not been determined. In the present PIOS study, the nascent ro-vibrational distribution of $O_2^+(A)$ has been determined by observing the $O_2^+(A-X)$ emission in the beam experiment.

In Figs. 2a) and 2c) are compared emission spectra obtained from the $He(2^3S)/O_2$ reaction in the beam and FA experiments. The $O_2^+(A-X)$ emission system from $v'=0-13$ and weak atomic oxygen lines are identified in both spectra. The maximum intensity of the beam spectrum shifts to blue and the fraction of the underlying continuum is large in comparison with the FA spectrum due to higher vibrational and rotational excitation. The ro-vibrational distribution in $O_2^+(A:v'=0-13)$ was determined by a computer simulation of the observed spectra, because a number of vibronic transitions are heavily overlapped. In Figs. 2c) and 2d) are shown the best fit FA and beam spectra, respectively, obtained assuming a single Boltzmann rotational distribution for each v' level. The ro-vibrational distributions thus obtained are given in Table IV together with reported FA data of Richardson and Setser[29] and RKR FC factors for $O_2(X) \rightarrow O_2^+(A)$ vertical ionization.

A comparison of our beam and FA data implies that the $O_2^+(A)$ ions are relaxed both vibrationally and rotationally in the FA conditions. It should be noted that the FA result of Richardson and Setser at a He pressure of 2 Torr[29] is more vibrationally excited than our FA result at 0.37 Torr and the beam one. The $O_2^+(A-X)$ emission resulting from the He_2^+/O_2 reaction in the He afterglow was much more vibrationally excited than that from the $He(2^3S)/O_2$ reaction.[19] Therefore, the most probable explanation of the high vibrational excitation in their FA data is the contribution of the He_2^+/O_2 charge-transfer reaction because of incomplete ion collection.

The nascent vibrational distribution has a peak at $v'=0$ and decreases rapidly with increasing v'. This differs markedly from FC factors for vertical ionization with a peak around $v'=7$. The rotational temperature decreases rapidly from 4200 K for $v'=0$ to 400 K for $v'=13$. The present PIOS data represent that 295 and 257 meV are deposited into vibration and rotation of $O_2^+(A)$ in process (6), respectively. This shows that 10.6 and 9.3% of the

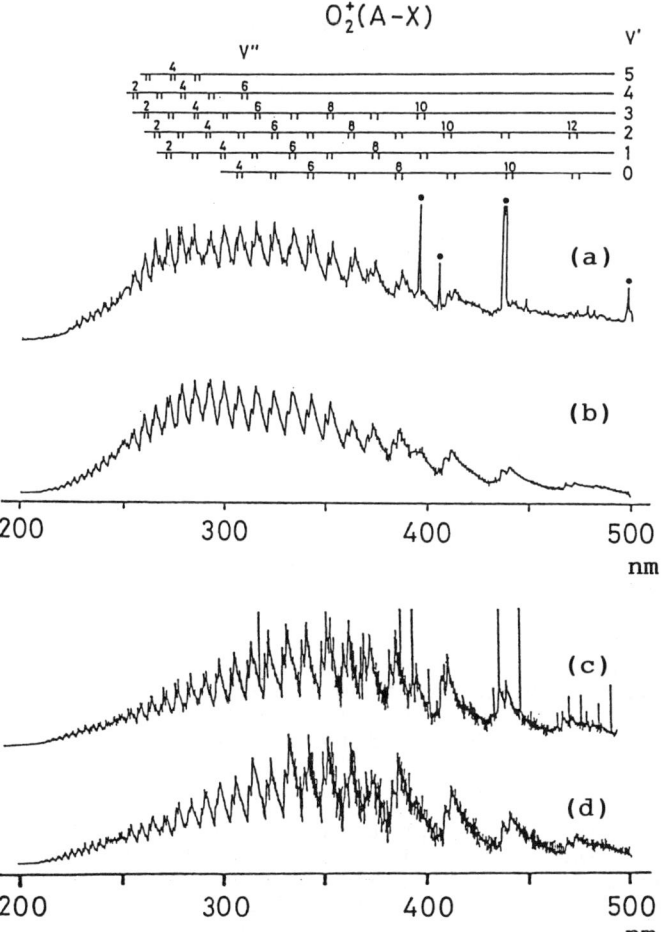

Fig. 2. $O_2^+(A-X)$ emissions produced from the $He(2^3S)/O_2$ Penning ionization.

(a) Beam: Obs. (c) FA: Obs.
(b) Beam: Calc. (d) FA: Calc.

total available energy is partitioned into vibration and rotation, respectively.

The following four cases are possible to explain the non-FC like vibrational population of $O_2^+(A)$; each case is shown in Fig. 3.

Case (a): (near) resonant excitation transfer: this process proceeds through excitation transfer from the covalent $^{1,3}A"$ $He(2^3S)-O_2$ entrance potentials to a (near) resonant Rydberg potential followed by autoionization into the exit ionic potential.

Case (b): pseudo-resonance excitation transfer via an ionic potential; this process proceeds through an avoided crossing from the covalent entrance potential to the $He^+-O_2^-$ ionic potential followed by a second avoided crossing onto an O_2^{**} potential, from which

Table IV. Rovibrational distributions of $O_2^+(A^2\Pi_u)$ produced from the $He(2^3S)/O_2$ Penning ionization.

		v'=0	1	2	3	4	5	6	7	8	9	10	11	12	13
Beam	$N_{v'}$	100	83	54	35	26	19	10	7.4	4.6	2.9	1.9	0.93	0.74	0.46
	T_R (K)	4200	3500	2800	2200	1700	1400	1000	800	600	600	600	400	400	400
FA[a]	$N_{v'}$	100	61	21	7.4	6.8	5.0	3.9	2.6	2.2	1.5	1.0	0.78	0.43	0.26
	T_R (K)	3100	2400	2000	1600	1300	1100	800	600	500	400	400	400	300	300
FA[b]	$N_{v'}$	100	94	63	47	22	22	---	---	---	---	---	---	---	---
FCF ($\times 10^{-3}$)		2.8	12	29	49	68	82	89	89	84	76	67	57	48	40

[a]This work: He pressure 0.37 Torr.
[b]Ref. 29: He pressure 2.0 Torr.

autoionization takes place.

Case (c): ionization via an ionic potential; this process results from an avoided crossing from the $He(2^3S)$-O_2 covalent potentials to the He^+-O_2^- ionic potentials followed by a direct autoionization into the product He-$O_2^+(A)$ potential.

Case (d): Penning ionization from an entrance covalent surface; a transition from the covalent into ionic configuration is symmetrically forbidden in $C_{\infty v}$ and C_{2v} symmetries for the $He(1s2s)$-O_2 quasi-molecule system.[33] Hence, the $He(2^3S)$-O_2 quasi-molecule system survives the crossing in the covalent conformation, if the symmetries is either $C_{\infty v}$ or C_{2v} when the crossing radius is reached.

In cases (a) and (b) Penning ionization occurs at rather long range where the exit surface is nearly flat, while it does at short range in cases (c) and (d) where the exit surface is repulsive. The latter processes probably play a significant role for the formation of $O_2^+(A)$ because the high rotational excitation is likely to result from a short-range repulsive interaction in the exit [He-$O_2^+(A) \longrightarrow He + O_2^+(A)$] channel. A strong underlying continuum probably due to a strongly attractive entrance potential is present in the PIES spectrum.[30-33] An additional evidence of the presence of an attractive entrance surface in process (7) has recently obtained by the dependence of partial cross section of the $O_2^+(A)$ on the reactant energy in a PIES study of Mitsuke et al.[34] On the basis of these facts, it is reasonable to assume that the dominant ionization

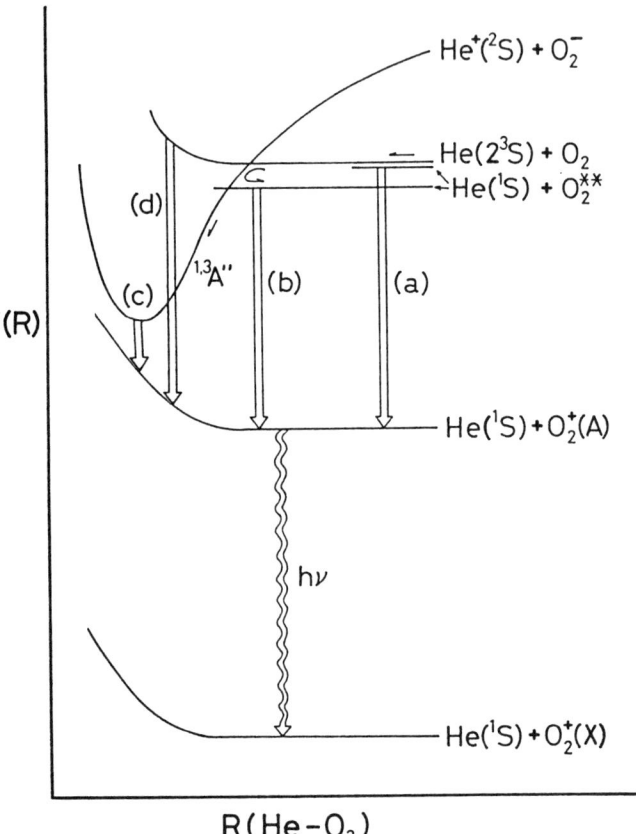

Fig. 3. Potential energy curves for the $He(2^3S)/O_2$ Penning ionization leading to $O_2^+(A^2\Pi_u)$ ions.

occurs via case (c) and a significant $V \rightarrow T,R$ transfer in the exit channel is the major determinant of the observed non-FC like vibrational distribution accompanied by high rotational excitation for low v' levels.

$He^+/H_2O, D_2O$ dissociative charge-transfer

Figs. 4a) and 4c) show emission spectra produced from the $He^+, He(2^3S)/H_2O$ reactions in the beam and FA experiments, respectively. The most prominent band in both spectra is the $OH^+(A^3\Pi-X^3\Sigma^-)$ system from $v'=0-3$. In addition, emissions from neutral fragments, $OH(A^2\Sigma^+-X^2\Pi)$ and H(Balmer series), are identified. In the beam experiment, the $OH^+(A-X)$ and $OH(A-X)$ bands disappeared almost completely by the ion trapping, while the H^* lines were independent of the ion trapping. From these facts and the energetics, the excited fragments are found to be formed through the following processes:

$He^+ + H_2O \longrightarrow OH^+(A^3\Pi) + H + He$, (11)
$\Delta H = -3.07$ eV,

$\longrightarrow OH(A^2\Sigma^+) + H + He$, (12)
$\Delta H = -1.88$ eV,

$He(2^3S) + H_2O \longrightarrow H(n=3-10) + OH + He$, (13)
$\Delta H = -2.64$ to -1.27 eV.

An outstanding feature of the FA spectrum is that the $OH(A-X)$ band is enhanced in comparison with the beam spectrum. This is due to the formation of $OH(A)$ through fast electron-ion recombination processes in the high pressure FA experiment.[35,36]

The reactions of He active species with D_2O in the beam experiment gave the corresponding isotopic band systems as shown in Fig. 5a). Based upon the energetis and the effect of ion trapping, $OD^+(A)$, $OD(A-X)$, and deuterium Balmer lines are found to be excited through the following processes:

$He^+ + D_2O \longrightarrow OD^+(A^3\Pi) + D + He$, (14)
$\Delta H = -2.92$ eV,

$\longrightarrow OD(A^2\Sigma^+) + D + He$, (15)
$\Delta H = -1.75$ eV,

$He(2^3S) + D_2O \longrightarrow D(n=3-10) + OD + He$, (16)
$\Delta H = -2.53$ to -1.16 eV.

This report focuses upon dissociative ionization (11) and (14). The rovibrational distributions of $OH^+(OD^+)$ in the $A^3\Pi$ state were determined by a computer simulation of the observed spectra. In the calculation twenty seven rotational branches were involved and the dependence of FC factors on the rotational quantum number was taken into consideration. The best fit spectra are

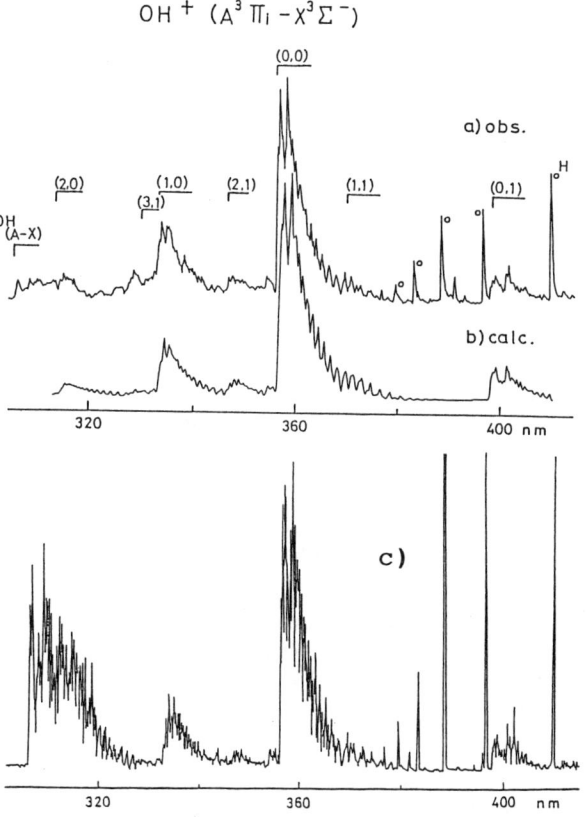

Fig. 4. Emission spectra of H_2O.
(a) Beam: Obs. (b) Beam: Calc.
(c) FA: Obs

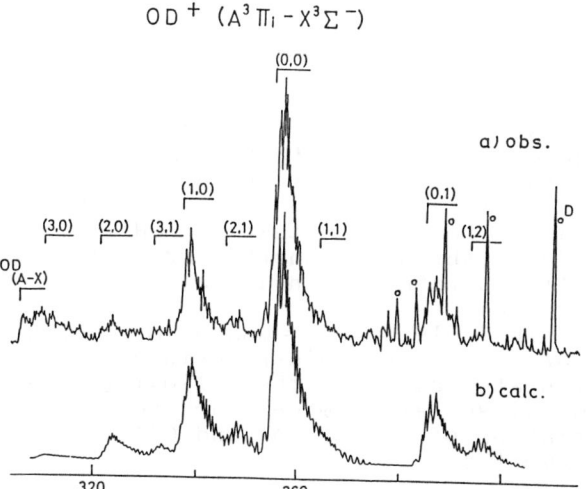

Fig. 5. Emission spectra of D_2O.
(a) Beam: Obs. (b) Beam: Calc.

shown in Fig. 4b) for OH⁺(A-X) and Fig. 5b) for OD⁺(A-X). The ro-vibrational distributions obtained in the beam and FA experiments are summarized in Table V along with the ICR data of Govers et al.[37] A comparison between the beam and FA data suggests that vibrational and rotational relaxation by collisions with the buffer He gas is slow during a rather long radiative lifetime of 2.5 μs. This result is contrasted with the cases of SiH⁺(A:τ:1.025 μs) and CH⁺(A:τ: 0.815 μs) which show significant vibrational and/or rotational relaxations in the He afterglow. The nascent vibrational distribution for v'=0-2 obtained in the present study is in good agreement with the ICR data. The $N_{v'}$ decreases exponentially with increasing v' with effective vibrational temperatures of 2860±500 K for OH⁺(A) and 2630±250 K for OD⁺(A). The fractions of the available energy deposited into rotation and vibration of OH⁺(OD⁺) were evaluated to be 4.8(5.5) % and 3.9(4.1)%, respectively, indicating that a large amount of the available energy is partitioned into the relative translational energy. No isotope effect is found in the energy disposal between the He⁺/H₂O and He⁺/D₂O reactions. This implies that the reaction dynamics is not controlled by a kinematic effect such as a mass combination of fragments.

The correlation diagram of H₂O⁺ with the states of (OH + H)⁺ is shown in Fig 6. The OH⁺(A) + H dissociation limit correlates to the steep repulsive first doubly excited ²A₁ state of H₂O⁺, too high in energy (~27 eV) to be populated by thermal energy He⁺(24.58 eV). Therefore, Govers et al.[37] predicted that distortion of H₂O⁺(²A₁) from C_{2v} to near C_s symmetry due to stretching of one O-H bond is necessary to lower the potential energy. This argument, coupled with the fact the excitation rate constants of OH⁺(A) and OD⁺(A) are small (4 × 10⁻¹² cm³s⁻¹),[37] leads us to predict that the OH⁺(A) is produced through a short range intimate collision rather than a long range FC type one.

In Table V are shown statistical prior vibrational distributions calculated assuming the following two-body dissociation model for the formation of the OH⁺(A) and OD⁺(A) fragments:

$$He^+ + H_2O \longrightarrow H_2O^{+*} + He, \quad (17)$$
$$H_2O^{+*} \longrightarrow OH^+(A) + H. \quad (18)$$

The prior vibrational distributions obtained from the relation $(1 - f_{v'})^{3/2}$ are more excited than the experimental

Table V. Ro-vibrational distributions of OH⁺(OD⁺) in the A³Π$_i$ state.

	v'	0	1	2	3
OH⁺	Obs.(Beam)	0.66	0.20	0.10	0.04
	Obs.(FA)	0.70	0.23	0.07	----
	Obs.(ICR)[a]	0.63	0.25	0.12	----
	Calc.(2-body)	0.30	0.26	0.23	0.21
	T_r (Beam)		1700±200 K		
	T_r (FA)		1200±100 K		
OD⁺	Obs.(Beam)	0.55	0.26	0.13	0.06
	Obs.(FA)	0.78	0.22	----	----
	Obs.(ICR)[a]	0.63	0.25	0.12	----
	Calc.(2-body)	0.29	0.26	0.24	0.22
	T_r (Beam)		1800±200 K		
	T_r (FA)		1400±100 K		

[a]Ref. 37.

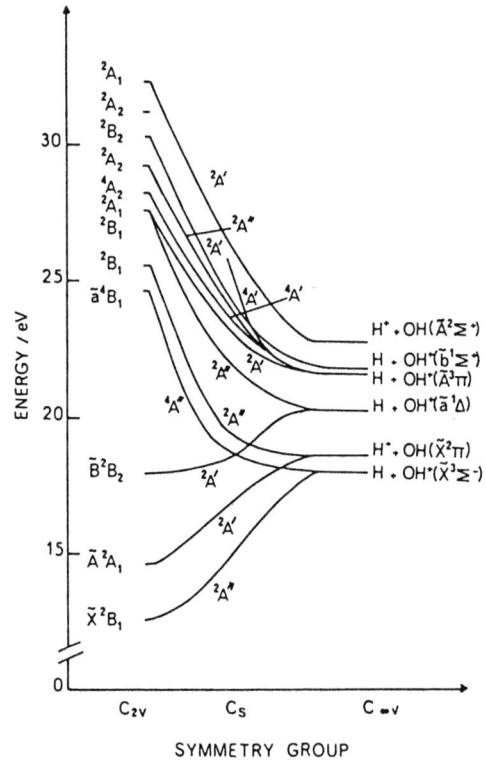

Fig. 6. Correlation diagram for dissociation of the lower excited states of H₂O⁺ into (H + OH)⁺ (reproduced from Ref. 37.)

ones. A similar disagreement was found for the rotational distributions between the prior distributions and the experimental ones. For example, rotational energy of OH⁺(A:v'=0) is predicted to be 1.23 eV from the relation $g_r = f_r/(1 - f_{v'})$. This value is much larger than the observed ones (0.15 eV). Large discrepancies between the statistical and experimental rovibrational distributions suggest that the internal energy distributions of the product fragments produced from the dissociation of precursor H_2O^{+*} (D_2O^{+*}) ions are not controlled by the statistical factor.

References

[1] E.E. Ferguson, F.C. Fehsenfeld, and A.L. Schmeltetopf, Adv. At. Mol. Phys. 5, 1 (1969).
[2] D.L. Albritton, At. Data Nucl. Data Tables 22, 1 (1978).
[3] M. Tsuji, in Techniques of Chemistry: Techniques for the Study of Ion Molecule Reactions, Eds. by J.M. Farrar and W. Saunders Jr., (John Wiley & Sons, New York 1988), pp 489-562.
[4] M. Tsuji, I. Murakami, and Y. Nishimura, J. Chem. Phys. 75, 5373 (1981).
[5] S. Yamaguchi, M. Tsuji, and Y. Nishimura, Chem. Phys. Lett. 138, 29 (1987).
[6] M. Tsuji, T. Mizuguchi, and Y. Nishimura, Can. J. Phys. 59, 985 (1981).
[7] Y. Nishimura, Rep. Res. Inst. Ind. Sci., Kyushu University, 80, 111 (1986).
[8] M. Tsuji, T. Mizuguchi, and Y. Nishimura, Chem. Phys. Lett. 84, 318 (1981).
[9] M. Tsuji, K. Shinohara, S. Nishitani, T. Mizuguchi, and Y. Nishimura, Can. J. Phys. 62, 353 (1984).
[10] I.R. Lambert, S.M. Mason, R.P. Tuckett, and A. Hopkirk, J. Chem. Phys. 89, 2675, 2683 (1988).
[11] M. Tsuji, J.P. Maier, H. Obase, and Y. Nishimura, Chem. Phys. 110, 17 (1986).
[12] H. Obase, M. Tsuji, Y. Nishimura, Chem. Phys. 57, 89 (1981).
[13] H. Obase, M. Tsuji, and Y. Nishimura, Chem. Phys. 74, 89 (1983).
[14] M. Tsuji, J.P. Maier, H. Obase, H. Sekiya, and Y. Nishimura, Chem. Phys. Lett. 137, 421 (1987).
[15] M. Tsuji, K. Tsuji, H. Fukutome, and Y. Nishimura, Chem. Lett. 673 (1977).
[16] H. Obase, M. Tsuji, and Y. Nishimura, Chem. Phys. Lett. 141, 133 (1987).
[17] M. Tsuji, K. Mizukami, H. Obase, and Y. Nishimura, J. Phys. Chem. 92, 1163 (1988).
[18] M. Endoh, M. Tsuji, and Y. Nishimura, Chem. Phys. 82, 67 (1983).
[19] M. Tsuji, H. Obase, M. Endoh, S. Yamaguchi, K. Yamaguchi, K. Kobarai, and Y. Nishimura, J. Chem. Phys. 89, 6753 (1988).
[20] H. Sekiya, M. Tsuji, and Y. Nishimura, Chem. Lett. 1997 (1986).
[21] H. Obase, M. Tsuji, and Y. Nishimura, J. Chem. Phys. 87, 2695 (1987).
[22] M. Tsuji and J.P. Maier, Chem. Phys. 126, 435 (1988).
[23] H. Sekiya, M. Tsuji, Y. Nishimura, J. Chem. Phys. 87, 325 (1987).
[24] S. Yamaguchi, M. Tsuji, and Y. Nishimura, J. Chem. Phys. 88, 3111 (1988).
[25] S. Yamaguchi, M. Tsuji, and Y. Nishimura, J. Chem. Phys. 87, 1637 (1987).
[26] S. Yamaguchi, M. Tsuji, H. Obase, H. Sekiya, and Y. Nishimura, J. Chem. Phys. 86, 4952 (1987).
[27] W.W. Robertson, J. Chem. Phys. 44, 2456 (1966).
[28] W.C. Richardson, D.W. Setser, D.L. Albritton, and A.L. Schmeltekopf, Chem. Phys, Lett. 12, 349 (1971).
[29] W.C. Richardson and D.W. Setser, J. Chem. Phys. 58, 1809 (1973).
[30] D.S.C. Yee, W.B. Stewart, C.A. McDowell, and C.E. Brion, J. Electron Spectrosc. 7, 93 (1975).
[31] H. Hotop and G. Hübler, J. Electron Spectrosc. 11, 101 (1977).
[32] H. Hotop, E. Kolb, and J. Lorenzen, J. Electron Spectrosc. 16, 213 (1979).
[33] O. Leisin, H. Morgner, and W. Müller, Z. Phys. A304, 23 (1982).
[34] K. Mitsute, T. Takami, and K. Ohno, J. Chem. Phys. submitted for publication.
[35] C.B. Collins and W.W. Robertson, J. Chem. Phys. 40, 701 (1964).
[36] A.J. Yencha and K.T. Wu, Chem. Phys. 32, 247 (1978).
[37] T.R. Govers, M. Gérard, and R. Marx, Chem. Phys. 15, 185 (1976).

MULTI-CHARGED ION/SLOW ELECTRON COLLISIONS IN COLD PLASMAS

L. Shmaenok

A.F. Ioffe Physico-Technical Institute, Leningrad, USSR

Experiments on electron/ion collisions in overcooled laser produced plasmas of Li and Be, expanding far from targets ($R \sim 1...10$ cm) are reviewed. Spectroscopic techniques including detection of VUV lines and laser excitation were used for an observation of three-particle recombination rate variation effects as well as for a determination of collisional transition rates.

INTRODUCTION

This report contains a review of experiments on interaction between electrons and ions being mainly in excited states in expanding laser produced plasmas. The experimental approach has appeared as a result of the previous spectroscopic investigation of the Be plasma at far distances ($R \gg 1$ cm) from the target [1,2]. In that works relative and absolute intensities of Be hydrogen- and helium-like ions at $R \sim 5...10$ cm were measured for the first time. The spectroscopic data combined with the evaluations of electron density were interpreted in the frames of the general description of the expanding (initially high-temperature) plasma [3,4]. The spectroscopic study of the laser produced plasma within the spatial zone $R \lesssim 1$ cm [5] as well as the mass-spectrometric measurements of the emitted ions parameters at the most far distances ($R \gtrsim 1$ m) [6] and the recombination processes numerical modelling [7,8] were taken into consideration. The main conclusions about expanding laser produced plasma properties in the spatial zone $R \sim 1...10$ cm, based on the experiments [1,2] are compatible with the total scope of data cited in [3-8]. These conclusions, which are important for the present topic, are following.

The plasma expanding in vacuum is caracterized at $R \sim 1$ cm by an electron temperature of several electronvolts, T_e falls down with distance rapidly to $\sim 0,1$ eV at $R \sim 5$ cm. The corresponding electron and ion densities are $\bar{N}_e = \bar{z}\bar{N}_i \sim 10^{16} - 10^{12}$ cm^{-3}. The average ion charge remains rather high ($\bar{z}_i > 2$) for the Be plasma and the concentrations of H- and He-like ions are comparable at such a low temperature. The expanding plasma (at least its leading part, containing multi-charged ions) is fully ionized thanks to a considerable difference in velocities of multicharged ions ($\sim 10^7$ cm s^{-1}) and neutrals ($\lesssim 10^6$ cm s^{-1}). The three-particle diffusive recombination of electrons and ions via highly excited states is the dominating process of collisions. As a result of this process excited states of ions are populated noticeably ($N^* \sim 10^8 - 10^9$ cm^{-3} at $R \sim 5$ cm). The three-particle recombination determines radiation spectrum features of an expanding laser produced plasma and the variation of ions concentrations as a function of R. All the mentioned parameters refer to the experimental conditions, which are ordinary for laboratory laser installations and were realized in the works [1,2,9,10,12]: neodimium (1,06 m) laser pulse energy 10 J, pulse duration 10-30 ns, focal power density $10^{12}-10^{13}$ W cm^{-3}.

SUPPRESSION OF THREE-PARTICLE RECOMBINATION IN INTERPENETRATING PLASMOIDS

An impressive evidence of the three-particle recombination dominating role has been obtained in the experiment with two crossing interpenetrating plasmas [9]. Initially the authors pursued in this work the goal to test an idea of some contribution of ion/ion collisions into the expanding plasma radiation. But the result appeared to be rather unexpected and exotic. In this experiment two laser beams were focussed onto the mutually perpendicular surfaces of two Be plates (see fig. 1,a) or onto a single plain Be target (fig. 1,b), the distance between the focal points in this case being d = 2...20 mm.

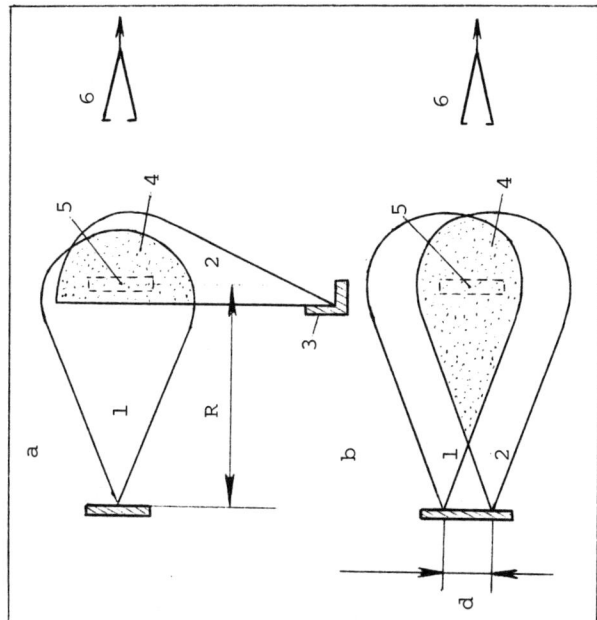

Figure 1. Crossing interpenetrating plasmoids. 1, 2 - expanding plasmas (plasmoids), 3 - collimator, 4 - interpenetration spatial region, 5 - spatial volume of spectroscopic observation, 6 - collector of ions.

The BeIII (2p-3d, 67,5 nm) and BeIV (2-3, 41,0 nm) lines emitted from the spatial volume at R = 6 cm were detected. The intensities of these lines were used as a measure of the three-particle recombination rates. The average charge of ions at R = 10 cm, $\bar{z}$, and its variation with experimental conditions $\bar{z}/\bar{z}_o$, where $\bar{z}$ - average ion charge of a single plasmoid, were measured by an ion collector technique [1].

In the case presented in fig. 1,a an interaction of the plasmas took place within a quite well localized region just at the distance R = 6 cm. It was found, that the radiation intensity (i.e. the recombination process rate) decreases by 15-20 times as compared to the values for two plasmas expanding independently. The spatial interaction region was increased considerably and drawn nearer to the target in the case 1,b. In the experiment with this geometry two effects were observed: an analogous decrease of the recombination process intensity and a noticeable increase of the average ion charge, i.e. the multi-charge ions yield. It is reasonable to compare the recombination intensity I_Σ ([1] - $cm^{-3}c^{-1}$) in the interpenetrating plasmas with the sum $(I_1 + I_2)$ for two plasmas expanding independently. Fig. 2 displays the dependences on the distance d between the focal points of the Be^{4+} and Be^{3+} ions relative recombination intensity $I_\Sigma/(I_1 + I_2)$ and of the relative average charge $\bar{z}/\bar{z}_o$.

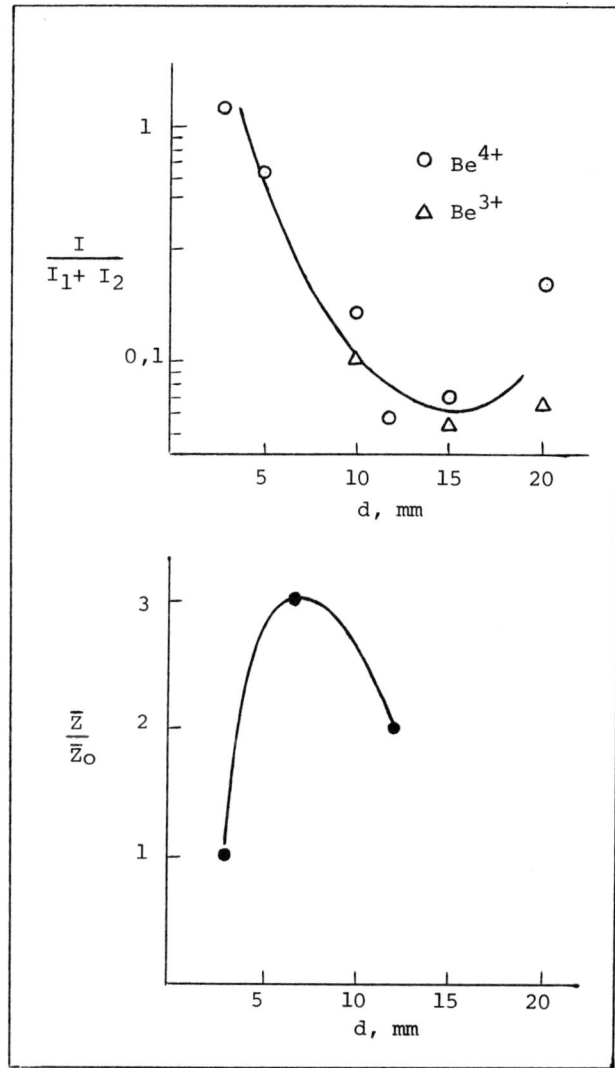

Figure 2. Recombination intensity (a) and average charge of ions (b) for different distances d between focal points R = 6 cm (a), 10 cm (b).

It is seen, that at some values of d the maximum suppression of recombination and the maximum enhancement of $\bar{z}$ are reached.

The effects discovered experimentally were explained by an increase of plasma electron temperature caused by a mutual friction of the plasmoids. Owing to the fact that T_e is less than a few eV at $R \geqslant 1$ cm a very small part of the ions kinetic energy ($E_k \sim 1$ keV [6]), transferred to the electron thermal energy via electron/ion collisions can be sufficient for a drastic fall of the three-particle recombination rate $\mathscr{æ}$, depending on T_e as $\mathscr{æ} \sim T_e^{-4.5}$. This qualitative suggestion has been confirmed by the numerical modelling [9], performed on the base of the laser produced plasma hydrodinamical description developed previously [8]. Leaving apart all the processes, taken into account in this model for the case of a single expanding plasma, we draw attention only to the term

$$Q_{ei} = N_e \nu_{ei} m_e u^2,$$

describing the process of heating electrons via collisions with ions of the other plasmoid, where ν_{ei} - electron/ion collisions frequency, u - interacting plasmoids relative velocity. The calculation show, that when the separation d between the focal points is not very small, d > 1 cm, the additional thermal energy is, as it was supposed, much less than the ions kinetic energy. The friction heating electrons at the distances $R \geqslant 1$ cm (fig. 1,b), according to the numerical model, leads to an increase of T_e from 1 eV to 15-20 eV, so the following expansion of the plasmas goes also at higher temperatures, preventing multi-charged ions (the initial average ion charge just after the laser pulse action is almost 4) from recombination. In the case 1, a considerable losses of multi-charge ions due to the recombination processes remain within the spatial region R < 6 cm, so no enhancement of the ion yield is observed. But an analogous friction heating within the volume of spectroscopic observation at R = 6 cm increases T_e by 1,5-2 times that diminishes the recombination rate approximately by 10 times.

It is not in place here to discuss practical applications of these results relating to possibilities of obtaining high concentrations of multi-charged ions from laser produced plasma sources or suppressing the recombination radiation of an expanding plasma for specific spectroscopic purposes. The experiment on interpenetrating expanding plasmas is probably the unique case of effecting and controlling the three particle recombination rate using a negligible portion of the plasmoids kinetic energy and ion/electron collisions.

DETERMINATION OF DEEXCITATION RATE USING SELECTIVE LASER PUMPING

The problem of direct measurements of collisional transition rates, especially for highly excited states of ions, remains to be as interesting as difficult. The technique of selective laser excitation, applied in many experiments with plasmas of different types, enabled to reach qualitatevely new results. This technique itself has been developed in details and now a success depends decisively on a good choice of plasma medium. The expanding laser produced plasma in the spatial region R = = 1...10 cm is attractive due to several pecularities. First of all the plasma is practically fully ionized, so it is possible not to bother about a contribution of undesirable collisions with neutrals when measuring electron collision transition rates. But at the same time obvious difficulties, connected with spatial nonuniformity and short lifetime of the plasma should be avoided.

A combination of the laser excitation technique with the spectroscopic approach to the expanding laser produced plasma investigation appeared in the work [10]. In this work the collisional deexcitation rate for the 2^3P state of He-like Li ions was measured. The geometry of the experiment is displaid in fig. 3.

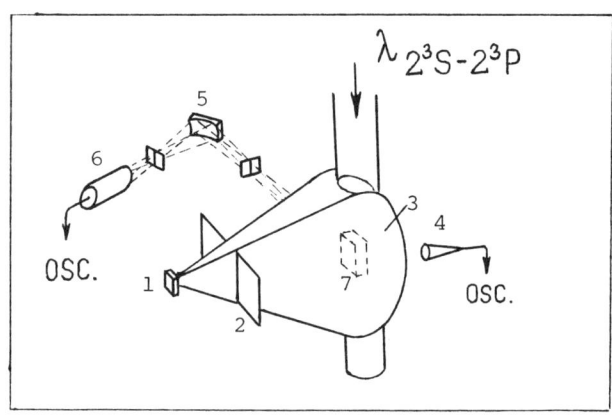

Figure 3. Geometry of experiment with laser excitation. 1 - Li target, 2 - collimators, 3 - plasmoid, 4 - ion collector, 5 - monochromator, 6 - photomultiplier, 7 - volume of observation

An expanding Li plasma was intersected at the distance R = 3 cm by a pulsed dye laser beam tuned to the LiII 2^3S-2^3P transition (λ = 548,5 nm, $\Delta\lambda$ = = 0,25 nm, Δt = 20 ns, P $\lesssim$ 10 kW cm^{-2}). The resonance fluorescence at the inverse transition and the plasma recombination radiation at this wavelength were detected using a Seya-Namioka monochromator supplied with a fast photomultiplier (time resolution 2 ns). It should be pointed out, that the width of the spatial region, from which the radiation was collected ($<$ 3 mm), was about ten times smaller than the dimensions of the plasmoid. So the plasma within that sensitive volume could be considered as being quite uniform.

The plasma electron density was measured as previously [1] using the ion collector technique. It is this element of the total experimental technology, which caused most unproper errors of the final physical results. The ion collector was mounted far from the sensitive spatial volume (ΔR = 7 cm) not to produce any disturbances in the plasma. So it was necessary to take into account the law of variation of N_e as a function of R.

The nonuniformity of the whole plasmoid causing some difficulties mentioned above was used at the same time successfully for observations of the resonant fluorescence from regions with different electron densities. The scheme of synchronization was adjusted for a variation of the pumping dye laser pulse delay relative to the initial moment of plasma expansion. The interval of obtained N_e values was $10^{13} - 10^{14}$ cm^{-3}.

Fig.5 displais the oscillograms of signals from the spectroscopic channel exit (a,b) and from the ion collector (c). In fig. 5,a a multiply increase of the plasma radiation intensity is seen, caused by the laser pumping of the upper level 2^3P. The oscillograms 5,b obtained at a more fast sweep demonstrate a variation of the 2^3P level relaxation time: from τ_γ = 44 ns at a small plasma density $N_e \approx 10^{13}$ cm^{-3} (pure radiative decay) to $\tau_{\gamma,e}$ = 30 ns in the plasma with $N_e \approx 10^{14}$ cm^{-3} (radiative and collisional decay). The ion collector signal (5,c) gives an information about the electron density at the collector location point (R = 10 cm) at different moments of expansion.

The 2^3P state collisional deexcitation rate was obtained from the expression, containing the values $\tau_{\gamma,e}$ and τ_γ, the latter being a well known constant,

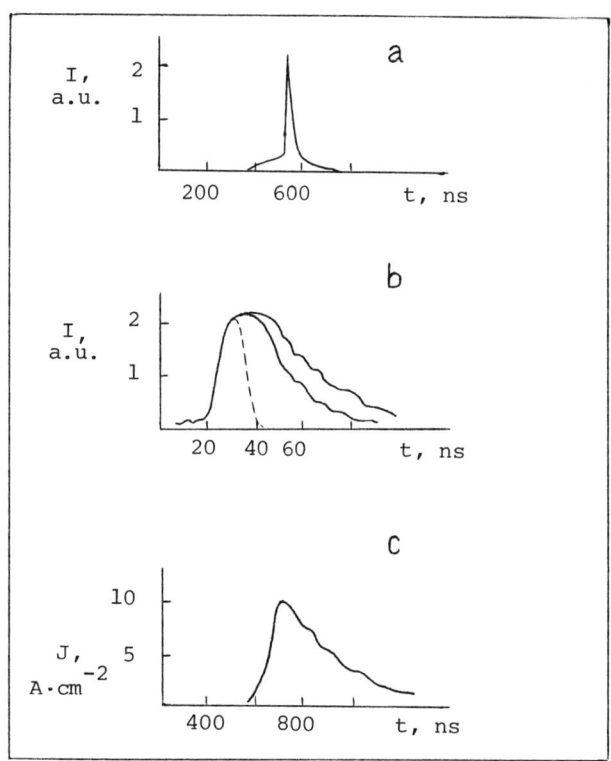

Figure 5. Typical oscillograms.
a,b - signals of spectroscopic channel detector (dashed line - pumping laser pulse), c - ion collector current.

$$\langle \sigma v \rangle = A_e/N_e = 1/N_e(1/\tau_{\gamma,e} - 1/\tau_\gamma)$$

It should be emphasised, that $\langle \sigma v \rangle$ was derived from the oscillograms analogous to that shown in fig. 5 in a single plasma flash.

The electron temperature T_e was found as previously [1] on the base of Gurevitch-Pitaevsky formula connecting intensity of the three-particle recombination process with T_e. The recombination intensity (number of recombination processes within 1 cm^3 at 1 s) was determined by measuring absolute intensities of lines, corresponding to low steps of the radiative cascade in Li. The plasma electron temperature appeared to be $\sim$ 0,2 eV. Thus the parameter $\beta = \Delta E/T_e$ which is considered in many theoretical works was $\beta \approx 10$ in the discussed case.

The value of $\langle \sigma v \rangle$ obtained in the work [10] for the Li$^+$ 2^3P state collisional relaxation rate is $\langle \sigma v \rangle$ = (5,0 $\pm$ 2,5) 10^{-7}cm^{3}s^{-1}. In spite of such a low precision of this result (caused by bad measurements of electron density) it is relevant to compare it with an attainable theoretical value. Such a value of $\langle \sigma v \rangle$, obtained in the frames of the Born-Coulomb approximation [11], is $3 \cdot 10^{-7} cm^3 s^{-1}$.

Completing the consideration of this experimental approach it is necessary to make a remark. To our opinion the method is quite acceptable for measuring deexcitation rates as it is possible to overcome difficulties connected with nonuniformity and nonstationarity of the expanding plasma. But it is also obvious that the ion collector technique applied to electron density measurements should be replaced by some reliable technique. Evaluations show an applicability of the Thomson scattering diagnostic to the expanding laser produced plasma with $T_e > 1$ eV and $N_e \sim 10^{15} - 10^{16} cm^{-3}$.

POPULATION DISTRIBUTION OF STATES IN H-LIKE Li IONS

An alternative approach to the investigation of collisions between excited ions and slow electrons, suggested first long ago [13] also has been used in the work on expanding laser produced plasma [12]. The essence of this approach is to measure relative populations of many levels depending considerably on the collisional transition probabilities and to compare them with a theoretical population distribution. Thus it is possible to test calculations of transition probabilities included in the theoretical model.

In the work [12] the population distribution of levels in Li^+ ions was studied. Owing to the fact that the laser produced plasma in the far distant zone $R > 1$ cm is transparent for at least nonresonant lines the experimental part of the work was reduced to precise measurements of LiIII lines intensities in the VUV region. It was important to detect all the lines to be measured in a single plasma flash. This problem was solved using a Seya-Namioka monochromator supplied with a MCP image intensifier and a photosensitive charged-coupled array. The exit slit of the spectrometer had been removed and the position-sensitive detector front surface was placed at the thoroidal grating focal point. The whole system enabled to obtain in one plasma flash a spectrum within the interval 20 nm containing the LiIII lines of the transitions (2 - n), n = 4...10.

The instrumental parameters of the spectral channel - wavelength dependances of grating reflectivity and detector sensitivity - were measured carefully. It is worth to dwell on the detector calibration. The spectral sensitivity of the detector has been found using a photoionization technique. In a separate experiment a low-pressure (10^{-4} Torr) photoionization chamber filled with Ar was mounted between the exit slit of the VUV monochromator and the detector. A laser produced plasma on a heavy (Ta) target served as a continuum source of radiation. So it was possible to compare the detector exit signal with the yield of Ar^+ ions at all the spectral points of interest. As the photoionization cross-sections for Ar are known precisely, the errors of the detector calibration did not exceed a few percents.

Relative populations of levels were obtained from the measurements of lines intensities on the assumption that the levels with different l but the same n are populated proportionally to statistical weights. These results are shown in fig. 6. Unfortunately the authors did not manage to measure correctly the relative population of the state with n = 3, which appeared to be very important for a comparison with the theoretical model.

All the spectral measurements were done for the spatial region at R = 2 cm.

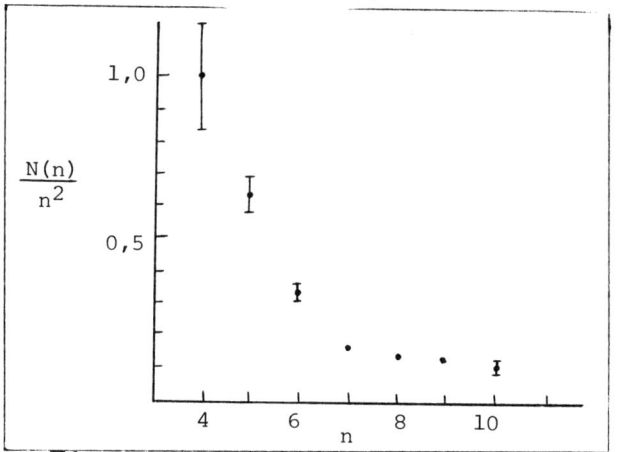

Figure 6. Obtained experimental values of Li^+ levels relative populations

An interpolation of data taken from different works [1, 5, 9] and available results of numerical calculations [7, 8] enabled to evaluate in advance the value of T_e, corresponding to the obtained population distribution, as $T_e \sim 1$ eV - essentially higher than in the work [10].

This temperature corresponds to $N_e \sim 5 \cdot 10^{15}$ cm^{-3}, derived as previously from ion collector measurements. The evaluation of $T_e \sim 1$eV appeared to be in agreement with the population distribution for the highest levels, presented in fig. 6. For the Li^{2+} ions at $T_e \sim 1$eV, $N_e \sim 10^{15}$ cm^{-3} populations of the states with n>8, according an analysis, should be determined only by a balance of the ionization and the three-particle recombination processes, i.e., should be in the equilibrium with the continuum. The relative populations of the n=8,9,10 states correspond just to the temperature ~1eV. The population distribution for the lowest levels (n=4,5) also differs considerably from that for the radiative cascade, so an application of the "I_r vs κ_r vs $T_e^{-4,5}$" dependence to determination of T_e (as in the works [1, 9, 10]) was not justified in this case.

The population distribution for the levels with n > 2 was calculated by numerically solving a system of balance equations

$$\frac{d}{dt} N_z(n) = -N_z(n)\, \Gamma_n + \sum_{m \neq n} N_z(m)\, \Gamma_{mn} + N_{z+1}\, R_n \quad (1)$$

where $N_z(n)$ = the population of the state n of the z-charged ions; Γ_n = the total decay probability for the state n, including the radiation decay probability A_n and the probabilities of the following processes: transitions caused by collisions with electrons $w_e(n) = \sum_{n' \neq n} w_{nn'}$ and ions $w_i(n)$ (the latter appeared to be negligible), ionization and recombination; Γ_{mn} = the probability of a transition from m to n; R_n - the probability of recombination to the state n of the z-charged ion.

For n < 8 the ionization probability is very small and can be excluded from the consideration. Thus for the levels n=2...8 only the collisions with electrons were taken into account. Finally owing to the fact that any excited state relaxes much faster than the plasma parameters N_e, T_e vary significantly, one can suppose that $\frac{dN(n)}{dt}=0$ for n>2. As a result the system (1) was reduced to a system of linear algebraic equations.

The transition cross-sections σ determining the transition probabilities $w_{nn'} = \langle \sigma v \rangle N_e$ were calculated in [12] in the Born-Coulomb approximation (formulas from [14] were used) as well as using a relatively new adiabatic model. This model originated from the novel approach to the problem of an electron in the field of two positively charged Coulomb centers [15], based on an investigation of analytical properties of potential curves in the complex plane of internuclear distances R. The theory, developed in [15] has been modified and suited to the case of a free slow electron with energy E, interacting with a bound excited one and resulted in the following formulas:

$$\sigma_{nn'} = \frac{1}{n^2} \sum_{\ell,m} \sigma_{n\ell m \to n'\ell'n'}$$

$$\sigma_{n\ell m \to n'\ell'm'} = \pi \rho_c^2$$

$$\exp\left(\frac{-4\Delta E_{1,2}\, J_m\, R_c}{v_1 + v_2} \right)$$

where $\rho_c = R_c^2 \left(1 + \frac{z}{R_c E} - \frac{E_{1,2}}{E}\right)$

$$v_{1,2} = v_0 \sqrt{1 - \frac{\rho^2}{R^2} + \frac{z}{R_c E} - \frac{E_{1,2}}{E}},$$

$$v_0 = \sqrt{\frac{2E}{m}}$$

$$R_c = \left[(\ell+\tfrac{1}{2})^2 + \frac{(m+1)^2}{2} + i(m+1)\right]$$

$$\cdot \sqrt{2(\ell+\tfrac{1}{2})^2 - \frac{(m+1)^2}{4}}$$

$$\cdot \frac{1}{(z+1)}.$$

The results of the population distribution calculations for $T_e = 1$eV and two N_e values: $N_e = 0,5 \cdot 10^{16}$ cm^{-3} $N_e = 1 \cdot 10^{16}$ cm^{-3} are shown in Fig. 7. It is seen that the maxima in the curves obtained with the Born-Coulomb transition rates are shifted towards larger n. It is connected with that this approximation results in larger values of transition rates. But the reached accuracy of the N_e measurements is not sufficient to discuss an applicability of both approaches: the experimental distribution appears to be in agreement with one of the theoretical curves if N_e varies by the factor 2. Future measurements of N_e as well as of the $N(3)/N(4)$ relation will help to complete this consideration.

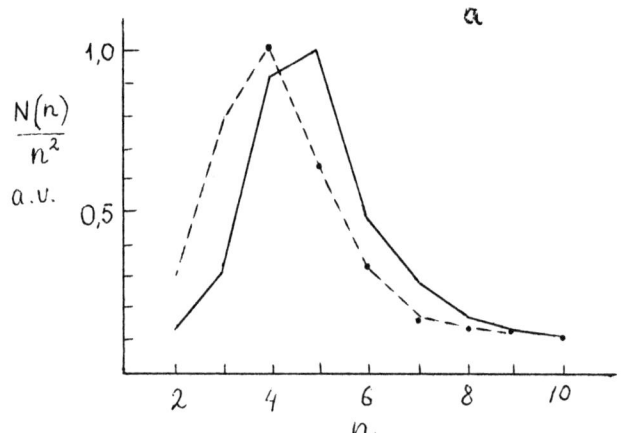

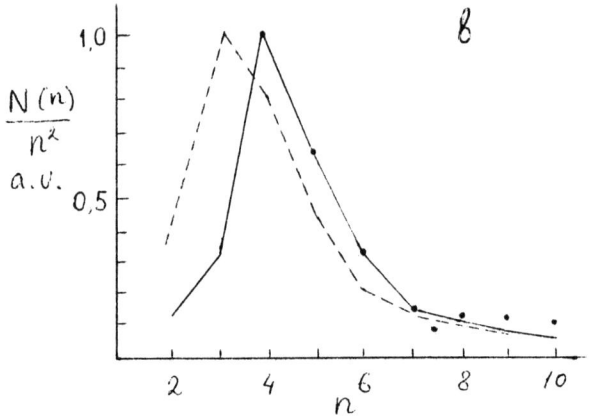

Fig. 7. Population distribution for Li^{2+} states at $T_e = 1eV$ and $N_e = 5\cdot10^{15}$ cm^{-3} (a), 1.10^{16} cm^{-3} (b); solid line - with the collision transition rates, obtained in the Born-Coulomb pproximation, dashed line - with the rates obtained within the adiabatic model; · - experimental results.

CONCLUSIONS

Spectroscopic experiments on the expanding laser produced plasma have brought much evidence of a dominating role of the three-particle recombination in the spatial region $R \gtrsim 1$cm, where $T_e \lesssim 1eV$, $N_e \lesssim 10^{16}$ cm^{-3}. It is possible to have an effect on and to control to some extent the recombination rate by mixing two plasmas and achieving via electron-ion collisions a transfer of a negligible part of the ion's kinetic energy to the electron's thermal energy.

Laser produced plasma is available for direct and indirect determination of collisional transition rates in the low energy region, where $\beta = \Delta E/T_e >> 1$. The nonuniformity and the instability of the laser produced plasma can be avoided in experiments and spectroscopic data are reliable. But local independent measurements of N_e are urgently needed, especially for testing the adiabatic model of electron/ion collisions.

ACKNOWLEDGMENTS

The author is thankful to Professor S. V. Bobashev, Drs. D. A. Simanovskii, S. Yu. Ovtchinnikov, S. V. Latyshev and to D. A. Mosesyan for many discussions.

REFERENCES

1. V. V. Afrosimov, S. V. Bobashev, A. V. Golubev, D. M. Simanovskii, L. A. Shmaenok, Sov. J. Phys. JETP 64, 284 (1986).

2. S. V. Bobashev, L. A. Shmaenok, in "Electronic and Atomic Collisions (ed. D. C. Lorents, W. E. Meyerhof, J. R. Peterson) (ICPEAC XIV Invited Talks), 1986, p. 479.

3. L. I. Gudzenko, S. I. Yakovlenko, Plasma Lasers (in Russian) Moscow, 1978, ch. 3.

4. Ya. B. Zel'dovitch, Yu. P. Raizer, Physics of Shock waves and High-Temperature Hydrodynamic Phenomena, ed. by W.D. Hayes, R. F. Probestein, Academic, New York, 1967.

5. V. A. Boiko, F. V Bunkin, V. K. Derzhiev, S. I. Yakovlenko, Isv. ANSSR, Ser. Fiz, 47, 10 (1983).

6. Yu. A. Bykovskii et al. Zh Eksp. Teor. Fiz. 60, 1306.

7. S. V. Latyshev, Preprint ITEP N 66, (1983).

8. S. V. Latyshev, I. V. Rudskoi, Sov. J. Plasma Phys. 11, 669 (1985).

9. S. V. Bobashev, S. V. Latyshev, I.V. Rudskoi, D. M. Simanovskii, L. A. Shmaenok, Sov. J. Plasma Phys. 13, 801 (1987).

10. S. V. Bobashev, D. M. Simanovskii, L. A. Shmaenok, Sov. Tech. Phys. Lett. 13, 249 (1987).

11. L. A. Vainshtein, I. I. Sobel'man, E. A. Yukov, Excitation of atoms and spectral line broadening (in Russian). Moscow, 1979.

12. S. V. Bobashev, S. Yu. Ovtchinnikov, D. M. Simanovskii, L. A. Shmaenck, X-Ray and Vacuum Ultraviolet Interaction Data Bases, Calculations and Measurements. Nancy Kerr del Grande, Ping Lee, J.A.R. Samson, David Y. Smith, Editors, Proc. SPIE 911, p. 67.

13. L. C. Johnson, E. Hinnov, Phys. Rev. 167, 143 (1969).

14. I. L. Beigman, A. M. Urnov, J. Quant. Spectr. Rad. Transf. 14, 1009 (1974).

15. S. Yu. Ovtchinnikov, E. A. Solovev. Sov. J. Phys. JETP 64, 280 (1986).

SPECTRAL DISTRIBUTION OF ELECTRONS EMITTED INTO THE CONTINUUM OF FAST PROJECTILES: THEORETICAL APPROACHES OF HIGHER ORDER IN COMPARISON WITH EXPERIMENT

D.H.Jakubaßa-Amundsen

Physics Section, University of Munich, 8046 Garching, Germany

A compilation of recent experimental data on the forward peak resulting from the capture of target electrons into the continuum of bare, partly stripped and neutral projectiles is presented. The impact-parameter dependence and the dependence of the peak shape on the projectile charge state as well as on the angular acceptance is considered. An interpretation is attempted within the second-order Born theory and the impulse approximation. Results from Monte Carlo calculations at lower impact energies are also included.

1. Introduction

Since its discovery, the forward peak (cusp) in the secondary electron spectrum has attracted great interest both experimentally and theoretically.[1] It consists of target or projectile electrons which are emitted into low-lying continuum states of the projectile, and hence appears in the laboratory frame at forward electron angles $\vartheta_f \approx 0$ and comprises electron momenta k_f in the vicinity of the collision velocity v. Recent coincidence experiments[2] have offered the possibility to separate the contributions from target electrons (CTC) and from projectile electrons (ELC). Although a very recent compilation of ELC data[3] calls for an improvement of the theoretical approaches beyond the customary first-order Born theory, I shall restrict myself to the CTC process since it is much more sensitive to higher-order effects than the electron loss process. Starting with the derivation of the second Born theory and the impulse approximation (IA) for structured projectiles (section 2), I shall consider CTC by bare projectiles and show how the dependence of the peak shape on the angular acceptance is related to the nonanalytic behaviour of the doubly differential CTC cross section at $\vec{k}_f = \vec{v}$ (section 3). It is further shown how the variation of the peak shape with projectile charge Z_P depends on the collision velocity. The influence of the collision dynamics on the peak formation is displayed with the help of a classical trajectory Monte Carlo (CTMC) calculation[4], and the impact-parameter dependence of the forward electrons is studied within the impulse approximation. In section 4, I consider electron capture by partly stripped projectiles and show that a description of the projectile in terms of a pointlike ionic charge is incorrect. The last section is devoted to CTC by neutral projectiles where the observation of a cusp-like peak[2] is a great challenge to theory.

2. Theory

For a theoretical description of charge transfer within a perturbative approach, one has to restrict oneself to energetic collisions where the velocity v exceeds the shell velocity of the electron in its initial target bound state, or where the ratio between target charge Z_T and projectile charge Z_P deviates largely from unity. In the following I shall treat the target as a quasi one-electron system, but will take the projectile electrons (as far as they exist) explicitly into account. The neglect of multiple target excitations is especially justified for the case $Z_P/Z_T \ll 1$.

In the semiclassical picture where the internuclear motion is described by a classical trajectory, the capture amplitude is given by (in atomic units $\hbar = e = m = 1$)

$$a_{fi} = -i \int dt \, \langle \psi_f^{(-)} | V_i | \varphi_i^T \phi_i^P \rangle \tag{2.1}$$

where φ_i^T is the wavefunction of the bound target electron and ϕ_i^P describes the electronic ground state of the projectile. The initial perturbation V_i is composed of the interaction V_{PT}^e of the projectile electrons with the target electron, the interaction V_{PT}^N of the projectile electrons with the target nucleus, and the interaction V_P between the projectile nucleus and the target electron. In the (prior) impulse approximation, valid for $Z_P \ll Z_T$, the exact scattering function $\psi_f^{(-)}$ is approximated by

$$|\psi_f^{(-)}\rangle = (1 + G^{(-)} V_f)|f\rangle \approx (1 + G_{IA}^{(-)} V_f)|f\rangle$$
$$|f\rangle = \sum_\lambda b_{f\lambda} |\psi_\lambda^{P+1}\rangle = \sum_{\lambda m} b_{f\lambda} |\tilde{\varphi}_{\lambda m}^P \phi_m^P\rangle \tag{2.2}$$

where $V_f = V_T$, the potential between the target electron and the target nucleus, and consequently $|f\rangle$ eigenstate to $H_f = H - V_T$ where H is the full electronic Hamiltonian ($H = H_P + T_e + V_P + V_T + V_{PT}^e + V_{PT}^N$; H_P describes the projectile electrons, T_e the one-electron kinetic energy). The state $|f\rangle$ can be represented in terms of eigenstates ψ_λ^{P+1} to the projectile plus one electron (in absence of the target nucleus) which again can be written as a superposition of electronic states ϕ_m^P of the projectile alone, where $\tilde{\varphi}_{\lambda m}^P$ is a one-electron scattering state. The full Greens function $G^{(-)} = (i\partial_t - H - i\varepsilon)^{-1}$ of the exact formulation is in the IA replaced by $G_{IA}^{(-)}$ corresponding to $H_{IA} = H_P + T_e + V_T + V_{PT}^N$. The second-order Born approximation would result if in $G^{(-)}$, H were replaced by $H_P + T_e$.

Introducing a complete set of eigenstates $|\phi_n^{PT} \vec{q}\rangle$ where $|\vec{q}\rangle$ is a one-electron plane wave of momentum $\vec{q}$ and ϕ_n^{PT} eigenstate to $H_{PT} = H_P + V_{PT}^N$, and going on-shell, one obtains

$$a_{fi}^{IA} = -i \int dt \int d\vec{q} \sum_n \langle f | \phi_n^{PT} \vec{q} \rangle \langle \phi_n^{PT} \varphi_{\vec{q}}^T | V_i | \varphi_i^T \phi_i^P \rangle \tag{2.3}$$

where $\varphi_{\vec{q}}^T$ is an electronic continuum target state. Expanding $\phi_n^{PT} = \sum_\mu a_{n\mu} \phi_\mu^P$ and making use of the orthogonality properties of the eigenstates, one arrives at

$$a_{fi}^{IA} = -i \int dt \int d\vec{q} \sum_{n\lambda} b_{f\lambda}^* \sum_m \langle \tilde{\varphi}_{\lambda m}^P | \vec{q} \rangle \, a_{nm}$$
$$\times \left[a_{ni}^* \langle \varphi_{\vec{q}}^T | V_P | \varphi_i^T \rangle + \sum_\mu a_{n\mu}^* \langle \phi_\mu^P \varphi_{\vec{q}}^T | V_{PT}^e | \varphi_i^T \phi_i^P \rangle \right] \tag{2.4}$$

The expansion coefficients a_{nk} and $b_{f\lambda}$ describe projectile excitations due to the interaction V_{PT}^N with the target nucleus:

$$a_{nk} = \langle \phi_k^P | (1 + G_{PT} V_{PT}^N) | \phi_n^P \rangle$$
$$b_{f\lambda} = \langle \psi_\lambda^{P+1} | (1 + G_f V_{PT}^N) | \psi_f^{P+1} \rangle \tag{2.5}$$

where G_{PT} corresponds to H_{PT} and G_f corresponds to H_f.

If excitation of the projectile is neglected ($a_{nk} = b_{nk} = \delta_{nk}$) the conventional form is obtained

$$a_{fi}^{IA} \approx -i \int dt \int d\vec{q} \, \langle \tilde{\varphi}_{fi}^P | \vec{q} \rangle$$
$$\times \langle \varphi_{\vec{q}}^T | (V_P + \langle \phi_i^P | V_{PT}^e | \phi_i^P \rangle) | \varphi_i^T \rangle \tag{2.6}$$

where $\tilde{\varphi}_{fi}^P$ is a one-electron scattering state to the projectile which is in the ground state ϕ_i^P.

In complete analogy, the second Born approximation reads

$$a_{fi}^{B2} \approx -i \int dt \langle \tilde{\varphi}_{fi}^P | (1 + V_T \int d\vec{q} | \vec{q} \rangle \frac{1}{i\partial_t - \frac{q^2}{2} + i\varepsilon} \langle \vec{q} |)$$
$$\times (V_P + \langle \phi_i^P | V_{PT}^e | \phi_i^P \rangle) | \varphi_i^T \rangle \tag{2.7}$$

In accordance with the omission of the internuclear potential in the transition matrix element, also the ground-state expectation value of V_{PT}^N (which only depends on the internuclear coordinate) has been omitted in (2.7).

The impulse approximation in its post form, suitable for $Z_P/Z_T \gg 1$, can be derived[5] in a similar way as (2.6) and contains an intermediate projectile eigenstate instead of the target state $\varphi_{\vec{q}}^T$.

3. CTC by bare projectiles

For bare projectiles, V_{PT}^e in (2.6) and (2.7) is zero and $\tilde{\varphi}_{fi}^P$ reduces to a Coulomb wave φ_f^P to the charge Z_P. In order to investigate the properties of the forward electrons it is instructive to look at the Fourier transform of φ_f^P at $\vec{\varkappa}_f = 0$ ($\vec{\varkappa}_f = \vec{k}_f - \vec{v}$ is the electron momentum in the projectile reference frame) which appears explicitly in the transition amplitude (2.6)

$$\varphi_{\vec{\varkappa}_f}^{*P}(\vec{s}) \rightarrow \frac{Z_P}{\pi^2} e^{\pi\eta_f/2} \Gamma(1 - i\eta_f) \frac{1}{s^4}$$

$$\times \exp(-2iZ_P[\cos\vartheta_s\cos\theta_f' + \sin\vartheta_s\sin\theta_f'\cos\varphi_s]/s),$$

$$\vec{\varkappa}_f \rightarrow 0 \qquad (3.1)$$

where θ_f' is the electron emission angle in the projectile frame, and $\eta_f = Z_P/\varkappa_f$.

Upon insertion into (2.6) it follows that the doubly differential cross section diverges like $\varkappa_f^{-1}$ due to the normalisation factor of the Coulomb wave. Furthermore, the dependence on θ_f' is nonanalytical because the integration variable $\vec{s} = \vec{q} - \vec{v}$ can attain the value zero[6]. Since the phase in (3.1) switches sign if one considers forward electrons with momentum below the peak ($k_f \lesssim v$, $\theta_f' = 180°$) and above the peak ($k_f \gtrsim v$, $\theta_f' = 0°$), the intensity of the emitted electrons is different on the two sides of the peak, leading to an asymmetric peak shape. This behaviour is, however, only visible in higher-order theories, because in the first-order OBK theory the phase information from (3.1) drops out when calculating $|a_{fi}|^2$. The asymmetry of the forward peak is clearly seen in the experiments as shown in Fig.1 in the case of O^{8+} + Ne collisions.[7]

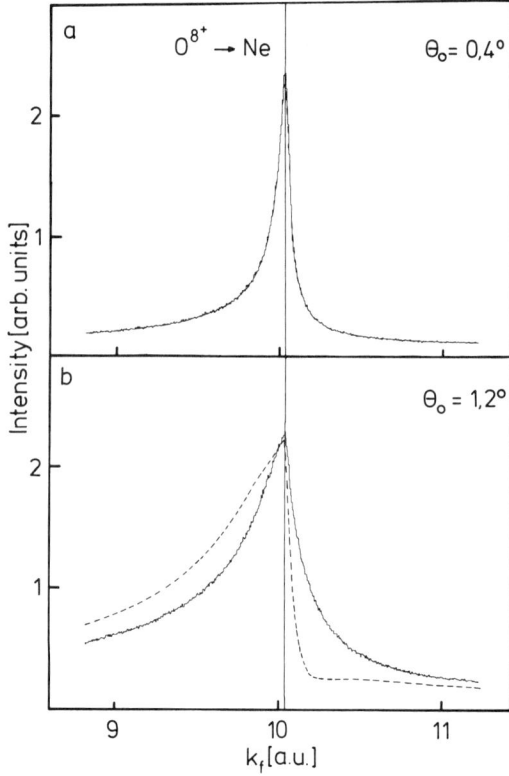

Fig.1. Cusp electron spectrum[7] from 40 MeV O^{8+} + Ne collisions at an energy resolution of 0.5% and $\Theta_o = 0.4°$ (a) and 1.2° (b). In (a), the fit by a 6-term expansion (eq. (3.2)) is indistinguishable from the data. (b) - - - - constructed "fit" with use of the same B_{lm} as for 0.4°

In order to compare theory with experiment, the differential cross section has to be averaged over the energy resolution ΔE_o and the angular resolution Θ_o of the detector. Conventionally, an expansion in terms of Legendre polynomials P_l and powers of $\varkappa_f$ is made[8]

$$\langle \frac{d^2\sigma}{dE_f d\Omega_f} \rangle_{\Theta_o, \Delta E_o} = \sum_{nl} B_{nl} \frac{1}{\Delta E_o} \int_{E_f-\Delta E_o/2}^{E_f+\Delta E_o/2} k_f \, dE_f$$

$$\frac{1}{1 - \cos\Theta_o} \int_0^{\Theta_o} \sin\vartheta_f' \, d\vartheta_f' \, \varkappa_f^{n-1} P_l(\cos\theta_f') \qquad (3.2)$$

in order to extract coefficients B_{nl} which are independent of the detector resolution. However, since the cross section is nonanalytic in θ_f' a truncation of the sum over l in (3.2) is in general not possible. From Table 1 it is seen that

θ_f'	N_1	N_2	N_3
0°	1	1	1
30°	1.02	1.01	1.05
60°	1.06	1.03	1.11
90°	1.15	1.09	1.28
120°	1.29	1.18	1.53
150°	1.42	1.27	1.79
180°	1.47	1.30	1.89

Table 1.
Cross section (in the projectile frame) for cusp electron emission at $\chi_f = 0$ within the transverse peaked prior IA. For p + He collisions with v = 4.4745, $d^2\sigma/d\varepsilon_f d\Omega_f' = N_1 \cdot 2.02 \cdot 10^5$ barn/keV·sr, with v = 6.328, $d^2\sigma/d\varepsilon_f d\Omega_f' = N_2 \cdot 6.43 \cdot 10^3$ barn/keV·sr, for He^{2+} + He collisions with v = 6.328, $d^2\sigma/d\varepsilon_f d\Omega_f' = N_3 \cdot 4.57 \cdot 10^4$ barn/keV·sr

the differential cross section in the projectile frame (which results upon multiplication by χ_f/k_f) increases weakly with Θ_f' near 0° and 180°, but rather strongly around 90°. A similar result as in the IA is also found with the second Born theory. The slope is the larger, the smaller the velocity v and the larger the projectile charge. Tentative considerations indicate that the nonanalyticity reveals itself in a singular behaviour of the higher derivatives, especially around 90°.

The failure of the conventional truncation of the series to 6 terms ($1 \leq 2$, $n \leq 1$) is readily seen if Θ_0 is varied in the experiment. In Fig.1a where $\Theta_0 = 0.4°$, the B_{1m} are fitted to the experimental spectrum. If these B_{1m} are used to construct the spectrum for e.g. $\Theta_0 = 1.2°$ (Fig. 1b) a clear discrepancy with the experimental data is found.

As a measure of the peak asymmetry, the half width at half maximum to the left (Γ_L) and to the right (Γ_R) of the line $k_f = v$ can be used. In Fig.2 the ratio Γ_L/Γ_R following from experiment and from the second Born theory is shown. The experimental decrease of the peak asymmetry for $\Theta_0 \to 0$ (which is another argument against a truncated series expansion) is qualitatively reproduced by theory. Note that although eq. (2.7) has

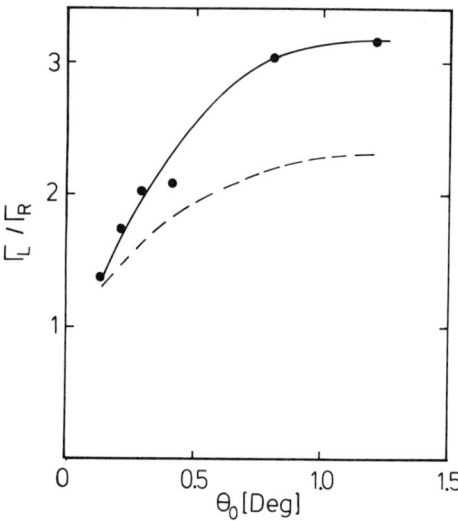

Fig.2. Ratio Γ_L/Γ_R for 40 MeV O^{8+} + Ne as a function of detector resolution[7]. Solid line, eye-guide to the data, dashed line, 2.Born

been evaluated without resorting to the asymptotic approximation of Shakeshaft[9], the second Born theory is not expected to give quantitative agreement with the data for systems with $Z_P \approx Z_T \approx v$.

The shape of the forward peak depends strongly on the system parameters like projectile charge and collision velocity. Fig.3 displays the cross section ratio for He^{2+} impact to proton impact on He at a rather low collision velocity of 2 au, and also the cross section ratio for O^{8+} impact to hydrogen impact on Ne at v = 10 au. The low-velocity data, which are well reproduced by a Monte Carlo calculation[10], scale approximately with Z_P^2 at small momenta of the ejected electron, but show an increase when $k_f > v$. This increase reflects a <u>decreasing</u> peak asymmetry with Z_P. On the other hand, the high-velocity data scale approximately with $Z_P^{2.3}$ but decrease when k_f increases beyond v. This behaviour is qualitatively explained by the (post) impulse approximation[11] (for a He target in order to obey the IA validity criterions) which scales like Z_P^3 and which leads to an asymmetry which <u>increases</u> according to Z_P/v. A similar difference in the v dependence of the asymmetry (an increase with v at small v but a

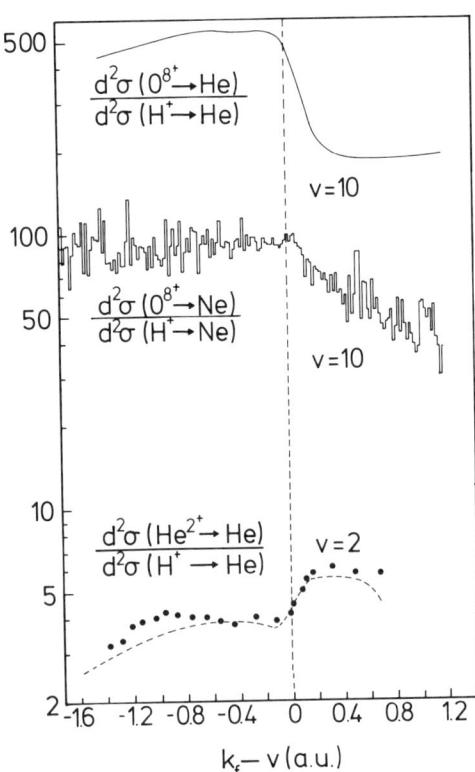

Fig.3.
(Top) Ratio of the doubly differential cross section for 2.5 MeV/N O^{8+} + Ne to H^+ + Ne collisions (exp: histogramm; $\vartheta_o \sim 0.8°$, $\Delta E_o/E_f = 0.5\%$) as a function of electron momentum relative to v; solid line, cross section ratio for 2.5 MeV/N O^{8+} + He to H^+ + He at $\vartheta_f = 1°$ within the post IA. (Bottom) Cross section ratio for 100 keV/N He^{2+} + He to H^+ + He. Data points, Bernardi et al, dashed line, CTMC calculation at $\vartheta_f = 5°$ (taken from Ref. 10)

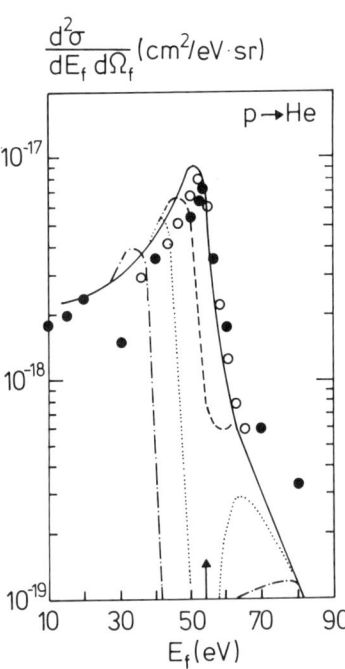

Fig.4.
Forward peak and its formation as a function of the internuclear distance R_o (where the CTMC calculation is stopped) for 100 keV p + He collisions at $\vartheta_f = 1°$. —·—·—, $R_o = 100$ au, ·······, $R_o = 500$ au, - - - -, $R_o = 3000$ au, ———, $R_o = 10^5$ au. Experiments: ●, Gibson and Reid[13], ○, Dahl[12] (taken from Ref.4)

decrease at large v) has been observed by Dahl[12].

A supplementary information about the cusp electrons can be extracted from an investigation of the collision dynamics. Especially suited is the Monte Carlo method, where the classical trajectories of the projectile can be followed. In Fig.4 is shown how the forward peak develops as the internuclear distance between the proton and the He target is increased after the encounter. Distances of the order of 10^5 au are necessary in order to give the correct energy spectrum which compares well with the experimental data of Gibson[13] and Dahl[12], pointing to strong post-collisional effects in medium-energy collisions. In quantum mechanical calculations, the information on the relevant projectile-target distances is contained in the impact parameter distribution. Fig.5 gives a comparison of the experimental data of Jagutzki et al[14] with the (prior) peaked impulse approximation[6] (which is scaled by the ratio between the unpeaked IA and the peaked IA for 1s-1s capture in the same collision at b = 0). For H^+ + Ne, theory reproduces experiment within the experimental uncertainty of the normalisation (~50%). A measurement of the b-distribution of δ-electrons emitted under zero degrees gives within the error bars the same shape as the b-distribution of the cusp electrons. This confirms the validity of the peaked IA which factorises into the ionisation probability times the squared normalisation constant of the final-state Coulomb wave.[6,15] For the more symmetric $^3He^{2+}$ + Ne collision, the IA gives poor results at an impact velocity as low as the electronic velocity of the

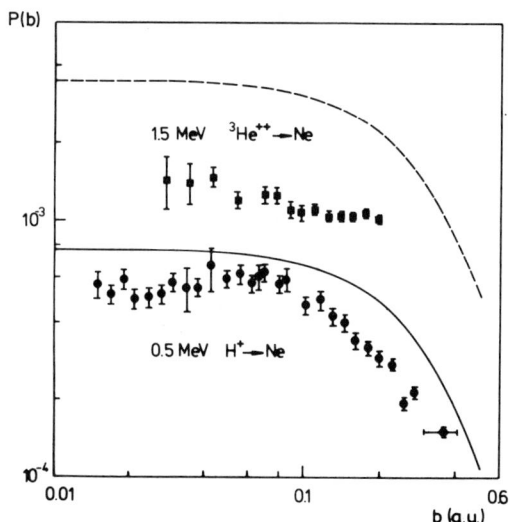

Fig.5.
Impact-parameter distribution of the cusp electrons within $\vartheta_f \leq 3°$ in 0.5 MeV/N H^+ + Ne and $^3He^{2+}$ + Ne collisions. The experimental data[14] are obtained by integrating the electron spectrum over the peak region. The calculations are performed with the fully peaked prior IA (integrated over the region $k_f = v \pm 0.1v$) and scaled down by a factor of 0.4 (solid line) and 0.23 (dashed line), respectively (see text)

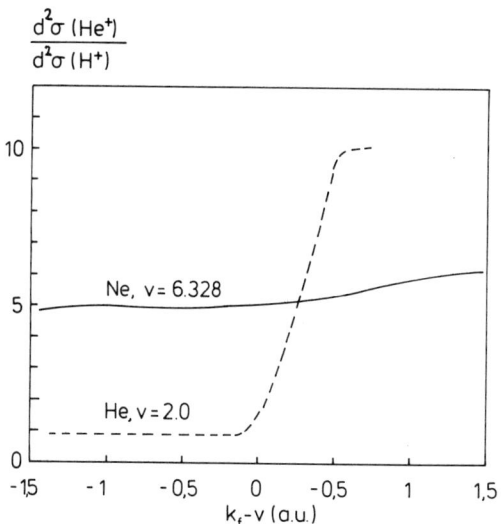

Fig.6.
Ratio of the doubly differential cross section for 100 keV/N He^+ + He to H^+ + He at $\vartheta_f = 5°$ within the Monte Carlo method[10] (dashed line) and for 1 MeV/N He^+ + Ne to H^+ + Ne for $\theta_0 = 1°$ within the fully peaked prior IA (solid line) as a function of electron momentum relative to v

target L-shell which yields the dominant contribution.

4. CTC by partly stripped projectiles

To simplify the theoretical description of the cusp electrons when the projectile carries electrons it is often assumed that the projectile acts like a pointlike particle of ionic charge. From the general theory (e.g. eq. (2.6)) it follows that the projectile field enters twofold, first directly as a transition operator in the matrix element, and second, implicitly in the final-state electronic wavefunction $\tilde{\varphi}_{fi}^P$. In Fig.6 is shown the doubly differential cross section ratio for He^+ + He relative to the H^+ + He collision system at an intermediate collision velocity of 2 au from a CTMC calculation when a static screened projectile field is used[10]. For small electron momenta, $k_f < v$, the ratio is approximately one which is conform with the picture of an ionic point charge. For $k_f > v$, however, He^+ is much more efficient in producing CTC cusp electrons than H^+. This is related to the fact that fast electrons require a large momentum transfer q, which is easier provided by the static screened He^+ potential which in the limit of $q \to \infty$ coincides with the He^{2+} field. Also shown in Fig.6 is the ratio for He^+ + Ne relative to H^+ + Ne in a fast collision (v = 6.328 au) from the (prior) peaked impulse approximation. A ratio of 4 is expected since the projectile field in the ionisation matrix element in (2.6) acts very much like a He^{2+} field for the large momenta required. The deviation of the ratio from 4 results from the consideration of the ionic field (including polarisation and exchange) in the calculation of the final-state electronic wavefunction.

5. CTC by neutral projectiles

For a short-range potential no cusp behaviour for the CTC electrons is expected as long as the projectile remains in its ground state. If the final-state electronic wavefunction is taken as a scattering eigenstate of the projectile (calcu-

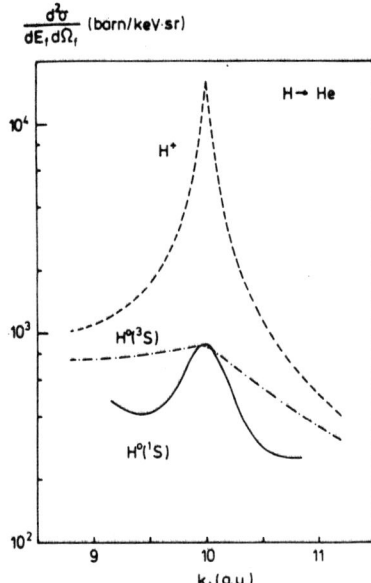

Fig.7.
Forward electron spectrum in 2.5 MeV H + He collisions for $\theta_0 = 0.5°$ within the fully peaked prior IA for different charge states of the projectile. - - - - , H^+, -.-.-.- , $H^0(^3S)$, ———, $H^0(^1S)$

lated from a Schrödinger-type equation including polarisation and exchange[16]) a forward peak is indeed obtained, although with a much larger width than in case of an ionic potential. Fig.7 displays the forward peak for neutral hydrogen projectiles in the spin singlet and spin triplet state in comparison with H^+ on a He target. The calculations are performed within the (prior) peaked impulse approximation where from the final-state wavefunction, only the $l = 0$ partial wave normalisation constant is required[17]. The existence of a peak for $\varkappa_f \to 0$ results from the attractive polarisation potential, and the enhancement of the peak for the singlet state is due to the exchange interaction.

Experiments, where the projectile charge state is measured in coincidence with the electrons in order to isolate the CTC contribution show, however, a cusp-like structure even for neutral projectiles.[2] Fig.8 displays the forward peak in 75 keV/N H^0 + Ar and He^0 + Ar collisions separa-

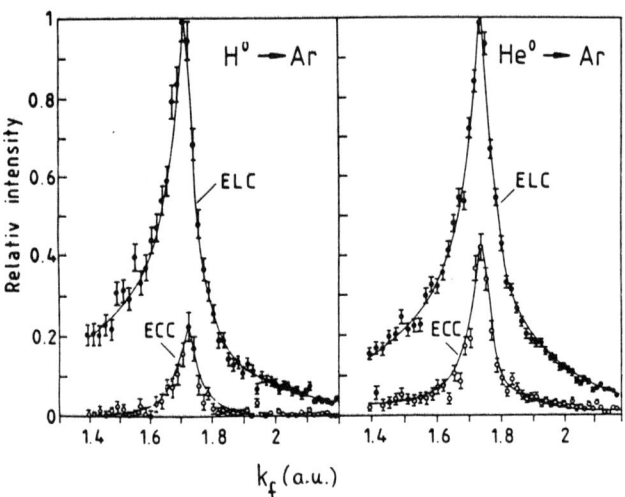

Fig.8.
Forward electron spectrum in 75 keV/N H^0 + Ar (left) and He^0 + Ar (right) collisions. Shown are the relative intensities for the electron loss contribution (ELC) and capture to continuum contribution (ECC) for $\theta_0 = 3.5°$. The solid lines are eye-guides to the data (taken from Ref.18)

ted into electron loss and electron capture contributions[18]. The fact that for the He^0 projectile, the cusp is even narrower than for a He^+ projectile has been tentatively explained in terms of a low-lying shape resonance[19]. In order to observe such a resonance which introduces a singularity into the final-state electronic wavefunction it is, however, necessary that the projectile is excited to a specific state prior to the CTC process. For He^0, the metastable 2^1S state (which may be present in the beam) can account for the occurrence of a resonance[20]. For neutral hydrogen on the other hand, it is difficult to imagine in which way a specific excited resonance-supporting projectile state can be sufficiently strong populated: From Fig.8 it follows that the peak intensity for CTC is about 20% of that for ELC. Assuming (optimistically) an equal transition probability for electron loss and electron capture by a projectile with fixed electronic configuration, the excitation probability of the projectile has to be as large as 0.2 !

In conclusion, it has been shown that the impulse approximation as well as the second Born theory are able to reproduce the features of the forward peak for energetic collisions with charged projectiles, such as the peak asymmetry and its dependence on velocity, projectile charge and angular acceptance. For neutral projectiles, the question of existence and interpretation of a cusp-like peak calls for further investigations.

Acknowledgments

The collaboration with W.Oswald (who provided the second Born results) and with the other members of the experimental group, H.-D.Betz and R.Schramm, is greatly acknowledged. I would also like to thank R.Koch and H.Schmidt-Böcking for providing the data on the b-distribution, and C.O.Reinhold for communicating the CTMC results prior to publication. Especially I would like to thank D.Berényi and his group for discussions and for their great hospitality during my visit in Debrecen.

References

1. Forward Electron Ejection in Ion-Atom Collisions, Lecture Notes in Physics, Vol. 213, eds. K.-O.Groeneveld, W.Meckbach and I.A. Sellin (Springer, Berlin 1984)
2. L.Sarkadi, J.Pálinkás, A.Kövér, D.Berényi and T.Vajnai, Phys.Rev.Lett. 62, 527 (1989)
3. H.Atan, W.Steckelmacher and M.W.Lucas, preprint (1989)
4. C.O.Reinhold and R.E.Olson, Phys.Rev. A39, 3861 (1989)
5. D.H.Jakubaßa-Amundsen, Int.J.Mod.Phys. A4, 769 (1989)
6. D.H.Jakubaßa-Amundsen, Phys.Rev. A38, 70 (1988)
7. W.Oswald, R.Schramm and H.-D.Betz, Phys.Rev. Lett. 62, 1114 (1989)
8. W.Meckbach, I.B.Nemirovsky and C.R.Garibotti, Phys.Rev. A24, 1793 (1981)
9. R.Shakeshaft and L.Spruch, Phys.Rev.Lett. 41, 1037 (1978)
10. C.O.Reinhold and D.R.Schultz, preprint (1989)
11. D.H.Jakubaßa-Amundsen, J.Phys. B16, 1761 (1983)
12. P.Dahl, J.Phys. B18, 1181 (1985)
13. D.K.Gibson and I.D.Reid, J.Phys. B19, 3265 (1986)
14. O.Jagutzki, R.Koch, A.Skutlartz and H.Schmidt-Böcking, Contributed Paper to the 3rd ECAMP, Bordeaux, Book of Abstracts p.654 (1989)
15. J.S.Briggs, J.Phys. B10, 3075 (1977)
16. H.Nakanishi and D.M.Schrader, Phys.Rev. A34, 1810 (1986)
17. D.H.Jakubaßa-Amundsen, J.Phys. B (in print)
18. A.Kövér, L.Sarkadi, J.Pálinkás, L.Gulyás, Gy.Szabó, T.Vajnai, D.Berényi, O.Heil, K.-O. Groeneveld, J.Gibbons and I.A.Sellin, preprint (1989)
19. C.R.Garibotti and R.O.Barrachina, Phys.Rev. A28, 2792 (1983)
20. R.O.Barrachina, preprint (1989)

MULTIPLE ELECTRON CAPTURE IN CLOSE ION-ATOM COLLISIONS

A.S. Schlachter,[a] J.W. Stearns,[a] K. H. Berkner,[a] E. M. Bernstein,[b]
M.W. Clark,[b] R.D. DuBois,[c] W.G. Graham,[d] T.J. Morgan,[e]
D.W. Mueller,[f] M.P. Stockli,[g] J.A. Tanis,[b] and W. T. Woodland[b]

a) Lawrence Berkeley Laboratory, Berkeley, California USA
b) Western Michigan University, Kalamazoo, Michigan USA
c) Pacific Northwest Laboratory, Richland, Washington USA
d) Queen's University, Belfast, Northern Ireland
e) Wesleyan University, Middletown, Connecticut USA
f) University of North Texas, Denton, Texas USA
g) Kansas State University, Manhattan, Kansas USA

Collisions in which a fast highly charged ion passes within the orbit of K electrons of a target gas atom are selected by emission of a K x-ray from the projectile or target. Measurement of the projectile charge state after the collision, in coincidence with the K x-ray, allows measurement of the charge-transfer probability during these close collisions. When the projectile velocity is approximately the same as that of target electrons, a large number of electrons can be transferred to the projectile in a single collision. The electron-capture probability is found to be a linear function of the number of vacancies in the projectile L shell for 47-MeV calcium ions in an Ar target.

I. INTRODUCTION

Charge transfer in ion-atom collisions is a fundamental process of considerable interest. The probability of capturing an electron in a collision is a function of the impact parameter for the collision, as well as a function of projectile velocity and charge state. Collisions in which a fast ion projectile passes close to a target atom, e.g., within the K or L shell, can be selected by measuring the angle through which the projectile is scattered in the collision: the projectile is scattered through a larger angle in a close collision than in a more distant one.[1-5] A less direct method of selecting close collisions is coincident detection of highly charged target recoil ions: a highly charged recoil ion implies that a close collision has taken place.[6-9] The impact-parameter dependence of inner-shell vacancy production has also been studied by measuring coincidences between charge transfer and x-ray emission.[10]

We have measured[11] cross sections for electron capture in coincidence with Ca or Ar K x-ray emission for 47-MeV Ca^{17+} + Ar, and showed that the cross section for capturing several electrons in a close collision can exceed that for capturing only one electron. The only previous report of the cross section for multiple-electron capture being larger than that for single-electron capture, determined by ion-x-ray coincidence, is the observation of double K-to-K transfer in collisions of bare 4.5-MeV/u Si ions in Ar.[12] The results of reference 11 were recently extended to measure capture of up to eight electrons, and the projectile velocity has been varied.[13] New results are presented here for varying the projectile charge state, which varies the number of initial vacancies in the projectile L shell.

II. APPARATUS AND METHOD

The apparatus and methods employed are similar[11,13,14] to those previously used. The apparatus is shown schematically in Fig. 1. A beam of fast Ca ions from the LBL SuperHILAC accelerator was passed through a thin foil, and ions in a given charge state were selected by a magnet. Calcium ions in charge states from 17+ (lithium-like) to 10+ (neon-like) were collimated and then passed through a differentially pumped gas cell. The magnetically-analyzed Ca ions were measured after the collision using a Faraday cup and electrometer for those ions which did not change charge in the target, while ions which captured one to eight electrons were counted with solid-state detectors. A Si(Li) x-ray detector viewed the gas cell. Coincidences between x-rays and projectiles which captured electrons were measured by starting time-to-digital or time-to-amplitude converters (TDCs or TACs) with an x-ray signal, and stopping them with delayed signals from the particle detectors. The data were corrected for absorption of x-rays in the detector window and for the detector solid angle. Errors were minimized by using a very thin target relative to all relevant electron-capture and -loss cross sections and by correcting for background effects. The relative uncertainties in the cross sections reported here are ±25%, the systematic uncertainty is ±30%, for an absolute uncertainty of approximately ±40%.

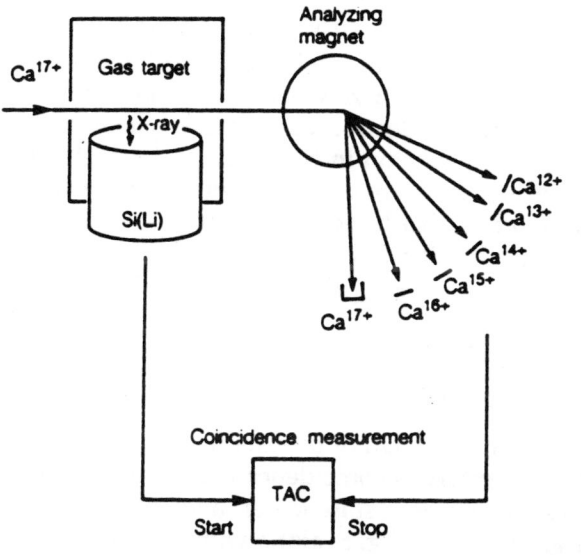

FIG. 1. Schematic diagram of the apparatus, shown for Ca^{17+} projectile; eight particle detectors were used (only five are shown). Each particle detector goes to a separate time-to-amplitude converter (TAC). The Si(Li) detector is used to detect x-rays.

III. CLOSE COLLISIONS OF 47-MeV Ca^{17+} IN Ar

A 47-MeV Ca ion has a velocity (1.5×10^9 cm/s) which is close to the velocity of the L-shell electrons of the Ar target (1.0×10^9 cm/s), and is slower than the K-shell electrons[15] of the projectile (4.2×10^9 cm/s) and target (3.4×10^9 cm/s). K-vacancy production in either partner is therefore expected to be appropriately described in the molecular-orbital picture. A schematic MO diagram for the relevant states of the Ca + Ar system is shown in Fig. 2. An Ar K electron can be transferred to the L shell of the Ca^{17+} ion at an internuclear separation of about the K-shell radius (approximately $0.08 a_0$) via $2p\sigma$-$2p\pi$ rotational coupling.[16] The resulting vacancy in the Ar K shell can be shared with the outgoing Ca ion by subsequent $2p\sigma$-$1s\sigma$ radial coupling[17] at a distance of several times the K-shell radius, giving rise to a Ca K vacancy. Simultaneous with the K-K transfer is the transfer of target L- and M-shell electrons to the projectile L shell, which has a high probability of occurrence over a range of several times the L-shell radius (see Fig. 2).

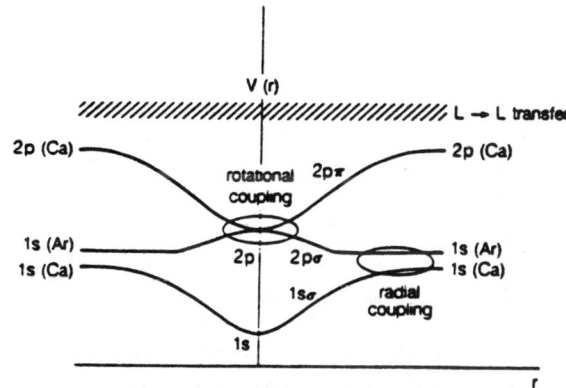

FIG. 2. Schematic molecular-orbital diagram for the Ca+Ar system showing the relevant levels. Rotational coupling creates an Ar K vacancy at small internuclear separations; radial coupling (vacancy sharing) at larger internuclear separations can lead to a Ca K vacancy. L-L transfer occurs over a still larger range of internuclear separations (indicated schematically by the cross hatching).

Cross sections for L- and M-shell electron capture coincident with a Ca K or Ar K x-ray are shown in Fig. 3 for 47-MeV Ca^{17+} in Ar. The cross sections for coincidence with an Ar K and Ca K x-ray have essentially the same shape, and in fact, coincide, if the cross sections for coincidence with an Ar K x-ray are shifted one charge state lower, as shown in Fig. 3. This shift is needed because an Ar K vacancy (Ar K x-ray emission) resulting from transfer of an Ar K electron to the Ca L shell appears as a capture event by the Ca ion (Fig. 4a). For the case of emission of a Ca K x-ray, however, an electron has been returned to the Ar by K-vacancy sharing (Fig. 4b), resulting in no net change of charge for the Ca ion. The cross sections are thus compared in terms of capture of target L- and M-shell electrons only.

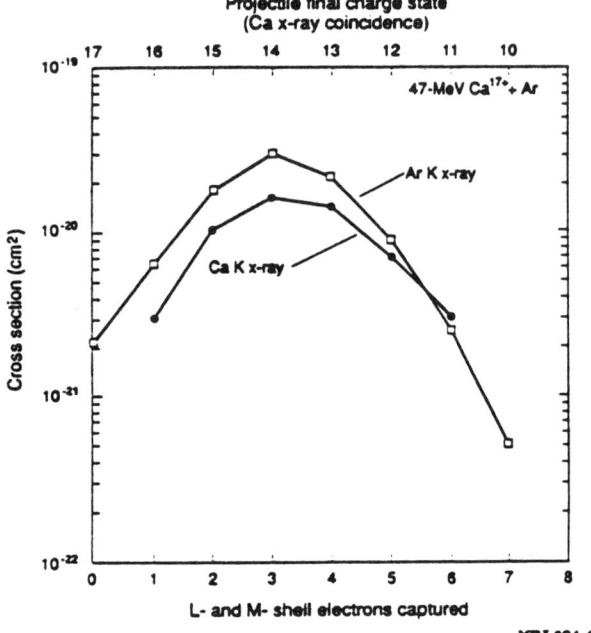

FIG. 3. Cross sections for electron capture in coincidence with an Ar or Ca K x-ray for 47-MeV Ca^{17+} in Ar, as a function of the number of L- and M-shell electrons captured. The upper scale shows the projectile final charge state for coincidence with a Ca K x-ray. The Ar K x-ray coincidence data have been shifted one charge state to the left to account for the promotion of an Ar K electron to the Ca L-shell for the case of emission of an Ar K x ray (see text).

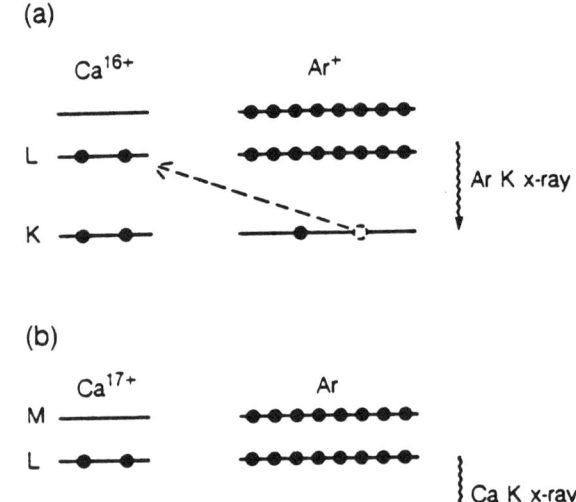

FIG. 4. Schematic diagram of energy levels for Ca^{17+} + Ar, showing (a) electron promotion and (b) vacancy sharing.

The electron-capture charge-state distribution can be described[18] by a binomial distribution if we assume that the L- and M-shell electrons are captured due to electron-nucleus interactions and that electron-electron interactions can be neglected, i.e., the electrons are captured without correlation. The binomial distribution for capture of n electrons with a probability p is written:

$$P_n = \binom{t}{n} p^n (1-p)^{t-n}$$

where t is the number of places into which electrons may be captured and $\binom{t}{n}$ is the binomial coefficient.

The measured relative probability distributions for electron capture in coincidence with an Ar K x-ray (shifted one charge state lower) and with a Ca K x-ray are shown in Fig. 5, along with a binomial distribution which has been fit to the experimental data for t=7, giving an electron-capture probability p = 0.45 for the Ar K x-ray data and p = 0.49 for the Ca K x-ray data. The good agreement of the measured charge-state distribution with a binomial distribution is consistent with independent electron-capture events, and argues against significant electron correlation effects.

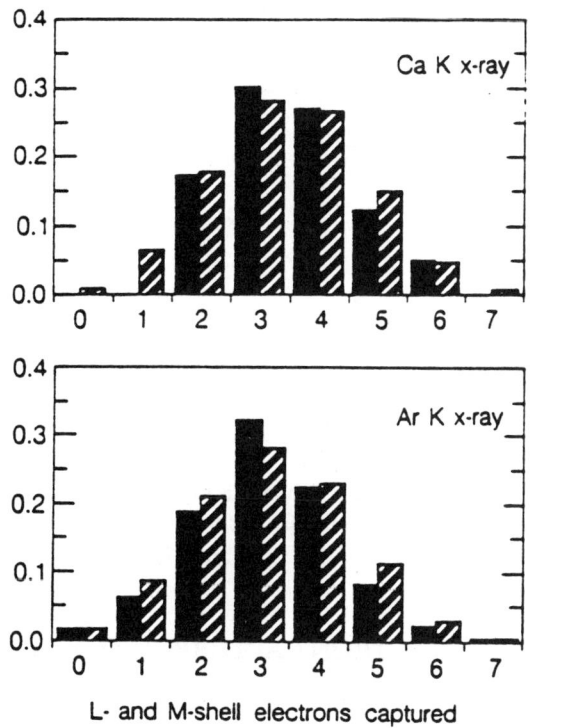

FIG. 5. Relative electron-capture probabilities for 47-MeV Ca^{17+} in Ar as a function of the number of L- and M-shell electrons transferred for coincidence with a Ca K or Ar K x-ray. The distribution for coincidence with an Ar K x-ray has been shifted to the left by one charge state (see text). Solid bars show the experimental results; shaded bars show the binomial distribution for an electron-capture probability of 0.47 (assuming 7 electrons can be captured). The experimental probabilities have been normalized by a factor determined by the fit to the binomial distribution.

Experimental K-vacancy-sharing probabilities, w_{exp}, can be calculated from the observed cross sections for electron capture in coincidence with a Ca or Ar K x ray, corrected by neutral-atom fluorescence yields. Agreement of experimental values, w_{exp}, with the theoretical[17] value $w_{th} = 0.30$ is good for capture of up to four electrons, where neutral-atom K binding energies have been used. The values of w_{exp} for a larger range of number of electrons captured shows that w_{exp} increases with increasing number of electrons captured (Fig. 6). Possible explanations for this dependence are that the K-vacancy-production process is not truly independent of the electron-capture mechanism for collisions in which many electrons are captured, or that the use of neutral-atom fluorescence yields is inadequate.

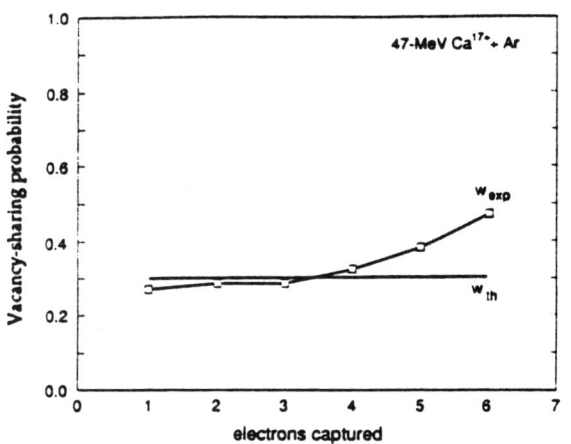

FIG. 6. K vacancy-sharing probabilities as a function of the number of L- and M-shell electrons captured. Experimental probabilities, w_{exp}, are calculated from the experimental results; w_{th} is the theoretical value.[17]

IV. CLOSE COLLISIONS OF 47-MeV Ca IONS IN Ar: VARYING THE NUMBER OF VACANCIES IN THE Ca L SHELL

Close collisions of 47-MeV Ca ions in Ar have been further investigated by varying the number of initial vacancies in the L shell of the Ca-ion; ions in charge states 10, 13, 15, 16, and 17 were used. Results are shown in Fig. 7, as a function of the number of L-shell electrons captured. Note that the abscissas are offset to emphasize that the cross-section distributions have the same shape. This suggests that cross sections for electron loss in coincidence with a Ca or Ar K x-ray might be appreciable.

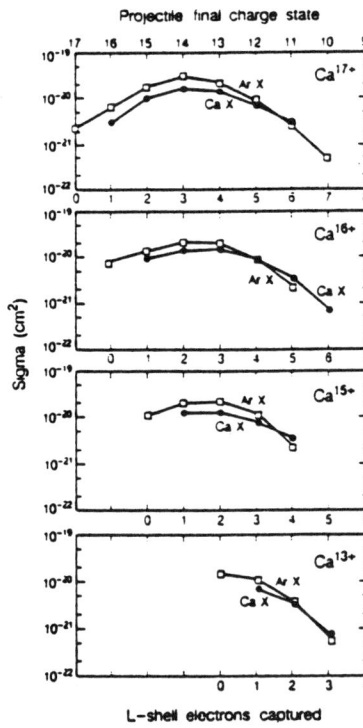

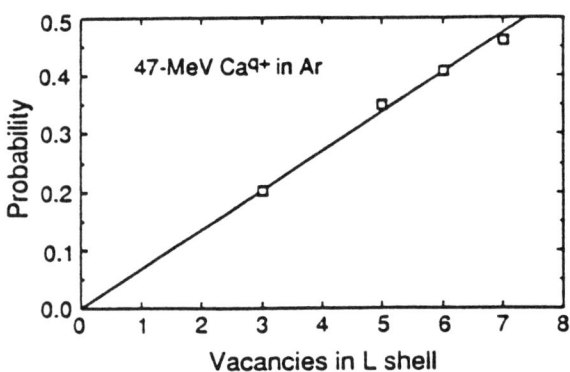

FIG. 8. Electron-capture probability as a function of the number of vacancies in the Ca L shell for close collisions of 47-MeV Ca^{q+} in Ar (q = 13, 15, 16, and 17).

FIG. 7. Cross sections for electron capture in coincidence with an Ar or Ca K x-ray for 47-MeV Ca^{q+} in Ar, for q = 13, 15, 16 and 17, as a function of the number of L-shell electrons captured. The upper scale shows the projectile final charge state for coincidence with a Ca K x-ray. The Ar K x-ray coincidence data have been shifted one charge state to the left.

The cross sections shown in Fig. 7 can be fit with a binomial distribution, as has been done in Fig. 5 for Ca^{17+} ions. The result in each case is an electron-capture probability. These electron-capture probabilities are shown in Fig. 8 as a function of the number of initial vacancies in the L shell of the Ca-ion projectile. This electron-capture probability is seen to be linear with the number of Ca-ion L-shell vacancies.

V. TOTAL ELECTRON CAPTURE: 47-MeV Ca IONS IN Ar

Total electron-capture cross sections for 47-MeV Ca^{17+} in Ar are shown in Fig. 9. A total cross section for capture of up to eight electrons in a single collision is observed. These cross sections, unlike those for close collisions (coincidence with a Ca or Ar K x-ray), which are also shown in Fig. 9, are a monotonically decreasing function of the number of electrons captured. We note that close collisions are only a small fraction of the total number of collisions, although this fraction is found to increase with increasing number of electrons captured, a result which might arise from greater interpenetration of the projectile and target L shells for collisions in which a greater number of electrons are captured.

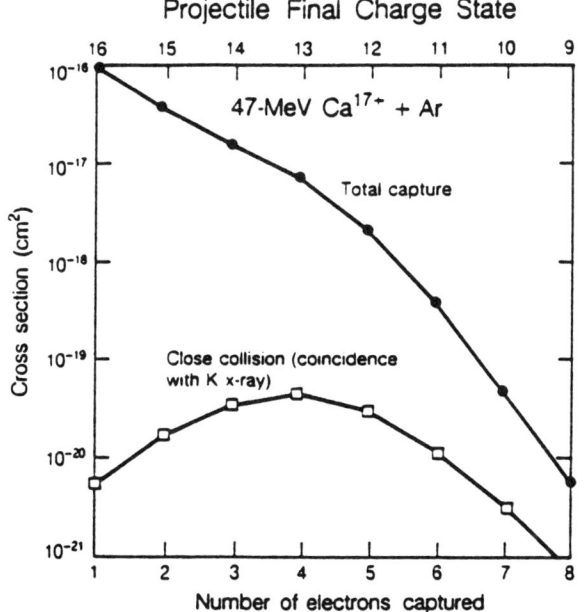

FIG. 9. Total and x-ray coincident (sum of Ar and Ca K x-rays) electron-capture cross sections for Ca^{17+} in Ar as a function of the number of electrons captured.

ACKNOWLEDGMENTS

This work was supported by the Director, Office of Energy Research, Office of Fusion Energy, of the U. S. Department of Energy under Contract No. DE-AC03-76SF00098, the U. S. Department of Energy, Office of Basic Energy Sciences, Division of Chemical Sciences, the Office of Health and Environmental Research under Contract No. DE-AC06-76RLO1830, the Western Michigan University Faculty Research and Creative Activities Support Fund, and the UKSERC.

REFERENCES

1. B. Rosner and D. Gur, *Phys. Rev. A* 15, 70 (1977).
2. E. N. Fuls, P. R. Jones, F. P. Ziemba, E. Everhart, *Phys. Rev.* 107, 704 (1957).
3. E. Everhart and Q. C. Kessel, *Phys. Rev.* 146, 16 (1966).
4. V. V. Afrosimov, Yu S. Gordeev, M. Panov, and N. V. Fedorenko, Zh. Eksp. Teor. Fiz. Pisma Red. 2, 291 (1965); *JETP Lett.* 2, 185.
5. Q. C. Kessel, *Phys. Rev. A* 2, 1881 (1970).
6. T. J. Gray, C. L. Cocke, and E. Justiniano, *Phys. Rev. A* 22, 849 (1980).
7. S. Kelbch, H. Schmidt-Böcking, J. Ullrich, R. Schuch, E. Justiniano, H. Ingwersen, and C. L. Cocke, *Z. Phys. A* 317, 9 (1984).
8. S. Kelbch, J. Ullrich, R. Mann, P. Richard, and H. Schmidt-Böcking, *J. Phys. B* 18, 323 (1985).
9. A. Müller, B. Schuch, W. Groh, E. Salzborn, H. F. Beyer, P. H. Mokler, and R. E. Olson, *Phys. Rev. A* 33, 3010 (1986).
10. H. J. Stein, H. O. Lutz, P. H. Mokler, K. Sistemich, and P. Armbruster, *Phys. Rev. Lett.* 24, 701 (1970).
11. A. S. Schlachter, E. M. Bernstein, M. W. Clark, R. D. DuBois, W. G. Graham, R. H. McFarland, T. J. Morgan, D. W. Mueller, K. R. Stalder, J. W. Stearns, M. P. Stockli, and J. A. Tanis, *J. Phys. B* 21, L291 (1988).
12. S. Andriamonje, J. F. Chemin, J. Roturier, B. Saboya, J. N. Scheurer, R. Gayet, A. Salin, H. Laurent, P. Aguer, and J. P. Thibaud, *Z. Phys. A* 317, 251 (1984).
13. A. S. Schlachter, J. W. Stearns, K. H. Berkner, E. M. Bernstein, M. W. Clark, R. D. DuBois, W. G. Graham, T. J. Morgan, D. W. Mueller, M. P. Stockli, J. A. Tanis, and W. T. Woodland, *Nucl. Instrum. Methods B* 40/41, 21 (1989).
14. J. A. Tanis, E. M. Bernstein, M. P. Stockli, W. G. Graham, K. H. Berkner, D. J. Markevich, R. H. McFarland, R. V. Pyle, J. W. Stearns, and J. E. Willis, *Phys. Rev. A* 29, 2232 (1984).
15. T. A. Carlson, C. W. Nestor, Jr., N. Wasserman, and J. D. McDowell, *Atomic Data* 2, 63 (1970).
16. K. Taulbjerg, J. S. Briggs, and J. Vaaban, *J. Phys. B* 9, 1351 (1976).
17. W. E. Meyerhof, *Phys. Rev. Lett.* 31, 1341 (1973).
18. R. L. Kauffman, J. H. McGuire, P. Richard, and C. F. Moore, *Phys. Rev. A* 8, 1233 (1973).

MOMENTUM TRANSFER BETWEEN PROJECTILE AND RECOIL ION IN FAST IONIZING PROTON – HELIUM COLLISIONS.

J. Ullrich[*], R.E. Olson[**], R. Dörner[*], H. Schmidt-Böcking[*]

[*] Institut für Kernphysik, Universität Frankfurt, D-6000 Frankfurt/Main, FRG
[**] Department of Physics, University of Missouri-Rolla, Rolla, Missouri 65401, USA

H^+ + He single ionization at 0.5 MeV has been investigated measuring the projectile deflection in a plane perpendicular to the beam direction in coincidence with the transverse momentum of the recoil-ion. The transverse momentum exchange between projectile, emitted electron and target nucleus is completely determined in single ionization collisions. Three scattering regions could be distinctly recognized that are dominated by proton – helium nucleus, proton – electron, or electron – helium nucleus interactions. On the basis of our experimental results, the peak structure in the relative fraction of double ionization at 0.9 mrad observed by Giese et al[6] is calculated to be due to two uncorrelated, subsequent scattering events of H^+ with both ionized He electrons.

Introduction

The investigation of single and double ionization of He by swift (MeV/u) singly charged projectiles is of fundamental interest in ion-atom collision physics and has recently attracted extensive experimental and theoretical attention, stimulated by surprising results for antimatter impact[1]. Although highly differential He ionization cross sections have been measured for e^- impact using a coincidence determination of the total momentum of the two outgoing electrons[2], most experimental studies for heavy projectile impact have concentrated on total cross sections. These experimental results can be theoretically described by quantum-mechanical[3] and classical[4,5] calculations. Only recently, two experimental studies[6,7] on the projectile scattering distribution for He single and double ionization have been reported and show unexpected structure in the scattering angle dependence: A sharp shoulder in the angular differential single ionization cross section was observed[7] for swift H^+ on He collisions at the maximum deflection of the H^+ by a stationary electron of 0.545 mrad and has been attributed to be due to projectile deflection on a "quasi-free" target electron. Giese et al[6] reported on a distinct peak in the relative fraction of He double ionization at a projectile laboratory scattering angle $\vartheta \approx 0.9$ mrad, which shows itself to be independent of the projectile velocity.

In order to elucidate the dynamics of ionizing collisions, we have measured doubly differential He ionization cross sections for 0.5 MeV H^+ impact in a ϑ regime between 0.1 and 2.0 mrad. As is shown in fig. 1, the projectile deflection in the plane perpendicular to the beam axis was determined in coincidence with the recoil-ion transverse momentum p_{Rx} and charge state. The apparatus is designed in such a way that the scattering plane in φ direction is defined by the beam axis and the recoil-ion apperture. Thus, for single ionization the complete 3-body transverse momentum exchange between projectile, emitted electron and recoiling target nucleus can experimentally be observed.

The experimental results show a strong influence of the emitted electron momentum on the heavy particle trajectories providing evidence that the use of a central scattering potential and a unique relation between impact parameter and ϑ is invalid to describe a major part of the ionization dynamics: H^+ deflections smaller than 0.5 mrad are found to be dominated by the scattering with the ejected electron, the recoil-ion momentum being completely independent of ϑ. On the basis of these results, the peak observed by Giese et al can be explained and calculated to be due to an uncorrelated double scattering event of the projectile with both ionized electrons.

The experimental results are compared to calculations made using the n-body classical trajectory Monte Carlo (nCTMC) approach[8]. Both target electrons are included in 3-dimensional calculations that incorporate all interactions except those between the electrons (i.e. the electrons are not correlated). The inclusion of the two electrons in the calculation allows both the projectile and the He nucleus to be properly screened from one another at both small and large internuclear distances. The nCTMC results give a total single ionization cross section of $3.34 \cdot 10^{-17}$ cm^2, slightly lower than the experimental value[9] of $(3.70 \pm 0.15) \cdot 10^{-17}$ cm^2. The calculated projectile angular differential cross sections are in quantitative agreement with Giese et al[6] and also display the shoulder at about 0.5 mrad observed by Kamber et al[7]. The nCTMC method explicitly includes the post collision interactions and the coupling between the electronic and nuclear motions.

Experiment

The experiments were performed at the 2 MV Van-de-Graaff accelerator of the Institut für Kernphysik, University of Frankfurt. The proton beam was collimated to a divergence of less than 0.08 mrad over a total collimation length of 5 m. A two-dimensional position-sensitive channelplate-detector with a resolution of less than 200 μm detected the scattered projectiles about 3 m downstream the target region. Therefore the overall angular resolution is about ± 0.06 mrad. The undeflected part of the beam was dumped on a mask of 0.5 mm diameter directly in front of the detector.

The recoil-ion transverse momentum (with respect to the beam axis) is measured by a time-of-flight technique[10] using a new spectrometer with a low temperature (30 K) He-target. The extended target region (a cylinder, coaxial to the beam direction of 5 mm diameter and a length of 40 mm with incoming and outgoing beam appertures of 1.0 mm diameter) was designed to be "free" of electric and magnetic fields. Hence, the recoiling target ions, produced along the beam axis, drift from the axis to the walls of the target-cylinder in a time interval Δt inversely proportional to the transverse velocity v_{Rx} they obtained in the collision. The recoil-ions leave the cylinder through a small apperture (Φ = 1 mm), they are accelerated, charge state analyzed, and detected by a two-dimensional position-sensitive channelplate detector. A coincidence between the scattered projectiles and the recoil-ions provides, after corrections for inherent flight-times in the apparatus and electronic processing times, the flight-time Δt of the recoil-ion. This can be transformed into the recoil ions transverse velocity, momentum or energy. Since the target region is extended, the recoil-ion scattering angle ζ is not determined but all ζ between +25° and 155° are accepted with the same solid angle. However, only the transverse component of the velocity is detected. Moreover, the apparatus design is such that the scattering plane is determined by the beam axis and the recoil-ion apperture. The target-gas pressure was about 10^{-3} hPa.

The energy transferred to the recoil-ion in a two-body collision with the projectile, which is scattered to a deflection angle of 0.1 mrad, is estimated to be 1.3 meV. From this value it becomes obvious that the Boltzmann thermal motion of the target atoms has to be reduced considerably to observe the very small momentum transfers of interest. Therefore, the He target-gas was cooled by a cryo-pump to a temperature of about 30 K, which was measured by Pt-resistors.

In order to avoid contact potentials, the cylindrically shaped target cell was constructed out of a single copper block and the recoil-ion apperture was covered on the outside by a copper grid (mesh width: 0.12 mm by 0.12 mm). In succeeding experiments the copper cell was gold-plated to circumvent possible potentials between the Cu basis material and oxides on the surface. Moreover, the outside mesh was removed and the inner surface was completely lined with either a gold or a copper grid in order to have a more "perfect" Faraday cage.

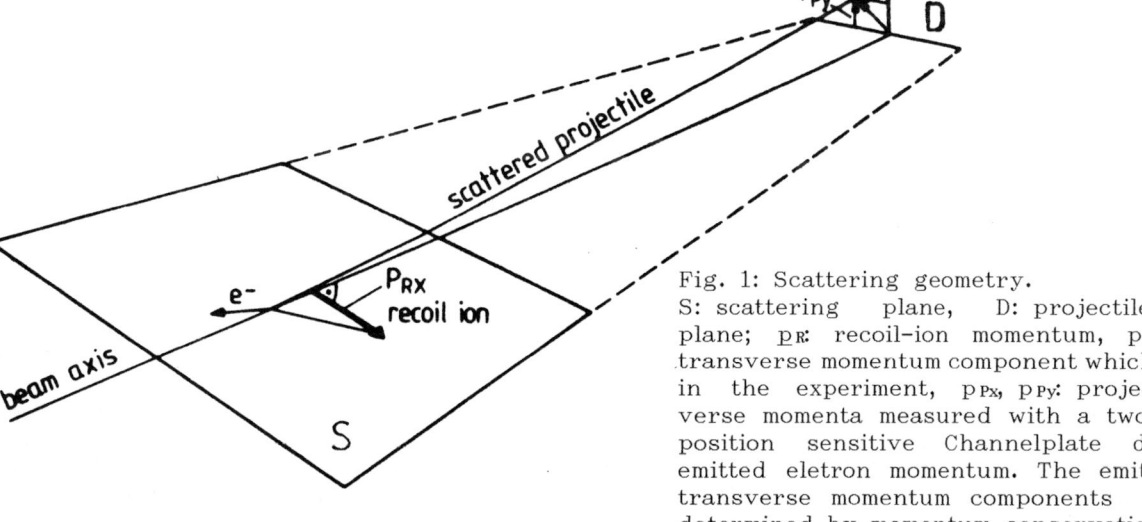

Fig. 1: Scattering geometry.
S: scattering plane, D: projectile detection plane; p_R: recoil-ion momentum, p_{Rx}: recoil-ion transverse momentum component which is measured in the experiment, p_{Px}, p_{Py}: projectile transverse momenta measured with a two dimensional position sensitive Channelplate detector. p_e: emitted eletron momentum. The emitted electron transverse momentum components p_{ex}, p_{ey} are determined by momentum conservation law: $p_{ex} = -(p_{Rx} + p_{Px})$, $p_{ey} = -p_{Py}$.

Before each experiment the target cell was cleaned in an ultrasonic bath. Identical results were obtained, within statistical error bars, for all three surface-grid combinations. Furthermore, in most cases, any contamination inside the target cell presents itself by a noticeable disturbance in the time-of-flight spectra. From our experimental tests, we estimate that the absolute accuracy of the recoil-ion's energy determination is ± 5 meV.

Results

In Fig.2a, the deflection (in mrad) of the protons in a plane perpendicular to the beam axis (x,y-plane) is shown for fixed momentum transfer to the singly charged He$^+$ recoil-ion in the x-direction of $0.9 \leq |p_{Rx}/p_0| \leq 1.1$ (p_0 is the incoming projectile momentum), which is equivalent to recoil-ion energies between 100 meV and 150 meV. In other words, illustrated in Fig.1, the projectile scattering on plane "D" is observed for a definite recoil-ion momentum p_{Rx}. For a 2-body collision between H$^+$ and the target nucleus ($p_{Px} = -p_{Rx}$) a projectile scattering into the shaded area (Fig.2a) in the plane "D" is expected. Zero deflection of the beam is marked by the cross. The overall angular resolution in two dimensions (diameter of the undeflected beam, position resolution of the detector, influence of thermal target motion) is indicated by the bars of the cross.

Instead of a definite scattering angle of the projectile into the indicated area, a large number of other projectile deflections can be observed. Since the experimental resolution is small compared to the observed deviation from the two-body scattering this broadening of the projectile angular distribution is due to the exchange of transverse momentum with the ejected target electron. Applying momentum conservation, the ejected electron transverse momentum can be obtained for each collision event from the distance of the detected projectile position to the shaded area. We note that the calculation of the projectile transverse deflection using any central scattering potential would lead to values indicated by the shaded area in the figure.

To allow for a quantitative comparison of the experimental results with nCTMC calculations, the distribution shown in Fig.2a is projected onto the x-axis, Fig.2b. Since no absolute experimental data could be obtained in this first measurement, the integral of the distribution is normalized on the theoretical results (open circles). Also indicated in the figure (shaded area) is the projectile momentum range which would correspond to the chosen recoil-ion momentum window for a two-body collision. The width of the measured distribution of about 1.1 mrad is consistent with the maximum possible proton deflection of 0.55 mrad in each direction from a binary encounter with a statio-

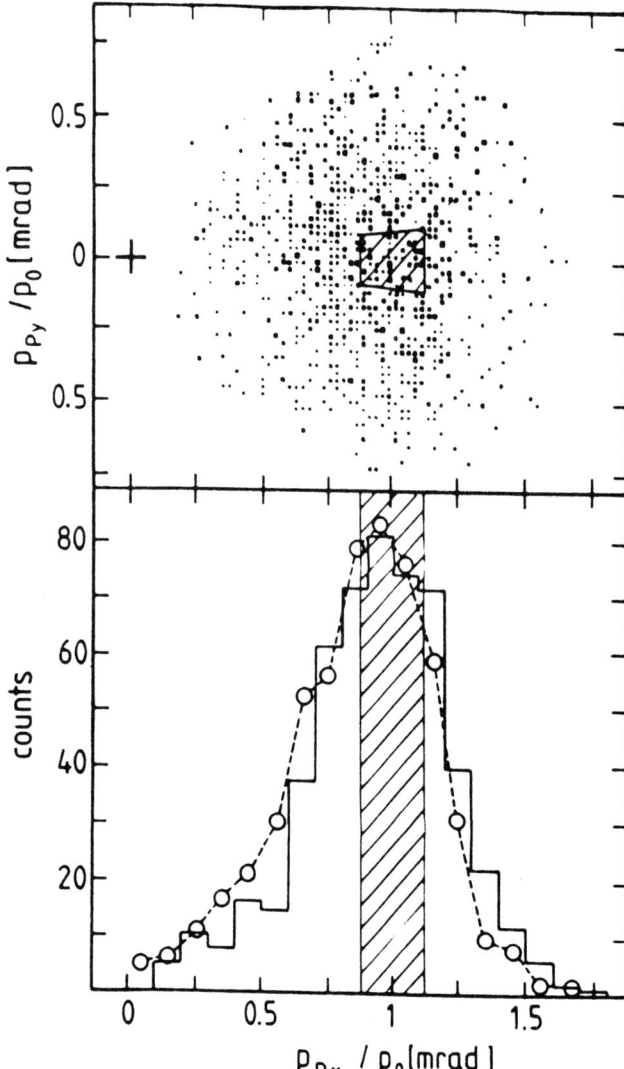

Fig. 2a: Projectile deflection in a plane perpendicular to the beam direction for fixed recoil-ion momentum of $0.9 \leq |p_{Rx}/p_0| \leq 1.1$ in 0.5 MeV singly ionizing H$^+$ - He collisions. The shaded area indicates the proton scattering angle expected from a two-body collision with the He-nucleus.

Fig. 2b: Projection of the proton scattering in Fig. 1a onto the x-axis (histogram). Open circles: result of nCTMC calculations. Shaded area: recoil-ion momentum window.

nary target electron and is in good agreement with the calculated one. The Compton profile of the He-electron before its ionization gives rise to a further broadening of the proton deflection which is reflected in both the theoretical and experimental results. As an unexpected result, the experimental as well as the theoretical mean momentum transfer to the proton is slightly smaller than that of the recoil-ion, indicating that the ionized electron is preferentialy ejected away from the recoiling He⁺ ion.

In Fig. 3a the mean transverse energy transferred to the recoiling target-ion is plotted versus the proton laboratory scattering angle ϑ. For a 2-body scattering-process between the projectile and the He-nucleus, as well as for any central scattering-potential, both final momenta will be identical which is indicated by the dashed line. In Fig. 3b the theoretical and the experimental results are divided by the 2-body momentum exchange value to visualize the strong observed deviation. Only for very close collisions $b \lesssim 0.1$ a_0 (leading to projectile deflections of more than ~1 mrad), which contribute only 3% to the total cross section, is the proton scattering dominated by the interaction with the He-nucleus. Even in this regime, the initial momentum distribution of the He target electron as well as that for the final ionized electron have a measurable influence on the proton trajectory, Fig. 2.

For H⁺ deflections smaller than 0.9 mrad, deviations from the expected 2-body proton - helium-nucleus scattering is observed. Around the maximum laboratory scattering angle for a proton from a free electron of ~0.55 mrad, the theoretical as well as experimental mean transverse recoil-ion energy is more than a factor of 2 below the 2-body value. (To allow for a quantitative comparison of the experimental data with theory at these very small momentum transfers, the theory was folded with the Boltzmann thermal motion distribution which is represented by the dotted line in the figure).

However, the mean recoil-ion energy does not drop to a value near zero as might have been expected, but saturates at a finite value for small projectile scattering angles. The experimental saturation value of about 15 meV is not determined by the target thermal motion velocity distribution, which only yields a small contribution of ~4 meV at 30 K. Another set of independent experimental data (bars in Fig.3b) obtained using a different target-cell grid combination as well as a slightly different target geometry (with a smaller recoil-ion detection solid angle and therefore larger statistical errorbars) yields identical results within statistical errors. A systematical error of not more than ±5 meV is estimated being due to uncertainties in the determination of the absolute gas-target temperature (±20 K),to residual stray fields and due to the uncertainty in the determination of the beam axis within the target cell.

Since the recoil-ion momentum is much larger than the one transferred to the proton, the 3-body momentum transfer is now primarily due to the final state interaction of the ionized electron with it's parent nucleus. The theoretical saturation value for the mean

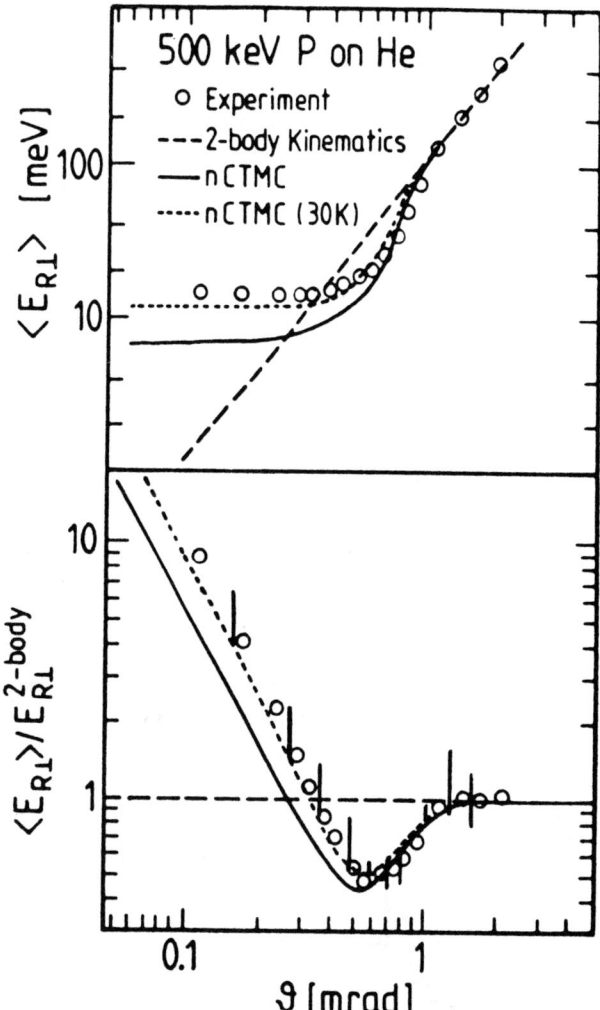

Fig. 3a: Mean transverse recoil-ion energy $\langle E_R \rangle$ (open circles) in dependence of the projectile scattering angle ϑ for singly ionizing 0.5 MeV H⁺ - He collisions. Full line: nCTMC calculation. Dotted line: nCTMC results folded with target thermal motion for the experimental target temperature of 30 K. Dashed line: results for a two-body collision between the proton and the He-nucleus.

Fig. 3b: Ratio of the recoil-ion transverse energy $\langle E_R \rangle$ to that expected in a 2-body collison E_R. Notation is the same as in Fig. 3a. Bars: Experimental results for different target material and geometry.

transverse recoil-ion energy of 6.5 meV for "zero-deflection" of the projectile is equivalent to a mean transverse energy of the ionized electron of 47 eV. The mean emitted electron energy, integrated over all emission angles, is calculated in the nCTMC approach to be 56 eV, which is in close agreement with the experimental value of 54 eV reported by Rudd et al[11]. Since the mean emission angle is about 65°, this corresponds to a calculated mean transverse electron energy of about 50 eV, which shows that the mean recoil-ion energy is completely compensated by that of the emitted electron. The saturation effect demonstrates the importance of the electron - nucleus interaction in the 3-body momentum exchange at projectile scattering angles smaller than 0.3 mrad.

The H$^+$ + He coincidence spectrum was also calculated for 1 MeV impact. Surprisingly, the saturation energy was not the same as at 0.5 MeV, which would be the case if it primarily reflected the electronic internal momentum distribution before the collision. We find the saturation energy to be ~50% larger at 1 MeV than at 0.5 MeV. The reason being that the calculated mean ionized electron energy increased to 62 eV from 55 eV (0.5 MeV), with the mean ejection angle also increasing to 70° from 65°. The saturation energy values are primarily determined by the ionized electron, recoil-ion momentum balance.

We have also calculated the recoil-ion spectra for 0.5 MeV antiproton-He collisions and find measurable differences in the recoil-ion spectra. Noteable, is that the recoil-ion saturation energy at small projectile scattering angles is ~50% greater for antiprotons than protons. This behavior is largely due to a greater average ionized electron scattering angle (i.e. closer to 90°) for antiproton versus proton impact on He.

The distinct peak in the ratio of double to total ionization at projectile deflections of about 0.9 mrad in 0.3 to 1.0 MeV proton He collisions (full dots in Fig.4 for 1 MeV proton impact) observed by Giese et al[6] can be explained on the basis of the results given here: the projectile scattering around 1.0 mrad is determined by only small impact parameter collisions for single ionization, while all impact parameters contribute for double ionization since an uncorrelated double scattering event of the proton from two target electrons can lead to a maximum deflection of the projectile of 1.1 mrad.

To simply prove our explanation we have convoluted the angular differential single ionization cross sections of Giese et al with themselves in two dimensions, which reflects an uncorrelated scattering event with the same deflection function. The obtained double ionization distribution has been normalized to the total double ionization cross section to account for the target thickness and is divided by the total ionization cross section to obtain the ratio. Still the magnitude of the ratio obtained by this simple model is to low by a factor of two for scattering angles larger than 0.2mrad. Uncertainties in reading the differential single ionisation cross section from figure 1 in Ref [6] as well as in the extrapolation of the experimental values to zero deflection yield small uncertainties in the calculated ratio, indicated in Fig.4 by the shaded area.

However, the overall shape of the calculated ratio $\sigma^{++}/(\sigma^{+} + \sigma^{++})$ always exhibits a distinct peak that is in shape and position in close agreement with the experimental findings. This result indicates that a complicated mechanisms like triple scattering, quantum interference processes or electron correlation effects may be not necessary to understand the observed peak. The ratio obtained in the forced-impulse technique of Reading et al[12] are given by the dashed line, there are no quantitative values from Vegh[13] who proposed a triple scattering event to give rise for the observed peak.

We have also calculated this ratio in the nCTMC approach and found agreement with the experimental data with a maximum at 0.9 mrad. This strongly supports the uncorrelated double collision mechanism to be the reason for the

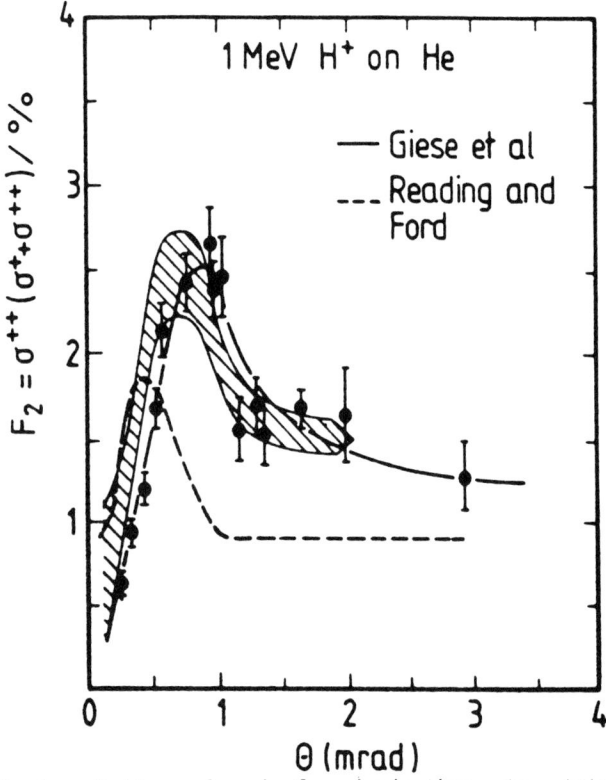

Fig.4: Ratio of single ionization to total ionization for 1 MeV H$^+$ on He in dependence of the scattering angle ϑ. Full dots: experimental results of Giese et al[6], full line: line to guide the eyes through experimental data, dashed area: calculated from differential cross sections for single ionization[6] asuming an uncorrelated double scattering event for the double ionisation normalization see text, dashed line: theoretical results of Reading and Ford[13].

structure since no electron-electron interaction has been included in our calculations.

Future Investigations

Very recently we investigated experimentally the double ionization process for 0.3, 0.5 and 1MeV p impact the data not being yet evaluated. The experimental methode should allow a detailed invetigation of the importance of dynamic and static correlation in the double ionization reaction.

Furthermore experiments are underway to measure the recoil ion emission angles for single and double ionization in dependence of the recoil ion energy. nCTMC calculations predict a nearly isotropic recoil-ion ejection, slightly being enhanced toward backwards angles for small recoil-ion energies and an emission around 90 degree for close collisions and high recoil energies.

The experimental technique enables to investigate single and double ionization of helium by electron, photon or antiparticle impact and therefore should help to get more complete knowledge of the importance of ionization mechanisms such as shake off or correlation effects. Experiments of that kind are under preparation. In addition, the experimental determination of the transverse recoil-ion energy in dependence of the recoil-ion charge state is also feasible for highly charged, heavy ion impact such as 500 MeV/u U^{92+} availible at the new cyclotron-storage ring-combination at GSI Darmstadt.

Conclusion

We have applied a new technique to measure the projectile scattering in a plane perpendicular to the beam direction in coincidence with the transverse momentum of singly ionized recoil-ions. For the first time the complete momentum balance between the projectile, the target-nucleus, and the ionized electron could experimentally determined for heavy particle impact. nCTMC calculations are found to be in good agreement with the experimental data. As a main conclusion of our work, we note that a central interaction potential is invalid to describe the heavy particle scattering for a major portion (97%) of ionizing collisions and as a consequence no unique relationship between impact parameter and projectile scattering angle exists. Moreover the proton scattering to about 1 mrad probes different impact parameter regions for the single and double ionization of He giving rise to the peak in the ratio. This underlines that the 3-body coupling between electronic and nuclear motion, which is usually ignored in theoretical approaches, is essential to understand the collision dynamics.

Acknowledgements

One of the authors (REO) would like to thank the von Humboldt foundation for support. The experimental work was supported by the Bundesministerium für Forschung und Technologie (BMFT: 060F/73).

References

1   L.H. Andersen, P. Hvelplund, H. Knudsen, P. Moller, K. Elsener, K-G. Rensfelt, and E. Uggerhoj, Phys. Rev. Lett. **57**, 2147 (1986).
2   H. Erhardt, K. Jung, G. Knoth, and P. Schlemmer, Z. Phys. D**1**, 3 (1986).
3   J.F. Reading and A.L. Ford A, Phys. Rev. Lett. **58**, 543 (1987).
4   S.J. Pfeifer, and R.E. Olson, Phys. Lett. **92A**, 175 (1982).
5   R.E. Olson, T.J. Gay, H.G. Berry, E.B. Hale, and V.D. Irby, Phys. Rev. Lett. **59**, 36 (1987).
6   J.P. Giese and E. Horsdal, Phys. Rev. Lett. **60**, 2018 (1988).
7   E.Y. Kamber, C.L. Cocke, S. Cheng, and S.L. Varghese, Phys. Rev. Lett. **60**, 2026 (1988) E.Y. Kamber, C.L. Cocke, S. Cheng, J.H. McGuire and S.L. Varghese, J. Phys. B **21**, L455 (1988).
8   R.E. Olson. Electronic and Atomic Collisions ed. by H.B. Gilbody, W.R. Newell, R.H. Read, and A.C.H. Smith, (Elsevier Science Pub), pp 271-285 (1988).
9   M.B. Shah and H.B. Gilbody, J. Phys. B **18**, 899 (1985)
10 J. Ullrich, H. Schmidt-Böcking, C. Kelbch, Nucl. Instr. Meth. A**268**, 216 (1988).
11 M.E. Rudd, Y.K. Kim, D.H. Madison, and J.W. Gallagher,Rev. Mod. Phys. **57**, 965 (1985)
12 J.F. Reading, A.L. Ford, and X. Fang, Phys. Rev. Lett. **62**, 245 (1989)
13 L. Vegh, J. Phys. B **22**, 135 (1989)

FIRST ATOMIC PHYSICS EXPERIMENTS WITH COOLED STORED ION BEAMS AT THE HEIDELBERG HEAVY-ION RING TSR

A. Wolf[a], V. Balykin[a,†], W. Baumann[a], J. Berger[a], G. Bisoffi[b], P. Blatt[a], M. Blum[b], A. Faulstich[a],
A. Friedrich[b], M. Gerhard[e], C. Geyer[b], M. Grieser[b], R. Grieser[e], D. Habs[a], H. W. Heyng[b], B. Hochadel[a],
B. Holzer[b], G. Huber[e], E. Jaeschke[b], M. Jung[b], A. Karafillidis[e], G. Kilgus[a], R. Klein[e], D. Krämer[b],
P. Krause[a], M. Krieg[e], T. Kühl[f], K. Matl[b], A. Müller[d], M. Music[a], R. Neumann[a,‡], G. Neureither[a],
W. Ott[b], W. Petrich[a], B. Povh[b], R. Repnow[b], S. Schröder[e], R. Schuch[c], D. Schwalm[a], P. Sigray[a,§],
M. Steck[b], R. Stokstad[a,¶], E. Szmola[a,‖], M. Wagner[d], B. Wanner[a], K. Welti[a] and S. Zwickler[a]

[a]Physikalisches Institut der Universität and Max-Planck-Institut für Kernphysik, Heidelberg, FRG
[b]Max-Planck-Institut für Kernphysik, Heidelberg, FRG
[c]Manne Siegbahn Institute (MSI), Stockholm, Sweden
[d]Institut für Kernphysik, Universität Giessen, FRG
[e]Institut für Physik, Universität Mainz, FRG
[f]Gesellschaft für Schwerionenforschung (GSI), Darmstadt, FRG

An overview of atomic physics experiments at the heavy ion Test Storage Ring (TSR) is given. Highly charged ions up to fully stripped silicon have been stored at energies between 4 and 12 MeV/u. The enhancement of the beam intensity by stacking, the beam lifetime, and electron cooling of these ion beams are discussed. Radiative and state-selective dielectronic recombination rates of hydrogen-like oxygen ions with free electrons from the electron cooler were measured. Beam noise spectra are being investigated with regard to collective effects caused by the Coulomb interaction in the cold ion beams. Resonance fluorescence from stored single-charged ions was observed using tunable narrow-band lasers. First indications of laser cooling in a storage ring were seen.

INTRODUCTION

Fast ion beams are commonly used for studying ionic and electronic collisions and for investigating highly ionized atomic systems. Also laser spectroscopy with fast ion beams has been performed successfully. New possibilities, as reviewed by Datz[1], are opened up for such experiments if the fast ions are kept on a circular orbit in a storage ring. The storage time, ranging from seconds up to hours, allows one to improve the phase-space density of the beam and to perform novel experiments which, in particular, profit from the long time of interaction of the ions with external fields or among each other.

At the Max-Planck-Institut für Kernphysik in Heidelberg, the heavy ion Test Storage Ring (TSR) came into operation[2] in 1988. Several atomic physics experiments under preparation for the TSR have been outlined recently[3]. In this presentation, an account of the first experimental results will be given, preceded by a short summary of the storage ring performance.

STORAGE RING LAYOUT

The beams for the TSR are supplied by a 13 MV tandem with a radiofrequency post accelerator, which yields ions of many elements in a large variety of charge states. Fully stripped ions up to Ni^{28+} at maximum energies of about 15 MeV/u are available. The storage ring with a circumference of 55 m, shown in Fig. 1, provides deflecting and focusing fields for particles with a magnetic rigidity up to 1.5 Tm. During the first runs the beam energy was between 4 and 12 MeV/u for highly charged ions ($^A[Z]^{q+}$, $q \approx Z$, $Z \leq 14$). The relatively heavy single-charged ions $^7Li^+$ and $^9Be^+$ were stored at energies up to 1.3 MeV/u, and protons at 21 MeV.

Between the magnetic elements four field-free straight sections, each of 5.2 m length, are available. One of these sections is occupied by the electron cooling device, where an intense electron beam can be merged with the ion beam over a length of 1.5 m. The process of electron cooling[4] is enabled by matching the mean electron velocity to that of the ions. The dynamical friction due to Coulomb collisions between electrons and ions presents a non-conservative external force, by which the phase-space density of the stored ion beam can be increased. This cooling method, discussed in more detail below, was for the first time applied to heavy ions at the TSR[5]. A short section at the first dipole magnet downstream of the electron cooler (experimental area I) is used for the detection of ions which change their charge state in the

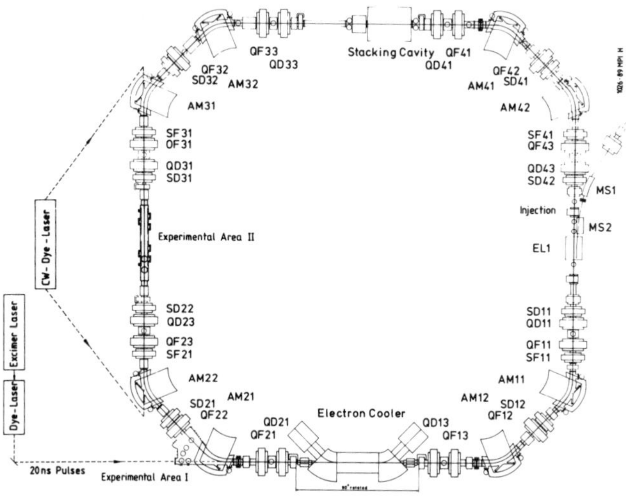

FIG. 1. TSR floor plan.

overlap region of electron and ion beam. Intense light pulses from a dye laser pumped by an excimer laser can be directed into this region for planned investigations[3] of laser-induced recombination.

Another straight section is dedicated to experimental installations (experimental area II) and presently used for laser spectroscopy and cooling of singly charged ions. As a further temporary installation in this section, a polarized atomic hydrogen target for nuclear physics experiments[6] is under construction.

STORAGE RING PERFORMANCE

Ion current and stacking. An important advantage of the storage ring is the possibility to increase the ion current by stacking ions from the accelerator over a large number of revolution periods. One class of stacking methods makes use of the fact that the phase space available for the stored ions is much larger than the phase-space volume occupied by the accelerator beam. External time-dependent fields can then be used to move the stored ions in phase space, thus creating free regions for injecting new ions and filling up the available volume. At TSR, two stacking methods of this class are used[7,8], one acting on the horizontal, the other on the longitudinal degree of freedom. The beam intensity obtained by these techniques is listed in the first two lines of Table I.

Additional stacking possibilities are opened up by using electron cooling. In contrast to external fields, the friction in the electron beam allows one to compress more and more ions into the same phase-space region. Results obtained at the TSR by combining electron cooling with the other stacking methods are also listed in Table I. At the given maximum currents the ion loss by further stacking compensated the feeding rate.

TABLE I. Beam currents obtained by combining stacking methods at TSR with 6 MeV/u C^{6+}. H: horizontal multiturn stacking, L: longitudinal (radiofrequency) stacking, EC: stacking by electron cooling.

Injected current	Stacking method H	L	EC	Stacking factor	Stored current	Number of ions
15 μA	•			40	0.6 mA	1.0×10^9
10 μA	•	•		700	6.9 mA	1.2×10^{10}
14 μA	•	•	•	1300	18 mA	3.0×10^{10}
2 μA	•		•	1400	2.7 mA	4.6×10^9

Ion beam lifetime. The lifetime of the stored ions is limited in most cases either by radiative recombination with free electrons in the electron cooler or by interactions with the rest gas. The TSR has been baked at 150°C in the beginning of 1989 and the average pressure obtained varied between 0.5 and 2×10^{-10} mbar. (In order to reach the design pressure of $< 1 \times 10^{-11}$ mbar, a 300°C bakeout is planned.) The typical residual gas composition was 93% of H_2, 6.7% of molecules containing C, N, and O atoms, and 0.3% of Ar. In Fig. 2 the predicted loss rates due to various processes are shown for ions with charge $q \approx Z$ as a function of Z. A pressure of 1×10^{-10} mbar, the given residual gas composition, and two beam energies typical for the runs in the first half of 1989 were used in the calculation. Details of the formulae used have been discussed elsewhere[3].

The measured inverse lifetimes plotted in Fig. 2 are in satisfactory agreement with the predicted loss rates. For fully stripped ions in the high-energy range (Fig. 2a) the recombination (REC) in the electron cooler determined the lifetime (5 h for 12 MeV/u C^{6+}). For the same ions at lower energies (Fig. 2b) electron capture (EC) in the rest gas became the limiting process (1 h for 6 MeV/u C^{6+}, 7 min for 6 MeV/u Si^{14+}). Very long lifetimes were obtained for protons at 21 MeV: 36 h with electron cooling,

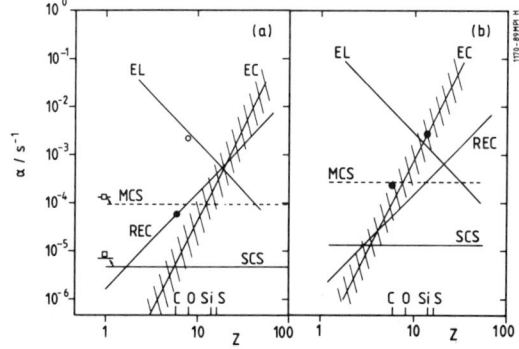

FIG. 2. Loss rates α for cooled ion beams in TSR as a function of the ion charge Z calculated for beam energies of 12 MeV/u (a) and 6 MeV/u (b). Measured loss rates for fully stripped ions (•), few-electron ions (9 MeV/u O^{7+}, o), and 21 MeV protons (□; upper point without cooling).

limited by single Coulomb scattering (SCS) only, and 3 h without cooling, limited by multiple Coulomb scattering (MCS). The loss rates related to Coulomb scattering often are negligible for the other beams stored in the TSR.

For highly charged ions with a few remaining electrons, the electron loss (EL) in rest gas collisions is predominant (6 min for 9 MeV/u O^{7+}, see Fig. 2a), except for very high Z. This process becomes much more frequent for the singly charged ions used in the laser experiments, in particular $^7Li^+$ and $^9Be^+$. These ions are much slower and the electrons are more weakly bound. The lifetime for such beams was about 15 s.

Electron cooling. In the first operation of electron cooling at the TSR, electron beam energies between 1 and 12 keV and currents up to 1 A were used. The electron beam had a diameter of 5 cm and was guided by a magnetic field in beam direction of a strength between 20 and 50 mT. In the typical case of 3×10^9 C^{6+} ions stored at 6 MeV/u, the momentum spread was reduced from 1.8×10^{-2} to the equilibrium of 2×10^{-4} (FWHM) within less than 3 s. The equilibrium momentum spread[9] was determined from the beam noise spectra discussed below. It depended on the beam intensity, the charge of the ions, and their energy as shown in Fig. 3. The increase of the momentum spread with the beam intensity is attributed to Coulomb collisions between the ions (intrabeam scattering), which heat the longitudinal motion in the rest frame on account of the transverse and also of the average translational energy. The lowest rest-frame energies in the longitudinal degree of freedom, calculated from the momentum spread, are indicated in Fig. 3.

An example for the reduction of the beam size by electron cooling is shown in Fig. 4. The spatial distribution of O^{6+} ions produced by charge-changing collisions in a beam of $\approx 4 \times 10^7$ O^{7+} ions at an energy of 9 MeV/u was measured. The width of these profiles roughly represents the size of the circulating beam. In the profiles for a cooled beam, which were dominated by radiative recombination in the electron cooler, the width was on the order of the detector aperture, corresponding to a radius of the circulating beam of roughly 1 mm. Taking into account the beam focusing, varying around the ring, this implies a divergence on the order of 0.3 mrad and a transverse ion energy in the rest frame on the order of 30 eV.

If the heating of the ion beam is minimized, for example by reducing the beam intensity, very low rest-frame temperatures can be obtained by electron cooling. Particularly low temperatures were reported for proton beams[10]. These possibilities are opened up by the experimental setup of a magnetically guided electron beam[11-13]. The typical transverse electron energy in such a beam is about 0.1 eV, corresponding to the cathode temperature. However, in the longitudinal direction

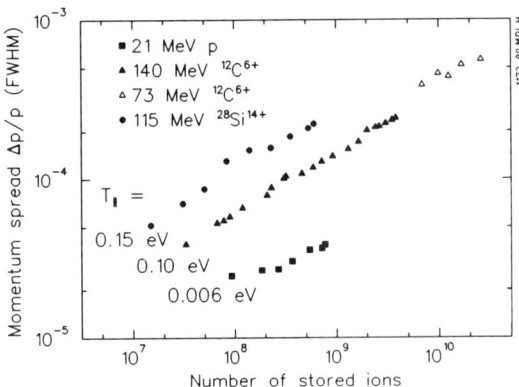

FIG. 3. Momentum spread of beams in TSR obtained by electron cooling.

a rest-frame energy down to 10^{-5} eV can be obtained because of the compression of the velocity distribution during the acceleration of the electrons. The friction force on an ion in the electron beam is to a large extent due to slow collisions, proceeding adiabatically with respect to the transverse cyclotron motion of the electrons. For these collisions, the effective electron temperature is given by that in the degree of freedom parallel to the magnetic field. Therefore, it is the longitudinal temperature of the electrons which determines the ultimate equilibrium temperature of the cooled ions. This illustrates that careful studies of the cooling and heating processes at the TSR could yield a further improvement of the beam temperature.

OVERVIEW OF EXPERIMENTS

Radiative recombination in the electron cooler. It has been recognized before[14] that the intense electron beam used for phase-space cooling, in connection with the stored intense and cold beam of highly charged ions, provides a favourable arrangement for studying electron-ion collisions. This also opens up the way to novel spectroscopic investigations of heavy ions. By detecting ions which capture or loose an electron in the electron cooler, recombination and electron impact ionization can be investigated. The installation of x-ray detectors close to the electron cooling device is planned for spectroscopy in connection with recombination and electron impact excitation.

In the first experiments at TSR with highly charged ions, the emphasis was on recombination processes with free electrons in the electron cooler. When the velocities of the electron and ion beams are matched during electron cooling, the radiative recombination of ions with free electrons at the rest-frame temperature of the electron beam is observed. The recombined ions are intercepted by a movable ion detector (Fig. 4a). A rate coefficient of $\alpha_r = (3.0 \pm 0.3) \times 10^{-10}$ cm^{3}s^{-1} was measured for the ra-

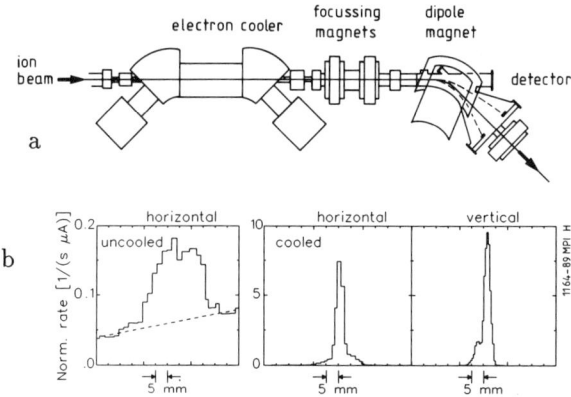

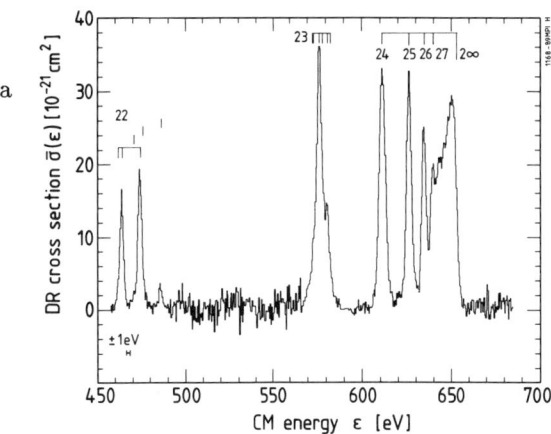

FIG. 4. Detection of O^{6+} ions formed by charge-changing collisions in an O^{7+} beam. (a) Scheme of experimental arrangement. (b) Profile of ions measured with a movable 2×2 mm detector. 'Uncooled': profile due to electron capture in the rest gas; 'cooled': profiles due to radiative recombination with free electrons.

diative recombination of O^{7+} ions, using a channel-plate detector of 18 mm diameter. Calculating α_r for bare ions[15] of charge $q = 7$, assuming a transverse thermal energy of 0.1 eV and a negligible longitudinal temperature of the electrons, one obtains 2.5×10^{-10} cm^3 s^{-1}, in fair agreement with the measurement for the hydrogen-like O^{7+}.

Dielectronic recombination. If ions with bound electrons are stored in the ring, state-selective measurements of dielectronic recombination (DR) can be performed using the electron cooler. The status of research on dielectronic recombination has been reviewed at this conference[16]. The ion investigated in a first measurement at TSR was O^{7+}, which can recombine via two-electron resonances of helium-like oxygen. Related to the three-body Coulomb problem, the positions and strengths of these resonances are of quite general interest and lend themselves to precise calculations. Moreover, the influence of electric stray fields on the resonance strengths is small in comparison to a number of earlier DR measurements.

The energy-integrated cross-sections of the investigated DR resonances are rather low, on the order of 10^{-20} cm^2 eV. The high sensitivity of the TSR experiment stems from the facts that intense electron and ion beams with a low energy spread are merged, and that the background from rest-gas electron capture is low because of the good vacuum. The O^{7+} ions were stored at an energy of 9 MeV/u and cooled by an electron beam of 4.9 keV. The center-of-mass (CM) electron energy ϵ of 460 eV, close to the $O^{6+}(2l\,2l')$ resonances, was reached by increasing the electron energy in the laboratory to about 9 keV. Spectra obtained by scanning the electron energy and measuring the recombination rate are shown

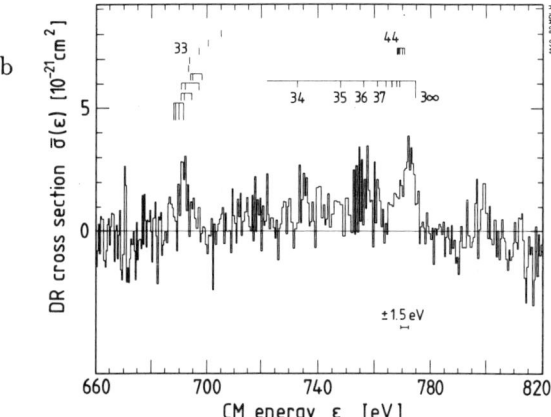

FIG. 5. Dielectronic recombination spectra of hydrogen-like oxygen. (a) Resonances due to $N = 2$ core excitation. (b) Resonances due to $N = 3$ core excitation.

in Fig. 5. The CM energy resolution was about 1 eV (FWHM). After a correction for the electron space-charge an absolute accuracy of ϵ better than ± 1 eV is estimated.

The spectra obtained in this experiment are discussed in more detail elsewhere[17]. Note that from the observed spectra, the energies and resonance strengths of all terms of the $2l\,2l'$ configuration can be determined and compared to theory. Detailed calculations can be checked against experiment for the complete Rydberg series $2l\,nl'$. A small peak is attributed to terms of the $3l\,3l'$ configuration according to Fig. 5b. Finally, the accuracy of the energy calibration is illustrated by comparing the measured 2∞ and 3∞ edge energies, 653.4(1) eV and 774.5(2) eV, to the hydrogenic calculation, yielding 653.7 eV and 774.7 eV.

Noise spectra and collective effects in cold ion beams. The observation of noise generated in a pick-up device by stochastic fluctuations of the beam current is a well established diagnostic method for the energy distribution of ion beams in a storage ring[18]. In the conventional high-energy rings, the ion beams are relatively hot

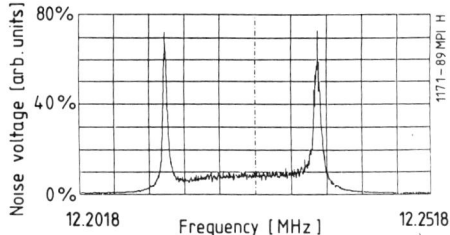

FIG. 6. Beam noise spectrum of 3×10^{10} C^{6+} ions cooled to a momentum spread of 5×10^{-4} (FWHM).

and therefore, the pulses induced in the pick-up by the individual ions add up independently so that the noise spectrum simply reflects the distribution function of the ion revolution frequency.

For the low beam temperatures obtained by electron cooling, however, the shape of the noise spectrum is determined by collective oscillations of the ions in the beam rest-frame[19,10]. A spectrum obtained at the TSR cooling an intense C^{6+} beam at 73 MeV is shown in Fig. 6. Around the 20th harmonic ($n = 20$) of the revolution frequency ω_r, a noise band, delimited by sharp maxima on both sides, is observed. The distance of the peaks from $n\omega_r$ (the center frequency of the band) is determined by the frequency $\omega_c(n)$ of the collective oscillation. This oscillation, caused by restoring forces due to the electromagnetic interaction among the ions, is similar to the density variations in a sound wave with a wavelength equal to the nth fraction of the storage ring circumference.

In the approximation of a weak coupling between the ions[19], the frequency $\omega_c(n)$ is related to their plasma frequency. At very low beam temperatures the ion beam will reach the limit of strong coupling, as expressed by the ratio of the Coulomb interaction energy at the average ion distance d to the thermal energy. Values $\gg 1$ of this ratio Γ lead to ordering and possibly to crystalline structures in the ion ensemble[20]. In this case, the noise spectrum will be of particular interest[3] since it is proportional to the dynamic structure function of the ion ensemble, transformed from the rest frame into the laboratory.

The present optimum of the *longitudinal* ion temperature for Si^{14+} (Fig. 3) and the ion distance calculated for an estimated beam radius of 1 mm ($d \approx 300$ μm) yield an ordering parameter of $\Gamma \approx 0.006$. In order to reach higher ordering parameters, efforts will be directed to the improvement[21] of electron cooling. In addition, the method of laser cooling promises even lower equilibrium beam temperatures in the longitudinal degree of freedom.

Laser spectroscopy and cooling. Resonance scattering of photons from a laser beam by moving ions gives rise to a light-pressure force with a very sharp dependence on the ion velocity[22,23]. Taking typical parameters for singly charged ion beams in the TSR, the natural line profile of the laser transition translates into a longitudinal velocity spread of $\Delta v/v \approx 10^{-7}$; only ions within this narrow range can strongly interact with the laser. For the situation depicted in Fig. 7, the ions are decelerated by the light-pressure force; the average relative change of their velocity is $\approx 10^{-8}$ per absorption process. Taking into account the lifetime of the excited state and the beam velocity, and assuming a laser intensity sufficient to saturate the transition, a resonant ion can perform one cycle of absorption and spontaneous re-emission within an average distance of about 0.1 m to 1 m. After several such cycles the ions are driven out of resonance by the light-pressure force, going to the low-velocity side of the resonance profile for antiparallel laser and ion beams. When the decelerated ions are kept in resonance by sweeping the laser frequency or by applying re-acceleration in an external field, the velocity profile of the ion beam can be compressed. The ultimate longi-

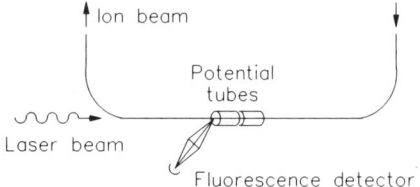

FIG. 7. Schematic setup for laser cooling at TSR.

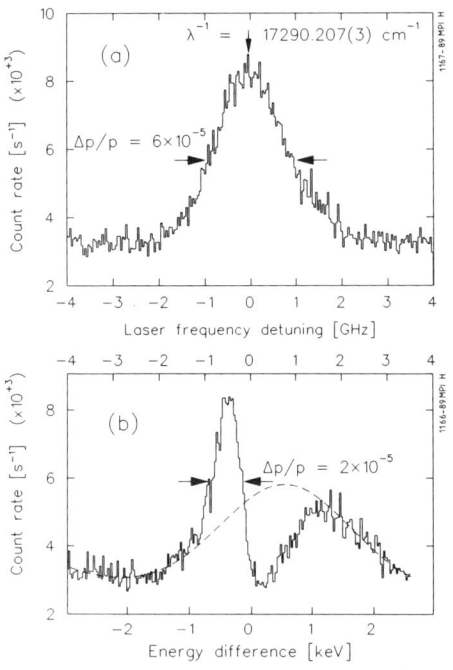

FIG. 8. Energy profiles of $^7Li^+$ ions interacting with resonant laser light in TSR. (a) Uncooled profile obtained by scanning the laser frequency. (b) Profile obtained by varying the 'retarding' voltage after the laser frequency had been scanned over the Doppler profile, as described in the text.

tudinal velocity spread of this cooling method is on the order of the natural width of the resonance profile (Δv), corresponding to a longitudinal thermal energy of about 10^{-7} eV.

Ions with suitable laser lines investigated at the TSR were ^{7}Li$^+$ and ^{9}Be$^+$. The ^{7}Li$^+$ ions[24] were excited from the metastable 1s2s $^3S_1$ ($F = \frac{5}{2}$) state (radiative lifetime 50 s) via a transition at a wavelength of 548 nm in the ion rest-frame. Approximately 10% of the ions coming from the accelerator are in this state. Observing the fluorescence intensity as a function of the laser wavelength, the longitudinal velocity distribution of the ions was measured (Fig. 8a). Using the Doppler formula, the average ion velocity can be determined with a high absolute precision ($\approx 1 \times 10^{-5}$) from the laser wavelength and the well known transition frequency of the ^{7}Li$^+$ ion.

The effect of the light-pressure force and indications of laser cooling were observed in a two-step experiment. First the laser resonance was moved over the Doppler profile (dashed line in Fig. 8b) from the high-velocity wing to its center. The ion velocities should be compressed into a narrow velocity interval on the low-velocity side of the laser resonance. With a second resonance, which had much weaker influence on the velocity distribution, the result of the first scan was probed. As shown in Fig. 7, the fluorescence of the ions was observed in a region where the ion energy could be varied slightly by an electric potential on the order of 1 kV. While the laser frequency was now kept unchanged, the second resonance was realized by varying this electric potential and measuring the fluorescence intensity as a function of its value. The signal is shown in Fig. 8b. For a slightly negative potential, corresponding to velocities just below the laser resonance, a strong increase of fluorescence within a velocity spread of 2×10^{-5} (FWHM) was observed. It is attributed to a fraction of the ions cooled by the laser. The width of the signal, much larger than expected from the natural line profile, is probably due to the relatively long scanning time and the delay (≈ 1 s) between the preparation and the probing of the velocity distribution; this will be avoided in later experiments.

As the laser only interacts with ions in the mentioned metastable state, a large fraction of the ^{7}Li$^+$ beam is not cooled. This was the main motivation to investigate also the laser interaction with stored ^{9}Be$^+$ ions[25], which can be excited from the ground state using a wavelength of 313 nm. Whereas all ions can be cooled in this system, the situation is more complex than in ^{7}Li$^+$ since optical pumping can occur. Thus, two adjacent hyperfine transitions must be driven simultaneously. In first experiments with ^{9}Be$^+$ this was achieved by Doppler tuning with the 'retarding' potential. Analyzing the time dependence of the fluorescence intensity, indications were observed that the ions are driven out of the resonance by the light pressure force.

We would like to acknowledge the skill and the effort of the technicians of the Max-Planck-Institut during the construction of the TSR and the preparation of the first experiments. This work has been funded by the German Federal Minister for Research and Technology (BMFT) under the contract numbers 06 HD 938 I, 0234 HD CI, 06 HD 852 I, and 06 MZ 458 I, and by the Gesellschaft für Schwerionenforschung (GSI), Darmstadt.

References

1. S. Datz, Nucl. Instrum. Methods **B24/25**, 3 (1987).
2. E. Jaeschke et al. in *Proc. European Particle Accelerator Conf.*, Rome, 1988, edited by S. Tazzari (World Scientific, Singapore, 1989), p. 365.
3. D. Habs et al., to be published in Nucl. Instrum. Methods.
4. G.I. Budker and A.N. Skrinsky, Sov. Phys. Usp. **21**, 277 (1978).
5. M. Steck et al., to be published in Nucl. Instrum. Methods.
6. W. Haeberli et al. in *Proc. 4th LEAR workshop*, Villars-sur-Ollon, 1987, edited by C. Amsler et al. (Harwood Acad. Publ., Chur, 1988), p. 195.
7. A. Noda and M. Grieser, MPI Heidelberg report V23 (1987).
8. D. Krämer et al., to be published in Nucl. Instrum. Methods.
9. M. Steck et al., to be published.
10. E.N. Dement'ev et al., Sov. Phys. Tech. Phys. **25**, 1001 (1980).
11. Ya.S. Derbenev and A.N. Skrinsky in *Physics Reviews*, edited by I.M. Khalatnikov, vol. 3, (Harwood Acad. Publ., Chur, 1981), p. 165.
12. A. Wolf, Phys. Scr. **T22**, 55 (1988).
13. N.S. Dikansky et al., Institute of Nuclear Physics Novosibirsk preprint 88-61.
14. R. Schuch, Nucl. Instrum. Methods **B24/25**, 11 (1987).
15. M. Bell and J.S. Bell, Part. Acc. **12**, 49 (1982).
16. Papers by A. Müller and by R. Schuch, these Proceedings.
17. G. Kilgus et al., submitted to Phys. Rev. Lett.
18. D. Boussard in: CERN 87-03, Geneva, 1987, p. 416.
19. V.V. Parkhomchuk and D.V. Pestrikov, Sov. Phys. Tech. Phys. **25**, 818 (1980).
20. *Proc. Workshop on Crystalline Ion Beams*, Wertheim, 1988, edited by R.W. Hasse, I. Hofmann and D. Liesen, GSI-89-10 (GSI, Darmstadt, 1989).
21. D. Habs et al., Phys. Scr. **T22**, 269 (1988).
22. T.W. Hänsch and A.L. Schawlow, Opt. Comm. **13**, 68 (1975).
23. S. Stenholm, Rev. Mod. Phys. **58**, 699 (1986), and references therein.
24. G. Huber et al. in *Proc. 9th Int. Conf. on Laser Spectroscopy*, Bretton Woods, 1989, to be published.
25. A. Faulstich et al. in same Proc. as Ref. 24.

---

†On leave from Academy of Sciences, Troizk, USSR
‡Now at GSI, Darmstadt, FRG
§On leave from MSI, Stockholm, Sweden
¶Senior Humboldt Awardee, on leave from Lawrence Berkeley Laboratory, Berkeley, USA
‖On leave from Technical University, Miskolc, Hungary

STATE-SELECTIVE ANGULAR-DIFFERENTIAL SINGLE-ELECTRON CAPTURE IN VERY SLOW Ar⁴⁺ - Ar COLLISIONS

C. Biedermann, H. Cederquist*, J. C. Levin, R. T. Short, L. Liljeby*, L. R. Andersson*,
H. Rothard**, K-O. Groeneveld**, C-S. O, C. R. Vane, J. P. Gibbons, S. B. Elston and I. A. Sellin

Department of Physics, University of Tennessee, Knoxville, Tennessee 37996-1200
and Oak Ridge National Laboratory, Oak Ridge, Tennessee 37831-6377
*Manne Siegbahn Institute for Physics, S-10405 Stockholm, Sweden
**Institut für Kernphysik der J. W. Goethe Universität, D-6000 Frankfurt am Main, FRG

> State-resolved angular distributions for single-electron capture in very-slow collisions (0.5-5 eV/amu) of Ar⁴⁺ and Ar have been measured with a new experimental apparatus. The energy-gain spectra show that single-electron capture populates the 4p level, while no capture to 4s is observed in this energy range. The 4p angular distribution has a single maximum close to 0° for energies larger than about 100 eV. At lower energies a second maximum develops and forward scattering is suppressed. The secondary maximum moves to larger scattering angles as the collision energy is lowered and below about 60 eV a third maximum appears. The positions of the primary and the secondary peaks as functions of the collision energy are reproduced by semiclassical calculations of the angular differential cross section.

In the past years there has been a growing interest in the physics of slow collisions between highly-charged ions and atoms. This is motivated by the urge to understand the details of the fundamental process of charge transfer reactions. One of the reasons why this problem has attracted so much attention lately is the importance of charge-exchange collisions in many laboratory and astrophysical plasmas. In the context of fusion research based on magnetic confinement of high temperature plasmas, these processes are relevant to energy loss mechanisms.[1] Even though the most important cooling mechanism in fusion plasma is loss of energetic photons by the capture of free electrons to highly-charged impurity ions, electron capture in slow collisions between highly-charged ions and neutral atoms contributes to cooling and is, moreover, important for fusion plasma diagnostics.[1] Information on near-resonant charge exchange is needed to get further insight into the interactions between the plasma and residual gas, which take place over a wide range of energies. Slow multiply-charged ions and neutral atoms may also coexist in astrophysical plasmas.[2] To estimate the abundance of elements and their ionization from astropysical spectra, a detailed understanding of electron-transfer mechanisms is needed.

The process of single-electron capture at low energy is fairly well understood at the level of total charge exchange cross sections and the final capture states of the projectile.[3-11] Therefore, experimental and theoretical activities have recently shifted towards detailed studies of more complicated processes such as, e.g., measurements and calculations of angular distributions of one- and two- electron processes in slow and very-slow collisions.[12-18] The merged-beam technique is well suited for measurements of absolute charge exchange cross sections at very low energy, since the scattering in the center-of-mass system is compressed in the laboratory. Due to this compression of scattering angle and the small velocity change of the projectile in the inelastic collision compared to the laboratory beam velocity, however, it is difficult to extract post-collisional kinematic effects. This technique has been used, for example, by Havener et.al.[17] to measure single-electron capture cross sections for N⁵⁺ and O⁵⁺ colliding with atomic hydrogen and deuterium at center-of-mass energies down to ~1 eV. So far, experiments involving state-selective angular distributions are rare. Until very recently[19], information on angular distributions has been obtained primarily for collision systems where the electron capture states are well known from complimentary experiments.[20-23]

In slow collisions, the relative velocity of the heavy particle is much smaller than the orbital velocity of the active electron. The capture of one or several electrons can thus be treated in the quasimolecular model. In this framework the collision complex is viewed as a diatomic molecule in which the internuclear separation R changes slowly compared to the motion of the electrons. When the collision partners are far apart, the energy eigenstates of the quasimolecule can be described by Stark-shifted atomic energy levels. As the nuclei approach each other, the potential shows a series of avoided crossings for nearby molecular states of the same spin and symmetry. The diabatic transition probability at an avoided crossing of adiabatic states depends on the collision velocity and the magnitude of states of the quasimolecule can be described by Stark-shifted atomic energy levels. As the nuclei approach each other, the potential shows a series of avoided crossings for nearby molecular states of the same spin and symmetry. The diabatic transition probability at an avoided crossing of adiabatic states depends on the collision velocity and the magnitude of the energy splitting. The capture probability is significant for a certain range of R-values, where adiabatic and diabatic transition probabilities are of the same order of magnitude. This is often referred to as the reaction window.[24] Closely related to this concept is the over-barrier model, where charge transfer corresponds to the transfer of an electron having sufficient energy to overcome the potential barrier between the projectile and the target nucleus.[25] The height of the barrier is given by the saddlepoint of the superimposed Coulomb potentials due to the positive ions. These two models and various extentions [3,11,26,27] are able to explain a variety of single-electron capture data[3,5,9,28] with reasonable agreement.

At very low velocities, i.e., where the inelasticity is of the same order of magnitude as the kinetic energy in the center-of-mass system, the locations of the active avoided crossings and the shape of interpotential curves have a greater influence on the nuclear trajectories. Since radial coupling at avoided crossings of adiabatic molecular states is expected to be the dominant electron transfer mechanism in both the slow and the very-slow collision regions, we hope to be able to reveal features of the electron-transfer process, which may be applied also at higher energies. At very low velocities it is possible to resolve, from measurements of angular distributions over a wide angular range and at various velocities, features of the interatomic potential and diabatic transition probabilities, which are hidden at higher velocities. In addition, investigations of this kind offer the possibility to test different theoretical approaches of charge exchange at very low velocities. Experimental results of state-selective angular distributions can serve as a guide to help determine in which situations a full quantum-mechanical treatment[10] of the collision system is necessary, or alternatively when a semiclassical treatment would be sufficient.

In this paper we present state-selective angular-differential measurements of single-electron capture of Ar^{4+}-Ar collisions at very low energies and compare them with results from a, somewhat preliminary, semiclassical calculation of the angular-differential cross section at laboratory collision energies 19, 40, 80 and 200 eV. As will be shown below, the comparison between theoretical and experimental results is quite favourable. The measurements required the development of an intense source of very-slow highly-charged ions, with good angular collimation, and a cooled gas target.

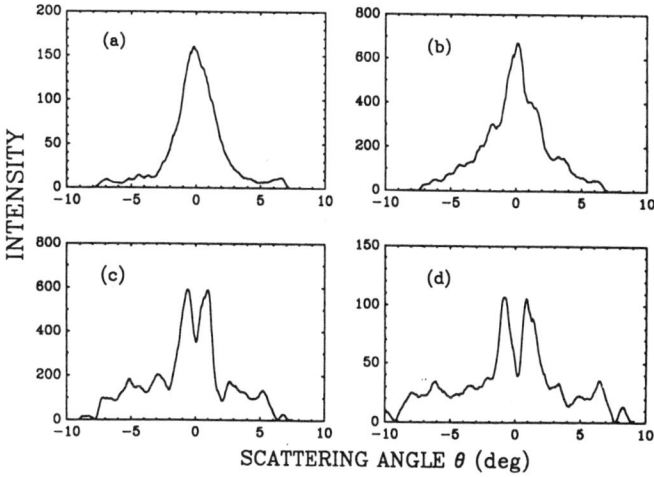

Fig. 1: *Angular distribution for single-electron capture to the 4p level of Ar^{3+} in Ar^{4+}-Ar collisions at (a) 200, (b) 80, (c) 40, (d) 19 eV. The small asymmetries are due to a non-uniform detector efficiency.*

The experimental setup consists of a differentially pumped recoil ion source, a Wien filter, an effusive gas jet and a 30° parallel-plate analyzer with a two-dimensional position-sensitive detector (PSD).[19] A detailed description of the low energy recoil-ion source and its performance is given elsewhere.[29] A 30 MeV Cl^{5+} beam from the Oak Ridge National Laboratory EN-12 tandem Van de Graaff accelerator is directed onto an Ar-filled gas cell. The Ar recoil ions are extracted to form a very-slow ion beam. In order to make energy analysis after the secondary collision with the gas jet meaningful, it is essential to collimate the heavy-ion pump beam tightly (~ 0.4 mm) and keep the extraction field low (≤ 0.5 V/cm). Outside the differential pump step, and after acceleration, the recoil ions enter the Wien filter, which separates the charge states of the beam. After the Wien filter, the beam is retarded to the collision energy. The very-slow beam has a diameter of 1 mm diameter at the target and is collimated to an angular divergence of ± 0.5°. The collision energy of the very-slow ion beam is given by the potential difference between the positions of the heavy-ion pump beam and the gas-jet target. The effusive gas jet is furnished by a concave glass-capillary array, which is focused onto the very-slow ion beam. To resolve structure in the angular distribution after the collision it is necessary to diminish the thermal velocity spread of the gas jet. The angular divergence of the drifting gas atoms is limited to less than 10° and the jet has a diameter of 1 mm at the intersection point. With the differentially pumped gas jet, a target pressure of 2 mTorr can be provided, while keeping the pressure in the main chamber at $5 \cdot 10^{-7}$ Torr. After the intersection of the very-slow ions with the gas-jet target, the charge and energy of the ions are analyzed with a 30° parallel plate electrostatic spectrometer. Since the analyzer is focusing only in one dimension, the energy-gain of the projectile can be obtained from the position along the direction of dispersion in the focal plane, while the position in the perpendicular direction gives information on the scattering angle. The energy and angular distributions in the focal plane are mapped onto a two-dimensional position-sensitive detector by two short acceleration fields.

The spectrometer is designed to have a relative energy resolution of 3%. Due to the energy spread of the very-slow ion beam, the measured energy resolution is ~4%. The angular acceptance of the present setup is fixed by the size of the micro-channel-plate and the resistive anode to about ±9°.

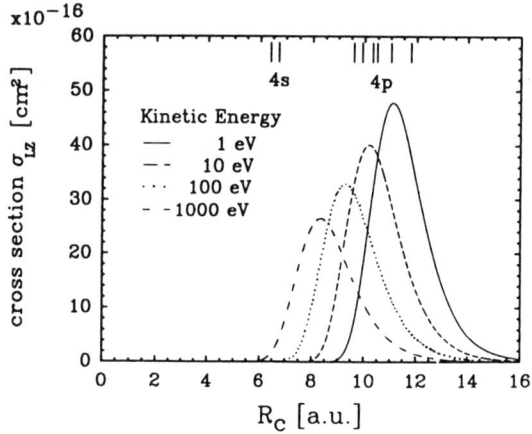

Fig. 2: *The reaction window for single-electron capture to the 4p level in Ar^{4+}-Ar collisions as a function of the laboratory kinetic energy. The cross section σ_{LZ} is calculated within the framework of a single crossing Landau-Zener model.*

In fig. 1 angular distributions are shown for single electron capture to the 4p level of Ar^{3+} for 200, 80, 40 and 19 eV laboratory energy of the incident Ar^{4+} projectile. The scale for the angular axis was calibrated from the vertical positions of the primary ion beam on the PSD along with the calculated trajectory length between target and detector. The angular distributions in fig. 1 are plotted for the energy-gain peak corresponding to capture into the Ar^{3+} 4p level (see fig. 1 in abstract of contributed papers, this conference). The energy calibration was obtained by recording the position on the PSD of the primary ion beam for a constant beam en-

ergy as a function of the analyzer voltage. Within the experimental resolution it is not possible to distinguish among the electron capture to the six different LS-terms, which are formed by adding a 4p electron to the Ar^{4+} $3p^2$ (^3P) core. These six unresolved states are spread within 1.6 eV excitation energy and are each separated by 0.1 to 0.5 eV. In energy-gain experiments of the same collision system at higher projectile energy (2180 eV), Giese et. al.[7] have found that electron transfer to the 4p and the 4s levels of Ar^{3+} takes place with a ratio of the cross sections >2. For the projectile energy of 800 eV, Puerta et. al.[8] have seen the 4p level dominating by a factor of ~5. Energy-gain spectra of single-electron capture in 160 to 400 eV Ar^{4+} on Ar collisions have been investigated by Hvelplund and Cederquist at Aarhus University.[30] As the projectile energy was lowered, the contribution of the 4p level dominated, and about 200 eV the energy-gain peak of 4s vanished in the background. Those results[30] agree with the present observation of capture to only the 4p level in the energy range 19 to 200 eV. This shift of intensity from the 4s to the 4p level of Ar^{3+} is in qualitative agreement with the expected shift of the reaction window towards larger crossing radii R_c for lower energies as seen in fig. 2. The cross section σ_{LZ} in fig. 2 is calculated within

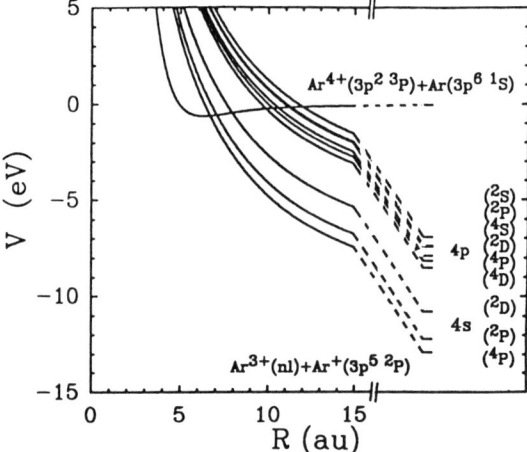

Fig. 3: *Potential energy diagram for the Ar^{4+}-Ar collision system. The potential energy curves of the incoming channel and the 4p and the 4s channels are plotted as functions of the internuclear distance R.*

the concept of a single-crossing Landau-Zener model[24] for capture of a 3p electron from Ar into the 4p level of Ar^{3+}, by integration of the transition probabilities over all impact parameters smaller than R_c. Evaluation of σ_{LZ} for various single-crossing distances R_c gives the reaction window. Only the coulomb repulsion and the energy of the final state are taken into account, and the semi-empirical expression for the adiabatic energy splitting H_{12} given by Olson and Salop[31] is modified by the factor proposed by Taulbjerg[32] for a multi-electron projectile.

The potential energy curve diagram of Ar^{4+}-Ar and Ar^{3+} (4p and 4s)-Ar^+ are shown as functions of the internuclear distance R in fig. 3. Polarization and penetration of the $3p^6$-shell of the neutral target atom by the Ar^{4+} projectile are included in the incoming channel. For the outgoing channels the major contribution to the potential energy is given by the coulomb interaction, from which the inelasticity of the reaction for the respective final state is subtracted. In addition, polarization and penetration of the Ar^+ $3p^5$ (^2P) term are included in the outgoing channels. The crossing of the adiabatic potential curves of the 4p levels with the incoming channel all lie in the range of internuclear distances between 9.6 and 11.9 au., while the crossing with the 4s levels occur at 6.3 to 7.6 au.

We have performed semiclassical calculations of angular-differential cross sections $\frac{d\sigma}{d\Omega}$ for single-electron capture to the 4p level of Ar^{3+} in Ar^{4+}-Ar collisions at laboratory energies 19, 40, 80 and 200 eV. The differential cross sections are calculated as sums of classical differential cross sections for the possible nuclear trajectories weighted by the corresponding mulitchannel Landau-Zener transition probability[24], as described by Andersson et. al.[33]. In the present work the model given in ref. 33 is extended to account for polarization, penetration and interference effects. With the potential curves of fig. 3 the classical deflection functions $\Theta(b)$, where b is the impact parameter, are calculated for each of the six possible final 4p (^{2s+1}L) terms. For example, for the 4p ^2S exit channel at 40 eV collision energy, we arrive at the deflection function of fig. 4. For other exit channels (^2P, ^4S, ^2D, ^4P or ^4D), which cross the incident channel at smaller internuclear distances, the deflection functions lie at smaller angles. We assume that the incident projectile is in its ground state $3p^2$ (^3P) and the ionic core of the recoiling target is in a $3p^5$ (^2P) state. Since it is not possible from the experimental resolution of the energy-gain spectrum to decide which of the final 4p levels are populated, all six possible 4p exit channels are considered. If the system behaves diabatically on the way in to the collision and adiabatically when crossing the 4p levels on the way out from the collision, the lower branch of the deflection function in fig. 4 is followed, giving rise to small scattering angles. For adiabatic transitions on the way into the collision, the projectile can encounter a repulsive wall at large R, which results in large scattering angles. The intensity at large scattering angles thus corresponds to the probability flux on the upper branches of the deflection function (fig. 4). The lower branch of the deflection function rises at small impact parameters, due to the assumed adiabaticity of the crossing of the incident channel with the $3p^2$ 4s (^2D) curve, producing a singularity in the classical cross section at the rainbow angle.[34] The multi-valued nature of the deflection function, allows the same scattering angle θ for several (up to seven for the ^2S exit channel) different impact parameters b. This leads to the possibility of interference effects, which in principle may result in oscillations in the angular-differential cross sections. As discussed by Ford and Wheeler[34] the interference phases are differences of action integrals along any two possible paths on the multi-valued deflection functions. This difference is proportional to the enclosed area within two of the branches of the deflection function between the rainbow angle and the observation angle. The variation of this area determines the frequency of the oscillations.

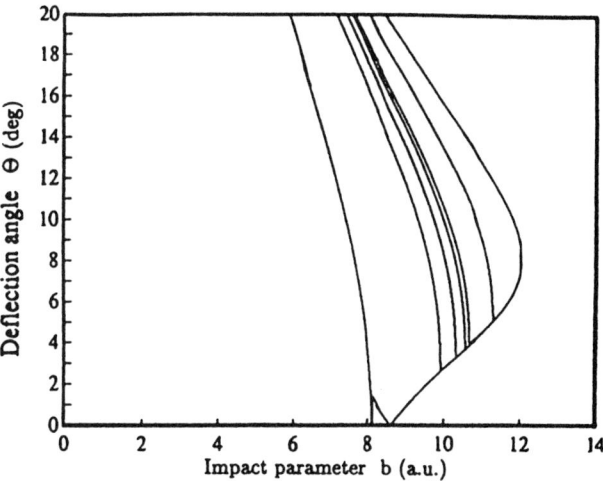

Fig. 4: *Laboratory deflection function for the process Ar^{4+} $3p^2$ (^3P) + Ar $3p^6$ (^1S) - Ar^{3+} $3p^2$ 4p (^2S) + Ar^+ $3p^5$ (^2P) at the laboratory collision energy 40 eV*

In the present calculation we used the semi-empirical expression for the adiabatic energy splitting given by Olson and Salop[31] H_{12}^{OS} and modified it for open shell projectiles colliding with multi-electron targets. The matrix element H_{12}^{OS} was developed to fit experimental single-electron capture cross sections for bare nuclei colliding with atomic hydrogen. For the present multi-electron collision system with an open shell projectile, the matrix element H_{12} is modified, allowing only transitions between the parts of the wave functions that have the correct angular properties. This modification is derived under the assumption of distant radial couplings between projectile and target with electron distributions well described by atomic wavefunctions. In the initial channel, the $Ar^{4+}(^3P)$ ion and the $Ar(^1S)$ atom can form a $^3\Sigma^-$ or a $^3\Pi$ state, of which the latter is doubly degenerate due to the two different possible directions of the projected total angular momentum. The calculation is divided in two parts, one for each symmetry of the initial channel. An angular-differential cross section is calculated for each of the six final 4p levels. These six results are then added incoherently to form a total 4p angular-differential capture cross section for each of the initial states $^3\Sigma^-$ and $^3\Pi$. At some crossings the final diabatic molecular curves are degenerate, which is taken into account by a modification of the multichannel Landau-Zener transition probabilities. The developments of the phase angles are calculated as differences of action integrals for each pair of branches in the deflection functions.

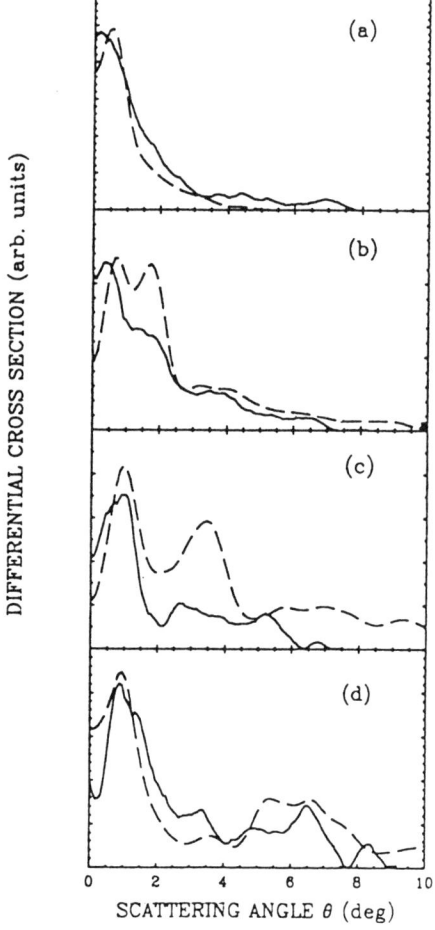

Fig. 5: *Calculated total differential cross section $\frac{d\sigma}{d\Omega}$ (dashed line) as functions of the laboratory scattering angle θ togehter with the experimental angular distribution (solid line) for (a) 200, (b) 80, (c) 40 and (d) 19 eV Ar^{4+} on Ar collisions.*

In fig. 5 the calculated total differential cross section $\frac{d\sigma}{d\Omega}$ (dashed line) is shown as a function of the scattering angle θ together with the measured angular distribution (solid line) for 19, 40, 80 and 200 eV collision energy. The angular distribution $\frac{d\sigma}{d\Omega}(\theta)$ is convoluted with a Gaussian response function with a FWHM of the peak of 0.5°.

The peak at small scattering angles is a rainbow resulting from the fact that many impact parameters in the lower branch of the deflection function can give rise to almost the same scattering angle. The angular position agrees fairly well with the scattering angle of the first maxima of the experimental angular distribution (solid line, fig. 5). If the 4s (^2D) level is assumed to be diabatic, however, and the 4s (^2S) level is taken as a repulsive wall, the rainbow angle is moved closer to zero degree at lower energies, in disagreement with the experimental data. It should be noted that while $\frac{d\sigma}{d\Omega}$ is calculated, the measured angular-differential cross sections are integrated over a range of scattering angles in the energy-dispersive plane, due to the focussing properties of the analyzer.

The calculated cross section $\frac{d\sigma}{d\Omega}$ is at very small angles θ extremly sensitive to the detailed shape of the potential curves, due to the $\sin\theta$ factor in the denominator of $\frac{d\sigma}{d\Omega}=\frac{b}{sin\theta}|\frac{db}{d\theta}|$. In the angular distribution for 19, 40 and 80 eV secondary peaks show up in the data as well as in the calculation. The secondary peaks move to larger scattering angles with lower energy. At 200 eV collision energy a second maximum in the angular distribution is not visible in the experiment or the calculation. Whereas the peak closest to the forward direction in each spectrum results from intensity on the lower path of the deflection functions, the secondary peaks are due to rainbow scattering corresponding to minima associated with the crossings of the incident channel with the 4p channels. A further result of the calculation, which is also visible in the reaction windows of fig. 2, is the dominating contribution of the innermost of the 4p channels to the cross section at the present collision energies. In the calculation of $\frac{d\sigma}{d\Omega}$, interference effects are included with an assumed phase offset of zero. The influence of the interference is relatively limited, since adding the intensity for all six outgoing channels tends to wash out any oscillatory structure. The third peak in the calculated angular distribution of 40 eV collision energy might be shifted in position due to the minimum next to it. This is at least partly caused by the semiclassical treatment of the collision. The intensity is dramatically reduced at the position corresponding to the scattering angle, where the two branches of the deflection function meet and $\frac{d\Theta}{db}$ becomes infinite. However, if quantum mechanical tunneling were included this singularity would be removed.

Results on inelasticities and angular distributions in very-slow collisions between highly-charged ions and neutral atoms have been presented. Important features of the experiment are a recoil-ion source of low energy spread producing an intense, collimated beam of very-slow highly-charged ions, a transversly cooled gas-jet target, and a line-focusing energy analyzer followed by a two-dimensional position-sensitive detector. The angular distributions of single electron capture to the 4p level of Ar^{3+} in Ar^{4+}-Ar collisions have been measured and calculated in the laboratory energy range 19 to 200 eV. The positions of the first and the second maxima near the forward direction are well reproduced for all energies, regardless of whether interference effects are included or not. In conclusion, the main structure of the angular-differential cross section for single-electron capture to the Ar^{3+} 4p level in Ar^{4+}-Ar collisions is qualitatively reproduced by the present semiclassical calculations in the energy range (0.5 - 5 eV/amu).

Work supported in part by the National Science Foundation and by the U. S. Department of Energy, Office of Basic Energy Sciences, Division of Chemical Sciences, under Contract No. DE-AC05-84OR21400 with Martin Marietta Energy Systems Inc., by the Deutsche Forschungsgesellschaft, the Bundesministerium für Forschung und Technologie / Bonn (FRG) and by the Swedish Natural Science Research Council (NFR).

References

[1] *Atomic and Molecular Processes in Controlled Thermonuclear fusion* eds. C. J. Joachain and D. E. Post (Plenum Press New York, 1983).
[2] A. Dalgarno and S. E. Butler, Comm. Atom. Mol. Phys. 7, 129 (1987).
[3] L. Liljeby, G. Astner, A. Bárány, H. Cederquist, H. Danared, S. Huldt, P. Hvelplund, A. Johnson, H. Knudsen and K-G. Rensfeldt, Physica Scripta 33, 310 (1986).
[4] H. Lebius, H. R. Koslowski and B. A. Huber, Z. Phys. D11, 53 (1989).
[5] C. Schmeissner, C. L. Cocke and R. Mann, W. Meyerhof, Phys. Rev A30, 1661 (1984).
[6] E. Y. Kamber and J. B. Hasted, J. Phys. B16, 3025 (1983).
[7] J. P. Giese, C. L. Cocke, W. Waggoner and L. N. Tunnell, S. L. Varghese, Phys. Rev. A34, 3770 (1986).
[8] J. Puerta, H-J. Kahlert, H. R. Koslowski, B. A. Huber, Nucl. Instr. & Meth. B9, 415 (1985).
[9] E. H. Nielsen, L. H. Andersen, A. Bárány, H. Cederquist, J. Heinemeier, P. Hvelplund, H. Knudsen, K. B. MacAdam and J. Sørensen, J. Phys. B18, 1789 (1985).
[10] J. Tan and C. D. Lin, Phys. Rev. A37, 1152 (1988).
[11] E. H. Niehaus, J. Phys, B19 2925 (1986).
[12] M. Barat, Comm. At. Mol. Phys. 21, 307 (1988).
[13] E. Y. Kamber, J. Phys. B21, 4185 (1988).
[14] H. Danared, A. Bárány, J. Phys. B19, 3109 (1986).
[15] C. L. Cocke, J. Physique 50, C1-19 (1989).
[16] H. Laurent, M. Barat, M. N. Gaboriaud, L. Guillemot, P. Roncin, J. Phys. B20, 6581 (1987).
[17] C. C. Havener, M. S. Huq, H. F. Krause, P. A. Schulz, R. A. Phaneuf, Phys. Rev. A39, 1725 (1989).
[18] B. Friedrich and Z. Hermann, Chem. Phys. Let. 107, 375 (1984).
[19] H. Cederquist, L. Liljeby, C. Biedermann, J. C. Levin, C-S. O, H. Rothard, K-O. Groeneveld, C. R. Vane, I. A. Sellin, Phys. Rev. A39, 4308 (1989).
[20] E. Y. Kamber, C. L. Cocke, J. P. Giese, J. O. K. Pedersen, W. Waggoner, Phys. Rev. A36, 5575 (1987).
[21] L. N. Tunnell, C. L. Cocke, J. P. Giese, E. Y. Kamber, S. L. Varghese, and W. Waggoner, Phys. Rev. A35, 3299 (1987).
[22] W. Waggoner, C. L. Cocke, L. N. Tunnell, C. C. Havener, F. W. Meyer and R. A. Phaneuf, Phys. Rev. A37, 2386 (1988).
[23] H. Schmidt-Böcking, M. H. Prior, R. Dörner, H. Berg, J. O. K. Pedersen, C. L. Cocke, M. Stockli, A. S. Schlachter, Phys. Rev. A37, 4640 (1988).
[24] M. Kimura, T. Iwai, Y. Kaneko, N. Kobayashi, A. Matsumoto, S. Ohtani, K. Okuno, S. Takagi, H. Tawara and S. Tsurubuchi, J. Phys. Japan 53, 2224 (1984).
[25] H. Ryufuku and K. Sasaki, T. Watanabe, Phys. Rev. A21, 745 (1980).
[26] A. Bárány, G. Astner, H. Cederquist, H. Danared, S. Huldt, P. Hvelplund, A. Johnson, H. Knudsen, L. Liljeby, K-G. Rensfeldt, Nucl. Instr. & Meth. B9, 397 (1985).
[27] P. Hvelplund, L. H. Andersen, A. Bárány, H. Cederquist, J. Heinemeier, H. Knudsen, K. B. MacAdam, E. H. Nielsen, J. Sørensen, Nucl. Instr. & Meth. B9, 421 (1985).
[28] H. Cederquist, L. H. Andersen, A. Bárány, P. Hvelplund, H. Knudsen, E. H. Nielsen, J. O. K. Pedersen and J. Sørensen, J. Phys. B18, 3951 (1985).
[29] H. Cederquist, C. Biedermann, J. C. Levin, C-S. O, I. A. Sellin, R. T. Short, Nucl. Instr. & Meth. B34, 243 (1988).
[30] P. Hvelplund and H. Cederquist (unpubl. work).
[31] R. E. Olson and A. Salop, Phys. Rev. A14, 579 (1976).
[32] K. Taulbjerg, J. Phys. B19 L367 (1986).
[33] L. R. Andersson, J. O. P. Pedersen, A. Bárány, J. P. Bansgaard, P. Hvelplund, J. Phys. B22, 1603 (1989).
[34] K. W. Ford and J. A. Wheeler, Ann. Phys. 7, 259 (1959).

l-STATE SELECTIVE CHARGE EXCHANGE CROSS SECTIONS FOR COLLISIONS OF He^{2+} ON ATOMIC AND MOLECULAR HYDROGEN

R. Hoekstra[*,**], F.J. de Heer[**] and R. Morgenstern[*]

[*]KVI, Rijksuniversiteit Groningen, Groningen, the Netherlands
[**]FOM-Institute for Atomic and Molecular Physics, Amsterdam, the Netherlands

We have now for the first time succeeded to determine absolute cross sections for electron capture into the quasi-degenerate $4l$ states of He^+ produced in collisions of He^{2+} on atomic and molecular hydrogen. The cross sections have been deduced from emission profile measurements along the beam axis. Measurements have been performed in the energy ranges of 2 - 13 keV/amu and 50 - 125 keV/amu.

INTRODUCTION

Electron capture processes in collisions of bare ions with atomic hydrogen are of fundamental interest[1] and play an important role in the diagnostics of fusion plasmas, e.g. in connection with the injection of atomic hydrogen beams[2,3]. The $\Delta n = 1$ transitions between high-n states, yielding light in the visible spectral region are particularly important for future plasma diagnostics[2,3], since they require only fibre optic connections between spectrometers and fusion reactor. We have now for the first time succeeded to measure the contributions of the different l states included in such a $\Delta n = 1$ transition. The previously measured emission cross sections for the CVI $8 \rightarrow 7$ and $7 \rightarrow 6$ transitions[4,5] and the HeII $4 \rightarrow 3$ transition[6] (with the numbers referring to the principal quantum numbers n involved in the transitions) can not be used directly for plasma diagnostics since in a tokamak the l states within one n shell are mixed[2,3]. Therefore it is important to deduce experimentally all the l state electron capture cross sections seperately. Due to the quasi-degeneracy of the l levels in the resulting hydrogenic ions, light emission from different l levels could until now not be distinguished and hence the measurements had to be compared with the sum of the various theoretical contributions. Comparing in this way experiments with the most elaborate atomic orbital[7] and molecular orbital[8,9] calculations, it was found that there was good agreement between theory and experiment for the dominant capture channels[4-6,10], whereas for the non-dominant high-n states there were considerable differences[4-6]. We have now concentrated our experimental work on determining the $4l$ cross sections in collisions of He^{2+} on H (the dominant channel is $n = 2$). The He^{2+} - H system has been choosen, because the HeII $4 \rightarrow 3$ is the most important for plasma physics, the number of l states is not too large and there are extended atomic orbital calculations[11] available for comparison.

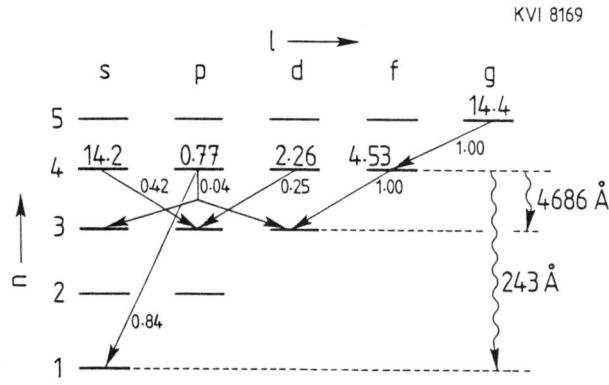

FIG. 1. Energy level scheme of HeII. The relevant transitions are shown together with the branching ratios and the lifetimes (in nsec) of upper states.

EXPERIMENTAL TECHNIQUE

To deduce the separate contributions of the $4l$ states to the $4 \rightarrow 3$ transitions we have exploited the fact that the lifetimes of the states are different. Fig.1. shows the energy level scheme of HeII together with the lifetimes and the branching ratios of the states relevant for our experiment. Radiation from shortlived levels, e.g. $4p$ is mainly concentrated on the target area, whereas radiation from longlived states, e.g $4s$ is still emitted downstream from the collision center, see fig.2. Hence the measurement of emission profiles along the beam axis gives information on the l states. This method,

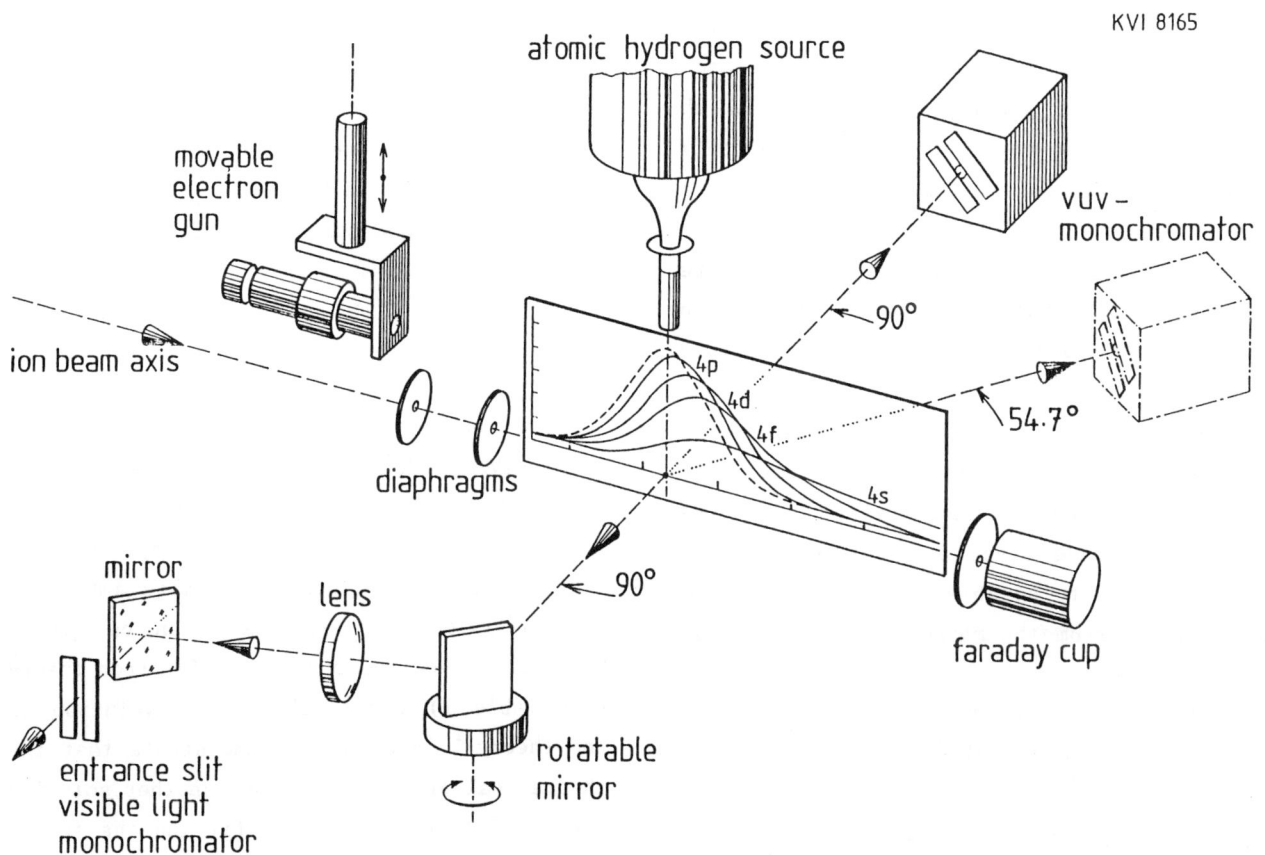

FIG. 2. Schematical view of the experimental set-up. Typical emission profiles of the $4l$ states are included. The target profile is indicated by the dashed curve.

often used in collision experiments on static targets has only once been used in combination with a beam target, namely to measure capture into $3l$ states in H^+ - Li collisions[12]. The case of He^{2+} colliding on atomic hydrogen is more intricate since (I) more l states have to be included, (II) the lifetimes of the states are closer to each other, (III) the target has two components, H and H_2. In the experimental set-up, schematically shown in fig.2 , the He^{2+} ions, produced with the ECR ion source installed at the KVI cross a partly dissociated hydrogen beam. The absolute density profiles of the atomic and molecular components of the beam produced by the radio-frequency source were determined by observation of electron impact induced atomic (Balmer-ß) and molecular radiation. The radiation was observed with a monochromator equipped with an imaging lens system which enabled the measurements of the target density along the beam axis. In this way[13] we found a dissociation of 75% in the center of the beam target. The HeII $4 \rightarrow 3$ emission profiles were observed perpendicular to the ion beam with the above mentioned spectrometer for visible light, while the $4p \rightarrow 1s$ emission was observed with a vuv spectrometer (both spectrometers are calibrated absolutely on wavelength and sensitivity[14]). The measurements have also to be performed on a pure H_2 target to be able to substract the H_2 contribution from the measurements on a mixed H/H_2 target.

Neglecting cascades the emission profiles, P_{4l} are described by

$$P_{4l}(z) = \frac{1}{v\tau_{4l}} \int_0^z T(z') e^{-(z'-z)/v\tau_{4l}} dz'$$

with v the velocity of the ions and τ_{4l} the lifetime of state $4l$, z the position along the beam axis and T(z) the target density profile. The measured signals, S(z) are equal to

$$S(z) = K \sum_l \beta(4l \rightarrow 3) P_{4l}(z) \sigma_{4l}$$

with K an absolute calibration constant, $\beta(4l \rightarrow 3)$ the branching ratio for transitions to $n = 3$ and σ_{4l} the electron capture cross sections. The signals have been measured at 25-30 positions along the beam axis, so it is possible to deduce the cross sections from a least square fit. To increase the accuracy of the fit results the contribution of the $4p$, which is comparably small due to the small branching ratio (fig.1.) has first been substracted from the signals. The σ_{4p} could be determined directly from the $4p \rightarrow 1s$ transition. Details of this work will be described in forthcoming article[15]. That article will also include the results for all Lyman transitions and the $5 \rightarrow 3$ transition on both atomic and molecular hydrogen.

RESULTS

Fig.3a and b show the emission profiles for 5 keV/amu He^{2+} colliding on molecular and atomic hydrogen. It can be seen that the measurements are well described by the fits and that the $4f$ is relatively more important for collisions on molecular than on atomic hydrogen. Since the results differ from theory (see table 1) fig.3c shows a convolution of the theory[11] in order to demonstrate that the theoretical prediction is not another solution for the fit. Comparing the fit and the theory it can be seen that the s, p and d are stronger populated than predicted. As a typical example of the comparison between experimental and theoretical $4l$ cross sections fig. 4 shows the results for the $4f$ state. It

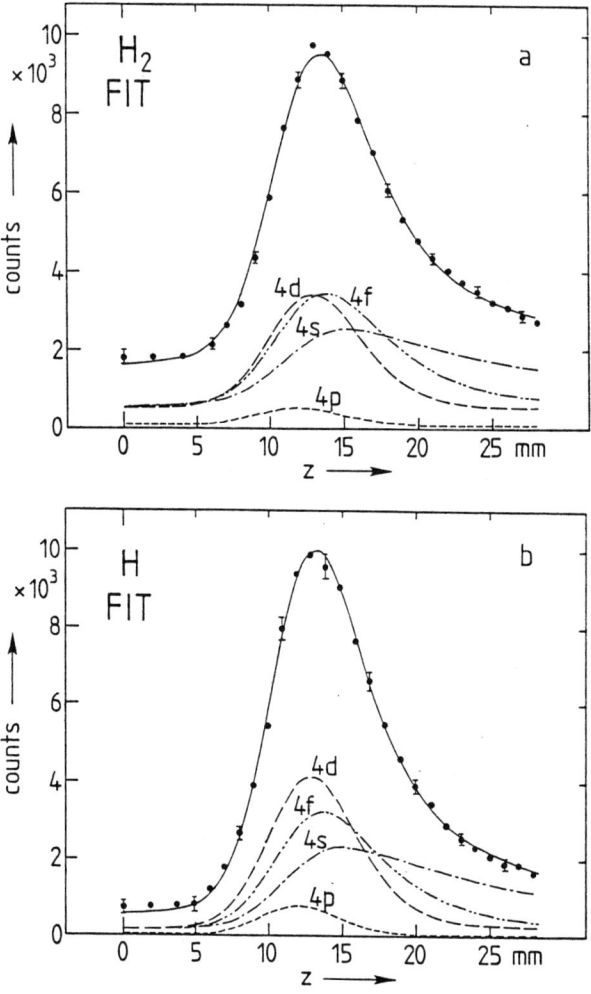

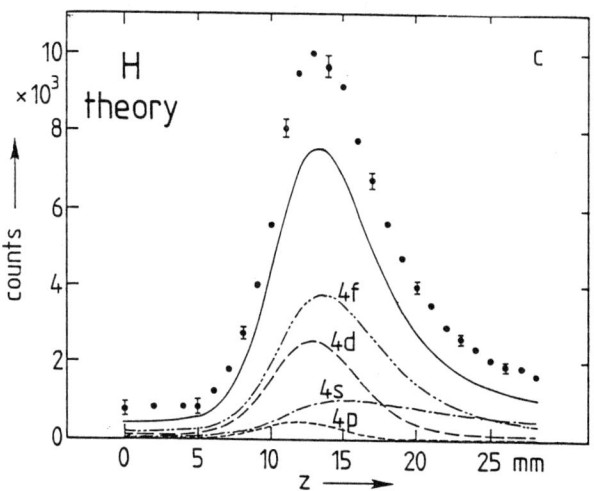

FIGS. 3.a-c. Emission profiles for 5keV/amu He^{2+} colliding on H_2 (fig.a) and H (figs.b and c). Figs.a and b show deconvolution results, whereas fig.c shows the convolution of the AO results[11].

can be seen that the experimental results increase smoothly with energy, whereas the theory shows a structure around for 4 keV/amu. Table 1 gives the results for the four impact energies in the range of 2 - 13 kev/amu at which there are theoretical and experimental results. However since the $4l$ cross sections are more than two orders of magnitude smaller than the dominant one for capture into $2p$ the theoretical calculations seem to describe the charge exchange processes to a fair extent. Furthermore it should be noted that the $4s$ cross section includes cascades since the effective lifetime of the cascades ($5g \rightarrow 4f \rightarrow 3d$) is close to the lifetime of the $4s$ state.

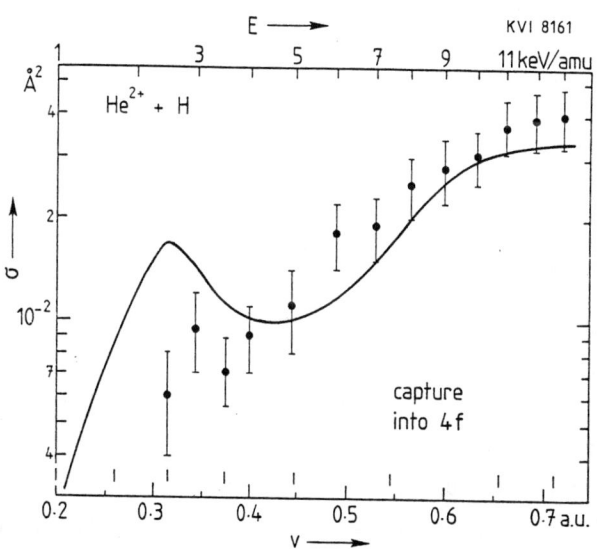

FIG. 4. Absolute electron capture cross sections for charge transfer into $He^+(4f)$ in collisions of He^{2+} on atomic hydrogen. Theoretical curve: atomic orbital calculation[11]

CONCLUSIONS

To conclude we have shown that it is possible to unraffle the $4l$ distribution in collisions of He^{2+} on molecular and atomic hydrogen by means of measuring emission profiles along the ion beam axis. In this way information relevant for fusion plasma diagnostics is obtained in the energy range of 2 - 13 keV/amu. However for that diagnostics not only these low energies are of importance but also higher energies. Therefore we have extended the measurements into the energy range of 50 - 125 keV/amu. These high energy beams have been produced by stripping a fast He^+ ion beam, which was extracted from the Van de Graaff accelerator installed at the FOM Institute for Atomic and Molecular Physics. The first results[16] in that energy region show, regarding the relative smallness of the $4l$ capture processes also a fair agreement with the extended atomic orbital calculations by Fritsch[11].

ACKNOWLEDGEMENT

The authors gratefully acknowledge the excellent technical support of J. Eilander and J. Sijbring. We would like to thank Dr. W. Fritsch (Hahn Meitner Institute Berlin) for communicating his theoretical results. This work is part of the research program of the Stichting voor Fundamenteel Onderzoek der Materie (FOM) with financial support by the Stichting voor Nederlands Wetenschappelijk Onderzoek (NWO). This research project is also supported financially by JET under contract JP7/9006.

Table 1. Experimental, σ_{ex} and theoretical atomic orbital[11], σ_{th} cross sections in 10^{-2} Å^2 for He^{2+} - H collisions. The relative errors are included in the table The absolute systematic error is 20%.

E(keV/amu)		2.5	3.5	5	10
$4s^*$	σ_{th}	1.4	1.6	1.2	3.1
	σ_{ex}	0.6	1.3	1.6	3.2
	$\Delta\sigma_{ex}$	0.2	0.3	0.4	0.6
$4p$	σ_{th}	1.1	1.4	2.1	4.8
	σ_{ex}	0.7	1.5	3.2	6.0
	$\Delta\sigma_{ex}$	0.3	0.4	0.5	0.7
$4d$	σ_{th}	1.6	0.8	2.3	3.0
	σ_{ex}	1.3	2.5	3.1	5.4
	$\Delta\sigma_{ex}$	0.4	0.5	0.6	1.0
$4f$	σ_{th}	1.4	1.6	1.2	3.1
	σ_{ex}	0.6	0.7	1.1	3.1
	$\Delta\sigma_{ex}$	0.2	0.2	0.3	0.6

* The $4s$ cross section includes cascades

REFERENCES

[1] R.K. Janev and H Winter, Phys. Rep. **117** 265 (1985).

[2] R.J. Fonck, D.S. Darrow and K.P. Jähnig, Phys. Rev. A **29** 3288 (1984).

[3] A. Boileau, M. von Hellermann, L.D. Horton, J. Spence and H.P. Summers, Plasma Physics and Controlled Fusion **31** 779 (1989).

[4] R. Hoekstra, D. Ćirić, A.N. Zinoviev, Yu.S. Gordeev, F.J. de Heer and R. Morgenstern, Z. Phys. D. **8** 57 (1988).

[5] R. Hoekstra, F.J. de Heer and R. Morgenstern, Phys. Scr. in press (1989).

[6] D. Ćirić, R. Hoekstra, F.J. de Heer and R. Morgenstern, *Invited Papers XV ICPEAC, Brighton, 1987*, edited by H.B. Gilbody, W.R. Newell, F.H. Read and A.C.H. Smith (North Holland, Amsterdam, 1988), p655.

[7] W. Fritsch and C.D. Lin, Phys. Rev. A **29** 3039 (1984).

[8] T.A. Green, E.J. Shipsey and J.C. Browne, Phys. Rev A **25** 1364 (1982).

[9] E.J. Shipsey, T.A. Green and J.C. Browne, Phys. Rev A **27** 821 (1983).

[10] R. Hoekstra, D. Čirič, F.J. de Heer and R. Morgenstern, Phys. Lett. **124** 73 (1987).

[11] W. Fritsch, Phys. Rev. A **38** 2664 (1988) and private communication.

[12] F. Aumayr, M. Fehringer and H. Winter, J. Phys. B **17** 4201 (1984).

[13] D. Čirič, D. Dijkkamp, E. Vlieg, A. de Boer and F.J. de Heer, J. Phys. B **18** 4745 (1985).

[14] D. Dijkkamp, D. Čirič, E. Vlieg, A. de Boer and F.J. de Heer, J. Phys. B **18** 4763 (1985).

[15] R. Hoekstra, F.J. de Heer, R. Morgenstern and W. Fritsch, to be published.

[16] J. Frieling, S. Kuppens, A.N. Zinoviev and D. Čirič, private communication.

Molecules

CHAOS IN MOLECULAR RYDBERG STATES

M. Lombardi*, P. Labastie**, M.C. Bordas** and M. Broyer**

*Laboratoire de Spectrométrie Physique, Université Joseph. Fourier de Grenoble. BP 87 38402 Saint Martin d'Hères France.
**Laboratoire de Spectrométrie Ionique et Moléculaire, Université Claude Bernard, 23 Boulevard. du 11 Novembre 1918. 69622 Villeurbanne. France

We discuss on the particular case of Rydberg states of a molecule the correspondence between classical chaos and quantum level distributions.

Introduction: We will use the study of Rydberg states of a diatomic molecule as an example to check some ideas on the correspondence between classical chaos and properties of the spectrum of a quantum system which has the same Hamiltonian. We will see that results agree with expectations in the semi-classical limit (h -> 0 or large quantum numbers), but that they do differ for understandable simple reasons in more quantal cases.

The hypothesis we want to check was put forward by nuclear physicists (O. Bohigas et al,[1]), and is based on qualitative arguments made by E.P. Wigner as early as 1956[2].

The idea of Wigner was, when analysing very complex spectra of highly excited nuclei, that if levels are not coupled their position should occur at random, as with a random number generator, while if they are coupled they should repell themselves. In the first case nothing prevents them to be very close, so that a simple computation shows that the nearest neighbour spacing distribution peaks at zero spacing (Poisson distribution). In the second case the repulsion make this distribution to peak at non zero spacing (Wigner distribution) (See Fig 1).

Bohigas et al elaborated on the following idea: what is the precise meaning of "uncoupled" levels? A coupling can be defined only if there is some kind of zeroth order Hamiltinian, which defines what are basic states which may be arbitrarily close, and if some kind of perturbing Hamiltonian couples the zeroth order states, making them to repell. Now zeroth order levels can be non accidentally very close, and cross if one can vary some parameter, only if they have different quantum number, i.e. if they belong to some kind of different symmetry class. They can be coupled by a perturbing Hamiltonian if this Hamiltonian has a lower symmetry,

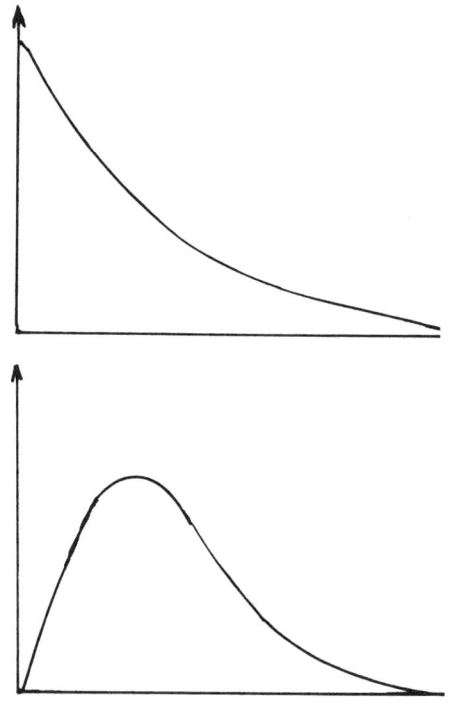

Figure 1: Nearest neighbour spacing distributions. Poisson (top), Wigner (bottom).

making some quantum numbers to disappear, and thus breaking down some selection rules which forbidded coupling.

In classical mechanics this situation corresponds exactly to the following one. A zeroth order Hamiltonian has enough first integrals or actions to be integrable (enough means a number n equal to the number of degrees of freedom of the system). Its motion is quasiperiodic, i.e. is the combination of n independant oscillations (it is perodic if the frequencies of these oscillations are commensurable). Now if a perturbation has lower symmetry, destroying at least one of the action integrals (quantum numbers in quantum mechanics), the motion becomes chaotic.

The hypothesis was thus that there is a correspondence between change in level statistics in quantum mechanis (transition from Poisson to Wigner nearest neighbour distribution), and transition to chaos in classical mechanics, both being due to some lowering of the symmetry of the system by a perturbation.

Method of Analysis. Before checking these ideas on a particular system, it is important to elucidate the relationship between quantum spectra and time evolution of a quantum system. The reasons are twofold:

(i) A practical reason for us is that we are basically spectroscopists of polyatomic molecules: in this case experimental spectra are frequently badly resolved, but we have shown[3] that it is possible even in this case to recover important features of the time evolution (or to make directly time measurements).

(ii) A theoretical reason is that classical chaos is basically a property of time behaviour of the systems so that physical insight on the correspondance is more likely to appear on the time evolution than on the spectrum, even if experimentally the spectrum comes first.

The correspondence is made through the usual formula which relates the spectrum intensity to the dipole autocorrelation function of the molecule.

$$I(\omega) = \int C(t) e^{-i\omega t} dt$$

with

$$C(t) = \langle d(0) d(t) \rangle = \Sigma \langle g | d | e \rangle^2 e^{-i(E_f - E_e)t}$$

One considers rather the square

$$|C(t)|^2 = \Sigma \langle e | D | e' \rangle^2 e^{-i(E_e - E_{e'})t}$$

with $\langle e | D | e' \rangle = \langle e | d | g \rangle \langle g | d | e' \rangle$

which depends only on the excited state if there is only one ground state (this can be obtained through various laser techniques).

This is basically a Fourier transform of the spectrum with a weight $\langle e | D | e' \rangle$ which is a Franck Condon factor.

One computes then the *average*

$$|C(t)|^2 = \iint \Sigma e^{-i(E_e - E_{e'})t} R_2(E_e, E_{e'}) dE_e dE_{e'}$$

where R_2 is some selected probability distribution to have a level at E_e and

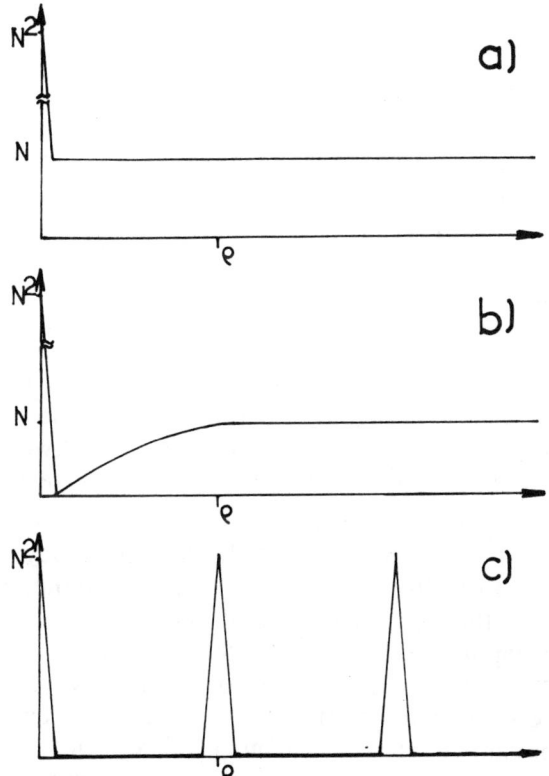

Figure 2. $|C(t)|^2$ (a) Poisson distribution. (b) GOE distribution. (c) picket fence.

another at E_e'. The important part of this distribution is the two level repulsion term $Y_2(E_e-E_e')$, whose Fourier transform $b_2(t)$ appear in the result of the computations.

These results are sketched on fig 2 for three simple limiting cases of level distribution:

(a) purely random (Poisson distribution),

(b) GOE or levels obtained by diagonalizing matrices with all terms, diagonal and off diagonal, randomly Gaussian distributed with the same order of magnitude, a model case for strong coupling which has been shown to give Wigner distribution of nearest neighbour spacings

(c) an equidistant set of levels (dubbed picket fence).

Notice that cases (a) and (c) correspond to two integrable cases, namely a set of uncoupled anharmonic oscillators, which has been shown to give Poisson nearest neighbour spacing distribution by Berry and Tabor[4], and a single harmonic oscillator, while (b) is supposed to correspond to a non integrable case.

The interpretation is the following. In case (a) there is a high fast peak at time zero (amplitude N^2, where N is the number of levels, width $1/\Delta$, where Δ is the energy span of the set of levels). This fast peak is followed by a small (N) long (time given by the width of the levels, usually due to some decay process) decay. In case (b) the coupling introduces after the fast peak a "correlation hole", i.e. a suppression of emission up to times corresponding to the reciprocal of the repulsion between the levels, whose shape is given by the "two level form factor" $b_2(t)$. In case (c) there are regular recurrences of amplitude N^2, at times given by the oscillator frequency. Transition from (a) to (b) can be thought of as the appearance of the hole corresponding to the level repulsion, which generates a more ordered sets of levels. Transition from (c) to (b) can be thought of as a broadening of all the recurrences which makes them to merge a broad plateau, leaving only a hole between the initial fast decay and the time where there was formerly the first recurrence.

This case corresponds to a lower ordering of the level set.

Rydberg Spectrum of Na_2. Fig 3 describes what is expected at first sight.

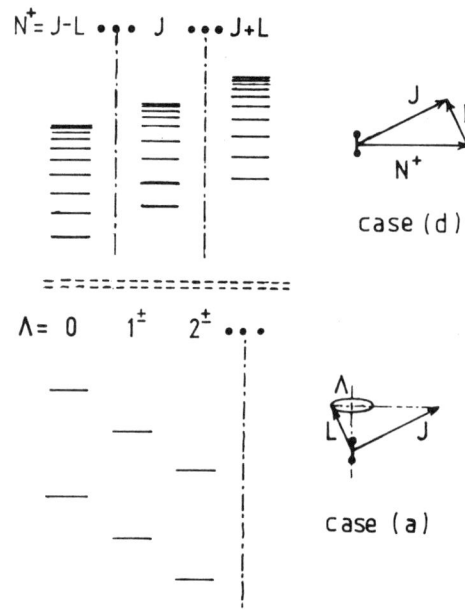

Figure 3. Left: spectrum of the Rydberg states of a molecule. Right: Hund's coupling cases.

At low energy the Rydberg electron is strongly coupled to the molecular core, giving rise to the so called Hund's (a) coupling case of spectroscopists, where the projection Λ of the angular momentum L of the electron on the molecular core axis is a good quantum number.

At high energy the electron is practically no more coupled to the core, giving Hund's case (d) coupling: Λ is no more a good quantum number, but N^+, the angular momentum of the core becomes a good quantum number. Coupling between core and Rydberg electron occurs only during short temporally well distant collisions. The Hund's case (d) is never truly attained for this reason.

In between, there is thus the breakdown of a quantum number, Λ, which should correspond in classical mechanics to a transition to chaos. In quantum mechanics, the breakdown of selection rules associated with this quantum number should produce an increase of complexity of the spectrum.

Experimentally however, it was observed after the onset of spectral congestion due to this phenomenon, that "clear" zone appear, where the number of lines drastically decrease[5].

The qualitative explanation was the following[5]. Clear zones appear when the splitting of the rotatinal levels of the core is equal to k times the separation between consecutive Rydberg levels. This means that every time the Rydberg returns to the core, the core has rotated an integral number of half turns, so that the electron sees it at the same position. The coupling is thus the same as at low energy where the core has no time to rotate between two collisions, so that the coupling returns to Hund's case (a) and the selection rules which simplify the spectrum reappear. At intermediate energies, to the contrary the electron sees the core at a rather random position, and this collision destabilizes its orbit if the interaction is strong enough.

We show in Fig. 4, the transition to chaos in one particular case, namely when sitting at the energy corresponding to the first resonance and increasing the coupling, i.e. the strength of the collision between the Rydberg electron and the core. A more complete description is given elsewhere[6]. In this figure one launches various classical orbits and plots on a sphere after each collision a point which represents the position of the angular momentum L of the electron in the molecular reference frame. This frame has Oz axis along the molecular core axis, and Ox axis along the molecular core angular momentum N^+. Points appear first on lines, which means that the motion is regular. One sees first the transition between Hund's case (d) at no coupling (L is fixed in laboratory frame, so that it seems to rotate around Ox axis in molecular frame), and Hund's case (a) at moderate coupling (The coupling produces a rotation around molecular Oz axis). When increasing further the coupling the conflict between these two axis of rotation produces the transition to chaos, which is evidenced by the points filling a surface instead of lines on the sphere.

The spectrum is computed accurately by using Multichannel Quantum Defect Theory[5]. One sees first clearly on Fig. 5 the onset of level repulsion when increasing coupling.

The time behaviour is more complex (Fig. 6). There is always a peaked structure. This structure correspond to the fact that the period of revolution of the electron on its orbit is rather constant. This is due to angular momentum selection rules and a proper choice of angular momentum L and N^+. Averaging on this peaked structure is equivalent to consider longer time evolution of the molecule, i.e. its average evolution after several collisions, In the no coupling case one has at long time a constant C(t) [2], which was expected after Berry and Tabor[4] for a Poisson spectrum. There is however a strong peak at short

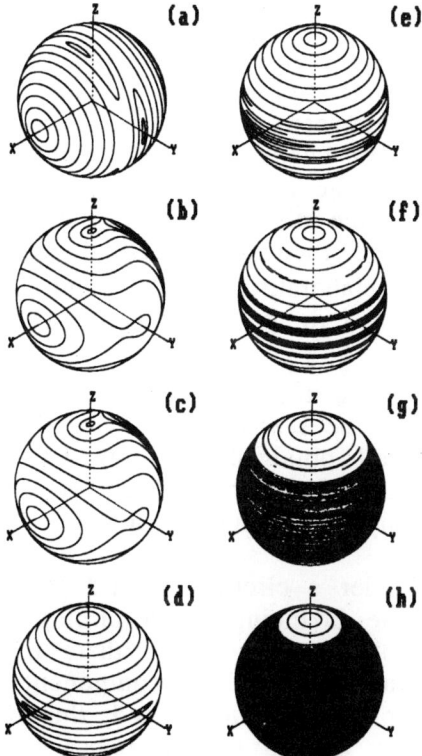

Figure 4. Transition to chaos when increasing coupling. (a)-(c) evolution from nearly Hund's case (d) to nearly Hund's case (a). (e)-(f): gradual onset on chaos when increasing further the coupling.

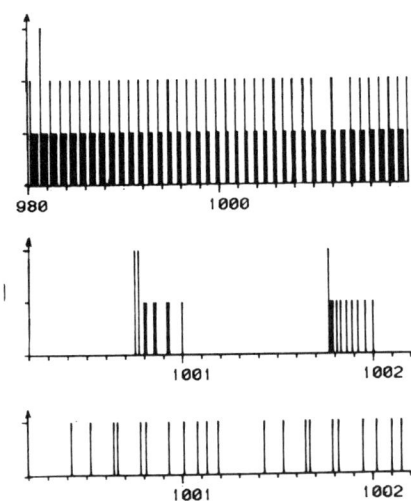

Figure 5. Spectrum of Rydberg molecule near a resonance. Top and medium (expanded); no coupling. Bottom: with coupling.

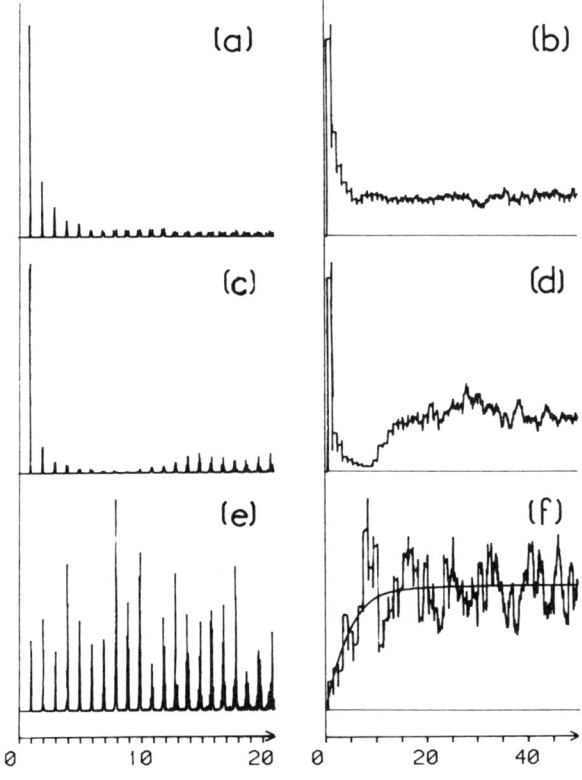

Figure 6 Time behaviour of a Rydberg molecule near the first resonance. Left: non averaged, Right averaged over one orbit time. Top to Bottom from no coupling to strong coupling.

time (not to be confused to the initial N^2 "fast decay" which occupy only channel 0 in these computations). This corresponds to the fact that the spectrum is not random but strongly clumped. The system is more like an harmonic oscillator of Fig 2c that to an ensemble of anharmonic oscillators of Fig. 2a. The evolution to an average constant at longer times is due to anharmonicities. It shows that one has not a generic situation, the anharmonic case, and an exceptional situation, the harmonic case, as stated by Berry and Tabor[4], but rather a time evolution between the two cases, the time constant for the transition between the two situations beeing the reciprocal of the anharmonicity. When increasing coupling transition to chaos in the classical system is paralleled as expected by a transition to the prediction of GOE (solid line) in the average (i.e. long term) time evolution.

Discussion. The predictions of the Bohigas et al[1] conjecture are thus nicely confirmed in this case, even if the distinction between short term stability and long term chaos introduces some complexity in the analysis. This is not too surprising since this conjecture has received since then a theoretical support by Berry[7], elaborating on the semi classical theory of Balian and Bloch[8] and Gutzwiller[9].

But it must be realized that this is valid only at the *semi-classical limit*, which is rather well obtained for high Rydberg states considered in the present analysis. This is easily understood on a very simple model.

Consider a circular billiard (i.e. a free particule bouncing on a circle). This is a separable system because there is a first integral, the vertical component of the angular momentum around the center of the circle, due to the invariance of the system by rotation around this axis. Levels corresponding to different values of this angular momentum do not repell, and can be arbitrarily close, giving a Poisson-like nearest neighbour spacing distribution. The corresponding classical motion is quasi periodic. Introduce now a fine granularity on the boundary. This has a drammatic effect on the classical system because trajectories which are initially very close but bounce off opposite sides of a grain separate rapidly. It has been shown that the system becomes chaotic in fact as soon as there is a convex part in the boundary, and that this is valid at any energy (i.e. velocity of the particule, which corresponds only to a time scaling). But quantum mechanics being a wave mechanics does not feel the granularity up to energies when the wavelength is of the order of magnitude of the grain size. It is only at this energy that level repulsion appear and that the spectrum becomes Wigner-like.

The conclusion is that quantum mechanics is intrinsically more stable than classical mechanis. This *ad hoc* model is rather peculiar, but one must be aware than when one increases even very moderately the number of the degrees of freedom, as in small polyatomic molecules, one has easily situations in which at relevant chemical energies the level density is very high, enabling precise level statistic analysis of the kind we have illustrated in Ryberg states, but the system is by no means semi-classical, because the level density is due to a combinatorial explosion of the number a ways to attain a given energy by distributing quanta among modes.

An other very important point to notice when attempting to apply this kind of ideas to non bound systems, like in collisions, is that classical chaos is a property of long time evolution (infinite time in theory) of the system. There is always, even in classical mechanics, time constants to attain such a situation, so that a chaotic system may well be regular if observed on too short times: think to the distinction between short time and long time behaviour in our molecular Rydberg case. Quantum mechanics introduce a further time limitation. Notice that for long times the two cases of Fig 2a and 2b which correspond to a regular and a chaotic system give the same (constant) limit. (Fig 2c, the harmonic oscillator is very peculiar because it is known that there is no wave packet diffusion so that the predictions of classical and quantum mechanics agree up to infinite time). The two, Fig 2a and Fig 2b situations merge at time corresponding to the reciprocal of the average level spacing, i.e. when Heisenberg's uncertainty principle forces the system to recognize that it is quantal.

References

1. O. Bohigas, M.J. Giannoni and C. Schmit, Phys. Rev. Lett., 52, 1 (1984)
2. E.P. Wigner, Oak Ridge National Laboratory, Report n° ORNL-2303, 1956 (unpublished) reprinted by C.E. Porter in Statistical Theories of Spectra : Fluctuations (Academic Press 1965, p. 139).
3. L. Leviandier, M. Lombardi, R. Jost and J.P. Pique. Phys. Rev. Lett. 56, 2449 (1986).
4. M.V. Berry and M. Tabor, Proc. Roy. Soc. London A 356, 375 (1977)
5. P. Labastie, M.C. Bordas, B. Tribollet and M. Broyer, Phys. Rev. Lett. 52, 1681 (1984)
   M.C. Bordas, M. Broyer, J. Chevaleyre, P. Labastie and S. Martin, J. Phys. (France) 46, 27 (1987)
6. M. Lombardi, P. Labastie, M.C. Bordas and M. Broyer. J. Chem. Phys. 89, 3479 (1988).
7. M.V.Berry, Proc. Roy. Soc. A 400, 229 (1985).
8. R. Balian and C. Bloch, Ann. Phys. 69, 76 (1972), Ibid 85, 514 (1974)
9. M.C. Gutzwiller, J. Math. Phys. 8, 1979 (1967), Ibid 10, 1004 (1969) Ibid, 11, 1791 (1970), Ibid 12, 343 (1971). Phys. Rev. Lett. 45, 150 (1980). Physica 7D, 341 (1983)

RECENT DEVELOPMENTS IN MOLECULAR ION RECOMBINATION RESEARCH.

J.B.A. MITCHELL

Dept. of Physics, University of Western Ontario,
London, Ontario, Canada. N6A 3K7.

A review of recent measurements of the recombination of H_2^+, HeH^+, N_2^+ and H_3^+ with electrons is presented.

INTRODUCTION.

Electron-ion recombination is a major collision process in the chemistry of plasmas. It can alter the state of ionization of the medium and lead to the formation of products which are physically and chemically distinct from the primary plasma constituents. Thus it plays an important role in the formation of interstellar molecules in the ionized interiors of interstellar clouds, it leads to population inversions in excimer laser discharges, to the diurnal variation of the electron density in the earth's ionosphere, and it results in the formation of reactive excited radicals in ion-etching plasmas. Despite its importance, it is a poorly understood phenomenon.

If we consider an ion such as O_2^+, formed in a variety of vibrationally excited states in a discharge, then the recombination proceeds rapidly via the transition of the electron + ion state to a repulsive neutral state which dissociates to form atomic oxygen pairs. The potential energy of the initial state is converted into the kinetic and potential energy of the products and the recombination is efficiently stabilized. This situation is illustrated in fig. 1.

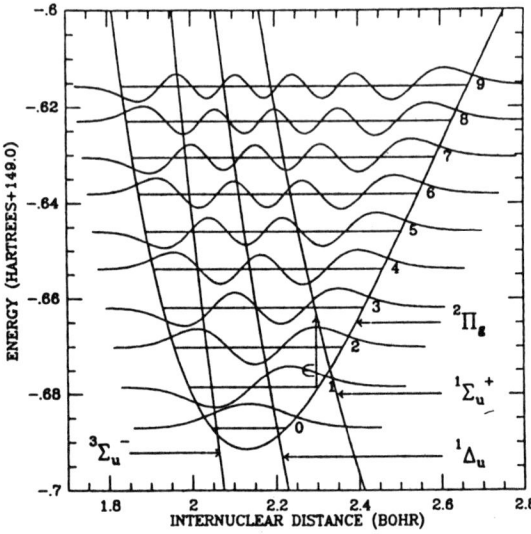

Fig. 1 - Potential energy states of O_2^+ and O_2^+ from reference 1.

The neutral states, through which the recombination proceeds are well characterized[1] and the process has been studied experimentally by a number of different techniques which have produced good mutual agreement in the measured value of the room temperature rate coefficient[2]. Even the branching ratios for the competing decay channels have been determined[3]. There are still questions regarding the agreement between experiment and theory but generally the situation appears to be well in hand.

Such is not the case for other ions however. If we examine the case of H^+_2, figure 2, we see that the dissociating state intersects the ion state in the vicinity of the v=1 level. This means that the direct recombination from the v=0 state should be considerably weaker and this allows alternative recombination channels to compete, thus introducing a considerable increase in complexity to the process. An example of a competing channel is the so-called "indirect mechanism" where the incoming electron can be captured via a vibronic coupling interaction into a vibrationally excited, autoionizing rydberg state which can subsequently predissociate via the repulsive $^1\Sigma^+_g$ state. When the electron energy is resonant with a suitable rydberg vibrational level, then the wave function for the system is concentrated in this bound configuration. Since however the vibronic coupling interaction is much weaker than the electronic coupling responsible for the direct capture, the recombination cross section exhibits deep window resonances at these energies. Such resonances have been calculated using Configuration Mixing[4] and Multi Channel Defect Theory[5,6] and a typical result for H^+_2 (v = 0) and H^+_2 (v=1) is illustrated in figure 3.

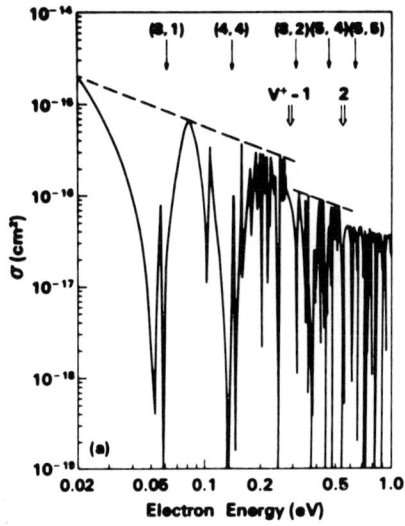

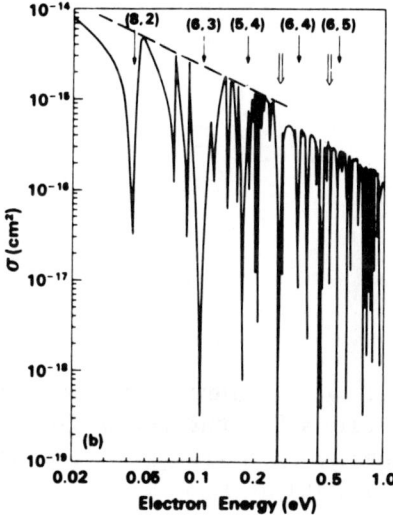

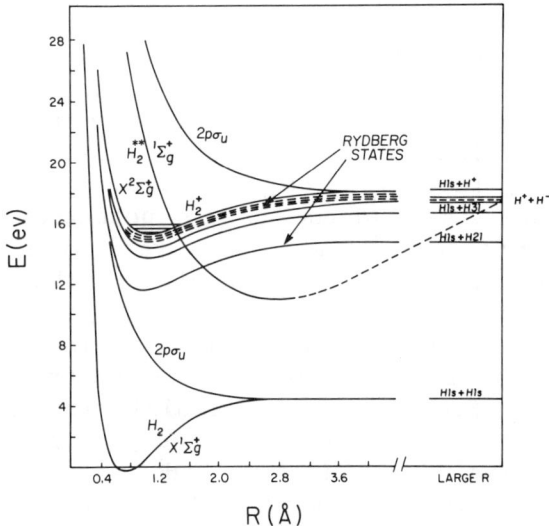

Fig. 2 - Potential energy states of H_2 and H^+_2.

Fig. 3 - Calculated cross sections for the dissociative recombination of
(a) H^+_2 (v = 0) and
(b) H^+_2 (v = 1). Ref. 6.

Recent high resolution merged beam experiments, using vibrationally selected ions formed in a trap ion source have identified these resonances as shown in fig. 4. Indeed they are very deep and extremely narrow. A theoretical analysis of these results is currently in progress in order to identify the states from which the resonances arise. A paper describing this work is currently in preparation[7].

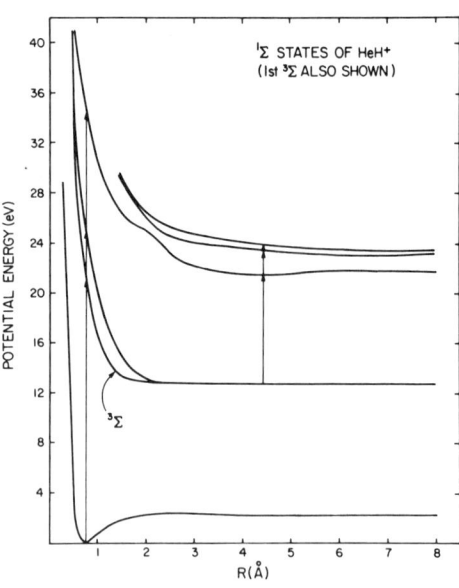

Fig. 5 - Potential energy curves for HeH$^+$ showing the absence of suitable crossings. From ref. 21.

The recombination would therefore appear to be ruled out and indeed in many published articles the rate has been assumed to be negligible. The results of a recent merged beam study[8] are illustrated in fig. 6 and it can be seen that in fact the recombination cross section is far from negligible. Complementary electron impact excitation measurements have shown that the ions were in their ground vibrational and electronic state. Clearly this is a situation where the recombination can only proceed through rydberg state capture although the mechanism for the stabilization of the recombination, i.e. the disbursement of the excess potential energy, has not as yet been identified. It is conceivable that capture can occur into a rydberg state which is subsequently predissociated by the ground state of the neutral HeH. (See fig. 5). Certainly it is known from spectroscopic studies[9,10,11], of the HeH* rydberg system that a number of these states are strongly predissociated. The complementary states for the HeD* are not predissociated and it will be

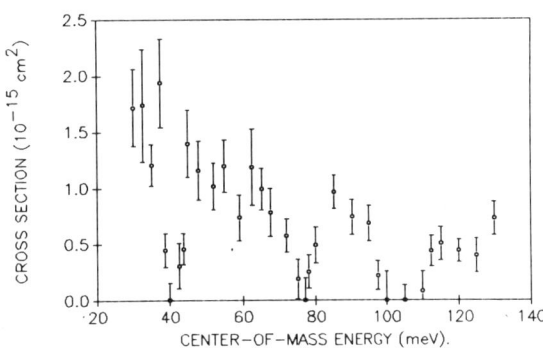

Fig. 4 - New experimental cross sections for the recombination of H$_2^+$. (Energy scale is subject to calibration.)

If we look at the case of HeH$^+$, the situation is even less clear. Because of the high excitation energies of the helium atom, this ion does not exhibit any curve crossing with a suitable dissociating state. (Fig. 5).

interesting to repeat the recombination measurement for HeD$^+$ to ascertain if this hypothesis is true or not.

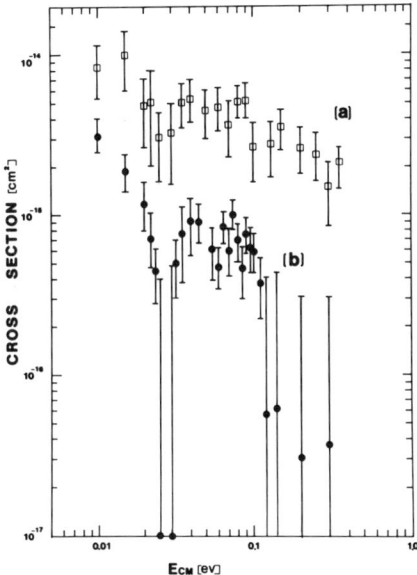

Fig. 6 - Cross sections for the recombination of HeH$^+$. From ref. 8.

The ability to produce beams of vibrationally cold molecules has led to a number of interesting results taken with the merged beam technique. A re-examination[12] of the recombination of N$^+_2$ has produced a surprising disagreement with previous measurements. This is a reaction which has had a stormy past[12] although measurements by Zipf[13] in 1980 had appeared to give the final word on the subject of whether the recombination rate depended upon the vibrational state or not. Zipf had used a laser induced fluorescence technique to track the decay of selected vibrational states of N$^+_2$ in an afterglow and found that in fact the decay rates for v=0, 1 and 2 were very similar. In a recent review however, Johnsen[14] disputed the conclusions of this study citing the interference of rapid charge exchange reactions with vibrationally excited nitrogen molecules in the apparatus as a means of altering the vibrational level populations in competition with recombination.

Our measurements of N$^+_2$ recombination taken with ions with varying internal energies are illustrated in fig. 7.

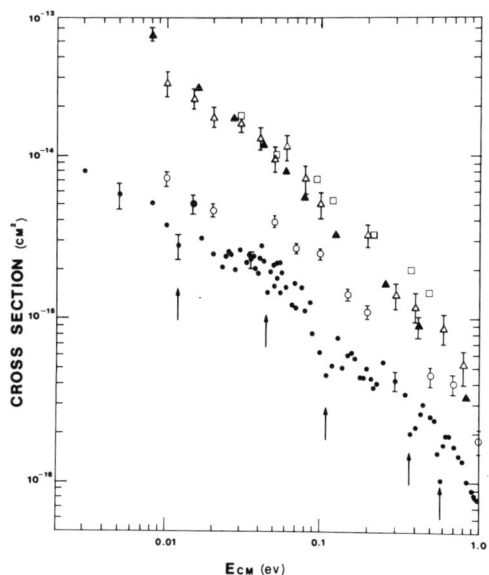

Fig. 7 - Cross sections for the dissociative recombination of H$^+_2$. ● v = 0, ○, , excited ions (ref. 22) ▲ ref. 23. From ref. 12.

It can be seen that for the case of N$^+_2$ (v=0), identified using electron impact excitation threshold measurements, the cross section is about a factor of 5 lower than previously accepted values. There is good reason to believe that the earlier studies were performed using ions which were not vibrationally relaxed. A fuller discussion of this is given by Noren et al[12]. Theoretical studies[15] of the curves intersecting the N$^+_2$ ground state in the vicinity of the v=0 level indicate that there is a good overlap with a number of states in this region that the cross section is reduced. Guberman[15] has argued that since capture into the G^3_g and $2^1\Sigma^+_g$ states involves a triple excitation, recombination via these states is less likely. The C' state leads to $^4S + ^2D$ atom pairs while the 3^3_u state dissociates to $^2D + ^2D$. Measurements of the decay channel branching ratios by Rowe et al[3] have shown that the $^2D + ^2D$ channel is favoured over the $^4S + ^2D$ channel. This suggests that the coupling to the C', channel is weak. Since the overlap of the 3^3_u state is less favourable with the v=0 level than with the v=1 state, this could explain the vibrational state dependance of the cross section found by Noren et al[12].

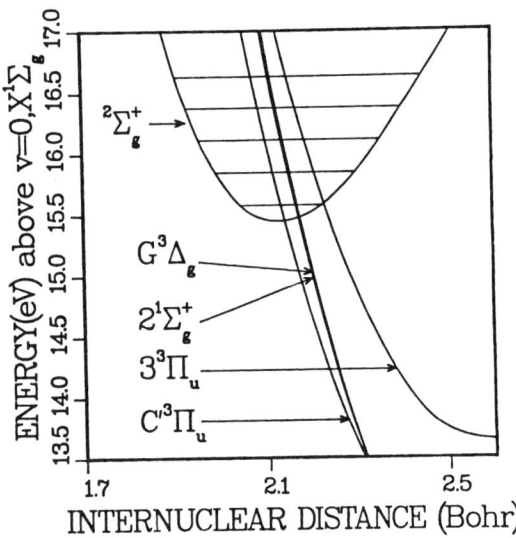

Fig. 8. - Potential energy curves for N_2 and N^+_2. From ref. 15.

Finally this report would be incomplete if mention was not made of the controversial H^+_3 recombination process. For many years this was accepted as a rapid reaction with a measured rate of $2\times10^{-7} cm^3 sec^{-1}$ at 300K. In 1984 however, Adams et al[16] published a new measurement, taken with the Flowing Afterglow Langmuir Probe technique which indicated that in fact the rate of recombination of H^+_3 ions in their ground vibrational states is a least an order of magnitude less than the above value. This result was supported by theoretical studies by Michels and Hobbs[17] and by Kulander and Guest[18]. Subsequent merged beam measurements using state selected H^+_3 ions confirmed that the rate for ions in v = 0 and possibly v = 1 was indeed $2\times10^{-8} cm^3 sec^{-1}$. In the meantime however, Adams and Smith had recalibrated their apparatus and had lowered the estimated rate to $2\times10^{-11} cm^3 sec^{-1}$.

A recent re-measurement of the excitation process used to determine the internal energies of the ions in the merged beam experiment has in fact confirmed that the ions were in the v=0 level only, (there is a sub-threshold resonance which had previously obscured the original measurements leading to ambiguity in the initial state determination). The Adams and Smith[19] calibration relied upon the fact that when helium was introduced into their apparatus, the measured rates for the decay of H^+_3, HeH^+ and He^+ were indistinguishable. Since the latter can only recombine radiatively with a calculated rate of $2\times10^{-11} cm^3 sec^{-1}$, and at the time, HeH^+ recombination was expected to have a similar rate, the rate for H^+_3 was also assigned this low value although the sensitivity of the FALP technique is insufficient to make this assignment definitive. In the light of the recent merged beam findings for HeH^+ described above however, it would appear that this calibration is based upon questionable assumptions. (It should be noted that attempts to measure a cross section for the recombination of He^+ using the merged beam apparatus, were unsuccessful since as expected, no signal could be detected arising from this process over a timescale of several hours.)

SUMMARY

The above discussions serve to illustrate that the recombination of molecular ions is a complex, beautiful subject. Many questions still remain concerning the identity of the recombination products, the excitation state of these products and the rates for ions where the locations of curve crossings are in dispute. Almost nothing is known about the recombination of complex hydrocarbon and cluster ions although FALP experiments[20] are making inroads into this area. All of these topics are of great importance in practical systems. Clearly there is wide scope for continued experimental and theoretical investigations into this fascinating phenomenon.

ACKNOWLEDGEMENTS

The author would like to thank the organizers of the XVIth ICPEAC for inviting this report, the United States Air Force Office of Scientific Research and the Canadian National Sciences and Engineering Research Council for supporting the merged beam work and my colleagues in the field of recombination research for many interesting and mostly convivial discussions.

REFERENCES.

1. S.L. Guberman in Physics of Ion-Ion and Electron-Ion Collisions (ed. F.Brouillard) Plenum, New York, 1983, p167.
2. E. Alge, N.G. Adams and D. Smith, J. Phys. B. 16, 1433, 1983.
3. B.R. Rowe and J.L. Queffelec in Dissociative Recombination: Theory, Experiment and Applications (Eds. J.B.A. Mitchell and S.L. Guberman) World Scientific, Singapore, 1989, p151.
4. A.P. Hickman, J. Phys. B. 20, 2091, 1987.
5. A. Giusti-Suzor, J.N. Bardsley and C. Derkits, Phys. Rev. A28, 682, 1983.
6. K. Nakashima, H. Takagi and H. Nakamura, J. Chem. Phys. 86,726,1986.
7. P. Van der Donk, F.B. Yousif, J.B.A. Mitchell and A.P. Hickman (In preparation, 1989).
8. F.B. Yousif and J.B.A. Mitchell, Phys. Rev. in press, 1989.
9. W. Ketterle, H. Figger and H. Walther, Phys. Rev. Lett. 55, 2941, 1985.
10. J.R. Peterson and Y.K. Bae, Phys. Rev. A34, 3517, 1986.
11. W.J. van der Zande, W. Koot, D.P. deBruijn and C. Kubach, Phys. Rev. Lett. 57, 1219, 1986.
12. C. Noren, F.B. Yousif and J.B.A. Mitchell, J. Chem Soc. (Faraday Trans. II) in press, 1989.
13. E.C. Zipf, Geophys. Res. Lett. 7, 645, 1980.
14. R. Johnson, Int. J. Mass. Spec. Ion Phys. 81, 67, 1987.
15. S.L. Guberman in Dissociative Recombination: Theory, Experiment and Applications (eds. J.B.A. Mitchell and S.L. Guberman) World Scientific, Singapore. 1989, p.45.
16. N.G. Adams, D. Smith and E. Alge, J. Chem. Phys. 81, 1778, 1978.
17. H.H. Michels and R.H. Hobbs, Ap.J. 286, L27, 1984.
18. K.C. Kulander and M.F. Guest, J. Phys. B. 12, L501, 1979.
19. N.G. Adams, D. Smith in Dissociative Recombination: Theory Experiment and Applications. (Eds. J.B.A. Mitchell and S.L. Guberman). World Scientific, Singapore, 1989, p.124.
20. N.G. Adams and D. Smith Chem. Phys. Lett.144, 11, 1988.
21. T.A. Green, H.H. Michels, J.C. Browne, J. Chem. Phys. 69, 101, 1978.
22. F.J. Mehr and M.A. Biondi, Phys. Rev. 181, 264, 1969.
23. P.M. Mul and J. Wm. McGowan, J. Phys. B. 12, 1591, 1979.

HIGH-RESOLUTION ELECTRON-MOLECULE SCATTERING USING SYNCHROTRON-GENERATED ELECTRON BEAMS

D. Field*, D.W. Knight**, S. Lunt***, G. Mrotzek***, J. Randell*, and J.P. Ziesel****

* School of Chemistry, University of Bristol, Bristol, UK
** Department of Science, Bristol Polytechnic, Bristol, UK
*** SERC Daresbury Laboratory, Warrington, UK
****LCAM, Université Paris-Sud, Orsay, France

A synchrotron photoionisation technique has been used in an electron spectrometer at the SERC-Daresbury Laboratory, to study electron-molecule scattering. We have shown that it is possible to achieve electron scattering at photon resolution down to an energy resolution of 3.5 meV for energies in a few hundred meV range. Electron scattering data in O_2 suggest an important contribution of the partial waves l = 4 and 6 which should give rise to rotational excitation of the molecule. Work in progress on electron scattering in NO and on vibrational excitation of the (100) and (001) modes near threshold in CO_2 benefits from the improved resolution.

INTRODUCTION

There are a number of processes in electron-molecule scattering the observation of which requires a very high energy resolution in the incident electron beam. To accurately determine the profile of long-lived resonances associated with negative-ion compound states, the electron energy width should be smaller than the natural width of the resonance[1,2]. Very steep scattering onset near threshold is predicted to occur in electronic, rovibrational or rotational excitation of molecules, especially when s-wave scattering is dominant[3]. Moreover in polyatomic molecules, some vibrational modes lie very close in energy. Attachment to yield negative ions near zero energy often follows a s-wave threshold law, resulting in a sharp cross-section[4,5]. It turns out that an energy width below 5 meV is thus required for detailed studies of electron scattering by molecules especially if we seek to observe rotational phenomena. The present contribution is concerned with the measurement of narrow resonances in O_2 and NO, and vibrational excitation near threshold in CO_2. These low-energy crossed-beam experiments are performed using a synchrotron photoionisation electron source with an energy width of a few meV and intensity around a picoamp; the energy range extends down to 100meV.

In the study of electron-molecule scattering, two experimental techniques have been traditionally employed. In the first, a beam of monoenergetic electrons is formed by emission from a hot filament and then energy selected by an electrostatic dispersive field[7,8]. The resolution obtained using 127° cylindrical sector or 180° hemispherical selectors is around 10 meV at best, in gas-phase experiments. Also, space charge effects limit both the resolution and the low-energy electron flux. In the second method, a pulse of electrons is formed and its energy is selected by time-of-flight measurement. In this case, the resolution is limited by timing uncertainties and is dependent on the electron energy; a resolution of 5 meV or less is obtainable at 250 meV or below but increases to 40 meV at 1 eV[9].

An alternative way towards very high resolution is by photoionisation of atoms. The initial electron resolution is then determined by the photon bandwidth and Doppler broadening. Gallagher and coworkers have succeeded in photoionising a beam of metastable ^{1}D barium atoms intracavity at 3250 Å using a He-Cd laser[10]. The resulting photoelectrons had 17 meV energy and scattering on helium has shown that the best electron beam resolution was about 2 meV. Besides using laser light, narrow bandwidth photon beams can be obtained by the monochromatisation of a broad photon source. The output of a Helium continuum lamp has been used by Chupka[11], Ajello and Chutjian[12] to photoionise rare gases and study the attachment of electrons in SF_6 by varying the wavelength with a normal incidence monochromator. The technique has been extended to study in situ electron attachment to a range of halogenated compounds[4,5], at energies of a few meV to a few tens of meV and energy resolution between 3 and 7 meV. We have chosen to use electron beams from synchrotron photoionisation of Ar to study electron-molecule scattering, first in a transmission experiment at the LURE-ACO storage ring in Orsay[13,14] then whith an electron spectrometer at the SERC-Daresbury Synchrotron Radiation Source[15,16].

EXPERIMENTAL

The Ar photoionization[17] was chosen because it presents a number of advantages with respect to the production of a monoenergetic electron beam: the cross section near the Ar^+ $^2P_{3/2}$ threshold (786.72Å) is dominated by a sharp autoionizing resonance Ar^* $3p^5(^2P_{1/2})$ 11s superimposed on a broader resonance Ar^* 9d and its value at peak is high at about 1.4×10^{-16} cm^2. By setting the photon wavelength on the 11s peak at 786.5Å and with a 0.2Å bandwidth, photoelectrons of 4.3meV kinetic energy with a FWHM energy spread of 4 meV can be produced inside a chamber filled with argon at ~10mtorr. Electrons can then be easily drawn-out by a weak uniform electric field through a small aperture and on passing through suitable optics can produce a focussed electron beam of variable energy.

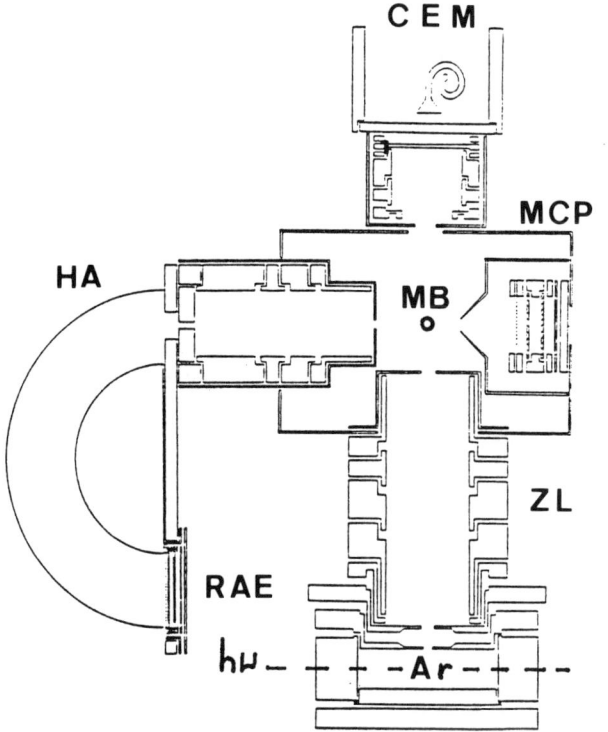

Fig.1 - Schematic diagram of the electron spectrometer: ZL is the zoom lens, MB is the supersonic molecular beam (into page), MCP are the microchannel plates, CEM is the channel electron multiplier, HA is the hemispherical analyzer and RAE is the resistive anode encoded strip.

A schematic diagram of the apparatus is shown in figure 1. The light from the VUV beam line 3.2 at Daresbury is monochromatised by a 5 m McPherson normal incidence monochromator and focussed at the center of the photoionisation source. The photoelectrons are accelerated and focussed by a 4-element zoom lens[18] onto a supersonic molecular beam (MB). Electrons scattered at right angles are detected either by the channelplates (MCP) with a large acceptance angle (±30°) or focussed by a zoom lens[18] at the entrance slit of a 180° hemispherical analyzer (HA) of radius R = 5 cm. The electrons are dispersed by the $1/R^2$ field between the two hemispheres and detected by a positional encoder device working in the risetime mode and composed of a pair of MCPs and a resistive anode. The range of electron kinetic energies simultaneously analyzed can be 1 eV and below. Unscattered electrons are collected at the channel electron multiplier (CEM). A number of factors discussed in more detail in ref. 13, could affect the energy resolution: i) Doppler motion of Ar in the source: due to the small photoelectron energy, the broadening ΔE_b is calculated to be 0.13 meV and thus negligible; ii) space charge across the photoionisation region[10] : the excess of positive ions creates a local potential field which contribute less that 0.2 meV; iii) extraction field: with a field of 0.4 V/cm the monochromator exit slit of 125µm set for 0.2 Å resolution with and 4:1 demagnifying mirror focussing on the center of the photoionisation region, the broadening is 1.2 meV and clearly, great care must be taken with the optical system; iv) the non-uniformity of surface potentials is minimized by cleanliness and aerodag (Matheson Co) coating and so far as possible by enlarging the dimensions so that the electron beam is far away from surfaces. Under typical operating conditions, we thus produce well-focussed incident beam currents of 1 picoamp for energies in the hundreds of meV range and 3-4 meV FWHM resolution.

The supersonic nozzle beam source (R. Campargue, CEN Saclay) operates by skimming in the zone of silence[19] where there is no interaction of the core of the jet with the background. Rotational cooling occurs by R-T collisions, mostly close to the nozzle. With a nozzle diameter of 300µ, a stagnation pressure around 1 bar and an optimum nozzle-skimmer distance of typically 15 mm, beams of local number density of more than 10^{13} cm^{-3} at the scattering center 69mm from the entrance to the nozzle are formed with rotational temperatures of ~20K for unseeded beams.

ELECTRON SCATTERING IN O_2.

The results in O_2 have been presented in detail elsewhere [15,16] and we will stress the main points only. The variation of the total scattering cross-section in the angular range 60-120°, measured with the MCP is shown in figure 2. The spectrum consists of a set of peaks superimposed on a rising background. These well-known resonances are associated with the vibrational sequence of the negative-ion O_2^- $^2\Pi_g$ (v',J'), formed by attachment of electrons with an even angular momentum ℓ to O_2 $^3\Sigma_g^-$(v"=0,J")[21,22]. The doublet structure of each peak is due to the spin-orbit splitting of each $^2\Pi$ vibrational level and has been observed in TOF transmission experiments [23,9] and in a recent derivative electron transmission experiment on a O_2 free jet[24]. Turning to figure 3 the lower curve is an expanded view of the O_2^- v'=6 resonance with the background subtracted; it was obtained with a pure O_2 molecular beam and experimental

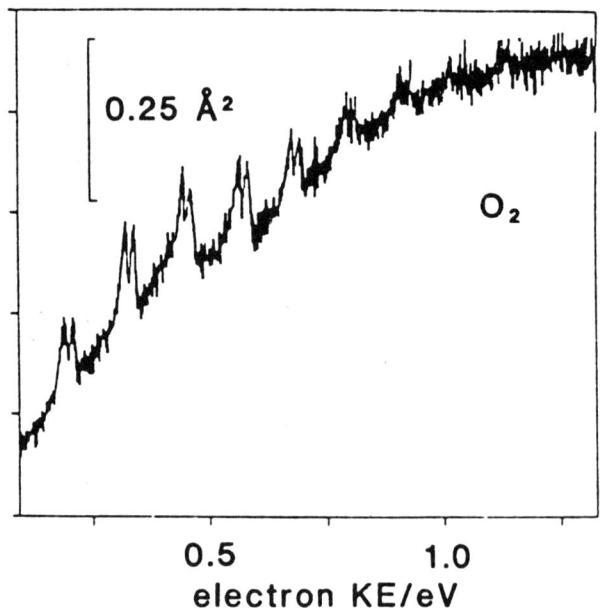

Fig.2 - Variation of the electron scattering cross section in O_2 at 90° in the 0.1 to 1.3 eV electron energy range. The lowest energy resonance shown is v'=5 of O_2^-; the v'=4 resonance lies at 91 meV[23].

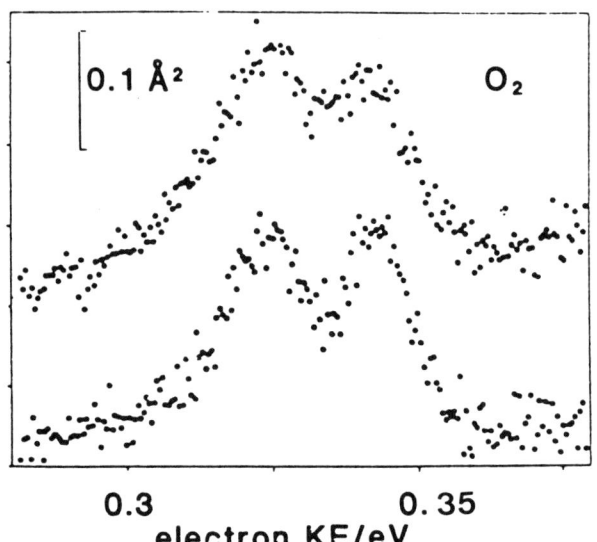

Fig.3- Two traces of the v'=6 resonance with a pure O_2 beam: upper trace : room temperature beam; lower trace : rotationally cold beam.

conditions where the beam should have rotational temperature of about 20K. The upper trace is the same resonance with an approximately room temperature molecular beam, produced by setting the nozzle-skimmer separation to 8 mm, conditions for which considerable scattering takes place in the vicinity of the skimmer The rotationally cooler beam experiment shows an apparent improvement in resolution which indicates that the doublet peaks are bands of rovibrational transitions and the greater the O_2 rotational population, the more the blurring of the doublet. This analysis[15,25] is supported by the observed narrowwidth of the resonance (fig.4) using a beam of O_2 30% seeded in He; due to the supercooling by He, the rotational temperature can then be 5K or less. The molecular constants of the O_2^- $^2\Pi_g$ state are determined from experiment: spin orbit coupling constant A = 18.75 ± 1 meV; vibrational constants ω_e = 135 ± 2meV; $\omega_e x_e$ = 1 ± 0.1meV. These values are in very good agreement with the other determinations[21,23,24]. Also the set of vibrational constants together with the v'=4 absolute energy[23] gives an electron affinity for O_2 close to the photodetachment experiment[26].

We have attempted to determine the lifetime of the O_2^- compound state by performing O_2^- v'=6 spectra simulation with: i) relative cross sections (O_2 (J") → O_2^- (J'))[25] ; ii) rotational populations in the supersonic beam using a two-temperature model[27], with the low temperature dominant; iii) gaussian electron energy distributions of width Γ_e; iv) lorentzian resonance lineshapes of witdth Γ_r. The convolution of rovibrational transitions with a given set of parameters gives a doublet envelope which may be compared with the experimental data.

The best global fit to the three sets of data (fig. 3-4) and other data[15,20] is obtained with an electron energy resolution Γ_e of 3.5 meV, a natural resonance linewidth Γ_r between 2 and 3 meV and contributions of the attaching electron partial waves ℓ= 2:4:6 in the ratio 1:0.5 : 0.25. The widths of the O_2^- resonances have been calculated[28-30] by theoretical analyses of the electron scattering data of Linder and Schmidt[20]. For the v' = 6 level, Parlant and Fiquet-Fayard[30] determine a width of 0.18 meV which is more than an order of magnitude smaller than the Γ_r value from our best fit. We could fit our data with the low calculated Γ_r only with contributions of the ℓ = 4 and 6 partial waves equal to that of ℓ = 2, which seems quite unrealistic. Recently, Noble et al[31] have performed R-matrix calculations of the width of the O_2^- $^2\Pi_g$ resonance for a

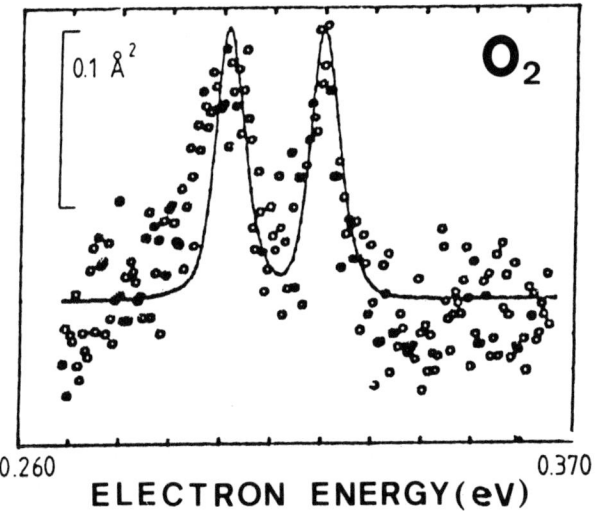

Fig.4 - The v'=6 resonance using O_2 seeded 30% in He : circles, experimental data; solid curve, simulation of data with the best set of parameters (see text).

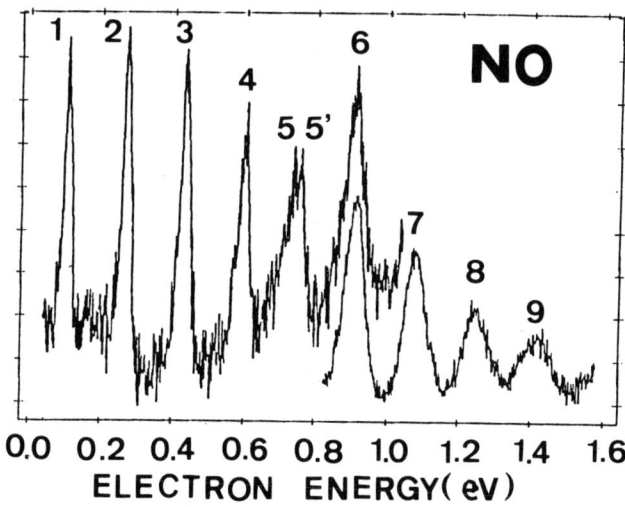

Fig.5 - Variation of the total electron scattering cross-section in NO at 90° measured on the MCP (see fig.1).

range of internuclear separations, using the adiabatic nuclei approximation; nine O_2 electronic states have been included in the model, in order to obtain a good CI wavefunction for the target. The calculated width is 2.6 meV, at the equilibrium distance of O_2. This distance is in the range of the v' = 5-6 levels of the O_2^- $^2\Pi_g$ experimental potential energy curve [26,30], and the results of the most sophisticated calculation to date are in good agreement with our value of 2-3 meV for the v'=6 resonance. A significant theoretical problem remains unanswered with regard to the large proportion of $\ell = 4$ and $\ell = 6$ that must be included in order to fit our data.

LOW-LYING RESONANT STATES IN NO

The lowest negative ion configuration in NO is NO^- $(2p\sigma)^2$ $(2p\pi)^4$ $(2p\pi)^2$ resulting in three resonant states $^3\Sigma^-$, $^1\Delta$, and $^1\Sigma^+$. The lowest state $^3\Sigma^-$ has a positive electron affinity of 24(+10, -5) meV determined by photodetachment[32]. In electron scattering on NO a resonant structure has been observed in the elastic cross-section with a good agreement on the peaks position[22] between various experiments. Burrow[33] has interpreted his data on elastic scattering at 180° in terms of the vibrational structures of NO^- $^3\Sigma^-$ below 0.7eV and NO^- $^1\Delta$ from this energy and up. The most complete results to date have been obtained in the electron spectroscopy experiment of Tronc et al[34]. The spectroscopic contants they have determined are ω_e=169meV and $\omega_e x_e$=1 meV for the $^3\Sigma^-$ state, ω_e = 185 meV and

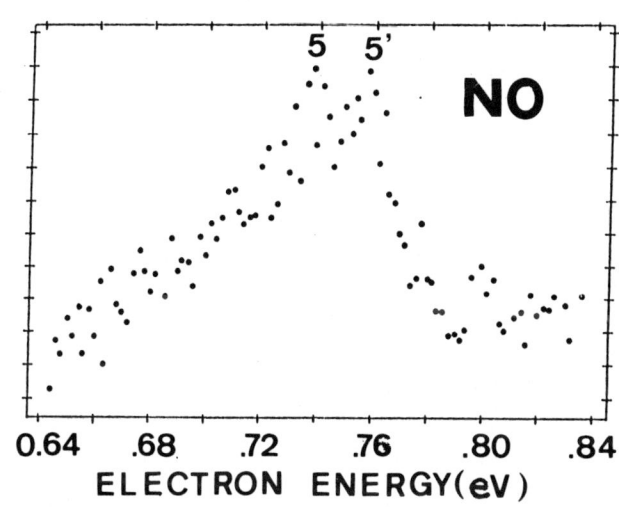

Fig.6 - Expanded spectrum of the NO^- resonance peaks around 747 meV showing clear splitting into two peaks.

T_e=0.75eV for the $^1\Delta$ state. The variation of the total cross-section which we have obtained between 50meV and 1.6eV is shown in fig.5. Scattered electrons are detected in the 60-120° angular range and the molecular beam is NO 15% seeded in helium. The peak energies and widths are listed in table 1 with the data of Tronc et al and the resonance linewidths Γ_r from the theoretical analysis of Teillet-Billy and Fiquet-Fayard[35]. The strong perturbation to peak

TABLE 1 - ELECTRON SCATTERING DATA IN NO

Peak	Energy/meV This work	a	Width/meV this work	a	Γ_r/meV[b] $^3\Sigma^-$		$^1\Delta$	
1	105	105	23		v' = 1	1		
2	271	275	29	40	v' = 2	5		
3	431	435	32	40	v' = 3	10		
4	599	595	32	40	v' = 4	15		
5	738	700	52				v' = 0	34
5'	758	740	52		v' = 5	21		
6	904	885	58	80	v' = 6	26	v' = 1	63
7	1077	1065	64	70	v' = 7	31	v' = 2	88
8	1248	1245	68		v' = 8	36	v' = 3	111
9	1415		85		v' = 9	41	v' = 4	131

a: elastic scattering experiment at 90° (ref. 34)
b: theoretical analysis of scattering data (ref.35)

---

heights and irregularities of spacings beyond the fourth resonance peak, together with the abrupt change in peak width, clearly indicate the involvement of more than one electronic state of NO⁻. The fifth peak is in fact a doublet (fig.6), indicating it is a superposition of two members of different vibrational series $^3\Sigma^-$ and $^1\Delta$, which begins to play a significant role above 700meV, as pointed out by Burrow, by Tronc et al and in agreement with the theoretical analysis[35]. The improved resolution in our experiment shows up in the experimental widths which increase with the vibrational quantum number and are then not instrument limited. There are strong discrepancies between the experimental widths and the calculated natural linewidths with a reversal between the $^3\Sigma^-$ and $^1\Delta$ states. More accurate ab initio calculations of the widths of the NO⁻ resonances in a range of internuclear separations are clearly needed[36].

NEAR THRESHOLD VIBRATIONAL EXCITATION IN CO_2.

Near threshold vibrational excitation of the (100), (010), and (100) modes have been studied experimentally by Stamatovic and Schulz[37] and Kochem et al[38]. The bending (010) and the asymmetric (001) modes are both optically allowed and the observed forward scattering[37] as well as the cross-section ratios are reasonably well described within the Born approximation. The differential Born formula, derived by Takayanagi[39] and Itikawa[40] using the infrared transition probability, reproduces well the shape of the (010) cross-section near threshold[38]. The agreement is not too satisfactory for the (001) cross-section, especially when the scattering angle increases and the electron energy gets closer to threshold. For the symmetric (100) excitation, the Born approximation fails to reproduce the isotropic angular distributions and the absolute magnitude of the cross-section[38]. The steep threshold in the symmetric-stretch excitation cross-section has been predicted by Morrison and Lane[41], and is due to the perturbation of the outgoing electron s-character wavefunction by a virtual state. The close coupling calculations of Whitten and Lane[42], and Kimura and Lane[44] and the virtual state model of Estrada and Domcke[43] are in good agreement with the experimental cross-section shape and magnitude[38].

We have studied the shape of the asymmetric (001) and symmetric (100) stretch excitation near-threshold with our high-resolution electron spectrometer. The electrons scattered at 90° by a pure CO_2 supersonic beam are energy-analyzed and detected on the resistive anode; the encoder window can be set up to detect either the whole range of energy losses simultaneously or a specific excitation process. The shape of the differential (001)cross-section at 90° is shown in figure 7; the onset is measured at 295 ± 25meV, which may be compared to the excitation energy of 291meV. The Born differential cross-section[40], calculated at the scattering angle of 90° is represented by the dashed line; for purposes of comparison the excitation curve has been moved arbitrarily to the excitation energy and the Born cross-section normalized to the maximum of the excitation curve. The shapes of the two curves near threshold are quite different, confirming the results at 39° and 54° of Kochem et al [37] who state that the Born approximation cannot explain the (001) experimental data. The steepness of the threshold

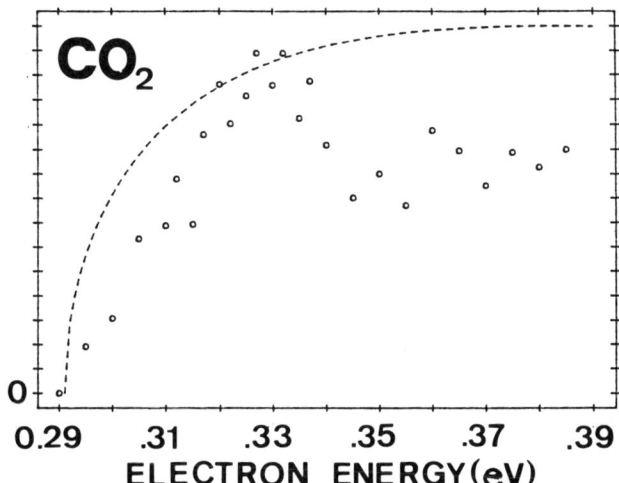

Fig.7 - Variation of the inelastic cross-section at 90° for excitation of the asymmetric stretch (001) in CO_2 : circles, experiment; dashed curve, differential Born cross-section.

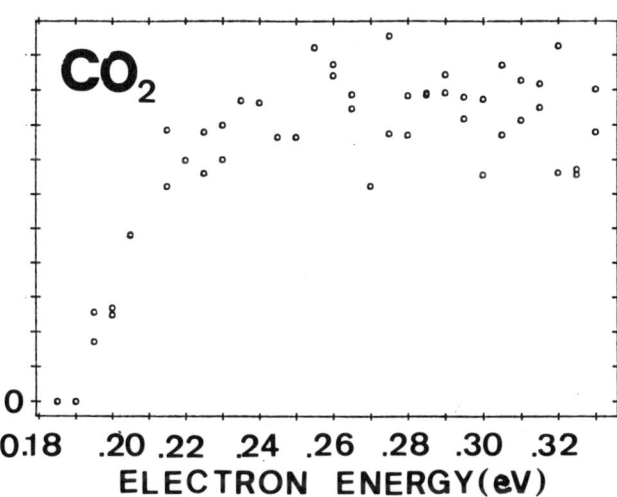

Fig.8 - Variation of the inelastic cross-section at 90° for excitation of the symmetric stretch (100) in CO_2.

peak in our data agrees with the two state vibrational close-coupling calculation of Whitten and Lane[42] for the (001) total cross-section near threshold. However we note that the forward scattering observed by Kochem et al rules out the s-wave scattering assumption which underpinned the theoretical model of Whitten and Lane.

The energy dependence of the (100) excitation curve at 90° shows a sharp onset at 190±25meV, within the limit of error equal to the excitation energy of 172meV. The steepness of the curve near threshold is comparable to the calculated cross-sections[42,44] where the rise takes place over 50meV approximately. It then flattens out from 250 to 330meV failing to reproduce the sharp threshold peak found in the calculations. Our measurements will be extended up to 1 eV for comparison with theory and previous data well above threshold.

CONCLUSION

The electron spectrometer at the SERC Daresbury Laboratory has shown that the synchrotron photoionisation technique is indeed effective for the production of usable high resolution electron beams. The instrument is capable of the detailed investigation of near-threshold vibrational excitation and the preliminary data on both CO_2 and NO reported here already present a significant challenge to theorists in this area. We intend to continue our work on NO and CO_2 to include further studies of elastic and inelastic scattering and their angular dependence.

References

1. R.J. Van Brunt and A.C. Gallagher, in Electronic and Atomic Collisions, Edited by G. Watel (North Holland, Amsterdam, 1978) pp. 129-142.
2. R.E. Kennerly, R.J. Van Brunt and A.C. Gallagher, Phys. Rev. A 23, 2430 (1981)
3. M.A. Morrison, in Advances in Atomic and Molecular Physics, edited by Sir D. Bates and B. Bederson (Academic, New York, 1987), vol. 24, pp.51-156.
4. A. Chutjian and S.H. Alajajian, J. Phys. B 18, 4159 (1985); Phys. Rev. A 35, 4512 (1987) and references therein.
5. S.H. Alajajian, M.T. Bernius and A. Chutjian, J. Phys. B 21, 4021, (1988).
6. L. Kerwin, P. Marmet and J.C. Carette, in Case Studies in Atomic Collision Physics, Edited by E.W. McDaniel and M.R.C. McDowell (North Holland, Amsterdam, 1969), vol. I, pp. 527-581.
7. J.A. Simpson, Rev. Sci. Instrum. 35, 1698 (1964); C.E. Kuyatt and J.A. Simpson, Rev. Sci. Instrum. 38, 103 (1967).
8. W. Raith, in Advances in Atomic and Molecular Physics, edited by D.R. Bates and B. Bederson (Academic, New York, 1976), vol. 12, pp.281-373.
9. J. Ferch, W. Raith and K. Shroder, J. Phys. B 13, 1481 (1980).
10. A.C. Callagher and G. York, Rev. Sci. Instrum. 45, 662 (1974).
11. W.A. Chupka, private communication (1974).
12. J.M. Ajello and A. Chutjian, J. Chem. Phys. 65, 5524 (1976); J. Chem. Phys. 71, 1079 (1979).
13. D. Field, J.P. Ziesel, P.M. Guyon and T.R. Govers, J. Phys. B 17, 4565 (1984).
14. J.P. Ziesel and D. Field, in Photophysics and Photochemistry above 6eV, edited by F. Lahmani (Elsevier, Amsterdam, 1985) pp. 221-225.
15. D. Field, G. Mrotzek, D.W. Knight, S. Lunt and J.P. Ziesel, J. Phys. B 21, 171 (1988).

16. D. Field, in Europhysics News, edited by the European Physical Society, **20**, 64 (1989).
17. K. Radler and J. Berkowitz, J. Chem. Phys. **70**, 221 (1979).
18. G. Martinez, M. Sancho and F.H. Read, J. Phys. E: Sci. Instrum. **16**, 631 (1983).
19. R. Campargue, J. Phys. Chem. **88**, 4466 (1984) and references therein.
20. G. Mrotzek, Thesis, University of Bristol, 1989.
21. F. Linder and H. Schmidt, Z. Naturf. **26A**, 1617 (1971)
22. G.J. Schulz, Rev. Mod. Phys. **45**, 423 (1973); in Principles of Laser Plasmas chap. II, edited by G. Bekefi (Interscience, New York, 1976)
23. J.E. Land and W. Raith, Phys. Rev. **A9**, 1592 (1974)
24. T.M. Stephen and P.D. Burrow, J. Phys. B **19**, 3167 (1986).
25. F. Fiquet-Fayard, J. Phys. B **8**, 2880, (1975).
26. R.J. Celotta, R.A. Bennet, J.L. Hall, M.W. Siegel and J. Levine, Phys. Rev. **A6**, 631 (1972).
27. S.P. Venkateshan, S.B. Ryali and J.B. Fenn, J. Chem. Phys. **77**, 2599, (1982).
28. F. Koike and T. Watanabe, J. Phys. Soc. Japan **34**, 1022 (1973).
29. F. Koike, J. Phys. Soc. Japan **35**, 1166 (1973).
30. G. Parlant and F. Fiquet Fayard, J. Phys. B **9**, 1617 (1976).
31. C.J. Noble, T.T. Scholz, C.J. Gillan and P.G. Burke, in Abstracts of Contributed papers, 16th ICPEAC, edited by A. Dalgarno, R.S. Freund, M.S. Lubell and T.B. Lucatorto, New York, 1989; C.J. Noble and P.G. Burke, to be published.
32. M.W. Siegel, R.J. Celotta, J.L. Hall, J. Levine and R.A. Bennet, Phys. Rev. **A6**, 607, (1972).
33. P.D. Burrow, Chem. Phys. Lett. **26**, 265, (1974).
34. M. Tronc, A. Huetz, M. Landau, F. Pichou and J. Reinhardt, J. Phys. B **8**, 1160 (1975)
35. D. Teillet-Billy and F. Fiquet-Fayard, J. Phys. B **10**, L111 (1977).
36. J. Tennyson and C.J. Noble, J. Phys. B **19**, 4025 (1986).
37. A. Stamatovic and G.J. Schulz, Phys. Rev. **188**, 213 (1969).
38. K.H. Kochem, W. Sohn, N. Hebel, K. Jung and H. Ehrhardt, J. Phys. B 18, 4455 (1985)
39. K. Takayanagi, Prog. Theoret. Phys. (Kyoto) Suppl.**40**, 216 (1967).
40. Y. Itakawa, Phys. Rev. **A 3**, 831 (1971).
41. M.A. Morrison and N.F. Lane, Chem. Phys. Lett. **66**, 527, (1979).
42. B.L. Whitten and N.F. Lane, Phys. Rev. **A26**, 3170, (1982).
43. M. Kimura and N.F. Lane, 1986, cited in ref. 3.
44. H. Estrada and W. Domcke, J. Phys. B **18**, 4469 (1985).

General

RESONANT RECOMBINATION AND AUTOIONIZATION IN ELECTRON-ION COLLISIONS

Alfred Müller

Strahlenzentrum der Justus-Liebig-Universität
D-6300 Giessen, Federal Republic of Germany

The occurence of resonances in elastic and inelastic electron-ion collisions is discussed. Resonant processes involve excitation of the ion with simultaneous capture of the initially free electron. The decay mechanism subsequent to the formation of the intermediate multiply excited state determines whether a resonance is found in recombination, excitation, elastic scattering, in single or even in multiple ionization. This review concentrates on resonances in the ionization channel. Correlated two-electron transitions are considered.

Introduction.

While resonant scattering of electrons from neutral atoms and molecules has been a subject of experimental research since several decades[1,2], resonances in electron-ion collisions have not directly been observed before 1970[3]. The first resonant electron-ion collision process studied theoretically was resonant photon-stabilized recombination. It had been suggested as a possible neutralization mechanism for ions in plasmas. Massey and Bates named this process "dielectronic recombination" (DR) and performed first theoretical calculations of rate coefficients[4]. Extended and more complete calculations by Burgess[5] established DR as an important fundamental collision process in any hot plasma. This also stimulated a longstanding interest in resonant electron-ion collision processes[6].

For the quantitative investigation of interactions between electrons and ions the technique of colliding beams[7] is most appropriate. Since 1961[8] this technique has been extensively used mostly in electron-impact ionization measurements[9]. It was soon recognized that indirect (or multi-step) mechanisms are of utmost importance in such collisions and a milestone for the whole field of research was set when LaGattuta and Hahn[10] calculated dominant contributions from excitation-autoionization and resonant-excitation-double-autoionization (REDA) in the electron-impact ionization of Fe^{15+} ions. Thus, experimentalists were confronted with a new challenge to find resonances in electron-ion collisions and measure cross sections not only for DR but also for REDA processes.

In 1983 first direct measurements of DR cross sections were published[11-14] and since then a number of colliding beams experiments on DR were carried out. An important new effect — the influence of an external electric field on the resonance strength — was experimentally established[15] and measurements were performed with multiply charged light ions[16].

Recently, several new breakthroughs were accomplished. DR of Ni^{26+} ions was observed in an electron-beam ion trap[17]. Cross sections for DR of multiply charged light ions were measured with greatly improved energy resolution[18] and very recently, the first measurements of cross sections for DR of hydrogen-like O^{7+} ions were carried out at a storage ring with an electron cooler[19].

During these same years, REDA and other resonant contributions to ionization of ions were inferred from experimental results[20-23], however, clearcut evidence for the presence of resonances in ionization cross sections could not be found although serious attempts were undertaken to reach this goal[24,25]. It was not until last year that successful experiments on the observation of narrow resonances in the electron-impact ionization of ions could be reported[26,27].

This paper reviews the work on resonances in electron-impact ionization of ions. An attempt is made to put the observations into a general frame and to clarify the connections between different resonance processes occuring in electron-ion collisions.

Fundamental processes.

Resonances in electron-ion collisions proceed through the capture of the incident electron by the ion. The excess energy is used up

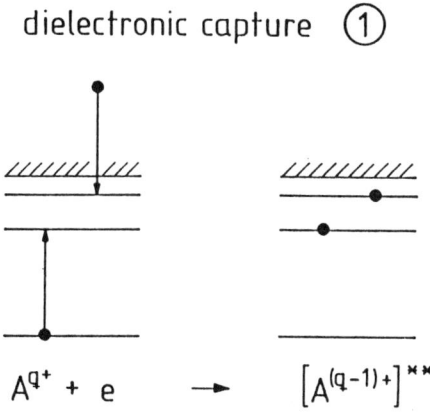

Figure 1. Simplified scheme for the process of dielectronic capture of a free electron by an ion and the resulting multiply excited state. Dielectronic capture is the first step in any electron-ion resonance process.

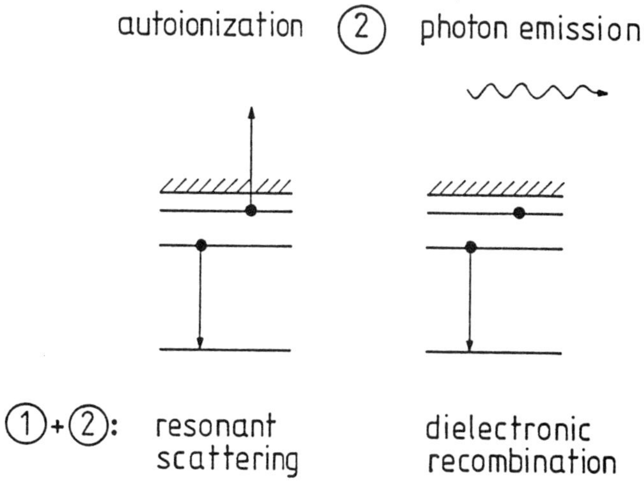

Figure 2. Decay processes of a doubly excited state. For light ions autoionization (Auger process) is generally the most likely decay mechanism. The net result of a dielectronic capture (Fig. 1) with subsequent autoionization is a resonant electron scattering from the parent ion. It is also possible to stabilize the doubly excited atom or ion by the emission of photons. A stabilizing transition into a non-autoionizing state is shown. The net result of dielectronic capture (Fig. 1) and subsequent stabilization by photon emission is called dielectronic recombination.

for an excitation of the electron core of the ion. Usually, it is one core electron that interacts with the projectile electron. The capture process can be represented by

$$e + A^{q+}(i) \rightarrow A^{(q-1)+}(j,nl) \qquad (1)$$

where $A^{q+}(i)$ is an ion in charge state q and in an initial electronic state i. By the interaction with the incident electron the ion electron core is excited to a state j and the initially free electron is bound in a subshell nl. (Of course excited and captured electrons are not distinguishable so that it could also be an electron from the core that is lifted into subshell nl). The resulting ion $A^{(q-1)+}(j,nl)$ with the charge state reduced by one unit is at least doubly excited. Fig. 1 presents a simple scheme of the process which can only happen if the energy of the free electron matches the resonance condition

$$E_{res} = E(j,nl) - E(i) \qquad (2)$$

Here, E_{res} is the electron energy at which the resonance occurs, $E(j,nl)$ and $E(i)$ are total energies of all electrons in the excited ion $A^{(q-1)+}(j,nl)$ and the parent ion $A^{q+}(i)$, respectively.

The process described by Eq. 1 is named differently in the literature, partly depending on the decay channel of the resonance observed by experiments. Widely used terms are resonant excitation, resonant recombination, dielectronic capture or even (incorrect) dielectronic recombination although the latter implies the subsequent stabilization of the ion $A^{(q-1)+}$ by photon emission in addition to the initial capture process. Each of the first three names has its merits and will probably further be used in the literature. The term dielectronic capture seems to be the most descriptive for the process shown by Fig. 1, however, it turned out that the excitation $i \rightarrow j$ in Eq. 1 can also involve two electrons which would require a term trielectronic capture. In this paper the term dielectronic (and were required trielectronic) capture will be used for the process described by Eq. 1.

Time-reverted dielectronic capture is the process of autoionization (or Auger electron emission). The probability for autoionization is usually described by the decay rate A_a. Due to the principle of detailed balance the cross section σ_c for dielectronic capture[28] is proportional to A_a.

$$\sigma_c = \frac{c}{E} \cdot \frac{g_f}{2g_i} \cdot \frac{A_a \cdot \Gamma(f)}{(E_{res} - E)^2 + \Gamma(f)^2/4} \qquad (3)$$

In this formula $c = 7.89 \cdot 10^{-5}$, σ_c is in cm^2,

E_{res} is in eV, E is the energy of the projectile electron (in units of eV), g_i and g_f are the statistical weights of the initial electronic state i and the final state f (which corresponds to (j,nl)), A_a is the autoionization rate (in units of s^{-1}) for time-reversed Eq. 1 and $\Gamma(f)$ is the total width (in units of eV) of the final state f. The cross section σ_c has Lorentzian shape.

The excited state (j,nl) formed by dielectronic capture is unstable. It can decay by autoionization as was discussed above

$$A^{(q-1)+}(j,nl) \rightarrow A^{q+}(k) + e \quad (4)$$

or by the emission of a photon

$$A^{(q-1)+}(j,nl) \rightarrow A^{(q-1)+}(p) + h\nu \quad (5)$$

The autoionization can lead to a state k in the ion A^{q+}, the photon emission to a state p in the ion $A^{(q-1)+}$. The states k and p can still be excited states submitted to further decay processes.

Fig. 2 shows simplified schemes for the decay of the intermediate excited state formed by dielectronic capture according to Fig. 1. For autoionization Fig. 2 shows the particular case where k = i. The net result of the combined processes Eq. 1 and Eq. 4 (with k = i) is a <u>resonant elastic scattering</u> of the electron e from the ion A^{q+}. When k describes an excited state of the A^{q+} ion one observes <u>resonant inelastic electron scattering</u>. The simplest case then is the one where state k can just decay by the emission of a photon. In an experiment looking at line radiation from the state k the combined processes Eq. 1 and Eq. 4 (with k ≠ i) would result in resonances in the photon-emission (excitation) cross section. The possibility that k is also an autoionizing state will be discussed later.

The cross section σ_{sc} for the combined two processes Eq. 1 and Eq. 4, i. e. resonant (elastic or inelastic) scattering, is given by the cross section for dielectronic capture multiplied with the branching ratio for a particular autoionization

$$\sigma_{sc} = \sigma_c \cdot \frac{\Gamma_a(j,nl \rightarrow k)}{\Gamma(f)} \quad (6)$$

Here, $\Gamma_a(j,nl \rightarrow k)$ is the rate of autoionizations leading from the state (j,nl) in the ion $A^{(q-1)+}$ to the state k in the ion A^{q+}. For elastic resonances k = i and $\Gamma_a = A_a$.

For radiative deexcitation Fig. 2 shows the particular case where p corresponds to a (non-autoionizing) configuration (i,nl). The net result of the combined processes Eq. 1 and Eq. 5 (with p non-autoionizing) is <u>dielectronic recombination</u>. The new charge state of the recombined ion $A^{(q-1)+}$ is then stabilized, though further emission of radiation is due from the state p with transitions finally ending in the ground state of the ion $A^{(q-1)+}$. The cross section σ_{DR} for the combined two processes Eq. 1 and Eq. 5 is given by

$$\sigma_{DR} = \sigma_c \cdot \frac{1}{\Gamma(f)} \cdot \sum_p \Gamma_r(j,nl \rightarrow p) \quad (7)$$

Here, $\Gamma_r(j,nl \rightarrow p)$ is the rate of radiative decays leading from the state (j,nl) in the ion $A^{(q-1)+}$ to a non-autoionizing state p in the ion $A^{(q-1)+}$. The sum extends over all non-autoionizing states p. In an experiment looking with low resolution at line radiation (the resonance line) from excited states r of the ion $A^{q+}(r)$ resonances in the cross section may be detected by the inclusion of dielectronic satellites arising from excited states $A^{(q-1)+}(r,nl)$ in the ion $A^{(q-1)+}$ produced in the course of a DR process (e.g. when r = j). By the presence of a spectator electron in the subshell nl the line radiation for the transition r → i is (slightly) shifted in wavelength but may be not resolved from the resonance line in a real experiment.

So far, resonances in elastic electron scattering, in stabilized electron-ion recombination and in electron-impact excitation of ions have been discussed. All these processes required not more than one electron bound in the parent ion. When the parent ion has at least two electrons an additional class of resonances becomes possible which can be seen in the channel of net single ionization of the ion. Fig. 3 shows possible schemes of the decay processes after a dielectronic capture event finally leading to ionization. For the sake of simplicity it is assumed that the parent ion has one electron in the lowest level (number 1) and (at least) one in the third level (number 3), i. e. a configuration (1,3). Dielectronic capture excites the deeply bound electron from level number 1 to level number 3 and (in the example) the initially free electron is captured also into level number 3. The resulting intermediate configuration (2,3,3) is displayed in Fig. 3. It can contribute to net ionization of the ion when two electrons are emitted from configuration (2,3,3). This is possible by double autoionization, i. e. two successive autoionization processes ejecting one electron each. In the left example of Fig. 3 it is the sequence (2,3,3) → (2,2) + e → (1) + e + e. Another possibility is the ejection of two electrons at a time which is also called auto-double-ionization or sometimes double-Auger process. This is sketched on the right side of Fig. 3 which shows the decay (2,3,3) → (1) + 2e where the electron in level 2 kicks out the other electrons when jumping into level 1.

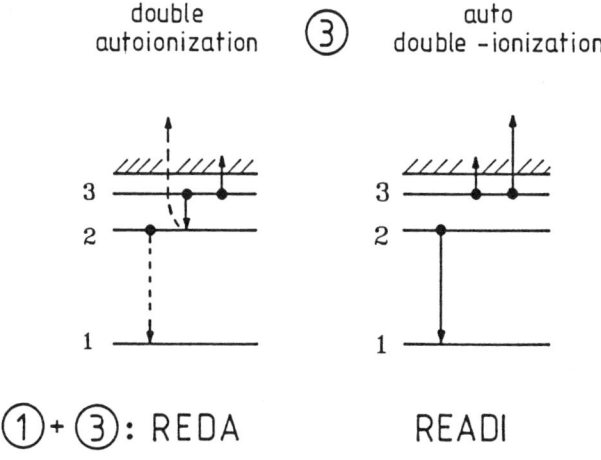

$$\sigma_{REDA} = \sigma_c \cdot \frac{1}{\Gamma(f)} \sum_{k,s} \Gamma_a(j,nl \to k) \cdot \frac{1}{\Gamma(k)} \Gamma_a(k \to s) \quad (9)$$

Here, $\Gamma(k)$ is the total width of the autoionizing intermediate state k of the ion A^{q+} and $\Gamma_a(k \to s)$ is the particular autoionization rate for the transition from state k of the ion A^{q+} to state s of the ion $A^{(q+1)+}$. The sum extends over all autoionizing intermediate states k and non-autoionizing states s. (In the case that also s is autoionizing even more complex mechanisms are possible and then Eq. 9 has to be modified).

When the intermediate multiply excited state formed by dielectronic capture decays by correlated emission of two electrons the whole mechanism is termed <u>resonant-excitation-auto-double-ionization</u> (READI or sometimes also READ). The cross section for a particular READI resonance is obtained by multiplying σ_c with the branching ratio for the particular two-electron emission process $\Gamma^{2e}/\Gamma(f)$

$$\sigma_{READI} = \sigma_c \cdot \Gamma^{2e}/\Gamma(f) \quad (10)$$

Beside REDA and READI there are even more resonant contributions to net single or multiple ionization possible which are mixed with photon emissions.

At the end of this chapter theoretical cross sections are presented as examples for resonant elastic scattering, DR and REDA. Fig. 4 shows σ_{RS} for resonant elastic scattering of electrons from C^{5+} ions in the energy range 265 eV - 285 eV. At these energies resonant intermediate states have configurations $2s^2$, $2s2p$ or $2p^2$. The cross section for each particular resonance was calculated from Eq. 6 (with k = i) by using published decay rates[29] from states with the above configurations. For direct comparison Fig. 5 shows σ_{DR} for dielectronic recombination of C^{5+} in the same energy region. The cross sections were calculated from Eq. 7 again by using the mentioned decay rates. Of course, the resonance positions are the same for σ_{RS} and σ_{DR}, however, the relative sizes of resonance peaks are different. In particular, σ_{RS} is much greater than σ_{DR} for C^{5+} which reflects the fact that autoionization rates usually exceed radiative rates for light ions by orders of magnitude. This changes when the atomic number of the ion is increased. In a real experiment the comparison of σ_{RS} and σ_{DR} would be complicated by the possible interference of resonant with non-resonant elastic scattering. So far, however, there is not a single experiment published yet on resonant elastic scattering of free electrons from ions, however, first results may be expected during the next years.

Figure 3. Decay mechanisms for triply excited states. Dielectronic capture (Fig. 1) may involve excitation of an inner-shell electron. The resulting configuration has an inner shell vacancy and partially filled outer shells. For simplicity only the minimum number of electrons required for a decay into the net-ionization channel is shown. When numbering the energy levels in the simplified scheme from the energetically lowest to the highest level the situation displayed in the example can be described by a set of "quantum numbers" (2,3,3). Such a state can decay by sequential or simultaneous emission of two electrons. The net result of dielectronic capture (Fig. 1) with two subsequent successive autoionizations is a single ionization of the parent ion. This resonant reaction path is called REDA (see text). The net result of dielectronic capture with subsequent emission of two correlated electrons is also a single ionization of the parent ion. The reaction path is also resonant. It is termed READI (see text).

In the notation used in Eqs. 1 and 4 the state k may be also autoionizing so that the ion $A^{q+}(k)$ can decay once again by the emission of an electron leading to the formation of an ion $A^{(q+1)+}$ in state s

$$A^{q+}(k) \to A^{(q+1)+}(s) + e \quad (8)$$

The mechanism proceeding via processes Eq. 1, Eq. 4 and Eq. 8 which lead to net single ionization of the parent ion after capture of the projectile electron is termed <u>resonant-excitation-double-autoionization</u> (REDA). The cross section for REDA can be written as

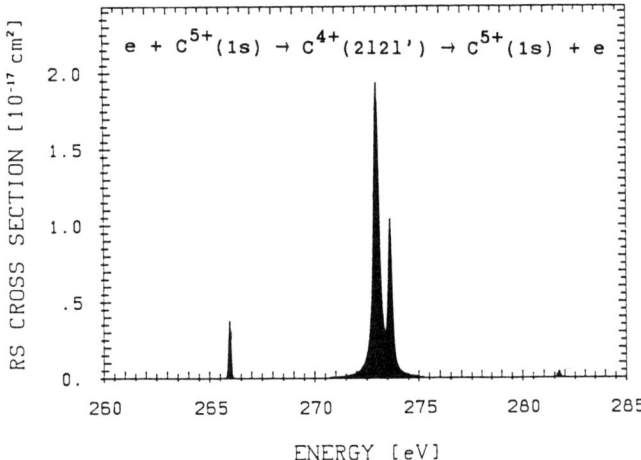

Figure 4. Cross section σ_{RS} for resonant scattering of electrons from C^{5+} ions calculated from Eq. 6 with k = i. Theoretical decay rates[29] for states with the intermediate configurations $2s^2$, $2s2p$ and $2p^2$ have been used. In a real experiment this cross section is influenced by non-resonant electron scattering. Interference may occur. The calculated resonances with natural widths between about 1 meV and 250 meV are convoluted with a Gaussian distribution of 0.1 eV FWHM.

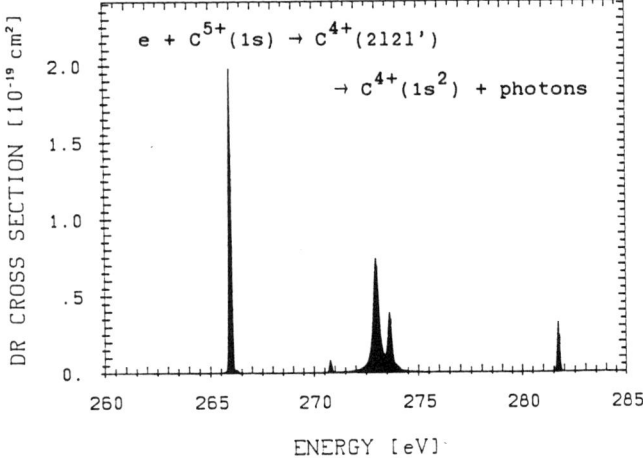

Figure 5. Cross section σ_{DR} for dielectronic recombination of electrons with C^{5+} ions calculated from Eq. 7. Theoretical decay rates[29] for states with the intermediate configurations $2s^2$, $2s2p$ and $2p^2$ have been used. The intermediate resonant states are the same as those contributing to the result shown in Fig. 4. Only the decay channel is different. Note the different scales of Figs. 4 and 5. The calculated resonances with natural widths between about 1 meV and 250 meV are convoluted with a Gaussian distribution of 0.1 eV FWHM.

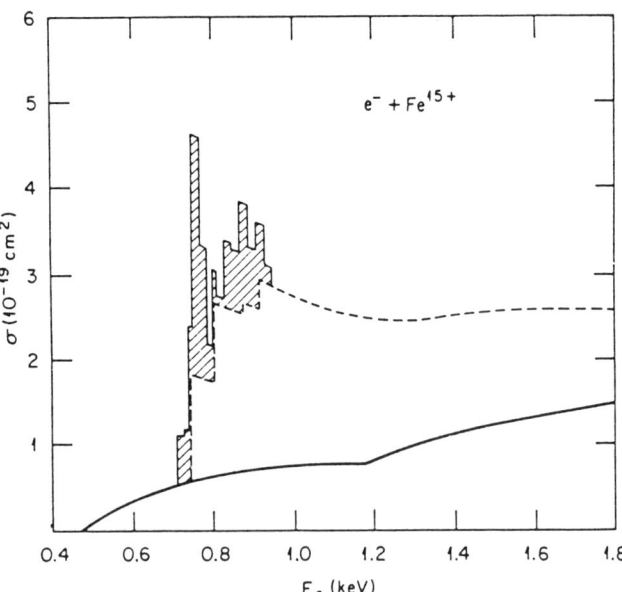

Figure 6: Theoretical cross section for electron-impact ionization of Fe^{15+} ions[10]. The solid curve is direct ionization of 3s or 2p electrons, the dashed curve is excitation-autoionization added to the direct part; the cross hatched area is resonant-excitation-double-autoionization (REDA) and added to the other two contributions.

Fig. 6 shows the benchmark calculation by LaGattuta and Hahn[10] on the ionization of sodium-like Fe^{15+}. The theoretical cross section comprises an estimate of direct ionization, calculated excitation-autoionization and REDA contributions where the resonances are averaged over energy bins of 20 eV width. The calculation shows dominant contributions from indirect ionization mechanisms and in particular REDA was predicted to dominate the cross section at energies around 750 eV. This work showed the potential importance of resonances in electron-impact ionization of ions and thus stimulated further experimental and theoretical research on REDA.

Resonances in electron-impact ionization of ions.

Direct single ionization is the only possible ionization mechanism when an electron collides with a one-electron parent ion. Cross sections for direct ionization can be represented by a smooth function of energy such as the Lotz formula[30] (σ roughly proportional to $1/E \cdot \ln E$). Any structure found in an electron-impact ionization cross section on top of such a smooth curve is due to indirect ionization mechanisms.

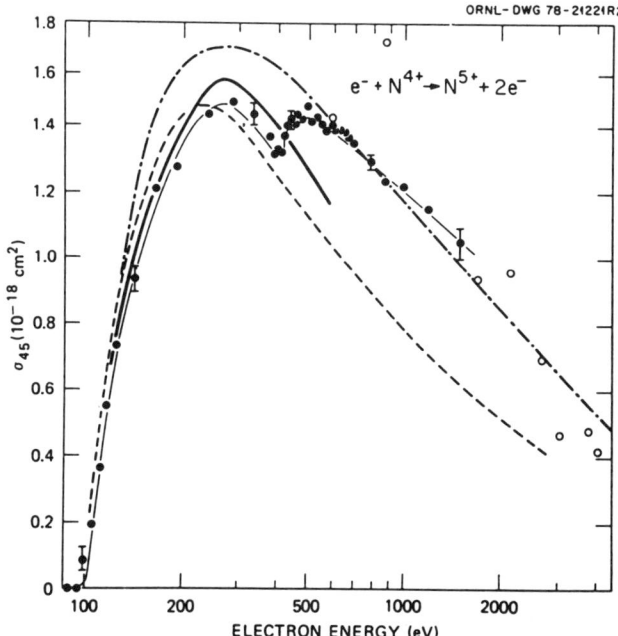

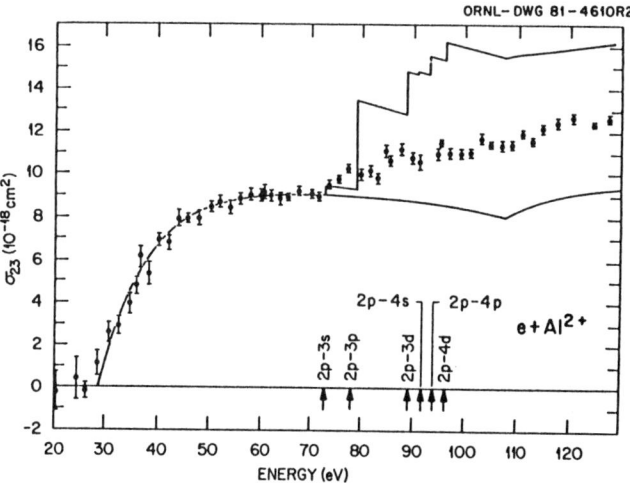

Figure 7. Cross section for electron-impact ionization of N^{4+} ions. Connected solid points are data of Crandall et al.[32]; open points are data of Donets and Ovsyannikov[42]; the solid curve is a Coulomb-Born calculation[43]; the dashed curve is scaled Coulomb Born[44]; the dot-dashed curve is from the semiempirical Lotz formula[30].

Figure 8. Electron impact ionization of Al^{2+} ions near threshold. Data are from Crandall et al.[34]. The solid curve is a distorted wave calculation of direct ionization by Younger[45], normalized to the experiment at 70 eV by multiplying Younger's results by 0.65. The distorted wave excitation of Griffin et al.[46] is added to the direct ionization calculation to give the upper solid curve. Arrows indicate center-of-gravity energies for excitation of 2p electrons to nl orbitals.

The first such mechanism clearly found in electron-ion collisions was that of excitation-autoionization[31]. An example is displayed in Fig. 7 which shows the cross section for single ionization of N^{4+} ions[32]. Beside direct ionization

$$e + N^{4+}(1s^2\, 2s) \rightarrow N^{5+}(1s^2) + 2e \qquad (11)$$

the step like increase at an energy of about 410 eV indicates a contribution from excitation of an inner shell electron and subsequent autoionization.

$$\begin{aligned} e + N^{4+}(1s^2\, 2s) &\rightarrow N^{4+}(1s\, 2s\, 2l) + e \\ &\rightarrow N^{5+}(1s^2) + 2e \end{aligned} \qquad (12)$$

In a later experiment[25] on O^{5+} ions (also Li-like) carried out with better counting statistics it was possible for the first time to see excitation of individual terms in the autoionizing 1s2s2l configurations. A serious attempt to find also READI resonances in that same experiment on ionization of O^{5+} ions failed due to the remaining experimental uncertainties. The first quantitative theoretical calculation of READI processes[33] (which had to deal with the correlated ejection of two electrons; see Fig. 3) gave contributions from READI which were 100 to 1000 times less than the cross section for direct ionization in Li-like ions, a result which appeared to make it impossible to ever see READI in these ions.

A better chance to see resonant processes in the ionization of ions could be inferred from the REDA hypothesis of LaGattuta and Hahn[10] (see Fig. 6). Indeed, the presence of REDA was inferred from measurements on Na-like Mg^+, Al^{2+} and Si^{3+} ions[34] although no single resonance had been resolved. Fig. 8 shows the measured electron-impact ionization cross section for Al^{2+} in the vicinity of the 2p → 3l excitation threshold together with calculations of direct ionization and excitation-autoionization contributions. The fact that there are measured points with their error bars (relative uncertainty) well above the calculated first excitation step ("snow on the stairs") lead Henry and Msezane[23] to a theoretical analysis that invoked the presence of REDA contributions. From a model calculation using close coupling methods they obtained a Rydberg series of REDA resonances between the thresholds for $2p^63s \rightarrow 2p^53s^2$ and $2p^63s \rightarrow 2p^53s3p$ excitations (see Fig. 9). The resonances belong to configurations of dielectronic-capture states

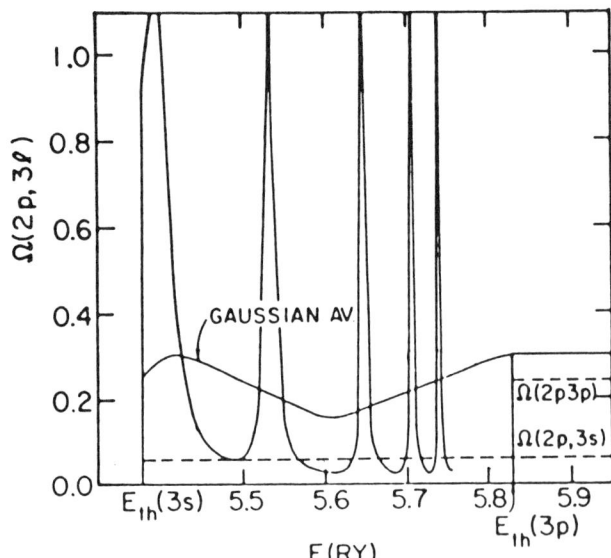

Figure 9. Collision strength Ω for electron-impact ionization of Na-like Al^{2+} ions from a model calculation of Henry and Msezane[23]. The dashed lines show collision strengths for $2p \to 3s$ and $2p \to 3p$ excitations. The solid lines are three-state close coupling model calculations and convoluted with a 2 eV FWHM Gaussian distribution. The resonances arise from $2p^53s3pnl$ Rydberg states with n = 3,4,5...

$2p^53s3pnl$ with n = 3 for the energetically lowest (and strongest) resonance seen just at the 3s excitation threshold. Double autoionization from these configurations is possible with a $2p^53s^2$ intermediate state. Below the excitation energy for a state $2p^53s^2$ REDA is not possible because then there is no suitable intermediate autoionizing state available. The energy resolution in the experiments on Al^{2+} (Fig. 8) was about 2eV and the convolution of the calculated REDA resonances (Fig. 9) with a Gaussian of 2 eV FWHM showed that none of the resonances could be separated in those measurements. Results and theoretical interpretations for Si^{3+} and Mg^+ are similar.

A new experimental approach[26] has been developed in Giessen to attack the question of intermediate autoionizing states formed in electron-ion collisions. The ion beam is crossed with an exceptionally intense electron beam[35] at a position of optimum beam overlap. In the present version the electron energy is changed about every 3 ms with a pause time of 300 μs needed to program the next energy and to generate necessary control signals for the measurement. An energy interval of 10 eV is divided into typically 256 steps so that one scan spanning 10 eV takes less than 1s. By

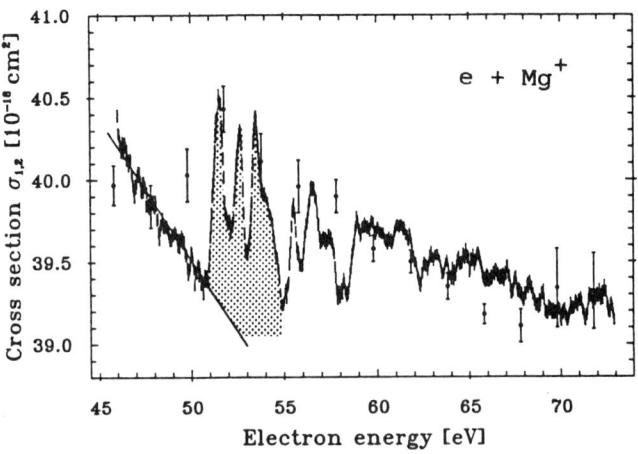

Figure 10. Cross sections for electron-impact ionization of Na-like Mg^+ ions. The Giessen scan data[37] are shown by bars indicating the statistical uncertainties. The open circles are data published by Crandall et al.[34]. The solid line represents direct ionization. The shaded REDA contributions are probably due to Rydberg states with configurations $2p^53s3pnl$, n=3,4,5, corresponding to the series shown in Fig. 9. This series is cut off at the low-energy end at an energy of 51 eV which is the lowest direct excitation threshold ($2p \to 3s$). Above this energy REDA processes are possible. Below this energy only READI processes can contribute (but are not resolved in this experiment). The original scan data were smoothed by using the running average (over 5 points) method.

repeating such scans thousands of times with simultaneous recording of signal counts, ion charge, electron charge and gate time accumulated for each energy and with subsequent normalization, fluctuations in the experimental conditions are averaged out and extremely high precision can be obtained which is only limited by counting statistics. The energy resolution $E/\Delta E$ in these experiments is of the order of 200. This resolution is made possible by compensating the substantial electron beam space charge with slow ions produced from background gas which is introduced into the collision region for this purpose[36].

The experimental technique briefly described above is exceptionally well suited for the observation of structure in ionization cross sections. Its potential to provide a new spectroscopic method for studying intermediate autoionizing states produced in electron-ion collisions is demonstrated already by the example shown in Fig. 10 which presents a scan measurement of the ionization cross section for Na-like Mg^+ ions[37] together with previous data[34].

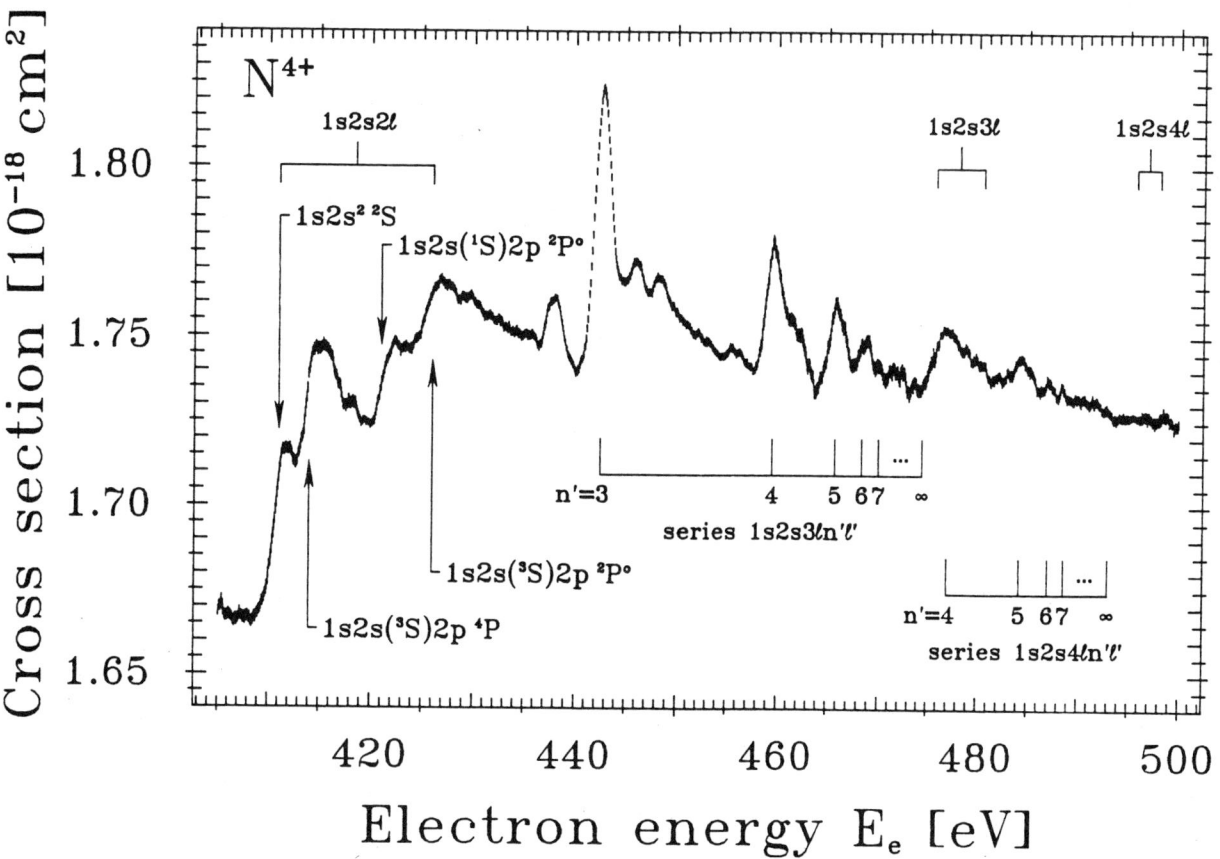

Figure 11. Cross sections for indirect contributions to electron impact ionization of Li-like N^{4+} ions[38]. The scan data were smoothed over bins of seven adjacent points of the original measurement. Assignments of states to the observed features are given on the basis of calculations using the code of Grant et al.[39]. Two series of Rydberg states involving $\Delta n = 2$ and $\Delta n = 3$ transitions can be recognized: 1s2s3lnl' (n = 3,4,5,6...) and 1s2s4lnl' (n = 4,5,...). The series limits 1s2s3l and 1s2s4l are at the thresholds for direct inner-shell excitations. The ranges of corresponding energy levels[47] are indicated.

The new measurements reveal even more structure than might have been expected simply on the basis of the model calculation[23] for a Na-like ion shown in Fig. 9. Apparently there are other resonance series and steps and dips in the cross section which make it difficult to see a clear excitation-autoionization feature. The resonance contributions (the snow) appear to be much higher than the excitation steps (the stairs) on top of which they are observed. One particular series of resonances between 51 eV and 55 eV is interpreted to be due to $2p^53s3pnl$ configurations; i.e. the Rydberg series corresponding to the one in Fig. 9. Peaks up to n = 6 are resolved, the higher Rydberg states lump together due to the limited energy resolution of the experiment. An upper limit of 0.3 eV to the energy spread is set by the width of the resonance observed at 55.5 eV. The strongest REDA peak observed makes up for about 3.5 % of the total ionization cross section at that particular energy. By this measurement a first experimental step is made towards the old challenge to see REDA processes in sodium-like Fe^{15+} ions.

Strong REDA resonances have also been found in the ionization of lithium-like ions[27,38] B^{2+}, C^{3+}, N^{4+}, O^{5+} and F^{6+}. A scan measurement on N^{4+} is shown in Fig. 11. Details in the vicinity of the excitation-autoionization threshold were studied (see also Fig. 7). Individual terms of the 1s2s2l excited configurations are resolved. In addition many resonance features are found.

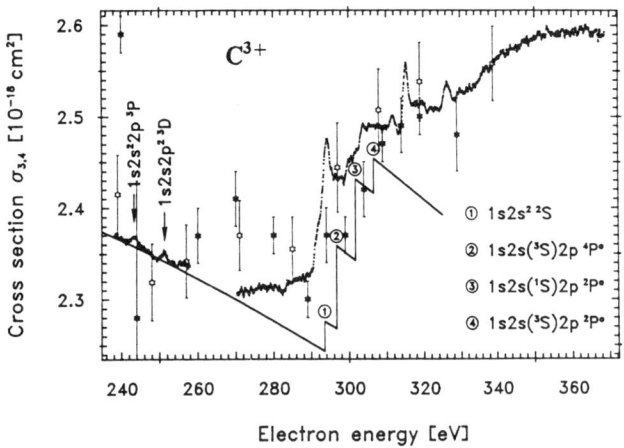

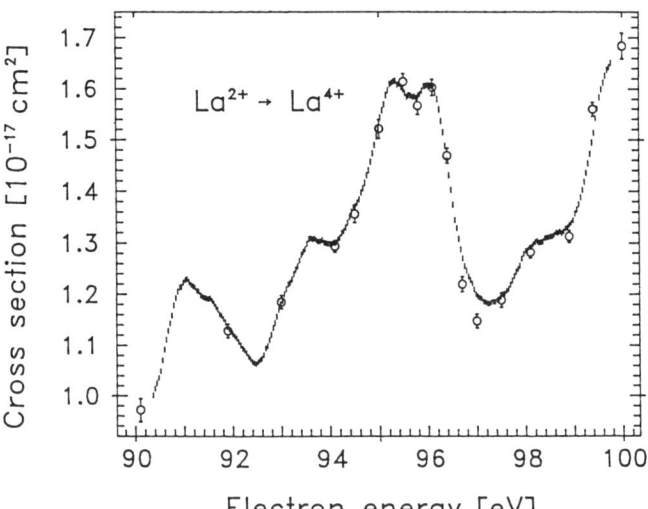

Figure 12. Cross sections for electron-impact ionization of Li-like C^{3+} ions[27]. Resonances due to the READI process (energy range 240 eV to 260 eV) are indicated. The solid line is the sum of the direct ionization (represented by 0.85 σ_{2s}; where σ_{2s} is Younger's[48] calculation) and the excitation-autoionization contributions calculated by Henry[47,49]. The calculated excitation thresholds slightly deviate from the measurements which are in good agreement with Auger spectroscopic data from beam-foil experiments[50]. The stars are absolute measurements by Crandall et al.[32] (solid symbols) and Müller et al.[27] (open symbols).

Figure 14. Cross sections for electron-impact double ionization of La^{2+} ions[26,40]. The scan data were smoothed over bins of 9 adjacent energies. They are represented by bars indicating the resulting statistical uncertainty. The open circles are independently absolute experimental data with their statistical uncertainty (two standard deviations). Note the absolute magnitude of the resonant contributions and their size compared to the total cross section (suppressed vertical scale).

Atomic-structure calculations using the code of Grant et al.[39] suggest the assignment of two Rydberg series to the dominant (REDA) resonances. The series 1s2s3lnl' (n = 3,4,5...) is associated with a $\Delta n = 2$ core transition in the ion, the series 1s2s4lnl' (n = 4,5,...) even with a $\Delta n = 3$ transition. This is remarkable since it was only recently that $\Delta n = 1$ transitions in dielectronic recombination could be quantitatively studied for the first time in a colliding-beams experiment[19].

Excitation-autoionization and resonance features are similar for all investigated ions along the Li-like iso-electronic sequence (B^{2+}, C^{3+}, N^{4+}, O^{5+} and F^{6+}). With increasing Z the indirect contributions decrease in absolute magnitude, but increase relative to direct ionization. The similarity is documented by Fig. 12 which shows indirect contributions found in the ionization of C^{3+} ions[27]. A new feature is seen about 50 eV below the lowest inner-shell excitation threshold. There are little peaks at about 240 eV to 250 eV. They can hardly be seen on the scale which had to be chosen, however, a closer look certifies that these peaks are real resonances. Since they are below the energetically lowest autoionizing state of C^{3+} ($1s2s^2$ ^2S) they must decay via simultaneous emission of two electrons to be visible in the ionization

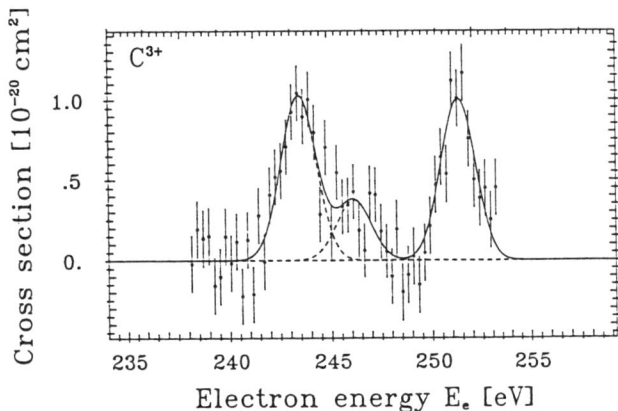

Figure 13. Contribution from READI processes to electron-impact ionization of Li-like C^{3+} ions[27]. From the data in Fig. 12 the solid line was subtracted. The original scan data with equidistant energy steps of 0.039 eV were combined to packets of 7 points to give 1 cross section every 0.27 eV with improved statistics. Three Gaussion distributions of 2 eV width are fitted to the data corresponding to states $1s2s^22p$ ^3P, $1s2s2p^2$ ^5P and $1s2s2p^2$ ^3D.

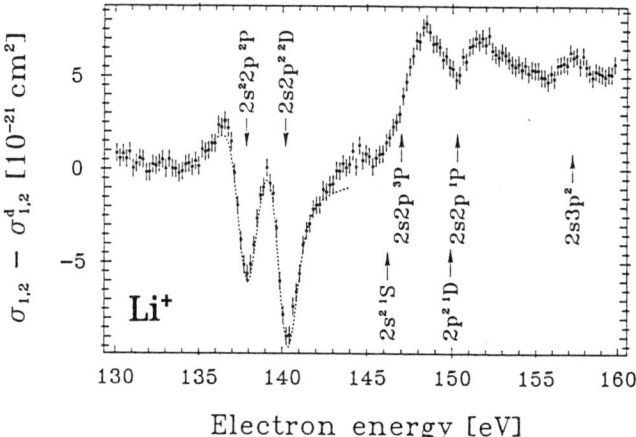

Figure 15. Details of the Li⁺ electron-impact ionization cross section which was measured[41] with 0.01 % statistics. The points were obtained by subtracting direct-ionization background $\sigma_{1,2}^d$ from scan data. The displayed error bars give the resulting statistical uncertainties. Energy levels and excitation thresholds for states with two vacancies in the K-shell of Li and Li⁺ are indicated by arrows. For the ^2P and ^2D resonances (three-electron states) the present experimental energy is indicated. For the doubly excited two-electron states energies from Auger spectroscopy[50] are given. On the basis of structure calculations using the code of Grant et al.[39] the resonance at 157.2 eV was suggested to be due to a configuration $2s3p^2$ (three-electron state). The dotted line represents the Fano q-parameter formula[51] fitted to the ^2P and ^2D resonances.

channel. So, these resonances are due to READI processes which had been predicted to be so unlikely compared to direct ionization. The precision of the scan measurements has made them visible. Basically three resonant states appear to be responsible for the READI peaks which have also been observed in B^{2+}, N^{4+} and O^{5+}: $1s2s^22p$ ^3P, $1s2s2p$ ^5P and $1s2s2p^2$ ^3D. The measurements for C^{3+} ions are displayed in Fig. 13 after subtraction of the cross section for direct ionization. Three Gaussian distributions were fitted to the data points. The widths of these distributions correspond to the energy resolution in the particular experiment (2 eV). Since it is not very difficult to calculate the cross section for dielectronic capture (Eq. 1) these measurements can serve to guide theory[33] in the calculation of correlated two-electron emission processes.

In a series of scan measurements on heavy metal ions[26,40] narrow capture resonances have been found even in multiple-ionization channels. Fig. 14 shows an example for capture resonances contributing to double ionization of La^{2+} ions.

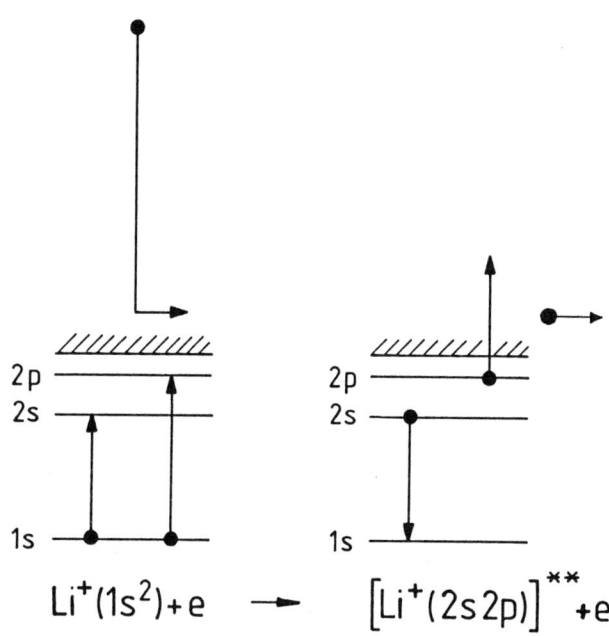

Figure 16. Simplified scheme for the process of electron-impact direct double excitation with subsequent Auger process seen in the ionization of He-like Li⁺ ions (Fig. 15).

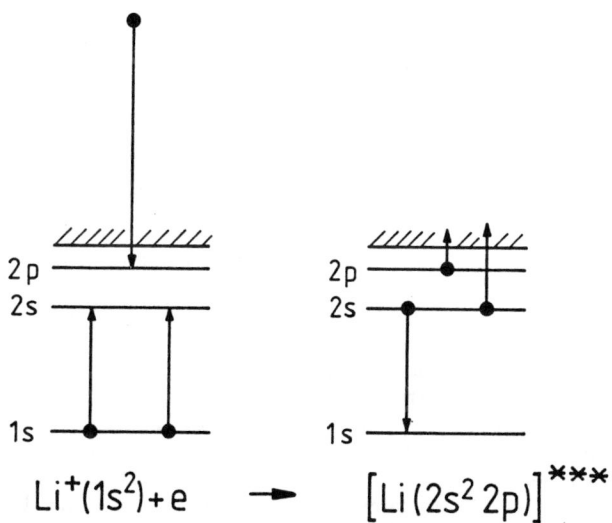

Figure 17. Simplified scheme for the process of trielectronic capture seen in the ionization of He-like Li⁺ ions (Fig. 15). Both K-shell electrons are excited at a time while the projectile electron is captured. The result is a triply excited Li atom which can decay by correlated auto-double-ionization (see Fig. 3) and thus contributes to the net-ionization channel.

In this case the intermediate multiply excited ion La^{1+} must decay by the emission of three electrons and one of the emission processes must involve an auto-double-ionization[40]. Still the observed resonance feature makes up for about one third of the total double ionization cross section. With respect to the term REDA for resonant contributions to net single ionization the new process was named RETA for resonant-excitation-triple-autoionization. Not enough with that, resonances were even found in net triple ionization e.g. of Ce^{1+} and Ce^{2+} ions which means that then four electrons have to be emitted after the dielectronic capture event[26].

An important experimental result concerning the resonances observed in heavy metal ions was the observation of branching into different decay channels from resonances produced through dielectronic capture by one parent ion species. The same initial resonances could be observed at identical energies in net single and in net double ionization or in net double and net triple ionization. This is similar to the situation shown in Figs. 4 and 5 where identical dielectronic capture resonances in an intermediate doubly excited C^{4+} ion once branch into the resonant-scattering channel (by autoionization) and once into the dielectronic recombination channel (by photon emission).

Many more experiments on resonances in the electron-impact ionization of ions have been carried out meanwhile by using the new energy-scanning technique. The format of this contribution does not allow to show more than a small percentage of the results which are now available. One recent measurement, however, still has to be discussed here. It is an experiment with He-like Li^{1+} ions[41]. As one might expect the total cross section for single ionization can be well represented by the Lotz formula[30] which describes only direct ionization processes. Excitation of one of the K-shell electrons would lead to a state which can only relax by radiative decay. Thus, there cannot be any conventional excitation-autoionization or resonant contributions to this cross section. And still, with a relative uncertainty of better than 0.01 %; i. e. a precision which is absolutely unprecedented in any colliding beams experiments, structure can be seen even in this cross section as evidenced by Fig. 14 which is obtained by subtracting direct ionization ($\sigma_{1,2}^d$) from a scan measurement. The features involve two-electron excitations. One is the direct double excitation of the two K-shell electrons into the L shell (see Fig. 16) and the other is double excitation of the two K-shell electrons into the L-shell with simultaneous capture of the projectile also into the L shell (see Fig. 17). In the first case a doubly excited state in the Li$^+$ ion is produced which can contribute to the measured cross section by a single Auger process. In the second case a triply excited intermediate Li atom is produced ("trielectronic capture") which can only be seen in the ionization channel when it decays by correlated two-electron emission (see Fig. 16). Thus, this process involves two correlated two-electron transitions at once. Nevertheless, it can clearly be seen in the ionization experiment. Interference of this resonant channel with direct single ionization is responsible for the observation of dips rather than peaks in the measured cross section.

Perspectives.

The results presented in this review article and in several other papers contributed to this volume clearly show that a new age of electron-ion collision studies has begun. With greatly improved experimental techniques and the new developments with heavy-ion cooler rings, the electron-beam ion trap and forthcoming merged-beams trochoidal-field-analyzer arrangements we are at the beginning of a fruitful and exciting new era of electron-ion collision studies.

Acknowledgements.

I want to thank Prof. E. Salzborn for his support during many years which made the work presented here possible. It has always been a pleasure to work with K. Tinschert and G. Hofmann and special thanks are due for their input to the ionization experiments. I am especially grateful to Prof. G. H. Dunn for longstanding fruitful collaboration and enlightening exchange of ideas. Prof. R. Becker's input into the development of the high-intensity electron gun used in the ionization experiments deserves a special acknowledgement. My thanks are devoted to many more colleagues with whom I am or have been collaborating on electron-ion collision studies. Support from Deutsche Forschungsgemeinschaft (DFG), Gesellschaft für Schwerionenforschung (GSI), Bundesministerium für Forschung und Technologie (BMFT), Max-Planck-Institut für Plasmaphysik (MPP) and NATO collaborative research grant (number RG 86/0510) is gratefully acknowledged.

References.
1. G. J. Schulz, Rev. Mod. Phys. 45, 378 (1973)
2. D. E. Golden, Advances in Atomic and Molecular Physics 14, 1 (1978)
3. D. S. Walton, B. Peart, and K. Dolder, J. Phys. B 3, L148 (1970)
4. H. S. W. Massey and D. R. Bates, Rep. Prog. Phys. 9, 62 (1942)
5. A. Burgess, Astrophys. J. 139, 776 (1964); 141, 1588 (1965)
6. Y. Hahn and K. J. LaGuttuta, Physics Reports 166, 195 (1988)

7. M. F. A. Harrison, in "Methods in Experimental Physics" (W. L. Fite and B. Bederson eds.), Vol. 7 B, Academic Press, New York 1968
8. K. T. Dolder, M. F. A. Harrison and P. C. Thonemann, Proc. R. Soc. A 264, 367 (1961)
9. K. Dolder and B. Peart, Advances in Atomic and Molecular Physics 22, 197 (1986)
10. K. J. LaGattuta and Y. Hahn, Phys. Rev. A 24, 2273 (1981)
11. D. S. Belić, G. H. Dunn, T. J. Morgan, D. W. Mueller, and C. Timmer, Phys. Rev. Lett. 50, 339 (1983)
12. J. B. A. Mitchell, C. T. Ng, J. L. Forand, D. P. Levac, R. E. Mitchell, A. Sen, D. B. Miko, and J. W. McGowan, Phys. Rev. Lett. 50, 335 (1983)
13. P. F. Dittner, S. Datz, P. D. Miller, C. D. Moak, P. H. Stelson, C. Bottcher, W. B. Dress, G. D. Alton and N. Nešković, Phys. Rev. Lett. 51, 31 (1983)
14. J. F. Williams, Phys. Rev. A 29, 2936 (1984)
15. A. Müller, D. S. Belić, B. D. DePaola, N. Djurić, G. H. Dunn, D. W. Mueller, and C. Timmer, Phys. Rev. Lett 56, 127 (1986); Phys. Rev. A 36, 599 (1987)
16. P. F. Dittner, Phys. Scr. T 22, 65 (1988)
17. D. A. Knapp, R. E. Marrs, M. A. Levine, C. L. Bennett, M. H. Chen, J. R. Henderson, M. B. Schneider and J. H. Scofield, Phys. Rev. Lett. 62, 2104 (1989)
18. L. H. Andersen, P. Hvelplund, H. Knudsen, and P. Kvistgaard, Phys. Rev. Lett. 62, 2656 (1989)
19. G. Kilgus, J. Berger, P. Blatt, E. Jaeschke, D. Habs, B. Hochadel, D. Krämer, G. Neureither, R. Neumann, D. Schwalm, M. Steck, R. Stokstad, E. Szmola, A. Wolf, R. Schuch, A. Müller, M. Wagner, to be published; see also A. Wolf, this Volume
20. D. C. Griffin, C. Bottcher, M. S. Pindzola, S. M. Younger, D. C. Gregory, and D. H. Crandall, Phys. Rev. A 29, 1729 (1984)
21. P. G. Burke, A. E. Kingston, and A. E. Thompson, J. Phys. B 16, L385 (1983)
22. M. S. Pindzola, C. Bottcher, and D. C. Griffin, J. Phys. B 20, 3535 (1987)
23. R. J. W. Henry and A. Z. Msezane, Phys. Rev. A 26, 2545 (1982)
24. D. C. Gregory, L. J. Wang, F. W. Meyer, K. Rinn, Phys. Rev. A 35, 3256 (1987)
25. K. Rinn, D. C. Gregory, L. J. Wang, R. A. Phaneuf, and A. Müller, Phys. Rev. A 36, 595 (1987)
26. A. Müller, K. Tinschert, G. Hofmann, E. Salzborn, and G. H. Dunn, Phys. Rev. Lett. 61, 70 (1988)
27. A. Müller, G. Hofmann, K. Tinschert, and E. Salzborn, Phys. Rev. Lett. 61, 1352 (1988)
28. M. J. Seaton and P. J. Storey, in "Atomic Processes and Applications" (P. G. Burke and B. L. Moiseiwitsch, eds.), North-Holland, Amsterdam, 1976), p. 133
29. L. A. Vainshtein and U. I. Safronova, Atomic Data and Nuclear Data Tables 21, 49 (1978)
30. W. Lotz, Z. Phys. 220, 466 (1969)
31. B. Peart and K. Dolder, J. Phys. B 8, 56 (1975)
32. D. H. Crandall, R. A. Phaneuf, B. E. Hasselquist, and D. C. Gregory, J. Phys. B 12, L249 (1979)
33. M. S. Pindzola and D. C. Griffin, Phys. Rev. A 36, 2628 (1987)
34. D. H. Crandall, R. A. Phaneuf, R. A. Falk, D. S. Belić, and G. H. Dunn, Phys. Rev. A 25, 143 (1982)
35. A. Müller, K. Huber, K. Tinschert, R. Becker, and E. Salzborn, J. Phys. B 18, 3011 (1985)
36. A. Müller, G. Hofmann, K. Tinschert, R. Sauer, E. Salzborn, and R. Becker, Nucl. Instrum. Methods Phys. Res. B 24/25, 369 (1987)
37. A. Müller, G. Hofmann, K. Tinschert, B. Weißbecker, and E. Salzborn, submitted to Z. Phys. D
38. G. Hofmann, A. Müller, K. Tinschert, E. Salborn, in preparation for Z. Phys. D
39. I. P. Grant, B. J. Mc Kenzie, and P. H. Norrington, Comp. Phys. Comm. 21, 207 (1980)
40. A. Müller, K. Tinschert, G. Hofmann, E. Salzborn, G. H. Dunn, S. M. Younger, and M. S. Pindzola, submitted to Phys. Rev.
41. A. Müller, G. Hofmann, B. Weißbecker, M. Stenke, K. Tinschert, and E. Salzborn, submitted to Phys. Rev. Lett.
42. E. D. Donets and V. P. Ovsyannikov, Joint Institute of Nuclear Research, Dubna, Report No. P7-10780 (1977)
43. D. L. Moores, J. Phys. B 11, 403 (1978)
44. L. B. Golden, and D. H. Sampson, J. Phys. B 10, 2229 (1977)
45. S. M. Younger, Phys. Rev. A 24, 1272 (1981)
46. D. C. Griffin, C. Bottcher, and M. S. Pindzola, Phys. Rev. A 25, 154 (1982)
47. D. H. Crandall, R. A. Phaneuf, D. C. Gregory, A. M. Howald, D. W. Mueller, T. J. Morgan, G. H. Dunn, D. C. Griffin, R. J. W. Henry, Phys. Rev. A 34, 1757 (1986)
48. S. M. Younger, Phys. Rev. A 22, 111 (1980)
49. R. J. W. Henry, J. Phys. B 12, L309 (1979)
50. M. Rødbro, R. Bruch and P. Bisgaard, J. Phys. B 12, 2413 (1979)
51. U. Fano, Phys. Rev. 124, 1866 (1961)

COLLISIONS IN, WITH AND ON CLUSTERS

J. McCombie and G. Scoles

Chemistry Department, Princeton University, Princeton, NJ 08544-1009

1.-Introduction. 2.-Brief review of cluster science. 3.-Collisions within small clusters.
4.-Collision with small clusters. 5.-Collisions in and on larger clusters. 6.-Outlook.

1. - Introduction

To give a precise definition of a cluster is not a simple task since the easy rule that the cluster be an atomic or molecular aggregate held together by weak van der Waals forces would exclude at least two classes of systems which are (in the first case) or should be (in the second case) commonly referred to as clusters. These are metallic clusters (hot or cold) and hot polyatomic molecules in which the internal energy is sufficient to put the molecule in a "structureless" fluxional state but insufficient to produce fragmentation.

Perhaps the following, and different, definition could be attempted. A cluster is an aggregate of atoms possessing a large number of isomers separated by energy barriers which are small with respect to the potential energy of the aggregate.

With the exception of dimers and trimers, which can be produced in reasonable quantities in a cell at equilibrium, the study of clusters has been made possible almost exclusively by the relatively recent development of beam production and detection techniques. However, due to the experimental difficulties connected with the production of intense monodispersed beams of neutral clusters, very few studies have been conducted so far which involve *collisions* with clusters in a collimated beam environment. In one of the few early examples Pauly and coworkers[1] measured the total integral cross section of Ar_2 dimers scattered by Ar, Ne and Kr atoms. In order to avoid significant contributions to the measured dimer signal arising from higher clusters undergoing dissociative ionization in the ion source (see below) the experiment had to be conducted at values of the source pressure low enough for the dimer cross section (measured without velocity dependence) to become independent from the pressure reading. This low-source-pressures extrapolation technique, which has been used several times since, is naturally not suitable for clusters with a number of monomers larger than two and, as we shall see later, has produced results, the interpretation of which frequently had to be revised.

In the present review we shall first very briefly discuss what is our present knowledge of atomic and molecular clusters and what are the principal sources of this knowledge about them. Afterwards we shall review three classes of collisional experiments carried out with cluster sizes varying from dimers to clusters containing a large number of monomers ($\sim 10^3$). In the first class the cluster beam (typically made of dimers) will be used only to prepare the colliding partners in a pre-determined configuration which will then be used as the starting point for a collisional event generated as the consequence of the electronic excitation or the photodissociation of one of the two partners.

In the second class of experiments a collimated beam of clusters is made to collide with another collimated beam and the angular distribution of the products of the collision is studied. For achieving meaningful results it is then mandatory that (at present) in this class of experiments the size of the clusters be kept relatively small (< ten monomers).

In the third and final class of experiments a beam composed of large clusters is made to collide with a (collimated or uncollimated) flux of molecules which attach to the clusters and, without deflecting them appreciably, proceed down the path of the beam together. In this case the cluster becomes a small reactor within which (or <u>on</u> which) further dynamic experiments can be carried out.

2. - Brief review of cluster science

The study of clusters has been an active field of research for many years both theoretically and experimentally. The focus of most of this activity has been the understanding of the evolution from atomic/molecular to bulk properties with increasing cluster size. To this end the tools of computer simulation and statistical mechanics have been extensively used. They have been joined over recent years by laser spectroscopy and molecular beams which have allowed much more specific information on structural, binding and reactive properties to be amassed. The collision-free environment provided by beams prevents further aggregation and hence clusters can be studied free of matrix effects. The price to pay is, of course, the loss of control on a very important parameter: the cluster's temperature.

Small Clusters

Apart from the initial cluster studies carried out largely with mass spectrometers and electron diffraction instruments, the majority of work in this field has been spectroscopic in nature and has dealt mostly with rather small clusters. Small clusters have been studied using a number of probes, principal among them are the spectroscopic tools of microwave, visible, UV and IR spectroscopy. Traditionally the gas phase molecular structures for binary and tertiary complexes (or clusters) have been obtained using microwave spectrosccopy.[2,3] In particular the techniques of molecular beam electric resonance and pulsed Fourier transform spectroscopy have allowed the rotational structure to be analyzed for a number of isotopically substituted complexes, and thus the structures obtained.

Infrared spectroscopy has the advantage that it is also applicable to non-polar species. In the near-infrared region of the spectrum the laser can be used to excite the vibrational degrees of freedom of the constituent monomer units of the cluster. However there also exist several far infrared techniques which are able to directly probe the vibrational degrees of freedom associated with the weak intermolecular bond, the interest in these low frequency vibrations arising from the fact that they depend directly on the details of the intermolecular potential surface. These techniques include intracavity absorption in free jets[4,5], molecular beam electric resonance[6] and FTIR spectroscopy.[7]

Larger clusters

Electron diffraction has played an important role for many years in molecular structure studies. By sampling a collimated supersonic molecular beam (a free jet is too inhomogeneous) with the technique of electron diffraction the Debye Scherrer rings obtained can give structural information about the microclusters present in the beam. Cluster sources have several advantages when obtaining pure Debye-Scherrer diffraction patterns, 1) the clusters are randomly oriented in the beam, thus forming a perfect powder; 2) they present a relatively narrow size distribution; 3) since there is no sample holder, coalescence is unlikely to occur and it becomes possible to study very small clusters (tens of molecules).

The first investigation of this type was carried out by Audit[8]. The technique has been used extensively since then by a number of groups. The work done can be basically divided into two groups, that done by the Orsay group [9-11] where the central core of converging nozzle expansions is studied and those carried out in the laboratories of Stein [12] and Bartell [13-16] where miniature Laval nozzles have been used which greatly slow down the expansion promoting nucleation at lower stagnation pressures and higher initial temperatures.

IR, visible, UV, and photoelectron spectroscopy have all been used to characterize large neutral clusters. The vast majority of such spectroscopic investigations have involved the generation of samples of clusters through supersonic expansions of condensable gases often seeded into carrier gases. Photoelectron spectrosccopy of van der Waals clusters gives direct information on the vertical ionization potentials for their ground and excited states, [17,18] However the usual problems exist in obtaining the photoelectron spectrum of a size selected cluster. The majority of the spectroscopic probing has been confined to laser induced fluorescence and to infrared spectroscopy using beam depletion techniques. A specific example of the use of IR spectroscopy[19,20] is discussed in detail later.

Mass spectrometry yielded the first direct evidence of cluster formation in supersonic jets and this method of detection has been very widely used to give detailed and extensive data on cluster size distributions. It has, as a technique, the advantages of high sensitivity, mass resolution and ease of combination with other experimental techniques. However it has the drawback that cluster ion distributions, as measured by mass spectrometry, are often severely distorted with respect to the original neutral cluster size distribution in the parent beam, the main reason for this is the extensive fragmentation of the clusters upon ionization.[21,22] Attempts to quantify relative concentrations of neutral van der Waals clusters have all shown that there are very substantial fragmentation effects with electron impact ionization, Penning ionization, fixed frequency VUV ionization and variable wavelength VUV ionization. Thus obviously fragmentation processes occupy a central, and problematic role in the ionizative detection of the clusters formed. The approach of some researchers to this problem involves labelling the neutral cluster so as to be able to discriminate different neutral contributions. Such methods include IR laser predissociation,[23] resonant two photon ionization [24] and the cluster beam-helium atomic scattering experiments of Buck and Meyer[25,26] (which are discussed at some length later in the text). This last cited technique means that the clusters are mass selected before the spectroscopic interrogation. Three other techniques have recently been introduced which may provide a method of interrogating mass selected neutral clusters. These are 1) the electromagnetic mass selection followed by the charge exchange technique of zu Pulitz and co workers[27,28] and 2) the spectroscopic retarded-ion-collection technique recently introduced by Knight and coworkers[29] and 3) several laser based ionization schemes which have been published within the last twelve months[30-34] and which will not be discussed here.

From the above it is clear that the "ideal" way to study neutral clusters has yet to be found, particularly for their basic vibrational properties, which are the key for the understanding of their dynamical behavior.

Since the application of neutron spectroscopy (which has contributed a great deal to our understanding of the vibrational properties of solids and liquids) to the study of clusters is hindered by substantial experimental difficulties a possible solution could, perhaps, be found in inelastic He scattering which is, in many ways, a technique very similar to neutron spectroscopy. While energy resolved He atom-cluster experiments could be designed and performed already with existing technology they would be very near to the limits imposed by the present state of the art. On the other hand, any improvements in the He detection, which are quite probable in view of the constant improvements in modern laser technology (see recent discovery of VUV lasing action in discharge generated plasmas[35,36]), are bound to make collisional spectroscopy of small and large clusters easily accessible and a very fruitful field for future research.

3. - Collisions within small clusters

Full control over the initial conditions of a reactive molecular collision and full analysis of the internal state of the products has been the goal of a great deal of experimental and theoretical activity in the last twenty years.[37-45] Ultimately one would like to control all parameters which define the initial geometric arrangement of the two gas phase reactants maintaining full control on their internal and external energies. However defining the impact parameter, b, has previously presented an intractable problem and effects due to different b's are generally deduced, in an averaged form, from the data. In the selective excitation of an atom-molecule (or two molecule) van der Waals complex, an electronically excited atom can be created in a known geometry, and with a known orbital symmetry with respect to the reaction coordinate, thus b is uniquely defined. In essence chemical reactions can then be studied within a narrow range of initial conditions limited only by the zero-point fluctuations of the weakly bonded van der Waals or hydrogen-bonded complexes. These conditions are now commonly referred to as precursor geometry limited (PGL) conditions.

In this type of experiment the van der Waals complex is prepared in the low temperature environment ($2 \rightarrow 10$ K) of a supersonic expansion. This technique is limited to systems where the ground state precursors do not react, or at least react very slowly, in the gas mixture before expansion. Ideally then the van der Waals bond is strong enough to provide alignment and yet weak enough to prevent significant chemical interaction between the two reagents prior to the desired reaction, thus providing an ideal precursor. Tunable laser radiation is then utilized to generate, selectively, an electronically excited state of the complex which can produce chemical products. Because the initial time is well defined by the optical excitation the time evolution of the complex can be studied.

Since a number of groups have used this technique, here the discussion will highlight the work of a few selected groups which have demonstrated different aspects of this type of experiment. No attempt will be made to achieve completeness in the coverage due to the obvious limitations of time and space.

The initial work performed using this method was carried out by Soep and coworkers in a series of experiments which test the basic premise of orbital stereospecificity in gas-phase reactions in a study of reactive excited states of van der Waals complexes. They performed the first study of this type on the harpoon reaction $Hg^* + Cl_2 \rightarrow HgCl$ $(B^2 \Sigma^+) + Cl$ [46] by freezing the reactants Hg and Cl_2 in a van der Waals complex ($Hg \cdots Cl_2$), exciting to the reactive state and characterizing the processes which occur after excitation of the ground state van der Waals complex by recording three types of spectra:

1) The fluorescence excitation spectrum in the vicinity of the transition of the free atom (for Hg this is the $^3P_1$ - $^1S_0$ transition). Here the total fluorescence is monitored as the UV laser wavelength is scanned. These spectra yields information on the regions of the potential surface which do not lead to product formation, or which produce Hg (6^3P_1) which subsequently fluoresces. 2) The action spectra in which a specific quantum state of a product is monitored with the probe laser as the exciting UV laser is scanned. This gives information about the regions of the upper state surface which produced the monitored product state. 3) Product spectra in which the UV laser is fixed in frequency and the probe laser is scanned to detect the product quantum states. These spectra illustrate how the product energy distribution depends on the particular portion of the upper surface initially populated in the excitation.

The results from this study indicated the direct spectroscopic observation of the charge transfer state $Hg^+ - Cl_2^-$, supporting the harpoon mechanism suggested as early as the 1930's by Polanyi [47] for the reactions between alkalis and halogens. The authors cite a model to account for these results whereby the accessed state (a mixed state) results from the avoided crossing of the covalent and ionic states, the covalent one bearing the oscillator strength. In the case of the $Hg \cdots Cl_2$ system the crossing between the covalent and ionic curve occurs at an internuclear distance of about 4 °A similar to the expected ground-state equilibrium distance of the van der Waals complex. Thus it was correctly predicted that the intermediate state from the crossing would be easily observed. Using this technique a number of harpoon type reactions,[48-50] have been studied. A review article appeared in 1987 which outlined the general features of the method for the systems studied so far. [51]

Concurrently Wittig and coworkers have introduced a different type of PGL technique using bimolecular systems, (CD······AB) where the initial electronic excitation of one of the components is sufficiently localized for the serial picture of the photodissociation of AB and subsequent reactive scattering of A with CD to be valid. The systems studied by them have included a) the reaction $H(^2S) + CO_2 (\chi\Sigma_g^+) \rightarrow OH (X^2\Pi) + CO_2$ ($X^1\Sigma$) (using the CO_2HBr [52,53] van der Waals precursor) b) the complementary work using the CO_2DBr complex[54] as a precursor for the reaction of deuterium with carbon dioxide, c) the D + OCS $\rightarrow$ OD ($X^2\Pi$) + CS ($X^1\Sigma$) or $\rightarrow$ SD ($X^2\Pi$) + CO ($X^1\Sigma^+$) using the SCODBr [55] complex, d) the use of CO_2H_2S [56] complexes to look at the OH distributions and three body interactions caused by simultaneous pairwise HS - H and H - OCO repulsions, e) the study of the H_2SClCN [57] complex, and most recently f) the reactions of hot H atoms with NO_2 in photoexcited N_2OHBr complexes[58] where several product channels were observed, namely: OH ($X^2\Pi$) + N_2 (k_1), OH ($A^2\Sigma$) + N_2 (k_2) and NH ($X^2\Sigma$) + NO ($X^2\Pi$) (k_3) under both bulk and PGL conditions. In both cases nascent NH is observed with low rotational excitation (bulk conditions $\approx$ 1000 cm^{-1}, PGL conditions $\approx$ 750 cm^{-1}) and very little vibrational excitation. No definite conclusions are drawn but the authors point to the possibility that the difference in rotational energies is due to the diminished internal excitation of HONO under PGL conditions caused by repulsive pairwise entrance channels interactions (Br-H and HNO_2). Of interest is the k_3/k_1 ($\approx$ 0.5) branching ratio that is observed under PGL conditions which points to the efficiency of the complex in promoting reaction via the high energy channel (ΔE=95 kcal mol^{-1}). A review article of this method as applied to bimolecular systems appeared in 1987.[59]

The discussion of this work is perhaps best carried out by focusing on the first two aforementioned systems which involve the specific case of atomic hydrogen (deuterium) reacting with carbon dioxide on the ground state PES via the endoergic process: $H(^2S) + CO_2 (\chi\Sigma_g^+) \rightarrow OH (X^2\Pi) + CO_2 (X^1\Sigma)$ and the corresponding reaction with $D(^2S)$. Wittig and coworkers point to two major difficulties to overcome in the study of these systems under PGL conditions. The first relates to the fact that most of the HBr (DBr) which is photodissociated is not complexed to CO_2, and therefore may cause the production of unwanted reaction products. This problem was minimized by using densities that would give no signal under 300K bulk conditions and short delay times between photolysis and probe lasers (the reaction is unimolecular with respect to the CO_2HBr complex and hence the OH deriving from the reaction does not vary with the delay on a nanosecond timescale). The second problem concerns contributions from higher clusters $(CO_2)_m(HBr)_n$, m+n > 2 since, although they can be discriminated against by the expansion conditions, they are conceivably more efficient in their production of OH than are the binary clusters. Under the experimental conditions used, the OH LIF signals are directly proportional to the $(CO_2HBr)^+$ ion signal and do not follow the higher cluster densities, and the LIF spectra were recorded where the binary cluster dominates.

The reaction of atomic hydrogen with CO_2 has been studied previously under bulk conditions[60,61]. It has been proposed that the reaction occurs on a potential surface of $^2A'$ symmetry corresponding to the ground state PES of the HOCO molecule. Matrix isolation studies of HOCO have determined the structure to be bent with an HOC angle of 120°.[62] The van der Waals bond of CO_2HBr complex is weak (~700 cm^{-1}) and thus there are large zero-point oscillations which can be described by χ_1 and χ_2, the respective angles that the HBr and CO_2 axes make with the axis connecting the HBr and CO_2 centers of mass. Average values for these angles were initially inferred from the known properties of CO_2HF and CO_2HCl and later from direct IR absorption measurements which indicated that the HBr and CO_2 axes are not aligned with the axis connecting the centers of mass of the two molecules but that the two molecular axis are instead more or less perpendicular with each other.

In order to calculate the distribution of reagent orientation the authors assumed the HBr and CO_2 to be rigid molecules in the CO_2HBr complex and that the motion of each about the axis connecting the CM's to be harmonic and uncoupled. The precursor geometry limited collision provides a very restricted range of parameters which are very different from the bulk.

The distributions deriving from the PGL reaction show less rotational excitation than the corresponding distributions derived from work in the bulk, whilst both the spin-orbit and the Λ doublet populations are similar in both cases. Neither distribution is dramatically non-statistical and the authors point to the possibility that the differences may be due to direct reaction deriving from the short HOCO lifetime and/or interactions between the OH product and the nearby Br which result in deactivation of the OH produced via HOCO decomposition.

Similar results were seen in the complementary work on the $D(^2S) + CO_2 (\tilde{X}\Sigma_g^+) \rightarrow OD (X^2\Pi) + CO_2 (X^1\Sigma)$ which observed the nascent OD $(X^2\Pi)$ rotational, vibrational, spin orbit and Λ doublet excitation distributions under both single-collision "bulk" conditions and from photolysis within a weakly bound CO_2DBr complex. The OH and OD LIF signal amplitudes do not differ markedly and Wittig et al point to the conclusion that tunneling is not very important in the PGL reaction of H and D with CO_2.

Zero point fluctuations are significant for hydrogen-bonded species and, as it is the zero-point amplitude of the lowest bound state of the complex which limits fundamentally the set of initial conditions available for a given precursor complex, this limits the production of significantly biased product state distributions. The aim of this work is to determine if, following excitation, the initial nonstatistical behavior of the nuclear motions is preserved throughout the reaction. Prinslow and Vaida [63] have recently investigated the photochemistry of two triatomic molecules, OCS and CS_2, and their respective van der Waals dimers using resonance enhanced multiphoton ionization, probing the competition between van der Waals and covalent bond chemistry and the energetic and kinetic factors affecting this competition.

In the case of OCS, excitation at 308 nm accesses the linear $^1\Sigma^+$ state and in the case of CS_2 excitation at 193 nm photon accesses the bent $^1B_2(^1\Sigma_u^+)$ state. In both cases the energy deposited is in excess of both the van der Waals bond and the covalent dissociative C - S bond, rendering both types of photochemical pathways energetically possible. Prinslow and Vaida examined the effect of dimer formation on the relative yields of OCS^+ and S_2^+ in order to establish the importance of the various photochemical pathways, the pathways for the photodissociation of the weak van der Waals bonds being $(OCS)_2 + h\upsilon \rightarrow OCS + OCS$ and $(CS_2)_2 + h\upsilon \rightarrow CS_2 + CS_2$ and those for photochemistry involving the covalent bonds being $(OCS)_2 + h\upsilon \rightarrow CO + CO + S_2$ and $(CS_2)_2 + h\upsilon \rightarrow C_2S_2 + 2S$ or $\rightarrow S_2 + 2CS$.

Their results demonstrate that, for both systems, despite the weaker van der Waals bond a large amount of covalent chemistry occurs in the dimers, a comparatively larger amount occurring in the case of the OCS dimer as compared to the CS_2 dimer. Prinslow and Vaida outline three rationalizations for these observations: 1) From studies of the homogeneous linewidth of the OCS $^1\Sigma^+$ and CS_2 $^1B_2(^1\Sigma_u^+)$ states at low temperatures[64,65] and from the homogeneous linewidth of the excitation they arrive at lifetimes of 0.01 and 0.02 ps for OCS and CS_2 respectively. Thus in the dimers for the $(CS_2)_2$ case there is a longer period of time in which energy can flow into a mode which will cause van der Waals dissociation before covalent dissociation, than there is in the case of $(OCS)_2$. 2) Suggesting that the dimer covalent chemistry occurs through the mechanisms $(OCS)_2 + 2h\upsilon \rightarrow CO + OCS_2 \rightarrow 2CO + S_2$, $(CS_2)_2 + h\upsilon \rightarrow CS \cdot CS + S \rightarrow S_2 + 2CS$ and $(CS_2)_2 + h\upsilon \rightarrow 2S + C_2S_2$ they point to the energetics of exchanging an S-S bond for a C-S bond, exothermic (0.7 eV) for OCS but endothermic (0.05 eV) for CS_2, as favouring covalent chemistry further in the case of the OCS dimer. 3) Finally they point to the proposed geometries of the dimers as a contributing factor. The proposed staggered parallel geometry of the OCS dimer [66] would mean that the excitation of the symmetric stretch in the transition to the linear $^1\Sigma^+$ state of OCS would not be disruptive to the van der Waals bond. In contrast the CS_2 dimer is thought to have a linear geometry and a transition to the bent $^1B_2(^1\Sigma_u^+)$ state will excite both the symmetric stretch and bending modes, both of which can disrupt the van der Waals bond.

It is not possible here to review adequately the bulk of work on collisions that has been carried out in the time resolved domain and so we shall just briefly mention the work carried out in collaboration by the groups of Bernstein and Zewail which also employs the "precursor dissociation" technique. Since, using this method, the zero time of the experiment can be established to be within the duration of the first laser pulse, which prepares the complex in a well defined state, the real-time dynamics of a bimolecular reaction can be probed by a second pulse, delayed in time with respect to the initiation pulse, which monitors the fragments in their different internal states. Previously Zewail et al. have performed a number of experimental investigations of time resolved state to state dynamics of molecular photofragmentation using picosecond spectroscopy. The main focus of these experiments was unimolecular reactions[67,68] although work was done on the vibrational predissociation of the phenol-benzene and cresol-benzene van der Waals dimer.[69] However the first experiment to exploit PGL conditions to study the real time dynamics of a bimolecular reaction used the van der Waals precursor molecule IH·····OCO[70] to study the lifetime of the collision complex HOCO for the reaction $H + CO_2 \rightarrow HO + CO$. The UV picosecond pump laser initializes the reaction by photolysing the HI component, ejecting a translationally hot H atom towards an O atom in CO_2. The probe laser tuned to a resonance wavelength of the free OH detects product formation. The rise time of the OH signal after deconvolution was found to be ~ 5 -15 ps. This represents the transient time decay of the collision complex HOCO formed from the $H + CO_2$ reaction at the given relative translational energy.

4. - Collisions with small clusters

While the literature on the interactions of clusters with electrons and photons is relatively abundant only a few experiments have been reported where a collimated beam of clusters collides with another collimated molecular beam and the angular distribution of the products is measured. We would like to discuss here two series of experiments of this type in some detail since both have been very important in enlarging the scope of experiments which are possible with condensed beams.

In the first type of experiment Herschbach and others have studied the exchange reactions between van der Waals dimers and atoms using crossed beam techniques. Reactions with higher (n>2) clusters also occur but the dimer reaction is resolved by adjusting experimental conditions (i.e. mild expansion conditions and a low electron bombardment voltage in the detector) and the exploitation of kinematic constraints which require that the velocity vectors of the product XA produced in the $X + A_2$ collisions fall on a narrow circular band in the center of mass coordinate system. Herschbach and coworkers[71] were the first to study van der Waals bond exchange in dimer - atom collisions using the molecular beam reactive scattering technique to study the angular and velocity distributions of XeAr from reactive scattering of crossed Xe and Ar_2 beams. The reactant beams intersected at 90° degrees and the direction and velocity of the products were measured by a rotatable mass spectrometer equipped with a time of flight analyzer. Typical source conditions were Xe at ~45 torr and 300K behind a nozzle of 200 μm diameter and Ar at ~75 torr and 120K behind a nozzle of 120 μm diameter. With less mild expansion conditions larger clusters became abundant in the argon beam and product distributions were distorted by fragmentation. However under these mild expansion conditions a "hole" in the XeAr distribution for θ<45 (θ measured from the initial Xe atom direction) was observed and a sequential impulse model based on pairwise hard-sphere interactions postulated.

In the second group of experiments Buck and coworkers through angular and velocity resolved atom cluster scattering have been able to carry out photodissociation experiments on mass selected neutral clusters, relying on the kinematically different behavior, upon collision, of clusters of different mass. This method of mass selection is independent of the detection mechanism and so overcomes and solves the problem of the unknown fragmentation of the clusters when detected by simple electron impact or photoionization.

In precis the experiment is carried out in a crossed molecular beam apparatus where the two beams are crossed at an intersection angle of 90°. The angular dependence of the scattered beam intensity is then measured by rotating the two differentially pumped source chambers relative to the scattering center and the fixed detector unit in the plane of the two crossing beams. The first beam is the cluster beam generated by expansion under conditions which favor the production of the desired oligomer, this cluster beam is then crossed by a He beam. Detection of the clusters is by electron impact ionization and mass analysis using a quadrupole mass filter and the velocity of the scattered particles is determined by time of flight analysis using the pseudorandom chopping method. At the widest angles only dimers can contribute to the detected signal while at smaller angles the higher cluster contributions can be separated for clusters containing up to 6 or 7 monomers. The conditions/assumptions which have to be fulfilled in this work are twofold: i) collisional dissociation should be neglected and ii) the angular and velocity resolution of the apparatus should be good enough to resolve the contributions of different cluster sizes.

This method has been used successfully by Buck and coworkers on a number of systems, to determine unambiguously the fragmentation process of clusters in the ion source[25,72], the cluster formation[26] in a supersonic expansion for clusters up to a maximum size n=10 and dynamical processes of clusters, such as the collisional energy transfer of the relatively tightly bound dimer $(NH_3)_2$,[73,74] and collisional dissociation processes such as $Ar_2 + He \rightarrow Ar + Ar + He$.[25,72] The first system for which this technique was exploited was the formation of Ar_n clusters where it was shown that, whilst the fragmentation of the dimers and trimers is always appreciable, it depends only slightly on the electron energy for energies between 30 and 100 eV.[25] Further work, done on the Ar_n system, addressed the question of how collisional dissociation influences the measured size distributions of these loosely bound van der Waals clusters.[72] This yielded somewhat different values for the fragmentation probability than was found in the earlier work [25] although the general trends remained the same. In summary large fragmentation probabilities were found for all Ar_n ($n \leq 6$) with a preference to form the Ar_2^+ dimer ion.

In a spectroscopic application of the same technique the vibrational predissociation of size selected ethylene clusters upon CO_2 laser irradiation has also been studied [75] and the pure dimer dissociation spectrum measured under conditions where any contamination from higher clusters could be excluded. The spectrum that Buck and coworkers observed exhibited two striking features compared to those obtained in other laboratories[76-78], a much broader total spectrum (FWHM = 32 cm^{-1}) and a pronounced structure indicating that the absorption profile is not homogeneously broadened. Buck et al point to the conclusion that in previous experiments, where either pulsed or continuous molecular beams were used, the effect of contributions from higher clusters was not taken into account. In this early work the authors drew attention to the fact that, due to the scattering with He atoms, the ethylene clusters contain on the average 29 meV of internal energy, and emphasize that further studies where the laser molecular beam interaction is placed before the scattering center (where the clusters are still cold) are necessary in order to determine the effect of this energy on the dissociation spectra. This work, the infrared photodissociation of selected internally cold ethylene clusters, was carried out by Huisken and Pertsch [79]. They found the frequency dependence of the dissociation spectra to be essentially the same for all investigated $(C_2H_4)_n$ clusters, in good agreement with all other work done on this system, and they also showed that both the broadening and the structure observed in the previous work were due to internal excitation. In this work the absorption profiles (n=2-5) are structureless and well described by Lorentzians with a FWHM of 12 cm^{-1} and a peak position around 953 cm^{-1}.

Further work on the $(C_2H_4)_2$ system (irradiating after the scattering center) was carried out by Buck and coworkers in conjunction with the Nijmegen group[80] combining the scattering analysis of the cluster beams with the use of a partially tunable waveguide laser to study the vibrational predissociation spectra of the dimers near the ν_7 absorption of the monomer. This work demonstrated very clearly that both the previously observed[81] narrow fine structure (with a linewidth of < 10 MHz) and the broad background are due to dimers, confirming two different timescales, τ=45 ns and τ=0.5 ps, for the dimer predissociation rates. The earlier work by the Nijmegen group[81] had given a lower limit for the predissociation of the ν_7 excited $(C_2H_4)_2$ of τ=45 ns.

In a similar experiment the infrared photodissociation spectra of $(CH_3OH)_n$, n=2-8 were measured near the absorption band of the C-O stretching frequency of the monomer.[82] The spectra of these oligomers are seen to vary from dimer to octamer. For instance the two-peak spectra of the hexamer is contrasted with the one-peak spectrum of the pentamer and is attributed to the existence in the former case of more than one isomer. Similarly to the work on ethylene clusters, earlier studies without size selection exhibited an unstructured spectrum.[83]

5. - Collisions in and on larger clusters.

We shall now move on to describe the work carried out by the authors' group which has been engendered by the development of a technique, referred to as the "pick-up" technique [84], which enables a guest species to be placed in or on a "host" cluster. This is done by exposing the supersonic beam of clusters to an effusive cross flux of the molecule of interest in the region between the nozzle and the skimmer. The guest molecule is then picked up by the clusters which, without appreciable deflection, after passing through the skimmer, are examined at some point downstream. This technique has been exploited in a series of experiments whereby the processes occurring after the collision of a molecule of the guest species, with either a pure cluster of noble gas atoms or with a cluster already seeded with a chromophore, have been studied.

The technique used to probe the clusters, after the collisional processes have occurred, employs infrared excitation of the chromophore and bolometric detection of the fragments evaporated from the cluster after the excited species has relaxed to its ground state. The infrared active clusters can be formed in three different ways: a) The conventional expansion technique, by which a very diluted (typically less than 1%) mixture of a guest species in Ar undergoes supersonic expansion from an ~30 μm diameter nozzle. b)

The surface deposition, or pick-up technique, in which a beam of Ar_n infrared inactive clusters, formed by supersonic expansion of pure argon, is crossed by an effusive flux of the guest species which is deposited on the cluster surface. c) A combination of a) and b) in which seeded infrared active clusters are formed by expansion of a dilute mixture of X/Ar, where X is the active chromophore, and the resulting infrared active beam is then crossed by an effusive flux of molecules which in turn attach themselves to the clusters.

Using technique c), a combination of conventional expansion and the pick-up technique, our group has carried out a study of the complex forming reaction between CH_3F and HCl[85,86] in medium-large argon clusters. An extensive series of measurements, aimed at the spectroscopic characterization of $(CH_3F)_m/Ar_n$ clusters (with m = 1-4 and n = ~10-10^3), were first carried out. These were carried out in view of the number of detailed studies on the vibrational relaxation of CH_3F in matrices[87-95] which make it of interest to examine the dependence of CH_3F infrared spectra on cluster size and to attempt to characterize the environment of the chromophore in the cluster.

Subsequently we undertook a series of experiments whereby the progress of the complex forming reaction between CH_3F, seeded into the cluster, and HCl, deposited by pick-up on its surface, was monitored as a reduction in the $(CH_3F)_m$ signal and the appearance of red shifted peaks due to the hydrogen bonded complexes $(CH_3F)_m(HCl)_p$. The appearance of this signal depends on the mobility of HCl in the cluster and provided us with a new way of monitoring molecular diffusional motions in this environment.

The main conclusions drawn from the results on complex formation by HCl pick-up are, that this technique makes possible the monitoring of chemical reactions between neutral species in a cluster environment and that the diffusion of molecules of that size is quite fast, even in relatively large clusters. We find that (for clusters containing up to approx. thousand argon atoms) the reaction between a CH_3F molecule in a cluster and a HCl molecule deposited on its surface occurs in a time scale shorter than 100 μs at temperatures which are estimated to be of the order of 30 to 40 K.

The use of the pick-up technique can also enable experiments to be carried out on the cluster surface.

The vibrational relaxation and photoevaporation of molecules adsorbed on the surface of solids is a problem which has received substantial attention [96-98]. The surface of a noble gas cluster is more irregular than the surface of an extended noble gas crystal, but does provide a viable source of alternative interest in its own right. "Surface" studies of this type have the added advantage of being less demanding with respect to the quality of the vacuum needed. Indeed since the surface is formed and the experiment is completed in a matter of a few hundred μseconds a vacuum of the order of 10^{-6} torr is sufficiently good while in standard surface science experiments a vacuum of at least 10^{-10} torr would be needed.

The first system on which a "surface" absorption line on argon clusters was seen was the SF_6/Ar system. These results have been reported by us in a series of publications.[84,99,100] The spectrum consists basically of three peaks. A peak at 937.9 cm-1 has previously been assigned to SF_6 residing in a matrix-like site in the solid argon cluster by comparison with the matrix isolation spectrum of SF_6 in Ar.[101] The cluster absorption is centered around the frequency of a matrix absorption doublet that disappears upon annealing of the sample and is red-shifted, with respect to the main absorption doublet, by about 1 cm-1. Since the red shift is proportional to the average polarizability of the solvent medium which is proportional to the Ar density we conclude that the environment hosting the SF_6 molecule in the cluster is more closely packed than in a well annealed solid.

A peak at 941.6 cm-1 has been assigned to the SF_6 sitting on the surface of the cluster by comparing the spectra of clusters produced with the seeding technique with those of clusters produced by means of the pick-up technique. In the spectrum of "seeded" clusters the "surface" absorption is less pronounced than in the case of "pick-up" clusters. At larger nozzle pressures the spectrum of pick-up clusters is clearly dominated by the "surface" absorption peak. The third peak (939.4 cm-1) which will not be discussed here, is located in exactly the same position as the absorption of SF_6 in a liquid Ar environment.

Recently we have also carried out a series of experiments using both the coexpansion and the pick-up technique on the CF_3Cl/Ar system. The initial results of this study are presented in ref. 102 where we demonstrate, for the first time, the presence of a surface dimer species.

Considering first the low pressure expansion spectra, at 30 psi there occurs a single broad absorption which is shifted slightly to the red of the gas phase absorption [103] (1108.4 cm-1). At this low expansion pressure the clusters contain fewer Ar atoms than are required for one solvation shell and hence the shift due to the solvent species is small.

In the range of coexpansion pressures 100-140 psi the spectra is dominated by one peak at 1099.3 cm-1 which is just to the blue of the frequency measured for CF_3Cl in liquid argon[104], and which we attribute to a very large site in which the molecule is only loosely solvated, i.e with a greater average CF_3Cl-Ar distance occurring in the cluster than in liquid Ar. To the best of our knowledge there is no spectrum recorded of CF_3Cl in a solid Ar matrix, however we would expect an annealed matrix site to give an absorption frequency lower than that in liquid Ar. At the medium expansion pressure of 200 psi the matrix peak is joined by an absorption at 1102 cm-1. This frequency suggests that the chromophore is not fully solvated although it is obviously not due to a decrease in average cluster size We attribute this peak to CF_3Cl sitting on or in the surface of these medium sized clusters. As the expansion pressure is increased the matrix peak diminishes whilst the surface peak grows in, indicating that for the larger clusters CF_3Cl is more stable at the cluster surface than in the loosely solvated matrix site. At 400 psi another absorption manifests itself at 1104.9 cm-1.

In a series of experiments where the pick-up method was used to study the effect on the spectrum of varying the density of chromophores in the pick-up beam, this second peak was assigned to the presence of a surface dimer (CF_3Cl)$_2$. In this study the background pressure of an effusive flux of CF_3Cl was varied from 1.75×10^{-5} to 6.0×10^{-5} torr for an expansion of pure argon at 325 psi and 400 psi, and the intensity of the monomer on the Ar surface and the dimer on the Ar surface recorded as a function of this effusive flux. The concentration dependence of the two peaks differs markedly, that of the monomer being linear whilst that of the dimer is roughly quadratic. The ratio of these two intensities (dimer/monomer) shows a linear dependence at low background pressures. This concentration behaviour indicates that the absorption at 1104.9 cm-1 is due to a $(CF_3Cl)_2/Ar_n$ species, or a "dimer". The position of this dimer peak, given that a complete Ar shell around the monomer shifts the frequency of the relevant transition by 9.1 cm-1, points to the fact that the dimer species resides on the surface of the cluster where it interacts with only a few Ar atoms. Furthermore, since the coexpansion spectra have shown that the larger clusters have a tendency to push the monomer outside the cluster and onto the surface, then, in order for this dimer peak to be assigned to a dimer residing inside the cluster, the process of dimerization would have to alter the Ar-CF_3Cl interaction sufficiently that dimer species formed on the surface of the cluster by monomer units combining would then find it energetically favourable to move inside the large Ar clusters. This dimer formation reaction on the surface of the cluster, while of of interest in itself, holds further potential since, by monitoring the monomer and the dimer surface peaks as a function of the pick-up gas flow we can also, with some assumptions, measure the <u>absolute</u> probability of picking up one molecule onto the cluster surface. Assuming a sticking coefficient this would allow us to determine the size of the clusters thus sampled in the beam. Work is presently underway to exploit this possibility.

6. - Outlook

While scattering experiments in, on and with clusters are still difficult there is little doubt that the continuous progress of molecular beam and laser techniques will produce a steady increase in the number of experiments of this type and in the detail of information extracted from them.

For the precursor geometry limited experiments we can expect the extension of this idea to other, more complex, oligomers and intensive application of time resolved techniques to problems of this type. It is interesting to note that PGL experiments in the purely inelastic regime (before the reactive threshold) have, to our knowledge, not yet been carried out. They present some difficulties connected with the presence of the large amounts of uncomplexed monomers in the beam but since the rovibrational states of these molecules, before the UV laser strikes, should be reactively unexcited, these difficulties may not be insurmountable.

The mass-selection-by-collision experiments also require a state of the art crossed beam scattering apparatus provided with TOF capabilities. While these machines are few and far between the experiment could be made relatively easier by the use of pulsed beams especially in cases when pulsed laser detection could be used advantageously over electron bombardment ionization mass spectrometry. Major advances in this area are to be expected as soon as tunable infrared c.w. lasers will be commercially available in the 100 to 1000 mW range which, we are told, may be around the corner. An experiment of this type, which is already feasible, but has not yet been carried out, is the spectroscopy of mass selected clusters carrying a visible chromophore. Because of the sensitivity of laser induced fluorescence detection it is possible that larger polymers could be studied in this way.

Future challenges in the area of collisions with large clusters include the study of (superfluid?) He and p-H_2 clusters, the determination of the sticking coefficients and the measurement of desorption probabilities for adsorbed species after excitation.

The field appears to be a rich one, dense of possible rewards for those who are not deterred by the technical challenges involved.

Acknowledgements.

This work was supported, in part, by the National Science Foundation (Grant CHE-8709572). It is a pleasure to acknowledge useful discussions with D. J. Levandier and S. Goyal who have also allowed us to discuss some of their results in advance of publication.

References
1. H. Vehmeyer, R. Feltgen, P. Chakraborti, M. Düker, F. Torello and H. Pauly, Chem. Phys. Lett. <u>42</u>, 597 (1976).
2. T. R. Dyke, Top. Curr. Chem. <u>120</u>, 86 (1984).
3. A. C. Legon and D. J. Millen, Chem. Rev. <u>86</u>, 635 (1986).
4. D. Ray, R. L. Robinson, D. -H Gwo, R. J. Saykally, J. Chem. Phys. <u>84</u>, 1171 (1986).
5. R. L. Robinson, D. -H Gwo, D. Ray and R. J. Saykally, J. Chem. Phys. <u>86</u>, 5211 (1987)
6. M. D. Marshall, A. Charo, H. O. Leung and W. Klemperer, J. Chem. Phys. <u>83</u>, 4924 (1985).
7. B. A Wofford, M. W. Jackson, J. W. Bevan, W. B. Olson and W. J. Lafferty, J. Chem. Phys. <u>84</u>, 6115 (1986).
8. P. Audit J. Phys. (Paris) <u>30</u>, 192 (1969).
9. J. Farges, M. F. deFeraudy, B. Raoult, and G. Torchet, J. Chem. Phys. <u>78</u>, 5067 (1983).
10. J. Farges, M. F. deFeraudy, B. Raoult, and G. Torchet, J. Chem. Phys. <u>84</u>, 3491 (1986).
11. J. Farges, M. F. deFeraudy, B. Raoult, and G. Torchet, in Large Finite Systems (Proceedings

of the 20th Jerusalem Symposium on Quantum Chemistry and Biochemistry, J. Jortner and B. Pullman, eds., D. Reidel, (1987).
12. J. W. Lee and G. D. Stein, J. Phys. Chem. 91, 2450 (1987).
13. L.S. Bartell, Chem. Rev. 86, 491(1986).
14. L. S. Bartell, E. J. Valente and J. C. Caillat J. Phys. Chem. 91, 2498 (1987).
15. L. S. Bartell Y. Z. Barshad J. Phys. Chem. 91, 2890 (1987).
16. E. J. Valente and L. S. Bartell J. Chem. Phys. 80, 1451 (1984).
17. P. M. Dehmer, S. T. Pratt and J. L. Dehmer, J. Phys. Chem.. 91, 593 (1987).
18. F. Carnovale, J. B. Peel and R. G. Rothwell, Physics and Chemistry of Small Clusters, p. 595, NATO ASI Series; Eds.: P. Jena, B. K. Rao, S. N. Khanna, Plenum Press: New York (1987).
19. T. E. Gough, M. Mengel, P. A. Rowntree and G. Scoles, J. Chem. Phys. 83, 4958 (1985).
20. T. E. Gough, D. G. Knight, P. A. Rowntree and G. Scoles, J. Phys. Chem. 90, 4026 (1986).
21. W. Henkes Z. Naturforsch. A17, 786 (1962).
22. A. Herrmann, S. Leutwyler, E. Schumacher, and L. Wöste, Helv. Chim. Acta. 61, 453 (1978).
23. J. Geraedts, S. Setiadi, S. Stolte and J. Reuss, Chem. Phys. Lett. 78, 277 (1981).
24. R. E. Smalley, L. Wharton and D. Levy, J. Chem. Phys. 66, 2750 (1977).
25. U. Buck and H. Meyer, Phys. Rev. Lett. 52(2), 109 (1984).
26. U. Buck and H. Meyer, Surf. Sci. 156, 275 (1985).
27. M. Arnold, J. Kowalski, G. zu Putlitz, T. Stehlin and F. Träger, Surf. Sci. 156, 149 (1985).
28. M. Arnold, J. Kowalski, G. zu Putlitz, T. Stehlin and F. Träger, Z. Physik, A322, 1179 (1985).
29. E. J. Bieske, M. W. Rainbird and A. E. W. Knight, J. Chem. Phys. 90, 2068 (1989).
30. M. Y. Hahn and R. Whetten, Phys. Rev. Lett. 61, 1190 (1988).
31. U. Even, N. Ben-Horin and J. Jortner, Phys. Rev. Lett. 62, 140 (1989).
32. J. Bösiger, R. Knochenmuss and S. Leutwyler, Phys. Rev. Lett. 62, 3058 (1989).
33. D. B. Smith and J. C. Miller, J. Chem. Phys. 90, 5203 (1989).
34. K. E. Schriver, M. Y. Hahn, J. L. Persson, M. E. LaVilla, R. L. Whetten, J. Phys. Chem. 93, 2869 (1989).
35. B. P. Stoicheff and T. Efthimiopulos, Am. Phys. soc. Bull. 34, 1686 (1989).

36. T. Efthmiopoulos, B. P. Stoicheff and R. I. Thompson Optics Lett. 14, 624 (1989).
37. R. B. Bernstein, Atom-Molecule Collision Theory. Plenum Press, New York (1979).
38. R. B. Bernstein, Chemical Dynamics via Molecular Beam and Laser Technique Oxford University Press. New York (1982).
39. S. Stolte, Ber. Bunsenges. Phys. Chem. 86, 413 (1982).
40. S. Stolte, In Atomic and Molecular Beam Methods Vol 1, Ed. G. Scoles. Oxford University Press, p. 631 (1988).
41. R. D. Levine and R. B.Bernstein Molecular Reaction Dynamics and Chemical Reactivity. Oxford University Press (1987).
42. M. A. Johnson, J. Allison and R. N. Zare, J. Chem. Phys. 85, 5723 (1986).
43. D. Parker, H. Jalink and S. Stolte, J. Phys. Chem. 91, 5427 (1987).
44. S. R. Ghandi, T. J. Curtiss, Q. X. Xu, S. E. Choi and R.B. Bernstein, Chem. Phys. Lett. 132, 6 (1986)
45. S. R. Ghandi, Q. X. Xu, T. J. Curtiss, and R.B. Bernstein, J. Phys. Chem. 91, 5437 (1987).
46. C. Jouvet, and B. Soep, Chem. Phys. Lett. 96, 426 (1983).
47. M. Polanyi Atomic Reactions. Williams and Norgate London (1932).
48. M. Boivineau, J. Le Calve, M. C. Castex, C. Jouvet, J. Chem. Phys. 84, 4712 (1986).
49. M. Boivineau, J. Le Calve, M. C. Castex, C. Jouvet, Chem. Phys. Lett. 130, 208 (1986).
50. W. H. Breckenridge, C. Jouvet, and B. Soep, J. Chem. Phys. 84, 1443 (1986).
51. C. Jouvet, M. Boivineau, M. C. Duval, and B. Soep, J. Phys. Chem. 91, 5416 (1987).
52. S. Buelow, G. Radhakrishnan, J. Catanzarite, and C. Wittig, J. Chem. Phys. 83, 444 (1985).
53. G. Radhakrishnan, S. Buelow, and C. Wittig, J. Chem. Phys. 84, 727 (1986).
54. S. Buelow, G. Radhakrishnan, and C. Wittig, J. Phys. Chem. 91, 5409 (1987).
55. D. Häusler, J. Rice, and C. Wittig, J. Phys. Chem. 91, 5413 (1987).
56. J. Rice, G. Hoffmann, and C. Wittig, J. Chem. Phys. 88, 2842 (1988).
57. J. de Juan, S. Callister, H. Reisler, G. A. Segal, and C. Wittig, J. Chem. Phys. 89, 1977 (1988).
58. G. Hoffmann, D. OH, H. Iams, and C. Wittig, Chem. Phys. Lett. 155, 356 (1989).
59. S. Buelow, M. Noble, G. Radhakrishnan, H. Reisler, C. Wittig and G. Hancock, J. Phys. Chem. 90, 1015 (1986).

60. C. R. Quick, Jr. and J. J. Tiee, Chem. Phys. Lett. 100, 223 (1983).
61. K. Kleinermanns and J. Wolfrum, Chem. Phys. Lett. 104, 157 (1984).
62. D. E. Milligan and M. E. Jacox, J. Chem. Phys. 54, 927 (1971).
63. D. A. Prinslow and V. Vaida, J. Phys. Chem. 93, 1836 (1989).
64. M. I. McCarthy, V. Vaida, J. Phys. Chem. in press,
65. R. J. Hemley, D. G. Leopold, J. L. Roebber, V. Vaida, J. Chem. Phys. 79, 5219 (1983).
66. Y. Ono, E. A. Osuch, C. Y. Ng, J. Chem. Phys. 74, 1645 (1981).
67. L. R. Khundkar, J. L. Knee and A. H. Zewail, J. Chem. Phys. 87, 77 (1987).
68. N. F. Scherer and A. H. Zewail, J. Chem. Phys. 87, 97 (1987).
69. J. L. Knee, L. R. Khundkar and A. H. Zewail, J. Chem. Phys. 87, 115 (1987).
70. N. F. Scherer, L. R. Khundkar, R. B. Bernstein and A. H. Zewail, J. Chem. Phys. 87, 1451 (1987).
71. D. R. Worsnop, S. J. Buelow, and D. R. Herschbach, J. Phys. Chem. 85, 3024 (1981).
72. U. Buck and H. Meyer, J. Chem. Phys. 84(9), 4854 (1986).
73. Z. Bacic, U. Buck, H. Meyer and R. Schinke, Chem. Phys. Lett 155, 47(1986)
74. U. Buck, H. Meyer, D. Nelson, Jr., G. Fraser and W. Klemperer, J. Chem. Phys. 88(5), 3028 (1988).
75. F. Huisken, H. Meyer, C. Lauenstein, R. Sroka and U. Buck, J. Chem. Phys. 84 (2), 1042 (1986).
76. M. P. Casassa, C. M. Western, and K. C. Janda, J. Chem. Phys. 81, 4950 (1984).
77. M. A. Hoffbauer, K. Liu, C. F. Giese and W. R. Gentry, J. Chem. Phys. 78, 5567 (1983).
78. J. Geraedts, Ph. D. thesis, University of Nijmegen, Nijmegen, (1983).
79. F. Huisken and T. Pertsch, J. Chem. Phys. 86, 106 (1987).
80. U. Buck, Ch. Lauenstein, A. Rudolph, B. Heijmen, C. Liedenbaum, S. Stolte and J. Reuss. Chem. Phys. Lett. 144, 167 (1988).
81. B. Heijmen, C. Liedenbaum, S. Stolte and J. Reuss. Z. Phys. D 6, 199 (1987).
83. M. A. Hoffbauer, C. F. Giese, W. R. Gentry, J. Phys. Chem. 88, 181 (1984).
82. U. Buck, X. Gu, C. Lauenstein and R. Rudolph, J. Phys. Chem. 92, 5561 (1988).
84. T. E. Gough, M. Mengel, P. A. Rowntree, G. J. Scoles, Chem. Phys. 83, 4958 (1985).
85. D. J. Levandier, J. McCombie, R. Pursel, G. Scoles, J. Chem. Phys. 86, 7239(1987)
86. D. J. Levandier, J. McCombie, R. Pursel, G. Scoles, Z. für Physik D, 10, 337 (1988).
87. V. A. Apkarian, E. Weitz, Chem. Phys. Lett. 76, 68 (1980).
88. W. Janiesch, V. A. Apkarian, E. Weitz, Chem. Phys. Lett. 85, 505 (1982).
89. L. Abouaf-Marguin, B. Gauthier-Roy, Chem. Phys. 51, 213 (1980).
90. B. Gauthier-Roy, L. Abouaf-Marguin and F. Legay, Chem. Phys. 46, 31(1980).
91. L. Abouaf-Marguin, B. Gauthier-Roy and F. Legay, Chem. Phys. 23, 443 (1977).
92. L. Young, C. B. Moore and J. Chem. Phys. 76, 5869 (1982).
93. B. Gauthier-Roy, C. Alamichel, A. Lecuyer and L. Abouaf-Marguin, J. Mol. Spectrosc. 88, 72 (1981).
94. L. H. Jones, B. I. Swanson, J. Chem. Phys. 76, 1634 (1982).
95. V. A. Apkarian, E. Weitz, J. Chem. Phys. 76, 5796 (1982).
96. H. J. Kreuzer, Faraday Discuss. Chem. Soc. 80, 265 (1985).
97. H. J. Kreuzer and Z. W. Gortel, Physisorption Kinetics, Springer Verlag: Berlin, 1985.
98. T. J. Chuang, Rep. Surf. Sci . 3, 1 (1983).
99. T. E. Gough, M. Mengel, P. Rowntree, and G. Scoles, Laser Applications in Chemistry SPIE 669, 129 (1986).
100. D.J. Levandier, M. Mengel, J. McCombie and G. Scoles Infrared Spectroscopy In and On Argon Clusters: Surface and Matrix Spectroscopy in the Gas Phase.in Proceedings of the CVIII Summer Course of the E. Fermi International School of Physics on the Chemical Physics of Atomic and Molecular Clusters, held in Varenna on Lake Como, from June 28 to July 7, 1988 (to be published).
101. B. I. Swanson and L. H. Jones, J. Chem. Phys. 74, 3205 (1981).
102. D. J. Levandier, S. Goyal, J. McCombie and G. Scoles - to be presented at the Faraday Discussion on Large Clusters, Warwick, Dec 1989.
103. H. Bürger, K. Burczk, R. Grassow and A. Ruoff, J. Mol. Spect., 93, 55 (1982).
104. T. D. Kolomiitsova, V. A. Kondaurov, S. M. Melikova and D. N. Shchepkin, J. Appl. Spec. 43 (3), 999 (1985).

QUANTUM MECHANICAL REACTIVE SCATTERING THEORY FOR SIMPLE CHEMICAL REACTIONS: RECENT DEVELOPMENTS IN METHODOLOGY AND APPLICATIONS

William H. Miller

Department of Chemistry, University of California, and
Materials and Chemical Sciences Division, Lawrence Berkeley
Laboratory, Berkeley, California 94720

It has recently been discovered that the S-matrix version of the Kohn variational principle is free of the "Kohn anomalies" that have plagued other versions and prevented its general use. This has made a major contribution to heavy particle reactive (and also to electron-atom/molecule) scattering which involve non-local (i.e., exchange) interactions that prevent solution of the coupled channel equations by propagation methods. This paper reviews the methodology briefly and presents a sample of integral and differential cross sections that have been obtained for the $H+H_2 \rightarrow H_2+H$ and $D+H_2 \rightarrow HD+H$ reactions in the high energy region (up to 1.2 eV translational energy) relevant to resonance structures reported in recent experiments.

I. Introduction

The description of chemical reactions at the most fundamental, state-to-state level has been a vision of physical chemists ever since molecular beam methodology[1] ushered in the modern reaction dynamics era over 30 years ago. From a theoretical perspective, the rigorous description of a bimolecular reaction is, of course, a problem in quantum mechanical scattering theory. Unfortunately, however, it is the most complicated kind of scattering process, namely a "rearrangement"[2] of the particles (here atoms), and this has made progress quite slow. In the 1970's there were many treatments of the <u>collinear</u> model of atom-diatom ($A+BC \rightarrow AB+C$) reactions but only a few attempts at rigorous calculations for the three-dimensional version of such reactions.[3] Almost all of the latter were for the prototype $H+H_2 \rightarrow H_2+H$ reaction,[3d] and here one singles out the work of Schatz and Kuppermann[3c] as the most successful for these early (1970's) calculations. The theoretical methodology employed for this, however, was sufficiently specialized to the $H+H_2$ system that it has not proved useful for extension to more general applications.

The 1980's have seen renewed activity in quantum reactive scattering calculations, spurred to some extent by the greater access academic researchers have had to supercomputer facilities and also by a new generation of state-to-state scattering experiments. At present there are several different categories of theoretical approaches that are being pursued for carrying out such calculations: the R-matrix propagation method developed by Light and co-workers,[4] methods based on the use of hyperspherical coordinates to describe the system of atoms,[5] methods based on the Faddeev equations, and the formulation of the problem given some years ago[6] by the author whereby the wavefunction is expressed as a coupled-channel expansion simultaneously in all arrangements, using the standard Jacobi coordinates of each arrangement. In addition to work at Berkeley using this latter approach, Truhlar, Kouri and co-workers[7] employ a methodology based on this formulation (which they emphasize also results from use of a particular "coupling scheme" in the coupled T-operator formalism developed in the 1970's by Kouri, Baer, <u>et al.</u>[8]) It is also useful to note that the formulation given in ref. 6, which was developed as a generalization of Burke and Taylor's[9] treatment of electron-atom scattering, is similar in spirit to the "resonating group method"[10] that was developed earlier to treat nuclear reactions.

The central difficulty with the formulation developed in ref. 6 is that the coupled-channel Schrödinger equation contains non-local <u>exchange</u> interactions which characterize the reaction. The only <u>general</u> way for dealing with exchange is to expand the coupled-channel radial functions in a basis set (of known functions) and determine the expansion coefficients via a variational principle. There are several different variational principles for scattering[2]: the

Kohn[11] (essentially the Rayleigh-Ritz) principle is the simplest - because it involves matrix elements only of the Hamiltonian operator - but it has in the past[12] suffered from "Kohn anomalies" (i.e., spurious, unphysical singularities) that have rendered it unsatisfactory. The Schwinger variational principle[2] has previously been shown[13] to yield reliable results for reactive scattering (and it has been widely used for electron-atom/molecule scattering[14]), and more recently the Newton variational principle[7] has also been successfully applied to the formulation of ref. 6. The latter two variational principles, however, require matrix elements involving the Green's function for a reference scattering problem, and this makes their application considerably more difficult than the Kohn principle.

It has thus been a major breakthrough to realize that the S-matrix version[15] of the Kohn variational principle is completely free of the "anomalies" that plague the K-matrix version that has traditionally been used. I.e., the results given by the Kohn variational principle are not invariant to how the scattering boundary conditions are applied (even with same basis set): applying it with standing wave boundary conditions to obtain the K-matrix, and then S via the Heitler damping equation, $(1+iK)(1-iK)^{-1}$, does not give the same S-matrix as applying it with incoming/outgoing wave boundary conditions to obtain the S-matrix directly. The latter (S-matrix) version is completely anomaly-free in a general and natural way, and though it may at first seem more difficult to apply because of the presence of complex basis functions (the incoming and outgoing waves), in practice this causes no additional effort.

The overall methodology[16] that results from applying the S-matrix version of the Kohn principle to the formulation of ref. 6 is extremely straight-forward quantum mechanics: one chooses basis functions, computes matrix elements of the Hamiltonian, and then performs a standard linear algebra calculation. Most of the intellectual input comes about, as is usually the case in quantum mechanics, in the first step, i.e., in choosing basis functions in the most efficient and useful manner. We note also that McCurdy, Rescigno, and Schneider[17] have made very significant advances in electron-molecule scattering by using this S-matrix version of the Kohn variational principle.

Section II briefly summarizes the basic working formulae for the S-matrix version of the Kohn variational principle. I have recently reviewed[16b,c] the basic ideas and methodology of this approach and discussed in detail why "Kohn anomalies" do not arise in this version. In this regard, another recent paper[18] has shown that the S-matrix Kohn method is free of anomalous (i.e., spurious, unphysical) singularities even in cases for which the Schwinger variational principle - which has been previously thought to be anomaly-free - actually does show anomalous singularities.

Integral and differential cross sections for the $H+H_2(v=j=0) \to H_2(v',\text{odd } j')+H$ and $D+H_2(v=j=0) \to HD(v',j')+H$ reactions are presented in Sections III and IV. These involve J (total angular momentum) values up to 24 and 31, respectively, in the partial wave sums to obtain converged results for the cross sections.

II. S-Matrix Version of the Kohn Variational Principle

All of the relevant features are illustrated by simple s-wave potential scattering. The methodology[16] will thus first be described with regard to this problem, and the generalization to multichannel rearrangement scattering given at the end.

The Hamiltonian is of the standard form

$$H = \frac{-\hbar^2}{2\mu} \frac{d^2}{dr^2} + V(r) , \qquad (2.1)$$

where $V(r) \to 0$ as $r \to \infty$. The S-matrix version of the Kohn variational approximation to the S-matrix (at energy E) can be stated as

$$S = \text{ext}[\tilde{S} + \frac{i}{\hbar} \langle \tilde{\psi}|H-E|\tilde{\psi}\rangle] , \qquad (2.2)$$

where $\tilde{\psi}(r)$ is a trial wavefunction that is regular at r=0 and has asymptotic form (as $r \to \infty$)

$$\tilde{\psi}(r) \sim -e^{-ikr}v^{-\frac{1}{2}} + e^{ikr}v^{-\frac{1}{2}}\tilde{S} \qquad (2.3)$$

where $v=\hbar k/\mu$ is the asymptotic velocity. (Note: The convention is used throughout this paper that the wavefunctions in the bra symbol $\langle\,|$ in bra-ket matrix element notation are not complex conjugated.) "ext" in Eq. (2.2) means that the quantity in square brackets is to be extremized by varying any parameters in $\tilde{\psi}(r)$. (Note that for a given trial function $\tilde{\psi}$, Eq. (2.2) may also be viewed as the distorted wave Born approximation, where $\tilde{\psi}$ is the distorted wave.)

A linear variational form is taken for the trial function $\tilde{\psi}(r)$,

$$\tilde{\psi}(r) = -u_0(r) + \sum_{\ell=1}^{N} u_\ell(r) c_\ell , \quad (2.4)$$

where $u_0(r)$ is a function that is regular at $r=0$ and has the asymptotic form (as $r \to \infty$)

$$u_0(r) \sim e^{-ikr} v^{-\frac{1}{2}} . \quad (2.5)$$

A simple choice for $u_0(r)$ is

$$u_0(r) = f(r) e^{-ikr} v^{-\frac{1}{2}} , \quad (2.6)$$

where $f(r)$ is a smooth cut-off function,

$$f(r) \to 0 , \quad r \to 0$$
$$f(r) \to 1 , \quad r \to \infty , \quad (2.7)$$

such as $f(r) = 1 - e^{-\alpha r}$. (More generally, $u_0(r)$ may be the (irregular) solution of some (e.g., long range) distortion potential that has asymptotic form Eq. (2.8), multiplied by a cut-off function to regularize it at $r=0$). The function $u_1(r)$ is

$$u_1(r) = u_0(r)^* , \quad (2.8)$$

and the basis functions $\{u_\ell(r)\}$, $\ell=2,\ldots,N$ are real, square-integrable functions. The coefficients $\{c_\ell\}$, $\ell=1,\ldots,N$ in Eq. (2.4) are the variational parameters in ψ.

With ψ of Eq. (2.4) substituted into Eq. (2.2) and the coefficients $\{c_\ell\}$ varied to extremize it, one obtains the following expression for the S-matrix

$$S = \frac{i}{\hbar} (M_{0,0} - \underset{\sim}{M}_0^T \cdot \underset{\approx}{M}^{-1} \cdot \underset{\sim}{M}_0) , \quad (2.9)$$

where $M_{0,0}$ is a 1x1 "matrix", $\underset{\sim}{M}_0$ a 1xN matrix, and $\underset{\approx}{M}$ an NxN matrix,

$$M_{0,0} = \langle u_0 | H-E | u_0 \rangle \quad (2.10a)$$

$$(\underset{\sim}{M}_0)_\ell = \langle u_\ell | H-E | u_0 \rangle \quad (2.10b)$$

$$(\underset{\approx}{M})_{\ell,\ell'} = \langle u_\ell | H-E | u_{\ell'} \rangle , \quad (2.10c)$$

for $\ell,\ell'=1,\ldots,N$, and where "T" denotes matrix transpose. Note that all matrix elements involving the unbounded basis functions u_0 and u_1 exist because

$$\lim_{r \to \infty} (H-E) \begin{Bmatrix} u_0(r) \\ u_1(r) \end{Bmatrix} = 0 . \quad (2.11)$$

Since the matrix $\underset{\approx}{M}$ of Eq. (2.10c) is complex-symmetric, there are no real values of E for which the matrix inverse in Eq. (2.9) is singular, and thus no "Kohn anomalies". In fact, the condition that Eq. (2.9) is singular, namely

$$\det(\underset{\approx}{M}) = \det [\langle u_\ell | H-E | u_{\ell'} \rangle] = 0 , \quad (2.12a)$$

$\ell,\ell'=1,\ldots,N$, is the secular equation for eigenvalues of the Schrödinger equation

$$(H-E)\psi(r) = 0 , \quad (2.12b)$$

with boundary condition (as $r \to \infty$)

$$\psi(r) \propto e^{ikr} ; \quad (2.12c)$$

i.e., Eq. (2.12a) is the expression that has been used before[19] for determining Siegert eigenvalues,[20] the complex energies that are the (physically correct) complex poles of the S-matrix which characterize the positions and widths of scattering resonances. Eq. (2.9) is thus singular only where it is supposed to be singular.

The S-matrix Kohn approach also allows one to identify a corresponding basis set approximation[15a] to matrix elements of the full outgoing wave Green's function $G^+(E) \equiv (E+i\epsilon-H)^{-1}$. This is

$$\langle a | G^+(E) | b \rangle = - \sum_{\ell,\ell'=1}^{N} \langle a | u_\ell \rangle (\underset{\approx}{M}^{-1})_{\ell,\ell'} \langle u_{\ell'} | b \rangle , \quad (2.15)$$

where $\underset{\approx}{M}$ is as above, Eq. (2.10c), and $|a\rangle$ and $|b\rangle$ are any square-integrable functions. Note that the complex-symmetric structure of the matrix $\underset{\approx}{M}$ is the same as that in complex scaling/coordinate rotation theory,[21-24] and for the same reasons. If the functions $|a\rangle$ and $|b\rangle$ are real, then Eq. (2.15) leads to a useful way for calculating matrix elements of the density of states operator,

$$\langle a|\delta(E-H)|b\rangle = -\pi^{-1}\,\text{Im}\langle a|G^+(E)|b\rangle. \quad (2.16)$$

In actual calculations for the S-matrix, Eq. (2.9), one does not wish to carry out numerical calculations with the complex symmetric matrix $\underset{\approx}{M}$. This can be avoided by the usual partitioning methods, so that Eq. (2.9) can be written in the equivalent form

$$S = \frac{i}{\hbar}(B - C \cdot B^{*-1} \cdot C), \quad (2.17)$$

where B and C are the 1×1 "matrices"

$$B = M_{0,0} - \underset{\sim}{M}_0^T \cdot \underset{\approx}{M}^{-1} \cdot \underset{\sim}{M}_0 \quad (2.18a)$$

$$C = M_{1,0} - \underset{\sim}{M}_0^{*T} \cdot \underset{\approx}{M}^{-1} \cdot \underset{\sim}{M}_0, \quad (2.18b)$$

when $M_{0,0}$, $\underset{\sim}{M}_0$, and $\underset{\approx}{M}$ are as before, Eq. (2.10), except that $\ell,\ell' = 2, \ldots, N$ (i.e., only the real basis functions), and

$$M_{1,0} = \langle u_0^*|H-E|u_0\rangle. \quad (2.18c)$$

Here the matrix $(\underset{\approx}{M})_{\ell,\ell'}$, $\ell,\ell' = 2, \ldots, N$ is real and symmetric, and thus more easily dealt with. (One can readily verify that a value of E for which $\det(\underset{\approx}{M}) = 0$ does not lead to a singularity in Eq. (2.17.)

Finally, for general multichannel rearrangement scattering,[6,25] let (q_γ, r_γ) denote the internal coordinates and radial scattering (i.e., translational) coordinate for arrangement γ; $\{\phi_n^\gamma(q_\gamma)\}$ are the asymptotic channel eigenfunctions for the internal degrees of freedom. Eqs. (2.17) and (2.18) generalize as follows

$$\underset{\approx}{S} = \frac{i}{\hbar}(\underset{\approx}{B} - \underset{\approx}{C}^T \cdot \underset{\approx}{B}^{*-1} \cdot \underset{\approx}{C}), \quad (2.19a)$$

where $\underset{\approx}{S}$, $\underset{\approx}{B}$, and $\underset{\approx}{C}$ are "small" square matrices, the dimension of the number of open channels, e.g., $\underset{\approx}{S} = [S_{n\gamma,n'\gamma'}]$, etc. $\underset{\approx}{B}$ and $\underset{\approx}{C}$ are given by

$$\underset{\approx}{B} = \underset{\approx}{M}_{0,0} - \underset{\approx}{M}_0^T \cdot \underset{\approx}{M}^{-1} \cdot \underset{\approx}{M}_0 \quad (2.19b)$$

$$\underset{\approx}{C} = \underset{\approx}{M}_{1,0} - \underset{\approx}{M}_0^{*T} \cdot \underset{\approx}{M}^{-1} \cdot \underset{\approx}{M}_0, \quad (2.19c)$$

where $\underset{\approx}{M}_{0,0}$ and $\underset{\approx}{M}_{1,0}$ are also "small" square matrices

$$(\underset{\approx}{M}_{0,0})_{n\gamma,n'\gamma'} = \langle u_{0n}^\gamma \phi_n^\gamma | H-E | u_{0n'}^{\gamma'} \phi_{n'}^{\gamma'}\rangle \quad (2.20a)$$

$$(\underset{\approx}{M}_{1,0})_{n\gamma,n'\gamma'} = \langle u_{0n}^{\gamma*} \phi_n^\gamma | H-E | u_{0n'}^{\gamma'} \phi_{n'}^{\gamma'}\rangle, \quad (2.20b)$$

$u_{0n}^\gamma(r_\gamma)$ is a function regular at $r_\gamma = 0$ and with asymptotic form (as $r_\gamma \to \infty$),

$$u_{0n}^\gamma(r_\gamma) \sim e^{-ik_{n\gamma}r_\gamma}/v_{n\gamma}^{1/2}.$$

$\underset{\approx}{M}$ is a "large" by "large" real symmetric matrix in the composite space, internal plus translational,

$$(\underset{\approx}{M})_{\ell n\gamma,\ell' n'\gamma'} = \langle u_{\ell n}^\gamma \phi_n^\gamma | H-E | u_{\ell' n'}^{\gamma'} \phi_{n'}^{\gamma'}\rangle, \quad (2.20c)$$

where $\{u_{\ell n}^\gamma(r_\gamma)\}$ is a square integrable basis (that need not depend on n — i.e., the same translational basis can be used for every channel). $\underset{\approx}{M}_0$ is a "large" by "small" rectangular matrix

$$(\underset{\approx}{M}_0)_{\ell n\gamma,n'\gamma'} = \langle u_{\ell n}^\gamma \phi_n^\gamma | H-E | u_{0n'}^{\gamma'} \phi_{n'}^{\gamma'}\rangle. \quad (2.20d)$$

Only open channels $\{n\gamma\}$ are included in the matrices $\underset{\approx}{M}_{00}$, $\underset{\approx}{M}_{1,0}$, and the "small" dimension of $\underset{\approx}{M}_0$, while open and closed channels are required in the matrix $\underset{\approx}{M}$ and the "large" dimension of $\underset{\approx}{M}_0$.

Eqs. (2.19) - (2.20) thus express the S-matrix for reactive scattering in an extremely straight-forward manner: one chooses basis functions, computes matrix elements of the Hamiltonian, and then does a standard linear algebra calculation.

III. $H + H_2(v=j=0) \to H_2(v',\text{odd}j') + H$

In this section and the next I survey some selected results obtained by Zhang and Miller[26-28] using this S-matrix Kohn approach. (The most complete summary of the methodology and its application to atom-diatom reactive in 3-d is given in ref. 28.) Integral cross sections are given by the standard formulae

$$\sigma_{v'j' \leftarrow vj}(E) = \pi [k_{vj}^2 (2j+1)]^{-1} \sum_{J=0} (2J+1)$$

$$\times \sum_{\ell',\ell} |S^J_{v'j'\ell', vj\ell}(E)|^2 , \quad (3.1)$$

and differential cross sections ($\equiv$ angular distributions) by

$$\sigma_{v'j'K' \leftarrow vjK}(\theta) = |(2ik_{vj})^{-1} \sum_J (2J+1) d^J_{K'K}(\theta)$$

$$\times \sum_{\ell',\ell} C(j'J\ell';K',-K') i^{\ell-\ell'} S^J_{v'j'\ell', vj\ell} C(jJ\ell;K,-K)|^2 ,$$

$$(3.2)$$

where J is the total angular momentum quantum number, $K \equiv m_j$ is the helicity, i.e., the projection quantum number for the diatom rotation with the relative translational velocity vector as the quantization axis. All other quantities in Eq. (3.1) and (3.2) are their usual selves (i.e., Clebsch-Gordan coefficients, Wigner rotation functions, etc.)

The calculations have been carried out for total energies ($E_{tot} = E_{trans} + \sim 0.27$ eV) up to 1.45 eV, and for the H+H$_2$ reaction this requires J values up to $\sim$20 to obtain convergence in the partial wave sums in Eqs. (3.1) and (3.2).

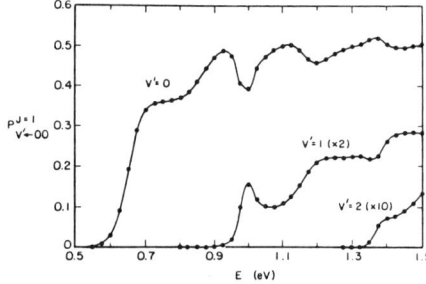

Figure 1. Reaction probability for the distinguishable atom reaction H+H$_2$(v$_1$=j$_1$=0)→H$_2$(v,all j)+H for total angular momentum J=1 as a function of total energy. The results for v'=1 and v'=2 have been multiplied by a factor of 2 and 10, respectively, for convenience in displaying them.

Fig. 1 first shows the energy dependence of the reaction probability for a single value of J,

$$P^J_{v' \leftarrow vj} \equiv \sum_{j'} \sum_{\ell',\ell} |S^J_{v'j'\ell', vj\ell}|^2 , \quad (3.3)$$

for the initial ground state v=j=0 of H$_2$. The most prominent feature is a resonance at $E \simeq 0.95$ eV due to a short-lived ($\sim$10-15 femtosec) collision complex.

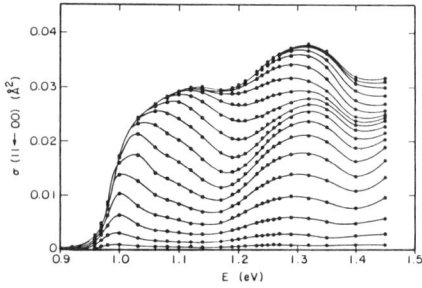

Figure 2. Integral cross sections for the properly symmetrized reaction H+H$_2$(vj)→H$_2$(v'j')+H for v=j=0, v'=j'=1, as a function of total energy. The various curves connecting the (calculated) points are for various values of J_{max}, the upper limit in the sum over J. The lowest curve is for J_{max}=0, the next one above it for J_{max}=1, and so forth, with the uppermost curve for J_{max}=18.

Next, Fig. 2 shows the integral cross section for the v=j=0 → v'=j'=1 reactive transition; the various curves correspond to various upper limits in the sum over J in Fig. (3.1). The uppermost curve is thus the converged result for the cross section. (Calculations with J_{max} up to 24 have subsequently been carried out to verify convergence.) The cross section is quite a smooth function of energy, showing no hint of the resonance at E = 0.95 eV seen in the reaction probability for a single small value of J (Fig. 1). This is what one would have expected[29] because of the large number of J values that contribute (incoherently) to the integral cross section.

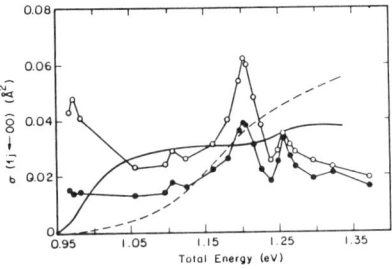

Figure 3. The points are the experimental integral cross sections of ref. 30 for the reaction H+H$_2$(vj)→H$_2$(v'j')+H for v=0, j averaged over 0 and 2, v'=1, j'=1 (open circles) and 3 (solid circles). The curves are the present theoretical cross sections for v=j=0, v'=1, j'=1 (solid curve) and 3 (dashed curve).

This result for the integral cross

section is in significant disagreement with the recent experimental report by Nieh and Valentini;[30] this is shown in Fig. 3, along with the present theoretical result. The magnitudes of the reactive cross sections are in excellent agreement, but the resonance-like structure reported experimentally is absent in the theoretical cross section. Recent independent calculations by Manolopoulos and Wyatt[31] have obtained results essentially identical to ours,[26] lending strong support to the belief that these theoretical calculations are the correct results for this potential energy surface,[32] which is thought to be extremely accurate. (Unpublished calculations[33] with another surface gives only modest differences.)

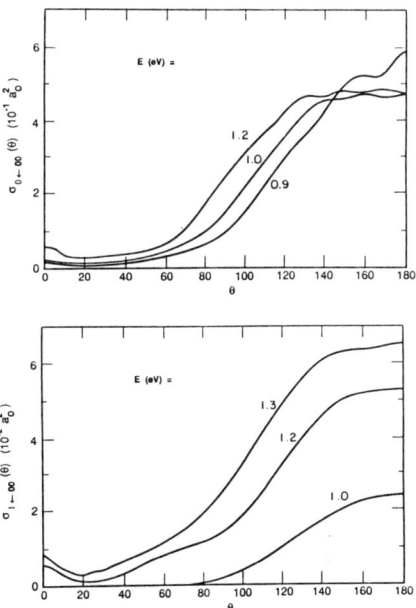

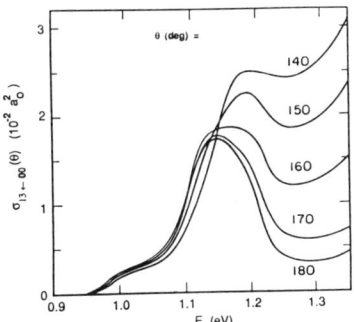

Figure 4. Differential cross section, at several fixed center-of-mass scattering angles θ for the reaction $H + H_2(v=j=0) \to H_2(v',j') + H$, as a function of total energy E, for the final state $(v',j') = (1,3)$.

We have suggested[26] that a partial resolution of this experimental-theoretical disagreement could be that the experiments are actually observing primarily back-scattered products rather than a true integral cross section. (The differential cross section at $\theta = 180°$ should have its dominant contribution from a few small values of J.) Thus both we[27] and Wyatt and Manolopoulos[31] have looked at the energy-dependence of the differential cross section at $\theta = 180°$ (and also other fixed angles). Fig. 4 shows this for the particular final state $v'=1, j'=3$. For $\theta = 180°$ there is indeed a structure in the energy dependence that seems roughly similar to the experimental structures in Fig. 3, but the similarity is at best only suggestive that this may be what the experiments have actually measured. Other suggestions[34] are that there may be non-linear optical effects in either the laser that photodissociates HI (to obtain fast H atoms) or in the CARS laser system used to observe the product $H_2(v'j')$.

Figure 5. Differential cross section for $H + H_2(v=j=0) \to H_2(v',j') + H$, summed over all odd values of j', for several values of the total energy E. (a) and (b) are for $v'=0$ and 1, respectively.

For completeness, Fig. 5a and b show the complete center-of-mass angular distributions for the reaction $v=j=0 \to v'=0$, all odd j' and $v'=1$, all odd j', respectively. As the energy increases, one sees the angular distribution in the backward direction flatten out (hard sphere-like behavior) and a small peak appear in the forward direction (stripping behavior).

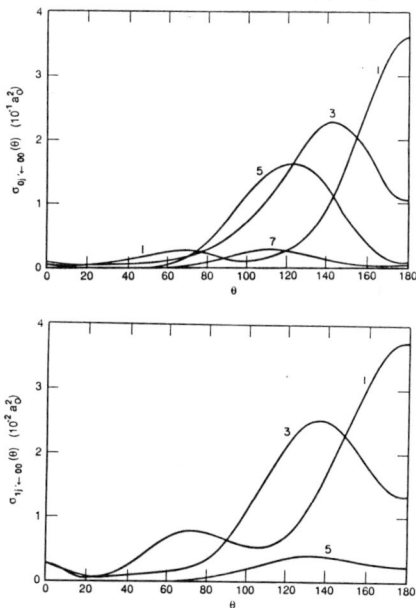

Figure 6. Differential cross section for

$H+H_2(v=j=0) \to H_2(v',j')+H$ for various values of j'. (a) is for $v'=0$ and $E=1.0$ eV, and (b) is for $v'=1$ and $E=1.3$ eV.

Finally, Fig. 6a and b show the angular distribution for specific final rotational states. The most significant feature here is strong coupling between the angular and internal state (here rotational state) distributions, i.e., the angular distribution is quite different for different final rotational states. The most probable scattering angle moves progressively away from the backward direction with increasing rotational excitation of the products.

IV. $D+H_2(v=j=0) \to HD(v',j')+H$

Similar calculations have been carried out for the $D+H_2$ reaction; here J values up to 34 were necessary in the partial wave sums to obtain convergence at the highest energies. The paper[28] reporting these results includes a comprehensive summary of the S-matrix Kohn methodology as it applies to atom-diatom reactions in full 3-d space. Here we give a selection of some of the results that were obtained.

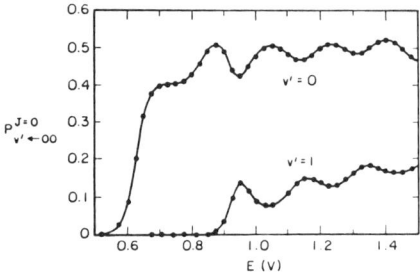

Figure 7. Reaction probability for $D+H_2(v=j=0) \to HD(v', \text{all } j')+H$ for $J=0$, as a function of total energy.

Fig. 7 shows the $J=0$ reaction probability from the ground state $v=j=0$ of H_2 to $v'=0$ and 1 (summed over final j' states) of HD, as a function of energy. One again sees a very broad resonance feature, here at $E \approx 0.95$ eV.

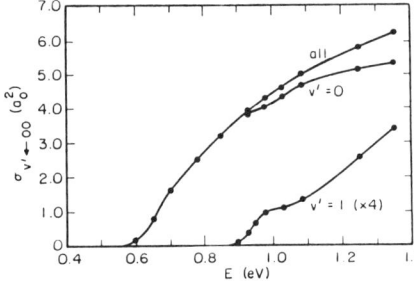

Figure 8. (Converged) integral cross sections for $D+H_2(v=j=0) \to HD(v', \text{all } j')$, as a function of energy.

In the integral cross section shown in Fig. 8, however, one sees that it has been almost totally washed out by the partial wave sum; there is only a slight inflection seen in the $v'=1$ cross section at $E \approx 0.95-1.0$ eV.

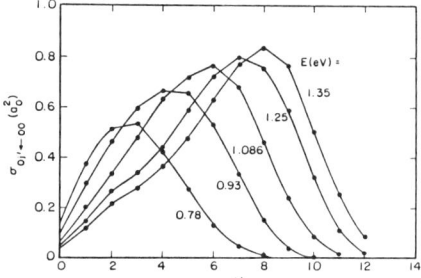

Figure 9. Integral cross sections for $D+H_2(v=j=0) \to HD(v'=0,j')+H$ as a function of j', for several values of E (eV).

Fig. 9 shows the integral cross section to specific final rotational states, i.e., for the transition $v=j=0 \to v'=0, j'$, at several values of energy. They all have a quasi-Boltzmann form,[35] $P(j) \propto (2j+1) \exp[-Bj(j+1)/kT_{eff}]$, where T_{eff} increases with E.

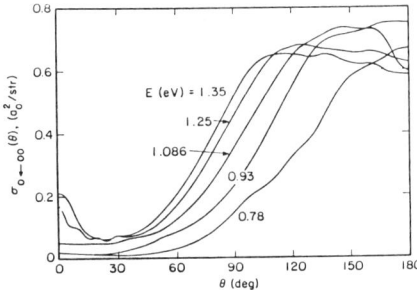

Figure 10. Differential cross sections for $D+H_2(v=j=0) \to HD(v'=0, \text{all } j')+H$, for several values of E (eV).

Fig. 10 shows the angular distribution (summed over final rotational states) for several values of E. As in Section III for $H+H_2$, one here sees the angular distribution flatten out in the backward direction as the energy increases and also the appearance of a stripping peak in the forward direction.

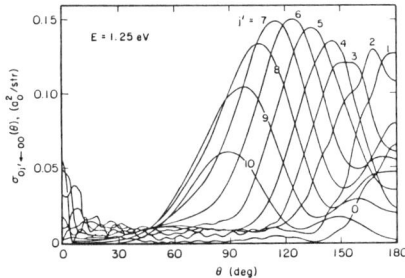

Figure 11. Differential cross section for $D+H_2(v=j=0) \to HD(v'=0,j')+H$, for various values of j', at $E = 1.25$ eV.

Finally, Fig. 11 shows the doubly differential cross section (i.e., differential in final scattering angle and final rotational state), and as in Section III one also sees here that the angular distribution depends very strongly on the final rotational state; j'=0 is strongly peaked in the backward direction (θ=180°), and the most probable scattering direction moves forward with increasing j', to θ≈90° for j'=10.

V. Concluding Remarks

The S-matrix version of the Kohn variational principle thus leads to a very straight-forward methodology for carrying out rigorous quantum mechanical calculations. The results described in Section III and IV, which are the first rigorous cross section calculations for these reactions in this high energy region (Schatz and Kuppermann[3c] earlier carried out rigorous calculations at lower energy), have employed the most simple minded version of the methodology. I.e., the basis set has been chosen to be of the simple direction product form, so that the same L^2 translational basis $\{u_\ell(r)\}$, $\ell=2, ..., N$ is used for each channel. The calculations could be made much more efficient - i.e., the basis set reduced in size - by fine tuning; e.g., it is clear that highly excited channels, which have low translational energy, need fewer translational functions than low energy channels (which have high translational energy). This, and many other tricks from the quantum chemistry of bound state eigenvalue calculational methodology, are available to enhance the applicability of this S-matrix Kohn approach to reactive scattering. I believe that the next few years will see a number of these developments and their application to more complex (though still relatively simple) chemical reactions.

Acknowledgment

This work has been supported by the National Science Foundation Grant CHE84-16345 and by the Director, Office of Energy Research, Office of Basic Energy Sciences, Chemical Sciences Division of the U.S. Department of Energy under Contract No. DE-AC03-76SF00090.

References

1. (a) For a survey of the early days of molecular beams in chemistry, see Molecular Beams, ed. J. Ross (Adv. Chem. Phys. 10, (1966)).
   (b) For a view of the current state of molecular beam chemistry, see Y. T. Lee, Chemical Scripta 27, 215 (1987); Science 236, 793 (1987).
2. See, for example, T.-Y. Wu and T. Ohmura, Quantum Theory of Scattering, Prentice-Hall, Englewood Cliffs, N.J., 1962, pp. 211 et. seq.
3. (a) G. Wolken and M. Karplus, J. Chem. Phys. 60, 351 (1974).
   (b) A. B. Elkowitz and R. E. Wyatt, J. Chem. Phys. 62, 2504 (1975); 63, 702 (1975).
   (c) G. C. Schatz and A. Kuppermann, J. Chem. Phys. 65, 4642, 4668 (1976).
   (d) For a review of H+H_2, see D. G. Truhlar and R. E. Wyatt, Ann. Rev. Phys. Chem. 27, 1 (1976).
4. (a) F. Webster and J. C. Light, J. Chem. Phys. 85, 4744 (1986).
   (b) F. Webster and J. C. Light, J. Chem. Phys. 90, 265 (1989).
   (c) F. Webster and J. C. Light, J. Chem. Phys. 90, 300 (1989).
5. (a) A. Kuppermann and P. G. Hipes, J. Chem. Phys. 84, 5962 (1986); P. G. Hipes and A. Kuppermann, Chem. Phys. Lett. 133, 1 (1987).
   (b) G. A. Parker, R. T Pack, B. J. Archer and R. B. Walker, Chem. Phys. Lett. 137, 564 (1987); R. T Pack and G. A. Parker, J. Chem. Phys. 87, 3888 (1987).
   (c) G. C. Schatz, Chem. Phys. Lett. 150, 92 (1988); 151 409 (1988).
   (d) M. Mishra, J. Linderberg and Y. Ohrn, Chem. Phys. Lett. 111, 439 (1984); J. Linderberg, Int. J. Quant. Chem. Symp. 19, 467 (1986); J. Linderberg and B. Vessal, Int. J. Quant. Chem. 31, 65 (1987).
   (e) J. M. Launay and B. Lepetit, Chem. Phys. Lett. 144, 346 (1988); 151, 287 (1988).
6. W. H. Miller, J. Chem. Phys. 50, 407 (1969).
7. (a) K. Haug, D. W. Schwenke, Y. Shima, D. G. Truhlar, J.Z.H. Zhang and D. J. Kouri, J. Phys. Chem. 90, 6757 (1986).
   (b) K. Haug, D. W. Schwenke, D. G. Truhlar, Y. Zhang, J.Z.H. Zhang and D. J. Kouri, J. Chem. Phys. 87, 1892 (1987).
   (c) D. W. Schwenke, K. Haug, D. G. Truhlar, Y. Sun, J.Z.H. Zhang and D. J. Kouri, J. Phys. Chem. 91, 6080 (1987).
   (d) J.Z.H. Zhang, D. J. Kouri, K. Haug, D. W. Schwenke, Y. Shima and D. G. Truhlar, J. Chem. Phys. 88, 2492 (1988).
   (e) M. Mladenovic, M. Zhao, D. G. Truhlar, D. W. Schwenke, Y. Sun and D. J. Kouri, Chem. Phys. Lett. 146, 358 (1988).

(f) Y. C. Zhang, J.Z.H. Zhang, D. J. Kouri, K. Haug, D. W. Schwenke and D. G. Truhlar, Phys. Rev. Lett. 60, 2367 (1988).
(g) M. Mladenovic, M. Zhao, D. G. Truhlar, D. W. Schwenke, Y. Sun and D. J. Kouri, J. Chem. Phys. 92, 7035 (1988).

8. M. Baer and D. J. Kouri, Phys. Rev. A 4, 1924 (1971). For a review of this approach to reactive scattering, see D. J. Kouri, in The Theory of Chemical Reaction Dynamics, Vol. I, ed. M. Baer, CRC Press, Boca Raton, FL, 1985, p. 163 et seq.
9. P. G. Burke and A. J. Taylor, Proc. Phys. Soc. (London) 88, 549 (1966).
10. For a review, see Y. C. Tang, M. LeMere and D. R. Thompson, Phys. Repts. 47, 167 (1978).
11. W. Kohn, Phys. Rev. 74, 1763 (1948). See also, T.-Y. Wu and T. Ohmura, Quantum Theory of Scattering, (Prentice-Hall, Englewood Cliffs, New Jersey, 1962), pp. 57-68.
12. (a) C. Schwartz, Phys. Rev. 124, 1468 (1961).
(b) Ann. Phys. (New York) 10, 36 (1961).
(c) R. K. Nesbet, Variational Methods in Electron-Atom Scattering Theory, (Plenum, New York, 1980), pp. 30-50.
13. (a) J. E. Adams and W. H. Miller, J. Chem. Phys. 83, 1505 (1979).
(b) M. R. Hermann and W. H. Miller, Chem. Phys. 109, 163 (1986).
14. (a) K. Takatsuka and V. McKoy, Phys. Rev. A 24, 2473 (1981); (b) ibid. 30, 1734 (1984); (c) T. L. Gibson, M. A. P. Lima, V. McKoy, and W. M. Huo, Phys. Rev. A 35, 2473 (1987).
15. (a) W. H. Miller and B. M. D. D. Jansen op de Haar, J. Chem. Phys. 86, 6213 (1987).
(b) This discovery was made earlier in nuclear physics; see M. Kamimura, Prog. Theor. Phys. Supplement 62, 236 (1977).
16. (a) J. Z. H. Zhang, S.-I. Chu and W. H. Miller, J. Chem. Phys. 88, 6233 (1988).
(b) W. H. Miller, Collect. Czech. Chem. Comm. 53, 1873 (1988).
(c) W. H. Miller, Comments At. Mol. Phys. 22, 115 (1988).
17. (a) C. W. McCurdy, T. N. Rescigno, and B. I. Schneider, Phys. Rev. A 36, 2061 (1987).
(b) C. W. McCurdy and T. N. Rescigno, Phys. Rev. A, submitted.
(c) T. N. Rescigno, C. W. McCurdy, and B. I. Schneider, Phys. Rev. Lett., submitted.
18. L. F. X. Gaucher and W. H. Miller, Israel J. Chem., in press.
19. A. D. Isaacson, C. W. McCurdy, and W. H. Miller, Chem. Phys. 34, 311 (1978).
20. A. J. F. Siegert, Phys. Rev. 56, 750 (1939).
21. (a) J. Nuttall and H. L. Cohen, Phys. Rev. 188, 1542 (1969).
(b) F. A. McDonald and J. Nuttall, Phys. Rev. C 6, 121 (1972).
(c) R. T. Baumel, M. C. Crocker, and J. Nuttall, Phys. Rev. A 12, 486 (1975).
22. (a) T. N. Rescigno and W. P. Reinhardt, Phys. Rev. A 8, 2828 (1973); 10, 158 (1974).
(b) B. R. Johnson and W. P. Reinhardt, ibid. 29, 2933 (1984).
23. C. W. McCurdy and T. N. Rescigno, Phys. Rev. A 21, 1499 (1980); 31, 624 (1985).
24. For reviews, see
(a) B. R. Junker, Adv. At. Mol. Phys. 18, 207 (1982).
(b) W. P. Reinhardt, Annu. Rev. Phys. Chem. 33, 223 (1982).
(c) Y. K. Ho, Phys. Rep. 99, 1 (1983).
25. (a) J. Z. H. Zhang and W. H. Miller, Chem. Phys. Lett. 140, 329 (1987).
(b) J. Z. H. Zhang and W. H. Miller, J. Chem. Phys. 88, 4549 (1988).
26. J. Z. H. Zhang and W. H. Miller, Chem. Phys. Lett. 153, 465 (1988).
27. J. Z. H. Zhang and W. H. Miller, Chem. Phys. Lett. (1989).
28. J. Z. H. Zhang and W. H. Miller, J. Chem. Phys. (1989).
29. G. C. Schatz, Ann. Rev. Phys. Chem. 39, 317 (1988).
30. J.-C. Nieh and J. J. Valentini, Phys. Rev. Lett. 60, 519 (1988).
31. D. E. Manolopoulos and R. E. Wyatt, Chem. Phys. Lett., in press.
32. (a) P. Siegbahn and B. Liu, J. Chem. Phys. 68, 2457 (1978).
(b) D. G. Truhlar and C. J. Horowitz, J. Chem. Phys. 68, 2466 (1978); 71, 1514(E) (1979).
33. S. M. Auerbach and W. H. Miller, to be published.
34. M. Shapiro, private discussion.
35. R. D. Levine and R. B. Bernstein, Molecular Reaction Dynamics and Chemical Reactivity, Oxford U. P., 1987, pp. 260-274.

HARPOONING AND CHEMISTRY AT SURFACES

Aart W. Kleyn

FOM-Institute for Atomic and Molecular Physics, Kruislaan 407,
1098 SJ Amsterdam, The Netherlands

Harpooning is a well known concept in gasphase collision chemical physics. For dissociative chemisorption at surfaces a similar concept is applicable. This is manifested by the formation of negative molecular ions in molecule-surface collisions. Electronic transitions may very well be the first step towards chemisorption of molecules and therefore chemistry in general at surfaces.

Harpooning reactions are well known chemical reactions in the gasphase, see e.g. refs. 1. The reactions proceed via an ionic intermediate, e.g. $K + Br_2 \rightarrow K^+ + Br_2^- \rightarrow KBr + Br$. The ionic intermediate can be formed at the crossing seam of the potential energy surfaces for the neutral (covalent) and ionic states at large separation of the reactants. The formation of free ion-pairs has been observed as soon as there is enough energy to form the pair in an (endothermic) harpooning reaction. This is clear from figure 1, where the total cross sections for the three major processes that can occur in $K + Br_2$ collisions: reaction, ion-pair formation and inelastic scattering are plotted. The figure is obtained from classical trajectory calculations that are in good agreement with all experimental data for this

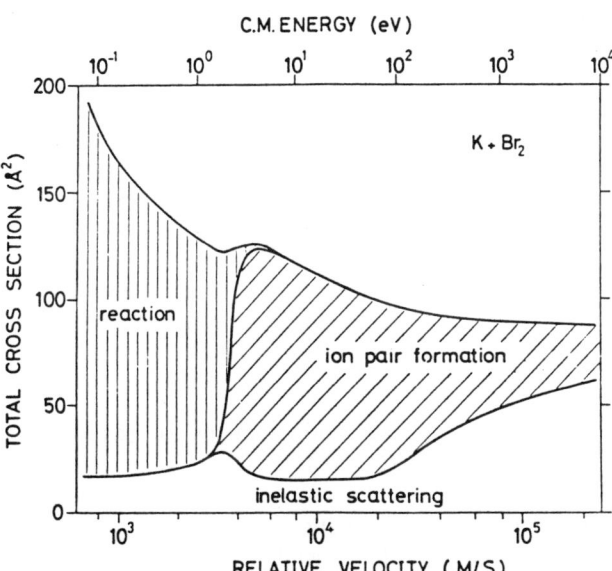

Figure 1. Total cross sections for the various nonelastic processes in $K + Br_2$ collisions as a function of the relative velocity and the CM-energy. From ref. 2.

system.[2] It is clear that ion-pair formation is a 'continuation' of the reaction with increasing energy: the mechanism is the same but at higher energies the 'bond' of the KBr molecule formed in the collision is immediately broken. From studies of ion-pair formation much information has been obtained for the understanding of the dynamics of harpooning reactions.[3] Similar, but more refined studies have been performed recently for charge transfer processes in the gasphase, see e.g. ref. 4.

Is the picture that electron transfer initiates reactive events in the gasphase relevant in surface science? It is good to note that dissociative chemisorption, i.e. $X_2 + 2M \rightarrow 2 XM = 2X_{ads} + 2M$, is the equivalent of the reactions discussed above. Here X_2 denotes a diatomic molecule to be adsorbed atomically at (metal) surface with atoms M. Another point to notice is that for harpooning in the gasphase the electron jump does enlarge the reactive cross section considerably with respect to gas-kinetic cross sections as is evident from figure 1, if one consideres that the gas kinetic cross section is on the order of 30 Å^2. In molecule-surface collisions leading to chemisorption this increase is irrelevant, because the molecule will hit the surface anyway. In surface collisions the importance of electron transfer is that it turns on an attractive force, that can initiate trapping, sticking, dissociation or even a reaction of the molecule, whereas otherwise elastic scattering occurs. By analogy to the gasphase studies on harpooning reactions these electron transfers, occurring possibly rather close to the surface, have also been termed harpooning in theoretical work by e.g. Gadzuk and Holloway, indicating that harpooning can occur in molecule-surface collisions.[5] Harpooning has also been seen for reactions of halogen molecules and alkali surfaces.[6]

For harpooning to occur we need a system for which dissociative chemisorption is preceded by formation of a molecular negative ion. This is the case in e.g. the oxygen silver interaction. Studies by several surface spectroscopies have indicated that formation of a negatively charged molecular ion is a precursor for the dissociative chemisorption of O_2 on Ag, see e.g. refs. 7. Campbell has summarized extensive studies in this system in an energy diagram reproduced in figure 2.[8] It is clear that three states exist at the surface, physisorbed O_2 bound by 0.2 eV, molecularly chemisorbed $O_2^{\partial-}$ bound by 0.5 eV ($1<\partial<2$), and atomically chemisorbed O^-, two atoms being bound by 1.7 eV. It is thought that dissociative chemisorption at thermal energies proceeds via two intermediates or precursors: O_2 and $O_2^{\partial-}$, see e.g. ref. 9.

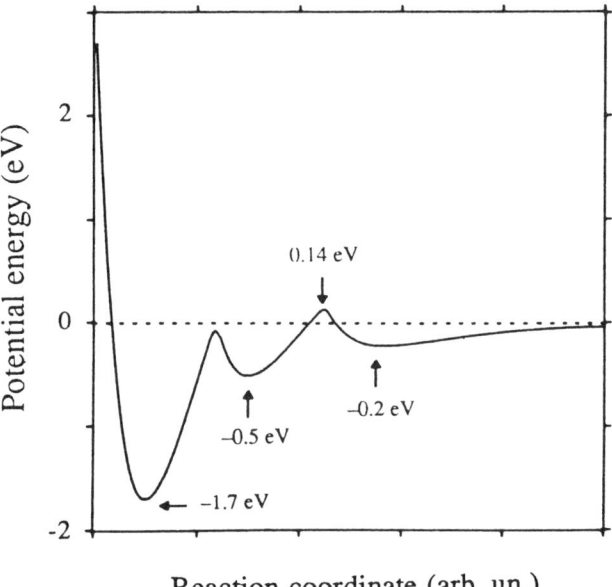

Figure 2. Potential energy diagram of the O_2/Ag(111) triple well system, as proposed by Campbell,[8] as a function of an arbitrary reaction coordinate.[9]

To study the dynamics of harpooning at surfaces one should perform experiments in which negative ion formation occurs, and in which the collision energy is sufficiently high to prevent subsequent sticking. Therefore we have studied O_2 scattering from Ag(111).[10-15] In the most recent experiments we use neutral O_2 beams with energies from 100-400 eV.[15] The beams are formed by charge exchange of O_2^+ on O_2. In earlier experiments also O_2^+ beams were used.[10-14] Energetic beams are necessary because the process of negative ion formation is endothermic by a few eV.

The result for scattering of a beam of O_2 with an energy E_i of 100eV from Ag(111) leading to negative ions is shown in fig. 3. The angle of incidence Θ_i measured from the surface normal is 70°. The figure shows the angular (Θ, measured from the surface normal) and final energy (E') spectrum of the negative ions. Essentially only one peak is visible at 80 eV in the specular direction. This peak can be attributed to O_2^-. The fact that the peak appears in the specular direction indicates that formation of O_2^- is a rather generally occurring process and the molecule does not have to impinge on special sites at the surface for charge transfer to occur. It also shows that O_2^- is not formed by sputtering, or because the molecule passes through the bulk. Assuming that for scattered O^- each of the atoms carries half of the translational energy of the corresponding O_2^-, we would expect a peak at half the energy of the O_2^- peak. This peak is barely visible indicating that dissociation does not occur under the given conditions, even though E_i is about 25 times the dissociation energy of O_2^-. By varying E_i and Θ_i we find

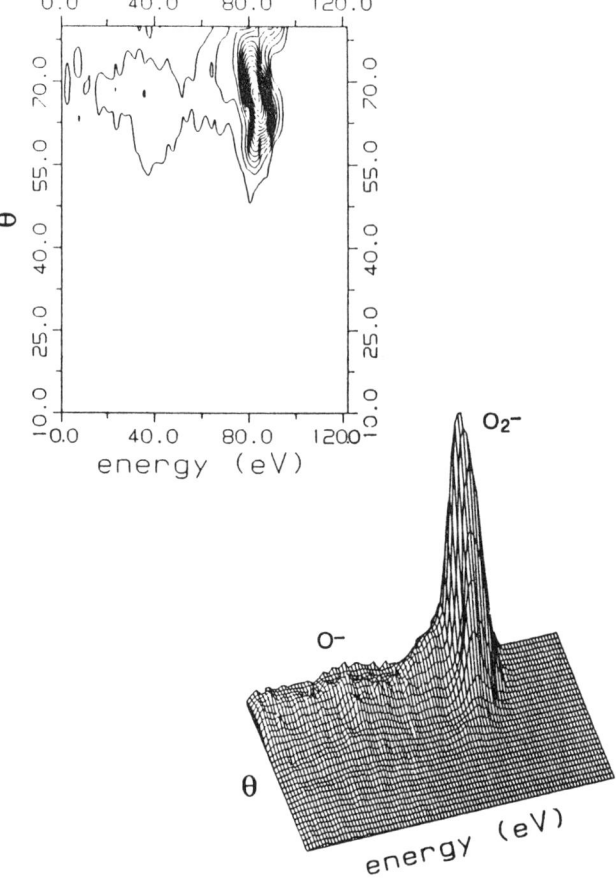

Figure 3. Intensity distribution for negative ions formed in glancing collisions ($\Theta_i = 70°$) of 100eV O_2 molecules with Ag(111). The measured intensity is plotted 3-dimensionally and in contour representation as a function of the final energy and polar scattering angle Θ. The axes are identical for both representations. From ref. 15.

that the O^- yield seems to scale with the normal component of the impact velocity, and can exceed the O_2^- yield.[10] Later in this report we will show that the cause of dissociation of O_2^- is most likely collision induced dissociation in an impulsive, mechanical collision between molecule and the repulsive wall of the surface. Dissociation does not occur due to an electronic transition to a steep repulsive potential, see e.g. refs. 16-17.

Having established that negative ions are formed in collisions at the surface leading to specular scattering, the next question concerns the yield of negative ions. In this case direct detection of both neutrals and ions is necessary. This has been performed in an apparatus, in which both ions and neutrals are directly counted by a microchannel plate. Consequently high particle energies (> 100 eV) are needed. To keep the normal component of the velocity low very glancing angles of incidence are used ($\Theta_i = 85°$). In the apparatus the time-of-flight (tof) of the particles from the surface to the detector can be measured.[18] The ions can be separated from the neutrals by postacceleration. A typical spectrum taken at $E_i = 270$ eV is shown in figure 4.[13] The large structure at 32 μs is due to O atoms and O_2 molecules, the latter forming the 'hat' on top of the much broader energy distribution of the O atoms. Also O_2^- ions are clearly observed and O^- ions are almost absent in the spectrum. The shape of the spectrum for the neutrals is attributed to the fact that two processes occur in the neutralisation of O_2^+. The atoms are formed in dissociative

probability to form O_2^- close to the classical turning point along the trajectory is much higher, but that most of the O_2^- reneutralizes in the exit channel or autodetaches when vibrationally hot.[13] By contrast the probability for forming O_2^+ is very low and positive ions are not detectable at grazing angles.

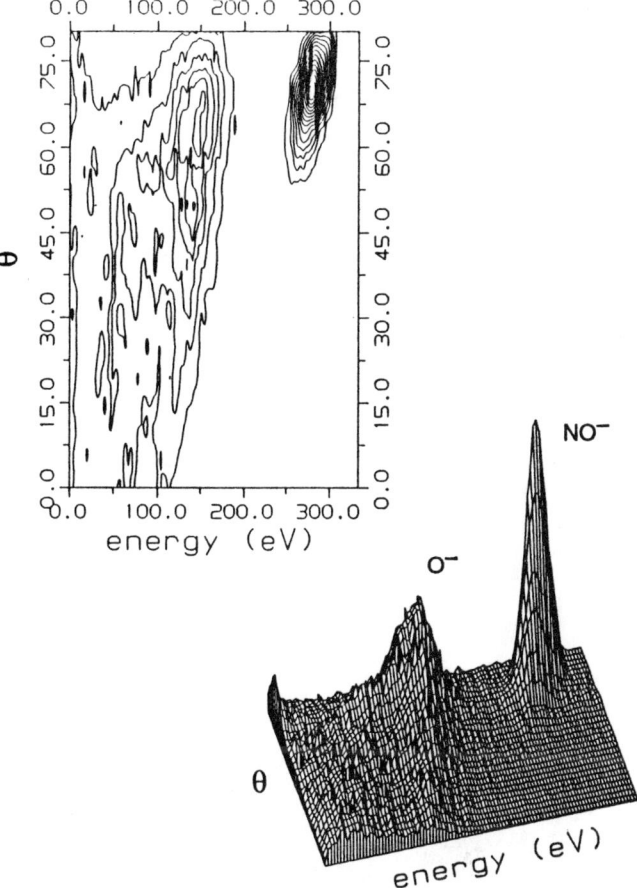

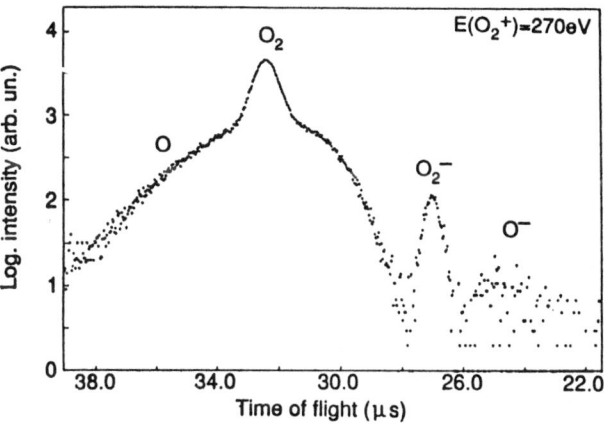

Figure 4. TOF distributions for neutrals and negative ions resulting from specular scattering of O_2^+ at $E_i = 270$ eV and $\Theta_i = 85°$. The intensity at the detector is plotted on a ^{10}log scale in arbitrary units as a function of the target-detector flight time in μs. From ref. 13.

neutralisation yielding a broad tof distribution; ground state O_2 molecules are formed by an Auger process, not affecting the tof distribution.[17] The molecules can pick up another electron whereas the atoms apparently cannot. From the yield of O_2^- with respect to O_2 we can see that the probability for charge transfer is at least a few percent. From the analysis of the data it has been concluded that the

Figure 5. Intensity distribution for negative ions formed in glancing collisions ($\Theta_i = 70°$) of 300eV NO^+ ions with Ag(111). The measured intensity is plotted 3-dimensionally and in contour representation as a function of the final energy and polar scattering angle Θ. The axes are identical for both representations. From ref. 15.

The formation of negative ions in molecule-surface collisions is by no means limited to the O_2/Ag system. In fact negative molecular ions have been seen for several systems: O_2^- from Ag(111), Ni(110) and Si(001),[10-12,14] CO_2^- from Ni(110),[12] I_2^- and several negative ions of organic molecules from diamond,[19] and NO^- from Ag(111).[15] The latter result is reproduced in figure 5. A beam of 300 eV NO^+ is incident at Ag(111) with $\Theta_i = 70°$. Clearly two peaks are observed in the specular direction, the one at low E' being attributed to O⁻ and the one at large E' to NO⁻. Similar experiments with O_2^+ or O_2 beams give almost indistinguishable results. This strongly suggests that also in case of the NO/Ag interaction harpooning can occur and negative ion intermediates can be formed. A similar conclusion has been reached on different grounds, namely by the analysis of vibrational excitation of NO at Ag(111), where a strong dependence on the surface temperature was observed and other evidence for vibrational excitation via an intermediate NO⁻ state.[20]

The observation of negative molecular ions and the potential energy diagram of figure 2 are reminiscent of the gasphase harpooning which is induced by a crossing of the relevant covalent (K + Br_2) and ionic (K^+ + Br_2^-) states of the system. These crossings are typically located at distances of several Å, outside the range of the repulsive (K + Br_2) potentials. The crossing is induced by the Coulomb interaction. In the case of a negative ion interaction with a metal surface the curve crossing needed could be induced by the electrostatic image potential.

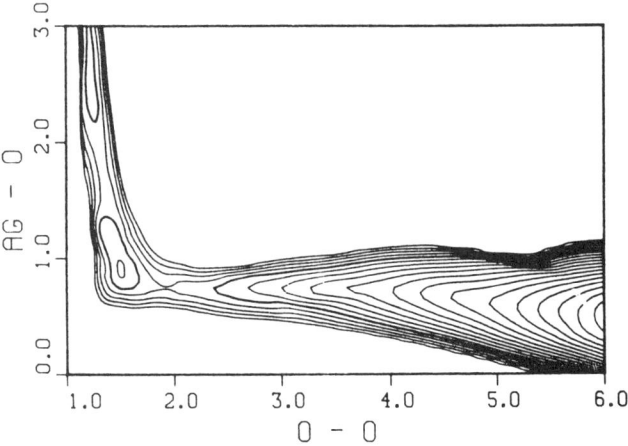

Figure 6. Calculated potential energy surface of O_2 chemisorbing and dissociating on Ag(110) as a function of the O-O separation and height above the surface (both in Å). The contour spacing is 0.1 eV. The zero of the energy scale (indicated by the heavy contours) is defined by the energy of Ag + O_2 at infinite separation. From ref. 21.

However, this is only one quarter as deep as the Coulomb potential, and consequently the neutral state of the system cannot be expected to cross the potential of the ionic (Ag^++ O_2^-) at a reasonably large molecule-surface distance. From the fact that a negatively charged $O_2^{\partial-}$ species is observed at the surface,[7] we conclude that extra stabilisation of the state occurs. This is confirmed by cluster calculations for the O_2/Ag system by Van den Hoek and Baerends.[21] Their computed potential energy hypersurface is shown in figure 6. The potential energy hypersurface shows the three states identified for the O_2/Ag system. The physisorption state is barely visible at R(Ag-O) = 2.5Å, which is not unreasonable since the cluster calculation at the SCF level used does not provide Van der Waals interactions. A molecularly chemisorbed state is clearly seen at r(O-O) = 1.5 Å. This O_2^- state has an equilibrium distance that is stretched with respect to ground state O_2 (r=1.2Å), the same being true for the free O_2^- ion (r=1.35Å). A very shallow and long barrier separates the molecularly chemisorbed state from the atomically chemisorbed state at R(Ag-O) = 0.5 Å. From the calculations it is clear that the adsorbed O_2^- correlates diabatically to free O_2^- far from the surface. However, the intramolecular potential of the adsorbed O_2^- is different from that of free O_2^-. The charge transfer is due to the formation of a (covalent) chemical bond to the surface. It is not exclusively induced by the electrostatic image potential.

At the conditions of figure 3 the normal velocity is about 0.1 Å.fs⁻¹. The vibrational period of O_2^- as observed in gasphase collisions is about 30 fs.[22] The molecular ion would for the conditions of figure 3 have to travel a distance normal to the surface of about 3 Å for a complete vibration. From the potential energy hypersurface in figure 6 it is clear that the O_2^- ions are formed close to the repulsive wall of the molecule-surface system and that the O-O repulsion is quite weak. Therefore one cannot expect it to vibrate considerably while the molecule approaches the surface. Because there is a small barrier to dissociation along the intermolecular separation, the molecular ion will not spontaneously dissociate during the collision. Thus so-called catalytic dissociation cannot be expected. This is confirmed by the results shown in figures 3 and 4, where no dissociation leading to O⁻ formation is observed at all. By contrast, dissociative neutralisation proceeds readily at these velocities because the relevant repulsive O_2 states are very steep in the Franck-Condon region.[16] Effective dissociation times may only be 20 fs, corresponding to distances of about 2 Å, which are easily passed in a dissociative neutralisation event.[17]

At higher perpendicular velocities of the incident $O_2^{(+)}$ dissociation of O_2^- is observed. This is clear from figure 5, which shows data for NO but the results for O_2 are very similar.[10,15] The dissociation seems not to be due to electronic transitions. As is clear from figure 6 the O⁻ + O⁻ system is only repulsive very close to the surface. Therefore the molecule needs a large time close to the surface to dissociate completely. Otherwise the intramolecular potential (of O_2 or O_2^-) will become bound again, if the molecule returns to the gasphase. Because the time available for molecular stretching on the repulsive O⁻ + O⁻ curve following electronic transitions only decreases when increasing the velocity, one would expect increasing dissociation or O⁻ + O⁻ formation with decreasing velocities. This is not observed experimentally. Therefore the dissociation is most likely collision induced, i.e. caused by the impulsive collision of the molecule with the repulsive wall of the surface. This is confirmed in classical trajectory calculations performed for scattering of O_2 from Ag(111) using ab-initio pairpotentials.[23] The potentials have been successfully tested for other systems.[24,25] In these calculations charge transfer is neglected. This has been done because the exact behaviour (both position and width) of the O_2^- level is not known and models using essentially only an image interaction may not be applicable. These models have been very successful in the explanation of formation of negative atomic ions at low workfunction surfaces, see e.g. refs. 26.

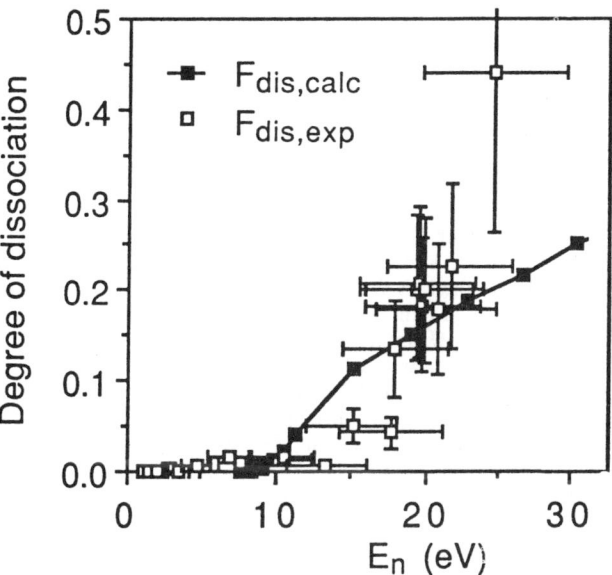

Figure 7. Degree of dissociation $F_{dis,exp}$ for the tof experiment and $F_{dis,calc}$ for the classical trajectory calculations as a function of $E_n = E_i \times \cos^2(\Theta_i)$. The line through the computed points was drawn to guide the eye. From ref. 13.

The onset of collision induced dissociation can be seen very well in tof experiments like the one shown in figure 4. When a reduced O⁻ yield in the specular direction is plotted a threshold appears as can be seen in figure 7. The x-axis shows the energy corresponding with the normal velocity ($E_n = E_i \times \cos^2(\Theta_i)$), the cosine factor being 0.0076. The threshold is consistent with the onset of collision induced dissociation in trajectory calculations, also shown in the figure. This indicates that collision induced dissociation occurs when E_n is about four times the dissociation energy of O_2^- and E_i is about 500 times this value. The normal component of the velocity is the only active component. This can very nicely be seen in figure 8,

where the average internal excitation of the O_2 molecule and the excitation of the solid is plotted as a function of the total energy, while keeping E_n constant. These quantities only decrease by less than a factor of 3 while the total energy increases by more than two orders of magnitude. In fact, it appears from the analysis by Van den Hoek and Kleyn that the strongest interaction takes place when parallel and normal velocity are comparable.[27] From these calculations it is clear that experiments at grazing angles of incidence can mimick experiments with low total energy but more normal angles of incidence very well. This is experimentally very convenient. The sites most effectively probed in specular scattering do depend on Θ_i. This could explain the different in the degree of dissociation for the negative ions at $\Theta_i = 70°$ and $\Theta_i = 85°$, if the charge transfer probability would be site dependend. Such a site dependence indeed has been observed recently in extended-Hückel calculations by Zonnevylle et al.[28]

Having demonstrated that harpooning transitions occur for the O_2/Ag(111) system, we now turn to the relevance to chemisorption for this system. For a slow O_2 molecule approaching the surface, the harpooning transition may take place along the way towards the surface, leading to chemisorption as O_2^-. However, from scattering experiments at near thermal energies it appears that harpooning is not very probable, because mainly direct inelastic scattering is seen and no desorption from an (activated) precursor like an O_2^- state.[25,29] This is in accord with the potential diagram in figure 2 proposed by Campbell, indicating that there is a barrier for formation of O_2^-. Also the low sticking probability of O_2 in the order of 10^{-6} indicates that the molecular negative ion state is not accessable for thermal molecules, because of a barrier. A hint of this barrier is seen in figure 6. The molecular beam studies on scattering and dissociative chemisorption of O_2 from Ag(111) indicate that the barrier cannot easily be overcome by an increase of translational energy.[9,25,29] The topology of the potential surface has to be more complex than shown in figure 6.

Only at values of E_n around 1 eV dissociative chemisorption and energy transfer in direct collisions increases, indicating that the O_2^- state is probed. At these energies there is enough time for subsequent dissociation of the O_2^- formed. Clearly the high translational and possibly also vibrational energy of the initial $O_2^{(+)}$ beams discussed in this report is sufficient to overcome the barrier(s), since the O_2^- relative yield is much larger than 10^{-6}. Dissociation is not observed in the high energy experiments, because the collision is too fast for dissociation to occur along the shallow repulsive wall between the O atoms as shown in figure 6 and discussed above. It is important to note that the O-O repulsion only exists at the surface. The absence of dissociation in fast collisions is corroborated by the fact that for a system where dissociative chemisorption does occur readily, O_2 on Ni(110), again little dissociation is seen in the high energy scattering experiment.[12] So we see that the high energy experiments on harpooning at surfaces probe intermediates or precursors towards chemisorption, that in some cases may be so short lived, that they cannot be identified by other means.

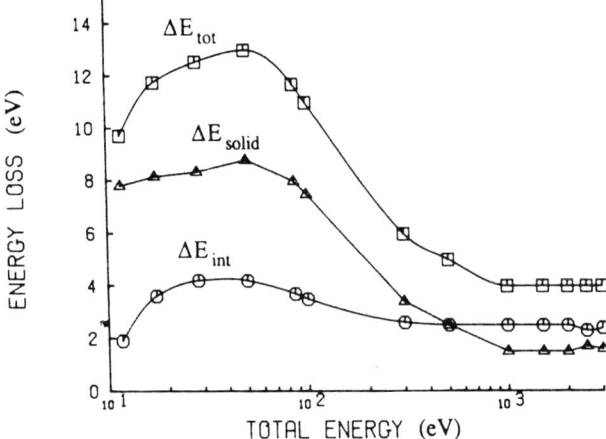

Figure 8. Total energy loss ΔE_{tot} (for non dissociated molecules), energy loss to the solid ΔE_{solid}, and energy loss to internal energy ΔE_{int} as a function of total incident energy with $E_n=11.7$ eV. From ref. 27.

We conclude that the importance of harpooning transitions in energy transfer processes at surfaces has been established experimentally and will add another dimension to these processes which traditionally are being thought of as collisions of hard spheres. We have shown that experiments carried out under grazing angles of incidence can explore chemical events happening at surfaces very well.

The work described in this progress report is the result of the work and collaboration of and stimulating discussions with Werner Heiland, Paul van den Hoek, Tom Horn, Uwe Imke, Edgar Kuipers, Joop Los, Akira Namiki, Pan Haochang, Heiner Rechtien, Paul Reijnen, Siegfried Schubert, Ken Snowdon and Marcel Spruit. Dolf de Vries is thanked for his careful reading of the manuscript. This work is part of the research program of FOM and is financially supported by NWO. Support from the Koninklijke Shell Laboratorium Amsterdam is also greatfully acknowledged.

REFERENCES

1. D.R.Herschbach, Adv.Chem.Phys. **10**, 319 (1966); "Alkali Halide Vapors, Eds. P. Davidovits and D.L. McFadden, Academic Press, New York, 1979.
2. J. Los and A.W. Kleyn, in: "Alkali Halide Vapors, Eds. P. Davidovits and D.L. McFadden, Academic Press, New York, 1979, p 275; C.W.A. Evers, Chem. Phys. **30**, 27.(1978)
3. A.W.Kleyn, J.Los and E.A.Gislason, Phys.Repts. **90**, 1 (1982).
4. E.A. Gislason and G. Parlant, Comm. At. Mol. Phys. **19**, 157.(1987).
5. J.W.Gadzuk, Comm.At.Mol.Phys. **16**, 219 (1985); S. Holloway, J. Vac. Sci. Tech. A **5**, 476 (1987).
6. D. Andersson, B. Kasemo and L. Walldén, Surf. Sci. **152/153**, 576 (1985); B. Kasemo and B.I. Lundqvist, Comm. At. Mol. Phys. **14**, 229 (1984).
7. C.Backx, C.P.M.de Groot and P.Biljoen, Surf.Sc. **104**, 300 (1981); C.T.Campbell, Surf.Sci. **173**, L641 (1986).
8. C.T.Campbell, Surf.Sci. **157**, 43 (1985).
9. M.E.M. Spruit and A.W. Kleyn, Chem. Phys. Lett. **159**, 342 (1989).
10. Pan Haochang, T.C.M.Horn and A.W.Kleyn, Phys.Rev.Lett. **57**, 3035 (1986); J. Elec. Spectr. Rel. Phen. **45**, 361 (1987); Pan Haochang, P.H.F. Reijnen, T.C.M. Horn and A.W. Kleyn, Radiation and Solids, in press.
11. P.H.F. Reijnen, A.W. Kleyn, U. Imke and K.J. Snowdon, Nucl. Instr. Meth. B33, 451 (1988).
12. S. Schubert, U. Imke, W. Heiland, K.J. Snowdon, P.H.F. Reijnen and A.W. Kleyn, Surf.Sci. Lett. **205**, L793 (1988).
13. P.H.F. Reijnen, P.J. van den Hoek, A.W. Kleyn, U. Imke and K.J. Snowdon, Surf.Sci., in press.
14. J.H. Rechtien, U. Imke, K.J. Snowdon, P.H.F. Reijnen, P.J. van den Hoek, A.W. Kleyn and A. Namiki, Surf.Sci., submitted
15. P.H.F. Reijnen and A.W. Kleyn, to be published.
16. W.J van der Zande, W. Koot, J.R. Peterson and J. Los, Chem. Phys. Lett. **140**, 175 (1987); Chem.Phys., in press.
17. K.J. Snowdon, B. Willerding and W. Heiland, Nucl. Inst. Meth. B **14**, 467 (1986); K.J. Snowdon, Nucl. Inst. Meth. B **33**, 365 (1988).
18. B. Willerding, H. Steininger, K.J. Snowdon and W. Heiland, Nucl. Instr. Meth. B **2**, 453 (1984).
19. A. Danon and A. Amirav, Phys. Rev. Lett. **61**, 2961 (1988); J. Chem. Phys. **86**, 4708 (1987); J. Phys. Chem. **93**, 5549 (1989).
20. C.T. Rettner, J. Kimman, F. Fabre, D.J. Auerbach and H. Morawitz, Surf. Sci. **192**, 107 (1987); D.M. Newns, Surf. Sci. **171**, 600 (1986); J.W. Gadzuk and S. Holloway, Phys. Rev. B **33**, 4298 (1986).
21. P.J. van den Hoek and E.J. Baerends, Surf. Sci. Lett. in press.
22. A.W. Kleyn, V.N. Khromov and J. Los, J. Chem. Phys. **72**, 5282 (1980); Chem. Phys. **52**, 65 (1980).
23. P.J. van den Hoek, T.C.M. Horn and A.W. Kleyn, Surf.Sci. Lett. **198**, L335 (1988).

24 T.C.M. Horn, Pan Haochang, P.J. van den Hoek and A.W. Kleyn, Surf.Sci. **201**, 573 (1988).
25 M.E.M. Spruit, P.J. van den Hoek, E.W. Kuipers, F.H. Geuzebroek and A.W. Kleyn,
Phys. Rev. B **39**, 3915 (1989);
Surf.Sci. **214**, 591 (1989).
26 R. Brako and D.M. Newns, Vacuum **32**, 39, (1982);
J.N.M. van Wunnik and J. Los,
Physica Scr. **T6**, 27 (1983);
J.J.C. Geerlings R. Rodink, J. Los and J.P. Gauyacq, Surf. Sci. **186**, 15 (1987).
27 P.J. van den Hoek and A.W. Kleyn, J. Chem. Phys., in press
28 M.C. Zonnevylle, R. Hoffmann, P.J. van den Hoek and R.A. van Santen, Surf. Sci. submitted.
29 M.E.M. Spruit, E.W. Kuipers, F.H. Geuzebroek and A.W. Kleyn, Surf.Sci. **215**, 421 (1989).

WAVEPACKET DYNAMICS AS A TOOL FOR CALCULATION OF AVERAGED PHOTOIONIZATION CROSS SECTIONS

William P. Reinhardt* and Daniel R. Kerner**

Department of Chemistry*, University of Pennsylvania
Philadelphia, PA 19104-6323 USA
and
Department of Chemistry**, University of Colorado and the
Joint Institute for Laboratory Astrophysics, University of Colorado and
National Institute for Standards and Technology
Boulder CO, 80309-0440 USA

Wavepackets may be used as the basis for computation of photoionization and photodetachment spectra, and have been of particular use in understanding classical-quantum correspondence when such processes take place in external fields. This follows from the fact that time propagation of an appropriate non-stationary state gives a time correlation function whose fourier transform is the desired spectrum. In this progress report we take advantage of the time-frequency fourier "uncertainly" relations to note that computation of low resolution spectra requires only short time propagation of the initial wave packet. This in turn necessitates only small L^2 expansion basies, and no specific representation of the (often infinite number of) continua representing open channels. These ideas are made specific, and it is shown how to obtain the averaged cross section with an arbitrary and prespecified averaging function. Illustrations are made to H, He and Li, where converged averaged photo-effect cross sections are seen to be easily computed, even in regions dominated by resonances.

Introduction

Atomic and molecular photo-effect spectra are usually calculated using stationary state wave functions to represent both the initial bound state, and the final continuum state[1]. In the case of photoionization of a highly stripped ion, many channels may be open, and due to the Feshbach type resonances associated with each threshold, such cross sections many be largely dominated by resonances, which, while individually narrow, may well be numerous enough to carry a large fraction of the oscillator strength. Such resonance structure is most conveniently summarized using the methods of multi-channel quantum defect theory[2] (MQDT), which necessitates theoretical calculation of a rather small number of parameters which may be thought of as parameterizing the spectrum.

However, for use in astrophysics or plasma physics, where what is wanted is a rate coefficient, rather than the beautifully detailed high resolution spectrum, cross sections are averaged, which often washes out all, or most, of the detail. In the case that tens to hundreds of open channels are present or multiple ionization is possible, considerable effort, both computational and intellectual, is required to carry out an MQDT analysis, and, if all that is wanted is a rate coefficient, one wonders as to whether a simpler and more cost effective procedure might not be available[3].

What is presented here is a method for *direct* calculation of *averaged* photoionization (or photodetachement) spectra. By *direct* is meant a method which does not produce the high resolution cross section at any intermediate stage of the computation. If the averaged cross section is a simpler and smoother function of frequency than the exact cross section we might hope that less work will be required to obtain it, as a less detailed description of the physics is required. We will see that this is indeed the case: the computational effort is, in fact, at least roughly inversely proportional to the resolution of the averaging function over a broad range of resolution. A simular situation arises in chemical rate theory, where if only an equilibrium or near equilibrium temperature dependent rate constant is desired, a Boltzmann average of the reaction cross section may be computed directly rather more efficiently than converged cross section itself[4].

How is it possible to compute an averaged cross section? For a number of years, Langhoff and co-workers have used the method of moments combined with Stieltjes imaging[5] to compute the smooth parts of atomic and molecular photo-effect cross sections. In this method, a small number of inverse power moments of the frequency dependent absorption spectrum are directly computed using only L^2 basis functions, and these power moments are used to construct a smooth oscillator strength distribution, which is often very close to the exact spectrum in the case that the latter has little structure. If all of the exact power moments are available the exact spectrum may be recovered. Thus computations using a "small" number of moments produce an averaged cross section, but the averaging function is, in fact, unknown. What is wanted here is a method which allows use of a prespecified averaging function, followed by convergence using an algorithm which directly produces the averaged cross section. Moment methods could be developed to this end; however, there is a more natural method utilizing the method of time-dependent quantum mechanics, and the familiar idea of "time-frequency" uncertainty when applied to fourier transform pairs. The organization of the remainder of this progress report is as follows: representation of the spectrum as the fourier transform of the dipole auto-correlation function & time- frequency uncertainty relations; spectral averages as convolutions illustrations for atomic hydrogen, helium and lithium; conclusions.

<u>Spectra and the Dipole Correlation Function.</u>
The usual (full resolution) photo-effect spectrum is given as

$$\sigma(\omega) \propto |<0|\mu|E>|^2, \quad (1)$$

which is the matrix element of the dipole transition operator, m, taken between the initial bound state $|0>$ and the final continuum state at energy $E = \hbar\omega + E_{bound}$, where E_{bound} is the initial energy of the state $|0>$. It is this quantity which is the object of usual computation schemes, the final state $|E>$ being obtained from scattering codes. It is computation of this final state, $|E>$, which we propose to avoid. Using standard quantum mechanical identities, Eqn(1) may be rewritten as[6], where now all constants have been included, a_0 being the Bohr radius:

$$\sigma(w) = 4\pi^2 a_0^2 \omega \, FT_\omega[<0|\mu(t).\mu(0)|>]. \quad (2)$$

where $FT_\omega[f(t)] = (2\pi)^{-1} \int \exp(i\omega t) f(t) dt$, is the usual fourier transform with respect to time, t, with the limits understood to be $-\infty$ to $+\infty$. It should be emphasized that Eqns(1) and (2) contain precisely the same physics

Computation of the spectrum, $\sigma(\omega)$, is thus equivalent to knowledge of the time dependent quantity $<0|\mu(t).\mu(0)|>$, where $\mu(t)$ is the Heisenberg form of the dipole operator, at time t. Stated in words, the spectrum is the fourier transform of the dipole auto correlation function. Physically, this relationship is what allows the experimental technique of *fourier transform spectroscopy*, wherein a system is pulsed at time t=0, and the subsequent transients of its dipole moment followed over time. In such an experiment collection of data for a short time yields a low resolution frequency spectrum, while long time data yields a high resolution spectrum, over the full frequency range. The relation between data collection time, Δt, and frequency resolution, $\Delta\omega$, is the usual transform uncertainty relation,

$$\Delta\omega \Delta t \approx 2\pi. \quad (3)$$

This simple result of fourier analysis correctly suggests that low resolution spectra should be obtained from short time dynamics. This is the topic of the following sub-section.

Before giving detailed consideration to the promised computation of average cross sections, we reinterpret the physical picture underlying the correlation function $<0|\mu(t).\mu(0)|0>$. Noting that

$$\mu(t) = e^{+iHt}\mu(0)e^{-iHt} \quad (4)$$

gives

$$<0|\mu(t)\cdot\mu(0)|0> = e^{+iEt}<\phi(0)|\phi(t)>, \quad (5)$$

where "E" is the energy of the initial state $|0>$, and acts as a fourier "shift" function, which sets the absolute frequency origin, and where

$$|\phi(0)> = \mu(0)|0>, \quad (6a)$$

and

$$|\phi(t)> = e^{-iHt}|\phi(0)>. \quad (6b)$$

If $|\phi(0)>$ thought of as an initial state, Eqn(6b) then tells us that $|\phi(t)>$ is this initial state time evolved to the later time t via the dynamics of the Hamiltonian H. The essential part of the correlation function, $<\phi(0)|\phi(t)>$, is thus interpreted as being the overlap of the initial wavepacket consisting of the initial bound state multiplied by the dipole operator, with the same packet allowed to time evolve to time t, in accord with the atomic dynamics. Computation of the correlation function thus amounts to propagation of a wave packet[7]

Two remarks are now in order: 1) As short time dynamics of wave packets is often given by classical dynamics, features in the spectrum may often be directly associated with underlying classical dynamics[8]; and, in particular, 2) oscillatory structures in spectra are associated with dynamics which causes the wave packet to leave its initial position and to then return. If such a recurrence is at time t_{recur}, the spectrum will inevitably show oscillations at frequency $2\pi/t_{recur}$ This latter observation has recently been exploited in the case of field induced oscillations in photoionization in magnetic[9], electric[10] and mixed[11] E and B fields. We now turn to the issue at hand: computation of average photoeffect spectra.

Averages as Convolutions

In what follows we assume that the desired average photoeffect cross section may be written as

$$\sigma^{ave}(G,\omega) = \int G(\omega-\omega')\sigma(\omega')d\omega' \quad (7)$$

where now G is the averaging function, and $\sigma(\omega)$ is the exact photoeffect cross section. Note that as G need not be even with respect to ω', that many types of useful averages may be written in this form. The key observation is that Eqn (7) is a convolution, and thus[12]

$$FT_t[\sigma^{ave}(G,\omega)] = FT_t[G(\omega)]FT_t[\sigma(\omega)] \quad (8a)$$

or in a usual notation, $\hat{G}(t) = FT_t[G(\omega)]$, we have

$$\hat{\sigma}^{ave}(G,t) = \hat{G}(t)\hat{\sigma}(t) \quad (8b)$$

Eqn(8) states that the time domain fourier transform of the average cross section is the product of the fourier transforms of the actual cross section and the averaging function. It is here that the role of the qualitative "uncertainty relation" of Equation (3) becomes clear: a sharp $G(\omega)$ with width $\Delta\omega$ will have transform $\hat{G}(t)$ with a time domain width $2\pi/\Delta\omega$; conversely, a broad $G(\omega)$ will have a narrow $G(t)$ which goes to zero in the (now assumed short) time $2\pi/\Delta\omega$. Eqn(8) thus explicitly tells us: if $\hat{G}(t)$ is strongly damped in time, wave packet dynamics only need be calculated up to the time when $\hat{G}(t)\hat{\sigma}(t)$ is no longer significant. Dynamics beyond this (often) very short time can have no effect whatsoever on the averaged spectrum. This idea is illustrated for some simple one, two and three electron system in the following section.

Illustrations using L^2 Basis Functions and Complex Scaled Atomic Hamiltonians.

In what follows, as the quantity for interest for computation is the correlation function $F(t)=<\phi(0)|\phi(t)>$, which is not precisely $\sigma(t)$, (due to the annoying presence of the factor "ω" preceding the FT in Eqn(2)), we first note that the cross section is

$$\sigma(\omega) = 4\pi^2 a_0^2 \, FT_\omega[2i(d/dt)\{e^{+iEt}F(t)\}] \quad (9a)$$

$$= 4\pi^2 a_0^2 FT_\omega[\hat{f}(t)] \quad (9b)$$

where $\hat{f}(t)$ is defined to be the quantity in square brackets in Eqn((9a)). In this notation,

$$\sigma^{ave}(G,\omega) = 4\pi^2 a_0^2 FT_\omega[\hat{G}(t)\hat{f}(t)] \quad (10)$$

and the average oscillator strength, which is

plotted in the following figures, is, even more simply.

$$df^{ave}(\omega)/d\omega = FT[\hat{G}(t)\hat{f}(t)]. \quad (11)$$

We are now ready to calculate average oscillator strength distributions for simple atomic systems. F(t), and thus f(t) are computed from

$$<\phi(0)|e^{-iHt}|\phi(0)>, \quad (12a)$$

which for short times may be replaced by

$$<\phi(0)|e^{-i\underline{H}t}|\phi(0)>, \quad (12b)$$

where now $\underline{H}$ is a finite dimensional matrix representation of H in an L^2 basis. Use of an L^2 basis is appropriate as the wave packet is L^2 at t = 0 and at all subsequent times.. Such an approximation can be valid for times up to $2\pi/\Delta E$ where ΔE is a typical spacing of the matrix eigenvalue discritization[13] of the electronic continuum (or continua). For times longer that this inverse level spacing artifical recurrences are induced by the discretization. Such recurrences due to continuum discretization[13] are easily damped, while those due to actual resonances and or bound states kept, by use of the method of complex scale transformations[14], which moves the eigenvalues corresponding to the discretized continua into the lower half complex energy plane, giving each an exponential damping. The desired complex scale transformation is effected by the change of variables r→rexp(iθ), where r is an interparticle distance. An atomic Hamiltonian is thus transformed as

$$H^{Atom}(\theta) = e^{-2i\theta} KE + e^{-i\theta} PE \quad (13)$$

where KE and PE are the potential and kinetic parts of the Hamiltonian, respectively[14]. The fact that wavepacket propagation is needed for only short times (provided that broad frequency averages are needed), combined with use of complex scaling, allows extraordinarily small expansions to be used.as shown in our first example.

H Atom Photoeffect at Low and Medium Resolution

In our first examples we consider the model problem of computing the average oscillator strength for the hydrogen atom. In these model computations the averaging function was taken as a Gaussian with half width W. Energies will be taken in Hartree atomic units, thus the IP of atomic hydrogen is at E = 1/2. Figures 1 and 2 show results for W = 1/8 in Hartree units (i.e. ca a 7 eV full width Gaussian slit function) which is quite broad compared to the spacings of bound levels and the distance between the first allowed transition (1s to 2p) and the onset of the continuum. The question to be answered here is can we, with a small basis, recover a converged and correct average oscillator strength distribution in a region where bound states (or resonances) carry a considerable fraction of that oscillator strength? In Figure 1, $|\hat{f}(t)|$, $\hat{G}(t)$ and the product $|\hat{G}(t)\hat{f}(t)|$ are shown. As expected $\hat{G}(t)$ is strongly damped and is essentially zero by 30 to 40 atomic time units. Thus the wave packet dynamics need only be followed for this ultrashort time: 1 atomic time unit is ca 2.4×10^{-17} sec.

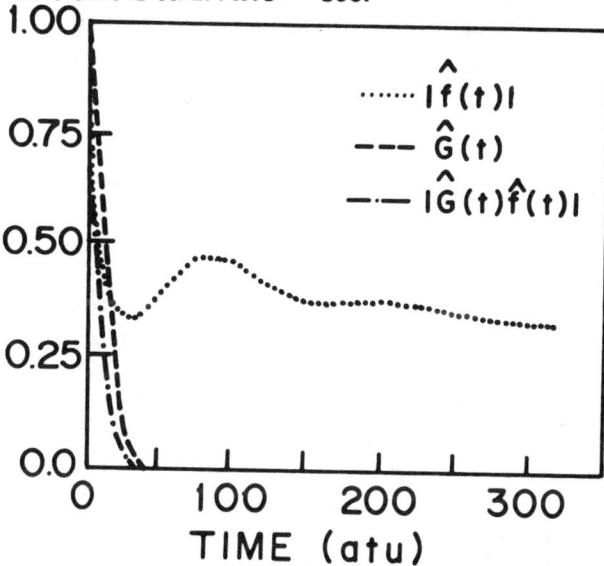

Figure 1. The functions $\hat{f}(t)$, derivative of the correlation function, $\hat{G}(t)$ the FT of the averaging function, and their product are shown. These converged results were obtained using only 4 Slater type basis functions. In this and in following figures only the absolute values of $\hat{f}(t)$ and $\hat{G}(t)\hat{f}(t)$ are shown, as these are complex functions.

Converged results with respect in increase in basis size and over a a reasonable range of rotation angle, were already obtained with four (4!) basis functions at a complex rotation angle of θ = 0.3 radians. Addition of further functions did not change the product G(t)f(t), or the average cross section shown in Figure 2.

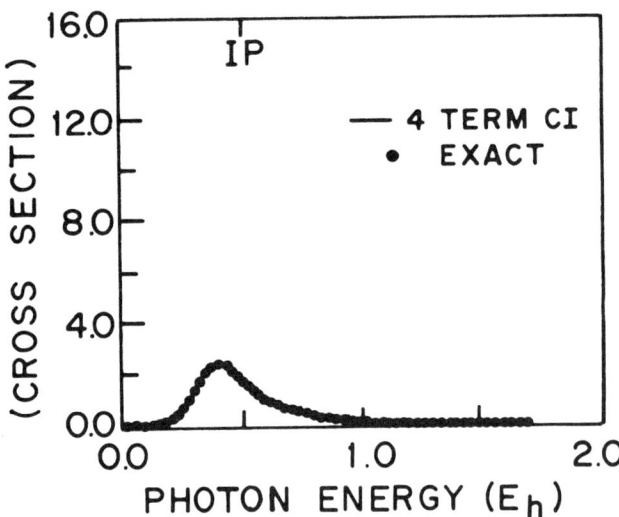

Figure 2. Averaged oscillator strength distribution for the H atom photoeffect, with a Gaussian slit function with half width of 1/8 Hartree atomic unit. The solid dots are the exact average distribution (which is proportional to the cross section), and the result of direct calculation using a 4 term basis is seen to be indistinguishable from the exact average.

As the averaging width becomes narrower, the time over which we need to propagate $\hat{f}(t)$ becomes longer, and thus the seize of the basis adequate to achieve convergence of $\hat{G}(t)\hat{f}(t)$ increases. In Figure 3 the time dependences of $\hat{f}, \hat{G}$ and $\hat{G}\hat{f}$ are shown where now G(w) is a Gaussian with width 1/64 Hartree unit. As expected from the relation of Eqn (3) the times of importance are now 8 times longer than if the case of 1/8 Hartree resolution illustrated in Figure 1.

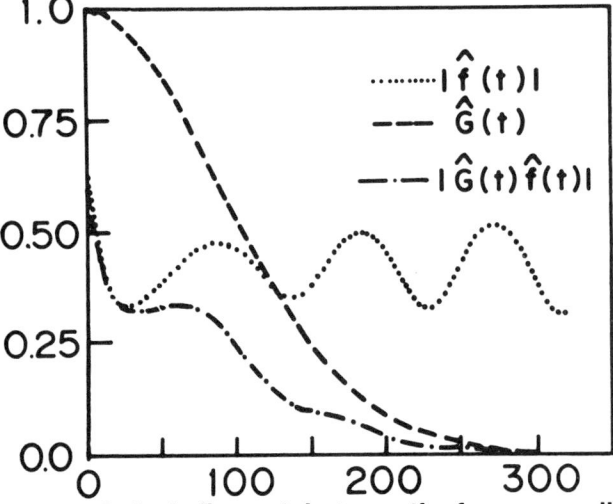

Figure 3. As in figure 1, but now the frequency slit width for averaging is 1/64 Hartree.

The corresponding average cross section, computed from the time evolution of Figure 3, which was obtained with eight (8) basis functions, is shown in figure 4.

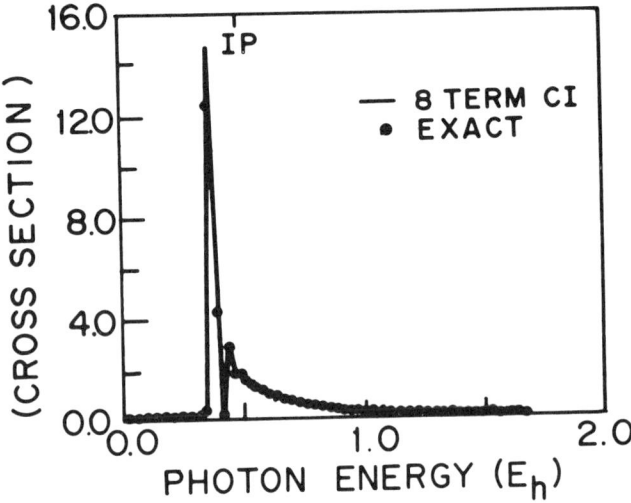

Figure 4. The average oscillator strength distribution for the photo-effect in atomic H as calculated using 8 basis functions. The desired resolution is 1/64 Hartree atomic units, and the directly calculated average cross section is not distinguishable from that calculated using the exact oscillator strength distribution suitably averaged.

It is evident from the results of Figure 4 that the concentration of oscillator strength corresponding to the bound part of the spectrum is correctly given, as is that portion corresponding to photo-ionization. This simple result suggests that correct averages over resonance dominated oscillator strength distributions may indeed be easily computed *without any of the resonance positions being explicitly determined, or indeed in the absence of a basis sufficient to accurately represent more than one or two...if any...of the individual resonances*[15].

Photoionization of Helium

Helium represents a more interesting challenge. Using relatively small configuration interaction basies can we represent the average contributions from the bound ^1P states, the 1skp continuum and from the 2s2p Feshbach resonance? Using a basis of 63 configurations, with a complex scaling of H^{Atom} of 0.3 radians (i.e. the same as for the H atom calculations!) the results of Figure 5 were obtained assuming a Gaussian averaging function of half width 0.04 Hartree atomic units.

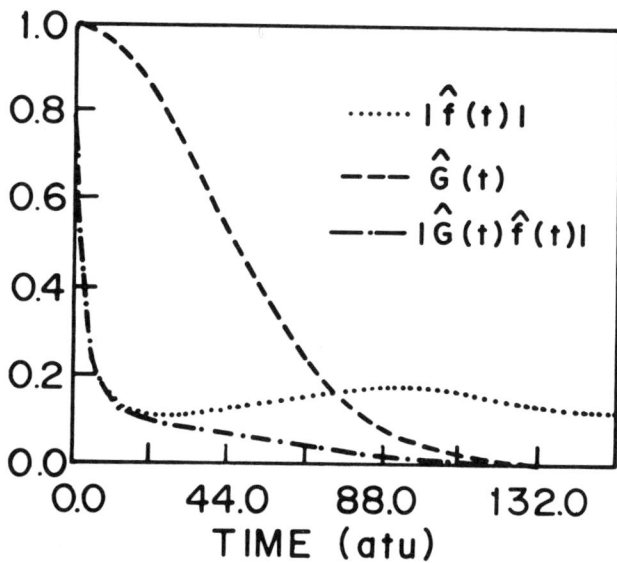

Figure 5. Time evolution of $\hat{f}(t)$, $\hat{G}(t)$ and $\hat{G}(t)\hat{f}(t)$ for the normal Helium photoeffect. Time evolution was calculated for a slit function of energy width 0.04 Hartree au, using a 63 term discretization of the two electron Hamiltonian. Transients were intentionally damped by use of complex scaling of the Hamiltonian, with a rotation angle of 0.3 rad.

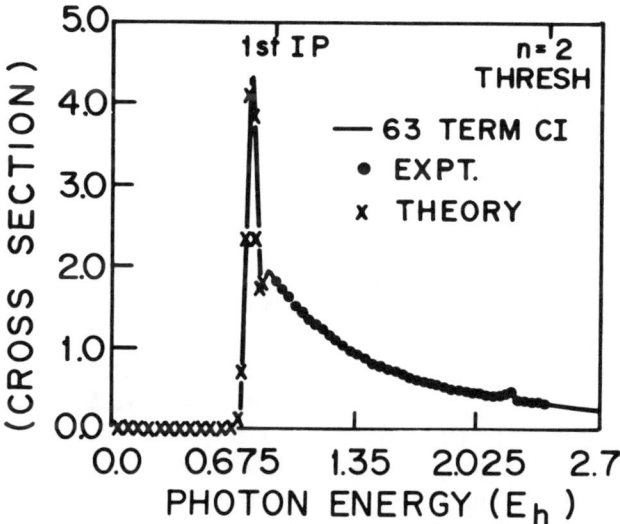

Figure 6. Oscillator strength distribution for the normal He photoeffect directly calculated at a resolution of 0.04 Hartrees. The solid curve is the present theoretical result. The xxx's were obtained by taking theoretical results of Kono and Hattori, and performing the appropriate average. The solid dots are the experimental results of Backx et. al. taken at the resolution of 0.04 Hartrees. The ability of theory to directly produce the observed low resolution spectrum is evident.

The average oscillator strength distribution is shown in Figure 6. What is evident from the results of Figure 6, is that the direct computation is able to easily reproduce the results of an experiment carried out at the same putative resolution, $\cong$ 1 eV. The averaged 2s2p $^1P^o$ Feshbach resonance is clearly seen in both the present calculation and in the experiment of Backx, Tol, Wright and Van der Wiel[16], as is the contribution from the bound transitions, as calculated by averaging the earlier (high resolution) theory of Kono and Hattori[17]. Similar results for the He isoelectronic series through O^{+6} were obtained[18].

Photoionization of Atomic Lithium

Three electron systems are far more computationally complex than two, and are thus more typical of real applications. We present here the first results[18] of the computation of the average oscillator strength distribution of atomic Lithium. In contrast to the work presented on H and He, these were calculated without the advantage of complex scaling of the Hamiltonian, and thus must be regarded as preliminary, as convergence was not attained, except for large W. The Li ^2S ground state was represented by a 176 term CI, and the ^2P spectrum by a 274 term expansion. As a measure of goodness of the initial and final state wave functions we note that 176 term expansion gave a ground state energy of -7.4763 Hartree units, to be compared with the exact non-relativistic value of -7.47807. The Thomas Reiche Kuhn sum rule was evaluated to be 2.9591, as opposed to the exact result of 3. The initial overlap $<0|\mu(0)\cdot\mu(0)|>$, was found to be 6.0511, while the exact value[19] is 6.085.

The basis expansions of the ground and ^2P spectrum thus meet at least minimal tests of goodness. The time evolution calculated using the 176 term approximate ground state and all 274 eigenfunctions of the ^2P spectrum in the calculation of exp(-i$\underline{H}$t), with no complex scaling, are shown in Figure 7. The corresponding frequency domain oscillator strength distribution is shown in Figure 8, where a qualitative comparison with experiment and prior theory can be made. The dots are the experimental results of Hudson and Carter[20] and Baker and Tomboulain[21]; the xxx's the appropriately averaged oscillator strengths of Caves and Dalgarno[22] In this case the averaging function was a Gaussian of half width 1.5 Hartree atomic units.

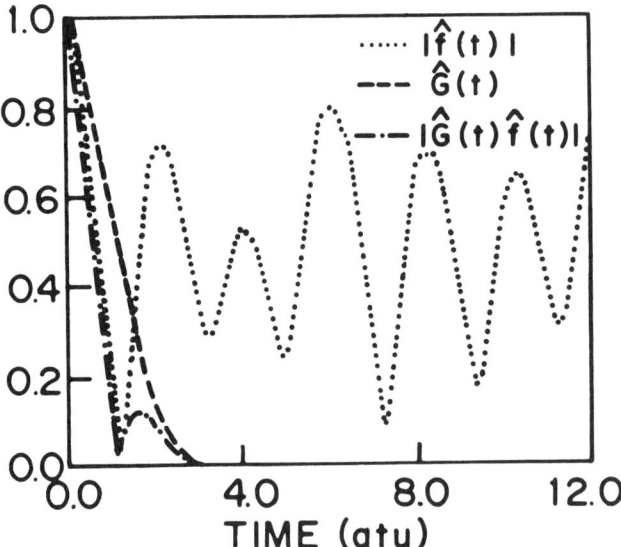

Figure 7. Time evolution of $\hat{F}(t)$, $\hat{G}(t)$ and the product $\hat{G}(t)\hat{f}(t)$ for a frequency half width of 1.5 Hartree atomic units. Much of the evident oscillatory structure in $\hat{f}(t)$ is due to the use of a finite basis, as complex scaling was not employed in this calculation. This necessitated use of the strongly damped $\hat{G}(t)$ shown. Were complex scaling used, more rapid and definitive converge would be expected.

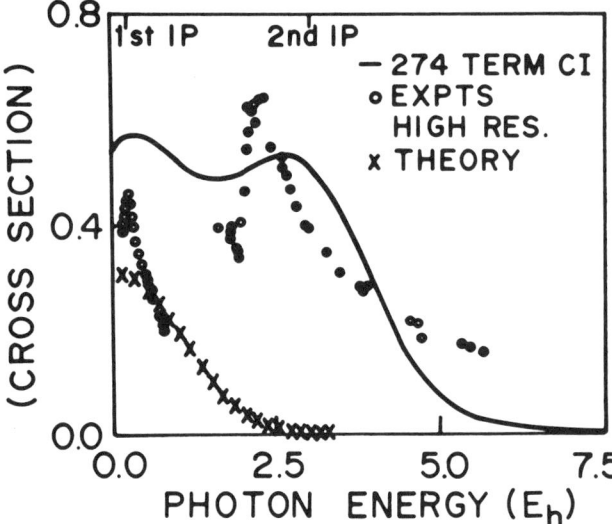

Figure 8. The solid line indicates the average oscillator strength distribution for the Li photoeffect. The distribution appears to be resonance dominated in the region just below the second IP. In this case the averaging slid width was 3 Hartrees, as necessitated by used of an unscaled Hamiltonian. Comparison with experiment, see text, is reasonable.

Summary and Discussion

In this brief Progress Report, we have focussed one a single aspect of wave packet dynamics. It is evident that short time propagation of wavepackets allows computation of appropriate correlation functions to allow direct computation of average cross sections, with a prespecified averaging function. One can expect that this should develop into a useful technique for production of rate coefficients for application in astrophysics, aeronomy and plasma physics, where the rich structures in the atomic physics are not necessarily of intrinsic interest.

It would be unfortunate, however, if limitations of time and space were to allow the impression that wavepackets were only of interest for such computations. Refs (9 through 11) give some indication of another type of application...to the spectroscopy of atoms in external fields. In both of the above cases wave packet dynamics is used to understand the results and model the dynamics of time independent experiments. However, and rather more recently, wave packets have been the subject of direct experimental observation[23] and appropriate theory is beginning to appear[24].

Acknowledgments

This work has been supported, in part, by grants from the US National Science Foundation. The suggestion, that it would be useful to be able to compute average cross sections directly if this could be easily accomplished, was made to one of the authors (WPR) by M. Seaton during an SRC sponsored visit to the Daresbury Laboratory (U.K.) and to University College London. We are most grateful for the opportunity to present these results at ICPEAC in New York City, and thank the organizers for their hospitality.

References

1) See, for example, R. J. W. Henry and L. Lipsky, Phys Rev 153,51(1967); P. G. Burke and W. D. Robb, Adv. At. Mol. Phys. 11, 143 (1975).
2) M. J. Seaton, Proc. Phys. Soc (Lon) 88,801(1966); Reports Prog. Phys. 46,167(1983).
3) M. J. Seaton, private communication.
4) W. H. Miller, S. D. Schwartz and J. W. Tromp, J. Chem. Phys. 79,4889(1983).
5) P. W. Langhoff, C. T. Corcoran, J. S. Sims, F Weinhold and R. M. Glover, Phys. Rev A 14, 1042(1976); W. P. Reinhardt, Computer Phys. Commun. 17,1(1979).
6) R. G. Gordon, Adv. Magnetic Resonance, 3, 1,(1968).
7) E. J. Heller, J. Chem. Phys. 62,1544(1975); *ibid* 68,2066,3891(1978).

8) M. J. Davis and E. J. Heller, J. Phys. Chem. 84, 1999(1980).
9) W. P. Reinhardt, J. Phys. B 16,L635(1983).
10) W. P. Reinhardt in *Atomic Exictation and Recombination in External Fields*, Eds M. H. Nayfeh and C. W. Clark, Gordon and Breach, New York, (1985), p 85.; H. C. Bryant et. al. Phys. Rev. Letts. 58,2412(1987)
11) W. P. Reinhardt, Am. Inst. Phys. Conf. Proc. 162,94(1987); C. Nessmann and W. P. Reinhardt, PHys. Rev. A 35,3269(1987).
12) P. M. Morse and H. Feshbach, *Methods of Theoretical. Phys*. McGraw-Hill, New York, 1953, p 464.
13) H. A. Yamani and W. P. Reinhardt, Phys. Rev. A11,1144(1975).
14) W. P. Reinhardt, Ann. Rev. Phys. Chem. 33,223(1982); B. R. Junker, Adv. At. Mol. Phys. 18,207(1982); Y. K. Ho, Phys. Reports 99,1 (1983).
15) The fact that information relating to oscillator strength distributions may be calculated in a bisis which cannot possibly represent the actual spectrum is at the heart of the methods discussed by A. Dalgarno and J. T. Lewis, Proc. Roy. Soc. (Lon) A233,70(1955).
16) C. Backx, R. R. Tol, G. R. Wright and M. J. Van der Wiel, J. Phys. B 8,2050(1975).
17) A. Kono and S. Hattori, Phys. Rev. A 29,2981(1984)
18) D. R. Kerner, Ph. D. Thesis, University of Colorado, unpublished, 1985 ,(JILA Thesis #128)
19) G. M. Stacey and A. Dalgarno, J. Chem. Phys. 48,2515(1968)
20) R. D. Hudson and V. L. Carter, J. Opt. Soc Am. 57,651(1967).
21) J. Baker and D. H. Tomboulian, Phys. Rev. 128,677(1962).
22) T. C. Caves and A. Dalgarno, J.Q.S.R.T. 12,1539(1972), which does not, however, include higher excitation channels.
23) J. A. Yeazell and C. R. Stroud, Jr., Phys. Rev. Letts 60,1494(1988); A. L. D. Noordam ten Wolde, A. Lagendijk and H. B. van Linden van den Heuvell, Phys. Rev. Letts. 61,2099(1988). ,
24) M. Nauenberg, Phys. Rev. A40,1133 (1989).

COLLISIONS OF RYDBERG ATOMS WITH NEUTRAL PARTICLES

V.S.Lebedev

P.N.Lebedev Physical Institute, Academy of Sciences of
the USSR, 53, Leninsky prospect, Moscow, 117924, USSR

The results of theoretical investigations of the excitation, quenching, ionization and broadening of the Rydberg atomic levels in inelastic collisions with ground-state atoms of buffer or parent gases are presented. The different physical mechanisms of these processes due to the scattering of the perturbing neutral particle on the quasifree electron and on the core of the Rydberg atom are considered. In particular the role of the potential and resonance scattering of the slow electron on the perturbing atoms in the quenching and broadening processes is discussed. The effects of the nonadiabatic transfer of the Rydberg electron energy to the kinetic energy of the colliding atoms and to the energy of the electronic shell of the quasimolecular ion are also considered. The results of calculations of the cross sections and the rate constants of the quenching, broadening and ionization processes for some physical systems are presented and are compared with available experimental data.

1. INTRODUCTION

Until recently the main part of experimental and theoretical research on collisions of Rydberg atoms with neutral particles has been devoted to the elastic and quasielastic processes. In particular the mixing of the highly excited states in the orbital angular moment $nl \to nl'$ as well as the broadening and shift of the Rydberg atomic levels at thermal collisions with ground-state atoms of the buffer gas have been actively investigated. In such processes the energy defect $\Delta \varepsilon_{if}$ between the initial ε_i and final ε_f levels of the Rydberg electron is very small or is equal to zero. The theory of these processes is now essentially complete (see Ref. 1, ch. 6, 7).

In the recent years many interesting experimental and theoretical results have been obtained for the processes of inelastic collisions of Rydberg atoms with neutral atomic particles. Such processes are of considerable interest for the atomic spectroscopy of highly excited states and for the kinetics of electronic energy relaxation in a gas or a low temperature plasma.

In this report we present the main results of the theory of inelastic transitions between the highly excited $nl \to n'l'$ (or hydrogenlike $n \to n'$) levels (1), direct (2a) and associative (2b) ionization of the Rydberg A(nl) atoms

$$A(nl) + B \longrightarrow \begin{cases} A(n'l') + B & (1) \\ A^+ + B + e & (2a) \\ BA^+ + e & (2b) \end{cases}$$

at thermal collisions with ground-state atoms of the buffer (B) or parent (A) gases. On the basis of this theory the processes of the quenching and broadening of the Rydberg atomic levels due to the inelastic collisions with perturbing atoms will be also discussed. The main part of the results presented here was obtained by the author jointly with V.S.Marchenko and was published in the papers [2-8].

The inelastic collisional processes (1,2) associated with the significant change of the Rydberg electron energy $\Delta \varepsilon_{if}$ may be defined by different physical mechanisms. For example, the ionization, quenching and broadening of the highly excited levels of the A(nl) atom may result from the potential or resonance scattering of the perturbing B atom on the quasifree electron e (see sec. 2.1-2.3). However in many cases the main role plays the competitive mechanisms due to the scattering of the perturber B on the core A^+ of the Rydberg $A(nl)$ atom (see sec. 3.1 and 3.2).

Below we consider some of the possible mechanisms of the inelastic excitation, quenching, broadening and ionization of the Rydberg levels. The analytical formulae for the cross sections and the rate constants of these processes along with the results of calculations for the various types of the colliding atoms will be presented.

2. MECHANISM OF THE PERTURBER-QUASI-FREE ELECTRON SCATTERING

2.1. Scattering length approximation

Consider first the traditional Fermi mechanism due to the potential scattering of the perturbing B atom on the quasifree electron e of the Rydberg $A(nl)$ atom. It is well known [1] that this mechanism plays the main role for the quasielastic $nl \to nl'$ transitions ($\Delta \varepsilon_{if} = 0$) between the highly excited states. The theory of inelastic $nl \to n'$ and $n \to n'$ transitions ($\Delta \varepsilon_{if} \neq 0$) was developed simultaneously by two different methods (see Ref. 5, 9, 10a). In paper 5 the perturbation theory with the Fermi pseudopotential $V = 2\pi L \delta(\vec{r} - \vec{R})$ and quasiclassical approximation were used. As a result the simple formula for the cross section of the $nl \to n'$ transition ($1 \ll n$) has been obtained

$$\sigma_{nl,n'} = \frac{4 L^2}{v^2 n'^3} \left\{ B_X(\tfrac{3}{2}, \tfrac{1}{2}) + \frac{\lambda}{2}\left[X - \ln(\frac{4}{\lambda^2 X}) \right] \right\} \quad (3)$$

Here $B_X(3/2, 1/2)$ is the incomplete beta function, $X = 1/(1 + \lambda^2/4)$, L is the scattering length, $\lambda = n|\Delta \varepsilon_{nl,n'}|/v$ is the parameter, and $\Delta \varepsilon_{nl,n'}$ is the energy defect between the initial nl-level ($\varepsilon_{nl} = -1/2(n - \delta_l)^2$, δ_l is the quantum defect) and hydrogenlike final n'-level ($\varepsilon_{n'} = -1/2 n'^2$). This equation (3) is applicable at the wide range of the $\Delta \varepsilon_{nl,n'} = |\delta_l + \Delta n|/n^3$ values (where $\Delta n = n' - n$).

For high quantum numbers $n \gg n_{max}$ the well known Omont result [11] for the cross section

$$\sigma_{nl,n'} = 2\pi L^2 / n^3 v^2 \quad (4)$$

of the quasielastic ($\Delta \varepsilon_{nl,n'} = 0$) process (1) directly follows from Eq.(3) as a special case ($\lambda \to 0$). The maximum of the inelastic cross section (3) is given by the formula

$$\sigma_{nl,n'}^{max} \approx 0.1 \, \sigma_{el} \, |\delta_l + \Delta n|^{3/2} v^{1/2} \quad (5)$$

at $n_{max} \approx 0.9(|\delta_l + \Delta n|/v)^{1/2}$ and $\lambda_{max} \approx 1.2$. Here $\sigma_{el} = 4\pi L^2$ is the total cross section for elastic electron-perturber scattering. The value of the cross section (5) is much lower than the corresponding cross section for the quasielastic $nl \to n$ transition ($\Delta \varepsilon = 0$) for the same n. At lower $n \ll n_{max}$ the decrease of the inelastic cross section of the $nl \to n'$ transitions with decreasing of the n value is described by the following expression ($\lambda \gg 1$)

$$\sigma_{nl,n'} \approx (4/3\pi) \sigma_{el} v \, n^3 / |\delta_l + \Delta n|^3 \quad (6)$$

The analytical result for the cross section of the inelastic transitions between the hydrogenlike levels $n \to n'$ is given by the formula [5]

$$\sigma_{nn'} = \frac{4 L^2}{v^2 n'^3} \left[\tan^{-1}(\frac{2}{\lambda}) - \frac{2\lambda(3\lambda^2 + 20)}{3(4 + \lambda^2)^2} \right] \quad (7)$$

where $\lambda = n|\Delta \varepsilon_{nn'}|/v \approx |\Delta n|/n^2 v$. The expressions analogous to Eq.(3) and (7) were obtained also (see Ref. 9, 10a) on the basis of the binary-encounter theory for the atomic formfactors and impulse approximation. Thus in the scattering length approximation the quasiclassical theory [5] and the binary-encounter theory [9,10a] are equivalent to each other. It should be noted also that the preliminary numerical calculations of the rate constants of the $n \to n'$ transitions based on the semiclassical theory were performed in papers [12,13].

The Eq.(3) for the cross sections of the $nl \to n'$ transitions was used [5] for the quantitative explanation of the existing experimental data on the inelastic quenching of ns-levels of Na and Rb Rydberg atoms ($\delta_s = 1.35$ and $\delta_s = 3.15$, respectively) in inert gases Ar and He. It was shown that because of the inelastic transitions with n-changing ($\Delta n \neq$

0 and $\Delta \varepsilon_{if} \neq 0$) the values of the quenching cross sections for ns-levels are considerably smaller as compared with the quasielastic quenching process of nd- and nf-levels ($\delta_d \approx 0$ and $\delta_f \approx 0$ for Na and Rb Rydberg atoms, respectively), see Fig. 1a and 1b.

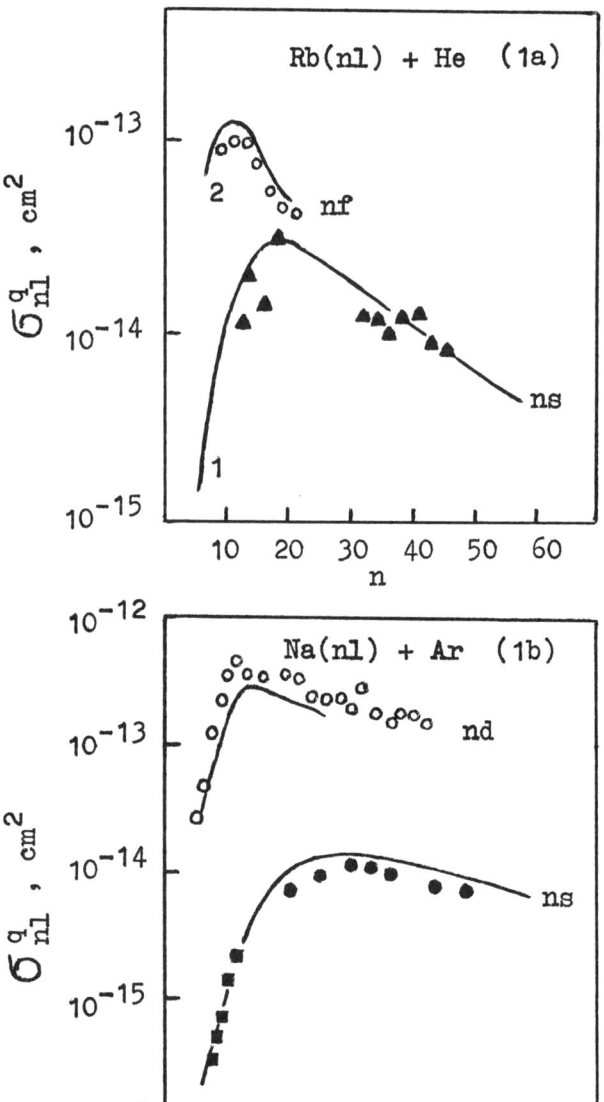

Fig.1. Cross sections σ_{nl}^q of the quenching of the Rydberg nl-levels for thermal collisions of Rb(nl) + He (1a) and Na(nl) + Ar (1b). Curves 1 are the results [5] of calculations for the inelastic quenching of the ns-levels(the main contribution is due to the transitions to the nearest energy levels ns→n', where n' = n - 3 for the Rb and n' = n-1 for the Na atoms); ▲,■, and ● are the experimental data [14-16]. Curves 2 are the results of calculations and ○ are the corresponding experimental data (see Ref. 1, ch. 6) for the quasielastic quenching of the nf-(Fig.1a) and nd-(Fig.1b) levels.

For the ionization process (2a) of the Rydberg atom by neutral atom, the theory [8] based on the impulse approximation and scattering length approximation leads to the following analytical expressions for the rate constants

$$K_{nl} = \frac{32\, L^2\, T}{\mu}\, n_{eff}\, \mathrm{EXP}\!\left(-\frac{1}{2\, n_{eff}^2\, T}\right)$$

$$K_n = \frac{2^{12}\, L^2\, T^2\, n^3}{\mu^2}\left(1 + \frac{1}{6\, n^2\, T}\right)\mathrm{EXP}\!\left(-\frac{1}{2n^2 T}\right) \tag{8}$$

Here T is the gas temperature. These equations are applicable at $v^{-1/2} \ll n \ll v^{-1}$ and are suitable for the estimation of the effectiveness of the Fermi mechanism of ionization and for the comparison with other competitive mechanisms. As follows from Eq.(8) and the results of calculation [8], the Fermi mechanism of ionization (2a) may be important for thermal energies T ~ 300-1000 K only at the range of high principal quantum numbers n (see below Fig. 5 and 7).

2.2. General expressions for inelastic cross sections

Consider now the general case when the amplitude $f_{eB}^{\beta,\beta'}(k,\theta)$ for elastic $\beta' = \beta$ or inelastic $\beta' \neq \beta$ scattering of the slow electron e by a perturbing B particle (atom or molecule) depends on the momentum k and the scattering angle θ. Here β and β' are the quantum numbers of the internal (vibrational-rotational) states of the perturbing particle with the energies ε_β and $\varepsilon_{\beta'}$. In this case the theory presented in sec. 2.1 becomes inapplicable. Thus the general formula relating the cross section $\sigma_{nl,n'}^{\beta,\beta'}$ of the inelastic nl→n' transition with the differential cross section $d\sigma_{eB}^{\beta,\beta'}/d\Omega = |f_{eB}^{\beta,\beta'}(k,\theta)|^2$ and the momentum distribution function $|G_{nl}(k)|^2$ of the Rydberg electron in the initial nl-state is required. This formula can be written in the form

$$\sigma_{nl,n'}^{\beta,\beta'} = \frac{\pi}{2^{1/2} v^2 n'^3} \int_{k_{min}}^{\infty} k^2 \, dk |G_{nl}(k)|^2 \cdot$$

$$\int_{-1}^{\nu_{max}} \frac{d(\cos\theta)}{(1-\cos\theta)^{1/2}} |f_{eB}^{\beta,\beta'}(k,\theta)|^2, \quad (9)$$

where $k_{min} = |\Delta\varepsilon_{nl,n'} + \Delta\varepsilon_{\beta\beta'}|/2v$ and $\nu_{max}(k) = 1 - 2k_{min}^2/k^2$. For the case of elastic electron-atom scattering ($\Delta\varepsilon_{\beta\beta'} = 0$) Eq.(9) was obtained in the paper [6a] within the framework of the impulse approximation. The analogous result was obtained also at $\Delta\varepsilon_{\beta\beta'} = 0$ on the basis of the semiclassical theory (see Ref. 10b). It should be noted that at asymptotic region of high $n \gg v^{-1/2}$ the result of Alekseev-Sobel'man [17] for the total cross section of the perturber-Rydberg atom scattering, directly follows from Eq.(9)(see Ref.6b).

The general formula [8] for the cross section of the ionization (2a) of the Rydberg A(nl) atom in collisions with perturbing B atom (or molecule) can be written as

$$\sigma_{nl}^{\beta,\beta'} = \frac{\pi}{V} \int_{Q_1}^{Q_2} dQ \left[Q - \frac{(|\varepsilon_{nl}| + \Delta\varepsilon_{\beta\beta'})}{V} - \frac{Q^2}{2\mu V} \right] \cdot$$

$$\int_{Q/2}^{\infty} k \, dk \, |G_{nl}(k)|^2 |f_{eB}^{\beta,\beta'}(k,Q)|^2, \quad (10)$$

$$Q_{1,2} = (2\mu E)^{1/2} \mp [2\mu(E - |\varepsilon_{nl}| - \Delta\varepsilon_{\beta\beta'})]^{1/2}$$

at the range of $v^{-1/2} \ll n \ll v^{-1}$. Here $Q = 2k\sin(\theta/2)$ is the momentum transfer and $E = \mu v^2/2$ is the kinetic energy of the colliding A(nl) and B atoms.

2.3. Resonance scattering

The general Eq.(9) may be used [6a,b] to investigate the role of the inelastic $nl \to n'$ transitions in the quenching and broadening of Rydberg atomic levels in the presence of a resonance in electron-perturber scattering. In the case of the narrow ($\Gamma_r \ll E_r$) ultra-low-energy ($E_r \sim 10^{-3}$ eV) resonances the preliminary asymptotic estimations of the broadening and shift cross sections have been obtained [18,19] on the basis of the Alekseev-Sobel'man theory.

Consider here the case of the ^{3}P-resonances on the quasidiscrete levels of the alkali-metal atoms Li, Na, K, Rb and Cs. At small energies $\varepsilon \sim 0.01-0.1$ eV the electron-atom scattering amplitude can be written as

$$|f_{eB}^{res}(k,\theta)|^2 \infty \frac{\gamma^2 \varepsilon^2 \cos^2\theta}{(\varepsilon-\varepsilon_o)^2 + \gamma^2 \varepsilon^3}, \quad (11)$$

where γ and ε_o are the ^{3}P-resonance parameters defining its energy E_r and width Γ_r. Let us substitute Eq.(11) into the general Eq.(9) and perform the integration over the scattering angle θ. Then we obtain the following result [6] for the ^{3}P-resonance contribution to the quenching σ_{nl}^q and broadening $\sigma_{nl}^{br} = \sigma_{nl}^q/2$ cross sections due to all inelastic $nl \to n'$ transitions

$$\sigma_{res}^{br}(nl) = \sum_{n'} \frac{63\pi\gamma^2}{2^{1/2} 20 v^2 n'^3} \int_{\varepsilon_{min}}^{\infty} |G_{nl}(\varepsilon)|^2$$

$$\frac{F_z(\varepsilon) \varepsilon^{5/2} d\varepsilon}{(\varepsilon-\varepsilon_o)^2 + \gamma^2 \varepsilon^3}, \quad (12)$$

where $\varepsilon = k^2/2$, $\varepsilon_{min} = |\Delta\varepsilon_{nl,n'}|^2/8v^2$, $F_z = 1 - (15Z + 20Z^3 - 12Z^5)/7$ and $Z = (\varepsilon/\varepsilon_{min})^{1/2}$.

The contribution of the potential electron-alkali atom scattering can be calculated independently on the basis of the scattering length approximation (see Eq.(3) and Ref. 6a,b).

The theory presented in sec.2.1-2.3 (see Eq.(3),(9) and (12)) has been used in papers 6a and 6b to explain available experimental data on the quenching [20] and broadening [21,22] processes of Rydberg atomic levels in alkali-metal vapours Rb(ns) + Rb, K(ns) + Rb and K(ns) + K (see Fig. 2). It was shown that at the range of $n \geq 25-30$ the dominant contribution to the cross sections of these processes is determined by inelastic $ns \to n'$ transitions to the nearest hydrogenlike n' levels ($n' = n - 2$

for K(ns) and n'=n-3 for Rb(ns) atoms; δ_s^K = 2.18 and δ_s^{Rb} = 3.15).

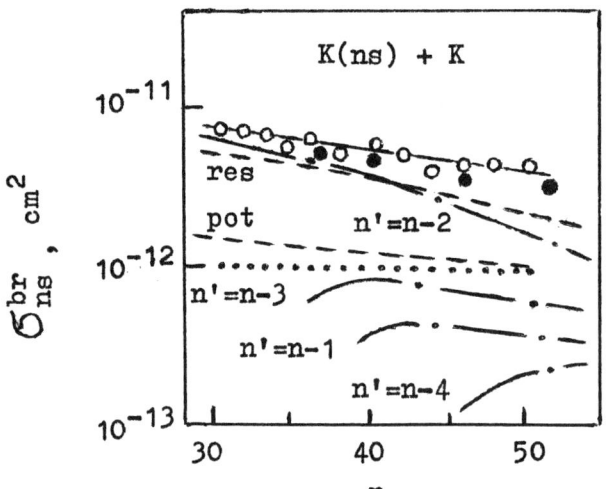

Fig.2. Cross sections σ_{ns}^{br} for the self-broadening process of K(ns)+ K(4s). Full curve is the result of calculation [6b] (at E= 510 K). Broken curves are the resonance and potential electron-perturber scattering contributions of all inelastic ns→n' transitions. Dotted curve is the elastic perturber-core scattering contribution. Dashed-dotted curves denote the contributions of the different inelastic ns→n' transitions. ● and ○ are the corresponding experimental data [21] and [22], respectively.

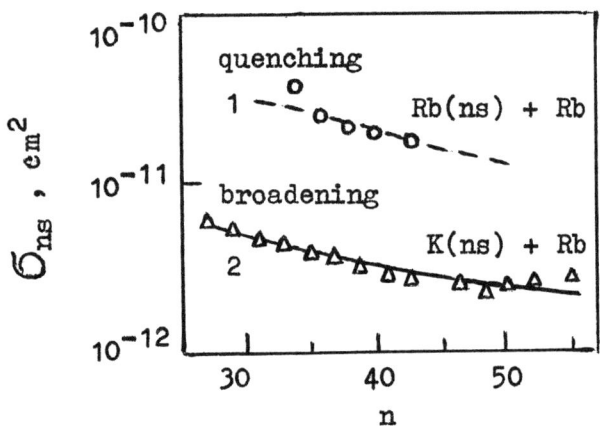

Fig.3. Cross sections for quenching σ_{ns}^q and broadening σ_{ns}^{br} of the Rydberg ns-levels of rubidium and potassium atoms at thermal collisions with Rb atoms. Curve 1 is the result of calculation [6a] of the quenching cross section for the Rb(ns)+Rb(5s) process. ○ are the experimental data [20]. Curve 2 is the broadening cross section for the K(ns)+Rb(5s) process. △ are the experimental data [21].

As follows from the results [6a,b] presented in Fig.2,3 the role of the resonance scattering in the quenching and broadening processes of the Rydberg atomic levels is greater than the potential electron-alkali atom scattering. Thus Eq.(12) may be used to determine the energies E_r and widths Γ_r of the ³P-resonances on the quasidiscrete levels of the alkali atoms from existing experimental data [20-22]. This determination must be performed only at high principal quantum numbers n ≳ 25-30, where the quasifree electron model is applicable. The ³P-resonance parameters obtained by this method from the monotonic component of the broadening cross sections of Rydberg atomic levels in alkali vapours are equal to E_r = 0.02 eV, Γ_r = 0.021 eV and E_r = 0.028 eV, Γ_r = 0.031 eV for K and Rb atoms, respectively. These values are in good agreement with Fabrikant data [23] obtained on the basis of the effective range theory. At the same time the data [24,25] essentially underestimate the E_r and Γ_r values (see [6a,b] for more details).

3. MECHANISMS OF THE PERTURBER-CORE SCATTERING

As follows from the results presented above, the cross sections of inelastic processes (1),(2) in the perturber-quasifree electron scattering mechanism decrease essentially with an increase of the transition frequency $\omega = \mathcal{E}_f - \mathcal{E}_i$. For this reason for the processes (1),(2) with large ω values the important role plays the alternative mechanisms which are due to the scattering of the perturbing B particle on the core A⁺ of the Rydberg A(nl) atom. In this case the quasimolecular BA⁺ + e system is formed with large orbital radius $r_e \sim 2n^2$ of the external electron e as compared to the internuclear R distance. The different types of the inelastic transitions in such systems can occur because of the interaction between the Rydberg electron e and the quasimolecular BA⁺ ion. In sec.3.1-3.2 we discuss the effects of the nonadiabatic transfer of the Rydberg electron energy to the kinetic energy of the nuclei or to the inner electrons energy.

3.1. Transitions within one electronic term

Consider first the processes of the excitation (deexcitation) and ionization of the Rydberg electron (1,2) due to the transitions within one electronic U(R) term of the quasimolecular BA^+ ion. In this case the direct exchange of the Rydberg electron energy with the kinetic energy of the colliding A^+ and B particles takes place. At lower frequency $\omega = |\varepsilon_f - \varepsilon_i| \ll v/b_{cap}$ the processes (1),(2) occur at large internuclear distances $b_{cap} \ll R \ll 2n^2$ (where b_{cap} and $\sigma_{cap} = \pi b_{cap}^2$ are respectively the impact parameter and cross section of the capture of the B atom by A^+ ion). In this region the inelastic $\sigma_{nn'}$ cross section and the total quenching cross section $\sigma_{nl}^q = \sum_{n'l'}' \sigma_{nl,n'l'}$ can be calculated in the shakeup model of the Rydberg electron based on the sudden perturbation theory. From this model the simple analytical expressions for quantum numbers $n \ll (\mu v/M_{A^+})^{-1/2}$ have been obtained in papers 4a, 6b

$$\sigma_{nn'} = \frac{4 g_{nn'} (\mu V / M_{A^+})^2}{3^{3/2} \pi n^5 n'^3 \omega_{nn'}^4} \sigma_{tr}^{BA^+}$$

$$\sigma_{nl}^q = (5/3)(\mu V / M_{A^+})^2 n^4 \sigma_{tr}^{BA^+} \quad (13)$$

Here $g_{nn'}$ is the Gaunt factor ($g_{nn'}=1$ in the Kramers approximation) and $\sigma_{tr}^{BA^+}$ is the transport cross section for elastic scattering of the perturbing B atom on the A^+ ion ($\sigma_{tr}^{BA^+} \approx 1.12\, \sigma^{BA^+}$ if the potential $U(R) = -C/R^4$ at $R \gg R_e$). These formulae generalize the preliminary estimations of V.A. Smirnov [26], who proposed the noninertial mechanism. The cross sections (13) increase rapidly (as n^4) with n increasing (see Fig. 4 at $n > 15$) and are the same order of magnitude as the $\sigma_{tr}^{BA^+}$ value at $n \sim n_0$ (where $n_0 \sim (\mu V/M_{A^+})^{-1/2} \approx 30-40$ for the light colliding atoms such as H(n) +He). At the same time for high principal quantum numbers n the cross sections of the $nl \to n'$ and $n \to n'$ transitions calculated in the quasifree electron model decrease rapidly (as n^{-3}) with n increasing (see Eq.(4)). For this reason at high $n \gtrsim 30-40$ the perturber-core scattering mechanism of the quenching and excitation (1) of Rydberg atomic levels is more effective than the traditional Fermi mechanism (see Fig. 4).

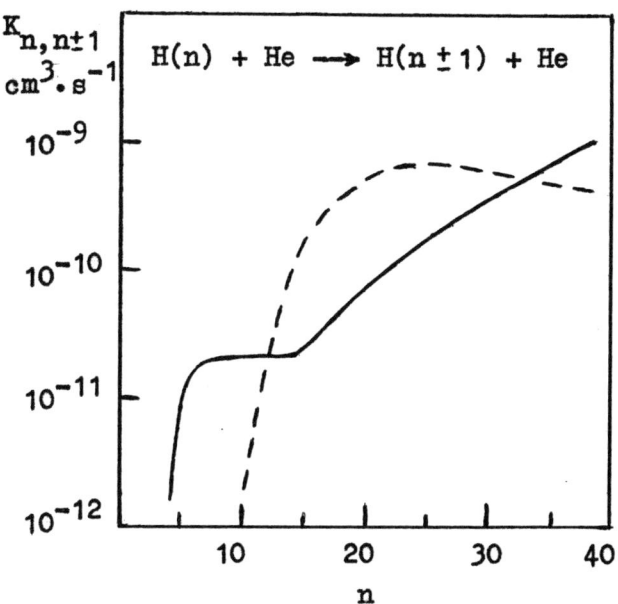

Fig.4. Rate constants $K_{n,n\pm 1}$ of the inelastic transitions $n \to n \pm 1$ in the hydrogen atom in collisions with helium atoms (T = 300 K). Full curve is the contribution of the perturber-core scattering mechanism at the range of $n \gtrsim n_0$ (results of calculation 4a). Broken curve is the contribution of the perturber-quasifree electron scattering mechanism (results of calculation 5).

At large frequency $\omega \gtrsim v/b_{cap}$ the excitation (deexcitation) and ionization processes (1) and (2) are realized at small internuclear distances $R \lesssim b_{cap} \ll 2n^2$. They are due to the interaction between the Rydberg electron e and the dipole moment D(R) of the quasimolecular BA^+ ion in the electronic state under consideration. Quantity D(R) takes into account both the contribution of the positive Coulomb A^+ center and the polarization of the electrons of the BA^+ ion in the collision of A^+ and B particles. This mechanism proposed in paper 3 is most effective, when the U(R) term has a deep well $E_0 = |U(R_e)|$ and a large value of the lower vibrational quantum ω_e (as for example HeH^+

ion). To calculate the cross sections and the rate constants of the processes (1) and (2), the stationary perturbation theory with dipole interaction $V = -\vec{D}(R)\cdot\vec{r}/r^3$ and the quasiclassical model for electronic and nuclear matrix elements was used (see Ref.4,8). The analytical expressions for the cross section of inelastic $n \rightarrow n'$ transition[4] and for the total cross section $\sigma_n^{ion} = \sigma_n^{d.i} + \sigma_n^{a.i}$ of direct (2a) and associative (2b) ionization[8] can be written in the following form

$$\sigma_{nn'} = \frac{8\pi\, d^2(\omega)}{3^{3/2}\, n^5\, n'^3\, \omega^2} \mathrm{EXP}\left(-\frac{2\omega}{\omega_e}\right)\sigma_{cap}^{BA^+},$$

$$\sigma_n^{ion} = \frac{16\pi\, d^2(|\varepsilon_n|)}{3^{3/2}\, n^3} E_2\left(-\frac{2|\varepsilon_n|}{\omega_e}\right)\sigma_{cap}^{BA^+}.$$

(14)

Here $E_2(Z)$ is the exponential integral function of the second order, $\omega = |\Delta n|/n^3$, $|\varepsilon_n| = 1/2\, n^2$. Quantity $d(\omega) \sim (dD/dR)_{R_\omega}\cdot \Delta R_\omega$ is determined by increment of the dipole moment of the BA^+ ion over the characteristic $\Delta R_\omega = |U'(R_\omega)/U''(R_\omega)|$ length of the $U(R)$ term in the vicinity of the distance R_ω, which gives the main contribution to the transition with frequency ω (see Ref. 4)

Expressions (14) take into account the large increase of the relative velocity $(v(R) = (2|U(R)|/\mu)^{1/2} \gg v(R=\infty)$, where $v(R=\infty) \equiv v = (2E/\mu)^{1/2}$) of the colliding A^+ and B particles in the deep potential $U(R)$. Due to this effect at the range of $\omega \gtrsim v/b_{cap}$ (i.e. $n \lesssim 15$ for $H(n)+He$) the behaviour of the cross sections of inelastic $n \rightarrow n'$ transitions (1) is changed radically as compared to the shakeup region $\omega \ll v/b_{cap}$ (see Fig.4). In particular in the mechanism considered the decrease of the $\sigma_{n,n-1}$ values at the range of $5 \lesssim n \lesssim 15$ does not occur. The strong decrease of the inelastic cross sections $\sigma_{n,n\pm 1} \sim \exp(-2\omega/\omega_e)$ takes place only at the adiabatic region $\omega \gg \omega_e$ (i.e. $n \lesssim 5$).

For this reason at $n \lesssim 10$ the perturber-core scattering mechanism under consideration plays the main role for the inelastic excitation and deexcitation $n \rightarrow n \pm \Delta n$ of the Rydberg atomic levels of hydrogen $H(n)$ in the buffer helium gas (see Fig.4).

The analogous calculations of the total rate constants $K_n^{ion}(T)$ of direct (2a) and associative (2b) ionization for thermal collisions of $H(n)$ and He atoms have been performed in paper [8]. As it may be seen from Fig.5 the values $K_n^{ion}(T)$ in the mechanism considered (full curve) do not depend practically on the gas temperature T and have the maximum $\sim 10^{-12}$ cm$^3\cdot$s^{-1}. Thus for the ionization process (2) of the Rydberg atoms ($H(n) + He$) the perturber-core scattering mechanism is more effective than the traditional Fermi mechanism at the range of $n \lesssim 20$-30.

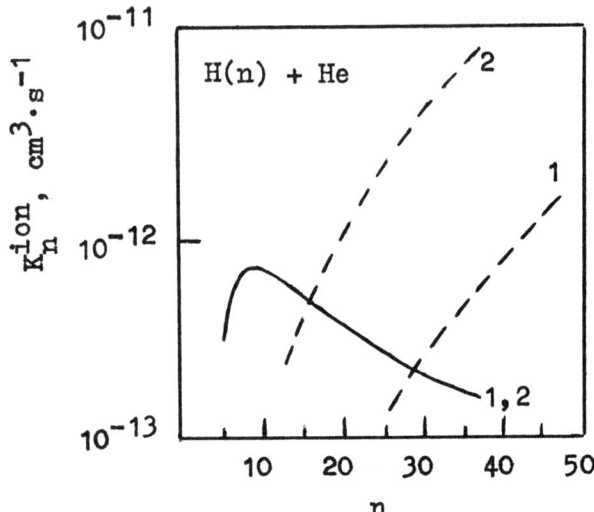

Fig.5. Total rate constants [8] of direct (2a) and associative (2b) ionization of Rydberg $H(n)$ atoms in collisions with He atoms at $T = 300$ K and $T = 1000$ K (curves 1 and 2, respectively). Full curve is the contribution of the perturber-core scattering mechanism. Broken curves correspond to the Fermi mechanism of ionization (see Eq.(8)).

3.2. Transitions between two different electronic terms.

In contrast to the mechanism discussed above the excitation (deexcitation) and ionization processes (1) and (2) may be effected by the energy exchange between the external and inner electrons of the quasimolecular $BA^+ + e$ system. In particular it is well known [27-29] that the ionization of the Rydberg alkali-metal $A(nl)$ atoms by the ground-state A atoms of the parent gas is due to the dipole transitions between the symmetrical and antisymmetrical electronic terms of the homonuclear quasimolecular A_2^+ ion. Recently for thermal collisions of different inert gas $A[n_op^5(^2P_{3/2})\,nl]$ and $B(^1S_o)$ atoms (such as $Xe[5p^5(^2P_{3/2})\,nl]$ and He, Ne, Ar, and Kr), a new effective mechanism of the quenching [7] and ionization [8] of Rydberg levels has been proposed. In this mechanism the the transition between two splitted $U_{j\Omega}(R)$ and $U_{j\Omega'}(R)$ terms of the quasimolecular BA^+ ion with different $\Omega = 3/2$ and $\Omega' = 1/2$ projections of the total electronic angular moment $j = 3/2$ on the internuclear $\vec{R}$ axis occurs. This transition exists mainly at the crossing R_ω point of the electronic terms $U_i(R) = U_{j\Omega}(R) + \mathcal{E}_i$ and $U_f(R) = U_{j\Omega'}(R) + \mathcal{E}_f$ of the compound $BA^+ + e$ system (where $\mathcal{E}_{i,f}$ is the energy of the external electron e in the discrete $\mathcal{E}_n = -1/2n^2$ or continuous $\mathcal{E} = k^2/2$ spectra). The excitation (1) and ionization (2) processes are due to the Coulomb interaction $V = \sum_x |\vec{r}_e - \vec{r}_x|^{-1}$ between the external electron $(\vec{r}_e)$ and inner electrons $(\vec{r}_x)$ (or vacancy $(\vec{r}_v)$) from the $n_op^5(^2P_{3/2})$-shell of the atomic A^+ core. The inner electrons from the ^1S_o closed shell of the inert gas perturbing B atom do not take part in the processes under consideration. The main role in this mechanism plays the quadrupole and short range parts of the Coulomb interaction V. This is due to the selection rules ($\Delta j = 0$; $\Delta \Omega = 1$ and $\Delta l = 0, \pm 2$; $\Delta m = -1$) in such $|j\Omega, nlm\rangle \longrightarrow |j\Omega', n'l'm'\rangle$ transitions.

Within the framework of the perturbation theory and quasiclassical approximation for the electronic and nuclear matrix elements the analytical expressions for the cross sections and the rate constants of the processes (1) and (2) have been obtained in the papers [7,8]. The simple analytical formula for the cross section of the inelastic $n \to n'$ transition is given by [7]

$$\sigma_{nn'} = \frac{8\pi^2 \gamma_p R_\omega^2 \,[1 - U_{j\Omega}(R_\omega)/E]^{1/2}}{25\, V\, n^5\, n'^3\, \Delta F(R_\omega)} \quad (15)$$

Here $\Delta F(R_\omega) = |dU_i/dR - dU_f/dR|_{R_\omega}$ is the difference of the derivatives of the Rydberg electronic terms (see Ref.2) of the $BA^+ + e$ quasimolecule at the crossing point R_ω given by the relation $\omega \equiv \Delta \mathcal{E}_{nn'} = \Delta U_{j\Omega, j\Omega'}(R_\omega)$, and

$$\gamma_p = (2^7/25\cdot 3^5)\left[\langle r_v^2\rangle_{n_op} + (3/14)\langle r_v^4\rangle_{n_op}\right]^2$$

is the parameter of the Rydberg electron-core interaction, which is defined by the multipole momenta $\langle r_v^s\rangle_{n_op}$ of vacancy over the Hartree-Fock radial wave functions of the n_op^5-shell (see Ref.7).

The analogous result for the total rate constant of direct (2a) and associative (2b) ionization of the Rydberg inert gas $A[n_op^5(^2P_{3/2})\,nl]$ atoms in collisions with the $B(^1S_o)$ atoms of the buffer inert gas can be written as [8]

$$K_n^{ion} = \frac{8\pi^2 \gamma_p}{25\, n^5}\int_0^{R_n} EXP\left[-\frac{U_{j\Omega}(R)}{T}\right]\Phi\left[\frac{U_{j\Omega'}(R)}{T}\right]R^2 dR \quad (16)$$

Here $\Delta U_{j\Omega, j\Omega'}(R_n) = 1/2n^2$; function $\Phi(Z) = 1$ at $Z > 0$ and $\Phi(Z) = (2/\pi^{1/2})\Gamma(3/2, Z)$ at $Z \leq 0$ (where $\Gamma(3/2, Z)$ is the incomplete gamma-function).

Discuss the results of calculations [7,8] for the quenching and ionization processes of the Rydberg xenon atoms at thermal collisions with He and Kr atoms. The rate constants $K_{nl}^q(T)$ and $K_{nl}^{ion}(T)$ with given values of the principal n and orbital l quantum numbers have a maximum for np-levels (l=1) and strongly decrease with l increasing. The corresponding values of the ionization rate constants are equal to $K_{np}^{ion} \sim 10^{-11} - 10^{-10}$ $cm^3 \cdot s^{-1}$ at the range of $n \sim 15-30$ and $T = 300-1000$ K (see Ref.8). For the case of equally populated of the Rydberg nlm sublevels at a given principal quantum number n the quenching and ionization rate constants are equal to $K_n^q \sim 10^{-10} - 10^{-11}$ $cm^3 \cdot s^{-1}$ at $n \sim 5-15$ (see Fig.6) and $K_n^{ion} \sim 10^{-12} - 10^{-13}$ $cm^3 \cdot s^{-1}$ at $n \sim 15-30$ (see Fig.7). These values exceed considerably the corresponding rate constants calculated on the basis of the quasifree electron model (see broken curves on Fig.6,7). Thus at the wide range of the principal quantum numbers n (most interesting for the experimental applications) the mechanism under consideration plays the main role in the quenching and ionization of Rydberg atoms in the buffer inert gas.

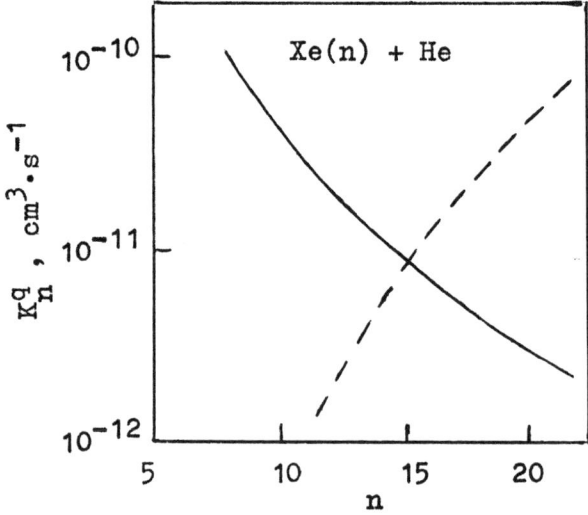

Fig.6. The rate constants [7] of the inelastic quenching (n → n - Δn, where Δn = 0,1,2,...) of the Rydberg levels of $Xe[5p^5(^2P_{3/2})n]$ atoms in the buffer helium gas (T = 300 K). Full curve is the contribution of the perturber-core scattering mechanism. Broken curve is the result of calculation in the quasifree electron model.

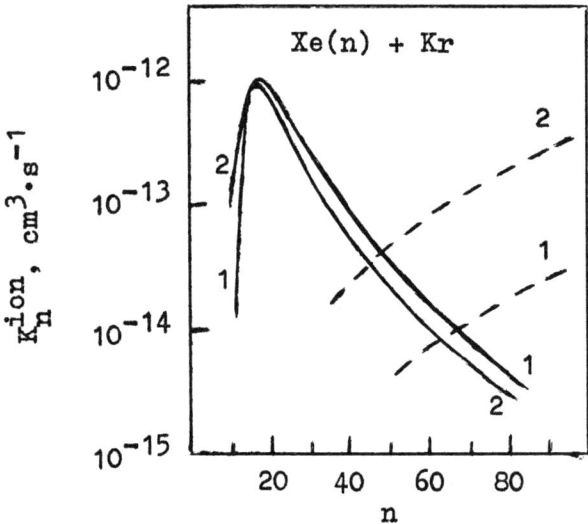

Fig.7. The rate constants [8] for the ionization of the Rydberg $Xe[5p^5(^2P_{3/2})n]$ atoms at thermal collisions with Kr atoms (curves 1 and 2 correspond to T= 300 K and T = 1000 K, respectively).

Full curves correspond to the perturber-core scattering mechanism, see Eq.(16). Broken curves are the results of calculations [8] in the quasifree electron model, see Eq.(8).

4. CONCLUSIONS

As follows from the results presented above the theoretical analysis of the inelastic collisions of the Rydberg atoms with neutral particles is more complicated problem in comparison with the elastic and quasielastic collisions. There are many competitive mechanisms which can play a significant role in the processes under consideration. The contributions of these mechanisms to the total cross sections of the processes (1),(2) depend on the type of colliding A(nl) and B atoms, their reduced mass μ and relative velocity v, the principal quantum number n, the orbital moment l and also on the transition frequency ω of the Rydberg electron.

REFERANCES

1. Rydberg States of Atoms and Molecules, ed. by R.F.Stebbings and F.B. Dunning (Cambridge University, Cambridge, New York, 1983)
2. L.I.Gudzenko and V.S.Lebedev, Proc. P.N.Lebedev Phys. Inst. Acad. of Sci. of the USSR, 120, 107 (1980)
3. V.S.Lebedev, V.S.Marchenko and S.I. Yakovlenko, Izv. Acad. of Sci. of the USSR, ser. Phys. 45, 2395(1981)
4. V.S.Lebedev and V.S.Marchenko, (a) Zh. Eksp. Teor. Fiz. 84, 1623(1983) [Sov. Phys.-JETP, 57, 946 (1983)]; (b) Khim. Fiz. 3, 210 (1984)[Sov. J. Chem. Phys. 3, 311 (1985)]; (c) J. Sov. Laser Research, 7, 489 (1986)
5. V.S.Lebedev and V.S.Marchenko, Zh. Eksp. Teor. Fiz. 88, 754 (1985) [Sov. Phys.-JETP, 61, 443 (1985)]
6. V.S.Lebedev and V.S.Marchenko, (a) Zh. Eksp. Teor. Fiz. 91, 428 (1986) [Sov. Phys.-JETP, 64, 251 (1986)]; (b) J. Phys. B, 20, 6041 (1987)
7. V.A.Ivanov, V.S.Lebedev and V.S. Marchenko, Pis. Zh. Tekh. Fiz. 14, 1575 (1988); Zh. Eksp. Teor. Fiz. 94, 86 (1988)
8. V.S.Lebedev, Preprint P.N.Lebedev Phys. Inst. Acad. of Sci. of the USSR, Moscow, № 150, p.1-56 (1989)
9. L.Petitjean and F.Gounand, Phys. Rev. A, 30, 2946 (1984)
10. Kaulakys, (a) J. Phys. B, 18, L167 (1985); (b) Zh. Eksp. Teor. Fiz.

$\underline{91}$, 391 (1986)
11. A.Omont, J. de Physique, $\underline{38}$, 1343 (1977)
12. D.R.Bates and S.P.Khare, Proc. Phys. Soc. $\underline{85}$, 231 (1965)
13. M.R.Flannery, Ann. Phys. $\underline{61}$, 465 (1970)
14. M.Hugon, B.Sayer, P.R.Fournier and F.Gounand, J.Phys. B. $\underline{15}$, 2391 (1982)
15. T.F.Gallagher and W.E.Cooke, Phys. Rev. A. $\underline{19}$, 2161 (1979)
16. J.Boulmer, J.F.Delpech, J.C.Gauthier and K.Safinya, J. Phys. B. $\underline{14}$, 4577 (1981)
17. V.A.Alekseev and I.I.Sobel'man, Zh. Eksp. Teor. Fiz. $\underline{49}$, 1274 (1965) [Sov. Phys.-JETP, $\underline{22}$, 882 (1966)]
18. M.Matsuzawa, J. Phys. B. $\underline{10}$, 1543 (1977)
19. B.P.Kaulakys, L.P.Presnyakov and P.D.Serapinas, Pis. Zh. Eksp. Teor. Fiz. $\underline{30}$, 60 (1979) JETP Lett. $\underline{30}$, 53 (1980)
20. M.Hugon, F.Gounand, P.R.Fournier and J.Berlande, J. Phys. B. $\underline{16}$, 2531 (1983)
21. H.Heinke, J.Lawrenz, K.Niemax and K.H.Weber, Z. Phys. A. $\underline{312}$, 329 (1983)
22. D.C.Thompson, E.Weinberger, G.X.Xu and B.P.Stoicheff, Phys. Rev. A, $\underline{35}$, 690 (1987)
23. I.I.Fabrikant, Opt. Spectrosk. $\underline{53}$, 223 (1982); J. Phys. B, $\underline{19}$, 1527 (1986)
24. B.P.Kaulakys, J. Phys. B, $\underline{15}$, L719 (1982)
25. Y.Rabin and F.Rebentrost, Opt. Commun. $\underline{40}$, 257 (1982)
26. V.A.Smirnov, Opt. Spectrosk. $\underline{37}$, 407 (1974)
27. A.Z.Devdariani, A.N.Klyucharev, A.V.Lazarenko and V.A.Sheverev, Pis. Zh. Tekh. Fiz. $\underline{4}$, 1013 (1978)
28. R.K.Janev and A.A.Mihajlov, Phys. Rev. A. $\underline{21}$, 819 (1980)
29. E.L.Duman and I.P.Shmatov, Zh. Eksp. Teor. Fiz. $\underline{78}$, 2116 (1980)

COLLISIONALLY INDUCED STOCHASTIC DYNAMICS OF FAST IONS IN SOLIDS

Joachim Burgdörfer

Dept. of Physics, University of Tennessee, Knoxville, TN 37996-1200, U.S.A.
and
Oak Ridge National Laboratory, Oak Ridge, TN 37831-6377, U.S.A.

Recent developments in the theory of excited state formation in collisions of fast highly charged ions with solids are reviewed. We discuss a classical transport theory employing Monte-Carlo sampling of solutions of a microscopic Langevin equation. Dynamical screening by the dielectric medium as well as multiple collisions are incorporated through the drift and stochastic forces in the Langevin equation. The close relationship between the extrinsically stochastic dynamics described by the Langevin equation and the intrinsic stochasticity in chaotic nonlinear dynamical systems is stressed. Comparison with experimental data and possible modification by quantum corrections are discussed.

I. INTRODUCTION

Ever since the seminal work by Bohr and Lindhard[1] on the evolution of the charge state and excitation state of fast atomic particles penetrating solids the investigation of the existence and the nature of the electronic excitation spectrum of swift ions has continued to draw considerable interest. A complex array of multiple scattering processes produces a variety of excited configurations not easily accessible by other means (e.g. photon excitation is limited by stringent selection rules). Despite the extensive application of the ion-solid interaction as a spectroscopic tool, a microscopic understanding of the dynamics of excitation process and of the evolution and transport of electrons accompanying fast ions is still rather poor. Many fundamental questions concerning the existence, the modes of formation, and the lifetimes of excited states in the dense medium are still left unanswered. The difficulties can be attributed to the complexity of the interaction process and to the fact that perturbations of excited states are sufficiently strong to preclude any perturbational treatment.

A review over recent experimental developments has been given by Chetioui[2] at the XV ICPEAC in Brighton. The probably most puzzling experimental observation is the abundant production of high-Rydberg[2-8] and low-energy continuum states[9] around fast highly charged projectiles penetrating thin foils. In particular, the observation of Rydberg states in high angular momentum states $\ell \gg 1$ is in sharp contrast to well known ion-atom collision processes which under otherwise similar conditions (collision velocity $v_p \gg 1$, charge state $q \gg 1$ with $q/v_p \simeq 1$) strongly favor low ℓ states. These findings, among many others, have posed an apparent paradox: On one hand, the formation of these states, observed only after the projectile has exited the solid, inside the medium is unlikely in view of the fact the characteristic radii $\langle r \rangle_{n\ell}$ exceeds the typical nearest neighbor spacing, d, in a solid[10] (Fig. 1). On the other hand, formation near or outside the exit surface should closely resemble single collisions (the "last-layer" hypothesis[11]) which, in turn, cannot account for the abundance of high-ℓ Rydberg states.

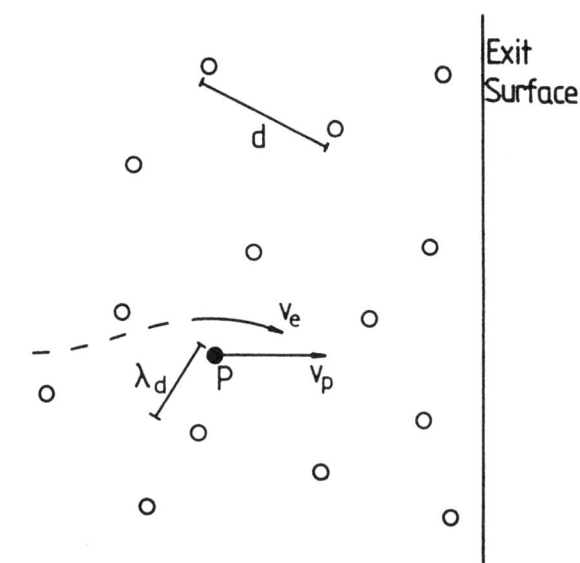

Fig. 1. Propagation of an electron associated the projectile P through the solid; d: nearest neighbor spacing, λ_D: dynamical screening length.

It has become clear that a resolution of this paradox requires a detailed study of the dynamical evolution of electronic states around swift ions inside the solid taking note of the fact that the observed final-state distributions bear some memory of the evolution in the bulk. More precisely, the final-state distribution is the result of the complex dynamical evolution of electronic charge cloud associated with the swift ion. Clearly, the transient wavepacket formed inside the solid has little in common with stationary atomic (or ionic) states. The build-up of a closely phase-space correlated, approximately isotachic electron flow in the proximity of the projectile is, however, a precursor event for the formation of post-foil Rydberg and convoy electron states.

A theoretical description is complicated by the intrinsic complexity of the underlying interaction processes. Viewing the evolution in terms of transient projectile-centered states one is faced with the problem of strong perturbations, both time-dependent and time-independent which preclude any perturbational treatment. Time-dependent perturbations result from multiple collisions of the electrons with target atoms in the solid. Time-independent perturbations are due to the fact that the interaction potential of the electron in the field of the swift projectile ion is not a pure Coulomb potential (or a Coulomb potential plus modifications by the ionic core for projectiles with partially filled inner shells at small distances) but completely altered at intermediate and large distances by the dynamical screening in the medium (the "wake field"[12]). In the following brief review we outline a classical transport theory employing stochastic dynamics. Even though highly desirable, the treatment of the evolution inside the solid in terms of a quantum mechanical transport theory appears to be a formidable task in view of the large number of coupled states, including those in the continuum, and of the nontrivial choice of an appropriate basis set that represents strongly perturbed atomic states in a dynamically screened projectile field. It is not at all obvious within which basis set of manageable size the evolution of the transient wavepacket can be represented with sufficient accuracy. The application of classical dynamics to Rydberg states relies on the limit of large quantum numbers (n, ℓ). The overall remarkably successful description of microwave ionization by classical dynamics[13] attests to its validity. However, noticeable deviations indicated that the applicability of the correspondence principle requires much more subtle considerations, in particular, concerning the density of dynamically accessible states[14]. Corrections due to quantum effects are therefore still an important and a largely open question.

Our approach is closely related to classical chaotic dynamics in Hamiltonian systems. The extrinsic stochasticity introduced via random forces into the Langevin equation mimics, under certain conditions, the intrinsic stochasticity in a chaotic system. Moreover, the non-separable wake potential induces irregular motion chaos even in absence of explicitly stochastic forces. Many tools developed in the context of classical and quantum chaos can be used to quantitatively characterize the electron transport problem. We will highlight the interplay of extrinsic and intrinsic stochasticity on the final-state distribution of ion-solid collisions.

II. PRIMARY EXCITATION: INNER SHELL PROCESSES AND THE SOURCE TERM FOR THE TRANSPORT PROBLEM

The excitation of highly charged ions in ion-solid collisions have been traditionally divided into two different regimes[2]: excitation of core states and excitation of highly excited states. While the dividing lines is far from being sharp, a qualitative distinction can be made according to the size of the orbit and, correspondingly, to its binding energy. Low-lying states with radii $\langle r \rangle_{n\ell} \ll d$ should closely resemble isolated atomic (or ionic) states. Likewise, the production of low-lying projectile states involving inner shells from both the target and projectile should resemble ion-atom collisions in the gas phase under otherwise identical conditions since on this length (or energy) scale solid state effect should be unimportant. The evolution of charge state and excited-state distributions in ion-solid collisions can therefore be described by rate equation models[15-17] (i.e., discrete master equations) with only a modest number of states included. The transition rates (W_{ij}) in the rate equation for the probability distribution P,

$$v_p \frac{dP_i}{dZ} = \sum_j (W_{ij} P_j - W_{ji} P_i) \qquad (1)$$

are assumed to be given in terms of atomic cross sections $\sigma_{j,i}$,

$$W_{ij} = v_p\, n\, \sigma_{j,i} \qquad (2)$$

where n is the effective target density. The channel indices (ij) in (1) refer to either different excited states of a given charge state or different charge states. Accordingly, $\sigma_{j,i}$ refer to the corresponding cross sections for excitation, ionization, electron capture ect. In Eq. (1) we have employed a straight-line constant velocity trajectory $Z = v_p t$ for the projectile neglecting energy loss and angular straggling in view of the small electron/projectile mass ratio. For the overall charge-state distributions as well as low-lying state populations this approach has proven to be remarkably successful. Limitations, however, have become apparent. Chetioui et al[2] have reported on a perturbation of the ℓ distribution predicted by (1) due to an effective field produced by the dynamical screening charge ("wake"). These post-collision-interaction (PCI) effects[18] can become important when long-range interactions persist on a time scale much larger than the collision time t_c of the primary collision event. Primary collision processes are characterized by a violent energy momentum transfer which require a close collision. The post-collision interaction time t_{PCI}, on the other hand, is determined by the range of the interaction and by the lifetime of the excited state which is limited by collisional and radiative destruction of the state. PCI effects have also been observed in experiments of resonant coherent excitation of channeled ions.[19]

For the formation of highly excited states validity of the master equation (1) breaks down. Not only atomic cross sections but also the states themselves are no longer well defined. A different description is needed to describe the evolution of the charge cloud in that region of the phase space. However, the primary excitation event involving a close collision associated with a large energy transfer is essentially unchanged. It is the subsequent evolution in the aftermath of this collision which is strongly influenced by condensed matter and collective effects. The primary excitation processes which include direct (projectile) excitation as well as charge transfer can be viewed as a source term for a charge cloud closely phase space correlated with the projectile ion which is the subject to modification due to a complex transport behavior.

III. CLASSICAL TRANSPORT THEORY

Within the framework of classical dynamics the electron transport problem is described by a phase-space master equation for the distribution function ρ (in $a.u.$),

$$v_p \frac{\partial}{\partial Z}\rho(\vec{r}, \vec{v}, Z) = (\hat{L} + \hat{R})\rho(\vec{r}, \vec{v}, Z) \quad (3)$$

where the classical Liouville operator

$$\hat{L} = -\vec{v} \cdot \vec{\nabla}_r + \vec{\nabla} V_p \cdot \vec{\nabla}_v \quad (4)$$

describes the phase space flow ("drift") due to the effective electron-projectile interaction V_p and $\hat{R}$ is the collision (integral) operator describing stochastic collisions of the electron in the solid. The phase-space coordinates $(\vec{r}, \vec{v})$ refer to the projectile frame. We use in (3) a one-electron approximation since the probability of producing a highly excited projectile state is small ($\lesssim 10^{-3}$) so that the simultaneous presence of a second electron in this phase space region is negligible. Electrons of the medium are well separated in phase space from the "active" electron since $v_p \gg v_F$ (v_F: Fermi velocity). The collision integral (relaxation operator) is given by

$$\hat{R}\rho = \int d^3\vec{q}\,[W(\vec{v}-\vec{q}, \vec{q})\rho(\vec{r}, \vec{v}-\vec{q}, Z) - W(\vec{v}, \vec{q})\rho(\vec{r}, \vec{v}, Z)] \quad (5)$$

where the transition rates W depends on both the momentum transfer $\vec{q}$ and the local momentum $\vec{v}$. The transition rates are proportional to the differential inverse mean free paths (DIMFP)[20] or momentum-differential cross sections. They include both elastic and inelastic scattering processes of the electron with target atoms and the electron gas of the medium. The assumption of a homogeneous ($\vec{r}$ independent) kernel is justified for homogeneous media and, to a lesser extent, for amorphous targets or projectile propagation along "random" directions in a solid. The Eqs. (3-5) represent a partial integro-differential equation in 6 + 1 dimensions and are not easily accessible to a direct numerical solution. A reduction to a Fokker-Planck equation is not applicable since the jump moments

$$\frac{d}{dt}\langle q_i^n \rangle = \int d^3q\, q_i^n\, W(v, q_i) \quad (i = x, y, z) \quad (6)$$

are not small. The latter follows from the fact that the large q limit of W displays a Rutherford tail $\sim q^{-4}$ thereby accentuating large momentum transfers.

We use an alternate strategy to solve Eq. (3-5) by employing "test particle discretization". We calculate trajectories of test particles determined by a stochastic equation of motion, the Langevin equation

$$\dot{\vec{v}} = -\vec{\nabla}V_p + \vec{F}(t) \quad (7)$$

and determine approximate phase space distributions by Monte Carlo (MC) sampling. A typical example of the effective projectile potential V_p used in the (7) is displayed in Fig. 2. Its characteristic features are a Coulomb trough at small distances and screening at large distances. In addition, an oscillatory structure ("wake") is trailing the ion. Since the drift term in (3) agrees with the deterministic part of the Langevin equation (7) the only non-trivial part in establishing a correspondence between (3) and (7) lies in the determination of an appropriate stochastic force $\vec{F}$. While such a construction is not unique, an obvious strategy is to optimize the agreement with the collision operator

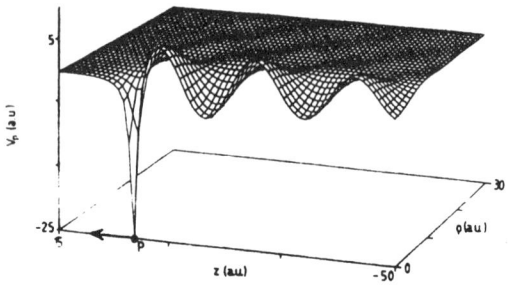

Fig. 2. Dynamical screening potential of S^{16+} in Al calculated in plasmon-pole approximation to the dielectric function ($v_P = 1 a.u.$).

for a finite number of jump moments. We describe stochastic force in terms of a sequence of impulsive momentum transfers ("kicks")

$$\vec{F} = \sum_{\alpha=1,2} \sum_i \Delta \vec{P}_i^\alpha\, \delta(t - t_i^\alpha) \quad (8)$$

where $\Delta \vec{P}_i^\alpha$ is the stochastic momentum transfer per collision at the time t_i^α. The determination of $\vec{F}(t)$ is thereby reduced to that of a stochastic sequence of pairs $(\Delta \vec{P}_i^\alpha, t_i^\alpha)$. The approximation of the collisional interactions of fast electrons with target atoms in terms of instantaneous momentum transfer is based on the observation that the interaction is short-ranged and determined by the static screening length in the medium (typically of the order of $1 a.u.$). The corresponding collision time $t_c \simeq 1/v_p$ is short compared to the orbital period $t_n = 2\pi\omega_n^{-1} = 4\pi(n/q)^2$ for $n \gg 1$. In (8) we have decomposed the stochastic sequence into two independent subsequences. One sequence ($\alpha = 1$) refers to elastic electron-target core scattering while the other ($\alpha = 2$) refers to inelastic electron-electron scattering. The terms (in)elastic refer to (non)-conservation of the kinetic energy of the scattered fast electron.

In accordance with the assumption of a homogeneous collision kernel the proper stochastic point process for the time intervals between subsequent events is Poissonian[21], i.e,

$$P(\Delta t^\alpha) = e^{-\Delta t^\alpha / \langle \Delta t^\alpha \rangle} \qquad (9)$$

where the mean time interval is given in terms of the corresponding mean free path (mfp) λ^α by

$$\langle \Delta t^\alpha \rangle = \lambda^\alpha / v_e \qquad (10)$$

It is straight-forward to modify (9) to take into account effects of the nearest-neighbor peak of the pair distribution function thereby relaxing the homogeneity assumption. First tests with different distribution functions have not shown significant modifications. The probability distribution function for the stochastic momentum transfer can be derived from the collision kernel as

$$P(\Delta \vec{P}^\alpha) = \frac{W(\vec{v}, \Delta \vec{P}^\alpha)}{\int d\vec{q}\, W(\vec{v}, \vec{q})}. \qquad (11)$$

The distribution function are, up to normalization constants, differential scattering cross sections associated with electron-core and electron-electron scattering. The distribution is three dimensional with both longitudinal ($\Delta \vec{P} \parallel \vec{v}_e$) and transverse components ($\Delta \vec{P} \perp \vec{v}_e$) with respect to the electron velocity in the target frame $\vec{v}_e = \vec{v}_p + \vec{v}$. For explicit calculations of (11) we have used the dielectric response approximation developed by Ashley et al for inelastic scattering[20] and a screened Coulomb potential for elastic scattering[22,23] employing a Firsov screening radius. One key advantage of the present formulation is that alternative and more sophisticated input can be used if necessary.

The distribution function $\rho(\vec{r}, \vec{v}, Z = 0)$ of initial conditions propagated according to Eq. (7) must take into account the fact that the population of excited projectile states occurs in a close collision at small impact parameters and accordingly, the phase space distribution should closely resemble binary ion-atom collisions. We choose therefore an initial phase space distribution which would yield the known behavior for excited state cross sections, $\sigma_{n\ell}$, if no further perturbation would occur along the trajectory. This choice consists of a classical uniform distribution in energy corresponding to a quantal n^{-3} law for excitation or charge transfer (and appropriate continuation into the projectile continuum) and an ℓ distribution peaked at small values $\ell \simeq O(1)$ whose detailed shape depends on charge state q, the dominant population mechanism (capture or excitation) and projectile velocity v_p. This guarantees that in the limit of zero path length or in the limit of small perturbations the final-state distribution under single collision conditions is recovered. The crucial difference to ion-atom collisions or the classical calculations of strongly perturbed atoms[24,25] is that the position space distribution is confined to small distances from the projectile ($r < r_c$) where r_c is determined by the dynamical screening length $\lambda_D = v_p/\omega_p$ (ω_p = plasma frequency) or the maximum distance below which sufficient momentum can be transferred in a primary collision event.

The initial conditions lie close to the pericenter of the orbit. Clearly, other choices are possible. For example, initial conditions can be taken from classical trajectory Monte-Carlo calculations (CTMC)[26] for binary ion-atom collision.

In order to relate the dynamical evolution in the bulk to the post-foil experimental observation, modifications due to the penetration of the exit surface must be taken into account. The sudden breakdown of the dynamical screening near the surface leads to a redistribution of the final-state population. The sudden switch from a screened Coulomb potential V_p to a bare potential V_p^0 closely resembles shakeup and shakedown processes in photoionization processes. The atomic orbitals of the old Hamiltonian (containing V_p) are redistributed among the eigenstates of the new Hamiltonian containing V_p^0. Shakeup and shakedown processes are easily incorporated in our classical formulation: The phase-space coordinates of the evolved orbit at the time of the passage through the surface $(r(t_s), v(t_s))$ are used to construct orbits in the bare Coulomb field V_p^0, projecting thereby, within a classical framework, orbits in a screened potential onto orbits of the bare potential.

The observed final-state distribution does not, in most cases of interest, correspond to an ensemble with well-defined distances of propagation Z between the point of primary excitation and the exit of solid, $\rho(\vec{r}, \vec{v}, Z)$, but to an ensemble average over different escape depths, i.e.

$$\tilde{\rho}(\vec{r}, \vec{v}) = \lim_{L \to \infty} \frac{1}{L} \int_0^L \rho(\vec{r}, \vec{v}, Z). \qquad (12)$$

We display all results for dynamical variables in terms of their asymptotic values in the Coulomb field using the projection of the phase space coordinates onto dynamical variables in the bare potential. This representation is motivated by the fact that only asymptotic (post-foil) values are experimentally accessible. Their values at a given distance Z inside the solid entering the calculation of the dynamical evolution are, however, different.

The mapping of the evolved phase space distribution onto the final-state distribution proceeds via standard binning techniques using semiclassical Bohr-Sommerfeld quantization rules. For example, for the ℓ distribution we use equally spaced bins centered around semiclassical eigenvalues $\ell + 1/2$. Similarly, the energy bins corresponding to an n-shell for hydrogenic final states extend from $\epsilon = -q^2/2(n - 1/2)^{-2}$ to $\epsilon = -q^2/2(n + 1/2)^{-2}$.

The conceptual advantage of a classical transport theory is that no a priori assumptions concerning the existence, structure, and lifetime of excited bound states inside the solid are required since the classical trajectories are well-defined irrespective of the properties of V_p and of the collision frequency. The space of classical phase space variables is automatically complete (within a given fixed number of degrees of freedom) so that no truncation errors occur. The price to pay is the neglect of intrinsic quantum corrections. In Sec. 6 we will briefly comment on their expected importance.

IV. COMPUTER SIMULATIONS
A. Free electron transport

The transport properties of monoenergetic free electrons through solids provide an important test[28] for the classical transport theory. In this case the Liouville operator describing the deterministic flow vanishes and the evolution of the energy and angular distribution is approximately separable. Slowing down and degradation (energy straggling)[29] as well as angular straggling belong to the few simple fundamental transport problems for which the resulting $1 + 1$ dimensional master equations (3) are solvable.[23] Approximate solutions for the master equations in energy and angle were first found by Landau[24] and by Goudsmit and Saunderson[22]. A comparison between very recent measurements for 1500 - 4800 eV electrons[23], a numerical solution of the master equation, and the ensemble solution of the Langevin equation[28] shows good agreement for the energy and the angular distribution. Since free electron transport corresponds to the $V_p \to 0$ limit of the full transport problem (Eq.(7)), this lends support to the applicability of classical stochastic dynamics. The master equation for free electron transport has been recently also applied[30,31] to convoy electrons in the field of the projectile ion. Its validity for convoy electrons is presently being tested by comparing distribution functions with and without the presence of V_p term in (7).[28]

B. Coulomb focussing and defocussing of convoy electrons

A portion of the flow of the electrons moving in close correlation with the projectile ion ends up in near-threshold states in the continuum ($\epsilon \simeq 0$) of the projectile. They are usually referred to as "convoy electrons"[32] and manifest themselves as a sharp peak ("cusp") in the forward spectrum. The occurrence of a "cusp" is an entirely classical phenomenon and is describable within a classical trajectory calculation.[33,34] Several yield measurements as a function of foil thickness[8,35,36] have indicated that convoy electrons display an enhanced mean free path, λ_c compared to that of isotachic free electrons, λ_f. While both a clear-cut definition of λ_c and a quantitative measurement are complicated by the sensitive dependence on the experimental resolution as well as on the simultaneously proceeding charge-state equilibration affecting inner shells (see Sec. 2) the underlying physical picture is deceivingly simple: the strong Coulomb force drags the charge cloud along and redirects scattered electrons into forward direction, thereby Coulomb focussing the "beam" and enhancing the effective mean free path.

The study of focussing effects with help of the Langevin equation has revealed a much more complex picture.[33] In Figs. 3 and 4, we compare the attenuation length l_f for free electrons and for convoy electrons l_c in the field of a highly charged sulfur ion S^{16+}. The attenuation length is defined in terms of an approximately exponential rate of depletion as a function of distance of propagation of a fixed small volume element in velocity space centered about the "cusp" velocity $\vec{v}_e \simeq \vec{v}_p$.

The comparison of the attenuation length for free electrons and for convoy electrons displays the two remarkable features: the attenuation of convoy electrons (curve a) is dramatically enhanced in apparent striking

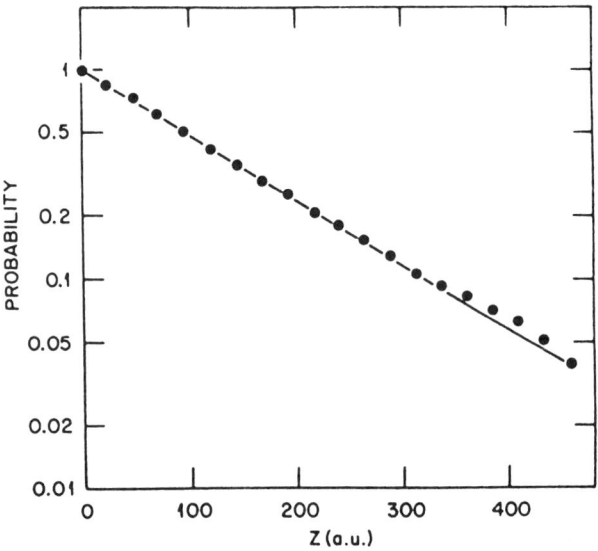

Fig. 3. Attenuation of free electrons with initial velocity $v_e = v_p = 12.5, v_\perp = 0$) in collection volume: $(v_p - 0.25 \leq v_{e\parallel} \leq v_p + 0.25; 0 \leq v_{e\perp} \leq 0.25)$

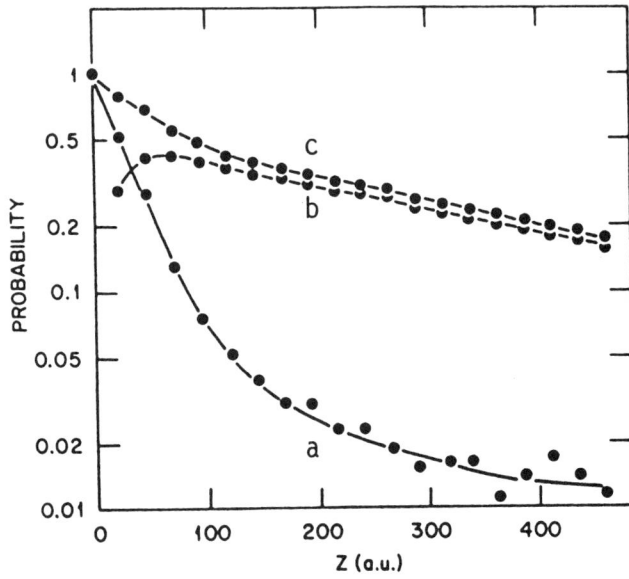

Fig. 4. Attenuation of convoy electrons ($\epsilon \simeq 0$) near a S^{16+} ion ($v_p = 12.5 a.u.$) in Al, collection volume as in Fig.3. a: convoy electrons; b. negative energy ("bound" states); c: total number of correlated electrons ($a + b$).

contradiction to the motion of Coulomb focussing ("Coulomb defocussing"). Furthermore, the rapid decay of the convoy electrons terminates and a long-time tail develops. Both effects are due to the changes of the dynamics in the presence of the (screened) Coulomb field. The interplay between the stochastic force and the Coulomb field introduces a nonlinear instability. After one collision the energy spread as seen in the "projectile" frame ($v_p = 12.5$, $v_\perp = 0$) is for a free electron

$$\Delta \epsilon = \Delta P^2 / 2 \qquad (13)$$

whereas for a Coulomb orbit we have

$$\Delta \epsilon = \vec{v}(\vec{r}) \cdot \vec{\Delta P} + \frac{\Delta P^2}{2} \qquad (14)$$

where $\vec{v}(\vec{r})$ is the local velocity on the Coulomb orbit. Since the local velocity becomes very large near the Coulomb singularity the first term in (14) dominates. The coupling of the externally supplied momentum to the local orbital momentum, $\vec{v}(\vec{r})$, rather than to the asymptotic momentum $\vec{v}(\vec{r} \to \infty) \simeq 0$ leads to an increased energy diffusion which manifests itself as a defocussing effect for convoy electrons. This mechanism possesses a close correspondence in the chaotic dynamics of the periodically perturbed ("kicked") hydrogen atom[37,38]. Indeed, the analogy reaches much further. The diffusive and stochastic dynamics we observe is not entirely due to the extrinsic stochasticity of the Langevin equation but is inherent in its deterministic counterpart (see Sec. 5). The origin of the long-time tails is the build-up of a significant "bound-state" population shown by curve b in Fig. 4. The term $\vec{v} \cdot \vec{\Delta P}$ is not positive-definite. Depending on the (pseudo) random orientation of the two vectors relative to each other, a collision results in either an energy gain or an energy loss. Therefore in nearly half of the trajectories the orbit starting with $\epsilon = 0$ visit regions of negative energy states, i.e. quasi-bound states. It should be stressed again that the energy refers to the asymptotic value in an unperturbed Coulomb field and not to the actual value in the dynamical screening potential. The stabilization of initially fragile ($\epsilon = 0$) orbits by multiple scattering may be called Coulomb "trapping". A similar trapping effect of recapturing unbound electrons has been observed for the periodically kicked hydrogen atom.[37]

Due to the trapping effect the decay curve for the total number of correlated electrons (curve c) has a small slope corresponding to an enhanced correlation length compared to that of the free electron attenuation. This enhancement can be considered a focussing effect. The long-time tail in the convoy yield (curve a) results from a delayed release of "trapped" electrons.

C. Angular momentum diffusion

The broad ℓ distribution accentuating high angular momentum Rydberg states observed in ion-solid collisions can be shown to be a consequence of multiple collisions described by the Langevin equation (7). In order to illustrate the stochastic dynamics of Rydberg orbits in detail we follow the evolution of a single state.

Figure 5 shows the projection of the phase-space distribution onto the (ℓ, ℓ_z) plane of an ensemble of initial orbits with $\ell_0 = 1$ in hydrogenic sulfur propagating at a speed $v_p = 12.5$ a.u. through carbon. All orbits lie initially on an energy shell which corresponds asymptotically to an $n = 32$ state. At each time, a subensemble

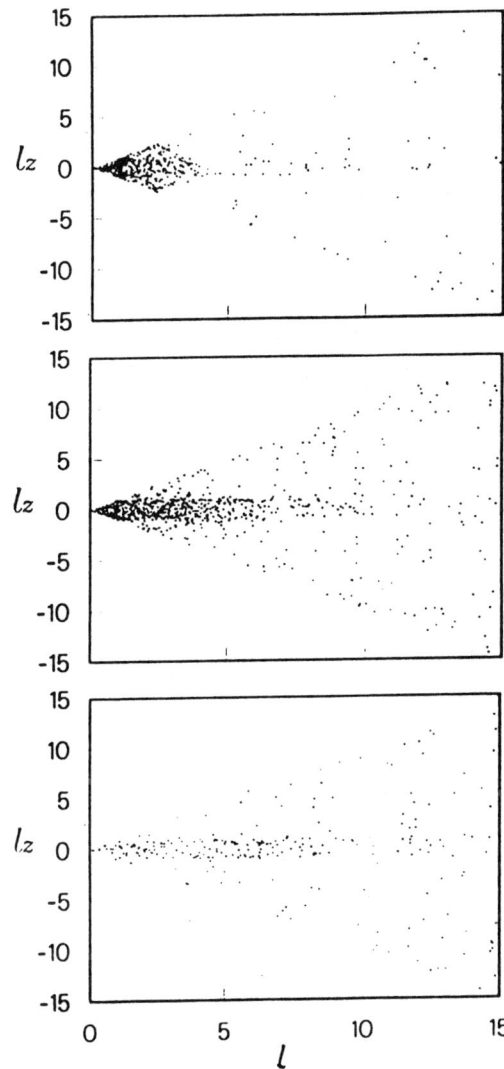

Fig. 5. Phase-space portrait of a subensemble of particles within the energy window ($-0.5 \leq \epsilon \leq 0.5 a.u.$), projected onto the $\ell - \ell_z$ plane. Initial state $S^{15+}(n_0 = 32, \ell_0 = 1)$ propagated through carbon foil at speed $v_p = 12.5 a.u.$ Frames from top to bottom: $t = 2.2 a.u., t = 7.6 a.u., t = 18.7 a.u..$

of orbits is selected for which the Coulomb energy lies close to threshold ($-0.5 \leq \epsilon \leq 0.5$). ℓ_z is the projection of the ℓ vector along the beam axis ($\hat{v}_p$), the axis of cylindrical symmetry of the problem. The spreading of the ℓ, the "ℓ diffusion", is clearly visible. The term "diffusion" should not be taken literally since the individual jump is not small compared to the characteristic orbital velocity. We also note that "scars" of the initial "magnetic substate" distribution ($-1 \leq \ell_z \leq 1$) survive for a long period of time because a significant fraction of stochastic collisions leads to excitation of longitudinal plasmons with $\Delta \ell_z = 0$. The cone-shaped boundary of the projected (ℓ, ℓ_z) phase space suggests that diffusion leads to a net drift into high-ℓ states.

A detailed quantitative test of this theory is complicated by the fact that most measurements are based on a cascade analysis[3-6] which allows only the determination of the global trends of the distribution $P(\ell)$ but does not permit a state-selective detection of the population of individual ℓ. Very recently, such a measurement has become available[7] for the 5ℓ population in autoionizing $C^{2+}(2p5\ell)$ states [Fig. 6(b)]. Comparison with the simultaneously measured ℓ distribution for both excitation $C^{2+} \to He$ and electron capture occur in the g-state population. ℓ diffusion has been observed also in the limit of adiabatic collisions[40] (i.e., $\Delta n = 0$). In the present case, the diffusion is strongly non-adiabatic. The growth in ℓ is strongly correlated with a growth in energy. This illustrated by the cross-correlation function $\langle \Delta \epsilon \Delta \ell \rangle$ (Fig. 7) for the C^{2+} data. The positive values indicate that growth in $\ell, \Delta \ell > 0$,

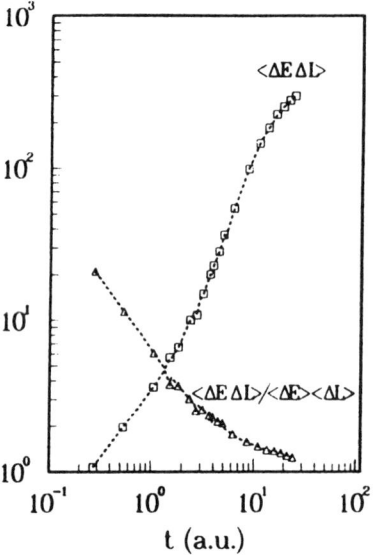

Fig. 7 Cross-correlation function $\langle \Delta \epsilon \Delta L \rangle$ and normalized $\langle \Delta \epsilon \Delta L \rangle / \langle \Delta \epsilon \rangle \langle \Delta L \rangle$ as a function of evolution time, parameters as in Fig. 6.

is positively correlated with a growth in energy. In the pictorial language of Fig. 10, we have a biased random walk along the diagonal towards the upper right corner in the $\epsilon - \ell$ plane. The decay of the normalized cross correlation function with an initial slope corresponding to an approximate power law t^{-1} indicates that the stochastic character of the perturbation destroys the long-term correlation between growth in $\Delta \epsilon$ and $\Delta \ell$.

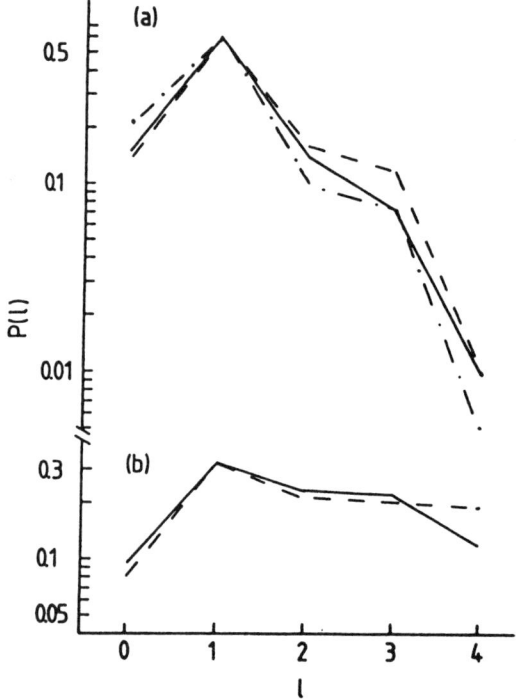

Fig. 6 ℓ distribution $P(\ell)$ in $n = 5\%$ of $C^{2+}(2p5\ell)$ ($v_p = 2.25 a.u.$). (a) - - -, experimental data for $C^{3+} + He \to C^{2+}(2p5\ell) + He^+$; — · — · —, experimental data for $C^{2+} + He \to C^{2+}(2p5\ell) + He$ (Ref. 7); —, simulated initial distribution for random walk. (b) - - -, experimental data (Ref. 7) for $C^+ \to$ carbon foil; —, escape-depth-averaged (steady state) solution of the Langevin equation.

$C^{3+} \to He$ in gas-phase collisions [Fig.6(a)] clearly displays the shift to high-ℓ states. Results for the ℓ distribution in $n = 5$ show astonishingly good agreement between the classical transport theory and the data.[39] The only statistically significant deviation, by a factor ≈ 1.5, from the experimental data appears to

V. RELATION TO DETERMINISTIC CHAOS

Our discussion so far has dealt with the dynamical evolution that is extrinsically stochastic since multiple scattering is a random process unless very special conditions such as channeling[19] in single crystals are met. Extrinsic stochasticity should be distinguished from intrinsic stochasticity which characterizes diffusive transport due to deterministic chaos. The Langevin equation (7) contains two deterministic dynamical systems as limiting cases which can be shown to be chaotic. This, in turn, suggests that the transport behavior we have observed is result of a complex interplay of extrinsic and intrinsic stochasticity.

One limiting case is found by switching off the random force ($\vec{F} = 0$) in Eq. (7). The resulting Newtonian equation of motion describes the motion of the electron in the wake potential (Fig. 1). This corresponds to a time-independent dynamical system with two degrees of freedom (the azimuthal angle is cyclic and L_z is a constant of motion because of the axial symmetry of the wake potential with respect to the beam axis ($\vec{v}_p$)). The wake potential is not separable and allows therefore for a divided phase space of regular and chaotic motion[41]. Restricting ourselves in the following only to the motion near the (distorted) Coulomb trough one

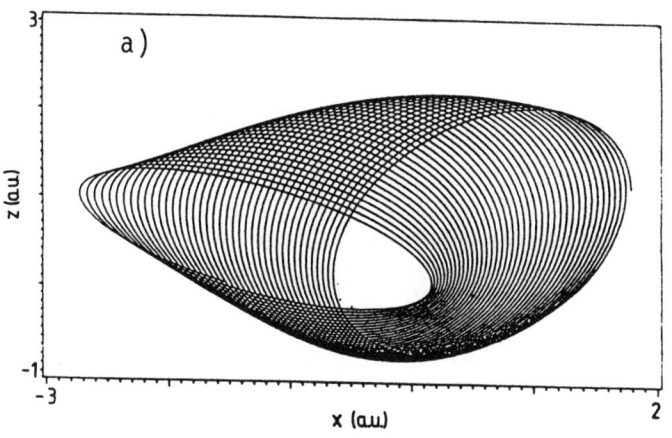

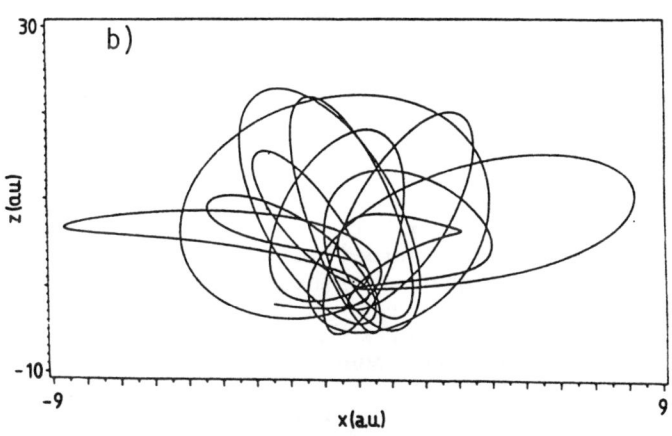

Fig. 8. Trajectory of an electronic orbit in S^{15+} ($v_p = 12.5 a.u.$) in Al with an energy corresponding to an unperturbed state with quantum number a) $n = 5$ and b) $n = 25$ state.

expects a transition from regular motion to chaotic motion with increasing energy since trajectories will explore regions of the potential where deviations from the pure Coulomb behavior become more pronounced. Figs. 8a and b show two trajectories for an energy corresponding to a low-lying state ($n = 5$) and a high-lying state ($n = 25$), both with $\ell_z = 1$. While Fig. 8a ($n = 5$) displays a precessing and "breathing" ellipse as expected for a weakly perturbed Coulomb potential Fig. 8b shows an irregular trajectory which lacks any quasi-periodic structure. The breathing mode of the ellipse in Fig. 8a corresponds to a change of eccentricity, or equivalently, of ℓ, since the dynamical screening potential breaks the spherical symmetry. It should be noted that in the limit of weak perturbation (i.e. low n) the precessional and breathing motion is the classical manifestation of the Stark effect due to the wake field discussed by Chetioui et al[2]. We have systematically investigated the evolution of the ℓ distribution due to the wake field as a function of the energy. An example is shown in Fig. 9 for an energy shell corresponding to an $n = 15$ state possessing a largely chaotic phase space. The evolution of ℓ as well as its rms fluctuation $\langle \Delta \ell \rangle = \langle (\ell - \langle \ell \rangle)^2 \rangle^{1/2}$ is shown as a function of time for an ensemble of 100 randomly chosen trajectories with $\ell_0 = 2$. The irregular dynamics ("mixing") manifests itself in the approach to a large equilibrium value for both ℓ and $\langle \Delta \ell \rangle$. We therefore

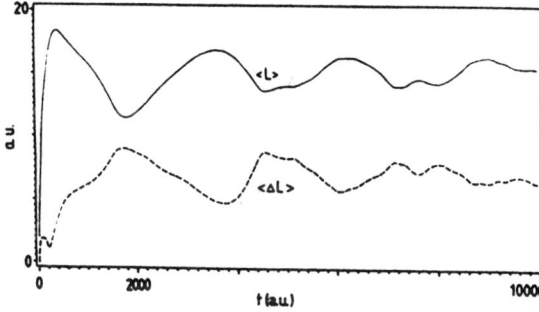

Fig. 9. Average value $\langle \ell \rangle$ and standard deviation from this average, $\langle \Delta \ell \rangle$, as a function of the evolution time for an ensemble of initial $\ell_0 = 2$ orbits in S^{15+} in Al ($v_p = 12.5$). The energy corresponds to an unperturbed $n = 15$ state.

find that in addition to multiple collisions the irregular motion in the wake potential can induce ℓ "diffusion" and growth in ℓ. Notice, however, that the characteristic time within which the deterministic ℓ diffusion occurs ($t \gtrsim 100$ a.u. corresponding to $Z \gtrsim 1200$ a.u.) is large compared to stochastic diffusion time (see Fig. 5). The important conclusion is that in the present case field mixing or, more generally wake potential mixing, is slow compared to collisional mixing.

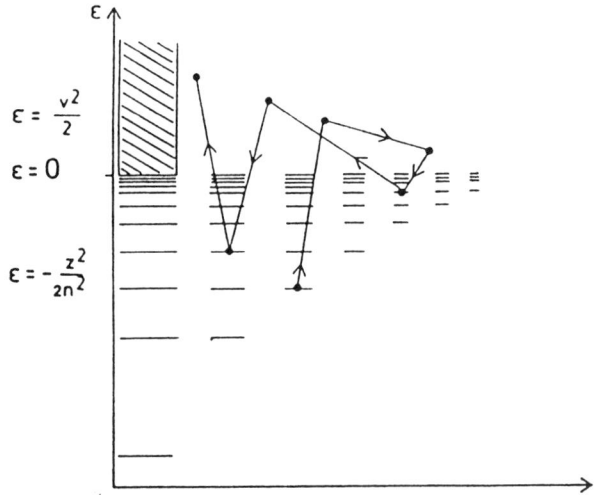

Fig. 10. Random walk in the $\epsilon - \ell$ plane of the phase space for an impulsively driven Rydberg atom.

The dynamically screened Coulomb potential possesses in addition to the Coulomb trough additional local minima which can support bound states ("wake riding electrons"). The latter can play an important role for the electronic evolution near an antiproton traversing a solid.[42]

The other limiting case results from replacing in (5) the dynamically screened potential V_p by a bare Coulomb potential V_p^0 and replacing the stochastic force by a deterministic periodic force

$$\vec{F}(t) = \sum_i (-1)^i \, \Delta \vec{P} \, \delta(t - iT/2) \qquad (15)$$

In this case we recover the Newtonian equation of motion for an impulsively driven hydrogen atom (or hydrogenic ion). This model has been previously studied [13,37,38,43] in connection with the microwave ionization of hydrogen. This connection can be made more explicit by Fourier analyzing the periodic force

$$\vec{F}(t) = \frac{4 \, \Delta \vec{P}}{T} \sum_{s=0}^{\infty} \cos(2s + 1)\omega t \qquad (16)$$

with $\omega = 2\pi/T$. The first term in (16), $s = 0$, agrees with the force exerted by a microwave field of field strength $F_o = 4 \, \Delta P/T$. The essential difference is the presence of an infinite number of higher harmonics which give rise to a large number of nonlinear resonances and to enhanced instability.

Choosing $\Delta \vec{P}$ as parallel to the $\hat{Z}$ axis the equation of motion possesses axial symmetry. Upon a canonical transformation Eq. (7) can be converted to a four dimensional nonlinear mapping[44]

$$\begin{pmatrix} n_{i+1} \\ \tau_{i+1} \\ \ell_{i+1} \\ \psi_{i+1} \end{pmatrix} = T \begin{pmatrix} n_i \\ \tau_i \\ \ell_i \\ \psi_i \end{pmatrix} \qquad (17)$$

where n_i is the classical action corresponding to the principal quantum number, τ_i is the conjugate angle variable (the mean anomaly), ℓ_i is the total angular momentum and ψ_i is its conjugate angle variable. The mapping (17) allows for a simple intuitive interpretation: The electron describes a (not necessarily random) "walk" in Coulomb state space (Fig. 10). The walk will be random when the dynamics becomes stochastic.

The stochasticity can be most conveniently measured with help of (positive) Lyapunov exponents β.

$$\beta = \lim_{k \to \infty} \lim_{\Delta X_0 \to 0} \frac{1}{k} \ln \frac{|T^{(k)}(\vec{X}_0 + \Delta \vec{X}_0) - T^{(k)}(\vec{X}_0)|}{|\Delta \vec{X}_0|} \qquad (18)$$

They measure the exponential divergence of nearby initial conditions. In (18), the number of iterations of the mapping is denoted by k. The vector of initial conditions is denoted by $\vec{X}_0$, the displacement of the nearby initial condition by $\Delta \vec{X}_0$. For the four-dimensional mapping (17) up to two positive Lyapunov exponents can exist. We focus on the largest exponent β_{max} which, for almost all $\Delta \vec{X}_0$, determines the overall rate of exponential separation. A trajectory is regular if $\beta_{max} = 0$ (within the numerical tolerance set in the following to 10^{-5}) and chaotic if $\beta_{max} > 0$. Both extrinsic (i.e. noise induced) and intrinsic stochasticity lead to positive Lyapunov exponents. One can therefore use β_{max} to study the interplay between the two. An illustrative example is shown in Fig.11. For an ensemble of $n = 100$, $\ell = 67.4$ orbits with randomized initial conditions for the Euler angles and τ the fraction of regular ($\beta_{max} = 0$) and ionized orbits have been determined as a function of the strength of the pulse, ΔP. The calculation was done for the periodically kicked hydrogen atom with a "kick" frequency $\omega = 3/2 \, \omega_a$ ($\omega_a = n^{-3}$ classical atomic frequency) and a stochastically perturbed hydrogen atom described by the Langevin equation. In the present case the frequency of the kicks was kept fixed but extrinsic stochasticity was introduced by a Gaussian distribution for the kick amplitude ΔP ("amplitude modulated noise") with a variance identical to that of the periodic perturbation. After 400 kicks we find that for small pulse heights ($\Delta P \leq 10^{-4}$) already a significant fraction of the phase space is intrinsically stochastic. The fraction of regular orbits is less than 20%. In the noisy counterpart almost all regular orbits are destroyed and the system proves to be completely stochastic. Near $\Delta P \geq 10^{-4}$ in either case all orbits are irregular and the difference between a deterministic and a random force ceases to be important. Note that, however, all orbits remain bound.

At higher values of $\Delta P \geq 2 \times 10^{-4}$ rapid ionization sets in. Choosing the 50% level as a threshold, the kicked hydrogen atom in an $n = 100$, $\ell = 67.4$ state ionizes at a periodic amplitude $\Delta P = 6 \times 10^{-4}$ and at

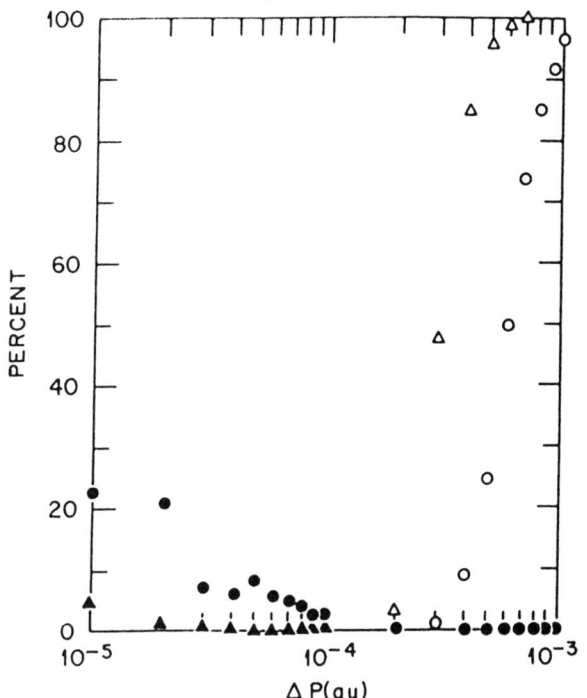

Fig. 11. Fractions of regular and ionized $n = 100, \ell = 67.4$ orbits after 400 pulses. solid symbols: percentage of regular orbits; open symbols: percentage of ionized atoms; circles: periodic perturbation; triangles: Gaussian amplitude modulated noise.

a noisy amplitude $\Delta P = 3 \times 10^{-4}$, i.e. the noise reduces the ionization threshold by approximately a factor of 2.

The observation of irregular trajectories and diffusive motion in the deterministic limiting cases of the Langevin equation provides clear evidence that diffusive transport is caused by both the randomness of the collisional interactions and the intrinsic dynamical instability. This implies that diffusive transport may also occur under ideal channeling conditions. The chaotic motion of charged particles in channels has recently been discussed by Kimball et al[45].

VI. OUTLOOK

We conclude this review by pointing to a few future direction of research: many experimental observation of foil excited atoms and ions are still not quantitatively accounted for. They include the long-time tails in the X ray cascades[46], the anisotropy in excited bound states and the high multipole content in the angular distribution of continuum electrons.[9] The classical transport theory should be able to address these problems.

On a more fundamental level, the question of quantum corrections to classical diffusive behavior is still open. Significant progress has been made for the microwave ionization problem. Casati et al[47] and Jensen et al[14] have shown that classical diffusive excitation is suppressed in the high-frequency region primarily because of the low density of accessible states. The saturation of diffusion in the quantum system, however, takes effect at longer times when the discrete structure of the energy spectrum becomes "visible". Since the characteristic times associated with electron transport in the solid are rather short corrections to the classical transport theory may be small. Blümel et al[48] have recently shown that noise can restore classical diffusion on an even larger time scale.

Finally, the intrinsic stochasticity observed in the "kicked" i.e. collisionally perturbed system is an example for chaotic dynamics in the continuum. The (quasi) bound states are, in fact, embedded in the continuum as illustrated by their rapid ionization (Fig. 11). The study of ion-solid collisions and, more generally, of atomic collisions provides the opportunity to study irregular dynamics in the continuum. The complexity of irregular scattering has only recently began to unfold.[48]

ACKNOWLEDGEMENTS

The author acknowledges many valuable contributions made by many colleagues to the work presented here. They include C. Bottcher, J. Gibbons, J. Müller, R. Ritchie, I. Sellin, and J. Wang. This work was supported in part by the National Science Foundation and by the U.S. Department of Energy, Office of Basic Energy Sciences, Division of Chemical Sciences, under Contract No. DE-AC05-84OR21400 with Martin Marietta Energy Systems, Inc.

REFERENCES

1. N. Bohr and L. Lindhard, Kgl. Danske Videnskab. Selskab.Mat. -Fys. Medd. **26**, No. 12 (1954).
2. A. Chetioui, J. Rozet, K. Wohrer, D. Vernhet, A. Touati and C. Stephan, in <u>Electronic and Atomic Collisions</u> (eds. H. Gilbody, W.Newell, F. Read, A. Smith) (North-Holland Publ. Co., Amsterdam, 1988) p. 309.
3. H. -D. Betz, D. Röschenthaler, and J. Rothermel, Phys.Rev. Lett. **50**, 34 (1983); H.D. Betz, in <u>Forward Electron Ejection in Ion Collisions</u>, edited by K.O. Groeneveld et al., Lecture Notes in Physics Vol. 213 (Springer-Verlag, Berlin, 1984), p. 115.
4. G. Dehmelt, A. Georgiadis, L.V. Gerdtell, D. Sträter,and P. van Brentano, Phys. Lett. **29A**, 193 (1982).
5. K. Dybdal, J. Sorensen, P. Hvelpund, and H. Knudsen, Nucl.Instr. Meth. Phys. Res., Sect. **B 13**, 581 (1986).
6. C. Can, R.J. Maurer, B. Bandong, and R.L. Watson, Phys.Rev. A **35**, 3244 (1987).
7. Y. Yamazaki et al., Phys. Rev. Lett. **61**, 2913 (1988).
8. G. Schiwietz, D. Schneider, and J. Tanis, Phys. Rev. Lett.**59**, 1561 (1987).
9. S.D. Berry, S.B. Elston, I.A. Sellin, M. Breinig, R.DeSerio, C.E. Gonzalez-Lepera, and L. Liljeby, J. Phys. B **19**, L149 (1986).
10. W. Brandt, in <u>Atomic Collisions In Solids</u>, edited by S. Datz, B. Appleton, and C. Moak (Plenum, New York, 1975), p. 761.

11. K. Dettmann, K.G. Harrison, M.W. Lucas, J. Phys. B **7**, 769 (1974).
12. J. Neufeld and R.H. Ritchie, Phys Rev. **98**, 1632 (1955).
13. P. Koch in Electronic and Atomic Collisions, Ref. 2, p. 501.
14. R. Jensen, Phys. Rev. A**30**, 386 (1984); R. Jensen, J. Leopold, and D. Richards, J. Phys. B**21**, L527 (1988); R. Jensen, S. Susskind, and M. Sanders, Phys. Rev. Lett. **62**, 1476 (1989).
15. H. Betz, Rev. Mod. Phys. **44**, 465 (1972).
16. C. Woods, N. Cowern, L. Bridwell, and C. Sofield, J.Phys. B**18**, 4113 (1985).
17. J. Rozet, A. Chetioui, P. Piquemal, D. Vernhet, K.Wohrer, C. Stephan, J. Phys. B**22**, 33 (1989).
18. J. Burgdörfer, Phys. Rev. A**24**, 1756 (1981); Nucl. Instr. Meth. **202**, 253 (1982). H.G. Berry, J. Dehaes, D. Neek and L. Sommerville, in Forward Electron Ejection in Ion Collisions, Lecture Notes in Physics Vol. 213 (Springer, Berlin, 1984), p. 150.
19. S. Datz et al., Phys. Rev. Lett. **40**, 843 (1978). S.Datz, J. de Physique (Paris) C1-335 (1979).
20. J. Ashley, C. Tung, and R. Ritchie, Surf. Sci. **81**,409 (1979); C. Martin, E. Arakawa, T. Callcott, and J. Ashley, J. Electron Spectrosc. Relat. Phenom. **35**, 307 (1985). J. Ashley, J. Cowan, R. Ritchie, V.E. Anderson, and J. Hoelz, Thin Films **60**, 361 (1979).
21. N. Van Kampen, Stochastic Processes in Physics and Chemistry, (North-Holland, Amsterdam, 1981).
22. S. Goudsmit and J. Saunderson, Phys. Rev. **57**, 24(1940); P. Sigmund and K. Winterbon, Nucl. Instr. Meth. **119**, 54l (1974).
23. S. Lencinas, J. Burgdörfer, J. Kemmler, O. Heil, K.Kroneberger, N. Keller, H. Rothard, and K.O. Groeneveld, submitted for publication to Phys. Rev. A.
24. J. Leopold and D. Richards, J. Phys. B**19**, 1125 (1986).
25. J. Delos, S. Knudsen, and D. Noid, Phys. Rev.A**28**, 7 (1983).
26. R. Abrines and I.C. Percival, Proc. Phys. Soc. (London)**88**, 861 and 873 (1966); R. Olson in Ref. 2, p. 271.
27. L. Landau, J. Physics (USSR) **8**, 201 (1944).
28. J. Burgdörfer and J. Gibbons, to be published.
29. K. Jacobi, Ph.D. thesis, TU Clausthal, W. Germany (1970, unpublished), see also H. Raether, Excitation of Plasmons and interband transitions by Electrons, Springer Tracts in Modern Physics, Vol. 88 (Springer, Berlin, 1980).
30. J. Kemmler, S. Lencinas, P. Koschar, O. Heil, H. Rothard, K. Kroneberger, G. Szabo, and K.O. Groeneveld, Nucl. Instr. Meth. B**33**, 317 (1988).
31. R. Barrachina, A. Goni, P. Focke, and W. Meckbach, Nucl. Instr. Meth. B**33**, 330 (1988).
32. M. Breinig et al., Phys. Rev. A**25**, 3015 (1982); K.O. Groeneveld, W. Meckbach, I. Sellin, and J. Burgdörfer, Comments At. Mol. Phys.**14**, 187 (1984).
33. J. Burgdörfer in Proceedings of the ThirdWorkshop on High Energy Ion-Atom Collisions, edited by D. Berenyi and G. Hock, Lecture Notes in Physics Vol. 294 (Springer-Verlag, Berlin, 1988), p.344.
34. C. Reinhold and R.E. Olson, Phys. Rev. A**39**, 3861 (1989).
35. I.A. Sellin, et al. in Forward Electron in Ion Collisions, Lecture Notes in Physics, Vol. 213 (Springer, Berlin, 1984), p. 109.
36. R. Schramm, P. Koschar, H.D. Betz, M. Burkhard, J.Kemmler, O. Heil, and K.O. Groeneveld, J. Phys. B**18**, L507 (1985).
37. T. Grozdanov and H.S. Taylor, J. Phys. B**20**, 3863 (1987).
38. J. Leopold and D. Richards, J. Phys. B**21**, 25(1988).
39. J. Burgdörfer and C. Bottcher, Phys. Rev. Lett. **61**, 2917 (1988).
40. I. Percival and D. Richards, J. Phys. B**12**, 7051 (1979).
41. J. Müller and J. Burgdörfer, to be published.
42. Y. Yamazaki et al., CERN Report No. CERN-PSCC 87-39 (unpublished), p. 108; CERN Report No. CERN-PSCC 87-40 (unpublished), p. 108; J.Burgdörfer, J. Wang and J. Müller, Phys. Rev. Lett. **62**, 1599 (1989).
43. A. Dhar, M. Nagaranjan, F. Izailev and R. Whitehead, J.Phys. B**16**, B17 (1983); A. Carnegie, J. Phys. B**17**, 3435 (1984).
44. J. Burgdörfer, Nucl. Instr. Meth. B (1989), in press.
45. J. Kimball, G. Petschel, and N. Cue, Nucl. Instr. Meth. B**33**, 53 (1988).
46. P. Richard, Phys. Lett. **45A**, 13 (1973).
47. G. Casati, B. Chirikov, D. Shepelyansky, I. Guarneri, Phys. Rep. **154**, 77 (1987).
48. R. Blümel, R. Graham. L. Sirko, V. Smilansky, H. Walther, and K. Yamada, Phys. Rev. Lett **62**, 341 (1989).
49. For an overview and early developments see, M. Berry and K. Mount, Rep. Prog, Phys. **35**, 315 (1972), more recent papers include: B. Eckart, Physica D**33**, 89 (1988); R. Blümel and U. Smilansky, Phys. Rev. Lett. **60**, 477 (1988).

OBSERVATIONS OF EXCITED H° ATOMS PRODUCED BY RELATIVISTIC H⁻ IONS IN CARBON FOILS

A.H. Mohagheghi[1], P.G. Harris[1], C.Y. Tang[1], H.C. Bryant[1],
J.B. Donahue[2], C.R. Quick[2], R.A. Reeder[2],
H. Sharifian[3], W.W. Smith[4], J.E. Stewart[5],
H. Toutounchi[6], T.C. Altman[7], D.C. Rislove[8]

[1] The University of New Mexico, Albuquerque, NM 87131
[2] Los Alamos National Laboratory, Los Alamos, NM 87545
[3] California State University, Long Beach, CA 90840
[4] The University of Connecticut, Storrs, CT 06268
[5] Western Washington University, Bellingham, WA 98225
[6] University of Mashhad, Mashhad, Iran
[7] Oswego High School, Oswego, NY 13126
[8] St. Mary's College, Winona, MN 55987

Studies of the production of H° in its various states by the passage of relativistic H⁻ ions through thin foils are under way at the linear accelerator at Los Alamos. Laser excitation and selective field ionization are used to isolate specific states. We present evidence that the excitation of Rydberg states by a foil at energies up to 800 MeV proceeds in a stepwise manner.

The Charles Addams Effect

The passage of matter through matter has been a topic of interest not only to scientists[1]. A study of the penetration of ionizing particles through solids was made very early in the modern era by Bohr[2]. Recently we have been measuring the yields of various excited states of H° resulting from directing beams of relativistic H⁻ ions at energies up to 800 MeV through thin carbon foils. For the foil thicknesses we have studied, ranging up to about 200 $\mu g/cm^2$, there is, as Charles Addams seemed to know, a chance that the loosely bound H⁻ will emerge unscathed from what must be a truly hostile environment. However, most of the H⁻ are stripped down to H°, distributed over its excited states as well as its ground state. Finally, as the foil thickness rises the fraction of fully-stripped protons inexorably must rise to unity. Skiers should avoid thick trees.

How does it feel to be an H⁻ ion passing through carbon?

Carbon atoms in a foil with a density of 2.2 gms/cm³ are spaced about 0.2 nm apart. A simple model[3] of the H⁻ ion predicts a characteristic radius of the outer electron also of about 0.2 nm. Thus an H⁻ ion passing through a foil could contain about 8 carbon atoms within its volume. How does an H⁻ remain intact?

The average field inside a carbon atom can be estimated from the Fermi-Thomas model to be of the order of 1.4×10^{10} V/cm. Fields transverse to the ion's motion are Lorentz enhanced by a factor of γ (1.85 at 800 MeV). The lifetime in such a strong D.C. field is of the order of 1.8×10^{-18} sec[4], which in turn is comparable to the time for an 800 MeV ion to pass through a carbon atom, 4×10^{-19} sec. The frequency with which carbon atoms are encountered is 2.5×10^{18} Hz.

The time to pass through a 100 $\mu g/cm^2$ foil is about one femtosecond for an 800 MeV H$^-$ ion.

An electron traveling with the speed of the ion would have a kinetic energy of 435 keV and would multiple scatter r.m.s. about 12 mrads[5], wandering some 60 Bohr radii from the proton with which it entered the foil. Indeed, it is difficult to imagine the atom remaining intact.

However, perhaps a glimmer of how an atom can pass through the foil unscathed can be obtained from the following consideration. At minimum ionizing a charged particle loses 1.78 MeV/(g/cm^2) passing through carbon[5]. Therefore, an electron at 435 keV (β = .842 corresponding to an 800 MeV H$^-$) passing through a 100 $\mu g/cm^2$ carbon foil would lose 178 eV of its energy. A proton would of course lose the same amount, but a much smaller fraction of its energy, and so, if the electrons and proton were not strongly coupled inside the foil, then the velocity of the electrons would be decreased more than the velocity of the proton. In fact, we can neglect the effect on the proton. The key question is "in the frame where the proton is at rest, how much kinetic energy would the electron have?"

The change in β in the lab frame, $\Delta\beta$, is given by

$$\Delta\beta = \Delta E / \beta \gamma^3 mc^2 \qquad (1)$$

where ΔE is the energy lost in the foil, β is v/c, $\gamma = (1 - \beta^2)^{-1/2}$ and mc^2 is the electron rest energy.

When we transform to the frame in which the proton is at rest, the relative velocity of the electron becomes $\gamma^2 \Delta\beta$. Therefore the kinetic energy of the electron in the proton's rest frame is

$$K.E. = (\Delta E)^2 / 2\beta^2 \gamma^2 mc^2, \qquad (2)$$

which for ΔE = 178 eV works out to be 12.9 meV.

Therefore we see that although the charged particles can deposit a relatively large amount of ionizing energy in the foil, the relative change in energy in the proton rest frame is much smaller than even the H$^-$ binding energy of 0.7542 eV.

Quantum Chaos?

It may be that an H$^-$ ion or H^0 atom perturbed by a femtosecond-long periodic jumble of carbon atoms must endure some sort of chaotic experience. According to Galvez et al.[6], when the scaled frequency of the perturbation, $n^3\omega$ (in atomic units), exceeds 2, quantum chaos sets in. The implication is that the scaled electric field required to ionize increases beyond the classical limit. In the cases we have treated, say n = 11 at 800 MeV, $n^3\omega$ = 4.5 x 10^5, which may also support the observation that atoms under such duress seem surprisingly durable, in spite of classical considerations mostly to the contrary.

Simple Theories

If one considers all H^0 atoms the same, regardless of the quantum state, then the H^0 yield, y, as a function of foil thickness, x, is given by[7]

$$y(x) = \frac{a}{b-c} [\exp(-cx) - \exp(-bx)] \qquad (3)$$

where, if ρ is the target number density and N_0 is the incident flux of H$^-$ ions, then

$$a = N_0 \rho \sigma_{-10},$$
$$b = \rho \sigma_{01},$$
and $$c = \rho (\sigma_{-10} + \sigma_{-11}),$$

where the cross sections are defined as follows for collisions with carbon atoms:

σ_{-10} for H$^-$ → H^0 + e,

σ_{-11} for H$^-$ → H$^+$ + 2e,

σ_{01} for H^0 → H$^+$ + e.

In this picture, the distribution of H° over its possible quantum states would be independent of x, the target thickness. That is, all the quantum states are made in each collision, with a probability which is independent of distance into the foil.

A different approach is to assume that the excitation process is diffusive. That is, to arrive at H°(n), one must first make H°(1), then H°(2) etc., up to n. In this case, if one considers a very thin foil for which the creation processes dominate, by the first model, we would expect, the flux of H°(1) should increase like x. H°(2) then should increase like x^2, and H°(n) like x^n. From this point of view, the distribution of excited states should vary with the thickness of the foil, with the higher n states peaking at the greater thicknesses. As we shall see, qualitatively, this statement agrees with observation.

Experimental Arrangement

Fig. 1 schematically exhibits our method for the study of yields from thin foils. A H⁻ beam, 1 or 2 mm diameter, whose energy can be as high as 800 MeV, is incident upon one of about 18 interchangeable foils of varying thickness ranging from 20 to 200 $\mu g/cm^2$. The first bending magnet after the foil sweeps target-produced electrons out of the beam, and by its motional electric field, strips high-lying Rydberg states. The surviving beam can now be probed with a laser beam in the interaction chamber using our standard technique[8]. With the fourth harmonic of the YAG we can continuously Doppler tune from 1.4 eV to 15.9 eV. The electron magnetic spectrometer bends electrons coming along the beam line through 90° into a scintillator. In addition, the spectrometer can be tuned to detect selectively high-lying Rydberg states in the beam by motional-field ionization. At 800 MeV, Rydberg states with principal quantum numbers higher than 10 can be detected in this way. A similar technique has been used with an electric field spectrometer by Zeitz, Kowallik and Schneider[9]. After passing through a fast ion chamber, the particle beam is stopped in a massive Faraday cup.

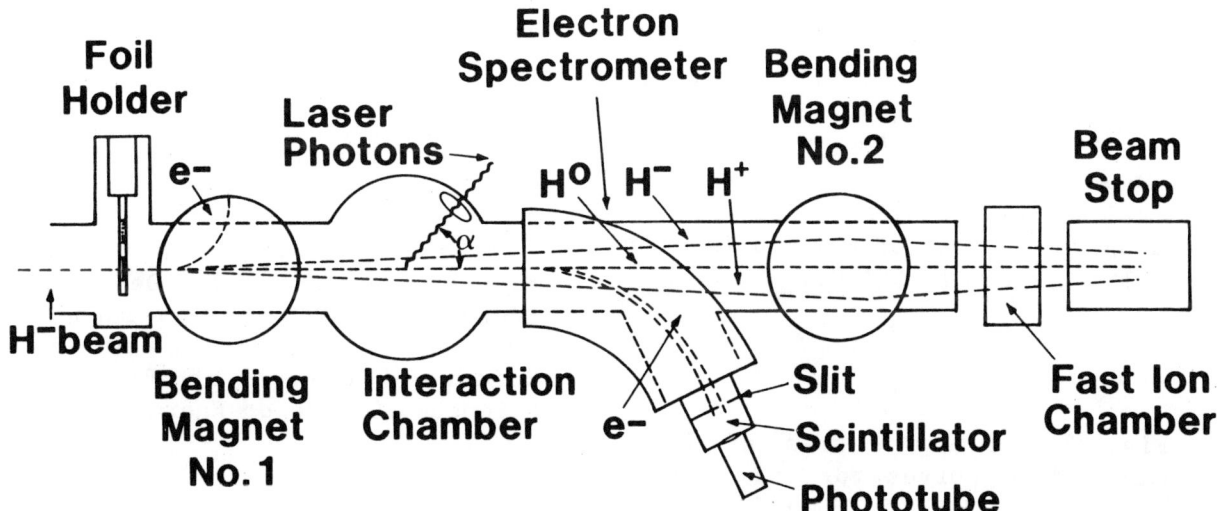

Fig. 1. Schematic of experimental arrangment in LAMPF beam line.

Some Preliminary Results

Fig. 2 shows relative yields of n=2, 3, and 4 produced by 581 MeV H⁻ on various carbon foils. These states were selectively laser-excited to n=13, which was detected by field ionization in the spectrometer. The yields are not normalized to each other since the relative laser intensities for the three measurements are not known. The three data sets were fit with eq.3, which we see is inadequate. Of interest to note is that the higher n states peak at greater thicknesses and that the onset of production also seems to increase in thickness with rising n, bespeaking a diffusive process as discussed above.

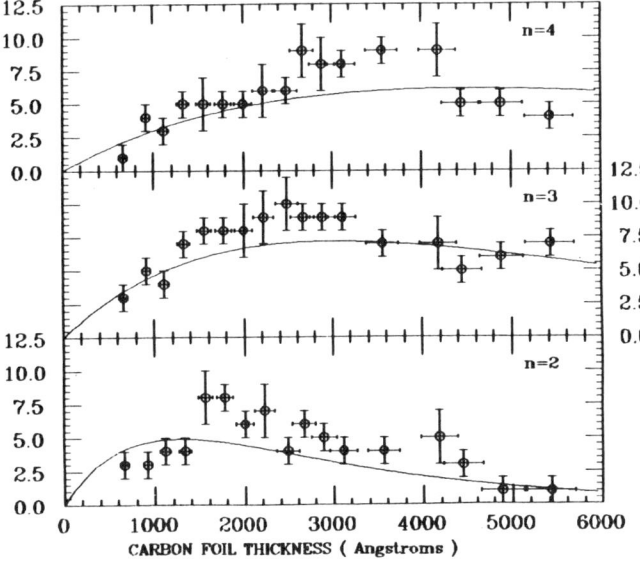

Fig. 2. Relative yields of n=2, 3, 4 (from bottom to top) of H° as a function of foil thickness at 581 MeV.

Fig. 3 shows a series of four relative yield curves for an 800 MeV H⁻ beam. Fig. 3A and 3B, n=1 and 2, were measured by laser excitation to n=11 followed by field ionization in the electron spectrometer which was tuned to receive only n=11. Fig 3C and 3D show the relative yields of n=10 and n=11 obtained by scanning the magnetic field of the spectrometer over the regions which selectively ionize these two states and bring the electron to the spectrometer slit. We notice the dramatic increase in the thickness at maximum yield as one goes from lower to higher Rydberg states.

Monte Carlo Simulations

We have tried to simulate the results of electron yield versus spectrometer setting using a Monte Carlo technique. Our preliminary result for a foil thickness of 70 $\mu g/cm^2$ and a beam energy of 581 MeV indicates that a $1/n$ distribution of Rydberg states with large ℓ enhancement[10,11] fits the data best, but much more modeling must be done, and fits made, before a clear picture of the distribution as a function of quantum numbers and foil thickness can be made.

Conclusion

Our data favor a production mechanism for excited H° atoms by a relativistic H⁻ beam passing through carbon foils which is, at least in part, diffusive. That is, there is evidence that the states are excited in a stepwise manner. The distribution function clearly cannot be represented by the product of a function of n times a function of the foil thickness, x.

This work was done under the auspices of the U.S. Department of Energy, the UNM portion supported in part by the Division of Chemical Sciences, Office of Basic Energy Sciences, Office of Energy Research.

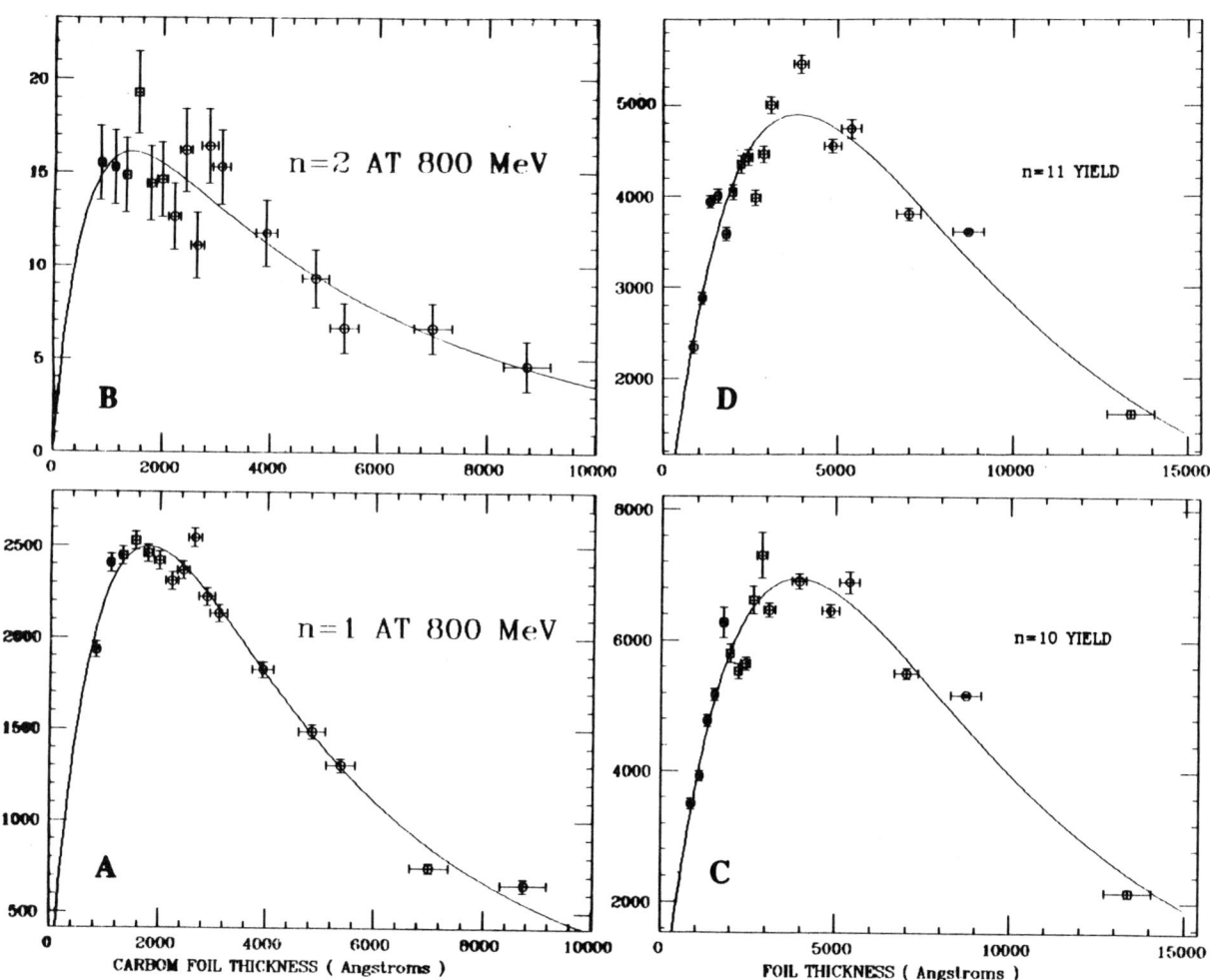

Fig. 3. Relative yields of H° states from an 800 MeV H⁻ beam as a function of foil thickness. Curves A, B, C, D correspond to n = 1, 2, 10, 11 respectively.

References

1. Charles Addams, The New Yorker, Jan. 13, 1940, p.13.
2. Niels Bohr, Phil. Mag. <u>30</u>, 581 (1915).
3. B.H. Armstrong, Phys. Rev. <u>131</u>, 1132 (1963).
4. A.J. Jason, D.W. Hudgings, O.B. van Dyck, IEEE <u>NS-28</u>, 2704 (1981).
5. Particle Properties Data Booklet April 1988 p. 95. See M. Aguilar-Benitez et al., Phys. Lett. <u>B204</u>, April 88.
6. E. J. Galvez et al., Phys. Rev. Lett. <u>61</u>, 2011 (1988). see also J.E. Bayfield et al., Phys. Rev. Lett. <u>63</u>, 364 (1989).
7. Eisenberg, M.D., "Target Thickness Data for various Electron Stripping Targets", LANL Memorandum, AT-4/84/157, July, 1984.
8. C.Y. Tang et al., Phys. Rev. A <u>39</u>, 6068 (1989).
9. W.-D. Zeitz, R. Kowallik and D. Schneider, Phys. Rev. A <u>39</u>, 43 (1989).
10. Y. Yamazaki et al., Phys. Rev. Letters <u>61</u>, 2913 (1988).
11. J. Burgdörfer and C. Bottcher, Phys. Rev. Letters <u>61</u>, 2917 (1988).

Classical ghosts in quantal microwave ionisation

D. Richards* and J. G. Leopold**
*Mathematics Faculty, Open University, Milton Keynes, MK7 6AA, UK.
**The Fritz Haber Molecular Dynamics Research Centre, Hebrew University, 91904 Jerusalem, Israel.

We provide a brief review of the response of an excited hydrogen atom to a microwave field, and a detailed discussion of the behaviour of a one-dimensional model in a high frequency field. In particular we show that although classical dynamics is incorrect for the parameters of experimental interest an examination of the classical behaviour is helpful in understanding the quantal behaviour. We show further that, for these parameters, the quantal probabilities fluctuate dramatically with the variation of most system parameters and attribute this non-classical behaviour to the dominant influence of relatively few, about ten, quasi-resonant states.

1 Introduction

The ionisation dynamics of an excited hydrogen atom is complicated and incompletely understood; rigorous result exist only for weak fields[1] and seem not to apply to the situations of experimental interest. Moreover there are many parameters defining an individual experiment or calculation so it is very difficult to obtain an overall picture of the system behaviour. The two main parameters defining the applied field are the field strength F_m and frequency Ω; these and the initial principal quantum number n_0 are the main system parameters, but it is convenient to measure these in classical scaled units $\Omega_0 = \Omega/\omega$, where $\omega = \mu e^4/(n_0\hbar)^3 = 6.576 \times 10^6 n_0^{-3}$ GHz is the unperturbed classical frequency, and F_0 the ratio of the applied field strength to the mean Coulomb field $F_0 = F_m(n_0\hbar)^4/\mu^2 e^6$. Classical ionisation probabilities satisfy scaling laws[2] and depend upon the parameters Ω, F_m and n_0 through only the two scaled variables Ω_0 and F_0. Other important parameters are the initial distribution of substates, the interaction time, the shape of the field envelope, $A(t)$ in equation 1 below, and the strength of the static field, if present.

There are currently two sets of experimental results needing different theoretical models. The experiments of Koch and co-workers[3] use an atomic beam of excited hydrogen atoms in a particular level, which can be chosen to be in the range $24 \leq n_0 \leq 90$, or even higher, but an approximate microcanonical distribution of substates. In these experiments the static field present in the cavity are usually small, although some experiments were performed with relatively strong static fields present[6]. Further, static fields are used outside the cavity for detection and to deflect beams and these impose upper limits on the principal quantum number. A full theoretical analysis requires three-dimensional dynamics: a quantal description of this system, without a drastic reduction in the basis size, is beyond the capabilities of modern computers: the numerical integration of Hamilton's equations, with careful regard of the Coulomb singularity, is however feasible on relatively modest computers. In the experiments of Bayfield and co-workers[4] an attempt is made to prepare the atoms in an extremal Stark-state, $n_e = n_2 - n_1 = 1 - n_0$, $n_2 = m = 0$, and to try to keep it from precessing[5] to $n_2 \neq 0$ states with a weak static field, $F_s \approx 1\,Vcm^{-1}$, so that levels up to $n \approx 160$ can exist in the waveguide.

Careful comparison of classical calculations and experimental results, mainly those of Koch[6,7,8], suggest that there are at least four ranges of scaled frequency in which the dynamics has different characteristics and needs different theoretical models. The boundaries of these regions are, of course, approximate and are given by the following inequalities. It must be emphasised that these boundaries are empirical and are based upon limited experience: typically the systems studied to date are exposed to between 100 and 300, and occasionally more, field oscillations, the field is switched on over many field cycles, about 40 or more, and the initial principal quantum number is in the range $30 < n_0 < 100$. Moreover in most experiments 'ionisation' means the sum of ionisation and excitation to states $n > (1.3 - 1.5)n_0$ in the interaction region.

1. **Low frequencies** $0 \leq \Omega_0 < 0.2$: in this region the experimental ionisation curves show a lot of structure not reproduced by classical simulations[8]. In[8] it is shown that this structure is reproduced fairly accurately by a $1d$ model using the adiabatic states defined in[9]. In this representation the quantum dynamics is well described by relatively few states and resonances between these states cause the observed structure. This mechanism cannot be reproduced by classical dynamics, hence the disagreement. As the parameter $C = n_0\Omega_0 F_0$ increases so does the required number of adiabatic states: for $C > 1.5$ many states are needed and although the quantum resonances persist there are so many that classical dynamics provides a good estimate of the mean quantal behaviour.

2. **Intermediate frequencies** $0.2 < \Omega_0 < 1.1$: here there is broad agreement between $3d$ classical simulations and experimental results[6], except at the low order resonances $\Omega_0 = r/s$, for integers $r < s$. Off resonance agreement is usually good for all field strengths: on resonance experiments[10] show that, for the quantum numbers currently accessible, the scaling laws breakdown as expected because the dynamics is dominated by island structures containing relatively few states. However, although detailed agreement is not good, it is clear that the real system, for principal quantum numbers $n_0 \sim 50$ is responding to a resonant field in a similar manner to the classical system.

3. **Medium frequencies** $1.1 < \Omega_0 < 2$: the only comparisons in this range are for the threshold fields $F_m(10)$ producing 10% ionisation. For both the $1d$[11,24] and the $3d$[7] dynamics the classical thresholds, on average, are too small, but the $3d$ comparisons suggests that the real atoms have a similar resonance structure to the classical atoms. Theoretical $1d$ calculations[12], which ignore the continuum, also suggest that the quantum dynamics retains many of the dynamical features seen in the classical dynamics. In this region $1d$ quantum dynamics provides reasonable estimates of $3d$ 10% threshold fields[21].

4. **High frequencies** $\Omega_0 > 2$: in this region two changes occur. First there is a clear divergence of the $3d$ classical and experimental threshold fields[7]; the classical thresholds, for long exposure times, decrease slowly with Ω_0, as $\Omega_0^{-1/3}$ according to resonance overlap theory[13], while experimental and quantal thresholds *increase* approximately linearly with Ω_0, see figure 1. Second, the $1d$ dynamics is dominated by the quasi-resonant (QR) states[14,21], that is those states whose energy differs from that of the initial state by an integer multiple of $\Omega \hbar$, see figure 5; this drastically reduces the effective density of states so making the classical and quantum dynamics quite different. In this high frequency region, for weak fields, the chaotic $1d$ classical dynamics is well described by diffusion in the action variable; because the quantum dynamics is dominated by the relatively fewer QR-states there is agreement between the two dynamics only when the initial principle quantum number n_0 is very large, and, according to[15], much larger than accessible in current experiments, see also equation 14 and[21].

The theory for the microwave ionisation of excited hydrogen atoms is difficult. The field strengths of interest are sufficiently strong to ionise the atom, the density of states is very large so there are very many coupled states, and the continuum needs to be included, although paradoxically it does not necessarily significantly affect the bound-state dynamics. For these reasons most quantal calculations have dealt with a $1d$ model, equation 1 below but with $F_s = 0$ and $A(t) = 1$; a full quantal $3d$ treatment is probably not possible on present computers. The numerical integration of Hamilton's equations is, however, relatively straight forward if regularisation is used to remove the Coulomb singularity[16].

The most detailed comparisons between theory and experiment are for the threshold fields. It has been shown that for $0.2 < \Omega_0 < 2.8$ the $1d$ and $3d$ classical thresholds agree reasonably well[7]; for $0.2 < \Omega_0 < 1$ these also agree well with the experimental results of Koch[6], but for $\Omega_0 > 1$ there is the systematic disagreement discussed above. For $0.4 < \Omega_0 < 1$ the $1d$ classical and quantal thresholds also agree[17]. For stronger fields, $F_m > F_m(10)$, the $1d$ and $3d$ dynamics is quite different, with the $1d$ ionisation probability increasing much more rapidly with F_m than the $3d$ probabilities because the $1d$ system emphasises the role of the Coulomb singularity: some comparisons are given in[8,18].

In the remainder of this article we concentrate on $1d$ systems as this allows direct comparison between classical and quantal probabilities.

2 One-dimensional dynamics

In this section we shall discuss some connections between the classical and quantum dynamics of the $1d$ system and shall concentrate on the high frequency region $\Omega_0 > 2$. For initial principal quantum numbers $n_0 \gg 1$ the Hamiltonian of this system can be accurately approximated by

$$H_1 = \frac{1}{2\mu}p^2 - \frac{e^2}{z} - z\left(F_s + F_m A(t)\sin\Omega t\right), \quad z \geq 0, \quad (1)$$

where $A(t)$ is the field envelope. Most comparisons between classical and quantal solutions have been performed using this Hamiltonian, but usually with $F_s = 0$ and $A(t) = 1$ which does not model any current experiment.

The Hamiltonian H_1 provides a good approximation to the dynamics of an atom prepared in an extremal Stark-state provided that $F_s \neq 0$ and $n_0 \gg 1$. If $F_s = 0$, or is too small, the Runge-Lenz vector precesses so all the n_e-states need to be included[5]; if n_0 is too small the wave function is insufficiently localised about the z-axis to ignore the off-axis motion; if there is no precession this error can, however, be corrected by modifying the matrix elements used in the time-dependent Schrödinger equation.

The static field has another crucial effect: it lowers the ionisation threshold so that there are only a finite number of bound states between n_0 and the ionisation limit $N_i \approx 161(F_s/\text{Vcm}^{-1})^{-1/4}$. For $F_s \sim 1\text{Vcm}^{-1}$ and $n_0 \sim 100$ the quantal dynamics involves about 100 states, which can be managed on most mainframe computers; the problem is that the continuum cannot be ignored. A relatively easy way of including this is described below.

It is generally accepted that classical and quantal dynamics are quite different in this high frequency region, but it is not generally realised that the reasons for this difference can be found by studying classical dynamics and that such studies also help perform efficient quantal calculations; the classical ghost in quantal ionisation.

In analysing the dynamics of H_1 it is important to note that the Coulomb singularity makes this system quite different from systems with smooth potentials encountered in similar molecular problems[18]. The effect of this singularity is most easily seen by using classical perturbation theory to compute the change in the action variable I during one field period: for this computation write H_1 in terms of the unperturbed angle-action variables (θ, I)

$$H_1 = E(I, F_s) - F_m A(t) \sum_{s=0}^{\infty} Z_s(I, F_s) \cos s\theta \sin \Omega t \quad (2)$$

where $E(I, F_s)$ is the unperturbed energy,

$$z(\theta, I) = \sum_{s=1}^{\infty} Z_s(I, F_s) \cos s\theta \quad (3)$$

and where the Fourier components Z_s are fairly complicated functions of I, F_s and s, see[19], but for $s \gg 1$ decrease as $s^{-5/3}$. On integrating Hamilton's equation over an unperturbed period $2\pi/\omega$, assuming $\Omega \gg \omega$ and using classical perturbation theory we obtain

$$\begin{aligned}\Delta I &\approx -\int_0^{2\pi/\omega} \frac{\partial H_1}{\partial \theta}, \quad \theta = \omega t + \theta_0 \\ &\approx \frac{\pi F_m r Z_r}{\omega} \cos r\theta_0, \quad r = [\Omega/\omega].\end{aligned} \quad (4)$$

where $[y]$ denotes the nearest integer to y. The change in energy is just $\Delta E = \omega \Delta I$. There are two important points to notice about this result. The first is that ΔE decays slowly with Ω, because Z_s decays slowly with s, which is a direct consequence of the Coulomb singularity. In other words, because the singularity causes infinite momentum, the motion needs all frequencies for its description so, no matter how large Ω, the field will always couple significantly with the motion. For frequencies accessible to current experiments there are only small departures from this result due to the eccentricity being slightly less than unity. The second point to note is that, when expressed in scaled units and putting $F_s = 0$ in Z_r, the energy change from level n is

$$\Delta E(n) \approx 2.58 F_0 n_0^{-2} \Omega_0^{-2/3} \cos r\theta_0, \quad (5)$$

which is *independent* of n; the significance of this will be discussed in the next section.

Heisenberg's correspondence principle[20] relates matrix elements to Fourier components,

$$\langle n|z|m \rangle \approx Z_{n-m}(I, F_s), \quad I = (n+m)\hbar/2, \quad (6)$$

so the slow decay of Z_s causes significant coupling if unperturbed Stark-states are used in the usual form of the time-dependent Schrödinger equation.

The Coulomb potential is also long ranged and as orbits extend a distance $a_0 n^2$, $Z_s \sim n^2$: but for $z \gg 1$ the applied field dominates and in this region there is almost no energy transfer, see figure 1 of[21]. All the dynamics occurs near the origin. There are three consequences of this. First, because $Z_s \approx \langle n|z|m \rangle \sim n^2$, the dipole gauge, although commonly used in calculations[22,23,24] is numerically exceedingly inefficient. Second, the compensated-energy Hamiltonian[2]

$$H_c = \frac{1}{2\mu} \left(p + \frac{F_m}{\mu \Omega} \cos \Omega t \right)^2 - \frac{e^2}{z} - zF_s \quad (7)$$

is almost constant over most of the orbit, so a basis comprising the set of states which diagonalise H_c – the compensated-energy-representation (CER) – is numerically far more efficient[21]. This is equivalent to using the momentum gauge. Finally, and perhaps most important, because all the interaction is concentrated in the region near $z = 0$ classical dynamics provides a good approximation over times $O(2\pi/\omega)$; only over longer times, when interference effects become important, does it fail.

This last feature gives two important results. First, in conjunction with semiclassical methods[14] we find that the transition amplitudes satisfy an approximate selection rule

$$S(n \to m) \approx \begin{cases} J_p(\mathcal{A}) & n - m = pr = p[\Omega/\omega], \\ 0 & \text{otherwise.} \end{cases} \quad (8)$$

where $\mathcal{A} = \pi F_m Z_r / (\omega \hbar)$ is the change in action during the time $2\pi/\omega$, and $J_p(z)$ is an ordinary Bessel function. Second, using the additional assumption that there is a negligible chance of capture back from the continuum to a bound state, we can use classical or semiclassical dynamics to obtain the decay rate[19] from a given level. A semiclassical approximation to the quantal evolution operator gives, after some manipulation, the ionisation probability per unperturbed period from a level n as[19]

$$P(n) = \sum_{k_m}^{\infty} J_k^2(\mathcal{A}), \quad \mathcal{A} = 2.58 F_0 n_0 \Omega_0^{-5/3}, \quad (9)$$

where

$$k_m = 1 + \left[\frac{-2e\sqrt{F_s} - E_n(F_s)}{\Omega \hbar} \right] \quad (10)$$

is the minimum number of photons required to reach the continuum from level n and each term in the sum represents a k-photon transition into the continuum from the level n. The classical limit $k_m \gg 1$ of this is simply

$$P^{cl}(n) = \begin{cases} (1/\pi)\cos^{-1}(k_m/\mathcal{A}), & \mathcal{A} > k_m \\ 0 & \mathcal{A} \leq k_m. \end{cases} \quad (11)$$

At the other extreme, $k_m = 1$ and $\mathcal{A} < 1$, $P(n)$ reduces to the Fermi Golden rule. These probabilities are used to estimate the decay rates for the time-dependent Schrödinger equation.

Thus we see that classical dynamics, although wrong in the long time limit, predicts the section rule 8 and provides a relatively painless way of incorporating the continuum into the time-dependent Schrödinger equation. This method of including the continuum should be contrasted with others of which there are basically three techniques. The first is to approximate the infinite dimensional vector space by one of finite dimension part of which overlaps the continuum, for example the 'boxed-continuum' method of [22] or the Sturmian basis of [24]; such methods fail for long times if there is significant ionisation. The second is to introduce decay factors into the time-dependent Schrödinger equation, for instance via complex (optical) potentials[25] or projection operator methods with approximate solutions to the continuum dynamics[23]; our method is a variation of this latter theme. Finally, there are the methods involving complex coordinate rotations[17], which should be treated with great caution when applied to the non-analytic 1d Coulomb problem.

3 High frequency behaviour

Using the CER-basis and introducing ionisation as described in the last section the time-dependent Schrödinger's equation can be integrated numerically relatively easily. In the first instance we show in Figure 1 the 10% threshold fields given by experiment[11] and our quantal model: the thresholds given here are for ionisation plus excitation to levels above $n = 130$. In these computations we take the field envelope to be $\sin\beta t$ with $\beta = \Omega/2N$, N being the number of field oscillations to which the atom is exposed, $n_0 = 98$ and $F_s = 0.875\text{Vcm}^{-1}$: typically $N \sim 120$.

There are two points to notice here. First that there is resonable agreement between theory and experiments, which gives us some confidence in the ionisation mechanism used. Second the theoretical results show more variation with Ω_0. The fluctuations in the quantal probabilities are seen very clearly in figure 2 where we compare the classical and the quantal ionisation probabilities as functions of the frequency. The statistical er-

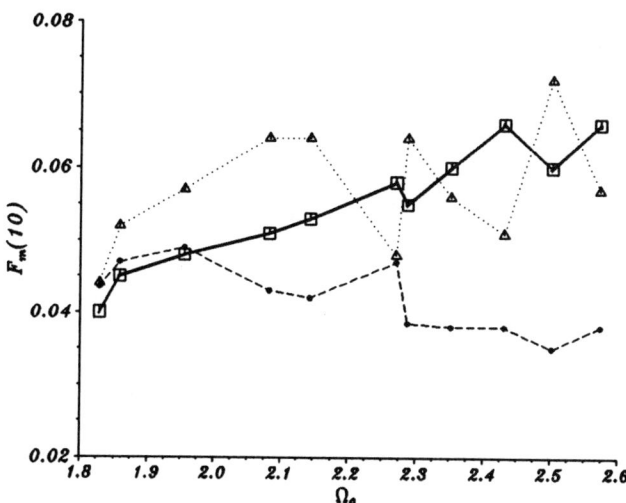

Figure 1: Figure showing the 10% threshold fields, $F_m(10)$ as functions of the scaled frequency: the experimental results □ are taken from [11] which are compared with the 1d quantal △ and classical ◇ theories. Here 'ionisation' is defined to be ionisation plus excitation to levels $n > 130$.

rors in the classical probabilities are less than ± 0.03. It is appropriate to use the Stark-shifted energy difference to define a scale frequency so in this graph we define $\Omega_1 = \Omega_0/n_0^3 \Delta E(n_0, n_0+1) = 1.083\Omega_0$, and it is seen that the classical system, curve (c), is more robust on- than off-resonance. The quantal probability is far more variable than the classical probability with order of magnitude changes with relatively small changes in Ω_1. Notice that the classical probabilities for *fixed* F_0 appear to be increasing, contrary to the statement made in section 1: this is entirely due to the relatively short exposure time. A more detailed comparison of the time-dependence of the probabilities is given in[19].

This erratic behaviour of the quantal probabilities is also seen in the variation of P_i with the static field, F_s, shown in figure 3. Again it is seen that the classical probabilities vary smoothly with F_s, in contrast to the erratic behaviour of the quantal probabilities. As far as we are aware the experimental ionisation probabilities also vary smoothly with F_s; we shall discuss a possible reason for this below. We find similar erratic variation at all frequencies and at higher field strengths.

On the other hand variations of both classical and quantal probabilities with F_0 are smooth and, in general, monotonic increasing. In figure 4 we show the ionisation curves for the scaled frequencies $\Omega_0 = 2.5$ and 2.6 corresponding to the local maximum and minimum at $\Omega_1 = 2.71$ and 2.82 of figure 2. Notice that that dramatic change in the quantal probabilities is not mirrored in the classical curves, and that P_i rises earlier and faster

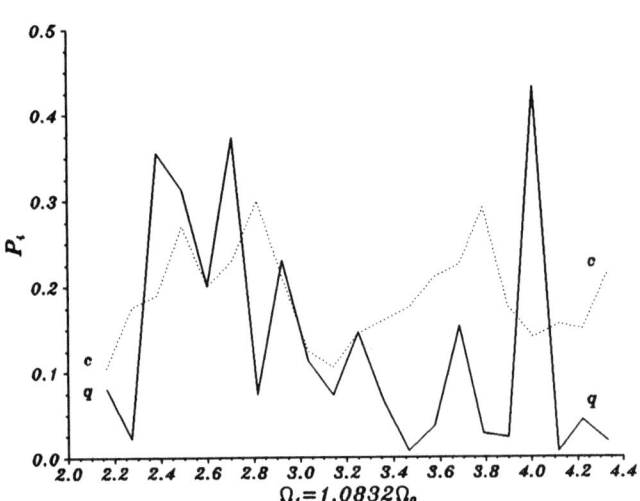

Figure 2: Classical, dotted curve c, and quantal, solid curve q, ionisation probabilities for the sinusoidal field envelope with 120 field oscillations $F_s = 1\text{Vcm}^{-1}$ and $n_0 = 100$. For these comparisons we changed the microwave field amplitude with frequency according to $F_0 = 0.025\Omega_0$, on order to allow for the decrease in P_i with increasing frequency. The meaning of Ω_1 is given in the text.

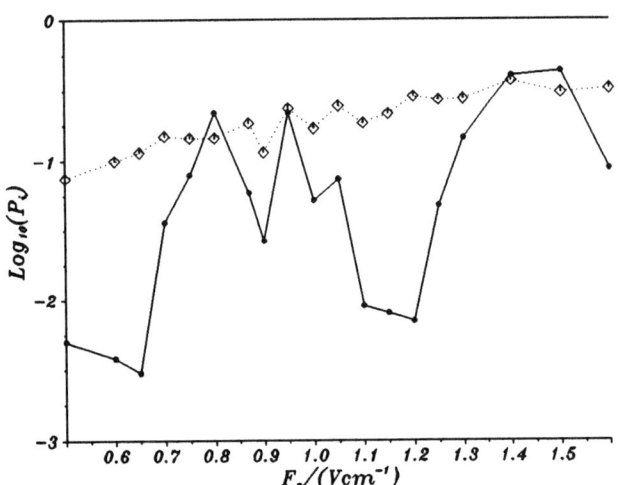

Figure 3: Graph showing the variation of the classical, curve $\diamond$, and quantal, $\bullet$, ionisation probabilities with the static field strength F_s in the case $\Omega_0 = 2.5736$, $F_0 = 0.057$ and $n_0 = 98$.

at $\Omega_0 = 2.5$; it is not clear what causes this robust behaviour at $\Omega_0 = 2.6$ but the level $n = 118$ is preferentially populated either because there are resonant transitions to it or because transitions to higher states are inhibited because the detuning parameters, discussed below equation 13, are large. We find that $P_i(F_0)$ is usually monotonically increasing, but that if ionisation is defined to be excitation above a certain level then the probability can have local maxima.

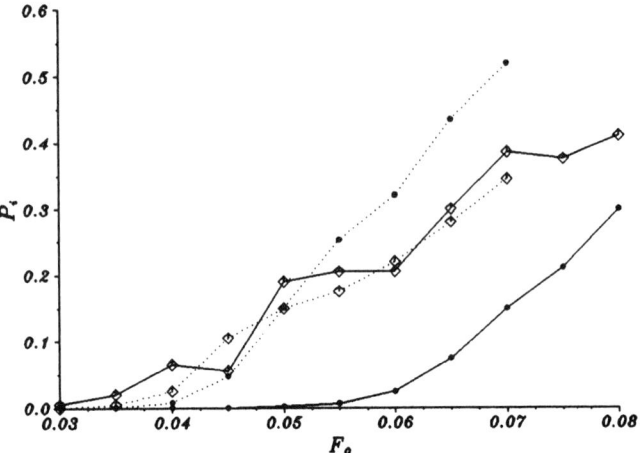

Figure 4: Graph showing the variation of the classical, curve $\diamond$, and quantal probabilities, $\bullet$, with the microwave field in the cases $\Omega_0 = 2.5$, dotted curve, and $\Omega_0 = 2.6$, solid curve, with $n_0 = 100$, $F_s = 1\text{Vcm}^{-1}$.

The reasons for this strange behaviour can be inferred from figure 5 in which we show the final classical and quantal probability distributions $P(n_0 \to n)$ and also the quantal distribution about half way through the cavity, when the microwave field is at its maximum strength. The final quantal probabilities are very strongly peaked on the QR-states, that is those states, n_R, for which the energy differences approximately equal the photon energy,

$$E_{n_{R+1}}(F_s) - E_{n_R}(F_s) \approx \Omega\hbar \quad R = \ldots -1, 0, 1, \ldots \quad (12)$$

with n_0 being the initial state. We find this behaviour to be ubiquitous for scaled frequencies $\Omega_0 > 2.5$; for weak fields it also occurs for $\Omega_0 > 1.5$. It is the consequence of the selection rule 8, derived using semiclassical methods. This selection rule is most pronounced when the field is switched off slowly, but it is present all through the cavity, as shown by the dashed curve, although the variation from maximum to minimum is not so dramatic.

When the parameters F_s and Ω_0 change the relative heights of the quantal peaks change dramatically, as would be expected from the above discussion. The classical probabilities, which for the purposes of comparison,

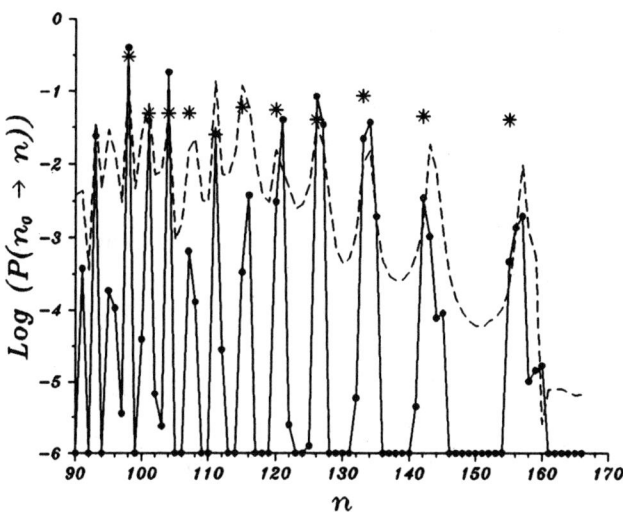

Figure 5: Probabilities, on a logarithmic scale, of excitation for the parameters $\Omega_0 = 2.5736$, $n_0 = 98$, $F_s = 1\text{Vcm}^{-1}$ and $F_0 = 0.06$ with 135 field cycles. The solid curve, •, is the final quantal probability; the dashed curve the quantal probability after 70 cycles. Stars depict the final classical probabilities.

are aggregated onto the QR-peaks, are relatively insensitive to changes in these parameters. As F_0 increases this picture does not change significantly.

It is important to remember that the occurrence of these peaks in the presence of the microwave field is basis-dependent; calculations performed in the dipole gauge using a basis of unperturbed wave functions do not show this structure[24], except for weak fields.

Figure 5 also provides a clue about the behaviour of the ionisation probability with the interaction time, for the classical energy transfer from level n, equation 5, gives a rough estimate of the level $n = n_i$ above which direct classical ionisation can occur. On making an approximation to $E(n, F_s)$ we find that

$$(n_i/n_0)^2 \approx \left[D + \sqrt{D^2 - (F_s/\text{Vcm}^{-2})(n_0/197)^4} \right]^{-1}$$
$$D = 2.6 F_0 \Omega_0^{-2/3} + (F_s/\text{Vcm}^{-2})^{-1/2} (n_0/189)^2.$$

For the parameters of figure 5 this gives $n_i \approx 126$ which is a typical value for the Bayfield experiments at those fields producing 10% ionisation. These relatively low values of n_i are readily accessible from the initial state, so we should expect the ionisation probability to depend significantly upon the interaction time. In particular, for field envelopes having a broad, flat maximum, as for instance in the Koch experiments, we should expect larger ionisation probabilities and hence smaller threshold fields, provided all other parameters remain the same. Theories providing estimates of $F_m(10)$ which ignore the shape of the field envelope clearly ignore much of the relevant physics of the process.

Figure 5 strongly suggests that the quantum dynamics is dominated by the QR-states, and that the other states play a relatively minor role. If this were true then the detuning parameters, defined by the equation

$$\frac{E_{n_{R+1}}(F_s) - E_{n_R}(F_s)}{E_{n_0}(F_s) - E_{n_0+1}(F_s)} = \Omega_1 + \alpha_R(F_s), \quad |\alpha_R| < \frac{1}{2}, \tag{13}$$

which are a measure of how near each QR-state is to resonance, would play an important role in the quantum dynamics. There are four reasons why we think that it is these few detuning parameters which are responsible for the erratic quantal behaviour.

1. The α_R depend only upon F_s and Ω_0 (or Ω_1) and are independent of F_0.

2. The α_R are discontinuous functions of F_s and Ω_0.

3. In all case we have examined in detail significant changes in P_i and $P(n_0 \to n)$ are associated with discontinuous changes in the detuning parameters.

4. If all states played an equivalent role then there would be too many detuning parameters for the discontinuous behaviour of a few to affect the probabilities.

Classical dynamics is also affected by the detuning parameters, but as it knows nothing of the selection rule 8 there are so many parameters that the variation in the classical probabilities is relatively smooth. Such behaviour is seen in figures 2 and 3, and also in Figure 5. These classical probabilities vary relatively smoothly with n_R and much of the observed variation is caused by Monte-Carlo statistical errors. For these quantum numbers, $n_0 \sim 100$, we find no agreement between classical and quantal probabilities as F_0 increases.

From these results we conclude that the quantum dynamics of the 1d system in a high frequency field is dominated by the QR-states and that this domination leads to the erratic behaviour seen in the figures. It seems reasonable to suppose that the classical limit can be reached only when n_0 is sufficiently large that the variation of the detuning parameters, α_R, over many, say 10 or more, QR-states is smooth; the condition for this is[21]

$$n_0 > 30\Omega_0^2, \quad \Omega_0 > 2.5, \quad (F_s = 0). \tag{14}$$

Finally, we note that the experimental results of Bayfield[11] vary relatively smoothly with all parameters. It is not absolutely clear why this should be so, but we note that in these experiments the substate selection is not perfect and that there is an almost uniform distribution of substates in the range $1 - n_0 \leq n_e < n_0/2$. If one makes the crude assumption that the n_e quantum number is constant then each set of n_e-substates will have an associated set of detuning parameters, $\alpha_R(n_e)$, dependent upon n_e, so the measured probability will be an average of the probabilities derived from these. Since there are many substates and since the the variation of P_i with α_R is so dramatic, it is easy to imagine that such an average would produce a smoother variation with F_s and Ω_0. In practice the n_e-substates are weakly coupled, but we should expect the same effect. It is a challenge for future experiments to observe this erratic quantal behaviour, which will probably be seen only with a very pure substate selection.

Acknowledgements

We thank the SERC for a visiting fellowship which made this work possible. We also thank Professor P M Koch for stimulating and maintaining our interest in this problem, and NATO for making visits to his Laboratory feasible.

References

1. Graffi S and Yajima K 1983 Comm Math Phys **89** 277

2. Leopold J G and Percival I C 1979 J Phys B **12** 709: Jones D A, Leopold J G and Percival I C 1980 J Phys B **13** 31

3. Koch P M 1982 J Phys (Paris) Colloq **43**:C2 187: Koch P M 1983 'Microwave ionisation of highly excited hydrogen atoms' in 'Electronic and Atomic Collisions' (Eds H B Gilbody, W R Newell, F H Read and A C H Smith, Elsevier Science): Koch P H and van Leeuwen K A H to be submitted to Phys Rev A

4. Bayfield J E 1987 'Studies of the sinusoidally driven weakly bound atomic electron in the threshold region for classically stochastic behaviour' in 'Quantum Measurements and Chaos' (Eds E R Pike and S Sarkar, Plenum)

5. Leopold J G and Richards D 1987 J Phys B **20** 2369

6. van Leeuwen K A H, v. Oppen G, Renwick S, Bowlin J B, Koch P M, Jensen R V, Rath O, Richards D and Leopold J G 1986 Phys Rev Lett **55** 2231

7. Galvez E J, Sauer B E, Moorman L, Koch P M and Richards D 1988 Phys Rev Lett **61** 2011

8. Richards D, Leopold J G, Koch P M, Galvez E J, van Leeuwen K A H, Moorman L, Sauer B E and Jensen R V 1989 J Phys B **22** 1307

9. Richards D 1987 J Phys B **20** 2171

10. Koch P M, Moorman L, Sauer B E, Galvez E J, van Leeuwen K A H, and Richards D, 'Experiments in Quantum Chaos: Microwave Ionisation of Hydrogen Atoms' (Proceedings of 20th EGAS Meeting, Graz, Austria 1987) 1989 Physica Scripta **T26**, 51

11. Bayfield J E, Casati G, Guarneri I and Sokol D W 1989 Preprint

12. Leopold J G and Richards D 1988 Phys Rev A **38** 2660 and J Phys B **21** 2179

13. Chirikov B V 1979 Phys Rep **52** 263

14. Jensen R V, Leopold J G and Richards D 1988 J Phys B **21** L527: Richards D, Leopold J G and Jensen R V 1989 B **22** 417

15. Jensen R V 1989 Phys Rev Lett **62** 1476

16. Leopold J G and Richards D 1985 J Phys B **18** 3369; Rath O and Richards D 1990 in preparation: Rath O and Richards D J 1988 Phys B **21** 555

17. Bardsley J N and Comella M J 1986 J Phys B **19** L565

18. Richards D 1989 'The Coulomb Potential and Microwave Ionisation' in proceedings of XVI ICPEAC satellite at Yale, July 1989

19. Leopold J G and Richards D 1989 In preparation

20. Percival I C and Richards D 1975 Adv Atom and Mol Phys **11** 1 (Eds D R Bates and B Bederson, Academic Press)

21. Leopold J G and Richards D 1989 J Phys B **22** 1931

22. Susskind S M and Jensen R V 1989 Phys Rev A **38** 2231

23. Blümel R and Smilansky U 1987 Z Phys D **6** 83

24. Casati G, Chirikov B V, Shepelyansky D L and Guarneri I 1987 Phys Rep **154** 77; Casati G, Guarneri I and Shepelyansky D L 1988 J Quantum Electronics **24** 1240

25. Brown R C and Wyatt R E 1986 J Phys Chem **90** 3590

RECENT PROGRESS IN ABOVE-THRESHOLD IONIZATION

P. H. Bucksbaum

AT&T Bell Laboratories,
Murray Hill, NJ 07974 USA

Above-threshold ionization spectra show very strong dependence on laser polarization. Experiments using sub-nanosecond linear, elliptical, and circularly polarized light show how the coupling between bound states and the wiggling electron continuum changes with polarization. Recently, these experiments have been extended to sub-picosecond time domains. The new results show that linearly polarized light induces ionization via Stark-shifted bound-state resonances, whereas circularly polarized light causes non-resonant ionization directly into the continuum.

1. INTRODUCTION

Above-threshold ionization, where atoms absorb more than the minimum number of photons during ionization in strong laser fields, is now a familiar part of the phenomenology of intense laser-matter interactions. However, ten years after its discovery,[1,2] and following several years of investigation,[1-3] we still do not understand how to perform calculations in the ATI regime, where the coulomb and laser fields are comparable in strength.

Part of the difficulty is comparing calculations, which are generally carried out using steady-state, plane-wave, and electric dipole approximations, to experiments using tightly focused ultra-short laser pulses. Experimental electron spectra are corrupted by ponderomotive energy shifts of the photoelectrons that must traverse the strong intensity gradients in the focus.[4]

These post-ionization energy shifts can be virtually eliminated by employing sub-picosecond laser pulses.[5] In such short light pulses, electrons have no time after ionization before the light turns off, to accelerate in the ponderomotive potential. Electrons are detected with the energy they received during the ionization process alone.

Sub-picosecond photoelectron spectra reveal that bound state resonances help to facilitate ATI, at least for linear polarization. These resonances produce a closely spaced substructure in the ATI peaks, which become visible once the free electron ponderomotive effects are reduced.[5]

This resonant structure in the ATI cross section complicates calculations. ATI with circularly polarized light presents a simpler system: angular momentum selection rules restrict the excited states that contribute to ionization, and allowed resonances have very weak multiphoton matrix elements, so that direct photoionization to the continuum should dominate the total rate. Recent measurements confirm this,[6] and point toward a new technique to study the final states in ATI.

2. PONDEROMOTIVE FORCES

Several ATI spectra at different intensities are shown in figure 1. These spectra are typical for ATI experiments carried out with pulses longer than a few picoseconds. The regular arrangement of peaks agrees with the n-photon version of Einstein's photoelectric formula:

$$E_{electron} = nh\nu - E_0, \quad (1)$$

where E_0 is the ionization potential in the absence of any A.C. Stark shift. The match is somewhat deceptive, since both the atoms and the electrons are shifted in energy by several eV at these intensities. Evidently, these shifts nearly cancel for figure 1.

The free electron shift may be calculated from the time-averaged kinetic energy due to classical wiggle motion of the electron in the laser field. This is called the "ponderomotive" energy[7]

$$U_P(x,t) = \frac{e^2 I(x,t)}{2\pi m_e c \nu^2}, \quad (2)$$

and equals 1 eV for an electron in a Nd:YAG laser field ($h\nu$=1.165eV) with an intensity of about $10^{13} W/cm^2$.

The ponderomotive energy depends linearly on intensity, so that it varies over the focused laser pulse. The spatial intensity variation gives rise to forces along the negative gradient of the wiggle energy. This time-averaged kinetic energy affects the electron's motion as a potential, the "ponderomotive potential."

These forces disrupt electron spectra, by scattering and accelerating electrons by as much as several eV in the fields necessary for ATI (See figure 2). In addition, temporal variations in the laser pulse can cause discernible energy gains or losses, known as "surfing".[8] Both ponderomotive surfing and scattering effects can be seen

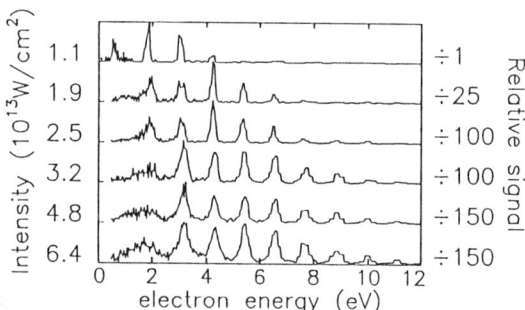

Figure 1: Photoelectron spectrum showing above-threshold ionization in xenon by 1.165 eV photons. The laser was linearly polarized and 100 psec in duration, focused to a 12 μm gaussian waist, with peak intensity shown on the left axis. The electrons were detected along the laser polarization direction, and energies were analyzed by time-of-flight. (From reference 3.)

clearly at the electron energies and laser intensities employed in ATI experiments (see figure 2).

Since a focused laser beam contains a continuous intensity distribution, we might expect the ponderomotive accelerations to produce a continuous spectrum of photoelectron energies in an ATI experiment. The reason that this does not happen is because the atomic levels undergo A.C. Stark shifts as well, which closely follow the free electron energy shifts. The situation is shown schematically in figure 3. In most ATI experiments, the ionization potential shifts with respect to the ground state by an amount nearly equal to U_P. In that case, the final state acceleration merely replenishes the electron energy to the value it would have had in the absence of any shifts. In other words, the Stark effect and the ponderomotive potential effectively hide each other.

3. CALCULATING ATI

ATI calculations fall into three broad categories: perturbation theories, analytical approximations, and numerical calculations.

Perturbation theories treat each ATI peak as a sum of discrete nth-order processes, where n is the number of photons absorbed in order to reach that energy from the ground state. These calculations become increasingly difficult as the intensity and number of photons become large. For most ATI, the peak electric field in the laser is on the order of several volts per angstrom, or not much different from the strength of the coulomb field that binds the valence electrons. Thus the nth order matrix element may have significant additional contributions from higher order processes, corresponding to stimulated emission and reabsorption of additional

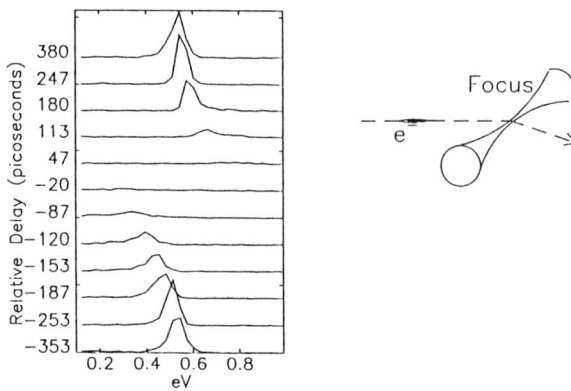

Figure 2: Scattering of electrons by light. 0.54 eV electrons were directed through the focus of a 100 psec laser pulse with peak ponderomotive potential of 8 eV. Those arriving early or late were undeflected. Electrons passing through the leading and trailing edges were accelerated or decelerated, respectively, due to the phenomenon of "surfing", described in the text. Electrons encountering the peak of the pulse were scattered away from the detector. (From reference 8.)

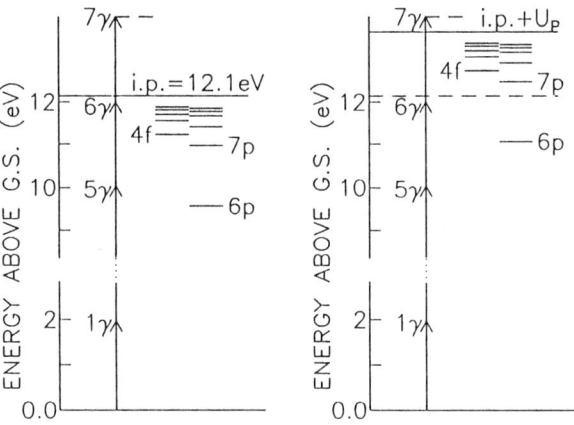

Figure 3. Energy level diagram for ionization in xenon by 2.0 eV photons. The figure on the left shows the unshifted energy levels. The lowest energy electrons emerge with approximately 1.9 eV of energy after absorbing seven photons. The figure on the right shows the relative A.C. Stark shifts for an intensity of $\approx 2 \times 10^{13}$ W/cm^2. The electron kinetic energy is substantially reduced; however, if the pulse length is greater than a few picoseconds, the ponderomotive potential energy of the photoelectron makes up the difference, so that electrons are detected at 1.9 eV.

photons.

Rather than treat the light as a time-dependent perturbation, one may directly solve Schrödinger's equation for the atom in a time-varying vector potential. The exact solution to a hydrogenic atom in the presence of a laser vector potential has not been found. However, several approximate methods have been applied to the problem, and these fall into two broad categories: analytic approximations, and numerical integrations of Schroedinger's equation. Numerical integrations are currently limited by the size and speed (and expense!) of computers, so we will not say more about them.[9] Analytic approximations have had some success, as we cite below.

3.1 Volkov States: The Free Electron in a Laser Field

If the continuum is dominated by the laser, so that coulomb field of the ion plays a negligible role, Schroedinger's equation can be written (in the $\mathbf{A} \cdot \mathbf{p}$ gauge)

$$\frac{[\mathbf{P} - \frac{e}{c}\mathbf{A}(\mathbf{x},t)]^2}{2m_e}\psi(\mathbf{x},t) = i\hbar\frac{\partial\psi(\mathbf{x},t)}{\partial t} \quad (5)$$

where $\mathbf{A}(\mathbf{x},t)$ is the classical (non-quantized) vector potential, which can be approximated as

$$\mathbf{A}(\mathbf{x},t) \approx \mathbf{A}(t) = \hat{\varepsilon} A_0 \sin\omega t \quad (6)$$

in the electric dipole approximation. The solutions, called Volkov states, are the quantum-mechanical versions of free wiggling electrons:[10]

$$\psi_{\text{Volkov}}(\mathbf{x},t) = \exp\left[\frac{i}{\hbar}\mathbf{p}\cdot\mathbf{x} - \frac{i}{\hbar}\frac{p^2}{2m_e}t \right.$$
$$- \frac{i}{\hbar}\frac{e^2 A_0^2}{4m_e c^2}t - \frac{i}{\hbar}\frac{e^2 A_0^2}{8\omega m_e c^2}\sin(2\omega t)$$
$$\left. + \frac{i}{\hbar}|\hat{\varepsilon}\cdot\mathbf{p}|\frac{eA_0}{m_e\omega c}\sin(\omega t)\right] \quad (7)$$

Like plane waves, the solutions are eigenstates of the *canonical* momentum with eigenvalue $\mathbf{p}$. The extra time-dependent phases are periodic energy fluctuations caused by the oscillating field. They correspond to the energy fluctuations in a classical wiggling electron, which modulate the energy at frequencies ω and 2ω. However, while classical wiggling electrons cover a continuous but finite distribution of energies, the Volkov state is an infinite superposition of plane wave energy eigenstates that are separated by integer multiples of $\hbar\omega$:

$$\psi_{\text{Volkov}}(\mathbf{x},t) = e^{\frac{i}{\hbar}(\mathbf{p}\cdot\mathbf{x} - \frac{p^2 t}{2m_e} - U_P t)}$$
$$\times \left[\sum_{n=-\infty}^{\infty}\sum_{m=-\infty}^{\infty} J_m(\frac{e^2 A_0^2}{8\hbar\omega m_e c^2}) \right.$$
$$\left. J_{n-2m}(|\hat{\varepsilon}\cdot\mathbf{p}|\frac{eA_0}{\hbar\omega m_e c})e^{in\omega t}\right] \quad (8)$$

Here J_m's are cylindrical Bessel functions, whose arguments are the coefficients of the oscillating terms in the Volkov phase.

This infinite series of eigenstates shows that Volkov states have an energy structure similar to the ATI final states of figure 1: they both consist of a series of energy peaks separated by $\hbar\omega$.

3.2 Circular Polarization

Volkov states can be incorporated into a scattering theory of photoionization originally put forward by Keldysh.[11] These have not been particularly successful. However, in 1986, H. Reiss suggested that certain Keldysh-type theories[12] proposed by him and by F. Faisal could accurately reproduce the circular polarized ATI spectra, with laser intensity as the only adjustable parameter.[13]

The Keldysh-Faisal-Reiss (KFR) calculations were much less successful for linear polarization. This may be due to the fact that the circularly polarized final states carry many units of angular momentum, and are therefore excluded from the vicinity of the ion core by a centrifugal barrier. Therefore, they may only sample regions of space where the Keldysh approximation is valid, i.e. regions where the ion potential is weak.

4. ATI IN ELLIPTICALLY POLARIZED LIGHT

4.1 Keldysh Prediction

The electron continuum may be studied in regions where the coulomb potential is relatively weak, by investigating ATI with elliptical polarization. Keldysh theories predict large modulations in the angular distributions as well as spectra for elliptically polarized light. Experiments show that the final states do not resemble free Volkov electrons in one crucial respect: the distributions do not display the required 4-fold symmetry, but are only symmetric with respect to rotations by π, as required by spatial isotropy. These asymmetries are partly explained by classical mechanics. After absorbing many elliptically polarized photons, an ATI photoelectron carries off several units of angular momentum. For example, the initial atom and final ion are both s states (as in He, for example,) the electron carries off about

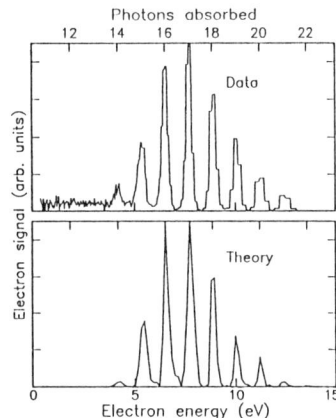

Figure 4. Top: ATI spectrum for circularly polarized 1064 nm light in xenon (from reference 14). Bottom: A computer simulation of the experiment, using ionization rates from the Keldysh theory of H. Reiss (reference 13), integrated over the measured temporal and spatial profile of the laser pulse.

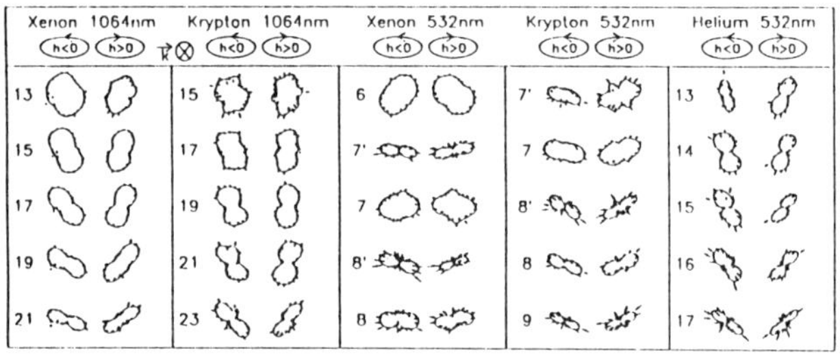

Figure 5. Comparison between data obtained with positive helicity light ($h=+0.82$), and data obtained under the same conditions, but with the helicity reversed ($h=-0.82$). The laser pulsewidth was 0.10 to 0.12 nsec. Xe 1064nm: $I_{peak}=4\times10^{13}$ W/cm^2; $P_{1/2}$ and $P_{3/2}$ final states were not resolved, so numbers indicate photons absorbed for the $P_{3/2}$ final state. Kr 1064nm: $I_{peak}=4\times10^{13}$ W/cm^2; $P_{3/2}$ final states only. Xe 532nm: $I_{peak}=1\times10^{13}$ W/cm^2; primed numbers designate photons absorbed to the final $P_{1/2}$ state; unprimed numbers designate the $P_{3/2}$ state. Kr 532nm: $I_{peak}=1.5\times10^{13}$ W/cm^2; primes mean the same as for Xe 532nm. Helium 532nm: $I_{peak}=1\times10^{14}$ W/cm^2. (From reference 15.)

$nh\hbar$ units of angular momentum, where h is the light helicity and n is the number of photons absorbed. The corresponding classical two-body system would have an impact parameter b of

$$b \approx \frac{nh\hbar}{\sqrt{2m_e(n\hbar\omega + E_G - U_P)}}. \qquad (9)$$

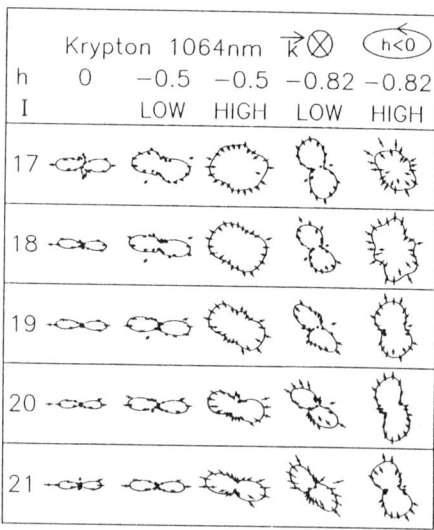

Figure 6. Azimuthal distributions for various retardations and laser intensities, for krypton photoionized by 1064 nm light. Numbers to the left of each row designate photons absorbed to the $P_{3/2}$ ion final state for each ATI peak. First column: linear polarization, $I_{peak}=2.5\times10^{13}$ W/cm^2. Second column: $h=-0.5$, $I_{peak}=2\times10^{13}$ W/cm^2. Third column: $h=-0.5$, $I_{peak}=4\times10^{13}$ W/cm^2. Fourth column: $h=-0.82$, $I_{peak}=2\times10^{13}$ W/cm^2. Fifth column: $h=-0.82$, $I_{peak}=4\times10^{13}$ W/cm^2. (From reference 15.)

Here E_G is the (negative) ground state energy. The quantum-mechanical manifestation of the classical impact parameter is the exclusion of the final state wave function from the ion by a centrifugal potential barrier.

This classical electron escapes via a hyperbolic trajectory, so that the asymptotic momentum deviates from the momentum at the point of closest approach by a substantial angle. For example, for 532 nm light with ($h=-0.82$), electrons ejected from helium after absorbing 14 photons would, in the classical theory, experience an angular deviation of −8° during the electrons transit out of the ion potential.

The quantum-mechanical origins of these deviations are phase shifts imposed on the outgoing electron wave, due to the static potential, bound state resonances, or continuum-continuum transitions.[16] KFR theory does not include these. Attempts to incorporate these shifts into ATI calculations have been successful in qualitative reproduction of the phase shifts shown above.[17]

5. ATI WITH SUB-PICOSECOND PULSES

5.1 Linear Polarization

As explained in the introduction, A.C. Stark shifts can be seen directly if very short laser pulses are used.[5] For pulses shorter than 1 psec, the ponderomotive potential is no longer conservative; in fact, the ponderomotive force acts for too short a time to significantly alter the electron's time-averaged momentum, so that the wiggle energy imposed by the field nearly disappears following the pulse. The top part of figure 7 shows spectra using 140 fsec linearly polarized 616nm pulses, which reveal that each ATI peak really consists of a "fine structure" of electrons with different energies, ionized when the laser intensity was at different values. In a long laser pulse, each of these electrons regains its energy deficit by accelerating out of the ponderomotive potential, and so all electrons in a given ATI peak appear with the same energy. For very short pulses, however, the fine structure is clearly visible, and has been associated with resonance enhancement with excited atomic states that are temporarily Stark shifted into resonance by the intense radiation.

The dominance of intermediate resonances complicates the business of calculating ATI. We would like know if systems exist for which high order multiphoton ionization is truly nonresonant. It now appears that at least one such system has been found: sub-picosecond photoionization of rare gases with *circularly polarized light*.

The bottom trace in figure 7 shows a xenon ati spectrum for 140 fsec 616 nm pulses with circular polarization. The main difference is the absence of resonances in the circularly polarized case. These allowed resonances are high lying states that are relatively circular, i.e. have their angular momentum L approaching their principal quantum number n. Thus they are quite far from the core. Evidently, transitions to continuum states is favored over transitions to these bound states. Even though continuum states require absorption of more photons and thus carry still more angular momentum, the continuum electrons can penetrate close to the ground state more easily because of their high kinetic energy. This argument is explained in more detail in reference 14, and gives very similar results to the KFR calculations for circular polarization, as in reference 13.

In any case, the absence of sharp resonances in the 140 fsec electron spectra is very good evidence that bound states play a minimal role in ATI with circular polarization. We may thus proceed to use these data as a test of theories such as the KFR model, which treat the ATI process as scattering between the ground state and Volkov states in the continuum.

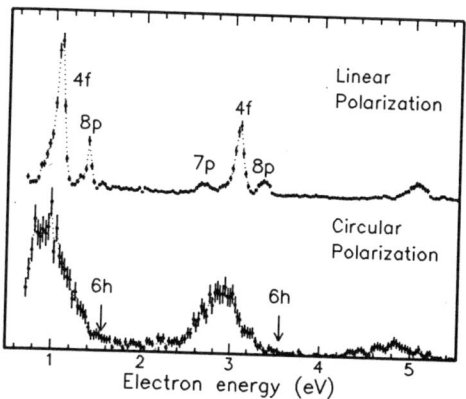

Figure 7. Top: Photoelectron spectrum in xenon for ionization by 140 fsec linearly polarized 616 nm pulses. The peaks correspond to the expected positions of six-photon resonant enhancements, provided that the resonant intermediate states shift by the full ponderomotive potential of a free electron in the laser field, as shown in figure 3. (From reference 6.) Bottom: Same conditions as the top spectrum, only now the laser is circularly polarized.

Figure 8 shows a series of circularly polarized ATI spectra for xenon, taken at different peak intensities. The gradual increase in the ionization potential is seen as a shift of the peaks to lower energies. In this figure, the solid line shows the energy that electrons would have after absorbing eight photons, if the atomic ionization potential did not shift at all. If, on the other hand, the shift is equal to the ponderomotive potential, then the positions will be those marked by the arrows.

For a pulse of known spatial and temporal shape, these data may be converted into photoelectron yields vs. intensity. Figure 9 shows such yield curves, obtained from some of the data in figure 8.

The main advantage of these yield curves over the calculations done previously to compare to the sub-nanosecond data, is that the shorter pulse spectra remove all ambiguity about the laser intensity. This is because the electron energy is tagged to the intensity where it was born in the ionization event.

6. CONCLUSIONS

These experiments have shown that above-threshold ionization is strongly affected by the state of the laser polarization. Previous experiments employing 0.1 nsec circular and elliptical polarization were useful in testing different approximate theories of the ionization process, but are ultimately limited by the inability to know the intensity of the ionizing region accurately. New above-threshold ionization experiments with sub-picosecond

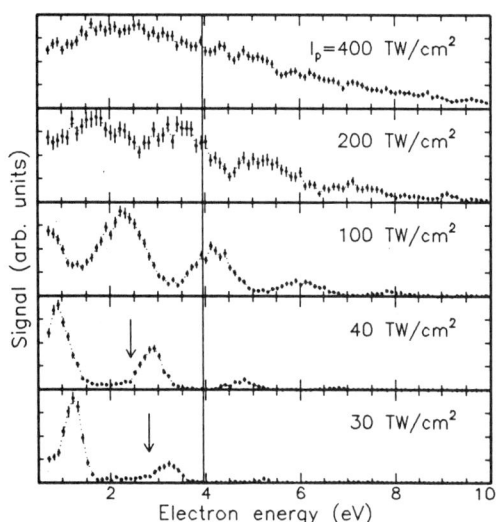

Figure 8. ATI for circular polarization in xenon, for several intensities. (From reference 6.) The solid line shows where the eight-photon peak would be expected for no light shift. (Long pulse ionization would show this, since the ponderomotive acceleration of the electrons following ionization would cancel the a.c. Stark shift of the atom, as described in the text.) The arrows show the highest shift expected for each spectrum, corresponding to the ponderomotive shift at the peak light intensity in the laser pulse.

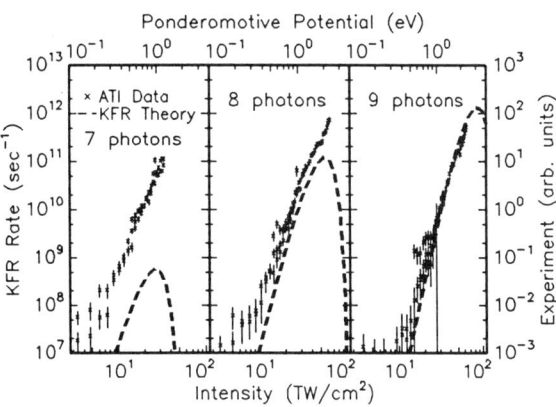

Figure 9 Experimentally determined yield curves for ATI in xenon by circularly polarized 616nm light (from reference 6).

616nm circularly polarized light overcome this problem. The new results show (1) ATI with circularly polarized light is truly nonresonant, unlike ATI using linear polarization; and (2) ionization spectra can be used to generate yield vs. intensity for direct comparison to theory.

All of the experimental work described here has been the result of collaborations at AT&T Bell Laboratories, involving at various times the efforts of M. Bashkansky, L. DiMauro, R.R. Freeman, M. Geusic, T.J. McIlrath, H. Milchberg, D.W. Schumacher, L. Van Woerkom, in addition to myself. I particularly thank Van Woerkom, Freeman, and Schumacher, for permitting me to discuss our most recent work on sub-picosecond circularly polarized ATI.

REFERENCES

1. P. Agostini, F. Fabre, G. Mainfray, G. Petite, and N. Rahman, Phys. Rev. Lett. **42**, 1127 (1979).

2. P. Kruit, J. Kimman, H. G. Muller, and J. J. van der Wiel, Phys. Rev. **A28**, 248 (1983).

3. T. J. McIlrath, P. H. Bucksbaum, R. R. Freeman, and M. Bashkansky, Phys. Rev. **A35**, 4611 (1987).

4. P.H. Bucksbaum, R.R. Freeman, M. Bashkansky, and T.J. McIlrath, Jour. Opt. Soc. Am. B **4**, 760 (1987).

5. R.R. Freeman, P.H. Bucksbaum, H. Milchberg, S. Darack, D. Schumacher, and M.E. Geusic, Phys. Rev. Letters **59**, 1092 (1987).

6. P.H. Bucksbaum, L. Van Woerkom, R.R. Freeman, and D.W. Schumacher, to be published.

7. L. S. Brown and T. W. B. Kibble, Phys. Rev. **133**, A705 (1965);

8. P. H. Bucksbaum, M. Bashkansky, and T. J. McIlrath, Phys. Rev. Lett. **58**, 349 (1987).

9. K.C. Kulander, Phys. Rev. A **38**, 778 (1988).

10. D. M. Volkov, Zeit. fur Physik **94**, 250 (1935).

11. L. V. Keldysh, Sov. Phys. JETP **20**, 1307 (1965).

12. F. H. M. Faisal, J. Phys. B **6**, L89 (1973); H. R. Reiss, Phys. Rev. A **22**, 1786 (1980).

13. H. R. Reiss, J. Phys. B **20**, L79 (1987).

14. P. H. Bucksbaum, M. Bashkansky, R. R. Freeman, T. J. McIlrath, and L. F. DiMauro, Phys. Rev. Lett. **56**, 2590 (1986).

15. M. Bashkansky, P.H. Bucksbaum, and D.W. Schumacher, Phys. Rev. Lett. **60**, 2458 (1988).

16. P. Lambropoulos and X. Tang, Phys. Rev. Lett **61**, 2506 (1988); H.G. Muller, G. Petite, and P. Agostini, Phys. Rev. Lett. **61**, 2507 (1988).

17. S. Basile, G. Ferrante, and F. Trombetta, Phys. Rev. Lett. **61**, 2435 (1988); J. Phys. B **21**, L539 (1988); J. Phys. B **21**, L377 (1988).

GENERATION OF VERY HIGH HARMONICS OF OPTICAL RADIATION IN RARE GASES

A.L'Huillier, L.A.Lompré, M.Ferray and G.Mainfray

Service de Physique des Atomes et des Surfaces
C.E.N. Saclay 91191 Gif–sur–Yvette, FRANCE

We review the main experimental and theoretical results on high–order harmonic generation in rare gas media exposed to strong laser fields. We present new experimental results obtained in xenon which help in understanding the influence of multiphoton ionization and also of propagation effects (phase matching) on the conversion efficiency.

Recent experiments[1-3] show the production of high–order harmonics in a rare gas medium, with photon energies much above the ionization energy. Up to the 33rd harmonic of a Nd–YAG laser (1064 nm) is thus observed in a 15 Torr pressure argon gas jet, at an intensity of 3×10^{13} W.cm^{-2}. These results raise two essential questions, which we shall try to address in the present report: (i) What is the influence of multiphoton ionization on the conversion efficiency ? Does harmonic generation occur preferably in the neutral medium or in an ionized medium ?
(ii) What is the role of propagation effects ? The conversion efficiency at a given harmonic frequency depends both on the single atom response (induced dipole moment) and on the many–atom response (through the phase matching conditions).Can one relate in a simple manner the experimental number of photons detected to a single–atom spectrum ?

Both questions are of the greatest interest in the perspective of the development of very intense lasers, which will enable the study of harmonic generation at very high laser intensity (above 10^{17} W.cm^{-2}).

An harmonic generation experiment consists in focusing an intense laser radiation into a rather dense rare gas medium (a few Torr) and in analyzing along the propagation axis the VUV light emitted during the interaction. In our case[3], we use the fundamental frequency of a mode–locked Nd–YAG laser (40 ps pulse width– 1064 nm wavelength), with a maximum energy of 1 GW at a 10 Hz repetition rate. The laser pulse is focused by a 200 mm–focal length lens over a 18 μm focal radius (confocal parameter b = 4.2 mm). The gaseous medium is provided by a pulsed gas jet[4] producing a well–collimated atomic beam (1 mm FWHM) with a 15 Torr pressure at 0.5 mm from the nozzle of the jet. The VUV light (from 350 nm to 10 nm) is analyzed along the laser axis by using a grating monochromator and detected by photomultipliers or a windowless electron multiplier at wavelengths below 120 nm.

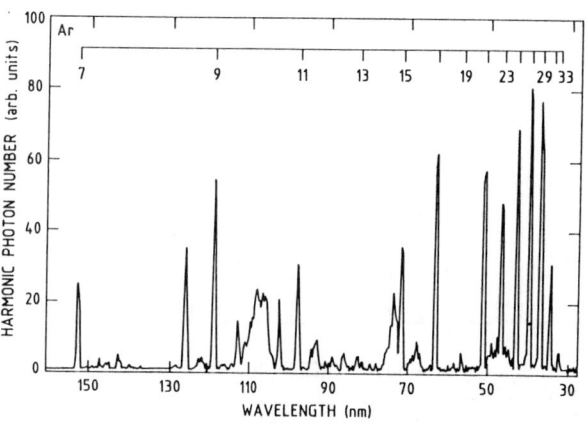

Figure 1. Photon spectrum obtained in Ar at 1064 nm, 4×10^{13} W.cm^{-2}, 15 Torr.

We show in Fig.1 a typical experimental photon spectrum obtained in Ar at 4×10^{13} W.cm^{-2}. The dominant features of this spectrum are the harmonics of the laser field from the 9th to the 33rd. Their widths reflect the resolution of our detection system (≈ 8 Å); their relative amplitudes are meaningless, since no corrections have been made for including the spectral efficiency of the detection system. The other peaks shown in Fig.1 are (besides some harmonics diffracted in the second order) fluorescence lines due to radiative decay from excited states of Ar, Ar$^+$, Ar^{2+}, probably populated through plasma recombination processes. Finally, there is a continuous background, possibly of the same origin as the fluorescence emission.

In Fig.2, we plot the number of photons detected (in a logarithmic scale) for each harmonic of the laser field, in Ar, at 3×10^{13} W.cm^{-2} and 15 Torr. The vertical scale is an estimate (within about one order of magnitude) of the absolute number of harmonic photons produced at each laser shot. We only observe *odd* harmonics (we did not however look for second order harmonic generation). This is to be expected for harmonic generation in an isotropic gaseous medium, with inversion symmetry. The distribution decreases first rather steeply from the third to the seventh harmonic; there is a plateau from the seventh to the 29th with a missing 13th harmonic; it decreases again steeply down to the 33rd harmonic. We also indicate in Fig.2 a photon energy scale in eV (the laser photon energy is 1.165 eV) and some characteristic ionization energies of Ar. The absence of the 13th harmonic is probably due to the influence of discrete excited states (inducing breaking of phase matching or reabsorption), since the energy of thirteen laser photons is just below the ionization limit, in a rather dense spectral region.

The harmonic distribution shown in Fig.2 represents the main physical result of these experiments on harmonic generation in rare gases. McPherson and coworkers[1] have obtained similar results, up to the 17th harmonic in neon and a distribution that decreases rapidly for the first orders and then much more slowly as the plateau in Fig.2, using an excimer laser (248 nm) at higher laser intensity ($\geq 10^{16}$ W.cm^{-2}). These results[1-3] have stimulated a number of theoretical works[5-8], either within the framework of perturbation theory (in hydrogen)[5] or going beyond it, using a Floquet method[6] (also in H) or performing time-dependent calculations[7-8]. Fig.3 shows, for example, the result obtained in hydrogen by Potvliege and Shakeshaft[6]. The square of the Fourier transform of the dipole moment is plotted in solid line as a function of the harmonic order. The dashed line indicates the result obtained in the framework of lowest-order perturbation theory: important non-perturbative effects appear beyond 10^{13} W.cm^{-2}.

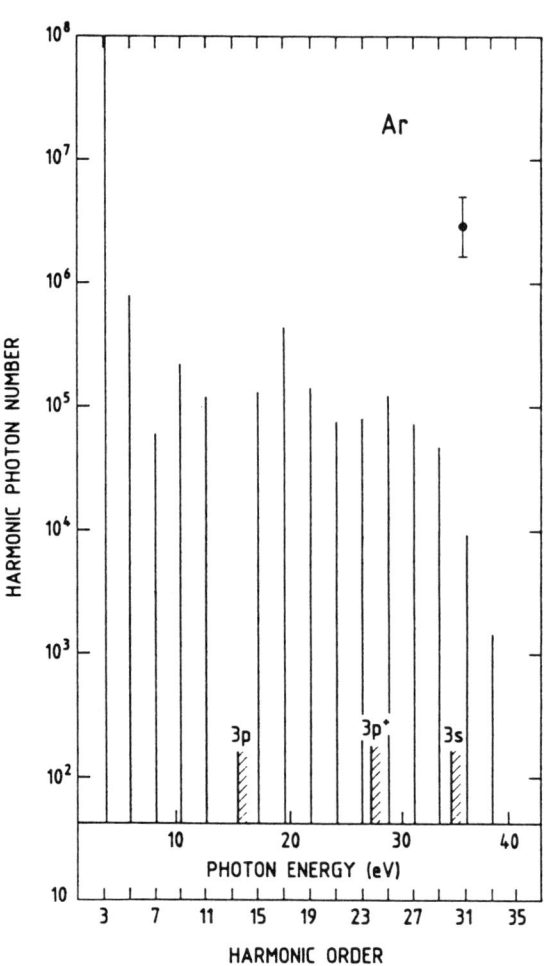

Figure 2. Harmonic intensity distribution in Ar at 3 10^{13} W.cm^{-2}

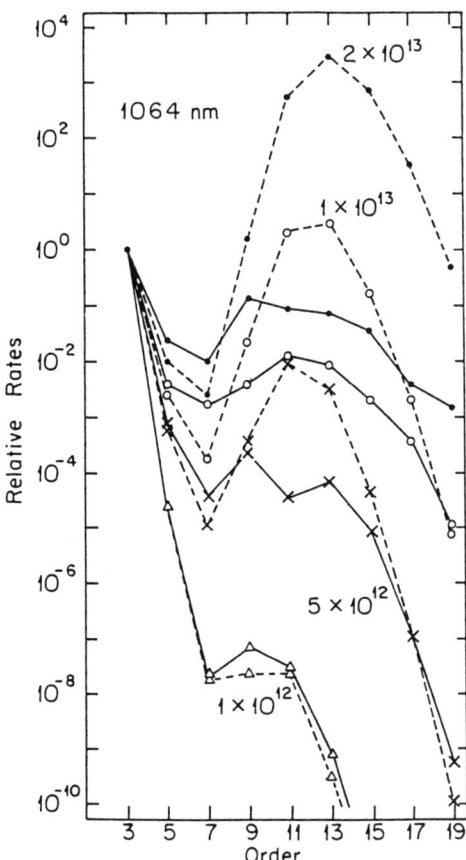

Figure 3. Relative harmonic generation rates in hydrogen[6].

In Fig.4, we show the result of Kulander and Shore[9] obtained in Xe, by numerically solving the time–dependent frozen–core Hartree–Fock equations. Besides the harmonics, there is a pronounced resonance and a continuous background.

If we assume that the number of photons that could be detected in an experiment is directly proportional to the rates shown in Figs.3,4, we see that both (non–perturbative) theoretical results (Figs.3,4) show a behavior quite similar to the experimental results (a rapid decrease, a plateau and a fall off). A comparison between experimental[3] and theoretical[7] results can be made in xenon : both results indeed agree quite well[7]. The important question that we would like to discuss in the present report is therefore the following : is it meaningful to compare in this way experimental and theoretical results or is the apparent good agreement accidental ? The theoretical results shown in Figs.3 and 4 include neither the ionization of the gaseous medium which occurs in the experiment, nor propagation effects (depending on the phase–matching conditions) which could significantly affect the overall harmonic distribution.

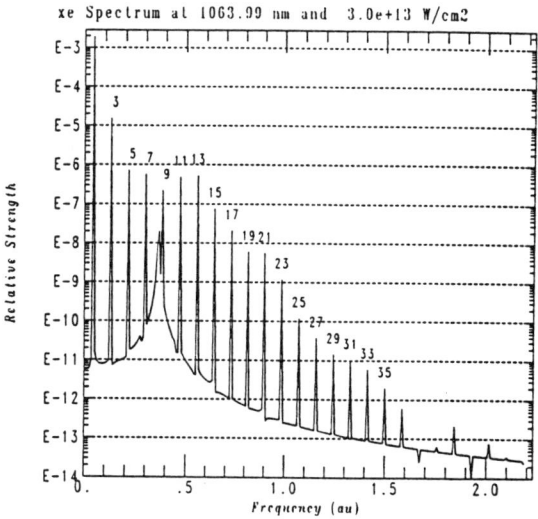

Figure 4. Relative harmonic generation rates in xenon[9].

Before presenting further experimental and theoretical works to clarify this question, let us briefly recall the main equations governing harmonic generation in an optically thick (neutral) gaseous medium. The starting point[9] is the propagation equation for the qth harmonic field (E_q) driven by the component oscillating at $q\omega$ of the nonlinear polarization (P_q).

$$\nabla_\perp^2 E_q + 2ik_q \frac{\partial}{\partial z} E_q = -\frac{4\pi}{c^2} q^2 \omega^2 P_q e^{-i\Delta k_q z}$$

where k_q is the wavevector of the qth harmonic field, c the speed of light and Δk_q ($=k_q-qk_1$) is the phase mismatch between E_q and P_q. Neglecting indirect (cascade) processes involving wave mixing with other harmonic fields, the *lowest–order* contribution to P_q is related to the laser field E_1 by

$$P_q = \mathcal{N}\chi^q E_1^q / 2^{q-1}$$

where χ^q denotes the qth order nonlinear susceptibility and $\mathcal{N}$ the atomic density. Assuming furthermore that the incident laser is a gaussian beam, the (analytical) solution of the propagation equation is also a gaussian beam with the same confocal parameter (b). The maximum intensity of the qth harmonic field is given by

$$I_q = \frac{2^{q+1} b^2 q^2 \pi^{q+3}}{\lambda^2 c^{q-1}} I^q |\chi^q|^2 \mathcal{N}^2 |F_q|^2$$

This expression involves two important terms, the square of the atomic nonlinear polarization at frequency $q\omega$, written here as $I^q|\chi^q|^2$ (perturbative picture), which characterizes the response of each individual atom to the radiation field; the phase matching factor F_q, defined by

$$F_q = \int_{-1/2}^{1/2} (1+2iz/b)^{1-q} e^{-iz\Delta k_q} \, 2dz/b$$

for a square atomic beam[11-12] of finite length (1). F_q describes the response of the whole medium and depends on macroscopic quantities such as the confocal parameter (b) or the atomic density ($\mathcal{N}$). Separating single–atom and many–atom contributions, the number of photons N_q detected in an experiment may be written as

$$N_q = K \, \{(2\pi I/c)^q |\chi^q|^2\} \, \{b^3 \mathcal{N}^2 |F_q|^2/\sqrt{q}\}$$

where K is a constant. This simple expression has the advantage of showing the important parameters governing harmonic generation, microscopic (χ^q) as well as macroscopic (b, $\mathcal{N}$). The quantity represented in Figs.3,4 is the non–perturbative generalization of the first parenthesis : $|d(q\omega)|^2 = (2\pi I/c)^q |\chi^q|^2$, where $d(q\omega)$ is the Fourier transform of the dipole moment. As already mentioned, two questions obviously require investigation: is the formalism valid for an ionized medium (or more precisely, a medium which becomes ionized as harmonics are generated) ? Is it possible to estimate the contribution of phase matching (last part of N_q) to the harmonic intensity distribution ?

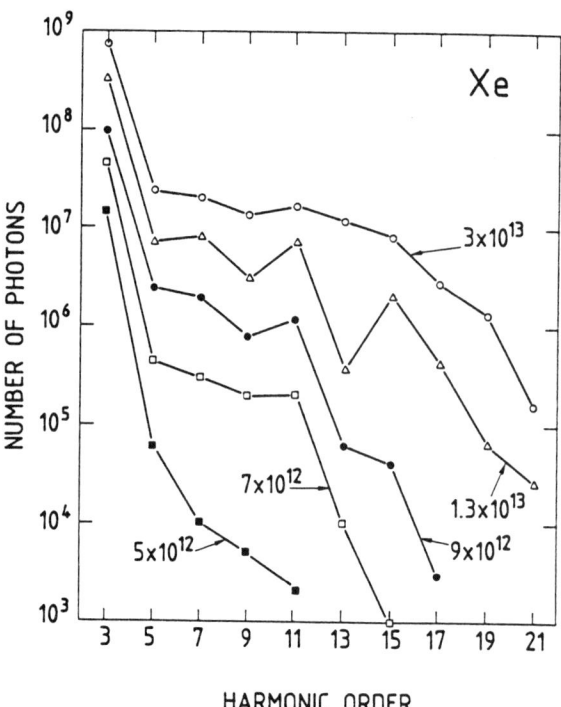

Figure 5. Harmonic intensity distribution in Xe at 15 Torr, at several laser intensities

Fig.5 shows the number of harmonic photons generated in xenon at several laser intensities from 5×10^{12} W.cm^{-2} to 3×10^{13} W.cm^{-2}. The saturation intensity for ionization, at which the ionization probability becomes close to unity and from which the number of ions created significantly deviates from a high power law (Refs.3,11) is estimated to 1.3×10^{13} W.cm^{-2}. As the intensity increases, a plateau appears (at 7×10^{12} W.cm^{-2}) and extends up to the saturation intensity. Beyond this intensity, the distribution becomes smoother, but the maximum order that can be detected remains constant (equal to 21). Most of the high–order harmonics are created at an intensity such that the medium is not or weakly ionized. This seems to rule out plasma effects[13] as responsible for the production of high–order harmonics of the laser field.

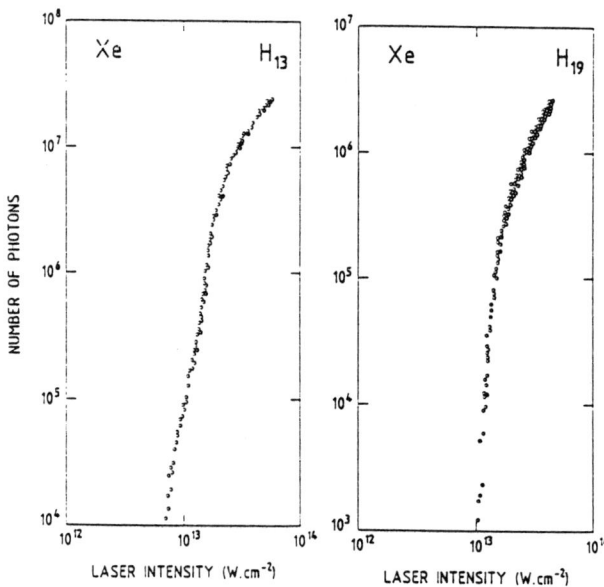

Figure 6. 13th and 19th harmonics as a function of the incident laser intensity.

In order to understand the influence of ionization on harmonic generation, we have studied the intensity dependences of all the harmonics, in particular in the vicinity of the saturation intensity. Figure 6 presents the number of photons at the 13th and at the 19th harmonic frequency in Xe as a function of the laser intensity in a double logarithmic plot. The first harmonics (3,5,7,9) closely follow a I^q power law before the onset of saturation. Beyond the saturation intensity I_s, they vary much more slowly. The 13th harmonic has a more complicated behavior, with a marked change of slope at an intensity below I_s, and which could have several explanations going from resonant or non-perturbative effects to phase matching or absorption (the energy of 13 photons is just above the 5p ionization threshold where the photoabsorption is maximum). Note that the anomalous behavior of the 13th harmonic is indeed reflected in the distributions shown in Fig.5. Finally the higher order harmonics (11, 15–21) approximately vary as I^{12} before saturation, with a pronounced bending at the onset of saturation (see Fig.6).

Multiphoton ionization therefore limits harmonic generation. There are two possible explanations for this limitation. First, the ionic nonlinear response (nonlinear susceptibilities) could be smaller than the atomic one. In this case, the saturation of the conversion efficiency would simply be due to the depletion of the nonlinear medium mainly responsible for harmonic generation. One may also think of breaking of phase matching conditions due to the presence of free electrons in the medium, introducing an additional (positive) phase mismatch (Refs.10,14). We have tried to model this latter effect and the result of the calculation is presented in Fig.7 for the third and fifth harmonic (together with the experimental results). The phase mismatch introduced by the presence of free electrons is equal to $2\pi e^2(q^2-1)\mathcal{N}_e/qmc\omega$ with

$$\mathcal{N}_e(r,z,t) = 1 - \exp(-\int_{-\infty}^{t} I^k(r,z,t)/I_s^k \, dt)$$

where k is the number of photons required for ionization. Since $\mathcal{N}_e$ (and Δk) depends on the transverse coordinate r, the propagation equation needs to be solved numerically. More details on this calculation will be presented elsewhere[15]. The agreement between the experimental and the theoretical results accounting for the presence of free electrons looks rather good. An important conclusion is that, below I_s, their influence is negligible. Above 7×10^{13} W.cm^{-2}, the I^3 or I^5 power law is recovered; the medium responsible for harmonic generation is then an ionic medium. (the intensity is so high that ions are created extremely rapidly at the beginning of the pulse, before harmonic generation takes place). Experimental results may be compared to theoretical results not taking into account ionization effects below saturation. Note that the deviation from the lowest order I^q power law for the high order harmonics below saturation could be due to non perturbative effects, but also to the presence of a few electrons in the medium which may already seriously affect phase matching.

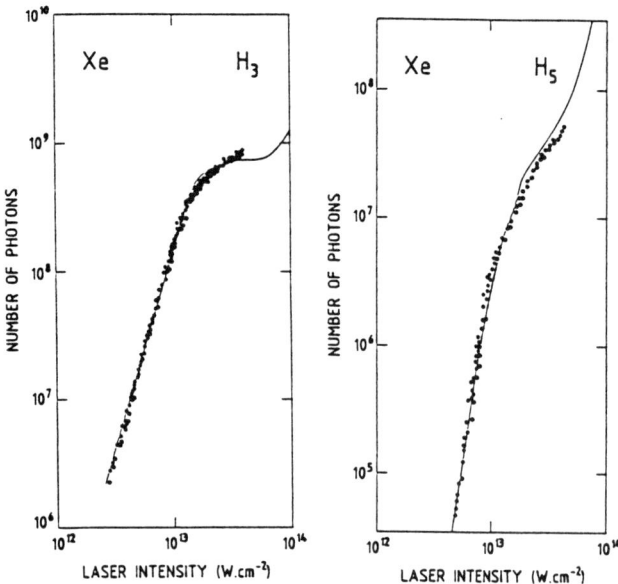

Figure 7. 3rd and 5th harmonics as a function of the incident laser intensity. The result of a calculation taking into account the presence of free electrons in the medium is shown in solid line.

The second question we would like to discuss is the role played by propagation effects. Harmonic generation depends on the individual response of an atom to the radiation field and also on the ability of the whole medium to assure proper phase matching of the generated radiation. We have tried to experimentally study the role of phase matching by varying the confocal parameter b from 1 to 6 mm (using different lenses to focus the laser beam). Some preliminary results[15] are shown in Fig.8. The number of harmonic photons detected at the saturation intensity in xenon (1.3×10^{13} W.cm^{-2}) is plotted as a function of the confocal parameter (we used three different focal lenses). The points follow a b^3 power law. In contrast, the background varies much more slowly with b.

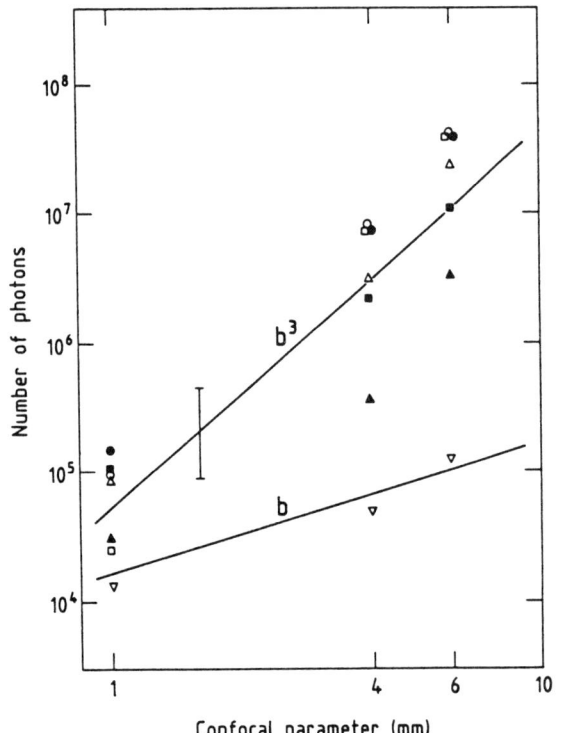

Figure 8. Number of photons as a function of the confocal parameter b; (□) fifth harmonic (H5); (o) H7; (Δ) H9; (●) H11; (▲) H13; (■) H15; (▽) background (at 74 nm).

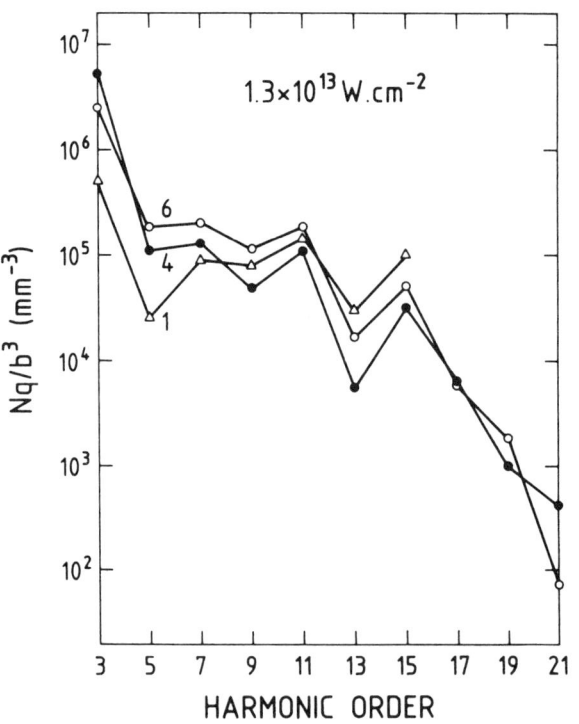

Figure 9. Relative harmonic distributions obtained at the saturation intensity. (Δ) f=75 mm; (●) f= 200 mm; (o) f=300 mm.

In Fig.9, we plot the number of harmonic photons detected at the saturation intensity 1.3×10^{13} W.cm^{-2}, divided by b^3.

We obtain extremely similar harmonic distributions (the difference between the relative distributions at the same laser intensity lies within our experimental error bar). We detect up to the 15th harmonic for b=1 mm, 21st harmonic at 4 mm and 23rd harmonic at 6 mm, but this difference may simply be attributed to a harmonic /background ratio which increases with b (see Fig.8), and which is barely above 1 for b=1mm and q>15. In conclusion, apart from the b^3 power law, these results do not show an important influence of phase matching effects. We obtain similar results whatever the focalization of the incident beam, as if the phase matching factor $|F_q|^2$ did not depend on b.

We have calculated the factor $|F_q|^2$, assuming a I^q power dependence for the qth harmonic[16], neglecting ionization of the medium, and using a 1 mm width Lorentzian distribution for representing the atomic density in the jet (Refs.3,11). The results of this calculation are shown in Fig.9 for a non-dispersive medium ($\Delta k=0$).

This figure does indeed show a variation of the phase matching factor as a function of b, which is not observed in the experiment. We have improved this calculation[15] by using Δk values obtained from a calculation of the dynamic polarizability of xenon[17]. The phase matching factor $|F_q|^2$ does not decrease as rapidly with q as for $\Delta k=0$ (the real part of Δk becomes negative above the threshold, thus favoring high order harmonic generation). However, the dependence on b remains similar (with a huge difference between b=1 mm and b= 4 or 6 mm, not observed in the experiment). Of course, this calculation lies upon several approximations (it neglects the ionization of the medium and assumes the weak field limit), which may not be valid in the present problem.

The interpretation of our experimental results and in particular the understanding of propagation of high-order harmonics in a nonlinear medium exposed to an intense laser field remains an open problem. Let us come back to the question raised at the beginning of this report : does the experimental harmonic distribution reflect the single-atom response to the nonlinear interaction ? The experimental results do not allow to give a definite answer to this question. The medium exposed to an intense infrared radiation field appears to behave as if the laser were a plane wave (or as if the effective interaction length were extremely small, much smaller than 1 mm, with $|F_q|^2$ independent of b).

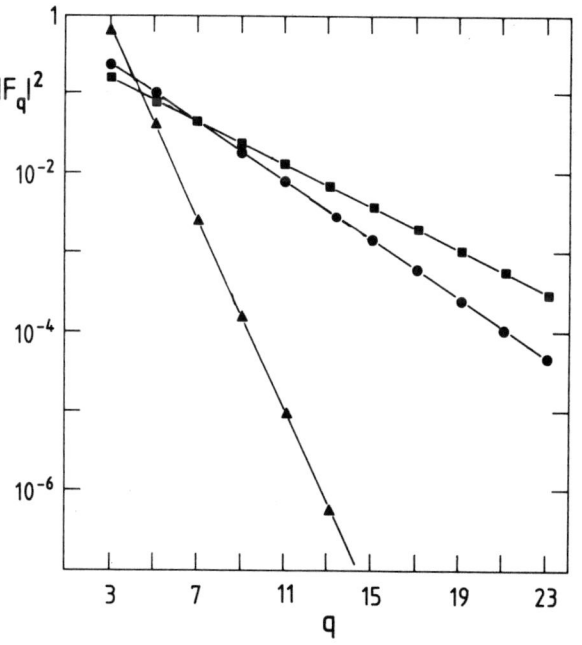

Figure 10. Phase matching function $|F_q|^2$ as a function of q, for different values of the confocal parameter; (▲) b= 1mm; (●) b= 4mm; (■) b= 6 mm; $\Delta k=0$

REFERENCES

1. A.McPherson, G.Gibson, H.Jara, U.Johann, T.S.Luk, I.McIntyre, K.Boyer and C.K.Rhodes J.Opt.Soc.Am.B **4**, 595 (1987)
2. M.Ferray, A.L'Huillier, X.F.Li, L.A.Lompré, G.Mainfray and C.Manus J.Phys.B **21**, L31 (1988)
3. X.F.Li, A.L'Huillier, M.Ferray, L.A.Lompré and G.Mainfray Phys.Rev.A**39**, 5751 (1989)

4. L.A.Lompré, M.Ferray, A.L'Huillier, X.F.Li and G.Mainfray, J.Appl.Phys. **63**, 1791 (1988)
5. Y.Gontier and M.Trahin IEEE, J.Quantum Electron. QE–18, 1137 (1982); R.M.Potvliege and R.Shakeshaft, Z.Phys.D **11**,93 (1989); B.Gao and A.F.Starace, Phys.Rev.A **39**, 4550 (1989); L.Pan, K.T.Taylor and C.W.Clark, Phys.Rev.A **39**, 4894 (1989)
6. R.M.Potvliege and R.Shakeshaft, Phys.Rev.A **40**, xxx (1989)
7. K.C.Kulander and B.W.Shore, Phys.Rev.Lett. **62**, 524 (1989)
8. J.H.Eberly, Q.Su and J.Javanainen, Phys.Rev.Lett. **62**, 881 (1989)
9. K.C.Kulander and B.W.Shore, to be published in J.Opt.Soc.Am.B (april 1990)
10. J.F.Reintjes, in "Nonlinear Optical Parametric Processes in Liquids and Gases", Academic Press, Inc. (1984); N.B.Delone and V.P.Krainov "Fundamentals of Nonlinear Optics of Atomic Gases", John Wiley and sons (1988)
11. A.Lago, G.Hilber and R.Wallenstein, Phys.Rev.A **36**, 3827 (1987)
12. A.L'Huillier, L.A.Lompré, G.Mainfray and C.Manus, J.Phys.B **16**, 1363 (1983)
13. R.L.Carman, C.K.Rhodes and R.F.Benjamin, Phys.Rev.A **24**, 2649 (1981); M.B.Isichenko and V.V.Yan'kov, Sov. Phys.JETP, **60**, 1101 (1984) [Zh. Eksp. Teor. Fiz. **87**, 1914 (1984)]
14. K.Myazaki and H.Kashiwagi, Phys. Rev.A, **18**, 635 (1978)
15. A. L'Huillier, L.A.Lompré, M.Ferray, X.F.Li, P.Monot and G.Mainfray, to be published
16. Non-perturbative behavior can also be included. See B.W.Shore and K.C.Kulander, J.Mod.Optics, in press.
17. A. L'Huillier, submitted to J.Phys.B Letters.

PHOTODETACHMENT IN STRONG OSCILLATING FIELDS

D. J. Larson, P. S. Armstrong, M. C. Baruch,
T. F. Gallagher, and T. Olsson

Department of Physics, University of Virginia
Charlottesville, VA 22901 USA

Photodetachment from negative chlorine ions has been studied in the presence of strong infrared and microwave fields. Detachment near the single photon threshold at 3.613 eV was observed by measuring the loss of ions or the production of neutrals resulting from illumination of the ions by ultraviolet light from a pulsed dye laser. Detachment rates with and without infrared light or microwaves were measured as a function of the ultraviolet wavelength. Infrared light from a pulsed Nd:YAG laser (λ=1064 nm) with a peak intensities up to 10^{11} W/cm^2 reduced the detachment cross section above threshold while producing detachment below threshold. The data are consistent with a model which includes a positive threshold shift and two-color, two-photon detachment. Microwaves with a frequency of 2.6 GHz and field strengths near 3 kV/cm produced detachment below threshold, non-zero detachment at threshold, and oscillations about the zero field cross section above threshold. Similar effects have been observed recently in experiments involving photodetachment from negative hydrogen ions in constant electric fields. No shift in the threshold due to the microwaves was evident.

Introduction

Two interesting effects observed in multiphoton ionization of atoms by intense laser fields are above threshold ionization (ATI), the absorption of more photons than the minimum number needed to reach the ionization limit, and an apparent increase in the ionization potential due to the extra ponderomotive or "wiggle" energy of an electron in an oscillating field.[1-4] A free electron in an oscillating electromagnetic field has a ponderomotive energy

$$U_p = e^2 F_0^2 / 4m\omega^2$$

where F_0 is the peak electric field at the position of the electron and ω is the frequency of the oscillation. This ponderomotive energy can be substantial compared to the photon or ionization energies. At the fundamental Nd:YAG laser frequency, the ponderomotive energy is approximately 1 eV at an intensity of 10^{13} W/cm^2. Despite large ponderomotive shifts, narrow electron energy peaks are observed in long-pulse ATI experiments. The electron's ponderomotive energy is converted into directed kinetic energy as it leaves the interaction region, compensating for the increased ionization potential and leaving the electron energy spectra unchanged, except for the absence of electrons with energies less than U_p. Thus the ionization potential shift does not enter directly into the position of electron energy peaks in such experiments. However, experiments done with pulses short compared to the time it takes the electron to leave the interaction region show structure that is consistent with the assumption that highly excited states, and presumably the ionization threshold, are shifted by an amount equal to the ponderomotive potential.[5]

This paper describes two experiments which directly probe the photodetachment threshold in negative chlorine ions in the presence of a strong infrared (YAG laser) or microwave (2.6 GHz) field. Negative ions have unique properties, due to the lack of a long range Coulomb interaction to bind the extra electron, which make them attractive atomic and molecular systems for the study of continuum thresholds in external fields. Experimentally, there is a clearly defined threshold which is not obscured by a converging series of bound excited states. Also, the detachment threshold can be reached with a small number of near visible photons. Theoretical analyses are simplified by the lack of bound excited states and by a continuum which in first approximation is equivalent to completely free electrons. Of course, experimental disadvantages of negative ions are that sample preparation is difficult and that achievable densities are much smaller than those used in experiments with neutral atoms. The present experiments use a Penning ion trap apparatus and an ion beam apparatus.

These experiments differ from most ATI experiments in that two frequencies of electromagnetic radiation are used. The threshold position is probed by a weak detaching beam in the presence of a much stronger perturbing field.

Photodetachment in a strong infrared field

In these measurements a Penning ion trap was used to contain negative chlorine ions which were illuminated by tunable ultraviolet light obtained by frequency doubling the output of a pulsed dye laser or pulsed dye amplifier. The ions were simultaneously illuminated by infrared light from the pulsed Nd:YAG laser, which was used to pump the pulsed dye laser. A schematic diagram of the ion trap and associated apparatus is shown in Fig. 1. Cl⁻ ions were produced in the trap by dissociative attachment of low energy electrons to CCl_4 gas which was leaked into the vacuum system at a pressure of approximately 3×10^{-9} Torr. After filling the trap with ions, the ion signal was measured by driving the axial motion and observing the currents induced on the ring electrode. Photodetachment was observed by measurement of the number of ions in the trap before and after 343-nm ultraviolet light pulses were sent through the trap. Photodetachment was measured alternately with and without focussed 1064-nm pulses from the YAG laser, which were overlapped in space and time with the ultraviolet light. Thus the ultraviolet light was used to probe the detachment cross section which was modified by the intense infrared light.

An important aspect of the experiment is the characterization of the detaching (ultraviolet) and perturbing (infrared) light pulses. The energy of the pulses was measured by a volume absorbing disc calorimeter. The temporal profiles of both the infrared and ultraviolet light pulses were measured using a fast photodiode and oscilloscope. The width of the infrared pulses was close to 9 nsec and the width of the ultraviolet pulses was about 3.5 nsec. An optical delay line was adjusted so that the ultraviolet pulse illuminated the ions near the peak of the infrared pulse. Focus spot sizes were obtained by measuring the light transmitted through a 6 μm slit which was stepped across the focus. The resulting focus intensity data were fitted to Gaussian distribution functions. For the data presented here, the diameters were near 200 μm for both the ultraviolet and infrared pulses. The maximum intensities of the infrared pulses were near 10^{11} W/cm²; for the ultraviolet pulses the maximum intensity was always less than 10^8 W/cm².

A preliminary version of this experiment[6] clearly showed a reduction in the detachment rate near threshold due to infrared light with peak intensities near 10^{10} W/cm². The data were consistent with a positive threshold shift of roughly 2 cm⁻¹ which is obviously smaller than the expected ponderomotive shift of approximately 8 cm⁻¹. The difference in the dynamic polarizabilities of the ground states of the negative ion and the neutral atom also should contribute to the shift. A recent evaluation of the ac polarizabilities for Cl and Cl⁻ shows that the Stark shifts are much less than the ponderomotive potential at 1064 nm, contributing a net shift of only 0.24 cm⁻¹ at 10^{10} W/cm², but they become comparable to or greater than the ponderomotive potential at higher frequencies[7].

The present experiment includes several improvements over the previous experiment including higher intensity, better characterization of the light pulses, and improved spectral resolution. An example of data obtained in the present measurements is shown in Fig. 2. Ion number data taken alternately with and without infrared light present are presented as a ratio of detachment rates with and without the infrared light. The general shape of this curve is easily understood. The ratio shoots up to very large values near threshold, because two-color, two-photon detachment involving one infrared photon and one ultraviolet photon keeps the cross section finite even below threshold when the infrared light is present. Just above threshold the detachment rate with infrared present is depressed, presumably due to a shift in the threshold to higher energies. Well above

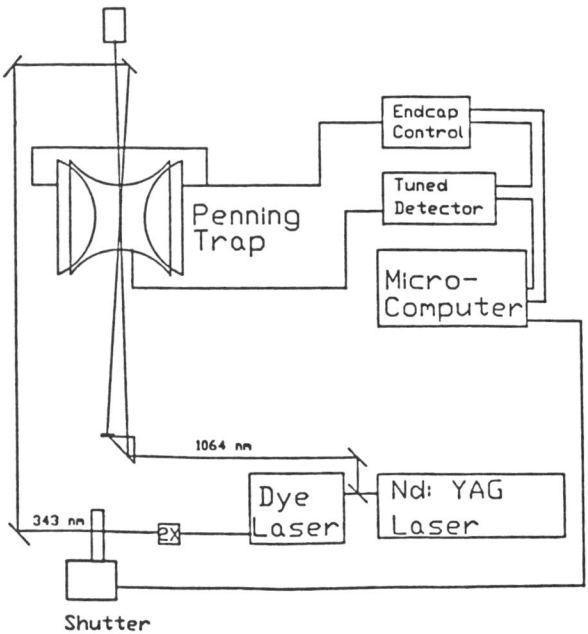

Fig. 1. Schematic diagram of the ion-trap apparatus. The ion trap is contained in a vacuum system with a base pressure below 1 nTorr.

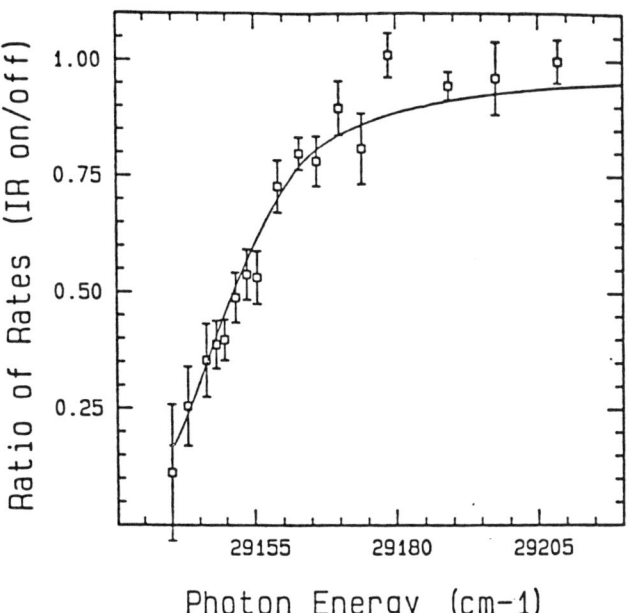

Fig. 2. The ratio of detachment rates with and without infrared light present plotted as a function of ultraviolet photon energy. The detachment threshold is at 29138 cm^{-1}. The solid line is the result of a fit to a model described in the text.

threshold, above the shifted threshold, the infrared has less and less effect, and the ratio approaches unity.

The solid line in Fig. 2 shows the results of a fit to the data by a model which includes two major elements, a threshold shift proportional to the infrared intensity and two-color, two-photon detachment. Two-color, two-photon detachment has been observed below threshold and, as expected, is found to be independent of ultraviolet photon energy, at least over a few tens of cm^{-1}. The model assumes a constant value for the two-photon cross section. Both the threshold shift and the two-photon detachment rate are proportional to the infrared intensity. Since the focus size of the infrared is very close to that of the ultraviolet, there is a substantial variation of the threshold shift and the two-photon detachment rate as a function of position in the beam. This has the effect of limiting the minimum value of the ratio and rounding off the curve. If the infrared illumination were uniform, the ratio should be small above threshold and independent of photon energy, up to the position of the shifted threshold.

Since the measured parameters for the spatial and temporal variation of the infrared and ultraviolet intensities are used in the model and the value of the two-photon cross section is inserted as a constant, the fit of the model to the data is a one parameter fit. This parameter is the proportionality constant between the threshold shift and the infrared intensity. For the data shown in Fig. 2 the peak infrared intensity is 8.4×10^{10} W/cm^2, corresponding to a peak ponderomotive energy of 72 cm^{-1}. The fit to the data gives a peak threshold shift of only 25 cm^{-1}.

A primary source of possible error in the experiment is in the determination of the intensity of the infrared light. Errors could occur in the determination of the pulse energy, the measurements of the spatial and temporal distributions, or, most likely, in the overlap of the infrared and ultraviolet beams. Since the infrared pulse is much longer than the ultraviolet, the results are not highly sensitive to small displacements in time. The spatial overlap is checked by measuring the detachment rate for a fixed ultraviolet wavelength a few cm^{-1} above threshold as a function of infrared beam position. At such a frequency the point of minimum detachment corresponds to maximum infrared intensity.

The two-color, two-photon cross section provides a check on the infrared intensity and overlap measurements. Like the threshold shift, the two-photon detachment rate should be proportional to the infrared intensity. An error in beam overlap would lead to a small two-photon detachment rate, and thus a small value for the cross section extracted from the data. A preliminary value for the two-color, two-photon cross section, derived from data taken below the single photon threshold, is 1.4×10^{-48} cm^{4}s. This is roughly thirty percent larger than the theoretical value for this cross section[8]. The agreement of the experimental and theoretical values for the two-photon cross section suggest that it is unlikely that the reason for the small apparent threshold shift is overlap or other problems with the real value of the infrared intensity.

Photodetachment in a strong microwave field

The size of the ponderomotive energy can be increased by increasing the intensity of the light. However, the presence of the $1/\omega^2$ term in the energy means that large ponderomotive energies can be achieved at relatively small intensities by reducing the frequency. At microwave frequencies the intensities that can be readily achieved are several orders of magnitude less than those attainable from pulsed laser sources, but the frequency difference more than compensates. We have carried out an experiment which is similar to the one described above except that the perturbing field is now at microwave frequencies. With a 1 kW pulsed microwave source feeding a cavity at 2.6 GHz, fields of 3 kV/cm, which should lead to ponderomotive energies of 15 eV, have been realized. The photodetachment cross section in the presence of such a field has been measured using an ion beam apparatus.

A schematic diagram of the apparatus used in these measurements is shown in Fig. 3. Negative chlorine ions are made from CCl$_4$ in a hot cathode source. The ions are accelerated to 1 keV and are

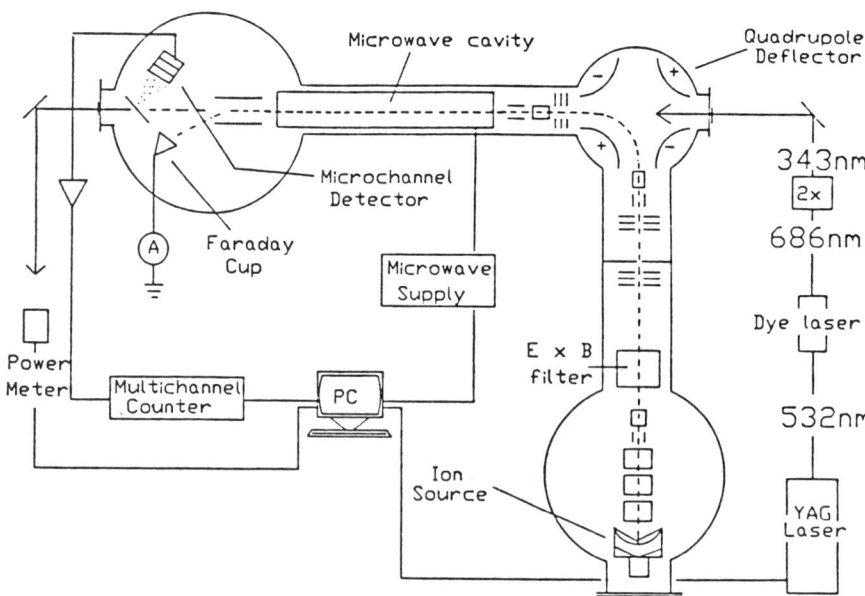

Fig. 3. Schematic diagram of the ion beam apparatus used for the microwave experiments.

sent through a number of focussing and deflecting elements, including an electric quadrupole deflector where the beam is deflected by 90°. The ions are then focussed into a 60 cm long microwave cavity operating on the TE_{101} mode. After exiting the cavity, the ions are deflected into a Faraday cup, while any neutrals impinge on a coated glass plate. Electrons produced by the collision of the fast neutral atoms with the glass plate are collected and detected using a microchannel-plate detector.

As in the infrared experiment, photodetachment is accomplished using a pulsed 343 nm beam obtained by frequency doubling the output of a pulsed dye laser. The laser beam is aligned collinear with the ion beam in the microwave cavity. The laser pulse length of a few nanoseconds is short compared to the microwave pulse and the time it takes the ions to traverse the microwave cavity, both of which are on the order of microseconds. Thus each photodetachment occurs with the ion in a region of microwave field with constant amplitude. However, photodetachment takes place all along the laser beam in microwave fields which vary from zero to the maximum amplitude. The microwave field is correlated with the position at which each neutral is created. Since all the atoms have the same velocity, the position of creation and thus the microwave field at the time of detachment is correlated with the time of arrival at the detector after the laser fires. A multichannel counter is used to separate the signals into 1 microsecond wide bins. Each bin is then associated with a position along the beam and thus a microwave field at the time of detachment. Care was taken to avoid counts due to electrons ejected from the glass plate by the ultraviolet light pulse.

In order to minimize the effects of ion beam and laser beam fluctuations in intensity or position, the microwaves were only pulsed on every other laser shot and detachment data with no microwaves present were collected as well as data with microwaves present. The laser was pulsed twenty times each second and the microwaves ten times a second. At each laser setting the ratio of neutrals produced with the microwaves on to neutrals produced with microwaves off was obtained. Below the zero field threshold, where there was no detachment signal without the microwaves, differences rather than a ratio were used. These data were then combined and plotted as a total cross section relative to the well verified Wigner law[9,10] cross section for detachment in zero field. An example of the data for one bin near the peak of the microwave field strength is presented in Fig. 4. Nonzero detachment below and at the zero field threshold and oscillations in the detachment about the zero field cross section above threshold are clearly evident. Any threshold shift is less than a few cm^{-1} and extremely small compared to a ponderomotive energy of 120,000 cm^{-1}.

The field induced structure in the detachment cross section is similar to that observed recently in experiments involving photodetachment from negative hydrogen ions in constant electric fields[11]. The two predominant features of the structure can be qualitatively understood by considering the potential

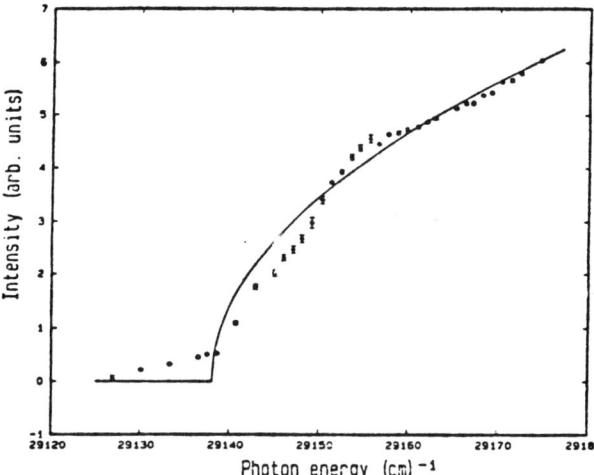

Fig. 4 Photodetachment cross section in the 2.6 GHz microwave field. The data are plotted relative to the zero field Wigner law cross section. The field amplitude for these data is about 3 kV/cm.

energy diagram in Fig. 5. The short range negative ion potential is depicted as a square well which is modified by the addition of a sloping potential in the presence of a static electric field. Below threshold detachment occurs due to tunneling through the barrier. The oscillations above threshold are due to interference between the part of the wavefunction corresponding to the electron's reflecting from the sloping potential and the part corresponding to the directly outgoing electron. Correspondingly, one can view the oscillations as stemming from the changing phase shift of the standing wave at the origin. Such oscillations have been seen in photoionization in a static electric field[12]. Also, the oscillations are similar the Frank-Condon oscillations observed in molecular photodissociation[13].

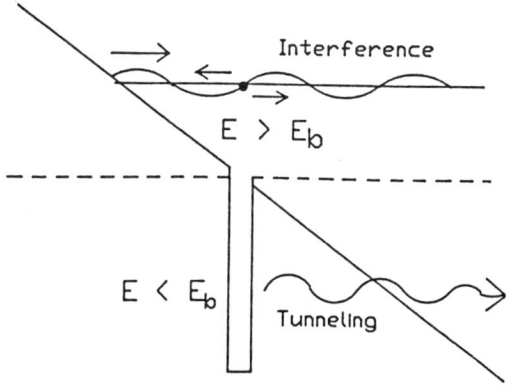

Fig. 5. Potential energy diagram for detachment in a static electric field. The short range negative ion potential is modeled as a square well and the electric field adds a sloping potential. The structure in the detachment cross section is due to tunneling below threshold and interference above threshold.

A quantitative expression for the photodetachment cross section in a constant electric field can be obtained by considering the transition of the electron from the short range negative ion ground state to a continuum which is modified by the presence of the constant electric field. The cross section near a threshold can be written as

$$\sigma = (D/F^{1/3}) \int_{-\infty}^{E} Ai^2(-E_z 2^{1/3}/F^{2/3}) \, dE_z$$

where F is the electric field strength, E is the energy of the detached electron, Ai is the Airy function, and D is a constant. Except for a factor of 1/k, this is the "modulating factor" given by Wong, Rau, and Greene [14]. Let us assume that detachment in the microwave field is detachment in a slowly varying field. In this case, the experimental cross section should be the cross section for a constant electric field averaged over a period of the microwave field. The data in Fig. 4 are consistent with such an assumption.

Discussion

These experiments on negative ion photodetachment in strong fields have yielded several interesting results. The observed shift of the detachment threshold due to perturbing light at 1064 nm is substantially smaller than the ponderomotive shift expected on the basis of ATI experiments. While this observation may be dependent upon the experimental conditions, the result is in at least qualitative agreement with an analogous measurement made on neutral Xe atoms[15]. In that experiment, the shifts in the levels of highly excited states of the atom were measured in the presence of perturbing light from a YAG laser. The measured shifts were only half the ponderomotive shift.

While the threshold shift observed at 1064 nm is roughly a factor of three smaller than the ponderomotive shift, any shift at 2.6 GHz is at least four orders of magnitude smaller than the ponderomotive energy of 120,000 cm^{-1}. Static field of the same magnitude, about 3kV/cm, produce effects similar to those observed in the microwave field. At low enough frequencies it is reasonable to expect the detachment cross section to be given by the time average of the cross section for detachment in a static field. In this connection it is interesting to note that ATI from neutral atoms in the low frequency limit can be treated in an exactly analogous fashion, as the liberation of an electron in an oscillating field[16-18]. This treatment predicts that all of the electrons should be ejected from the field with energies in excess of the ponderomotive energy. This prediction is borne out by experiment, and it is clear, at least in the microwave ATI experiment with Rydberg atoms[17], that the energy to surmount the pondermotive potential comes from the microwave field. Presumably the same must be true for negative ion photodetachment in microwave fields. If we analyzed the energies of the photodetached electrons we should see energies greater than or equal to the ponderomotive energy.

In order to develop a complete understanding of such effects it is necessary to make a connection between low frequencies, where the field picture can be used for ATI and threshold shift experiments, and high frequencies, where ATI is considered in terms of the photon picture. Valuable information should be obtained from threshold shift experiments at intermediate frequencies, from ATI experiments with negative ions, and from further consideration of the dynamics of atomic systems in strong external fields.

References

[1] P. Agostini, F. Fabre, G. Mainfray, G. Petite, and N.K. Rahman, Phys. Rev. Lett. 42, 1127 (1979).

[2] P. Kruit, J. Kimman, H.G. Muller, and M.J. van der Wiel, Phys. Rev. A 28, 248 (1983).

[3] H.G. Muller, A. Tip, and M.J. van der Wiel, J. Phys. B 16, L679 (1983).

[4] R.R. Freeman, T.J. McIlrath, P.H. Bucksbaum, and M. Bashkansky, Phys. Rev. Lett. 57, 3156 (1986).

[5] R.R. Freeman, P.H. Bucksbaum, H. Milchberg, S. Darack, D. Schumacher, and M.E. Geusic, Phys. Rev. Lett. 59, 1092 (1987).

[6] R. Trainham, G.D. Fletcher, N.B. Mansour, and D.J. Larson, Phys. Rev. Lett. 59, 2291 (1987).

[7] M. Kutzner, H.P. Kelly, D.J. Larson, and Z. Altun, Phys. Rev. A 38, 5107 (1988).

[8] M. Crance, Private Communications (1989).

[9] E.P. Wigner, Phys. Rev. 73, 1002 (1948).

[10] W.C. Lineberger, B.W. Woodward, Phys. Rev. Lett. 25, 424 (1970).

[11] J.E. Stewart, H.C. Bryant, P.G. Harris, A.H. Mohaghegdi, J.B. Donahue, C.R. Quick, R.A. Reeder, V. Yuan, C.R. Hummer, W.W. Smith, Stanley Cohen, Phys. Rev. A. 38, 5628 (1988).

[12] R.R. Freeman, N.P. Economou, G.C. Bjorklund, K.T. Lu, Phys. Rev. Lett. 41, 1463 (1978).

[13] Photons and Continuum States of Atoms and Molecules, J. Tellinghuisen, ed. N.K. Rahman, G. Guidotti, M. Allegrini, Springer Verlag Berlin, 1987.

[14] Hin-Yiu Wong, R.R. Rau, Chris H. Greene, Phys. Rev. A 37, 2393 (1988).

[15] D. Normand, L.A. Lompre, A. L'Huillier, J. Morellec, M. Ferray, J. Lavancier, G. Mainfray, and C. Manus, J. Opt. Soc. Am. B (to be published).

[16] Multiphoton Processes, H.B. van Linden van den Heuvell, H.G. Muller, ed. S.J. Smith.

[17] T.F. Gallagher, Phys. Rev. Lett. 61, 2304 (1988).

[18] P.B. Corkum, N.H. Burnett, F. Brunel, Phys. Rev. Lett. 62, 1259 (1989).

ATOMIC PHYSICS IN SURFACE STUDIES: AN OVERVIEW

F. B. Dunning

Department of Space Physics and Astronomy
Rice University, Houston, Texas 77251

In recent years a number of novel techniques have been devised to probe the geometric, electronic and magnetic properties of surfaces that make use of apparatus and methodologies similar to those employed in atomic collision studies. A selection of these techniques is reviewed to illustrate the varied information regarding surface properties that can be derived from studies of the scattering of electron, ion and neutral-atom beams at a surface. The data presented highlight a variety of interesting surface phenomena and suggest a number of future research opportunities.

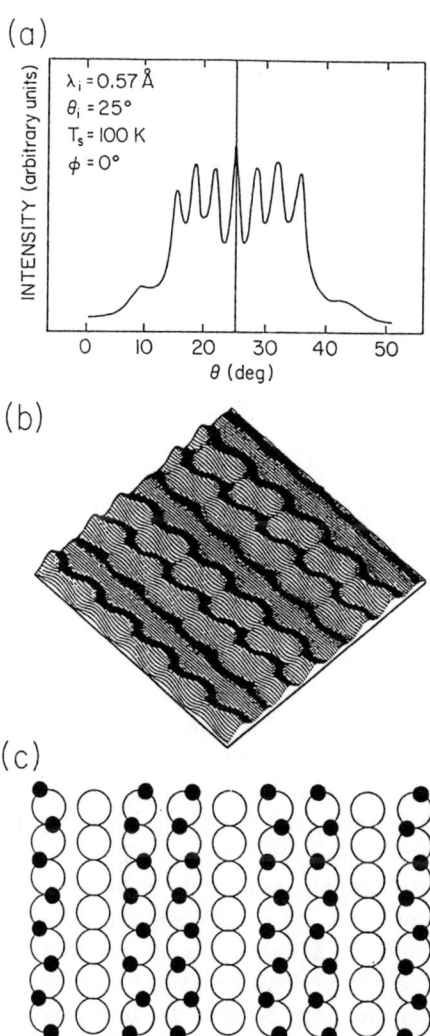

Fig. 1. Helium diffraction at a Ni(110) surface with two-thirds of a monolayer of adsorbed hydrogen. (a) In-plane helium diffraction spectrum. (b) The best fit corrugation function. (c) Hard-sphere model of the adsorbate structure.

In recent years a variety of novel techniques have been developed to probe the geometric, electronic and magnetic properties of clean surfaces and of adsorbed overlayers. A number of these approaches make use of apparatus and methodology similar to that employed in atomic collision studies. The purpose of the present article is to demonstrate the opportunities that exist in the use of particle beams as surface probes by providing examples of the information that can be derived from studies of the scattering of electron, ion and neutral atom beams at a surface. These examples also highlight a variety of interesting surface phenomena and suggest a number of possible avenues for future research.

With the development of supersonic nozzle beam sources it is now possible to produce highly-collimated rare gas atom beams with high intensity and narrow velocity distributions.[1] In the case of helium, surface scattering is predominantly elastic and, because the deBroglie wavelengths are on the order of an Angstrom, diffraction effects dominate. Thus, for an ordered surface, the size and orientation of the surface unit cells can be deduced from measurements of the angular positions of the diffracted beams using a movable mass spectrometer. Analysis of the intensities of the diffracted beams yields the surface corrugations, from which the geometrical arrangement of atoms at the surface may be deduced. Because the incident and diffracted beams are both neutral, this technique can be equally well applied to insulator, semiconductor and metal surfaces.

The capabilities of this method are illustrated in Fig. 1 which shows data obtained at a Ni(110) surface with two thirds of a monolayer of adsorbed hydrogen.[1,2] A typical in-plane helium diffraction scan is presented in Fig. 1a and contains a number of diffraction peaks. Such diffraction patterns result from the periodic modulation of the repulsive part of the atom-surface potential in the direction parallel to the surface. The classical turning point for an incident atom depends on whether it is incident on top of or between surface atoms. The locus of classical turning points defines a scattering surface that maps a surface electron density contour and is termed the corrugation function. Given a particular corrugation function it is then possible, under certain assumptions, to calculate the corresponding diffracted beam intensities. In analyzing experimental data the inverse problem is encountered, namely to calculate the corrugation function that provides the best fit to a particular data set. This is undertaken using a combination of Fourier techniques and trial and error, and the corrugation function that provides the best fit to the Ni(110)/H data is shown in Fig. 1b. The adatom configuration can be inferred directly from this function as each pronounced maxima corresponds to an adsorbed hydrogen atom. A hard sphere model of the corresponding adsorbate structure is presented in Fig. 1c. The hydrogen atoms form a series of ordered double zig-zag chains in which the zig-zag and zag-zig configurations alternate, each separated by a row of nickel atoms, resulting in a c(2x6) phase.

The data in Fig. 1 highlight two particular capabilities of diffraction methods. Light adsorbates can be readily studied, even on heavy substrates, and large unit cells comprising many atoms can be analyzed. (The c(2x6) phase just discussed comprises eight adatoms per unit cell.) This latter capability has, for example, recently been used to advantage in the study of long-range structures on Si(001) surfaces.[3] Diffraction experiments can, however, provide information not only on the regular arrangement of scatterers but also, through the form and intensity of the diffuse scattering background, on deviations from order.[4]

When a beam of helium atoms is reflected from a fluid surface the elastic peak in the energy distribution of the scattered atoms is weakly inelastic. This broadening of the reflected energy distribution with respect to the incident energy distribution is related to the lateral diffusive motions of surface atoms. Thus high-resolution time-of-flight studies of "quasi-elastic" helium scattering can be used to investigate self-diffusion at a surface. This technique has recently been employed in a study of surface melting of Pb(100) which revealed that atoms on the surface attain liquid-like mobilities ~50K below the bulk melting point.[5]

Inelastic atomic scattering may also result from excitation of low-lying vibrational modes associated with atoms or molecules adsorbed on a surface. Measurements of the inelastic energy loss (or gain) on scattering from a surface can therefore be used to determine, with high resolution, the corresponding vibrational frequencies. Shifts in the vibrational frequencies of molecules upon adsorption provide information on the nature of the surface bonding. Changes in frequency with increasing surface coverage point to the presence of adsorbate-adsorbate interactions or

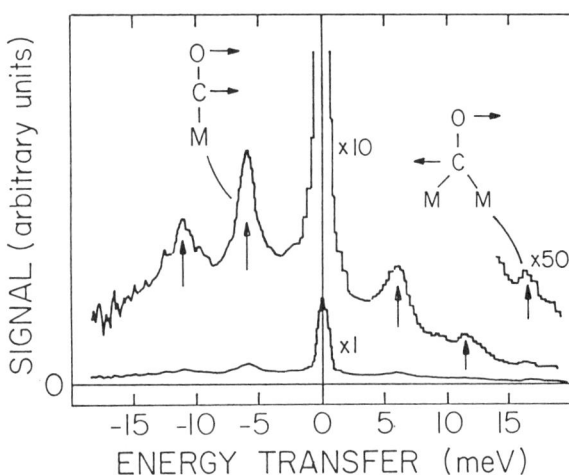

Fig. 2. Inelastic helium scattering at a Pt(111) surface with ~0.1 monolayer of CO. ($\theta_i = 29°$, $\theta_s = 61°$, $T_s = 300K$.) The insets indicate the vibrational motions associated with the observed features.

coverage-induced changes in the adsorbate-substrate interaction.

Inelastic helium scattering data obtained at a Pt(111) surface with 0.1 monolayer of CO are shown in Fig. 2.[6] At such low coverages CO does not form an ordered overlayer and the effects observed must be associated with vibrations of the isolated molecule. The energy gain and loss peaks ~6 meV to the right and left of the central elastic feature are assigned to CO vibrations. The features corresponding to energy transfers of ~ ± 11 meV are attributed to the first overtone. A small feature is also observed at an energy transfer of ~17 meV. These data, and data for the higher energy vibrational modes obtained using electron energy loss spectroscopy (to be discussed later), may then be used to carry out a normal mode analysis and derive a complete set of force constants for CO adsorbed at both on-top and bridge sites. On the basis of this analysis, the 6 meV feature apparent in Fig. 2 is assigned to hindered parallel translation at an on-top site and the 17 meV feature to hindered rotation at a bridge site, and these motions are illustrated in the figure.

Excited-state atom beams can also be used to investigate surface properties. In particular, experiments in several laboratories have demonstrated that metastable-atom deexcitation spectroscopy (MDS) provides a valuable probe of surface electronic structure.[7] In MDS, a thermal-energy beam of rare gas metastable atoms is directed at the surface of interest and the energy distribution of electrons ejected from the surface as a result of metastable atom deexcitation is measured. Because the internal energy of the incident metastable atoms is well defined, analysis of the ejected

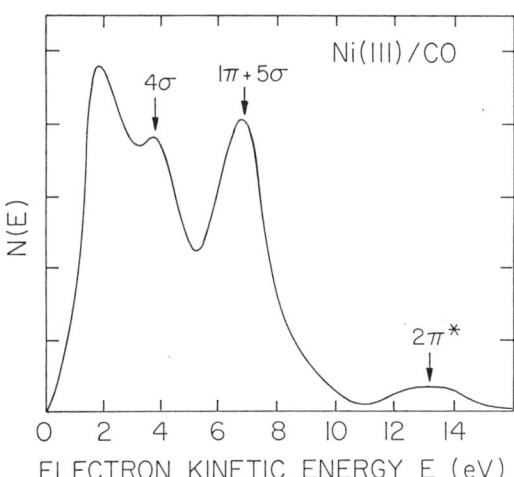

Fig. 3. Energy distribution of electrons ejected from a CO-covered room-temperature Ni(111) surface by incident He(2^1S) metastable atoms.

electron distribution provides information on the local density of electronic states at the surface. The ejected electron energy distribution observed when He(2^1S) metastable atoms are deexcited at a room-temperature Ni(111) surface with a saturation coverage of CO is shown in Fig. 3.[8] A number of features are observed that can be associated with electron emission from particular molecular orbitals. The highest energy feature results from emission from the $2\pi^*$ antibonding orbital which is populated by "back donation" of electrons from the metal. Detailed interpretation of such data, however, requires a thorough understanding of the mechanisms that give rise to electron ejection. Recent work has demonstrated that the use of spin-labelling techniques, specifically the use of electron-spin-polarized He(2^3S) metastable atoms coupled with spin analysis of the ejected electrons, can illuminate directly the dynamics of metastable atom-surface interactions.

The electron-ejection processes that can occur when a He(2^3S) atom is deexcited at a clean metal surface[9] are illustrated schematically in the energy-level diagrams shown in Fig. 4. If the work function of the surface is sufficiently large, then an incident 2^3S atom first undergoes resonant ionization (RI) in which the excited 2s electron tunnels into an unfilled level above the Fermi surface in the metal, as indicated by the wavy arrow in Fig. 4(a). The resulting He$^+$ ion continues toward the surface where it is neutralized by a conduction electron from the metal, the released energy being imparted to a second (Auger) electron in the metal which may, if the energy transferred is sufficiently large, be ejected from the surface. This two-electron process is termed Auger neutralization (AN). At low work-function surfaces RI cannot occur because there are no vacant levels of appropriate energy within the metal. In this situation He(2^3S) atoms are deexcited via the Auger deexcitation (AD) process diagramed in Fig. 4(b). In this process an electron from the metal tunnels into the helium ground atomic state and the energy released is communicated to the 2s electron which is ejected. The presence of adsorbed layers on a metal surface can also inhibit RI by preventing a good overlap between the 2s wave function and vacant states within the metal. In this situation deexcitation again occurs by AD. Because the electron tunneling to the helium ground state originates in the adsorbed layer, this process is similar to gas phase Penning ionization and is frequently termed surface Penning ionization (SPI).

The nature of the ejection process operative under a particular set of surface conditions can be probed directly, as is evident from Figs. 4(a) and 4(b), by spin polarizing the incident He(2^3S) atoms and measuring the polarizations of the ejected electrons. In RI + AN the ejected electron originates in the surface in which case, for nonmagnetic surfaces, any detected polarization is a measure of the correlation in spin orientation between the ejected and neutralizing electrons. In AD or SPI it is the 2s atomic electron that is ejected with polarization equal to that of the incident He(2^3S) atom beam.

Ejected-electron energy distributions and energy-resolved polarizations (normalized to unit incident metastable atom polarization) measured for a chemically cleaned (but otherwise untreated) Cu(100) surface, an atomically clean Cu(100) surface, and a Cu(100) surface with a potassium adlayer are shown in Figs. 5a, 5b, and 5c respectively.[10] The marked differences apparent in the

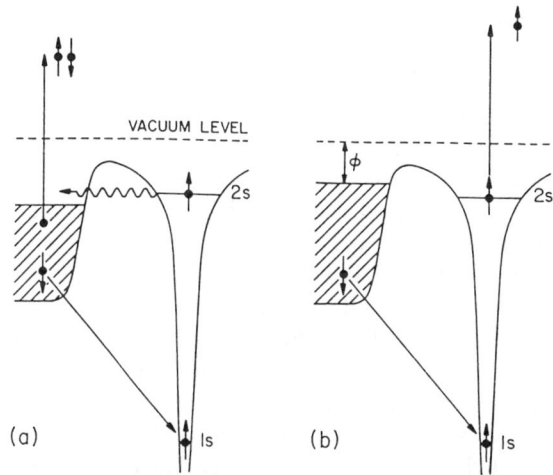

Fig. 4. Schematic diagrams of the electron-ejection processes that can occur when He(2^3S) metastable atoms are deexcited at a clean metal surface.

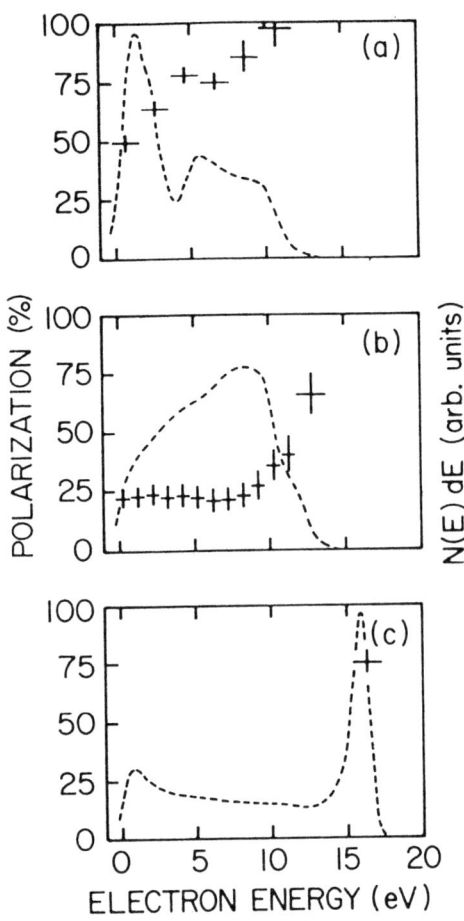

Fig. 5. Ejected electron energy distributions (---) and energy resolved polarizations (+), normalized to unit incident metastable atom polarization, for He(2^3S) deexcitation at (a) a chemically-cleaned Cu(100) surface, (b) a clean Cu(100) surface and (c) a Cu(100) surface with potassium adlayer.

observed energy and polarization distributions suggest that in each case different ejection processes are operative.

The polarization of electrons ejected from the chemically cleaned surface is large (Fig. 5a) indicating that, as might be expected, electron ejection results from SPI. The ejected-electron polarization, however, is on average, less than unity and is energy dependent. This can be explained in terms of the generation of (unpolarized) secondary electrons by electrons produced in the metastable atom-surface interaction that are initially directed into the surface. Because secondary-electron distributions peak at low energies, the relative importance of secondary electrons is expected to increase with decreasing ejected-electron energies, leading to a decrease in the net electron polarization.

In contrast, the polarization of electrons ejected from clean metal surfaces is low. The low polarization suggests that deexcitation occurs via RI + AN, as expected for metal surfaces with large work functions. The measured polarizations are, however, nonzero and, given the possibility of secondary-electron production, must represent a lower bound to the true polarization of electrons generated in direct metastable atom-surface interactions. Thus, the present results require that the two electrons involved in the AN process tend to have antiparallel spins, an effect that is most pronounced when both electrons originate near the Fermi surface, i.e., at the highest ejected-electron energies. The data are, however, consistent with the Pauli principle which suggests a higher probability for finding electrons of antiparallel spins in close proximity.

Deposition of a potassium adlayer on the clean Cu(100) surface results in a marked decrease in the surface work function and in the appearance of a sharp high-energy feature in the ejected-electron energy distribution (Fig. 5c) that has been ascribed to the potassium 4s band. The measured polarization of these electrons is large, indicating that they result from AD. Their polarization is, however, less than unity suggesting degradation due to secondary-electron production. (The other possibility, namely contributions from RI + AN, appears unlikely at these energies.)

The data in Fig. 5 show that use of spin-labelling techniques can provide significant new information regarding the dynamics of metastable atom-surface interactions that will permit more detailed interpretation of MDS spectra and significantly enhance the power of this spectroscopy. Use of spin-polarized metastable atoms also permits the study of surface ferromagnetism. At a ferromagnetic surface the local densities of majority- and minority-spin states are different. This results in differences between the total ejected electron yields obtained when the incident metastable atoms are polarized parallel and antiparallel to the surface magnetization and this asymmetry is related to the surface ferromagnetic order. This technique, termed spin-polarized metastable atom deexcitation spectroscopy (SPMDS), has been used to explore Ni(110) surface magnetism, and its dependence on temperature and the presence of controlled adsorbates.[11] Indeed, the asymmetry was observed to disappear at even submonolayer coverages of CO pointing to the extreme surface specificity of SPMDS, which results because the incident thermal-energy metastable atoms do not penetrate the surface and interact only with the outermost atomic layer. In contrast, exposure of a ferromagnetic Fe(110) surface to oxygen changes the sign of the SPMDS asymmetry, suggesting that the adsorbed oxygen is ferromagnetic at monolayer coverage.

Surface magnetic properties can also be investigated using electron capture spectroscopy (ECS).[12] The basic process in ECS is the capture of one or two electrons during grazing-incidence reflection of fast (~150 keV) deuterons at the target surface. Because the distance of closest approach of the ions to the surface is on the order of only one or two Angstroms, this technique again probes only the density of electronic states at the topmost surface layer.

Consider initially one-electron capture. If the surface is ferromagnetic electron capture will result in the formation of deuterium atoms having a net electron spin polarization. This initial electron polarization is partially converted to nuclear polarization by the hyperfine interaction. Determination of this nuclear polarization by measuring the angular distribution of α-particles emitted in the reaction T(d,n)α thus provides a measure of the net spin polarization of the captured electrons. Because electrons captured by different deuterons originate at widely different points on the surface, observation of a nonzero spin polarization indicates the presence of long-range ferromagnetic order at the surface. The temperature dependence of the spin polarization of electrons captured at a V(100)p(1x1) surface is shown in Fig. 6.[13] The data clearly demonstrate the existence of long-range ferromagnetic order at the surface, even though Vanadium is paramagnetic in the bulk, and point to a surface Curie temperature of ~540K. Such marked differences between bulk and surface magnetic properties are not uncommon[14] and their explanation provides a continuing challenge to both theorists and experimenters. Monolayer-level films grown epitaxially on suitable substrates also exhibit novel magnetic properties that can be investigated by taking advantage of the extreme surface specificity of spectroscopies such as ECS or SPMDS. Since thin epitaxial films may be grown in which the atomic spacings and/or structure are not those characteristic of bulk samples of the same material, such studies can delineate the influence of these parameters, and parameters such as film thickness and the nature of the substrate, on magnetic properties.

In the case of two electron capture, the only stable state of D^- is the $1s^2$ ^1S state. D^- ions can therefore only be formed by capture of electrons with *opposite* spins. Furthermore, the characteristic length within which an incident deuteron captures these electrons is small, ~10-20Å. Thus the captured electrons must originate in the same "local" surface region and two-electron capture will be strongly suppressed by the presence of local short-range magnetic order at a surface. The reduction in the D^- to D^+ ratio in the reflected beam, relative to that observed for a nonmagnetic sample such as copper, therefore provides a measure of the short-range ferromagnetic order existing at a surface on an atomic scale. ECS data show that, at many surfaces, short-range order persists far above the Curie temperature.[12]

Electron capture can result in the population of excited states. If the captured electrons have a net spin polarization this introduces an anisotropy in the total angular momentum of the excited state that can be measured by observing the polarization of the radiation emitted in subsequent spontaneous decay.[15] This approach has recently been adopted, using He^+, N^+ and Ar^+ ion beams, to investigate Fe(110) surface magnetism.

As evident from Fig. 6, the net spin polarization of electrons captured at a magnetized surface can be large. Indeed, experiments on a room-temperature Ni(110) surface indicate that all the captured electrons have essentially the same (minority) spin[12] This suggests that multi-electron capture at a magnetized surface can be employed to produce beams of atoms or ions in high spin states or, because of subsequent hyperfine coupling, in states of high nuclear polarization.[16]

A number of spin-sensitive techniques have also been developed to probe surface magnetism that use incident electron beams. One of the most versatile of these is scanning electron microscopy with polarization analysis (SEMPA)[17,18] This takes advantage of the fact that high energy electrons incident on a magnetized surface eject low energy secondary electrons that are polarized, and whose polarization reflects the local surface magnetization. Thus, if the secondary electrons are generated by the tightly-focused electron beam in a scanning electron microscope, it is possible by rastering the beam and measuring the polarization of the ejected secondary electrons to build up an image of the sample magnetization. With an appropriate spin analyzer, the components of ejected-electron spin polarization both in the plane of, and perpendicular to, the sample surface can be separately measured thereby obtaining an image of the vector magnetization. Further, the same

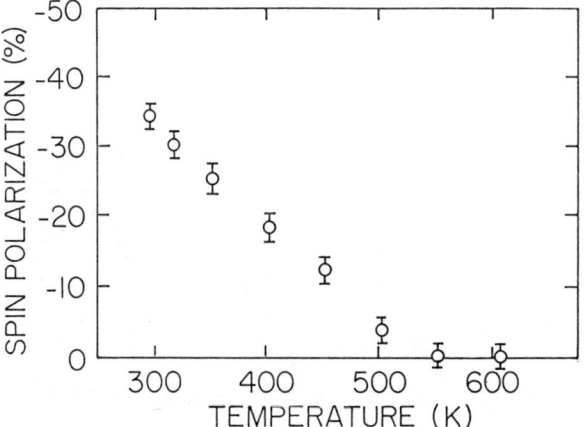

Fig. 6. Temperature-dependence of the spin polarization of electrons captured by fast deuterons during grazing-angle reflection at a V(100) p(1x1) surface.

Fig. 7. (a) High resolution polarization image recorded from an Fe-3% Si crystal. The gray scale indicates the component of magnetization along an axis in the surface plane. (b) The corresponding intensity image.

electrons that are used to obtain the magnetization image also generate the secondary electron intensity topographic image, and both images are built up simultaneously.

High-resolution polarization and intensity images recorded from an Fe-3% Si crystal are shown in Fig. 7.[18] The gray scale in Fig. 7a corresponds to one of the in-plane components of magnetization. Several different domains are clearly evident and the dagger-shaped domain in the center of the image appears to have domain walls that are pinned at the surface defects indicated by the arrows in Fig. 7b. As illustrated by Fig. 7, SEMPA is ideally suited to investigate the effect of surface defects or, through the addition of an Auger microprobe, surface chemical composition on magnetic microstructure. The high spatial resolution achievable with electron microscopy will also permit study of spin rotation in domain walls. Furthermore, because secondary electron escape depths are quite small, SEMPA can also be used to characterize the magnetic properties of films only a few monolayers thick.[19]

Electronic excitation of adsorbed molecules by an incident electron beam (or photons) can lead to the desorption of both atomic and molecular ions and of neutral species. A number of studies have demonstrated that ions desorb in directions that mirror the original orientation of the bond that was ruptured by the excitation.[20] This suggests that desorption proceeds via the excitation of highly repulsive states that give rise to a strong repulsive interaction in the direction of the breaking bond. The trajectories of the liberated ions are, however, influenced by image charge effects resulting in an increase in the polar angle of emission, although such effects do not change the azimuthal angle of emission. Nonetheless, measurements of the angular distributions of desorbed ions can provide very direct information about the geometrical arrangement of adsorbed molecules and this technique is referred to as ESDIAD (electron stimulated desorption ion angular distribution). Because ESDIAD is sensitive to local molecular structure, long-range order in the overlayer is not required for determination of adsorbate geometry.

The spatial distribution of H^+ ions ejected from NH_3 chemisorbed on Ni(110) at 85K is shown in Fig. 8a.[20] The data were acquired digitally using a position sensitive detector and are corrected for background effects associated with the soft x-rays that also result from electron impact on the target surface. The center of the figure corresponds to ion emission normal to surface and the polar angle of emission increases with increasing distance from this point.

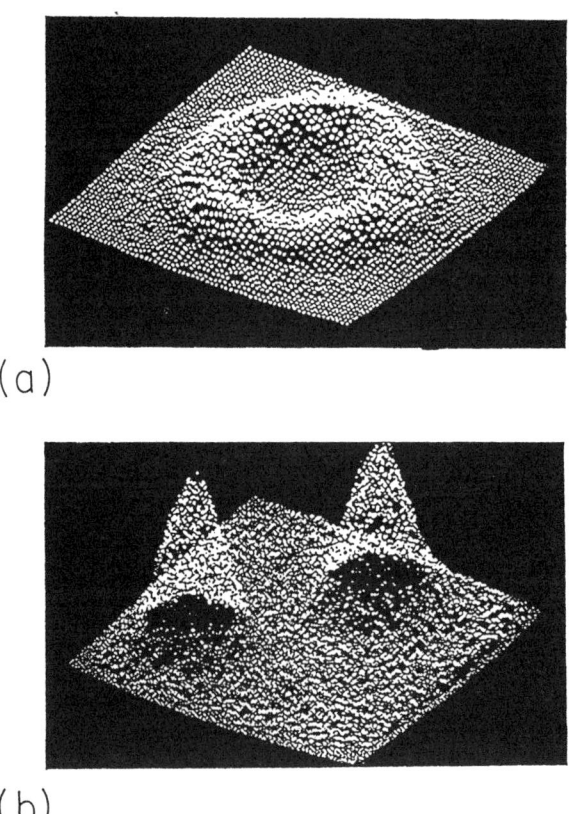

Fig. 8. Spatial distribution of H^+ ions ejected from (a) NH_3 and (b) NH_2 chemisorbed on Ni(110).

The H$^+$ ions are emitted in a cone giving rise to the observed "halo" pattern. This ion distribution indicates that the N-H bonds are all oriented at the same polar angle, i.e., the NH$_3$ stands upright from the surface, but that there is no preferred azimuthal bond angle. Fig. 8b shows the angular distribution of H$^+$ ions ejected from NH$_2$ chemisorbed on Ni(110). The H$^+$ angular distribution exhibits two distinct lobes of emission parallel to the [001] surface direction (i.e., perpendicular to the rows of Ni atoms) demonstrating that the NH$_2$ is rigidly locked relative to the Ni(110) lattice.

ESDIAD can also be used to investigate the interaction between two different species coadsorbed on the same surface. This illustrated by the data in Fig. 9 which show the changes in the ESDIAD patterns observed as a Ni(110) surface having an initial coverage of a quarter monolayer of NH$_3$ is exposed to increasing amounts of CO.[21] In the absence of CO, (Fig. 9a) the H$^+$ halo pattern just discussed is observed. As the CO coverage is increased the ammonia halo is preferentially attenuated along the [110] surface direction resulting in a twin-arc H$^+$ pattern. At the highest CO coverages a central O$^+$ ion beam is apparent. The observed changes in the H$^+$ distribution clearly show the existence of a significant interaction between the adsorbed NH$_3$ and CO molecules. The preferential ejection of H$^+$ ions along the [001] surface direction is explained by postulating a weak NH$_3$ ⋯ CO interaction that orients a hydrogen atom in an NH$_3$ molecule toward a neighboring CO molecule on the same Ni ridge. Given that such H$_2$NH ⋯ CO complexes can be oriented in either direction along a particular ridge, torsional oscillations will give rise to the observed twin-arc H$^+$ pattern. The data in Figs. 8 and 9 demonstrate the remarkable power of ESDIAD in investigating chemical bonding at surfaces and this technique has been used to study a wide variety of adsorbate/substrate systems.

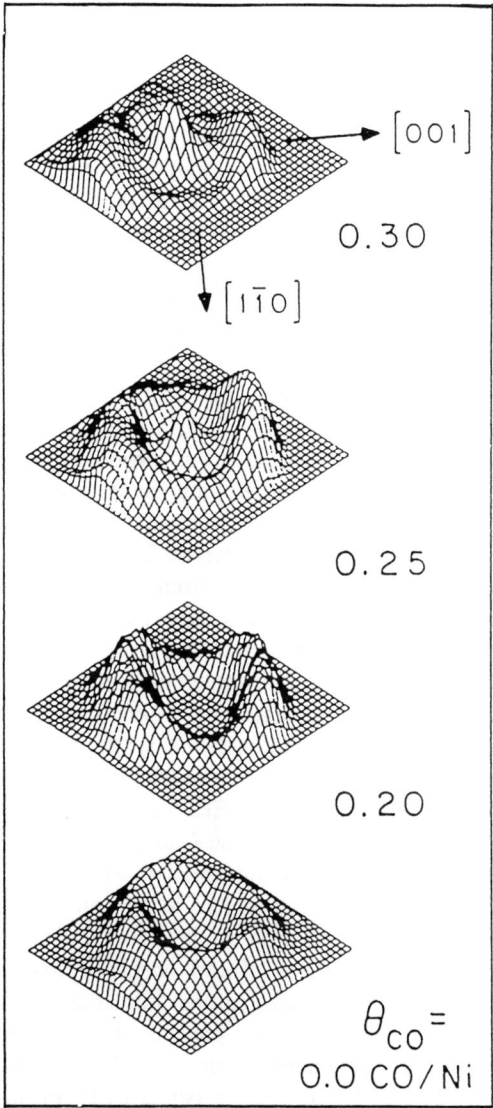

Fig. 9. Variation in the angular distribution of ions desorbed from a Ni(110) surface (T$_S$ = 84K) with a quarter monolayer of adsorbed NH$_3$ caused by increasing exposure to CO.

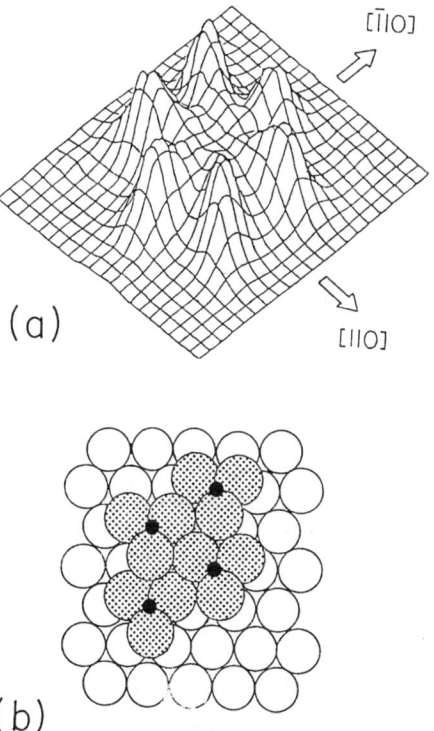

Fig. 10. (a) Angular distribution of F$^-$ ions ejected from a PF$_3$ overlayer on Ru(0001). (b) Proposed structure for the overlayer.

ESDIAD has recently been extended to include measurement of the angular distribution of negative ions ejected from a surface.[22] The major challenge in such studies is to separate the relatively weak negative ion signal from the much larger secondary electron signal and this is accomplished by using a pulsed electron beam in conjunction with time-of-flight techniques. The angular distribution of F^- ions ejected from a PF_3 overlayer on Ru(0001) is shown in Fig. 10. The surface was prepared by dosing to saturation with PF_3 at 110K followed by annealing to ~270K. F^- emission only occurs in certain specific directions pointing to the existence of an ordered overlayer. The proposed structure for this overlayer is indicated in Fig. 10b: the PF_3 molecular axis is tilted ~16° from the surface normal. Negative-ion ESDIAD can thus provide structural information that complements that available from positive-ion ESDIAD and comparisons between such data obtained at the same surface can illuminate the mechanisms of ion formation.

As just discussed, molecules adsorbed on a surface are frequently aligned in specific directions relative to the surface lattice. Thus surfaces provide a novel means to study collision processes involving oriented molecules. One such approach is illustrated schematically in Fig. 11 which shows two molecules that are coadsorbed on a surface and that are aligned in the same plane. If the surface is irradiated with UV light one of the adsorbed molecules may undergo photodissociation liberating, say, atom C. Atom C is then incident on the second molecule and may react with it. This reaction, however, involves a restricted range of collision energies (determined by the UV wavelength), collision angles and impact parameters. Thus, by changing the reactant alignments by, for example, changing the

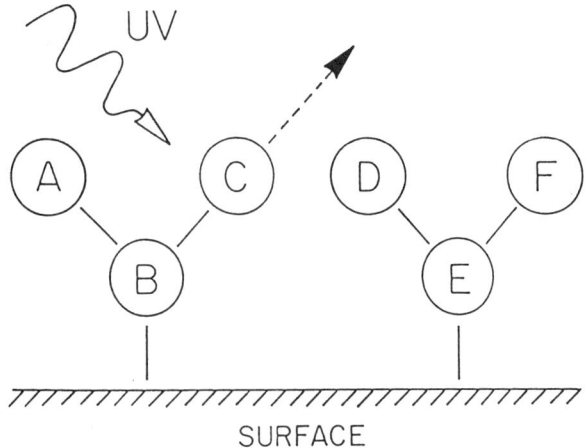

Fig. 11. Schematic diagram showing the basic principle of surface aligned photochemistry.

adsorbate coverage or substrate, it is in principle possible to investigate how a particular reaction depends on these factors. This technique, termed surface aligned photochemistry, has been applied in studies involving H_2S molecules physisorbed on LiF(001).[23] The interaction of photodissociated hydrogen atoms with neighboring H_2S molecules was observed to result in the formation of H_2 molecules having a bimodal velocity distribution suggesting the existence of at least two preferred orientations for H_2S molecules adsorbed on LiF.

Studies of the inelastic energy losses suffered by electrons scattered from surfaces provide an important probe of the physical and chemical properties of surfaces, and this technique is referred to as electron energy loss spectroscopy (EELS).[24] Typical EELS data obtained for β_1H on W(100) are presented in Fig. 12.[25] For specular scattering (Fig. 12a), a single large inelastic energy loss feature is evident at ~130 meV that results from excitation of the symmetric stretch vibrational mode indicated in the figure. (The hydrogen atoms are adsorbed at bridge sites between adjacent tungsten atoms.) This feature is associated with dipole scattering and results from the (long-range) interaction between the electric field produced by the incident electron, which (taking into account the image charge) is normal to the surface, and the dipole moment associated with the adsorbed molecule. Because of the dipole selection rule, however, such dipole scattering probes only those vibrational modes with dynamic moments perpendicular to the surface. This restriction can be removed, and vibrational modes parallel to the surface observed, by studying inelastic scattering at angles well removed from specular. Such data are shown in Fig. 12b and reveal the existence of additional energy-loss features at 80 and 160meV that are associated with the wag and asymmetric stretch modes diagrammed in the figure. This second scattering regime, termed impact scattering, is analogous to inelastic scattering from gas phase molecules and, with its less restrictive selection rules, greatly increases the power of EELS as a tool for surface analysis. (The feature at 260 meV also evident in Fig. 12 is an overtone of the symmetric stretch mode.)

EELS data can be interpreted at several levels. For example, surface composition can be inferred directly from the heights and energies of the observed loss peaks. Characteristic vibrational signatures immediately identify which atoms or molecules are present on the surface and their approximate concentrations. Indeed, whether a particular molecule adsorbs dissociatively or not can frequently be determined simply from the observed energy losses. Much additional information regarding bond distances, surface interaction potentials, structural configurations of surface and near-surface atoms, long-range interactions, etc., is, however, also available from EELS

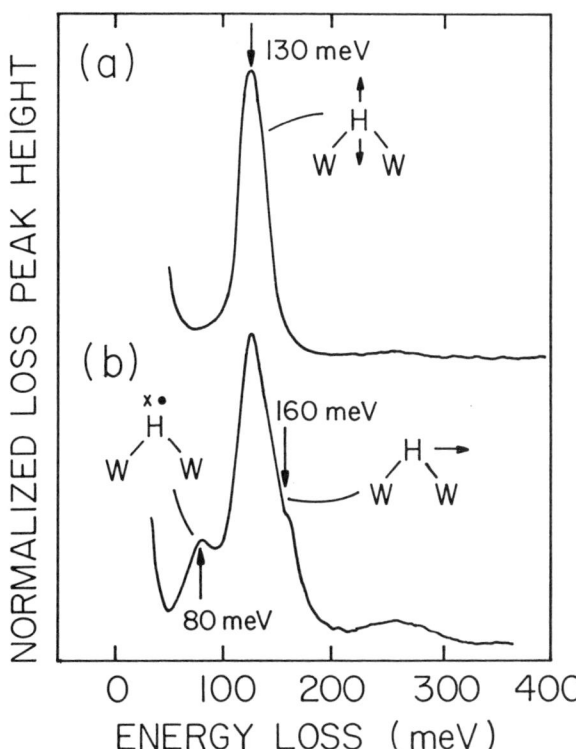

Fig. 12. EELS data for β_1H on W(100). (a) Specular scattering ($\theta_i = \theta_s = 30°$). (b) Off-specular scattering ($\theta_i = 30°$, $\theta_s = 5°$). The insets indicate the vibrational motions that give rise to each energy loss feature.

data. This can be obtained through detailed comparisons between experimental data, including vibrational energy loss line shapes and the energy- and angle-dependence of the loss features, and the results of calculations undertaken using different assumed lattice dynamical models. This approach has been applied to a number of adsorbate/substrate systems.[26]

With recent improvements in instrumentation, it is now possible to record EELS data on a millisecond time scale. Such TREELS (time-resolved EELS) experiments have been used to explore adsorption/desorption kinetics.[27,28] The time-development of the population of CO molecules chemisorbed on an initially clean Cu(100) surface during and following dosing with a 150-msec wide pulsed molecular beam of CO is shown in Fig. 13.[27] In this work the EELS spectrometer was tuned to the CO vibrational band and data were recorded with ~5 msec time resolution. The CO coverage was maintained sufficiently low that the EELS signal intensity was directly proportional to the coverage. The chemisorption of CO on Cu(100) is nondissociative and completely reversible. Thus, for low CO coverages, the adsorption and desorption

kinetics can be described by a simple first order model (i.e., the rate of desorption is proportional to the surface coverage). The decay of the CO signal therefore provides a direct measure of the mean surface residence time τ_r, which for the data in Fig. 13 is ~650 msec. Measurements of τ_r as a function of substrate temperature can be used, with the aid of an Ahrenius plot, to determine the activation energy of desorption. TREELS has also been used to investigate through temperature programmed reaction studies, the decomposition of molecules adsorbed on a surface.[27,29] In addition, the role of metastable precursor states in adsorption processes and of intermediate states in surface-mediated chemical reactions, can be probed.

Recent experiments have demonstrated that by use of a spin-polarized primary beam and/or spin analysis of the scattered electrons it is possible to study single particle (electron-hole pair) excitation at magnetized surfaces. Such spin-polarized EELS (SPEELS) experiments show that the probability for inelastic scattering can be strongly spin dependent and the technique has been applied to investigate low-lying electronic excitations of the Stoner continuum.[30]

EELS apparatus can also be used to measure the angular distribution of electrons elastically scattered from adsorbed molecules. Such scattering has recently been investigated in a series of model calculations.[31,32] These calculations embody a number of simplifying assumptions, for example they do not take into account the structure of the surface atomic layer or the effects of multiple scattering. The angular characteristics of the calculated elastic differential scattering cross sections do, however, depend

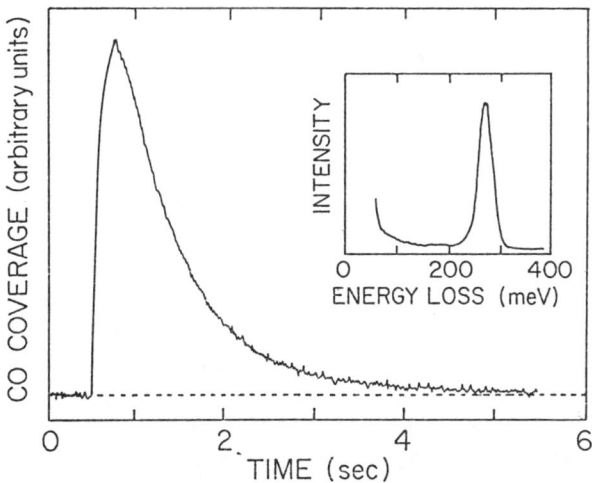

Fig. 13. Time-development of the CO coverage on an initially clean Cu(100) surface (T_s=193K) during and following dosing with a 150-msec wide pulsed molecular beam of CO. The inset shows the CO vibrational band used to monitor the coverage.

markedly on the assumed orientation of the adsorbed molecules. This suggests that a combination of differential electron scattering cross section measurements and electron scattering calculations can be used to obtain information concerning the orientation of molecules adsorbed on a surface.

As is evident from the above discussion, a wealth of information regarding surface geometric, electronic and magnetic structure can be derived from studies of the scattering of electron, ion and neutral-atom beams at a surface. New techniques for the application of particle beams in surface studies continue to be devised and it is certain that these will reveal many new and exciting surface phenomena in the future.

The research by the author and his colleagues described in this article is supported by the Division of Materials Science, Office of Basic Energy Sciences, U. S. Department of Energy, the National Science Foundation, and the Robert A. Welch Foundation.

[1] K. H. Reider, *Contemp. Phys.*, **26**, 559 (1985)
[2] K. H. Reider, *Phys. Rev. B.*, **27**, 7799 (1983).
[3] D. M. Rohlfing, J. Ellis, B. J. Hinch, W. Allison and R. F. Willis, *Surface Science*, **207**, L955 (1989).
[4] W. A. Schlup and K. H. Reider, *Phys. Rev. Letts.*, **56**, 73 (1986).
[5] J. W. M. Frenken, J. P. Toennies, and Ch. Wöll, *Phys. Rev. Letts.*, **60**, 1727 (1988).
[6] A. M. Lahee, J. P. Toennies, and Ch. Wöll, *Surface Science*, **177**, 371 (1986).
[7] See, for example, J. Arais, C. P. Hanrahan, R. M. Martin and H. Metiu, *Surface Science*, **165**, L95 (1986); W. Sesselmann, B. Woratschek, G. Ertl, J. Küppers and H. Haberland, *ibid*, **146**, 17 (1984); G. Boiziau, A. Garot, R. Nuvolone, and J. Roussel, *ibid*, **91**, 313 (1980); Y. Harada, H. Ozaki, K. Ohno, and K. Kajiwara, *ibid*, **147**, 356 (1984).
[8] F. Bozso, J. Arias, J. T. Yates, Jr., R. M. Martin, and H. Metiu, *Chem. Phys. Letts.*, **94**, 243 (1983).
[9] H. D. Hagstrum in *Electron and Ion Spectroscopy of Solids*, edited by L. Furmans, J. Vennick and W. Dekeyser (Plenum, New York, 1978).
[10] M. W. Hart, M. S. Hammond, F. B. Dunning, and G. K. Walters, *Phys. Rev. B*, **39**, 5488 (1989).
[11] M. Onellion, M. W. Hart, F. B. Dunning, and G. K. Walters, *Phys. Rev. Letts.*, **52**, 380 (1984).
[12] C. Rau, *Applic. of Surface Science.*, **13**, 310 (1982).
[13] C. Rau, C. Liu, A. Schmalzbauer, and G. Xing, *Phys. Rev. Letts.*, **57**, 2311 (1986).
[14] C. Rau and S. Eichner, *Phys. Rev. B*, **34**, 6347 (1986).
[15] H. Winter, H. Hagedorn, R. Zimny, H. Nienhaus, and J. Kirschner, *Phys. Rev. Letts.*, **62**, 296 (1989).
[16] J. S. Helman, C. Rau, and C. F. Bunge, *Phys. Rev. A.*, **27**, 562 (1983).
[17] R. J. Celotta, and D. T. Pierce, *Science*, **234**, 249 (1986).
[18] D. T. Pierce, J. Unguris and R. J. Celotta, *MRS Bulletin, XII*, 19 (1988).
[19] J. L. Robins, R. J. Celotta, J. Unguris, D. T. Pierce, B. T. Jonker, and G. A. Prinz, *Appl. Phys. Letts.*, **52**, 1918 (1988).
[20] M. J. Dresser, M. D. Alvey, and J. T. Yates, Jr., *Surface Science*, **169**, 91 (1986).
[21] M. J. Dresser, A.-M. Lanzillotto, M. D. Alvey, and J. T. Yates, Jr., *Surface Science*, **191**, 1 (1987).
[22] A. L. Johnson, S. A. Joyce and T. E. Madey, *Phys. Rev. Letts.*, **61**, 2578 (1988).
[23] I. Harrison, J. C. Polanyi, and P. A. Young, *J. Chem. Phys.*, **89**, 1498 (1988).
[24] J. L. Erskine, *CRC Crit. Rev. in Solid State and Mats. Sci.*, **13**, 311 (1987).
[25] W. Ho, R. F. Willis, and E. W. Plummer, *Phys. Rev. B.*, **21**, 4202 (1980).
[26] See, for example, R. L. Strong, and J. L. Erskine, *Phys. Rev. B.*, **31**, 6305 (1985).
[27] L. H. Dubois, T. H. Ellis, and S. D. Kevan, *J. of Electron. Spect. and Rel. Phen.*, **39**, 27 (1986).
[28] H. Froitzheim, U. Köhler, and H. Lammering, *Phys. Rev. B*, **34**, 2125 (1986).
[29] L. J. Richter and W. Ho, *J. Chem. Phys.*, **85**, 2569 (1985).
[30] See, for example, D. L. Abraham, and H. Hopster, *Phys. Rev. Letts.*, **62**, 1157 (1989).
[31] S. Nagano, Z.-P. Luo, H. Metiu, W. M. Huo, M. A. P. Lima and V. McKoy, *Surface Science.*, **186**, L548 (1987).
[32] S. Nagano, Z.-P. Luo, H. Metiu, W. M. Huo, and V. McKoy, *J. Chem. Phys.*, **88**, 7970 (1988).

Hydrogen-surface electron transition rates

P. Nordlander
Department of Physics and Astronomy, Rutgers University
Piscataway, New Jersey 08855-0849
and
J.C. Tully
Bell Laboratories, Murray Hill, NJ 07974
(31 July 1989)

A theoretical method for the calculation of the broadening and shifts of atomic hydrogen levels near surfaces is presented. The calculation employs a technique based on complex scaling, whereby the energies of the atomic resonances are obtained directly from the Schroedinger equation without the explicit construction of continuum states. An application to hydrogen on aluminium surfaces shows that the lifetimes of the atomic states can be much longer than previous calculations have predicted. We also present calculations for hydrogen outside impurity covered surfaces. It is shown that both the shifts and widths of the atomic resonances depend sensitively on the microscopic properties of the surface.

I. Introduction

Charge transfer processes play an important role in surface science. Many surface reactions such as dissociation and sticking are believed to depend crucially on the probability for an electron to transfer from the surface into excited atomic or molecular levels.[1,2]

Electron transfer processes also play an important role in quenching certain dynamical responses that could be induced by electronic transitions. One example of this is stimulated desorption.[3,4] The stimulated desorption process can occur if an atom is electronically excited into an antibonding state sufficiently longlived to enable the atom to acquire sufficient kinetic energy to leave the surface.[5,6,7] In the same way, photochemistry at surfaces depends on the enhanced reactivity of electronically excited species and the time the photon induced excitation survives.[8]

Excited atomic states near metals can decay through many different channels.[9] The dominant decay mechanism at typical physisorption distances is the resonant tunneling mechanism, where an electron in an occupied atomic state tunnel through the surface potential barrier into the metal.[9]

The purpose of the present paper is present a description of how the tunneling rates between atoms and surfaces can be calculated. We will present results for the shift and the broadening of lowest excited states of hydrogen on a clean metal and outside a metal surface covered with various impurities. A comparison between hydrogen outside a potassium and chlorine covered metal surface reveal that both the shifts and the broadening of the atomic levels as a function of H-surface separation depend sensitively on the details of the impurity induced changes of the surface electronic structure.

II. Theory

In this section the theoretical background necessary to describe how the tunneling rates between an atom and a metal surface can be calculated. Atomic levels shift and broaden in the vicinity of a metal surface. The origin of these effects are changes in the electron potential around an atom due to the presence of a surface. It is therefore important to model this potential accurately. Section A is a decription of how the surface electron potential can be calculated from density functional methods. Once the surface potential has been specified, the width and energy of the atomic resonances can be calculated using scattering theory methods. In section B, our method for calculating the shifts and widths of atomic levels is presented.

A. Surface electron potential

In the following we will assume a one-electron description of the surface and atomic electrons. Atomic units will be used throughout the text except when otherwise indicated. The coordinate system will be cylindrical with the positive z-axis oriented towards vacuum. The coordinate origin, z=0, corresponds to the actual location of the surface. The radial coordinate ρ refers to a surface normal through the atom. Upper case letters will be used to describe the coordinates of the atom and lower case letters refers to the electron coordinates.

In order to describe this process we invoke the jellium model of the surface.[10] The jellium approximation amount to treating the conduction electron in the solid as an electron gas. The positive background in the solid is smeared out into a constant attractive potential. The surface is modeled by abruptly truncating this potential at z=0. The jellium model is entirely specified by the valence electron density and the corresponding r_s describing the average distance between the electrons. In what follows we will consider jellium with $r_s=2.0$ corresponding to aluminum.

In the jellium model, the total potential for the electron at coordinates (ρ,z) in the presence of a proton at distance Z from the surface can be written as

$$V^{eff}(\rho,z;Z) = V_0^s(z) - \frac{1}{r} + \Delta V_A^s(\rho,z;Z) \quad (2.1)$$

The first part of the potential describes the bare electron-surface interaction. This potential can be calculated within the non-local density functional scheme. In the present case we have adopted the weighted density approximation.[11,12]

This particular many body approach describes both the image interaction and the potential in the bulk. For large z, $V_0^s \longrightarrow \frac{-1}{4(z-z_{im})}$, where z_{im} is the image plane defined as the first moment of the charge distribution induced by an external electric field.[13] The $\Delta V_A^s(\rho,z;Z)$ term describes how the bare surface electron potential is modified when an adsorbate is present. For large z this term approaches $\frac{1}{\sqrt{\rho^2+(z+Z-2z_{im})^2}}$. For intermediate distances this term is estimated using a linear response approach. The electron charge induced by an external pertubation is distributed in a thin layer around the surface. The thickness of this layer Δ as well as z_{im} depends on the r_s of the metal and has been calculated within the local density approximation.[13]

From the proton induced surface electron charge the induced electrostatic potential as well as exchange correlation potential can be calculated using Poisson's equation and a proper exchange correlation functional.

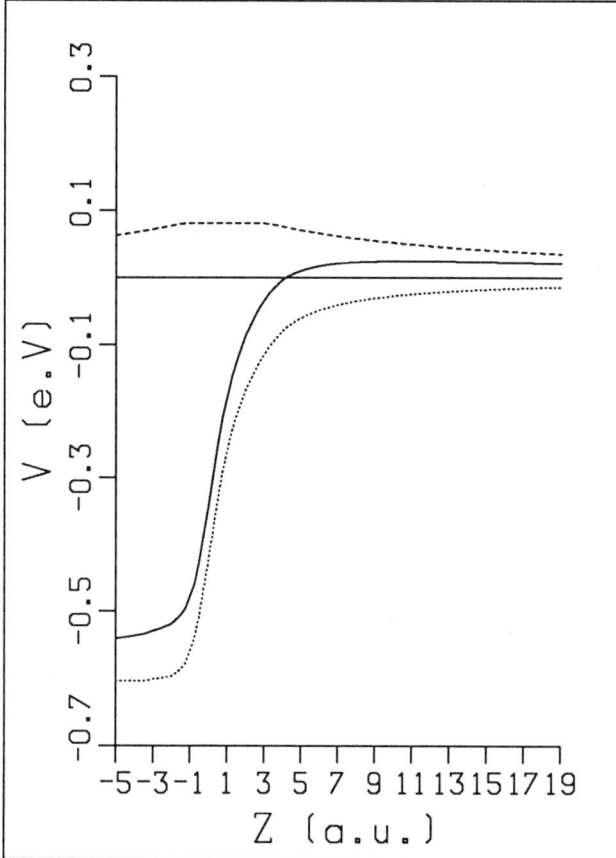

FIG. 1. The different contribution to the electron potential outside a metal surface. The hydrogen atom is placed at Z=10 a.u. The dotted line is the bare surface potential V_0^s, the dashed line is the proton induced potential, ΔV_a^s and the solid line is the total potential.

In Fig. 1, we show the electron potential for an adsorbate outside a jellium surface modeled using the present many body approach. We note that the bare surface potential, V_0^s provides a force towards the surface on the electron while the proton induced potential, ΔV_A^s is repulsive and therefore reduces the tunneling rates. In order to properly describe the interaction between the atom and the surface, it is important to use an accurate description of the surface potential.[14]

We note that close to the atom, the total surface induced electron potential is repulsive. The hydrogen states will thus shift upwards. This means that it will be easier to ionize an atom close to the surface. From the figure it is also clear why the atomic states will broaden, i.e. become resonances. Inside the surface the electron potential is attractive and electrons can tunnel between the atom and the surface.

1. Impurity covered surfaces

Alkali and halogen coadsorption on metal surfaces can dramatically increase or reduce the catalytic activity of transition metal surfaces. This phenomenon, referred to as catalytic promotion or poisoning, has been the subject of numerous experimental and theoretical treatments. The origin of this effect is not well understood. It has been proposed[15] that the effect could be due to the strong electrostatic fields induced in the vicinity of electro positive or electro negative chemisorbed atoms. Such fields may shift the levels of chemisorbed atoms and thus alter their reactivity. Coadsorption of alkali atoms on metal surfaces is also often used as a means of changing the work function of a metal.

In the context of understanding the charge transfer dynamics between atoms and real surfaces, it therefore appears important to investigate the microscopic interaction between atoms and impurity covered metal surfaces.

In order to model these types of systems, calculations of the shift and broadening of atomic levels will be performed for hydrogen levels outside a potassium and a chlorine chemisorbed on jellium. Due to the reduced symmetry of the two atom situation, the hydrogen will be always be placed along the surface normal through the chemisorbed impurity so that cylindrical symmetry of the potential is maintained.

The chemisorbed potassium is modeled using a K pseudopotential[16] and a negative image

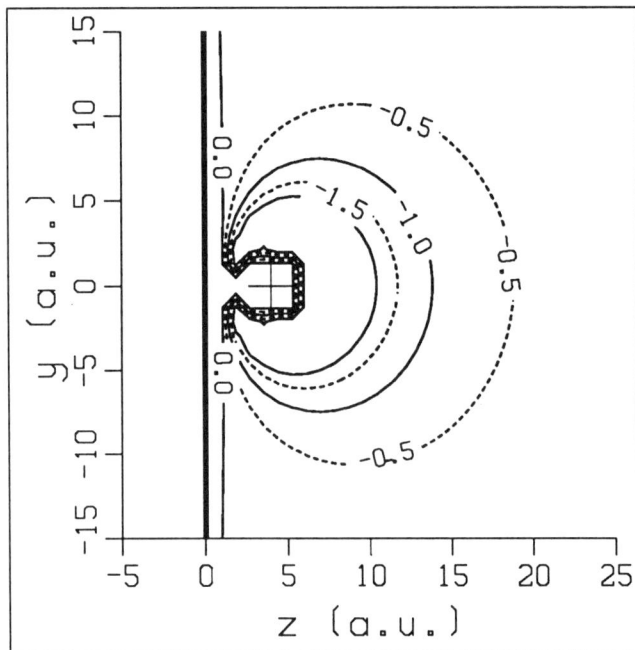

FIG. 2. Contour plot of the K induced electrostatic potential. The K, is placed at 4.0 a.u. outside the jellium edge. The energy unit is eV.

charge located at $2z_{im} - Z_K$.

The chemisorption distance Z_K of potassium on Al jellium is 4 a.u. Due to the electro positivity of the potassium, the induced electrostatic potential is negative in the surface region. In fig. 2, a contour plot of the induced electrostatic potential is shown. The present modeling of the alkali induced electrostatic fields quantitatively agrees with the results from ab initio calculations.[15]

Halogen atoms are electronegative and upon chemisorption on a free electron metal, an electron is transfered to the chlorine. The induced electrostatic potential is therefore positive outside the surface. In the present paper we use the self consistently calculated induced electrostatic potential[15] for a chlorine atom chemisorbed at a distance 2.625 from the Al surface to model the Cl impurity.

The impurity induced changes in the surface exchang correlation potential are neglected since these are confined to the immidiate vicinity of the impurity, and therefore not expected to influence the shift or witdth of the hydrogen resonances. The calculated impurity induced potential is added to the expression (2.1) for the total potential V^{eff}.

It is clear that both the shifts of atomic level and the tunneling rates will be influenced by such impurities. In the next section a calculation of these quantities will be presented.

B. Solution of the Schroedinger equation

In order to calculate the energies of the atomic resonances, the Schroedinger equation for the electrons must be solved

$$[-\nabla^2 + V^{eff}(\rho, z; Z)]\psi = \epsilon\psi. \qquad (2.2)$$

by imposing so-called Siegert boundary conditions:

$$\Psi(r) \longrightarrow e^{ik_R r + k_I r} f(\Omega) \qquad (2.3)$$

where k_I is positive. The energy is related to the complex wavenumber, k through $\epsilon = -\frac{1}{2}(k_R + ik_I)^2$. The energy thus becomes complex, $\epsilon = \epsilon_R - i\epsilon_I$. The real part of the energy ϵ_R describes the energy of the level and the imaginary part, ϵ_I describes the width of the resonance.

It can be seen that these boundary conditions diverge at infinity. This is the case because the number of electrons has to be conserved. The time evolution of the resonance state is

$$|\Psi(r,t)|^2 = e^{-2\epsilon_I t}|\Psi(r,0)|^2 \qquad (2.4)$$

At infinite time the integral over all space of this expression has to be finite. A convenient solution to the Schrodinger equation is provided by the so-called complex scaling method.[17,18] The idea here is to introduce a complex variable substitution in the radial coordinate r,

$$r \to e^{i\theta} r. \qquad (2.5)$$

Upon this variable transformation the resonance boundary condition is changed to

$$\Psi(r) \to e^{ik_R^{eff}(\theta)r + k_I^{eff}(\theta)r} f(\Omega) \qquad (2.6)$$

where

$$\begin{aligned} k_R^{eff} &= k_R \cos\theta + k_I \sin\theta \\ k_I^{eff} &= k_I \cos\theta - k_R \sin\theta \end{aligned} \qquad (2.7)$$

If θ is chosen larger than $\arctan\frac{k_I}{k_R}$ this boundary condition goes to zero for large r. This means that the resulting Hamiltonian can be diagonalized using an integrable basis. The advantage of simpler boundary conditions is at the expense of having to invert a complex nonhermitian Hamiltonian. This lengthens the computation time somewhat but is not a serious problem. The wave functions are expanded in a finite basis set consisting of generalized Laguerre polynomials. The Hamiltonian is then diagonalized. The accuracy of the calculations can be checked by investigating the dependence of the calculated eigenvalues on the parameter θ. For a complete set of basis functions there should be no θ dependence provided $\theta > \arctan\frac{k_I}{k_R}$. Further details of the method has been published elsewhere.[14]

III. Results for H outside metals

In figure 3 we show how the lowest excited hydrogen levels shift and broaden with distance from an Al surface(r_s=2). All levels shift upwards and become broader with decreasing

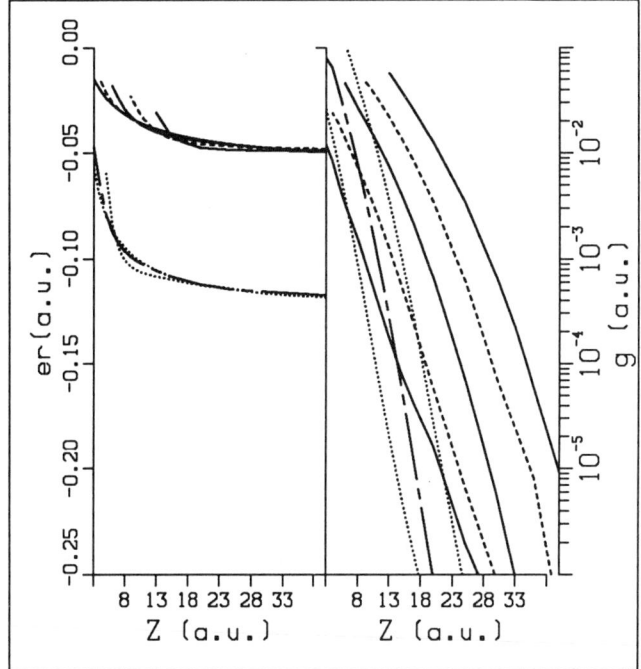

FIG. 3. Calculated energy shifts, to the left and broadening to the right of atomic hydrogen levels as function of distance outside a metal surface. The dotted lines are the H(n=2,m=0) states. The chaindotted line is the H(n=2,m=1) state. The solid lines are the H(n=3,m=0) states and the dashed lines are the H(n=3,m=1) states.

atom-surface separation. We note that the degeneracy of the n=2 and n=3 states is lifted. The n=2 states are initially four-fold degenerate. As the atom comes close to the surface three distinct states are formed that are derived approximately from the atomic Stark states. These states are labeled $\psi_{2s+2p_z}, \psi_{2p_{x,y}}$ and ψ_{2s-2p_z}, and are oriented towards, parallel and away from the surface respectively. The different levels show a complicated behaviour with distance. The state oriented towards the surface increases its energy fastest due to the overlap with the surface electrons and eventually crosses the two other states. We note that the different states have very different lifetimes. The state oriented towards the surface, ψ_{2s+2p_z}, has about two orders of magnitude larger width than the state oriented towards vacuum ψ_{2s-2p_z}. The reason for the large differences in lifetimes is that tunneling is exponential in distance and therefore strongly dependent on the orientation of the electronic state. The n=3 levels are initially nine-fold degenerate. The surface lifts some of the degeneracy and six distinct states are formed. At a given distance the most longlived of the n=3 states is three orders of magnitude narrower than the broadest level.

The calculated widths of the hydrogen levels on the clean metal are several order of magnitudes smaller than previous estimates of these widths. This finding is due to our use of realistic(non diverging) surface potential and the poper consideration of the intra atomic hybridisation.[19]

In Fig. 4, we compare the calculated shifts as a function of distance for H(n=2,n=3) states outside the impurity covered surface. The energy shift as function of distance behave very differently for chlorine and potassium impurities. It can be seen that in the latter case the hydrogen levels shift downwards toward the surface. This downshift occurs because the potassium induced electrostatic potential is larger than the conventional image potential which shift the levels upwards as apparent in fig. 3. We also note that the splittings between the different n=2 and n=3 states is strongly increased. These effects are induced by the strong dipole field induced by the potassium. Such large downshift of atomic levels near chemisorbed impurities means that charge transfer will be facilitated.

The effects of the chlorine impurity is the opposite. From fig. 4, it can be seen that the atomic levels shift upwards faster with decreasing atom-surface separation than for the clean metal.

In fig. 5, we show a comparison of the widths

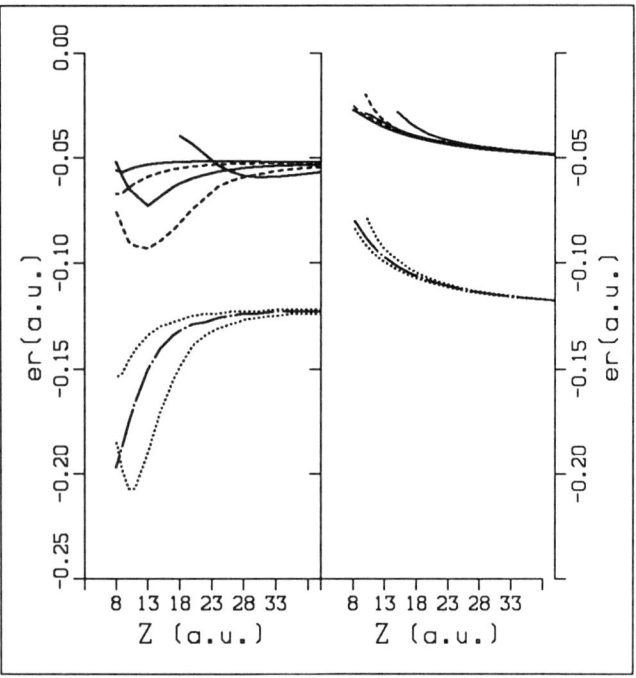

FIG. 4. A comparison between the shifts of the lowest excited hydrogen levels outside impurity covered surfaces. The left figure shows H/K/Al and the right figure shows the result for H/Cl/Al. The notation is as in fig. 3

of the H(n=2,n=3) states outside a clean Al, outside a chemisorbed potassium on Al and outside a chemisorbed chlorine on Al. It can be seen that the widths increase when potassium is present on the surface.

This is due to the induced electrostatic field which tend to lower the potential barrier between the atom and the surface. For the chlorine impurity the effect is reversed and the width at a given distance is reduced slightly. This happens because the dipole field from the Cl is positive and increase the surface barrier.

These effects must be included for a proper discussion of the charge transfer processes between atoms moving through alkali or halogen covered surfaces.

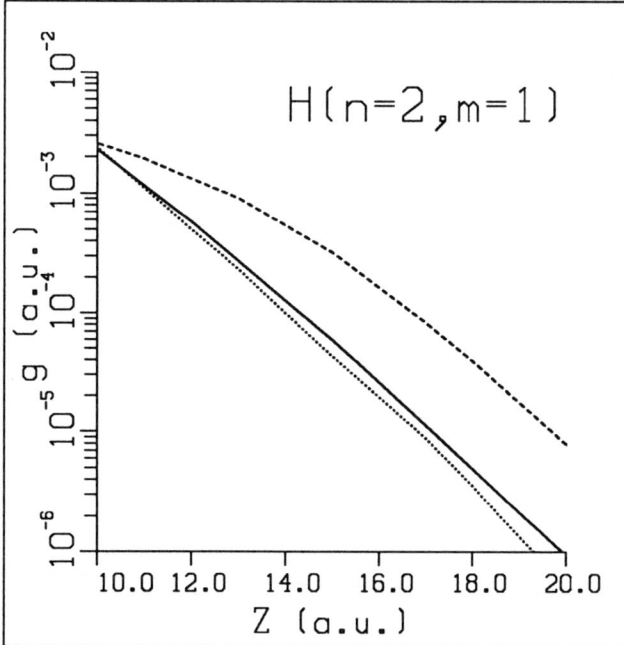

FIG. 5. Calculated width as function of distance Z for the H(n=2,m=1) level outside an Al surface. The solid line refers to the clean metal, the dashed curve is outside the K atom, and the dotted curve refers to the Cl covered metal.

IV. Discussion and Conclusion

We have shown how accurate calculations of atomic level shifts and broadening can be performed for realistic surface potentials. The proposed method is nonperturbative and gives the resonance energies directly from the surface potential. This is a clear advantage compared to perturbative techniques that require the explicit construction of the continuum states of the metal.

In a previous application it has been demonstrated that the calculated widths can be combined with a time dependent nonadiabatic theory to describe charge transfer processes in ion-surface scattering experiments.[20]

The finding of narrrow widths of atomic resonances near surfaces means that charge transfer reaction only will be possible relatively close to the surface. As demonstrated in this paper the atomic level shifts in this region can be non image like. In particular when impurities are present on the surface.

The finding of strongly downshifted hydrogen levels near alkali atoms on surfaces means that these impurities can serve as neutralisation centers. In recent stimulated desorption experiments of hydrogen from potassium covered Pt, it has been observed that neutral excited hydrogen can desorb from even for workfunctions larger than the energy of the atomic state. The experiments thus provide direct evidence for the downshift of the desorbing hydrogen electronic states in the vicinity of potassium impurities.[21]

The formalism is presently extended to treat more complicated atom-surface systems. In particular the formation of negative ions in atom surface scattering experiments is of interest.

The finding of low tunneling rates for resonant tunneling makes it important to reconsider other forms of charge transfer processes such as Auger decay. Such processes have sometimes been neglected since they have been assumed to have lower probability than the resonant tunneling process.[22]

V. Acknowledgements

Acknowledgement is made to the Donors of the Petroleum Research Fund, administred by the American Chemical Society, for the partial support of this research.

REFERENCES

[1] P. Sjovall, B. Hellsing, K. E. Keck, and B. Kasemo, J. Vac. Sci. Tech. A5, 1065 (1987).

[2] J. K. Norskov, D. M. Newns, and B. I. Lundqvist, Surf. Sci. 80, 179 (1979).

[3] N. H. Tolk, J. C. Tully, J. S. Kraus, W. Heiland, and S. H. Neff, Phys. Rev. Lett. **41**, 643 (1978).

[4] T. E. Madey, Science **234**, 316 (1986).

[5] D. Menzel and R. Gomer, J. Chem. Phys. **41**, 3311 (1964).

[6] P. E. Redhead, Can. J. Phys. **42**, 886 (1964).

[7] P. J. Feibelman and M. L. Knotek, Phys. Rev. B **18**, 531 (1978).

[8] W. Ho, Comments Cond. Mat. Phys. **13**, 293 (1988).

[9] H. D. Hagstrum, J. Vac. Sci. Tech. **12**, 7 (1975).

[10] N. D. Lang, Solid State Physics **28**, 225 (1985).

[11] O. Gunnarsson and R. O. Jones, Phys. Scr. **21**, 394 (1980).

[12] S. Ossicini, C. M. Bertoni, and P. Gies, Europhys. Lett. **1**, 661 (1986).

[13] N. D. Lang, Phys. Rev. B **7**, 3541 (1973).

[14] P. Nordlander and J. C. Tully, Submitted to Phys. Rev. B (1989).

[15] N. D. Lang, S. Holloway, and J. K. Norskov, Surf. Sci. **150**, 24 (1985).

[16] J. N. Bardsley, Case Studies in Atomic Physics **4**, 299 (1974).

[17] W. P. Reinhardt, Ann. Rev. Phys. Chem. **33**, 223 (1982).

[18] B. R. Junker, Adv. At. Mol. Phys. **18**, 207 (1982).

[19] P. Nordlander and J. C. Tully, Phys. Rev. Lett. **61**, 990 (1988).

[20] P. Nordlander and J. C. Tully, Surf. Sci **211**, 207 (1989).

[21] P. D. Johnson, A. J. Viescas, P. Nordlander, and J. C. Tully, To be published (1989).

[22] N. N. Nedeljkovic, R. K. Janev, and V. Y. Lazur, Phys. Rev. B. **38**, 3088 (1988).

Symposium:
Correlated Transfer/Excitation
and Autoionization

GENERAL CONSIDERATIONS FOR THE "SYMPOSIUM ON CORRELATED TRANSFER/EXCITATION AND AUTOIONIZATION"

J. A. Tanis

Department of Physics, Western Michigan University, Kalamazoo, MI 49008, USA

This paper discusses the general aspects of correlated transfer/excitation and autoionization. The close relationship between dielectronic recombination (DR) and resonant transfer and excitation (RTE) is reviewed and competing processes are considered. Several "electron-ion like" aspects of ion-atom collisions are identified and their significance discussed.

Recombination of ions in collisions with free electrons involves fundamental interaction mechanisms and is a subject of considerable interest. Dielectronic recombination[1] (DR) is a resonant process in which electron capture is accompanied by simultaneous excitation of the ion due to the electron-electron interaction followed by radiative stabilization. DR is, in fact, the principal mechanism by which free electrons recombine with ions. DR has been identified as an energy-loss mechanism in magnetically confined nuclear fusion plasmas since impurity ions (such as C, O, Fe, etc.) in the plasma can recombine by this mechanism.[2] The cross sections and spectral properties of the DR process are also of interest in plasma diagnostics[3] and in the study of astrophysical plasmas.[1] Therefore, DR has been the subject of intense experimental and theoretical investigations. However, measurements[4-6] of cross sections for DR have proven to be a formidable task since either crossed-beam or merged-beam techniques are required, and only recently have measurements become available which test the DR theory[2] in sufficient detail. It is worth noting that until very recently, DR has been measured only for transitions involving $\Delta n=0$ excitation, i.e., 2s-->2p, 3s-->3p, etc. Exceptions to this are the work[7] of the EBIT group at Livermore and the work of the Aarhus and Heidelberg groups as reported in this symposium.

A mechanism closely related to DR occurs in ion-atom collisions when projectile excitation is accompanied by simultaneous capture of a weakly bound target electron, a process known as resonant transfer and excitation[8] (RTE). RTE and DR proceed through an inverse Auger transition and, hence, are resonant processes as shown in Fig. 1. If the

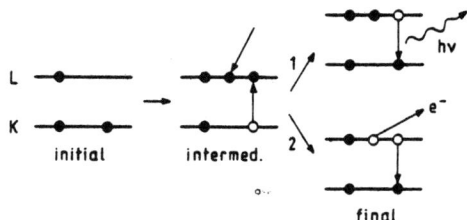

Figure 1. Schematic of the DR (or RTE) process showing resonant formation of the doubly excited intermediate state followed by decay via photon emission (1) or electron emission (2).

projectile velocity is large compared to the target electron velocity, then RTE can be treated in the impulse approximation.[9] Interestingly enough, the first suggestion of the possibility of RTE comes from Shakeshaft and Spruch.[10] In talking about electron capture at high velocities these authors make the comment "... another process could conceivably dominate in some energy range. One might imagine--this is a point we have not checked--that ... the dominant capture process might be through <u>dielectronic recombination</u>."

In the last few years a number of experimental[8,11,12] and theoretical studies[2,13] of RTE have clearly demonstrated the close relationship between RTE and DR. The RTE studies, done principally for 1s-->2p excitation, i.e., $\Delta n=1$, have shown that the measured RTE cross sections are in good agreement with calculations based on theoretical DR cross sections. Because of this close relationship between RTE and DR, most of the detailed tests of the DR theory have come from these RTE measurements, but only for excitation involving $\Delta n=1$ transitions, i.e., 1s-->2p and 2p-->3d.

The intermediate excited state which is formed in the RTE or DR processes can decay by photon emission or electron emission (Auger decay), and the entire reaction is described by the equation

$$A^{q+} + B \longrightarrow [A^{(q-1)+}]^{**} + B^+$$

$$\longrightarrow A^{(q-1)+} + h\nu \quad (1a)$$

$$\longrightarrow A^{q+} + e^- \quad (1b)$$

In the case of radiative decay, RTE is completely analogous to dielectronic recombination (DR) except that for DR the captured electron is initially free instead of bound.

In an ion-atom collision one must allow as well for the possibility of electron-nucleus interactions as indicated in Fig. 2. In the subsequent discussion we will denote electron-electron interactions by (e-e) and electron-nucleus interactions by (e-n). It is also noted that nucleus-nucleus interactions (n-n) can occur, but these interactions involve nuclear physics and will not be considered further here. If the electron capture and projectile excitation occur in a single encounter due to separate (e-n) interactions, i.e.,

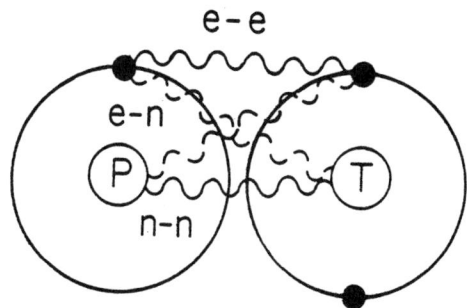

Figure 2. Schematic showing the possible interactions between colliding ions (atoms) each having electrons.

electron capture (projectile nucleus and target electron) and projectile excitation (target nucleus and projectile electron), then the process is a nonresonant two-step process which has been given the name nonresonant transfer and excitation (NTE).[14]

A formal theoretical treatment of simultaneous charge transfer and excitation in single collisions, incorporating separate amplitudes for the resonant and nonresonant contributions has been developed by Feagin, Briggs, and Reeves.[15] In this formal treatment of capture and excitation in single collisions, it is concluded that, to first order, RTE is equivalent to dielectronic recombination averaged over the electron momentum distribution of the target electrons as first proposed by Brandt.[9] In this simple first order theory of Brandt it is assumed that the incident ion velocity is much greater than the target electron velocity, thereby approximating a collision between an ion and a free electron. In this case the cross section for RTE is given by

$$\sigma_{RTE} \sim \sigma_{DR} \Sigma J_i(p_{iz}), \quad (2)$$

where σ_{DR} is the dielectronic recombination cross section, and $J_i(p_{iz})$ is the probability for finding a particular target electron with momentum component p_{iz} along the beam axis. Since the electron momentum distributions of all atoms have been tabulated[16], the cross sections for RTE can be calculated from this model if the DR cross sections are known. Of course, the calculation of the DR cross sections is, in general, difficult, but results[2] for several ions have been reported to date.

To study DR or RTE, these particular interaction mechanisms need to be isolated from competing reaction channels. Two basic techniques exist: (a) coincidence methods in which electron capture is associated with the appropriate decay process (photon or Auger electron), and (b) high resolution x-ray or Auger measurements in which specific intermediate states are identified. Experimentally, the combined process of electron capture and projectile excitation in a single collision has been identified and investigated by measuring the energy dependence of (a) coincidences between deexcitation photons associated with projectiles[8] which have captured an electron, and (b) high-resolution measurements of Auger electron emission associated with capture events.[17]

As noted above, for high-velocity ion-atom collisions in which the ion velocity is much greater than the bound target electron velocity, i.e., $v_{ion} \gg v_{elec}$, these bound electrons can be considered "quasi-free" and the impulse approximation should be valid. In this case "electron-ion like" aspects of ion-atom collisions can become manifest. The "electron-ion like" processes of interest to this symposium are: (1) RTE[8] (resonant transfer and excitation) in which electron capture is accompanied by simultaneous projectile excitation via an (e-e) interaction, (2) eeE (electron-electron excitation)[18] which is analogous to electron-impact excitation and occurs via an (e-e) interaction, and (3) CTE[19] or 2eTE[20] (continuum transfer excitation) which involves projectile excitation via an (e-e) interaction and charge transfer via an (e-n) interaction.

Resonant transfer and excitation (RTE) and its relationship to dielectronic recombination has already been discussed above. A typical RTE result[21] is shown in Fig. 3 for S^{13+} + He

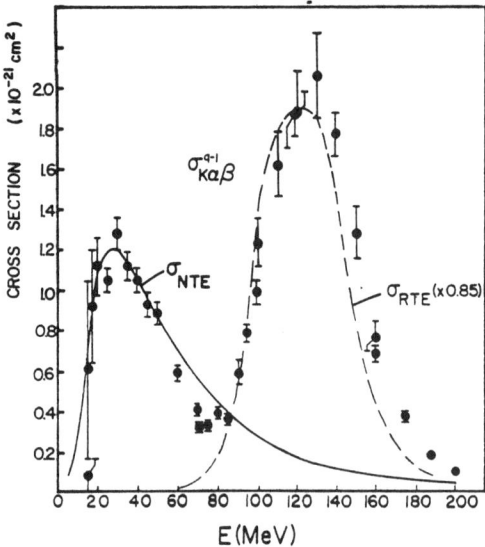

Figure 3. Cross section $\sigma_{K\alpha\beta}^{q-1}$ for K x-ray emission coincident with single electron capture in S^{13+} + He collisions (from Ref. 21). The dashed curve is the calculated RTE cross section multiplied by 0.85 and the solid curve is the calculated NTE cross section normalized to the data (see Ref. 21).

collisions in which K x rays were measured in coincidence with single electron capture. This work clearly shows that RTE closely approximates DR, the main difference being the momentum distribution of the target electrons in the case of RTE and, furthermore, that effects due to (e-e) interactions, i.e., RTE, can be separated from effects due to (e-n) interactions, i.e., NTE. The discrepancy between theory and experiment at high energies is noted in which the theory falls off faster than the data. This same general trend has been observed for lithium-like calcium ions as well. Attempts to explain this discrepancy through the contribution of 1s-->nln'l' (n,n' ≥ 3) transitions or through a core field effect have proven unsuccessful.[22] The two-step NTE process is seen to occur with a probability comparable to that for RTE at least for this collision system. Calculations[23] indicate, however, that the importance of NTE should decrease relative to RTE as the atomic number of the projectile increases.

For continuum transfer excitation[19,20] (CTE or 2eTE) the excitation is due to a direct (e-e) interaction (similar to electron impact excitation) and the capture process is due to an (e-n) interaction as shown schematically in Fig. 4. These

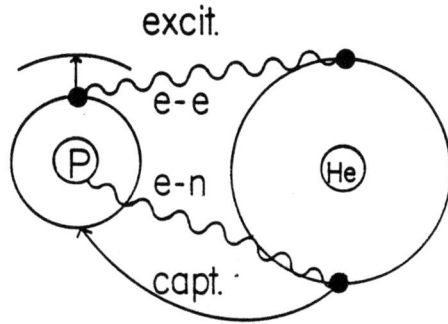

Figure 4. Schematic of the CTE (or 2eTE) process.

excitation and capture interactions are independent. The CTE mechanism was first proposed[17] in 1985 as a possible explanation for resolving the high energy discrepancy between theory and experiment for RTE cross sections (see Fig. 3). The first experimental evidence for 2eTE was recently reported by Schulz et al.[20] In this work, involving the high resolution measurements of Auger electron emission as a function of beam energy for F^{8+} + H_2 collisions, a maximum was found in the Auger emission for an incident energy of 29 MeV which is well above any maximum which could be due to RTE. However, by assuming the maximum resulted from an (e-e) interaction, giving rise to a threshold behavior broadened by the Compton profile of the target electrons, accompanied by an independent electron transfer event (e-n) in the same collision the maximum could be explained. These measurements provided the first evidence for the importance of electron-impact excitation via the (e-e) interaction in an ion-atom collision.

If the CTE (2eTE) process involving two separate interactions takes place, then it would be expected that the (e-e) part of the process, as shown in Fig. 5,

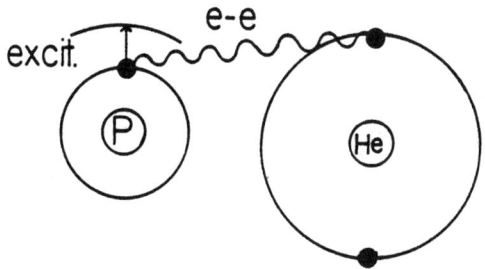

Figure 5. Schematic of the eeE process.

should occur by itself since this latter process represents a first order effect. Here the excitation, due strictly to the (e-e) interaction, should again exhibit a threshold behavior (corresponding to 1s-->2p excitation, for example) broadened by the Compton profile of the target electrons. This ion-atom analog of electron-impact excitation has recently been observed and has been dubbed electron-electron excitation (eeE).[18] In these high resolution measurements of Auger electron emission from the $(1s\,2s\,2p)^4P$ state for F^{6+} + H_2, He collisions, a broad threshold was observed at an incident energy of 25 MeV corresponding to 1s-->2p excitation. The broadening of the threshold was consistent with that expected from the Compton profiles of H_2 and He, respectively.

Generally, 1s-->2p excitation will be dominated by (e-n) interactions. The $(1s\,2s\,2p)^4P$ state was chosen for these observations because it cannot be produced by (e-n) interactions since a spin flip is required. The formation of this state proceeds instead through an (e-e) interaction involving electron exchange. These results thus provide the first direct evidence for the analog of electron-impact excitation in ion-atom collisions.

Very recently, Hahn and Ramadan[24] have presented a unified picture of (e-n) and (e-e) interactions in ion-atom collisions. To describe a general interaction involving a one-electron ion with a two-electron target they use the notation

$$(a+e_1) + (b+e_2+e_3) \rightarrow (a+e_1+e_2)^{**} + (b+e_3)$$
$$\underbrace{\qquad}_{\text{ion}} \quad \underbrace{\qquad\qquad}_{\text{target}}$$

where a and b represent nuclei and e_1, e_2, e_3 represent electrons. Three modes of transfer excitation, namely, RTE, NTE and CTE, are considered. Schematically, these processes involve the following interactions:

RTE: (e-e)   correlation
     [V_{12}]

NTE: (e-n) + (e-n) independent
     [V_{1b} and V_{2a}]

CTE: (e-e) + (e-n) independent
     [V_{13} and V_{2a}]

where the V's represent appropriate interaction potentials.

Calculations of this unified picture were applied to the S^{13+} + He data of Ref. 21 (see Fig. 3). By including contributions from these three possible modes of transfer excitation, general overall agreement with the data is obtained. In particular, by including the CTE contribution, some of the discrepancy at high energies between the data and the RTE theory is removed. Because of the difficulty of the calculations and the approximations which are necessary, additional theoretical work is needed to determine the relative importance of CTE compared to RTE.

In summary, dielectronic recombination (DR) and its ion-atom analog resonant transfer and excitation (RTE) are considered and discussed in this symposium. In addition to RTE, two other "electron-ion like" aspects of ion-atom collisions are considered, namely, continuum transfer excitation (CTE or 2eTE) and electron-electron excitation (eeE). It is found that these electron-ion like aspects can not only be isolated and identified, but that they can, in fact, be quite important to understanding the dynamics of ion-atom collision processes. Electron-ion like processes can involve <u>direct</u> interactions, as in the case of eeE and CTE, or <u>correlated</u> interactions as in the case of RTE. The role and significance of electron-ion like interactions in ion-atom collisions has only been recognized in recent years and it is clear that further experimental and theoretical work is needed to better define and understand the interplay of (e-e) and (e-n) interactions in these collisions.

The author is deeply indebted to the many colleagues with whom he has worked closely for several years. This work was supported in part by the U.S. Department of Energy, Office of Basic Energy Sciences, Division of Chemical Sciences.

References
1. A. Burgess, Astrophys. J. <u>139</u>, 776 (1964); and <u>141</u>, 1588 (1965).
2. Y. Hahn and K. J. LaGattuta, Phys. Rpt. <u>166</u>, 196 (1988), and references therein.
3. M. Bitter, S. von Goeler, S. Cohen, K. W. Hill, S. Sesnic, F. Tenney, J. Timberlake, U. I. Safronova, L. A. Vainshtein, J. Dubau, M. Loulergue, F. Bely-Dubau, and L. Steenman-Clark, Phys. Rev. A<u>29</u>, 661 (1984).
4. D. S. Belic, G. H. Dunn, T. J. Morgan, D. W. Mueller, and C. Timmer, Phys. Rev. Lett. <u>50</u>, 339 (1983).
5. P. F. Dittner, S. Datz, P. D. Miller, C. D. Moak, P. H. Stelson, C. Bottcher, W. B. Dress, G. D. Alton, N. Neskovic, and C. M. Fou, Phys. Rev. Lett. <u>51</u>, 31 (1983).
6. J. B. A. Mitchell, C. T. Ng, J. L. Forand, D. P. Levac, R. E. Mitchell, A. Sen, D. B. Miko, and J. Wm. McGowan, Phys. Rev. Lett. <u>50</u>, 335 (1983).
7. D. A. Knapp, R. E. Marrs, M. A. Levine, C. L. Bennett, M. H. Chen, J. R. Henderson, M. B. Schneider, and J. H. Scofield, Phys. Rev. Lett. <u>62</u>, 2104 (1989).

8. J. A. Tanis, Nucl. Instrum. Meth. Phys. Res. A262, 52 (1987), and references therein.
9. D. Brandt, Phys. Rev. A27, 1314 (1983).
10. R. Shakeshaft and L. Spruch, Rev. Mod. Phys. 51, 369 (1979).
11. M. Clark, D. Brandt, J. K. Swenson, and S. M. Shafroth, Phys. Rev. Lett. 54, 544 (1985).
12. S. Reusch, P. H. Mokler, R. Schuch, E. Justiniano, M. Schulz, A. Müller, and Z. Stachura, Nucl. Instrum. Meth. Phys. Res. B23, 137 (1987).
13. D. J. McLaughlin and Y. Hahn, Phys. Lett. 88A, 394 (1982); D. J. McLaughlin and Y. Hahn, Phys. Lett. 112A, 389 (1985); G. Omar and Y. Hahn, Phys. Rev. A35, 918 (1987).
14. P. L. Pepmiller, P. Richard, J. Newcomb, R. Dillingham, J. M. Hall, T. J. Gray, and M. Stöckli, IEEE Trans. Nucl. Sci. NS-30, 1002 (1983); and Phys. Rev. A31, 734 (1985).
15. J. M. Feagin, J. S. Briggs, and T. M. Reeves, J. Phys. B17, 1057 (1984).
16. F. Biggs, L. B. Mendelsohn, and J. B. Mann, Atom. Data Nucl. Data Tables 16, 201 (1975).
17. J. K. Swenson, Y. Yamazaki, P. D. Miller, H. F. Krause, P. F. Dittner, P. L. Pepmiller, S. Datz, and N. Stolterfoht, Phys. Rev. Lett. 57, 3042 (1986).
18. T. J. M. Zouros, D. H. Lee, and P. Richard, Phys. Rev. Lett. 62, 2261 (1989).
19. Y. Hahn, unpublished, 1985.
20. M. Schulz, J. P. Giese, J. K. Swenson, S. Datz, P. F. Dittner, H. F. Krause, H. Schöne, C. R. Vane, M. Benhenni, and S. M. Shafroth, Phys. Rev. Lett. 62, 1738 (1989).
21. J. A. Tanis, E. M. Bernstein, M. W. Clark, W. G. Graham, R. H. McFarland, T. J. Morgan, B. M. Johnson, K. W. Jones, and M. Meron, Phys. Rev. A31, 4040 (1985).
22. Y. Hahn, Second US-Mexico Symp. on Atomic and Molecular Physics: Two-Electron Phenomena, Cocoyoc, Mexico (1986); published in Notas de Fisica, eds. I. Alvarez, C. Cisneros, J. de Urquijo and T. J. Morgan (Instituto de Fisica, UNAM, Cuernavaca, Mor., Mexico, 1987) vol. 10, no. 2, pp. 91-104.
23. T. M. Reeves, J. M. Feagin, and E. Merzbacher, 14th Int. Conf. on the Physics of Electronic and Atomic Collisions, Palo Alto, California, 1985, Abstracts of Contributed Papers, p. 392.
24. Y. Hahn and H. Ramadan, Nucl. Instrum. Meth. Phys. Res. (1989), in press.

TRANSFER AND EXCITATION WITH HEAVY PROJECTILES AND TARGETS

W. G. Graham

Physics Department, Queen's University, Belfast BT7 1NN, Northern Ireland

It has been shown that projectile excitation and electron capture can occur together in a single encounter between an ion projectile and a target electron. The capture and excitation events can be correlated (Resonant Transfer and Excitation (RTE)) or uncorrelated (Nonresonant Transfer and Excitation (NTE) and Uncorrelated Transfer and Excitation (UTE)). In collisions of intermediate Z projectiles with H_2 and He there is now generally good agreement between experiment and theory but a few existing discrepancies are discussed. There is as yet no theoretical work for heavy projectiles but experimental work is now underway. In heavy targets it appears that the uncorrelated processes dominate over RTE.

Introduction

Recently, there have been considerable advances in our understanding of the processes occurring in collisions between ions and atoms, largely due to the realization that, even at high energies where the collision times are very short ($\sim 10^{-18}$ s), more than one event can take place in a single collision. In particular it has been shown that projectile excitation and electron capture can occur together in a single encounter between an ion projectile and a target atom. This results in the formation of an intermediate doubly excited state which subsequently decays by either photon or electron emission, i.e.,

$$A^{q+} + B \longrightarrow [A^{(q-1)+}]^{**} + B^+ \quad (1)$$
$$\longrightarrow A^{(q-1)+} + h\nu$$
$$\longrightarrow A^{q+} + e^-$$

The capture and excitation events can be correlated or uncorrelated. The correlated process was first suggested by Shakeshaft and Spruch[1] and subsequently confirmed experimentally by Tanis et al.[2,3] and involves an electron-electron interaction between a projectile electron and a (weakly-bound) target electron similar to an inverse Auger transition. Resonant formation of the intermediate state occurs for incident ion velocities equal to the Auger electron velocities. This correlated process is called Resonant Transfer and Excitation (RTE) and has been shown to be analogous to dielectronic recombination (DR)[4] where in DR the captured electron is initially free. RTE has now been widely observed in both the photon and electron decay channels.[2,3,5-7] It has been shown[8] that the RTE cross section can be calculated from the DR cross section by folding in the Compton profile ($\Sigma J_i(P_{iz})$) of the target electron, i.e.,

$$\sigma(RTE) = \sigma(DR) \sum_1 J_i(P_{iz}) \quad (2)$$

There are other combinations of excitation and capture events which can lead to <u>uncorrelated</u> transfer and excitation in a single collision. In nonresonant transfer and excitation (NTE)[7] the capture involves a target electron-projectile nucleus interaction while the excitation is through a target nucleus-projectile electron interaction. The cross section for NTE will depend on the separate probability for capture and excitation. It can be shown that

$$\sigma(NTE) = P_L^{capt}(0)\, \sigma_K^{excit} \qquad (3)$$

where $P_L^{capt}(0)$ is the capture probability at small (zero) impact parameter and σ_K^{excit} is the K-shell excitation cross section.[8,9]

In what was initially called CTE[10] or 2eTE[11] and is now known as uncorrelated transfer and excitation (UTE) two electrons in the target participate separately, one interacting with the projectile nucleus and leading to capture, a second with a projectile electron resulting in excitation. The theory for UTE is now just developing.[12] It is important to note that, since this is a direct excitation, the threshold for UTE is at the projectile electron excitation energy, i.e., the high energy edge of the RTE resonance.

Here these correlated and uncorrelated processes will be discussed in the context of measurements with highly stripped heavy ions where photon emission, in the x-ray region, is the dominate decay mode of the intermediate state.

Apparatus

The experimental measurements discussed here were carried out principally at the Lawrence Berkeley Laboratory using the SuperHILAC and Bevalac facilities and at the Brookhaven National Laboratory, using the tandem Van de Graaff facility. The experimental techniques used at the SuperHILAC and Brookhaven have been described in detail elsewhere.[2,3]

In the Bevalac experiment, the total single electron capture cross section was measured in carbon foils. The collimated beam from the Bevalac was passed through the foil target and the magnetically separated charge state components were detected in a position sensitive detector. The charge state fraction was measured as a function of foil thickness to obtain the cross section and to check for single collision conditions.

Results and Discussion

a) <u>Intermediate Z projectiles in light targets</u>. The most direct evidence for both NTE and RTE has been found in studies of collisions of S^{13+} with He.[13] As shown in Figure 1, two maxima are observed in the energy dependence of the cross sections for K x-ray production in coincidence with electron capture, $\sigma_{K\alpha\beta}^{q-1}$. The high energy maximum is consistent with that expected for RTE. The individual maxima associated with each of the intermediate states are smeared out by the contribution from the momentum distribution of the target electrons. The assignment of this maximum as RTE is confirmed by theoretical RTE calculations, determined using the simple first order theory of Brandt[8] and using the DR calculations of McLaughlin and Hahn[14] (Figure 1). The maximum near 30 MeV is attributed to nonresonant transfer and excitation (NTE). Calculated NTE cross sections[15] based on equation 3 are also shown in Figure 1.

In Figure 2 the energy dependence of σ_{Kx}^{q-1} in $Ca^{17+} + H_2$ collisions[16] is shown. The peaks are assigned to RTE, NTE being expected at substantially lower energies. Here the contributions from the intermediate states can be resolved into two groups, one centered around 200 MeV corresponding to the production of intermediate states with quantum numbers n=2,2, another centered around 260 MeV corresponding to the production of n=2,≥3 states. Also shown in Figure 2 are very recent intermediate coupling calculations of Badnell.[17] These and the calculations of Bhalla and Karim[18] appear to be in excellent agreement with the experimental measurements. Some discrepancies between theory and experimental are apparent in both Figures 1 and 2, and are also

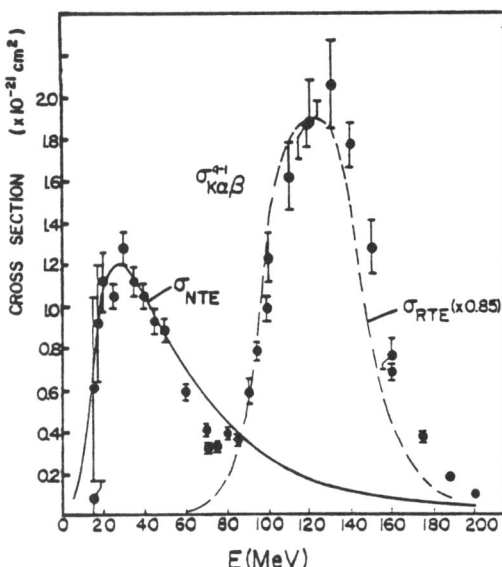

Figure 1. Cross sections for K x-ray emission in coincidence with single electron capture for S^{13+} in He. The dashed curve is the calculated RTE cross section multiplied by 0.85 and the solid curve is the calculated NTE cross section normalized to the experimental data at 30 MeV (Ref. 13).

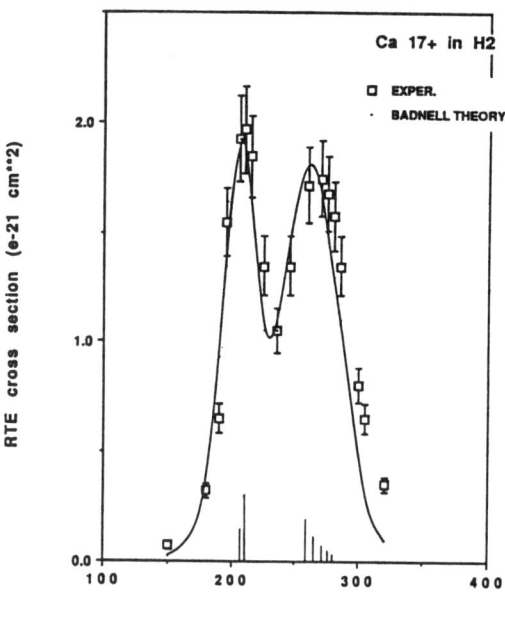

Figure 2. Cross section for K x-ray emission in coincidence with single electron capture for Ca^{17+} in H_2 (Ref. 16). The solid curve is the calculated RTE cross sections of Badnell (Ref. 17). The vertical bars indicate the positions of the DR resonance states.

observed in measurements and calculations with other charge states and projectile species[16] and in L-shell RTE.[19] At the high energy edge of the RTE resonance there is a consistent trend for the experimental measurements to lie above the theoretical calculations, while at the lower energy edge there is a trend for the experimental measurements to lie below the theoretical calculations. Hence, it is apparent that peak positions tend to be at higher energies in the experimental results than in the theoretical calculations. There are some suggestions as to the nature of these discrepancies. They seem to be consistent with an energy difference between theory and experiments.[20] If the theoretical calculations in Figure 2 are shifted up in energy by 2.5 MeV, then the peak heights match and the low energy discrepancy disappears although there is still a shortfall in the theoretical values at the high energy edge of the resonance. Such an energy adjustment could be required if the theoretical calculations failed to account correctly for the binding energy of the target electron.

Recently a careful study of the energy dependence of σ_{Kx}^{q-1} in the region of the n=2,2 resonance for Fe^{23+} in H_2 has been made.[21] Preliminary results indicate good agreement between theory and experiment on the high energy edge of this resonance but at the low energy edge the experimental values are consistently lower than the theoretical calculations.

It has been suggested[10,11] that the high energy discrepancy is due to the contribution of UTE to the σ_{Kx}^{q-1} cross section. As mentioned previously this would be at the high energy edge of the RTE resonance but the threshold will be smeared out by the target electron momentum distribution.

b) <u>Heavy projectiles</u>. The projectile atomic number (Z) dependence of transfer and excitation has been studied both theoretically and experimentally only over a very limited range, approximately $8 \leq Z \leq 32$.[2,3,6-7] For $Z \leq 14$ the DR cross section has been found to be fairly independent of the projectile Z.[22] Calculations of RTE cross sections for Li-like projectiles using equation 2 and assuming the DR cross section remains constant with Z show[23] that the increasing energy separation between the different intermediate states leads to the increased resolution and decreasing cross sections as Z increases.[23] As shown in Figure 3, extending a similar calculation to U^{89+} in H_2 indicates that the cross sections for producing several of the individual intermediate states can be completely resolved.[24] The U^{89+} in H_2 system is interesting since the resonances are expected at very high energy where the total single electron capture cross section is very small and dominated by radiative electron capture (REC).[25] It has been shown that in systems such as Ca^{17+} in H_2 the RTE contribution can be seen in the total electron capture cross section.[26] The predicted energy dependence of total electron capture for U^{89+} in H_2,[24] shown in Figure 3, implies that a relatively simple measurement of total electron capture should allow the determination of the RTE cross section for U^{89+} in H_2.

The magnitude of the DR cross sections (the basis of RTE calculations) for high Z projectiles is difficult to determine. The Z scaling of DR and RTE cross sections has been discussed by several authors.[27,28] They predict a maximum in the DR cross section at about Z=35, implying a maximum in the RTE cross section at about Z=30 due to the Z dependence of the Compton profile. At very high Z's the RTE cross section is expected to decrease as Z^3. These scaling rules rely on values and scalings of the radiative and Auger decay widths over a wide range of projectiles and charge states and may be uncertain. Theoretical calculations with very heavy projectiles are just becoming feasible.

In the meantime, experimental

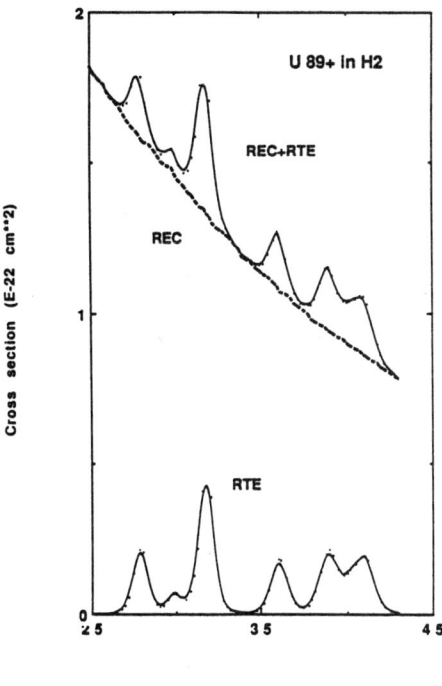

Figure 3. Calculated RTE cross sections for U^{89+} in H_2 (Ref. 24). The magnitude of the cross section is obtained by normalizing an RTE Z scaling (Ref. 28) to RTE calculations at Z=26 (Ref. 22). The dashed line is the calculated REC cross section (Ref. 24). Also shown is the calculated total single electron capture cross section (i.e., $\sigma_{rec} + \sigma_{RTE}$).

measurements of total single electron capture for U^{89+} in carbon foils have been made[25] in the energy range 26 to 43 GeV where the $1s2s2p^2$ resonance is expected. These measurements imply that the RTE cross section decreases above Z~35 and while a determination of the RTE cross section was not possible, an upper limit for the RTE cross section and DR cross section for the $1s2s2p^2$ intermediate state formation for U^{89+} could be inferred.[29]

c) <u>Heavy targets.</u> Measurements[30] of σ_{Kx}^{q-1} for Ca^{17+} in Ar are shown in Figure 4, along with measurements in H_2[16] and theoretical calculations of the RTE cross section in Ar. There is a large discrepancy between the calculated RTE cross section and σ_{Kx}^{q-1} for Ar. Similar results were found for Ne and Kr targets.

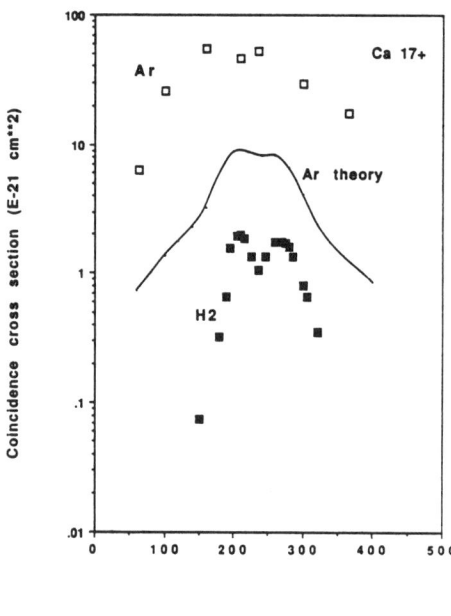

Figure 4. Cross sections for K x-ray emission in coincidence with single electron capture for Ca^{17+} in Ar and H$_2$. The curve is the calculated RTE cross section for an Ar target.

As mentioned previously in light targets there has been good agreement between theory and experiment. It therefore appears that in heavy targets electron transfer and excitation is dominated by NTE.

From a consideration of equation 3 and the fact that, in the energy range where the RTE resonance is expected, measured[26,30] K x-ray excitation cross sections in Ar are approximately 40 times larger than in H$_2$ and the electron capture cross sections are about three orders of magnitude larger it would be expected that NTE would be more pronounced in heavier targets. At this time the role of UTE is uncertain and it will only contribute at the highest energies in the present measurements. The increasing number of electrons in the targets would be expected to increase the probability of UTE since this involves two target electrons.

Conclusion

In general for intermediate Z projectiles in light targets there is now excellent agreement between the magnitude of theoretically and experimentally determined cross sections at the RTE maxima. There are some discrepancies between theory and experiment at the low and high energy edges of the RTE resonance and hence in the peak positions. It has been suggested that the high energy discrepancy is due to a contribution from uncorrelated transfer and excitation. Scaling to heavy projectiles is uncertain and more theoretical work is required to complement preliminary and planned experiments with U^{89+}. It would appear, for heavy targets, that, at least for intermediate Z projectiles, transfer and excitation in a single collision is dominated by the uncorrelated process, NTE.

Acknowledgments

I would like to thank the many people who have been involved in obtaining the measurements shown here. Gene Bernstein, Mark Clark and John Tanis have made major contributions to all aspects of this work. Important contributions and assistance has been provided by K. H. Berkner, R. D. DuBois, B. M. Johnson, K. W. Jones, M. Meron, R. J. McDonald, R. H. McFarland, T. J. Morgan, J. R. Mowat, D. W. Mueller, A. Müller, C. S. Ogelsby, A. S. Schlachter and J. W. Stearns. I would also like to thank Dr. N. R. Badnell for providing his theoretical RTE calculations prior to their publication. The author was supported by the United Kingdom Science and Engineering Research Council.

References

1. R. Shakeshaft and L. Spruch, Rev. Mod. Phys. **51**, 369 (1979).
2. J. A. Tanis, E. M. Bernstein, W. G. Graham, M. Clark, S. M. Shafroth, B. M. Johnson, K. W. Jones and M. Meron, Phys. Rev. Lett. **49**, 1325 (1982).

3. J. A. Tanis, E. M. Bernstein, W. G. Graham, M. P. Stöckli, M. Clark, R. H. McFarland, T. J. Morgan, K. H. Berkner, A. S. Schlachter and J. W. Stearns, Phys. Rev. Lett. 53, 2551 (1984).
4. A. Burgess, Astrophys. J. 139, 776 (1964) and 141, 1588 (1965).
5. A. Itoh, T. J. M. Zorros, D. Schneider, M. Stettner, W. Zeitz and N. Stolderfoht, J. Phys. B18, 4581 (1985).
6. J. K. Swenson, Y. Yamazaki, P. D. Miller, M. F. Krause, P. F. Dittner, P. L. Pepmiller, S. Datz and N. Stolderfoht, Phys. Rev. Lett. 57, 3042 (1986).
7. M. Schulz, E. Justiniano, R. Schuch, P. H. Mokler and S. Reusch, Phys. Rev. Lett. 58, 1724 (1987).
8. D. Brandt, Phys. Rev. A27, 1314 (1983).
9. P. L. Pepmiller, P. Richard, J. Newcomb, R. Dillingham, J. M. Hall, T. J. Gray and M. Stöckli, IEEE Trans. Nucl. Sci. NS30, 1002 (1983) and Phys. Rev. A31, 734 (1985).
10. Y. Hahn, unpublished (1985)
11. M. Schulz, J. P. Giese, J. K. Swenson, S. Datz, P. F. Dittner, H. F. Krause, H. Shöne, C. R. Vane, M. Benhenni and S. M. Shafroth, Phys. Rev. Lett. 62, 1738 (1989).
12. Y. Hahn and H. Ramadan, Nucl. Instrum. Meth. Phys. Res. (1989), in press.
13. J. A. Tanis, E. M. Bernstein, M. W. Clark, W. G. Graham, R. H. McFarland, T. J. Morgan, B. M. Johnson, K. W. Jones and M. Meron, Phys. Rev. A31, 4040 (1985).
14. D. J. McLaughlin and Y. Hahn, Phys. Lett. A88, 394 (1982).
15. T. L. Abee, Nucl. Instr. and Meth., 214, 89 (1983) and D. Brandt, ibid. pg. 93.
16. J. A. Tanis, E. M. Bernstein, M. W. Clark, W. G. Graham, R. H. McFarland, T. J. Morgan, J. R. Mowat, D. W. Mueller, A. Müller, M. P. Stöckli, K. H. Berkner, P. Gohil, R. J. McDonald, A. S. Schlachter and J. W. Stearns, Phys. Rev. A34, 2543 (1986).
17. N. R. Badnell, J. Phys. B (in press).
18. C.P. Bhalla and K. R. Karim, Phys. Rev. A39, 6060 (1989).
19. E. M. Bernstein, M. W. Clark, J. A. Tanis, K. H. Berkner, R. J. McDonald, A. S. Schlachter, J. W. Stearns, W. G. Graham, R. H. McFarland, T. J. Morgan, J. R. Mowat, D. W. Mueller and M. P. Stöckli, J. Phys. B20, L505 (1987).
20. E. Kanter (private communication).
21. J. A. Tanis, E. M. Bernstein, M. Clark, R. D. DuBois, W. G. Graham, T. J. Morgan, A. S. Schlachter and M. P. Stöckli (in preparation).
22. D. McLaughlin and Y. Hahn, Phys. Rev. A31, 1926 (1985).
23. J. A. Tanis, E. M. Bernstein, C. S. Oglesby, W. G. Graham, M. Clark, R. H. McFarland, T. J. Morgan, M. P. Stöckli, K. H. Berkner, A. S. Schlachter, J. W. Stearns, B. M. Johnson, K. W. Jones and M. Meron, Nucl. Instr. and Meth. B10/11, 128 (1985).
24. W. G. Graham, E. M. Bernstein, M. W. Clark and J. A. Tanis, Phys. Lett. A125, 134 (1987).
25. W. E. Meyerhof, R. Anholt, J. Eichler, H. Gould, Ch. Munger, J. Alonso, P. Theiberger and H. E. Wegner, Phys. Rev. A32, 3291 (1985).
26. W. G. Graham, E. M. Bernstein, M. W. Clark, J. A. Tanis, K. H. Berkner, P. Gohil, R. J. McDonald, A. S. Schlachter, J. W. Stearns, R. H. McFarland, T. J. Morgan and A. Müller, Phys. Rev. A33, 3591 (1986).
27. Y. Hahn, Phys. Rev. A22, 2896 (1980).
28. S. Reusch, GSI Report 88-19 (September 1988).
29. W. G. Graham, K. H. Berkner, E. M. Bernstein, M. W. Clark, H. Crawford, B. Feinberg, M. Flores, L. Greiner, R. J. McDonald, P. M. Mokler, T. J. Morgan, W. Rathbun, A. S. Schlachter and J. A. Tanis, 16th Int. Conf. on Physics of Electronic and Atomic Collisions, New York, New York (1989), Abstracts of Contributed Papers, p. 547 and in preparation.
30. M. W. Clark et al., J. de Physique 48, C9-203 (1987).

RESONANT PROCESSES IN ATOMIC COLLISIONS AND A UNIFIED VIEW

Yukap Hahn

Department of Physics, University of Connecticut, Storrs, CT 06268 USA

Resonant states of ions are copiously produced in violent electron-ion and ion-atom collisions when inner-shell electrons are excited or excitation of the ion is followed by electron capture. Various resonant processes are inter-related by unitarity, analyticity and impulse approximation, so that their cross section data can be correlated. The recent progress made in dielectronic recombination and transfer-excitation is discussed. A resonance model for the pair line production in heavy ion collisions is examined and the predicted spectrum is presented.

1. INTRODUCTION

Electron-ion and ion-atom collision data are needed in modelling laboratory and astrophysical plasmas, and much experimental and theoretical efforts are being expended to generate a complete set of reliable rate coefficients for the various processes which can take place inside plasmas. The required theoretical task is enormous and computer-intensive, while direct experimental measurements are often difficult and time-consuming. It is therefore useful if relationships among the various rates can be found, which then may be used to correlate scanty available data. A unified picture of both electron-ion (e-I) and ion-atom (I-A) collision processes was described previously[1] where the indirect resonance mode dominates over the direct counterparts. Thus, the 'RESONANCE CUBE' was defined, in Fig. 1, where the front four are the e-I processes and the back four are for I-A. The resonant e-I processes are inter-related by analyticity (QDT) and unitarity (ω/ξ branching ratios), while the e-I vs I-A processes can be connected in impulse approximation (IMA). Thus, for example, dielectronic recombination (DR) and resonant-transfer-excitation followed by x-ray emission (RTEX) are related by IMA, and Auger ionization and RTEA are connected by QDT/IMA. Theoretical analyses are routinely carried out[2] following the paths that are shown in the cube.

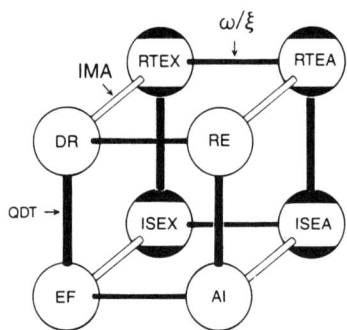

RESONANCE CUBE

Fig. 1. Resonance cube that summarizes various e-I and I-A resonant processes.
RE=resonant excitation, EF= excitation fluorescence. ISEA= inner-shell excitation, followed by Auger electron emission, RTEX=resonant transfer excitation followed by x-ray emission.

In this report, we discuss some of the latest works on DR in e-I collisions and transfer-excitation (TE) in I-A collisions, all of which proceed through the intermediate resonance states. As the third topic, we summarize a resonance model similar to Auger ionization (AI), which is used to explain the observed line pair production in heavy ion collision, and the predicted spectrum is compared with experiment.

2. DIELECTRONIC RECOMBINATION

We discuss three recent works on DR which are of special significance: (i) The DR cross sections for the metastable C(4+) and O(6+) of the He isoelectronic sequence have been measured by Andersen et al [3], involving 2s→2p intrashell excitations ($\Delta n_t=0$). The theoretical calculation[4] is compared with experiment in Fig. 2. for O(6+). A small fraction of the ion beam was in the metastable 1s2s ^{1}S and ^{3}S states, of 20-30%. We used this fraction and the field enhancement ratio r_F as parameters in generating the result in Fig. 2. Two things are noted; first, the cross section for the ^{1}S → ^{1}P is extremely large, of the order of 10^{-14} cm^2, and secondly there is a major discrepancy around 2.5 and 6.8 eV. Configuration interaction and intermediate coupling effects are insufficient to explain this. On the other hand, the C(4+) data are very well explained by the same theoretical calculation.

Fig. 2. DR for metastable O(6+) ions. The experimental data are from ref. 3 (solid curve), and the theory is from ref. 4 (histogram).

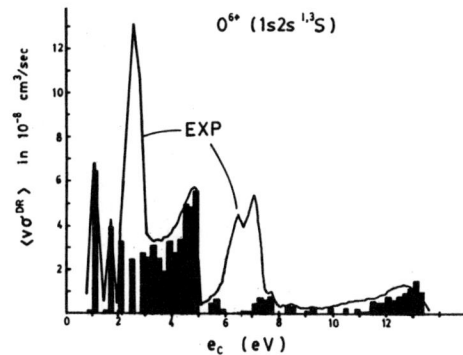

(ii) Preliminary data on DR for C(3+) have been reported[5], which, in addition to sizable electric field enhancement[2], suggest possible contributions coming from the continuum capture[6] in the $2p_{1/2}$ channel. This effect was suggested in ref. 6 in an attempt to resolve the discrepancy in the captured state distribution in the case of Mg(+) DR[7].

(iii) A DR anomaly was found for ions with N=19 electrons[8]. Contrary to a smooth behavior of the DR rates in the core charge Z_c expected for ions of given isoelectronic sequence, Sc(2+) showed an unusual dip in the rates as compared to that for the neighboring ions Ca(+) and Ti(3+). A typical result is shown in Fig. 3. For N ≳ 18, electron-electron correlation can be very important and the excited states are densely distributed, so that the usual distinction between the intra- and inter-shell excitation transitions starts to disappear.

Fig. 3. The DR cross sections for the four ions Ca, Sc, Ti and Fe of the K isoelectronic sequence, with 19 electrons.

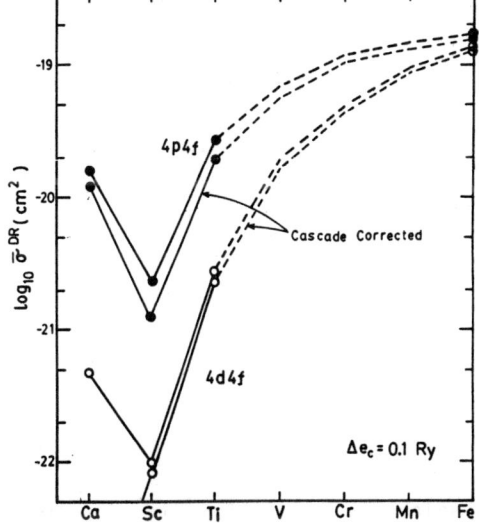

3. TRANSFER EXCITATION

Two modes of the TE process have been known since 1982, the resonant RTE and the nonresonant NTE, both of which produce resonant intermediate states of the projectile[9,10]. Subsequently, these states decay by either x-ray emission (TEX) or Auger electron emission (TEA). More recently, a third TE process, the uncorrelated TE (UTE), was observed[11] in the collision of the $F^{6+} + H_2$ system. Such a process was postulated earlier[12] in order to explain the apparent discrepancy found in the RTEX cross sections at high energies[13]. A comprehensive theory of TE has been formulated recently[14] in which all three modes (NTE, RTE and UTE) are incorporated. Thus, for a collision system defined schematically as

$$(a + 1) + (2 + 3 + b) \longrightarrow (a + 1 + 2)^{**} + (3 + b)^*$$
$$\quad A \qquad\qquad B \qquad\qquad\qquad A' \qquad\qquad B'$$

the TEX amplitude for example is given by

$$T^{TEX} = \sum_d \langle f | D | d \rangle \frac{1}{E - E_d + i\Gamma(d)} \langle d | V_i | i \rangle$$

where

$$V_i = \underbrace{V_{12}}_{RTE} + \underbrace{V_{13}}_{UTE} + \underbrace{V_{1b}}_{NTE} + (V_{2a} + V_{3a} + V_{ab})$$

excitation→

The exchange effect is neglected here for simplicity, but should be included in the actual calculation. The three modes are dominant in different collision energies, for projectiles with $Z_c \gtrsim 10$, but can interfere in the overlapping regions.

<u>3A. RTE</u>. The resonant mode of TE is mediated by V_{12} in V_i; that is, electron 2 excites the projectile A and at the same time is captured to form A'. The cross section is given in impulse approximation[15]

$$\sigma^{RTEX}(i \to d) = \sigma^{DR}(i \to d) \cdot W_B(p_z) \cdot (\Delta e_c / K_i)$$

where the resonance state d decays by x-ray emission. W_B is the Compton profile for the target B, and e_c is the kinetic energy of electron 2 in the rest frame of the ion A. For the DR cross section, we have

$$\sigma^{DR}(i \to d) = \sigma_e^{ECT-CAP}(i \to d) \cdot \omega(d), \quad \omega(d) = \text{fluorescence yield}.$$

For other modes of decay, we have RTEA[16] and also RTEXX[17], RTEXA, RTEAA, etc. The connection between DR and RTEX has been well established[9,15]

<u>3B. NTE</u>. The nonresonant mode is represented by V_{1b} in V_i; the projectile A is excited by the target core b, while one of the target electrons, say electron 2, is captured to form an intermediate state d. The cross section is given by[14,18,19]

$$\sigma^{NTEX} = \int \sigma_N^{ECT}(p) \cdot P_{CAP}(p) \, dp \cdot W_B(p) \cdot \omega(d)$$

where σ_N^{ECT} is the excitation cross section of A by core b, and P_{CAP} is the electron capture probability which is evaluated in closure approximation[14]

$$P_{CAP}(i \text{---} d) = 2\pi \left| \langle d | V_{2a} \cdot e^{i\vec{p} \cdot \vec{r}_{2a}} | c \rangle \right|^2 \cdot \left(\frac{1}{e_{A'} - \bar{e}_c} \right)^2 .$$

$\sigma_N^{ECT}(i \to d)$ is calculated in plane-wave Born approximation at high energies, but for low $E_L \lesssim 30$ MeV a correction to impulse approximation is needed, as[14]

$$\sigma_N^{ECT} \longrightarrow \sigma_N^{ECT} \cdot \exp(-2\Delta_{fi} \gamma_o / e_c) \cdot F_o$$

where the adiabatic correction factor includes two parameters γ_o and F_o. Δ_{fi} is the excitation energy.

<u>3C. UTE</u>. The uncorrelated TE is identified[14] with the term V_{13} in V_i; for a target system with more than one electrons, there is a third possibility of forming the intermediate resonance state d in which electron 3 excites the projectile as electron 2 is captured. The cross section is given by[14]

$$\sigma^{UTEX} = \int \sigma_e^{ECT}(p) \cdot P_{CAP} \cdot W_B(p) \, dp \, \omega(d)$$

where P_{CAP} is the same as that in NTEX. The electronic excitation cross section σ_e^{ECT} is related to σ (ECT-CAP/e) of RTEX by analytic continuation of one of the bound states to continuum. We now compare the theoretical predictions with experimental data.

3D. Results. (i) TE for S^{13+} + He. This is the first successful and complete calculation of the TE cross sections for the all three modes. The result is given[20] in Fig. 4 and compared with experimental data of Tanis et al[9]. The excitation cross section was first evaluated and the parameters γ_o and F_o determined using the experimental points (Fig. 4a). Next, this σ(ECT/N) is used in the evaluation of the NTEX cross section by explicitly calculating P_{CAP}. The UTEX cross section contains the same P_{CAP}, so that only the σ(ECT/e) needs to be evaluated. This was done for many partial waves and intermediate states d, all in angular momentum average procedure[2]. The agreement is excellent. We note that, in order to obtain the NTEX peak position correctly, the low energy behavior of the excitation cross section requires the adiabaticity correction.

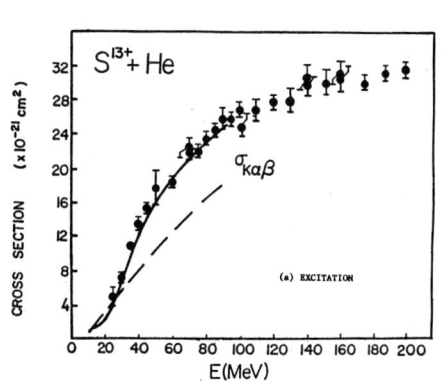
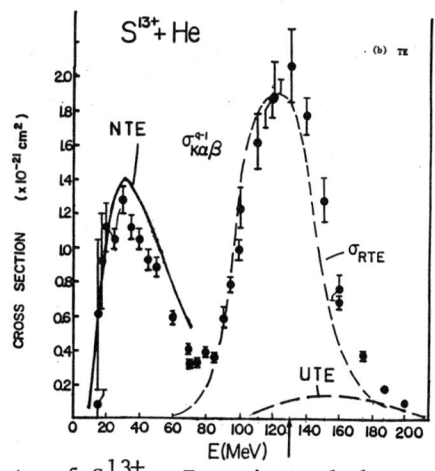

Fig. 4. (a) The excitation cross section for the 1s2s2p state of S^{13+}. Experimental data are from ref. 9, and the solid curve is the adiabaticity-corrected cross section. (b) The three modes of the TE processes are compared with experiment of ref. 9. The RTEX curve is from ref. 21.

(ii) More recently, the F^{6+} + H_2 system was studied[22] in the process ISEA for the excitation of d=1s2s2p, without electron transfer. The theoretical calculation[23] again fits the data reasonably well, both in magnitude and the location of the maximum. This process is closely related to UTEA in that the excitation is caused by one of the target electrons. A similar study was also reported[24] on the F^{8+} + H_2 system.

(iii) A very accurate measurement of the RTEX cross section was carried out sometime ago[25] for the system Nb(31+) + H_2, in which the 2p electron excitation is involved. The agreement with theory[26] of RTEX was good, except a large and clearcut discrepancy in the high energy tail. The UTEX cross section has been calculated recently[23] and the result is compared in Fig. 5 with experiment. Because of the high quality of the experiment in this high energy region, this is the most definitive case yet of the new UTE process.

Figure 5. The UTEX cross section for Nb^{31+} + H_2, together with the RTEX cross section from ref. 26. Experimental data are from ref. 25. The dotted curve is the UTEX, and the solid curve shows the total TEX in this region.

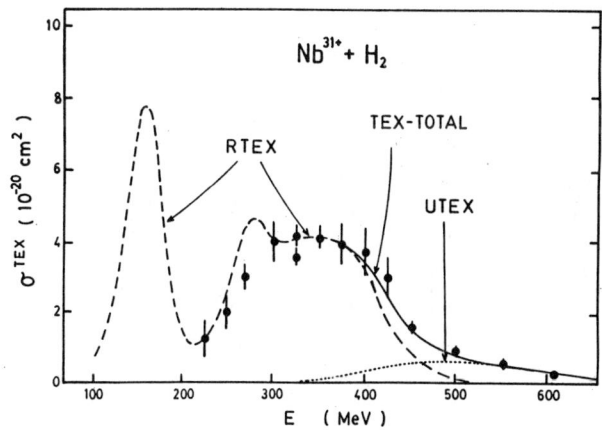

4. LINE PAIR PRODUCTION IN HEAVY ION COLLISIONS

Since the first experimental observation[27,28] of electron-positron pair line production in heavy ion collisions more than six years ago, many theoretical speculations have been put forward as to the origin of such phenomenon, some exotic and others highly speculative. None of them, however, have satisfactorily explained all the refined experimental measurements[29,30]. Experimentally, for the colliding ions with $Z_A \cong Z_B \gtrsim 80$, the e_+e_- pairs are copiously produced. The kinetic energy distribution of the ejected positron is smooth, with a maximum around 350 keV. Superimposed on this are several narrow resonance-like peaks at E_+ = 220, 320, 380, and 420 keV, etc. In particular, the pairs are produced back-toback.

Recently, a dynamical model was proposed[31] in which these sharp peaks are assumed to be produced by the decay of resonant states formed by doubly excited pairs, in complete analogy to the Auger ionization process discussed in sec. 2. A similar picture in the context of molecular collision was described earlier by Lichten[32], but without concrete predictions. The effective one-center energy diagram is given in Fig. 6, where the critical Z are indicated. Assuming sizable correlations among the pairs, we can then predict the pair energies at each Z_{cr}. The predicted spectrum of the model is summarized in Table I, and the suggested strengths and energies are compared with the observed values. Noting that the field produced by the two approaching ions at their midpoint is zero, while the net electric potential is very large, the model seems to explain all the constraints of the experiments. However, details of the dynamics are yet to be worked out.

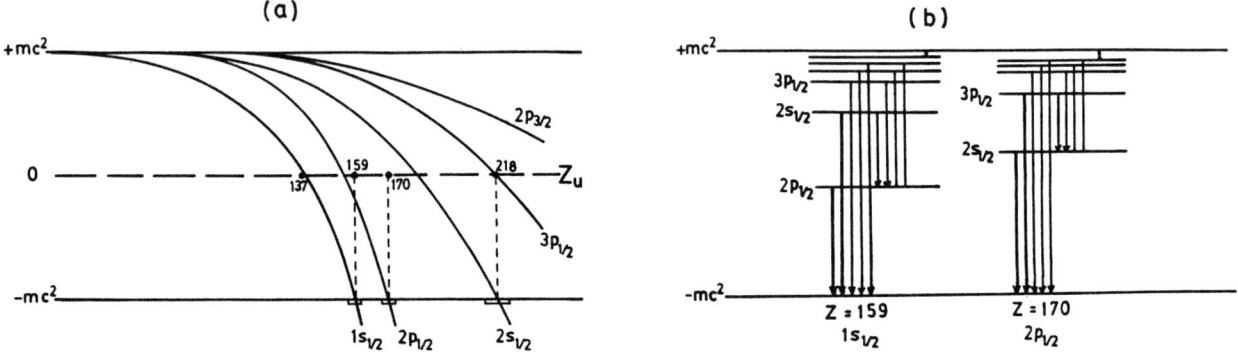

Fig. 6. Effective single-center energy levels for the two approaching heavy ions, as deduced approximately in ref. 31.

Table I. The transitions which are assumed to be responsible for the pair line production are summarized, together with the positron energies and suggested line strengths.

Transitions	Z_u	Pair energies	Strength	E_+-Theory	E_+-Exp.	(d) states	Z_u	Trans. Energies	Strength	E_+-Theory	E_+-Exp.
$2p_{1/2}$ - $1s_{1/2}$	$<Z_{cr}(2p_{1/2})$	430 keV	***	220 keV	220 keV	$2p_{1/2} + 2p_{1/2}$	$<Z_{cr}(2p_{1/2})$	860 keV	x	--- keV	--- keV
$2s_{1/2}$ - $1s_{1/2}$		730	***	370	380	$2p_{1/2} + 2s_{1/2}$		1160	*	70	
$3p_{1/2}$ - $1s_{1/2}$		870	**	440		$2p_{1/2} + 3p_{1/2}$		1300	*	140	
$2p_{3/2}$ - $1s_{1/2}$		930	*	470		$2p_{1/2} + 2p_{3/2}$		1360	*	170	
$2s_{1/2}$ - $2p_{1/2}$		300	*	150	165(?)	$2s_{1/2} + 2s_{1/2}$		1460	**	220	220
$3p_{1/2}$ - $2p_{1/2}$		440	*	220	230(?)	$2s_{1/2} + 3p_{1/2}$		1600	*	290	
$2p_{3/2}$ - $2p_{1/2}$		500	*	250	280(?)	$2s_{1/2} + 2p_{3/2}$		1660	*	340	
$3p_{1/2}$ - $2s_{1/2}$		140	*	70		$3p_{1/2} + 3p_{1/2}$		1740	**	360	360(?)
:		200	*	100		$3p_{1/2} + 2p_{3/2}$		1800	*	390	
		60	*	30		$2p_{3/2} + 2p_{3/2}$		1860	**	420	420
$2s_{1/2}$ - $2p_{1/2}$	$<Z_{cr}(2s_{1/2})$	610	****	310	320	:					
$3p_{1/2}$ - $2p_{1/2}$		830	***	420	410	$2s_{1/2} + 2s_{1/2}$	$<Z_{cr}(2s_{1/2})$	1220	**	100	(?)
$2p_{3/2}$ - $2p_{1/2}$		920	**	460	460(?)	$2s_{1/2} + 3p_{1/2}$		1440	*	210	
:		940	*	470	500(?)	$2s_{1/2} + 2p_{3/2}$		1530	*	260	
$3p_{1/2}$ - $2s_{1/2}$		220	*	110		$3p_{1/2} + 3p_{1/2}$		1660	***	320	320
$2p_{3/2}$ - $2s_{1/2}$		310	*	160	170(?)	$3p_{1/2} + 2p_{3/2}$		1750	*	370	
:		330	*	170		$2p_{3/2} + 2p_{3/2}$		1840	**	410	420
		90	*	50		:					
		110	*	60							
		20	*	10							

5. CONCLUSIONS

The resonant states formed in collisions of electrons and ions with a target system can provide a coherent picture of diverse reaction mechanisms and allows correlation of atomic data needed for plasma research. Some of the recent experimental data on DR are yet to be explained, as discussed in sec. 2, while the TE theory seems to contain for the first time all the essential physics, as described in secs. 3 and 4. But, much more refined calculations are now needed. In so far as the model for pair production is concerned, it is a drastic departure from the quasi-static picture adopted in many of the earlier work. A detailed calculation of the formation of the double-pair excited states and their decay widths is desired.

This work is supported in part by a DOE grant.

References

1. Y. Hahn, Comments Atom. Molec. Phys. $\underline{19}$, 99 (1987) and $\underline{13}$, 103 (1983)
2. Y. Hahn, Adv. Atom. Molec. Phys. $\underline{21}$, $\overline{123}$ (1985); Y. Hahn and K. LaGattuta, Phys. Rep. $\underline{166}$,195(1988)
3. L. H. Andersen, P. Hvelplund, H. Knudsen and P. Kvistgaad, Phys. Rev. Lett. $\underline{62}$, 2656 (1989)
4. Y. Hahn and R. Bellantone, Phys. Rev. A (RC), submitted
5. A. Young, J. Kohl, et al (private communication)
6. I. Nasser and Y. Hahn, Phys. Rev. $\underline{A36}$, 4704 (1987)
7. A. Muller et al, Phys. Rev. Lett. $\underline{56}$, 127 (1986); D. S. Belic et al, ibid $\underline{50}$, 339 (1983)
8. G. Omar, A. Moussa and Y. Hahn, Phys. Rev. A (BR), to be published.
9. J. A. Tanis, Nuclear Inst. Methods $\underline{A262}$, 52 (1987); J. Tanis et al, Phys. Rev. Lett., $\underline{49}$,1325(1982)
10. M. Clark et al, Phys. Rev. Lett. $\underline{54}$, 544 (1985)
11. E. Justiniano et al, ICPEAC Proc. 1987, p477; M. Schulz et al, Phys. Rev. $\underline{A38}$, 5454 (RC) (1988)
12. Y. Hahn, 1985 (unpublished)
13. Y. Hahn, Second US-Mexico Symposium, Cocoyoc, 1985
14. Y. Hahn, Phys. Rev. A (to be published)
15. D. Brandt, Phys. Rev. $\underline{A27}$, 1314 (1983)
16. J. Swenson et al, Phys. Rev. Lett. $\underline{57}$, 74 (1986); Y. Hahn, Phys. Lett. $\underline{A119}$, 293 (1986)
17. M. Schulz et al, Phys. Rev. Lett. $\underline{58}$, 1734 (1987); D. McLaughlin and Y. Hahn, Phys. Rev.$\underline{A38}$,531
18. J. M. Feagin et al, J. Phys. $\underline{B17}$, 1057 (1984)
19. D. Brandt, Nuclear Inst. Methods $\underline{214}$, 93 (183)
20. Y. Hahn and H. Ramadan, Nuclear Inst. Methods (to be published)
21. D. McLaughlin and Y. Hahn, Phys. Lett. $\underline{A88}$, 394 (1981)
22. T. J. Zurous et al, Phys. Rev. Lett. $\underline{62}$, 2261 (1989)
23. Y. Hahn and H. Ramadan, Phys. Rev. A (RC)
24. M. Schulz et al, Phys. Rev. Lett. $\underline{62}$, 1738 (1989)
25. J. Tanis et al, J. Phys. $\underline{B20}$, L505 (1987)
26. G. Omar et al, Phys. Rev. $\underline{A36}$, 576 (1987)
27. T. Cowan et al, Phys. Rev. Lett. $\underline{54}$, 1761 (1985)
28. M. Clemente et al, Phys. Lett. $\underline{B137}$, 41 (1984)
29. ICPEAC, 1985; J. Reinhardt et al, p389, J. Schweppe et al, p405, T. E. Cowan, p369
30. P. Kienle, Atomic Physics, Vol. 10. H. Narumi and I. Shimamura eds.(Elsevier, 1987)
31. Y. Hahn, Phys. Rev. Lett. (submitted)
32. W. Lichten and A. Robatino, Phys. Rev. Lett. $\underline{54}$, 781 (1985)

RECOMBINATION BETWEEN FREE ELECTRONS AND MULTIPLY CHARGED IONS

Preben Hvelplund

Institute of Physics, University of Aarhus
DK-8000 Aarhus C, Denmark

During the last couple of years, a group at the University of Aarhus has been working on an experiment aiming at measurements of recombination rates in collisions between multiply charged ions and electrons. The first results from this effort have recently been obtained. In this communication, I will discuss the experimental method used and the experimental results merely with respect to what may be learned from the result of the increased experimental resolution. Since several review papers on recombination have appeared during the past two years, a brief introduction to the field will suffice here.

INTRODUCTION

When atomic systems make a transition from one bound state to another, energy is normally emitted as photons or, if doubly excited states are involved, also as kinetic energy of Auger electrons. Likewise, when free bound transitions (recombination) are involved, excess energy can be carried away in the form of a photon or by another electron which can be either free or initially bound to the ion.

When the excess energy is carried away in the form of a photon (coupling to the radiation field), one talks of radiative-recombination reactions,

$$e^- + A^{q+} \rightarrow A^{(q-1)+} + h\nu \quad (RR) . \quad (1)$$

When a second electron is involved, we have either dielectronic recombination when this electron is initially bound to the ion, and the intermediate state decays radiatively,

$$e^- + A^{q+} \rightarrow A^{(q-1)+**} \rightarrow A^{(q-1)+*} h\nu \quad (DR) \quad (2)$$

or terminary recombination when this electron is initially free,

$$e^- + e^- + A^{q+} \rightarrow e^- + A^{(q-1)+} \quad (TR) . \quad (3)$$

The DR process is resonant, which means that only free electrons with discrete, well defined energies may become bound. Typical cross sections are in the order of 10^{-17}-10^{-18} cm^2. Both RR and TR are nonresonant processes, and the cross sections for these processes diverge as the relative energy approaches zero. A typical cross-section value for RR is about 10^{-20} cm^2 at a relative energy ~1 eV. The TR recombination rate is proportional to the square of the electron density, and this process is relevant only at high electron densities.

The cross sections for RR was first calculated by Stobbe[1] in 1930 and later in an analytical form by Bethe and Salpeter[2]. DR was first suggested as a recombination process by Sayers in 1939 and considered in a quantitative form by Massey and Bates[3]. Three decades later, Bates et al.[4] treated the TR process for hydrogenic plasmas and showed that this recombination mechanism is dominating at very low energies and high electron densities.

The recombination process mentioned above has become of interest to collision physicists within the last decade. At the Berlin ICPEAC in 1983, a hot-topic session was devoted to DR, and several laboratories reported in 1983 on DR measurement[5-7]. This experimental activity started new theoretical efforts represented by, e.g., papers by LaGattuta and Hahn[8] and Griffin, Pindzola, and Bottcher[9]. For recent review papers within the field of DR, the readers are referred to those of Dunn[10] and Hahn and LaGattuta[11].

Recent experimental improvements, which have led to the production of intense monoenergetic electron beams ('electron coolers')[12] for phase-space compression in storage rings, initiated further experimental efforts within the field of ion-electron recombination. With such an electron target, it is possible to obtain relative energy spreads as low as ~0.1 eV in a merged-beam experiment. Electron coolers are already in operation at several storage rings, and recombination results are about to come out in the near future.

In the experiment discussed here, an electron cooler is used in connection with a tandem accelerator. Both DR and RR measurements with very high energy resolution have been obtained. The experimental results, particularly new findings originating as a result of the improved energy resolution, are discussed in relation to available theoretical models.

THE MERGED-BEAM EXPERIMENT

The electrons are accelerated in an electron gun with Pierce geometry, and three solenoids and two toroids guide the electrons from the gun through the interaction region to the collector, see Fig. 1. The electron current is space-charge-limited with a typical value of about 10 mA. The beam diameter in the interaction region is about 1 cm. The energy spread of the electrons in such a beam is normally of the order of 0.15 eV.

Figure 1 shows the experimental set-up schematically. An ion and an electron beam are merged in a 1m long interaction region. The projectiles, which have captured an electron, are identified after electrostatic separation from the main beam. This field sets an upper limit n_{max} for the main quantum number of

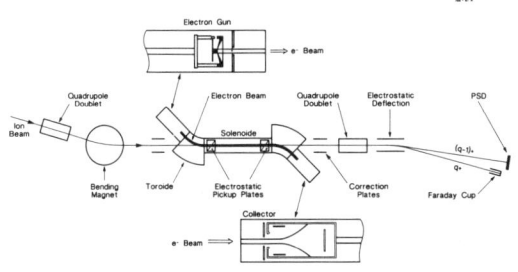

Figure 1

states which survives through the analyzer and accordingly contribute to the signal.

The actual energy of the electrons is given by the formula (in the laboratory frame

$$E_e = (V_c + V_s)e , \qquad (4)$$

where V_c is the cathode potential, V_s is the space-charge potential in the electron beam, and e is the elementary charge. In the rest frame of the ion, the energy of the electron, E_r is obtained as

$$|E_r| = E_e + \frac{m}{M}E_i - \sqrt{E_i E_e \frac{m}{M}} , \qquad (5)$$

where E_i is the ion energy and m and M are the electron mass and ion mass, respectively. We use the convention that negative E_r corresponds to $v_e < v_i$ and positive E_r corresponds to $v_e > v_i$, where v_e and v_i are the electron and ion laboratory velocity, respectively.

We choose to represent our data in the form of the rate coefficient defined as

$$\langle v\sigma \rangle = \frac{N^{(q-1)+}}{N^{q+}} \frac{v_i}{\varrho_e L} , \qquad (6)$$

where v is the electron velocity in the ion frame, ϱ_e is the electron density, L the interaction length, and v_i is the ion velocity. The combined uncertainty of $\langle v\sigma \rangle$ is estimated to be ±20%.

To be able to compare the experimentally obtained rate coefficient, Eq. (6), with theory, the electron-velocity distribution $f(\vec{v})$, as seen in the rest frame of the ion, must be known. The relation between the cross section σ and the rate coefficient is

$$\langle v\sigma \rangle = \int \sigma(v) v f(\vec{v}) d\vec{v} \qquad (7)$$

We choose to write $f(\vec{v})$ as a product of the two Maxwell distributions,

$$f(\vec{v}) = \frac{m}{2\pi k T_\perp} e^{-m v_\perp^2 / 2kT_\perp}$$

$$\cdot \sqrt{\frac{m}{2\pi k T_\parallel}} e^{-m(v_\parallel - \Delta)^2 / 2kT_\parallel} \qquad (8)$$

where $v_\perp$ and $v_\parallel$ are the electron-velocity components perpendicular and parallel to the ion beam, respectively, and Δ is the detuning velocity given by $\frac{1}{2}m\Delta^2 = E_r$. $T_\perp$ and $T_\parallel$ are the two temperatures which characterize the relative motion of the electrons and ions. From fits to several resonances (for details, see Ref.13), we obtain $kT_\perp = 0.135 \pm 0.010$ eV and $kT_\parallel = 10 \times 10^{-4} \pm 5 \times 10^{-4}$ eV for the two temperatures. As expected, due to the kinematic reduction of the longitudinal energy spread, $kT_\parallel \ll kT_\perp$. There is no kinematic reduction of the transverse energy spread, and $kT_\perp$ is close to that expected from a cathode at a temperature of 1000 K.

ELECTRON-ION RECOMBINATION

As is clear from the introduction, it is possible to measure radiative as well as dielectronic recombination with our experimental setup. An approximate analytical expression for RR was derived by Bethe and Salpeter[2] and is given for all principal quantum numbers n by

$$\sigma_n = 2.1 \times 10^{-22} \frac{E_0^2}{(nE(E_0 + n^2 E))} \text{cm}^2 , \qquad (9)$$

where E_0 is the ground-state (1s) binding energy, and E is the electron energy in the rest frame. If one uses this cross section in formula (7) and assume the velocity distribution from Eq. (8), the total RR rate reads[14]

$$\alpha_{RR} = \frac{3.02}{\sqrt{kT}} q^2 \left\{ \ln\frac{11.32q}{\sqrt{kT}} + 0.14 \left(\frac{kT}{q^2}\right)^{1/3} \right\} \cdot 10^{-13}$$

$$\text{cm}^3/\text{sec} \qquad (10)$$

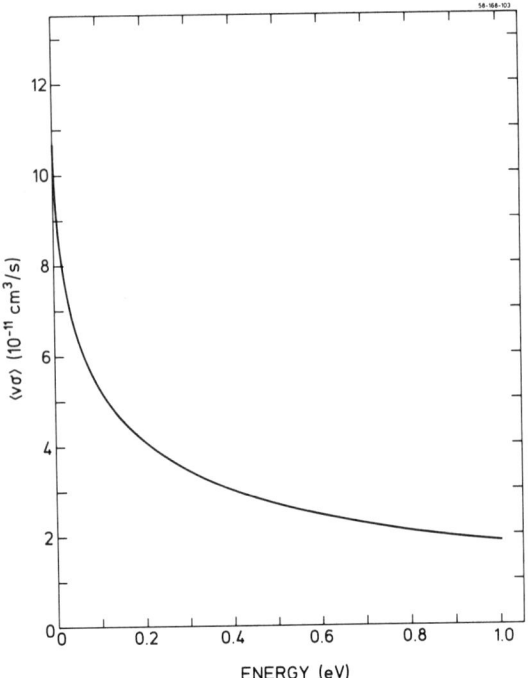

Figure 2

In Fig. 2, α_{RR} is shown as a function of energy for a q=6 ion interacting with the electron beam described earlier, ($kT_\perp = 0.15$ eV, $kT_\parallel = 2 \times 10^{-3}$ eV).

To gain insight in the physics of DR, let us follow the discussion as given by Müller et al.[15]. Let the cross section for DR into a given state be described as

$$\sigma = \sigma_c [A_r / (A_a + A_r)] , \qquad (11)$$

where σ_c is the cross section for electron capture into the intermediate state (see Eq. (2)). $A_r/(A_a + A_r)$ where A_r and A_a are transition rates for radiation and autoionization, respectively, is the fraction of the intermediate doubly excited state which undergoes radiative stabilization. From the principle of detailed balance, it follows that σ_c is proportional to A_a, and the cross section for capture into a state represented by quantum numbers n,l can be written as

$$\sigma_{n,l} = \sigma_0 2(2l+1) \frac{A_a(n,l) A_r(n,l)}{A_a(n,l) + A_r(n,l)} , \qquad (12)$$

where σ_0 is a constant and $2(2l+1)$ is the statistical weight of the final n,l states. It is argued by Müller et al.[15] that for low-to-moderate n and for low l, where $A_a \gg A_r$, one has $\sigma_{nl} (2l+1)A_r$. For high l's, A_a decreases rapidly, and one can define a maximum l state (l_c) which effectively contributes to the DR cross section.

The total cross section for capture into a given state can now be written as

$$\sigma_n \approx \sum_{l=0}^{l_c} \sigma_{n,l} = \sigma_0 2(l_c+1)^2 A_r \ . \quad (13)$$

While this imposes no limitation for small n's, where $l_{max}=n-1 > l_c$, for moderate n's, possible contributions from l's with $l_c \leq l \leq n-1$ are suppressed because of the strong decrease of A_a with l.

An electric field in the interaction region will cause a mixing of the wave functions corresponding to high l states, with these corresponding to low l states which have a stronger coupling to the continuum. This has the effect of increasing l_c, i.e., more states effectively participating in the recombination process when a field is applied to the collision region.

DR EXPERIMENTS WITH C^{3+}

In Fig. 3 is shown the DR spectrum for the process

$$C^{3+}(1s^2 2s)+e^- \rightarrow C^{2+}(1s^2 2pnl) \\ \rightarrow C^{2+}(1s^2 2snl) \ . \quad (14)$$

The theoretical cross sections are being obtained from distorted-wave calculations by Griffin et al.[16]. The cross sections were integrated over small energy bins to yield an energy-averaged cross section $\bar{\sigma}$,

$$\bar{\sigma}_{\varepsilon_0} = \frac{1}{\Delta\varepsilon}\int_{\varepsilon_0 - \Delta\varepsilon/2}^{\sigma_0 + \Delta\varepsilon/2} \sigma(\varepsilon) d\varepsilon \ , \quad (15)$$

where $\Delta\varepsilon$ is the energy-bin size of 0.0005 Hartrees. To obtain the rate coefficient, we calculated

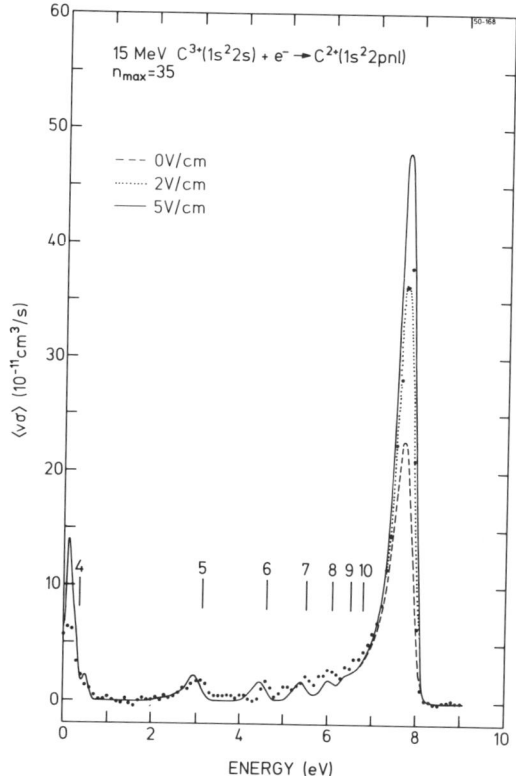

Figure 3

$$\alpha_{DR} = \int_0^\infty \sigma(E)\sqrt{\frac{2E}{m}} \frac{dN}{dE}(E)dE \ , \quad (16)$$

where the density dN/dE is given as

$$\frac{dN}{dE} = \frac{v}{m} \int_\Omega f(\vec{v}) d\Omega \quad (17)$$

with $f(\vec{v})$ defined in Eq. (8). We used $kT_\parallel = 10^{-3}$ eV and $kT_\perp = 0.135$ eV.

The distorted-wave calculation[16] was performed with different electric fields in the interaction region. The electric field perpendicular to the beam direction is due to the space-charge potential. As can be seen from Fig. 3, only the high-n part of the spectrum is influenced by the field, as was also argued in the previous section. It should be noted that the very good energy resolution in our experiment allows a comparison between theory and experiment for individual peaks, where electric fields play no rôle as well as for the high-n peak which is strongly influenced by even small fields.

It is natural to divide this spectrum into three parts, namely the n=4 peak, the peaks from n=5 to ~10, and the high-n peak which actually terminates at n=35 because of field stripping in the analyzer field. Good agreement between experiment and theory is observed for n= 5-10. This is an important observation because it is a case where field effects can be ignored, as discussed earlier. The apparent disagreement for n=4 ($1s^2 2p4l$ intermediate resonances) should not be taken too seriously at present since the cross section is very sensitive to the not very well known energy of the levels, in particular so close to threshold. On the other hand, a better understanding of this discrepancy will lead to valuable information, either about precise energies of doubly excited states or errors in the DR theory. For the high-n peak, experiment and theory are in agreement for a field of about 2-3 V/cm. The precise field strength is difficult to obtain because the exact ion-beam profile is not known in the interaction region. If we assume that the ion-beam profile is Gaussian, with a width of 2 mm (2σ), we obtain an average field of 1.8 V/cm.

The information gained from Fig. 3 is probably typical for any three-electron system, but so far, we have only done measurements with C^{3+} and O^{5+} (Ref. 13).

DR EXPERIMENTS WITH C^{4+}

The DR spectrum for C^{4+} is shown in Fig. 4. The observed structures are due to the presence of metastables ($1s2s(^3S)$ and $1s2s(^1S)$) in the incoming beam. It was argued in Ref. 13 that the beam composition at the target is $1s^2(^1S)$: 70%, $1s2s(^3S)$: 30%, and $1s2s(^1S)$: ~1%.

In Fig. 5 is shown a schematic level diagram for the relevant states in C^{4+} and C^{3+}. As seen from the figure, there are up to five different excitations for the He-like ions. The DR resonances are expected at relavetive energies E_n according to

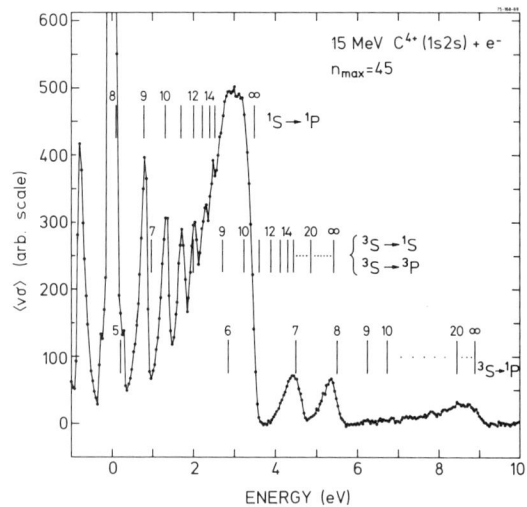

Figure 4

$$E_n = \Delta E - \frac{q^2 \cdot 13.6}{n^2} \qquad (18)$$

where ΔE is the core-excitation energy relative to one of the five possible excitations. In Fig. 4, the expected positions of the DR resonances according to Eq. (18) are shown. It is seen that the signal related to the $1s2s(^1S) \rightarrow 1s2p(^1P)$ excitation dominates. This is due to the large radiative transition rate A_r for the doubly excited intermediate state $1s2p(^1P)nl$.

The $1s2s(^1S)$ and $1s2p(^1P)$ core states are almost energy-degenerated. The $1s2p(^3P)$ state lies just above the $1s2s(^1S)$ state, the energy difference being only 0.0008 eV. Thus, in prin-

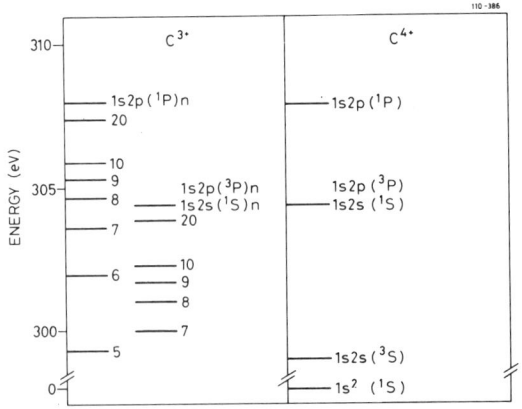

Figure 5

ciple, dielectronic recombination involving very high Rydberg states is possible for the $^1S \rightarrow ^3P$ excitation. According to Eq. (18), n should then be larger than about 160. Such high Rydberg states are ionized in the analyzer field, and consequently, this DR process will not be observed.

For the ^3S initial core state, three excitations are possible. It is immediately seen that these series are weaker than the strong $^1S \rightarrow ^1P$ series, and only the series limits (large n values) are observed in the spectra. The two transitions $^3S \rightarrow ^1S$ and $^3S \rightarrow ^3P$ appear as only one series due to the small energy difference.

The large peak around 4.5 eV cannot be explained based only on one of the transitions mentioned above. However, it is seen that the peak appears where the n=7 resonance of the $^3S \rightarrow ^1P$ series nearly coincides with the n=14 and 15 resonances of the $^3S \rightarrow S^1, ^3P$ series. Such extraordinarily large DR signals have also been observed for He-like O^{6+} (Ref. 17) and have been interpreted as being due to configuration interaction (CI)^{17,18}. The near-degeneracy of the states will cause them to interact so that the different core configurations mix. For the present case, this means that the ^1S and ^3P cores mix with the ^1P core. Since the ^1P core very quickly makes the radiative transition to the core ground state (large A_r), an enhanced DR signal results.

For states above the limits for the 1s2s ^1S and 1s2p ^3P, the population of the 1s2p ^1Pnl states is much reduced. This is not because the configuration interaction ceases to operate, but because Auger decay into the ^1S and ^3P channels now becomes possible with a reduction of the DR channels as a consequence.

NOTE ON RR EXPERIMENTS

Radiative recombination has now been measured for C^{6+} ions in our laboratory, and the results will be published shortly.

ACKNOWLEDGEMENTS

This work was supported by the Carlsberg Foundation and Ib Henriksen's Foundation. The author would like to thank his coworkers and students, in particular L.H. Andersen and H. Knudsen, for collaboration on this project.

REFERENCES

1) M. Stobbe, Ann. Physik 7, 661 (1939)
2) H. Bethe and E. Salpeter, in: *Handbuch der Physik*, Vol. 35, p. 88 (Heidelberg, New York: Springer, 1957)
3) H.S.W. Massey and D.R. Bates, Rep. Prog.Phys. 9, 62 (1942)
4) D.R. Bates, A.E. Kingston, and R.W.P. McWhirter, Proc.Roy.Soc. 267A, 197 (1962)
5) D.S. Belić, G.H. Dunn, T.J. Morgan, D.W. Müller, and C. Timmer, Phys. Rev.Lett. 50, 339 (1983)
6) J.B.A. Mitchell, C.T. Ng, J.L. Forand, D.P. Levac, R.E. Mitchell, A. Sen, D.B. Miko, and J.W. McGowan, Phys. Rev.Lett. 50, 335 (1983)
7) P.F. Dittner, S. Datz, P.D. Miller, C.D. Moak, P.H. Stelson, C. Bottcher, W.B. Dress, G.D. Alton, and N. Nešković, Phys.Rev.Lett. 51, 31 (1983)
8) K. LaGattuta and Y. Hahn, J.Phys.B. 15, 2101 (1982)
9) D.C. Griffin, M.S. Pindzola, and C. Bottcher, Phys.Rev.A 31, 568 (1985)
10) G.H. Dunn, in: *Atomic Processes in Electron-Ion and Ion-Ion Collisions*, ed. F. Brouillard (Plenum, New York, 1986) p. 93
11) Y. Hahn and K.J. LaGattuta, Phys. Rep. 166, 195 (1988)
12) A. Wolf, Physica Scripta T22, 55 (1988)
13) L.H. Andersen and J. Bolko, to be published in Phys.Rev.A (1989)
14) M. Bell and J.S. Bell, Part.Acc. 12, 49 (1982)
15) A. Müller et al., Phys.Rev.A 36 (1987)
16) D.C. Griffin, M.S. Pindzola, and P. Krylstedt, to be published in Phys. Rev.A (1989)
17) L.H. Andersen, H. Knudsen, P. Hvelplund, and P. Kvistgaard, Phys.Rev. Lett. 62, 2656 (1989)
18) K. Taulbjerg and J. Macek, Phys.Rev. Lett. 62, 2766 (1989)

RTE OF HYDROGEN-LIKE AND LITHIUM-LIKE IONS

R. Schuch[1], E. Justiniano[2], M. Schulz[3], P.H. Mokler[4], S. Reusch[4], S. Datz[3], P.F. Dittner[3], J. Giese[3], P.D. Miller[3],
H. Schoene[3], T. Kambara[5], A. Müller[6], Z. Stachura[7], R. Vane[3], A. Warczak[8], G. Wintermeyer[9],

[1] Manne Siegbahn Institute of Physics, S-104 05 Stockholm, Sweden
[2] Department of Physics, East Carolina University, Greenville, NC 27858-4353, USA
[3] Oak Ridge National Laboratory, Oak Ridge, TN 37831-6377, USA
[4] GSI, D-6100 Darmstadt 11, FRG
[5] RIKEN, Wako, Saitama, Japan
[6] Strahlenzentrum d. Universität, D-6300 Gießen, FRG
[7] Inst. of Nuclear Physics, 31-342 Krakow, Poland
[8] Inst. of Physics, University, 30-059 Krakow, Poland
[9] Physikalisches Institut d. Universität, D-6900 Heidelberg, FRG

Recent results on electron capture with simultaneous excitation of a projectile electron (RTE) are reviewed. These processes are identified by measuring coincidences between two K-x-rays or between one K-x-ray and an ion with its charge decreased by one unit, for hydrogen-like or lithium-like ions, respectively. In the dependence of the cross section on the beam energy it was seen, that this process is resonant and behaves very much like dielectronic recombination. In this work we present and discuss results from measurements of resonant transfer and excitation in collisions of lithium-like F, hydrogen-like S, lithium-, and hydrogen-like Ge with H_2. Emphasis is given to investigating the population of very high n states.

1. INTRODUCTION.

In Resonant Transfer and Excitation (RTE) or Dielectronic Recombination (DR) the capture of an electron, that has in the ion frame of reference the kinetic energy of an Auger electron from the recombined ion, occurs simultaneously with the excitation of an electron originally bound to the projectile. Following a collision where RTE (or DR) occurs a hydrogen-like projectile is left in a doubly excited state. In the case of lithium-like or more electron ions a singly excited intermediate state may be populated, but as a general rule the reaction also proceeds through intermediate doubly excited states. In Fig.1 the RTE process is schematically shown for H-like and Li-like ions. Depending on the branching ratios for the decay of the intermediate excited state it may either autoionize, causing the ion charge state to remain as before the collision, or it may emit one, two or more x rays, in which case the charge state changes by one unit. Quite often in the literature the two branches are referred to as RTEA or RTEX, respectively.

For DR or RTE to occur with hydrogenlike ions the electron bound in the 1s state has to change its main quantum number N. The captured electron populates states with main quantum numbers $n \geq N$. For a coarse characterization of the transition one often uses the Auger notation, e.g. KLM for an electron excited from the K to the L shell and another electron captured into n=3 (M shell). With ions having more than one bound electron, also intra shell transitions e.g. in the L shell can occur when the electron is captured. Up to now in RTE processes with H, He and Li-like ions $\Delta N = 1$ transitions have been detected [1-4]. Whereas in the first generation of experiments on DR with merged or crossed electron-ion beams [5-8] $\Delta N = 0$ transitions were measured. Only recently [9,10] also $\Delta N \geq 1$ transitions were detected in DR measurements.

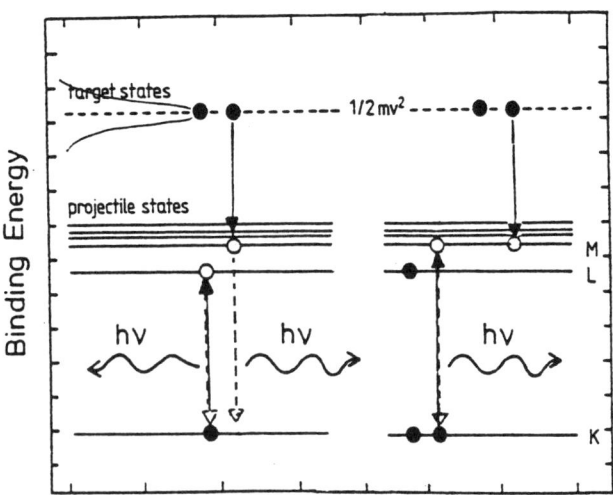

Figure 1. Schematics of DR and RTE for H-like and Li-like ions.

The difference between RTE and DR is that in RTE the ion captures the electron from a bound state of a target atom whereas in DR the ion recombines with a free electron. So the two processes could provide an important link between ion-atom and electron-ion collisions. A rather simple relationship between RTE and DR cross sections (σ_{RTE} and σ_{DR}, respectively) can be found within the impulse approximation [11].

$$\sigma_{RTE} = \int \sigma_{DR}(p_{iz}) J(p_{iz}) dp_{iz}$$

with

$$J(p_{iz}) = \int\int |\Psi_i(p_i)|^2 dp_{ix} dp_{iy}$$

the Compton profile representing the probability to find a target electron i with the z-component of its momentum between p_{iz} and $p_{iz} + dp_{iz}$. Exploiting the fact that the Compton profile varies only very weakly over the width of a resonance in σ_{DR} and using the relationship

$$\epsilon = \sqrt{\frac{2E}{M}} p_{iz} + (\frac{m_e}{M}) E$$

(E and M are the projectile ion energy and mass, respectively) for the electron energy in the projectile frame, one has:

$$\sigma_{RTE} = \sqrt{\frac{M}{2E}} \sum_i J(p_{iz}) \int \sigma_{DR} d\epsilon$$

RTE could exploit the high electron densities available in atomic targets to study low cross section resonant transfer which are not reachable with the present luminosities of electron-ion crossed or merged beam techniques. There are other interesting applications of RTE, such as the population of a metastable state and the observation of its decay by two photon coincidences[12], or the attempts to detect the simultaneous electron capture and nuclear excitation[13].

In ion-atom collisions RTE competes with the non-resonant transfer and excitation process (NTE). In NTE, the target electron is captured and simultaneously an electron bound to the projectile is excited by the target nucleus. The energy dependence for this reaction is thus governed by both that for excitation as well as by that for capture and could therefore also show a maximum at certain projectile energy. Useful estimates may be obtained through the method described by Brandt[14]. In this model for NTE the mutually independent probabilities for the two processes (capture and excitation) are factorized for a certain impact parameter (b) and then their product is integrated over b. From this approximation one finds that in the collision sytems discussed here the NTE cross section peaks at much lower collision energies than those we shall describe. This model also predicts NTE and RTE to show very different energy dependences. NTE cross sections should rise steeply at low energies, reach a maximum and then decrease very slowly for higher energies. The shape of the peak is much more asymmetric and the maximum is also much broader than in case of RTE. This behaviour and also the position of the peak at low energy has been observed for S^{13+} + He collisions[2].

For the case of isolated resonances in the intermediate state $(N, L; n, l)$ the DR cross section can be approximated as:

$$\sigma_{DR} = \sigma_0 \Gamma(N, L) \Sigma_l 2(2l+1) \frac{A_a(n,l) A_r(n,l)}{A_a(n,l) + A_r(n,l)}$$

Where σ_0 is a cross section containing all the constants including the energy bin, Γ is the Lorentz profile for the decay of the state (N,L;n,l), $A_a(n,l)$ and $A_r(n,l)$ are the autoionizing and radiative decay rates of this state. For high n-states (Rydberg states) the radiative rate $A_r(n,l)$ is independent of n,l and determined only by N,L. The autoionization rate $A_a(n,l)$ varies with n^{-3}. For low n,l one has $A_a \gg A_r$ and $\sigma_{DR} \sim \Sigma_l 2(2l+1) A_r(n,l) \sim n^2$. Therefore, σ_{RTE} for e.g. the KLn series should increase for low n and start to decrease for higher n. Since $A_r \sim Z^4$ this n-dependence of the RTE cross section should vary with the ion nuclear charge. We will discuss this dependence later in the comparison of the shapes in the measured RTE excitation functions for different ions. Also this formula for σ_{DR} will be used together with the Compton profile in the above given formula to calculate the values for σ_{RTE} in this work.

2. EXPERIMENTS.

The schematic setup for the two types of RTE experiments are shown in Fig. 2. The ion beam is in general poststripped and analysed to provide the Li-like or H-like charge state. Following collimation a charge state purifying system is often used before the beam entered the gas cell. This is done to eliminate charge changed particles, which undergo capture in the residual gas of the beam line. This is important in the experiments which detect charge changed ion - x ray coincidences. In the x-x coincidence setup with H-like ions this is not as crucial because ions which capture an electron prior to the gas cell do not contribute to either the true or the random coincidences. The exception to this rule being projectiles originally in the ^3S metastable state. In the experiment described the fraction of ions in this state was found to be negligible.

In the experiments reported here a H_2 gas target was used. The reason for this choice of target system is the comparatively narrow width of its Compton profile. The targets were typically about 10 mm long and multistage differentially pumped. Dependences of the true coincidence yield on the target pressure were recorded in order to find the optimum target density ($\sim 0.1 - 1$ mT cm) which allows maximum count rate under single collision conditions.

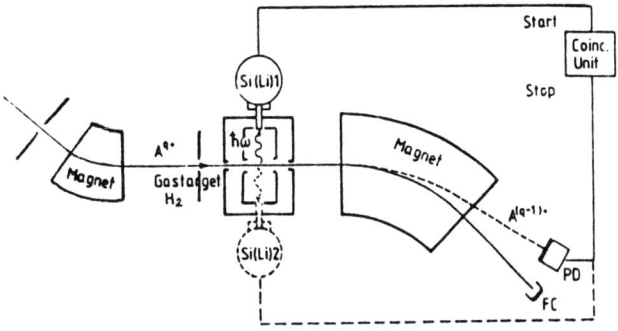

Figure 2. Typical experimental set-up for RTE measurements with Li-like (full line) and H-like (dashed line) ions.

In the experiment with F projectiles the x rays were detected by proportional counters and the charge changed particles by a ceratron. The correction for detection efficiency in this case is described below. In the cases of heavier projectiles the x rays were detected by Si(Li) x-ray detectors and the particles by surface barrier detectors (PD in Fig. 2). The efficiency of these detectors is close to unity and the respective corrections are straightforward. The x-ray resolution achieved with the Si(Li) detectors varied from about 160 eV for S K x rays to about 320 eV for the Ge K x rays. In all cases this was enough to resolve the K_α and K_β x rays. X-ray transitions from higher n states as well as hypersatellite transitions were obtained by fitting the spectra.

The intensity of the true coincident K x rays was obtained for different ion beam energies. The measurements were performed for small steps in the ion energy covering the RTE resonance, i.e. the ion velocity was varied so as to correspond to the electron-ion relative velocity needed to obtain the associated DR resonances. In order to normalize the different runs a Faraday cup (FC in Fig. 1) was used.

3. RTE WITH LITHIUM-LIKE IONS.

Much data has been obtained to date on RTE with Li-like ions. Here we report on data for F^{6+} and Ge^{29+} ions. The experiment with F^{6+} ions was performed at the EN tandem accelerator at Oak Ridge National Laboratory[15]. Because of the low energy of the F K x rays, two gas proportional counters with 3.5 μm Mylar windows were chosen for their detection. The value of the x-ray - particle coincidence efficiency was determined from capture of F^{9+} ions in H_2 and the assumption that 90% of the produced F^{8+} ions emit an x ray. These x rays, however, have a somewhat higher energy (827 eV) than those from RTE of F^{6+} (710 eV). So a small correction of the absorption was further applied (the yields were multiplied with a constant factor of 2). The results from this measurement are shown in Fig. 3 in comparison with the predictions of two calculations. In both of these a detailed analysis of the formation of highly excited intermediate states were performed. The dashed curve is taken from ref. 16 and is devided by a factor 2. This calculation includes $1s^22s + e^- \rightarrow 1s2s2pnl$ transitions. The full curve is from ref.17 and is devided by 2.5. It contains $1s^22s + e^- \rightarrow 1s2sNLnl$ ($n \leq 9, N \leq 3$) transitions.

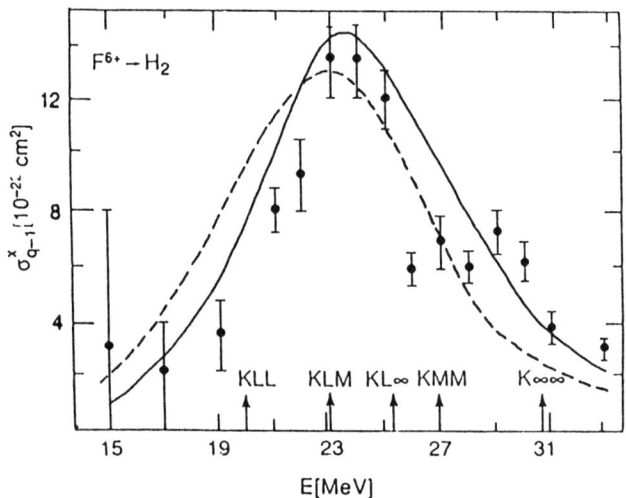

Figure 3. Comparison between measured and calculated RTE yields for Li-like F ions colliding with H_2 (see text).

At the UNILAC at GSI Darmstadt coincidences between K x rays and charge changed ions were detected for a Li-like Ge^{29+} beam in the energy range between 12 and 18.5 MeV/amu^{12} colliding with H_2. In this excitation function (see Fig. 4) a structure showing three clearly separated peaks may be seen. The first peak coincides well with the expected position (see arrows in Fig. 4) of the KLL resonances. According to these energies the second peak is due to KLM and the third peak to KLn ($n \geq 4$) resonances. The dotted line represents a calculation by McLaughlin and Hahn[16]. It reproduces rather well the observed structure in the excitation function of the K x ray - Ge^{28+} ion coincidences. Only for the high n-states it overestimates somewhat the measured cross section.

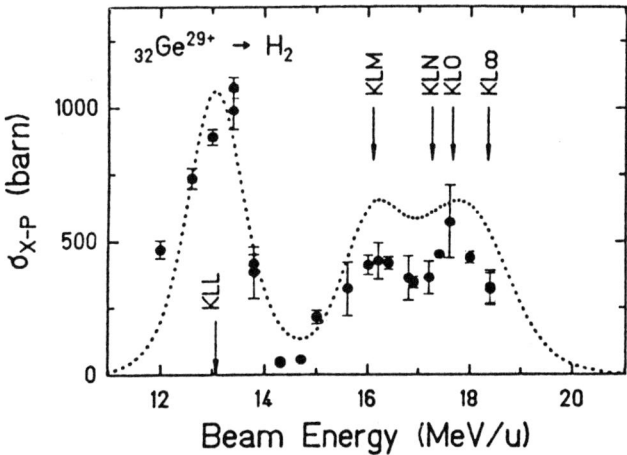

Figure 4. A comparison between the measured and calculated RTE cross section for Ge^{29+} on H_2 (see text).

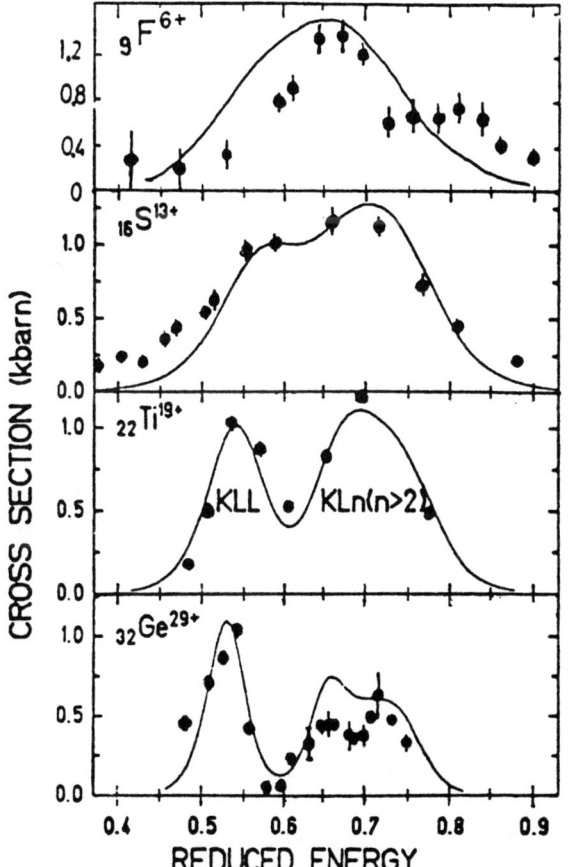

Figure 5. Measured and calculated RTE cross sections for Li-like ions as function of the reduced beam energy.

In order to compare the measured excitation functions for different Li-like projectiles we scale the ion energy according to $\eta = \frac{m_e}{m}\frac{E}{E_K}$ (E_K is the K-shell binding energy) [18]. In Fig. 5 we include not only experimental data but also theoretically predicted cross sections, which were obtained by scaling calculations from refs. 16-18. The experimental results for S^{13+} are from ref. 2 and were actually measured with a He target. Therefore they show a somewhat larger width than the calculated distribution. The experimental results for Ti^{19+} have been given in ref. 4. In this figure we see, first of all, that the maximum of the RTE cross section varies only very slowly with the nuclear charge Z. However, one sees a clear variation of the shape of the resonances with Z.

The structure of the KLL, KLM and higher states becomes more and more resolved with increasing Z. This is due to the fact that the transition energy scales approximately with Z^2. The Compton profile width increases therefore with Z, and as the subshells split approximately with Z^4, increases the separability with Z^3. Furtheron can be seen from Fig. 5, that the KLL contribution is weak for low Z, compared to KLM and higher resonances, and becomes more pronounced for heavier ions. The reason for this variation with Z is ascribed to the relative sizes of the radiative and autoionization rates as well as to the change of the radiative rate with Z, mentioned earlier in this work.

4. RTE WITH HYDROGEN-LIKE IONS.

The x-ray x-ray coincidence measurements for S^{15+} on H_2 were done for the first time at the MP tandem of the MPI for Nuclear Physics in Heidelberg [3]. In these measurements and also in all the others presented here no clear indication of $\Delta N \geq 2$ transitions (i.e. KMM and higher resonances) were seen. To search for reactions of this type the ion energy range was extended drastically in a subsequent experiment at the Holifield tandem accelerator at Oak Ridge National Laboratory[19]. The two sets of data are combined in Fig. 6 which shows the cross sections for time correlated emission of two K_α x rays (K_α-K_α), of a K_α and a K_β x ray (K_α-K_β), and of two K_β or K_γ and higher x rays ($K_{\beta\gamma}$-$K_{\beta\gamma}$). The expected positions of the resonance energies are indicated by arrows. Also shown in Fig. 6 are dashed lines and a cross representing results of calculations by McLaughlin and Hahn[16]. The calculation agrees rather well with the data in the KLL maximum of K_α-K_α. At higher energy it also shows a second maximum in K_α-K_α present in the experimental data as well. The origin of this maximum is attributed to cascading transitions from high n-states. At higher energies (above the KMM resonance energy of 163 MeV) also coincidences between two K_β and higher x rays become apparent as one may see in $K_{\beta\gamma}$-$K_{\beta\gamma}$ (Fig. 6). This is evidence for $\Delta N \geq 2$ transitions in RTE. The cross section for this process is about one order of magnitude smaller than of K_α-K_α. At 175 MeV McLaughlin and Hahn[16] predict a cross section for K_β-K_β coincidences, that is about a factor of 3 smaller than found experimentally. But it should be noted that this calculated value considers coincidences between the K_β transitions only.

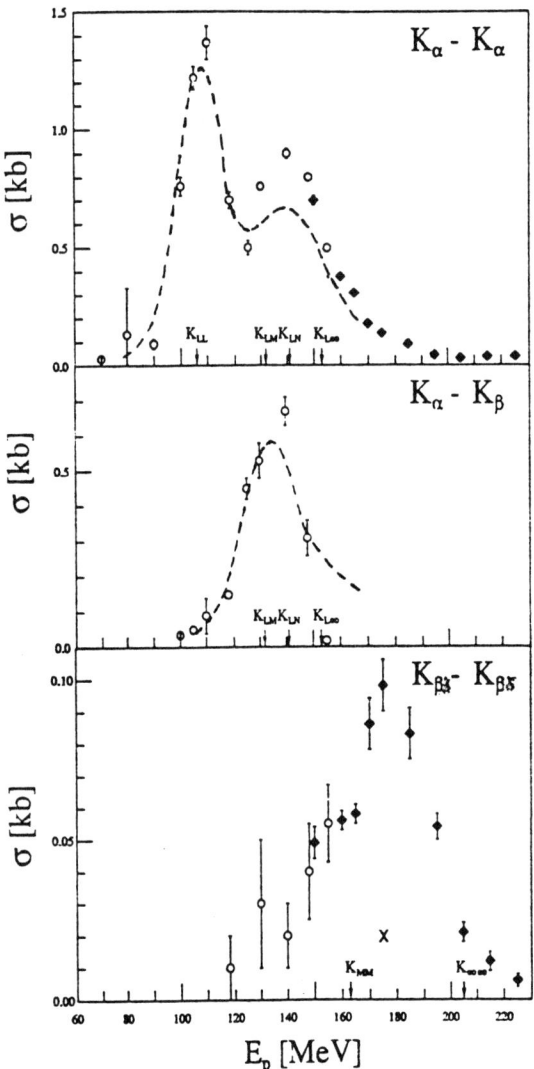

Figure 6. RTE cross section for S^{15+} on H_2 from coincidences between different K-x-ray satelite lines. Full line and cross represents a prediction by McLaughlin and Hahn[16].

The heaviest hydrogen-like ion for which RTE has been observed up to now is Ge^{31+}. This has been done by x-ray x-ray coincidences at the UNILAC of GSI Darmstadt[12]. In Fig. 7 an event plot of the true coincident x rays for 16.5 MeV/amu Ge^{31+} incident ions is shown. Each event represents the decay of a doubly excited state in Ge^{30+}. Besides the K_α-K_α coincidences (both x rays at about 10.5 keV) one recognizes K_α-K_β coincidences (one x ray at about 10.5 keV, the other at about 12.5 keV).

At this energy the K_β-K_β coincidences are obviously absent, but one observes three ridges (indicated by the dashed lines in Fig. 7). Two of them are for a two photon decay where one of them was a 10.5 keV x ray, the other was continuous. The diagonal ridge comes from the emission of two continuous x rays with a sum energy of 10.5 keV, which is the 2 E1 decay of the $1s2s^1S_0$ state. This state is populated through a hypersatellite K_α transition from the $2s2p^1P_1$ state. So primarily this state is populated which feeds the $1s2s^1S_0$ state. By observing the 2 E1 decay and scanning the energy, the resonant population of a single state, the $^1P_1$ state, can be observed.

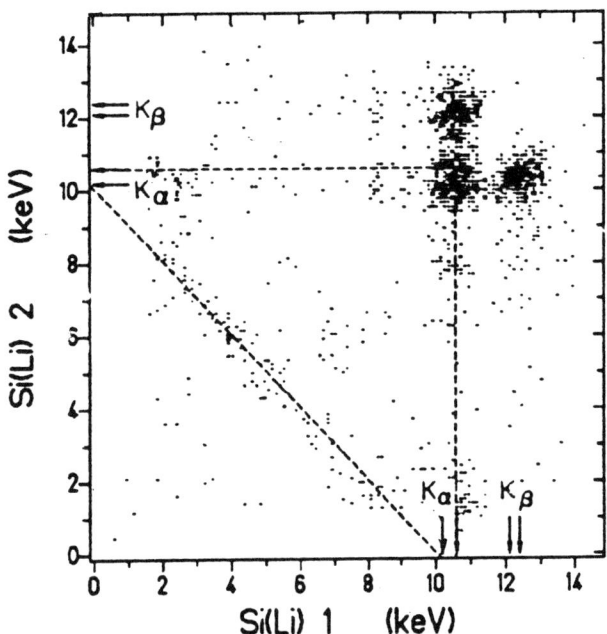

Figure 7. A two dimensional x-ray - x-ray true coincidence spectrum from 16.5 MeV/amu Ge^{31+} on H_2.

The excitation function of all the K-x-ray K-x-ray true coincidences is shown in Fig. 8 as full dots. The resonance peaks are clearly visible at positions indicated by arrows at KLL and KLM. Taking only K_α-K_β x rays (triangles) one observes a peak - as expected from such a resonant process - only at and above the KLM position. The dashed line and the dotted line are calculated up to N=2, n=3 for the total K-K coincidences and the K_α-K_β

coincident x rays, respectively. For this calculation the Compton profile of H_2[20], the tabulated autoionization rates[21], and a Dirac-Fock programm[22] to get the energies and radiative transition rates were used. The agreement in absolute value at the KLL resonance is very good. The energy position seems to be slightly shifted to lower energies. Some cross section seems to be missing at the KLM resonance, which could be explained by the limited number of states taken into account.

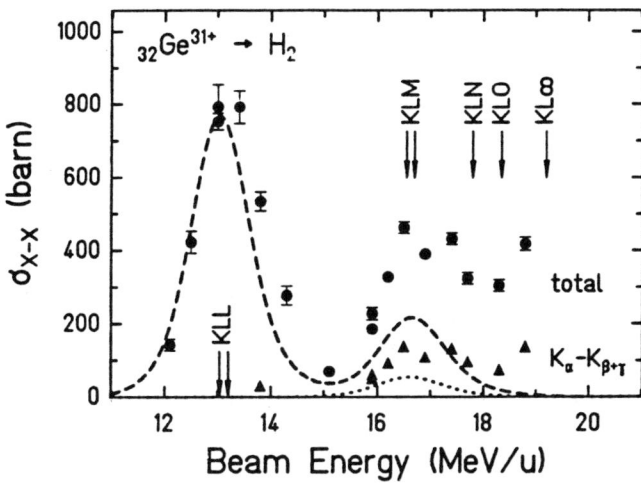

Figure 8. A comparison between the measured and calculated RTE cross section for Ge^{31+} on H_2 (see text).

5. CONCLUSION.

We conclude that RTE is reasonably well understood in terms of dielectronic recombination with bound target electrons, by treating them quasi free in an impulse approximation. There are some discrepancies still in the calculated absolute value of the cross section for light ions and for cascade contributions from high n states.

References.

[1] J.A. Tanis, E.M. Bernstein, W.G. Graham, M.P. Stöckli, M. Clark, R.H. McFarland, T.J. Morgan, K.H. Berkner, A.S. Schlachter, and J.W. Stearns, Phys. Rev. Lett. **53**, 2551 (1984).

[2] J.A. Tanis, Nucl. Instrum. Methods **A262**, 52, (1987).

[3] M. Schulz, E. Justiniano, R. Schuch, P.H. Mokler, and S. Reusch, Phys. Rev. Lett. **58**, 1734 (1987).

[4] S. Reusch, P.H. Mokler, R. Schuch, E. Justiniano, M. Schulz, A. Müller, and Z. Stachura, Nucl. Instrum. Methods Phys. Res. **B23**, 137 (1987).

[5] D.S. Belić, G.H. Dunn, T.J. Morgan, D.W. Mueller, and C. Timmer, Phys. Rev. Lett. **50**, 339 (1983).

[6] P.F. Dittner, S. Datz, P.D. Miller, C.D. Moak., P.H. Stelson, C. Bottcher, W.B. Dress, G.D. Alton, N. Nesković, and C.M. Fou, Phys. Rev. Lett. **51**, 31 (1983).

[7] J.B.A. Mitchell, C.T. Ng, J.L. Forand, D.P. Levac, R.E. Mitchell, A. Sen, D.B. Miko, and J.W. McGowan, Phys. Rev. Lett. **50**, 335 (1983).

[8] L.H. Anderson, P. Hvelplund, H. Knudson, P. Kvistgaard, Phys. Rev. Lett. **62**, 2656, (1989).

[9] D.R. Knapp, R.E. Marrs, M.A. Levine, C.L. Bennett, M.H. Chen, J.R. Henderson, M.B. Schneider, J.H. Scofield, Phys. Rev. Lett. **62**, 3104, (1989).

[10] G. Kilgus et al. submitted to Phys. Rev. Lett. and A. Wolf et al. in this volume.

[11] D. Brandt, Phys. Rev. **A27**, 1314 (1983).

[12] P.H. Mokler, S. Reusch, Th. Stöhlker, R. Schuch, M. Schulz, G. Wintermeyer, Z. Stachura, A. Warczak, A. Mueller, T. Kambara, Rad. Effects to be published

[13] N. Cue, J-C Poizat, J. Remillieux, Europhys. Lett. to be publ.

[14] D. Brandt, Nucl. Instrum. Methods Phys. Res. Sect. **A214**, 93 (1983).

[15] M. Schulz, R. Schuch, S. Datz, E.L. Justiniano, P.D. Miller, H. Schöne, Phys. Rev. **A38**, 5454, (1988).

[16] D.J. McLaughlin and Y. Hahn, Phys. Rev. **A38**, 531 (1988) and private communication.

[17] C.P. Bhalla, K.R. Karim, Phys. Rev. **A39** 6060, (1989).

[18] P.H. Mokler, S. Reusch, Z. Phys. **D8**, 393, (1988).

[19] R. Schuch, E. Justiniano, M. Schulz, S. Datz, J. Giese, H.F. Krause, R. Vane, P.F. Dittner, S. Shafroth, to be published

[20] P. Eisenberger, Phys. Rev. **A2**, 1678, (1970).

[21] L.A. Vainshtein, U.I. Safronova, At. Data and Nucl. Data Tables **25**, 311, (1980).

[22] I.P. Grant et al. Comp. Phys. Comm. **21** 207, (1980).

OBSERVATION OF ELECTRON-ELECTRON INTERACTION IN COLLISIONS OF O^{5+} IONS WITH H_2 TARGETS

T.J.M. Zouros, D.H. Lee, and P. Richard

J.R. Macdonald Laboratory, Kansas State University,
Manhattan, Kansas 66506, USA

Using high resolution projectile electron spectroscopy at zero degrees, we have determined projectile Auger electron spectra from collisions of Li-like O^{5+} beams with H_2 targets at energies between 4 - 32 MeV. The strongest populated Li-like $(1s2s2p)$ states resulting from $1s \to 2p$ excitation and Be-like $(1s2s2p^2)$ states resulting from Transfer-Excitation (TE) were resolved. The production of $(1s2s2p^2)^3D$ states by Resonance Transfer-Excitation was clearly observed and compared to theory. The production of $(1s2s2p)^4P$ and $^2P_+$ states above ~ 12 MeV was found to increase sharply with projectile energy, indicating the onset of projectile excitation due to an interaction with one of the target electrons.

Extensive efforts have been made, in the past few years, to understand electron-electron interactions by studying a variety of processes occuring in collisions of electron beams with ions.[1] Various types of crossed-beam or merged-beam arrangements have been used yielding important information about electron-impact excitation (eIE) and ionization of ions.[1]

An alternate route for obtaining information about electron-electron interactions, particularly for inner-shells could possibly be provided by the study of energetic ion-atom collisions. Projectile electrons could be excited or ionized by the Coulomb interaction with a target electron. Competing processes such as Coulomb interaction of the projectile electron with the target nucleus could be minimized by utilizing low-Z targets such as H_2 or He.

We have recently used high resolution projectile electron spectroscopy at zero degrees, to determine[2,3] projectile Auger electron spectra from collisions of various Li-like beams with H_2 and He targets at energies between 0.25 - 2 MeV/u. In particular, we have identified various inner shell Li-like and Be-like Auger states resulting from electron Excitation or TE of the projectile, thus obtaining state-selective information on the energy dependence of the various cross sections. In this contribution, we report on our recent measurements of the $O^{5+} + H_2$ collision system, where we observe clear evidence of projectile electron - target electron interactions as manifested by Resonance Transfer-Excitation (RTE)[3,4] and electron-electron Excitation (eeE).[2]

The measurements were performed using the Kansas State University $0°$ tandem electron spectrometer.[5] Experimental details have been reported[2,3,5] previously and are not presented here. In Fig. 1 are shown selected electron spectra obtained at various projectile energies. The normalized yields are displayed after subtraction of background continuum electrons and transformation to the projectile rest frame. The double differential cross sections in Fig. 1 are obtained by normalizing to the known Ne K Auger cross section[6] for 3 MeV p + Ne collisions. Various lines of the KLL Auger spectrum are resolved and identified in the overall electron spectra. Cross sections were extracted by fitting the observed lines to Lorentzian or Fano line shapes of natural width folded with the Gaussian-like response function of the exit analyzer.

On the low energy side of the electron spectra can be seen (Fig. 1) the doubly-excited Li-like $O^{5+}(1s2s2p)$ states formed by $1s \to 2p$ excitation in the collision

$$O^{5+}(1s^22s) + H_2 \to O^{5+}(1s2s2p) + H_2(?) \quad (Excitation)$$
$$\downarrow$$
$$O^{6+}(1s^2) + e^- \quad (Auger\ decay)$$

On the high energy side of the electron spectra can be seen the doubly-excited Be-like $O^{4+}(1s2s2p^2)$ states formed by TE in the collision

$$O^{5+}(1s^22s) + H_2 \to O^{4+}(1s2s2p^2) + H_2^+(?) \quad (TE)$$
$$\downarrow$$
$$O^{5+}(1s^22s) + e^- \quad (Auger\ decay)$$

in which there is an excitation of the projectile 1s electron to the 2p orbital and a simultaneous capture of a single target electron into the $O^{4+}(1s2s2p^2)$ state.

In both cases autoionizing states are formed and decay by ejecting an Auger electron which is detected with high resolution at $0°$ with respect to the beam direction.[7] The final state of the target is unknown since it was not determined in this measurement.

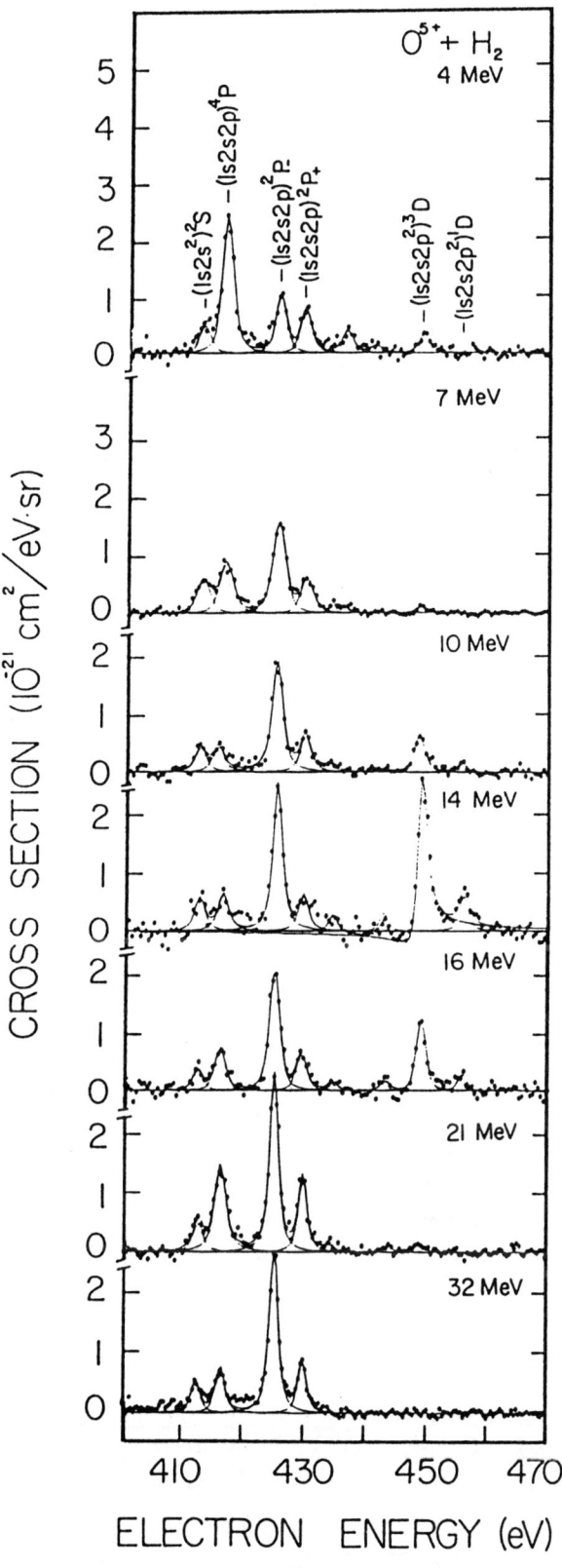

Figure 1. Normalized zero-degree electron spectra (projectile rest frame) produced in $O^{5+}(1s^22s) + H_2$ collisions at various projectile energies.

Excitation and TE production cross sections are displayed as a function of projectile energy in Fig. 2. Only the cross sections for the production of the $(1s2s2p)^4P$, $(1s2s2p)^2P_+$ and $(1s2s2p^2)^3D$ states are shown. The total production cross sections are obtained by multiplying the $0°$ differential cross sections by 4π and dividing by the calculated Auger yields.[2] In addition, corrections due to the long lifetime of the metastable ^4P_J state affecting its measured yield were also included.[2] For all three cases, the observed peaking of the measured cross sections with projectile energy is due to electron-electron interactions between the active projectile electron and either of the two target electrons.

In these energetic collisions, the collision time is fast relative to the orbiting time of the target electrons and the impulse approximation model[4] can be applied. In this model, the target electron can be considered to be practically free carrying into the collision its momentum distribution due to its orbital motion around the target nucleus. Thus, in the projectile frame, the target electron is seen to approach the projectile with an energy ϵ, where

$$\epsilon(p_z) = \frac{m}{M}K - E_t + \sqrt{\frac{2K}{M}}p_z \qquad (1)$$

K is the projectile energy, p_z is the target electron momentum component due to its orbital motion around the target along the beam direction (z-axis), E_t is the target electron binding energy (15.5 eV for H_2) and m and M are the electron and projectile masses respectively. In this approximation, we can obtain an ion-atom cross section by using the ion-free-electron cross section (either calculated or measured in a merged electron-ion beam experiment) and integrating over the initial momentum distribution of the approaching quasi-free target electron. Thus,

$$\sigma_{ion-bound\,e^-}(K) = \int \sigma_{ion-e^-}(\epsilon(p_z))J(p_z)dp_z \qquad (2)$$

The experimentally determined Compton profile of H_2 was used[2] for $J(p_z)$. The ion-free-electron cross sections for excitation, σ_{ion-e^-} were taken from calculated[8] electron impact excitation (eIE) cross sections in collisions with O^{5+} ions. The ion-free-electron cross sections for Transfer-Excitation were taken from calculated[9] dielectronic capture cross sections in collisions with O^{5+} ions. Upon integrating over the initial momentum distribution, $J(p_z)dp_z$, of the incoming target electron, we obtain the ion-atom cross sections known in the literature as electron-electron excitation (eeE)[2] or Resonance Transfer-Excitation (RTE)[4] cross sections for the excitation and TE processes respectively. The results of calculations using Eq. 2 are also included in Fig. 2 for comparison. Good agreement is observed in the energy dependence over the entire high energy region. In absolute magnitude, the data are found to be larger than the calculation by a factor of ~ 2.5 probably reflecting the known non-isotropic emission of the observed Auger electrons.

The RTE process is well known and has been previously studied for a variety of collision systems.[10] For more details we refer the reader to our recent article Ref. [3]. However, the eeE process has only recently been directly observed over the usually dominant projectile electron-target nucleus interaction (enE).[2]

In the production of the $(1s2s2p)^4P$ state, the excitation of the projectile electron by interaction with the target nucleus (enE) cannot occur to first order, since it is forbidden by spin-flip considerations.[11] However, the ^4P state can be formed by eeE, if electron exchange is included. The sharp rise of the excitation cross section above 12 MeV reflects the underlying sharp threshold for eIE.[2] Finally, a Transfer-Loss process (TL), involving the transfer of a target electron to the 2p projectile orbital with the simultaneous loss of a projectile 1s electron, could also result in a net $1s \to 2p$ excitation of the projectile.[11] TL is expected to be less important at high velocities due to the rapid fall-off of the capture cross section with increasing projectile energy. Thus, only eeE remains a likely contributor at energies near and above the threshold projectile energy.

In the production of the $(1s2s2p)^2P_+$ state also shown in Fig. 2, the $1s \to 2p$ projectile excitation can proceed via all three competing processes, i.e. enE, eeE and TL. Again, TL is expected to be small above 12 MeV and the only processes left will be enE and eeE. This is consistent with the measured total cross sections given in Fig. 2. The increase of the $^2P_+$ cross section above 12 MeV is superimposed on a rather flat "background" characteristic of enE.

In conclusion, we have observed both RTE and eeE in $O^{5+} + H_2$ collisions. By relating ion-atom excitation to electron-impact excitation via Eq. (2), ion-atom measurements could possibly provide cross sections for inner-shell electron-impact excitation of ions for which there are presently almost no measurements.

This work was supported by the Division of Chemical Sciences, Office of Basic Energy Sciences, Office of Energy Research, U.S. Department of Energy.

References

1. R.A. Phaneuf, in Physics of Electron-Ion and Ion-Ion Collisions, ed. F. Brouillard (Plenum, New York, 1986) p. 117 and references therein.

2. T.J.M. Zouros, D.H. Lee, and P. Richard, Phys. Rev. Let. 62, 2261 (1989).

3. T.J.M. Zouros, D.H. Lee, P. Richard, J.M. Sanders, J.L. Shinpaugh, S.L. Varghese, K.R. Karim and C.P.

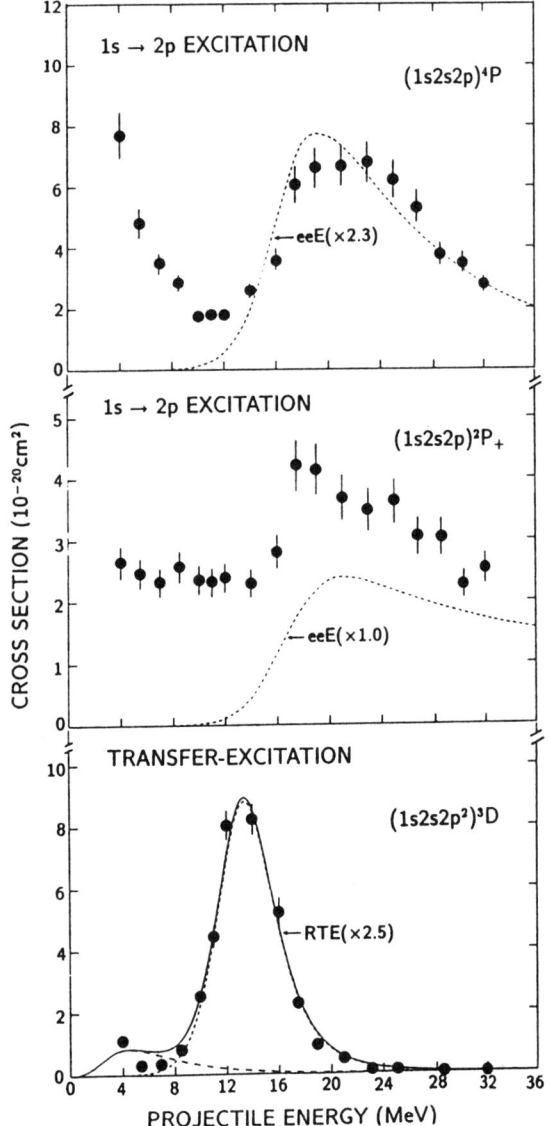

Figure 2. Data for $O^{5+} + H_2$. Dashed lines are calculations using ion-electron cross sections and the impulse approximation.

4. D. Brandt, Phys. Rev. A$\underline{27}$, 1314 (1983).

5. D.H. Lee, T.J.M. Zouros, J.M. Sanders, J.L. Shinpaugh, T.N. Tipping, S.L. Varghese, B.D. DePaola and P. Richard, Nucl. Instr. & Meth. in Phys. Res. B$\underline{40/41}$, 1229 (1989).

6. C.Y. Woods, R.L. Kaufman, K.A. Jamison, N. Stolterfoht and P. Richard, Phys. Rev. A $\underline{13}$, 1358 (1976).

7. A. Itoh, T. Schneider, G. Schiwietz, Z. Roller, H. Platten, G. Nolte, D. Schneider, and N. Stolterfoht, J. Phys. B$\underline{16}$, 3965 (1983).

8. S.J. Goett and D.H. Sampson, At. Dat. & Nucl. Dat. Tables $\underline{29}$, 535 (1983).

9. Y. Hahn, Phys. Lett. A 119, 293 (1986); ibid., Phys. Rev. A 1989 to be published.

10. J.A. Tanis, E.M. Bernstein, W.G. Graham, M. Clark, S.M. Shafroth, B. M. Johnson, K.W. Jones, and M. Meron, Phys. Rev. Lett. $\underline{49}$, 1325 (1982); M. Clark, D. Brandt, J.K. Swenson, and S.M. Shafroth, Phys. Rev. Lett. $\underline{54}$, 544 (1985); P.L. Pepmiller, P. Richard, J. Newcomb, J. Hall, T.R. Dillingham, Phys. Rev. A$\underline{31}$, 734 (1985); J.K. Swenson, Y. Yamazaki, P.D. Miller, H.F. Krause, P.F. Dittner, P.L. Pepmiller, S. Datz, and N. Stolterfoht, Phys. Rev. Lett. $\underline{57}$, 3042 (1986) and references therein.

11. N. Stolterfoht, P.D. Miller, H.F. Krause, Y. Yamazaki, J.K. Swenson, R. Bruch, P.F. Dittner, P.L. Pepmiller, and S. Datz, Nucl. Instrum. & Meth. in Phys. Res. B$\underline{24/25}$, 168 (1987).

Symposium:
Collisions With Cold Particles

MEASUREMENTS ON VERY LOW-ENERGY ION/ATOM-MOLECULE COLLISIONS

Gordon H. Dunn,* M.M Schauer, and S.R. Jefferts

Joint Institute for Laboratory Astrophysics, University of Colorado and National Institute of Standards and Technology
Boulder, Colorado 80309-0440

Data were obtained on very-low energy ion-molecule collisions for the first time about eight years ago when the cooled Penning trap was introduced for the purpose of such studies. Several complementary techniques have been introduced since then. In this paper, the various techniques are briefly described. A few results are presented, illustrating some kinds of physics accessible through such studies.

Introduction

For over three decades, the study of ion-molecule reactions has played a central role in the understanding of planetary atmospheres, flames, laboratory plasmas, reaction kinetics, and interstellar cloud chemistry. Yet, it is relatively recently[1] that these collisions could be studied at temperatures below 80 K.

Much motivation for carrying the temperatures to the range down to 10 K comes from the fact that molecular astrophysics[2] in the dark interstellar clouds typically involves temperatures below 50 K. Also, as is the case with many areas of physics, when one looks into new regions of parameter space, interesting physical phenomena present themselves. Such has been the case as laboratory collision energies have been reduced to cover the 1 - 7 meV range (10 - 80 K), which we refer to as "very-low energies" to distinguish the energies (in some cases down to nano-electronvolts) now being achieved in atom and ion traps by laser cooling.

The techniques that have emerged for the studies during the past eight years are the cooled Penning ion trap at JILA[1,3] the cooled selected ion drift tube at Heidelberg,[4] the CRESU (Cinetique de Reactions en Encoulement Supersonique Uniforme) supersonic expansion method at Meudon,[5] the free jet flow reactor technique at Tucson,[6] and the merged-guided beam/molecular beam and cooled RF-trap methods at Frieburg.[7] In this short paper we will briefly describe these techniques, with greatest emphasis on the Penning trap method used in this laboratory, and we will describe a sample of results to demonstrate physics which can be accessed in this energy range.

Since, at these low energies, much of the interesting physics occurs when an intermediate complex is formed, before describing the methods, it is useful to review some of the simple kinetics involved in such collisions. Some of the features provide a basis for com-paring the various techniques and noting how they complement one another. Consider the collision of A^+ and B which, upon collision, form a complex $(A \cdot B)^+$ with the excess initial kinetic energy distributed to internal modes of the complex:

$$A^+ + B \underset{A_d}{\overset{k_f}{\rightleftarrows}} (A \cdot B)^+ \overset{A_p}{\to} C^+ + D \ . \quad (1)$$

Here k_f is the rate coefficient for the forward reaction to form the complex, A_d is the dissociation rate of the complex to the original particles (resulting in elastic scattering), and A_p is the rate for the complex's proceeding to the products C^+ and D. Clearly, the rate coefficient for forming the products is

$$k_p = k_f \left(\frac{A_p}{A_p + A_d} \right) \ . \quad (2)$$

Very often the back dissociation rate is much larger than the product stabilization rate. Note also that A_d and k_f are related by detailed balancing through a function involving the phase space ratio, so that

$$k_p \approx \frac{k_f}{A_d} A_p = V(T) A_p \quad (3)$$

where $V(T)$ is a strongly increasing function with decreasing temperature T.

If we allow for the possible presence of a third body M in Eq. (1), then we have in addition to the process Eq. (1)

$$M + A^+ + B \underset{A_d}{\overset{k_f}{\rightleftarrows}} (A^+ \cdot B) + M \overset{k_s}{\to} C^+ + D + M \quad (4)$$

so that

$$k_p = k_F \left\{ \frac{A_p}{A_d + A_p + k_s[M]} + \frac{k_s[M]}{A_d + A_p + k_s[M]} \right\} (5)$$

and with the same approximation of $A_d \gg A_p$ and inserting $V(T)$ we have

$$k_p = V(T) A_p + \frac{k_F}{A_d} k_s [M] \ . \quad (6)$$

Thus, at appropriate densities [M], the three body stabilization rate can dominate. Furthermore, by varying [M] and making other measurements or theoreti-

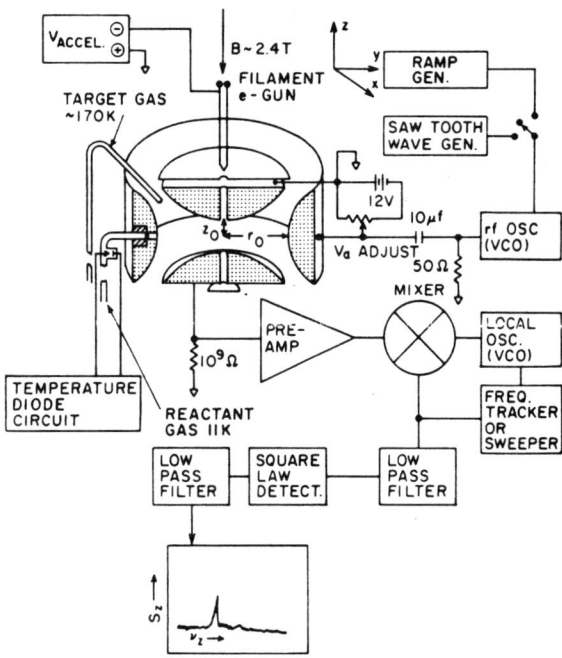

Fig. 1 Schematic of JILA Trap Apparatus

cal estimates of k_f and k_s, one can get information about the magnitude of A_d. If one is in an appropriately low enough range of [M] to insure linearity, one may possibly extrapolate to zero [M] to obtain k_p for pure binary collisions. The various techniques cover the range of using very high [M] to negligibly low [M], and thus provide a range of information about the collisions.

Experimental Methods

Cooled Penning Trap. A schematic of the hyperbolic Penning trap is shown in Fig. 1, borrowed from Ref. 3 which describes the technique in some detail. A target gas from which ions are made by electron impact is introduced through a tube which can be warmed with a heater to allow gas to pass. The trap itself is anchored through a thermal impedance to a flange held at 4 K, and the temperature of the trap can then be varied by a heater on the trap between about 9 and 70 K. Reactant gas is introduced into the side of the trap through a tube held at the temperature of the trap, so that the gas is at the temperature of the trap. Ions are detected by the image currents induced in the end caps of the trap at the frequency of oscillation in the harmonic well, and detection is made mass specific by selective heating at the cyclotron resonance frequency. After making an ion sample, the ions cool by evaporation to about V/13, where V is the depth of the trapping well. Further cooling can take place by momentum transfer in gas collisions or radiative cooling to the electronic circuit. The first is relied upon, since the latter is very slow for ions. Of course, in principle, ions could be cooled by laser cooling, but this is not employed in these experiments. The interaction temperature is then a combination of the gas and ion temperatures and the magnetron drift of the ion cloud about the magnetic field axis. The latter offers another parameter which can be used to vary the temperature of interaction over a limited range.

Since one depends upon the temperature of the environment to help establish the reaction temperature, reactant gases are pretty well limited to H_2 and He for the lowest temperatures noted. Reactant gas densities are usually low [10^7 cm^{-3}] so that Eq. (2) [or (3)] applies directly as opposed to Eq. (5) [or (6)]. Sensitivity is very high, and rates down to 10^{-15} cm^3 s^{-1} can be measured, but this limit could be increased by increasing the trapping magnetic field.

One will recognize that another version of the Penning trap in the form of ICR (Ion Cyclotron Resonance) cells are in wide use for ion-molecule and other studies throughout the chemical community, with instruments being commercially available. It is not usual to cool these instruments to the low temperatures discussed here, and typically the particles cool only by evaporation before reacting so that equivalent temperatures are often several hundreds to tens of thousands of degrees (fractions of eV to several eV).

Cooled Selected Ion Drift Tube (SIDT). In this apparatus used at Heidelberg ions from an electron impact ion source are mass selected with an RF quadrupole, and enter a drift tube which is cooled to temperatures down to 18 K. Since densities are generally high [10^{16} to 10^{17} cm^{-3}] there is rapid thermalization of the interaction temperatures, and product ions are detected on a channel electron multiplier after acceleration from the exit of the drift tube and mass selection with another quadrupole mass filter. Since the apparatus operates at high densities, Eqs. (5) and (6) apply and one can get some information about A_d but must extrapolate over a large range to get information on the binary rate coefficient. The sensitivity is moderate, and rate coefficients in the range to 10^{-13} cm^3 s^{-1} have been measured. As with the Penning device, since the gases must be in contact with walls at low temperatures, the range of gases accessible for study at the lowest temperatures is very limited. Unfortunately this apparatus is no longer in use.

Supersonic Expansion Jet. The uniform supersonic expansion technique for cooling was applied for ion-molecule studies by the group at Meudon.[5] Measurements are made under thermal conditions in the isentropic core of a uniform supersonic jet, avoiding the problems of temperature and density gradients associated with so-called free expansion jets. A high energy electron beam with variable distance from the skimmer hole leading to the mass spectrometer is the source of ionization. Product ions are monitored versus the neutral reactant flow rate and for different distances from the e-beam. Decrease of the reactant ion leads to the rate coefficient. Densities are typically about 10^{15} cm^{-3}, so similar to the previous method, some information is obtained about the complex lifetime, but extrapolation over a long range is required to deduce binary rate coefficients. An important advantage of the technique is the fact that neutral collision partners are not limited to non-condensibles at the low temperatures studied. Without question, the Meudon group using this technique has produced more studies and data on low temperature collisions than any other. However, sensitivity is relatively low, and rates lower than 10^{-12} cm^{-3} cannot readily be measured. A free expansion technique has recently been used at Arizona[6] for detailed rate studies around 2 K.

Slow Merged Ion and Molecular Beams (SMIMB). A very recent innovation has been the introduction in Freiburg[7] of the merging of very slow ($\approx$50 meV) ion beams (using the "guided" beam approach) with molecular beams from pulsed supersonic nozzles. This is an extension of the familiar merged beams approach which takes advantage of the fact that particle beams traveling in the laboratory with high velocities may still have a very small *relative* velocity (down to $\approx$1 meV), and there is also a kinematic shrinking of the interaction energy spread. The difficulty in the past has been that the merged beams typically had energies in the kilovolt range, and any slight difference in direction of the travel of the beams (non-zero angle) gave a high relative energy component. With the slow guided ion beams and molecular beams, angles can be tens of degrees and still contribute less than milli-electronvolts to the interaction energy. The method has not yet been widely applied for this type of study, but it promises to be a winner, since it can give a broad energy range, broad spectrum of collision partners, very good energy resolution, and lends itself readily to such things as laser preparation of ion states or laser diagnosis of products. The only apparent drawback is the unfavorably low sensitivity. Target densities are quite low, [M] $\approx 10^{11}$ cm^{-3}, so binary collisions will be quite competitive with three-body collisions, and extrapolation (if needed at all) to binary rates is over a very small range.

Guided Beam/Temperature Variable Trap. Another very promising technique recently instituted at Freiburg[7] involves using the guided beam technology to introduce low energy ions into a temperature variable radio frequency ion trap. This RF trap is different from the familiar Paul trap and consists of a stack of rings with alternate rings at opposite ends of an rf coil. The ring stack is tied to a cryogenically cooled thermal bath, and the reactant gas is in the trap cell. The trap has been cooled only to liquid nitrogen temperature (78 K) so far, but changes are being implemented to cool to 20 K. At the lower temperature, the method will be limited to non-condensible target gases (He and H_2), but it has extremely high sensitivity (rate coefficients less than 10^{-16} cm^2 s^{-1} have been measured), and densities are in the very flexible range of 10^{12} to 10^{13} cm^{-3} where information on both the binary and ternary rates can be obtained.

Examples of Results

In the brief space of this paper it is impossible to sample the various types of interesting results obtained. One of the more interesting phenomena is that of radiative association which was observed with the ion trap and has more recently been studied using the Freiburg trap. For much of the work, the results of the two methods are consistent, but for $CH_3^+ + H_2 \rightarrow CH_5^+ + h\nu$ there is a serious discrepancy, the Freiburg results yielding a much smaller rate coefficient, albeit at 78 K. This gives cause to reexamine the JILA results for the possibility of parent gas contamination in the measurement. Radiative association has been reviewed recently by Bates and Herbst[8] and won't be dealt with further here.

Another interesting class of process is that which involves quantum mechanical tunneling through a potential

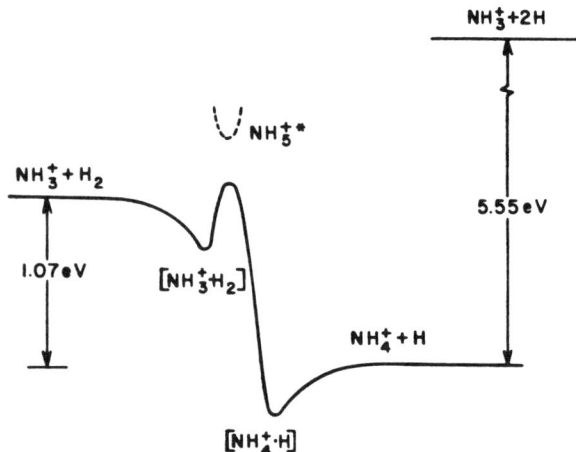

Fig. 2 Schematic reaction coordinate diagram

barrier after formation of the intermediate reaction complex $(A \cdot B)^+$. An example of such a collision is shown in the cartoon of Fig. 2 which shows a proposed reaction coordinate diagram for the abstraction process $NH_3^+ + H_2 \rightarrow (NH_3 \cdot H_2)^+ \rightarrow NH_4^+ + H$, as based upon results obtained using the ion trap technique,[9] the cooled selected ion drift tube,[10] and the higher temperature SIFT method of Adams and Smith.[11] The results are shown in Fig. 3. The complex, formed by the ion-induced dipole potential lives long enough for tunneling to occur through the barrier. The results show the expected isotope dependence associated with tunneling: The greatest rate is for the heavy isotope (greatest density of states for the complex) and the lightest tunneling particle. The next largest is again the light tunneler but the lighter ion; then the heavy ion - heavy tunneler; and finally the light ion - heavy tunneler. The decrease from high temperatures occurs as one descends further and further below the top of the barrier, then there is a minimum as complex formation and tunneling take over. These collisions play an important role in proposed models for ammonia formation in the interstellar medium (ISM).

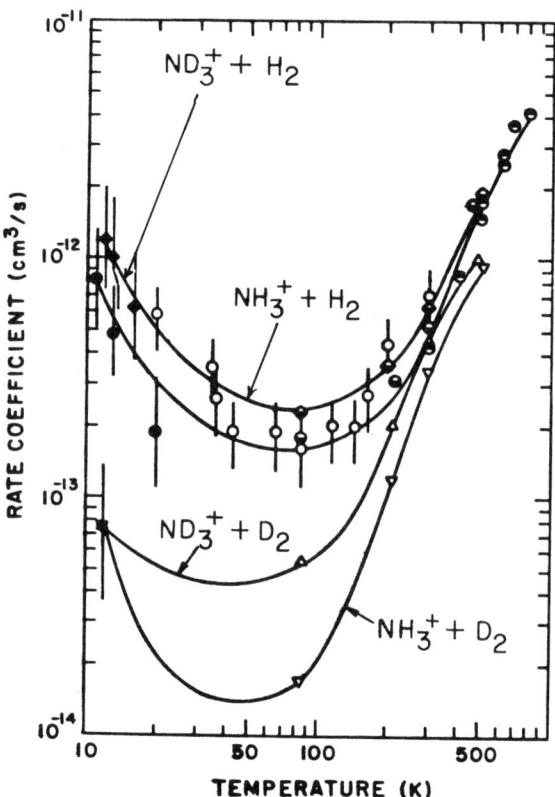

Fig. 3 Ammonium formation reaction from Ref. 9. Lines for visual purposes only. ◆, ●, ▲, ▼ from Ref. 9; ◇, ⊙, △, ▽ from Ref. 11; o from Ref. 10.

A simpler collision system which apparently involves two barriers is that of $He^+ + H_2 \rightarrow He + H^+ + H$. This is a very exoergic process which doesn't correlate with the product state without encountering barriers. The first barrier along the reaction coordinate is only about 3 meV high and is at large distances, while the second may be as much as about 1 eV high, but it is narrow and can be tunneled from the ion-induced dipole potential well complex. Results, which are a composite of ion trap[12] and cooled SIDT[13,14] measurements are shown in Fig. 4. One observes the decline from higher temperatures as less and less of the particles can negotiate the large barrier, a minimum is reached as complex formation and tunneling become

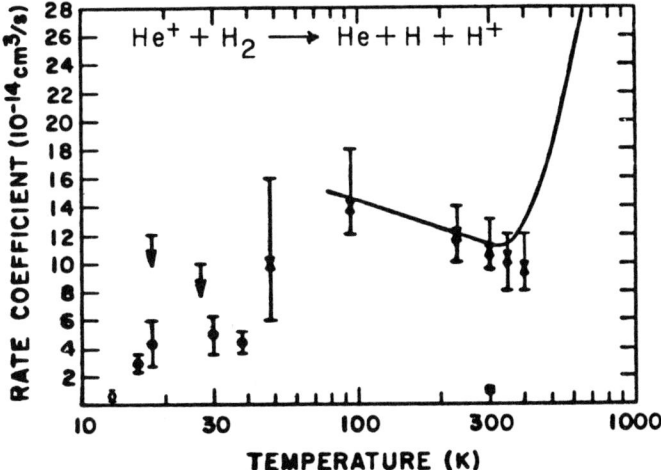

Fig. 4 Dissociative charge transfer. Solid curve, Ref. 14; triangle pairs, Ref. 13; Solid circles, Ref. 12; arrows are upper limits from Ref. 13; Open circle, Ref. 15.

competitive, and finally the rate coefficient again decreases as the 3 meV barrier excludes reactants from coming together. This tiny barrier keeps He^+ - H_2 collisions from being the main means by which He^+ is destroyed and its ionization energy transferred to the ISM. Rather, the next most plentiful molecule, CO, will be the primary collision partner for He^+ loss.

Summary and Acknowledgments

Five complementary methods have been described which have grown up over the past eight years by which one can study ion-molecule reactions at very low energies. Limitations and advantages have been noted. Some results which graphically illustrate interesting physical processes have been briefly recounted, and importance to astrochemistry has been mentioned.

The ion-trap work at JILA has been supported in part by the National Science Foundation Grant No. PHY86-04504 to the University of Colorado.

References

*Staff Member, Quantum Physics Division, NIST.

1. J. A. Luine and G. H. Dunn, Book of Abstracts, XII ICPEAC, Gatlinberg, Tennessee (1981), p. 1035.
2. For an introduction to the literature, see *Molecular Astrophysics: State of the Art and Future Directions*, eds. G. Diercksen, W. Huebner, P. Langhoff (Reidel, Dordrecht, 1985).
3. S. E. Barlow, J. A. Luine and G. H. Dunn, Int. J. Mass. Spectrom. Ion Proc. 74, 97 (1986).
4. H. Böhringer and F. Arnold, Int. J. Mass Spectrom. Ion Phys. 49, 61 (1983).
5. B. R. Rowe et al., J. Chem. Phys. 80, 241 (1984).
6. M. Hawley and M. A. Smith, J. Am. Chem. Soc. (in press, 1989).
7. D. Gerlich and G. Kaefer, 5th Internat'l Swarm Seminar, Birmingham, UK, 1987. Also, Symp. on Atomic and Surface Physics, La Plagne, France, 1988.
8. D. R. Bates and E. Herbst, in *Rate Coefficients in Astrochemistry*, T. J. Millar and D. A. Williams, eds. (Kluwer, Boston, 1988), p. 17.
9. S. E. Barlow and G. H. Dunn, Int. J. Mass Spectrom. Ion Proc. 80, 227 (1987).
10. H. Böhringer, Chem. Phys. Lett. 122, 185 (1985).
11. N. G. Adams and D. H. Smith, Int. J. Mass Spectrom. Ion Proc. 61, 133 (1984).
12. M. M. Schauer, S. R. Jefferts, S. E. Barlow and G. H. Dunn, J. Chem. Phys. 91 (#8, 1989).
13. H. Böhringer and F. Arnold, J. Chem. Phys. 84, 1459 (1986).
14. R. Johnsen, A. Chen and M. A. Biondi, J. Chem. Phys. 72, 3085 (1980).
15. S. E. Barlow, Ph.D. thesis, Univ. of Colorado, Boulder, 1983.

THEORY OF ULTRACOLD ATOMIC COLLISIONS IN OPTICAL TRAPS

P. S. Julienne

B268 Physics, National Institute of Standards and Technology, Gaithersburg, MD 20899 USA

Ultracold collisions of neutral atoms can now be studied in the laboratory in the temperature range below 0.001 K. Rate coefficients for ultracold collisions of excited states are decreased by spontaneous emission during the long duration of the collision, but can be dramatically modified by intense laser effects on collision dynamics. Generalized MCQDT is used to develop criteria for the onset of quantum threshold behavior, which generally begins at temperatures above the Doppler cooling limit of light traps. Threshold Penning ionization of He metastable atoms in a light trap is predicted to occur with a rate coefficient larger than $5 \times 10(-11)$ cm^3/sec.

1. INTRODUCTION

Laser cooling and atom trapping methods have been developed in recent years[1-6] so that collisions of neutral atoms can now be studied at ultracold temperatures, T < 0.001 K, near and probably below[7] the Doppler cooling limit $T_D = \hbar/2\tau k_B$, where $k_B T_D$ is comparable to the natural linewidth $\hbar/\tau$ of the cooling transition (τ = lifetime). The cross section for associative ionization (AI) of two excited Na(3p) atoms has been measured to be nearly 10^{-13} cm^2 at 750 μK.[8] Velocity selected atomic beam methods have also been used to measure this cross section at 60 mK.[9] Collisional processes leading to loss of trapped atoms has been reported for Na and Cs traps,[10,11] and predicted for Na traps.[12,13] Ultracold photoassociation spectroscopy should permit high resolution studies of excited molecular states.[14] It should be possible to observe Penning ionization collisions of the metastable rare gas species Ne* (and presumably also He*)[15] which can also be cooled and trapped.[5] The study of ultracold collisions is now an exciting new area of atomic physics, because of the following novel effects associated with the low collision energy and the long time and distance scales of these collisions: (1) Excited state collisions are modified by spontaneous emission due to the long collision time scale. (2) Collision rates can be modified or controlled by external optical and/or magnetic fields. (3) The long DeBroglie wavelength results in quantum threshold behavior which is not described semiclassically.

2. EXCITED STATE SPONTANEOUS EMISSION

Assume that the atoms are excited by a low intensity cooling laser whose red detuning from resonance is on the order of the natural linewidth $\hbar/\tau$ of the transition. The response of the individual atoms to photo-induced forces determine the temperature. If two like atoms are colliding, either atom or both can be excited by the laser. Therefore, collisional processes involving two ground state atoms, g+g, one ground and one excited atoms, g+e, or two excited atoms, e+e, can occur. The observed process of associative ionization (AI) of two Na(3p) atoms is a good example of the last case.

Due to the very low kinetic energy of ultracold collisions ($k_B T = 9 \times 10^{-8}$ eV or $k T_B/h$ = 21 Mhz at 1 mK), the long range interatomic potential is very important in determining collision dynamics. The long range potentials of the diatomic quasimolecule formed in a collison vary as the van der Waals $1/R^6$, resonant dipole-dipole $1/R^3$, and quadrupole-quadrupole $1/R^5$ interactions for the respective cases g+g, g+e, and e+e. The dominant potential at long range is the g+e C_3/R^3 potential. Figure 1 shows the well-known retarded nonrelativistic $^1\Sigma_u$ and $^1\Pi_u$ potentials for S+P interactions.[16] The respective ordinate and abscissa are shown in reduced units of natural linewidth and $R_\omega = \lambda_\omega/2\pi$, where λ_ω is the wavelength of the cooling transition. Note that the interaction potential is on the order of one natural linewidth when R is on the order of R_ω. Therefore, when the atoms are much closer together than R_ω, $R \ll R_\omega$, the quasimolecule is detuned from resonance with the laser, and neither the g+e nor the e+e excited states of the molecule can be excited by the laser. Thus, if AI is observed when the atoms are excited by weak near-resonant radiation, we know that the two excited atoms must

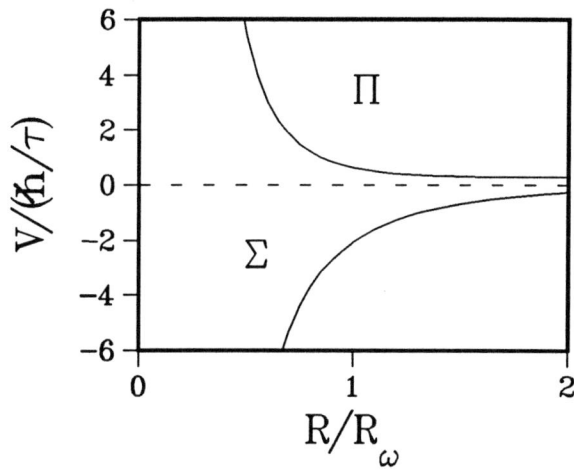

Figure 1. Σ and Π ungerade molecular potentials between S and P atoms of the same species.

have been produced when their internuclear separation was on the order of R_ω or more.

Associative ionization only occurs if the collision brings the two atoms together on the doubly excited molecular state to a separation R near the equilibrium internuclear distance R_e of the molecular ion. The time required to do this is $\tau_c \approx R_\omega/v$. Typical collision velocities at "normal" and ultracold energies are 10^5 cm/sec and 100 cm/sec respectively. If we rewrite these as 10000 Å/nsec and 10 Å/nsec and use τ = 8 nsec for the Na_2 molecule, we see that τ_c is much longer than τ for ultracold conditions, but much shorter than τ for "normal" collision energy. Thus, in an ultracold collision, the spontaneous decay of the e+e quasimolecular state as the atoms move between R_ω and R_e will dramatically decrease the effective rate coefficient of the AI process, probably by several orders of magnitude. The reason why AI of Na(3p) atoms is observed at all at 750 μK is explained in the next section as due to intense laser effects.

A general conclusion we draw from this kind of argument is that effective collision rate coefficients for excited state processes are expected to be strongly modified by excited state spontaneous emission if T is much less than some characteristic temperature T_S. This temperature may be estimated by requiring that the collision time $\tau_c \approx R_\omega/v_S$ be comparable to the molecular lifetime τ. This condition is equivalent to saying that the Doppler shift due to the relative motion during a collision is comparable to the natural linewidth.[17] Table 1 shows T_S as well as T_D for several cooling transitions. Since $T_D \ll T_S$, we generally expect excited state processes to be modified by spontaneous emission in optical and magneto-optical traps.

A real challenge for theory is to develop new formulations which properly account for the role of the complex optical pumping process which occurs during the long range approach of the two cold atoms where one or more absorption-emission cycles occur in the quasimolecule. Since the process of "preparing" the excited atoms for the collision is not decoupled from the collision itself, the validity of the usual cross section formulation using close coupled scattering methods is called into question.

3. ROLE OF INTENSE LASER FIELDS

A strong laser field can induce large modifications in effective rate coefficients for ultracold collisions, since the dipole interaction $\vec{E}\cdot\vec{\mu}$ can be large compared to $k_B T$.[18] In fact, this is why AI of Na(3p) atoms is observed with a large cross section in the optical trap of the NIST experiment, since we have just shown above that weak field excitation should result is a small cross section. Recall that the NIST hybrid optical trap consists of two time cycles: one in which a weak near-resonant cooling laser is on and one in which a very intense off-resonant trap laser is on. The intensity ratio between the two cycles is 30000:1. In a new time resolved experiment,[9,19] an ion signal due to AI is only observed during the trap cycle, not during the cooling cycle, although the excited state density, as measured by atomic fluorescence, is about the same during both cycles. Since the cross section originally reported actually represents an average over both cycles of the trap,[8] this cross section must be corrected for the fact that AI only occurs when the intense laser is on.

TABLE 1

Species	λ_ω(nm)	T_D	T_S	n	T_Q
H + H	120	1.8mK	7mK	6	20K
Hs(^{3}S)+He(^{3}S)	1080	40μK	1mK	6	100mK
Cs + Cs	852	130μK	200mK	6	330μK
Na + Na	589	250μK	60mK	6	10mK
Na + Na(3p)	589	250μK	60mK	3	$<10^{-9}$K
Na(3p)+Na(3p)	589	250μK	60mK	5	740μK

In order to understand the qualitative effects of the intense laser on the collision dynamics, it is necessary to set up the Hamiltonian for the system atoms + field + interaction. The adiabatic dressed molecule (ADM) potentials obtained by diagonalizing this Hamiltonian as a function of R can be used to explain the basic physics of the ultracold collision in an intense field.[18] The asymptotic atomic states are the field-dressed atomic states in which ground and excited states are mixed by the intense laser. All ADM potentials vary at long range as $1/R^3$ because of this mixing. The interaction may be attractive or repulsive, depending on the angle θ between the internuclear axis $\vec{R}$ and the laboratory quantization axis $\hat{z}$ defined by the laser polarization. A range of θ exists for which an attractive interaction path leading to AI is possible for both the long range $1/R^3$ and the shorter range $1/R^5$ quadrupole-quadrupole potentials. The energy of the field dressed atomic state which correlates with the e+e state at short range is shifted by twice the single atom light shift, or about 50 $k_B T$ in the NIST trap. As the atoms come together along an attractive path, the atoms remain strongly coupled to the excitation laser until they reach a distance R_I, dependent on laser intensity I, where the molecule becomes decoupled from the radiation field due to the R-dependent detuning of the molecular potential. For the NIST trap, $R_I \approx 0.1 \times R_\omega$. Since the atoms are accelerated to an energy $\approx 50\ k_B T$ near R_I, equal to the amount of light shift lost when the atoms become decoupled from the light field, the two excited atoms can readily travel with no emission losses from R_I to R_e, where AI occurs. Furthermore, the $1/R^3$ long range potential is effective in capturing a wider range of partial waves than a $1/R^5$ potential. Therefore, theory predicts[18] and experiment confirms[9,19] that the effective rate coefficient for AI is large during the intense laser cycle of the trap but is much smaller during the cooling laser cycle.

We see that ultracold collisions offer the prospect that some degree of laser control over effective excited state collision rate coefficients will be possible by varying the laser parameters. It should be possible to explore such effects in atomic beams as well as atom traps. Theory also predicts that rate coefficients should be sensitive to atomic orientation and direction of approach. Such effects may be more prominent for ultracold collisions than "normal" collisions, since few partial waves contribute. Such effects are averaged in cell-like trap experiments, but can be studied in beam experiments.

4. QUANTUM THRESHOLD EFFECTS

Although quantum threshold laws are well established,[20-22] the generalized form of multi-channel quantum defect theory (MCQDT)[23-25] offers some fresh insights and practical tools for exploring the threshold effects which we may now expect to observe in ultracold neutral atom collisions. The MCQDT gives a rigorous treatment of threshold behavior with a physical picture that is remarkably similar in spirit to the picture used by Bethe[20] in 1935 to derive the basic threshold law for inelastic exothermic processes, namely $\sigma \propto 1/v$, where v is entrance channel velocity. The origin of quantum threshold behavior is the mismatch which exists between the short range wavefunction with local DeBroglie wavelength,

$$\lambda(E,R) = 2\pi/k(E,R) = h/[2\mu(E-V(R))]^{1/2}, \quad (1)$$

and the asymptotic entrance channel wave with DeBroglie wavelength, $\lambda_\infty = 2\pi/k_\infty$, that is long compared to the size of the interatomic potential.

The close coupled wavefunction is written is some basis $|\gamma\rangle$ as

$$\Psi(E,R) = |\gamma\rangle F(E,R) / R . \quad (2)$$

In the MCQDT analysis the multichannel wavefunction in the "inner" zone where molecular interactions lead to inelastic scattering events is represented as

$$F(E,R) = f(E,R) + g(E,R)Y(E,R) , \quad (3)$$

where f and g are <u>diagonal</u> matrices whose elements, $f_{nn}(E,R)$ and $g_{nn}(E,R)$ are the independent solutions of the 1-dimensional Schrodinger equation for the reference potential $V_n(R)$. The reference potentials are <u>chosen</u> for our convenience in analyzing Ψ, but observable quantities are independent of our choice. The reference solutions can be written as

$$f(E,R) = \alpha(E,R)\sin\beta(E,R) \quad (4a)$$
$$g(E,R) = \alpha(E,R)\cos\beta(E,R) . \quad (4b)$$

In the WKB-assisted version of the theory,
$$\alpha(E,R) = 1/[k(E,R)]^{1/2} , \quad (5)$$

is the WKB normalization and $\beta(E,R)$ is the WKB phase. The MCQDT matrix Y develops in this inner zone due to the nonadiabatic interactions among reference states. It is independent of the <u>asymptotic</u> boundary conditions, is slowly varying with energy and continuous across asymptotic thresholds, and is defined for both asymptotic open and closed channels.

The wavefunction in the asymptotic zone,

$$F(E,R) = s(E,R) + c(E,R)T(E) \quad (6)$$

may be represented in terms of the diagonal matrices of the reference functions $s_n(E,R)$ and $c_n(E,R)$, which have standard scattering boundary conditions and are normalized so as to let us represent the plane and scattered wave asymptotically:

$$s(E,R) = \sin(k_\infty R - \pi\ell/2 + \xi)/k_\infty^{1/2} \quad (7)$$

$$c(E,R) = \cos(k_\infty R - \pi\ell/2 + \xi)/k_\infty^{1/2} \quad . \quad (8)$$

The complete S matrix is represented as

$$\mathbf{S}(E) = e^{i\xi(E)}[1+i\mathbf{T}(E)][1-i\mathbf{T}(E)]^{-1}e^{i\xi(E)} \quad (9)$$

$$\mathbf{T}(E) = \mathbf{C}(E)^{-1}[\mathbf{Y}(E)^{-1} - \tan\lambda(E)]^{-1}\mathbf{C}(E)^{-1} \quad , \quad (10)$$

where $\mathbf{Y}(E)$ is the asymptotic value of $\mathbf{Y}(E,R)$ and the diagonal matrices $\mathbf{C}(E)$ and $\tan\lambda(E)$ contain the parameters introduced by the theory to match f and g to s and c for each reference potential:

$$f(E,R) = \mathbf{C}(E) \, s(E,R) \quad (11)$$

$$g(E,R) = \mathbf{C}(E)^{-1}[c(E,R) - \tan\lambda(E)\mathbf{C}(E)^2 s(E,R)] . \quad (12)$$

For collision energies $E \gg E_Q$ well above threshold, $\mathbf{C}(E) \to 1$, $\tan\lambda \to 0$, and the two sets of reference functions are identical: f = s, g = c. This is the case when the WKB validity criterion,

$$d\lambda(E,R)/dR \ll 1 \quad , \quad (13)$$

applies for all R between the inner and asymptotic zones, and a WKB connection can be made between the inner and outer solutions. However, as $E \to 0$ at threshold, $E \ll E_Q$, there will always be some range of R between the inner and outer zones where the WKB approximation fails to apply (for $1/R^n$ potentials with $n \geq 3$), and $\mathbf{C}(E)$ and $\tan\lambda$ have the following characteristic threshold behavior for $\ell = 0$ (s waves):

$$\mathbf{C}(E)^{-2} \to A_0 k_\infty \quad (14)$$

$$\tan\lambda(E) \to -\cot\nu(0) \quad , \quad (15)$$

where $\nu(0)$ is the threshold value of the MCQDT quantum number function and A_0 is an inelastic scattering length parameter.

The essential features of the threshold cross section in the MCQDT picture can be interpreted in terms of the $\mathbf{C}(E)^{-1}$ function, which matches the wavefunctions f and s having respective inner zone WKB normalization and asymptotic zone scattering normalization. Although f and s are equivalent well above threshold, $s = \mathbf{C}^{-1}f$ is seen to vanish in the inner zone as $k_\infty^{1/2}$ as threshold is approached. A typical example of $\mathbf{C}(E)^{-2}$ for s, p, and d waves is shown in Fig. 2. This reduction in inner zone amplitude of the function s with scattering boundary conditions

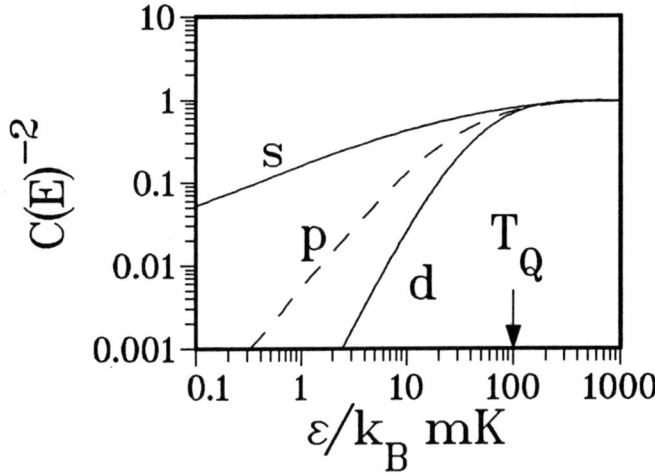

Figure 2. MCQDT matching function versus collision energy for a model $\text{He}(^3S) + \text{He}(^3S)$ $^1\Sigma_g^+$ potential. The p wave does not actually contribute to Penning ionization, due to homonuclear symmetry.

causes S matrix elements for exothermic inelastic processes to vanish also as $k_\infty^{1/2}$ for s waves.

The cross section for an inelastic, exothermic transition from initial channel i to final channel f can be written as

$$\sigma(E, i \to f) = (\pi/k_\infty^2) \sum (2\ell+1) P_\ell(E, i \to f) \quad , \quad (16)$$

where the opacity function $P \leq 1$ depends on $\mathbf{Y}$, $\mathbf{C}$, and $\tan\lambda$, and is proportional to $\mathbf{C}(E)^{-2}$ for the entrance channel. Near threshold only s waves contribute to σ, which is seen from Eq. (14) to diverge as $1/k_\infty$. In this limit, the rate coefficient for the reaction, $K(T, i \to f) = \langle \sigma(E, i \to f) v \rangle$, is seen to approach a temperature-independent constant,

$$K(0, i \to f) = (hA_0'/2\mu) P^0(i \to f) . \quad (17)$$

The inner zone opacity P^0, which expresses the physics of the inner zone interactions, depends on $\mathbf{Y}$ only and is insensitive to energy in the threshold region. If the nonadiabatic couplings expressed by $\mathbf{Y} \neq 0$ are not too large, and if there is not a bound state too close to threshold, A_0' is the scattering length parameter of the entrance channel $\mathbf{C}(E)^{-2}$ function in Eq. (14). On the other hand, if a bound state exists sufficiently close to threshold, and the nonadiabatic couplings are large enough that its width or shift is comparable to its binding energy, then A_0' is different from A_0 in a way that depends on $\mathbf{Y}$, $\mathbf{C}$, and $\tan\lambda$. In any case, if we write (17) as

$$K(0, i \to f) = 1 \times 10^{-11} (A_0'/\mu) P^0 \text{ cm}^3 \text{sec}^{-1}, \quad (18)$$

where A_0' is in Bohr and μ is in atomic mass units, we see that rate coefficients for allowed ultracold processes (i.e., those with an intrinsic opacity near unity) can easily be in the range $10^{-12} - 10^{-10}$ cm^{3}sec^{-1}. This range is similar to that for "allowed" processes at normal temperatures.

A simple criterion for the switching energy E_Q where threshold behavior begins can be derived by choosing E_Q to be the highest energy for which a WKB connection can be made for a $1/R^n$ potential, using the condition $d\lambda(E_Q,R)/dR \le 1/2$ at all R between inner and asymptotic zones. The resulting $T_Q = E_Q/k_B$ is shown in Table 1 for several examples. We see that T_Q is larger than T_D for ground state collisions. Therefore, a general conclusion is that semiclassical theory is not expected to be accurate for ground state collisions in traps operating near T_D. However, T_Q is so low for collisions on a $1/R^3$ potential that semiclassical theory should be useful for such collisions.

We have carried out model calculations to illustrate features of threshold behavior. One is for the Penning ionization of He(^{3}S) metastable atoms,

$$\text{He}(^3\text{S}) + \text{He}(^3\text{S}) \rightarrow \text{He} + \text{He}^+ + e \quad . \quad (19)$$

This has a large room temperature rate coefficient, 10^{-9}cm^{3}sec^{-1} and has been observed in a single beam with T as low as 20 K.[25] A magneto-optical trap operating at or below 100 μK should be feasible for these metastable atoms.[15] It is desirable to have an estimate of the threshold rate coefficient in order to estimate the importance of collisional loss processes in such a trap. Collisions at 100 μK occur in the quantum threshold regime, since $T_Q = 100$ mK. Of the three potentials which correlate with the asymptotic He(^{3}S) atoms, $^1\Sigma_g^+$, $^3\Sigma_u^+$, and $^5\Sigma_g^+$, only the $^1\Sigma_g^+$ gives rise to allowed s wave Penning ionization in the ^{4}He isotope. By making small variations in the known $^1\Sigma_g^+$ potential, we estimate that A_0 is not less than about 30 Bohr. Fig. 2 illustrates model calculations of the $C(E)^{-2}$ function for s, p, and d waves for this system. If m and m' represent the Zeeman projection quantum numbers of the electron spin on the laboratory quantization axis, the only nonvanishing threshold rate coefficients K(m,m') are K(0,0) and K(1,-1), which we estimate to have a magnitude not less than 5×10^{-11}cm^{3}sec^{-1}

(We use an opacity of 1/3, which accounts for the probability of being on the $^1\Sigma_g^+$ potential, given the projection quantum numbers). The upper bound to the cross section at 100 μK due to unitarity of the S matrix is 10^{-9} cm^{3}sec^{-1}. Therefore, rapid threshold Penning ionization of unpolarized samples of He(^{3}S) metastables is predicted. The rate coefficient K(1,1) for spin-polarized atoms is expected to be orders of magnitude smaller, since the process is spin-forbidden.

We have also done a model calculation to test the validity of semiclassical theory for collisions on a $1/R^3$ potential.[12] We have estimated the rate coefficient for the process

$$\text{Na} + \text{Na}(3p^2P_{3/2}) \rightarrow \text{Na} + \text{Na} + h\nu \quad , \quad (20)$$

where the emission is due to a free-free transition of the Na$_2$ molecule and is red shifted from the resonance transition energy. If the red shift is enough, then the two ground state product atoms will have enough kinetic energy to escape the trapping potential, and this process results in loss of atoms from the trap. We have carried out for 1 mK collisions both quantal and semiclassical calculations of the emission spectrum and loss rate coefficient for various trap depths. As expected due to the very low T_Q for $1/R^3$ potentials (Table 1), the quantal and semiclassical calculations are in reasonable

Figure 3. Quantal (oscillatory) and semiclassical (smooth) emission spectrum for the $0_u^+ \rightarrow 0_g^+$ transition versus ground state separation energy for $\epsilon/k_B = 1.4$ mK collisions of Na + Na($^2P_{3/2}$).

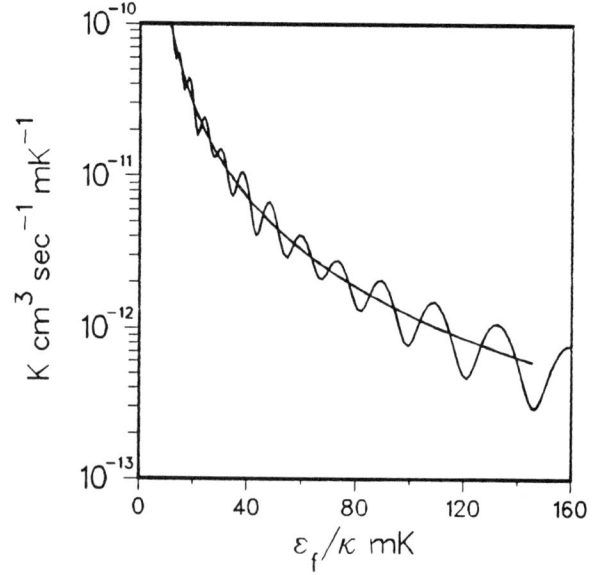

agreement. An example comparing the two spectra, uncorrected for spontaneous emission flux loss, is shown in Fig. 3. In accordance with the discussion in Section II, spontaneous emission causes a strong modification of the effective rate coefficient for trap loss. We estimate for process (20) at 1 mK a loss rate coefficient of 5×10^{-11} cm^{3}sec^{-1} for a 10 mK deep trap and 2×10^{-12} cm^{3}sec^{-1} for a 500 mK deep trap.[26]

5. ACKNOWLEDGEMENT

This work was supported in part by a grant from the Air Force Office of Scientific Research.

6. REFERENCES

1. J. V. Prodan, A. Migdall, W. D. Phillips, I. So, H. J. Metcalf, and J. Dalibard, Phys. Rev. Lett. **54**, 992(1985).

2. W. Ertmer, R. Blatt, J. L. Hall, and M. Zhu, Phys. Rev. Lett. **54**, 996(1985).

3. S. Chu, L. Holberg, J. E. Bjorkholm, A. Cable, and A. Ashkin, Phys. Rev. Lett. **55**, 49(1985).

4. E. L. Raab, M. Prentiss, A. Cable, S. Chu, and D. E. Pritchard, Phys. Rev. Lett. **59**, 2631(1987).

5. F. Shimizu, K. Shimizu, and H. Takuma, Phys. Rev. A **39**, 2758(1989).

6. D. Sesko, C. Fan, and C. E. Wieman, J. Opt. Soc. Am. B **5**, 1225(1988).

7. P. D. Lett, R. N. Watts, C. I. Westbrook, W. D. Phillips, P. L. Gould, and H. J. Metcalf, Phys. Rev. Lett. **61**, 169(1988).

8. P. L. Gould, P. D. Lett, P. S. Julienne, W. D. Phillips, H. R. Thorsheim, and J. Weiner, Phys. Rev. Lett. **60**, 788(1988).

9. J. Weiner, this volume.

10. M. Prentiss, A. Cable, J. E. Bjorkholm, S. Chu, E. L. Raab, and D. E. Pritchard, Opt. Lett. **13**, 452(1988).

11. T. Walker, this volume.

12. P. S. Julienne, S. H. Pan, H. R. Thorsheim, and J. Weiner, in _Advances in Laser Science_, Vol. 3, edited by A. C. Tam, J. L. Gole, and W. C. Stwalley, AIP Conference Proceedings 172 (American Institute of Physics, New York, 1988), p.308.

13. J. Vigue, Phys. Rev. A**34**, 4476 (1986).

14. H. R. Thorsheim, J. Weiner, and P. S. Julienne, Phys. Rev. Lett. **58**, 2420(1987).

15. A. Aspect, private communication(1989).

16. W. J. Meath, J. Chem. Phys. **48**, 227(1968).

17. J. Cooper and D. N. Stacey, J. Phys. B**7**, 2143 (1974).

18. P. S. Julienne, Phys. Rev. Lett. **61**, 698(1988).

19. P. L. Gould, P. D. Lett, R. N. Watts, C. I. Westbrook, P. S. Julienne, W. D. Phillips, H. R. Thorsheim, and J. Weiner, in _Atomic Physics 11_, edited by S. Haroche, J. C. Gay, and G. Grynberg (World Scientific, Singapore, 1989).

20. H. Bethe, Phys. Rev. **47**, 747(1935).

21. E. P. Wigner, Phys. Rev. **73**, 1002(1948).

22. L. M. Delves, Nuclear Physics **8**, 358(1958).

23. C. H. Greene, A. R. P. Rau, and U. Fano, Phys. Rev. A **26**, 2441(1982).

24. (a) F. H. Mies, J. Chem. Phys. **80**, 2514(1984); (b) F. H. Mies and P. S. Julienne, J. Chem. Phys. **80**, 2526 (1984).

25. M. Muller, W. Bussert, M.-W. Ruf, H. Hotop, and W. Meyer, Phys. Rev. Lett. **59**, 2279(1987).

26. H. R. Thorsheim, Thesis, Department of Physics, University of Maryland, (1989).

Theoretical treatment of the associative ionization reaction between laser excited sodium atoms : energy dependence and anisotropy effects

Anne HENRIET[+] and Françoise MASNOU-SEEUWS

Laboratoire des Collisions Atomiques et Moléculaires, Bât. 351, Université Paris-Sud
91405 ORSAY Cedex France

The first order MQDT treatment proposed by Giusti (1980) for dissociative recombination is applied to the present problem. Two electron calculations for the intermediate Rydberg states of Na_2 are used to define a diabatic representation, where dissociative doubly-excited states Q_j are interacting with monoexcited Rydberg series, converging to the Na_2^+ ground state X. The ionization probability is controlled by the strength of this electronic coupling and by the position of the crossing between the Q_j and the X curves. The cross sections may thus differ considerably from one molecular symmetry to another, explaining the strong anisotropy observed experimentally. However the long range population sharing between the various doubly-excited curves may be dependent on the collision velocity, so that the conclusions obtained at room temperature should not be extrapolated to ultracold collisions

Introduction

The associative ionization (AI) reaction between two atoms :

$$A + B \rightarrow AB^+ + e \quad (1)$$

is a simple example for the formation of a chemical bond. The central importance of AI, or of the reverse reaction, which is dissociative recombination (DR), for the gas phase chemistry has long been recognized. For instance, in 1947, Bates and Massey[1] could explain the daily variation of the electron density in the earth's ionosphere by considering the balance between the photoionisation of the atmospheric molecules by the solar UV radiation and the dissociative recombination of the molecular ions. However, the theoretical treatment of AI and DR still remains in many respects an open problem. The major difficulty lies in the treatment of bound-continuum interactions : in the initial state, all the electrons are bound while the relative motion of the two atoms is free ; in the final state one electron is free while the vibrational motion of the two nuclei is constrained inside a potential well.

In the case where both atoms are excited, much experimental investigation has been devoted to the reaction between laser excited alkali atoms, especially :

$$Na(3p) + Na(3p) \rightarrow Na_2^+ + e . \quad (2)$$

The measurements concern the total cross-section at room temperature [2,3] as well as the variation of this cross-section as a function of the polarization of the laser light exciting the atoms in particular combinations of $|Jm_J>$ sublevels [4,5].

The process of associative ionization may also have some influence on experiments involving laser cooled atoms. The occurrence of reaction (1) may lead to an important loss of atoms in a trap, as the molecular ions are not generally sensitive to the trapping radiation. Besides, the presence of charged particles may strongly disturb the trapped atoms. The need for a good theoretical estimate of the cross-section for the reaction (2) therefore exists.

As discussed by Julienne [6], collisions between ultra-cold atoms are controlled by the long range part of the potentials, so that at a given collision energy E the cross-section may be written :

$$\sigma(E) = \pi b_{max}^2 \overline{P(E)} \quad (3)$$

In (3), b_{max} is the maximum impact parameter for which the reaction zone is attained, and P(E) is an average probability for the reaction (2) to take place in the inner region. Indeed at threshold the Na_2^+ product cannot be formed in a vibrational state v excited beyond v=3, so that the two colliding atoms have to reach the region R < 8 au of internuclear distances. In this region the autoionization of the Na_2 molecule may take place, leaving the molecular ion in a vibrational level v ≤ 3, while the free electron evacuates the excess energy. However, the autoionization probability differs markedly according to the various channels, so that P(E) should be considered as an average. The aim of the present paper is to discuss the current state of the knowledge concerning P(E).

One of the main difficulties in treating the associative ionization problem is the large number of channels that are involved. We know that of the twelve molecular potential curves correlated to the Na(3p) + Na(3p) dissociation limit, ten may contribute to the reaction (2). Such curves are embedded in the Rydberg series converging to the various Na(3s) + Na(nl) limits : the (3s + 5s) and (3s + 4d) limits are respectively 92 meV under and 70 meV above the doubly-excited limit. Turning now to the reaction region, the full treatment of the autoionization problem for a given molecular symmetry requires in principle the consideration of an infinite number of states, as molecular Rydberg series are involved. We shall see that this complexity can greatly be reduced by use of a multichannel quantum defect (MQDT) approach[7].

On the other hand, the correlation to the separated atoms limit still raises some problems.

2. Treatment of the Na2 molecule : adiabatic representation

We first compute the potential energy curves of the Na_2 molecule : as the final state of the reaction (2) involves Na_2^+, it is also necessary to describe the molecular ion. Due to the weak bonding energy of the sodium valence electron compared to the core electrons, a model potential treatment should be satisfactory [8]. The Na_2^+ ion is then treated as an effective one electron system, and the Na_2 molecule as an effective two electron system. The main difference with the related H_2^+ and H_2 problems arises from the presence of two polarisable cores. We treat the core-polarization effects via effective operators[9], the interaction between one electron and one core being represented by a parametric potential fitted to the sodium experimental spectrum [10]. Fine structure effects are neglected in this treatment.

In a first step, the Na_2^+ orbitals χ_p are evaluated accurately by solving the one-electron Schrödinger equation in a space defined by a large basis set of generalized Slater orbitals in prolate spheroïdal coordinates[11]. The potential curves for the ground and first excited states of the Na_2^+ molecular ion are displayed in Fig. 1.

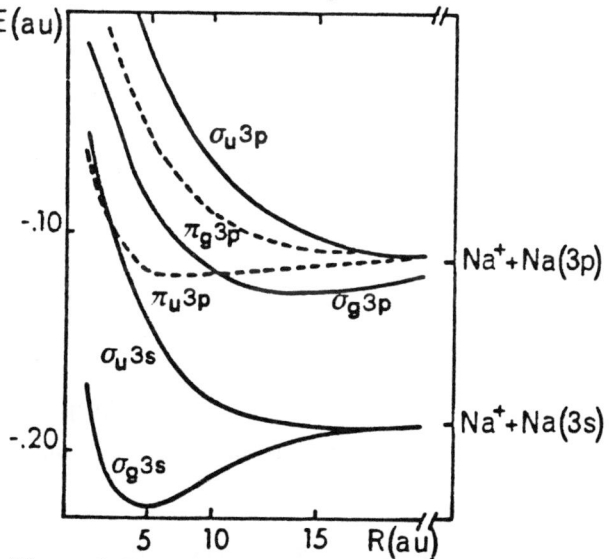

Figure 1. Potential curves for Na_2^+.

We may note two main differences with the H_2^+ case :
i) the equilibrium distance for the ground state is large ($R_e = 6.8$ au) and therefore the energy splitting between the ground state $\sigma_g 3s$ bonding orbital and the first antibonding orbital is small (2.22 eV at R_e).

ii) four excited orbitals, $\sigma_g 3p$, $\pi_u 3p$, $\sigma_u 3p$ and $\pi_g 3p$, which are correlated to the $Na^+ + Na(3p)$ dissociation limit, have at short internuclear distances energies close to the $\sigma_u 3s$ orbital.

We next solve the Na_2 problem by expanding the two electron wavefunction in antisymmetrized products of two Na_2^+ orbitals χ_p and χ_q :

$$\mathcal{F}(1,2) = \sum_i a_i \left\{ \chi_p^i(1) \chi_q^i(2) \pm \chi_p^i(2) \chi_q^i(1) \right\} \quad (3)$$

We have chosen to perform a numerical evaluation of the bielectronic integrals[12]. An expansion on 30 configurations for the Σ symmetry, 20 for the π symmetry, is sufficient to obtain, by diagonalization of the full hamiltonian $\mathcal{H}(1,2,R)$, the various adiabatic potential curves of Na_2[14] with an accuracy which can be assessed by comparing to spectroscopic data : the first excited states are well known, and accurate data are also available for the intermediate Rydberg states of Σ and Δ symmetry[13].

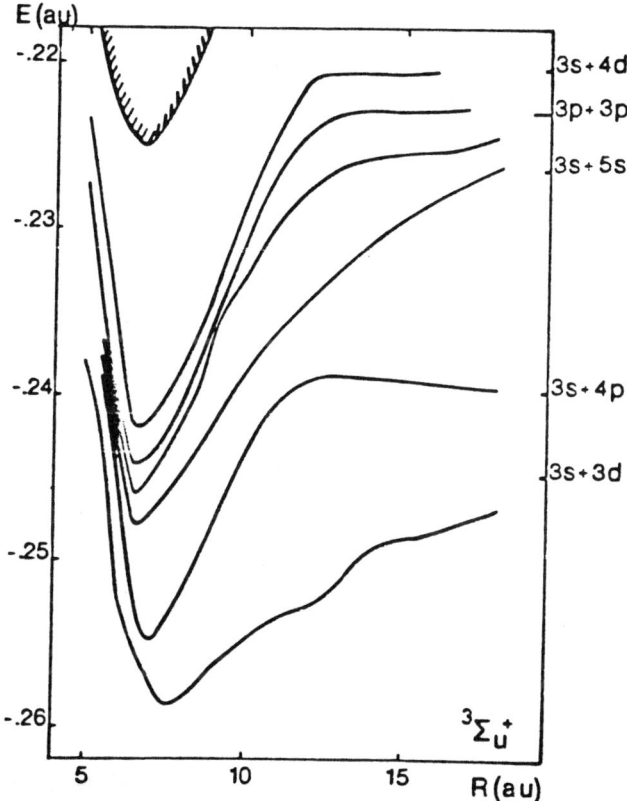

Figure 2. Adiabatic potential curves of $^3\Sigma_u^+$ symmetry for the Na_2 molecule.

The conclusion[14] is that the energies are presently determined to within 200 cm^{-1}(24 meV) and the position of the minima to within 0.1 Å. Moreover, an independent check of the calculations is the good agreement obtained with the potential curves computed by Jeung[15] up to the 3s + 5s dissociation limit using an *ab initio* pseudo-potential treatment with a gaussian orbitals basis set.

are involved. We shall see that this complexity can greatly be reduced by use of a multichannel quantum defect (MQDT) approach[7].

On the other hand, the correlation to the separated atoms limit still raises some problems.

2. Treatment of the Na₂ molecule : adiabatic representation

We first compute the potential energy curves of the Na_2 molecule : as the final state of the reaction (2) involves Na_2^+, it is also necessary to describe the molecular ion. Due to the weak bonding energy of the sodium valence electron compared to the core electrons, a model potential treatment should be satisfactory [8]. The Na_2^+ ion is then treated as an effective one electron system, and the Na_2 molecule as an effective two electron system. The main difference with the related H_2^+ and H_2 problems arises from the presence of two polarisable cores. We treat the core-polarization effects via effective operators[9], the interaction between one electron and one core being represented by a parametric potential fitted to the sodium experimental spectrum [10]. Fine structure effects are neglected in this treatment.

In a first step, the Na_2^+ orbitals χ_p are evaluated accurately by solving the one-electron Schrödinger equation in a space defined by a large basis set of generalized Slater orbitals in prolate spheroïdal coordinates[11]. The potential curves for the ground and first excited states of the Na_2^+ molecular ion are displayed in Fig. 1.

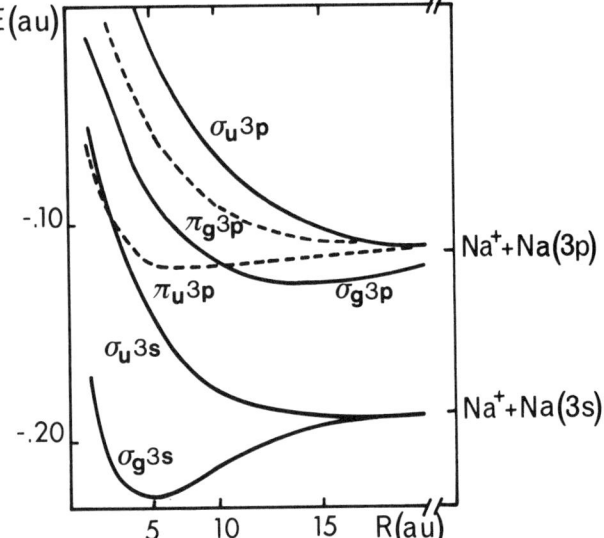

Figure 1. Potential curves for Na_2^+.

We may note two main differences with the H_2^+ case :

i) the equilibrium distance for the ground state is large ($R_e = 6.8$ au) and therefore the energy splitting between the ground state $\sigma_g 3s$ bonding orbital and the first antibonding orbital is small (2.22 eV at R_e).

ii) four excited orbitals, $\sigma_g 3p$, $\pi_u 3p$, $\sigma_u 3p$ and $\pi_g 3p$, which are correlated to the $Na^+ + Na(3p)$ dissociation limit, have at short internuclear distances energies close to the $\sigma_u 3s$ orbital.

We next solve the Na_2 problem by expanding the two electron wavefunction in antisymmetrized products of two Na_2^+ orbitals χ_p and χ_q :

$$\mathcal{F}(1,2) = \sum_i a_i \left\{ \chi_p^i(1) \chi_q^i(2) \pm \chi_p^i(2) \chi_q^i(1) \right\} \quad (3)$$

We have chosen to perform a numerical evaluation of the bielectronic integrals[12]. An expansion on 30 configurations for the Σ symmetry, 20 for the π symmetry, is sufficient to obtain, by diagonalization of the full hamiltonian $\mathcal{H}(1,2,R)$, the various adiabatic potential curves of Na_2[14] with an accuracy which can be assessed by comparing to spectroscopic data : the first excited states are well known, and accurate data are also available for the intermediate Rydberg states of Σ and Δ symmetry[13].

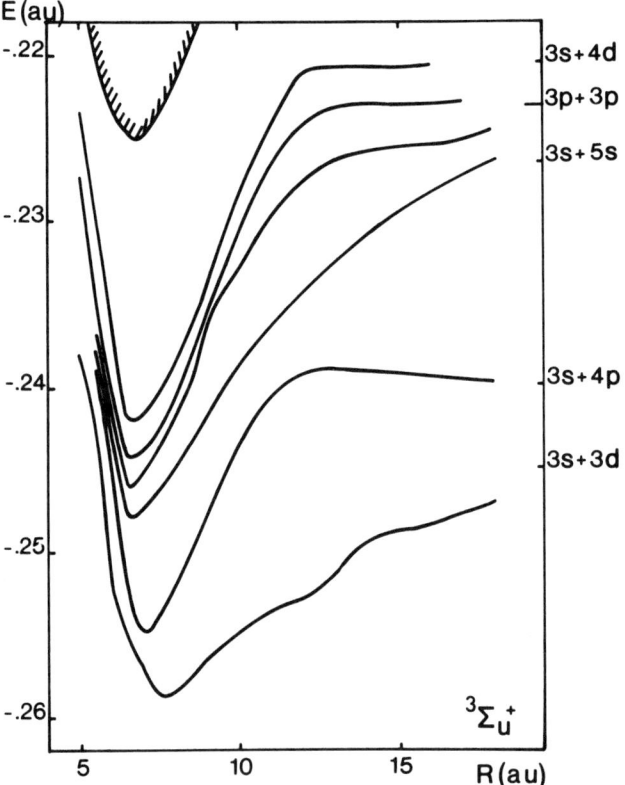

Figure 2. Adiabatic potential curves of $^3\Sigma_u^+$ symmetry for the Na_2 molecule.

The conclusion[14] is that the energies are presently determined to within 200 cm⁻¹ (24 meV) and the position of the minima to within 0.1 Å. Moreover, an independent check of the calculations is the good agreement obtained with the potential curves computed by Jeung[15] up to the 3s + 5s dissociation limit using an *ab initio* pseudo-potential treatment with a gaussian orbitals basis set.

For some symmetries (see Fig.2) the adiabatic potential curves display many avoided crossings, the Rydberg series being strongly perturbed by a doubly-excited state. It appears therefore more convenient to use a diabatic representation to treat the molecular problem.

3. Treatment of the Na$_2$ molecule : diabatic representation

An alternative procedure, first introduced by O'Malley[16] to treat autoionizing atomic states, consists in diagonalizing the two electron hamiltonian $\mathcal{H}(1,2,R)$ by dividing the configuration space, for a given molecular symmetry, in two subspaces :

i) in one subspace, hereafter named $\mathcal{P}$, we consider configurations such that one electron stays on the bonding $\sigma_g 3s$ Na$_2^+$ orbital. Partial diagonalization in $\mathcal{P}$ yields regular Rydberg series converging to the Na$_2^+$ ground state X. For all the states $|P_i^k\rangle$ of a series k, the ionization energies are obtained by the Rydberg formula :

$$\langle P_i^k | \mathcal{H}(1,2,R) | P_j^k \rangle - E_X(R) = \delta_{ij} \mathcal{R}/(n_i^k - \mu_k(R))^2 \quad (4)$$

In (4), $E_X(R)$ is the energy of the Na$_2^+$ ground state, $\mathcal{R}$ is the Rydberg constant, n_i^k an integer number increasing by one unit from the state $|i\rangle$ to the next upper one, and $\mu_k(R)$ is indeed independent of the index i. Such states correspond to a Rydberg electron moving in the field of a ground state Na$_2^+$ core.

ii) in another subspace, hereafter named $\mathcal{Q}$, we consider doubly-excited configurations, such that both electrons occupy excited Na$_2^+$ orbitals. Partial diagonalization in $\mathcal{Q}$ yields doubly-excited potential curves, corresponding to Rydberg series converging to an excited state of Na$_2^+$ or more generally to a linear combination of such excited states.

In order to avoid an incorrect dissociation, we have modified the subspace $\mathcal{Q}$ into $\mathcal{Q}'$ by orthogonalization to the lowest state obtained in (i). For most symmetries, this procedure causes a minor modification of the Q_j states. An example of such states is displayed in Fig.3 for the $^3\Sigma_u^+$ symmetry.

After this block diagonalization, the electronic hamiltonian $\mathcal{H}(1,2,R)$ has non-zero matrix elements between a state of $\mathcal{P}$ and a state of $\mathcal{Q}$. As the states of $\mathcal{P}$ are grouped in Rydberg series, there exists[17] a scaling law for such matrix elements :

$$\langle P_i^k | \mathcal{H}(1,2,R) | Q_j \rangle = (n_i^k - \mu_k(R))^{-3/2} \mathcal{V}_{jk}(R) \quad (5)$$

where the reduced interaction $\mathcal{V}_{jk}(R)$ may be considered as a constant along a Rydberg series. Examples for this scaling law are displayed in table 1.

Table 1. Variation of the reduced interaction between the second doubly-excited state of $^3\Sigma_u^+$ symmetry and two levels of the "f" Rydberg series in the subspace P.

R = 5au	$n_i - \mu_i(R)$	3.90	4.90
	$\mathcal{V}_{jk}(R)$	0.048	0.046
R = 6au	$n_i - \mu_i(R)$	3.88	4.89
	$\mathcal{V}_{jk}(R)$	0.044	0.042
R = 7au	$n_i - \mu_i(R)$	3.80	4.81
	$\mathcal{V}_{jk}(R)$	0.049	0.044
R = 8au	$n_i - \mu_i(R)$	3.78	4.78
	$\mathcal{V}_{jk}(R)$	0.048	0.043

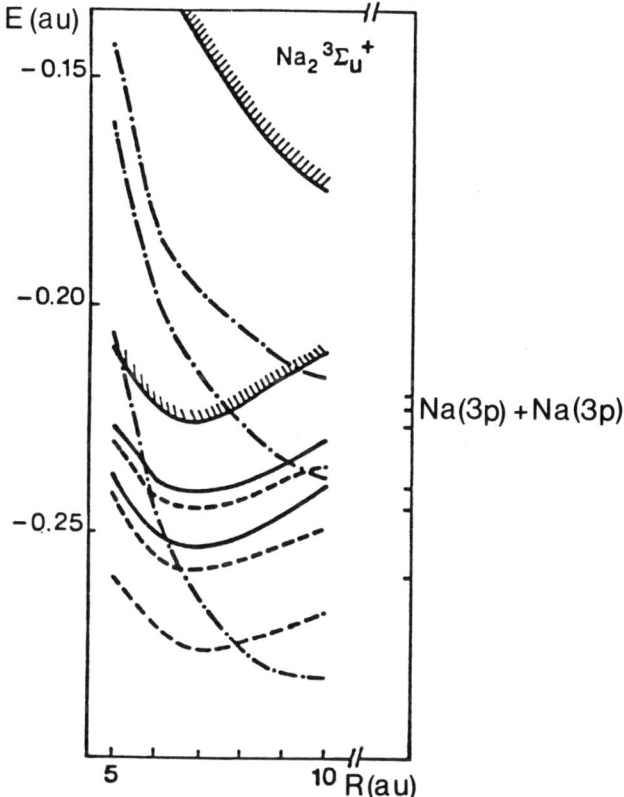

Figure 3. Diabatic potential curves for Na$_2$ $^3\Sigma_u^+$. dashed lines, "p" Rydberg series ; solid lines "f" Rydberg series ; dash-dotted line, doubly excited states. The shadowed curves correspond to the ground and first excited states of Na$_2^+$.

The associative ionization can therefore be understood as a two electron process, where in the initial state one electron occupies an excited antibonding Na$_2^+$ orbital, the other electron occupying a Rydberg orbital. *The transfer of the* *inner electron towards the ground state Na_2^+ orbital yields enough energy (>2.2eV at equilibrium) to enable the second electron to escape.*

We have considered separately two groups of doubly-excited states, hereafter labelled A and B. The group A contains the second doubly-excited state of each symmetry which, for internuclear distances R > 10 au, is mainly described by configurations containing two of the four orbitals correlated to Na^+ + Na (3p). The group B contains the first excited state of each symmetry, which for R > 10 au is mainly described by a configuration involving the $\sigma_u 3s$ orbital. The corresponding potential curves are displayed in Figs.4 and 5 respectively.

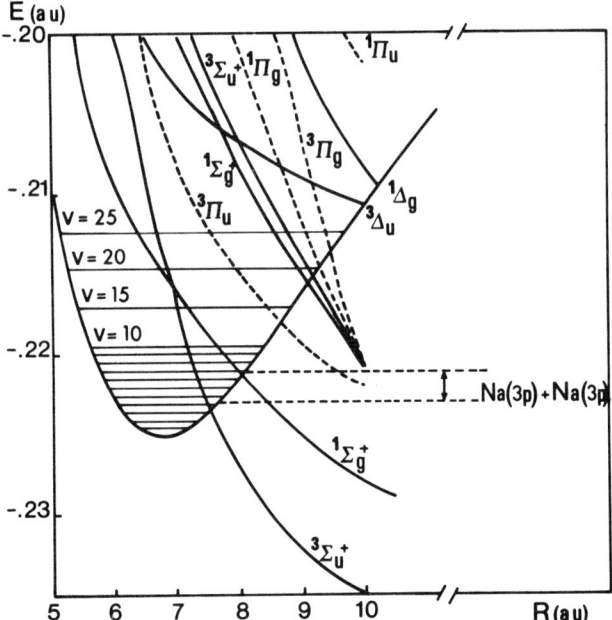

Figure 4. Doubly-excited states of Na_2 correlated to the 3p+3p dissociation limit.

4. Treatment of the dynamical problem : the autoionization region

Following Giusti[7], the autoionisation of a molecule can be described in the framework of a MQDT treatment. Let us consider the ionization of the doubly-excited state Q_j resulting into a continuum state, with energy ε, corresponding to the extrapolation of the Rydberg series k :

$$(Na_2^{**})_j^l \rightarrow (Na_2^+ (X^2\Sigma_g^+), v, l) + e(\varepsilon))_k . \quad (6)$$

We assume that the angular momentum $\hbar\sqrt{l(l+1)}$ of the heavy particle motion is conserved. The cross-section may be evaluated, in the weak coupling approximation, by :

$$\sigma_{jk}(E) = \frac{\pi}{2\mu E} r \sum_{l=0}^{l_{max}} (2l+1) a_{jk}^l (E) . \quad (7)$$

In Eq(7), E is the collision energy, r the multiplicity ratio between the final and the initial state, l the partial wave and $a_{jk}^l (E)$ a dimensionless quantity such that :

$$a_{jk}^l (E) = \sum_{v=0}^{v_{max}} \frac{4 |\xi_{vl}^{jk}|^2}{[1 + \sum_w |\xi_{wl}^{jk}|^2]^2} \quad (8)$$

$$\xi_{vl}^{jk} = \pi \int \chi_{vl}(R) \, \nu_{jk}(R) \, f_j^l(E,R) \, dR . \quad (9)$$

In (8), the summation is over the vibrational levels of Na_2^+ which may be populated through the reaction (6) at energy E. In (9), $\chi_{vl}(R)$ is the vibrational wavefunction in the final state, $\nu_{jk}(R)$ the reduced electronic interaction defined in Eq(5) and $f_j^l(E,R)$ the radial wavefunction associated with the relative motion of the two atoms, interacting with the potential $Q_j(R)$, for given energy E and angular momentum $\hbar\sqrt{l(l+1)}$. As the quantity $\nu_{jk}(R)$ is a slowly varying function of R (see table 1), the autoionization process finally depends upon quantities which can be estimated as :

$$\xi_{vl}^{jk} = \pi \, \overline{\nu_{jk}} \int \chi_{vl}(R) \, f_j^l(E,R) \, dR . \quad (10)$$

In (10) $\overline{\nu_{jk}}$ is an average value and the integral is merely the *overlap* between the radial wavefunctions describing the nuclear motion in the initial and in the final state.

The autoionization probability is thus controlled both by the strength of the electronic coupling and by the Franck-Condon factor for the initial and final nuclear wavefunctions. Such factors depend markedly upon the relative position of the Na_2^+ ground state potential curve X and of the doubly-excited curve Q_j. *Even in the case of a strong electronic coupling, the autoionization process can hardly take place for any value of the vibrational quantum number v when the two curves are not crossing. It is also negligible for the vibrational numbers v $\leq$ 3 when the crossing is located at internuclear distances R > 8 au.*

We may therefore predict a strong selectivity of the autoionization process at low collision energy E in favour of molecular symmetries leading to a doubly-excited diabatic curve presenting a crossing with the Na_2^+ ground state potential curve close to its minimum.

If we consider the group A of states, displayed in Fig.4, we see that one of the two $^3\Sigma_u^+$ curves, hereafter labelled Q_A, crosses the X curve at R = 7.5 au, whereas a $^1\Sigma_g^+$ curve, hereafter labelled Q_B, crosses the X curve at R = 8.1 au. The autoionization via a $^3\Pi_u$ state requires a collision energy E $\geq$ 0.14 eV. We may therefore conclude that for E < 0.14

eV, the autoionization proceeds preferentially via Σ molecular states. For a collision energy E < 0.054 eV, the $Q_A \, {}^3\Sigma_u^+$ channel is the only possible one. Such a conclusion is in agreement with the experimental findings[4,5], and with the analysis of Jones and Dahler[18] (see below), provided that we may assume an adiabatic correlation of the Q_A state to a state of the separated atoms with two σ orbitals. The ${}^3\Sigma_u^+$ states of the group A are constructed mainly with the $(\sigma_g 3p \, \sigma_u 3p)$ and $(\pi_g 3p \, \pi_u 3p)$ configurations. There is a crossing at R = 13 au between the energy curves of those two configurations, the first one being lower at large distances. The comparison with the experimental results could indicate that this crossing is adiabatic.: however, a correct treatment of the population sharing from the states of the separated atoms to the molecular states must be preformed. In the case of the ${}^1\Sigma_g^+$ curve, the discussion is even more complicated, as four configurations $(\sigma_g 3p)^2 \, (\sigma_u 3p)^2 \, (\pi_g 3p)^2$ and $(\pi_u 3p)^2$ should be considered.

At threshold, when in the group A of states only the ${}^3\Sigma_u^+$ Q_A state can be populated, it is possible to discuss the energy behaviour of the autoionization cross-section (7). When the collision energy E becomes negligible compared to the potential energy $Q_j(R) - Q_j(\infty)$ in the region R < 8 au where the integral (9) is computed, the nuclear wavefunction $f_j^l(R)$ is the product of an energy independent function by a $E^{-1/8}$ normalization factor. The energy dependence of the $a_{jk}^l(E)$ through the Franck-Condon factors is then weak. The number $v_{max}+1$ of vibrational levels that can be populated stays equal to 4. The energy variation of the cross-section in (7) is controlled by the E^{-1} factor and by the energy dependence of the maximum partial wave l_{max}. The behaviour predicted by Julienne (see (3)) is then verified. However, assuming a statistical population of the Q_A state of $\frac{1}{12}$ from the Na(3p) + Na(3p) limit, and noting that the r factor in (7) is $\frac{4}{3}$, we find, when the Q_A curve is arbitrarily extrapolated to the separated atom limit, a cross section of 12 Å^2 at a collision energy E of 0.03 eV and 18 Å^2 at 0.05 eV, while the experimental cross section[2,3] is of the order of 1 Å^2. In our calculations l_{max} is the maximum rotational number for a given vibrational level of the molecular ion. The presence of a small long-range barrier would considerably change the results.

We also predict a selective population of the levels v=2 and v=3 which contain 65% of the total population at 0.03 and 0.05 eV and more than 90% at 0.01 eV when the level v=4 can no longer be populated.

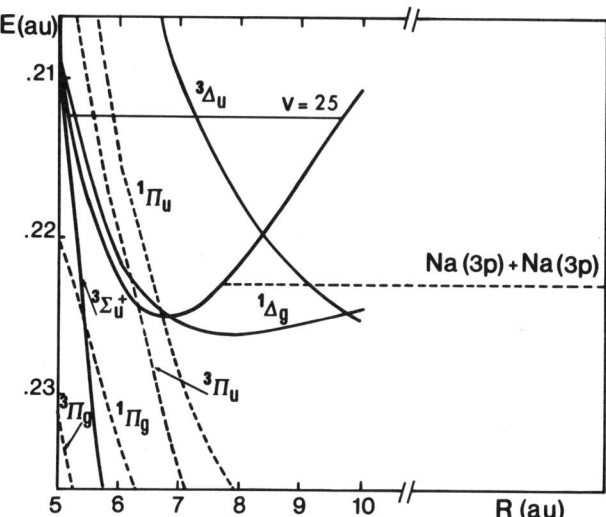

Figure 5. Doubly-excited states of Na$_2$ corresponding for each symmetry to the lowest state in the subspace Q. They are correlated at infinity to an ionic curve. The ${}^1\Sigma_g^+$ curve is lower and not visible in the figure.

Considering now the group B of states (see Fig.5) we see that provided they may be populated from the Na(3p) + Na(3p) dissociation limit, the ${}^1\Delta_g, {}^3\Delta_u, {}^1\Pi_u$ and ${}^3\Pi_u$ doubly-excited states could also contribute to the ionization process. For instance, at short distances, the ${}^3\Delta_u$ state is mainly represented by the $(\pi_g 3p \, \pi_u 3p)$ configuration, which has a lower energy than the $(\sigma_u 3s \, \delta_g 3d)$ configuration. Although this situation is reversed at large distances, we should note that the states of the group B should be correlated to ionic Na$^+$-Na$^-$ states, the Na$^-$ ion being excited for Π and Δ states. Due to the small value of the electron affinity for such states[19], ionic-covalent crossings could be expected in the region 9-10 au.

A detailed treatment of the correlation between the atomic representation at large R and the diabatic molecular representation at small R is therefore required

5. Discussion

It is thus possible to treat the autoionization of the Na$_2$ molecule by considering the interaction of dissociative doubly-excited states with mono-excited Rydberg series. Several points of this treatment still need to be refined :
i) the potential curves are required to better accuracy. *Ab initio* calculations by W. Meyer and colleagues in Kaiserslautern offer hope for an improvement in the future.
ii) the MQDT evaluation of the **K** matrix should be extended to second order, including the indirect ionization process via vibrational autoionization of a monoexcited Rydberg state :

Na(3p)+Na(3p)→(Na$_2$)** →Na$_2^*$(v$_1$,l) → Na$_2^+$ (v,l) + e   (11)

Such work is presently in progress.

Nevertheless, the qualitative conclusion that a $^3\Sigma_u^+$ doubly-excited state is mainly responsible for the autoionization process is consistent with the experimental finding that, for collision energies $E \simeq 0.1\text{-}0.3$ eV, the reaction (2) is favoured when both atoms are prepared in a $|\frac{3}{2}\frac{1}{2}\rangle$ Zeeman sublevel. Indeed, assuming a sudden decoupling of the spin at large internuclear distances, the $|Jm_J\rangle$ states decouple into $|Lm_L\rangle|Sm_S\rangle$ states so that the $|\frac{3}{2}\frac{3}{2}\rangle$ preparation of both atoms yields only $\pi 3p$ orbitals, whereas the $|\frac{3}{2}\frac{1}{2}\rangle$ preparation may yield two $\sigma 3p$ orbitals. By analysis of the variation of the ion signal as a function of the polarisation angle when the atoms are excited by linearly polarized laser light, Jones and Dahler[18] have shown that the contribution of a $^3\Sigma_u^+$ (σ^2) state is dominant, and that $^1\Sigma_g^+$, $^1\Pi_u$ and $^3\Pi_u$ states are also likely to contribute to the ionization process. Moreover Wang et al[20] have interpreted their experimental cross-sections by considering the ionization via a $^3\Delta_u$ channel.

However, at lower collision energies, the hypothesis of a sudden decoupling of the spin is no longer justified. The connection between the atomic states representation in the asymptotic region and the diabatic molecular representation in the autoionization region has to be correctly treated, including spin-orbit effects. Long range crossings between the potential curves are predicted both by Bussery and Aubert-Frecon[21] and by Kowalczyk[22] and could contribute to the sharing of the population. AI cross-sections should then be written as a sum over various channels:

$$\sigma = \sum_j \alpha_j(E) \, \pi (b_{max}^j)^2 P_j(E) \qquad (12)$$

with $b_{max}^j = l_{max}^j / \sqrt{2\mu E}$

where P_j is the autoionization probability for the doubly-excited state Q_j, b_{max}^j the maximum impact parameter for this channel, and $\alpha_j(E)$ the population of this state. *At low collision energy, both α_j and b_{max}^j might depend upon long range effects.*

We may conclude that while the molecular autoionization seems to be a reasonably understood problem, future work on the associative ionization reaction between two alkalis, in connection with the experiments involving ultra cold atoms, *should focus on the treatment of the population transfer from the atomic states to the molecular autoionizing states.*

[1]Present address : Laboratoire des interactions ioniques et moléculaires. Université d'Aix-Marseille 1 - Centre Saint Jérôme, 13397 MARSEILLE Cedex 13

Acknowledgements

The authors wish to acknowledge a stimulating collaboration with A Giusti-Suzor and O.Dulieu as well as discussions with H.A.J. Meijer, R.Morgenstern, J. Weiner and R. Mc Carroll. They thank S.Sandmeier for her help in preparing the manuscript

References

1. D.R. Bates and M.S.W. Massey, Proc.Roy.Soc.(London) A192 1 (1947)
2. J. Huennekens and A. Gallagher, Phys.Rev.A28 1276 (1983)
3. R. Bonanno, J. Boulmer and J. Weiner, Phys.A 28 604 (1983)
   R. Bonanno, J. Boulmer and J. Weiner, Comments At.Mol.Phys. 16 109 (1985)
4. J.G. Kircz, R. Morgenstern, and G. Nienhuis, Phys.Rev.Lett. 48 610 (1982)
5. M.X. Wang, J. Keller, J. Boulmer and J. Weiner, Phys.Rev.A 34 4497 (1986)
6. P.S. Julienne, Phys.Rev.Lett. 61 698 (1988)
7. A. Giusti, J.Phys.B 13 3867 (1980)
8. A. Dalgarno, Atomic physics, vol.4 (Plenum Press, New York 1975) pp 325-334
9. C. Bottcher and A. Dalgarno, Proc. Roy Soc Lond. A340 187 (1974)
10. M. Klapisch, Comput.Phys.Commun.2 239 (1971)
11. A. Henriet, J.Phys.B.18 3085 (1985)
12. A. Henriet, C. Le Sech and F. Masnou-Seeuws, Chem.Phys.Lett.158 389 (1989)
13. A.J. Taylor, K.M. Jones and A.L. Schawlow, Opt.Commun. 39 47 (1981)
14. A.Henriet and F. Masnou-Seeuws, J.Phys.B.21 L339, (1988) and J.Phys.B. in press (1989)
15. G.H. Jeung, Phys.Rev.A35 26, (1987)
16. T. O'Malley, Adv.At.Mol.Phys. 7 223 (1971)
17. M.J. Seaton, Rep.Prog.Phys. 46 167 (1983)
18. D.M. Jones and JS Dahler, Phys.Rev.A35 3688 (1987)
19. O. Dulieu, Z.Phys.D : Atoms,.Molecules and Clusters 13 17 (1989)
20. M.X. Wang, J. Keller, J. Boulmer and J. Weiner, Phys.Rev.A35 934 (1987)
21. B. Bussery and M. Aubert-Frecon, J.Chem.Phys.82 3224 (1985) and private communication
22. P. Kowalczyk, J.Phys.B.17 817 (1984)

COLLISIONAL LOSS MECHANISMS IN LIGHT-FORCE ATOM TRAPS

T. G. Walker, D. W. Sesko, C. Monroe, and C. Wieman

Joint Institute for Laboratory Astrophysics, University of Colorado and National Institute for Science and Technology, and Department of Physics, University of Colorado, Boulder, CO 80309-0440

Evidence for atom-atom collisional loss mechanisms in spontaneous-force optical traps is presented. Ground-state–ground-state collisions involving hyperfine state changes dominate the loss at low trap-laser intensities. When the trap depth is greater than the hyperfine energy, however, the ground state collisions are ineffective, and collisions involving excitation of atom pairs by the trapping light dominate the loss of atoms from the trap. The addition of a "catalysis laser" allows a study of these collisions as a function of detuning and shows the novel effects of spontaneous emission of light during the collisions. A simple model proposed by Gallagher and Pritchard for the latter loss mechanism is explained and compared to the experimental results.

Traps for neutral atoms have recently been made with depths of $h\Delta_T \sim 1K$ and atom temperatures of $T < 10^{-3}K$.[1,2] Under these conditions the dominant collisional limitations on the achievable atom densities are inelastic collision mechanisms which transfer sufficient kinetic energy for the atoms to leave the trap. The low temperatures in these traps adds novel features to these collisions, such as spontaneous emission during the collision[3] and modification of the collision dynamics by the presence of the relatively weak laser field.[4] At room temperature, radiative lifetimes (10^{-8}s) are much longer than collision times (10^{-12}s), and the effects of strong light fields are only important for intensities obtainable in pulsed lasers.[5] In contrast, at millikelvin temperatures and below, collision times are comparable to radiative lifetimes. Thus we expect studies of collisions in traps to reveal new and interesting features not amenable to conventional techniques.

We have studied the collisional loss rates for Cs atoms in such a trap. We attribute the losses to hyperfine-state changing collisions between two ground-state atoms (when the trap depth is less than the ground-state hyperfine energy) and to collisions involving excitation of atom pairs by the light fields. We present data which shows the importance of spontaneous emission of light on the latter mechanism.

Our trap is of the Zeeman-tuned spontaneous-force type,[2] which is static and has both trapping and damping forces. The trapping light is produced by a diode laser stabilized by optical feedback and locked 5Mhz to the red of the $6S_{1/2}(F=4) - 6P_{3/2}(F'=5)$ transition. For the experiments described here, adequate loading of the trap was obtained by simply trapping the low-velocity tail of the velocity distribution in a

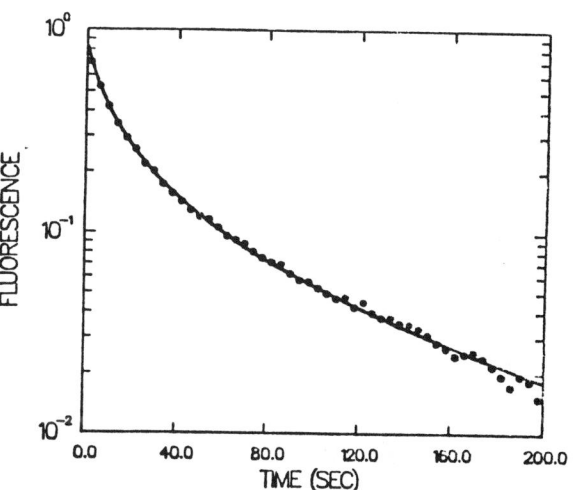

Figure 1: Decay of the fluorescence from the trapped atoms as a function of the time after the atomic beam is blocked. The solid line is the fit to the equation $dF/dt = -\beta_F F^2 - \Gamma F$.

thermal Cs atomic beam. In this way we were able to load 4×10^4 atoms into the trap, with resulting densities of $4 \times 10^9 cm^{-3}$. Although we have loaded as many as 4×10^8 atoms into our trap, with more than 4×10^4 the spatial distribution of the atoms depends on the number in the trap.[6]. Thus to obtain reliable data it was necessary to work with the relatively small number of atoms. After loading the atoms into the trap, we blocked the atomic beam and observed the fluorescence from the trapped atoms (which is proportional to the Cs density n) as it decreased with time. As shown in Fig. 1, the decay of the density was well-characterized by the equation $dn/dt = -\beta n^2 - \Gamma n$, where Γ is the loss rate due to collisions with the background gas in the vacuum chamber, and β is the rate coefficient for loss of atoms due to collisions between atoms in the trap. Similar behavior was previously observed by Prentiss et al.,[7] but their studies of the mechanisms involved were inconclusive. For more experimental details, see Ref.[8] and Ref.[9]. In this paper we will concentrate on the mechanisms responsible for β.

Fig. 2 shows the measured trap-loss rate coefficient as a function of the intensity of the trapping light. We

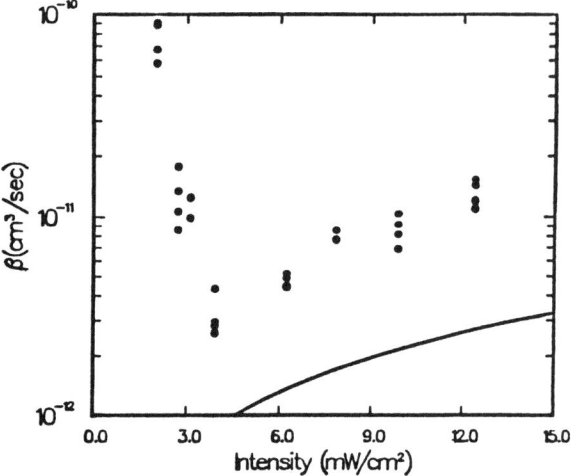

Figure 2: Dependence of the measured trap-loss rate coefficient β on the light intensity in the trap. The solid line is the prediction of the Gallagher-Pritchard model.

account for this unusual intensity dependence as follows. For low intensities ($I < 4\text{mW}/\text{cm}^2$) β decreases rapidly with increasing intensity. We believe this shows the suppression of the loss of atoms caused by ground-state collisions as the depth of the trap is increased. Since our trap operates on the upper ($F = 4$) hyperfine level of the ground state, one of the atoms of a colliding pair may change to the $F = 3$ level during the collision, converting the hyperfine energy (.5K for Cs) to kinetic energy. If the trap depth Δ_T is less than the hyperfine energy, the atoms leave the trap. When the trap becomes sufficiently deep, the atoms which are accelerated by these collisions are not removed from the trap. This indentification is supported by a calculation of the expected rate coefficient for this process:

$$\beta_{HF} = k_L P_{HF} = \frac{9\pi}{4}\sqrt{\frac{3T}{\mu}}\left(\frac{6C_6}{T}\right)^{1/3} P_{HF} \quad (1)$$
$$= P_{HF} \times 2.2 \times 10^{-10} \text{cm}^3/\text{s}$$

at $T = 300\mu\text{K}$ (the measured temperature of our trap). This is the Langevin rate coefficient[10] multiplied by the probability P_{HF} of changing the hyperfine state. Here μ is the reduced mass of the Cs pair and $-C_6/R^6 = -4600\text{eV}\text{Å}^6/R^6$ is the energy[11] of two ground-state Cs atoms separated a distance R. A value of $P_{HF} = .3$ would reproduce our results. It is clear that these collisions are analogous to the corresponding phenomena at higher temperatures. We will therefore devote the rest of this paper to the more interesting excited-state collisions. We note, however, that the suppression of the loss due to hyperfine-state changing collisions was only obtained with careful alignment of the trapping lasers as well as improvement of the spatial quality of the lasers. With poor alignment or spatial profiles, the trap loss was independent of intensity, being dominated by the hyperfine-state changing collisions.

Fig. 2 shows that once the contribution of ground-state collisions to β has been suppressed by increasing the trap depth ($I > 4\text{mW}/\text{cm}^2$), β begins to increase with increasing intensity. This suggests that a process involving one excitation of a Cs-Cs pair by the laser is dominating the trap loss. Gallagher and Pritchard[3] have proposed a simple model for such a process which is explained in Fig. 3.

To begin, we note that the conventional concept of an excited state collision beginning at $R \sim \infty$ is invalid at the temperatures under consideration here. At $300\mu\text{K}$ a Cs atom travels only 100Å per atomic lifetime. Thus in order for the two atoms to reach small internuclear separations where $> \hbar\Delta_T \sim 1\text{K}$ of energy transfer can take place, the atoms must begin close enough that the excited state potential can accelerate them into small R before radiating. We conclude that instead of considering the collision as beginning with one atom excited at essentially infinite separation from the unexcited atom, we must consider the collision as starting with the atoms unexcited at some internuclear separation R_0. They absorb a photon at R_0, and then are accelerated towards each other by the relatively strong excited state potential. If they reach the small R region without radiating, they have some probability $P_{\Delta E}$ of having sufficient energy transfer occur to leave the trap. We find it useful to think of this entire process in four steps, as shown in Fig. 3.

Following Gallagher and Pritchard, and referring to Fig. 3, we calculate the collisional loss rate by calculat-

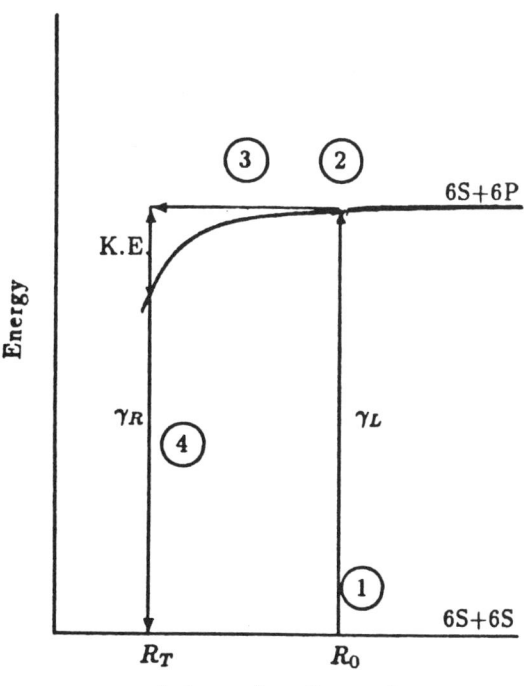

Figure 3: The Gallagher-Pritchard model involves four steps. 1) Two Cs atoms at internuclear separation R_0 are excited by the laser field to the first excited state. 2) They are accelerated by the R^{-3} potential. 3) If they reach the region of small R without radiating, and 4) radiate at R_T or closer, for example, they pick up sufficient kinetic energy to leave the trap.

ing the density of atom pairs at $R = R_0$, multiplying by the excitation rate of the pairs at R_0, multiplying by the probability of reaching small R without radiating and finally multiplying by the probability of energy transfer:

$$\beta n^2 = 2 \int_0^\infty \frac{n^2 4\pi R_0^2 dR_0}{2} \mathcal{R}(R_0, \Delta_L) \times \exp\left(-t(R_0)/\tau_M\right) P_{\Delta E}. \qquad (2)$$

The first factor of 2 is necessary because two atoms leave the trap per "incident," $n^2 4\pi R_0^2/2 dR_0$ is the density of atom pairs between R_0 and $R_0 + dR_0$, $\mathcal{R}(R_0, \Delta_L)$ is the excitation rate for pairs at $R = R_0$ due to the laser detuned Δ_L from the atomic transition, $\tau_M = \tau_A/2$ is the natural lifetime of the molecular state, and the exponential factor gives the probability of reaching small R without radiating.

The excitation rate is computed semi-classically, assuming a Lorentzian lineshape for the molecular excitation rate:

$$\mathcal{R}(R_0, \Delta_L) = \frac{I}{hc/\lambda} \frac{\lambda^2/2\pi}{1 + 4(\Delta - \Delta_L)^2/\Delta_M^2} \qquad (3)$$

where I is the intensity, $\Delta = -C_3/hR^3$, and $\Delta_M = (2\pi \tau_M)^{-1}$ is the molecular linewidth. The potential curve is taken to be of the form $-C_3/R^3$, which neglects the fine[12] and hyperfine structure. With this assumption the time for an initially stationary pair at R_0 to reach $R = 0$ is

$$t(R_0) \approx .746 \sqrt{\frac{\mu R_0^5}{2 C_3}}. \qquad (4)$$

The probability of energy transfer $P_{\Delta E}$ is calculated as follows. There are two processes which can transfer the $> h\Delta_T$ kinetic energy required for the atoms to leave the trap: radiative redistribution of light, and fine-structure changing transitions. The radiative redistribution process requires the atoms to radiate at internuclear separations less than $R_T = (C_3/h\Delta_T)^{1/3} = 105$Å, where the kinetic energy picked up by the atoms exceeds the trap depth. Since the nuclear kinetic energy is conserved after radiating, the atoms leave the trap. The probability of radiating once having reached R_T is

$$P_R = \frac{2\sqrt{2} C_3^{1/3} m^{1/2}}{5 \tau_M (h\Delta_T)^{5/6}} = .027 \qquad (5)$$

for Cs.

The probability for fine-structure changing can be estimated from measurements at high temperatures.[14] The cross-section for fine-structure changing collisions has been measured to be 20Å^2 at 350K compared to the cross-section, $\pi R^2 \simeq 200$Å^2, to be expected for reaching the fine-structure recoupling region ($R \sim 8$Å). This implies a probability of $P_{\Delta J} \sim 0.1$ for changing fine-structure state at room temperature. In the Landau-Zener approximation,[15] $P_{\Delta J} = 2p(1-p)$, where $p = \exp(-2\gamma)$ and γ is the adiabaticity parameter. At 350K the data give $\gamma = 1.47$. Since γ is inversely proportional to the velocity, we then estimate

$$\gamma = 1.47 \sqrt{\frac{E_{FS}/2 + k_B \times 350\text{K}}{E_{FS}/2}} = 2.01$$

at $T = 0$K, where $E_{FS} = 555$cm^{-1} is the Cs fine structure splitting. This value of γ gives $P_{\Delta J} = .035$. The total probability of gaining enough energy to leave the trap is then $P_{\Delta E} = P_{\Delta J} + P_R = .062$.

The result (2) should properly be averaged over the different potential curves, including the fine and hyperfine structure, as well as averaging over the initial velocity distribution of the atoms. An additional effect considered by Gallagher and Pritchard are the multiple traversals of the small R region when the probability of spontaneous emission is small, but for the data presented here this has little effect. The initial velocity distribution has been included in the solid line of Fig. 2, but a single potential curve has been assumed. The resulting prediction of Equation (2) is low by about a factor of 4.5. In view of the large number of approximations and simplifications used, this is a quite reasonable result. Also, since the time to reach small R is sensitive to the potential at R_0, the model is quite sensitive to the value of the C_3 coefficient.

An important feature of Equation (2) is that it simplifies greatly when the detuning is much greater than the molecular linewidth, and the relevant interesting physics is simpler to understand. At large detunings, the excitation rate is strongly peaked about the laser frequency while the other factors in Equation (2) vary slowly. This is the quasi-static limit, and, again ignoring hyperfine structure, a simple equation for β results:

$$\beta = \frac{I\lambda^3 P_{\Delta E} C_3 \Delta_M}{12\pi\hbar^2 c \Delta_L^2} \exp\left(-(\Delta_\tau/\Delta_L)^{5/6}\right) \quad (6)$$

where $\Delta_\tau = 246$ MHz (for Cs) is the detuning at which the time to reach $R = 0$ is equal to the molecular lifetime. Equation (6) predicts that for large laser detunings β will decrease as Δ_L^{-2}. Physically, this is because of the decreasing density of atom pairs at smaller internuclear separations. With smaller detuning β increases until the detuning comes close to the atomic line. Then the probability of spontaneous emission before reaching the small R region becomes important, and the trap-loss coefficient again begins to decrease.

In order to check this behavior, we irradiated the trapped atoms with an additional laser ("catalysis laser") and observed the resulting change in β. Care was taken that the catalysis laser did not affect the position or the size of the atom cloud, and therefore did not affect the trap depth. The experimental results for β as a function of the detuning of the catalysis laser are shown in Fig. 4. Outside the excited-state hyperfine structure, where the assumption of a single R^{-3} potential function is best, the predictions of Equation (6) agree with the data to better than a factor of 2. Inside the hyperfine structure, we see the qualitative behavior predicted by (6), but quantitative comparisons are made difficult due to the hyperfine structure. Nevertheless, the data clearly shows the importance of spon-

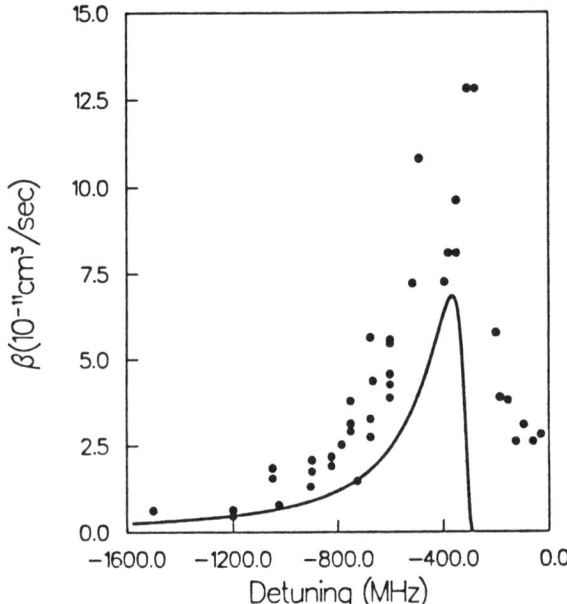

Figure 4: Dependence of the measured trap-loss rate coefficient β on the detuning of the catalysis laser (intensity 24mW/cm^2). The solid line is the prediction of the Gallagher-Pritchard model, neglecting the excited state hyperfine structure. For this purpose the center of gravity of the hyperfine structure has been chosen as the zero of detuning.

taneous emission during these collisions. The conventional concepts of excited-state collisions must be revised in order to understand the data of Fig. 4.

Since the deBroglie wavelength of a Cs atom at 300μK is 140Å, which is of the order of the range of the potential, one might wonder why the assumption of classical motion is valid. The reason for this is that the relevant deBroglie wavelength is determined by the wavelength for the motion in the excited state, which is greatly reduced by the steep potential, except for small detunings. Thus the main effect of the finite deBroglie wavelength is to broaden the absorption lineshape.

It is also of interest to investigate any intensity saturation properties of the trap-loss coefficient. As Gallagher and Pritchard point out, the line broadening due to atomic motion should dramatically increase the saturation intensity for these molecules over the saturation intensity for isolated Cs atoms. We have attempted to measure the saturation intensity by looking for a saturation in β as the intensity of the catalysis laser was varied, but observed no saturation. Fig. 5 shows the data for 600MHz detuning, with no discernable saturation at intensities of up to 50mW/cm^2. We have made similar measurements at a variety of other detunings with similar results. We note that this measurement is made difficult by the need to not affect the trap parameters with the catalysis laser. For every detuning we find that the trap is affected before saturation becomes important.

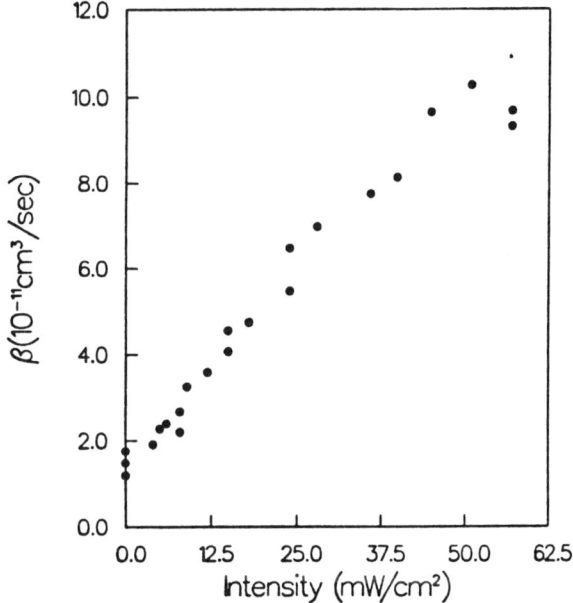

Figure 5: The trap-loss coefficient β as a function of the intensity of the catalysis laser shows no saturation characteristics.

Although the Gallagher-Pritchard model gives satisfactory agreement with experiment, there are a number of issues which need to be addressed. First of all, effects of multiple excitations and deexcitations by the trapping lasers on the collision dynamics have been ignored.[4] This is likely to effect the predictions in the small-detuning case where the initial velocity distribution is important. Secondly, the myriad of potential curves which exist have likewise been ignored. Since the time required to reach small R depends sensitively on the potential at R_0, the details of the potential curves are of importance. A third simplification is the assumption of the Lorentzian excitation rate (3). This neglects any velocity dependences and line broadening. Again, this is not expected to affect the trap-loss coefficient at large detunings where only the integrated cross-section is important, but at small detunings where the trap-loss is dominated by absorption in the wings of the lineshape these effects may be expected to be significant.

We note that one other group[7] has observed collisional loss in a Na trap, but concluded that the rate coefficient was independent of intensity. They rejected ground-state–ground state, ground-state–excited-state and excited-state–excited-state collisions as possible mechanisms for the observed trap loss, and concluded that the loss rate depended only on the total atomic density. We cannot reconcile their results with ours, except to note that only upon very careful alignment of the trapping lasers and improvement of the spatial profile of the lasers were we able to achieve sufficient trap-depths to suppress the hyperfine-changing collisions. With poorer beam quality or alignment we observe only the hyperfine-changing collisions, which are independent of laser intensity. In addition, if too many atoms are in the trap, the contraction of the cloud as the number of atoms decreases makes the interpretation of the experiments difficult.

There are some important implications of this work for the design and operation of this and similar optical traps. In order to minimize the trap loss and therefore maximize the acheivable density, the optimal trap would operate out of the absolute ground level of the atom. This would eliminate the troublesome ground-state loss mechanism which for our trap requires about 4mW/cm^2 of laser power to suppress. Thus if the trap uses the ground state of the atom, the light intensity can be reduced, which also reduces the excited-state loss mechanism. This also reduces the sensitivity of the loss rate to alignment.

In conclusion, we have identified the important trap-loss mechanisms for current spontaneous-force light traps. In particular, we have found that both ground-ground and ground-excited state collisions are important in these traps. These collisions will be present in other types of traps as well. The novel effects of spontaneous emission during these collisions have been observed for the first time and we have shown that the collisions can be reliably predicted by the model of Gallagher and Pritchard.[3]

This work was supported by the Office of Naval Research and the National Science Foundation.

References

[1] D. Pritchard, E. Raab, V. Bagnato, C. Wieman, and R. Watts, Phys. Rev. Lett. **57**, 310 (1986).

[2] E. L. Raab, M. G. Prentiss, A. E. Cable, S. Chu, and D. E. Pritchard, Phys. Rev. Lett **59**, 2631 (1987).

[3] A. Gallagher and D. Pritchard, JILA preprint.

[4] P. Julienne, Phys. Rev. Lett. **61**, 698 (1988).

[5] P. D. Kleiber, K. Burnett, and J. Cooper, Phys. Rev. Lett **47**, 1595 (1981).

[6] D. Sesko, T. Walker, and C. Wieman, to be published.

[7] M. Prentiss, A. Cable, J. E. Bjorkholm, S. Chu, E, L. Raab, and D. E. Pritchard, Opt. Lett. **13**, 452 (1988).

[8] D. Sesko, T. Walker, C. Monroe, A. Gallagher, and C. Wieman, JILA preprint.

[9] D. Sesko, C. G. Fan, and C. Wieman, JOSA B **5**, 1225 (1988).

[10] T. Su and M. T. Bowers, in *Gas Phase Ion Chemistry*, ed. M. T. Bowers, (Academic Press, London, 1979).

[11] A. Dalgarno and W. D. Davidson, Adv. Atom. Mol. Phys. **2**, 1 (1966).

[12] M. Movre and G. Pichler, J. Phys. B **10**, 2631 (1977).

[13] M. Abramowitz and I. Stegun, Natl. Bur. Std. (U. S.), Appl. Math. Ser. **55** (1964).

[14] A. Gallagher, private communication.

[15] N. F. Mott and H. S. W. Massey, *The Theory of Atomic Collisions*, (Oxford, Clarendon Press, 1965).

ATOMIC HYDROGEN: GAS AND SURFACE COLLISIONS FOR $T \to 0$

J.T.M. Walraven

Van der Waals Laboratorium, Universiteit van Amsterdam, Valckenierstraat 67, 1018 XE Amsterdam, The Netherlands

> Hydrogen atoms are very well suited to study elastic and inelastic collisions at very low energies. Atom-atom collisions are being studied in the gas phase, where the lowest temperatures are achieved in magnetic traps. Atom-surface collisions are studied by observing the reflection of a beam of H-atoms from the surface of liquid Helium.

1 Introduction

At the end of the nineteen-seventies it became possible to create high purity samples of gaseous atomic hydrogen under cryogenic conditions. Polarization of the electron spins in a high magnetic field causes the atoms to interact pairwise via the non-binding $b\text{-}^3\Sigma_u^+$ potential and results in a stabilized gas if the temperature is sufficiently low to avoid spin flips. Further, massive surface-catalyzed recombination can be strongly suppressed by covering all surfaces exposed to the gas with a film of liquid helium which offers the weakest conceivable Van der Waals attraction to the hydrogen atoms. With the availability of these samples, intensive research has revealed fascinating new properties. The light mass and the weak triplet interaction lead to a positive internal energy down to the absolute zero of temperature and make hydrogen into *the only* quantum gas. This gas has been studied at densities ranging from 10^9 to 10^{18} atoms/cm^3 and at temperatures between 1 mK and 1 K. At present a wealth of information is available about the system although a major goal of the research, the observation of collective quantum effects such as Bose-Einstein Condensation (BEC) in the bulk gas or the Kosterlitz-Thouless Transition (KTT) in the adsorbed gas still awaits realization.

Comprehensive reviews, providing a fairly complete summary of the field, are written by Greytak and Kleppner[1] and by Silvera and Walraven.[2] Zero-field magnetic resonance[3] and the low-temperature hydrogen maser[4] have been summarized by Hardy and collaborators, spin-waves by Freed[5] and Lee[6] and the dynamics of surface adsorption and thermal accommodation by Berkhout and Walraven.[7]

In this paper emphasis is put on collisional phenomena between hydrogen atoms confined in magnetic traps and on collisions of cold hydrogen atoms with the surface of liquid helium. Before introducing the trapping experiments (section 4), some of the basic properties of spin-polarized hydrogen are summarized in section 2 and 3. The results of the trapping experiments are presented

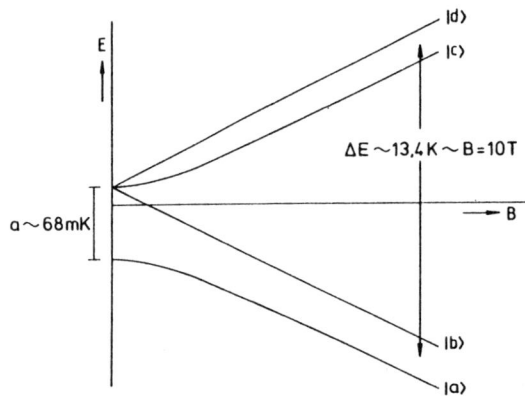

Figure 1: The hyperfine structure of the 1S manifold of the hydrogen atom.

in section 5. In section 6 it is shown that in the low energy limit of atom-surface scattering the probability of inelastic scattering is vanishing. This leads to a quantum reflection regime which is experimentally illustrated in section 7 by focussing a beam of hydrogen atoms with the aid of a liquid-helium-coated concave mirror.

2 Fundamentals

To characterize atomic hydrogen in experiments one usually refers to the state of polarization of the gas. To introduce the notation the hyperfine structure of the 1S manifold of hydrogen is shown in Fig.1. By convention the ground state hyperfine levels are labeled a, b, c and d in order of increasing energy. The b and d states are pure spin states, the a and c states are hyperfine mixed linear combinations of up and down spin states:

$$\begin{aligned}
|a> &= \sin\theta|\uparrow\Downarrow> - \cos\theta|\downarrow\Uparrow>, \\
|b> &= |\downarrow\Downarrow>, \\
|c> &= \cos\theta|\uparrow\Downarrow> + \sin\theta|\downarrow\Uparrow>, \\
|d> &= |\uparrow\Uparrow>,
\end{aligned} \qquad (1)$$

where

$$\tan 2\theta \equiv a/[\hbar(\gamma_e + \gamma_p)B] \, . \quad (2)$$

Here a is the hyperfine constant, γ_e and γ_p the electron and proton gyromagnetic ratios, $2\pi\hbar$ the Planck constant and B the applied magnetic field. The simple arrows ↑ and ↓ refer to the magnetic quantum number of the electron spins and crossed arrows ⇑ and ⇓ to that of the proton spins. Unpolarized gas is referred to as H, up or down electron-spin-polarized gases as H↑ and H↓ respectively. Further one distinguishes the doubly (both electron and proton spin) polarized gases, consisting predominantly of b-state (H↓⇑) or d-state (H↑⇑) atoms. An analogous notation is used for deuterium. Sometimes it is convenient to label the atoms by the direction of the force caused by a magnetic field gradient. For this purpose the terminology high-field-seekers (for H↓) and low-field-seekers (for H↑) is used.

The interaction between two H-atoms depends on the spin states of the atoms. Thus, the four 1S hyperfine states give rise to 16 potential curves, 11 of which are distinct in zero field (for D-D 22 out of 36). However, for most practical purposes the hyperfine interaction may be neglected and a description in terms of the X-$^1\Sigma_g^+$ and b-$^3\Sigma_u^+$ potentials is adequate. These potentials are calculated to high precision by Kolos and Wolniewicz.[8,9]

3 Stability of the gas phase

The remarkable stability of spin-polarized hydrogen gas at low temperature ($T \leq 0.5\,\text{K}$) arises from the nonbonding character of the triplet potential. Although curve crossing to the singlet potential can lead to resonance recombination this channel is unimportant in most low temperature experiments. The efficiency of room temperature recombination is due to the accessibility of several efficient resonances. At low temperatures resonances are not important. The ($v = 14$, $J = 4$) vibrational-rotational state is the only molecular state that gives rise to an observable resonance within a thermal band of energies from the dissociation limit. This is a bound state and therefore beyond cut-off in all experiments with H↑. In experiments with H↓ resonance recombination to the (14,4) state may only result from a-a collisions. The $|ab>$ pair state has a vanishing spin-projection on all para-molecular states and b-b collisions have a pure triplet character. However, also resonance recombination arising from a-a collisions may be eliminated (suppressed exponentially) by increasing the excitation energy of the resonance with a modest magnetic field ($B \geq 4\,\text{T}$). In the absence of resonance recombination, molecule formation can only result from direct three-body collisions, which at a given density are much less frequent than the two-body events. These three-body processes have been studied extensively but are not further discussed in the present context. Radiative recombination in a two-body collision is negligible as homonuclear diatomic molecules have no permanent dipole moment.

It is interesting to compare the stability of the hydrogens with that of laser cooled spin-polarized alkalis. The number of vibrational- rotational levels rapidly grows with the mass of the atoms, making it increasingly difficult to avoid the resonances. For deuterium two resonances, (21,0) and (21,1), have to be avoided, either by careful selection of magnetic field and temperature or by using doubly polarized gas. For the alkalis the singlet-resonances become virtually impossible to avoid, leaving only double polarization as a practical means to suppress molecule formation. This was realized more than 20 years ago by Bernheim[10] and Kastler[11] and first experimentally demonstrated by Alzetta et al.[12,13] using optical pumping. Vigué pointed to an important difference between hydrogen and the alkali atoms.[14] The larger polarizability and the larger mass makes that in addition to the singlet potential the $^3\Sigma_u^+$ potential also supports vibrational levels, 9 in the case of ^{6}Li$_2$ and 20 for Cs$_2$. Hence the gas has to be cooled to below the temperature corresponding to the energy of the lowest relevant orbiting resonance above the dissociation threshold. As the exact location of these resonances in the continuum is not known, at present no reliable statements concerning the low temperature stability of the alkalis may be made. However in view of the typical spacing between the resonances it may be estimated that for Li the temperature has to be reduced to at least well below 60 mK and for Cs to below 200 μK. For H such an estimate yields a temperature of 20 K. In view of the availability of powerful optical cooling methods for the alkalis these extremely low temperatures appear to be within experimental reach.

4 Trapping experiments

Most experiments with spin-polarized hydrogen have been performed with surface confined samples of H↓ in a liquid He coated environment. As mentioned in the introduction in such environments surface-catalyzed recombination may be strongly suppressed. However, for a really radical elimination of surface recombination one has to rely on surface-free confinement methods. This perspective leads to interesting parallels between trapping of hydrogen and trapping of alkalis or meta-stable inert gases, although it should be emphasized that the experimental realizations differ widely. Hydrogen can be loaded in a trap by purely conventional cryogenic techniques and is unique in this respect. It is not clear whether this is even possible for deuterium. For any other atomic system one

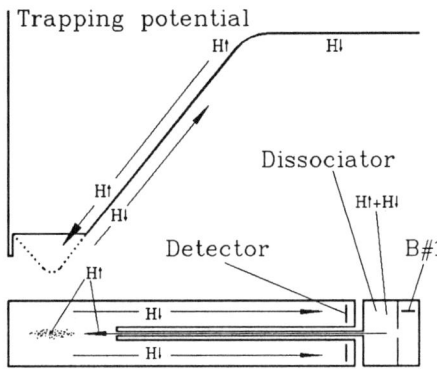

Figure 2: Principle of the trapping experiment of Van Roijen et al.. The H↑-atoms are driven to the B-field minimum. The sample is studied by observing the H↓-atoms ejected from the trap by relaxation.

has to rely on optical manipulation.

Several interesting concepts were analyzed to construct electromagnetic bottles for H-gas. The type of trap employed in the hydrogen experiments is the static minimum-B-field trap, proposed by Hess.[15,16] This type of trap is also used at NBS[17] and MIT[18] to magnetically trap optically cooled Na atoms. The static trapping concept only works for H↑ as trapping of H↓ would require a local magnetic field maximum which is not a solution of the Maxwell equations.[19] This is very unfortunate since H↓ is much stabler against electron-spin flips due to the absence of sufficient thermal energy to populate the c and d levels in a high magnetic field. To trap H↓ dynamical trapping methods have to be used. Interesting AC schemes were proposed for this purpose by Lovelace et al.[20] and Agosta et al.[21,22] but so far they were not put into practice.

For the trapping experiments in Amsterdam a static trap was constructed with a non-zero B field minimum. The principle of the experimental arrangement is illustrated in Fig.2. Both the sample cell and the effective trapping potential are shown. As in all experiments with cryogenic hydrogen all surfaces are coated with a film of liquid helium. The solid line represents the trapping field at the wall of the sample cell. The dashed line corresponds to the field along the symmetry axis. The curves are seen to coincide everywhere except in the trapping region and reflect the magnetic trapping potential. In addition the surface adsorption potential is shown which is due to the helium on the wall at the left hand side of the sample cell. The effective 'depth' of our trap is $\epsilon_{tr}/k_B = 0.92\,\text{K}$. This value is only slightly smaller than the binding energy of H on the surface of liquid ^{4}He, $\epsilon_a/k_B \approx 1\,\text{K}$, and much larger than the adsorption energy on a ^{3}He/^{4}He mixture, $\epsilon_a/k_B \approx 0.4\,\text{K}$. In comparison to the wall-confined arrangement of the experiments with H↓ this yields (for equal bulk densities) an enormous suppression of the density of adsorbed gas (σ). This is best seen from the low density expression for the surface adsorption isotherm in a form relating σ to the density in the center of the trap (n_0),

$$\sigma = n_0 \lambda_{th} \exp[(\epsilon_a - \epsilon_{tr})/k_B T] , \qquad (3)$$

where $\lambda_{th} = (2\pi\hbar^2/mk_B T)^{1/2}$ is the thermal de Broglie wavelength. Only for hydrogen the traps can be made sufficiently deep to make the exponent in equation 3 negative, implying that thermal activation is required to populate the surface states. Hence thermal equilibrium drives the atoms to the trap rather than to the surface.

In Amsterdam Van Roijen et al.[23] studied the loading and relaxational decay of H↑ down to 80 mK. The H is produced in an RF dissociator operated at $T \approx 600\,\text{mK}$ and situated in the high field region of our setup. The H↑ is guided towards the trapping region where the density will build up until a steady state is reached. The actual trapping of the H↑ results from interatomic collisions in which potential energy due to the trapping potential is converted into kinetic energy which is carried off by the walls of the sample cell. Only a few collisions per atom are sufficient to establish a thermal density distribution of the H↑ in the trapping region of the cell. Note that the H↑ atoms cannot escape the low field region due to the presence of walls and the high field barrier. The H↓ also generated in the discharge is continuously removed by forced recombination on a helium-free bolometer chip (B#1), mounted near the dissociator in the high field region. To study the trapped H↑ we measure the flux of H↓ atoms which are ejected from the trap due to dipole-dipole relaxation and spin-exchange in the sample. Detection is done with a helium free thermal detector ("pumping plate")[28] positioned at the high-field end of the sample cell to collect the ejected atoms. With this arrangement 4×10^{13} atoms could be collected in the trap at 100 mK, which corresponds to a density $n_0 = 3 \times 10^{14}\,\text{atoms/cm}^3$ in the trap center.

Trapping of H↑ was first demonstrated at MIT by Hess et al.,[24] who loaded up to 5×10^{12} atoms at temperatures as low as $T = 40\,\text{mK}$. These authors used an experimental arrangement that differs from the one shown schematically in Fig.2 in that low-field-seekers may escape from the trap on the left-hand-side if their energy is sufficiently high to overcome a magnetic barrier. Hess et al. demonstrated that this process results in cooling of the gas remaining in the trap. This technique is known as evaporative cooling. The gas was studied by measuring the quantity of gas remaining in the trap after a certain holding time by a magnetic resonance method. It was found that the gas could be cooled

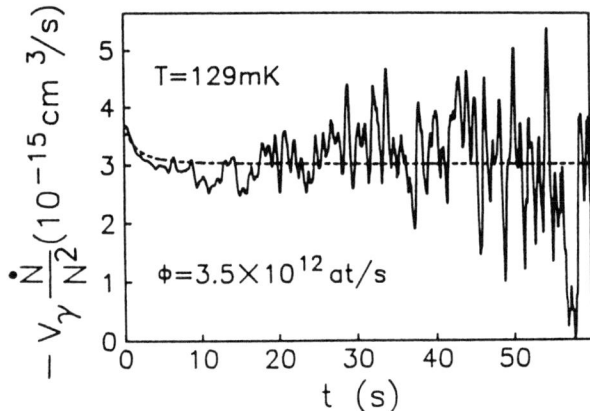

Figure 3: Typical decay curve plotted as $-V_\gamma \dot{N}/N^2$ versus time; ϕ is the H↑ filling flux and the dashed line represents a simulation.

to a temperature well below the wall temperature. In more recent experiments with this setup a central density $n_0 = 7.6 \times 10^{12}\,\text{cm}^{-3}$ was reached and temperatures below 1 mK.[25]

5 Gas collisions

In this section we consider the implications of the magnetic relaxation processes for the life time of a sample of H↑. The dominant decay mechanisms for trapped H↑ were calculated Lagendijk et al.[26] and by Stoof et al..[27] Binary collisions lead to relaxation of atoms in the c and d states to the a or b state. Spin-exchange is the most efficient process in low field, but in the s-wave scattering regime it only occurs if two atoms in the c state collide. Other collisions between low-field seekers, i.e. c-d and d-d collisions, can lead to relaxation through the magnetic dipolar interaction between the electronic spins. The electron-proton and proton-proton contributions to the dipolar interaction are entirely negligible. In low field ($B < 0.2\,\text{T}$), as in the center of the minimum-B-field traps, the spin-exchange rate is considerably faster than the dipolar rate. Hence the gas will be quickly depleted of c state atoms, leaving the gas in the doubly polarized d state: H↑↑.

To study the relaxation of H↑ to the high field seeking hyperfine states Van Roijen et al.[23] plot their data as $-V_\gamma \dot{N}/N^2$ versus time as shown in Fig.3. $N(t)$ is the total number of atoms in the trap at a given time t, obtained by integrating the observed flux $\dot{N}$ from t to ∞. V_γ is the effective volume of the sample defined by

$$V_\gamma \equiv V_{1e}^2/V_{2e}, \qquad (4)$$

where

$$V_{me} \equiv \int [\frac{n(\mathbf{r})}{n_0}]^m d\mathbf{r} \qquad (5)$$

is the effective volume associated with the density distribution to the mth power. For the Van Roijen trap $V_\gamma T_g^{-5/2} \approx 180\,\text{cm}^3\text{K}^{-5/2}$. In plotting the data the gas temperature T_g is set to be equal to the wall temperature T_w. For temperatures above 100 mK the initial decay tends to be faster than the decay expected for dipole-dipole relaxation alone. This enhanced initial decay (EID) is observed during up to 8 s after switching off the discharge. This effect is believed to arise from rapid spin-exchange relaxation. For $T < 100\,\text{mK}$ a reduced initial decay (RID) is observed for up to 15 s. The RID is attributed to loss of thermal contact between gas and cell walls.

Using the theoretical results of Stoof et al.[27] $\dot{N}(t)$ may be calculated by averaging the field and temperature dependent relaxation rates over a thermal density distribution. The contribution due to an individual process may be written as

$$\dot{N} = (\gamma_0 G_0/V_\gamma) N_{h_1} N_{h_2} \qquad (6)$$

where G_0 is the rate constant of the process, evaluated for the conditions at the center of the trap, and

$$\gamma_0 \equiv \frac{1}{V_{2e}} \int \frac{G(\mathbf{r})}{G_0}[\frac{n(\mathbf{r})}{n_0}]^2 d\mathbf{r} \qquad (7)$$

includes all effects due the field dependence of the rate. The effective volume V_γ accounts for all effects associated with the spatial distribution of the gas. N_{h_i} refers to the total number of atoms in the trap in hyperfine state h_i. The correction factor γ_0 varies substantially with temperature. For the dipolar relaxation $\gamma_0 \equiv \gamma_{dd}$ is calculated to increase from 2.1 to 3.5 for T_g increasing from 80 to 225 mK. For spin-exchange $\gamma_0 \equiv \gamma_{se}$ decreases from 0.28 to 0.08 over the same temperature range. Calculating the decay curves we find that the dominance of the spin-exchange terms in the rate equations leads to nuclear polarization and to an EID-period after which the polarization reaches a steady state. In Fig.3 the calculated EID is denoted by the dashed curve. We find that the asymptotic decay is described to $-V_\gamma \dot{N}/N^2 \equiv \gamma_{dd} G_{dd}$, where $G_{dd} \equiv 2G_{ddaa}^d + G_{ddac}^d + G_{ddad}^d$ in the notation of Lagendijk et al..[26] The results are shown as a function of temperature in Fig.4. The open circles refer to ^{4}He covered walls, the solid circles to the data with a monolayer coverage of ^{3}He. Note that the observed decay rate does not depend significantly on the surface chosen as is to be expected for trapped gas. The dashed curve is obtained if the theoretical model of Stoof et al. is evaluated for the field at the center of the trap. Taking into account the field average by including the γ_{dd}-factor one finds the drawn curve which is in good agreement with

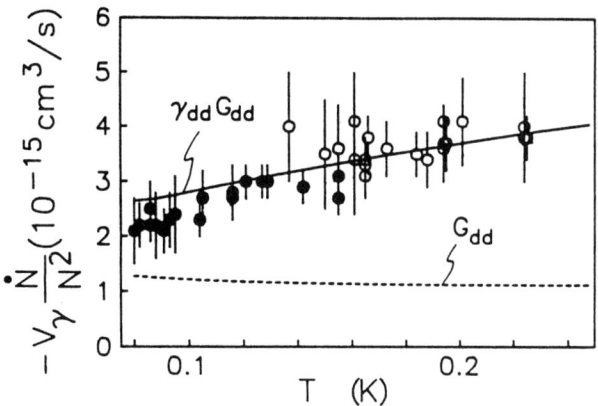

Figure 4: Experimental results for the dipole-dipole rate constant as a function of temperature.

the data considering the presence of a systematic error of order 17%.

Thus far only the inelastic scattering channels were mentioned. The elastic collisions are important to establish thermal equilibrium in the gas. For hydrogen these processes are straightforward as for the conditions in the trap simple s-wave scattering with a constant cross-section is dominant. For deuterium however something special is expected to happen from a scattering point of view. Just taking into account the Van der Waals interaction the elastic cross-section should vanish as T^2 as is readily verified within PWBA for p-wave scattering. However, for temperatures below 1 mK the atoms will start to become sensitive for the long range tail of the dipolar interaction between the atoms. A PWBA analysis shows here that for these extremely low temperatures the $1/r^3$ tail gives rise to a constant cross-section as in the boson case. It would be extremely interesting if this 'dipolar gas phase' could be observed.

6 Surface collisions

We next address the low-energy limit of atom-surface scattering. Hydrogen atoms are particularly well suited to study this limit as their mass is the smallest of any atomic system, so that the low energy behavior is observed at a relatively high temperature. In particular the phenomenon of quantum reflection at low temperature will be discussed.

To observe specular reflection one requires both conservation of the atomic momentum parallel to the surface, and of the absolute value of the normal momentum of the atoms, i.e. the wave vectors $k_\parallel$ and $|k_\perp|$ should be conserved. To conserve $k_\parallel$ one requires translational invariance, which is satisfied to a sufficient level if the surface roughness is small in comparison to the de Broglie wavelength which may be associated with the normal momentum of the atom. Crystalline structure should be avoided as it gives rise to diffraction. Clearly this condition is best satisfied for light atoms with small kinetic energy or at grazing incidence. To conserve $|k_\perp|$ one should choose a surface which is difficult to excite and preferably free of excitations.

The requirements mentioned in the previous paragraph are well satisfied for H-atoms and the surface of liquid ^{4}He below 0.5 K. In experiments with spin-polarized hydrogen we are used to work with H-atoms of typically 200 mK which corresponds to $\lambda_{th} = 39$ Å. The elementary excitations of the surface of liquid helium are the ripplons. They are characterized by a dispersion relation of the form[29,30]

$$\omega_q^2 = [g_e q + \frac{\gamma}{\rho_0} q^3] \tanh qd \qquad (8)$$

where q is the ripplon wave vector and d the thickness of the helium film. Equation 8 consists of two terms. The term linear in q corresponds to a gravity-like restoring force with effective acceleration $g_e = g + 3\alpha/d^4$, where g is the gravity acceleration proper and $3\alpha/d^4$ is the acceleration due to the Van der Waals interaction with the substrate. The q^3 term describes the dispersion due to capillarity for a liquid with surface tension γ and mass density ρ_0. The tanh function describes the change in dispersion from a wave in a thick layer of liquid, $\lambda \ll d$, to a wave in a thin layer of liquid, $\lambda \gg d$. For a liquid helium film of thickness $d = 125$ Å capillarity dominates the dispersion for a ripplon wavelength of less than 6000 Å. The thermal wavelength of the ripplons is approximately 100 Å for $T = 200$ mK. This means that ripplons with a wavelength of less than 100 Å are not thermally excited under typical experimental conditions. Consequently the only roughness on this length scale originates from the zero-point fluctuations of the surface, which are calculated to have a rms amplitude of approximately 2 Å.[29]

Non-specular reflection of H-atoms from the surface of liquid helium covering a 'flat' substrate implies inelastic scattering. The theory of these inelastic processes is discussed by various authors in the literature.[30-36] The scattering may be into a surface bound state followed by subsequent re-evaporation or direct, i.e. without adsorption. Below 500 mK both processes are accompanied by emission (or absorption in the case of the direct process) of a *single* ripplon. Phonon excitation is known to be negligible at these temperatures.[31] In linear response the adsorption rate is given by[32]

$$\Gamma = \frac{2\pi}{\hbar} \sum_q \frac{\hbar q}{2\rho_0 \omega_q}[1 + n(\omega_q)] \cdot$$
$$\cdot |<B|\frac{\partial U_q}{\partial z}|k_\perp>|^2 \delta(E_i - E_f), \quad (9)$$

where $|B>$ is the surface bound state. Apart from the overlap matrix element which contains the normal derivative of the (q- dependent) adsorption potential and the δ-function included for energy conservation, equation 9 contains the atom-ripplon coupling strength which depends on the thermal occupation $n(\omega_q)$ of the ripplons. For $T \ll \epsilon_a/k_B T \approx 1\,\text{K}$ the low temperature limit is reached and the energy conservation is dominated by the large adsorption energy that has to be accommodated. Momentum conservation then yields that adsorbed atom and excited ripplon should have equal and opposite momentum. This ripplon is calculated to have wavelength of 48 Å and only the q vectors corresponding to this value enters in the sum over final states in equation 9. As these ripplons have a wavelength clearly shorter than the wavelength of the thermal ripplons, their thermal occupation will be negligible ($n(\omega_q) \approx 0$) and the adsorption is purely induced by the zero-point fluctuations of the liquid surface. One easily verifies that the only temperature dependence that enters equation 9 arises from the temperature dependence of the average normal velocity of the atoms. As a result the adsorption rate should fall as $\Gamma \sim T$ for $T \to 0$. Dividing by the wall collision rate shows that the sticking probability should vanish as $T^{1/2}$. Experimentally the sticking probabilities (s) was found to scale as $sT^{-1} = 0.33\,\text{K}^{-1}$ with temperature, for $0.2 < T < 0.5$, were derived from a capillary flow experiment by Berkhout et al.[37] and from a measurement of thermal accommodation by Helffrich et al..[38] Thus far the temperatures have not been sufficiently low to observe the $T^{1/2}$ behavior.

7 Focussing of H-beam

A very nice demonstration of quantum reflection due to a vanishing probability of surface excitation was realized by Berkhout et al.,[39] who measured the specular reflectivity of a helium coated hemispheric mirror with the experimental cell shown in Fig.5. In this experiment the buffer volume V_B is filled with hydrogen to a density of 10^{14} atoms/cm^3. The gas expands through an orifice and is detected on a pumping plate H-flux detector[28] which also keeps the secondary volume free of hydrogen. The exponential decay time of the flux escaping the buffer volume is measured. If the center of the mirror coincides with the exit orifice of the buffer volume V_B the beam of atoms expanding through the diaphragm is reflected back into the buffer volume, causing the exponential decay time to increase. If the dimensions and temperature of the cell are accurately known one may calculate the loss factor χ representing the probability that an atom is *not* scattered back into the buffer volume

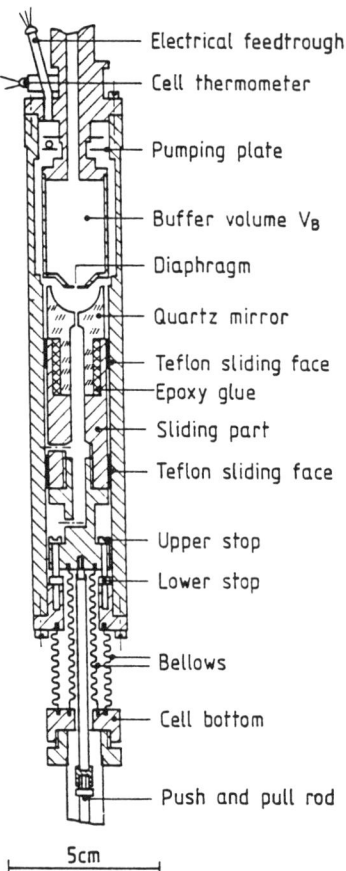

Figure 5: The experimental cell used in the mirror reflection experiment of Berkhout et al.. A hemispheric quartz mirror coated with a layer of superfluid helium may be moved up and down with a push and pull mechanism to adjust the focus of the beam expanding from the diaphragm.

$$\chi \equiv \frac{\tau_0}{\tau} \approx -\frac{\tau_0}{N}\frac{dN}{dt} \ . \qquad (10)$$

Here N is the total number of atoms in the buffer volume and $\tau_0 \equiv 4V_B/\bar{v}A$ the first-order decay time in the absence of the mirror, $\bar{v} = [8k_BT/\pi m]^{1/2}$ is the average thermal velocity of the atoms and A is the cross-sectional area of the orifice. In Fig.6 the loss factor is shown as a function of the vertical position of the mirror. With the mirror far from focussing conditions the loss is found to approach unity as in the absence of the mirror. As seen from the figure the loss from the buffer volume may be reduced by a factor 4-5 by proper adjustment of the mirror. This probably represents the first demonstration ever of focussing of an atomic beam by means of a mirror. In the absence of lateral misalignment and spherical aberration the loss at the minimum of the position scan corresponds to the deviation from perfect reflectivity of the mirror.

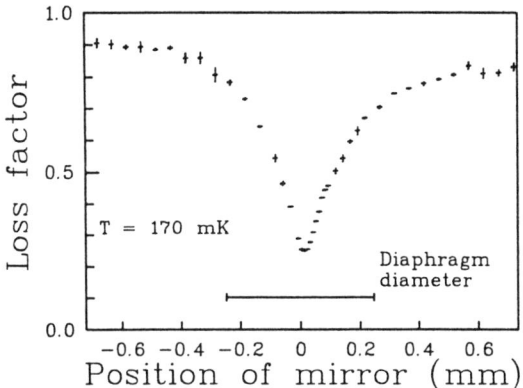

Figure 6: The loss factor as a function of the vertical mirror position as observed by Berkhout *et al.*.

Plotting the loss factor at the minimum of the scan curve as a function of temperature one obtains the results shown in Fig.7. The main results are the solid triangles. According to ref.[39] this is best described by the solid curve which results from four independent loss factors, i=1-4, accumulated according to

$$\chi = 1 - \prod_{i=1}^{i=4}(1-\chi_i) \qquad (11)$$

The loss contributions are adsorption at normal incidence ($\chi_1 T^{-1} = 0.5\,\mathrm{K}^{-1}$), direct inelastic scattering ($\chi_2 T^{-2} = 0.5(1)\,\mathrm{K}^{-2}$), lateral misalignment ($\chi_3 = 0.2$) and other geometrical factors such as spherical aberration ($\chi_4 = 0.042$). No significant effect was observed upon adding up to 0.1% ^{3}He (solid square and open circles). This quantity should be sufficient to assure full monolayer coverage. This is remarkable as such a layer changes the adsorption energy by a factor 2.5 and the surface tension by a factor 2. A similar lack of effect was observed by ref..[37] If the helium film thickness is reduced as much as possible the effective reflectivity decreases, see crosses in Fig.7. This may be explained by the presence of static roughness due to the substrate.

Acknowledgements

Much of the material presented in this paper was collected in relation to the PhD studies of Raymond van Roijen and Jaap Berkhout whose contributions I would like to emphasize here. Further I wish to acknowledge the contributions of Erik Wolters, Jom Luiten, Irwan Setija, Tom Hijmans and our guests Simo Jaakkola and Takao Mizusaki. The work benefitted from the technical expertise of Otto Höpfner and Michiel Groeneveld. The stimulating environment presented of the Spectroscopy

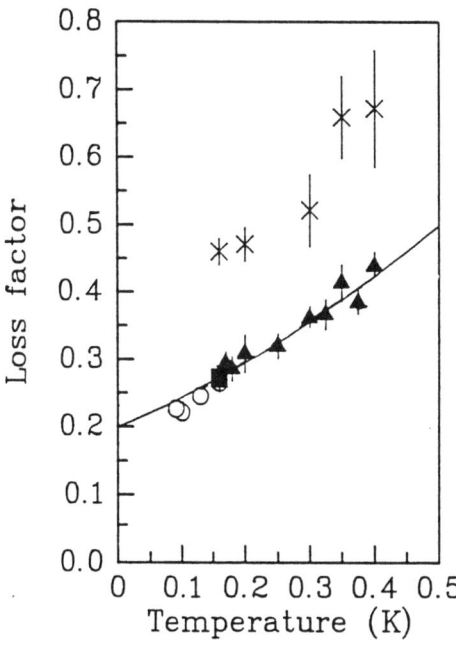

Figure 7: Measured loss factors at the minimum of the scan curves as a function of temperature.

of condensed matter group of Ad Lagendijk and Rudolf Sprik is warmly acknowledged.

This work is part of the research program of the Stichting voor Fundamenteel Onderzoek der Materie (FOM), which is financially supported by the Nederlandse Organisatie voor wetenschappelijk onderzoek (NWO).

References

[1] T.J. Greytak and D. Kleppner, *New Trends in Atomic Physics*, G. Grynberg and R. Stora (Eds.), Elsevier, Amsterdam, 1984, Vol.II, 1125.

[2] I.F. Silvera and J.T.M. Walraven, Prog. Low Temp. Phys. **X**, 139 (1986).

[3] W.N. Hardy, M. Morrow, R. Jochemsen and A.J. Berlinsky, (LT-16), Physica **109 & 110B**, 1964 (1982).

[4] W.N. Hardy, M.D. Hürlimann and R.W. Cline, (LT-18), Jap. J. Appl. Phys. **26**, 2065 (1987).

[5] J.H. Freed, Ann. Phys. (France) **10**, 901 (1985).

[6] D.M. Lee, (LT-18), Jap. J. Appl. Phys. **26**, 1841 (1987).

[7] J.J. Berkhout and J.T.M. Walraven, *Spin Polarized Quantum Systems*, S. Stringari (Ed.) World Scientific, Singapore, 1989, p. 201.

[8] W. Kolos and L. Wolniewicz, J. Chem. Phys. **43**, 2429 (1965); Chem. Phys. Lett. **24**, 457 (1974); J. Mol. Spectrosc. **54**, 303 (1975).

[9] W. Kolos, K. Szalewicz, and H.J. Monkhorst, J. Chem. Phys. **84**, 3278 (1986).

[10] R. Bernheim, *Optical Pumping. An Introduction*, W.A. Benjamin, New York (1965), p.64.

[11] A. Kastler, Acta Phys. Pol. **XXXIV**, 693 (1968).

[12] G. Alzetta, A. Gozzini, and L. Moi, C.R. Acad. Sci. Paris, **274**, 39 (1972).

[13] M. Allegrini, G. Alzetta, P. Bicchi, S. Gozzini, and L. Moi, Ann. Phys. (France) **10**, 883 (1985).

[14] J. Vigué, Phys. Rev. A **34**, 4476 (1986).

[15] H.F. Hess, Bul.Am.Phys.Soc. **30**, 854 (1985).

[16] H.F. Hess, Phys. Rev. B **34**, 3476 (1986).

[17] A.L. Migdall, J.V. Prodan, W.D. Phillips, T.H. Bergeman, and H.J.Metcalf, Phys. Rev. Lett. **54**, 2596 (1985)

[18] V.S. Bagnato, G.P. Lafyatis, A.G. Martin, E.L. Raab, R.N. Ahmad-Bitar, and D.E. Pritchard, Phys. Rev. Lett. **58**, 2194 (1987).

[19] W.H. Wing, Prog. Quant. Electron. **8**, 181 (1984).

[20] R.V.E. Lovelace, C. Mehanian, T.J. Tommila, and D.M. Lee, Nature **318**, 30 (1985)

[21] C.C. Agosta and I.F. Silvera, *Spin Polarized Quantum Systems*, S. Stringari (Ed.) World Scientific, Singapore, 1989, p. 254.

[22] C.C. Agosta, I.F. Silvera, H.T.C. Stoof and B.J. Verhaar, Phys. Rev. Lett. **62**, 2361 (1989).

[23] R. Van Roijen, J.J. Berkhout, S. Jaakkola, and J.T.M. Walraven, Phys. Rev. Lett. **61**, 931 (1988).

[24] H.F. Hess, G.P. Kochanski, J.M. Doyle, N. Masuhara, D. Kleppner, and T.J. Greytak, Phys. Rev. Lett. **59**, 672 (1987).

[25] N. Masuhara, J.M. Doyle, J.C. Sandberg, D. Kleppner, and T.J. Greytak Phys. Rev. Lett. **61**, 935 (1988).

[26] A. Lagendijk, I.F. Silvera en B.J. Verhaar, Phys. Rev. B **33**, 626 (1986).

[27] H.T.C. Stoof, J.M.V.A. Koelman, and B.J. Verhaar, Phys. Rev. B **38**, 4688 (1988).

[28] J.J. Berkhout, O.H. Höpfner, E.J. Wolters, and J.T.M. Walraven, (LT-18), Jap. J. Appl. Phys. **26**, 231 (1987).

[29] M.W. Cole, Phys. Rev. B **2**, 4239 (1970); Phys. Rev. A **1**, 1838 (1970).

[30] B.W. Statt, Phys. Rev. B **32**, 7160 (1985).

[31] Y. Kagan and G.V. Shlyapnikov, Phys. Lett. A **95**, 309 (1983).

[32] D.S. Zimmerman and A.J. Berlinsky, Can. J. Phys. **61**, 508 (1983).

[33] V.V. Goldman, Phys. Rev. Lett. **56**, 612 (1986).

[34] T.W. Hijmans and G.V. Shlyapnikov, Phys. Lett. A, accepted for publication.

[35] E. Tiesinga, H.T.C. Stoof and B.J. Verhaar, to be published.

[36] Yu. Kagan, N.A. Glukhov, B.V. Svistunov, and G.V. Shlyapnikov, to be published.

[37] J.J. Berkhout, E.J. Wolters, R. van Roijen, and J.T.M. Walraven, Phys. Rev. Lett. **57**, 2387 (1986).

[38] J. Helffrich, M.P. Maley, M. Krusius, and J.C. Wheatley, Phys. Rev. B **34**, 6550 (1986).

[39] J.J. Berkhout, O.J. Luiten, I.D. Setija, T.W. Hijmans, T. Mizusaki, and J.T.M. Walraven, Phys. Rev. Lett. **63**, 1689 (1989).

EXPERIMENTS IN COLD AND ULTRACOLD COLLISIONS

J. Weiner

Department of Chemistry and Biochemistry
University of Maryland
College Park, Maryland 20742

Two-particle and light-field interactions characterize collision physics in the cold and ultracold kinetic energy regimes. Cross sections increasing by orders of magnitude and acute sensitivity to light field intensity and polarization distinguish this new domain from conventional thermal environments. This contribution reports results from collisions within optical traps and atomic beams.

1. Introduction

As kinetic energy decreases below 1 K, many conventional ideas concerning heavy particle collisions are turned upside down. The deBroglie wavelength becomes orders of magnitude longer than the range of the chemical bonding force, the translational Maxwell-Boltzmann distribution assumes a width comparable to the natural width of excited atomic states, radiative lifetimes become long compared to collision times; and weak, long-range interactions control the probability of inelastic events. The present paper reports on the first few exploratory experiments in this largely uncharted territory. We define collisions between 1K and 1mK as "cold" and below 1 mK as "ultracold". In the cold regime alignment and orientation of the weakly interacting atomic populations control the collision probability, while in the ultracold regime optical field state dressing plays an indispensable role. At this writing ultracold collisions have been studied in optical traps and cold collisions within an atomic beam.

2. Associative Ionization in an Optical Trap

Associative ionization (AI) between two resonantly excited Na atoms in an optical trap presented itself as a likely candidate for a first experiment since the process was known to occur[1] and the fraction of excited states in a dipole trap is always near saturation (about 30%). In collaboration with the group of W.D. Phillips at N.I.S.T., therefore, we searched for the dimer ions resulting from

$$Na(3p) + Na(3p) \longrightarrow Na_2^+ + e \quad (1)$$

using some of the atomic cooling and trapping techniques developed by them and others over the last several years[2]. The optical trap design specific to this experiment has already been described in some detail elsewhere[3], and we will only summarize it briefly here. A sodium atomic beam is first laser cooled by using a tapered magnetic field technique to compensate for the decreasing Doppler shift as the atoms decelerate. At the end of the tapered solenoid some fraction of the cooled atoms drift up to the zone where six orthogonal laser beams intersect to form a volume of "laser molasses" from which the dipole trap is loaded. Trapping the atoms from molasses serves to concentrate the density to about 10^{10} atoms cm^{-3} thereby increasing the collision frequency to an observable level. The trap consists of two counterpropagating, circularly polarized laser beams focused to a waist of about 100 μm with about 40 mW power in each beam. The laser beams are chopped in time to avoid standing-wave heating, and the foci are spatially separated by about 5 cm. Trapping and cooling cycles alternate to prevent the atoms from boiling out of the trap, and both the molasses and trap are situated between field plates which accelerate charged particles onto a multiplier detector mounted vertically above. A time-of-flight measurement determines ion mass, confirming that only Na_2^+ is produced in the trap. Variation of the ion signal intensity with density confirms a quadratic dependence characteristic of a bimolecular collision. Using this experimental setup we determined (see ref. 3) the absolute rate constant (and the corresponding cross section) for process (1) at 0.75 mK. The result is $K=1.5\times10^{-11}$ $cm^3 sec^{-1}$.

This value is not too different from those measured at conventional temperatures, but the corresponding cross section $\sigma=8.6\times10^{-14}$ cm^2 is about three orders of magnitude greater. Increasing cross sections at very low kinetic energy are to be generally expected because the deBroglie wavelength of the reduced mass increases inversely as the velocity.

Another curious consequence of this regime is the long duration of a collision compared to spontaneous emission. Julienne[4] as pointed out that the characteristic time for collision between $^2S_{1/2}$ and $^2P_{3/2}$ Na atoms at 1 mK is about 30 nsec -- four radiative lifetimes of the corresponding Na_2 molecule. Thus associative ionization between two excited Na atoms can only take place if stimulated absorption can maintain population in the excited state over a sufficient fraction of the total collision time. As the two atoms approach, long-range interaction will shift their levels out of resonance with the laser field, and the excited state population will begin to deplete at the spontaneous emission rate. Relative strength of atomic coupling to the laser field compared to collisional interaction will determine the fraction of excited-state population available for associative ionization. In ref. 4 Julienne has worked out a four-state model for this coupling competition for the conditions of the molasses (weak field) and trap (strong field) appropriate to the experiment described above. The model predicts that the ratio of the AI rate constants with the trap on and off will be about 10^4. In order to test this prediction ion production was measured synchronously with the trapping and cooling cycle, and the results shown in fig (1).

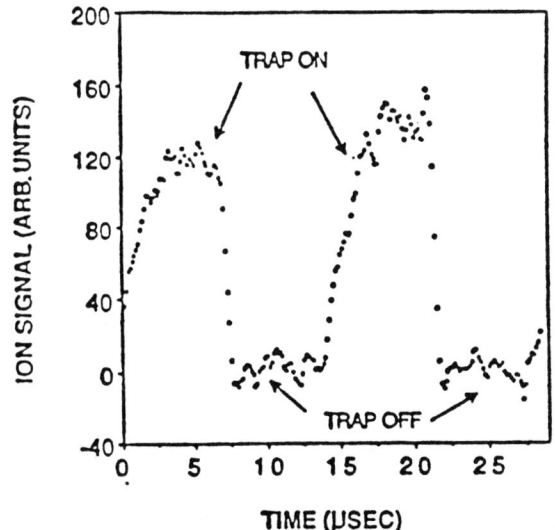

Fig 1. Ion intensity modulation with optical trap switching.

Julienne's qualitative prediction of marked enhancement in the strong-field AI rate appears to be correct, but experimental data are not yet precise enough to test the model quantitatively.

3 <u>Associative Ionization in a Single Beam</u>

Although optical traps achieve temperatures at which kT approaches the natural width of an atomic line, as containers for collision experiments they are essentially cells of very cold gas in which the distribution of collision directions is isotropic. Atomic beams, however, provide a principal laboratory axis and are therefore useful for studying the effects of orientation and alignment in collision problems. Cold collisions can even be studied in a beam without recourse to "standard" cooling or trapping techniques. Consider a highly collimated atomic beam crossed by a cw, monomode laser propagating in the opposite direction and tuned to the red of the atomic line rest frequency. The Doppler shift of a narrow velocity class will just compensate for the laser detuning, producing a narrow velocity distribution of excited atoms. Collisions between these excited atoms can result in associative ionization just as it does in the optical trap. Analysis[5] hows that the average velocity is 11.9 m sec^{-1}, corresponding to a "temperature" ($T=2E/3k_B$) of 65 mK. Although this collision energy cannot be considered "ultracold", it is cold enough so that normally negligible long-range interactions dominate the reaction probability, and the effect of circular and linear polarization as well as laser field intensity on collision cross section can be studied systematically. We have just begun to exploit the possibilities of cold collisions in beams, and the results reported here must be regarded as preliminary. Nevertheless, they serve to illustrate possibilities of the technique. Figure (2) is schematic of the atomic beam apparatus which can be used in either crossed- or single-beam mode. For the experiments reported here one of the beams was flagged off. The obvious first experiment was to measure the cross section for Na(3p) + Na(3p) AI at a temperature intermediate between the ultracold value (T=0.75 mK, σ=8.6x10^{-14} cm^2) and that measured[6] under conventional alkali vapor conditions (T=650K, σ=1.0x10^{-16} cm^2). Two counterpropagating laser beams are tuned such that the laser frequency is resonant with the Doppler blue-shifted transition in one direction while resonant with the Doppler

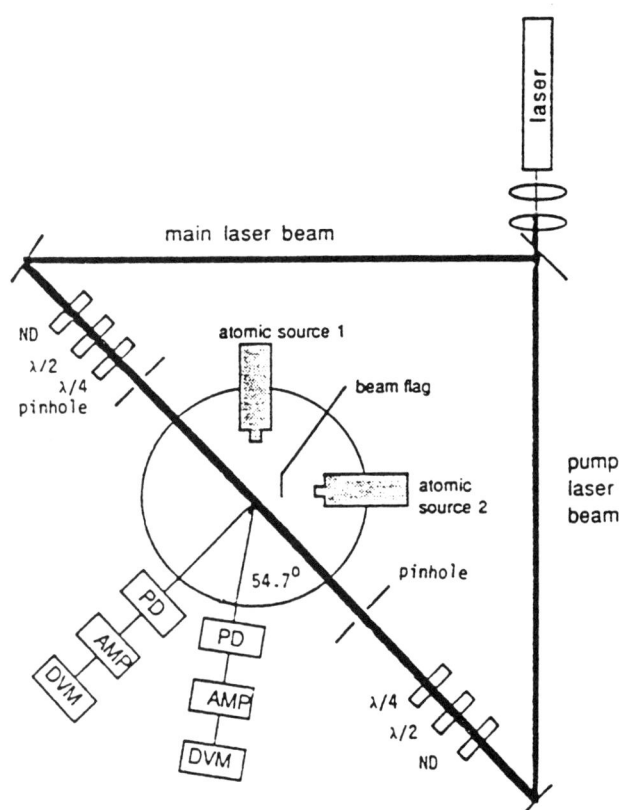

Fig 2. Crossed-beam apparatus used to measure cold collision properties.

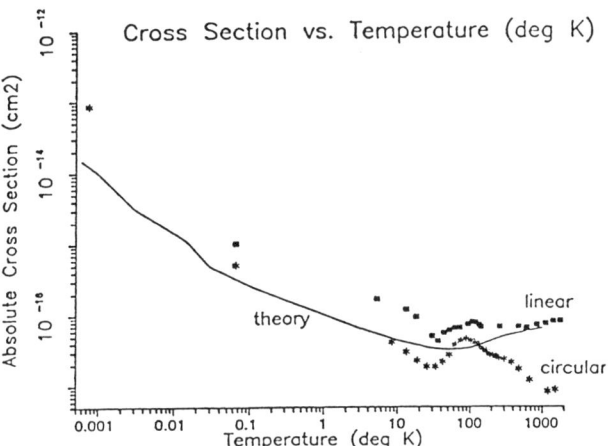

Fig 3. Cross section for process (1) as a function of collision "temperature" over six orders of magnitude--theory and experiment.

red-shifted $_{12}$ transition in the other. As pointed out almost 15 years ago[7], his two-level, single-frequency technique maximizes excited-state population and avoids intensity-dependent optical pumping effects even when the laser light is linearly polarized. The rate constant using circular polarization was measured to be 6.1×10^{-13} cm^{3}sec^{-1} corresponding to a cross section of 5.1×10^{-16} cm^2. Light polarized linearly in the collision plane yields a rate constant greater by a factor of 2.0. The next experiment excites a single atomic beam with two velocity groups from two independently tunable lasers. The relative velocity of collision between the two groups can then be varied over a range of "temperature" from about 4 K to 300 K. Figure (3) shows a plot of the associative ionization cross section vs. temperature over a range of six orders of magnitude together with a theoretical curve calculated from a semi-classical model recently published by Geltman[8]. The theory predicts cross sections decreasing with decreasing temperature to a minimum at 75 K due to the closing of entrance channels with repulsive quadrupole-quadrupole interaction terms. Only channels with a quadrupole orientation leading to π states remain open. As the temperature further decreases, the cross section rises again as the effects of a decreasing angular momentum barrier and increasing deBroglie wavelength begin to dominate. The experimental results follow the same trends but the minimum appears closer to 40 K. The theory does not take account of state-dressing effects that are essential for understanding collision dynamics at ultracold temperatures; and, therefore lack of agreement with the optical trap results at 0.75 mK is not surprising.

The single-beam technique can also be used to study the effects of optical-field state dressing, but a narrow velocity class must be isolated in one of the Na ground state hyperfine levels so as to avoid an increased velocity dispersion by power broadening. We have used a setup originally suggested earlier[9] in which the atomic beam is crossed by the laser beam three times: the first pass optically pumps all the population into F=1 of the ground state Na atom, the second pass selects a narrow velocity group by excitation to Na($3p^2P_{3/2}$) followed by decay to F=2 in the ground state. At this point only the optically selected velocity class is present in the F=2. The third pass again excites this velocity class to the upper level from which AI takes place. In this third step, however, the intensity and polarization of the laser field can be varied to observe the effect on the rate constant. The transition is power broadened but the velocity class is not. Figure (4) shows some recent results employing this technique. Polarization-dependent intensity effects on the AI cross section are clearly evident, but to date a definitive interpretation has not been carried out. The observed behavior may arise either from competition between optical and collisional coupling, leading to increased "survival" of the excited-state population (against spontaneous emission) during the incoming branch of the collision, or to variation in the atomic excited-state density

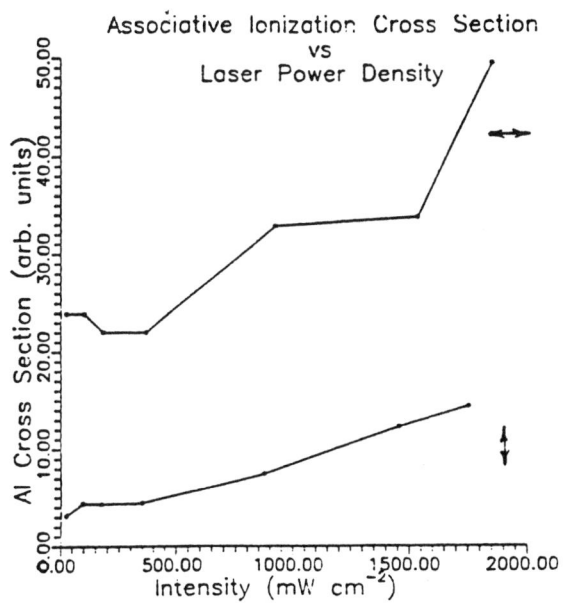

Fig 4. AI cross section as a function of intensity in single-beam experiments. Note that for polarization parallel and perpendicular to the atomic beam axis, cross sections increase markedly with laser intensity.

matrix (due to optical pumping) as laser intensity increases. The hope is that these new results will inspire a redoubled theoretical effort, and that a detailed analysis of intensity and polarization effects in the very-cold collision regime will soon be forthcoming.

References

1. J. Weiner, F. Masnou-Seeuws, and A. Giusti-Suzor, Adv. At. Mol. Opt. Phys. **26** (in press).

2. For an early review of "The Mechanical Effects of Light" see JOSA B **2** 1705 (1985) and for a more recent account, JOSA B **6** No. 11 (to be published).

3. P.L. Gould, P.D. Lett, P.S. Julienne, W.D. Phillips, H.R. Thorsheim, and J. Weiner, Phys. Rev. Lett. **60**, 788 (1988).

4. P.S. Julienne, Phys. Rev. Lett. **61**, 698 (1988).

5. H.A. Thorsheim, Y. Wang, and J. Weiner (to be published).

6. J. Huennekens and A. Gallagherer, Phys. Rev. A **28**, 1276 (1983) and R. Bonanno, J. Bouler, and J. Weiner, Comments At. Mol. Phys. **16**, 109 (1985).

7. G.M. Carter, D.E. Pritchard, and T.W. Ducas, Appl. Phys. Lett. **27**, 498 (1975).

8. S. Geltman, J. Phys. B; At. Mol. Opt. Phys. **21**, L375 (1988).

9. F. Schuda and C.R. Stroud, Optics Comm. **9**, 14 (1973) and D.G. Steel and R.A. McFarlane, Optics Lett. **8**, 33 (1983).

Symposium:
Collisions Involving Positrons

INTRODUCTION TO THE SYMPOSIUM ON COLLISIONS INVOLVING POSITRONS

J.W. Humberston
Department of Physics and Astronomy
University College London, Gower Street, London WC1E 6BT, U.K.

The study of positron collision physics has increased so substantially over the past several years that the subject now fully deserves this special symposium at ICPEAC. Sheldon Datz, in his keynote address, made reference to the increasing proportion of contributed papers at successive ICPEACs devoted to 'exotic' topics, and the majority of these have involved positrons in some form. This year the number of papers on positron collisions has risen to more than thirty, in addition to the papers presented in this symposium.

As well as the direct representation of the subject at ICPEAC, a series of small specialised satellite workshops on positron collision physics was established in 1981 with meetings at York University, Toronto (1981), Royal Holloway College, Egham (1983), Wayne State University, Detroit (1985), University College London (1987) and, most recently, at the Goddard Space Flight Center, Greenbelt (1989). References to the proceedings of the first four workshops are given at the end of this introduction.

A brief summary of the contents of the latest workshop has been prepared by R.J. Drachman (Goddard Space Flight Center) for the Symposium, but the full proceedings will be published separately. Of particular interest at the Workshop were the talks on various astrophysical aspects of positrons associated with the observations of electron-positron annihilation radiation coming from the direction of the galactic centre.

Theoretical investigations of positron collisions with atoms and molecules have been carried out for many years, even before the availability of experimental results with which to make comparisons. Sometimes results for positrons were obtained almost as an afterthought in the principal investigation of electron collisions, all that was required being a change of sign of the charge of the incident particle and the neglect of exchange. However, it soon became apparent that several of the standard methods of approximation developed for electron scattering do not work so well for positrons, particularly at low positron energies where strong electron-positron correlations in the form of virtual, or real, states of positronium interfere significantly with the usual central field approximations of atomic physics. These correlations require the use of very flexible trial wave functions with explicit dependence on the electron-positron coordinate if accurate results are to be obtained. Such wave functions, used in conjunction with some form of variational method, have yielded very accurate results for positron scattering from atomic hydrogen, helium and, recently, molecular hydrogen. For larger atoms, where such detailed allowance for correlations is no longer feasible, some form of polarised orbital method seems to provide reasonably satisfactory results.

The most significant progress in positron collision physics has been made since the development of monoenergetic low energy positron beams, first introduced in the early 1970s. Initially the beam intensities were so low that only total scattering cross sections could be measured, and, then, only with rather large statistical errors. Beam intensities have increased steadily since then and it is now possible to measure various partial, and even differential, cross sections.

The availability of increasingly good experimental data has stimulated further associated theoretical investigations and A.D. Stauffer describes recent results for the noble gases and alkali atoms from the group at York University, Toronto.

The system which has attracted most theoretical attention over many years is, not surprisingly, positron scattering by atomic hydrogen, and essentially exact results have been obtained for some of the low energy scattering parameters. Nevertheless, the system continues to attract the interest of theoreticians, and A.S. Ghosh from I.A.C.S., Calcutta discusses the use of the close coupling approximation with the inclusion of ground state positronium in a calculation of the low energy elastic scattering phase shifts.

Even with the more intense positron beams now available, it has been exceedingly difficult to measure any cross sections for positron scattering by atomic hydrogen because of the low density of atomic hydrogen in the target gas cell. The first such measurement, of the ionization of atomic hydrogen has, however, recently been made in Bielefeld and a report on this work is given by W. Raith. Further results of improved accuracy should be available in the near future when this experiment is transferred to the intense positron beam facility at Brookhaven.

Comparisons between the scattering cross sections for positrons and electrons from the same target provide one of the most interesting aspects of atomic collision physics, hopefully providing further insight into the roles of exchange and positronium formation. One curious feature of the comparison is the merging of the two total cross sections at much lower energies than the energies at which individual partial cross sections merge. These and other related matters are discussed by W.E. Kauppila of Wayne State University.

Monoenergetic positron beams have usually been used to study positron scattering from individual atoms or molecules in a gas target. Such beams are also now being used to investigate the processes which occur when positrons with a few tens of keV enter a solid and slow down. P.J. Schultz of the University of Western Ontario describes recent work in this field.

Positron collision physics has made rapid advances in the past twenty years to the stage where the range, and quality, of experiments is beginning to approach that for electrons. Such improvements are, in turn, challenging the theoreticians to provide more detailed and accurate results for various scattering parameters. Further significant developments in both areas can be expected in the near future.

I should like to thank Prof. R.P. McEachran and his colleagues Profs. J.W. Darewych and A.D. Stauffer for their hospitality at York University, Toronto where this introduction was written.

References

1. Proceedings of the International Conference on Positron Scattering and Annihilation in Gases, York University, Toronto, Canada, 1981. Can. J. Phys. **60**, pp; 461-617 (1982).
2. 'Positron Scattering in Gases', edited by J.W. Humberston and M.R.C. McDowell (Plenum, New York, 1984) Proceedings of the NATO Advanced Workshop on Positron Scattering in Gases, Royal Holloway College, Egham, U.K., 1983.
3. 'Positron (Electron)- Gas Scattering', edited by W.E. Kauppila, T.S. Stein and J.M. Wadehra (World Scientific, Singapore, 1986) Proceedings of the Third International Workshop on Positron (Electron)- Gas Scattering, Wayne State University, Detroit, U.S.A., 1985.
4. 'Atomic Physics with Positrons', edited by J.W. Humberston and E.A.G. Armour (Plenum, New York, 1987). Proceedings of the NATO Advanced Research Workshop on Atomic Physics with Positrons, University College London, London, U.K., 1987.

THE WORKSHOP ON ANNIHILATION IN GASES AND GALAXIES

Richard J. Drachman

Laboratory for Astronomy and Solar Physics
Goddard Space Flight Center
Greenbelt, MD 20771, USA

This is a brief report on the Workshop held in Greenbelt from July 19 to July 21, 1989, as a Satellite of ICPEAC XVI. It was the sixth in an informal biennial series designed to emphasize the interaction of positrons with atoms in non-condensed materials.

INTRODUCTION

It has become a tradition to hold a positron workshop as a satellite of the biennial meeting of ICPEAC at a location fairly convenient to the main meeting. To some extent this was a reaction to the fact that the large alternate-year conferences on positron annihilation had become dominated by the field of positron interactions in condensed matter. The organizers of the first of these workshops simply wished to re-emphasize the interesting problems involving "gaseous positronics" as Sir Harrie Massey liked to call it. In addition, they wanted a gathering of a manageable size, where lively discussions and personal interactions would not be inhibited by sheer numbers of participants.

Each of the subsequent workshops has followed this pattern, aiming for an attendance of under 100. Several have also featured some unique variation on the theme of positrons in gases; one year the comparison of positrons with electrons was emphasized, and one year atomic physics with positrons was central. Because this year's workshop was held under the auspices of NASA's Goddard Space Flight Center it seemed appropriate to devote time to positron astrophysics and antimatter, along with the more traditional topics. Since quite a few of the traditional topics presented at the Workshop are also being covered here at ICPEAC, I will concentrate here on reporting the less familiar sorts of work discussed in Greenbelt.

ASTROPHYSICS

Reuven Ramaty (Goddard) reviewed the theoretical and experimental situation with respect to annihilation radiation from the direction of the Galactic Center:

Definite observations of annihilation gamma rays have been made repeatedly, mainly from balloon-borne instruments but also (twice) from a satellite. The line resulting from two-photon (singlet) annihilation has the laboratory energy, apparently not significantly Doppler shifted. In addition, there is a strong indication of the three-photon continuum from triplet annihilation; since this comes almost entirely from positronium (Ps) annihilation it is possible to conclude that the Ps fraction is about 90%. This kind of data is important in constraining the possible physical conditions existing in the annihilation region.

Each successive observation up to 1981 measured a different intensity for the radiation, and after 1981 the next few balloon flights failed to see anything! If the various observations are in fact comparable in terms of sensitivity and angular resolution this rapid variation is strong evidence that the Galactic Center source is extremely compact, no larger than 1/2 light year.

Combined with its very large luminosity (if it is actually at the indicated distance and not just a foreground object) this small size calls for an unusual kind of source, perhaps a massive black hole. Then its time dependence could indicate the irregularity of the distribution of infalling matter or perhaps the motion of clouds of matter drifting across a beam of intense radiation from the black hole in which the positrons are created and then annihilated.

It is important to remember that the atomic physics relevant to the production of gamma rays has to do with the slowing down and subsequent annihilation of positrons in a very low density environment, which we will assume to be composed mainly of neutral atomic hydrogen. Positrons are injected into this medium at some high energy and slow down at first by inelastic scattering and ionization. After a while (below about 100 eV) Ps formation begins to compete. Calculations show that about 95% of all positrons injected form Ps in flight with a mean energy of about 30 eV. Three-fourths of these contribute to the continuum, the rest to the Ps annihilation line with a width of over 6 keV, just due to the Doppler effect. In addition, the positrons that thermalize without forming Ps eventually annihilate with slowly moving H atoms; the width of their annihilation line is only 1.3 keV due to the velocity distribution of the electrons in the atom. (Since all of these contribute to the two-photon peak they are more important than indicated by their 5% share of the population.) This seems to be inconsistent with the observed width of about 3 keV, but if the fraction of thermalized positrons is increased somewhat (consistent with laboratory simulations) agreement may be achieved. Partial ionization of the medium accelerates the thermalization and can account for the observations.

Marvin Leventhal (AT&T Bell Labs) discussed two new balloon flights made during 1988. After seven years the Galactic Center (GC) source of annihilation radiation has now been observed again at about the same intensity as before it turned off in 1981! Eleven hours of data were obtained on flights over Australia in May and October with the gamma-ray spectrometer pointed at the GC; the instrument has a 17° field of view. In addition, seven hours of data were obtained during the second flight with the instrument pointed 25° away from the GC but in the Galactic Plane. The flux for the off-center pointing was significantly lower, and this fact supports the idea that the dominant annihilation-line emission does come from a source concentrated in the GC. The line width was 3.6 ± 0.5 keV. This is the first true measurement of this width and was possible because of the excellent energy resolution (1.8 keV FWHM) of the array of seven germanium detectors cooled to 90K by liquid nitrogen. The relatively good angular resolution produced by active NaI shields and aperture holes above the detectors will be improved in the future (to about 1°) by the use of a coded aperture system. There is an indication of significant 3-photon continuum radiation in the data, perhaps at the 90% level, but further analysis will be needed in order to subtract the background correctly. (A more recent balloon observation, made in May, 1989, reported that the GC source has once again turned off!)

Gerald Share (Naval Research Lab) discussed measurements of Galactic positron annihilation made as an unexpected bonus by SMM, the Solar Maximum Mission. This satellite was originally launched to observe the Sun during its active phase, but it became famous as the first satellite to be repaired in orbit by astronauts in the Shuttle; it has enjoyed a long and productive post-operative life. The NRL group managed to make a series of annual observations of the GC using SMM. The satellite remains pointed at the Sun but it also sees the GC in the course of the motion of the earth during the year. By a fairly complicated analysis it is possible to separate out the radiation coming from the GC, and in this way the time history of the annihilation gamma rays has been studied. Almost no variation in intensity is seen by the NRL group, in apparent contradiction to the rapid and strong variation reported by Leventhal and others. (Fig.1 shows this time variation with the SMM data marked "S".) Share seemed willing to accept the results of the balloon observations, and the conclusion is that SMM is probably measuring a diffuse component of the radiation rather than the compact source. This is reasonable, since the SMM angular resolution is very broad. The last word on this subject may not yet have been spoken, but compromise seems in the wind.

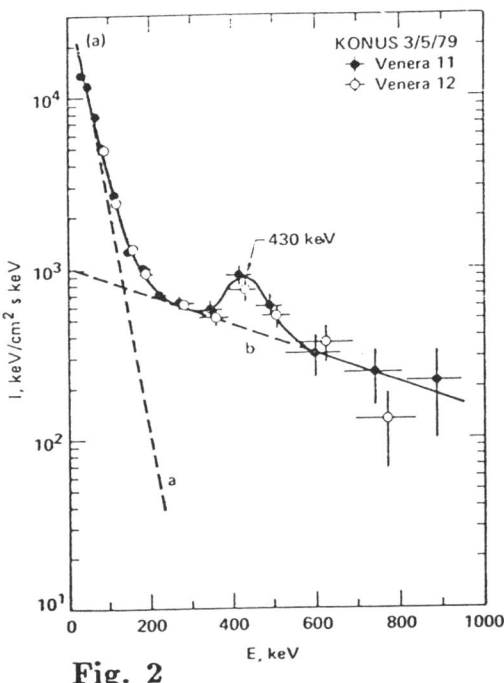

Fig. 2

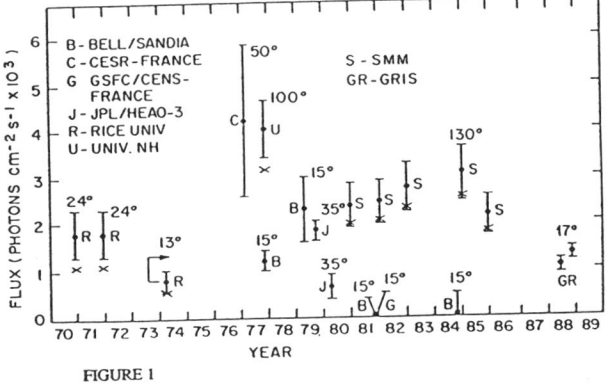

FIGURE 1

Alice Harding (Goddard) gave an interesting review of a strange class of celestial sources called gamma-ray bursts (GRBs.) These are transient sources of gamma rays with almost none of their emission below 1 keV. They last usually for only a few seconds, are not identified with any known astronomical objects, and almost never have been seen to recur (over the 15-year period since they were first observed.) The famous burst of March 5, 1979 had a rise time of less than 0.2 ms; the source size must have been smaller than 60 km. Some GRBs (Fig.2) have interesting peaks in their spectra lying near 400 keV, and they might represent red-shifted annihilation radiation. If so, their red shifts are consistent with the gravitational red shift at the surface of a neutron star (for certain neutron-matter equations of state.) It is thus necessary to re-analyze the old familiar positron processes in the strong magnetic field environment of these exotic condensed objects. In this context "strong" means 10^{12} gauss, and one might guess that fairly unusual physical processes might take place. For example, single photon annihilation and creation in vacuum can take place with the magnetic field taking up the recoil momentum. The e^+-e^- pairs must be in the discrete Landau states appropriate in the magnetic field, and this effect can cause remarkable resonant behavior. The question is whether this type of exotic physics can reasonably result in the kind of spectrum observed. Simulations incorporating this physics do in fact show an annihilation feature above the continuum for a variety of parameters, although the models for annihilation in GRBs are still not very sophisticated. The experimental situation should improve soon with the launch of GRO, the

Gamma-Ray Observatory. It will have greatly improved sensitivity and time resolution as short as 10μs, and may finally decide whether positron annihilation is actually occurring in GRBs.

Floyd Stecker (Goddard) gave a detailed theoretical talk on the problem of "dark" or "missing" mass, its possible resolution in terms of particle physics, and possible observational consequences. There are many indications that the Universe must contain much more matter than is represented in stars and other visible astronomical objects. For one thing, the rotation of galaxies would not be consistent with Kepler's law if the only gravitating matter in the galaxies were the visible matter. Second, the virial theorem applied to clusters of galaxies also seems to demand additional invisible matter. Finally, there are fairly convincing reasons connected with the closure of the Universe that lead to the same conclusion. Many astronomers look for sub-luminous stars and other normal types of invisible objects to make up the deficit, but others predict that there may be new classes of weakly interacting elementary particles that are able to balance the accounts. Stecker argued that some of these may annihilate in new ways, giving rise to unique radiation signatures that may be observable. For example, a characteristically hard spectrum in the 100 MeV region could serve as a signature of χ particle annihilation.

ANTIMATTER

Robert Tjoelker (Harvard) filled in on short notice for Gerald Gabrielse and discussed their experiments on isolating, trapping, and cooling antiprotons. The $\bar{p}$ source is a CERN accelerator, the antiparticles are first accumulated in a storage ring at E>5 MeV and then slowed by collisions to about 3 keV. They are then stored in a Penning trap. (This involves a magnetic field that contains the particles along two dimensions and electric fields that close the ends of the trap.) The $\bar{p}$s are allowed into the trap which is then closed (in 3 ns) before they they can escape. Collisions with electrons which have been cooled by synchrotron radiation then complete the thermalization. Some 12000 $\bar{p}$s can be stored in one load, their lifetime in the trap is about 50 hours, and they can be allowed to leak out gradually in order to measure their temperature. At present, the inertial mass of the $\bar{p}$ is being measured by coupling the cyclotron motion to the axial oscillations in the trap. This is intended to test further the TCP theorem that predicts equality between particle and antiparticle masses. Equality has been checked so far to 2 parts per million.

John Humberston (London) discussed the most basic bit of neutral antimatter--the antihydrogen atom. It is, like the antiproton discussed above, of importance for testing fundamental invariance principles. Since it is neutral, it may be useful in investigating the gravitational interaction between matter and antimatter, where electrical neutrality is essential. In addition, the atomic interaction between matter and antimatter, leading to annihilation, is of theoretical and (as we will see below) engineering importance.

The two most often discussed processes for producing this exotic atom from its constituents are radiative capture

$$e^+ + \bar{p} \rightarrow \bar{H} + \gamma$$

and charge exchange

$$Ps + \bar{p} \rightarrow \bar{H} + e^-.$$

The former has a much smaller cross-section than the latter process, but the two colliding particles can be stored and recycled. The latter is of considerable theoretical interest and may also be a practical production method. The nice point about calculating the charge exchange cross-section is that it can be simply related to the cross-section for Ps formation

$$e^+ + H \to Ps + p$$

by application of charge conjugation and time reversal invariance. Humberston then reviewed the existing calculations of this process, including his own definitive ones at low energies. In addition, certain less accurate calculations involving excited states were discussed, and a total antihydrogen cross section of at least $20\pi a_0^2$ can be estimated. With expected currents of the initial particles the production rate of antihydrogen atoms would be a few per second.

David Morgan (Livermore) gave a talk that gracefully spanned the gulf between understanding the antiproton-matter interaction and considering possible engineering applications. The latter include interplanetary and interstellar rocket propulsion (requiring the production and storage of antimatter in gram amounts) and the more modest goals of antiproton cancer therapy and three-dimensional imaging of materials including the human body.

Most of the lecture was concerned with the fundamentals of antimatter annihilation following slowing down in matter. The most interesting physics, perhaps, concerns what happens to a proton of low energy when it encounters a hydrogen atom. The annihilation rates at low- and sub-eV energies are dominated by a rearrangement reaction in which the antiproton becomes bound to a nucleus, while the electron absorbs the energy, most likely through ionization. An adiabatic treatment follows the $\bar{p}$ along a classical trajectory through the atom under the influence of the induced dipole (R^{-4}) potential. If it reaches the critical radius of $0.639a_0$ there is a very high (80%) probability of rearrangement, capture, and subsequent annihilation of the two heavy particles mainly into pi mesons. The accuracy of the calculation is increased by the fact that the inner turning point of the antiproton's path is a discontinuous function of the impact parameter. From the critical impact parameter the cross-section can be computed.

BOUND STATES AND RESONANCES

Yew Kam Ho (Louisiana State) gave a review of the techniques for calculating bound states and resonances involving positron-containing systems and reported on the most recent results. In particular, he discussed the "polyelectron" systems Ps^- (Ps^+) and Ps_2. The former has been known to be bound for many years, and recent variational calculations have obtained its binding energy to an accuracy of one part in 10^{12}, along with its lifetime against annihilation. An advantage of the variational method is that without additional work one can look for scattering resonances; these appear as "stabilized" eigenvalues above the e-Ps threshold. In this way Ho has found several resonances lying below higher thresholds of Ps that correspond exactly to the well-known resonances in e-H scattering. The widths of these resonances can be calculated with the complex-rotation method: the three-body Hamiltonian is analytically continued by a certain simple transformation on the coordinates of the particles, after which diagonalization is carried out as usual. The resulting energy spectrum has some isolated complex eigenvalues of the form

$$Z = E - i\Gamma/2$$

where E is the (real) energy and Γ is the width of the resonance. In this way the existence of resonances indicated by stabilized energies can be verified. Ho has carried out these types of calculations to quite high accuracy.

The Positronium molecule Ps_2 has also been known to be particle stable, but its exact binding energy has proven to be more elusive. Ho reported what is probably the best variational result for the binding energy, using a clever trick to adapt a previously existing computer program. A well-verified program to treat the PsH system ($2e^-, e^+, p$) was modified by replacing the mass of the proton by that of a positron. For simplicity, only the electron system was explicitly anti-symmetrized; as expected, the positron part of the wave function separated into singlet and triplet parts, and the lowest energy could be found quite accurately. In this system too there are resonances; in this case they are physically described as quasi-bound Coulomb states of an electron and a Ps ion.

Some years ago the Ps ion gained added respectability when Allen Mills (AT&T Bell Labs) produced it by double electron pickup by a positron and measured its annihilation lifetime. At the Workshop he gave a progress report on new experiments which should be able to measure the lifetime with enough precision to challenge the theorists.

For a number of years there has been the suspicion that bound states or at least long-lived resonances must exist in systems containing a large molecule and a positron, based on the very large observed annihilation rates. The argument was that such a large value must really represent a fairly large cross-section for forming the complex, after which annihilation occurs with almost 100% efficiency. Two new types of experiments were reported at the Workshop that strengthen that conclusion. Cliff Surko (San Diego) conducted experiments on storing positrons in a Penning trap, allowing them to enter and then lose enough energy by collision with background nitrogen gas that they cannot exit. The first experiments were disappointing; confinement times were several orders of magnitude shorter than expected. It turned out that the particles were not escaping but being annihilated at rates much too high for N_2; impurity molecules in trace amounts were responsible. Further introduction of hydrocarbon molecules in controlled amounts led to the clear conclusion that resonances were occurring, and a simple model of the resonance formation was applied.

D. L. Donohue (Oak Ridge) and his colleagues expanded on this work in an interesting way. They constructed a high-quality mass spectrometer to analyze the ions resulting from removal of an electron from these large molecules by annihilation. Initial results were encouraging. They showed strong evidence for e^+ attachment in toluene, followed by annihilation in which almost no fragment ions were produced.

CONCLUSION

In addition to the particular topics I singled out for discussion, there were many other interesting invited and contributed papers. Gottfried Spicher (Bielefeld) and his colleagues presented results on the first e^+-H crossed-beam experiment, something we have all been waiting for. Jack Straton (Kansas State) presented the first results of a heroic calculation of Ps formation in e^+-H^- collisions using the Fock-Tani method. Alkali atoms were featured, both experimentally and theoretically. David Schrader (Marquette) discussed positrons in liquids, illustrating the contrast with gases. Leonard Roellig (CUNY) described some experiments on Ps-surface interactions. And there were more.

The field seems as active as ever, and I look forward to the next Workshop, to be held in 1991 in Sydney, Australia.

LOW ENERGY POSITRON HYDROGEN ATOM SCATTERING USING CCA

A.S.Ghosh, M.Mukherjee and Madhumita Basu

Department of Theoretical Physics
Indian Association for the Cultivation of Science
Jadavpur, Calcutta 700032, INDIA

The article reports low order phase shifts using CCA with different coupling schemes. The phase shifts are found to be systematically improved with the increase of target states. The effect of Ps formation is found to be significant in the energy range considered. The CCA predictions including the rearrangement channel are found to be encouraging.

Close coupling or pseudostate close coupling approximations are found to be successful in predicting at least the elastic scattering parameters. On the other hand, literature reveals that the elastic phase shifts for positron-atom scattering at low incident energies obtained by employing close coupling approximation (CCA) are not very encouraging when compared with variational predictions. This is evident from Fig.1. The elastic s-wave phase shifts using six (1s,2s,2p,3s,3p,3d)-state CCA[1] and four (1s,2s,2p-2p,3d) pseudostate CCA[2] are compared with exact variational numbers of Bhatia et al[3]. The results of McEachran and Fraser[1] are very poor. The inclusion of pseudostates in the coupling scheme although improves the results to a large extent but still the difference between the pseudostate CCA and variational predictions is appreciable. The situation in the case of elastic p- and d-wave phase shifts are not very much different. The conclusion that CCA is unsuitable for studying positron-atom scattering is based on these predictions. However, in most of the CCA calculations (Ghosh et al[4]), the rearrangement channel is not taken into account.

In positron-atom scattering, long- and short-range correlations are of vital importance. In the framework of CCA the main effect of distortion of the atom may be included by retaining the (O)-orbitals of the target in the coupling scheme. The inclusion of the positronium (Ps) formation channel in the coupling scheme accounts for the important attractive short-range effect on the elastic channel. Ps-formation channel as either a real or virtual process has no parallel in the electron scattering. The inclusion of Ps formation channel in the coupling has been emphasised by Burke et al[5], Fon and Gallahar[2] and Mukherjee et al[6].

Consider the importance of the long- and short-range correlation, we employ the following coupling schemes
 i) H(1s,2s,2p),Ps(1s)
 ii) H(1s,2s,2p),Ps(1s)
 iii) H(1s,2s,2p,3d),Ps(1s)
The pseudostates H(2p) and H(3d) are taken from Damburg and Karule[7]. However we are aware that the pseudostates employed may not be very suitable at low energies. The use of these pseudostates are qualified by considering the fact

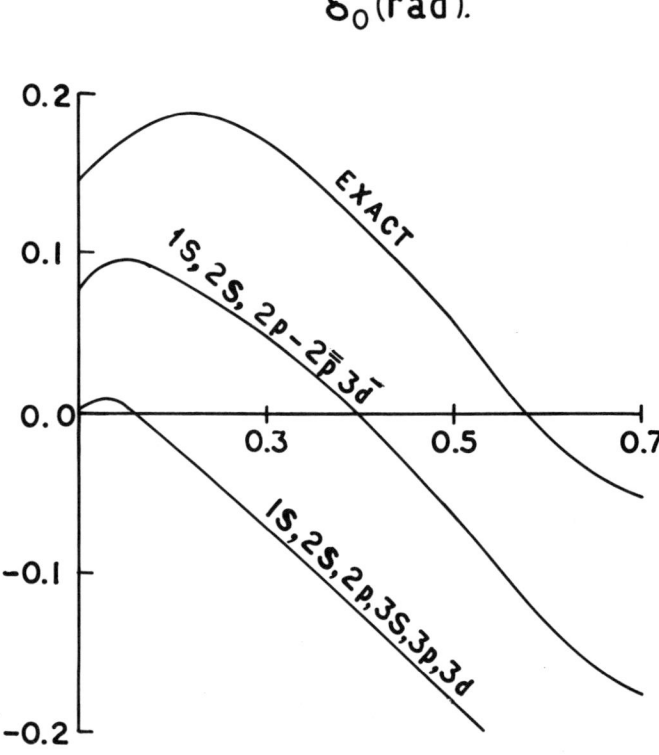

Figure 1. s-wave phase shifts.

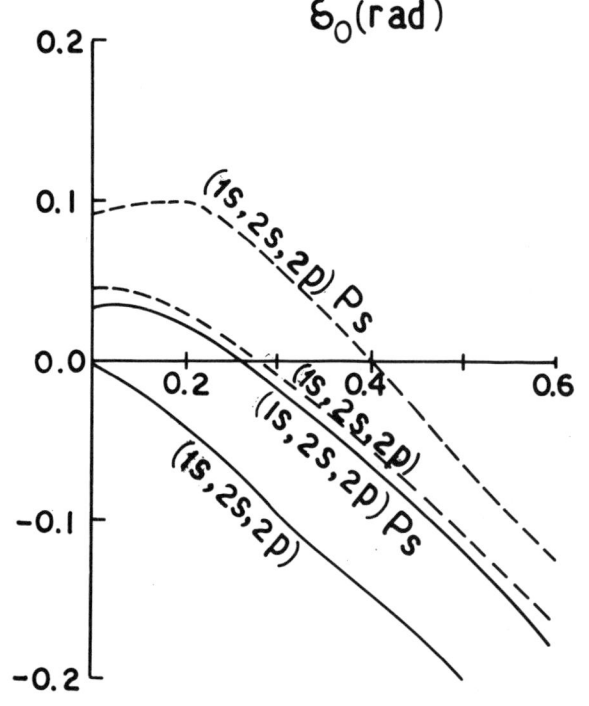

Figure 2. s-wave phase shifts.

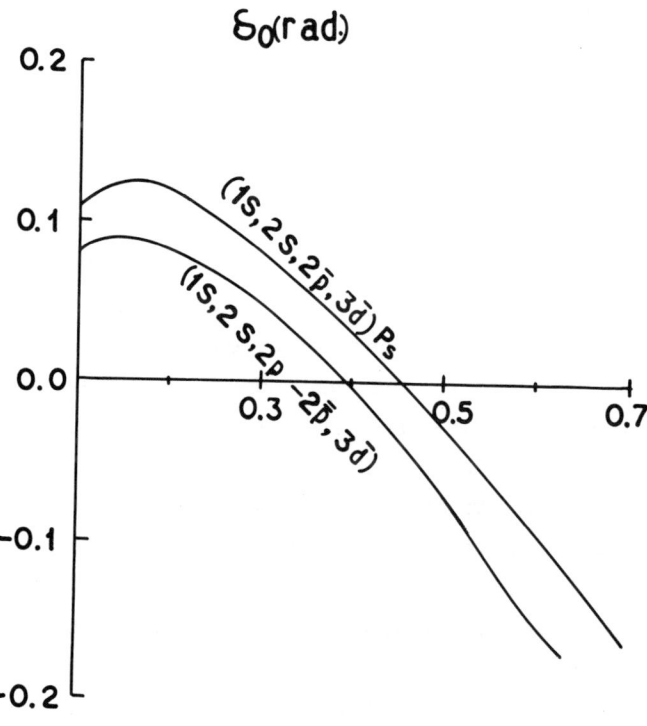

Figure 3. s-wave phase shifts.

that the proper choice of the coupling scheme instead of mere choice of pseudostates is our motivation.

In Figs. 2-3, the s-wave phase shifts below the Ps-formation threshold are shown using different coupling schemes. The effect of Ps-formation can be estimated by comparing the two results with and without inclusion of Ps formation channel in the coupling scheme. The inclusion of Ps-formation channel influence the s-wave phase shifts very significantly. The elastic phase shifts (model iii) are enhanced by about 50/ over those of corresponding coupling scheme neglecting Ps-formation channel. The enhancement of s-wave phase shifts with the inclusion of Ps-formation channel assures that the effect of the rearrangement channel on the elastic scattering is short-range and attractive in nature. Moreover, the effect of rearrangement channel in e$^+$-atom scattering is of key importance.

Fig. 4 presents the s-wave phase shifts using our three models. The varia-

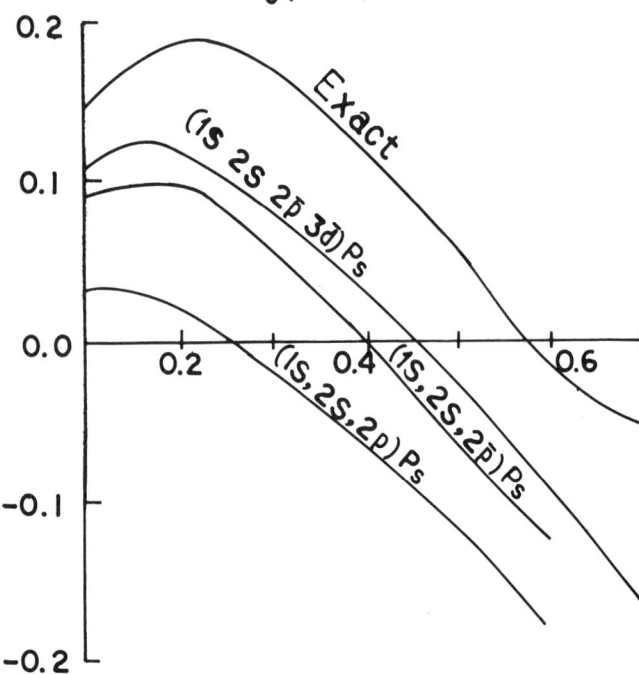

Figure 4. s-wave phase shifts.

tional predictions of Bhatia et al[3] are compared with the present results. The present s-wave phase shifts are approaching towards exact variational numbers with the increase of target states in the coupling scheme. The effect of added target states can be judged from the present three results. The present results using model (iii) (H(1s,2s,2p,3d), Ps) are improved by about 50/ over the existing best CCA predictions (Fon and Gallahar[2]) and are in fair agreement with the rigorous lower bound results of Drachman[8]. However, the present best results still differ appreciably from those of Bhatia et al[3].

Three sets of present p-wave phase shifts are plotted in Fig.5. The inclu-

sion of Ps-formation enhances the p-wave

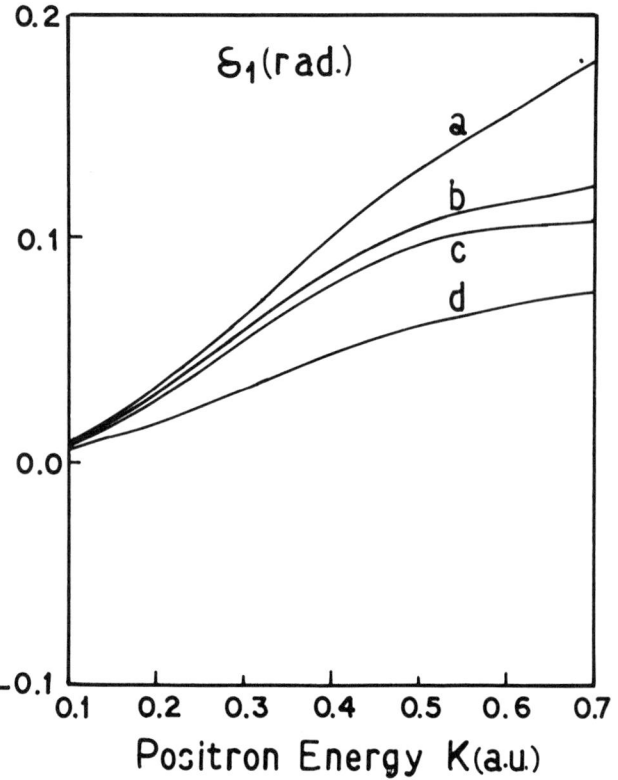

Figure 5. p-wave phase shifts. a: Variational results[9], b: 1s,2s,2p,3d,ps, c: 1s,2s,2p,ps, d: 1s,2s,2p,ps

phase shifts very appreciably as in s-wave. The relative importance of the addition of 2p and 3d target states are evident (Fig.5.). The agreement between the present p-wave phase shifts (model iii) and exact variational prediction (Bhutia et al[9]) is better than that in the case of s-wave.

The present d-wave phase shifts are plotted in Fig.6 along with the reliable predictions of Winick and Reinherdt[10]. Here also, we have found that the effect of Ps formation is very appreciable. The present d-wave phase shift (model iii) are in excellent agreement with those of Winick and Reinherdt. In other words, we conclude present d-wave results are reliable.

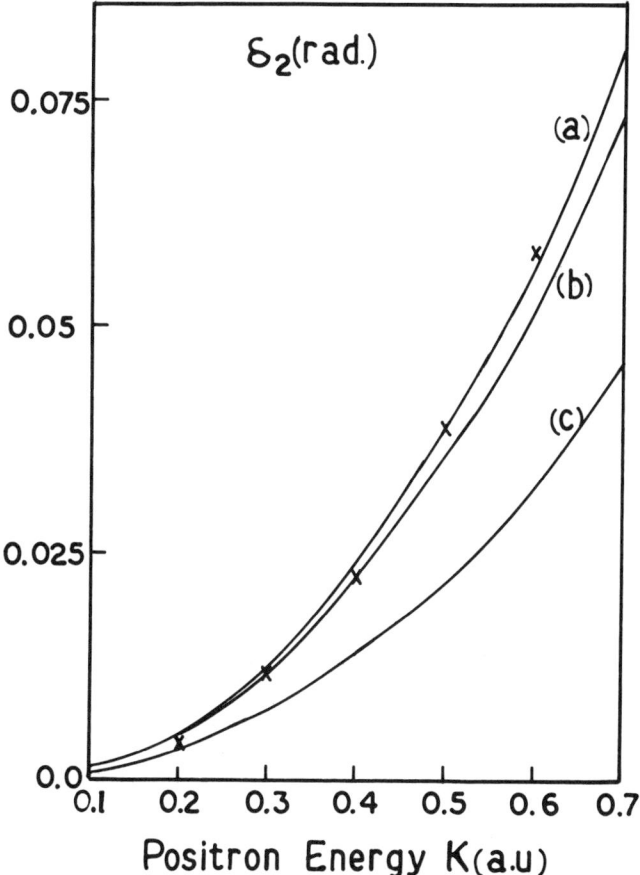

Figure 6. d-wave phase shifts. a: 1s,2s,2p,3d,ps, b: 1s,2s,2p,ps, c: 1s,2s,2p,ps. X: Winick and Reinherdt.

The e^+-H scattering process has been studied using CCA with three different coupling schemes which explicitly include the Ps formation channel. The present s-wave phase shifts below the Ps-pick up threshold, are found to be systematically improved as we increase the target states. The effect of Ps formation channel is found to be very significant in the energy range considered. Our best (1s,2s,2p,3d,ps) s-wave phase shifts, though different from the exact variational numbers, are appreciably better than the best pseudostate CCA predictions of Fon and Gallaher and are in fair agreement with rigorous lower bound results of Drachman[8].

With the increase of angular momentum, the differences between variational predictions and present CCA results decrease. The present d-wave elastic phase shifts coalesce with those of Wincik and Reinherdt which are considered to be most exact. One can obtain accurate elastic integrated and differential cross section at low incident energies by combining the variational results (2) and the present CCA results (2).

In the present calculation, we are not very careful in choosing the pseudostates. Moreover, we have not included any ns pseudostates. Perkins[10] pointed out inclusion of ns state will improve the s-wave phase shifts appreciably. Calculation with the above refinements will be worth trying to justify the use of CCA in investigating e^+-H scattering

References

[1] R.P. McEachran and P.A. Fraser, Proc. Phys. Soc. B6, 369 (1965).

[2] W.C. Fon and D.F. Gallahar, J. Phys. B5, 943 (1972)

[3] A.K. Bhatia, A. Temkin, R.J. Drachman and H. Eiserike, Phys. Rev. A3, 1328 (1971)

[4] A.S. Ghosh, N.C. Sil, and P. Mandal, Phys. Rep. 87, 313 (1982)

[5] P.G. Burke, H.M. Schey and K. Smith, Phys. Rev. 129, 1258 (1963).

[6] M. Mukherjee, M. Basu and A.S. Ghosh, Phys. Rev. A38, 1079 (1988).

[7] R.J. Damburg and E.M. Karule, Proc. Phys. Soc. 90, 637 (1967).

[8] R.J. Drachman, Phys. Rev. 173, 191 (1968).

[9] A.K. Bhatia, A. Temkin and H. Eiserike, Phys. Rev. A9, 219 (1974).

[10] J. R. Winick and W. P. Reinherdt, Phys. Rev. A$\underline{18}$, 925 (1978).

[11] J. F. Perkins, Phys. Rev. $\underline{173}$, 164 (1968).

MEASUREMENTS OF POSITRON AND ELECTRON TOTAL AND ELASTIC SCATTERING BY ATOMS

W.E. Kauppila and T.S. Stein

Department of Physics and Astronomy, Wayne State University
Detroit, Michigan, USA

Recent measurements of positron and electron total scattering by alkali metal atoms (sodium and potassium) and differential elastic scattering by the inert gases are discussed. These measurements are helping to provide information on how positron and electron total scattering cross sections (Q_t's) by a given atomic target may be the same, while the separate scattering channels contributing to Q_t may still be quite different for positrons and electrons.

1. INTRODUCTION

A comparison of cross sections for the scattering of positrons and electrons by helium as shown in Fig. 1 provides a good representative background of the differences and similarities that occur for the scattering of positrons and electrons by atoms. The Q_t measurements of Kauppila et al.[1,2] and Stein et al.[3] for positrons (+) and for electrons (o), which have been made in the same experimental system, revealed a merging of the Q_t's (to within 2%) for energies above 200 eV, while at lower energies the electron Q_t becomes as much as 100X larger than the positron Q_t. The lowest threshold energies for inelastic scattering are indicated by the labelled arrows, which are positronium (Ps) formation for positrons and atomic excitation (exc.) for electrons. As a result, it is seen that the largest difference in the Q_t's occur in the purely elastic scattering energy region while the Q_t's merge where elastic and all possible inelastic channels are open.

A comparison of the above Q_t measurements in the elastic scattering energy region shows excellent agreement for electrons with the variational calculations of Nesbet[4] (———), and very good agreement for positrons with a theoretical result (heavy ———) obtained by Wadehra et al.[5] (using the s-wave phaseshifts of Humberston,[6] the p-wave phaseshifts of Humberston and Campeanu,[7] the lowest set of d-wave phaseshifts of Drachman,[8] and higher phaseshifts up to l = 20 from the Born approximation expression of O'Malley et al.[9]) when the inability of the experiment to discriminate against all small angle elastic scattering is considered.[5] The low energy electron Q_t measurements of Buckman and Lohmann[10] extend the excellent agreement with Nesbet[4] down to even lower energies.

At higher energies (>100 eV) the distorted wave second Born approximation (DWSBA) calculations of Dewangan and Walters[11] for positrons (heavy ———) and electrons (heavy − −) do not predict a merging to within 2% of the Q_t's until E = 2000 eV. It is interesting that at 200 eV where the positron and electron Q_t measurements agree to within 2% the DWSBA calculation predicts that the elastic scattering cross section (Q_{el}) for electrons (thin − −) should be 2.4 times as large as for positrons (thin ———). This latter comparison of measured Q_t's and theoretical Q_{el}'s for positrons and electrons at 200 eV suggests that the total

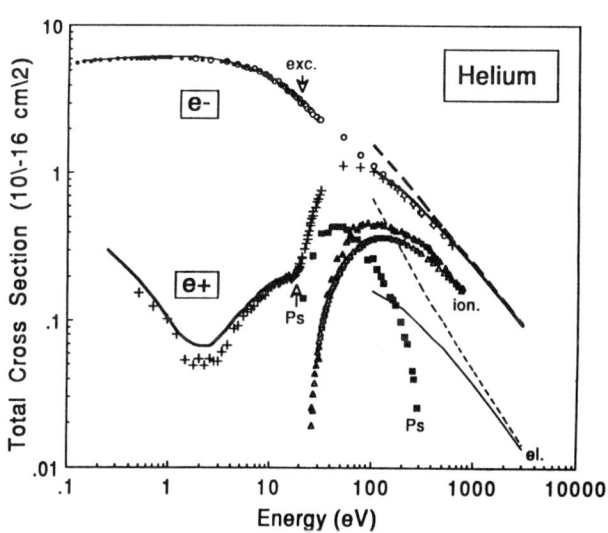

FIG. 1. Comparisons of cross sections for positron and electron scattering by helium.

inelastic cross section (Q_{inel}) for positrons at 200 eV must be larger than for electrons. One important difference between positron and electron scattering by the same target is that positronium formation can only occur for positron scattering. Measurements of Q_{Ps} by Fromme et al.[12] (solid squares) show that Ps formation accounts for about 50% of the positron Q_t between 20-30 eV, but then Q_{Ps} falls rapidly above 100 eV becoming <few% of Q_t above 300 eV. As a result, it is not expected that Ps formation by itself will account for the above-mentioned difference that may be expected between the Q_{inel}'s for positrons and electrons. Relative measurements of the direct ionization cross section for positron-He scattering (closed triangles) by Fromme et al.,[12] when normalized above 750 eV to the electron Q_{ion} measurements (open triangles) of Montague et al.,[13] show that the positron Q_{ion} is larger than that for electrons in the intermediate energy region.

Some other general features observed in comparisons of positron and electron scattering by many atoms (and also molecules) can be seen in Fig. 2 where measured Q_t's for positron and electron scattering by the inert gas atoms[1-3,10,14-17] are displayed. The lowest energy inelastic thresholds are again labelled by the arrows corresponding to atomic excitation for electrons and Ps formation for positrons. For positron scattering there are significant increases in Q_t after the Ps formation thresholds with Q_t reaching an intermediate energy maximum that does not seem to be directly related to Q_{el} (as one might expect from the shape of Q_{el} below the Ps formation threshold), while for electron scattering no noticeable changes in the shapes of the Q_t curves are apparent as the energy is increased through the lowest atomic excitation threshold with the result being that the Q_t maxima for electrons are associated with Q_{el}.

2. TOTAL SCATTERING BY Na AND K

A current set of investigations in our laboratory is to measure Q_t's for positrons and electrons scattering from alkali metal atoms using the experimental setup[18] shown in Fig. 3, where a beam transmission experiment is used to deduce Q_t from the attenuation of the projectile beam as it passes through a heated gas scattering cell containing the alkali vapor. The positron beam for this work (having an intensity of >100/sec and a FWHM energy width <0.1 eV) is obtained from the radioactive decay of a van de Graaff accelerator produced carbon-11 positron source.

Alkali metal atoms are of particular interest for investigation because they are hydrogen-like, they have large atomic polarizabilities, and they have low ionization potentials (which are less than the 6.8 eV

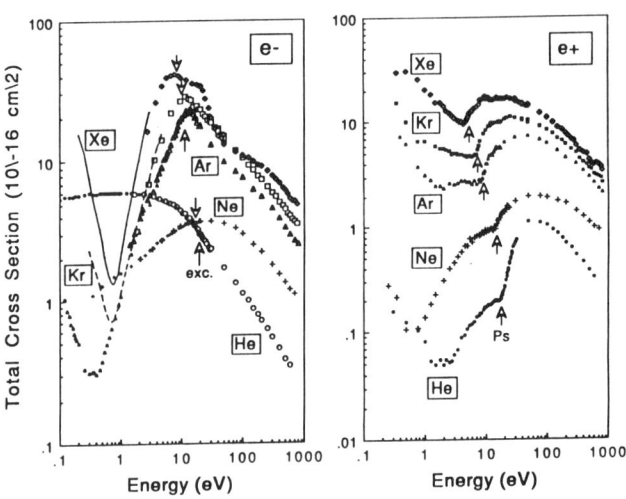

FIG. 2. Comparisons of total cross section measurements for positron and electron scattering by inert gases.

binding energy of Ps in its ground state). As a result of their ionizaton potentials, a positron of arbitrarily small energy interacting with an alkali atom may form Ps.

The most recent comparison Q_t measurements obtained[19] for positrons and electrons scattering from Na and K are displayed in Figs. 4 and 5. The most intriguing aspect of these comparison measurements is that the positron and electron Q_t's for each of these target atoms are remarkably close down to even the lowest

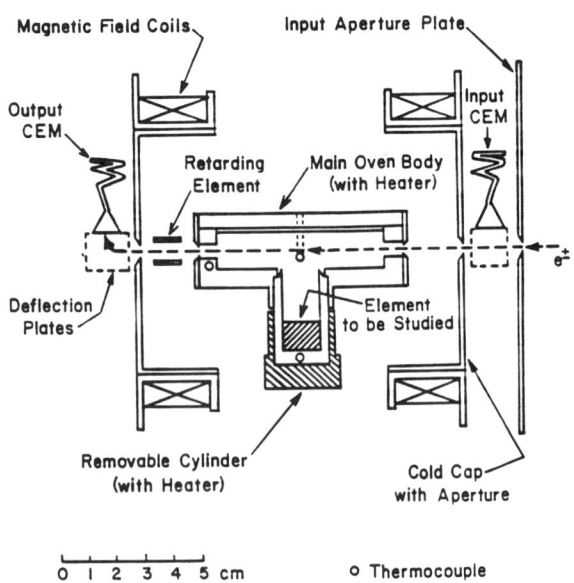

FIG. 3. Experimental setup for measuring Q_t's for positron and electron scattering by alkali metal atoms. (From Ref. 18)

energies of investigation with the positron Q_t's becoming noticeably larger, which is in marked contrast to the situation for the inert gases where the electron Q_t's are appreciably larger at low energies.

Insight can be obtained into the somewhat anomalous behavior for the comparisons of the positron and electron Q_t's for the alkali atoms by considering the partial cross sections that are expected to be the most significant contributors to the Q_t's. Following the lead of a theoretical analysis by Walters[20] to deduce Q_t's for electron scattering by Na and K, the electron results referred to as "Total, Th+Exp" in Figs 4 and 5 represent the sum of the four partial cross sections shown where the elastic ("El.") and ionization ("Ion.") cross sections were selected from existing theoretical and experimental results by Walters,[20] and the cross sections for resonance excitation ("Res."), 3s-3p for Na and 4s-4p for K, and discrete excitations other than resonance excitation ("Other Exc.") are those measured by Phelps and Lin[21] for Na and by Phelps et al.[22] for K. The "Total, Ward" results shown for positron scattering by Na and K are those obtained by Ward et al.[23,24] using five state close-coupling calculations that include the cross sections for elastic scattering, resonance excitation ("Res. Exc."), and a few other discrete excitations (3s-4s, 3d, 4p for Na and 4s-5s, 3d, 5p for K), but do not include the cross sections for Ps formation and ionization, which are both expected to be rather small above 10 eV. The Ps formation cross section values from the distorted-wave approximation calculation of Guha and Mandal[25] are shown in Figs. 4 and 5, while the ionization cross sections for positrons should be close to (and perhaps slightly larger than) those for electrons.

At low energies the Q_t measurements of Kwan et al.[19] are expected to be too small due to an inability to discriminate against all small angle elastic scattering (which for the alkalis is very pronounced), while at higher energies the measurements should be quite close to the actual Q_t's because the experiment can discriminate 100% against all inelastic scattering. As a result of these considerations there does seem to be meaningful agreement between the measurements of Kwan et al.[19] and the "Total" cross sections obtained by adding various partial cross sections. On the basis of the above information it seems reasonable to expect that the positron Q_t curves may rise above the electron curves at lower energies due to the relatively small contributions to the Q_t's of the elastic cross sections, which are predicted to be slightly larger for electrons, and the large contributions of inelastic processes, which are predicted to have larger cross sections for positrons.

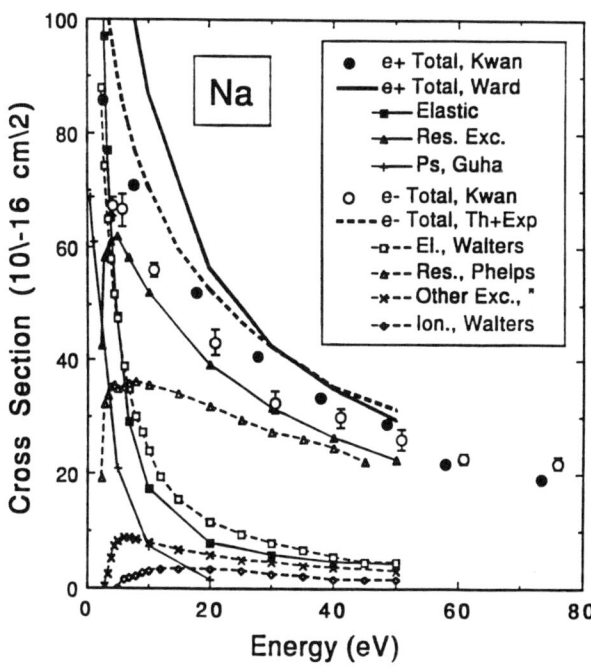

FIG. 4. Comparisons of cross sections for positron and electron scattering by Na.

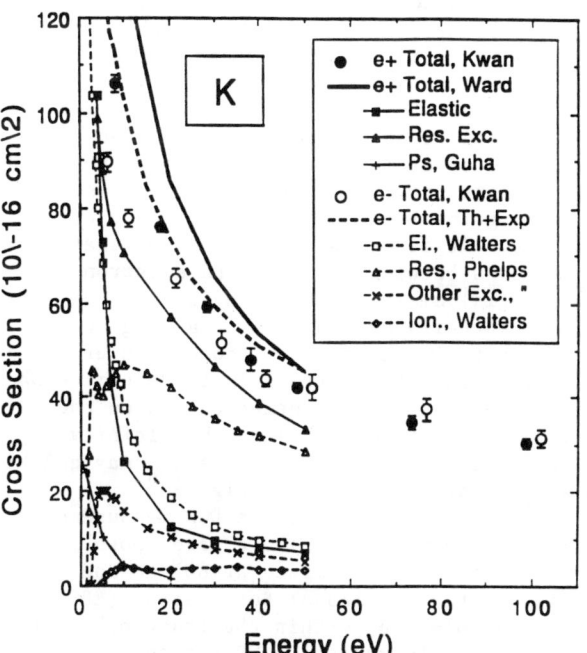

FIG. 5. Comparisons of cross sections for positron and electron scattering by K.

A possible explanation for the observed near merging of the Q_t's for positron and electron scattering by Na and K, discussed by Stein et al.,[26] relates to a theoretical analysis by Dewangan[27] of higher order Born amplitudes calculated in the closure approximation, which have been shown[28,29] to imply that if electron exchange can be ignored for electrons, and the closure approximation is valid, then a merging of the positron and electron Q_t's can occur at energies considerably lower than the asymptotic energies at which the first Born approximation is valid. These arguments also provide a possible explanation[28,30] for the "premature" merging of the Q_t curves for positron and electron scattering by He.

3. DIFFERENTIAL ELASTIC SCATTERING BY INERT GAS ATOMS

Another set of investigations currently underway in our laboratory is to measure relative elastic differential cross sections (DCS's) for the scattering of positrons and electrons by inert gas atoms. The basic experimental setup (see Fig. 6) and approach are the same as that used by Hyder et al.[31] where a projectile positron (or electron) beam is crossed with an atom beam effusing from a capillary array. Two movable channeltron electron multipliers with electric field retarding elements are used to detect positrons (electrons) elastically scattered at angles of 30-134° from the incident beam. The positron beam for this experiment originates from a 160 millicurie sodium-22 radioactive source with a tungsten backscattering moderator, which gives a beam having an intensity >100,000/sec at 100 eV and a FWHM energy width of about 2 eV.

There are several reasons for studying the differential scattering of positrons (and electrons) by atoms. It is well known that DCS measurements provide a more sensitive test of scattering theories than the integrated cross sections. It is of interest to investigate comparisons between positron and electron differential elastic scattering at high energies where eventually one may expect a merging of their scattering behavior. Another interesting investigation for positrons is to determine what effect the onset of Ps formation has on the elastic scattering channel in the vicinity of the Ps formation threshold where Q_{Ps} may quickly become as large as Q_{el} (see Fig. 2).

Some initial positron-He DCS measurements our group has made[32] at 200 eV are shown in Fig. 7 where they are normalized at 60° to and compared with the eikonal Born series (EBS) method calculation (within the framework of the optical model formalism) of Byron and Joachain.[33] It is seen that within the statistical uncertainties (represented by the

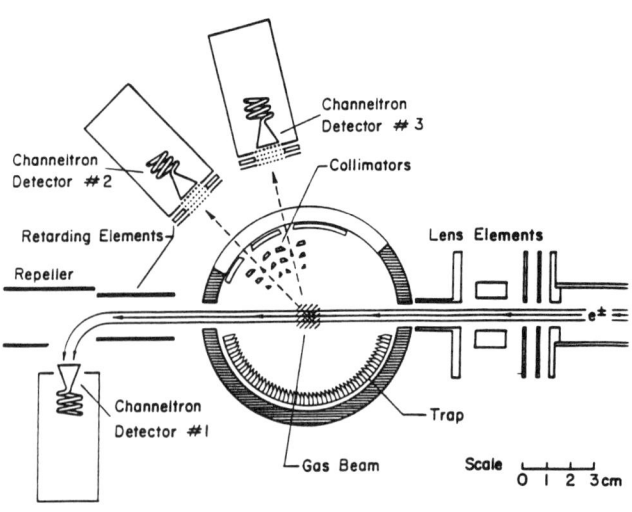

FIG. 6. Experimental setup for DCS measurements. (From Ref. 32)

error bars) the agreement in shape between theory and experiment is good. In the case of the corresponding electron DCS measurements our group has found very good shape agreement with the same EBS calculation of Byron and Joachain.[33] It is noteworthy that the EBS calculation of Byron and Joachain,[33] and the DWSBA calculation of Dewangan and Walters[11] for the positron (and electron) elastic DCS's are very close to being the same, which when

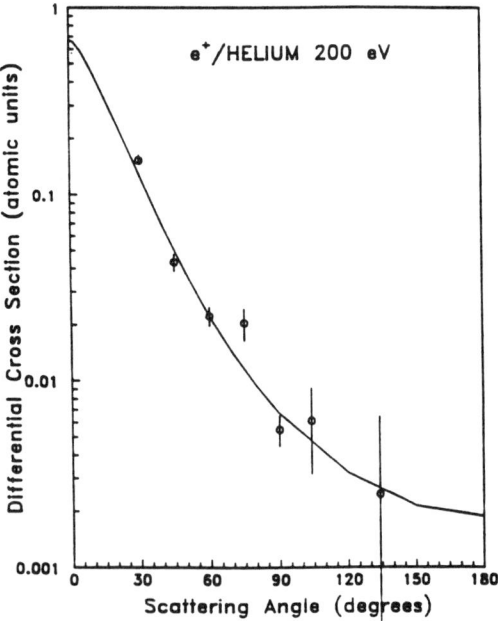

FIG. 7. Elastic DCS results for positron scattering from He at 200 eV. (From Ref. 32)

combined with the information in Figs. 1 and 7 is interesting because this tends to support the predicted large difference in positron and electron Q_{el}'s at 200 eV while the Q_t's are merged to within 2%. This could be consistent with the "closure approximation" arguments of the preceding section.

For positron energies just below the Ps formation thresholds for several of the inert gases our group[32,34] has obtained results (solid circles in Fig. 8) in quite good qualitative agreement with the polarized orbital calculations of McEachran et al.[35-37] (solid line), as shown in Fig. 8 for Ar at 8.7 eV along with prior measurements of Coleman and McNutt[38] (+) and Floeder et al.[39] (open squares) and a calculation by Montgomery and LaBahn[40] (- -). For positron energies above the Ps formation thresholds the DCS measurements[32,34,39] become appreciably different from the polarized orbital calculations, as shown for Ar at 30 eV in Fig. 8. It is to be noted that the polarized orbital calculations of McEachran et al.[35-37] do not include any consideration for the effect of inelastic scattering channels on their elastic scattering calculations. Therefore, it seems that inelastic scattering may be having an effect on the elastic scattering channel. Optical potential calculations made by Bartschat et al.[41] (- -) at 30 eV, where they have considered the "absorption" effects of ten inelastic excitation channels (but not Ps formation or ionization) on the elastic channel, indicate that these "absorption effects" tend to remove the structure (maxima and minima) in the positron DCS curves, which in this respect is consistent with the positron DCS measurements shown in Fig. 8 at 30 eV. The exact shape of the results of Smith et al.,[34] however, differs noticeably from the calculations of Bartschat et al.[41] suggesting that their calculation is incomplete, which may relate to their not including the absorption effect of Ps formation on their calculation. It is relevant to this discussion that the Q_t results in Fig. 2 and the Q_{Ps} measurements of Fornari et al.[42] indicate that Q_{Ps} increases rapidly after the Ps formation threshold becoming more than 50% of Q_t at 20 eV. In the corresponding case of electron differential elastic scattering by Ar the measureements of Smith et al.[34] (solid circles in Fig. 8) are in reasonable agreement with the measurements of Williams[43] (- -) at 8.7 eV, and Williams and Willis[44] (- -) and Srivastava et al.[45] (open squares) at 30 eV. A polarized orbital calculation by McEachran and Stauffer[46] (——) for the electron elastic DCS (where no allowance is made for absorption effects) is seen in Fig. 8 to be in good agreement with experiment at 8.7 and 30 eV, which indicates that absorption effects do not play an important role in electron scattering by Ar.

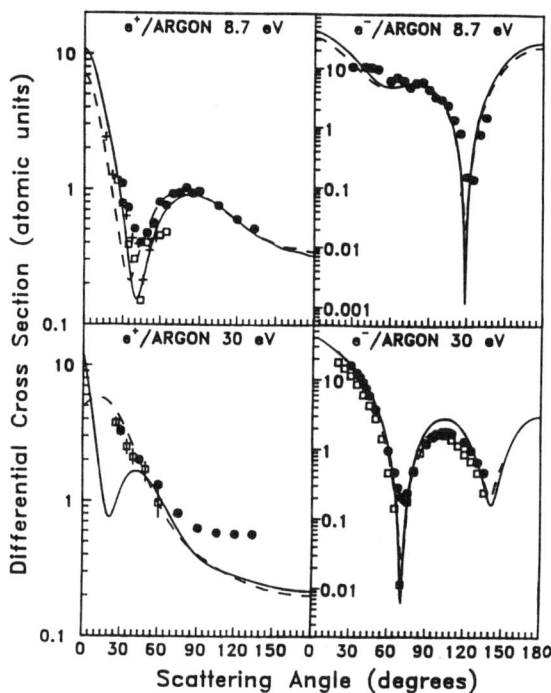

FIG. 8. Elastic DCS results for positron and electron scattering from Ar. (From Ref. 34)

ACKNOWLEDGEMENT

We gratefully acknowledge the support of the National Science Foundation for our research program.

REFERENCES

1. W.E. Kauppila, T.S. Stein, G. Jesion, M.S. Dababneh, and V. Pol, Rev. Sci. Instrum. 48, 822 (1977).
2. W.E. Kauppila, T.S. Stein, J.H. Smart, M.S. Dababneh, Y.K. Ho, J.P. Downing, and V. Pol, Phys. Rev. A 24, 725 (1981).
3. T.S. Stein, W.E. Kauppila, V. Pol, and G. Jesion, Phys. Rev. A 17, 1600 (1978).
4. R.K. Nesbet, Phys. Rev. A 20, 58 (1979).
5. J.M. Wadehra, T.S. Stein, and W.E. Kauppila, J. Phys. B 14, L783 (1981).
6. J.W. Humberston, Adv. At. Mol. Phys. 15, 101 (1979).
7. J.W. Humberston and R.I. Campeanu, J. Phys. B 13, 4907 (1980).
8. R.J. Drachman, Phys. Rev. 144, 25 (1966).
9. T.F. O'Malley, L. Spruch, and L. Rosenberg, J. Math. Phys. 2, 491 (1961).
10. S.J. Buckman and B. Lohmann, J. Phys. B 19, 2547 (1986).
11. D.P. Dewangan and H.R.J. Walters, J. Phys. B 10, 637 (1977).

12. D. Fromme, G. Kruse, W. Raith, and G. Sinapius, Phys. Rev. Lett. 57, 3031 (1986).
13. R.G. Montague, M.F.A. Harrison, and A.C.H. Smith, J. Phys. B 17, 3295 (1984).
14. M.S. Dababneh, W.E. Kauppila, J.P. Downing, F. Laperriere, V. Pol, J.H. Smart, and T.S. Stein, Phys. Rev. A 22, 1872 (1980).
15. M.S. Dababneh, Y.-F. Hsieh, W.E. Kauppila, V. Pol, and T.S. Stein, Phys. Rev. A 26, 1252 (1982).
16. A. Salop and H.H. Nakano, Phys. Rev. A 2, 127 (1970).
17. K. Jost, P.G.F. Bisling, F. Eschen, M. Felsmann, and L. Walther, Proc. ICPEAC-13 Abstracts, p. 91 (1983).
18. T.S. Stein, R.D. Gomez, Y.-F. Hsieh, W.E. Kauppila, C.K. Kwan, and Y.J. Wan, Phys. Rev. Lett. 55, 488 (1985).
19. C.K. Kwan, W.E. Kauppila, R.A. Lukaszew, S.P. Parikh, T.S. Stein, Y.J. Wan, S. Zhou, and M.S. Dababneh (to be published).
20. H.R.J. Walters, J. Phys. B 9, 227 (1976).
21. J.O. Phelps and C.C. Lin, Phys. Rev. A 24, 1299 (1981).
22. J.O. Phelps, J.E. Solomon, D.F. Korff, C.C. Lin, and E.T.P. Lee, Phys. Rev. A 20, 1418 (1979).
23. S.J. Ward, M. Horbatsch, R.P. McEachran, and A.D. Stauffer, J. Phys. B 21, L611 (1988).
24. S.J. Ward, M. Horbatsch, R.P. McEachran, and A.D. Stauffer, J. Phys. B 22, 1845 (1989).
25. S. Guha and P. Mandal, J. Phys. B 13, 1919 (1980).
26. T.S. Stein, M.S. Dababneh, W.E. Kauppila, C.K. Kwan, and Y.J. Wan, in "Atomic Physics with Positrons", edited by J.W. Humberston and E.A.G. Armour, NATO ASI Series B, Vol. 169, pp. 251-263 (Plenum, 1987).
27. D.P. Dewangan, J. Phys. B 13, L595 (1980).
28. H.R.J. Walters, Phys. Rep. 116, 1 (1984).
29. F.W. Byron, Jr., C.J. Joachain, and R.M. Potvliege, J. Phys. B 15, 3915 (1982).
30. H.R.J. Walters, J. Phys. B 21, 1893 (1988).
31. G.M.A. Hyder, M.S. Dababneh, Y.-F. Hsieh, W.E. Kauppila, C.K. Kwan, M. Mahdavi-Hezaveh, and T.S. Stein, Phys. Rev. Lett. 57, 2252 (1986).
32. W.E. Kauppila, S.J. Smith, C.K. Kwan, and T.S. Stein, to appear in Proceedings of the "Workshop on Annihilation in Gases and Galaxies", to be published as a NASA Conference Publication (NASA Goddard Space Flight Center, July, 1989).
33. F.W. Byron, Jr. and C.J. Joachain, Phys. Rev. A 15, 128 (1977).
34. S.J. Smith, W.E. Kauppila, C.K. Kwan, and T.S. Stein, Abstracts of Contributed Papers, ICPEAC XVI, p. 403 (New York, 1989).
35. R.P McEachran, A.G. Ryman, and A.D. Stauffer, J. Phys. B 11, 551 (1978).
36. R.P McEachran, A.D. Stauffer, and L.E.M. Campbell, J. Phys. B 13, 1281 (1980).
37. R.P McEachran, A.G. Ryman, and A.D. Stauffer, J. Phys. B 12, 1031 (1979).
38. P.G. Coleman and J.D. McNutt, Phys. Rev. Lett. 42, 1130 (1979).
39. K. Floeder, P. Honer, W. Raith, A. Schwab, G. Sinapius, and G. Spicher, Phys. Rev. Lett. 60, 2363 (1988).
40. R.E. Montgomery and R.W. LaBahn, Can J. Phys. 48, 1288 (1970); and private communication.
41. K. Bartschat, R.P. McEachran, and A.D. Stauffer, J. Phys. B 21, 2789 (1988).
42. L.S. Fornari, L.M. Diana, and P.G. Coleman, Phys. Rev. Lett. 51, 2276 (1983).
43. J.F. Williams, J. Phys. B 12, 265 (1979).
44. J.F. Williams and B.A. Willis, J. Phys. B 8, 1670 (1975).
45. S.K. Srivastava, H. Tanaka, A. Chutjian, and S. Trajmar, Phys. Rev. A 23, 2156 (1981).
46. R.P McEachran and A.D. Stauffer, J. Phys. B 16, 4023 (1983).

POSITRON-IMPACT IONIZATION OF ATOMIC HYDROGEN

Wilhelm Raith, Björn Olsson, Günther Sinapius, Wolfgang Sperber
and Gottfried Spicher

Fakultät für Physik, Universität Bielefeld, D-4800 Bielefeld 1
Federal Republic of Germany

> In a crossed-beam experiment the impact-ionization cross section, integrated over all angles and energy partitions <u>but excluding positronium formation</u>, has been measured for positrons as well as electrons in the energy region of 30 to 600 eV. Below 400 eV the positron cross section is larger than the electron one, at 40 eV by about a factor of 2. Absolute cross sections are obtained by comparing the electron data with literature values. Our cross sections for positron-impact ionization lie above all theoretical predictions; within the errors they barely agree with the distorted-wave polarized-orbital calculations of Ghosh et al. for the model in which the scattered positron is described by a plane wave.

Of all positron-atom scattering problems the e^+-H interaction can be calculated with the highest accuracy and experimental tests are, therefore, of fundamental importance. In addition, the e^+-H interaction is of great astrophysical interest. Historically, the laboratory experiments on positron-atom scattering began with simple gas-target transmission experiments and only recently progressed to crossed-beam experiments with noble gases.[1,2] For the very first crossed-beam e^+-H measurements we chose the impact-ionization cross section because of its distinctive signature consisting of a formed H^+ ion and a scattered positron. By detecting ion and positron in time correlation the positronium (Ps) formation (in which the positron annihilates) is excluded. By measuring the ion flight time, atomic and molecular ions can be distinguished and the impact ionization of H and H_2 be determined simultaneously. The H_2 is a part of the incompletely dissociated hydrogen beam and the main component of the residual gas in the scattering chamber.

Fig.1 shows schematically the experimental arrangement. The hydrogen atoms emerge from a Slevin type RF-discharge source (Leisk Ltd.). The positrons originate in a 170 MBq ^{22}Na radioactive source. The moderator consists of two highly annealed tungsten meshes. With reversed optical potentials the secondary electrons from the moderator are used for the electron measurements.

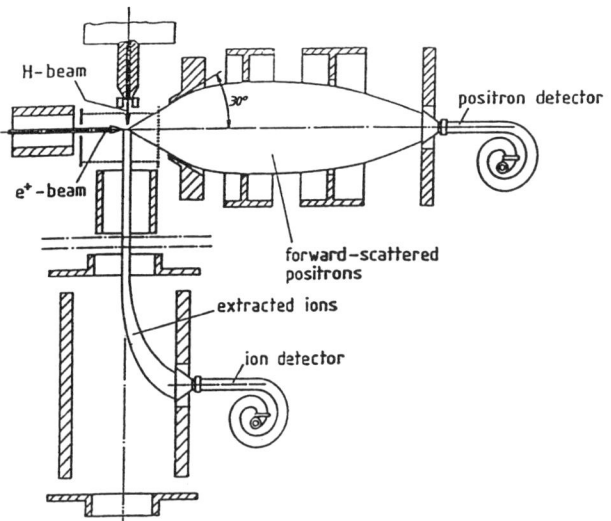

Figure 1. Layout of the experiment

The channel electron multiplier (CEM) No.1 detects the unscattered beam positrons (electrons) and those scattered into a forward cone of 30° apex half-angle. The ions are extracted from the interaction region by a weak electric field. They move along the atomic beam into the detector region where they are electrostatically bent toward the CEM No.2. Typical counting rates for CEM 1 and 2 are 3000 s^{-1} and 10 s^{-1}, respectively. Only about 0.1% of the CEM 2 counts are time-correlated to e$^+$(e$^-$)-detection events occurring within the preceding 4 μs. With "inverted timing" we measure the time between an ion-detection event of CEM 2 and the <u>delayed</u> positron(electron)-detection event of CEM 1. A typical time-distribution spectrum is shown in Fig.2. Clearly visible are the two peaks associated with H$^+$ and H$_2^+$ ions. The "prompt" peak is thought to result from photons originating at the anode of CEM 1 and getting into the funnel of CEM 2, perhaps after wall reflections. This peak provides a useful time mark. The background is attributed to uncorrelated CEM 2 events mainly due to Lyman-α photons from the discharge tube. Only after shielding the discharge tube and blackening all internal surfaces was this background low enough to permit measurements. From the peak signal, corrected for background, we determine the relative impact-ionization cross sections at a given initial energy of the positrons(electrons).

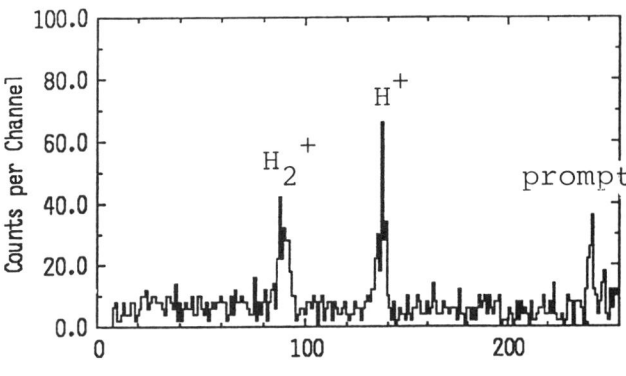

Figure 2. Typical time-correlation spectrum. Time interval = 4μs. Start of MCA by response of CEM 2; Stop by response of CEM 1, delayed for 4μs.

Table 1 lists the relevant ionization processes, the cross-section symbols used in this paper and the respective threshold energies. We measure $\sigma_{Ion}(H)$ and $\sigma_{Ion}(H_2)$ for positron(electron) impact. The Ps-formation processes with cross sections $\sigma_{Ps}(H)$, $\sigma_{Ps}(H_2)$ and $\sigma_{Ps,Diss}(H_2)$ lead to uncorrelated ions which are not measured here. The dissociative cross section $\sigma_{Ion,Diss}(H_2)$ contributes, in principle, a correlated H$^+$ ion. However, in

Process	Cross Section	Threshold
$e^+ + H \rightarrow e^+ + e^- + H^+$	$\sigma_{Ion}(H)$	13.6 eV
$e^+ + H_2 \rightarrow e^+ + e^- + H_2^+$	$\sigma_{Ion}(H_2)$	15.4 eV
$e^+ + H \rightarrow Ps + H^+$	$\sigma_{Ps}(H)$	6.8 eV
$e^+ + H_2 \rightarrow Ps + H_2^+$	$\sigma_{Ps}(H_2)$	8.6 eV
$e^+ + H_2 \rightarrow Ps + H + H^+$	$\sigma_{Ps,Diss}(H_2)$	11.1 eV
$e^+ + H_2 \rightarrow e^+ + e^- + H + H^+$	$\sigma_{Ion,Diss}(H_2)$	17.9 eV

Table 1. The relevant processes for the positron-impact ionization of atomic and molecular hydrogen

tests with 100 eV electrons or positrons and a pure H_2 beam (discharge turned off) we could not detect any H^+ ions. Apparently, most dissociative ionization events proceed via highly excited states and yield ions of high kinetic energy which we collect with low efficiency. For electrons at 100 eV their contribution to our H^+ peak is less than 1% although the cross section $\sigma_{Ion,Diss}^-(H_2)$ amounts to 7% of $\sigma_{Ion}^-(H_2)$.[3]

The most serious disadvantage of our experimental arrangement is the limitation of the positron (electron) detection to scattering angles of less than 30°. From studies of differential electron-impact ionization, however, it is known that most ionization events are associated with small momentum transfers and correspondingly small scattering angles of the primary electrons. This behavior is typical for higher energies but does not apply to ionization near threshold. By taking electron data as a function of the energy and comparing them with literature values it is possible to find out how the restriction to small angles affects our measurements.

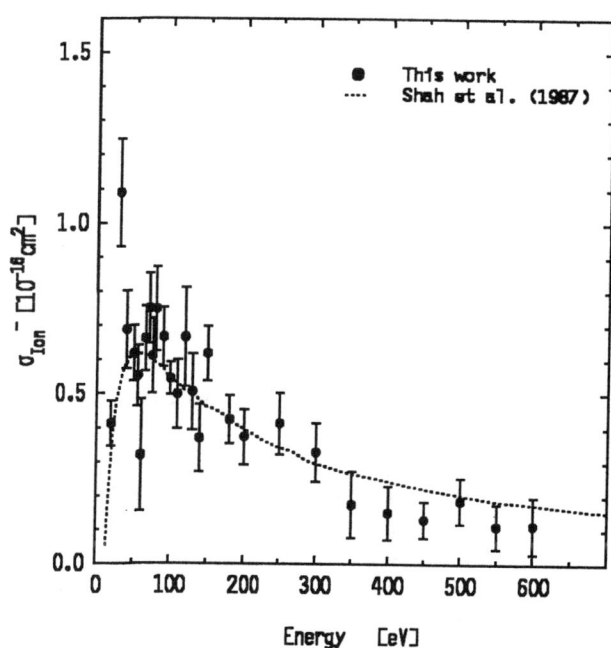

Figure 3. Electron-impact ionization of atomic hydrogen. Our data compared with and normalized to the experimental results of Shah et al.

Fig.3 shows our electron ionization cross section $\sigma_{Ion}^-(H)$, normalized to the values of Shah et al.[4] at 100 eV. The good agreement over the whole range of measurement indicates that we detect nearly all the electrons scattered in the ionization processes. The also-measured cross section $\sigma_{Ion}^+(H_2)$ agrees very well with the results of Fromme et al.[5] obtained with an entirely different set-up.

The cross sections σ_{Ion}^+ and σ_{Ion}^- should merge at sufficiently high energy: the influence of exchange effects in electron scattering and Ps formation in positron scattering decreases with increasing energy and, ultimately, the First Born Approximation (FBA) should become valid for both. It is of great interest to test experimentally at what energy the merging occurs. Our data accumulated thus far are consistent with equal positron and electron cross sections above 500 eV.

By normalizing our electron cross section $\sigma_{Ion}^-(H)$ to the experimental results of Shah et al. (which have an absolute uncertainty of 7%) we obtain an absolute scale for our positron data, $\sigma_{Ion}^+(H)$. The latter is plotted in Fig.4. Compared with the maximum of the electron cross section (Fig.3) our positron maximum is about twice as high. All theoretical prediction[6-9] lie more or less below the experimental values (Fig.5) but give approximately the right shape of the curve and the maximum at the right energy.

Ghosh et al.[6] calculated the total impact ionization cross section up to 58 eV by using a distorted-wave polarized-orbital method. They found a strong dependence on the chosen final-channel wave function. Their results obtained with a plane wave for the scattered positron, which are close to their FBA results, are shown in Fig.5. They agree with our data within the statistical and normalization errors. (However, their results obtained with a Coulomb wave for the scattered positron (not shown in Fig.5) lie considerably lower.)

The predictions of Mukherjee et al.[7] (their model DCPE which yields the highest cross-section maximum) lie lower. These authors implemented some corrections which were used by Campeanu et al.[10] for positron-impact ionization of helium and gave good agreement with the experimental results of Fromme et al.[11] The Monte-Carlo results of Ohsaki et al.[8] and

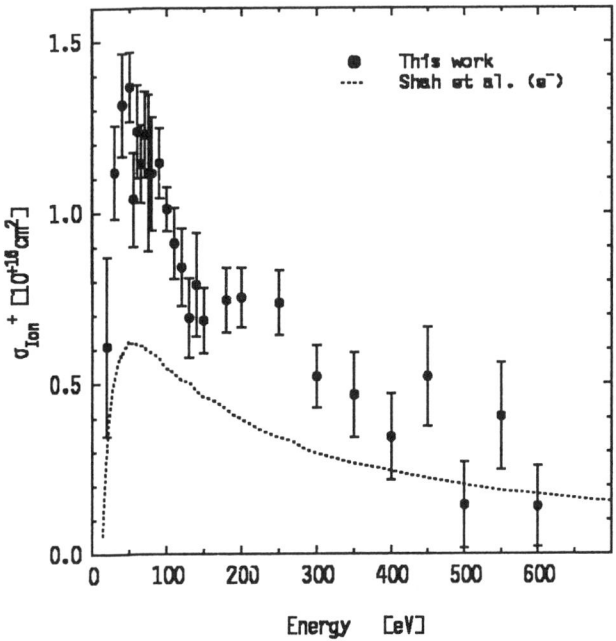

Figure 4. Positron-impact ionization of atomic hydrogen. Our data compared with the <u>electron</u> cross sections of Shah et al.

Wetmore and Olson[9] lie definitely outside the error margin of our measurements.

The work on e^+-H scattering reported here has been limited by the very low signal rate of less than 0.01 s^{-1} which prohibited more ambitious measurements. The main experimental problem was the reduction of the background from the hydrogen source due to Lyman-α photons and charged particles. The data presented here were obtained in automated around-the-clock data-taking over 100 days. The tune-up for finding the first correlation peaks was rather difficult and took several months.

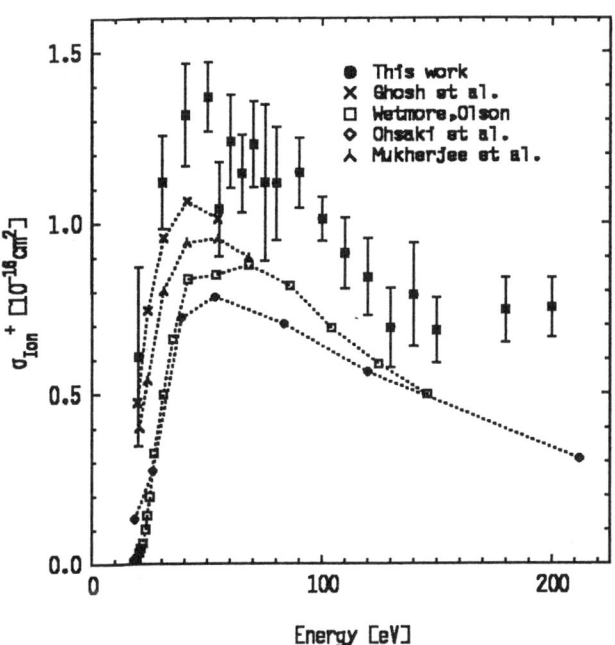

Figure 5. Our data up to 200 eV, compared with theoretical predictions.

In the near future this experiment will be moved to the high-current positron source of the Brookhaven National Laboratory (BNL).[1,2] We will perform further e^+-H measurements in collaboration with other physicists, in particular Dr. Kelvin Lynn (BNL), who developed the unique BNL positron source and Dr. Arnim Schwab (Bielefeld), who is already working at BNL. By repeating the reported measurements with higher positron currents it will be possible to determine in what energy range the large-angle scattering can no longer be neglected. The higher current will also permit measurement of <u>differential</u> ionization by detecting positrons scattered only within a small angular interval. With the use of a quadrupole mass spectrometer, set at mass one, we can measure the sum of the two cross sections $\sigma_{Ion^+}(H) + \sigma_{Ps}(H)$ and then separate $\sigma_{Ps}(H)$. The use of the mass spectrometer will considerably improve the signal-to-noise ratio of the ion detection. Measurements on elastic and inelastic e^+-H scattering will follow.

Starting April this year, our work has been supported by the Bundesminister für Forschung und Technologie (Grant 03-RA2BIE). Before that, this work had been supported by the Deutsche Forschungsgemeinschaft (Grant RA 297/6-3).

References

1. G.M. Hyder, M.S. Dababneh, Y.-F. Hsieh, W.E. Kauppila, C.K. Kwan, M. Mahdavi-Hezaveh, and T.S. Stein, Phys.Rev.Lett. <u>57</u>, 2252 (1986)

2. K. Floeder, P. Höner, W. Raith, A. Schwab, G. Sinapius, and G. Spicher, Phys.Rev.Lett. <u>60</u>, 2363 (1988)

3. D. Rapp, P. Englander-Golden and D.D. Briglia, J.Chem.Phys. <u>42</u>, 4081 (1965)

4. M.B. Shah, D.S. Elliott, and H.B. Gilbody, J.Phys.B <u>20</u>, 3501 (1987)

5. D. Fromme, G. Kruse, W. Raith, and G. Sinapius, J.Phys.B. <u>21</u>, L261 (1988)

6. A.S. Ghosh, P.S. Majumdar, M. Basu, Can.J.Phys. <u>63</u>, 621 (1985)

7. A. Ohsaki, T. Watanabe, K. Nakanishi, K. Iguchi, Phys. Rev.A <u>32</u>, 2640 (1985)

8. A.E. Wetmore, R.E. Olson, Phys.Rev. <u>34</u>, 2822 (1986)

9. K.K. Mukherjee, N.R. Singh, P.S. Mazumdar, J.Phys.B <u>22</u>, 99 (1989)

10. R.I. Campeanu, R.P. McEachran, and A.D. Stauffer, J.Phys.B <u>20</u>, 1635 (1987)

11. D. Fromme, G. Kruse, W. Raith, and G. Sinapius, Phys.Rev. Lett. <u>57</u>, 3031 (1986)

12. K.G. Lynn, A.P. Mills, Jr., L.O. Roellig, and M. Weber, in <u>Electronic and Atomic Collisions</u>, D.C. Lorentz, W.E. Meyerhof, J.R. Peterson (eds.), Elsevior Science Publishers B.V., 227 (1986)

SLOWING DOWN OF POSITRONS IN SOLIDS

Peter J. Schultz, L.R. Logan, W.N. Lennard and G.R. Massoumi

Department of Physics, The University of Western Ontario
London, Ontario, CANADA N6A 3K7

When monoenergetic positrons enter a solid they scatter and lose energy via processes similar to those for electrons. Theoretical details of these processes have been well established for decades, but experimental results using low energy positron beams are only now becoming available for comparison. We review the theoretical results for elastic and inelastic scattering of positrons and the predictions that follow for backscattering, inner-shell ionization, energy loss and stopping profiles. Emphasis is given to specific comparisons with calculations for electrons, and recent experimental results in each of these areas are shown.

1.0 Introduction

The interaction of a monoenergetic positron (e^+) with a solid surface is in many ways different from that for the electron (e^-), which is precisely why variable-energy positron beams have recently been receiving considerable attention. In order to use positron-beam techniques for quantitative, analytical studies[1] it is important to know where positrons stop in the solid, how they subsequently diffuse, and how they are affected by electronic and structural properties of the solid.

Similar to electrons, incident positron directions are randomized through Mott (relativistic Coulomb) scattering, and the energy is lost via energy transfer to the bound electrons and by radiative processes (bremsstrahlung). For positrons below ≈100 keV radiative losses are negligible[2]. Energy loss down to a few hundred eV takes on the order of 10^{-12} s, independent of kinetic energy T, at which point the positron is near its final position in the solid. The final stages of thermalization for a positron involve plasmon scattering (≈10–100 eV), electron-hole creation (≈0.1–10 eV), and phonon scattering (≈0–0.01 eV).

The purpose of the present paper is to review the present understanding of positron scattering, energy-loss, and eventual stopping in solids. We develop the discussion around previous theoretical work, emphasizing the differences between positrons and electrons. We also include more recent experimental studies which systematically compare some of the interactions for monoenergetic positrons and electrons in the "low" energy region, ≈0–50 keV.

A general review of the application of variable-energy positrons to studies of solid surfaces and thin films can be found in ref. 1, and a more detailed discussion of positron stopping in solids is given in ref. 2. Theoretical aspects of positron-solid interactions are discussed in refs. 3–6.

2.0 Elastic Scattering

Elastic scattering cross sections for both electrons (σ^-) and positrons (σ^+) from a central Coulomb field of strength Ze were calculated exactly by Mott[7], and presented by several authors[3] as simpler expansions in αZ ($\alpha = e^2/\hbar c$). The cross sections predicted for keV electrons are always larger than (or equal to) those for positrons, as shown for an aluminum target in Fig. 1. The primary reason for this is the fact that the electron exchange-correlation causes the atomic potential to be somewhat larger in the tails for electrons than for positrons. The differences tend to be largest at

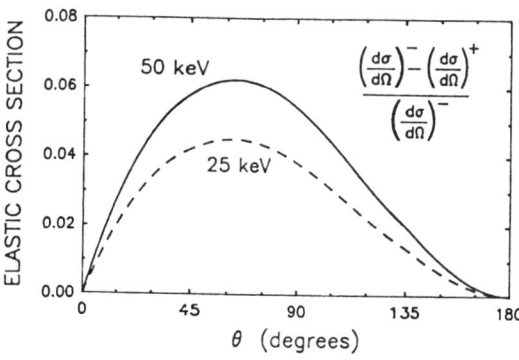

Figure 1. Relative difference of elastic scattering cross sections for electrons and positrons as a function of scattering angle, θ. (From ref. 2)

intermediate angles, although these events are relatively rare. Nevertheless, it is primarily those medium to large angle scattering events which lead to backscattering. The more frequent small angle scattering events lead to a broadening of the incident beam direction[8].

2.1 Backscattering

Large angle elastic (single) scattering, or plural scattering at intermediate angles, ultimately leads to some fraction of the incident beam being backscattered from the solid. The backscattered fraction, η, is a function of material (Z, density and thickness) and of incident particle energy. Differences in the elastic cross sections result in significantly different predictions for η^- (e^-) and η^+ (e^+), as illustrated in Fig. 2. A great deal of experimental evidence for electron backscattering at higher energies has confirmed the shape and (in most cases) the magnitude of the predictions in the figure[9-10].

In the low energy range of present interest experimental data for electron backscattering are sparse (see ref. 2 for a complete list). One area which has so far not received much attention is full doubly-differential (i.e. angle and energy) measurements of the backscattered particle distributions. Preliminary data from our present study are shown in Fig. 3, showing the expected decrease of the most probable energy with increasing scattering angle. The fitting function is known to be incorrect, and more recent data acquired to higher statistics clearly show the appropriate tail projecting towards zero energy. All data were aquired using an optimized surface barrier detector and using 30 keV electrons at normal incidence. These results will be compared with comparable measurements with monoenergetic positrons in the same apparatus.

There are even fewer experimental results for monoenergetic positrons than for electrons. One indirect measurement for silicon[11] revealed no difference (to within 1%) between η^- and η^+ at ≈45 keV, which does not agree with the theoretical predictions that η^- should be at least 10% larger than η^+ (Fig. 2). Other measurements of positron backscattering[12], shown in Fig. 4, support the theoretical prediction that $\eta^+<\eta^-$ for any given material. The figure also shows a recent theoretical prediction[13] based on a simple Kronig—Penny model potential, which suggests that there may not be a monatonic dependence of η on atomic number, as is usually suggested[14]. The data of Baker and Coleman shown in Fig. 4 lend qualitative support to this calculation, since the coefficient for positron backscattering η^+ from Cu ($Z=29$) is anomalously high by comparison with that for Ag ($Z=47$). In spite of decades of research (particularly for electrons), experimental backscattering data are not sufficiently precise or consistent to allow a credible comparison with theory. The study of electron and positron backscattering differences will continue to be an active area of research.

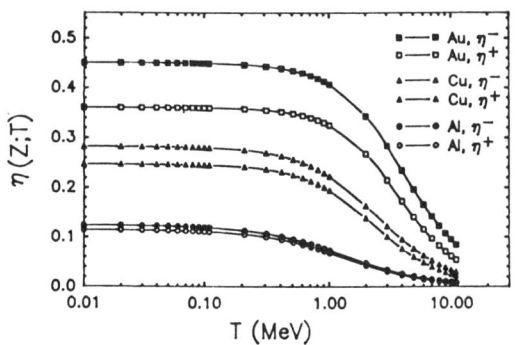

Figure 2. Backscattering coefficients, η, calculated for electrons and positrons. (From ref. 2)

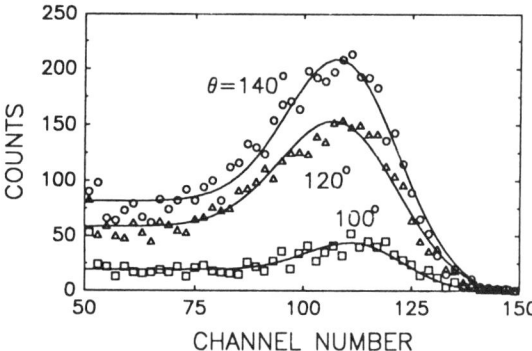

Figure 3. Doubly differential distributions of 30 keV incident electrons scattered from a Au target. The incident beam energy corresponds to channel 119, and the detector resolution is ≈4.8 keV (19 channels).

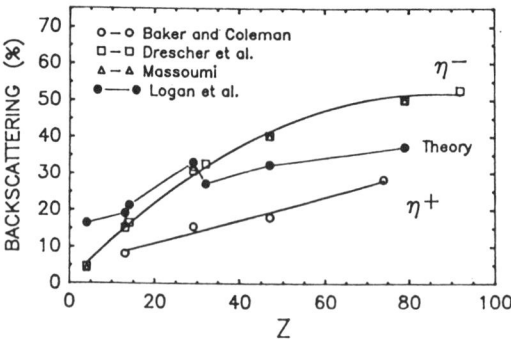

Figure 4. Backscattering coefficients versus atomic number for electrons and positrons at $T=25$ keV. Data are from Baker and Coleman[12], Drescher et al.[25], and Massoumi (this work). Theory is from Logan et al.[13].

3.0 Inelastic Scattering

Scattering of electrons by free electrons is described by Møller's cross section[15], and for positrons the appropriate relativistic cross section was derived by Bhabha[16]. The differences between Møller and Bhabha cross sections for kinetic energies $T=25$ keV and $T=50$ keV are illustrated in Fig. 5. The data are plotted as a ratio to a common prefactor, $\chi/T\epsilon^2$, where ϵ is the fractional energy transfer, q/T, and $\chi=(2\pi e^4)/(mv^2)$.

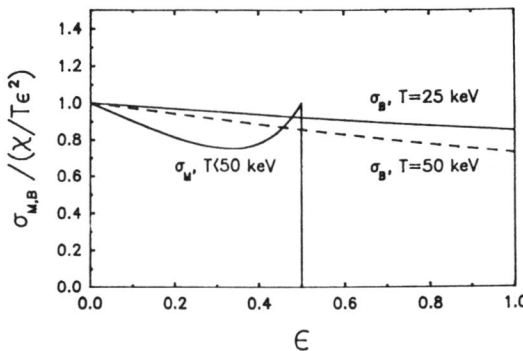

Figure 5. Differential inelastic cross sections for electron–electron (Møller, σ_m) and positron–electron (Bhabha, σ) scattering. (From ref. 2)

Because of the indistinguishability of electrons, the outgoing electron with the higher energy in a binary collision is always defined as the primary electron. This means that the maximum energy transfer possible in Møller scattering is $\epsilon_m = 1/2$. Inelastic scattering at these energies is much stronger for positrons than for electrons due to this cutoff, and the inherently larger Bhabha cross section. This results in larger energy straggling and mean energy loss for positrons than for electrons. At higher energies (>345 keV) the Møller cross section dominates[3].

One effect that is often neglected in calculations of energy loss and inelastic scattering is channeling[17]. Channeling is the phenomenon whereby energetic particles are guided by the semi-continuous potential of a highly ordered crystalline solid. The assumption that classical channeling effects are not important is reasonable for electrons, which interact too strongly with the ion cores at these low energies (<100 keV). In contrast to this, positrons have been shown to have a relatively high probablility of channeling in this energy region[18]. Channeling may be potentially important for stopping distributions of positrons, leading to 2–dimensional depth distributions which have distinct lobes in high symmetry directions.

3.1 Inner–shell ionization

One of the clearest examples of the difference between Møller and Bhabha scattering which has been observed experimentally is for inner–shell ionization. These events involve the largest energy transfers in the slowing down process, although they tend to be rare and therefore less important in stopping than the more numerous outer shell (or soft) collisions.

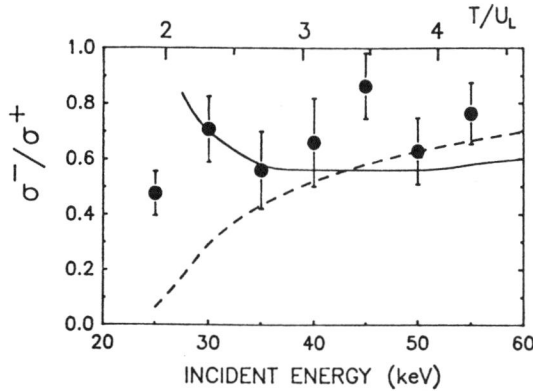

Figure 6. Au L–shell ionization cross section for electrons relative to that for positrons. (From ref. 19)

In Fig. 6 we show results for L–shell ionization of a thin (40 μg/cm^2) Au target using monoenergetic electrons and positrons between 25 and 55 keV[19]. The data are plotted as the ratio of the total cross sections, and show the first directly measured evidence that the Bhabha cross section is greater than the Møller cross section at these energies. Kolbenstvedt (1967) developed a theory for estimating K–shell ionization by electrons which separated the contributions for impact parameters greater than and less than the shell radius. The large impact parameter projectiles contribute to the total ionization only through the electric field, which is similar to a radiation field for $v \to c$. The close collision effects are estimated by integrating the Møller cross section over all energy transfers from the mean ionization potential, $\epsilon = U/T$, to $\epsilon_m = 1/2$. This calculation was extended to positron ionization using the Bhabha cross section, where $\epsilon_m = 1$, and applied to the L–shell in order to obtain the predicted cross section ratio shown by the dashed line in Fig. 6.

The most significant source of the discrepancy between the Kolbenstvedt type of theory and the data shown in Fig. 6 is the distortion of the incoming projectiles due to the Coulomb field of the nucleus. Fig. 7 shows a schematic representation of both e$^-$ and e$^+$ trajectories through the electron cloud of an atom, from which two significant "Coulomb" effects can be inferred. First, the average electron separation from the nucleus will be less than that for positrons, which will

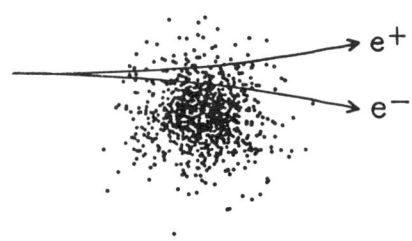

Figure 7. Classical trajectories are used to calculate projectile wavefunction overlap with atomic electron wavefunctions, shown for the radially symmetric Cu K–shell. (From ref. 2)

increase the overlap of the incident projectile with the atomic electron distribution and therefore increase σ^- relative to σ^+. Secondly, and more important, the electron will speed up as it approaches the nucleus, whereas the positron will slow down. Several authors have derived semiempirical expressions for total ionization cross sections[20]. At low velocities these all show a strongly positive dependence on energy, and thus the velocity shift should also increase σ^- relative to σ^+. Logan et al. (refs. 2, 19, and unpublished) have included these effects in calculations of total ionization cross sections for electrons and positrons. Their method involves the calculation of an ionization probability over a set of classically determined electron trajectories. The local ionization probability along such a path is weighted by the atomic electron probability density function. The contribution to the total ionization cross section due to close collisions is then obtained by summing over all classical paths in three dimensions. The solid curve in Fig. 6 shows their model, which fits the L–shell data reasonably well.

Electron/positron comparisons have also been made for K–shell ionization of Ag and Cu[21-23]. These data are shown in Fig. 8 together with the Kolbenstvedt–type prediction (dashed line). The influence of the Coulomb field of the nucleus on the relative ionization cross sections is much more pronounced for the K–shell. The theoretical calculations described above are much less successful in reproducing the measurements than they are for the L–shell data.

3.2 Energy loss

The collisional stopping power due to energy transfer from incident projectiles to bound atomic electrons is calculated using the Bethe–Bloch formalism, which separates energy transfers into two classes depending on whether ϵ is above or below a limiting value, $\epsilon_1 = q_1/T$. The limit, q_1, is chosen to be large compared to the (outer shell) atomic electron binding energies of the stopping medium. In addition, impact parameters associated with small energy transfer collisions ($\epsilon < \epsilon_1$) are required to be large compared to atomic dimensions.

The average energy loss per unit path length, or stopping power (dE/ds), for small energy transfers is calculated directly from the Bethe–Bloch collision cross–section. For large energy transfers the atomic electrons are regarded as free and at rest, and the appropriate differential inelastic cross–section is integrated over all possible energy transfers. The total stopping power is the sum of these two terms, and it is larger for positrons than for electrons by $\approx 6-10\%$ in the low energy region of present interest. This is reflected in Fig. 9, where we show the relative difference for positrons and electrons[2-4]. The crossover at ≈ 345 keV is due to the fact that the Møller cross section becomes larger than the Bhabha cross section at high energies, as

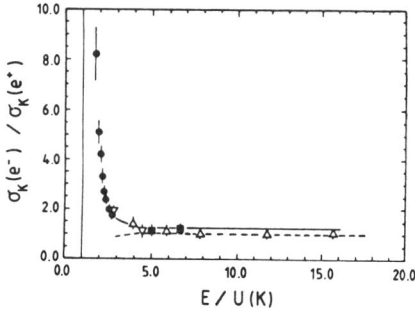

Figure 8: K–shell ionization cross section for electrons relative to positrons. Data shown are for Cu (∇ Schultz and Campbell[22]; ■ Ebel et al.[23]) and Ag (Δ Ito et al.[21]; ● Ebel et al.[23]). (From ref. 23)

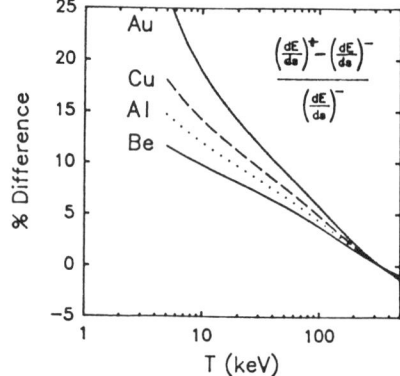

Figure 9. Difference in mean energy loss for positrons and electrons. (From ref. 2)

mentioned above. Extensive tables for stopping power versus energy for both positrons and electrons in various materials are compiled in ref. 4.

Experimental studies of stopping are normally done by measuring the energy distribution of particles transmitted through foils that are thin compared to the particle range. This distribution is characterized by a peak corresponding to the most probable energy loss, and a tail extending to lower energy. In discussing energy loss distributions, the individual collisions are usually classified as soft, intermediate or hard, depending on whether the energy transfer involved is less than, near to, or greater than a critical value, ϵ_c ($\epsilon_c \approx 0.005$, $i.e.$, energy transfer $\approx 0.005\,T$). The soft and intermediate collisions determine the most probable energy loss, since the hard collisions are so infrequent. Thus, the different maximum energy transfers (ϵ_m) for electrons and positrons have little influence. This has been demonstrated experimentally for 38–56 keV electrons and positrons passing through a 1.5 mg/cm^2 Be foil[11], as well as for much lower energy (7 keV) electrons and positrons as shown in Fig. 10.

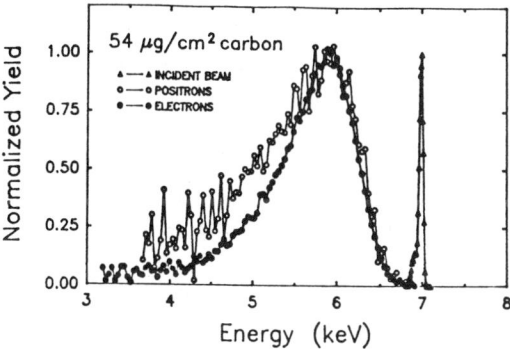

Figure 10. Energy loss distributions for 7 keV electrons and positrons incident on a C target. (From ref. 2)

The high loss part of the distribution is caused primarily by the hard collisions, and the FWHM is determined primarily by the intermediate collisions. Differences in the shapes of the distributions for energy loss greater than the most probable are, therefore, related to differences in the corresponding Møller and Bhabha cross sections for energy transfers near $\epsilon = \epsilon_c$. These differences are clearly evident in the FWHM due to the fact that energy loss straggling depends on the square of the energy transfer. The mean energy loss shows somewhat less difference since the stopping power depends only linearly on energy[2].

For the 7 keV incident particles shown in Fig. 10, the above considerations would lead us to predict $\approx 5\%$ larger FWHM for positrons than for electrons. We observe an effect that is even larger in these preliminary data, for which we have no explanation at this time. More precise and systematic experiments designed to address these questions at low energies will be conducted using a high resolution energy analyzer presently being constructed.

3.3 Stopping profiles

Experimental and theoretical research concerning scattering and energy loss processes is largely motivated by a fundamental interest, particularly since experiments can (now) often be readily compared for both electrons and positrons. However, a very important aspect of this research is the need to provide accurate stopping or implantation profiles for monoenergetic positrons, critical for applications such as the near surface studies and depth profiling described in ref. 1.

So far the most detailed investigation of positron stopping profiles in the literature is the Monte Carlo study reported by Valkealahti and Nieminen[6]. Fig. 11 shows the end points of 5 keV positrons incident on semi–infinite Al, calculated for 1000 particle histories down to a termination energy of 20 eV. The distribution of positrons versus depth (z) in the sample can be described using a Makhovian function, which is very nearly a derivative of a Gaussian. The mean depth of the positron stopping distribution is related to incident energy through a simple power law, and both experimental data[24-25] and Monte Carlo simulations[6] support the same somewhat suprising conclusion: Low energy positrons penetrate deeper into the solid than electrons, in spite of the higher cross section for inelastic scattering. This indicates the importance of the elastic scattering cross section (which is larger for electrons) in randomizing the directions before a significant fraction of the energy is lost.

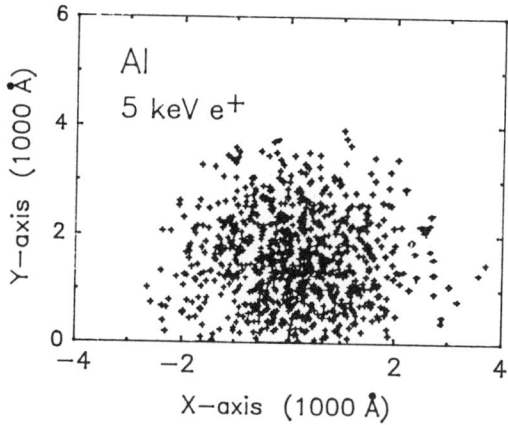

Figure 11. Monte Carlo calculation of 5 keV incident positrons stopping in aluminum. (From ref. 6)

4.0 Conclusions

The slowing-down of monoenergetic, low energy (<100 keV) positrons which enter a solid surface has been discussed, emphasizing the differences between positron and electron interactions. The total elastic scattering cross section is larger for electrons than for positrons, which results in larger predicted backscattering coefficients. Experimental data for positron and electron backscattering are not as convincing, and the results of both on-going and future studies will be required to establish the relative fractions.

Inelastic scattering cross sections, contrary to elastic, are significantly larger for positrons than for electrons in this energy regime. This inequality has been confirmed by studies of inner-shell ionization, but energy-straggling distributions measured for equivelocity positrons and electrons are still somewhat ambiguous.

References

[1] P.J. Schultz and K.G. Lynn, Rev. Mod. Phys. 60, 701 (1988).
[2] P.J. Schultz, L.R. Logan, W.N. Lennard, and G.R. Massoumi, Scanning Microscopy, Suppl. 4 (to be published).
[3] F. Rohrlich and B.C. Carlson, Phys. Rev. 93, 38, (1954).
[4] International Commision on Radiation Units and Measures, Report No. 37: Stopping powers for electrons and positrons (1984).
[5] R.M. Nieminen and J. Oliva, Phys. Rev. B 22, 2226 (1980).
[6] S. Valkealahti and R.M. Nieminen, Appl. Phys. A 35, 51 (1984).
[7] N.F. Mott, Proc. Roy. Soc. A 124, 426 (1929); ibid A 135, 429 (1932).
[8] P. Sigmund and K.B. Winterbon, Nucl. Instrum. Meth. 119, 541 (1974).
[9] T. Tabata, R. Ito, and S. Okabe, Nucl. Instrum. Meth. 94, 509 (1971).
[10] V.A. Kuzminikh, I.A. Tsekhanovski, and S.A. Vorobiev, Nucl. Instrum. Meth. 118, 269 (1974).
[11] W.N. Lennard, P.J. Schultz, and G.R. Massoumi, Nucl. Instrum. Meth. B 33, 128 (1988).
[12] J.A. Baker and P.G. Coleman, J. Phys. C: Solid State Phys. 21, L875 (1988).
[13] L.R. Logan, M.G. Cottam, P.J. Schultz and H.H. Jorch, p.300 in: Positron Annihilation, Proc. of the 8th International Conf., Gent, Belgium, 1988, World Scientific Publ., Singapore (1989).
[14] E.H. Darlington, J. Phys. D: Appl. Phys. 8, 85 (1975).
[15] C. Møller, Annalen der Physik 14, 531 (1932).
[16] H.J. Bhabha, Proc. Roy. Soc. A 154, 195 (1936).
[17] J. Lindhard, Dan. Vidensk. Selsk. Mat.—Fys. Medd. 34, No. 14 (1965).
[18] P.J. Schultz, L.R. Logan, T.E. Jackman and J.A. Davies, Phys. Rev. B 38, 6369 (1988).
[19] W.N. Lennard, P.J. Schultz, G.R. Massoumi and L.R. Logan, Phys. Rev. Lett. 61, 2428 (1988).
[20] M. Gryzinski, Phys. Rev. A 138, 336 (1965); other references contained in C.J. Powell, Rev. Mod. Phys. 48, 33 (1976).
[21] S. Ito, S. Shimizu, T. Kawaratani and K. Kubota, Phys. Rev. A 22, 407 (1980).
[22] P.J. Schultz and J.L. Campbell, Phys. Lett. 112A, 316 (1985).
[23] F. Ebel, W. Faust, C. Hahn, M. Rückert, H. Schneider, A. Singe and I. Tobehn, Phys. Lett. A, in press (1989).
[24] A.P. Mills, Jr., and R.J. Wilson, Phys. Rev. A 26, 490 (1982).
[25] A.Y. Vyatskin and V.Y. Khramov, Radio Tek. Electron. Phys. 21, 1931 (1976).
[26] H. Drescher, L. Reimer, and H. Seidel, Z. Angew. Phys. 29, 331 (1970).

THEORETICAL CALCULATIONS OF POSITRON COLLISIONS WITH ATOMS

A. D. Stauffer†, K. Bartschat‡, R. I. Campeanu†, M. Horbatsch†, R. P. McEachran†, L. A. Parcell§ and S. J. Ward¶

†Department of Physics, York University, Toronto, Canada
‡Department of Physics, Drake University, Des Moines, Iowa, USA
§School of Mathematics, Physics, Computing and Electronics, Macquarie University, Sydney, Australia
¶Department of Physics and Astronomy, University of Tennessee, Knoxville, Tennessee, USA

We review calculations carried out for positron scattering at low and intermediate energies from the noble gases as well as from the alkali atoms. Processes considered are elastic scattering (differential and integrated cross sections), excitation and ionization. Emphasis is placed on comparison with experimental results.

INTRODUCTION

It is obvious from the experimental papers in this symposium that we are rapidly approaching the situation where any scattering experiment that can be carried out with electrons will also be able to be done with positrons. Thus measurements of total cross sections are giving way to measurements of elastic differential cross sections.[1-5] Ionization cross sections are being measured as well as those for positronium formation.[6-9] In addition, initial attempts have been made to determine excitation cross sections.[10-11] We have attempted to keep pace with all these experimental advances by concurrently carrying out theoretical calculations at low and intermediate energies for the processes being measured.

This review will cover our recent work in positron scattering from the noble gases and from alkali atoms. Since these two types of atoms require rather different calculational techniques we shall deal with them separately.

THE NOBLE GASES

The noble gases are characterized by a tightly bound, closed shell structure. The excitation thresholds occur at relatively large energies and the polarizabilities are small. Thus below the first excitation threshold (which is positronium formation) methods which take into account the static and polarization interactions between the incident positron and the atom, for example the polarized-orbital method, produce results which are in good agreement with the experimental data for the integrated cross sections for helium, neon, argon, krypton and xenon.[12-15] For instance, they predicts a Ramsauer minimum in the integrated cross section for helium and neon but not for the heavier noble gases. These methods also work well in this energy range for the differential cross sections which have so far been measured in neon and argon, producing a distinct minimum in the differential cross section.[3,5,16-19]

However, in the inelastic region these methods becomes increasingly unreliable as they continue to predict a minimum in the differential cross section at all energies. This is not the case with the experimental data where the minimum disappears as the energy is increased sufficiently far above the inelastic thresholds.[3,5] If the effects of the inelastic channels on the elastic channel are taken into account, for example by using second order methods such as optical potentials or the eikonal Born series, this minimum disappears and the results are consistent with the experimental data for argon.[20,21] This is illustrated in figure 1 where we show the differential cross section for the scattering of positrons from argon at 30 eV. However, in the elastic scattering regime the minimum is present.[3]

For neon the picture is less clear since the theoretical results are similar to the case of argon while the experimental data suggests that the minimum persists well above the inelastic thresholds.[5] In figure 2 we display the results for the differential cross section for positron–neon scattering at 20 eV. Absolute measurements for these cross sections, which are in the process of being determined experimentally, should help in assessing the accuracy of the theoretical methods.[22]

In the case of ionization, only integrated cross sections have been measured to date for positron helium scattering.[7-9] Since the two outgoing particles are distinct, exchange is not present and relatively simple models of this process are effective. Representing the scattered and ejected particles by distorted waves yields very good agreement with the latest experimental measurements as is shown in figure 3.[23,24] We are in the process of calculating differential scattering cross sections for ionization. The production of accurate differential cross sections will be a much more stringent test of theory than integrated cross sections.

So far the measurements of the excitation cross sections for positron scattering from the noble gases have produced results which are somewhat ambiguous.[10] We have attempted to reproduce these results theoretically for helium and neon using a distorted wave approxima-

© 1990 American Institute of Physics

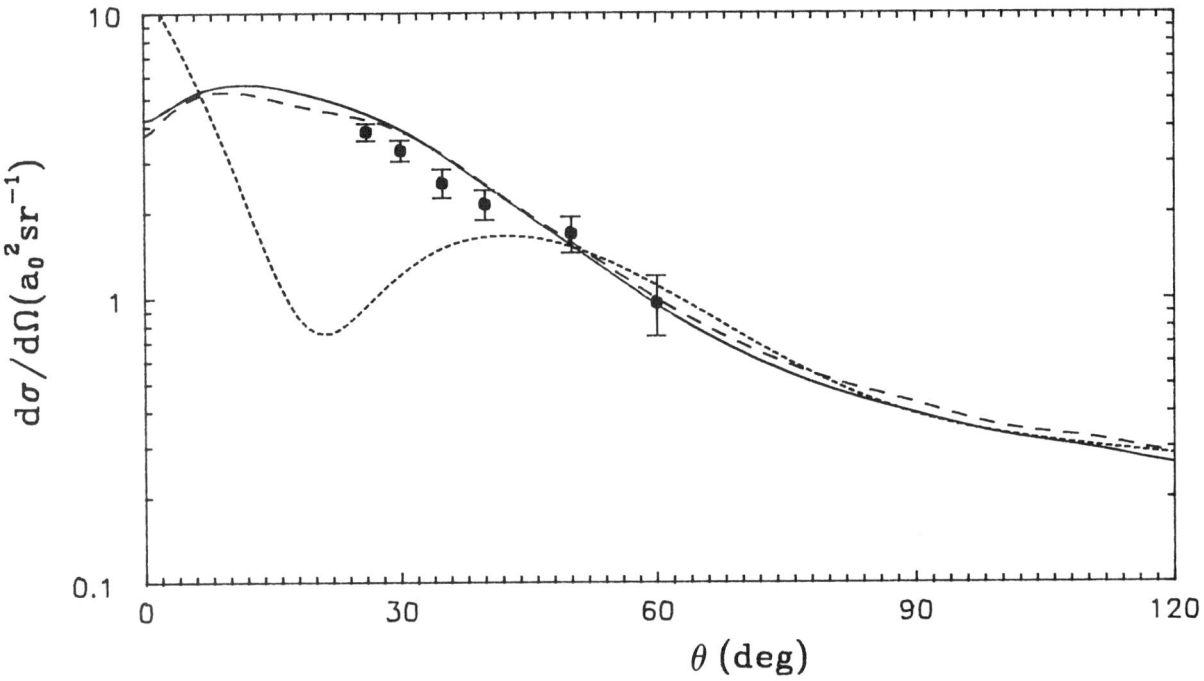

FIG. 1. Differential cross sections for positron scattering from argon at 30 eV. (———), ten-state optical potential using free-wave Green's functions[20]; (— — —), ten-state optical potential using numerical Green's functions; (- - -), polarized-orbital approximation[16]; •, experimental data normalized at 60°.[3]

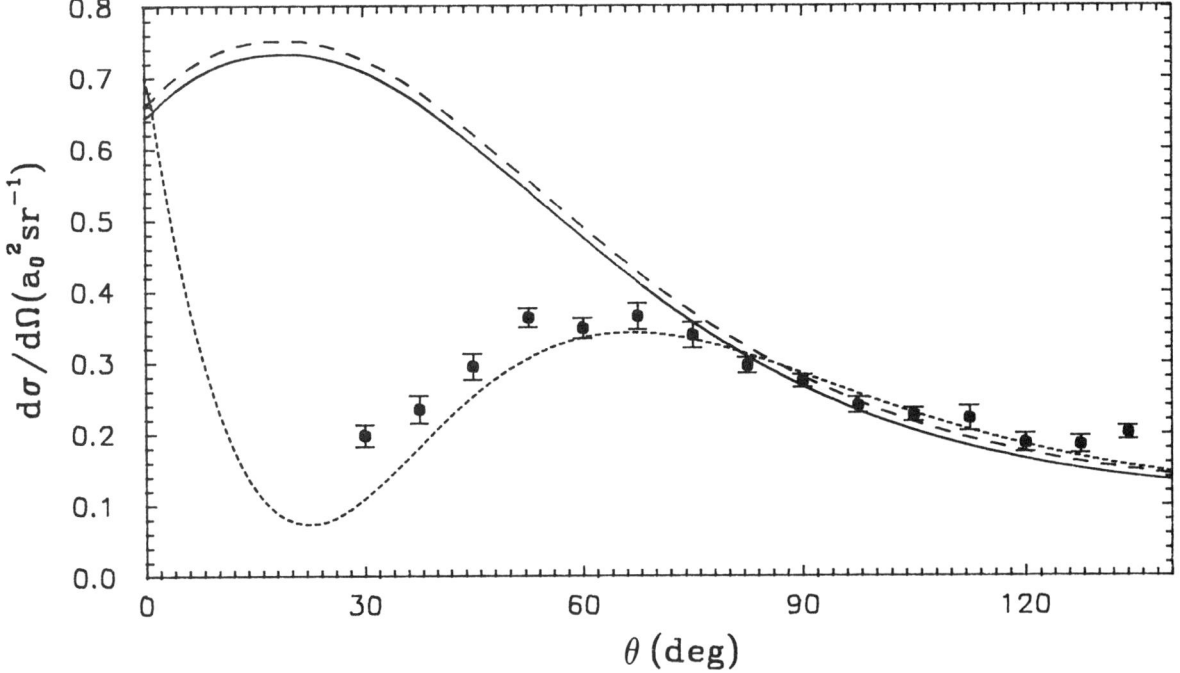

FIG. 2. Differential cross sections for positron scattering from neon at 20 eV. (———), fourteen-state optical potential using free-wave Green's functions; (— — —), fourteen-state optical potential using numerical Green's functions; (- - -), polarized-orbital approximation[16]; •, experimental data normalized at 90°.[5]

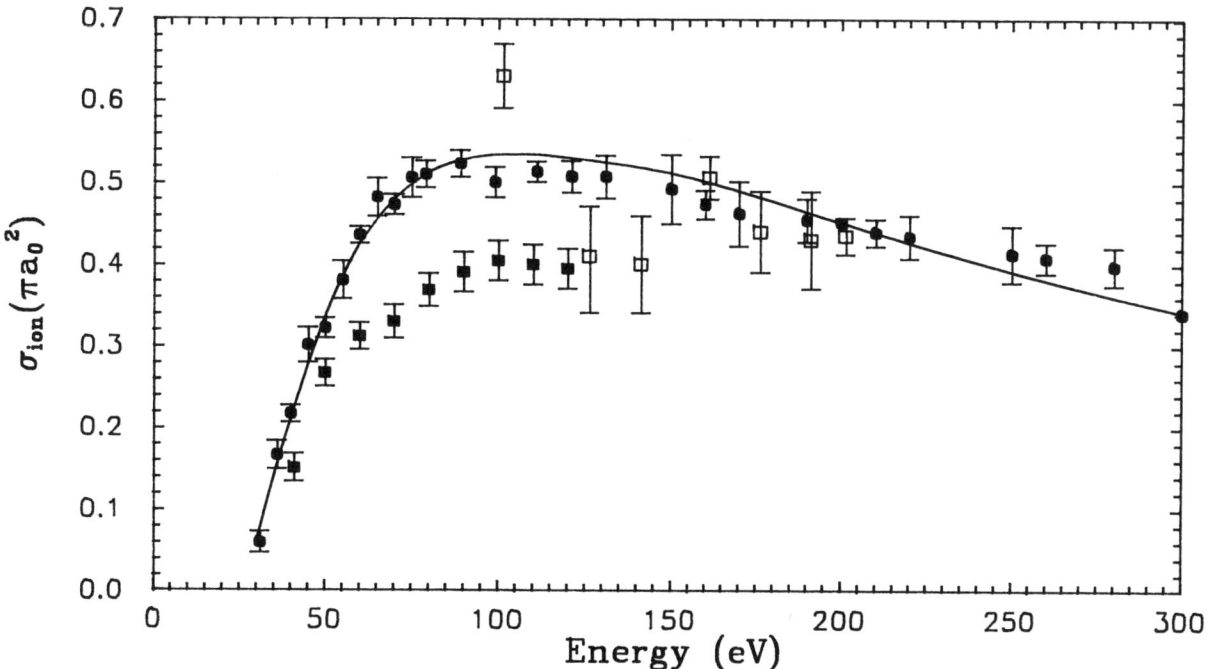

FIG. 3. Cross section for the ionization of helium by positron impact. Theory: (———), model DCPT3[23]. Experiment: •, Fromme et al;[7] □, Diana et al;[8] ■, Suoeka.[9]

tion but there remain substantial differences between our calculations and the experimental data.[25,26] Clearly this is an area which requires more work, both theoretically and experimentally.

THE ALKALIS

The alkali atoms are characterised by a loosely bound electron outside a closed shell. The polarizabilities of these atoms are large and the excitation thresholds lie at very low energies. Thus it is necessary to take into account both elastic and excitation channels at all scattering energies. In addition the positronium formation channel is open at zero scattering energy since the ionization potential of the alkalis is less than the binding energy of positronium.

The close-coupling approximation has been employed to calculate cross sections for positron scattering from lithium, sodium, potassium and rubidium where up to the first five atomic states have been included in the eigenfunction expansion.[27–31] As yet positronium formation has only been included in calculations which do not take account of the excited states of the atom.[32–35] Positronium formation is difficult to deal with theoretically since it requires a two-centre formulation to treat it accurately. The hot topics talk in this symposium deals in greater detail with this problem.[36]

So far only total cross sections have been measured for these atoms and the experimental technique cannot detect positrons elastically scattered through small angles in the forward direction.[37–40] Since the differential cross sections for elastic scattering are highly peaked in the forward direction, the experimental mesurements miss a large proportion of the elastic cross section. In order to take this situation into account we have produced an effective total cross section which corresponds directly to the experimental data. By this method we have obtained results which are in good agreement with the experimental data available to date. Effective total cross sections for positron scattering from sodium and potassium are given in figures 4 and 5 respectively along with the corresponding experimental results and a higher energy calculation using a modified Glauber approximation.[41–43]

In our close-coupling calculations[29,44] we have found the first evidence of resonances in positron–alkali scattering. This resonant behaviour is associated with the various inelastic thresholds and has resonant widths of up to 130 meV. Figure 6 shows the partial integrated cross section for the elastic scattering channel in sodium. A narrow S-wave resonance is visible along with a much broader P-wave resonance. There is a good possibility of being able to detect this broad resonances experimentally even with existing experimental setups.[22] In analogy with electron–alkali scattering and given the large polarizabilities of these atoms, such resonant behaviour is not unexpected. However, experimental confirmation of such behaviour would be a breakthrough and would settle a long standing question of whether resonances actually exist in positron–atom scattering.

648 Calculations of Positron Collisions

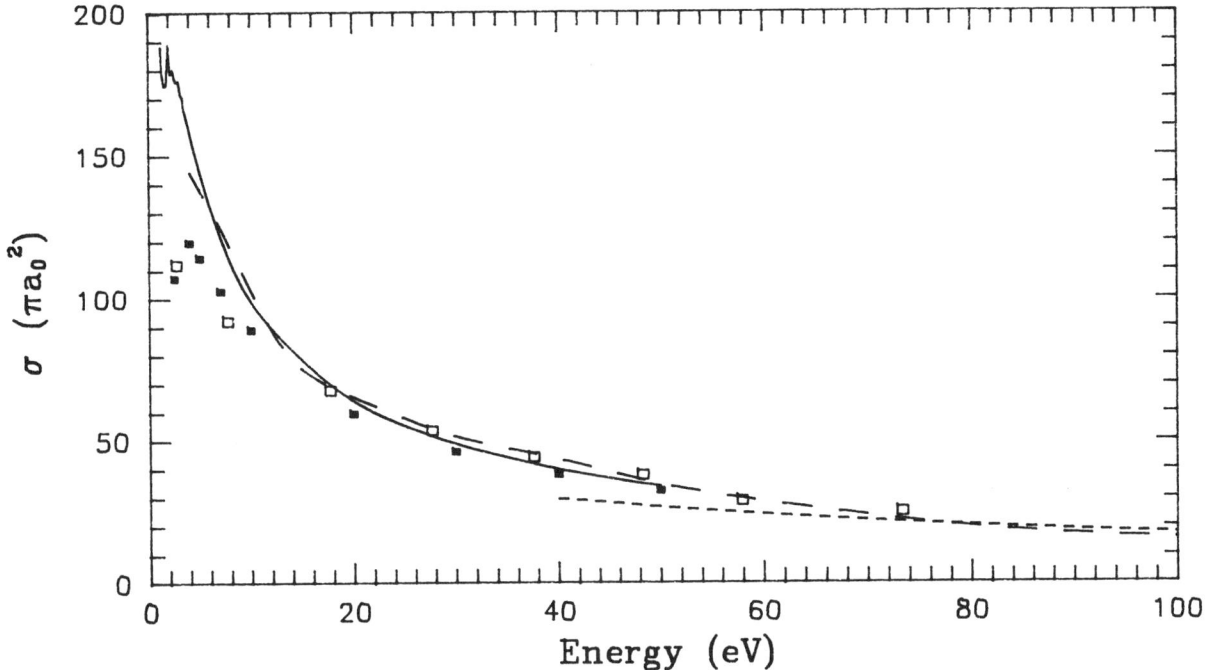

FIG. 4. The total cross section for e^+–Na scattering: (– – –), 4-state close-coupling approximation (CCA), Sarkar et al;[31] (———), 5-state CCA, Ward et al;[28] ■, effective 5-state CCA, Ward et al;[28] (- - -), core-corrected modified Glauber approximation, Gien;[42] □, experimental data, Kwan et al.[39]

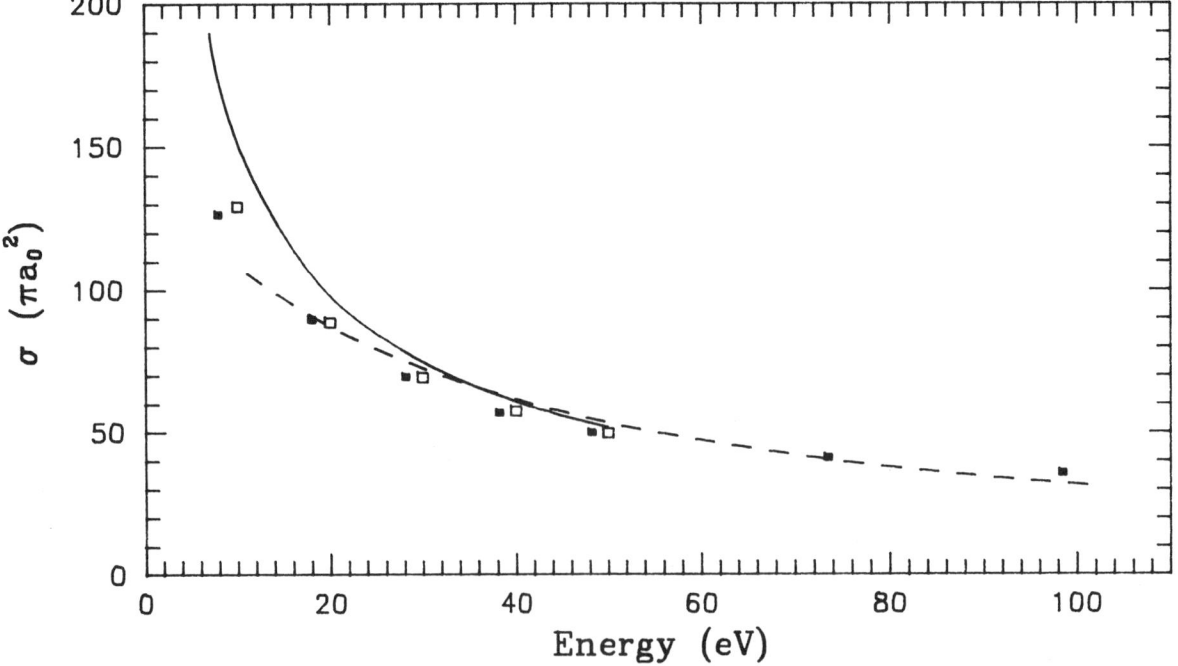

FIG. 5. The total cross section for e^+–K scattering: (———), 5-state close-coupling approximation (CCA), Ward et al;[28] ■, effective 5-state CCA, Ward et al;[28] (- - -), core-corrected modified Glauber approximation, Gien;[43] □, experimental data, Stein et al.[37,38]

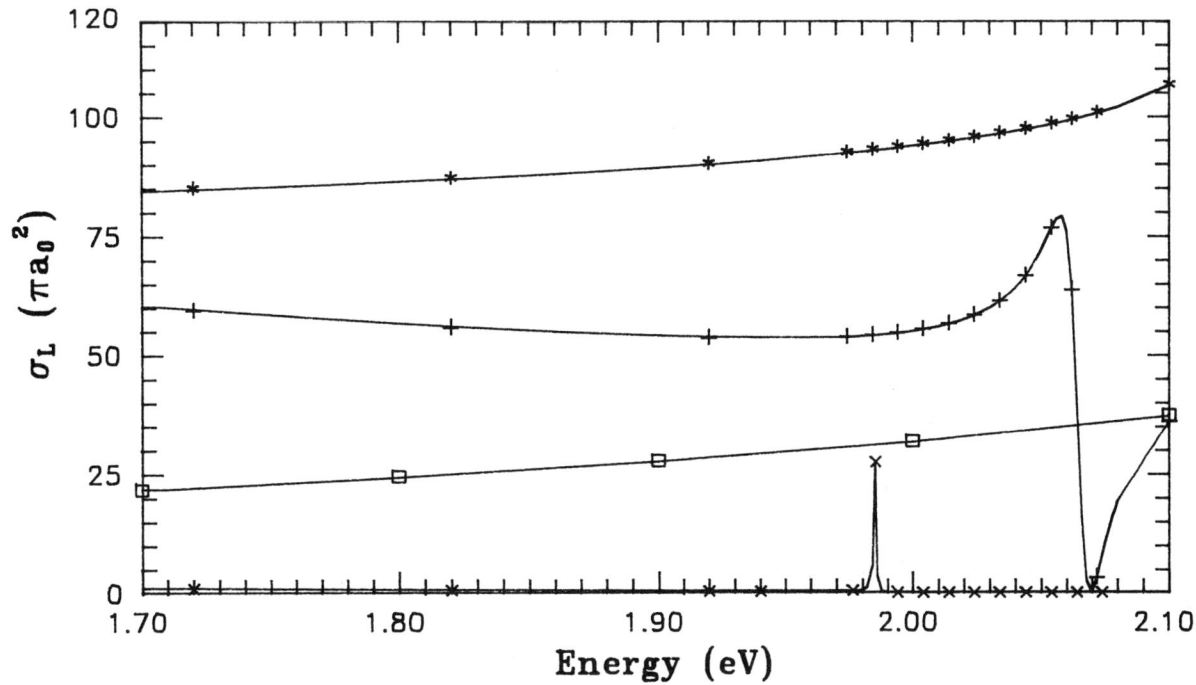

FIG. 6. Partial wave elastic cross sections (in πa_0^2) for $L = 0$, ×; $L = 1$, +; $L = 2$, * and $L = 3$, □, e^+–Na scattering in the (3s–3p–4s–3d–4p) model potential close-coupling approximation.

In addition, we have found evidence for bound states of the positron–alkali system since the s-wave phase shifts start at a positive multiple of π. While we have not included the positronium channel in our calculations, we do not expect it to make a major change to the values of our cross section, except perhaps at very low energies. However, it is not clear at present what effect positronium formation will have on the position and widths of the resonances we have found or even if it will surpress them altogether. Furthermore we may find that the bound states, for which we have found evidence, may not be stable against dissociation into positronium and an alkali ion. This is clearly an area of high priority and we are attempting to carry out calculations which take it into account explicitly.

CONCLUSIONS

As experimental results have become more particular and detailed, theoretical calculations have kept pace and the agreement between theory and experiment has been quite satisfactory for the most part. We look forward to increasingly sophisticated experiments which will provide evermore stringent tests of our theoretical models. Since these models are ones that are also used for electron scattering calculations, the existence of this alternate set of data will be particularly useful in assessing the accuracy of these models and differences between electron and positron scattering will give us further insights into the physics involved in scattering phenomena.

ACKNOWLEDGMENTS

We wish to thank Professors W. E. Kauppila and T. S. Stein for sending us their experimental data in advance of publication. This research was supported in part by grants from the Natural Sciences and Engineering Research Council of Canada and by the Deutsche Forschungsgemeinschaft.

---

[1] W. E. Kauppila, this symposium.
[2] G. M. A. Hyder, M. S. Dababneh, Y.-F. Hsieh, W. E. Kauppila, C. K. Kwan, M. Mahdavi-Hezaveh and T. S. Stein, Phys. Rev. Lett. **57**, 2252, (1986).
[3] K. Floeder, P. Höner, W. Raith, A. Schwab, G. Sinapius and G. Spicher, Phys. Rev. Lett. **60**, 2363, (1988).
[4] P. G. Coleman and J. D. McNutt, Phys. Rev. Lett. **42**, 1130, (1979).

[5] W. E. Kauppila, S. J. Smith, C. K. Kwan and T. S. Stein, *Proc. Workshop on Annihilation of Gases and Galaxies*, to be published.

[6] W. Raith, this symposium.

[7] D. Fromme, G. Kruse, W. Raith and G. Sinapius, Phys. Rev. Lett. **57**, 3031, (1986).

[8] L. M. Diana, L. S. Fornari, S. C. Sharma, P. K. Pendleton and P. G. Coleman, *Proc. 7th Int. Conf. on Positron Annihilation*, ed. P. C. Jain, R. M. Singru and K. P. Gopinathan (Singapore: World Scientific), (1985).

[9] O. Sueoka, J. Phys. Soc. Japan **51**, 2381, (1982).

[10] P. G. Coleman, J. T. Hutton, D. R. Cook and C. A. Chandler, Can. J. Phys. **60**, 584, (1982).

[11] O. Sueoka, J. Phys. Soc. Japan **51**, 3757, (1982).

[12] R. P. McEachran, D. L. Morgan, A. G. Ryman and A. D. Stauffer, J. Phys. B. **10**, 663, (1977) and J. Phys. B. **11** 951, (1978).

[13] R. P. McEachran, A. G. Ryman and A. D. Stauffer, J. Phys. B. **11**, 551, (1978).

[14] R. P. McEachran, A. G. Ryman and A. D. Stauffer, J. Phys. B. **12**, 1031, (1979).

[15] R. P. McEachran, A. D. Stauffer and L. E. M. Campbell, J. Phys. B. **13**, 1281, (1980).

[16] R. P. McEachran and A. D. Stauffer, *Proc. of the Third Int. Workshop on Positron (Electron)-Gas Scattering*, ed. W. E. Kauppila, T. S. Stein and J. M. Wadhera (Singapore: World Scientific) p. 122, (1986).

[17] H. Nakanishi and D. M. Schrader, Phys. Rev. A **34**, 1823, (1986).

[18] S. K. Datta, S. K. Mandal, P. Khan and A. S. Ghosh, Phys. Rev. A **34**, 633, (1985).

[19] S. N. Nahar and J.M. Wadehra, Phys. Rev. A **35**, 2051, (1987).

[20] K. Bartschat, R. P. McEachran and A. D. Stauffer, J. Phys. B. **21**, 2789, (1988).

[21] C. J. Joachain and R. M. Potvliege, Phys. Rev. A **35**, 4873, (1987).

[22] W. E. Kauppila and T. S. Stein, private communication.

[23] R. I. Campeanu, R. P. McEachran and A. D. Stauffer, J. Phys. B. **20**, 1635, (1987).

[24] R. I. Campeanu, D. Fromme, G. Kruse, R. P. McEachran, L. A. Parcell, W. Raith, G. Sinapius and A. D. Stauffer, J. Phys. B. **20**, 3557, (1987).

[25] L. A. Parcell, R. P. McEachran and A. D. Stauffer, J. Phys. B. **16**, 4259, (1983).

[26] L. A. Parcell, R. P. McEachran and A. D. Stauffer, J. Phys. B. **20**, 2307, (1987).

[27] S. J. Ward, M. Horbatsch, R. P. McEachran and A. D. Stauffer, J. Phys. B. **21**, L611, (1988).

[28] S. J. Ward, M. Horbatsch, R. P. McEachran and A. D. Stauffer, J. Phys. B. **22**, 1845, (1988).

[29] S. J. Ward, M. Horbatsch, R. P. McEachran and A. D. Stauffer, Nucl. Instrum. Methods (1989), in press.

[30] P. Khan, S. Dutta and A. S. Ghosh, J. Phys. B. **20**, 2927, (1987).

[31] K. P. Sarkar, M. Basu and A. S. Ghosh, J. Phys. B. **21**, 1649, (1988).

[32] S. Guha and P. J. Mandal, J. Phys. B. **13**, 1919, (1980).

[33] P. J. Mandal and S. Guha, J. Phys. B. **13**, 1337, (1980).

[34] S. Guha and B. C. Saha, Phys. Rev. A **21**, 564, (1980).

[35] S. Guha and A. S. Ghosh, Phys. Rev. A **23**, 743, (1981).

[36] A. S. Ghosh, this symposium.

[37] T. S. Stein, R. D. Gomez, Y.-F. Hsieh, W. E. Kauppila, C. K. Kwan and Y. J. Wan, Phys. Rev. Lett. **55**, 488, (1985).

[38] T. S. Stein, M. S. Dababneh, W. E. Kauppila, C. K. Kwan, and Y. J. Wan, *Atomic Physics with Positrons*, ed. J. W. Humberston and E. A. G. Armour (New York: Plenum) p. 251, (1988).

[39] C. K. Kwan, W. E. Kauppila, R. A. Likaszew, S. P. Parikh, T. S. Stein, Y. J. Wan and M.S. Dababneh, Phys. Rev. A, (1989), submitted.

[40] T. S. Stein, C. K. Kwan, W. E. Kauppila, R. A. Likaszew, S. P. Parikh, Y. J. Wan and M. S. Dababneh, *Proc. Workshop on Annihilation of Gases and Galaxies*, to be published.

[41] T. T. Gien, Phys. Rev. A **35**, 2026, (1987).

[42] T. T. Gien, Chem. Phys. Lett. **139**, 23, (1987) (Erratum **142**, 575).

[43] T. T. Gien, J. Phys. B. **22**, L129, (1989).

[44] S. J. Ward, M. Horbatsch, R. P. McEachran and A. D. Stauffer, J. Phys. B. **22**, (1989), submitted for publication.

Symposium: Supercomputational Collision Physics

SUPERCOMPUTERS AND THE FUTURE OF COMPUTATIONAL ATOMIC SCATTERING PHYSICS

Stephen M. Younger

X-Division, Los Alamos National Laboratory, Los Alamos, NM 87545 USA

The advent of the supercomputer has opened new vistas for the computational atomic physicist. Problems of hitherto unparalleled complexity are now being examined using these new machines, and important connections with other fields of physics are being established. This talk briefly reviews some of the most important trends in computational scattering physics and suggests some exciting possibilities for the future.

Spectacular advances in computer technology made over the past two decades are opening new vistas for science in general, and scattering physics in particular. The formation of supercomputer centers around the world has made these machines available to many scientists, and has allowed workers at widely dispersed locations to collaborate on programs and calculations based at a single remote facility. As the speed and memory of computers continues to increase, a new breed of physicist is emerging, a hybrid between the traditional analytic theorist and the experimentalist. Problems which are too complex for analytic approaches, such as those involving many coupled equations or a very large number of independent variables, can now be solved numerically. By varying the parameters of the calculation, the behavior of very complex systems can be simulated starting from only fundamental assumptions. In some cases this new type of physics is leading to a rethinking of fundamental issues in the underlying science itself.

The following paragraphs will briefly examine some of the advances in supercomputer technology which have brought about this new thinking and some of the scattering physics problems that are being approached by means of large scale calculations. I will also present some connections being forged between atomic physics and other areas such as elementary particle and solid state physics.

The rapid advance of computer capability is illustrated in Figure 1, which shows the log of the number of numerical calculations that the machine can perform per second plotted against the year in which it became available.[1] This curve demonstrates that computer speed is increasing exponentially at a rate of an order of magnitude every decade. In the early years these impressive increases were achieved by more sophisticated central processor designs and by reducing the signal transit time between components by means of more compact assembly. Recently, advanced architectures such as vector processing machines have been developed which can simultaneously process whole arrays of numbers. Another major development is the construction of massively parallel machines, which contain hundreds or thousands of independent processors connected by a common memory. Each of the individual processors solves a predetermined small part of the overall problem, with some form of integration of the results occurring later on. A challenge associated with many-processor machines, however, is the special programming which is sometimes required to make full use of the machines' capabilities. Most large physics computer codes are written in the Fortran language, essentially a single processor, sequential language. Indeed, some theoretical methods used in computational scattering physics were developed with this programming in

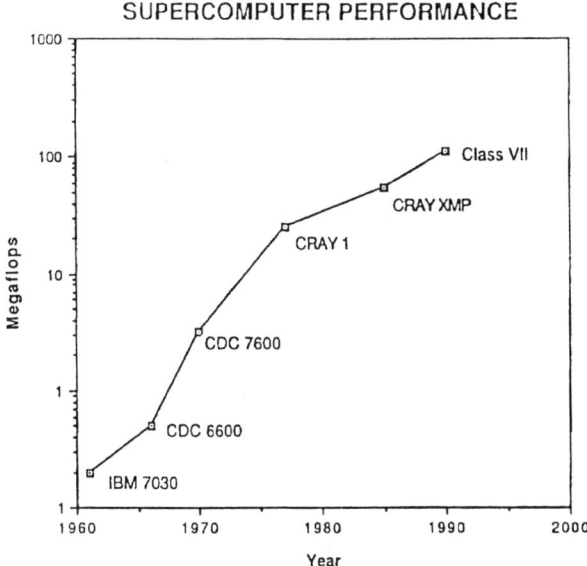

Figure 1. Supercomputer speed measured in millions of floating point (numerical) operations per second.

mind. To make full use of the capabilities of massively parallel processing machines may require not only a rewrite of the codes, but a reconsideration of the physics problem that is under study.

What all of this means to the atomic physicist can be illustrated by the example of electronic structure calculations. In 1930 it took Douglas Hartree many months of laborious hand calculations to compute the electronic structure of an atom such as rubidium. By 1960 such a calculation could be performed overnight on a digital computer, although considerable effort was required to master the primitive programming languages available at that time. By 1970 Hartree-Fock structure calculations were routine, and standard programs were available. Today, using a Cray computer or its equivilant, the atomic structure of rubidium can be computed in about one second.

It is important to note, however, that even these prodigious advances in computational power are themselves inadequate for a brute force attack on any but the simplist problems in atomic physics. The large number of independent variables in a complex atom requires the use of an approximation. The challenge is to maximize the return on a computational investment given the technology available.

Increases in computer speed have been matched by similar improvements in memory. Indeed, the "core" or direct access memory which can be rapidly accessed by programs is expanding seemingly without limit. Even smaller machines often have huge "virtual" memories made available by clever programming and special machine instructions. Problems that just a few years ago required a large mainframe computer can now be comfortably handled by an advanced desktop machine. Moreover, problems that exceeded the theorists' most diligent economies can now fit on the newest generation of supercomputers, opening new avenues of research.

The range of topics which fall under the heading "atomic scattering physics" is very broad. High energy heavy ion accelerators can produce close collisions between high-Z nuclei in which super-critical fields are reached. At the opposite extreme, slow collisions with surfaces are of interest both as probes of the complex surface layer and as practical means of "atomic engineering" in microelectronic fabrication. The talks which comprise this symposium look at a variety of topics in scattering physics, each of which exploits supercomputer technology for the solutions of hitherto unapproachable problems.

Electron-atom collisions are among the more simple collision processes to describe theoretically, in that the atomic nucleus provides a center of symmetry. Nevertheless, the complexity of atomic structure can make an accurate calculation very difficult. In the relatively straightforward distorted wave approximation the target wavefunction is frozen and the continuum wave is computed in an approximate scattering potential. The approximate scattering potential. The interaction between the scattering electron and the target can then be computed via perturbation theory to yield a cross section. The distorted wave approximation breaks down when there is significant interaction between the continuum electron and the target.

In more sophisticated approximations, the interaction of the scattering electron and the target atom is described by a compound state represented in terms of an expansion over single atom states plus some type of distorted continuum wave. Two major computational methods have been developed to treat electron scattering from complex atoms: the close coupling approximation and the R-matrix method. In the close coupling approximation a set of coupled equations for low lying states of the atom + continuum complex are solved to model the total wavefunction. Because the computational labor increases rapidly with the number of states, considerable care is used in defining the basis states so as to reproduce accurate optical oscillator strengths, energies, etc. In expansion methods problems in convergence can occur which effectively limit the number of basis states, making the final answer somewhat dependent on the cleverness with which the basis is chosen. Calculations with 9-15 states are now commonplace, although rigorous proof of the convergence of the expansion remains an open question. In particular, the difficulty of treating continuum components continues to be a problem.

The R-matrix method is based on a diagonalization of the target+scattering electron Hamiltonian in a basis set defined within some finite volume surrounding the atom. The boundary of this volume is chosen such that the electron-atom interaction has reached its asymptotic form, so that the solution in the exterior region is known analytically. The diagonalization process produces a coupled wavefunction which can then be used in scattering cross section calculations. As with any finite basis expansion, the accuracy of the R-matrix method depends on the completeness of the basis set.

Both the close coupling approximation and the R-matrix approximation have been applied to problems in electron excitation and photoionization, but run into difficulties when applied to electron ionization or other processes containing two or more continuum electrons in the final state. The interaction of multiple continuum electrons represents higher order terms in a Born-type expansion. Here the problem is more than obtaining computational resources sufficient to handle the complexity of interacting long range orbitals. Rather there is the lack of a fundamental theory to guide the calculations. More information on the close coupling and R-matrix methods can be found in invited lectures included in this volume.

The collision of two atoms or ions introduces an additional level of complexity owing to the loss of a single center of symmetry. At low scattering energies the trajectories of the atoms can be significantly perturbed by electronic interactions, requiring a self-consistent treatment of ion and electronic motion. At higher energies one may employ an approximate form of the interatomic interaction to determine the trajectories and concentrate on electronic transitions such as excitation, ionization, or charge transfer. To describe such processes one can employ much of the language and methodology of quantum chemistry to describe level crossings and other resonant phenomena important for the determination of cross sections. A finite basis expansion of the total wavefunction based on nucleus-centered states is usually performed, sometimes with the addition of auxiliary functions located between the nuclei or at other strategic locations. As with any basis expansion method, the two-electron Coulomb operator with its four orbital components causes the size of the basis set size. For any but the simplest atoms one can rapidly fill a supercomputer memory. Kulander et al.[2] among others has developed a time dependent Hartree-Fock approximation for the description of atomic and molecular scattering.

With the completion of the heavy ion accelerator at GSI Darmstat atomic physicists have the opportunity to observe very close collisions of heavy nuclei containing no or only a few electrons. Clearly any "atomic" states which might exist during such collisions will be strongly influenced by the changing nuclear configuration as well as by large relativistic and QED effects. During close collisions "quasi-molecular" states can occur in which bound electrons are shared among the nearby nuclei. During the collision of two heavy nuclei it is possible for the local electrostatic field to exceed the critical value, i.e., that which would correspond to a nucleus with Z=137. In the supercritical field regime, pair production is possible, and has been observed. At the present time there is still debate concerning the physical mechanism for this pair production, since it is sometimes observed to occur after the nuclei have moved sufficiently apart to drop the local electrostatic field below the critical level. Modeling such phenomena requires a robust methodology which is capable of describing the three dimensional motion of the nuclear configurations as a function of time. Bottcher et al.[3] have applied basis spline techniques to such problems with exciting results. In these calculations space is divided into a large number of discrete volume elements, the physics in which is determined by the usual equations of QED. A variety of methods can be used to propagate solutions in time so that overall rate coefficients can be determined.

Atomic collisions with surfaces is a subject of considerable practical importance for the semiconductor industry. In the drive for ever more sophisticated microelectronics "atomic engineering" is being performed wherein monoatomic layers are deposited on surfaces to perform special functions. With the advent of advanced electron microscopes, we can map the surfaces of materials down to the level of individual atoms, providing fascinating images of the subject under study. As one might expect, the theoretical treatment of atom-surface interactions is exceptionally complex, owing both to the discontinuous nature of the surface itself and the complex interactions possible between the surface and a scattering atom. Garcia et al.[4] have developed a method which they have applied to atomic scattering from hard surfaces and from "jellium." They are able to describe the time-dependent distortion of the atomic wavefunction as it interacts with the surface and is repelled. Figure 2 shows the result of one of their calculations of a hydrogen atom being reflected from a hard barrier.

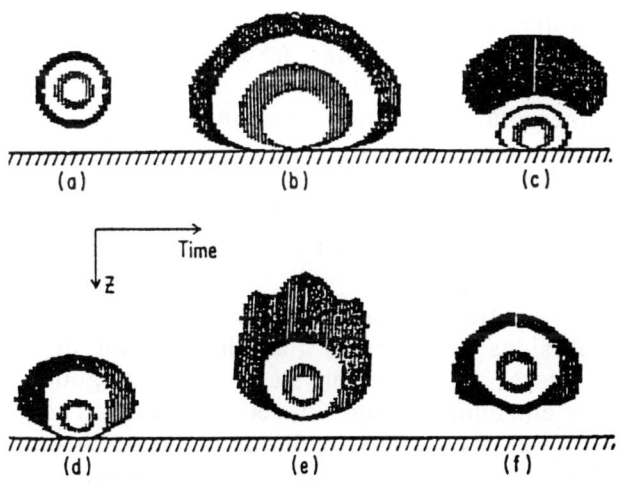

Figure 2. Density contours of the electronic wavefunction of a hydrogen atom scattering off of a hard barrier. The initial energy of the atom was 25 eV.(Ref. 4.)

With more computer power it is possible to construct a computational model consisting of several atomic layers with some type of realistic boundary condition to model the deeper material. The incident atom can then be projected against the surface at different angles and impact parameters to generate probabilities for sticking, reflection, and transmission into the solid.

A fascinating practical problem which borrows heavily from quantum chemistry is that of scattering in a dense liquid or solid. Here the complication of many simultaneous interactions is present. In a liquid, there is the additional complexity of a disordered asymptotic boundary. Recent calculations performed with an approximate time-dependent Hartree-Fock code have demonstrated that many-atom interactions in liquids soften the interatomic interaction compared to what one would observe in binary collisions. The origin of this softening is the self-consistent redistribution of the electronic charge density during close collisions. Although such calculations which treat 30 or more atoms can take many thousands of minutes even on a supercomputer, they demonstrate that new phenomena await the computational theorist, and that with advances in computer hardware, problems of great complexity can be handled in a more less direct fashion.[5]

Midway between few-atom collisions and liquids lies the domain of clusters. Cluster physics has undergone a tremendous boom in recent years as new experimental techniques have been developed both to produce and examine the properties of clusters with well characterized numbers of atoms. On the theoretical side, Car and Parrinello[6] have developed an efficient computational theory for clusters based on the local density approximation. The electronic wavefunction is expanded over a plane wave basis set and the time dependent Schrodinger equation is solved for the motion of the nuclei. By imparting a velocity distribution to the nuclei they were able to study structural transitions in small clusters as a function of temperature. A particularly interesting finding was that at high temperatures the density of states increases dramatically, i.e. there are a very large number of nearly degenerate bound configurations. Hence rather than "melting" of the sample into a true disordered liquid state, they identified transitions into a large but countable number of discrete bound configurations.

Increasing the temperature to where a plasma forms brings another class of problems. In contrast to the low temperature case, where only the ground state or some set of low lying states is populated, in a plasma the electronic population may be spread over many hundreds or thousands of energy levels. The level populations are controlled by a complex series of excitation, ionization, and recombination processes which involve ions, electrons, and photons. To model an atom embedded in a plasma it is necessary to have detailed rate coefficients for all of these processes for all of the ionization stages expected to be important for the kinetics. This information is fed into a rate equation solver which yields the populations of individual levels. Such a procedure is challenging even for simple atoms. For complex atoms, such as rare earths, it becomes impractical owing to the many thousands of levels involved. Here one may employ a statistical approximation which deals with entire arrays of transitions rather than the individual lines themselves. Klapisch et al[7] have developed such an unresolved transition array (UTA) approximation which has been successfully applied to a wide class of problems in laser interactions and plasma physics.

The immense complexity of rare earth and other heavy element spectra and the need to develop useful approximations is one indicator that increased computational power alone is increased computational power alone is not the answer to every atomic physics problem. Equally important is the development of realistic theoretical structures which efficiently treat the dominant interactions in a way that will allow solutions to be obtained in a reasonable amount of time. As we approach even more complicated problems, such as the simulation of macromolecular interactions, we will be forced either to break the calculation into manageable parts or to use one or more simplifying assumptions in the calculation. Thus, far from reducing physics to mere computation, the development of computational atomic physics will require a new level of creativity to model a wide range of phenomena in pure and applied research. Theoretical approximations and numerical algorithms developed for past generations of single processor computers may not be the most suitable ones for the next generations of multi-processor machines. For example, rather than fretting over the completeness of highly optimized basis sets used in expansion methods, it may be easier to use a many-body perturbation theory approach which computes a very large number of correlation diagrams automatically given a reasonably realistic starting Hamiltonian.

As atomic physicists begin to harness the power of supercomputers for their own interests, they are also noting the similarities between their problems and those from a wide variety of other scientific disciplines. The connection between atomic and molecular scattering and quantum chemistry is an old example. More recently, however, recognition has been increasing that the methods developed for dense plasma interactions also have application to the quark gluon plasma expected to be formed in ultrarelativistic nuclear collisions. Similarly, workers who are approaching the liquid state from a traditional atomic-molecular level are deriving guidance from the statistical mechanics community. In the study of electronic phenomena occurring during the collision of heavy nuclei one must explicitly consider the time dependent development of nuclear shapes during the collision.

The recognition that numerical recipes and algorithms can be applied to a wide variety of problems in seemingly disparate fields of physics, chemistry, and material science is opening new vistas to the computational physicist. In effect, the computer is becoming a theoretical laboratory where numerical experiments can be conducted over a wide range of physical parameters. Exciting similarities between phenomena at the eV and GeV range are being uncovered, revealing common mechanisms operating over unprecedented ranges of time, space, and energy.

Acknowledgement

This work performed under the auspices of the United States Department of Energy by the Lawrence Livermore National Laboratory under contract No. W-7405-ENG-48.

References

1. F.H. McMahon, *The Livermore Fortran Kernels: A Computer Test of the Numerical Performance Range*, Lawrence Livermore National Laboratory Report UCRL-53745 (1986).
2. D. Tiszauer and K.C. Kulander, Phys. Rev. A **29**, 2909 (1984).
3. C. Bottcher and M.R. Strayer, Nucl. Instr. and Meth. **B31**, 122 (1988).
4. K.J. Schafer, J.D. Garcia, and N.-H. Kwong, Phys. Rev. B **31**, 122 (1988).
5. S.M. Younger, A.K. Harrison, and G. Suguyama, Phys. Rev. A 1989.
6. R. Car and M. Parrinello, Phys. Rev. Lett. **55**, 2471 (1985).
7. M. Klapisch, this volume.

Electron Correlation in the Continuum

Christopher Bottcher *and* Michael R. Strayer

Oak Ridge National Laboratory, Oak Ridge, Tennessee 37831, USA

We consider a class of problems, notably double ionization, which require accurate descriptions of correlation in both the initial and final states. Methods are presented for representing correlated wavefunctions on a basis spline lattice, and for calculating bound-continuum transition probabilities.

I. Introduction: problems and approaches

Double photoionization

$$h\nu + \text{He} \longrightarrow e + e + \text{He}^{++} \qquad (1)$$

has long been recognized as the prototype of correlated atomic processes, in that it is obviously forbidden in a truly independent particle description. Thus it has become the test bench for theoretical methods which attempt to go beyond independent particle approximations. Interest has intensified with the advent of experiments which replace the photon by a fast projectile, in particular a highly charged ion. We shall give a preliminary report on a new approach to these problems, which we expect to improve on previous theories in two respects. First, the initial and final states are calculated in a consistent framework, and secondly, the angular and energy distributions of the outgoing electrons are extracted.

Let us briefly review the difficulties in the way of calculating the cross section of a process such as (1). If the photon or projectile is described by perturbation theory, we have to evaluate a matrix element

$$M_{fi} = <\Psi_f|D|\Phi_i> \qquad (2)$$

where Φ_i is the initial (bound) state, Ψ_f the final (continuum) state and D is a dipole matrix element, say

$$D = z_1 + z_2 . \qquad (3)$$

More generally, D would be a sum of rather complicated multipoles, but the principle is unaltered. At this level of approximation, only the final state wavefunction Ψ_f poses any difficulties. If the excitation mechanism is a slow collision, final state and post-collision effects enter to make the calculation formidably more difficult.

The leading difficulty in constructing Ψ_f for particles with Coulomb interactions at large distances has long been that of specifying the asymptotic boundary conditions. Much progress has recently been made on the boundary conditions [1,2], but other problems remain. Most variational or R-matrix prescriptions for calculating the short-range part of a scattering wavefunction with two unbound particles suffer from divergences, traceable to the free-particle Green function. Thus a complete calculation of Ψ_f is some distance in the future. However the matrix element M_{fi} should be calculable without a complete knowledge of Ψ_f, particularly in the asymptotic region. Successful descriptions of bound-continuum processes have been carried out by many authors using bound state techniques to describe the short-range part of Ψ_f. The idea in a rigorous form dates from Ref. 3, and it has of course been applied in numerous papers since then [4].

II. Schrödinger equation for the three-body problem

Since we wish to consider methods which have the possibility of converging to an accurate solution, we shall represent the dependence on the internuclear distance explicitly, rather than through configuration interaction. In this section we present the analysis needed for an accurate numerical representation of a two-electron wavefunction of given orbital angular momentum, spin and parity.

Consider two electrons moving in the field of an infinitely heavy nucleus of charge C. If the electrons have position vectors $\vec{r_1}, \vec{r_2}$, the (non relativistic)

Hamiltonian is given by

$$H = -\frac{1}{2}(\nabla_1^2 + \nabla_2^2) + \left(\frac{1}{r_{12}} - \frac{C}{r_1} - \frac{C}{r_2}\right), \quad (4)$$

where $r_{12} = \left(r_1^2 + r_2^2 - 2r_1 r_2 \cos\vartheta\right)^{\frac{1}{2}}$ and ϑ is the angle between $\vec{r_1}$ and $\vec{r_2}$.

The *exact* nonrelativistic wavefunction for a state of total spin S and total orbital angular momentum L can be expanded as follows,

$$\Psi(\vec{r_1}, \vec{r_2}) = \Theta_{SM_S} \sum_{\ell=\varpi}^{L} \psi_\ell(r_1, r_2, \vartheta) \\ \times \mathcal{Y}_{LM_L}(\ell, L + \varpi - \ell) \quad (5)$$

where Θ is a spin eigenfunction, and $\mathcal{Y}_{LM_L}(\ell, \ell')$ are the usual coupled spherical harmonics. The switch $\varpi = 0, 1$ dictates the overall parity: $\Pi = (-1)^{L+\varpi}$. Then it is straightforward to derive a set of coupled equations satisfied by the functions ψ_ℓ, whose arguments are the three "dynamical" variables r_1, r_2, and ϑ,

$$[\mathsf{h}(\ell, L + \varpi - \ell) - E]\psi_\ell = \\ \sum_{\ell'}[\mathsf{v}_{\ell\ell'}^{(1)} + \mathsf{v}_{\ell\ell'}^{(2)}]\psi_{\ell'} \quad . \quad (6)$$

The diagonal Hamiltonian has the obvious form

$$\mathsf{h}(\ell_1, \ell_2) = T[r_1] + T[r_2] + T[\vartheta] \\ + \left[\frac{\ell_1(\ell_1+1)}{2r_1^2} + \frac{\ell_2(\ell_2+1)}{2r_2^2}\right], \quad (7)$$

where $-2T[r] = \partial^2/\partial r^2$ and

$$-2T[\vartheta] = \left(\frac{1}{r_1^2} + \frac{1}{r_2^2}\right)\frac{1}{\sin\vartheta}\frac{\partial}{\partial\vartheta}\left(\sin\vartheta\frac{\partial}{\partial\vartheta}\right). \quad (8)$$

The coupling terms are given by

$$\mathsf{v}_{\ell\ell'}^{(j)} = \frac{\mathcal{Z}_{\ell\ell'}^{(j)}(\vartheta)}{r_j^2} \frac{1}{\sin\vartheta}\frac{\partial}{\partial\vartheta} \quad (9)$$

for $j = 1, 2$. General formulae for the coefficients $\mathcal{Z}$ as polynomials in $\cos\vartheta$ have been obtained from angular momentum theory. In addition to these equations, the functions ψ_ℓ possess symmetries under interchange of r_1 and r_2, depending on S, L and ϖ. The analysis can be applied to three particles of arbitrary mass, if the coordinates are appropriately redefined.

III. Discretization on a lattice

We choose to represent the equations (6) on a discrete lattice of points in r_1, r_2, ϑ space by means of the Basis Spline Collocation technique. This method has been applied with success to the Schrodinger and Dirac equations in three cartesian coordinates in the context of ion-atom collisions, so its extension to a similar equation in another three-dimensional space is a natural one. The earlier Finite Element method was extensively applied to two electron systems, using the time dependent wavepacket approach [5]. Since the theory of the BSC method has been described in several papers [6], we only state the underlying postulates and final working equations here, for completeness.

In a collocation method, an operator equation $L[\psi] = 0$ is discretized by expanding $\psi(\vec{x})$ in a finite basis $u_k(\vec{x}), k = 1, \ldots, K$, where the coefficients are determined by setting $L[\sum \psi^k u_k] = 0$ at N collocation points $\vec{\xi}_\alpha, \alpha = 1, \ldots, N$. In practice each function is localized around a small subset of these points, and the coefficients ψ_k are eliminated in favour of the values of the solution ψ_α at the points $\vec{\xi}_\alpha$. The resulting equations have the structure of finite difference algorithms, while drawing upon the full power of modern functional methods. We choose localized piecewise continuous polynomials, known as *basis splines*, as overall the most powerful and flexible interpolating functions yet devised.

The solutions of (6) are thus expanded in products of splines,

$$\psi_\ell(r_1, r_2, \vartheta) = \sum_{ijk} c_{ijk} u_i(r_1) u_j(r_2) w_k(\vartheta), \quad (10)$$

and the collocation principle is applied as described to obtain the equations satisfied by the vector of $\psi_\ell(r_{1\alpha}, r_{2\beta}, \vartheta_\gamma)$ on the collocation lattice. Operators become matrices in this space. The potential energy, and indeed any local function of r_1, r_2, ϑ, is diagonal in the indices α, β, γ. Each term in the kinetic energy is a sparse block matrix, e.g. $T[r_1]$ has the structure $t_{\alpha,\alpha'}\delta_{\beta,\beta'}\delta_{\gamma,\gamma'}$. This extreme sparseness gives the collocation method an advantage, particularly in three dimensions, over other formulations, such as that of Galerkin. We want the Hamiltonian to be sparse, since we do not wish to store its full matrix representation. Rather, all algorithms are broken down into canonical operations of the form: matrix × vector ⟶ vector, where the matrix need only be represented implicitly by its non zero blocks.

Before proceding to the subject of linear operator methods, some technical aspects of the collocation methods are worthy of comment. As implied by (10), different bases are used for the radial and angular coordinates. To obtain accurate wavefunctions, these bases must satisfy precise boundary conditions which accommodate the cusp conditions at the nucleus, and when the two electrons are close together. The radial functions $u_i(r) \sim r - Cr^2$, in the metric of (7). Thus we arrange that u_i is a polynomial in r, $= 0$ at the origin. Near $\vartheta = 0$, the expansion $\psi \sim a + b\vartheta + \cdots$ fits the two-body cusp: $\psi \sim c(1 + \frac{1}{2}r_{12})$, while near $\vartheta = \pi$, the correct expansion is $\psi \sim a' + b'(\pi - \vartheta)^2$. These conditions are satisfied if $w_k(\vartheta)$ is a polynomial in the variable ϑ as opposed to $\cos\vartheta$, and $\neq 0$ at $\vartheta = 0, \pi$.

Finally, we note that the collocation representation of the Hamiltonian is usually not self-adjoint, i.e. the kinetic energy matrices are not symmetric. They do satisfy the more general criterion of factorizing into a product $\mathbf{T} = \mathbf{S}^{-1}\tilde{\mathbf{T}}$ of Hermitian matrices, $\mathbf{S}$ being positive definite, a sufficient condition to ensure that the spectrum of $\mathbf{T}$ is real. A practical consequence is that normal and adjoint solutions must always be carried along together, e.g.

$$(\mathbf{H} - E)\mathbf{\Psi} = 0, \quad (\mathbf{H}^T - E)\tilde{\mathbf{\Psi}} = 0, \quad (11)$$

where norms and inner products always have the form $\tilde{\mathbf{\Psi}}^T \mathbf{\Psi}$. Intuitively one may think of the adjoint vector as carrying a quadrature weight.

IV. Linear operator methods

To obtain the initial and final states in (2), we use linear algebraic techniques taken over from the ion-atom problem with little modification [6,7]. We shall use the customary notation of quantum mechanics, provided it is understood that operators are matrices, kets are vectors and bras are vectors in adjoint space. Bound state solutions are calculated by damped relaxation methods. The solution of $(H - E_n)\Phi_n = 0$ is calculated by iterating

$$\Phi^{(I+1)} = \Phi^{(I)} + \mathcal{D}(H - E^{(I)})\Phi^{(I)}, \quad (12)$$

where $E = <H>$ and $\mathcal{D} = \lambda(1 + T/\mu)^{-1}$. In practice we construct an approximate $\mathcal{D}$ separable in the spaces of r_1, r_2, ϑ. The constants λ, μ can always be chosen to guarantee rapid convergence. If excited states are required, one enforces orthogonality to all lower-lying states by projecting at each iteration. Indeed any desired constraint can be incorporated in this method, a notable example being Feshbach projection to calculate autoionizing states. Suppose ω_q are the bound states of the one body Hamiltonian embedded in (7), e.g. that of He^+. The Feshbach projection operators are

$$P_q = |\omega_q(1)><\omega_q(1)| + (1 \leftrightarrow 2). \quad (13)$$

If we project out the $1s$ state, by enforcing $P_{1s}\Phi = 0$, we can use (12) to calculate autoionizing states, e.g. $He(2s^2)$.

Turning now to the calculation of (2), we shall reformulate the problem slightly. Let $M_{fi} = M_\lambda(E)$ where f is a continuum state of energy E, distinguished by other quantum numbers λ. Then

$$D\Phi_i = \sum_\lambda \int dE \, M_\lambda(E)\Psi_\lambda(E). \quad (14)$$

Bound state contributions to (14) are omitted for clarity; we note that they are easy to calculate explicitly and subtract out. Now we introduce the Gaussian filter operator,

$$F(E - H, \Delta) = (\sqrt{\pi}\Delta)^{-1/2} \exp\left[-\frac{(E - H)^2}{2\Delta^2}\right], \quad (15)$$

so normalized that $\int F^2 dE = 1$. Applying this operator to (15), elementary quantum mechanics shows that

$$<\chi(E, \Delta)|\chi(E, \Delta)> = \sum_\lambda \int d\epsilon F(E - \epsilon, \Delta)^2 \times |M_\lambda(\epsilon)|^2, \quad (16)$$

where

$$\chi(E, \Delta) = F(E - H, \Delta)D\Phi_i. \quad (17)$$

For reasonably small values of Δ, we can write simply

$$<\chi(E, \Delta)|\chi(E, \Delta)> \simeq \sum_\lambda |M_\lambda(E)|^2. \quad (18)$$

The filter considered here is a functional of the full Hamiltonian: any other operator could replace H, leading to projections on a different spectrum. If the individual λ terms could not be disentangled, this would not be a very interesting result: the sum is usually dominated by single ionization leaving the ion in its ground state. Fortunately we can do better, but it is necessary to look more closely at the properties of χ.

For an $E > 0$, (7) defines a normalizable wavepacket of spatial extent $a \sim E^{1/2}/\Delta$. Thus for small Δ (such that $a < r_q$, the size of a bound state), χ can be represented over most of space by its asymptotic form

$$\chi \sim \sum \omega_q(1) f_q^{(-)}(2)$$
$$+ \sum \omega_q(2) f_q^{(-)}(1) + \chi_C, \qquad (19)$$

where ω_q are bound states of the ion as before, and χ_C is a superposition of true doubly ionized states. Thus if $\chi_q = P_q \chi$, in the notation of (13),

$$<\chi_q|\chi_q> = |M_q^{(1)}|^2, \qquad (20)$$

the contribution from single ionization, leaving the ion in state q. The term χ_C in (19) is obviously given by $\prod(1 - P_q)\chi$, whence

$$<\chi_C|\chi_C> = \sum_\lambda |M_\lambda^{(2)}|^2, \qquad (21)$$

the true double ionization contribution.

We emphasize that these calculations are valid for a lattice larger that a bound atom, though finite. We also recall that the spline basis can represent continuum states on such a lattice very well. Our procedure should be contrasted with the incorrect one of projecting (14) before filtering, i.e. calculating $FP_q D\Phi_i$ instead of $P_q FD\Phi_i$. Since P_q is a functional of the independent electron part of the Hamiltonian, it does not commute with F. This is an aspect of the well recognized fact [8] that projection from a configuration interaction representation by itself cannot distinguish single from double ionization. In summary, projection as in (20) is valid because the Gaussian filter picks out the asymptotic region. In particular, (20) does not contain contributions from autoionizing states. The filter plays a role in lattice calculations analogous to that of the Green function in Hilbert space.

We have finally to deconvolute different contributions to the double ionization contained in the wavepacket $|\chi_C|^2$. The analysis of a wavepacket of two unbound electrons was discussed in Ref. 5. Let us express $|\chi_C|^2$ in the hyperspherical coordinate system R, α, ϑ, where $r_1 = R\cos\alpha, r_2 \sin\alpha$. Consider the probability in a cone defined by $\alpha_0 - \Delta\alpha/2 < \alpha < \alpha_0 + \Delta\alpha/2$, $\vartheta_0 - \Delta\vartheta/2 < \vartheta < \vartheta_0 + \Delta\vartheta/2$, and $0 < R < R_{max}$. If the electrons are only weakly interacting in most of the space, this probability (divided by $\Delta\alpha\Delta\vartheta$) is the amplitude associated with the two-electron state α_0, ϑ_0; in other words, the electrons are ejected at a relative angle ϑ and with energies E_1, E_2 such that $E_1/E_2 = \tan^2 \alpha_0$. The proof of this result from scattering theory is rather lengthy, and our remarks here are only intended as a heuristic justification.

V. Numerical calculations

The numerical implementation of (17) was described in detail in Ref. 9. There we evaluated the Gaussian by a splitting technique: letting $Z = (E-H)^2/\Delta^2$ we write $\exp(-Z) = [\exp(-Z/M)]^M$, where M is sufficiently large that $\exp(-Z/M)$ can be expanded in a Taylor series. Though effective, this is a very slow procedure, so we have replaced the Gaussian by a rational function

$$F = \mathcal{N}\left(1 + \frac{Z}{M}\right)^{-M}, \qquad (22)$$

where $\mathcal{N}$ is the normalizing factor. The operator in (22) is inverted by a damped relaxation method, similar to (12): to solve $(1 + Z/M)\xi = \eta$ we iterate

$$\xi^{(I+1)} = \xi^{(I)} + \mathcal{D}^2 \rho^{(I)}$$
$$\rho^{(I)} = \eta - (1 + Z/M)\xi^{(I)}. \qquad (23)$$

Notice that $\mathcal{D}$ is squared to cancel H^2 in Z. Our revised procedure is hundreds of times faster than the old.

The methods outlined in Section IV are novel, but fortunately they can be tested thoroughly. The use of the filter to extract matrix elements (16) has been applied to a variety of soluble single channel problems [9]. We have carried out similar tests for the multichannel cases of interest here. A crucial issue is the validity of the projection technique underlying (19)-(21). If the postulate (19) is valid the calculated amplitudes should be independent of Δ, which they are to a good approximation. We can also check sum rules, such as

$$\int dE \sum_\lambda |M_\lambda(E)|^2 = <\Phi_i|D^2|\Phi_i>. \quad (24)$$

In general our calculated total amplitudes for double ionization and single ionization with simultaneous excitation are in accord with earlier accurate calculations [10,11], and results will be presented elsewhere. The differential amplitudes for double ejection show strong correlations between the two electrons as predicted by the Wannier theory [1,5], and these correlations persist to energies well above threshold.

VI. Conclusions and prospects

We have presented a unified description of scattering processes involving electronic correlation in both bound and continuum states. Many other processes can be treated by the same methods. A simple example is provided by double excitation, usually leading to an autoionizing state

$$\mathcal{Z} + \text{He} \longrightarrow \mathcal{Z}^* + \text{He}(qq') \quad (25)$$

where $\mathcal{Z}$ stands for any fast projectile, and the final state of Helium is labelled by two single electron orbital designations q, q' (e.g. $2s, 2p$). The quantum numbers of the final state in (24) including its allignment can be determined experimentally. Theoretical interest focusses on the competition between single and multiple step mechanisms [12].

Another class of problems involves the decay of autoionizing states in external fields. As before, we require bound-continuum matrix elements,

$$\mathcal{V}_\lambda(E) = <\Psi_\lambda(E)|H - E + D|\Phi_{qq'}> \quad (26)$$

where D is the operator of the external field. The autoionizing state $\Phi_{qq'}$ is calculated by projection, and the continuum state $\Psi_\lambda(E)$ describes the outgoing electrons. The angular distribution of these electrons is of great experimental interest [13], but has been difficult to predict because of its extreme sensitivity to correlation in the wavefunctions.

Time dependent processes can also be addressed, insofar as they can properly be described by a limited number of angular momentum eigenstates. A problem of pedagogical interest is to monitor the decay of autoionizing states in time, a process which remains intuitively mysterious, in spite of its long history.

Acknowledgements

This research was supported by the U.S. Department of Energy, Office of Basic Energy Sciences, Division of Chemical Sciences. Oak Ridge National Laboratory is operated by Martin Marietta Energy Systems under Contract No. DE-AC05-84OR21400 with the U.S. Department of Energy.

References.
1. C. Bottcher, Adv. Atom. Molec. Phys. **25**, 303 (1989).
2. P. L. Altick, Phys. Rev. **A25**, 128 (1982).
3. Y. M. Chan and A. Dalgarno, Proc. Phys. London **85**, 277 (1965).
4. H. S.Taylor and A. U. Hazi, Phys. Rev. **A14**, 2071 (1976).
5. C. Bottcher, Adv. Atom. Molec. Phys. **20**, 241 (1985).
6. C. Bottcher and M. R. Strayer, Ann. Phys. (N.Y.) **175**, 64 (1987).
7. C. Bottcher, M. R. Strayer, A. S. Umar and P. G. Reinhardt, Phys. Rev. **A**, in press (1989).
8. A. L. Ford and J. F. Reading, J. Phys. **B21**, L685 (1988).
9. C. Bottcher, M. R. Strayer, A. S. Umar and V. E. Oberacker, Phys. Rev. **C37**, 2487 (1988).
10. S. L. Carter and H. P. Kelly, Phys. Rev. **A24**, 170 (1981).
11. M. R. H. Rudge, J. Phys. **B21**, 1887 (1988).
12. J. P. Giese, M. Schultz, J. K. Swenson, H. Schöne, S. L. Varghese, C. R. Vane, M. Benhenni, P. F. Dittner and S. Datz, contributed paper no. 484, *this conference* (1989).
13. J. K. Swenson, C. C. Havener, N. Stolterfoht, K. Sommer and F. W. Meyer, Phys. Rev. Lett. **63**, 35 (1989).

ION-METAL AND ION-ATOM COLLISIONS: INSTANT REPLAYS AND MEAN-FIELD THEORIES*

J. D. Garcia[†], N. H. Kwong[†], and K. J. Schafer[††]

[†] Department of Physics, University of Arizona, Tucson, AZ, 85721
[††] Lawrence Livermore National Laboratory, Livermore, CA, 94550

In this paper, we describe the results of our general long-term programmatic goal of investigating the strengths and weaknesses of time-dependent mean-field theories for collisions. We have made some progress in: (a) obtaining a better formulation of the theory, which has the exact full Schrödinger equation as one limit and permits appropriate classical treatment of heavy particles correctly coupled to the quantally treated electrons; (b) restructuring our numerical treatment to make it fully three-dimensional, improve accuracy and decrease cycle time, so that larger problems more in keeping with the mean-field concept can be treated; and (c) incorporating the electrons in the conduction band of a metal into our quantal treatment, making possible the description of collisions of atoms and ions with solids. Numerical results for protons traversing a thin metallic foil, among other examples, are presented and discussed.

The collision dynamics of aggregates of charged particles presents an interesting juxtaposition of circumstances that make theoretical descriptions difficult. One well-known feature is the slow decrease of the potential at large distances, or equivalently, the singularity at small momentum transfers, which shows up in a singularity in the phase of the wave function. This causes difficulties for cases involving more than two particles. On the other hand, for a system with zero net charge as one partner in a collision, the net effective interaction is highly nonlinear and of short duration. Despite this, many of us believe that an appropriate set of simple analytical models can be devised, albeit different ones for different sets of circumstances, which will more or less accurately describe these results. Finding such formulations presents an interesting challenge. We have approached this problem by investigating the appropriateness of time-dependent mean-field theory for providing an accurate, detailed general description, then using those results for developing special simpler, hopefully analytical models in various regimes. Of course, the time-dependent mean-field formulation is already an idealized model of the full Schrödinger equation. It may be the simplest accurate description possible for some circumstances. In any case, our direct numerical approach has the advantage of being able to produce stop-action pictures of the collision in progress, in addition to ultimately producing the standard predictions in terms of probabilities, cross sections, etc. Here we describe our progress to date and the prognosis for the future.

We use a variational principle to obtain the equations of motion, starting from the full Schrödinger equation for all particles. The correct coupling of the motion of the heavy particles to the electronic motion that emerges has as central features the conservation of total energy, momentum, and angular momentum. Details of the derivation can be found elsewhere.[1] The results are a set of classical Hamiltonian equations for the heavy particles, coupled to the time-dependent Hartree-Fock equations for the electrons:

$$\vec{\dot{X}}_j = \vec{P}_j / M_j$$

$$\vec{\dot{P}}_j = - \sum_{k \ne j} \vec{\nabla}_{X_j} V_{nn}(X_j, \vec{X}_k)$$

$$- \sum_\alpha \langle \phi_\alpha(\vec{x}) | \vec{\nabla}_{X_j} V_{en}(\vec{x}, \vec{X}_j) | \phi_\alpha(\vec{x}) \rangle$$

$$i\hbar \frac{\partial \phi_\alpha(\vec{x})}{\partial t} = \hat{\mathcal{H}} \phi_\alpha(\vec{x}) + V_{exch}$$

where

$$\hat{\mathcal{H}} = \vec{p}^2/2m + \sum_k V_{en}(\vec{x}, \vec{X}_k) + V_{ee}$$

$$V_{exch} = - \sum_\beta \phi_\beta(\vec{x}) \langle \phi_\beta(\vec{x}') | 1/|\vec{x}-\vec{x}'| | \phi_\alpha(\vec{x}') \rangle$$

$$V_{ee} = \sum_\beta \langle \phi_\beta(\vec{x}') | 1/|\vec{x}-\vec{x}'| | \phi_\beta(\vec{x}') \rangle$$

$$V_{en}(\vec{x}, \vec{X}_k) = - Z_k / |\vec{x} - \vec{X}_k|$$

$$V_{nn} = + Z_k Z_j / |\vec{X}_k - \vec{X}_j|$$

Here, the coordinates of the heavy particles are denoted by capital letters and those of the electrons by lower case letters, as is customary.

The variational principle yields different equations depending upon the restrictions imposed on the form of the total wave function. If no restrictions are imposed, the full Schrödinger equation results. The above results arise from assumption of a WKB form for the wave functions of the heavy particles.

We solve this set of coupled equations numerically on a three-dimensional grid, using a basis-spline collocation method originated by Bottcher and Strayer.[2]

The electrons in the solid are treated using the successful "jellium" model for the conduction band, in which the electrons occupy states in a potential provided by a frozen distribution of positive charge. This adds a corresponding potential to the above equations. We have adapted the basis-spline method to this situation by reformulating it to accommodate periodic boundary conditions. This enables us to describe the behavior of electrons in a three-dimensional sheet of metal, i.e., a foil of finite thickness in one dimension. We then start an atom or ion at some large distance from the slab and evolve the entire system self-consistently in time. Our time-evolution technique takes advantage of the fact that the potential energy matrix is diagonal in configuration space. We use an operator splitting technique as follows: instead of expanding the time-evolution operator in a power series, we use a representation correct to third order

$$\psi(t+\Delta t) = e^{-i\hat{\mathcal{H}}\Delta t} \psi(t) = e^{-i(T+V)\Delta t} \psi(t)$$

$$= e^{-iV\Delta t/2} e^{-iT\Delta t} e^{-iV\Delta t/2} \psi(t) + O(\Delta t^3)$$

The kinetic energy piece of the operator, $e^{-iT\Delta t}$, is carried out by using an intermediate (time-independent) kinetic-energy basis set. The potential energy part is simply a phase factor in configuration space. The primary advantage is that this entire operator is now vectorizable on the Cray-2. This gives us a running-time savings of about a factor of 10. Though explicit, this scheme is manifestly unitary and thus stable.

Our initial tests of the time-dependent mean-field approach involved collisions of atoms and ions including only a small number of electrons, so that we could compare our results with other theoretical treatments. Figures 1 and 2 show the results of application of this technique to He^{++}-He collisions and muon-hydrogen collisions (details of this latter calculation can be seen in our paper[3] at this conference; these are both finite-difference, cylindrical grid calculations). In Fig. 1, the process being described is a two-electron transfer from the helium atom to the α-particle; this is of interest because time-dependent mean-field theories track best the one-electron processes occurring. The fact that our complete results, depicted by the crossed circles, agree well with experiment and bridge the gap between atomic and molecular basis-expansion results is significant.

In Fig. 2, the results of our calculation of muon capture by hydrogen are shown and compared to other calculations. In this calculation, the muon is treated semiclassically, as is the proton, while the electron is quantal. The results, listed as CQC (classical-quantal coupling), elucidate the non-adiabatic nature of this process at high muon incident energies. Other results shown include those from classical trajectory Monte Carlo calculations (CTMC), from adiabatic ionization theory (AI), and some approximate cross sections from a treatment in which the muon is described by a moving Gaussian wave packet, labeled TDHF.

Our previous papers[4] indicate that the time-dependent mean-field procedures provide reliable guides to <u>inclusive</u> probabilities. These tests are continuing.

These systems, however, did not truly have enough electrons to justify the mean-field concept. Nevertheless, our results are seen to be quite satisfactory when compared to other theories and to experiment. In Fig. 3, we show a time sequence of a proton traversing a metal foil with the same electron density as aluminum. In this calculation we are following 59 wave functions in space and time in a unit cell of the solid, properly coupled to the motion of the proton. We have been able to examine the charge capture, energy loss and time dependence of the electron density in the metal for incident proton velocities from 1/2 to 8 times the Fermi velocity. Typically, such a calculation takes 1-2 CPU hours on a Cray-2.

The proton in Fig. 3 was initially moving in the direction perpendicular to the foil surface at a speed equal to the Fermi velocity. The legend above the induced density plot gives the position of the proton at the time of the frame. It can be seen from the first density plot that the proton pulls electron density out of the metal surface (i.e., captures charge via the tail of the electronic wave function) even before it reaches the surface at -10 a.u. As it traverses the foil, it carries with it the electrons that fall into the Coulomb potential well and excites density waves that can be seen to radiate from the proton. As it exits the rear surface, it again is able to carry some electron density out of the foil with it, leaving an enhanced induced density at the surface, which quickly dissipates. At higher proton speeds, the conduction electrons do not have time to adjust to the motion of the proton, so the pre-capture as well as the final capture decrease rapidly with increasing proton speed.

Our results for energy loss to the conduction electrons seem to agree with the earlier results of Lindhard. Additional studies of these phenomena, including direct comparisons of our results with linear response theory, are currently under investigation.

We plan a variety of applications of our three-dimensional computer program, which will include atom-laser, solid-laser interactions, as well as other ion-atom collisions.

It appears that the self-consistent mean-field technique can provide a quite feasible, conceptually simple approach for exploring strong interaction, nonlinear non-adiabatic processes.

References

*Supported in part by National Science Foundation Grant No. PHY87-11139

1. N. H. Kwong, J. Phys. B <u>20</u>, L647 (1987).

2. C. Bottcher and M. R. Strayer, Ann. Phys. (NY) <u>175</u>, 64 (1987).

3. N. H. Kwong, J. D. Garcia and J. S. Cohen, XVI ICPEAC, New York.

4. J. D. Garcia, Nucl. Instrum. Methods <u>A240</u>, 552 (1985).

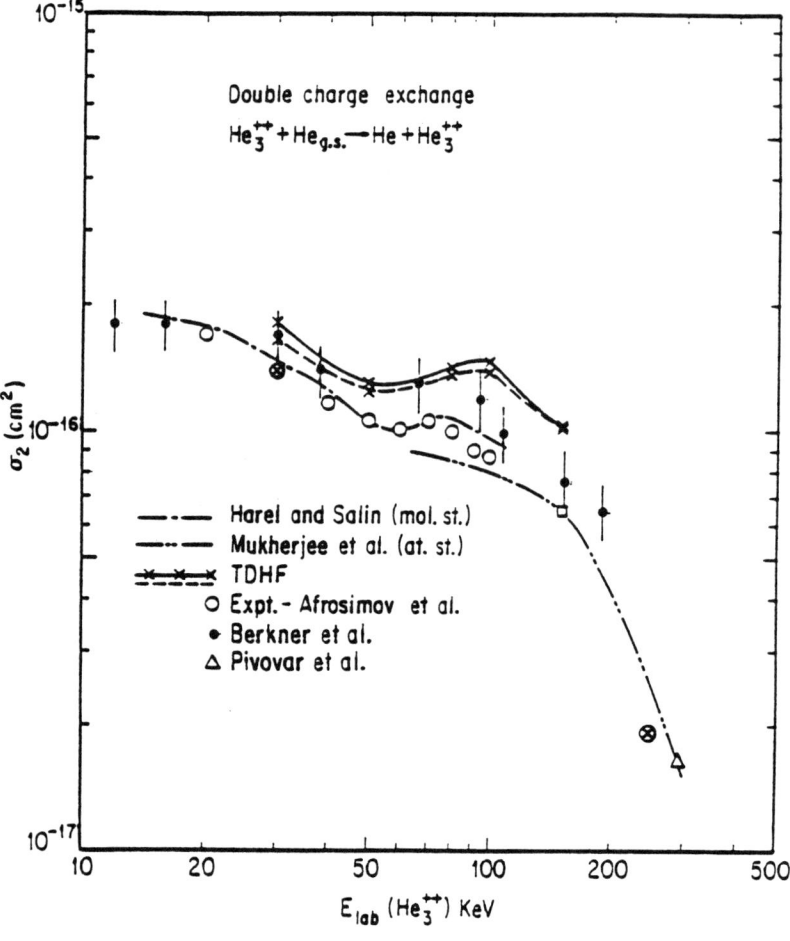

Figure 1. Capture of two electrons by an α-particle colliding with a helium atom. The crossed circles are the TDHF results including Coriolis forces. See Ref. 4 for more details.

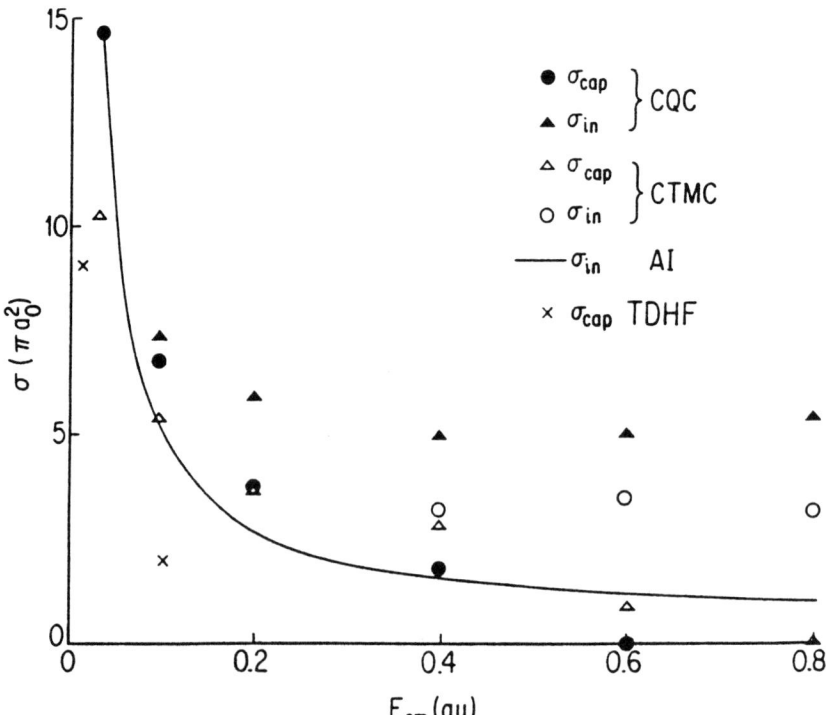

Figure 2. Muon capture cross sections and hydrogen inelastic excitation cross sections for various energy muons impinging upon a ground-state hydrogen atom. CQC represents our current results; CTMC are classical trajectory Monte Carlo results; AI is the adiabatic ionization theory. TDHF is an approximate result with a quantal treatment of the muon as well.

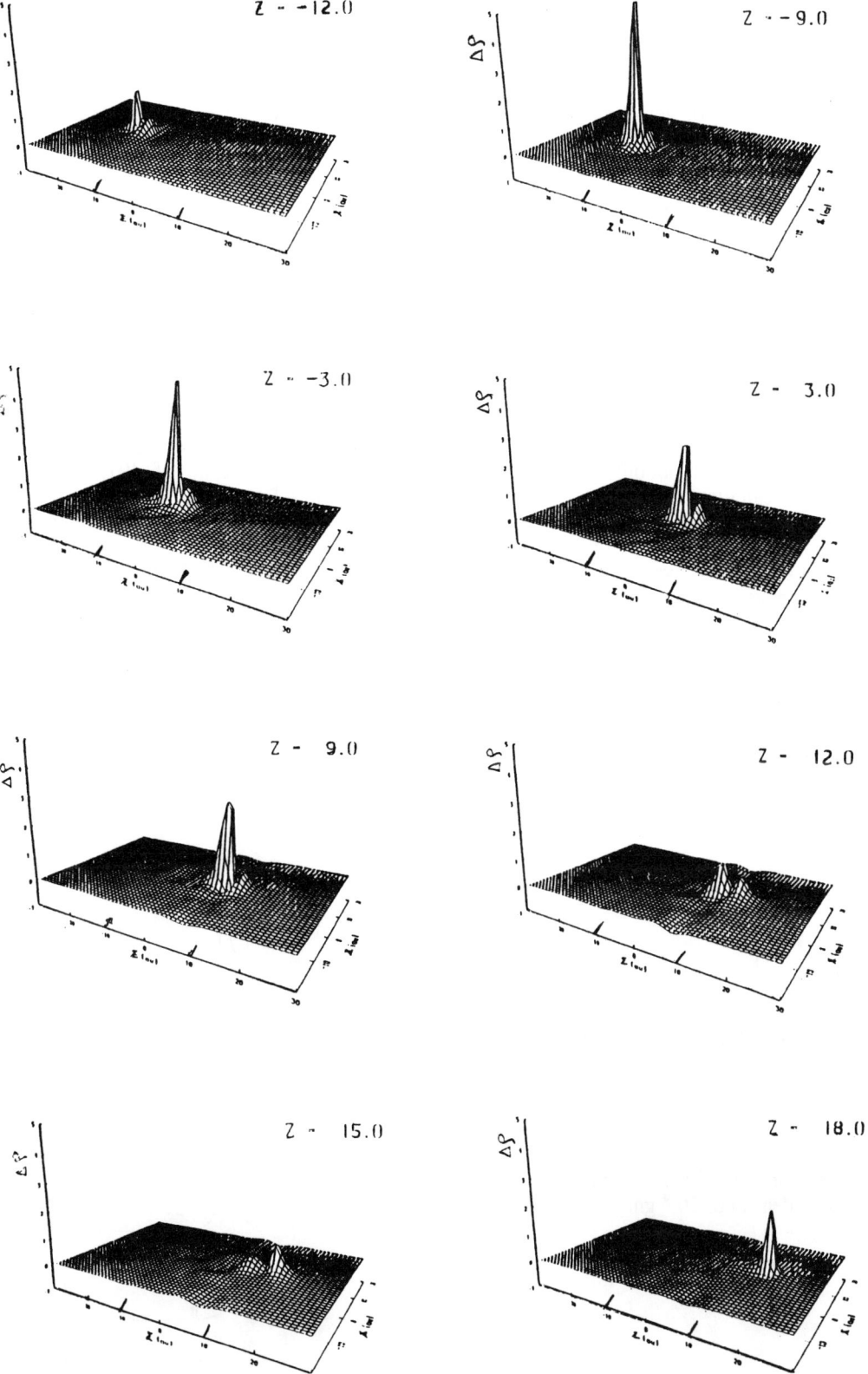

Figure 3. The charge density $\Delta\rho$ induced by a proton traversing an aluminum foil. The foil extends from -10 a.u. to +10 a.u., as marked on the axis. The initial proton speed was equal to the Fermi velocity, perpendicular to the foil. Charge capture can be seen in the last frames as the proton emerges from the foil.

LARGE SCALE CALCULATIONS OF ELECTRON-ATOM/ION PROCESSES

K T Taylor

Department of Mathematics, Royal Holloway and Bedford New College, University of London, UK

A number of 'large-scale' calculations using supercomputers are discussed including the generation of atomic data in the Opacity Project; the diagonalisation of large matrices on the CRAY X-MP/48, and application in state-of-the-art calculations on electron-helium scattering; the novel topic of Rydberg states of diatomic molecules in a magnetic field. The use of linked arrays of transputers is considered, not only in relation to large existing FORTRAN codes but also with new codes employing extremely advantageous algorithms.

The Opacity Project: 'Large-scale' in the title will be given a number of distinct interpretations in this paper. In the Opacity Project it is an apt descriptor not only of the size of many of the individual calculations but is also appropriate in considering the overall volume of data produced. This project[1], co-ordinated and inspired by M J Seaton of University College London has, over the past 5 years, involved collaboration amongst groups in the United Kingdom, the United States, France, West Germany and Venezuela in the calculation of atomic radiative data relevant to a much improved understanding of stellar envelope opacities.

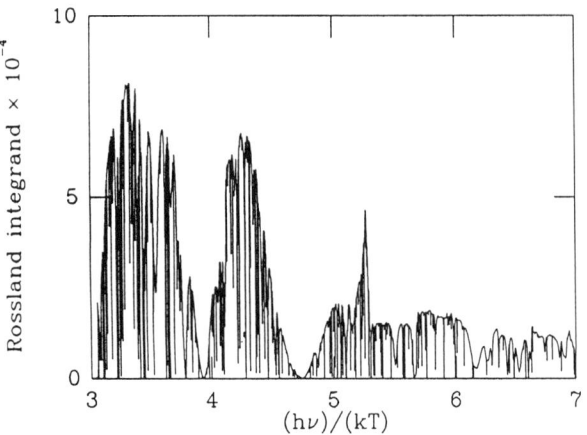

Figure 1. The Rosseland integrand for carbon at a temperature of $10^{4.5}$ K and density of 10^{-8} gm cm^{-3}

The Rosseland mean opacity K_R is defined in a reciprocal manner:

$$\frac{1}{K_R} = \int_0^\infty \frac{1}{K_\nu} R(\nu)\, d\nu$$

where K_ν is the monochromatic opacity and $R(\nu)$ is a weighting function. Integration is over the full frequency spectrum. Figure 1 is a plot[2] of the above integrand for carbon, for typical conditions of temperature and density and where all ionisation stages have been accounted for. This involves *ab initio* calculation of 491 energy levels, oscillator strengths for 3360 discrete lines, and photoionisation cross sections from 307 distinct initial states all to an accuracy better than 10%. Such work differs markedly in scale from most calculations in atomic physics, whose end result is, for example, one or two energy levels or photoionisation cross sections. Even with a supercomputer the choice of algorithm is a crucial factor. Adoption of the R-matrix approach[3] in treating the basic electron-ion scattering problem has allowed the efficient generation of such data by employing basis set techniques in the complicated multi-electronic part of the problem.

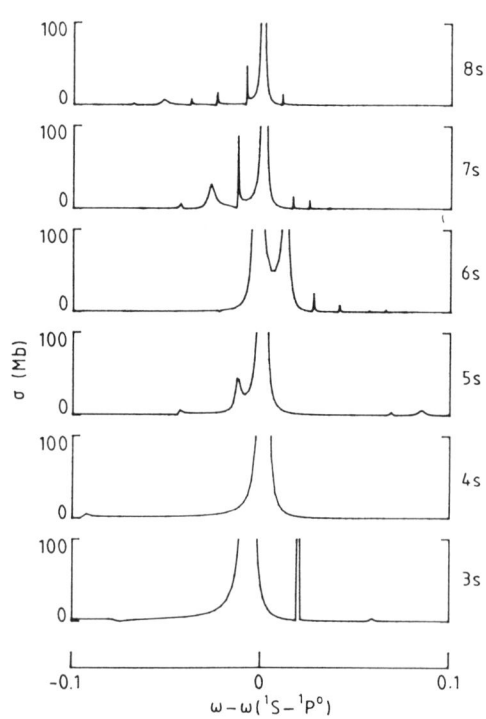

Figure 2. Photoionisation cross sections for C$^+$.

In Figure 1 the pronounced notches are the most obvious features. These are due to resonances in the photoionisation cross sections caused by photoexcitation of the core (PEC resonances). Figure 2 displays calculated photoionisation cross sections[4] from C^+ initial states $2s^2ns$ ^2S with n=3,8 where PEC resonances carry the bulk of the cross section, viz:

$2s^2$ (^1S) ns + hν -> 2s2p ($^1P^o$) ns -> $2s^2$ ^1S + e.

It should be noted that such important contributions to the cross sections (and corresponding large oscillator strengths in discrete transitions) are completely missed out in calculations within a hydrogenic model.

In summary, radiative data are being calculated for each member of every isoelectronic sequence up to that of magnesium, and for all ionisation stages of cosmically abundant elements heavier than this. Transitions out of all levels involving an electron with a principal quantum number less than 11 are required. Most data are in hand with the small remainder to be available by the end of 1989.

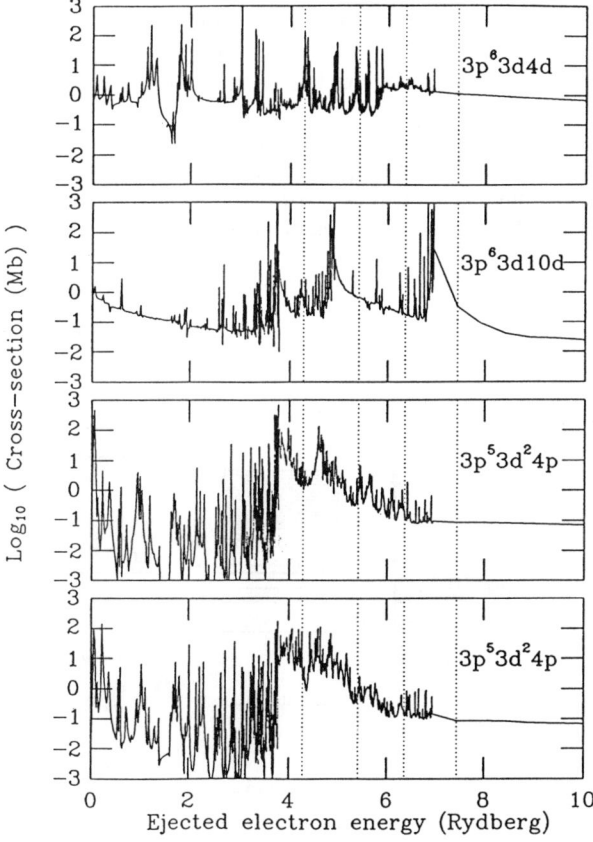

Figure 3. Photoionisation cross sections for Fe^{6+}.

It is worth concluding this section with the example of Fe^{6+} which requires a calculation[5] typical of those for the heavier ions. In this case it was necessary to include the lowest 31 levels of Fe^{7+} in the scattering process. The calculation has yielded 1834 bound levels for Fe^{6+}, 33960 oscillator strengths in Fe^{6+} and phoionisation cross sections from 515 separate initial states. Thus this single ion yields more data than arises from *all* ionisation stages of carbon. Figure 3 displays a sample of 4 photoionisation cross sections from initial states of ^1P symmetry. The detailed resonance structure is typical of all 515 cross sections.

Matrix Diagonalisation on the CRAY X-MP/48: In this section 'large scale' is an appropriate description of the order of matrix to be diagonalised. The R-matrix method[3] allows a basis set representation of the electron plus target atom or ion Hamiltonian over the multi-electronic portion of space immediately surrounding the nucleus. Using an orthogonal basis set yields an ordinary eigenvalue problem for the Hamiltonian matrix produced. In contrast to most other basis set approaches, *all* the eigenvalues and *all* the eigenvectors must be found. Since due to symmetry, only the upper triangle need be retained, a matrix of order up to 3,500 can be stored in the 7MW fast memory common to the 4 processors on the CRAY X-MP/48. Clearly it will be a very inefficient use of such a machine if a program uses only *one* of the processors and that processor demands *all* the memory. For improved throughput also, it is important that such a job be multi-tasked where possible. Sawey and Berrington[6] have carefully scrutinised the Householder reduction to tri-diagonal form, the Sturm sequence method for the eigenvalues and the inverse iteration computation of the eigenvectors. They have found that the latter two computations can be carried out entirely in parallel as also can the most time consuming portions of the tri-diagonalisation step. The following timings have been achieved using a test matrix whose diagonal elements increase from unity in steps of one and whose off-diagonal elements are all equal to 0.1. (This mimics quite closely the structure of Hamiltonian matrices produced in real calculations.)

Processor	Time (min : sec)
1	12 : 05
2	12 : 27
3	11 : 57
4	14 : 39
Total	51 : 08

Thus throughput is achieved in less than 20 minutes rather than in about an hour if only one processor is employed. The new program is vectorised where possible and typically achieves 70 Mflops (millions of floating point operations per second) on each processor.

The new program has been exploited by Sawey et al[7] in electron hydrogen and electron helium calculations using the R-matrix method. The general procedure in a given calculation has been to include in the close-coupling expansion *all* states of the target atom until a certain value of principal quantum number n for the outer electron of this target is reached. For electron hydrogen scattering a 21 state calculation (all states n ≤ 6) is just possible within the CRAY X-MP/48 and involves Hamiltonian matrices up to order 3680. For the electron helium case, including all states with n ≤ 5, gives rise to a 29 state calculation and Hamiltonian matrices of order up to 3363. To illustrate the importance of this latter calculation, Figure 4 displays a typical collision strength, that for excitation of He(2^3S) from the ground state, and compares it with results from previous 5 state (n ≤ 2)[8], 11 state (n ≤ 3)[9] and 19 state (n ≤ 4)[10] calculations. (The upward pointing arrows on the energy axis indicate positions of other excited states of the atom that couple into the process.) The expected trend is apparent, namely, that the more states of the target atom included the higher the electron impact energy for which the results can be considered reasonable. In addition, one gains extra real resonance structure in the neighbourhood of each additional state of the atom that is coupled in. In general, the authors consider the results reliable up to the threshold energy for the highest target state explicitly included.

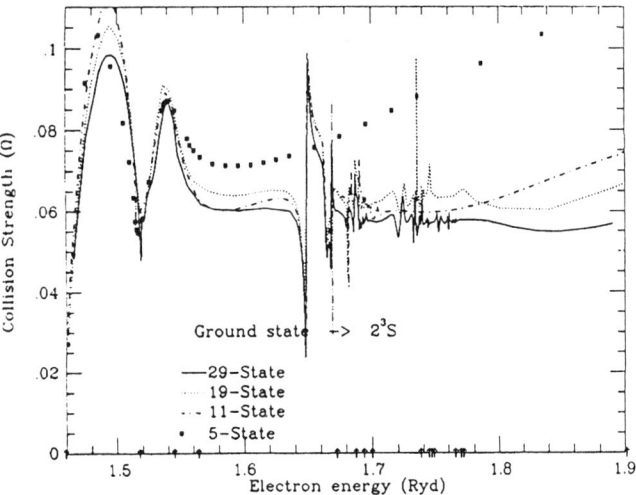

Figure 4. Electron impact excitation collision strength from He(1^1S) to He(2^3S).

Unfortunately the size of such calculations grows rather more quickly than just the increase in number of states would indicate. The persistent growth in radial extent of excited target states demands ever larger internal regions that must be spanned by an R-matrix basis.

H$_2$ in a magnetic field: Over the past ten years the H atom in a laboratory strength magnetic field (say, about 5 tesla) has been studied extensively[11] both theoretically and experimentally. The Hamiltonian in atomic units for this simple system is given by

$$H = -\frac{1}{2}\nabla^2 - \frac{1}{r} + \frac{1}{2}\beta^2 r^2 \sin^2\theta$$

where β measures the magnetic field in atomic units (4.7 tesla corresponds to a β of 10^{-5}), r is the radial co-ordinate of the electron and θ the angle between its position vector and the field direction. For such a laboratory strength field the electron must be in a high Rydberg (or continuum) state to have its wavefunction appreciably altered from its field-free form. Such states are typically of several thousand Bohr in radial extent and thus can readily be described as 'large scale'. The same term applies to the latest basis set approach[12] (involving approximately 200,000 basis members) used on a CRAY 2 machine in stepping upwards as far as possible towards the ionisation threshold.

Although no more difficult than H to handle in the laboratory, H$_2$ has been neglected by the experimentalists. Its field-free spectrum is markedly richer than that of H, reflecting not only the electronic structure of the inner ionic core (as in a non-hydrogenic atom) but also that energy goes into rotational and vibrational motions. A simple energy level diagram in Figure 5, indicates, that a series converging on the lowest N=0 rotational level of the residual H$_2^+$ core can be repeatedly perturbed by interloping members from other series converging on higher rotational levels of the ion. The magnetic field will alter the detailed interaction between the levels, but, more importantly, since for any given energy the interloping levels have a substantially smaller radial extent, it is reasonable to suppose them to behave rather differently in the magnetic field.

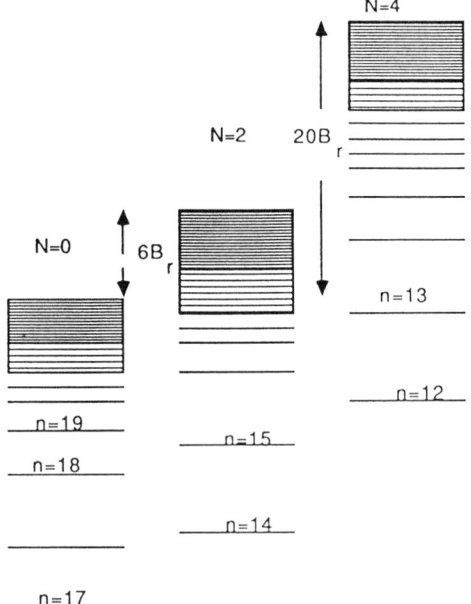

Figure 5. Schematic energy level diagram for field-free H$_2$.

Monteiro and Taylor[13] have recently extended the R-matrix method developed by O'Mahony and Taylor[14] to handle discrete levels of non-hydrogenic atoms, so that these interesting questions about the H_2 spectrum could be addressed. This method rests on identifying an inner region where the magnetic contributions to the Hamiltonian can be neglected and an outer region where although magnetic terms are crucial, only one electron is present. Treatment of the outer region involves a basis set approach and for H_2 using a CRAY X-MP *all* eigenvalues and *all* eigenvectors have been found in the banded generalised eigenvalue problem, with matrices of order up to several thousand. The technical details in H_2 are complicated by the strong electrostatic coupling between the inner portion of the Rydberg electron wavefunction and the residual ion core, whereas the outer portion is entirely decoupled from the core by the magnetic field. This leads to a non-trivial linear Zeeman effect for such states. In contrast to the atomic case, the only good magnetic quantum number over all space is that for the complete molecule rather than that for its Rydberg electron.

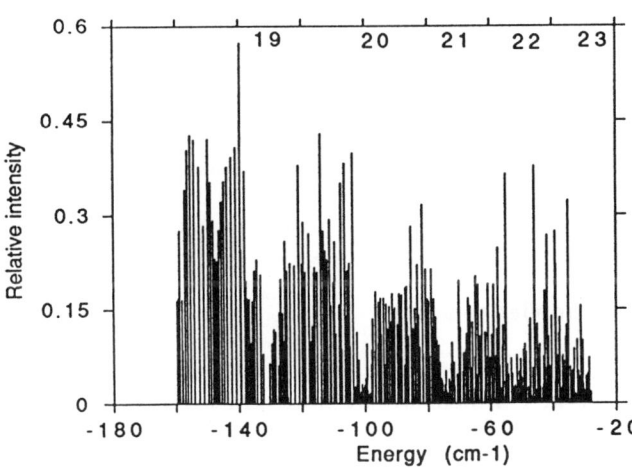

Figure 6. Photoabsorption spectrum for H_2 in a magnetic field of 4.7 tesla.

Figure 6 displays the calculated spectrum for photoabsorption by the ground state of H_2 in a magnetic field of 4.7 tesla using linearly polarised light so that the upper state has overall magnetic quantum number $M = 0$. The energy axis is measured from zero at the field-free $N = 0$ threshold. The spectrum is regular at the lower end, but as in the case of H, develops irregularity as the ionisation threshold is approached. The novelty in H_2 is that the irregularity appears confined to bands, with low intensity $N=2$ lines filling the notches between them. The full significance of this result in connection with ideas of chaos in the classical mechanics of this system has not, as yet, been fully explored, but it appears that in H_2 there is the opportunity to study interacting regular and irregular spectra, clearly not possible in the simple H system.

Transputers and their Application: Over the past year atomic physics groups at Daresbury Laboratory, Durham University and Queen's University Belfast have had Meiko M10 Computing Surfaces installed. In the Queen's and Durham instances the surface consists of 20 Inmos T800 transputers each of which is attached to a dedicated 4 Mbytes of RAM. Thus 'large scale' here can be applied to the number of processors running in parallel. Each transputer has a microprocessor, a 64-bit floating point unit, fast serial interconnecting links to allow communication with its fellows, a dynamic memory interface and a timer *all placed on a single chip*! This is illustrated schematically in Figure 7.

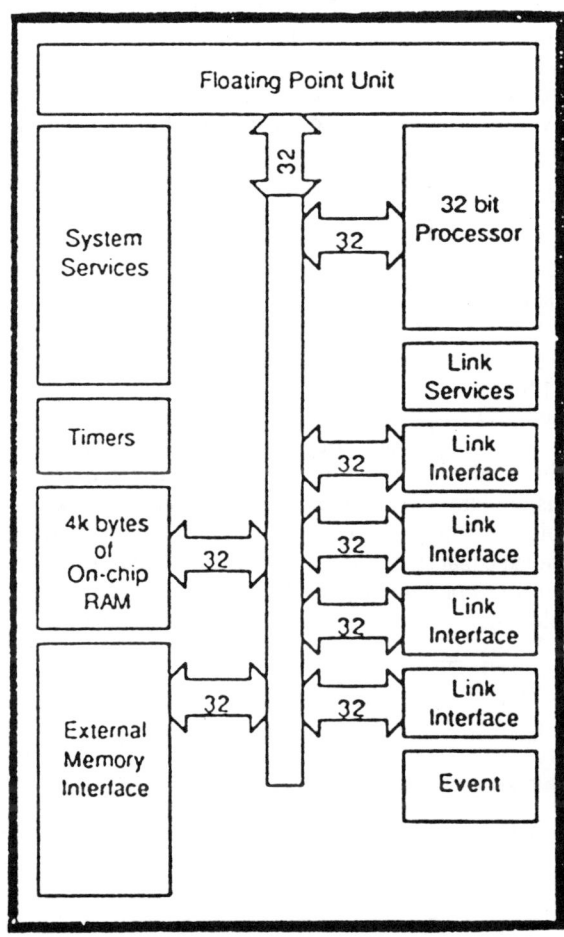

Figure 7. Schematic diagram of the Inmos T800 transputer.

Although the Meiko is a multi-processor machine the dedicated memory per processor makes it fundamentally different from an entirely shared memory multi-processor machine such as the CRAY X-MP. This in turn means that algorithms, such as the matrix diagonalistion described above, which are especially suited to the X-MP, will not be so easy to implement nor will they run so advantageously on the Meiko. On the other hand many algorithms, already in use in large FORTRAN codes, can readily exploit the dedicated memory machine. A difficulty, though, is that the transputer has its own programming language, occam-2. It would be clearly impractical to re-write FORTRAN codes that have taken many years to develop, and frequently access specialised FORTRAN libraries, in this new language. Allan, Heck and Zureck[15] have overcome this obstacle by developing an occam 'harness' known as Fortnet. Written in occam-2, Fortnet allows a series of concurrent buffer processes which pass data into and out of the FORTRAN serial processes under them (distributed over slave transputers), and ship data between these serial processes and to the front-end filing system. For greatest simplicity of both design and operation, the transputers have been linked as a dual daisy chain, that is, links are placed only between nearest neighbour transputers on a one-dimensional chain.

A number of electron scattering calculations have already been got underway. The aim of work by Whelan et al[16] is to calculate double and single differential cross sections as well as total cross sections for electron impact ionisation of atoms. In this work, the progression beyond the fundamental triple differential cross section requires time-consuming integrations over the ejected and scattered electrons' angles as well as over the partition of escape energy between them. The approach taken to parallelising the program has been to distribute the inner of a set of nested DO loops over all the available processors except one. This turns out equivalent to calculating the triple differential cross section at the whole set of differing collision parameters and summing on the last processor to obtain the integral cross section. The speedup factor achieved equals the number of processors minus one. Figure 8 illustrates preliminary results achieved by Whelan et al[17]. employing a Coulomb Projected Born approximation.

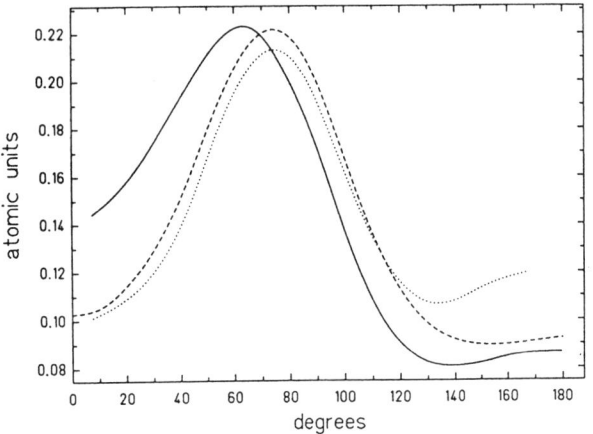

Figure 8. Double differential cross section for electron impact ionisation of H.

The double differential cross section (with integration over all angles of the scattered electron) for electron impact ionisation of atomic hydrogen with an incident electron energy of 250 eV and an ejected electron energy of 5 eV is plotted against ejected electron angle (dashed line) in Figure 8. Comparison is made with a pseudo-state calculation by Curran and Walters[18] (dotted line) and a standard first Born result (full line).

An algorithm for application in the multi-photon absorption problem and well-suited to the Meiko implementing the Fortnet harness is as follows[19]. At a typical step in the calculation, each processor solves a set of approximately 300 linear algebraic equations with a single column vector right-hand-side. The single column vector solution obtained by each processor forms the right-hand-side column vector for nearest neighbour processors at the next step. The matrix of coefficients held in the RAM dedicated to a given processor is the same from one step to the next, but completely different to that on any other processor. Such an algorithm clearly exploits a distributed-memory machine. Its other advantages are : all processors work on the same size matrices and thus remain in step; the amount of data transferred between processors is relatively small and takes little time; the algorithm can exploit any number of processors and indeed increasing the number would be important in carrying out convergence studies.

Conclusion: There is a role in large-scale electron atom/ion collision calculations for both the shared-memory and distributed-memory, multi-processor machines. For certain algorithms, and especially with the advent of vector processing registers, it is likely that the transputer will prove the 'faster' machine.

Acknowledgements

The author would like to thank R J Allan, K A Berrington, L Heck, T S Monteiro, M Sawey, M J Seaton, P J Storey and C T Whelan for supplying results in advance of publication.

References

1. M J Seaton, J Phys B:At Molec Phys,**20**,6363 (1987)
2. Figure 2 represents preliminary work of the Opacity Project and has been kindly supplied by M J Seaton.
3. K A Berrington, P G Burke, K Butler, M J Seaton, P J Storey, K T Taylor and Yu Yan, J Phys B:At Molec Phys,**20**,6379 (1987)
4. Yu Yan and M J Seaton, J Phys B:At Molec Phys, **20**, 6409 (1987)
5. P J Storey and H E Saraph, J Phys B:At Molec Opt Phys to be submitted.
6. M Sawey and K A Berrington, Comput Phys Comm to be submitted.
7. M Sawey, K A Berrington, P G Burke and A E Kingston, Abs. XVI International Conference on the Physics of Electronic and Atomic Collisions, New York 1989 and J Phys B:At Molec Opt Phys to be submitted.
8. K A Berrington, P G Burke and A L Sinfailam, J Phys B:At Molec Phys, **8**, 1459 (1975)

9. K A Berrington, P G Burke, L C G Freitas and A E Kingston, J Phys B:At Molec Phys, **18**, 4135 (1985)
10. K A Berrington and A E Kingston, J Phys B:At Molec Phys, **20**, 6631 (1987)
11. For recent reviews, see K T Taylor, M H Nayfeh and C W Clark (Eds.) ; *Atomic Spectra and Collisions in External Fields* Plenum Press Inc New York - London 1988
12. G Zeller and G Wunner, private communication
13. T S Monteiro and K T Taylor, J Phys B: At Molec Cpt Phys **22** L191(1989) and J Phys B: At Molec Opt Phys submitted.
14. P F O'Mahony and K T Taylor, Phys Rev Lett **57**, 2931 (1986).
15. R J Allan, L Heck and S Zureck, Comput Phys Comm submitted.
16. C T Whelan, H R J Walter, J Hanssen and R M Dreizler, Z Phys D in press.
17. C T Whelan, H R J Walters, R J Allan, J Hanssen and R M Dreizler, in preparation.
18. E P Curran and H R J Walters, J Phys B:At Molec Phys,**20**, 1105 (1987)
19. K T Taylor, J Phys B: At Molec Opt Phys to be submitted.

Author Index

A

Adams, N. G., 90, 325
Altman, T. C., 487
Amusia, M. Ya, 201
Andersson, L. R., 384
Andresen, P., 337
Andriamonje, S., 309
Anne, R., 309
Armstrong, P. S., 513
Ast, H. J., 258

B

Balykin, V., 378
Bárány, A., 246
Bartschat, K., 645
Baruch, M. C., 513
Basu, M., 622
Baumann, W., 378
Becker, U., 162
Beijerinck, H. C. W., 317
Bemis, Jr., C. E., 157
Berger, J., 378
Berkner, K. H., 366
Bernstein, E. M., 366
Biedermann, C., 384
Bisoffi, G., 378
Bizau, J. M., 224
Blatt, P., 378
Blum, M., 378
Bonfert, J., 130
Bordas, M. C., 398
Bottcher, C., 658
Broyer, M., 398
Bryant, H. C., 487
Bucksbaum, P. H., 499
Bürgdorfer, J., 476

C

Campeanu, R. I., 645
Carré, B., 224
Cederquist, H., 384
Chevallier, M., 309
Clark, M. W., 366
Claytor, N., 157
Cohen, C., 309
Cubaynes, D., 224

D

Datz, S., 2, 562
de Castro Faria, N. V., 309
de Heer, F. J., 390
Dittner, P. F., 562
Donahue, J. B., 487

Dörner, R., 372
Drachman, R. J., 616
Dreizler, R. M., 258
DuBois, R. D., 366
Dunn, G. H., 574
Dunning, F. B., 519
Dural, J., 309

E

Elston, S. B., 384

F

Fainstein, P. D., 264
Farizon-Mazuy, B., 309
Faulstich, A., 378
Feinberg, B., 157
Ferray, M., 505
Field, D., 410
Friedrich, A., 378

G

Gaillard, M. J., 309
Gallagher, T. F., 513
Garcia, J. D., 663
Genre, R., 309
Gerhard, M., 378
Geyer, C., 378
Ghosh, A. S., 622
Gibbons, J. P., 384
Giese, J., 562
Gomez del Campo, J., 157
Gould, H., 157
Graf, H., 130
Graham, W. G., 366, 544
Grieser, M., 378
Grieser, R., 378
Groeneveld, K-O., 384

H

Habs, D., 378
Hage-Ali, M., 309
Hahn, Y., 550
Harris, P. G., 487
Henne, A., 258
Henriet, A., 586
Herd, C. R., 90
Heyng, H. W., 378
Hirayama, T., 82
Hochadel, B., 378
Hoekstra, R., 390
Holzer, B., 378

Horbatsch, M., 645
Huber, G., 378
Huillier, A. L., 505
Humberston, J. W., 614
Hüwel, L., 337
Hvelplund, P., 556

J

Jaeschke, E., 378
Jakubassa-Amundsen, D. H., 358
Jefferts, S. R., 574
Joachain, C. J., 68
Julienne, P. S., 580
Jung, M., 378
Justiniano, E., 562

K

Kambara, T., 562
Karafillidis, A., 378
Kauppila, W. E., 627
Kelley, M. H., 103
Kerner, D. R., 458
Kilgus, G., 378
Kingston, A. E., 137
Kirsch, R., 309
Klein, R., 378
Kleyn, A. W., 451
Knight, D. W., 410
Krause, P., 378
Krämer, D., 378
Krieg, M., 378
Kühl, T., 378
Kwong, N. H., 663

L

L'Hoir, A., 309
Labastie, P., 398
Larson, D. J., 513
Lebedev, V. S., 466
Lee, D. H., 568
Lennard, W. N., 639
Leopold, J. G., 492
Levin, J. C., 176, 384
Liljeby, L., 384
Logan, L. R., 639
Lombardi, M., 398
Lompré, L. A., 505
Lüdde, H. J., 258
Ludemann, C. A., 157
Lunt, S., 410

M

Madison, D. H., 149
Mainfray, G., 505
Manson, S. T., 189
Masnou-Seeuws, F., 586
Massoumi, G. R., 639
Matl, K., 378
McCombie, J., 430
McEachran, R. P., 645
McGuire, J. H., 280
Miller, P. D., 562
Miller, W. H., 442
Mitchell, J. B. A., 404
Mittleman, M. H., 184
Mohagheghi, A. H., 487
Mokler, P. H., 562
Monroe, C., 593
Montemayor, V., 299
Morgan, L. A., 96
Morgan, T. J., 366
Morgenstern, R., 390
Mory, J., 309
Moulin, J., 309
Mrotzek, G., 410
Mueller, D. W., 366
Mukherjee, M., 622
Müller, A., 378, 418, 562
Music, M., 378

N

Nakel, W., 130
Neumann, R., 378
Neumark, D. M., 33
Neureither, G., 378
Newell, W. R., 122
Nordlander, P., 529

O

O, C-S., 384
Olson, R. E., 372
Olsson, B., 633
Olsson, T., 513
Ott, W., 378

P

Parcell, L. A., 645
Pegg, D. J., 233
Petrich, W., 278
Poizat, J. C., 309
Ponce, H., 264
Povh, B., 378

Q

Quéré, Y., 309
Quick, C. R., 487

R

Raith, W., 633
Randell, J., 410
Reeder, R. A., 487
Reinhardt, W. P., 458
Remillieux, J., 309
Repnow, R., 378
Reusch, S., 562
Richard, P., 568
Richards, D., 492
Rislove, D. C., 487
Rivarola, R. D., 264
Rothard, H., 384

S

Salzborn, E., 290
Schafer, K. J., 663
Schartner, K-H., 215
Schauer, M. M., 574
Schiwietz, G., 299
Schlachter, A. S., 366
Schmaus, D., 309
Schmidt, V., 241
Schneider, D., 299
Schoene, H., 562
Schröder, S., 378
Schuch, R., 378, 562
Schulz, M., 562
Schultz, P. J., 639
Schwalm, D., 378
Scoles, G., 430
Sellin, I. A., 384
Sesko, D. W., 593
Sharifian, H., 487
Shmaenok, L., 350
Shmidt-Böcking, H., 372
Short, R. T., 384
Sigray, P., 378
Sinapius, G., 633
Skogvall, B., 299
Smith, D., 90, 325
Smith, W. W., 487
Sperber, W., 633
Spicher, G., 633
Stachura, Z., 562
Stary, C., 258
Stauffer, A. D., 645
Stearns, J. W., 366
Steck, M., 378
Stein, T. S., 627
Stewart, J. E., 487
Stockli, M. P., 366
Stokstad, R., 378
Stolterfoht, N., 299
Straton, J. C., 280
Strayer, M. R., 658
Suzuki, H., 82
Szmola, E., 378

T

Takayanagi, K., 49
Takayanagi, T., 82
Tang, C. Y., 487
Tanis, J. A., 366, 538
Taulbjerg, K., 273
Taylor, K. T., 668
Toulemonde, M., 309
Toutounchi, H., 487
Tsuji, M., 342
Tully, J. C., 529

U

Ullrich, J., 372

V

Vane, C. R., 157, 384
Vane, R., 562
Voges, H., 337

W

Wagner, M., 378
Walker, T. G., 593
Walraven, J. T. M., 599
Wanner, B., 378
Ward, S. J., 645
Warzcak, A., 562
Weiner, J., 607
Welge, K. H., 16
Welti, K., 378
Wieman, C., 593
Williams, J. F., 115
Wintermeyer, G., 562
Wodtke, A. M., 337
Wolf, A., 378
Woodland, W. T., 366
Wuilleumier, F. J., 224

Y

Younger, S. M., 652

Z

Ziesel, J. P., 410
Zouros, T. J. M., 568
Zwickler, S., 378

AIP Conference Proceedings

		L.C. Number	ISBN
No. 125	Capture Gamma-Ray Spectroscopy and Related Topics – 1984 (Internat. Symposium, Knoxville)	84-73303	0-88318-324-2
No. 126	Solar Neutrinos and Neutrino Astronomy (Homestake, 1984)	84-63143	0-88318-325-0
No. 127	Physics of High Energy Particle Accelerators (BNL/SUNY Summer School, 1983)	85-70057	0-88318-326-9
No. 128	Nuclear Physics with Stored, Cooled Beams (McCormick's Creek State Park, Indiana, 1984)	85-71167	0-88318-327-7
No. 129	Radiofrequency Plasma Heating (Sixth Topical Conference, Callaway Gardens, GA, 1985)	85-48027	0-88318-328-5
No. 130	Laser Acceleration of Particles (Malibu, California, 1985)	85-48028	0-88318-329-3
No. 131	Workshop on Polarized ^{3}He Beams and Targets (Princeton, New Jersey, 1984)	85-48026	0-88318-330-7
No. 132	Hadron Spectroscopy–1985 (International Conference, Univ. of Maryland)	85-72537	0-88318-331-5
No. 133	Hadronic Probes and Nuclear Interactions (Arizona State University, 1985)	85-72638	0-88318-332-3
No. 134	The State of High Energy Physics (BNL/SUNY Summer School, 1983)	85-73170	0-88318-333-1
No. 135	Energy Sources: Conservation and Renewables (APS, Washington, DC, 1985)	85-73019	0-88318-334-X
No. 136	Atomic Theory Workshop on Relativistic and QED Effects in Heavy Atoms	85-73790	0-88318-335-8
No. 137	Polymer-Flow Interaction (La Jolla Institute, 1985)	85-73915	0-88318-336-6
No. 138	Frontiers in Electronic Materials and Processing (Houston, TX, 1985)	86-70108	0-88318-337-4
No. 139	High-Current, High-Brightness, and High-Duty Factor Ion Injectors (La Jolla Institute, 1985)	86-70245	0-88318-338-2
No. 140	Boron-Rich Solids (Albuquerque, NM, 1985)	86-70246	0-88318-339-0
No. 141	Gamma-Ray Bursts (Stanford, CA, 1984)	86-70761	0-88318-340-4
No. 142	Nuclear Structure at High Spin, Excitation, and Momentum Transfer (Indiana University, 1985)	86-70837	0-88318-341-2
No. 143	Mexican School of Particles and Fields (Oaxtepec, México, 1984)	86-81187	0-88318-342-0
No. 144	Magnetospheric Phenomena in Astrophysics (Los Alamos, 1984)	86-71149	0-88318-343-9
No. 145	Polarized Beams at SSC & Polarized Antiprotons (Ann Arbor, MI & Bodega Bay, CA, 1985)	86-71343	0-88318-344-7

No. 146	Advances in Laser Science–I (Dallas, TX, 1985)	86-71536	0-88318-345-5
No. 147	Short Wavelength Coherent Radiation: Generation and Applications (Monterey, CA, 1986)	86-71674	0-88318-346-3
No. 148	Space Colonization: Technology and The Liberal Arts (Geneva, NY, 1985)	86-71675	0-88318-347-1
No. 149	Physics and Chemistry of Protective Coatings (Universal City, CA, 1985)	86-72019	0-88318-348-X
No. 150	Intersections Between Particle and Nuclear Physics (Lake Louise, Canada, 1986)	86-72018	0-88318-349-8
No. 151	Neural Networks for Computing (Snowbird, UT, 1986)	86-72481	0-88318-351-X
No. 152	Heavy Ion Inertial Fusion (Washington, DC, 1986)	86-73185	0-88318-352-8
No. 153	Physics of Particle Accelerators (SLAC Summer School, 1985) (Fermilab Summer School, 1984)	87-70103	0-88318-353-6
No. 154	Physics and Chemistry of Porous Media—II (Ridge Field, CT, 1986)	83-73640	0-88318-354-4
No. 155	The Galactic Center: Proceedings of the Symposium Honoring C. H. Townes (Berkeley, CA, 1986)	86-73186	0-88318-355-2
No. 156	Advanced Accelerator Concepts (Madison, WI, 1986)	87-70635	0-88318-358-0
No. 157	Stability of Amorphous Silicon Alloy Materials and Devices (Palo Alto, CA, 1987)	87-70990	0-88318-359-9
No. 158	Production and Neutralization of Negative Ions and Beams (Brookhaven, NY, 1986)	87-71695	0-88318-358-7
No. 159	Applications of Radio-Frequency Power to Plasma: Seventh Topical Conference (Kissimmee, FL, 1987)	87-71812	0-88318-359-5
No. 160	Advances in Laser Science–II (Seattle, WA, 1986)	87-71962	0-88318-360-9
No. 161	Electron Scattering in Nuclear and Particle Science: In Commemoration of the 35th Anniversary of the Lyman-Hanson-Scott Experiment (Urbana, IL, 1986)	87-72403	0-88318-361-7
No. 162	Few-Body Systems and Multiparticle Dynamics (Crystal City, VA, 1987)	87-72594	0-88318-362-5
No. 163	Pion–Nucleus Physics: Future Directions and New Facilities at LAMPF (Los Alamos, NM, 1987)	87-72961	0-88318-363-3
No. 164	Nuclei Far from Stability: Fifth International Conference (Rosseau Lake, ON, 1987)	87-73214	0-88318-364-1
No. 165	Thin Film Processing and Characterization of High-Temperature Superconductors	87-73420	0-88318-365-X

No. 166	Photovoltaic Safety (Denver, CO, 1988)	88-42854	0-88318-366-8
No. 167	Deposition and Growth: Limits for Microelectronics (Anaheim, CA, 1987)	88-71432	0-88318-367-6
No. 168	Atomic Processes in Plasmas (Santa Fe, NM, 1987)	88-71273	0-88318-368-4
No. 169	Modern Physics in America: A Michelson-Morley Centennial Symposium (Cleveland, OH, 1987)	88-71348	0-88318-369-2
No. 170	Nuclear Spectroscopy of Astrophysical Sources (Washington, D.C., 1987)	88-71625	0-88318-370-6
No. 171	Vacuum Design of Advanced and Compact Synchrotron Light Sources (Upton, NY, 1988)	88-71824	0-88318-371-4
No. 172	Advances in Laser Science–III: Proceedings of the International Laser Science Conference (Atlantic City, NJ, 1987)	88-71879	0-88318-372-2
No. 173	Cooperative Networks in Physics Education (Oaxtepec, Mexico 1987)	88-72091	0-88318-373-0
No. 174	Radio Wave Scattering in the Interstellar Medium (San Diego, CA 1988)	88-72092	0-88318-374-9
No. 175	Non-neutral Plasma Physics (Washington, DC 1988)	88-72275	0-88318-375-7
No. 176	Intersections Between Particle and Nuclear Physics (Third International Conference) (Rockport, ME 1988)	88-62535	0-88318-376-5
No. 177	Linear Accelerator and Beam Optics Codes (La Jolla, CA 1988)	88-46074	0-88318-377-3
No. 178	Nuclear Arms Technologies in the 1990s (Washington, DC 1988)	88-83262	0-88318-378-1
No. 179	The Michelson Era in American Science: 1870–1930 (Cleveland, OH 1987)	88-83369	0-88318-379-X
No. 180	Frontiers in Science: International Symposium (Urbana, IL 1987)	88-83526	0-88318-380-3
No. 181	Muon-Catalyzed Fusion (Sanibel Island, FL 1988)	88-83636	0-88318-381-1
No. 182	High T_c Superconducting Thin Films, Devices, and Application (Atlanta, GA 1988)	88-03947	0-88318-382-X
No. 183	Cosmic Abundances of Matter (Minneapolis, MN 1988)	89-80147	0-88318-383-8
No. 184	Physics of Particle Accelerators (Ithaca, NY 1988)	87-07208	0-88318-384-6
No. 185	Glueballs, Hybrids, and Exotic Hadrons (Upton, NY 1988)	89-83513	0-88318-385-4

No. 186	High-Energy Radiation Background in Space (Sanibel Island, FL 1987)	89-083833	0-88318-386-2
No. 187	High-Energy Spin Physics (Minneapolis, MN 1988)	89-083948	0-88318-387-0
No. 188	International Symposium on Electron Beam Ion Sources and their Applications (Upton, NY 1988)	89-084343	0-88318-388-9
No. 189	Relativistic, Quantum Electrodynamic, and Weak Interaction Effects in Atoms (Santa Barbara, CA 1988)	89-084431	0-88318-389-7
No. 190	Radio-frequency Power in Plasmas (Irvine, CA 1989)	89-045805	0-88318-397-8
No. 191	Advances in Laser Science–IV (Atlanta, GA 1988)	89-085595	0-88318-391-9
No. 192	Vacuum Mechatronics (First International Workshop) (Santa Barbara, CA 1989)	89-045905	0-88318-394-3
No. 193	Advanced Accelerator Concepts (Lake Arrowhead, CA 1989)	89-045914	0-88318-393-5
No. 194	Quantum Fluids and Solids—1989 (Gainesville, FL, 1989)	89-81079	0-88318-395-1
No. 195	Dense Z-Pinches (Laguna Beach, CA, 1989)	89-46212	0-88318-396-X
No. 196	Heavy Quark Physics (Ithaca, NY, 1989)	89-81583	0-88318-644-6
No. 197	Drops and Bubbles (Monterey, CA, 1988)	89-46360	0-88318-392-7
No. 198	Astrophysics in Antarctica (Newark, DE, 1989)	89-46421	0-88318-398-6
No. 199	Surface Conditioning of Vacuum Systems (Los Angeles, CA, 1989)	89-82542	0-88318-756-6
No. 200	High T_c Superconducting Thin Films: Processing, Characterization, and Applications (Boston, MA, 1989)	90-80006	0-88318-759-0
No. 201	QED Stucture Functions (Ann Arbor, MI, 1989)	90-80229	0-88318-671-3
No. 202	NASA Workshop on Physics From a Lunar Base (Stanford, CA, 1989)	90-55073	0-88318-646-2
No. 203	Particle Astrophysics: The NASA Cosmic Ray Program for the 1990s and Beyond (Greenbelt, MD, 1989)	90-55077	0-88318-763-9
No. 204	Aspects of Electron-Molecule Scattering and Photoionization (New Haven, CT, 1989)	90-55175	0-88318-764-7

APR 16 1991